SOLUTIONS TO EXERCISES

CHEMISTRY

THE CENTRAL SCIENCE

14TH EDITION

SOLUTIONS TO EXERCISES

Roxy Wilson
University of Illinois, Urbana-Champaign

CHEMISTRY

THE CENTRAL SCIENCE

14TH EDITION

BROWN | LeMAY
BURSTEN | MURPHY
WOODWARD | STOLTZFUS

 Pearson

Courseware Portfolio Manager: Terry Haugen

Managing Producer, Science: Kristen Flatham

Product Marketing Manager: Elizabeth Bell

Content Producer, Science: Beth Sweeten

Full Service Vendor: Cenveo® Publisher Services

Main Text Cover Designer: Jeff Puda

Supplement Cover Designer: 17th Street Studios

Buyer: Maura Zaldivar-Garcia

1 17

ISBN-10: 0-13-455224-5

ISBN-13: 978-0-13-455224-8

Contents

Introduction

Chemistry: The Central Science, 14th edition, contains over 2700 end-of-chapter exercises. Considerable attention has been given to these exercises because one of the best ways for students to master chemistry is by solving problems. Grouping the exercises according to subject matter is intended to aid the student in selecting and recognizing particular types of problems. Within each subject matter group, similar problems are arranged in pairs. This provides the student with an opportunity to reinforce a particular kind of problem. There are also a substantial number of general exercises in each chapter to supplement those grouped by topic. Integrative exercises, which require students to integrate concepts from several chapters, are a continuing feature of the 14th edition. Answers to the odd-numbered topical exercises plus selected general and integrative exercises, about 1150 in all, are provided in the textbook. These appendix answers help to make the textbook a useful self-contained vehicle for learning.

This manual, **Solutions to Exercises in Chemistry: The Central Science, 14th edition**, was written to enhance the end-of-chapter exercises by providing documented solutions. The manual assists the instructor by saving time spent generating solutions for assigned problem sets and aids the student by offering a convenient independent source to check their understanding of the material. Most solutions have been worked in the same detail as the in-chapter sample exercises to help guide students in their studies.

To reinforce the *Analyze, Plan, Solve, Check* problem-solving method used extensively in the text, this strategy has also been incorporated into the Solutions Manual. Solutions to most red topical exercises and selected Additional and Integrative exercises feature this four-step approach. We strongly encourage students to master this powerful and totally general method.

When using this manual, keep in mind that the numerical result of any calculation is influenced by the precision of the numbers used in the calculation. In this manual, for example, atomic masses and physical constants are typically expressed to four significant figures, or at least as precisely as the data given in the problem. If students use slightly different values to solve problems, their answers will differ slightly from those listed in the appendix of the text or this manual. This is a normal and a common occurrence when comparing results from different calculations or experiments.

Rounding methods are another source of differences between calculated values. In this manual, when a solution is given in steps, intermediate results will be rounded to the correct number of significant figures; however, unrounded numbers will be used in subsequent calculations. By following this scheme, calculators need not be cleared to re-enter rounded intermediate results in the middle of a calculation sequence. The final answer will appear with the correct number of significant figures. This may result in a small discrepancy in the last significant digit between student-calculated answers and those given in this manual. Variations due to rounding can occur in any analysis of numerical data.

The first step in checking your solution and resolving differences between your answer and the listed value is to look for similarities and differences in problem-solving methods. Ultimately, resolving the small numerical differences described above is less important than understanding the general method for solving a problem. The goal of this manual is to provide a reference for sound and consistent problem-solving methods in addition to accurate answers to text exercises.

Extraordinary efforts have been made to keep this manual as error-free as possible. All exercises were worked and proofread by at least three chemists to ensure clarity in methods and accuracy in mathematics. The ongoing work and advice of Dr. Richard Helmich, Ms. Rene Musto, and Dr. Christopher Musto continue to be invaluable to this project. In any written work as technically challenging as this manual, typos and errors inevitably creep in, despite our combined efforts. Please help us find and eliminate them. We hope that both instructors and students will find this manual accurate, helpful, and instructive.

Roxy B. Wilson, Ph.D.
1829 Maynard Dr.
Champaign, IL 61822
rbwilson@illinois.edu

1 Introduction: Matter, Energy, and Measurement

Visualizing Concepts

1.1 *Pure elements* contain only one kind of atom. Atoms can be present singly or as tightly bound groups called molecules. *Compounds* contain two or more kinds of atoms bound tightly into molecules. *Mixtures* contain more than one kind of atom and/or molecule, not bound into discrete particles.

   (a) pure element: i

   (b) mixture of elements: v, vi

   (c) pure compound: iv

   (d) mixture of an element and a compound: ii, iii

1.2 After a *physical change*, the identities of the substances involved are the same as their identities before the change. That is, molecules retain their original composition. During a *chemical change*, at least one new substance is produced; rearrangement of atoms into new molecules occurs.

   Diagram (b) represents a chemical change because the molecules after the change are different than the molecules before the change.

1.3 (a) Brass is composed of two different kinds of atoms, so it is a mixture. The mixture appears homogeneous under an optical microscope, so it is a homogeneous mixture.

   (b) Because brass is a homogeneous mixture, it is a solution. We usually think of solutions as liquids, but they can be solids, liquids, or gases.

1.4 Refer to Sample Exercise 1.5.

   (a) $\text{density} = \dfrac{\text{mass}}{\text{volume}}$; $\text{mass} = \text{density} \times \text{volume}$

   $$\text{mass(Al)} = \frac{2.70\,\text{g}}{\text{cm}^3} \times 196\,\text{cm}^3 \times \frac{1\,\text{kg}}{1000\,\text{g}} = 0.5292 = 0.529\,\text{kg}$$

   $$\text{mass(Ag)} = \frac{10.49\,\text{g}}{\text{cm}^3} \times 196\,\text{cm}^3 \times \frac{1\,\text{kg}}{1000\,\text{g}} = 2.056 = 2.06\,\text{kg}$$

   (b) The amount of work done to lift each sphere to a height of 2.2 m can be calculated using Equation 1.1, $w = F \times d = m \times g \times d$. Use mass in kg and distance in meters so the final unit for work is joules, J.

$$w(Al) = 0.5292 \text{ kg Al} \times \frac{9.8 \text{ m}}{s^2} \times 2.2 \text{ m} = 11.41 = \frac{11 \text{ kg-m}^2}{1 \text{ s}^2} = 11 \text{ J}$$

$$w(Ag) = 2.056 \text{ kg Ag} \times \frac{9.8 \text{ m}}{s^2} \times 2.2 \text{ m} = 44.33 = \frac{44 \text{ kg-m}^2}{s^2} = 44 \text{ J}$$

(The results have two significant figures because the height had only two significant figures.)

Note: This is the first exercise where "intermediate rounding" occurs. In this manual, when a solution is given in steps, the intermediate result will be rounded to the correct number of significant figures. However, the **unrounded** number will be used in subsequent calculations. The final answer will appear with the correct number of significant figures. That is, calculators need not be cleared and new numbers entered in the middle of a calculation sequence. This may result in a small discrepancy in the last significant digit between student-calculated answers and those given in the manual. These variations occur in any analysis of numerical data.

For example, in this exercise the mass of the Ag sphere, 2.056×10^3 g, is rounded to 2.06×10^3 g, but 2.056×10^3 g is retained in the subsequent calculation of work, 44 J. In this case, the unrounded and rounded masses both lead to the same rounded value for work. In other exercises, the correctly rounded results of the two methods may not be identical.

(c) Less work is done on the Al sphere, so its potential energy increases by a smaller amount than the potential energy of the Ag sphere.

(d) No. The two spheres do not have the same kinetic energy when they touch the ground. Because the Ag sphere has greater mass, it gains more potential energy when it is lifted. At the point of impact, all potential energy has been converted to kinetic energy. Because of its greater potential energy, the Ag sphere has greater kinetic energy when it hits the floor.

1.5   Filtration. When brewing a cup of coffee, hot water contacts the coffee grounds and dissolves components of the coffee bean that are water-soluble. This creates a heterogeneous mixture of undissolved coffee bean solids and liquid coffee solution; this mixture is separated by filtration. Undissolved grounds remain on the filter paper and liquid coffee drips into the container below.

1.6   (a) time   (b) mass   (c) temperature   (d) area   (e) length

(f) area   (g) temperature   (h) density   (i) volume

1.7   Density is the ratio of mass to volume. For a sphere, size is like volume; both are determined by the radius of the sphere.

(a) For spheres of the same size or volume, the denominator of the density relationship is the same. The denser the sphere, the heavier it is. A list from lightest to heaviest is in order of increasing density and mass. The aluminum sphere (density = 2.70 g/cm^3) is lightest, then nickel (density = 8.90 g/cm^3), then silver (density = 10.49 g/cm^3).

(b)   For cubes of equal mass, the numerator of the density relationship is the same. The denser the sphere, the smaller its volume or size. A list from smallest to largest is in order of decreasing density. The platinum sphere (density = 21.45 g/cm^3) is smallest, then gold (density = 19.30 g/cm^3), then lead (density = 11.35 g/cm^3).

1.8   Results (bullet holes) that are close to each other are *precise*. Results that are close to the "true value" (the bull's-eye) are *accurate*.

    (a)   The results on target A are precise but not accurate. The individual shots landed close to each other, but not close to the bull's-eye. The results on target B are both precise and accurate. The individual shots landed close together and in the center of the target. The results on target C are neither precise nor accurate. The individual shots are scattered around the target.

    (b)   The precise grouping on A is high and to the right of center. To improve accuracy, the sighting mechanism on the gun should be adjusted down and slightly left from its current position.

          To improve the results on C, someone who can produce a precise grouping must shoot the student's target rifle. Then, the position of the sighting mechanism can be adjusted to produce an accurate shot. After adjusting the sight, the student needs more practice to improve precision.

1.9   (a)   7.5 cm. There are two significant figures in this measurement; the number of cm can be read precisely, but there is some estimating (uncertainty) required to read tenths of a centimeter. Listing two significant figures is consistent with the convention that measured quantities are reported so that there is uncertainty in only the last digit.

    (b)   The speed is 72 mi/hr (inner scale, two significant figures) or 115 km/hr (outer scale, three significant figures). Both scales are read with certainty in the "hundreds" and "tens" place, and some uncertainty in the "ones" place. The km/hr speed has one more significant figure because its magnitude is in the hundreds.

1.10   (a)   Volume = length × width × height. Because the operation is multiplication, the dimension with fewest significant figures (sig figs) determines the number of sig figs in the result. The dimension "2.5 cm" has 2 sig figs, so the volume is reported with 2 sig figs.

    (b)   Density = mass/volume. Because the operation is division, again the datum with fewer significant figures determines the number of sig figs in the result. While mass, 104.72 g, has 5 sig figs, volume [from (a)] has 2 sig figs, so density is also reported to 2 sig figs.

1.11   Given: masses of six jelly beans, mass of full jar, mass of empty jar. Find: number of jelly beans in the jar. The total mass of jelly beans is the mass of the jar full minus the mass of the jar empty. The mass of an "average" jelly bean is the average of the six masses. Then, the number of jelly beans is the total mass of beans divided by the average mass of a single bean.

Total mass of beans = 2082 g – 653 g = 1429 g

Average mass of a bean = (3.15 + 3.12 + 2.98 + 3.14 + 3.02 + 3.09) / 6 = 3.08 g

Number of beans = 1429 g total / 3.08 g per bean = 463.96 = 464 beans

By applying the significant figure rules for addition and subtraction, the total mass of beans has 0 decimal places and thus 4 significant figures. The mass of an average bean has 2 decimal places and 3 sig figs. The number of beans then has 3 sig figs, by the rules for multiplication and division. This makes sense, because we expect an integer number of beans in the jar. (Note that using an unrounded average bean mass of 3.0833 g predicts the number of beans to be 463.46, which rounds to 463 beans. The difference in these two values shows uncertainty in the last significant figure of the number of beans, as we expect in an experimental result.)

1.12   Compounds are pure substances, they have constant composition and properties throughout. The agate stone cannot be a compound, because materials with different properties appear as irregular rings in the stone. Ellen is correct.

Classification and Properties of Matter (Sections 1.2 and 1.3)

1.13   (a)   heterogeneous mixture

(b)   homogeneous mixture (If there are undissolved particles, such as sand or decaying plants, the mixture is heterogeneous.)

(c)   pure substance

(d)   pure substance

1.14   (a)   homogeneous mixture

(b)   heterogeneous mixture (particles in liquid)

(c)   pure substance

(d)   heterogeneous mixture

1.15   (a)   S   (b) Au   (c) K   (d) Cl   (e) Cu   (f) uranium

(g)   nickel   (h) sodium   (i) aluminum   (j) silicon

1.16   (a)   C   (b) N   (c) Ti   (d) Zn   (e) Fe   (f) phosphorus

(g)   calcium (h) helium   (i) lead   (j) silver

1.17 $A(s) \rightarrow B(s) + C(g)$

Substances A and C are definitely compounds; B is probably a compound. When solid carbon is burned in excess oxygen gas, the two elements combine to form a gaseous compound, carbon dioxide. Clearly substance C is this compound. Because C is produced when A is heated in the absence of oxygen (from air), both the carbon and the oxygen in C must have been present in A originally. A is, therefore, a compound composed of two or more elements chemically combined. Without more information on the chemical or physical properties of B, we cannot determine absolutely whether it is an element or a compound. However, few if any elements exist as white solids, so B is probably also a compound.

1.18   Gold, Au, and "fool's gold," FeS_2, are similar in appearance and physical state, so these two properties are not useful for discriminating between the two substances. Density and melting point would both help determine if the nugget is gold. (For these two substances, density is easiest to measure and is definitive.)

1.19   *Physical properties*: silvery white (color); lustrous; melting point = 649 °C; boiling point = 1105 °C; density at 20 °C = 1.738 g/cm^3; pounded into sheets (malleable); drawn into wires (ductile); good conductor. *Chemical properties*: burns in air to give intense white light; reacts with Cl_2 to produce brittle white solid.

1.20   (a)   *Physical properties*: melting point = 420 °C; hardness = 2.5 Mohs; density = 7.13 g/cm^3 at 25 °C. *Chemical properties*: granules react with dilute sulfuric acid to produce hydrogen gas; at elevated temperatures, reacts slowly with oxygen gas to produce ZnO.

       (b)   From the photo, we see that zinc is a shiny dark gray solid at atmospheric conditions, (physical state, color). These are physical properties.

1.21   (a)   chemical   (b) physical   (c) physical   (d) chemical   (e) chemical

1.22   (a)   chemical

       (b)   physical

       (c)   physical (The production of H_2O is a chemical change, but its *condensation* is a physical change.)

       (d)   physical (The production of soot is a chemical change, but its *deposition* is a physical change.)

1.23   Distillation. A solution of sugar and water is a homogeneous mixture that cannot be separated by filtration. Distillation takes advantage of the much lower boiling point of water.

1.24   (a)   Chemical. A substance with new chemical properties is produced when the two clear liquids are mixed. The product of physical mixing would be another colorless solution.

       (b)   Of the three methods, filtration is the most convenient way to separate an insoluble solid from a liquid.

The Nature of Energy (Section 1.4)

1.25   (a)   *Plan.* $E_k = 1/2\,mv^2$; m = 1200 kg; v = 18 m/s; 1 kg-m^2/s^2 = 1 J
             Solve. $E_k = 1/2 \times 1200\ kg \times (18)^2\ m^2/s^2 = 1.944 \times 10^5 = 1.9 \times 10^5$ J

       (b)   1 cal = 4.184 J; $1.944 \times 10^5\ J \times \dfrac{1\ cal}{4.184\ J} = 4.646 \times 10^4$ cal $= 4.6 \times 10^4$ cal

       (c)   As the automobile brakes to a stop, its speed (and hence its kinetic energy) drops to 0. This "lost" kinetic energy is mostly converted to heat. (The heat shows up in the brake parts, tires, and road.) It is not converted to some form of potential energy.

1.26   (a)   *Analyze.* Given: mass and speed of ball.   Find: kinetic energy.

             Plan. Because $1\ J = 1\ kg\text{-}m^2/s^2$, convert oz to kg and mph to m/s to obtain E_k in J.

             Solve. $5.13\ oz \times \dfrac{1\ lb}{16\ oz} \times \dfrac{1\ kg}{2.205\ lb} = 0.14541 = 0.145$ kg

$$\frac{95.0 \text{ mi}}{1 \text{ hr}} \times \frac{1.6093 \text{ km}}{1 \text{ mi}} \times \frac{1000 \text{ m}}{1 \text{ km}} \times \frac{1 \text{ hr}}{60 \text{ min}} \times \frac{1 \text{ min}}{60 \text{ sec}} = 42.468 = 42.5 \text{ m/s}$$

$$E_k = 1/2 \, mv^2 = 1/2 \times 0.14541 \text{ kg} \times \left(\frac{42.468 \text{ m}}{1 \text{ s}}\right)^2 = \frac{131 \text{ kg-m}^2}{1 \text{ s}^2} = 131 \text{ J}$$

Check. $1/2(0.15 \times 1600) \approx 1/2(160+80) \approx 120 \text{ J}$

(b) Kinetic energy is related to velocity squared (v^2); if the speed of the ball decreases to 55.0 mph, the kinetic energy of the ball will decrease by a factor of $(55.0/95.0)^2$. (The conversion factors to m/s apply to both speeds and will cancel in the ratio.) The numerical multiplier is $(55/95)^2 = 0.335$. The kinetic energy decreases by approximately a factor of 3.

(c) As the ball hits the catcher's glove, its speed (and hence its kinetic energy) drops to 0. Most of the energy is converted to heat, which is transferred to the glove and the hand inside. If the catcher's arm (and shoulder) is considered a spring, some kineic energy is transferred to potential energy of the arm assembly, which recoils while catching the ball. [The event can also be considered an inelastic collision between the ball and the arm assembly, with the combined masses (arm+ ball) having the same kinetic energy and the arm doing work to decelerate the (arm + ball).]

1.27 (a) Kinetic energy; the particles move apart.

(b) Potential energy decreases. The greater the separation prior to release, the smaller the electrostatic repulsion and potential energy.

1.28 (a) Increases. The magnitude of the electrostatic attractive potential energy gets smaller as the distance between the particles increases, but the sign is negative. The potential energy increases because it becomes less negative. From another perspective, energy must be supplied to separate particles that are attracted to each other. This energy leads to an increase in the potential energy of the particles.

(b) Increases. Work must be done to pump water in opposition to the force of gravity. This work increases the potential energy of the water.

(c) Increases. Energy must be supplied to break a chemical bond. The potential energy of the atoms increases.

1.29 *Analyze/Plan.* Use results from Solution 1.4 and energy relationships discussed in Section 1.4 to solve for kinetic energy and velocity.

Solve. From Solution 1.4 we have the work required to lift the spheres. This is equal to the potential energy of the spheres at 2.2 m. As a sphere hits the floor, all potential energy is changed to kinetic energy. The kinetic energy of the Al sphere is then 11 J.

$E_k = 11.41 = 11 \text{ J}$; mass(Al) $= 0.5292 = 0.529$ kg; $E_k = 1/2 \, mv^2$; $v = (2 E_k / m)^{1/2}$;

$$v(\text{Al}) = \left(2 \times \frac{11.41 \text{ kg-m}^2}{s^2} \times \frac{1}{0.5292 \text{ kg}}\right)^{1/2} = 6.567 = 6.6 \text{ m/s}$$

6

1.30 $E_k = 44.33 = 44$ J; mass(Ag) = 2.056 kg = 2.06 kg; $E_k = 1/2\, mv^2$; $v = (2\, E_k\, /m)^{1/2}$

$$v(Ag) = \left(2 \times \frac{44.33\ \text{kg-m}^2}{\text{s}^2} \times \frac{1}{2.056\ \text{kg}} \right)^{1/2} = 6.567 = 6.6\ \text{m/s}$$

Note that calculation shows the speeds of the Al and Ag spheres at impact to be equal, as stated in Exercise 1.4(d).

Units and Measurement (Section 1.5)

1.31   (a) 1×10^{-1} (b) 1×10^{-2} (c) 1×10^{-15} (d) 1×10^{-6} (e) 1×10^{6}

        (f) 1×10^{3} (g) 1×10^{-9} (h) 1×10^{-3} (i) 1×10^{-12}

1.32   (a) $2.3 \times 10^{-10}\ \text{L} \times \dfrac{1\ \text{nL}}{1 \times 10^{-9}\ \text{L}} = 0.23\ \text{nL}$

        (b) $4.7 \times 10^{-6}\ \text{g} \times \dfrac{1\ \mu\text{g}}{1 \times 10^{-6}\ \text{g}} = 4.7\ \mu\text{g}$

        (c) $1.85 \times 10^{-12}\ \text{m} \times \dfrac{1\ \text{pm}}{1 \times 10^{-12}\ \text{m}} = 1.85\ \text{pm}$

        (d) $16.7 \times 10^{6}\ \text{s} \times \dfrac{1\ \text{Ms}}{1 \times 10^{6}\ \text{s}} = 16.7\ \text{Ms}$

        (e) $15.7 \times 10^{3}\ \text{g} \times \dfrac{1\ \text{kg}}{1 \times 10^{3}\ \text{g}} = 15.7\ \text{kg}$

        (f) $1.34 \times 10^{-3}\ \text{m} \times \dfrac{1\ \text{mm}}{1 \times 10^{-3}\ \text{m}} = 1.34\ \text{mm}$

        (g) $1.84 \times 10^{2}\ \text{cm} \times \dfrac{1\ \text{m}}{1 \times 10^{2}\ \text{cm}} = 1.84\ \text{m}$

1.33   (a) $^\circ\text{C} = 5/9\ (^\circ\text{F} - 32\ ^\circ)$; $5/9\ (72 - 32) = 22\ ^\circ\text{C}$

        (b) $^\circ\text{F} = 9/5\ (^\circ\text{C}) + 32\ ^\circ$; $9/5\ (216.7) + 32 = 422.1\ ^\circ\text{F}$

        (c) $\text{K} = ^\circ\text{C} + 273.15$; $233\ ^\circ\text{C} + 273.15 = 506\ \text{K}$

        (d) $^\circ\text{C} = 315\ \text{K} - 273.15 = 41.85 = 42\ ^\circ\text{C}$; $^\circ\text{F} = 9/5\ (41.85\ ^\circ\text{C}) + 32 = 107\ ^\circ\text{F}$

        (e) $^\circ\text{C} = 5/9\ (^\circ\text{F} - 32)$; $5/9\ (2500 - 32) = 1371 = 1400\ ^\circ\text{C}$

           $\text{K} = 1371\ ^\circ\text{C} + 273.15 = 1644 = 1600\ \text{K}$

        (f) $^\circ\text{C} = 0\ \text{K} - 273.15 = -273.15\ ^\circ\text{C}$; $^\circ\text{F} = 9/5\ (-273.15\ ^\circ\text{C}) + 32 = -459.67\ ^\circ\text{F}$

           (assuming 0 K has infinite sig figs)

1.34   (a) $^\circ\text{C} = 5/9\ (87\ ^\circ\text{F} - 32\ ^\circ) = 31\ ^\circ\text{C}$

        (b) $\text{K} = 25\ ^\circ\text{C} + 273.15 = 298\ \text{K}$; $^\circ\text{F} = 9/5\ (25\ ^\circ\text{C}) + 32 = 77\ ^\circ\text{F}$

        (c) $^\circ\text{C} = 5/9\ (400\ ^\circ\text{F} - 32\ ^\circ) = 204.444 = 200\ ^\circ\text{C}$

           $\text{K} = ^\circ\text{C} + 273.15 = 204.444\ ^\circ\text{C} + 273.15 = 477.59 = 500\ \text{K}$

        (d) $^\circ\text{C} = 77\ \text{K} - 273.15 = -196.15 = -196\ ^\circ\text{C}$

           $^\circ\text{F} = 9/5\ (-196.15\ ^\circ\text{C}) + 32 = -321.07 = -321\ ^\circ\text{F}$

1.35   (a) $\text{density} = \dfrac{\text{mass}}{\text{volume}} = \dfrac{40.55 \text{ g}}{25.0 \text{ mL}} = 1.62 \text{ g/mL or } 1.62 \text{ g/cm}^3$

(The units cm^3 and mL will be used interchangeably in this manual.)

Tetrachloroethylene, 1.62 g/mL, is more dense than water, 1.00 g/mL; tetrachloroethylene will sink rather than float on water.

(b) $25.0 \text{ cm}^3 \times 0.469 \dfrac{\text{g}}{\text{cm}^3} = 11.7 \text{ g}$

1.36   (a) $\text{volume} = \text{length}^3 \text{ (cm}^3\text{); density} = \text{mass/volume (g/cm}^3\text{)}$

$\text{volume} = (1.500)^3 \text{ cm}^3 = 3.375 \text{ cm}^3$

$\text{density} = \dfrac{76.31 \text{ g}}{3.375 \text{ cm}^3} = 22.61 \text{ g/cm}^3 \text{ osmium}$

(b) $125.0 \text{ mL} \times \dfrac{1 \text{ cm}^3}{1 \text{ mL}} \times \dfrac{4.51 \text{ g}}{1 \text{ cm}^3} = 563.75 = 564 \text{ g titanium}$

(c) $0.1500 \text{ L} \times \dfrac{1 \text{ mL}}{1 \times 10^{-3} \text{ L}} \times \dfrac{0.8787 \text{ g}}{1 \text{ mL}} = 131.8 \text{ g benzene}$

1.37   (a) $\text{calculated density} = \dfrac{38.5 \text{ g}}{45 \text{ mL}} = 0.86 \text{ g/mL}$

The substance is probably toluene, density = 0.866 g/mL.

(b) $45.0 \text{ g} \times \dfrac{1 \text{ mL}}{1.114 \text{ g}} = 40.4 \text{ mL ethylene glycol}$

(c)   A 100 mL graduated cylinder such as the one in Figure 1.21 usually has 1-mL markings. One can read the volume with certainty to the nearest mL, and estimate tenths of a mL. The volume calculated in (b) has uncertainty in the tenths place, so a graduated cylinder like this will provide the appropriate accuracy of measurement.

(d) $(5.00)^3 \text{ cm}^3 \times \dfrac{8.90 \text{ g}}{1 \text{ cm}^3} = 1.11 \times 10^3 \text{ g (1.11 kg) nickel}$

1.38   (a) $\dfrac{21.95 \text{ g}}{25.0 \text{ mL}} = 0.878 \text{ g/mL}$

The tabulated value has four significant figures, while the experimental value has three. The tabulated value rounded to three figures is 0.879. The values agree within one in the last significant figure of the experimental value; the two results agree. The liquid could be benzene.

(b) $15.0 \text{ g} \times \dfrac{1 \text{ mL}}{0.7781 \text{ g}} = 19.3 \text{ mL cyclohexane}$

(c) $r = d/2 = 5.0 \text{ cm}/2 = 2.5 \text{ cm}$

$V = 4/3 \pi r^3 = 4/3 \times \pi \times (2.5)^3 \text{ cm}^3 = 65.4498 = 65 \text{ cm}^3$

$65.4498 \text{ cm}^3 \times \dfrac{11.34 \text{ g}}{\text{cm}^3} = 7.4 \times 10^2 \text{ g}$

(The answer has two significant figures because the diameter had only two significant figures.)

1.39 $36 \text{ billion metric tons} \times \dfrac{1 \times 10^9 \text{ metric tons}}{1 \text{ billion metric tons}} \times \dfrac{1000 \text{ kg}}{1 \text{ metric ton}} \times \dfrac{1000 \text{ g}}{1 \text{ kg}} = 3.6 \times 10^{16} \text{ g}$

The metric prefix for 1×10^{15} is peta, abbreviated P.

$3.6 \times 10^{16} \text{ g} \times \dfrac{1 \text{ Pg}}{1 \times 10^{15} \text{ g}} = 36 \text{ Pg}$

1.40   (a)   The wafers have the same diameter as the boule, so the question becomes "how many 0.75 mm wafers can be cut from the 2 m boule?"

$\dfrac{2.0 \text{ m}}{\text{boule}} \times \dfrac{1 \text{ mm}}{1 \times 10^{-3} \text{ m}} \times \dfrac{1 \text{ wafer}}{0.75 \text{ mm}} = 2667 = 2.7 \times 10^3 \text{ wafers}$

As a practical matter, the thickness of the cutting blade reduces the actual number of disks that can be produced, so the real number is something less than 2667 wafers. Perhaps a more realistic answer, to 2 sig figs, is 2.6×10^3 wafers.

(b)   Calculate the volume of the wafer in cm^3. $V = r^2 h$

$r = \dfrac{d}{2} = \dfrac{300 \text{ mm}}{2} \times \dfrac{1 \text{ cm}}{10 \text{ mm}} = 15 \text{ cm}; \quad h = 0.75 \text{ mm} \times \dfrac{1 \text{ cm}}{10 \text{ mm}} = 7.5 \times 10^{-2} \text{ cm}$

$V = \pi r^2 h = \pi (15 \text{ cm})^2 (7.5 \times 10^{-2} \text{ cm}) = 53.0144 = 53 \text{ cm}^3$

Density = mass / V;   mass = density × V

$\dfrac{2.33 \text{ g}}{\text{cm}^3} \times 53.0144 \text{ cm}^3 = 123.52 = 1.2 \times 10^2 \text{ g}$

1.41   *Analyze.* Given: heat capacity of water = 1 Btu/lb-°F Find: J/Btu

Plan. heat capacity of water = $\dfrac{1 \text{ cal}}{\text{g-°C}}; \quad \dfrac{\text{cal}}{\text{g-°C}} \rightarrow \dfrac{\text{J}}{\text{g-°C}} \rightarrow \dfrac{\text{J}}{\text{lb-°F}} \rightarrow \dfrac{\text{J}}{\text{Btu}}$

This strategy requires changing °F to °C. Because this involves the magnitude of a degree on each scale, rather than a specific temperature, the 32 in the temperature relationship is not needed. 100 °C = 180 °F; 5 °C = 9 °F

Solve. $\dfrac{1 \text{ cal}}{\text{g-°C}} \times \dfrac{4.184 \text{ J}}{\text{cal}} \times \dfrac{453.6 \text{ g}}{\text{lb}} \times \dfrac{5 \text{ °C}}{9 \text{ °F}} \times \dfrac{1 \text{ lb-°F}}{1 \text{ Btu}} = 1054 \text{ J/Btu}$

1.42   (a)   *Analyze.* Given: 1 kwh; 1 watt = 1 J/s; 1 watt-s = 1 J. Find: conversion factor for joules and kwh.

Plan. kwh → wh → ws → J

Solve. $1 \, kwh \times \dfrac{1000 \, w}{1 \, kw} \times \dfrac{60 \, min}{h} \times \dfrac{60 \, s}{min} \times \dfrac{1 \, J}{1 \, w\text{-}s} = 3.6 \times 10^6 \, J$

$1 \, kwh = 3.6 \times 10^6 \, J$

(b)   *Analyze.* Given: 100 watt bulb. Find: heat in kcal radiated by bulb or person in 24 hr.

Plan. 1 watt = 1 J/s; 1 kcal = 4.184×10^3 J; watt → J/s → J → kcal.   *Solve.*

$100 \, watt = \dfrac{100 \, J}{1 \, s} \times \dfrac{60 \, sec}{min} \times \dfrac{60 \, min}{hr} \times 24 \, hr \times \dfrac{1 \, kcal}{4.184 \times 10^3 \, J} = 2065 = 2.1 \times 10^3 \, kcal$

24 hr has 2 sig figs, but 100 watt is ambiguous. The answer to 1 sig fig would be 2×10^3 kcal.

Uncertainty in Measurement (Section 1.6)

1.43   Exact: (b), (d), and (f) (All others depend on measurements and standards that have margins of error, e.g., the length of a week as defined by the Earth's rotation.)

1.44   Exact: (b), (e) (The number of students is exact on any given day.)

1.45   (a)   3   (b) 2   (c) 5   (d) 3   (e) 5   (f) 1 [See Sample Exercise 1.7 (c)]

1.46   (a)   4   (b) 3   (c) 4   (d) 5   (e) 6   (f) 2

1.47   (a) 1.025×10^2 (b) 6.570×10^2 (c) 8.543×10^{-3}
             (d) 2.579×10^{-4} (e) -3.572×10^{-2}

1.48   (a) 7.93×10^3 mi   (b) 4.001×10^4 km

1.49   (a)   14.3505 + 2.65 = 17.0005 = 17.00 (For addition and subtraction, the minimum number of decimal places, here two, determines decimal places in the result.)

(b)   952.7 − 140.7389 = 812.0

(c) $(3.29 \times 10^4)(0.2501) = 8.23 \times 10^3$ (For multiplication and division, the minimum number of significant figures, here three, determines sig figs in the result.)

(d) $0.0588 / 0.677 = 8.69 \times 10^{-2}$

1.50   (a) $[320.5 - 6104.5/2.3] = -2.3 \times 10^3$ (The intermediate result has two significant figures, so only the thousand and hundred places in the answer are significant.)

(b) $[285.3 \times 10^5 - 0.01200 \times 10^5] \times 2.8954 = 8.260 \times 10^7$ (Because subtraction depends on decimal places, both numbers must have the same exponent to determine decimal places/sig figs. The intermediate result has 1 decimal place and 4 sig figs, so the answer has 4 sig figs.)

(c) $(0.0045 \times 20{,}000.0)$ + (2813×12) $= 3.4 \times 10^4$
    2 sig figs /0 dec pl   2 sig figs /first 2 digits

(d)   863 $\times$ [1255 $-$ (3.45×108)] $= 7.62 \times 10^5$
    (3 sig figs /0 dec pl)

    3 sig figs $\times$ [0 dec pl/3 sig figs]   = 3 sig figs

1.51   The mass 21.427 g has 5 significant figures.

1.52   The volume in the graduated cylinder is 19.5 mL. Liquid volumes are read at the bottom of the meniscus, so the volume is slightly less than 20 mL. Volumes in this cylinder can be read with certainty to 1 mL, and with some uncertainty to 0.1 mL, so this measurement has 3 sig figs.

Dimensional Analysis (Section 1.7)

1.53   In each conversion factor, the old unit appears in the denominator, so it cancels, and the new unit appears in the numerator.

(a)   mm $\to$ nm: $\dfrac{1 \times 10^{-3} \text{ m}}{1 \text{ mm}} \times \dfrac{1 \text{ nm}}{1 \times 10^{-9} \text{ m}} = 1 \times 10^6 \text{ nm/mm}$

(b)   mg $\to$ kg: $\dfrac{1 \times 10^{-3} \text{ g}}{1 \text{ mg}} \times \dfrac{1 \text{ kg}}{1000 \text{ g}} = 1 \times 10^{-6} \text{ kg/mg}$

(c)   km $\to$ ft: $\dfrac{1000 \text{ m}}{1 \text{ km}} \times \dfrac{1 \text{ cm}}{1 \times 10^{-2} \text{ m}} \times \dfrac{1 \text{ in}}{2.54 \text{ cm}} \times \dfrac{1 \text{ ft}}{12 \text{ in}} = 3.28 \times 10^3 \text{ km/ft}$

(d) $\text{in}^3 \to \text{cm}^3$: $\dfrac{(2.54)^3 \text{ cm}^3}{1^3 \text{ in}^3} = 16.4 \text{ cm}^3/\text{in}^3$

1.54   In each conversion factor, the old unit appears in the denominator, so it cancels, and the new unit appears in the numerator.

(a) $\mu\text{m} \to \text{mm}$: $\dfrac{1 \times 10^{-6} \text{ m}}{1 \ \mu\text{m}} \times \dfrac{1 \text{ mm}}{1 \times 10^{-3} \text{ m}} = 1 \times 10^{-3} \text{ mm/}\mu\text{m}$

(b)   ms $\to$ ns: $\dfrac{1 \times 10^{-3} \text{ s}}{1 \text{ ms}} \times \dfrac{1 \text{ ns}}{1 \times 10^{-9} \text{ s}} = 1 \times 10^6 \text{ ns/ms}$

(c)   mi $\to$ km: 1.6093 km/mi

(d) $\text{ft}^3 \to \text{L}$: $\dfrac{(12)^3 \text{ in}^3}{1 \text{ ft}^3} \times \dfrac{(2.54)^3 \text{ cm}^3}{1 \text{ in}^3} \times \dfrac{1 \text{ L}}{1000 \text{ cm}^3} = 28.3 \text{ L/ft}^3$

1.55   (a) $\dfrac{15.2 \text{ m}}{\text{s}} \times \dfrac{1 \text{ km}}{1000 \text{ m}} \times \dfrac{60 \text{ s}}{1 \text{ min}} \times \dfrac{60 \text{ min}}{1 \text{ hr}} = 54.7 \text{ km/hr}$

(b) $5.0 \times 10^3 \text{ L} \times \dfrac{1 \text{ gal}}{3.7854 \text{ L}} = 1.3 \times 10^3 \text{ gal}$

(c) $151\,\text{ft} \times \dfrac{1\,\text{yd}}{3\,\text{ft}} \times \dfrac{1\,\text{m}}{1.0936\,\text{yd}} = 46.025 = 46.0\,\text{m}$

(d) $\dfrac{60.0\,\text{cm}}{\text{d}} \times \dfrac{1\,\text{in}}{2.54\,\text{cm}} \times \dfrac{1\,\text{d}}{24\,\text{hr}} = 0.984\,\text{in/hr}$

1.56   (a) $\dfrac{2.998 \times 10^8\,\text{m}}{\text{s}} \times \dfrac{1\,\text{km}}{1000\,\text{m}} \times \dfrac{1\,\text{mi}}{1.6093\,\text{km}} \times \dfrac{60\,\text{s}}{1\,\text{min}} \times \dfrac{60\,\text{min}}{1\,\text{hr}} = 6.707 \times 10^8\,\text{mi/hr}$

(b) $1454\,\text{ft} \times \dfrac{1\,\text{yd}}{3\,\text{ft}} \times \dfrac{1\,\text{m}}{1.0936\,\text{yd}} = 443.18 = 443.2\,\text{m}$

(c) $3{,}666{,}500\,\text{m}^3 \times \dfrac{1^3\,\text{dm}^3}{(1 \times 10^{-1})^3\,\text{m}^3} \times \dfrac{1\,\text{L}}{1\,\text{dm}^3} = 3.6665 \times 10^9\,\text{L}$

(d) $\dfrac{242\,\text{mg cholesterol}}{100\,\text{mL blood}} \times \dfrac{1\,\text{mL}}{1 \times 10^{-3}\,\text{L}} \times 5.2\,\text{L} \times \dfrac{1 \times 10^{-3}\,\text{g}}{1\,\text{mg}} = 12.58 = 13\,\text{g cholesterol}$

1.57   (a) $5.00\,\text{days} \times \dfrac{24\,\text{hr}}{1\,\text{day}} \times \dfrac{60\,\text{min}}{1\,\text{hr}} \times \dfrac{60\,\text{s}}{1\,\text{min}} = 4.32 \times 10^5\,\text{s}$

(b) $0.0550\,\text{mi} \times \dfrac{1.6093\,\text{km}}{\text{mi}} \times \dfrac{1000\,\text{m}}{1\,\text{km}} = 88.5\,\text{m}$

(c) $\dfrac{\$1.89}{\text{gal}} \times \dfrac{1\,\text{gal}}{3.7854\,\text{L}} = \dfrac{\$0.499}{\text{L}}$

(d) $\dfrac{0.510\,\text{in}}{\text{ms}} \times \dfrac{2.54\,\text{cm}}{1\,\text{in}} \times \dfrac{1 \times 10^{-2}\,\text{m}}{1\,\text{cm}} \times \dfrac{1\,\text{km}}{1000\,\text{m}} \times \dfrac{1\,\text{ms}}{1 \times 10^{-3}\,\text{s}} \times \dfrac{60\,\text{s}}{1\,\text{min}} \times \dfrac{60\,\text{min}}{1\,\text{hr}} = \dfrac{46.6\,\text{km}}{\text{hr}}$

Estimate: $0.5 \times 2.5 = 1.25$; $1.25 \times 0.01 \approx 0.01$; $0.01 \times 60 \times 60 \approx 36\,\text{km/hr}$

(e) $\dfrac{22.50\,\text{gal}}{\text{min}} \times \dfrac{3.7854\,\text{L}}{\text{gal}} \times \dfrac{1\,\text{min}}{60\,\text{s}} = 1.41953 = 1.420\,\text{L/s}$

Estimate: $20 \times 4 = 80$; $80/60 \approx 1.3\,\text{L/s}$

(f) $0.02500\,\text{ft}^3 \times \dfrac{12^3\,\text{in}^3}{1\,\text{ft}^3} \times \dfrac{2.54^3\,\text{cm}^3}{1\,\text{in}^3} = 707.9\,\text{cm}^3$

Estimate: $10^3 = 1000$; $3^3 = 27$; $1000 \times 27 = 27{,}000$; $27{,}000/0.04 \approx 700\,\text{cm}^3$

1.58   (a) $0.105\,\text{in} \times \dfrac{2.54\,\text{cm}}{\text{in}} \times \dfrac{1 \times 10^{-2}\,\text{m}}{\text{cm}} \times \dfrac{1\,\text{mm}}{1 \times 10^{-3}\,\text{m}} = 2.667 = 2.67\,\text{mm}$

(b) $0.650\,\text{qt} \times \dfrac{1\,\text{L}}{1.057\,\text{qt}} \times \dfrac{1\,\text{mL}}{1 \times 10^{-3}\,\text{L}} = 614.94 = 615\,\text{mL}$

(c) $\dfrac{8.75\,\mu m}{s} \times \dfrac{1\times10^{-6}\,m}{1\,\mu m} \times \dfrac{1\,km}{1\times10^3\,m} \times \dfrac{60\,s}{1\,min} \times \dfrac{60\,min}{1\,hr} = 3.15\times10^{-5}\,km/hr$

(d) $1.955\,m^3 \times \dfrac{(1.0936)^3\,yd^3}{1\,m^3} = 2.55695 = 2.557\,yd^3$

(e) $\dfrac{\$3.99}{lb} \times \dfrac{2.205\,lb}{1\,kg} = 8.798 = \$8.80/kg$

(f) $\dfrac{8.75\,lb}{ft^3} \times \dfrac{453.59\,g}{1\,lb} \times \dfrac{1\,ft^3}{12^3\,in^3} \times \dfrac{1\,in^3}{2.54^3\,cm^3} \times \dfrac{1\,cm^3}{1\,mL} = 0.140\,g/mL$

1.59 (a) $31\,gal \times \dfrac{4\,qt}{1\,gal} \times \dfrac{1\,L}{1.057\,qt} = 1.2\times10^2\,L$

Estimate: $(30 \times 4)/1 \approx 120\,L$

(b) $\dfrac{6\,mg}{kg\,(body)} \times \dfrac{1\,kg}{2.205\,lb} \times 185\,lb = 5\times10^2\,mg$

Estimate: $6/2 = 3;\ 3 \times 180 = 540\,mg$

(c) $\dfrac{400\,km}{47.3\,L} \times \dfrac{1\,mi}{1.6093\,km} \times \dfrac{1\,L}{1.057\,qt} \times \dfrac{4\,qt}{1\,gal} = \dfrac{19.9\,mi}{gal}$

$(2\times10^1\,mi/gal$ for 1 sig fig)

Estimate: $400/50 = 8;\ 8/1.6 = 5;\ 5/1 = 5;\ 5 \times 4 \approx 20\,mi/gal$

(d) $200\,cups \times \dfrac{1\,lb}{50\,cups} \times \dfrac{1\,kg}{2.205\,lb} = 1.81\,kg$

(2 kg for 1 sig fig)

Estimate: $1\,lb = 50\,cups,\ 4\,lb = 200\,cups;\ 4\,lb \approx 2\,kg$

1.60 (a) $1257\,mi \times \dfrac{1\,km}{0.62137\,mi} \times \dfrac{charge}{225\,km} = 8.99\,charges$

Because charges are integral events, 9 total charges are required. The trip begins with a full charge, so 8 additional charges during the trip are needed.

(b) $\dfrac{14\,m}{s} \times \dfrac{1\,km}{1\times10^3\,m} \times \dfrac{1\,mi}{1.6093\,km} \times \dfrac{60\,s}{1\,min} \times \dfrac{60\,min}{1\,hr} = 31\,mi/hr$

(c) $450\,in^3 \times \dfrac{(2.54)^3\,cm^3}{1\,in^3} \times \dfrac{1\,mL}{1\,cm^3} \times \dfrac{1\times10^{-3}\,L}{1\,mL} = 7.37\,L$

(d) $2.4\times10^5\,barrels \times \dfrac{42\,gal}{1\,barrel} \times \dfrac{4\,qt}{1\,gal} \times \dfrac{1\,L}{1.057\,qt} = 3.8\times10^7\,L$

1.61   14.5 ft × 16.5 ft × 8.0 ft = 1914 = 1.9×10^3 ft^3 (2 sig figs)

$$1914 \text{ ft}^3 \times \frac{(1 \text{ yd})^3}{(3 \text{ ft})^3} \times \frac{(1 \text{ m})^3}{(1.0936)^3 \text{ yd}^3} \times \frac{10^3 \text{ dm}^3}{1 \text{ m}^3} \times \frac{1 \text{ L}}{1 \text{ dm}^3} \times \frac{1.19 \text{ g}}{\text{L}} \times \frac{1 \text{ kg}}{1000 \text{ g}} = 64.4985 = 64 \text{ kg air}$$

Estimate: 1900/27 ≈ 60; (60 × 1)/1 ≈ 60 kg

1.62   10.6 ft × 14.8 ft × 20.5 ft = 3216.04 = 3.22×10^3 ft^3

$$3216.04 \text{ ft}^3 \times \frac{(1 \text{ yd})^3}{(3 \text{ ft})^3} \times \frac{(1 \text{ m})^3}{(1.0936 \text{ yd})^3} \times \frac{48 \text{ μg Co}}{1 \text{ m}^3} \times \frac{1 \times 10^{-6} \text{ g}}{1 \text{ μg}} = 4.4 \times 10^{-3} \text{ g Co}$$

1.63   Strategy: 1)   Calculate volume of gold (Au) in cm^3 in the sheet

2)   Mass = density × volume

3)   Change g → troy oz and $

$$100 \text{ ft} \times 82 \text{ ft} \times \frac{(12)^2 \text{ in}^2}{1 \text{ ft}^2} \times 5 \times 10^{-6} \text{ in} \times \frac{(2.54)^3 \text{ cm}^3}{1 \text{ in}^3} = 96.75 = 1 \times 10^2 \text{ cm}^3 \text{ Au}$$

$$96.75 \text{ cm}^3 \text{ Au} \times \frac{19.32 \text{ g}}{1 \text{ cm}^3} \times \frac{1 \text{ troy oz}}{31.1034768 \text{ g}} \times \frac{\$1654}{\text{troy oz}} = \$99,399 = \$1 \times 10^5$$

(Strictly speaking, the datum 100 ft has 1 sig fig, so the result has 1 sig fig.)

1.64   A wire is a very long, thin cylinder of volume, V = π r^2 h, where h is the length of the wire and π r^2 is the cross-sectional area of the wire.

Strategy: 1)   Calculate total volume of copper in cm^3 from mass and density

2)   h (length in cm) $= \dfrac{V}{\pi r^2}$

3)   Change cm → ft

$$150 \text{ lb Cu} \times \frac{453.6 \text{ g}}{1 \text{ lb Cu}} \times \frac{1 \text{ cm}^3}{8.94 \text{ g}} = 7610.7 = 7.61 \times 10^3 \text{ cm}^3$$

$$r = d/2 = 7.50 \text{ mm} \times \frac{1 \text{ cm}}{10 \text{ mm}} \times \frac{1}{2} = 0.375 \text{ cm}$$

$$h = \frac{V}{\pi r^2} = \frac{7610.7 \text{ cm}^3}{\pi (0.375)^2 \text{ cm}^2} = 1.7227 \times 10^4 = 1.72 \times 10^4 \text{ cm}$$

$$1.7227 \times 10^4 \text{ cm} \times \frac{1 \text{ in}}{2.54 \text{ cm}} \times \frac{1 \text{ ft}}{12 \text{ in}} = 565 \text{ ft}$$

(too difficult to estimate)

Additional Exercises

1.65   (a)   A gold ingot is pure metallic gold, a pure substance.

       (b)   A cup of coffee is a solution if there are no suspended solids (coffee grounds). It is a heterogeneous mixture if there are grounds. If cream or sugar is added, the homogeneity of the mixture depends on how thoroughly the components are mixed.

       (c)   A wood plank is a heterogeneous mixture of various cellulose components. The different domains in the mixture are visible as wood grain or knots.

1.66   (a)   A hypothesis is more likely to eventually be shown to be incorrect. A hypothesis is a possible explanation based on preliminary experimental data. A theory may be more general, and has a significant body of experimental evidence to support it.

       (b)   A <u>law</u> reliably predicts the behavior of matter, while a <u>theory</u> provides an explanation for that behavior.

1.67   According to the law of constant composition, any sample of vitamin C has the same relative amount of carbon and oxygen; the ratio of oxygen to carbon in the isolated sample is the same as the ratio in synthesized vitamin C.

$$\frac{2.00 \text{ g O}}{1.50 \text{ g C}} = \frac{x \text{ g O}}{6.35 \text{ g C}}; \quad x = \frac{(2.00 \text{ g O})(6.35 \text{ g C})}{1.50 \text{ g C}} = 8.47 \text{ g O}$$

1.68   (a)   (l) $\rightarrow$ (g)

       (b) $°F = 9/5 \, (°C) + 32 \,°$; $9/5 \, (12) + 32 = 53.6 = 54 \,°F$

       (c) $0.765 \text{ g/mL} \times 103.5 \text{ mL} = 79.178 = 79.2$ g ethyl chloride

1.69   (a)   I. $(22.52 + 22.48 + 22.54)/3 = 22.51$

             II. $(22.64 + 22.58 + 22.62)/3 = 22.61$

             Based on the average, set I is more accurate. That is, it is closer to the true value of 22.52%.

       (b)   Average deviation = Σ | value-average |/3

             I. | 22.52 – 22.51 | + | 22.48 – 22.51 | + | 22.54 – 22.51 |/3 = 0.02

             II. | 22.64 – 22.61 | + | 22.58 – 22.61 | + | 22.62 – 22.61 |/3 = 0.02

             Based on average deviations, the two sets display the same precision, even though set I is more accurate. [According to Section 1.5, standard deviation is a measure that is often used to determine precision. Using the formula for calculating standard deviation given in Appendix A.5, the values for the two sets are 0.03 and 0.03, respectively. The standard deviations of the two sets are also the same, confirming that the two sets are equally precise.]

1.70   (a)   Inappropriate. The circulation number indicates a precision of one part per million, 1 ppm. First, the term "circulation" is not specific, and can include paid subscriptions, news stand sales and other means of distribution. Even if we consider paid subscriptions only, it is unlikely that the number of paid subscriptions is know to a precision of 1 ppm.

(b)   Inappropriate. The population number implies a precision of 1 part per million, 1 ppm. Census data are taken only once every ten years and cannot directly measure every citizen even in the year the census is taken. Population fluctuates daily, so this precision is not reasonable.

(c)   Appropriate. The percentage has three significant figures. In a population as large as that of the United States, the number of people named Brown can surely be counted by census data or otherwise to a precision of three significant figures.

(d)   Inappropriate. Letter grades are posted at most to two decimal places and three significant figures (if plus and minus modifiers are quantified). The grade-point-average, obtained by addition and division, cannot have more decimal places or significant figures than the numbers being averaged.

1.71   (a)   volume   (b) area   (c) volume   (d) density

      (e)   time   (f) length   (g) temperature

1.72   (a) $\dfrac{m}{s^2}$ (b) $\dfrac{kg\text{-}m}{s^2}$ (c) $\dfrac{kg\text{-}m}{s^2} \times m = \dfrac{kg\text{-}m^2}{s^2}$

      (d) $\dfrac{kg\text{-}m}{s^2} \times \dfrac{1}{m^2} = \dfrac{kg}{m\text{-}s^2}$ (e) $\dfrac{kg\text{-}m^2}{s^2} \times \dfrac{1}{s} = \dfrac{kg\text{-}m^2}{s^3}$

      (f) $\dfrac{m}{s}$ (g) $kg \times \left(\dfrac{m}{s}\right)^2 = \dfrac{kg\text{-}m^2}{s^2}$

1.73   (a) $2.4 \times 10^5 \text{ mi} \times \dfrac{1.609 \text{ km}}{1 \text{ mi}} \times \dfrac{1000 \text{ m}}{1 \text{ km}} = 3.862 \times 10^8 = 3.9 \times 10^8 \text{ m}$

      (b) $2.4 \times 10^5 \text{ mi} \times \dfrac{1.609 \text{ km}}{1 \text{ mi}} \times \dfrac{1 \text{ hr}}{350 \text{ km}} \times \dfrac{60 \text{ min}}{1 \text{ hr}} \times \dfrac{60 \text{ s}}{1 \text{ min}} = 4.0 \times 10^6 \text{ s}$

      (c) $3.862 \times 10^8 \text{ m} \times 2 \times \dfrac{1 \text{ s}}{3.00 \times 10^8 \text{ m}} = 2.574 = 2.6 \text{ s}$

      (d) $\dfrac{29.783 \text{ km}}{s} \times \dfrac{1 \text{ mi}}{1.6093 \text{ km}} \times \dfrac{60 \text{ s}}{1 \text{ min}} \times \dfrac{60 \text{ min}}{1 \text{ hr}} = 6.6624 \times 10^4 \text{ mi/hr}$

1.74   (a)   Baking powder is not a pure substance. It is a mixture of basic and acidic compounds that, in the presence of water, react to form CO_2 which causes baked goods to rise.

      (b)   Lemon juice is not a pure substance. It is a mixture of water, citric acid, and other natural flavors. Its exact composition depends on the characteristics of the lemon that produces the juice.

      (c)   Propane is a nearly pure substance. Propane itself is odorless, but tank propane contains trace amounts of an odor-producing substance so that leaks can easily be detected.

      (d)   Aluminum foil is a nearly pure substance.

      (e)   Ibuprofen itself is a pure substance. Ibuprofen tablets probably contain the active ingredient and other binders, drying agents and coatings.

(f)   Bourbon whiskey is not a pure substance. It is a mixture of water, alcohol, and flavoring compounds.

(g)   Helium gas is a pure substance.

(h)   Clear water pumped from a deep aquifer is a nearly pure substance. All natural water (not distilled or deionized) contains trace minerals.

1.75   (a) $575 \text{ ft} \times \dfrac{12 \text{ in}}{1 \text{ ft}} \times \dfrac{2.54 \text{ cm}}{1 \text{ in}} \times \dfrac{10 \text{ mm}}{1 \text{ cm}} \times \dfrac{1 \text{ quarter}}{1.55 \text{ mm}} = 1.1307 \times 10^5 = 1.13 \times 10^5 \text{ quarters}$

        (b) $1.1307 \times 10^5 \text{ quarters} \times \dfrac{5.67 \text{ g}}{1 \text{ quarter}} = 6.41 \times 10^5 \text{ g (641 kg)}$

        (c) $1.1307 \times 10^5 \text{ quarters} \times \dfrac{1 \text{ dollar}}{4 \text{ quarters}} = \$28,268 = \$2.83 \times 10^4$

        (d) $\$16,213,166,914,811.11 \times \dfrac{1 \text{ stack}}{\$28,268} = 5.7355 \times 10^8 = 5.74 \times 10^8 \text{ stacks}$

1.76   (a) $\dfrac{\$1950}{\text{acre-ft}} \times \dfrac{1 \text{ acre}}{4840 \text{ yd}^2} \times \dfrac{3 \text{ ft}}{1 \text{ yd}} \times \dfrac{(1.094 \text{ yd})^3}{(1 \text{ m})^3} \times \dfrac{(1 \text{ m})^3}{(10 \text{ dm})^3} \times \dfrac{(1 \text{ dm})^3}{1 \text{ L}} =$

        $\$1.583 \times 10^{-3}/\text{L or } 0.1583 \text{ ¢/L} \quad (0.158 \text{ ¢/L to 3 sig figs})$

        (b) $\dfrac{\$1950}{\text{acre-ft}} \times \dfrac{1 \text{ acre-ft}}{2 \text{ households-year}} \times \dfrac{1 \text{ year}}{365 \text{ days}} \times 1 \text{ household} = \dfrac{\$2.671}{\text{day}} = \dfrac{\$2.67}{\text{day}}$

1.77   Select a common unit for comparison, in this case the kg.

        $1 \text{ kg} > 2 \text{ lb}, 1 \text{ L} \approx 1 \text{ qt}$

        $5 \text{ lb potatoes} < 2.5 \text{ kg}$

        $5 \text{ kg sugar} = 5 \text{ kg}$

        $1 \text{ gal} = 4 \text{ qt} \approx 4 \text{ L}; 1 \text{ mL H}_2\text{O} = 1 \text{ g H}_2\text{O}; 1 \text{ L} = 1000 \text{ g}, 4 \text{ L} = 4000 \text{ g} = 4 \text{ kg}$

        The order of mass from lightest to heaviest is 5 lb potatoes < 1 gal water < 5 kg sugar.

1.78   There are 347 degrees between the freezing and boiling points on the oleic acid (O) scale and 100 degrees on the celsius (C) scale. Also, 13 °C = 0 °O. By analogy with °F and °C,

        $^\circ\text{O} = \dfrac{100}{347}(^\circ\text{C} - 13) \quad \text{or} \quad ^\circ\text{C} = \dfrac{347}{100}(^\circ\text{O}) + 13$

        These equations correctly relate the freezing point (and boiling point) of oleic acid on the two scales.

        f.p. of H_2O: $^\circ\text{O} = \dfrac{100}{347}(0 \text{ }^\circ\text{C} - 13) = -3.746 = -4 \text{ }^\circ\text{O}$

1.79 The most dense liquid, Hg, will sink; the least dense, cyclohexane, will float; H_2O will be in the middle.

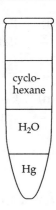

1.80 Density is the ratio of mass and volume. For samples with the same volume, in this case spheres with the same diameter, the denser ball will have a greater mass. The heavier ball, the red one on the right in the diagram is more dense.

1.81 The mass of water in the bottle does not change with temperature, but the density (ratio of mass to volume) does. That is, the amount of volume occupied by a certain mass of water changes with temperature. Calculate the mass of water in the bottle at 25 °C, and then the volume occupied by this mass at −10 °C.

(a) 25 °C: $1.50 \text{ L } H_2O \times \dfrac{1000 \text{ cm}^3}{1 \text{ L}} \times \dfrac{0.997 \text{ g } H_2O}{1 \text{ cm}^3} = 1.4955 \times 10^3 = 1.50 \times 10^3 \text{ g } H_2O$

−10 °C: $1.4955 \times 10^3 \text{ g } H_2O \times \dfrac{1 \text{ cm}^3}{0.917 \text{ g } H_2O} \times \dfrac{1 \text{ L}}{1000 \text{ cm}^3} = 1.6309 = 1.63 \text{ L}$

(b) No. If the soft-drink bottle is completely filled with 1.50 L of water, the 1.63 L of ice **cannot** be contained in the bottle. The extra volume of ice will push through any opening in the bottle, or crack the bottle to create an opening.

1.82 mass of toluene = 58.58 g – 32.65 g = 25.93 g

volume of toluene = $25.93 \text{ g} \times \dfrac{1 \text{ mL}}{0.864 \text{ g}} = 30.0116 = 30.0 \text{ mL}$

volume of solid = 50.00 mL − 30.0116 mL = 19.9884 = 20.0 mL

density of solid = $\dfrac{32.65 \text{ g}}{19.9884 \text{ mL}} = 1.63 \text{ g/mL}$

1.83 $V = 4/3 \, \pi \, r^3 = 4/3 \, \pi \, (28.9 \text{ cm})^3 = 1.0111 \times 10^5 = 1.01 \times 10^5 = 1.01 \times 10^5 \text{ cm}^3$

$1.0111 \times 10^5 \text{ cm}^3 \times \dfrac{19.3 \text{ g}}{\text{cm}^3} \times \dfrac{1 \text{ lb}}{453.59 \text{ g}} = 4302 = 4.30 \times 10^3 \text{ lb}$

No, the thief is unlikely to be able to carry the sphere without assistance. It weighs 4300 pounds, more than two tons.

1.84 $1.00 \text{ gal battery acid} \times \dfrac{4 \text{ qt}}{1 \text{ gal}} \times \dfrac{1000 \text{ mL}}{1.0567 \text{ qt}} \times \dfrac{1.28 \text{ g}}{\text{mL}} = 4845.3 = 4.85 \times 10^3 \text{ g battery acid}$

   $4.8453 \times 10^3 \text{ g battery acid} \times \dfrac{38.1 \text{ g sulfuric acid}}{100 \text{ g battery acid}} = 1846 = 1.85 \times 10^3 \text{ g sulfuric acid}$

1.85   (a) $\dfrac{40 \text{ lb peat}}{14 \times 20 \times 30 \text{ in}^3} \times \dfrac{1 \text{ in}^3}{(2.54)^3 \text{ cm}^3} \times \dfrac{453.6 \text{ g}}{1 \text{ lb}} = 0.13 \text{ g/cm}^3 \text{ peat}$

   $\dfrac{40 \text{ lb soil}}{1.9 \text{ gal}} \times \dfrac{1 \text{ gal}}{4 \text{ qt}} \times \dfrac{1.057 \text{ qt}}{1 \text{ L}} \times \dfrac{1 \times 10^{-3} \text{ L}}{1 \text{ mL}} \times \dfrac{1 \text{ mL}}{1 \text{ cm}^3} \times \dfrac{453.6 \text{ g}}{1 \text{ lb}} = 2.5 \text{ g/cm}^3 \text{ soil}$

   No. Volume must be specified to compare mass. The densities tell us that a certain volume of peat moss is "lighter" (weighs less) than the same volume of top soil.

   (b) $1 \text{ bag peat} = 14 \times 20 \times 30 = 8.4 \times 10^3 \text{ in}^3$

   $15.0 \text{ ft} \times 20.0 \text{ ft} \times 3.0 \text{ in} \times \dfrac{12^2 \text{ in}^2}{\text{ft}^2} = 129{,}600 = 1.3 \times 10^5 \text{ in}^3 \text{ peat needed}$

   $129{,}600 \text{ in}^3 \times \dfrac{1 \text{ bag}}{8.4 \times 10^3 \text{ in}^3} = 15.4 = 15 \text{ bags (Buy 16 bags of peat.)}$

1.86 $8.0 \text{ oz} \times \dfrac{1 \text{ lb}}{16 \text{ oz}} \times \dfrac{453.6 \text{ g}}{\text{lb}} \times \dfrac{1 \text{ cm}^3}{2.70 \text{ g}} = 84.00 = 84 \text{ cm}^3$

   $\dfrac{84 \text{ cm}^3}{50 \text{ ft}^2} \times \dfrac{1^2 \text{ ft}^2}{12^2 \text{ in}^2} \times \dfrac{1^2 \text{ in}^2}{2.54^2 \text{ cm}^2} \times \dfrac{10 \text{ mm}}{1 \text{ cm}} = 0.018 \text{ mm}$

1.87   The total solar flux, the power provided by the sun, is the average solar flux times the area of the disc in sunlight at any instant. However, because the earth rotates, each disc is in sunlight for only half of the day, so only half of this flux can be collected.

   $\dfrac{680 \text{ W}}{\text{m}^2} \times 1.28 \times 10^{14} \text{ m}^2 \times 1/2 = 4.352 \times 10^{16} = 4.4 \times 10^{16} \text{ W}$

   Collection is 10% efficient, so only 10% of this power is available, $4.4 \times 10^{15} \text{ W}$

   $15 \text{ TW} \times \dfrac{1 \times 10^{12} \text{ W}}{1 \text{ TW}} = 15 \times 10^{12} \text{ W needed}$

   $\dfrac{15 \times 10^{12} \text{ W needed}}{4.352 \times 10^{15} \text{ W available}} \times 100 = 0.3447 = 0.34\% \text{ Earth's surface needed}$

1.88   Warren's first hypothesis was that the bacterium he observed was involved in causing ulcers. This was a radical hypothesis in 1979, when it was commonly accepted that the acidic environment of the stomach would not support microorganisms. Strong evidence was needed to change the opinion and practice of the medical community.

To test his hypothesis, Warren first collected many more tissue samples, from patients with and without ulcers. He established that healthy tissue did not contain the bacterium, while many of the ulcerated samples did. From there, with Marshall and other collaborators, he established two different ways to diagnose the infection (determine the presence of the bacterium in patients with potential ulcers), and learned to culture (grow in the laboratory) the new bacterium so that its effects could be studied. Finally, in patients diagnosed and treated for bacterial ulcers, he demonstrated that the recurrence of ulcers was rare.

This information was obtained from the autobiography of J. Robin Warren, http://www.nobelprize.org/nobel_prizes/medicine/laureates/2005/warren-autobio.html

1.89 $45.23 \text{ g ethanol} \times \dfrac{1 \text{ cm}^3}{0.789 \text{ g ethanol}} = 57.3257 = 57.33 \text{ cm}^3$, volume of cylinder

$$V = \pi r^2 h; \, r = (V/\pi h)^{1/2} = \left[\frac{57.3257 \text{ cm}^3}{\pi \times 25.0 \text{ cm}} \right]^{1/2} = 0.854338 = 0.854 \text{ cm}$$

$$d = 2\,r = 1.71 \text{ cm}$$

1.90   (a)   Let x = mass of Au in jewelry

9.85 – x = mass of Ag in jewelry

The total volume of jewelry = volume of Au + volume of Ag

$$0.675 \text{ cm}^3 = x \text{ g} \times \frac{1 \text{ cm}^3}{19.3 \text{ g}} + (9.85 - x) \text{ g} \times \frac{1 \text{ cm}^3}{10.5 \text{ g}}$$

$$0.675 = \frac{x}{19.3} + \frac{9.85 - x}{10.5} \quad \text{(To solve, multiply both sides by (19.3)(10.5))}$$

$$0.675 \, (19.3)(10.5) = 10.5 \, x + (9.85 - x)(19.3)$$

$$136.79 = 10.5 \, x + 190.105 - 19.3 \, x$$

$$-53.315 = -8.8 \, x$$

x = 6.06 g Au; 9.85 g total – 6.06 g Au = 3.79 g Ag

$$\text{mass \% Au} = \frac{6.06 \text{ g Au}}{9.85 \, g \text{ jewelry}} \times 100 = 61.5\% \text{ Au}$$

(b) $24 \text{ carats} \times 0.615 = 15 \text{ carat gold}$

1.91   The separation with distinctly separated red and blue spots is more successful. The procedure that produced the purple blur did not separate the two dyes. To quantify the characteristics of the separation, calculate a reference value for each spot that is

$$\frac{\text{distance traveled by spot}}{\text{distance traveled by solvent}}$$

If the values for the two spots are fairly different, the separation is successful. (One could measure the distance between the spots, but this would depend on the length of paper used and be different for each experiment. The values suggested above are independent of the length of paper.)

1.92 (a) False. Air and water are not elements. Air is a homogeneous mixture of gases and pure water is a compound.

(b) False. Mixtures can contain any number of pure substances, either elements or compounds or both.

(c) True.

(d) True.

(e) False. When yellow stains in a kitchen sink are treated with bleach water, a chemical change occurs.

(f) True.

(g) False. The number 0.0033 has the same number of significant figures as 0.033.

(h) True. (In a conversion factor, the quantity in the numerator is equal to the quantity in the denominator, so the overall numerical value is one.)

(i) True.

1.93 The densities are:

carbon tetrachloride (methane, tetrachloro) – 1.5940 g/cm^3

hexane – 0.6603 g/cm^3

benzene – 0.87654 g/cm^3

methylene iodide (methane, diiodo) – 3.3254 g/cm^3

Only methylene iodide will separate the two granular solids. The undesirable solid (2.04 g/cm^3) is less dense than methylene iodide and will float; the desired material is more dense than methylene iodide and will sink. The other three liquids are less dense than both solids and will not produce separation.

1.94 (a) $10.0 \text{ mg} \times \dfrac{1 \times 10^{-3} \text{ g}}{1 \text{ mg}} \times \dfrac{1 \text{ cm}^3}{0.20 \text{ g}} = 0.050 \text{ cm}^3 = 0.050 \text{ mL volume}$

(b) $10.0 \text{ mg} \times \dfrac{1 \times 10^{-3} \text{ g}}{1 \text{ mg}} \times \dfrac{1242 \text{ m}^2}{1 \text{ g}} = 12.42 = 12.4 \text{ m}^2 \text{ surface area}$

(c) 7.748 mg Hg initial – 0.001 mg Hg remain = 7.747 mg Hg removed

$\dfrac{7.747 \text{ mg Hg removed}}{7.748 \text{ mg Hg initial}} \times 100 = 99.99\% \text{ Hg removed}$

(d) 10.0 mg "spongy" initial + 7.747 mg Hg removed = 17.747 = 17.7 mg after exposure

2 Atoms, Molecules, and Ions

Visualizing Concepts

2.1   (a)   Like charges repel and opposite charges attract, so the sign of the electrical charge on the particle is negative.

         (b)   The greater the magnitude of the charges, the greater the electrostatic repulsion or attraction. As the charge on the plates is increased, the bending will increase.

         (c)   As the mass of the particle increases and speed stays the same, linear momentum (mv) of the particle increases and bending decreases. (See **A Closer Look**: The Mass Spectrometer.)

2.2   (a) $\% \text{ abundance} = \dfrac{\# \text{ of mass number} \times \text{particles}}{\text{total number of particles}} \times 100$

             12 red ^{293}Nv particles

             8 blue ^{295}Nv particles

             20 total particles

             $\% \text{ abundance} \,^{293}\text{Nv} = \dfrac{12}{20} \times 100 = 60\%$

             $\% \text{ abundance} \,^{295}\text{Nv} = \dfrac{8}{20} \times 100 = 40\%$

         (b)   Atomic weight (AW) is the same as average atomic mass.

             Atomic weight (average atomic mass) = $\sum$ fractional abundance $\times$ mass of isotope

             AW of Nv = 0.60(293.15) + 0.40(295.15) = 293.95 amu

             (Because % abundance was calculated by counting exact numbers of particles, assume % abundance is an exact number. Then, the number of significant figures in the AW is determined by the number of sig figs in the masses of the isotopes.)

2.3   In general, metals occupy the left side of the chart, and nonmetals the right side.

         metals: red and green   *nonmetals*: blue and yellow

         alkaline earth metal: red   *noble gas*: yellow

2.4   Because the number of electrons (negatively charged particles) does not equal the number of protons (positively charged particles), the particle is an ion. The charge on the ion is 2–.

Atomic number = number of protons = 16. The element is S, sulfur.

Mass number = protons + neutrons = 32

$^{32}_{16}S^{2-}$

2.5   In a solid, particles are close together and their relative positions are fixed. In a liquid, particles are close but moving relative to each other. In a gas, particles are far apart and moving. Most ionic compounds are solids because of the strong forces among charged particles. Molecular compounds can exist in any state: solid, liquid, or gas.

Because the molecules in *ii* are far apart, *ii* must be a molecular compound. The particles in *i* are near each other and exist in a regular, ordered arrangement, so *i* is likely to be an ionic compound.

2.6   Formula: IF_5 Name: iodine pentafluoride

Because the compound is composed of elements that are all nonmetals, it is molecular.

2.7   See Figure 2.17. yellow box: 1+ (group 1A); blue box: 2+ (group 2A)

black box: 3+ (a metal in Group 3A); red box: 2– (a nonmetal in group 6A);

green box: 1– (a nonmetal in group 7A)

2.8   Cations (red spheres) have positive charges; anions (blue spheres) have negative charges. There are twice as many anions as cations, so the formula has the general form CA_2. Only $Ca(NO_3)_2$, calcium nitrate, is consistent with the diagram.

2.9   These two compounds are isomers. They have the same chemical formula, C_4H_9Cl, but different arrangements of atoms. That is, they have different chemical structures. In the first isomer, the Cl atom is bound to the second C atom from the left. In the second isomer, the Cl atom is bound to the right-most C atom.

2.10   (a)   In the absence of an electric field, there is no electrostatic interaction between the oil drops and the apparatus, so the rate of fall of the oil drops is determined solely by the force of gravity. In the presence of an electric field, there is electrostatic attraction between the negatively charged oil drops and the positively charged plate, as well as electrostatic repulsion between the negatively charged oil drops and the negative plate. These electrostatic forces oppose the force of gravity and change the rate of fall of the drops.

         (b)   Each individual drop has a different number of electrons associated with it. The greater the accumulated negative charge on the drop, the greater the electrostatic forces between the oil drop and the plates. If the combined electrostatic forces are greater than the force of gravity, the drop moves up.

The Atomic Theory of Matter and the Discovery of Atomic Structure
(Sections 2.1 and 2.2)

2.11   (a)   ratio of masses $= \dfrac{0.727 \text{ g O}}{0.273 \text{ g C}} = 2.663 = 2.66$

      (b)   ratio of masses $= \dfrac{0.571 \text{ g O}}{0.429 \text{ g C}} = 1.331 = 1.33$

      (c)   The two mass ratios are related by a factor of 2. In the first compound, CO_2, twice as much O is bound to one gram of C as in the second compound. The empirical formula of the second compound is then CO.

2.12   (a)   6.500 g compound – 0.384 g hydrogen = 6.116 g sulfur

      (b)   *Conservation of mass*

2.13   (a) $\dfrac{17.60 \text{ g oxygen}}{30.82 \text{ g nitrogen}} = \dfrac{0.5711 \text{ g O}}{1 \text{ g N}};\ 0.5711 / 0.5711 = 1.0$

              $\dfrac{35.20 \text{ g oxygen}}{30.82 \text{ g nitrogen}} = \dfrac{1.142 \text{ g O}}{1 \text{ g N}};\ 1.142 / 0.5711 = 2.0$

              $\dfrac{70.40 \text{ g oxygen}}{30.82 \text{ g nitrogen}} = \dfrac{2.284 \text{ g O}}{1 \text{ g N}};\ 2.284 / 0.5711 = 4.0$

              $\dfrac{88.00 \text{ g oxygen}}{30.82 \text{ g nitrogen}} = \dfrac{2.855 \text{ g O}}{1 \text{ g N}};\ 2.855 / 0.5711 = 5.0$

      (b)   These masses of oxygen per one gram nitrogen are in the ratio of 1:2:4:5 and thus obey the *law of multiple proportions*. Multiple proportions arise because atoms are the indivisible entities combining, as stated in Dalton's theory. Because atoms are indivisible, they must combine in ratios of small whole numbers.

2.14   (a)   1: $\dfrac{3.56 \text{ g fluorine}}{4.75 \text{ g iodine}} = 0.749$ g fluorine/1 g iodine

             2:   $\dfrac{3.43 \text{ g fluorine}}{7.64 \text{ g iodine}} = 0.449$ g fluorine/1 g iodine

             3:   $\dfrac{9.86 \text{ g fluorine}}{9.41 \text{ g iodine}} = 1.05$ g fluorine/1 g iodine

      (b)   To look for integer relationships among these values, divide each one by the smallest. If the quotients aren't all integers, multiply by a common factor to obtain all integers.

         1: $0.749 / 0.449 = 1.67;\ 1.67 \times 3 = 5$

         2: $0.449 / 0.449 = 1.00;\ 1.00 \times 3 = 3$

         3: $1.05 / 0.449 = 2.34;\ 2.34 \times 3 = 7$

         The ratio of g fluorine to g iodine in the three compounds is 5:3:7. These are in the ratio of small whole numbers and, therefore, obey the *law of multiple proportions*. This integer ratio indicates that the combining fluorine "units" (atoms) are indivisible entities.

2.15   Electrons, in the form of cathode rays, were discovered first. Neutrons were discovered last.

2.16   (a)   Beta rays have the mass and charge of an electron. If the unknown particle is a proton, it will be deflected in the opposite direction as a beta ray because protons and beta rays have opposite electrical charges.

           (b)   In an electric field, lighter particles are deflected by a greater amount than heavier ones. Protons have larger mass than beta rays (electrons) so they would be deflected by a smaller amount.

2.17   *Analyze.* We are given the diameters of a gold atom and its nucleus, and a gold foil that is two atoms thick. What fraction of alpha particles in Rutherford's experiment are deflected at large angles?

        Plan. In order to be deflected at a large angle, an alpha particle must directly strike a gold nucleus. Assume that the gold atoms in a single row touch. Consider the cross-sectional area of the gold foil exposed to the beam of alpha particles. Calculate the percentage of this area occupied by the nucleus. But, there are two rows of gold particles, offset relative to one another (Figure 2.9). Assume each alpha particle has two chances to hit a gold nucleus, so the fraction deflected at large angles is twice the ratio of areas. [This approach ignores empty space in the arrangement of gold atoms, which is about 9% of the total cross-sectional area.]

        Solve.

$$\text{fraction of alpha particles deflected at large angles} = \frac{\text{area of Au nucleus}}{\text{area of Au atom}} \times 2$$

        The cross-sectional area of a spherical atom is a circle. Area = πr^2

$$\text{fraction deflected at large angles} = \frac{\pi[r(\text{nucleus})]^2}{\pi[r(\text{atom})]^2} \times 2$$

$$\text{fraction deflected at large angles} = \frac{(1.0 \times 10^{-4}\ \text{Å})^2}{(2.7\ \text{Å})^2} \times 2 = 2.7 \times 10^{-9}$$

        That is, 1 out of approximately 365 million alpha particles is deflected at a large angle.

2.18   (a)   The droplets carry different total charges because there may be 1, 2, 3, or more electrons on the droplet.

           (b)   The electronic charge is likely to be the lowest common factor in all the observed charges.

           (c)   Assuming this is so, we calculate the apparent electronic charge from each drop as follows:

                A: $1.60 \times 10^{-19} / 1 = 1.60 \times 10^{-19}\ \text{C}$
                B: $3.15 \times 10^{-19} / 2 = 1.58 \times 10^{-19}\ \text{C}$
                C: $4.81 \times 10^{-19} / 3 = 1.60 \times 10^{-19}\ \text{C}$
                D: $6.31 \times 10^{-19} / 4 = 1.58 \times 10^{-19}\ \text{C}$

                The reported value is the average of these four values. Because each calculated charge has three significant figures, the average will also have three significant figures.

                $(1.60 \times 10^{-19}\ \text{C} + 1.58 \times 10^{-19}\ \text{C} + 1.60 \times 10^{-19}\ \text{C} + 1.58 \times 10^{-19}\ \text{C}) / 4 = 1.59 \times 10^{-19}\ \text{C}$

The Modern View of Atomic Structure; Atomic Weights (Sections 2.3 and 2.4)

2.19   (a) $1.35\,\text{Å} \times \dfrac{1 \times 10^{-10}\,\text{m}}{1\,\text{Å}} \times \dfrac{1\,\text{nm}}{1 \times 10^{-9}\,\text{m}} = 0.135\,\text{nm}$

              $1.35\,\text{Å} \times \dfrac{1 \times 10^{-10}\,\text{m}}{1\,\text{Å}} \times \dfrac{1\,\text{pm}}{1 \times 10^{-12}\,\text{m}} = 1.35 \times 10^2 \text{ or } 135\,\text{pm} \ (1\,\text{Å} = 100\,\text{pm})$

    (b)   Aligned Au atoms have **diameters** touching. $d = 2r = 2(1.35\,\text{Å}) = 2.70\,\text{Å}$

              $1.0\,\text{mm} \times \dfrac{1\,\text{m}}{1000\,\text{mm}} \times \dfrac{1\,\text{Å}}{1 \times 10^{-10}\,\text{m}} \times \dfrac{1\,\text{Au atom}}{2.70\,\text{Å}} = 3.70 \times 10^6 \text{ Au atoms}$

    (c) $V = 4/3\,\pi r^3. \ \ r = 1.35\,\text{Å} \times \dfrac{1 \times 10^{-10}\,\text{m}}{1\,\text{Å}} \times \dfrac{100\,\text{cm}}{\text{m}} = 1.35 \times 10^{-8}\,\text{cm}$

              $V = (4/3)(\pi)(1.35 \times 10^{-8})^3\,\text{cm}^3 = 1.03 \times 10^{-23}\,\text{cm}^3$

2.20   (a) $r = d/2; \ r = \dfrac{2.7 \times 10^{-8}\,\text{cm}}{2} \times \dfrac{1\,\text{Å}}{1 \times 10^{-8}\,\text{cm}} = 1.35 = 1.4\,\text{Å}$

              $r = \dfrac{2.7 \times 10^{-8}\,\text{cm}}{2} \times \dfrac{1\,\text{m}}{100\,\text{cm}} = 1.35 \times 10^{-10} = 1.4 \times 10^{-10}\,\text{m}$

    (b)   Aligned Rh atoms have **diameters** touching. $d = 2.7 \times 10^{-8}\,\text{cm} = 2.7 \times 10^{-10}\,\text{m}$

              $6.0\,\mu\text{m} \times \dfrac{1 \times 10^{-6}\,\text{m}}{1\,\mu\text{m}} \times \dfrac{1\,\text{Rh atom}}{2.7 \times 10^{-10}\,\text{m}} = 2.2 \times 10^4 \text{ Rh atoms}$

    (c) $V = 4/3\,\pi r^3; \ r = 1.35 \times 10^{-10} = 1.4 \times 10^{-10}\,\text{m}$

              $V = (4/3)[(\pi(1.35 \times 10^{-10})^3]\,\text{m}^3 = 1.031 \times 10^{-29} = 1.0 \times 10^{-29}\,\text{m}^3$

2.21   (a)   proton, neutron, electron

    (b)   proton = +1, neutron = 0, electron = –1

    (c)   The neutron is most massive. (The neutron and proton have very similar masses.)

    (d)   The electron is least massive.

2.22   (a)   False. The nucleus has most of the mass **but occupies very little** of the volume of an atom.

    (b)   True.

    (c)   False. The number of electrons in a neutral atom is equal to the number of **protons** in the atom.

    (d)   True.

2.23   (a)   5 protons, 5 neutrons, 5 electrons. Every neutral ^{10}B atom has 5 protons and 5 neutrons. The mass number of this B atom is 10, so it has $(10 - 5) = 5$ neutrons.

    (b) $^{11}_{6}\text{C}$. Adding a proton increases the atomic number of the atom to 6 and the mass number to 11. The element with atomic number 6 is carbon.

(c) $^{11}_{5}$B. Adding a neutron increases the mass number by 1, but does not change the identity of the atom.

(d)   The atom in part (c) is an isotope of ^{10}B. Isotopes are atoms of the same element with different masses.

2.24   (a)   29 protons, 34 neutrons, 29 electrons. Every neutral Cu atom has 29 protons and 29 electrons. The mass number of this Cu atom is 63, so it has (63 − 29) = 34 neutrons.

       (b) $^{63}_{29}$Cu^{2+}. Removing electrons produces a positively charged ion; it does not change the atomic number or mass number of the atom.

       (c) $^{65}_{29}$Cu. A copper atom with 36 neutrons has a mass number of (29 + 36) = 65.

2.25   (a)   *Atomic number* is the number of protons in the nucleus of an atom. *Mass number* is the total number of nuclear particles, protons plus neutrons, in an atom.

       (b)   The mass number can vary without changing the identity of the atom, but the atomic number of every atom of a given element is the same.

2.26   (a) $^{31}_{16}$X and $^{32}_{16}$X are isotopes of the same element, because they have identical atomic numbers.

       (b)   These are isotopes of the element sulfur, S, atomic number = 16.

2.27   p = protons, n = neutrons, e = electrons

       (a) ^{40}Ar has 18 p, 22 n, 18 e   (b) ^{65}Zn has 30 p, 35 n, 30 e

       (c) ^{70}Ga has 31 p, 39 n, 31 e   (d) ^{80}Br has 35 p, 45 n, 35 e

       (e) ^{184}W has 74 p, 110 n, 74 e   (f) ^{243}Am has 95 p, 148 n, 95 e

2.28   (a) ^{32}P has 15 p, 17 n   (b) ^{51}Cr has 24 p, 27 n

       (c) ^{60}Co has 27 p, 33 n   (d) ^{99}Tc has 43 p, 56 n

       (e) ^{131}I has 53 p, 78 n   (f) ^{201}Tl has 81 p, 120 n

2.29

Symbol	^{79}Br	^{55}Mn	^{112}Cd	^{222}Rn	^{207}Pb
Protons	35	25	48	86	82
Neutrons	44	30	64	136	125
Electrons	35	25	48	86	82
Mass no.	79	55	112	222	207

2.30

Symbol	^{112}Cd	^{96}Sr	^{87}Sr	^{81}Kr	^{235}U
Protons	48	38	38	36	92
Neutrons	64	58	49	45	143
Electrons	48	38	38	36	92
Mass No.	112	96	87	81	235

2.31   (a) $^{196}_{78}$Pt   (b) $^{84}_{36}$Kr   (c) $^{75}_{33}$As   (d) $^{24}_{12}$Mg

2.32   Because the two nuclides are atoms of the same element, by definition they have the same number of protons, 54. They differ in mass number (and mass) because they have different numbers of neutrons. ^{129}Xe has 75 neutrons and ^{130}Xe has 76 neutrons.

2.33   (a) $^{12}_{6}$C

  (b)   Atomic weights are really average atomic masses, the sum of the mass of each naturally occurring isotope of an element times its fractional abundance. Each B atom will have the mass of one of the naturally occurring isotopes, whereas the "atomic weight" is an average value. The naturally occurring isotopes of B, their atomic masses, and relative abundances are:

   ^{10}B, 10.012937, 19.9%; ^{11}B, 11.009305, 80.1%.

2.34   (a)   12 amu

  (b)   The atomic weight of carbon reported on the front-inside cover of the text is the abundance-weighted average of the atomic masses of the two naturally occurring isotopes of carbon, ^{12}C and ^{13}C. The mass of a ^{12}C atom is exactly 12 amu, but the atomic weight of 12.011 takes into account the presence of some ^{13}C atoms in every natural sample of the element.

2.35   Atomic weight (average atomic mass) = Σ fractional abundance × mass of isotope

   Atomic weight = 0.6917(62.9296) + 0.3083(64.9278) = 63.5456 = 63.55 amu

2.36   Atomic weight (average atomic mass) = Σ fractional abundance × mass of isotope

   Atomic weight = 0.7215(84.9118) + 0.2785(86.9092) = 85.4681 = 85.47 amu

   (The result has 2 decimal places and 4 sig figs because each term in the sum has 4 sig figs and 2 decimal places.)

2.37   (a)   In Thomson's cathode ray tube, the charged particles are electrons. In a mass spectrometer, the charged particles are positively charged ions (cations).

  (b)   The x-axis label (independent variable) is atomic mass (or particle mass) and the y-axis label (dependent variable) is signal intensity.

  (c)   The Cl^{2+} ion will be deflected more. The greater the charge on the positive ion, the larger its interaction with the electric and magnetic fields. (For this reason, the x-axis label of a mass spectrum is usually mass-to-charge ratio of the particles.)

2.38   (a)   True.

(b)   False. The height of each peak in the mass spectrum is directly proportional to the relative abundance of the isotope.

(c)   True.

2.39   (a)   Average atomic mass = 0.7899(23.98504) + 0.1000(24.98584) + 0.1101(25.98259)

= 24.31 amu

(b)

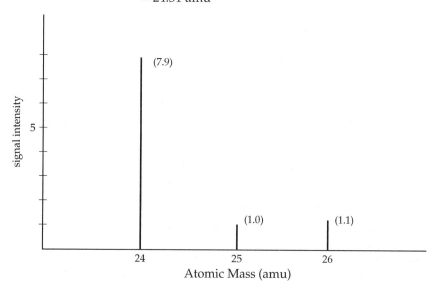

The relative intensities of the peaks in the mass spectrum are the same as the relative abundances of the isotopes. The abundances and peak heights are in the ratio ^{24}Mg: ^{25}Mg: ^{26}Mg as 7.8 : 1.0 : 1.1.

2.40   (a)   Three peaks: ^{1}H – ^{1}H, ^{1}H – ^{2}H, ^{2}H – ^{2}H

(b) ^{1}H – ^{1}H = 2(1.00783) = 2.01566 amu

^{1}H – ^{2}H = 1.00783 + 2.01410 = 3.02193 amu

^{2}H – ^{2}H = 2(2.01410) = 4.02820 amu

The mass ratios are 1 : 1.49923 : 1.99845 or 1 : 1.5 : 2.

(c) ^{1}H – ^{1}H is largest, because there is the greatest chance that two atoms of the more abundant isotope will combine.

^{2}H – ^{2}H is the smallest, because there is the least chance that two atoms of the less abundant isotope will combine.

The Periodic Table, Molecules and Molecular Compounds, and Ions and Ionic Compounds (Sections 2.5, 2.6, and 2.7)

2.41   (a)   Cr, 24 (metal)   (b) He, 2 (nonmetal)   (c) P, 15 (nonmetal)

(d)   Zn, 30 (metal)   (e) Mg, 12 (metal)   (f) Br, 35 (nonmetal)

(g)   As, 33 (metalloid)

2.42   (a)   lithium, 3 (metal)   (b)   scandium, 21 (metal)

        (c)   germanium, 32 (metalloid)   (d)   ytterbium, 70 (metal)

        (e)   manganese, 25 (metal)   (f)   antimony, 51 (metalloid)

        (g)   xenon, 54 (nonmetal)

2.43   (a)   K, alkali metals (metal)   (b)   I, halogens (nonmetal)

        (c)   Mg, alkaline earth metals (metal)   (d)   Ar, noble gases (nonmetal)

        (e)   S, chalcogens (nonmetal)

2.44   C, carbon, nonmetal; Si, silicon, metalloid; Ge, germanium, metalloid; Sn, tin, metal; Pb, lead, metal

2.45   (a) C_4H_{10} is the molecular formula for both compounds. For the molecular formula, count the total number of each kind of atom in the structural formula.

        (b) C_2H_5. Starting with the molecular formula, divide subscripts by any common factors to determine the simplest ratio of atom types in the molecule. In this example the common factor for both molecules is 2.

        (c)   Structural. In this example, the molecules are *structural isomers* and only the structural formulas allow us to determine that the molecules are different.

2.46   (a)   Benzene, C_6H_6; acetylene, C_2H_2. For the molecular formula, count the total number of each kind of atom in the ball and stick representations.

        (b)   Benzene, CH; acetylene, CH. Starting with the molecular formula, divide subscripts by any common factors to determine the simplest ratio of atom types in the molecule. In this example the common factor for benzene is 6 and for acetylene is 2.

2.47   From left to right, the molecular and empirical formulas are: N_2H_4, NH_2; N_2H_2, NH; NH_3, NH_3

2.48   No. Two substances with the same molecular and empirical formulas can be isomers. They are not necessarily the same compound.

2.49   (a) $AlBr_3$ (b) C_4H_5 (c) C_2H_4O

        (d) P_2O_5 (e) C_3H_2Cl (f) BNH_2

2.50   A molecular formula contains all atoms in a molecule. An empirical formula shows the simplest ratio of atoms in a molecule or elements in a compound.

        (a)   molecular formula: C_6H_6; empirical formula: CH

        (b)   molecular formula: $SiCl_4$; empirical formula: $SiCl_4$ (1:4 is the simplest ratio)

        (c)   molecular: B_2H_6; empirical: BH_3

        (d)   molecular: $C_6H_{12}O_6$; empirical: CH_2O

2.51   (a)   6   (b) 10   (c) 12

2.52   (a)   6   (b) 6   (c) 9

2.53   (a) C_2H_6O H—C—O—C—H   (b) C_2H_6O H—C—C—O—H

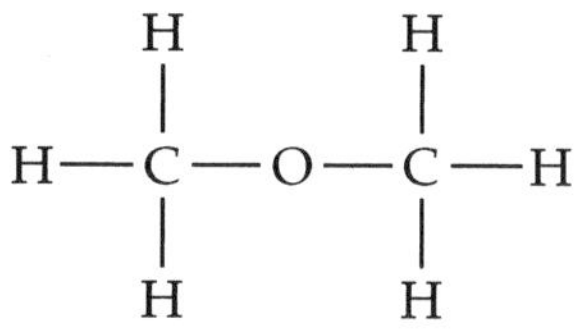

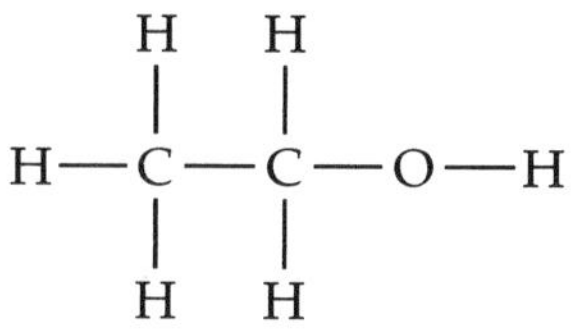

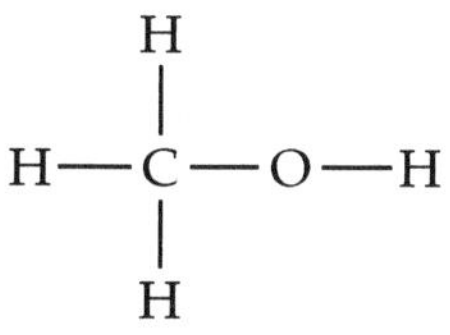

     (c) CH_4O H—C—O—H   (d) PF_3 F—P—F

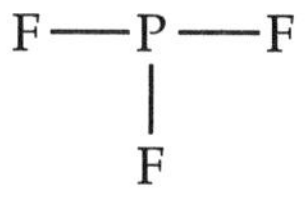

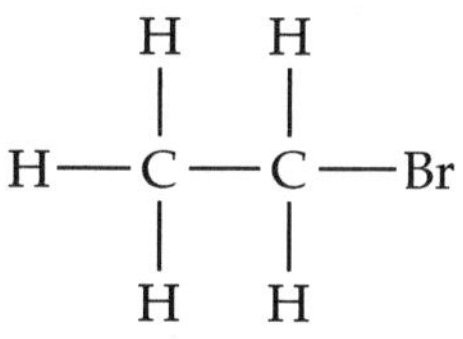

2.54   (a) C_2H_5Br H—C—C—Br   (b) C_2H_7N H—C—N—C—H

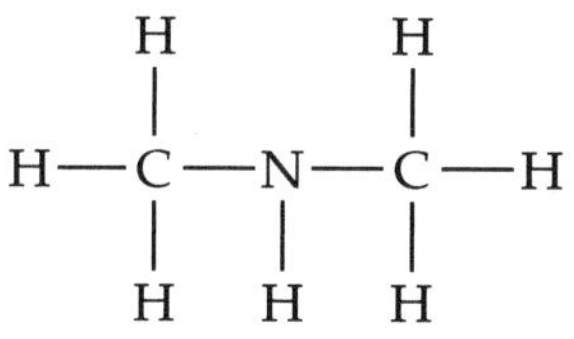

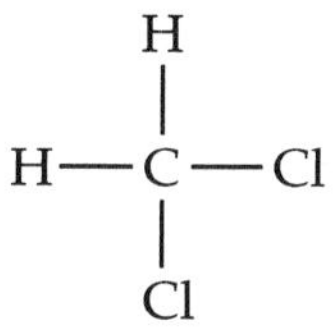

     (c) CH_2Cl_2 H—C—Cl   (d) NH_2OH H—N—O

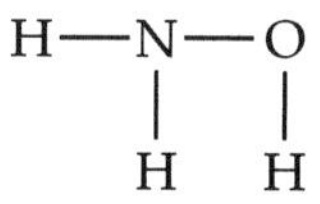

2.55

Symbol	$^{59}Co^{3+}$	$^{80}Se^{2-}$	$^{192}Os^{2+}$	$^{200}Hg^{2+}$
Protons	27	34	76	80
Neutrons	32	46	116	120
Electrons	24	36	74	78
Net Charge	3+	2−	2+	2+

2.56

Symbol	$^{31}P^{3-}$	$^{79}Se^{2-}$	$^{119}Sn^{4+}$	$^{197}Au^{3+}$
Protons	15	34	50	79
Neutrons	16	45	69	118
Electrons	18	36	46	76
Net Charge	3−	2−	4+	3+

2.57   (a) Mg^{2+} (b) Al^{3+} (c) K^+ (d) S^{2-} (e) F^-

2.58   (a) Ga^{3+} (b) Sr^{2+} (c) As^{3-} (d) Br^- (e) Se^{2-}

2.59   (a) GaF_3, gallium(III) fluoride   (b) LiH, lithium hydride

     (c) AlI_3, aluminum iodide   (d) K_2S, potassium sulfide

2.60   (a) ScI_3 (b) Sc_2S_3 (c) ScN

2.61   (a) $CaBr_2$ (b) K_2CO_3 (c) $Al(CH_3COO)_3$ (d) $(NH_4)_2SO_4$ (e) $Mg_3(PO_4)_2$

2.62   (a) $CrBr_3$ (b) Fe_2O_3 (c) Hg_2CO_3 (d) $Ca(ClO_3)_2$ (e) $(NH_4)_3PO_4$

2.63

Ion	K^+	NH_4^+	Mg^{2+}	Fe^{3+}
Cl^-	KCl	NH_4Cl	$MgCl_2$	$FeCl_3$
OH^-	KOH	NH_4OH^*	$Mg(OH)_2$	$Fe(OH)_3$
CO_3^{2-}	K_2CO_3	$(NH_4)_2CO_3$	$MgCO_3$	$Fe_2(CO_3)_3$
PO_4^{3-}	K_3PO_4	$(NH_4)_3PO_4$	$Mg_3(PO_4)_2$	$FePO_4$

*Equivalent to $NH_3(aq)$.

2.64

Ion	Na^+	Ca^{2+}	Fe^{2+}	Al^{3+}
O^{2-}	Na_2O	CaO	FeO	Al_2O_3
NO_3^-	$NaNO_3$	$Ca(NO_3)_2$	$Fe(NO_3)_2$	$Al(NO_3)_3$
SO_4^{2-}	Na_2SO_4	$CaSO_4$	$FeSO_4$	$Al_2(SO_4)_3$
AsO_4^{3-}	Na_3AsO_4	$Ca_3(AsO_4)_2$	$Fe_3(AsO_4)_2$	$AlAsO_4$

2.65   Molecular (all elements are nonmetals):

     (a) B_2H_6 (b) CH_3OH (f) $NOCl$ (g) NF_3

     Ionic (formed by a cation and an anion, usually contains a metal cation):

     (c) $LiNO_3$ (d) Sc_2O_3 (e) $CsBr$ (h) Ag_2SO_4

2.66   Molecular (all elements are nonmetals):

     (a) PF_5 (c) SCl_2 (h) N_2O_4

     Ionic (formed from ions, usually contains a metal cation):

     (b) NaI (d) $Ca(NO_3)_2$ (e) $FeCl_3$ (f) LaP (g) $CoCO_3$

Naming Inorganic Compounds; Some Simple Organic Compounds (Sections 2.8 and 2.9)

2.67   (a) ClO_2^- (b) Cl^- (c) ClO_3^- (d) ClO_4^- (e) ClO^-

2.68   (a)   selenate   (b) selenide   (c) hydrogen selenide (biselenide)

     (d)   hydrogen selenite (biselenite)

2.69   (a)   calcium, 2+; oxide, 2–   (b) sodium, 1+; sulfate, 2–

     (c)   potassium, 1+; perchlorate, 1–   (d) iron, 2+; nitrate, 1–

     (e)   chromium, 3+; hydroxide, 1–

2.70   (a)   copper, 2+; sulfide, 2–   (b)   silver, 1+; sulfate, 2–

         (c)   aluminum, 3+; chlorate, 1–   (d)   cobalt, 2+; hydroxide, 1–

         (e)   lead, 2+; carbonate, 2–

2.71   (a)   lithium oxide   (b)   iron(III) chloride (ferric chloride)

         (c)   sodium hypochlorite   (d)   calcium sulfite

         (e)   copper(II) hydroxide (cupric hydroxide)   (f)   iron(II) nitrate (ferrous nitrate)

         (g)   calcium acetate   (h)   chromium(III) carbonate (chromic carbonate)

         (i)   potassium chromate   (j)   ammonium sulfate

2.72   (a)   potassium cyanide   (b)   sodium bromite

         (c)   strontium hydroxide   (d)   cobalt(II) telluride (cobaltous telluride)

         (e)   iron(III) carbonate (ferric carbonate)

         (f)   chromium(III) nitrate (chromic nitrate)

         (g)   ammonium sulfite   (h)   sodium dihydrogen phosphate

         (i)   potassium permanganate   (j)   silver dichromate

2.73   (a) $Al(OH)_3$ (b) K_2SO_4 (c) Cu_2O (d) $Zn(NO_3)_2$

         (e) $HgBr_2$ (f) $Fe_2(CO_3)_3$ (g) $NaBrO$

2.74   (a) Na_3PO_4 (b) $Zn(NO_3)_2$ (c) $Ba(BrO_3)_2$ (d) $Fe(ClO_4)_2$

         (e) $Co(HCO_3)_2$ (f) $Cr(CH_3COO)_3$ (g) $K_2Cr_2O_7$

2.75   (a)   bromic acid   (b)   hydrobromic acid   (c)   phosphoric acid

         (d) $HClO$ (e) HIO_3 (f) H_2SO_3

2.76   (a) HI (b) $HClO_3$ (c) HNO_2

         (d)   carbonic acid   (e)   perchloric acid   (f)   acetic acid

2.77   (a)   sulfur hexafluoride   (b)   iodine pentafluoride   (c)   xenon trioxide

         (d) N_2O_4 (e) HCN (f) P_4S_6

2.78   (a)   dinitrogen monoxide   (b)   nitrogen monoxide   (c)   nitrogen dioxide

         (d)   dinitrogen pentoxide   (e)   dinitrogen tetroxide

2.79   (a) $ZnCO_3, ZnO, CO_2$ (b) HF, SiO_2, SiF_4, H_2O (c) SO_2, H_2O, H_2SO_3

         (d) PH_3 (e) $HClO_4, Cd, Cd(ClO_4)_2$ (f) VBr_3

2.80   (a) $NaHCO_3$ (b) $Ca(ClO)_2$ (c) HCN

         (d) $Mg(OH)_2$ (e) SnF_2 (f) CdS, H_2SO_4, H_2S

2.81   (a)   A hydrocarbon is a compound composed of the elements hydrogen and carbon only.

        (b)

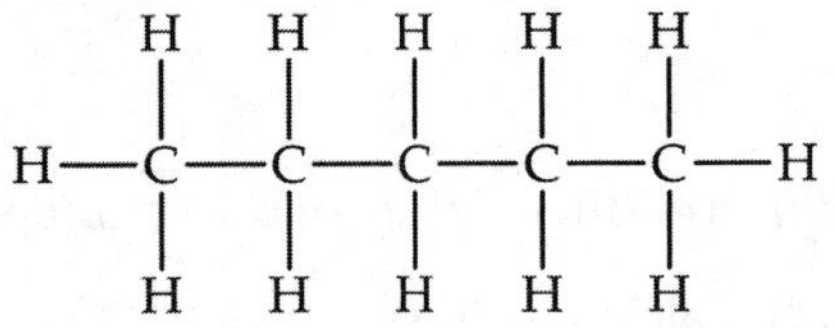

molecular and empirical formulas: C_5H_{12}

2.82   (a)   Isomers are molecules with the same molecular formula, but different structural formulas. Isomers have the same number and kinds of atoms, but these atoms are arranged in different ways.

        (b)   Butane and pentane are both capable of existing in isomeric forms. There is more than one way to arrange the four C atoms and ten H atoms of butane, and more than one way to arrange the five C atoms and twelve H atoms of pentane. There is only one way to arrange the two C atoms and six H atoms of ethane and only one way to arrange the three C atoms and eight H atoms of propane.

2.83   (a)   A *functional group* is a group of specific atoms that are constant (arranged the same way) from one molecule to the next.

        (b)   The characteristic alcohol functional group is an –OH. Another way to say this is that whenever a molecule is called an alcohol, it contains the –OH group.

        (c)

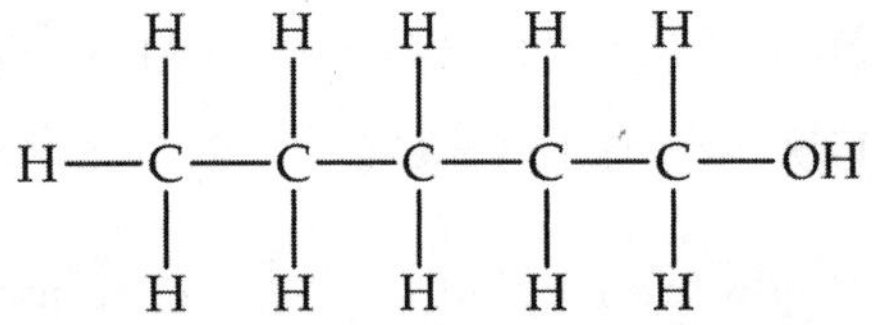

2.84   (a)   The *suffix -ol* indicates that an organic compound contains an –OH group. Organic compounds that contain the –OH functional group are called alcohols. Ethanol and propanol contain –OH groups (and are alcohols.)

        (b)   The *root prop-* indicates that a molecule contains three carbon atoms. Propane and propanol contain three carbon atoms.

2.85   (a)

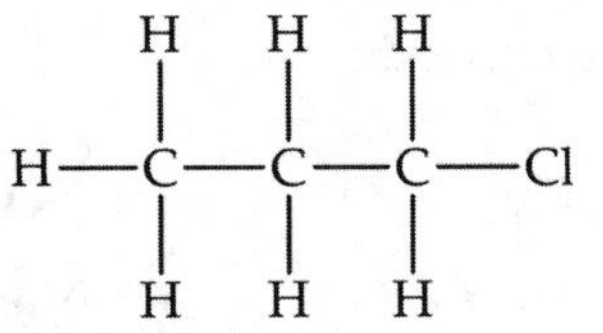

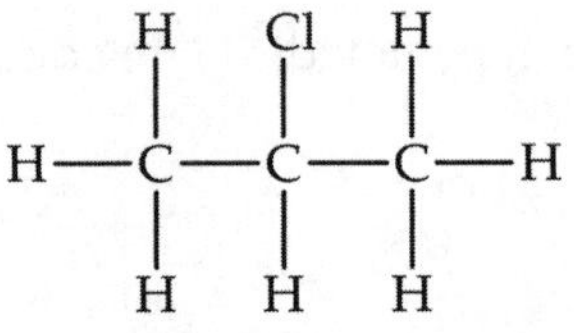

        (b)   1-chloropropane   2-chloropropane

2.86

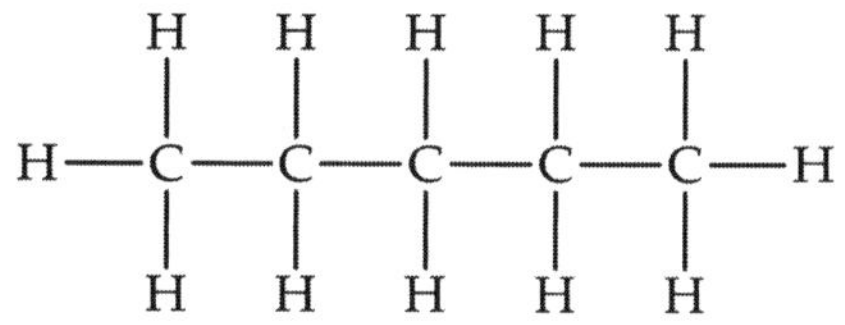

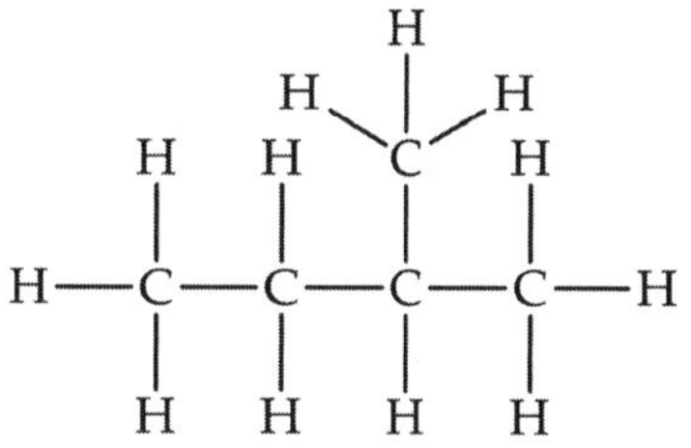

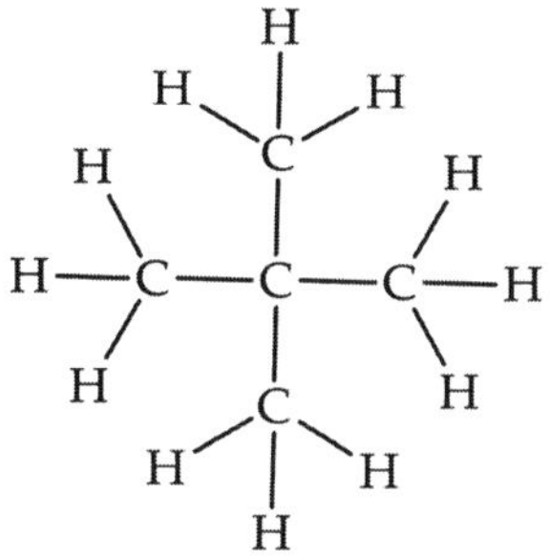

Additional Exercises

2.87 (a) Droplet D would fall most slowly. It carries the most negative charge, so it would be most strongly attracted to the upper (+) plate and most strongly repelled by the lower (–) plate. These electrostatic forces would provide the greatest opposition to gravity.

(b) Calculate the lowest common factor.

A: $3.84 \times 10^{-8} / 2.88 \times 10^{-8} = 1.33; 1.33 \times 3 = 4$

B: $4.80 \times 10^{-8} / 2.88 \times 10^{-8} = 1.67; 1.67 \times 3 = 5$

C: $2.88 \times 10^{-8} / 2.88 \times 10^{-8} = 1.00; 1.00 \times 3 = 3$

D: $8.64 \times 10^{-8} / 2.88 \times 10^{-8} = 3.00; 3.00 \times 3 = 9$

The total charge on the drops is in the ratio of 4:5:3:9. Divide the total charge on each drop by the appropriate integer and average the four values to get the charge of an electron in warmombs.

A: $3.84 \times 10^{-8} / 4 = 9.60 \times 10^{-9}$ wa

B: $4.80 \times 10^{-8} / 5 = 9.60 \times 10^{-9}$ wa

C: $2.88 \times 10^{-8} / 3 = 9.60 \times 10^{-9}$ wa

D: $8.64 \times 10^{-8} / 9 = 9.60 \times 10^{-9}$ wa

The charge on an electron is 9.60×10^{-9} wa

(c) The number of electrons on each drop are the integers calculated in part (b). A has $4 \, e^-$, B has $5 \, e^-$, C has $3 \, e^-$ and D has $9 \, e^-$.

(d) $$\frac{9.60 \times 10^{-9} \text{ wa}}{1 \, e^-} \times \frac{1 \, e^-}{1.60 \times 10^{-16} \text{ C}} = 6.00 \times 10^7 \text{ wa/C}$$

2.88   (a) ^{3}He has 2 protons, 1 neutron, and 2 electrons.

        (b) ^{3}H has 1 proton, 2 neutrons, and 1 electron.

^{3}He: $2(1.672621673 \times 10^{-24}\ g) + 1.674927211 \times 10^{-24}\ g + 2(9.10938215 \times 10^{-28}\ g)$

$$= 5.021992 \times 10^{-24}\ g$$

^{3}H: $1.672621673 \times 10^{-24}\ g + 2(1.674927211 \times 10^{-24}\ g) + 9.10938215 \times 10^{-28}\ g$

$$= 5.023387 \times 10^{-24}\ g$$

Tritium, ^{3}H, is more massive.

        (c)   The masses of the two particles differ by $0.0014 \times 10^{-24}\ g$. Each particle loses 1 electron to form the +1 ion, so the difference in the masses of the ions is still 1.4×10^{-27}. A mass spectrometer would need precision to $1 \times 10^{-27}\ g$ to differentiate ^{3}He$^+$ and ^{3}H.

2.89   (a)   Calculate the mass of a single gold atom, then divide the mass of the cube by the mass of the gold atom.

$$\frac{197.0\ amu}{gold\ atom} \times \frac{1\ g}{6.022 \times 10^{23}\ amu} = 3.2713 \times 10^{-22} = 3.271 \times 10^{-22}\ g/gold\ atom$$

$$\frac{19.3\ g}{cube} \times \frac{1\ gold\ atom}{3.271 \times 10^{-22}\ g} = 5.90 \times 10^{22}\ Au\ atoms\ in\ the\ cube$$

        (b)   The shape of atoms is spherical; spheres cannot be arranged into a cube so that there is no empty space. The question is, how much empty space is there? We can calculate the two limiting cases, no empty space and maximum empty space. The true diameter will be somewhere in this range.

No empty space: volume cube/number of atoms = volume of one atom

$V = 4/3\pi\ r^3$; $r = (3\ V/4\ \pi)^{1/3}$; $d = 2r$

$$volume\ of\ cube = (1.0 \times 1.0 \times 1.0) = \frac{1.0\ cm^3}{5.90 \times 10^{22}\ Au\ atoms} = 1.695 \times 10^{-23}$$

$$= 1.7 \times 10^{-23}\ cm^3$$

$r = [3\ (1.695 \times 10^{-23}\ cm^3)/4\ \pi]^{1/3} = 1.6 \times 10^{-8}\ cm$; $d = 2r = 3.2 \times 10^{-8}\ cm$

Maximum empty space: Assume atoms are arranged in rows in all three directions so they are touching across their diameters. That is, each atom occupies the volume of a cube, with the atomic diameter as the length of the side of the cube. The number of atoms along one edge of the gold cube is then

$(5.90 \times 10^{22})^{1/3} = 3.893 \times 10^7 = 3.89 \times 10^7$ atoms/1.0 cm.

The diameter of a single atom is $1.0\ cm/3.89 \times 10^7$ atoms $= 2.569 \times 10^{-8}$

$$= 2.6 \times 10^{-8}\ cm.$$

The diameter of a gold atom is between $2.6 \times 10^{-8}\ cm$ and $3.2 \times 10^{-8}\ cm$ (2.6 – 3.2 Å).

(c)   Some atomic arrangement must be assumed, because none is specified. The solid state is characterized by an orderly arrangement of particles, so it isn't surprising that atomic arrangement is required to calculate the density of a solid. A more detailed discussion of solid-state structure and density appears in Chapter 11.

2.90   (a)   In arrangement A, the number of atoms in 1 cm^2 is just the square of the number that fit linearly in 1 cm.

$$1.0 \text{ cm} \times \frac{1 \text{ atom}}{4.95 \text{ Å}} \times \frac{1 \times 10^{10} \text{ Å}}{1 \text{ m}} \times \frac{1 \text{ m}}{100 \text{ cm}} = 2.02 \times 10^7 = 2.0 \times 10^7 \text{ atoms/cm}$$

$$1.0 \text{ cm}^2 = (2.02 \times 10^7)^2 = 4.081 \times 10^{14} = 4.1 \times 10^{14} \text{ atoms/cm}^2$$

(b)   In arrangement B, the atoms in the horizontal rows are touching along their diameters, as in arrangement A. The number of Rb atoms in a 1.0 cm row is then 2.0×10^7 Rb atoms. Relative to arrangement A, the vertical rows are offset by 1/2 of an atom. Atoms in a "column" are no longer touching along their vertical diameter. We must calculate the vertical distance occupied by a row of atoms, which is now less than the diameter of one Rb atom.

Consider the triangle shown below. This is an isosceles triangle (equal side lengths, equal interior angles) with a side-length of 2d and an angle of 60°. Drop a bisector to the uppermost angle so that it bisects the opposite side.

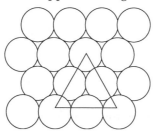 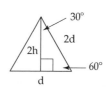

The result is a right triangle with two known side lengths. The length of the unknown side (the angle bisector) is 2h, two times the vertical distance occupied by a row of atoms. Solve for h, the "height" of one row of atoms.

$(2h)^2 + d^2 = (2d)^2; 4h^2 = 4d^2 - d^2 = 3d^2; h^2 = 3d^2/4$

$h = (3d^2/4)^{1/2} = (3(4.95 \text{ Å})^2/4)^{1/2} = 4.2868 = 4.29 \text{ Å}$

The number of rows of atoms in 1 cm is then

$$1.0 \text{ cm} \times \frac{1 \text{ row}}{4.2868 \text{ Å}} \times \frac{1 \times 10^{10} \text{ Å}}{1 \text{ m}} \times \frac{1 \text{ m}}{100 \text{ cm}} = 2.333 \times 10^7 = 2.3 \times 10^7$$

The number of atoms in a 1.0 cm^2 square area is then

$$\frac{2.020 \times 10^7 \text{ atoms}}{1 \text{ row}} \times 2.333 \times 10^7 \text{ rows} = 4.713 \times 10^{14} = 4.7 \times 10^{14} \text{ atoms}$$

Note that we have ignored the loss of "1/2" atom at the end of each horizontal row. Out of 2.0×10^7 atoms per row, one atom is not significant.

(c)   The ratio of atoms in arrangement B to arrangement A is then 4.713 × 10^{14} atoms/4.081 × 10^{14} = 1.155 = 1.2:1. Clearly, arrangement B results in less empty space per unit area or volume. If extended to three dimensions, arrangement B would lead to a greater density for Rb metal.

2.91   (a)   diameter of nucleus $= 1 \times 10^{-4}$ Å; diameter of atom $= 1$ Å

$V = 4/3 \, \pi \, r^3 ; r = d/2; r_n = 0.5 \times 10^{-4}$ Å; $r_a = 0.5$ Å

volume of nucleus $= 4/3 \, \pi \, (0.5 \times 10^{-4})^3$ Å^3

volume of atom $= 4/3 \, \pi \, (0.5)^3$ Å^3

volume fraction of nucleus $= \dfrac{\text{volume of nucleus}}{\text{volume of atom}} = \dfrac{4/3 \pi (0.5 \times 10^{-4})^3 \, \text{Å}^3}{4/3 \pi (0.5)^3 \, \text{Å}^3} = 1 \times 10^{-12}$

diameter of atom $= 5$ Å, $r_a = 2.5$ Å

volume fraction of nucleus $= \dfrac{4/3 \pi (0.5 \times 10^{-4})^3 \, \text{Å}^3}{4/3 \pi (2.5)^3 \, \text{Å}^3} = 8 \times 10^{-15}$

Depending on the radius of the atom, the volume fraction of the nucleus is between 1×10^{-12} and 8×10^{-15}, that is, between 1 part in 10^{12} and 8 parts in 10^{15}.

    (b)   mass of proton $= 1.0073$ amu

1.0073 amu $\times 1.66054 \times 10^{-24}$ g/amu $= 1.6727 \times 10^{-24}$ g

diameter $= 1.0 \times 10^{-15}$ m, radius $= 0.50 \times 10^{-15}$ m $\times \dfrac{100 \text{ cm}}{1 \text{ m}} = 5.0 \times 10^{-14}$ cm

Assuming a proton is a sphere, $V = 4/3 \, \pi \, r^3$.

density $= \dfrac{g}{cm^3} = \dfrac{1.6727 \times 10^{-24}}{4/3 \pi (5.0 \times 10^{-14})^3 \, cm^3} = 3.2 \times 10^{15}$ g/cm^3

2.92   The integer on the lower left of a nuclide is the atomic number; it is the number of protons in any atom of the element and gives the element's identity. The number of neutrons is the mass number (upper left) minus atomic number.

    (a)   As, 33 protons, 41 neutrons

    (b)   I, 53 protons, 74 neutrons

    (c)   Eu, 63 protons, 89 neutrons

    (d)   Bi, 83 protons, 126 neutrons

2.93   (a) ^{6}Li, 3 protons, 3 neutrons; ^{7}Li, 3 protons, 4 neutrons

    (b)   Average atomic weight of sample $= \Sigma$ fractional abundance $\times$ mass of isotope

Av. atomic weight $= 0.01442(6.015122) + 0.98558(7.016004) = 7.001571 = 7.002$ amu

2.94   (a) $^{16}_{8}$O, $^{17}_{8}$O, $^{18}_{8}$O,

    (b)   All isotopes are atoms of the same element, oxygen, with the same atomic number ($Z = 8$), 8 protons in the nucleus and 8 electrons. Elements with similar electron arrangements have similar chemical properties (Section 2.5). Because the 3 isotopes all have 8 electrons, we expect their electron arrangements to be the same and their chemical properties to be very similar, perhaps identical. Each has a different number of neutrons (8, 9, or 10), a different mass number (A = 16, 17, or 18) and thus a different atomic mass.

2.95   Atomic weight (average atomic mass) = Σ fractional abundance × mass of isotope

Atomic weight = 0.014(203.97302) + 0.241(205.97444) + 0.221(206.97587) +

$$0.524(207.97663) = 207.22 = 207 \text{ amu}$$

(The result has 0 decimal places and 3 sig figs because the fourth term in the sum has 3 sig figs and 0 decimal places.)

2.96   (a)   The 68.926 amu isotope has a mass number of 69, with 31 protons, 38 neutrons and the symbol $^{69}_{31}\text{Ga}$. The 70.925 amu isotope has a mass number of 71, 31 protons, 40 neutrons, and symbol $^{71}_{31}\text{Ga}$. (All Ga atoms have 31 protons.)

         (b)   The average mass of a Ga atom is 69.72 amu. Let x = abundance of the lighter isotope, 1–x = abundance of the heavier isotope. Then x(68.926) + (1–x)(70.925) = 69.72; x = 0.6028 = 0.603, ^{69}Ga = 60.3%, ^{71}Ga = 39.7%.

2.97   (a)   There are 24 known isotopes of Ni, from ^{51}Ni to ^{74}Ni.

         (b,c)   The five most abundant isotopes (b) and their natural abundances (c) are

^{58}Ni, 57.935346 amu, 68.077%

^{60}Ni, 59.930788 amu, 26.223%

^{62}Ni, 61.928346 amu, 3.634%

^{61}Ni, 60.931058 amu, 1.140%

^{64}Ni, 63.927968 amu, 0.926%

Data from *Handbook of Chemistry and Physics*, 74th edition [Data may differ slightly in other editions.]

2.98   (a)   A Br_2 molecule could consist of two atoms of the same isotope or one atom of each of the two different isotopes. This second possibility is twice as likely as the first. Therefore, the second peak (twice as large as peaks 1 and 3) represents a Br_2 molecule containing different isotopes. The mass numbers of the two isotopes are determined from the masses of the two smaller peaks. Because 157.836 ≈ 158, the first peak represents a $^{79}\text{Br}—^{79}\text{Br}$ molecule. Peak 3, 161.832 ≈ 162, represents a $^{81}\text{Br}—^{81}\text{Br}$ molecule. Peak 2 then contains one atom of each isotope, $^{79}\text{Br}—^{81}\text{Br}$, with an approximate mass of 160 amu.

         (b)   The mass of the lighter isotope is 157.836 amu/2 atoms, or 78.918 amu/atom. For the heavier one, 161.832 amu/2 atoms = 80.916 amu/atom.

         (c)   The relative size of the three peaks in the mass spectrum of Br_2 indicates their relative abundance. The average mass of a Br_2 molecule is

0.2569(157.836) + 0.4999(159.834) + 0.2431(161.832) = 159.79 amu.

(Each product has four significant figures and two decimal places, so the answer has two decimal places.)

(d) $\dfrac{159.79 \text{ amu}}{\text{avg. Br}_2 \text{ molecule}} \times \dfrac{1 \text{ Br}_2 \text{ molecule}}{2 \text{ Br atoms}} = 79.895 \text{ amu}$

(e)   Let x = the abundance of ^{79}Br, 1 – x = abundance of ^{81}Br. From (b), the masses of the two isotopes are 78.918 amu and 80.916 amu, respectively. From (d), the mass of an average Br atom is 79.895 amu.

$$x(78.918) + (1 - x)(80.916) = 79.895, \ x = 0.5110$$
$$^{79}\text{Br} = 51.10\%, \ ^{81}\text{Br} = 48.90\%$$

2.99   (a)   Five significant figures. ^{1}H^{+} is a bare proton with mass 1.0073 amu. ^{1}H is a hydrogen atom, with 1 proton and 1 electron. The mass of the electron is 5.486×10^{-4} or 0.0005486 amu. Thus the mass of the electron is significant in the fourth decimal place or fifth significant figure in the mass of ^{1}H.

(b)   Mass of ^{1}H = 1.0073 amu   (proton)

           <u>0.0005486 amu</u>   (electron)

           1.0078 amu   (We have not rounded up to 1.0079 because 49 < 50 in the final sum.)

$$\text{Mass \% of electron} = \frac{\text{mass of e}^{-}}{\text{mass of }^{1}\text{H}} \times 100 = \frac{5.486 \times 10^{-4} \text{ amu}}{1.0078 \text{ amu}} \times 100 = 0.05444\%$$

2.100   (a)   an alkali metal: K   (b) an alkaline earth metal: Ca   (c) a noble gas: Ar

        (d)   a halogen: Br   (e) a metalloid: Ge   (f) a nonmetal in 1A: H

        (g)   a metal that forms a 3+ ion: Al   (h) a nonmetal that forms a 2– ion: O

        (i)   an element that resembles Al: Ga

2.101   (a) $^{266}_{106}$Sg has 106 protons, 160 neutrons and 106 electrons

        (b)   Sg is in Group 6B (or 6) and immediately below tungsten, W. We expect the chemical properties of Sg to most closely resemble those of W.

2.102   Strontium is an alkaline earth metal, similar in chemical properties to calcium and magnesium. Calcium is ubiquitous in biological organisms, humans included. It is a vital nutrient required for formation and maintenance of healthy bones and teeth. As such, there are efficient pathways for calcium uptake and distribution in the body, pathways that are also available to chemically similar strontium. Harmful strontium imitates calcium and then behaves badly when the body tries to use it as it uses calcium.

2.103   Calculate the volume of the penny, use density to calculate mass and price to calculate the value of copper in the penny.

V = $\pi \, r^2 h$; d = 19 mm, r = d/2 = 9.5 mm; h = 1.5 mm

$$V = \pi \times (9.5 \text{ mm})^2 \times 1.5 \text{ mm} \times \frac{1 \text{ cm}^3}{(10)^3 \text{ mm}^3} = 0.4253 = 0.43 \text{ cm}^3$$

$$0.4253 \text{ cm}^3 \times \frac{8.9 \text{ g}}{\text{cm}^3} \times \frac{1 \text{ lb}}{453.6 \text{ g}} \times \frac{\$2.40}{\text{lb}} = \$0.02003 = \$0.020$$

That is, the copper in each penny is worth two pennies!

2.104   Calculate the volume of the coin, use density to calculate mass and price to calculate the value of silver in the coin.

$V = \pi r^2 h$; d = 41 mm, r = d/2 = 20.5 mm; h = 2.5 mm

$$V = \pi \times (20.5 \text{ mm})^2 \times 2.5 \text{ mm} \times \frac{1 \text{ cm}^3}{(10)^3 \text{ mm}^3} = 3.3006 = 3.3 \text{ cm}^3$$

$$3.3006 \text{ cm}^3 \times \frac{10.5 \text{ g}}{\text{cm}^3} \times \frac{\$0.51}{\text{g}} = \$17.675 = \$18$$

Wow! The silver in each Silver Eagle dollar coin is worth \$18.

2.105   (a)   chlorine gas, Cl_2: ii   (b) propane, C_3H_8: v   (c) nitrate ion, NO_3^- : i

(d)   sulfur trioxide, SO_3: iii   (e) methylchloride, CH_3Cl: iv

2.106   (a)   nickel(II) oxide, 2+   (b) manganese(IV) oxide, 4+

(c)   chromium(III) oxide, 3+   (d) molybdenium(VI) oxide, 6+

2.107

Cation	Anion	Formula	Name
Li^+	O^{2-}	Li_2O	Lithium oxide
Fe^{2+}	PO_4^{3-}	$Fe_3(PO_4)_2$	Iron(II) phosphate
Al^{3+}	SO_4^{2-}	$Al_2(SO_4)_3$	Aluminum sulfate
Cu^{2+}	NO_3^-	$Cu(NO_3)_2$	Copper(II) nitrate
Cr^{3+}	I^-	CrI_3	Chromium(III) iodide
Mn^+	ClO_2^-	$MnClO_2$	Manganese(I) chlorite
NH_4^+	CO_3^{2-}	$(NH_4)_2CO_3$	Ammonium carbonate
Zn^{2+}	ClO_4^-	$Zn(ClO_4)_2$	Zinc perchlorate

2.108   (a)   Empirical formula, CH_3

The empirical and molecular formulas of propane are C_3H_8. Propane has two more H atoms than cyclopropane, so the empirical and molecular formulas are different.

(b)   The solid wedges indicate bonds from C atoms to H atoms that are above the plane of the page; the dashed wedges show bonds from C atoms to H atoms that are behind the plane of the page.

(c)   To illustrate chlorocyclopropane, replace any one of the H atoms on cyclopropane with a Cl atom. There are no isomers of chlorocyclopropane, because a structure with a Cl atom at any one of the six positions can be rotated into the original structure.

2.109   (a)   perbromate ion   (b)   selenite ion

         (c) AsO_4^{3-} (d) $HTeO_4^-$

2.110   Carbonic acid: H_2CO_3; the cation is H^+ because it is an acid; the anion is carbonate because the acid reacts with lithium hydroxide to form lithium carbonate.
Lithium hydroxide: LiOH; lithium carbonate: Li_2CO_3

2.111   (a)   sodium chloride   (b) sodium bicarbonate (or sodium hydrogen carbonate)

         (c)   sodium hypochlorite   (d) sodium hydroxide

         (e)   ammonium carbonate   (f) calcium sulfate

2.112   (a)   potassium nitrate   (b) sodium carbonate   (c) calcium oxide

         (d)   hydrochloric acid   (e) magnesium sulfate   (f) magnesium hydroxide

2.113   (a)   CaS, $Ca(HS)_2$ (b) HBr, $HBrO_3$ (c) AlN, $Al(NO_2)_3$ (d) FeO, Fe_2O_3

         (e) NH_3, NH_4^+ (f) K_2SO_3, $KHSO_3$ (g) Hg_2Cl_2, $HgCl_2$ (h) $HClO_3$, $HClO_4$

2.114   In the nucleus. The strong force holds the protons together against the repulsive electrostatic force.

3 Chemical Reactions and Reaction Stoichiometry

Visualizing Concepts

3.1 Reactant A = blue, reactant B = red

Overall, 4 blue A_2 molecules + 4 red B atoms $\rightarrow$ 4 A_2B molecules

Because 4 is a common factor, this equation reduces to equation (a).

3.2 (a) There are four CH_3OH molecules in the products box. CO is the only source of C atoms for the reaction, so there must be four CO molecules in the reactants box.

(b) $CO + 2 H_2 \rightarrow CH_3OH$

3.3 (a) There are twice as many O atoms as N atoms, so the empirical formula of the original compound is NO_2.

(b) No, because we have no way of knowing whether the empirical and molecular formulas are the same. NO_2 represents the simplest ratio of atoms in a molecule but not the only possible molecular formula.

3.4 The box contains 4 C atoms and 10 H atoms, so the empirical formula of the hydrocarbon is C_2H_5.

3.5 (a) *Analyze.* Given the molecular model, write the molecular formula.

Plan. Use the colors of the atoms (spheres) in the model to determine the number of atoms of each element.

Solve. Observe 2 gray C atoms, 5 white H atoms, 1 blue N atom, 2 red O atoms. $C_2H_5NO_2$

(b) *Plan.* Follow the method in Sample Exercise 3.9. Calculate formula weight in amu and molar mass in grams.

2 C atoms = 2(12.0 amu) = 24.0 amu

5 H atoms = 5(1.0 amu) = 5.0 amu

1 N atoms = 1(14.0 amu) = 14.0 amu

2 O atoms = 2(16.0 amu) = <u>32.0 amu</u>

75.0 amu

Formula weight = 75.0 amu, molar mass = 75.0 g/mol

(c) *Plan.* The molar mass of a substance provides the factor for converting grams to moles (or moles to grams).

Solve.

Because the mass of glycine has 4 significant figures, use a molar mass of glycine that has at least 4 significant figures. Using molar masses of the component elements from the periodic chart, the molar mass of glycine is

[2(12.0107) + 5(1.00794) + 14.0067 + 2(15.9994)] = 75.0666 = 75.067 g/mol

$$100.0 \text{ g glycine} \times \frac{1 \text{ mol glycine}}{75.067 \text{ g glycine}} = 1.3321 = 1.332 \text{ mol glycine}$$

(d) *Plan.* Use the definition of mass % and the results from parts (a) and (b) above to find mass % N in glycine.

Solve. $\text{mass \% N} = \dfrac{\text{g N}}{\text{g C}_2\text{H}_5\text{NO}_2} \times 100$

Assume 1 mol $C_2H_5NO_2$. From the molecular formula of glycine [part (a)], there is 1 mol N/mol glycine.

$$\text{mass \% N} = \frac{1 \times (\text{molar mass N})}{\text{molar mass glycine}} \times 100 = \frac{14.0 \text{ g}}{75.0 \text{ g}} \times 100 = 18.7\%$$

3.6 *Analyze.* Given: 4.0 mol CH_4. Find: mol CO and mol H_2

Plan. Examine the boxes to determine the CH_4:CO mol ratio and CH_4:H_2O mole ratio.

Solve. There are 2 CH_4 molecules in the reactant box and 2 CO molecules in the product box. The mole ratio is 2:2 or 1:1. Therefore, 4.0 mol CH_4 can produce 4.0 mol CO. There are 2 CH_4 molecules in the reactant box and 6 H_2 molecules in the product box. The mole ratio is 2:6 or 1:3. So, 4.0 mol CH_4 can produce 12:0 mol H_2.

Check. Use proportions. 2 mol CH_4/2 mol CO = 4 mol CH_4/4 mol CO;

2 mol CH_4/6 mol H_2 = 4 mol CH_4/12 mol H_2.

3.7 *Analyze.* Given a box diagram and formulas of reactants, answer questions about the reaction mixture in the box.

Plan. Write and balance the chemical equation. Determine combining ratios of elements and decide on limiting reactant. Decide the maximum number of NH_3 molecules that can be produced and the number of leftover reactant molecules.

Solve.

(a) $N_2 + 3 H_2 \longrightarrow 2 NH_3$

(b) H_2 is the limiting reactant. There are 4 N_2 molecules and 9 H_2 molecules in the box. According to the chemical reaction, each N_2 molecule requires 3 H_2 molecules for complete reaction. If all N_2 molecules were to react, 12 H_2 molecules would be required. There are only 9 H_2 molecules, so H_2 is the limiting reactant.

(c) 6 NH_3 molecules. Because H_2 is the limiting reactant, the amount of H_2 available determines the amount of NH_3 produced. Three H_2 molecules produce 2 NH_3 molecules, so 9 H_2 molecules produce 6 NH_3 molecules.

(d) One N_2 molecule is left over. The 9 H_2 molecules react with 3 N_2 molecules, leaving one N_2 molecule unreacted. All H_2 molecules are used up.

3.8   (a) 8 NO_2 molecules can be produced. The overall chemical reaction is

$$2\,NO + O_2 \rightarrow 2\,NO_2.$$

There are 8 NO molecules and 5 O_2 molecules in the box. Each NO molecule reacts with 1 O atom (1/2 of an O_2 molecule) to produce 1 NO_2 molecule. Eight NO molecules react with 8 O atoms (4 O_2 molecules) to produce 8 NO_2 molecules. One O_2 molecule doesn't react (is in excess).

(b) NO is the limiting reactant, because it limits (determines) the amount of product that can be produced. It is completely used up if the reaction goes to completion.

(c) If the yield is 75%, 6 NO_2 molecules, 2 NO molecules and 2 O_2 molecules are present. A 100% yield is 8 NO_2 molecules. 75% of that is 6 NO_2 molecules. Formation of 6 NO_2 molecules requires 6 NO molecules and 6 O atoms or 3 O_2 molecules. This leaves 2 NO molecules and 2 O_2 molecules unreacted.

Chemical Equations and Simple Patterns of Chemical Reactivity
(Sections 3.1 and 3.2)

3.9   (a) False. We balance chemical equations as we do because mass must be conserved.

(b) True. Mass is conserved.

(c) False. Subscripts in chemical formulas cannot be changed when balancing an equation. Changing a subscript changes the identity of a compound, which changes the overall reaction.

3.10 (a) $CaO(s) + H_2O(l) \rightarrow Ca(OH)_2(aq)$

(b) The only way to write a balanced equation with CaOH(aq) as a product is to include $OH^-(aq)$ as a second product. Otherwise, the ratio of elements in the product is never the same as the ratio of elements in the reactants.

$$CaO(s) + H_2O(l) \rightarrow CaOH(aq) + OH^-(aq)$$

But, if CaOH is a neutral compound, this equation violates the principle of charge balance. The equation above cannot be the correct balanced equation for the reaction of calcium oxide with water.

3.11 (a) $2\,CO(g) + O_2(g) \rightarrow 2\,CO_2(g)$

(b) $N_2O_5(g) + H_2O(l) \rightarrow 2\,HNO_3(aq)$

(c) $CH_4(g) + 4\,Cl_2(g) \rightarrow CCl_4(l) + 4\,HCl(g)$

(d) $Zn(OH)_2(s) + 2\,HNO_3(aq) \rightarrow Zn(NO_3)_2(aq) + 2\,H_2O(l)$

3.12   (a) $6 \, Li(s) + N_2(g) \rightarrow 2 \, Li_3N(s)$

        (b) $TiCl_4(l) + 2 \, H_2O(l) \rightarrow TiO_2(s) + 4 \, HCl(aq)$

        (c) $2 \, NH_4NO_3(s) \rightarrow 2 \, N_2(g) + O_2(g) + 4 \, H_2O(g)$

        (d) $2 \, AlCl_3(s) + Ca_3N_2(s) \rightarrow 2 \, AlN(s) + 3 \, CaCl_2(s)$

3.13   (a) $Al_4C_3(s) + 12 \, H_2O(l) \rightarrow 4 \, Al(OH)_3(s) + 3 \, CH_4(g)$

        (b) $2 \, C_5H_{10}O_2(l) + 13 \, O_2(g) \rightarrow 10 \, CO_2(g) + 10 \, H_2O(g)$

        (c) $2 \, Fe(OH)_3(s) + 3 \, H_2SO_4(aq) \rightarrow Fe_2(SO_4)_3(aq) + 6 \, H_2O(l)$

        (d) $Mg_3N_2(s) + 4 \, H_2SO_4(aq) \rightarrow 3 \, MgSO_4(aq) + (NH_4)_2SO_4(aq)$

3.14   (a) $Ca_3P_2(s) + 6 \, H_2O(l) \rightarrow 3 \, Ca(OH)_2(aq) + 2 \, PH_3(g)$

        (b) $2 \, Al(OH)_3(s) + 3 \, H_2SO_4(aq) \rightarrow Al_2(SO_4)_3(aq) + 6 \, H_2O(l)$

        (c) $2 \, AgNO_3(aq) + Na_2CO_3(aq) \rightarrow Ag_2CO_3(s) + 2 \, NaNO_3(aq)$

        (d) $4 \, C_2H_5NH_2(g) + 15 \, O_2(g) \rightarrow 8 \, CO_2(g) + 14 \, H_2O(g) + 2 \, N_2(g)$

3.15   (a) $CaC_2(s) + 2 \, H_2O(l) \rightarrow Ca(OH)_2(aq) + C_2H_2(g)$

        (b) $2 \, KClO_3(s) \overset{\Delta}{\rightarrow} 2 \, KCl(s) + 3 \, O_2(g)$

        (c) $Zn(s) + H_2SO_4(aq) \rightarrow H_2(g) + ZnSO_4(aq)$

        (d) $PCl_3(l) + 3 \, H_2O(l) \rightarrow H_3PO_3(aq) + 3 \, HCl(aq)$

        (e) $3 \, H_2S(g) + 2 \, Fe(OH)_3(s) \rightarrow Fe_2S_3(s) + 6 \, H_2O(g)$

3.16   (a) $SO_3(g) + H_2O(l) \rightarrow H_2SO_4(aq)$

        (b) $B_2S_3(s) + 6 \, H_2O(l) \rightarrow 2 \, H_3BO_3(aq) + 3 \, H_2S(g)$

        (c) $4 \, PH_3(g) + 8 \, O_2(g) \rightarrow P_4O_{10}(s) + 6 \, H_2O(g)$

        (d) $2 \, Hg(NO_3)_2(s) \overset{\Delta}{\rightarrow} 2 \, HgO(s) + 4 \, NO_2(g) + O_2(g)$

        (e) $Cu(s) + 2 \, H_2SO_4(aq) \rightarrow CuSO_4(aq) + SO_2(g) + 2 \, H_2O(l)$

Patterns of Chemical Reactivity (Section 3.2)

3.17   (a)   NaBr. When a metal reacts with a nonmetal, an ionic compound forms. The combining ratio of the atoms is such that the total positive charge on the metal cation(s) is equal to the total negative charge on the nonmetal anion(s). Determine the formula by balancing the positive and negative charges in the ionic product.

        (b)   The product is a solid at room temperature. All ionic compounds are solids.

        (c)   In the balanced chemical equation, the coefficient in front of the product is two.

               $2 \, Na(s) + Br_2(l) \rightarrow 2 \, NaBr(s)$

3.18   (a) $O_2(g)$. Combustion is reaction with oxygen.

       (b)   The products are $CO_2(g)$ and $H_2O(l)$.

       (c)   The sum of the balanced coefficients is 11. (Remember that the coefficient in front of $C_3H_6O(l)$ is one.)

$$C_3H_6O(l) + 4\,O_2(g) \rightarrow 3\,CO_2(g) + 3\,H_2O(l)$$

3.19   (a) $Mg(s) + Cl_2(g) \rightarrow MgCl_2(s)$

       (b) $BaCO_3(s) \xrightarrow{\Delta} BaO(s) + CO_2(g)$

       (c) $C_8H_8(l) + 10\,O_2(g) \rightarrow 8\,CO_2(g) + 4\,H_2O(l)$

       (d) CH_3OCH_3 is C_2H_6O. $C_2H_6O(g) + 3\,O_2(g) \rightarrow 2\,CO_2(g) + 3\,H_2O(l)$

3.20   (a) $2\,Ti(s) + O_2(g) \rightarrow 2\,TiO(s)$ [or $Ti(s) + O_2(g) \rightarrow TiO_2(s)$]

       (b) $2\,Ag_2O(s) \xrightarrow{\Delta} 4\,Ag(s) + O_2(g)$

       (c) $2\,C_3H_7OH(l) + 9\,O_2(g) \rightarrow 6\,CO_2(g) + 8\,H_2O(l)$

       (d) $2\,C_5H_{12}O(l) + 15\,O_2(g) \rightarrow 10\,CO_2(g) + 12\,H_2O(l)$

3.21   (a) $2\,C_3H_6(g) + 9\,O_2(g) \rightarrow 6\,CO_2(g) + 6\,H_2O(g)$ combustion

       (b) $NH_4NO_3(s) \rightarrow N_2O(g) + 2\,H_2O(g)$ decomposition

       (c) $C_5H_6O(l) + 6\,O_2(g) \rightarrow 5\,CO_2(g) + 3\,H_2O(g)$ combustion

       (d) $N_2(g) + 3\,H_2(g) \rightarrow 2\,NH_3(g)$ combination

       (e) $K_2O(s) + H_2O(l) \rightarrow 2\,KOH(aq)$ combination

3.22   (a) $PbCO_3(s) \rightarrow PbO(s) + CO_2(g)$ decomposition

       (b) $C_2H_4(g) + 3\,O_2(g) \rightarrow 2\,CO_2(g) + 2\,H_2O(g)$ combustion

       (c) $3\,Mg(s) + N_2(g) \rightarrow Mg_3N_2(s)$ combination

       (d) $C_7H_8O_2(l) + 8\,O_2(g) \rightarrow 7\,CO_2(g) + 4\,H_2O(g)$ combustion

       (e) $2\,Al(s) + 3\,Cl_2(g) \rightarrow 2\,AlCl_3(s)$ combination

Formula Weights (Section 3.3)

3.23   *Analyze.* Given molecular formula or name, calculate formula weight.

       Plan. If a name is given, write the correct molecular formula. Then, follow the method in Sample Exercise 3.5.   *Solve.*

       (a) HNO_3: $1(1.0) + 1(14.0) + 3(16.0) = 63.0$ amu

       (b) $KMnO_4$: $1(39.1) + 1(54.9) + 4(16.0) = 158.0$ amu

       (c) $Ca_3(PO_4)_2$: $3(40.1) + 2(31.0) + 8(16.0) = 310.3$ amu

       (d) SiO_2: $1(28.1) + 2(16.0) = 60.1$ amu

       (e) Ga_2S_3: $2(69.7) + 3(32.1) = 235.7$ amu

       (f) $Cr_2(SO_4)_3$: $2(52.0) + 3(32.1) + 12(16.0) = 392.3$ amu

       (g) PCl_3: $1(31.0) + 3(35.5) = 137.5$ amu

3.24   Formula weight in amu to 1 decimal place.

     (a) N_2O: FW = 2(14.0) + 1(16.0) = 44.0 amu

     (b) C_6H_5COOH: 7(12.0) + 6(1.0) + 2(16.0) = 122.0 amu

     (c) $Mg(OH)_2$: 1(24.3) + 2(16.0) + 2(1.0) = 58.3 amu

     (d) $(NH_2)_2CO$: 2(14.0) + 4(1.0) + 1(12.0) + 1(16.0) = 60.0 amu

     (e) $CH_3CO_2C_5H_{11}$: 7(12.0) + 14(1.0) + 2(16.0) = 130.0 amu

3.25   *Plan.* Calculate the formula weight (FW), then the mass % oxygen in the compound.   *Solve.*

     (a) $C_{17}H_{19}NO_3$:   FW = 17(12.0) + 19(1.0) + 1(14.0) + 3(16.0) = 285.0 amu

$$\% \text{ O} = \frac{3(16.0) \text{ amu}}{285.0 \text{ amu}} \times 100 = 16.842 = 16.8\%$$

     (b) $C_{18}H_{21}NO_3$: FW = 18(12.0) + 21(1.0) + 1(14.0) + 3(16.0) = 299.0 amu

$$\% \text{ O} = \frac{3(16.0) \text{ amu}}{299.0 \text{ amu}} \times 100 = 16.054 = 16.1\%$$

     (c) $C_{17}H_{21}NO_4$: FW = 17(12.0) + 21(1.0) + 1(14.0) + 4(16.0) = 303.0 amu

$$\% \text{ O} = \frac{4(16.0) \text{ amu}}{303.0 \text{ amu}} \times 100 = 21.122 = 21.1\%$$

     (d) $C_{22}H_{24}N_2O_8$:   FW = 22(12.0) + 24(1.0) + 2(14.0) + 8(16.0) = 444.0 amu

$$\% \text{ O} = \frac{8(16.0) \text{ amu}}{444.0 \text{ amu}} \times 100 = 28.829 = 28.8\%$$

     (e) $C_{41}H_{64}O_{13}$:   FW = 41(12.0) + 64(1.0) + 13(16.0) = 764.0 amu

$$\% \text{ O} = \frac{13(16.0) \text{ amu}}{764 \text{ amu}} \times 100 = 27.225 = 27.2\%$$

     (f) $C_{66}H_{75}Cl_2N_9O_{24}$:   FW = 66(12.0) + 75(1.0) + 2(35.5) + 9(14.0) + 24(16.0) = 1448.0 amu

$$\% \text{ O} = \frac{24(16.0) \text{ amu}}{1448.0 \text{ amu}} \times 100 = 26.519 = 26.5\%$$

3.26   (a) C_2H_2: FW = 2(12.0) + 2(1.0) = 26.0 amu

$$\% \text{ C} = \frac{2(12.0) \text{ amu}}{26.0 \text{ amu}} \times 100 = 92.3\%$$

     (b) $HC_6H_7O_6$: FW = 6(12.0) + 8(1.0) + 6(16.0) = 176.0 amu

$$\% \text{ H} = \frac{8(1.0) \text{ amu}}{176.0 \text{ amu}} \times 100 = 4.5\%$$

     (c) $(NH_4)_2SO_4$: FW = 2(14.0) + 8(1.0) + 1(32.1) + 4(16.0) = 132.1 amu

$$\% \text{ H} = \frac{8(1.0) \text{ amu}}{132.1 \text{ amu}} \times 100 = 6.1\%$$

     (d) $PtCl_2(NH_3)_2$: FW = 1(195.1) + 2(35.5) + 2(14.0) + 6(1.0) = 300.1 amu

$$\% \text{ Pt} = \frac{1(195.1) \text{ amu}}{300.1 \text{ amu}} \times 100 = 65.01\%$$

(e) $C_{18}H_{24}O_2$: FW = 18(12.0) + 24(1.0) + 2(16.0) = 272.0 amu

$$\% \ O = \frac{2(16.0) \ amu}{272.0 \ amu} \times 100 = 11.8\%$$

(f) $C_{18}H_{27}NO_3$: FW = 18(12.0) + 27(1.0) + 1(14.0) + 3(16.0) = 305.0 amu

$$\% \ C = \frac{18(12.0) \ amu}{305.0 \ amu} \times 100 = 70.8\%$$

3.27   *Plan.* Follow the logic for calculating mass % C given in Sample Exercise 3.6.   *Solve.*

(a) C_7H_6O: FW = 7(12.0) + 6(1.0) + 1(16.0) = 106.0 amu

$$\% \ C = \frac{7(12.0) \ amu}{106.0 \ amu} \times 100 = 79.2\%$$

(b) $C_8H_8O_3$: FW = 8(12.0) + 8(1.0) + 3(16.0) = 152.0 amu

$$\% \ C = \frac{8(12.0) \ amu}{152.0 \ amu} \times 100 = 63.2\%$$

(c) $C_7H_{14}O_2$: FW = 7(12.0) + 14(1.0) + 2(16.0) = 130.0 amu

$$\% \ C = \frac{7(12.0) \ amu}{130.0 \ amu} \times 100 = 64.6\%$$

3.28   (a) CO_2: FW = 1(12.0) + 2(16.0) = 44.0 amu

$$\% \ C = \frac{12.0 \ amu}{44.0 \ amu} \times 100 = 27.3\%$$

(b) CH_3OH: FW = 1(12.0) + 4(1.0) + 1(16.0) = 32.0 amu

$$\% \ C = \frac{12.0 \ amu}{32.0 \ amu} \times 100 = 37.5\%$$

(c) C_2H_6: FW = 2(12.0) + 6(1.0) = 30.0 amu

$$\% \ C = \frac{2(12.0) \ amu}{30.0 \ amu} \times 100 = 80.0\%$$

(d) $CS(NH_2)_2$: FW = 1(12.0) + 1(32.1) + 2(14.0) + 4(1.0) = 76.1 amu

$$\% \ C = \frac{12.0 \ amu}{76.1 \ amu} \times 100 = 15.8\%$$

Avogadro's Number and the Mole (Section 3.4)

3.29   (a)   False. A mole of horses contains four moles of horse legs.

(b)   True.

(c)   False. The mass of one mole of water is 18.0 g.

(d)   True. Electrically neutral NaCl is composed of Na^+ cations and Cl^- anions.

3.30   (a)   <u>exactly</u> 12 g   (b) 6.0221421×10^{23}, Avogadro's number

3.31   *Plan.* Because the mole is a counting unit, use it as a basis of comparison; determine the total moles of atoms in each given quantity.   *Solve.*

23 g Na contains 1 mol of atoms

0.5 mol H_2O contains (3 atoms $\times$ 0.5 mol) = 1.5 mol atoms

6.0×10^{23} N_2 molecules contains (2 atoms $\times$ 1 mol) = 2 mol atoms

3.32   42 g $NaHCO_3$ (molar mass = 84 g) contains (6 atoms $\times$ 0.5 mol) = 3 mol atoms

1.5 mol CO_2 contains (3 atoms $\times$ 1.5 mol) = 4.5 mol atoms

6.0×10^{24} Ne atoms contains (1 atoms $\times$ 10 mol) = 10 mol atoms

3.33   *Analyze.* Given: 160 lb/person; Avogadro's number of people, 6.022×10^{23} people. Find: mass in kg of Avogadro's number of people; compare with mass of Earth.

Plan. people $\rightarrow$ mass in lb $\rightarrow$ mass in kg; mass of people/mass of Earth

Solve. 6.022×10^{23} people $\times \dfrac{160\ lb}{person} \times \dfrac{1\ kg}{2.2046\ lb} = 4.370 \times 10^{25} = 4.37 \times 10^{25}$ or 4.4×10^{25} kg

$\dfrac{4.370 \times 10^{25}\ kg\ of\ people}{5.98 \times 10^{24}\ kg\ Earth} = 7.31$ or 7.3

One mole of people weighs 7.3 times as much as Earth.

Check. This mass of people is reasonable because Avogadro's number is large.

Estimate: 160 lb $\approx$ 70 kg; $6 \times 10^{23} \times 70 = 420 \times 10^{23} = 4.2 \times 10^{25}$ kg

3.34   321 million = $321 \times 10^6 = 3.21 \times 10^8$ people

$\dfrac{6.022 \times 10^{23}\ \cent}{3.21 \times 10^8\ people} \times \dfrac{\$1}{100\ \cent} = \dfrac{\$6.022 \times 10^{21}}{3.21 \times 10^8\ people} = 1.876 \times 10^{13} = \$1.88 \times 10^{13}$ gift / person

$17.419 trillion = \$1.7419 \times 10^{13} = \$1.74 \times 10^{13}$ debt; $\dfrac{\$1.876 \times 10^{13}\ gift}{1.7419 \times 10^{13}\ debt} = 1.1011 = 1.10$

Each person would receive an amount that is 1.10 times the dollar amount of the national debt.

3.35   (a)   *Analyze.* Given: 0.105 mol sucrose, $C_{12}H_{22}O_{11}$. Find: mass in g.

           Plan. Use molar mass (g/mol) of $C_{12}H_{22}O_{11}$ to find g $C_{12}H_{22}O_{11}$.

           Solve. molar mass = 12(12.0107) + 22(1.00794) + 11(15.9994) = 342.296 = 342.30

           0.105 mol sucrose $\times \dfrac{342.30\ g}{1\ mol} = 35.942 = 35.9$ g $C_{12}H_{22}O_{11}$

           Check. 0.1(342) = 34.2 g. The calculated result is reasonable.

(b)   *Analyze.* Given: mass. Find: moles. *Plan.* Use molar mass of $Zn(NO_3)_2$.

Solve. molar mass = 1(65.39) + 2(14.0067) + 6(15.9994) = 189.3998 = 189.40

$$143.50 \text{ g Zn(NO}_3)_2 \times \frac{1 \text{ mol}}{189.40 \text{ g Zn(NO}_3)_2} = 0.75766 \text{ mol Zn(NO}_3)_2$$

Check. $140/180 \approx 7/9 = 0.78$ mol

(c)   *Analyze.* Given: moles. Find: molecules. *Plan.* Use Avogadro's number.

Solve. $1.0 \times 10^{-6} \text{ mol CH}_3\text{CH}_2\text{OH} \times \frac{6.022 \times 10^{23} \text{ molecules}}{1 \text{ mol}} = 6.022 \times 10^{17}$

$6.0 \times 10^{17} \text{ CH}_3\text{CH}_2\text{OH molecules}$

Check. $(1.0 \times 10^{-6})(6 \times 10^{23}) = 6 \times 10^{17}$

(d)   *Analyze.* Given: mol NH_3. Find: N atoms.

Plan. mol $NH_3 \rightarrow$ mol N atoms $\rightarrow$ N atoms

Solve. $0.410 \text{ mol NH}_3 \times \frac{1 \text{ mol N atoms}}{1 \text{ mol NH}_3} \times \frac{6.022 \times 10^{23} \text{ atoms}}{1 \text{ mol}}$

$= 2.47 \times 10^{23}$ N atoms

Check. $(0.4)(6 \times 10^{23}) = 2.4 \times 10^{23}$.

3.36   (a)   molar mass = 1(112.41) + 1(32.07) = 144.48 g

$$1.50 \times 10^{-2} \text{ mol CdS} \times \frac{144.48 \text{ g}}{1 \text{ mol}} = 2.17 \text{ g CdS}$$

(b)   molar mass = 1(14.01) + 4(1.008) + 1(35.45) = 53.49 g/mol

$$86.6 \text{ g NH}_4\text{Cl} \times \frac{1 \text{ mol}}{53.49 \text{ g}} = 1.6190 = 1.62 \text{ mol NH}_4\text{Cl}$$

(c) $8.447 \times 10^{-2} \text{ mol C}_6\text{H}_6 \times \frac{6.02214 \times 10^{23} \text{ molecules}}{1 \text{ mol}} = 5.087 \times 10^{22} \text{ C}_6\text{H}_6 \text{ molecules}$

(d) $6.25 \times 10^{-3} \text{ mol Al(NO}_3)_3 \times \frac{9 \text{ mol O}}{1 \text{ mol Al(NO}_3)_3} \times \frac{6.022 \times 10^{23} \text{ O atoms}}{1 \text{ mol}}$

$= 3.39 \times 10^{22}$ O atoms

3.37   *Analyze/Plan.* See 3.35 for stepwise problem-solving approaches.   *Solve.*

(a) $(NH_4)_3PO_4$ molar mass = 3(14.007) + 12(1.008) + 1(30.974) + 4(16.00) = 149.091

$= 149.1$ g/mol

$$2.50 \times 10^{-3} \text{ mol (NH}_4)_3\text{PO}_4 \times \frac{149.1 \text{ g (NH}_4)_3\text{PO}_4}{1 \text{ mol}} = 0.373 \text{ g (NH}_4)_3\text{PO}_4$$

(b) $AlCl_3$ molar mass = 26.982 + 3(35.453) = 133.341 = 133.34 g/mol

$$0.2550 \text{ g AlCl}_3 \times \frac{1 \text{ mol}}{133.34 \text{ g AlCl}_3} \times \frac{3 \text{ mol Cl}^-}{1 \text{ mol AlCl}_3} = 5.737 \times 10^{-3} \text{ mol Cl}^-$$

(c) $C_8H_{10}N_4O_2$ molar mass $= 8(12.01) + 10(1.008) + 4(14.01) + 2(16.00) = 194.20$

$$= 194.2 \text{ g/mol}$$

$$7.70 \times 10^{20} \text{ molecules} \times \frac{1 \text{ mol}}{6.022 \times 10^{23} \text{ molecules}} \times \frac{194.2 \text{ g } C_8H_{10}N_4O_2}{1 \text{ mol caffeine}}$$

$$= 0.248 \text{ g } C_8H_{10}N_4O_2$$

(d) $\dfrac{0.406 \text{ g cholesterol}}{0.00105 \text{ mol}} = 387 \text{ g cholesterol/mol}$

3.38   (a) $Fe_2(SO_4)_3$ molar mass $= 2(55.845) + 3(32.07) + 12(16.00) = 399.900 = 399.9 \text{ g/mol}$

$$1.223 \text{ mol } Fe_2(SO_4)_3 \times \frac{399.9 \text{ g } Fe_2(SO_4)_3}{1 \text{ mol}} = 489.077 = 489.1 \text{ g } Fe_2(SO_4)_3$$

(b) $(NH_4)_2CO_3$ molar mass $= 2(14.007) + 8(1.008) + 12.011 + 3(15.9994) = 96.0872$

$$= 96.087 \text{ g/mol}$$

$$6.955 \text{ g } (NH_4)_2CO_3 \times \frac{1 \text{ mol}}{96.087 \text{ g } (NH_4)_2CO_3} \times \frac{2 \text{ mol } NH_4^+}{1 \text{ mol } (NH_4)_2CO_3} = 0.1448 \text{ mol } NH_4^+$$

(c) $C_9H_8O_4$ molar mass $= 9(12.01) + 8(1.008) + 4(16.00) = 180.154 = 180.2 \text{ g/mol}$

$$1.50 \times 10^{21} \text{ molecules} \times \frac{1 \text{ mol}}{6.022 \times 10^{23} \text{ molecules}} \times \frac{180.2 \text{ g } C_9H_8O_4}{1 \text{ mol aspirin}} = 0.449 \text{ g } C_9H_8O_4$$

(d) $\dfrac{15.86 \text{ g diazepam}}{0.05570 \text{ mol}} = 284.7 \text{ g diazepam/mol}$

3.39   (a) $C_6H_{10}OS_2$ molar mass $= 6(12.01) + 10(1.008) + 1(16.00) + 2(32.07) = 162.28$

$$= 162.3 \text{ g/mol}$$

(b)   *Plan.* mg $\to$ g $\to$ mol   *Solve.*

$$5.00 \text{ mg allicin} \times \frac{1 \times 10^{-3} \text{ g}}{1 \text{ mg}} \times \frac{1 \text{ mol}}{162.3 \text{ g}} = 3.081 \times 10^{-5} = 3.08 \times 10^{-5} \text{ mol allicin}$$

Check. 5.00 mg is a small mass, so the small answer is reasonable.

$(5 \times 10^{-3})/200 = 2.5 \times 10^{-5}$

(c)   *Plan.* Use mol from part (b) and Avogadro's number to calculate molecules.

$$\textit{Solve.} \ 3.081 \times 10^{-5} \text{ mol allicin} \times \frac{6.022 \times 10^{23} \text{ molecules}}{\text{mol}} = 1.855 \times 10^{19}$$

$$= 1.86 \times 10^{19} \text{ allicin molecules}$$

Check. $(3 \times 10^{-5})(6 \times 10^{23}) = 18 \times 10^{18} = 1.8 \times 10^{19}$

(d)   *Plan.* Use molecules from part (c) and molecular formula to calculate S atoms.

$$\textit{Solve.} \ 1.855 \times 10^{19} \text{ allicin molecules} \times \frac{2 \text{ S atoms}}{1 \text{ allicin molecule}} = 3.71 \times 10^{19} \text{ S atoms}$$

Check. Obvious.

3.40   (a) $C_{14}H_{18}N_2O_5$ molar mass = 14(12.01) + 18(1.008) + 2(14.01) + 5(16.00)

$$= 294.30 \text{ g/mol}$$

     (b) $1.00 \text{ mg aspartame} \times \dfrac{1 \times 10^{-3} \text{g}}{1 \text{ mg}} \times \dfrac{1 \text{ mol}}{294.3 \text{ g}} = 3.398 \times 10^{-6} = 3.40 \times 10^{-6} \text{ mol aspartame}$

     (c) $3.398 \times 10^{-6} \text{ mol aspartame} \times \dfrac{6.022 \times 10^{23} \text{molecules}}{1 \text{ mol}} = 2.046 \times 10^{18}$

$$= 2.05 \times 10^{18} \text{ aspartame molecules}$$

     (d) $2.046 \times 10^{18} \text{ aspartame molecules} \times \dfrac{18 \text{ H atoms}}{1 \text{ aspartame molecule}} = 3.68 \times 10^{19} \text{ H atoms}$

3.41   (a)   *Analyze.* Given: $C_6H_{12}O_6$, 1.250×10^{21} C atoms. Find: H atoms.

         Plan. Use molecular formula to determine number of H atoms that are present with 1.250×10^{21} C atoms.   *Solve.*

$$\frac{12 \text{ H atoms}}{6 \text{ C atoms}} = \frac{2 \text{ H}}{1 \text{ C}} \times 1.250 \times 10^{21} \text{ C atoms} = 2.500 \times 10^{21} \text{ H atoms}$$

         Check. $(2 \times 1 \times 10^{21}) = 2 \times 10^{21}$

     (b)   *Plan.* Use molecular formula to find the number of glucose molecules that contain 1.250×10^{21} C atoms.   *Solve.*

$$\frac{1 \text{ } C_6H_{12}O_6 \text{ molecule}}{6 \text{ C atoms}} \times 1.250 \times 10^{21} \text{ C atoms} = 2.0833 \times 10^{20}$$

$$= 2.083 \times 10^{20} \text{ } C_6H_{12}O_6 \text{ molecules}$$

         Check. $(12 \times 10^{20}/6) = 2 \times 10^{20}$

     (c)   *Plan.* Use Avogadro's number to change molecules $\rightarrow$ mol.   *Solve.*

$$2.0833 \times 10^{20} \text{ } C_6H_{12}O_6 \text{ molecules} \times \frac{1 \text{ mol}}{6.022 \times 10^{23} \text{ molecules}}$$

$$= 3.4595 \times 10^{-4} = 3.460 \times 10^{-4} \text{ mol } C_6H_{12}O_6$$

         Check. $(2 \times 10^{20})/(6 \times 10^{23}) = 0.33 \times 10^{-3} = 3.3 \times 10^{-4}$

     (d)   *Plan.* Use molar mass to change mol $\rightarrow$ g.   *Solve.*

         1 mole of $C_6H_{12}O_6$ weighs 180.0 g (Sample Exercise 3.9)

$$3.4595 \times 10^{-4} \text{ mol } C_6H_{12}O_6 \times \frac{180.0 \text{ g } C_6H_{12}O_6}{1 \text{ mol}} = 0.06227 \text{ g } C_6H_{12}O_6$$

         Check. $3.5 \times 180 = 630$; $630 \times 10^{-4} = 0.063$

3.42   (a) $3.88 \times 10^{21} \text{ H atoms} \times \dfrac{19 \text{ C atoms}}{28 \text{ H atoms}} = 2.63 \times 10^{21} \text{ C atoms}$

     (b) $3.88 \times 10^{21} \text{ H atoms} \times \dfrac{1 \text{ } C_{19}H_{28}O_2 \text{ molecule}}{28 \text{ H atoms}} = 1.3857 \times 10^{20}$

$$= 1.39 \times 10^{20} \text{ } C_{19}H_{28}O_2 \text{ molecules}$$

(c) 1.3857×10^{20} $C_{19}H_{28}O_2$ molecules $\times \dfrac{1 \text{ mol}}{6.022 \times 10^{23} \text{ molecules}} = 2.301 \times 10^{-4}$

$$= 2.30 \times 10^{-4} \text{ mol } C_{19}H_{28}O_2$$

(d) $C_{19}H_{28}O_2$ molar mass $= 19(12.01) + 28(1.008) + 2(16.00) = 288.41 = 288.4 \text{ g/mol}$

$$2.301 \times 10^{-4} \text{ mol } C_{19}H_{28}O_2 \times \dfrac{288.4 \text{ g } C_{19}H_{28}O_2}{1 \text{ mol}} = 0.0664 \text{ g } C_{19}H_{28}O_2$$

3.43   *Analyze.* Given: g C_2H_3Cl/L.   Find: mol/L, molecules/L.

Plan. The /L is constant throughout the problem, so we can ignore it. Use molar mass for g → mol, Avogadro's number for mol → molecules.   *Solve.*

$$\dfrac{2.0 \times 10^{-6} \text{ g } C_2H_3Cl}{1 \text{ L}} \times \dfrac{1 \text{ mol } C_2H_3Cl}{62.50 \text{ g } C_2H_3Cl} = 3.20 \times 10^{-8} = 3.2 \times 10^{-8} \text{ mol } C_2H_3Cl/L$$

$$\dfrac{3.20 \times 10^{-8} \text{ mol } C_2H_3Cl}{1 \text{ L}} \times \dfrac{6.022 \times 10^{23} \text{ molecules}}{1 \text{ mol}} = 1.9 \times 10^{16} \text{ molecules/L}$$

Check. $(200 \times 10^{-8})/60 = 2.5 \times 10^{-8} \text{ mol}$

$$(2.5 \times 10^{-8}) \times (6 \times 10^{23}) = 15 \times 10^{15} = 1.5 \times 10^{16}$$

3.44 $25 \times 10^{-6} \text{ g } C_{21}H_{30}O_2 \times \dfrac{1 \text{ mol } C_{21}H_{30}O_2}{314.5 \text{ g } C_{21}H_{30}O_2} = 7.95 \times 10^{-8} = 8.0 \times 10^{-8} \text{ mol } C_{21}H_{30}O_2$

$$7.95 \times 10^{-8} \text{ mol } C_{21}H_{30}O_2 \times \dfrac{6.022 \times 10^{23} \text{ molecules}}{1 \text{ mol}} = 4.8 \times 10^{16} \, C_{21}H_{30}O_2 \text{ molecules}$$

Empirical Formulas from Analyses (Section 3.5)

3.45   (a)   *Analyze.* Given: moles. Find: empirical formula.

Plan. Find the **simplest ratio of moles** by dividing by the smallest number of moles present.

Solve. 0.0130 mol C / 0.0065 = 2

           0.0390 mol H / 0.0065 = 6

           0.0065 mol O / 0.0065 = 1

The empirical formula is C_2H_6O.

Check. The subscripts are simple integers.

(b)   *Analyze.* Given: grams. Find: empirical formula.

Plan. Calculate the moles of each element present, then the simplest ratio of moles.

Solve. $11.66 \text{ g Fe} \times \dfrac{1 \text{ mol Fe}}{55.85 \text{ g Fe}} = 0.2088 \text{ mol Fe}; \; 0.2088/0.2088 = 1$

$5.01 \text{ g O} \times \dfrac{1 \text{ mol O}}{16.00 \text{ g O}} = 0.3131 \text{ mol O}; \; 0.3131/0.2088 \approx 1.5$

Multiplying by two, the integer ratio is 2 Fe : 3 O; the empirical formula is Fe_2O_3.

Check. The subscripts are simple integers.

(c)   *Analyze.* Given: mass %. Find: empirical formulas.

Plan. Assume 100 g sample, calculate moles of each element, find the simplest ratio of moles.

Solve. $40.0 \text{ g C} \times \dfrac{1 \text{ mol C}}{12.01 \text{ g C}} = 3.33 \text{ mol C};\ 3.33/3.33 = 1$

$6.7 \text{ g H} \times \dfrac{1 \text{ mol H}}{1.008 \text{ mol H}} = 6.65 \text{ mol H};\ 6.65/3.33 \approx 2$

$53.3 \text{ g O} \times \dfrac{1 \text{ mol O}}{16.00 \text{ mol O}} = 3.33 \text{ mol O};\ 3.33/3.33 = 1$

The empirical formula is CH_2O.

Check. The subscripts are simple integers.

3.46   (a)   Calculate the simplest ratio of moles.

0.104 mol K / 0.052 = 2

0.052 mol C / 0.052 = 1

0.156 mol O / 0.052 = 3

The empirical formula is K_2CO_3.

(b)   Calculate moles of each element present, then the simplest ratio of moles.

$5.28 \text{ g Sn} \times \dfrac{1 \text{ mol Sn}}{118.7 \text{ g Sn}} = 0.04448 \text{ mol Sn};\ 0.04448/0.04448 = 1$

$3.37 \text{ g F} \times \dfrac{1 \text{ mol F}}{19.00 \text{ g F}} = 0.1774 \text{ mol F};\ 0.1774/0.04448 \approx 4$

The integer ratio is $1 \text{ Sn} : 4 \text{ F}$; the empirical formula is SnF_4.

(c)   Assume 100 g sample, calculate moles of each element, find the simplest ratio of moles.

$87.5\% \text{ N} = 87.5 \text{ g N} \times \dfrac{1 \text{ mol N}}{14.01 \text{ g}} = 6.25 \text{ mol N};\ 6.25/6.25 = 1$

$12.5\% \text{ H} = 12.5 \text{ g H} \times \dfrac{1 \text{ mol}}{1.008 \text{ g}} = 12.4 \text{ mol H};\ 12.4/6.25 \approx 2$

The empirical formula is NH_2.

3.47   *Analyze/Plan.* The procedure in all these cases is to assume 100 g of sample, calculate the number of moles of each element present in that 100 g, then obtain the ratio of moles as smallest whole numbers.   *Solve.*

(a) $10.4 \text{ g C} \times \dfrac{1 \text{ mol C}}{12.01 \text{ g C}} = 0.866 \text{ mol C};\ 0.866/0.866 = 1$

$27.8 \text{ g S} \times \dfrac{1 \text{ mol S}}{32.07 \text{ g S}} = 0.867 \text{ mol S};\ 0.867/0.866 \approx 1$

$61.7 \text{ g Cl} \times \dfrac{1 \text{ mol Cl}}{35.45 \text{ g Cl}} = 1.74 \text{ mol Cl};\ 1.74/0.866 \approx 2$

The empirical formula is $CSCl_2$.

(b) $21.7 \text{ g C} \times \dfrac{1 \text{ mol C}}{12.01 \text{ g C}} = 1.81 \text{ mol C}; \ 1.81 / 0.600 \approx 3$

$9.6 \text{ g O} \times \dfrac{1 \text{ mol O}}{16.00 \text{ g O}} = 0.600 \text{ mol O}; \ 0.600 / 0.600 = 1$

$68.7 \text{ g F} \times \dfrac{1 \text{ mol F}}{19.00 \text{ g F}} = 3.62 \text{ mol F}; \ 3.62 / 0.600 \approx 6$

The empirical formula is C_3OF_6.

(c)   The mass of F is [100 g total – (32.79 g Na + 13.02 Al)] = 54.19 g F

$32.79 \text{ g Na} \times \dfrac{1 \text{ mol Na}}{22.99 \text{ g Na}} = 1.426 \text{ mol Na}; \ 1.426 / 0.4826 \approx 3$

$13.02 \text{ g Al} \times \dfrac{1 \text{ mol Al}}{26.98 \text{ g Al}} = 0.4826 \text{ mol Al}; \ 0.4826 / 0.4826 = 1$

$54.19 \text{ g F} \times \dfrac{1 \text{ mol F}}{19.00 \text{ g F}} = 2.852 \text{ mol F}; \ 2.852 / 0.4826 \approx 6$

The empirical formula is Na_3AlF_6.

3.48   See Solution 3.47 for stepwise problem-solving approach.

(a) $55.3 \text{ g K} \times \dfrac{1 \text{ mol K}}{39.10 \text{ g K}} = 1.414 \text{ mol K}; 1.414 / 0.4714 \approx 3$

$14.6 \text{ g P} \times \dfrac{1 \text{ mol P}}{30.97 \text{ g P}} = 0.4714 \text{ mol P}; \ 0.4714 / 0.4714 = 1$

$30.1 \text{ g O} \times \dfrac{1 \text{ mol O}}{16.00 \text{ g O}} = 1.881 \text{ mol O}; \ 1.881 / 0.4714 \approx 4$

The empirical formula is K_3PO_4.

(b) $24.5 \text{ g Na} \times \dfrac{1 \text{ mol Na}}{22.99 \text{ g Na}} = 1.066 \text{ mol Na}; \ 1.066 / 0.5304 \approx 2$

$14.9 \text{ g Si} \times \dfrac{1 \text{ mol Si}}{28.09 \text{ Si}} = 0.5304 \text{ mol si}; \ 0.5304 / 0.5304 = 1$

$60.6 \text{ g F} \times \dfrac{1 \text{ mol F}}{19.00 \text{ g F}} = 3.189 \text{ mol F}; 3.189 / 0.5304 \approx 6$

The empirical formula is Na_2SiF_6.

(c)   The mass of O is [100 g total – (62.1 g C + 5.21 g H + 12.1 g N)] = 20.59 = 20.6 g O

$62.1 \text{ g C} \times \dfrac{1 \text{ mol C}}{12.01 \text{ g C}} = 5.17 \text{ mol C}; \ 5.17 / 0.864 \approx 6$

$5.21 \text{ g H} \times \dfrac{1 \text{ mol H}}{1.008 \text{ g H}} = 5.17 \text{ mol O}; \ 5.17 / 0.864 \approx 6$

$12.1 \text{ g N} \times \dfrac{1 \text{ mol N}}{14.01 \text{ g N}} = 0.864 \text{ mol N}; \ 0.864 / 0.864 = 1$

$20.6 \text{ g O} \times \dfrac{1 \text{ mol O}}{16.00 \text{ g O}} = 1.29 \text{ mol O}; \ 1.29 / 0.864 \approx 1.5$

Multiplying by 2, the empirical formula is $C_{12}H_{12}N_2O_3$.

3.49   *Analyze.* Given: mass% F; empirical formula XF_3 implies 3:1 ratio of mol F to mol X.
        Find: atomic mass (AM) of X.

        Plan. Calculate mol F. This is 3 times mol X. mol X = 35 g X/AM X.
        mol F/3 = 35 g X/AM X. Solve for AM X.

        Solve. Mol F = 65/19.0 = 3.421 = 3.4; mol X = 3.421/3 = 1.14035 = 1.1

        1.14035 mol X = 35 g X/AM X; AM X = 35 g X/1.14035 mol X = 30.69 = 31 g/mol

        The element is likely phosphorus and the compound is then PF_3.

3.50   Follow the logic in Solution 3.49. Match the calculated atomic mass to that of an element.

        mol Cl = 75.0/35.453 = 2.1155 = 2.12; mol X = 2.1155/4 = 0.52887 = 0.529

        0.52887 mol X = 25.0 g X/AM X; AM X = 25.0 g X/0.52887 mol X = 47.271 = 47.3 g/mol

        The element with atomic mass closest to 47.3 is **Ti**, atomic mass = 47.867 g/mol.

3.51   *Analyze.* Given: empirical formula, molar mass. Find: molecular formula.

        Plan. Calculate the empirical formula weight (FW); divide FW by molar mass (MM) to
        calculate the integer that relates the empirical and molecular formulas. Check. If
        FW/MM is an integer, the result is reasonable.   *Solve.*

        (a)   FW CH_2 = 12.0 + 2(1.01) = 14.0 $\dfrac{MM}{FW} = \dfrac{84.0}{14.0} = 6.00 = 6$

                The subscripts in the empirical formula are multiplied by 6. The molecular
                formula is C_6H_{12}.

        (b)   FW NH_2Cl = 14.01 + 2(1.008) + 35.45 = 51.48. $\dfrac{MM}{FW} = \dfrac{51.5}{51.5} = 1$

                The empirical and molecular formulas are NH_2Cl.

3.52   (a)   FW HCO_2 = 12.01 + 1.008 + 2(16.00) = 45.0 $\dfrac{MM}{FW} = \dfrac{90.0}{45.0} = 2$

                The molecular formula is $C_2H_2O_4$.

        (b)   FW C_2H_4O = 2(12.0) + 4(1.01) + 16.0 = 44.0 $\dfrac{MM}{FW} = \dfrac{88.0}{44.0} = 2.00 = 2$

                The molecular formula is $C_4H_8O_2$.

3.53   *Analyze.* Given: mass %, molar mass. Find: molecular formula.

        Plan. Use the plan detailed in Solution 3.47 to find an empirical formula from mass %
        data. Then use the plan detailed in Solution 3.51 to find the molecular formula. Note
        that some indication of molar mass must be given, or the molecular formula cannot be
        determined. *Check.* If there is an integer ratio of moles and MM/ FW is an integer, the
        result is reasonable.   *Solve.*

        (a)   92.3 g C × $\dfrac{1 \text{ mol C}}{12.01 \text{ g C}}$ = 7.685 mol C; 7.685/7.639 = 1.006 ≈ 1

                7.7 g H × $\dfrac{1 \text{ mol H}}{1.008 \text{ g H}}$ = 7.639 mol H; 7.639/7.639 = 1

                The empirical formula is CH, FW = 13.

                $\dfrac{MM}{FW} = \dfrac{104}{13} = 8$; the molecular formula is C_8H_8.

(b) $49.5 \text{ g C} \times \dfrac{1 \text{ mol C}}{12.01 \text{ g C}} = 4.12 \text{ mol C}; \quad 4.12 / 1.03 \approx 4$

$5.15 \text{ g H} \times \dfrac{1 \text{ mol H}}{1.008 \text{ g H}} = 5.11 \text{ mol H}; \quad 5.11 / 1.03 \approx 5$

$28.9 \text{ g N} \times \dfrac{1 \text{ mol N}}{14.01 \text{ g N}} = 2.06 \text{ mol N}; \quad 2.06 / 1.03 \approx 2$

$16.5 \text{ g O} \times \dfrac{1 \text{ mol O}}{16.00 \text{ g O}} = 1.03 \text{ mol O}; \quad 1.03 / 1.03 = 1$

Thus, $C_4H_5N_2O$, FW = 97. If the molar mass is about 195, a factor of 2 gives the molecular formula $C_8H_{10}N_4O_2$.

(c) $35.51 \text{ g C} \times \dfrac{1 \text{ mol C}}{12.01 \text{ g C}} = 2.96 \text{ mol C}; \quad 2.96 / 0.592 = 5$

$4.77 \text{ g H} \times \dfrac{1 \text{ mol H}}{1.008 \text{ g H}} = 4.73 \text{ mol H}; \quad 4.73 / 0.592 = 7.99 \approx 8$

$37.85 \text{ g O} \times \dfrac{1 \text{ mol O}}{16.00 \text{ g O}} = 2.37 \text{ mol O}; \quad 2.37 / 0.592 = 4$

$8.29 \text{ g N} \times \dfrac{1 \text{ mol N}}{14.01 \text{ g N}} = 0.592 \text{ mol N}; \quad 0.592 / 0.592 = 1$

$13.60 \text{ g Na} \times \dfrac{1 \text{ mol Na}}{22.99 \text{ g Na}} = 0.592 \text{ mol Na}; \quad 0.592 / 0.592 = 1$

The empirical formula is $C_5H_8O_4NNa$, FW = 169 g. Because the empirical formula weight and molar mass are approximately equal, the empirical and molecular formulas are both $NaC_5H_8O_4N$.

3.54 Assume 100 g in the following problems.

(a) $75.69 \text{ g C} \times \dfrac{1 \text{ mol C}}{12.01 \text{ g C}} = 6.30 \text{ mol C}; \quad 6.30 / 0.969 = 6.5$

$8.80 \text{ g H} \times \dfrac{1 \text{ mol H}}{1.008 \text{ g H}} = 8.73 \text{ mol H}; \quad 8.73 / 0.969 = 9.0$

$15.51 \text{ g O} \times \dfrac{1 \text{ mol O}}{16.00 \text{ g O}} = 0.969 \text{ mol O}; \quad 0.969 / 0.969 = 1$

Multiply by 2 to obtain the integer ratio 13:18:2. The empirical formula is $C_{13}H_{18}O_2$, FW = 206 g. Because the empirical formula weight and the molar mass are equal (206 g), the empirical and molecular formulas are $C_{13}H_{18}O_2$.

(b) $58.55 \text{ g C} \times \dfrac{1 \text{ mol C}}{12.01 \text{ g C}} = 4.875 \text{ mol C}; \quad 4.875 / 1.956 \approx 2.5$

$13.81 \text{ g H} \times \dfrac{1 \text{ mol H}}{1.008 \text{ g H}} = 13.700 \text{ mol H}; \quad 13.700 / 1.956 \approx 7.0$

$27.40 \text{ g N} \times \dfrac{1 \text{ mol N}}{14.01 \text{ g N}} = 1.956 \text{ mol N}; \quad 1.956 / 1.956 = 1.0$

Multiply by 2 to obtain the integer ratio 5:14:2. The empirical formula is $C_5H_{14}N_2$; FW = 102. Because the empirical formula weight and the molar mass are equal (102 g), the empirical and molecular formulas are $C_5H_{14}N_2$.

(c) $59.0 \text{ g C} \times \dfrac{1 \text{ mol C}}{12.01 \text{ g C}} = 4.91 \text{ mol C}; \quad 4.91 / 0.550 \approx 9$

$7.1 \text{ g H} \times \dfrac{1 \text{ mol H}}{1.008 \text{ g H}} = 7.04 \text{ mol H}; \quad 7.04 / 0.550 \approx 13$

$26.2 \text{ g O} \times \dfrac{1 \text{ mol O}}{16.00 \text{ g O}} = 1.64 \text{ mol O}; \quad 1.64 / 0.550 \approx 3$

$7.7 \text{ g N} \times \dfrac{1 \text{ mol N}}{14.01 \text{ g N}} = 0.550 \text{ mol N}; \quad 0.550 / 0.550 = 1$

The empirical formula is $C_9H_{13}O_3N$, FW = 183 amu (or g). Because the molar mass is approximately 180 amu, the empirical formula and molecular formula are the same, $C_9H_{13}O_3N$.

3.55 (a) *Analyze.* Given: mg CO_2, mg H_2O Find: empirical formula of hydrocarbon, C_xH_y

Plan. Upon combustion, all $C \rightarrow CO_2$, all $H \rightarrow H_2O$.

mg $CO_2 \rightarrow$ g $CO_2 \rightarrow$ mol C; mg $H_2O \rightarrow$ g H_2O, mol H

Find simplest ratio of moles and empirical formula. *Solve.*

$5.86 \times 10^{-3} \text{ g CO}_2 \times \dfrac{1 \text{ mol CO}_2}{44.01 \text{ g CO}_2} \times \dfrac{1 \text{ mol C}}{1 \text{ mol CO}_2} = 1.33 \times 10^{-4} \text{ mol C}$

$1.37 \times 10^{-3} \text{ g H}_2O \times \dfrac{1 \text{ mol H}_2O}{18.02 \text{ g H}_2O} \times \dfrac{2 \text{ mol H}}{1 \text{ mol H}_2O} = 1.52 \times 10^{-4} \text{ mol H}$

Dividing both values by 1.33×10^{-4} gives C:H of 1:1.14. This is not "close enough" to be considered 1:1. No obvious multipliers (2, 3, 4) produce an integer ratio. Testing other multipliers (trial and error!), the correct factor seems to be 7. The empirical formula is C_7H_8.

Check. See discussion of C:H ratio above.

(b) *Analyze.* Given: g of menthol, g CO_2, g H_2O, molar mass. Find: molecular formula.

Plan/Solve. Calculate mol C and mol H in the sample.

$0.2829 \text{ g CO}_2 \times \dfrac{1 \text{ mol CO}_2}{44.01 \text{ g CO}_2} \times \dfrac{1 \text{ mol C}}{1 \text{ mol CO}_2} = 0.0064281 = 0.006428 \text{ mol C}$

$0.1159 \text{ g H}_2O \times \dfrac{1 \text{ mol H}_2O}{18.02 \text{ g H}_2O} \times \dfrac{2 \text{ mol H}}{1 \text{ mol H}_2O} = 0.012863 = 0.01286 \text{ mol H}$

Calculate g C, g H and get g O by subtraction.

$0.0064281 \text{ mol C} \times \dfrac{12.01 \text{ g C}}{1 \text{ mol C}} = 0.07720 \text{ g C}$

$$0.012863 \text{ mol H} \times \frac{1.008 \text{ g H}}{1 \text{ mol H}} = 0.01297 \text{ g H}$$

mass O = 0.1005 g sample – (0.07720 g C + 0.01297 g H) = 0.01033 g O

Calculate mol O and find integer ratio of mol C: mol H: mol O.

$$0.01033 \text{ g O} \times \frac{1 \text{ mol O}}{16.00 \text{ g O}} = 6.456 \times 10^{-4} \text{ mol O}$$

Divide moles by 6.456×10^{-4}.

$$\text{C:} \frac{0.006428}{6.456 \times 10^{-4}} \approx 10; \quad \text{H:} \frac{0.01286}{6.456 \times 10^{-4}} \approx 20; \quad \text{O:} \frac{6.456 \times 10^{-4}}{6.456 \times 10^{-4}} = 1$$

The empirical formula is $C_{10}H_{20}O$.

$$\text{FW} = 10(12) + 20(1) + 16 = 156; \quad \frac{M}{FW} = \frac{156}{156} = 1$$

The molecular formula is the same as the empirical formula, $C_{10}H_{20}O$.

Check. The mass of O wasn't negative or greater than the sample mass; empirical and molecular formulas are reasonable.

3.56　　(a)　*Plan.* Calculate mol C and mol H, then g C and g H; get g O by subtraction.

Solve.

$$6.32 \times 10^{-3} \text{ g CO}_2 \times \frac{1 \text{ mol CO}_2}{44.01 \text{ g CO}_2} \times \frac{1 \text{ mol C}}{1 \text{ mol CO}_2} = 1.436 \times 10^{-4} = 1.44 \times 10^{-4} \text{ mol C}$$

$$2.58 \times 10^{-3} \text{ g H}_2\text{O} \times \frac{1 \text{ mol H}_2\text{O}}{18.02 \text{ g H}_2\text{O}} \times \frac{2 \text{ mol H}}{1 \text{ mol H}_2\text{O}} = 2.863 \times 10^{-4} = 2.86 \times 10^{-4} \text{ mol H}$$

$$1.436 \times 10^{-4} \text{ mol C} \times \frac{12.01 \text{ g C}}{1 \text{ mol C}} = 1.725 \times 10^{-3} \text{ g C} = 1.73 \text{ mg C}$$

$$2.863 \times 10^{-4} \text{ mol H} \times \frac{1.008 \text{ g H}}{1 \text{ mol H}} = 2.886 \times 10^{-4} \text{ g H} = 0.289 \text{ mg H}$$

mass of O = 2.78 mg sample – (1.725 mg C + 0.289 mg H) = 0.77 mg O

$$0.77 \times 10^{-3} \text{ g O} \times \frac{1 \text{ mol O}}{16.00 \text{ g O}} = 4.81 \times 10^{-5} \text{ mol O. Divide moles by } 4.81 \times 10^{-5}.$$

$$\text{C:} \frac{1.44 \times 10^{-4}}{4.81 \times 10^{-5}} \approx 3; \quad \text{H:} \frac{2.86 \times 10^{-4}}{4.81 \times 10^{-5}} \approx 6; \quad \text{O:} \frac{4.81 \times 10^{-5}}{4.81 \times 10^{-5}} = 1$$

The empirical formula is C_3H_6O.

(b)　*Plan.* Calculate mol C and mol H, then g C and g H. In this case, get N by subtraction. *Solve.*

$$14.242 \times 10^{-3} \text{ g CO}_2 \times \frac{1 \text{ mol CO}_2}{44.01 \text{ g CO}_2} \times \frac{1 \text{ mol C}}{1 \text{ mol CO}_2} = 3.2361 \times 10^{-4} \text{ mol C}$$

$$4.083 \times 10^{-3} \text{ g H}_2\text{O} \times \frac{1 \text{ mol H}_2\text{O}}{18.02 \text{ g H}_2\text{O}} \times \frac{2 \text{ mol H}}{1 \text{ mol H}_2\text{O}} = 4.5316 \times 10^{-4} = 4.532 \times 10^{-4} \text{ mol H}$$

$$3.2361 \times 10^{-4} \text{ mol C} \times \frac{12.01 \text{ g C}}{1 \text{ mol H}} = 3.8866 \times 10^{-3} \text{ g C} = 3.8866 \text{ mg C}$$

$$4.532 \times 10^{-4} \text{ mol H} \times \frac{1.008 \text{ g H}}{1 \text{ mol H}} = 0.45683 \times 10^{-3} \text{ g H} = 0.4568 \text{ mg H}$$

mass of N = 5.250 mg sample – (3.8866 mg C + 0.4568 mg H) = 0.9066

= 0.907 mg N

$$0.9066 \times 10^{-3} \text{ g N} \times \frac{1 \text{ mol N}}{14.01 \text{ g N}} = 6.47 \times 10^{-5} \text{ mol N}. \text{ Divide moles by } 6.47 \times 10^{-5}.$$

$$\text{C: } \frac{3.24 \times 10^{-4}}{6.47 \times 10^{-5}} \approx 5; \quad \text{H: } \frac{4.53 \times 10^{-4}}{6.47 \times 10^{-5}} \approx 7; \quad \text{N: } \frac{6.47 \times 10^{-5}}{6.47 \times 10^{-5}} = 1$$

The empirical formula is C_5H_7N, FW = 81. A molar mass of 160 ± 5 indicates a factor of 2 and a molecular formula of $C_{10}H_{14}N_2$.

3.57   *Analyze.* Given: mass of H_2O and mass of CO_2 from combustion of 0.165 g of valproic acid; molar mass of valproic acid. Find: empirical and molecular formulas of valproic acid.

Plan. Calculate mol C and mol H, then g C and g H; get g O by subtraction.

Solve.

$$0.403 \text{ g CO}_2 \times \frac{1 \text{ mol CO}_2}{44.01 \text{ g CO}_2} \times \frac{1 \text{ mol C}}{1 \text{ mol CO}_2} = 9.157 \times 10^{-3} = 9.16 \times 10^{-3} \text{ mol C}$$

$$0.166 \text{ g H}_2O \times \frac{1 \text{ mol H}_2O}{18.02 \text{ g H}_2O} \times \frac{2 \text{ mol H}}{1 \text{ mol H}_2O} = 0.018424 = 0.0184 \text{ mol H}$$

$$9.157 \times 10^{-3} \text{ mol C} \times \frac{12.01 \text{ g C}}{1 \text{ mol C}} = 0.10998 \text{ g C} = 0.110 \text{ g C}$$

$$0.018424 \text{ mol H} \times \frac{1.008 \text{ g H}}{1 \text{ mol H}} = 0.01857 \text{ g H} = 0.0186 \text{ g H}$$

mass of O = 0.165 g sample – (0.10998 g C + 0.01857 g H) = 0.03645 = 0.036 g O

$$0.03645 \text{ g O} \times \frac{1 \text{ mol O}}{16.00 \text{ g O}} = 2.278 \times 10^{-3} = 2.3 \times 10^{-3} \text{ mol O}. \text{ Divide moles by } 2.278 \times 10^{-3}.$$

$$\text{C: } \frac{9.157 \times 10^{-3}}{2.278 \times 10^{-3}} \approx 4; \quad \text{H: } \frac{0.018424}{2.278 \times 10^{-3}} \approx 8; \quad \text{O: } \frac{2.278 \times 10^{-3}}{2.278 \times 10^{-3}} = 1$$

The empirical formula is C_4H_8O, FW = 72. A molar mass of 144 g/mol indicates a factor of two and a molecular formula of $C_8H_{16}O_2$.

3.58   *Analyze/Plan.* Calculate theoretical mass% C and H in propenoic acid, $C_3H_4O_2$, and experimental mass% C and H in the combustion sample. Compare the theoretical and experimental mass percents to determine if the sample is propenic acid. (This is a common method in modern chemical analysis.)

Solve. The molar mass of $C_3H_4O_2$ is [3(12.0107) + 4(1.00794 + 2(15.9994))] = 72.0627 = 72.06 g/mol

$$\% \text{ C(theo)} = \frac{3(12.01) \text{ g}}{72.06 \text{ g}} \times 100 = 50.0\%$$

61

$$\% \text{ H(theo)} = \frac{4(1.008) \text{ g}}{72.06 \text{ g}} \times 100 = 5.60\%$$

$$0.374 \text{ g CO}_2 \times \frac{1 \text{ mol CO}_2}{44.01 \text{ g CO}_2} \times \frac{1 \text{ mol C}}{1 \text{ mol CO}_2} = 8.498 \times 10^{-3} = 8.50 \times 10^{-3} \text{ mol C}$$

$$0.102 \text{ g H}_2\text{O} \times \frac{1 \text{mol H}_2\text{O}}{18.02 \text{ g H}_2\text{O}} \times \frac{2 \text{ mol H}}{1 \text{ mol H}_2\text{O}} = 0.01132 = 0.0113 \text{ mol H}$$

$$8.498 \times 10^{-3} \text{ mol C} \times \frac{12.01 \text{ g C}}{1 \text{ mol C}} = 0.10206 \text{ g C} = 0.102 \text{ g C}$$

$$0.01132 \text{ mol H} \times \frac{1.008 \text{ g H}}{1 \text{ mol H}} = 0.01141 \text{ g H} = 0.0114 \text{ g H}$$

$$\% \text{ C(exptl)} = \frac{0.10206 \text{ g C}}{0.275 \text{ g sample}} \times 100 = 37.1\%$$

$$\% \text{ H(exptl)} = \frac{0.01141 \text{ g H}}{0.275 \text{ g sample}} \times 100 = 4.15\%$$

Clearly theoretical and experimental mass percents don't match. The unknown liquid is not propenoic acid.

3.59   *Analyze.* Given 2.558 g $Na_2CO_3 \cdot xH_2O$, 0.948 g Na_2CO_3. Find: x.

Plan. The reaction involved is $Na_2CO_3 \cdot xH_2O(s) \rightarrow Na_2CO_3(s) + xH_2O(g)$.

Calculate the mass of H_2O lost and then the mole ratio of Na_2CO_3 and H_2O.

Solve. g H_2O lost = 2.558 g sample – 0.948 g Na_2CO_3 = 1.610 g H_2O

$$0.948 \text{ g Na}_2\text{CO}_3 \times \frac{1 \text{ mol Na}_2\text{CO}_3}{106.0 \text{ g Na}_2\text{CO}_3} = 0.00894 \text{ mol Na}_2\text{CO}_3$$

$$1.610 \text{ g H}_2\text{O} \times \frac{1 \text{ mol H}_2\text{O}}{18.02 \text{ g H}_2\text{O}} = 0.08935 \text{ mol H}_2\text{O}$$

The formula is $Na_2CO_3 \cdot \underline{\textbf{10}} \text{ H}_2\text{O}$.

Check. x is an integer.

3.60   The reaction involved is $MgSO_4 \cdot xH_2O(s) \rightarrow MgSO_4(s) + xH_2O(g)$. First, calculate the number of moles of product $MgSO_4$; this is the same as the number of moles of starting hydrate.

$$2.472 \text{ g MgSO}_4 \times \frac{1 \text{ mol MgSO}_4}{120.4 \text{ g MgSO}_4} \times \frac{1 \text{ mol MgSO}_4 \cdot x\text{H}_2\text{O}}{1 \text{ mol MgSO}_4} = 0.02053 \text{ mol MgSO}_4 \cdot x \text{ H}_2\text{O}$$

Thus, $\dfrac{5.061 \text{ g MgSO}_4 \cdot x\text{H}_2\text{O}}{0.02053} = 246.5 \text{ g/mol} = \text{FW of MgSO}_4 \cdot x\text{H}_2\text{O}$

FW of $MgSO_4 \cdot xH_2O$ = FW of $MgSO_4$ + x(FW of H_2O).

246.5 = 120.4 + x(18.02). x = 6.998. The hydrate formula is $MgSO_4 \cdot \underline{7}H_2O$.

Alternatively, we could calculate the number of moles of water represented by weight loss: (5.061 – 2.472) = 2.589 g H_2O lost.

$$2.589 \text{ g H}_2\text{O} \times \frac{1 \text{ mol H}_2\text{O}}{18.02 \text{ g H}_2\text{O}} = 0.1437 \text{ mol H}_2\text{O}; \quad \frac{\text{mol H}_2\text{O}}{\text{mol MgSO}_4} = \frac{0.1437}{0.02053} = 7.000$$

Again the correct formula is $MgSO_4 \cdot \underline{7}H_2O$.

Quantitative Information from Balanced Equations (Section 3.6)

3.61 $Na_2SiO_3(s) + 8\,HF(aq) \rightarrow H_2SiF_6(aq) + 2\,NaF(aq) + 3\,H_2O(l)$

    (a)   *Analyze.* Given: mol Na_2SiO_3. Find: mol HF. *Plan.* Use the mole ratio 8 HF : 1 Na_2SiO_3 from the balanced equation to relate moles of the two reactants.

         Solve. $0.300\ \text{mol } Na_2SiO_3 \times \dfrac{8\ \text{mol HF}}{1\ \text{mol } Na_2SiO_3} = 2.40\ \text{mol HF}$

         Check. Mol HF should be greater than mol Na_2SiO_3.

    (b)   *Analyze.* Given: mol HF. Find: g NaF. *Plan.* Use the mole ratio 2 NaF : 8 HF to change mol HF to mol NaF, then molar mass to get NaF.   *Solve.*

         $0.500\ \text{mol HF} \times \dfrac{2\ \text{mol NaF}}{8\ \text{mol HF}} \times \dfrac{41.99\ \text{g NaF}}{1\ \text{mol NaF}} = 5.25\ \text{g NaF}$

         Check. $(0.5/4) = 0.125;\ 0.13 \times 42 > 4$ g NaF

    (c)   *Analyze.* Given: g HF Find: g Na_2SiO_3.

         Plan. $\text{g HF} \rightarrow \text{mol HF}\ \left(\dfrac{\text{mol}}{\text{ratio}}\right) \rightarrow \text{mol } Na_2SiO_3 \rightarrow \text{g } Na_2SiO_3$

         The mole ratio is at the heart of every stoichiometry problem. Molar mass is used to change to and from grams.   *Solve.*

         $0.800\ \text{g HF} \times \dfrac{1\ \text{mol HF}}{20.01\text{g HF}} \times \dfrac{1\ \text{mol } Na_2SiO_3}{8\ \text{mol HF}} \times \dfrac{122.1\ \text{g } Na_2SiO_3}{1\ \text{mol } Na_2SiO_3} = 0.610\ \text{g } Na_2SiO_3$

         Check. $0.8\ (120/160) < 0.75$ mol

3.62 $4\,KO_2 + 2\,CO_2 \rightarrow 2\,K_2CO_3 + 3\,O_2$

    (a) $0.400\ \text{mol } KO_2 \times \dfrac{3\ \text{mol } O_2}{4\ \text{mol } KO_2} = 0.300\ \text{mol } O_2$

    (b) $7.50\ \text{g } O_2 \times \dfrac{1\ \text{mol } O_2}{32.00\ \text{g } O_2} \times \dfrac{4\ \text{mol } KO_2}{3\ \text{mol } O_2} \times \dfrac{71.10\ \text{g } KO_2}{1\ \text{mol } KO_2} = 22.2\ \text{g } KO_2$

    (c) $7.50\ \text{g } O_2 \times \dfrac{1\ \text{mol } O_2}{32.00\ \text{g } O_2} \times \dfrac{2\ \text{mol } CO_2}{3\ \text{mol } O_2} \times \dfrac{44.01\ \text{g } CO_2}{1\ \text{mol } CO_2} = 6.88\ \text{g } CO_2$

3.63   (a) $Al(OH)_3(s) + 3\,HCl(aq) \rightarrow AlCl_3(aq) + 3\,H_2O(l)$

    (b)   *Analyze.* Given mass of one reactant, find stoichiometric mass of other reactant and products.

         Plan. Follow the logic in Sample Exercise 3.16. Calculate mol $Al(OH)_3$ in 0.500 g $Al(OH_3)_3$ separately, because it will be used several times.

         Solve. $0.500\ \text{g } Al(OH)_3 \times \dfrac{1\ \text{mol } Al(OH)_3}{78.00\ \text{g } Al(OH)_3} = 6.410 \times 10^{-3} = 6.41 \times 10^{-3}\ \text{mol } Al(OH)_3$

         $6.410 \times 10^{-3}\ \text{mol } Al(OH)_3 \times \dfrac{3\ \text{mol HCl}}{1\ \text{mol } Al(OH)_3} \times \dfrac{36.46\ \text{g HCl}}{1\ \text{mol HCl}} = 0.7012 = 0.701\ \text{g HCl}$

         $6.410 \times 10^{-3}\ \text{mol } Al(OH)_3 \times \dfrac{1\ \text{mol HCl}}{1\ \text{mol } Al(OH)_3} \times \dfrac{133.34\ \text{g } AlCl_3}{1\ \text{mol } AlCl_3} = 0.8547$

                                           $= 0.855\ \text{g } AlCl_3$

(c) 6.410×10^{-3} mol $Al(OH)_3 \times \dfrac{3 \text{ mol } H_2O}{1 \text{ mol } Al(OH)_3} \times \dfrac{18.02 \text{ g } H_2O}{1 \text{ mol } H_2O} = 0.3465 = 0.347$ g H_2O

(d)   Conservation of mass: mass of products = mass of reactants

reactants: $Al(OH)_3$ + HCl, 0.500 g + 0.701 g = 1.201 g

products: $AlCl_3 + H_2O$, 0.855 g + 0.347 g = 1.202 g

The 0.001 g difference is due to rounding (0.8547 + 0.3465 = 1.2012). This is an excellent *check* of results.

3.64   (a) $Fe_2O_3(s) + 3\, CO(g) \rightarrow 2\, Fe(s) + 3\, CO_2(g)$

(b) 0.350 kg $Fe_2O_3 \times \dfrac{1000 \text{ g}}{1 \text{ kg}} \times \dfrac{1 \text{ mol } Fe_2O_3}{159.688 \text{ g } Fe_2O_3} = 2.1918 = 2.19$ mol Fe_2O_3

2.1918 mol $Fe_2O_3 \times \dfrac{3 \text{ mol } CO}{1 \text{ mol } Fe_2O_3} \times \dfrac{28.01 \text{ g } CO}{1 \text{ mol } CO} = 184.17 = 184$ g CO

(c) 2.1918 mol $Fe_2O_3 \times \dfrac{2 \text{ mol } Fe}{1 \text{ mol } Fe_2O_3} \times \dfrac{55.845 \text{ g } Fe}{1 \text{ mol } Fe} = 244.80 = 245$ g Fe

2.1918 mol $Fe_2O_3 \times \dfrac{3 \text{ mol } CO_2}{1 \text{ mol } Fe_2O_3} \times \dfrac{44.01 \text{ g } CO_2}{1 \text{ mol } CO_2} = 289.38 = 289$ g CO_2

(d)   reactants: 350 g Fe_2O_3 + 184.17 g CO = 534.17 = 534 g

products: 244.80 g Fe + 289.38 g CO_2 = 534.18 = 534 g

Mass is conserved.

3.65   (a) $Al_2S_3(s) + 6\, H_2O(l) \rightarrow 2\, Al(OH)_3(s) + 3\, H_2S(g)$

(b)   *Plan.* g A → mol A → mol B → g B. See Solution 3.61 (c).   *Solve.*

14.2 g $Al_2S_3 \times \dfrac{1 \text{ mol } Al_2S_3}{150.2 \text{ g } Al_2S_3} \times \dfrac{2 \text{ mol } Al(OH)_3}{1 \text{ mol } Al_2S_3} \times \dfrac{78.00 \text{ g } Al(OH)_3}{1 \text{ mol } Al(OH)_3} = 14.7$ g $Al(OH)_3$

Check. $14 \left(\dfrac{2 \times 78}{150} \right) \approx 14(1) \approx 14$ g $Al(OH)_3$

3.66   (a) $CaH_2(s) + 2\, H_2O(l) \rightarrow Ca(OH)_2(aq) + 2\, H_2(g)$

(b) 4.500 g $H_2 \times \dfrac{1 \text{ mol } H_2}{2.016 \text{ g } H_2} \times \dfrac{1 \text{ mol } CaH_2}{2 \text{ mol } H_2} \times \dfrac{42.10 \text{ g } CaH_2}{1 \text{ mol } CaH_2} = 46.99$ g CaH_2

3.67   (a)   *Analyze.* Given: mol NaN_3. Find: mol N_2.

Plan. Use mole ratio from balanced equation.   *Solve.*

1.50 mol $NaN_3 \times \dfrac{3 \text{ mol } N_2}{2 \text{ mol } NaN_3} = 2.25$ mol N_2

Check. The resulting mol N_2 should be greater than mol NaN_3, (the N_2:NaN_3 ratio is > 1), and it is.

(b)   *Analyze.* Given: g N_2 Find: g NaN_3.

Plan. Use molar masses to get from and to grams, mol ratio to relate moles of the two substances.   *Solve.*

10.0 g $N_2 \times \dfrac{1 \text{ mol } N_2}{28.01 \text{ g } N_2} \times \dfrac{2 \text{ mol } NaN_3}{3 \text{ mol } N_2} \times \dfrac{65.01 \text{ g } NaN_3}{1 \text{ mol } NaN_3} = 15.5$ g NaN_3

Check. Mass relations are less intuitive than mole relations. Estimating the ratio of molar masses is sometimes useful. In this case, 65 g NaN_3/28 g $N_2 \approx 2.25$. Then, $(10 \times 2/3 \times 2.25) \approx 15$ g NaN_3. The calculated result looks reasonable.

(c) *Analyze.* Given: vol N_2 in ft^3, density N_2 in g/L. Find: g NaN_3.

Plan. First determine how many g N_2 are in 10.0 ft^3, using the density of N_2. Then proceed as in part (b).

Solve.

$$\frac{1.25 \text{ g}}{1 \text{ L}} \times \frac{1 \text{ L}}{1000 \text{ cm}^3} \times \frac{(2.54)^3 \text{ cm}^3}{1 \text{ in}^3} \times \frac{(12)^3 \text{ in}^3}{1 \text{ ft}^3} \times 10.0 \text{ ft}^3 = 354.0 = 354 \text{ g N}_2$$

$$354.0 \text{ g N}_2 \times \frac{1 \text{ mol N}_2}{28.01 \text{ g N}_2} \times \frac{2 \text{ mol NaN}_3}{3 \text{ mol N}_2} \times \frac{65.01 \text{ g NaN}_3}{1 \text{ mol NaN}_3} = 548 \text{ g NaN}_3$$

Check. 1 $ft^3 \sim 28$ L; 10 $ft^3 \sim 280$ L; 280 L $\times 1.25 \sim 350$ g N_2

Using the ratio of molar masses from part (b), $(350 \times 2/3 \times 2.25) \approx 525$ g NaN_3

3.68 $2 C_8H_{18}(l) + 25 O_2(g) \rightarrow 16 CO_2(g) + 18 H_2O(l)$

(a) $1.50 \text{ mol C}_8\text{H}_{18} \times \dfrac{25 \text{ mol O}_2}{2 \text{ mol C}_8\text{H}_{18}} = 18.75 = 18.8 \text{ mol O}_2$

(b) $10.0 \text{ g C}_8\text{H}_{18} \times \dfrac{1 \text{ mol C}_8\text{H}_{18}}{114.2 \text{ g C}_8\text{H}_{18}} \times \dfrac{25 \text{ mol O}_2}{2 \text{ mol C}_8\text{H}_{18}} \times \dfrac{32.00 \text{ g O}_2}{1 \text{ mol O}_2} = 35.0 \text{ g O}_2$

(c) $15.0 \text{ gal C}_8\text{H}_{18} \times \dfrac{3.7854 \text{ L}}{1 \text{ gal}} \times \dfrac{1000 \text{ mL}}{1 \text{ L}} \times \dfrac{0.692 \text{ g}}{1 \text{ mL}} = 39,292 = 3.93 \times 10^4 \text{ g C}_8\text{H}_{18}$

$3.9292 \times 10^4 \text{ g C}_8\text{H}_{18} \times \dfrac{1 \text{ mol C}_8\text{H}_{18}}{114.2 \text{ g C}_8\text{H}_{18}} \times \dfrac{25 \text{ mol O}_2}{2 \text{ mol C}_8\text{H}_{18}} \times \dfrac{32.00 \text{ g O}_2}{1 \text{ mol O}_2} = 137,627 \text{ g}$

$$= 1.38 \times 10^5 \text{ g O}_2$$

(d) $3.9292 \times 10^4 \text{ g C}_8\text{H}_{18} \times \dfrac{1 \text{ mol C}_8\text{H}_{18}}{114.2 \text{ g C}_8\text{H}_{18}} \times \dfrac{16 \text{ mol CO}_2}{2 \text{ mol C}_8\text{H}_{18}} \times \dfrac{44.01 \text{ g CO}_2}{1 \text{ mol O}_2} = 121,139 \text{ g}$

$$= 1.21 \times 10^5 \text{ g CO}_2$$

3.69   (a) *Analyze.* Given: dimensions of Al foil. Find: mol Al.

Plan. Dimensions $\rightarrow$ vol $\xrightarrow{\text{density}}$ mass $\xrightarrow[\text{mass}]{\text{molar}}$ mol Al

Solve. $1.00 \text{ cm} \times 1.00 \text{ cm} \times 0.550 \text{ mm} \times \dfrac{1 \text{ cm}}{10 \text{ mm}} = 0.0550 \text{ cm}^3 \text{ Al}$

$0.0550 \text{ cm}^3 \text{ Al} \times \dfrac{2.699 \text{ g Al}}{1 \text{ cm}^3} \times \dfrac{1 \text{ mol Al}}{26.98 \text{ g Al}} = 5.502 \times 10^{-3} = 5.50 \times 10^{-3} \text{ mol Al}$

Check. $2.699/26.98 \approx 0.1$; $(0.055 \text{ cm}^3 \times 0.1) = 5.5 \times 10^{-3}$ mol Al

(b) *Plan.* Write the balanced equation to get a mole ratio; change mol Al $\rightarrow$ mol $AlBr_3 \rightarrow$ g $AlBr_3$.

Solve. $2 Al(s) + 3 Br_2(l) \rightarrow 2 AlBr_3(s)$

$5.502 \times 10^{-3} \text{ mol Al} \times \dfrac{2 \text{ mol AlBr}_3}{2 \text{ mol Al}} \times \dfrac{266.69 \text{ g AlBr}_3}{1 \text{ mol AlBr}_3} = 1.467 = 1.47 \text{ g AlBr}_3$

Check. $(0.006 \times 1 \times 270) \approx 1.6$ g $AlBr_3$

65

3.70   (a)   *Plan.* Calculate a "mole ratio" between nitroglycerine and total moles of gas produced. $(12 + 6 + 1 + 10) = 29$ mol gas; 4 mol nitro: 29 total mol gas.   *Solve.*

$$2.00 \text{ mL nitro} \times \frac{1.592 \text{ g}}{\text{mL}} \times \frac{1 \text{ mol nitro}}{227.1 \text{ g nitro}} \times \frac{29 \text{ mol gas}}{4 \text{ mol nitro}} = 0.10165 = 0.102 \text{ mol gas}$$

        (b) $$0.10165 \text{ mol gas} \times \frac{55 \text{ L}}{\text{mol}} = 5.5906 = 5.6 \text{ L}$$

        (c) $$2.00 \text{ mL nitro} \times \frac{1.592 \text{ g}}{\text{mL}} \times \frac{1 \text{ mol nitro}}{227.1 \text{ g nitro}} \times \frac{6 \text{ mol N}_2}{4 \text{ mol nitro}} \times \frac{28.01 \text{ g N}_2}{1 \text{ mol N}_2} = 0.589 \text{ g N}_2$$

3.71   *Analyze/Plan.* We are given heat produced by combustion of one mole of ethanol and a mass of ethanol to burn. Change mass ethanol to moles ethanol and multiply by kJ/mol.

        Solve. The molar mass of CH_3CH_2OH is $[2(12.0107) + 6(1.00794 + 15.9994] = 46.06844 = 46.07$ g/mol

$$235.0 \text{ g } CH_3CH_2OH \times \frac{1 \text{ mol } CH_3CH_2OH}{46.068 \text{ g } CH_3CH_2OH} \times \frac{1367 \text{ kJ}}{1 \text{ mol } CH_3CH_2OH} = 6973.3 = 6.973 \times 10^3 \text{ kJ}$$

3.72   The molar mass of $CH_3(CH_2)_6CH_3$ or C_8H_{18} is $[8(12.0107) + 18(1.00794)] = 114.22852 = 114.2$ g/mol

$$1.000 \text{ gal } C_8H_{18} \times \frac{3.7854 \text{ L}}{1 \text{ gal}} \times \frac{1000 \text{ mL}}{1 \text{ L}} \times \frac{0.692 \text{ g}}{1 \text{ mL}} = 2619.5 = 2.62 \times 10^3 \text{ g } C_8H_{18}$$

$$2.6195 \times 10^3 \text{ g } C_8H_{18} \times \frac{1 \text{ mol } C_8H_{18}}{114.23 \text{ g } C_8H_{18}} \times \frac{5470 \text{ kJ}}{1 \text{ mol } C_2H_5OH} = 125,437 = 1.25 \times 10^5 \text{ kJ}$$

        [Three sig figs in the density dictate 3 sig figs in the result.]

Limiting Reactants (Section 3.7)

3.73   (a)   The *limiting reactant* determines the maximum number of product moles resulting from a chemical reaction; any other reactant is an *excess reactant*.

        (b)   The limiting reactant regulates the amount of products, because it is completely used up during the reaction; no more product can be made when one of the reactants is unavailable.

        (c)   Combining ratios are molecule and mole ratios. Since different molecules have different masses, equal masses of different reactants will not have equal numbers of molecules. By comparing initial moles, we compare numbers of available reactant molecules, the fundamental combining units in a chemical reaction.

3.74   (a)   *Theoretical yield* is the maximum amount of product possible, as predicted by stoichiometry, assuming that the limiting reactant is converted entirely to product.

               Actual yield is the amount of product actually obtained, less than or equal to the theoretical yield. *Percent yield* is the ratio of (actual yield to theoretical yield) $\times$ 100.

(b)   No reaction is perfect. Not all reactant molecules come together effectively to form products; alternative reaction pathways may produce secondary products and reduce the amount of desired product actually obtained, or it might not be possible to completely isolate the desired product from the reaction mixture. In any case, these factors reduce the actual yield of a reaction.

(c)   No, 110% actual yield is not possible. Theoretical yield is the maximum possible amount of pure product, assuming all available limiting reactant is converted to product, and that all product is isolated. If an actual yield of 110% is obtained, the product must contain impurities which increase the experimental mass.

3.75   (a) $2\,C_2H_5OH + 6\,O_2 \rightarrow 4\,CO_2 + 6\,H_2O$

The equation above corresponds to the contents of the diagram shown in Exercise 3.75, but does not have the simplest ratio of coefficients. Divide all integer coefficients by 2 to obtain $C_2H_5OH + 3\,O_2 \rightarrow 2\,CO_2 + 3\,H_2O$

(b) C_2H_5OH is the limiting reactant. According to the balanced equation, two molecules of C_2H_5OH require six molecules of O_2 for reaction. In the box there are two molecules of C_2H_5OH and seven molecules of O_2. O_2 is present in excess and C_2H_5OH limits.

(c)   If the reaction goes to completion, there will be four molecules of CO_2, six molecules of H_2O, zero molecules of C_2H_5OH (the limiting reactant is completely consumed), and one molecule of O_2 (excess reactant).

3.76   (a) $C_3H_8 + 5\,O_2 \rightarrow 3\,CO_2(g) + 4\,H_2O$

(b) O_2 is the limiting reactant. According to the balanced equation, one molecule of C_3H_8 requires five molecules of O_2 for reaction. The three C_3H_8 molecules in the box would require fifteen molecules of O_2 for complete reaction, but only ten O_2 molecules are present. C_3H_8 is present in excess and O_2 limits.

(c)   If the reaction goes to completion, there will be six molecules of CO_2, eight molecules of H_2O, one molecule of C_3H_8 (excess reactant), and zero molecules of O_2 (the limiting reactant is completely consumed).

3.77   *Analyze.* Given: 1.85 mol NaOH, 1.00 mol CO_2. Find: mol Na_2CO_3.

Plan. Amounts of more than one reactant are given, so we must determine which reactant regulates (limits) product. Then apply the appropriate mole ratio from the balanced equation.

Solve. The mole ratio is 2 NaOH: 1 CO_2, so 1.00 mol CO_2 requires 2.00 mol NaOH for complete reaction. Less than 2.00 mol NaOH are present, so NaOH is the limiting reactant.

$$1.85\text{ mol NaOH} \times \frac{1\text{ mol Na}_2\text{CO}_3}{2\text{ mol NaOH}} = 0.925 \text{ mol Na}_2\text{CO}_3 \text{ can be produced}$$

The Na_2CO_3:CO_2 ratio is 1:1, so 0.925 mol Na_2CO_3 produced requires 0.925 mol CO_2 consumed. (Alternately, 1.85 mol NaOH × 1 mol CO_2/2 mol NaOH = 0.925 mol CO_2 reacted). 1.00 mol CO_2 initial – 0.925 mol CO_2 reacted = 0.075 mol CO_2 remain.

Check.	2 NaOH(s)	+	CO_2(g)	→	Na_2CO_3(s)	+	H_2O(l)
initial	1.85 mol		1.00 mol		0 mol		
change (reaction)	–1.85 mol		–0.925 mol		+0.925 mol		
final	0 mol		0.075 mol		0.925 mol		

Note that the "change" line (but not necessarily the "final" line) reflects the mole ratios from the balanced equation.

3.78 $0.500 \text{ mol Al(OH)}_3 \times \dfrac{3 \text{ mol H}_2\text{SO}_4}{2 \text{ mol Al(OH)}_3} = 0.750 \text{ mol H}_2\text{SO}_4$ needed for complete reaction

Only 0.500 mol H_2SO_4 available, so H_2SO_4 limits.

$0.500 \text{ mol H}_2\text{SO}_4 \times \dfrac{1 \text{ mol Al}_2(\text{SO}_4)_3}{3 \text{ mol H}_2\text{SO}_4} = 0.1667 = 0.167 \text{ mol Al}_2(\text{SO}_4)_3$ can form

$0.500 \text{ mol H}_2\text{SO}_4 \times \dfrac{2 \text{ mol Al(OH)}_3}{3 \text{ mol H}_2\text{SO}_4} = 0.3333 = 0.333 \text{ mol Al(OH)}_3$ react

$0.500 \text{ mol Al(OH)}_3$ initial $- 0.333 \text{ mol react} = 0.167 \text{ mol Al(OH)}_3$ remain

3.79 $3 \text{ NaHCO}_3(\text{aq}) + \text{H}_3\text{C}_6\text{H}_5\text{O}_7(\text{aq}) \rightarrow 3 \text{ CO}_2(\text{g}) + 3 \text{ H}_2\text{O}(\text{l}) + \text{Na}_3\text{C}_6\text{H}_5\text{O}_7(\text{aq})$

(a)   *Analyze/Plan.* Abbreviate citric acid as H_3Cit. Follow the approach in Sample Exercise 3.19. *Solve.*

$1.00 \text{ g NaHCO}_3 \times \dfrac{1 \text{ mol NaHCO}_3}{84.01 \text{ g NaHCO}_3} = 1.190 \times 10^{-2} = 1.19 \times 10^{-2} \text{ mol NaHCO}_3$

$1.00 \text{ g H}_3\text{C}_6\text{H}_5\text{O}_7 \times \dfrac{1 \text{ mol H}_3\text{Cit}}{192.1 \text{ g H}_3\text{Cit}} = 5.206 \times 10^{-3} = 5.21 \times 10^{-3} \text{ mol H}_3\text{Cit}$

But $NaHCO_3$ and H_3Cit react in a 3:1 ratio, so 5.21×10^{-3} mol H_3Cit require $3(5.21 \times 10^{-3}) = 1.56 \times 10^{-2}$ mol $NaHCO_3$. We have only 1.19×10^{-2} mol $NaHCO_3$, so $NaHCO_3$ is the limiting reactant.

(b) $1.190 \times 10^{-2} \text{ mol NaHCO}_3 \times \dfrac{3 \text{ mol CO}_2}{3 \text{ mol NaHCO}_3} \times \dfrac{44.01 \text{ g CO}_2}{1 \text{ mol CO}_2} = 0.524 \text{ g CO}_2$

(c) $1.190 \times 10^{-2} \text{ mol NaHCO}_3 \times \dfrac{1 \text{ mol H}_3\text{Cit}}{3 \text{ mol NaHCO}_3} = 3.968 \times 10^{-3}$

$= 3.97 \times 10^{-3} \text{ mol H}_3\text{Cit react}$

$5.206 \times 10^{-3} \text{ mol H}_3\text{Cit} - 3.968 \times 10^{-3} \text{ mol react} = 1.238 \times 10^{-3}$

$= 1.24 \times 10^{-3} \text{ mol H}_3\text{Cit remain}$

$1.238 \times 10^{-3} \text{ mol H}_3\text{Cit} \times \dfrac{192.1 \text{ g H}_3\text{Cit}}{\text{mol H}_3\text{Cit}} = 0.238 \text{ g H}_3\text{Cit remain}$

3.80 $4 \text{ NH}_3(\text{g}) + 5 \text{ O}_2(\text{g}) \rightarrow 4 \text{ NO(g)} + 6 \text{ H}_2\text{O(g)}$

(a)   Follow the approach in Sample Exercise 3.19.

$2.00 \text{ g NH}_3 \times \dfrac{1 \text{ mol NH}_3}{17.03 \text{ g NH}_3} = 0.11744 = 0.117 \text{ mol NH}_3$

$$2.50 \text{ g O}_2 \times \frac{1 \text{ mol O}_2}{32.00 \text{ g O}_2} = 0.07813 = 0.0781 \text{ mol O}_2$$

$$0.07813 \text{ mol O}_2 \times \frac{4 \text{ mol NH}_3}{5 \text{ mol O}_2} = 0.06250 = 0.0625 \text{ mol NH}_3 \text{ required}$$

More than 0.0625 mol NH_3 is available, so O_2 is the limiting reactant.

(b) $0.07813 \text{ mol O}_2 \times \dfrac{4 \text{ mol NO}}{5 \text{ mol O}_2} \times \dfrac{30.01 \text{ g NO}}{1 \text{ mol NO}} = 1.8756 = 1.88 \text{ g NO produced}$

         $0.07813 \text{ mol O}_2 \times \dfrac{6 \text{ mol H}_2\text{O}}{5 \text{ mol O}_2} \times \dfrac{18.02 \text{ g H}_2\text{O}}{1 \text{ mol H}_2\text{O}} = 1.6894 = 1.69 \text{ g H}_2\text{O produced}$

(c) $0.11744 \text{ mol NH}_3 - 0.0625 \text{ mol NH}_3 \text{ reacted} = 0.05494 = 0.0549 \text{ mol NH}_3 \text{ remain}$

         $0.05494 \text{ mol NH}_3 \times \dfrac{17.03 \text{ g NH}_3}{1 \text{ mol NH}_3} = 0.93563 = 0.936 \text{ g NH}_3 \text{ remain}$

(d)   mass products = 1.8756 g NO + 1.6894 g H_2O + 0.9356 g NH_3 remaining = 4.50g products

         mass reactants = 2.00 g NH_3 + 2.50 g O_2 = 4.50 g reactants

         (For comparison purposes, the mass of excess reactant can be either added to the products, as above, or subtracted from reactants.)

3.81   *Analyze.* Given: initial g Na_2CO_3, g $AgNO_3$. Find: final g Na_2CO_3, $AgNO_3$, Ag_2CO_3, $NaNO_3$

         Plan. Write balanced equation; determine limiting reactant; calculate amounts of excess reactant remaining and products, based on limiting reactant.

         Solve. $2 \text{ AgNO}_3(aq) + \text{Na}_2\text{CO}_3(aq) \rightarrow \text{Ag}_2\text{CO}_3(s) + 2 \text{ NaNO}_3(aq)$

$$3.50 \text{ g Na}_2\text{CO}_3 \times \frac{1 \text{ mol Na}_2\text{CO}_3}{106.0 \text{ g Na}_2\text{CO}_3} = 0.03302 = 0.0330 \text{ mol Na}_2\text{CO}_3$$

$$5.00 \text{ g AgNO}_3 \times \frac{1 \text{ mol AgNO}_3}{169.9 \text{ g AgNO}_3} = 0.02943 = 0.0294 \text{ mol AgNO}_3$$

$$0.02943 \text{ mol AgNO}_3 \times \frac{1 \text{ mol Na}_2\text{CO}_3}{2 \text{ mol AgNO}_3} = 0.01471 = 0.0147 \text{ mol Na}_2\text{CO}_3 \text{ required}$$

$AgNO_3$ is the limiting reactant and Na_2CO_3 is present in excess.

	$2 \text{ AgNO}_3(aq)$	+	$\text{Na}_2\text{CO}_3(aq)$	$\rightarrow$	$\text{Ag}_2\text{CO}_3(s)$	+	$2 \text{ NaNO}_3(aq)$
initial	0.0294 mol		0.0330 mol		0 mol		0 mol
reaction	−0.0294 mol		−0.0147 mol		+0.0147 mol		+0.0294 mol
final	0 mol		0.0183 mol		0.0147 mol		0.0294 mol

         0.01830 mol Na_2CO_3 × 106.0 g/mol = 1.940 = 1.94 g Na_2CO_3

         0.01471 mol Ag_2CO_3 × 275.8 g/mol = 4.057 = 4.06 g Ag_2CO_3

         0.02943 mol $NaNO_3$ × 85.00 g/mol = 2.502 = 2.50 g $NaNO_3$

         Check. The initial mass of reactants was 8.50 g, and the final mass of excess reactant and products is 8.50 g; mass is conserved.

3.82   *Plan.* Write balanced equation; determine limiting reactant; calculate amounts of excess reactant remaining and products, based on limiting reactant.

Solve. $H_2SO_4(aq) + Pb(C_2H_3O_2)_2(aq) \rightarrow PbSO_4(s) + 2\,HC_2H_3O_2(aq)$

$$5.00 \text{ g } H_2SO_4 \times \frac{1 \text{ mol } H_2SO_4}{98.09 \text{ g } H_2SO_4} = 0.05097 = 0.0510 \text{ mol } H_2SO_4$$

$$5.00 \text{ g } Pb(C_2H_3O_2)_2 \times \frac{1 \text{ mol } Pb(C_2H_3O_2)_2}{325.3 \text{ g } Pb(C_2H_3O_2)_2} = 0.015370 = 0.0154 \text{ mol } Pb(C_2H_3O_2)_2$$

1 mol H_2SO_4 : 1 mol $Pb(C_2H_3O_2)_2$, so $Pb(C_2H_3O_2)_2$ is the limiting reactant.

0 mol $Pb(C_2H_3O_2)_2$, $(0.05097 - 0.01537) = 0.0356$ mol H_2SO_4, 0.0154 mol $PbSO_4$,

$(0.01537 \times 2) = 0.0307$ mol $HC_2H_3O_2$ are present after reaction

0.03560 mol $H_2SO_4 \times 98.09$ g/mol = 3.4920 = 3.49 g H_2SO_4

0.01537 mol $PbSO_4 \times 303.3$ g/mol = 4.6619 = 4.66 g $PbSO_4$

0.03074 mol $HC_2H_3O_2 \times 60.05$ g/mol = 1.8460 = 1.85 g $HC_2H_3O_2$

Check. The initial mass of reactants was 10.00 g; and the final mass of excess reactant and products is 10.00 g; mass is conserved.

3.83   *Analyze.* Given: amounts of two reactants. Find: theoretical yield.

Plan. Determine the limiting reactant and the maximum amount of product it could produce. Then calculate % yield.   *Solve.*

(a) $30.0 \text{ g } C_6H_6 \times \dfrac{1 \text{ mol } C_6H_6}{78.11 \text{ g } C_6H_6} = 0.3841 = 0.384 \text{ mol } C_6H_6$

$65.0 \text{ g } Br_2 \times \dfrac{1 \text{ mol } Br_2}{159.8 \text{ g } Br_2} = 0.4068 = 0.407 \text{ mol } Br_2$

Because C_6H_6 and Br_2 react in a 1:1 mole ratio, C_6H_6 is the limiting reactant and determines the theoretical yield.

$0.3841 \text{ mol } C_6H_6 \times \dfrac{1 \text{ mol } C_6H_5Br}{1 \text{ mol } C_6H_6} \times \dfrac{157.0 \text{ g } C_6H_5Br}{1 \text{ mol } C_6H_5Br} = 60.30 = 60.3 \text{ g } C_6H_5Br$

Check. $30/78 \sim 3/8$ mol C_6H_6. $65/160 \sim 3/8$ mol Br_2. Because moles of the two reactants are similar, a precise calculation is needed to determine the limiting reactant. $3/8 \times 160 \approx 60$ g product

(b) $\% \text{ yield} = \dfrac{42.3 \text{ g } C_6H_5Br \text{ actual}}{60.3 \text{ g } C_6H_5Br \text{ theoretical}} \times 100 = 70.149 = 70.10\%$

3.84   (a) $C_2H_6 + Cl_2 \rightarrow C_2H_5Cl + HCl$

$125 \text{ g } C_2H_6 \times \dfrac{1 \text{ mol } C_2H_6}{30.07 \text{ g } C_2H_6} = 4.157 = 4.16 \text{ mol } C_2H_6$

$255 \text{ g } Cl_2 \times \dfrac{1 \text{ mol } Cl_2}{70.91 \text{ g } Cl_2} = 3.596 = 3.60 \text{ mol } Cl_2$

Because the reactants combine in a 1:1 mole ratio, Cl_2 is the limiting reactant. The theoretical yield is:

$$3.596 \text{ mol } Cl_2 \times \frac{1 \text{ mol } C_2H_5Cl}{1 \text{ mol } Cl_2} \times \frac{64.51 \text{ g } C_2H_5Cl}{1 \text{ mol } C_2H_5Cl} = 231.98 = 232 \text{ g } C_2H_5Cl$$

(b) $\%$ yield $= \dfrac{206 \text{ g } C_2H_5Cl \text{ actual}}{232 \text{ g } C_2H_5Cl \text{ theoretical}} \times 100 = 88.8\%$

3.85   *Analyze.* Given: g of two reactants, $\%$ yield. Find: g S_8.

Plan. Determine limiting reactant and theoretical yield. Use definition of $\%$ yield to calculate actual yield.   *Solve.*

$$30.0 \text{ g } H_2S \times \frac{1 \text{ mol } H_2S}{34.08 \text{ g } H_2S} = 0.8803 = 0.880 \text{ mol } H_2S$$

$$50.0 \text{ g } O_2 \times \frac{1 \text{ mol } O_2}{32.00 \text{ g } H_2S} = 1.5625 = 1.56 \text{ mol } O_2$$

$$0.8803 \text{ mol } H_2S \times \frac{4 \text{ mol } O_2}{8 \text{ mol } H_2S} = 0.4401 = 0.440 \text{ mol } O_2 \text{ required}$$

Because there is more than enough O_2 to react exactly with 0.880 mol H_2S, O_2 is present in excess and H_2S is the limiting reactant.

$$0.8803 \text{ mol } H_2S \times \frac{1 \text{ mol } S_8}{8 \text{ mol } H_2S} \times \frac{256.56 \text{ g } S_8}{1 \text{ mol } S_8} = 28.231 = 28.2 \text{ g } S_8 \text{ theoretical yield}$$

Check. $30/34 \approx 1$ mol H_2S; $50/32 \approx 1.5$ mol O_2. Twice as many mol H_2S as mol O_2 are required, so H_2S limits. $1 \times (260/8) \approx 30$ g S_8 theoretical.

$$\% \text{ yield} = \frac{\text{actual}}{\text{theoretical}} \times 100; \quad \frac{\% \text{ yield} \times \text{theoretical}}{100} = \text{actual yield}$$

$$\frac{98\%}{100} \times 28.231 \text{ g } S_8 = 27.666 = 28 \text{ g } S_8 \text{ actual}$$

3.86 $H_2S(g) + 2 \text{ NaOH}(aq) \rightarrow Na_2S(aq) + 2 H_2O(l)$

$$1.25 \text{ g } H_2S \times \frac{1 \text{ mol } H_2S}{34.08 \text{ g } H_2S} = 0.03668 = 0.0367 \text{ mol } H_2S$$

$$2.00 \text{ g NaOH} \times \frac{1 \text{ mol NaOH}}{40.00 \text{ g NaOH}} = 0.0500 \text{ mol NaOH}$$

By inspection, twice as many mol NaOH as H_2S are needed for exact reaction, but mol NaOH given is less than twice mol H_2S, so NaOH limits.

$$0.0500 \text{ mol NaOH} \times \frac{1 \text{ mol } Na_2S}{2 \text{ mol NaOH}} \times \frac{78.05 \text{ g } Na_2S}{1 \text{ mol } Na_2S} = 1.95125 = 1.95 \text{ g } Na_2S \text{ theoretical}$$

$$\frac{92.0\%}{100} \times 1.95125 \text{ g } Na_2S \text{ theoretical} = 1.7951 = 1.80 \text{ g } Na_2S \text{ actual}$$

Additional Exercises

3.87   (a) $CH_3COOH = C_2H_4O_2$. At room temperature and pressure, pure acetic acid is a liquid. $C_2H_4O_2(l) + 2\ O_2(g) \rightarrow 2\ CO_2(g) + 2\ H_2O(l)$

       (b) $Ca(OH)_2(s) \rightarrow CaO(s) + H_2O(g)$

       (c) $Ni(s) + Cl_2(g) \rightarrow NiCl_2(s)$

3.88 $C_2H_5OH(l) + 3\ O_2(g) \rightarrow 2\ CO_2(g) + 3\ H_2O(g)$

       $C_3H_8(g) + 5\ O_2(g) \rightarrow 3\ CO_2(g) + 4\ H_2O(g)$

       $CH_3CH_2COCH_3(l) + 11/2\ O_2(g) \rightarrow 4\ CO_2(g) + 4\ H_2O(l)$

       In a combustion reaction, all H in the fuel is transformed to H_2O in the products. The reactant with most mol H/mol fuel will produce the most H_2O. C_3H_8 and $CH_3CH_2COCH_3$ (C_4H_8O) both have 8 mol H/mol fuel, so 1.5 mol of either fuel will produce the same amount of H_2O. 1.5 mol C_2H_5OH will produce less H_2O.

3.89   The formulas of the fertilizers are NH_3, NH_4NO_3, $(NH_4)_2SO_4$ and $(NH_2)_2CO$. Qualitatively, the more heavy, non-nitrogen atoms in a molecule, the smaller the mass % of N. By inspection, the mass of NH_3 is dominated by N, so it will have the greatest % N, $(NH_4)_2SO_4$ will have the least. In order of increasing % N:

       $(NH_4)_2SO_4 < NH_4NO_3 < (NH_2)_2CO < NH_3$.

       Check by calculation:

       $(NH_4)_2SO_4$: FW = 2(14.0) + 8(1.0) + 1(32.1) + 4(16.0) = 132.1 amu

       % N = [2(14.0)/132.1] × 100 = 21.2%

       NH_4NO_3: FW = 2(14.0) + 4(1.0) + 3(16.0) = 80.0 amu

       % N = [2(14.0)/80.0] × 100 = 35.0%

       $(NH_2)_2CO$: FW = 2(14.0) + 4(1.0) = 1(12.0) + 1(16.0) = 60.0 amu

       % N = [2(14.0)/60.0] × 100 = 46.7% N

       NH_3: FW = 1(14.0) + 3(1.0) = 17.0 amu

       % N = [14.0/17.0] × 100 = 82.4 % N

3.90   (a) $0.500\ g\ C_9H_8O_4 \times \dfrac{1\ mol\ C_9H_8O_4}{180.2\ g\ C_9H_8O_4} = 2.7747 \times 10^{-3} = 2.77 \times 10^{-3}\ mol\ C_9H_8O_4$

       (b) $0.0027747\ mol\ C_9H_8O_4 \times \dfrac{6.022 \times 10^{23}\ molecules}{1\ mol} = 1.67 \times 10^{21}\ C_9H_8O_4\ molecules$

       (c) $1.67 \times 10^{21}\ C_9H_8O_4\ molecules \times \dfrac{9\ C\ atoms}{1\ C_9H_8O_4\ molecule} = 1.50 \times 10^{22}\ C\ atoms$

3.91   (a)   *Analyze.* Given: diameter of CdSe sphere (dot), density of CdSe. Find: mass of dot.

            Plan. Calculate volume of sphere in cm^3, use density to calculate mass of the sphere (dot).

            Solve. $V = 4/3\pi r^3$; $r = d/2$

$$\text{radius of dot} = \frac{2.5 \text{ nm}}{2} \times \frac{1 \times 10^{-9} \text{ m}}{1 \text{ nm}} \times \frac{1 \text{ cm}}{1 \times 10^{-2} \text{ m}} = 1.25 \times 10^{-7} \text{ cm}$$

$$\text{volume of dot} = (4/3) \times \pi \times (1.25 \times 10^{-7})^3 = 8.1812 \times 10^{-21} = 8.2 \times 10^{-21} \text{ cm}^3$$

$$8.1812 \times 10^{-21} \text{ cm}^3 \times \frac{5.82 \text{ g CdSe}}{\text{cm}^3} = 4.7615 \times 10^{-20} = 4.8 \times 10^{-20} \text{ g/dot}$$

   (b)   *Plan.* Change g CdSe to mol Cd using molar mass, then mol Cd to atoms Cd using Avogadro's number.   *Solve.*

$$4.7615 \times 10^{-20} \text{ g CdSe} \times \frac{1 \text{ mol CdSe}}{191.385 \text{ g CdSe}} \times \frac{1 \text{ mol Cd}}{1 \text{ mol CdSe}} \times \frac{6.0221 \times 10^{23} \text{ Cd atoms}}{\text{mol Cd}}$$

$$= 149.82 = 150 \text{ Cd atoms}$$

   (c) $\text{volume of dot} = (4/3) \times \pi \times (3.25 \times 10^{-7})^3 = 1.4379 \times 10^{-19} = 1.4 \times 10^{-19} \text{ cm}^3$

$$1.4379 \times 10^{-19} \text{ cm}^3 \times \frac{5.82 \text{ g CdSe}}{\text{cm}^3} = 8.3688 \times 10^{-19} = 8.4 \times 10^{-19} \text{ g/dot}$$

   (d) $8.3688 \times 10^{-19} \text{ g CdSe} \times \frac{1 \text{ mol CdSe}}{191.385 \text{ g CdSe}} \times \frac{1 \text{ mol Cd}}{1 \text{ mol CdSe}} \times \frac{6.022 \times 10^{23} \text{ Cd atoms}}{\text{mol Cd}}$

$$= 2633.3 = 2.6 \times 10^3 \text{ Cd atoms}$$

   (e)   We can calculate the number of 2.5-nm dots required to make a 6.5-nm dot using either the ratio of the volumes of the dots, or the ratio of the number of CdSe units in the two dots.

$$\frac{1.4379 \times 10^{-19} \text{ cm}^3}{8.1812 \times 10^{-21} \text{ cm}^3} = 17.5758 \text{ 2.5-nm dots}; \quad \frac{2633.3 \text{ CdSe units}}{149.82 \text{ CdSe units}} = 17.5764 \text{ 2.5-nm dots}$$

            The number of 2.5-nm dots required to make a 6.5-nm dot and the number of CdSe units left over are integer numbers. We have included extra figures in the ratio calculations to show that the results are amazingly close. Both ratios indicate that 18 of the smaller dots are needed to produce one 6.5 nm dot. There will be a few CdSe units left over, slightly less than half of one dot. Taking the average of the "leftovers,"

            0.424 small dot x 150 CdSe units/small dot = 63.6 = 64 CdSe units left over

3.92   (a) $\dfrac{5.342 \times 10^{-21} \text{ g}}{1 \text{ molecule penicillin G}} \times \dfrac{6.0221 \times 10^{23} \text{ molecules}}{1 \text{ mol}} = 3217 \text{ g/mol penicillin G}$

   (b)   1.00 g hemoglobin (hem) contains 3.40×10^{-3} g Fe.

$$\frac{1.00 \text{ g hem}}{3.40 \times 10^{-3} \text{ g Fe}} \times \frac{55.85 \text{ g Fe}}{1 \text{ mol Fe}} \times \frac{4 \text{ mol Fe}}{1 \text{ mol hem}} = 6.57 \times 10^4 \text{ g/mol hemoglobin}$$

3.93   *Plan.* Assume 100 g, calculate mole ratios, empirical formula, then molecular formula from molar mass.   *Solve.*

$$68.2 \text{ g C} \times \frac{1 \text{ mol C}}{12.01 \text{ g C}} = 5.68 \text{ mol C}; \; 5.68/0.568 \approx 10$$

$$6.86 \text{ g H} \times \frac{1 \text{ mol H}}{1.008 \text{ g H}} = 6.81 \text{ mol H}; \; 6.81/0.568 \approx 12$$

$$15.9 \text{ g N} \times \frac{1 \text{ mol N}}{14.01 \text{ g N}} = 1.13 \text{ mol N}; \; 1.13/0.568 \approx 2$$

$$9.08 \text{ g O} \times \frac{1 \text{ mol O}}{16.00 \text{ g O}} = 0.568 \text{ mol O}; \; 0.568/0.568 = 1$$

The empirical formula is $C_{10}H_{12}N_2O$, FW = 176 amu (or g). Because the molar mass is 176, the empirical and molecular formula are the same, $C_{10}H_{12}N_2O$.

3.94   *Plan.* Assume 1.000 g and get mass O by subtraction.   *Solve.*

(a) $$0.7787 \text{ g C} \times \frac{1 \text{ mol C}}{12.01 \text{ g C}} = 0.06484 \text{ mol C}$$

$$0.1176 \text{ g H} \times \frac{1 \text{ mol H}}{1.008 \text{ g H}} = 0.1167 \text{ mol H}$$

$$0.1037 \text{ g O} \times \frac{1 \text{ mol C}}{16.00 \text{ g O}} = 0.006481 \text{ mol O}$$

Dividing through by the smallest of these values we obtain $C_{10}H_{18}O$.

(b)   The formula weight of $C_{10}H_{18}O$ is 154. Thus, the empirical formula is also the molecular formula.

3.95   Because all the C in the vanillin must be present in the CO_2 produced, get g C from g CO_2.

$$2.43 \text{ g CO}_2 \times \frac{1 \text{ mol CO}_2}{44.01 \text{ g CO}_2} \times \frac{12.01 \text{ g C}}{1 \text{ mol C}} = 0.6631 = 0.663 \text{ g C}$$

Because all the H in vanillin must be present in the H_2O produced, get g H from g H_2O.

$$0.50 \text{ g H}_2\text{O} \times \frac{1 \text{ mol H}_2\text{O}}{18.02 \text{ g H}_2\text{O}} \times \frac{2 \text{ mol H}}{1 \text{ mol H}_2\text{O}} \times \frac{1.008 \text{ g H}}{1 \text{ mol H}} = 0.0559 = 0.056 \text{ g H}$$

Get g O by subtraction. (Because the analysis was performed by combustion, an unspecified amount of O_2 was a reactant, and thus not all the O in the CO_2 and H_2O produced came from vanillin.) 1.05 g vanillin – 0.663 g C – 0.056 g H = 0.331 g O

$$0.6631 \text{ g C} \times \frac{1 \text{ mol C}}{12.01 \text{ g C}} = 0.0552 \text{ mol C}; \; 0.0552 / 0.0207 = 2.67$$

$$0.0559 \text{ g H} \times \frac{1 \text{ mol H}}{1.008 \text{ g H}} = 0.0555 \text{ mol C}; \; 0.0555/0.0207 = 2.68$$

$$0.331 \text{ g O} \times \frac{1 \text{ mol O}}{16.00 \text{ g O}} = 0.0207 \text{ mol O}; \; 0.0207/0.0207 = 1.00$$

Multiplying the numbers above by **3** to obtain an integer ratio of moles, the empirical formula of vanillin is $C_8H_8O_3$.

3.96   *Plan.* Because different sample sizes were used to analyze the different elements, calculate mass % of each element in the sample.

      i.   Calculate mass % C from g CO_2.

      ii.   Calculate mass % Cl from AgCl.

      iii.   Get mass % H by subtraction.

      iv.   Calculate mole ratios and the empirical formulas.

    Solve.

      i. $3.52 \text{ g CO}_2 \times \dfrac{1 \text{ mol CO}_2}{44.01 \text{ g CO}_2} \times \dfrac{1 \text{ mol C}}{1 \text{ mol CO}_2} \times \dfrac{12.01 \text{ g C}}{1 \text{ mol C}} = 0.9606 = 0.961 \text{ g C}$

           $\dfrac{0.9606 \text{ g C}}{1.50 \text{ g sample}} \times 100 = 64.04 = 64.0\% \text{ C}$

      ii. $1.27 \text{ g AgCl} \times \dfrac{1 \text{ mol AgCl}}{143.3 \text{ g AgCl}} \times \dfrac{1 \text{ mol Cl}}{1 \text{ mol AgCl}} \times \dfrac{35.45 \text{ g Cl}}{1 \text{ mol Cl}} = 0.3142 = 0.314 \text{ g Cl}$

           $\dfrac{0.3142 \text{ g Cl}}{1.00 \text{ g sample}} \times 100 = 31.42 = 31.4\% \text{ Cl}$

      iii.   % H = 100.0 − (64.04% C + 31.42% Cl) = 4.54 = 4.5% H

      iv.   Assume 100 g sample.

           $64.04 \text{ g C} \times \dfrac{1 \text{ mol C}}{12.01 \text{ g C}} = 5.33 \text{ mol C}; \quad 5.33/0.886 = 6.02$

           $31.42 \text{ g Cl} \times \dfrac{1 \text{ mol Cl}}{35.45 \text{ g Cl}} = 0.886 \text{ mol Cl}; \quad 0.886/0.886 = 1.00$

           $4.54 \text{ g H} \times \dfrac{1 \text{ mol H}}{1.008 \text{ g H}} = 4.50 \text{ mol H}; \quad 4.50/0.886 = 5.08$

      The empirical formula is probably C_6H_5Cl.

      The subscript for H, 5.08, is relatively far from 5.00, but C_6H_5Cl makes chemical sense. More significant figures in the mass data are required for a more accurate mole ratio.

3.97   The mass percentage is determined by the relative number of atoms of the element times the atomic weight, divided by the total formula mass. Thus, the mass percent of bromine in $KBrO_x$ is given by $0.5292 = \dfrac{79.91}{39.10 + 79.91 + x(16.00)}$.

    Solving for x, we obtain x = 2.00. Thus, the formula is $KBrO_2$.

3.98   (a)   Let AW = the atomic weight of X.

          According to the chemical reaction, moles XI_3 reacted = moles XCl_3 produced

          $0.5000 \text{ g XI}_3 \times 1 \text{ mol XI}_3 / (\text{AW} + 380.71) \text{ g XI}_3$

                      $= 0.2360 \text{ g XCl}_3 \times \dfrac{1 \text{ mol XCl}_3}{(\text{AW} + 106.36) \text{ g XCl}_3}$

          0.5000 (AW + 106.36) = 0.2360 (AW + 380.71)

          0.5000 AW + 53.180 = 0.2360 AW + 89.848

          0.2640 AW = 36.67; AW = 138.9 g

    (b)   X is lanthanum, La, atomic number 57.

3.99 $O_3(g) + 2\,NaI(aq) + H_2O(l) \rightarrow O_2(g) + I_2(s) + 2\,NaOH(aq)$

(a) $5.95 \times 10^{-6}\ mol\ O_3 \times \dfrac{2\ mol\ NaI}{1\ mol\ O_3} = 1.19 \times 10^{-5}\ mol\ NaI$

(b) $1.3\ mg\ O_3 \times \dfrac{1 \times 10^{-3}\ g}{1\ mg} \times \dfrac{1\ mol\ O_3}{48.00\ g\ O_3} \times \dfrac{2\ mol\ NaI}{1\ mol\ O_3} \times \dfrac{149.9\ g\ NaI}{1\ mol\ NaI}$

$= 8.120 \times 10^{-3} = 8.1 \times 10^{-3}\ g\ NaI = 8.1\ mg\ NaI$

3.100 $2\,NaCl(aq) + 2\,H_2O(l) \rightarrow 2\,NaOH(aq) + H_2(g) + Cl_2(g)$

Calculate mol Cl_2 and relate to mol H_2, mol NaOH.

$1.5 \times 10^6\ kg \times \dfrac{1000\ g}{1\ kg} \times \dfrac{1\ mol\ Cl_2}{70.91\ g\ Cl_2} = 2.115 \times 10^7 = 2.1 \times 10^7\ mol\ Cl_2$

$2.115 \times 10^7\ mol\ Cl_2 \times \dfrac{1\ mol\ H_2}{1\ mol\ Cl_2} \times \dfrac{2.016\ g\ H_2}{1\ mol\ H_2} = 4.26 \times 10^7\ g\ H_2 = 4.3 \times 10^4\ kg\ H_2$

$4.3 \times 10^7\ g \times \dfrac{1\ metric\ ton}{1 \times 10^6\ g\ (1\ Mg)} = 43\ metric\ tons\ H_2$

$2.115 \times 10^7\ mol\ Cl_2 \times \dfrac{2\ mol\ NaOH}{1\ mol\ Cl_2} \times \dfrac{40.0\ g\ NaOH}{1\ mol\ NaOH} = 1.69 \times 10^9 = 1.7 \times 10^9\ g\ NaOH$

$1.7 \times 10^9\ g\ NaOH = 1.7 \times 10^6\ kg\ NaOH = 1.7 \times 10^3\ metric\ tons\ NaOH$

3.101 $2\,C_{57}H_{110}O_6 + 163\,O_2 \rightarrow 114\,CO_2 + 110\,H_2O$

molar mass of fat $= 57(12.01) + 110(1.008) + 6(16.00) = 891.5$

$1.0\ kg\ fat \times \dfrac{1000\ g}{1\ kg} \times \dfrac{1\ mol\ fat}{891.5\ g\ fat} \times \dfrac{110\ mol\ H_2O}{2\ mol\ fat} \times \dfrac{18.02\ g\ H_2O}{1\ mol\ H_2O} \times \dfrac{1\ kg}{1000\ g} = 1.1\ kg\ H_2O$

3.102   (a)   *Plan.* Calculate the total mass of C from g CO and g CO_2. Calculate the mass of H from g H_2O. Calculate mole ratios and the empirical formula.   *Solve.*

$0.467\ g\ CO \times \dfrac{1\ mol\ CO}{28.01\ g\ CO} \times \dfrac{1\ mol\ C}{1\ mol\ CO} \times 12.01\ g\ C = 0.200\ g\ C$

$0.733\ g\ CO_2 \times \dfrac{1\ mol\ CO_2}{44.01\ g\ CO_2} \times \dfrac{1\ mol\ C}{1\ mol\ CO_2} \times 12.01\ g\ C = 0.200\ g\ C$

Total mass C is $0.200\ g + 0.200\ g = 0.400\ g\ C$.

$0.450\ g\ H_2O \times \dfrac{1\ mol\ H_2O}{18.02\ g\ H_2O} \times \dfrac{2\ mol\ H}{1\ mol\ H_2O} \times \dfrac{1.008\ g\ H}{1\ mol\ H} = 0.0503\ g\ H$

(Because hydrocarbons contain only the elements C and H, g H can also be obtained by subtraction: 0.450 g sample – 0.400 g C = 0.050 g H.)

$0.400\ g\ C \times \dfrac{1\ mol\ C}{12.01\ g\ C} = 0.0333\ mol\ C;\ \ 0.0333/0.0333 = 1.0$

$0.0503\ g\ H \times \dfrac{1\ mol\ H}{1.008\ g\ H} = 0.0499\ mol\ H;\ \ 0.0499/0.0333 = 1.5$

Multiplying by a factor of 2, the empirical formula is C_2H_3.

(b)   Mass is conserved. Total mass products – mass sample = mass O_2 consumed.

0.467 g CO + 0.733 g CO_2 + 0.450 g H_2O – 0.450 g sample = 1.200 g O_2 consumed

(c)   For complete combustion, 0.467 g CO must be converted to CO_2.

$2 CO(g) + O_2(g) \rightarrow 2 CO_2(g)$

$0.467 \text{ g CO} \times \dfrac{1 \text{ mol CO}}{28.01 \text{ g CO}} \times \dfrac{1 \text{ mol } O_2}{2 \text{ mol CO}} \times \dfrac{32.00 \text{ g } O_2}{1 \text{ mol } O_2} = 0.267 \text{ g } O_2$

The total mass of O_2 required for complete combustion is
1.200 g + 0.267 g = 1.467 g O_2.

3.103 $N_2(g) + 3 H_2(g) \rightarrow 2 NH_3(g)$

Determine the moles of N_2 and H_2 required to form the 3.0 moles of NH_3 present after the reaction has stopped.

$3.0 \text{ mol } NH_3 \times \dfrac{3 \text{ mol } H_2}{2 \text{ mol } NH_3} = 4.5 \text{ mol } H_2 \text{ reacted}$

$3.0 \text{ mol } NH_3 \times \dfrac{1 \text{ mol } N_2}{2 \text{ mol } NH_3} = 1.5 \text{ mol } N_2 \text{ reacted}$

mol H_2 initial = 3.0 mol H_2 remain + 4.5 mol H_2 reacted = 7.5 mol H_2

mol N_2 initial = 3.0 mol N_2 remain + 1.5 mol N_2 reacted = 4.5 mol N_2

In tabular form:	$N_2(g)$	+	$3 H_2(g)$	$\rightarrow$	$2 NH_3(g)$
initial	4.5 mol		7.5 mol		0 mol
reaction	–1.5 mol		–4.5 mol		+3.0 mol
final	3.0 mol		3.0 mol		3.0 mol

(Tables like this will be extremely useful for solving chemical equilibrium problems in Chapter 15.)

3.104   All of the O_2 is produced from $KClO_3$; get g $KClO_3$ from g O_2. All of the H_2O is produced from $KHCO_3$; get g $KHCO_3$ from g H_2O. The g H_2O produced also reveals the g CO_2 from the decomposition of $KHCO_3$. The remaining CO_2 (13.2 g CO_2– g CO_2 from $KHCO_3$) is due to K_2CO_3 and g K_2CO_3 can be derived from it.

$4.00 \text{ g } O_2 \times \dfrac{1 \text{ mol } O_2}{32.00 \text{ g } O_2} \times \dfrac{2 \text{ mol } KClO_3}{3 \text{ mol } O_2} \times \dfrac{122.6 \text{ g } KClO_3}{1 \text{ mol } KClO_3} = 10.22 = 10.2 \text{ g } KClO_3$

$1.80 \text{ g } H_2O \times \dfrac{1 \text{ mol } H_2O}{18.02 \text{ g } H_2O} \times \dfrac{2 \text{ mol } KHCO_3}{1 \text{ mol } H_2O} \times \dfrac{100.1 \text{ g } KHCO_3}{1 \text{ mol } KHCO_3} = 20.00 = 20.0 \text{ g } KHCO_3$

$1.80 \text{ g } H_2O \times \dfrac{1 \text{ mol } H_2O}{18.02 \text{ g } H_2O} \times \dfrac{2 \text{ mol } CO_2}{1 \text{ mol } H_2O} \times \dfrac{44.01 \text{ g } CO_2}{1 \text{ mol } CO_2} = 8.792 = 8.79 \text{ g } CO_2 \text{ from } KHCO_3$

13.20 g CO_2 total – 8.792 CO_2 from $KHCO_3$ = 4.408 = 4.41 g CO_2 from K_2CO_3

$4.408 \text{ g } CO_2 \times \dfrac{1 \text{ mol } CO_2}{44.01 \text{ g } CO_2} \times \dfrac{1 \text{ mol } K_2CO_3}{1 \text{ mol } CO_2} \times \dfrac{138.2 \text{ g } K_2CO_3}{1 \text{ mol } K_2CO_3} = 13.84 = 13.8 \text{ g } K_2CO_3$

100.0 g mixture – 10.22 g $KClO_3$ – 20.00 g $KHCO_3$ – 13.84 g K_2CO_3 = 55.9 g KCl

3.105   (a) $2\,C_2H_2(g) + 5\,O_2(g) \rightarrow 4\,CO_2(g) + 2\,H_2O(g)$

         (b)   Following the approach in Sample Exercise 3.18,

$$10.0\text{ g }C_2H_2 \times \frac{1\text{ mol }C_2H_2}{26.04\text{ g }C_2H_2} \times \frac{5\text{ mol }O_2}{2\text{ mol }C_2H_2} \times \frac{32.00\text{ g }O_2}{1\text{ mol }O_2} = 30.7\text{ g }O_2\text{ required}$$

Only 10.0 g O_2 are available, so O_2 limits.

         (c)   Because O_2 limits, 0.0 g O_2 remain.

Next, calculate the g C_2H_2 consumed and the amounts of CO_2 and H_2O produced by reaction of 10.0 g O_2.

$$10.0\text{ g }O_2 \times \frac{1\text{ mol }O_2}{32.00\text{ g }O_2} \times \frac{2\text{ mol }C_2H_2}{5\text{ mol }O_2} \times \frac{26.04\text{ g }C_2H_2}{1\text{ mol }C_2H_2} = 3.26\text{ g }C_2H_2\text{ consumed}$$

10.0 g C_2H_2 initial − 3.26 g consumed = 6.74 = 6.7 g C_2H_2 remain

$$10.0\text{ g }O_2 \times \frac{1\text{ mol }O_2}{32.00\text{ g }O_2} \times \frac{4\text{ mol }CO_2}{5\text{ mol }O_2} \times \frac{44.01\text{ g }CO_2}{1\text{ mol }CO_2} = 11.0\text{ g }CO_2\text{ produced}$$

$$10.0\text{ g }O_2 \times \frac{1\text{ mol }O_2}{32.00\text{ g }O_2} \times \frac{2\text{ mol }H_2O}{5\text{ mol }O_2} \times \frac{18.02\text{ g }H_2O}{1\text{ mol }H_2O} = 2.25\text{ g }H_2O\text{ produced}$$

Integrative Exercises

3.106   *Plan.* Write a balanced chemical equation for the synthesis of GaAs, assuming the reaction is homogeneous and occurs in the gas phase. Use given masses of each reactant, along with molar masses, to find moles of each reactant. Use mole ratios from the equation determine limiting reactant and mass of GaAs produced. Use density to change mass from part (a) to volume of GaAs, then calculate area of the thin film produced in part (c).   *Solve.*

         (a) $Ga(CH_3)_3(g) + AsH_3(g) \rightarrow GaAs(s) + 3\,CH_4(g)$

Determine the moles of $Ga(CH_3)_3(g)$ and $AsH_3(g)$ available for reaction. The molar masses are 114.827 g/mol and 77.9454 g/mol, respectively.

$$450.0\text{ g }Ga(CH_3)_3 \times \frac{1\text{ mol }Ga(CH_3)_3}{114.827\text{ g }Ga(CH_3)_3} = 3.91894 = 3.919\text{ mol }Ga(CH_3)_3$$

$$300.0\text{ g }AsH_3 \times \frac{1\text{ mol }AsH_3}{77.9454\text{ g }AsH_3} = 3.84885 = 3.849\text{ mol }AsH_3$$

According to the balanced equation, $Ga(CH_3)_3$ and AsH_3 react in a 1:1 mole ratio. We have nearly equal moles of the two reactants, but there are fewer moles of AsH_3. The amount of AsH_3 determines the amount of GaAs that can be made.

$$3.84885\text{ mol }AsH_3 \times \frac{1\text{ mol GaAs}}{1\text{ mol }AsH_3} \times \frac{144.645\text{ g GaAs}}{\text{mol GaAs}} = 556.717 = 556.7\text{ g GaAs}$$

         (b) $Ga(CH_3)_3$ is present in excess. Because of the 1:1 reacting ratio, the amount of $Ga(CH_3)_3$ left over is

(3.91894 mol available − 3.84885 mol reacted) = 0.07009 mol $Ga(CH_3)_3$ left over

         (c) $$556.717\text{ g GaAs} \times \frac{1\text{ cm}^3}{5.32\text{ g}} \times \frac{1}{40\text{ nm}} \times \frac{1\text{ nm}}{1 \times 10^{-7}\text{ cm}} = 26{,}161{,}513 = 2.62 \times 10^7\text{ cm}^2$$

The area of the 40-nm thick GaAs film is 2.62×10^7 cm^2.

3.107   (average yield)^{11} = 0.05; average yield = $(0.05)^{1/11}$ = 0.7616 = 0.8; 80% average yield

Think of two reactions in a sequence that each has a yield of 50%. Relative to the amount of starting material, the first product is 50% of the maximum possible product. The second product is 50% of the maximum relative to the product of the first reaction. Relative to the amount of starting material, the quantity of final product is 25% of the maximum possible product. That is,

(0.5)(0.5) = 0.25   or   (average yield)^2 = 0.25

Extending this example to eleven reactions in a sequence, the net yield, relative to the initial amount of starting material, is 5%.

(average yield)^{11} = 0.05; average yield = 80%

3.108   *Plan.* Volume cube $\xrightarrow{\text{density}}$ mass $CaCO_3 \rightarrow$ moles $CaCO_3 \rightarrow$ moles O $\rightarrow$ O atoms

       Solve. $(2.005)^3 \text{ in}^3 \times \dfrac{(2.54)^3 \text{ cm}^3}{1 \text{ in}^3} \times \dfrac{2.71 \text{ g CaCO}_3}{1 \text{ cm}^3} \times \dfrac{1 \text{ mol CaCO}_3}{100.1 \text{ g CaCO}_3} \times \dfrac{3 \text{ mol O}}{1 \text{ mol CaCO}_3}$

                         $\times \dfrac{6.022 \times 10^{23} \text{ O atoms}}{1 \text{ mol O}} = 6.46 \times 10^{24} \text{ O atoms}$

3.109   (a)   *Plan.* volume of Ag cube $\xrightarrow{\text{density}}$ mass of Ag $\rightarrow$ mol Ag $\rightarrow$ Ag atoms

           Solve. $(1.000)^3 \text{ cm}^3 \text{ Ag} \times \dfrac{10.5 \text{ g Ag}}{1 \text{ cm}^3 \text{ Ag}} \times \dfrac{1 \text{ mol Ag}}{107.87 \text{ g Ag}} \times \dfrac{6.022 \times 10^{23} \text{ atoms}}{1 \text{ mol}}$

                           $= 5.8618 \times 10^{22} = 5.86 \times 10^{22} \text{ Ag atoms}$

    (b)   1.000 cm^3 cube volume, 74% is occupied by Ag atoms

        0.74 cm^3 = volume of 5.86×10^{22} Ag atoms

        $\dfrac{0.7400 \text{ cm}^3}{5.8618 \times 10^{22} \text{ Ag atoms}} = 1.2624 \times 10^{-23} = 1.3 \times 10^{-23} \text{ cm}^3 / \text{Ag atom}$

        Because atomic dimensions are usually given in Å, we will show this conversion.

        $1.2624 \times 10^{-23} \text{ cm}^3 \times \dfrac{(1 \times 10^{-2})^3 \text{ m}^3}{1 \text{ cm}^3} \times \dfrac{1 \text{ Å}^3}{(1 \times 10^{-10})^3 \text{ m}^3} = 12.62 = 13 \text{ Å}^3 / \text{Ag atom}$

    (c) $V = 4/3\pi r^3; r^3 = 3V/4\pi; r = (3V/4\pi)^{1/3}$

        $r_A = (3 \times 12.62 \text{ Å}^3 / 4\pi)^{1/3} = 1.444 = 1.4 \text{ Å}$

3.110   (a)   *Analyze.* Given: gasoline = C_8H_{18}, density = 0.692 g/mL, 20.5 mi/gal, 225 mi. Find: kg CO_2.

        Plan. Write and balance the equation for the combustion of octane. Change mi $\rightarrow$ gal octane $\rightarrow$ mL $\rightarrow$ g octane. Use stoichiometry to calculate g and kg CO_2 from g octane.

Solve. $2 C_8H_{18}(l) + 25 O_2(g) \rightarrow 16 CO_2(g) + 18 H_2O(l)$

$$225 \text{ mi} \times \frac{1 \text{ gal}}{20.5 \text{ mi}} \times \frac{3.7854 \text{ L}}{1 \text{ gal}} \times \frac{1 \text{ mL}}{1 \times 10^{-3} \text{ L}} \times \frac{0.692 \text{ g octane}}{1 \text{ mL}} = 2.8751 \times 10^4 \text{ g}$$

$$= 28.8 \text{ kg octane}$$

$$2.8751 \times 10^4 \text{ g } C_8H_{18} \times \frac{1 \text{ mol } C_8H_{18}}{114.2 \text{ g } C_8H_{18}} \times \frac{16 \text{ mol } CO_2}{2 \text{ mol } C_8H_{18}} \times \frac{44.01 \text{ g } CO_2}{1 \text{ mol } CO_2} = 8.8638 \times 10^4 \text{ g}$$

$$= 88.6 \text{ kg } CO_2$$

Check. $\left(\dfrac{225 \times 4 \times 0.7}{20} \right) \times 10^3 = (45 \times 0.7) \times 10^3 = 30 \times 10^3 \text{ g} = 30 \text{ kg octane}$

$\dfrac{44}{114} \approx \dfrac{1}{3}$; $\dfrac{30 \text{ kg} \times 8}{3} \approx 80 \text{ kg } CO_2$

(b)   *Plan.* Use the same strategy as part (a).   *Solve.*

$$225 \text{ mi} \times \frac{1 \text{ gal}}{5 \text{ mi}} \times \frac{3.7854 \text{ L}}{1 \text{ gal}} \times \frac{1 \text{ mL}}{1 \times 10^{-3} \text{ L}} \times \frac{0.69 \text{ g octane}}{1 \text{ mL}} = 1.1754 \times 10^5$$

$$= 1 \times 10^2 \text{ kg octane}$$

$$1.1754 \times 10^5 \text{ g } C_8H_{18} \times \frac{1 \text{ mol } C_8H_{18}}{114.2 \text{ g } C_8H_{18}} \times \frac{16 \text{ mol } CO_2}{2 \text{ mol } C_8H_{18}} \times \frac{44.01 \text{ g } CO_2}{1 \text{ mol } CO_2}$$

$$= 3.624 \times 10^5 \text{ g} = 4 \times 10^2 \text{ kg } CO_2$$

Check. Mileage of 5 mi/gal requires ~4 times as much gasoline as mileage of 20.5 mi/gal, so it should produce ~4 times as much CO_2. 90 kg CO_2 [from (a)] × 4 = 360 = 4×10^2 kg CO_2 [from (b)].

3.111   Structural isomers, like 1-propanol and 2-propanol, have the same number and kinds of atoms, but different arrangements of these atoms. Because molecular weight is the sum of atomic weights, and number and kinds of atoms are the same, the molecular weights of structural isomers are the same. Again, because number and kinds of atoms are the same, percent composition and therefore combustion analysis results will be the same. Physical properties, like boiling point and density, are influenced by structure as well as molecular weight, and are different for structural isomers.

The properties (a) boiling point and (d) density will distinguish between 1-propanol and 2-propanol. This is confirmed by comparing these properties from either Wolfram Alpha (WA) or the CRC Handbook of Chemistry and Physics (CRC).

Compound	Boiling Point (WA)	Boiling Point (CRC)	Density (WA)	Density (CRC)
1-propanol	97 °C	97.4 °C	0.804 g/cm³	0.8035 g/cm³
2-propanol	82 °C	82.4 °C	0.785 g/cm³	0.7855 g/cm³

3.112　(a)　$S(s) + O_2(g) \rightarrow SO_2(g); SO_2(g) + CaO(s) \rightarrow CaSO_3(s)$

　　　(b)　If the coal contains 2.5% S, then 1 g coal contains 0.025 g S.

$$\frac{2000.0 \text{ tons coal}}{\text{day}} \times \frac{2000 \text{ lb}}{1 \text{ ton}} \times \frac{1 \text{ kg}}{2.2046 \text{ lb}} \times \frac{1000 \text{ g}}{1 \text{ kg}} \times \frac{0.025 \text{ g S}}{1 \text{ g coal}} \times \frac{1 \text{ mol S}}{32.065 \text{ g S}}$$

$$\times \frac{1 \text{ mol } SO_2}{1 \text{ mol S}} \times \frac{1 \text{ mol CaO}}{1 \text{ mol } SO_2} \times \frac{56.077 \text{ g CaO}}{1 \text{ mol CaO}} \times \frac{1 \text{ kg CaO}}{1000 \text{ g CaO}} =$$

$$79,327.49 = 7.9 \times 10^4 \text{ kg CaO or } 7.9 \times 10^7 \text{ g CaO}$$

　　　(c)　$1 \text{ mol CaO} = 1 \text{ mol } CaSO_3$

$$7.9327 \times 10^7 \text{ g CaO} \times \frac{1 \text{ mol CaO}}{56.077 \text{ g CaO}} \times \frac{1 \text{ mol } CaSO_3}{1 \text{ mol CaO}} \times \frac{120.14 \text{ g } CaSO_3}{1 \text{ mol } CaSO_3}$$

$$= 1.6995 \times 10^8 = 1.7 \times 10^8 \text{ g } CaSO_3$$

This corresponds to about 190 tons of $CaSO_3$ per day as a waste product.

3.113　(a)　*Plan.* Calculate the kg of air in the room and then the mass of HCN required to produce a dose of 300 mg HCN/kg air.　　*Solve.*

$$12 \text{ ft} \times 15 \text{ ft} \times 8.0 \text{ ft} = 1440 = 1.4 \times 10^3 \text{ ft}^3 \text{ of air in the room}$$

$$1440 \text{ ft}^3 \text{ air} \times \frac{(12 \text{ in})^3}{1 \text{ ft}^3} \times \frac{(2.54 \text{ cm})^3}{1 \text{ in}^3} \times \frac{0.00118 \text{ g air}}{1 \text{ cm}^3 \text{ air}} \times \frac{1 \text{ kg}}{1000 \text{ g}} = 48.12 = 48 \text{ kg air}$$

$$48.12 \text{ kg air} \times \frac{300 \text{ mg HCN}}{1 \text{ kg air}} \times \frac{1 \text{ g}}{1000 \text{ mg}} = 14.43 = 14 \text{ g HCN}$$

　　　(b)　$2 \text{ NaCN}(s) + H_2SO_4(aq) \rightarrow Na_2SO_4(aq) + 2 \text{ HCN}(g)$

The question can be restated as: What mass of NaCN is required to produce 14 g of HCN according to the above reaction?

$$14.43 \text{ g HCN} \times \frac{1 \text{ mol HCN}}{27.03 \text{ g HCN}} \times \frac{2 \text{ mol NaCN}}{2 \text{ mol HCN}} \times \frac{49.01 \text{ g NaCN}}{1 \text{ mol NaCN}} = 26.2 = 26 \text{ g NaCN}$$

　　　(c)　$$12 \text{ ft} \times 15 \text{ ft} \times \frac{1 \text{ yd}^2}{9 \text{ ft}^2} \times \frac{30 \text{ oz}}{1 \text{ yd}^2} \times \frac{1 \text{ lb}}{16 \text{ oz}} \times \frac{454 \text{ g}}{1 \text{ lb}} = 17,025$$

$$= 1.7 \times 10^4 \text{ g } CH_2CHCN \text{ in the room}$$

50% of the carpet burns, so the starting amount of CH_2CHCN is
$0.50(17,025) = 8513 = 8.5 \times 10^3 \text{ g}$

$$8513 \text{ g } CH_2CHCN \times \frac{50.9 \text{ g HCN}}{100 \text{ g } CH_2CHCN} = 4333 = 4.3 \times 10^3 \text{ g HCN possible}$$

If the actual yield of combustion is 20%, actual g HCN = 4333(0.20) = 866.6 $= 8.7 \times 10^2$ g HCN produced. From part (a), 14 g of HCN is a lethal dose. The fire produces much more than a lethal dose of HCN.

3.114   (a) $N_2(g) + O_2(g) \rightarrow 2\,NO(g); \; 2\,NO(g) + O_2(g) \rightarrow 2\,NO_2(g)$

(b)   1 million $= 1 \times 10^6$

$$19 \times 10^6 \text{ tons } NO_2 \times \frac{2000\,\text{lb}}{1\,\text{ton}} \times \frac{453.6\,\text{g}}{1\,\text{lb}} = 1.724 \times 10^{13} = 1.7 \times 10^{13} \text{ g } NO_2$$

(c)   *Plan.* Calculate g O_2 needed to burn 500 g octane. This is 85% of total O_2 in the engine. 15% of total O_2 is used to produce NO_2, according to the second equation in part (a).

Solve. $2\,C_8H_{18}(l) + 25\,O_2(g) \rightarrow 16\,CO_2(g) + 18\,H_2O(l)$

$$500 \text{ g } C_8H_{18} \times \frac{1\,\text{mol } C_8H_{18}}{114.2\,\text{g } C_8H_{18}} \times \frac{25\,\text{mol } O_2}{2\,\text{mol } C_8H_{18}} \times \frac{32.00\,\text{g } O_2}{\text{mol } O_2} = 1751 = 1.75 \times 10^3 \text{ g } O_2$$

$$\frac{1751\,\text{g } O_2}{\text{total g } O_2} = 0.85; \; 2060 = 2.1 \times 10^3 \text{ g } O_2 \text{ total in engine}$$

2060 g O_2 total $\times 0.15 = 309.1 = 3.1 \times 10^2$ g O_2 used to produce NO_2. One mol O_2 produces 2 mol NO. Then 2 mol NO react with a second mol O_2 to produce 2 mol NO_2. Two mol O_2 are required to produce 2 mol NO_2; one mol O_2 per mol NO_2.

$$309.1 \text{ g } O_2 \times \frac{1\,\text{mol } O_2}{32.00\,\text{g}} \times \frac{1\,\text{mol } NO_2}{1\,\text{mol } O_2} \times \frac{46.01\,\text{g } NO_2}{1\,\text{mol } O_2} = 444.4 = 4.4 \times 10^2 \text{ g } NO_2$$

3.115   (a) $Fe_2O_3(s) + 2\,Al(s) \rightarrow Al_2O_3(s) + 2\,Fe(l)$

(b) $$500.0 \text{ g } Fe_2O_3 \times \frac{1\,\text{mol } Fe_2O_3}{159.688\,\text{g } Fe_2O_3} \times \frac{2\,\text{mol } Al}{1\,\text{mol } Fe_2O_3} \times \frac{26.9815\,\text{g } Al}{1\,\text{mol } Al} = 168.964$$

$$= 169.0 \text{ g } Al$$

(c) $$1.00 \times 10^4 \text{ kJ} \times \frac{1\,\text{mol } Fe_2O_3}{852\,\text{kJ}} \times \frac{159.688\,\text{g } Fe_2O_3}{1\,\text{mol } Fe_2O_3} = 1874.27 = 1.87 \times 10^3 \text{ g } Fe_2O_3$$

(d)   Heat is a reactant in the reverse reaction.

3.116   (a)   Represent the hexyl group, $CH_3CH_2CH_2CH_2CH_2CH_2-$, as C_6H_{13}. The Ugi product is quite complex. Moving from left to right, we can write $C_6H_{13}CON(C_6H_{13})C(C_6H_{13})_2CON(H)C_6H_{13}$. The parentheses indicate hexyl groups that are not in the backbone of the molecule. The molecular formula of the product is $(C_6H_{13})_5C_3HN_2O_2$ or $C_{33}H_{66}N_2O_2$. The molecular formula of the product is convenient for calculating molar mass in part (b).

$$(C_6H_{13})_2CO + C_6H_{13}NH_2 + C_6H_{13}COOH + C_6H_{13}NC \rightarrow$$

$$C_6H_{13}CON(C_6H_{13})C(C_6H_{13})_2CON(H)C_6H_{13}$$

(b) $$435.0 \text{ g } C_6H_{13}NH_2 \times \frac{1\,\text{mol } C_6H_{13}NH_2}{101.190\,\text{g } C_6H_{13}NH_2} \times \frac{1\,\text{mol } C_{33}H_{66}N_2O_2}{1\,\text{mol } C_6H_{13}NH_2} \times$$

$$\frac{522.889 \text{ g } C_{33}H_{66}N_2O_2}{1\,\text{mol } C_{33}H_{66}N_2O_2} = 2247.82 = 2248 \text{ g } C_{33}H_{66}N_2O_2$$

4 Reactions in Aqueous Solution

Visualizing Concepts

4.1   *Analyze.* Correlate the formula of the solute with the charged spheres in the diagrams.

Plan. Determine the electrolyte properties of the solute and the relative number of cations, anions, or neutral molecules produced when the solute dissolves.

Solve. Li_2SO_4 is a strong electrolyte, a soluble ionic solid that dissociates into separate Li^+ and SO_4^{2-} when it dissolves in water. There are twice as many Li^+ cations as SO_4^{2-} anions. Diagram (c) represents the aqueous solution of a 2:1 electrolyte.

4.2   *Analyze/Plan.* Correlate the neutral molecules, cations, and anions in the diagrams with the definitions of strong, weak, and nonelectrolytes.   *Solve.*

(a)   AX is a nonelectrolyte because no ions form when the molecules dissolve.

(b)   AY is a weak electrolyte because a few molecules ionize when they dissolve, but most do not.

(c)   AZ is a strong electrolyte because all molecules break up into ions when they dissolve.

4.3   *Analyze/Plan.*   From the molecular representations, write molecular formulas for the compounds. Using Table 4.2 and molecular formulas (there are no ionic compounds in this exercise), classify the compounds as strong acid, strong base, weak acid, weak base (NH_3), or nonelectrolyte. Strong acids and bases are strong electrolytes, weak acids and bases are weak electrolytes.   *Solve.*

(a)   HCOOH. The molecule has a –COOH group; it is a weak acid and weak electrolyte (it is not one of the strong acids listed in Table 4.2).

(b) HNO_3. The molecule is a strong acid (Table 4.2) and a strong electrolyte.

(c) CH_3CH_2OH. The molecule is neither an acid nor a base; it is a nonelectrolyte.

4.4   Statement (b) is most correct. Statement (a) is incorrect because, at equilibrium, the chemical reactions are ongoing. Statement (c) is incorrect because the concentration of the product (or reactant) is changing until equilibrium is reached, but not at equilibrium. Equilibrium is a state of dynamic constancy.

4.5   *Analyze/Plan.* From the names and/or formulas of the three possible solids, determine which exhibits the described solubility properties. Use Table 4.1.

Solve. The three possible compounds are $BaCl_2$, $PbCl_2$, and $ZnCl_2$. $PbCl_2$ does not dissolve in water to give a clear solution, so it can be eliminated. Of the remaining possibilities, Ba^{2+} has a sulfate precipitate, but Zn^{2+} does not. The compound is indeed $BaCl_2$.

4.6   *Analyze.* Given the formulas of some ions, determine whether these ions ever form
        precipitates in aqueous solution. *Plan.* Use Table 4.1 to determine if the given ions can
        form precipitates. If not, they will always be spectator ions.   *Solve.*

(a) Cl^- can form precipitates with Ag^+, Hg_2^{2+}, and Pb^{2+}.

(b) NO_3^- never forms precipitates, so it is always a spectator.

(c) NH_4^+ never forms precipitates, so it is always a spectator.

(d) S^{2-} usually forms precipitates.

(e) SO_4^{2-} can form precipitates with Sr^{2+}, Ba^{2+}, Hg_2^{2+}, and Pb^{2+}.

        Check. NH_4^+ is a soluble exception for sulfides, phosphates, and carbonates which
        usually form precipitates, so all rules indicate that it is a perpetual spectator.

4.7   *Analyze/Plan.* Given three metal powders and three 1 *M* solutions, use Table 4.5, the
        activity series of metals, to find a scheme to distinguish the metals.

        Solve. In the activity series, any metal on the list can be oxidized by the ions of elements
        below it. The nitric acid solution contains H^+(aq). This solution will oxidize and thus
        dissolve Zn(s) and Pb(s), which appear above H_2(g) on the list. Platinum, Pt(s), is
        distinguished by its lack of reaction with nitric acid.

        To distinguish between Zn and Pb, use a metal ion that occurs between them on the list.
        We have such an ion, Ni^{2+}(aq) in the nickel nitrate solution. Ni^{2+}(aq) will oxidize and
        thus dissolve Zn(s), which is above it on the list. Ni^{2+}(aq) will not oxidize or dissolve
        Pb(s), which is below it on the list.

        To summarize, Pt(s) will neither be oxidized by nor dissolve in any of the three
        available solutions. Pb(s) is oxidized by and will dissolve in the nitric acid solution, but
        not the nickel nitrate solution. Zn(s) is oxidized by and will dissolve in both nitric acid
        and nickel nitrate solutions.

4.8   In a redox reaction, one reactant loses electrons and a different reactant gains electrons;
        electrons are transferred. Acids ionize in aqueous solution to produce (donate)
        hydrogen ions (H^+, protons). Bases are substances that react with or accept protons
        (H^+). In an neutralization reaction, protons are transferred from an acid to a base. We
        characterize redox reactions by tracking electron transfer using oxidation numbers. We
        characterize neutralization reactions by tracking H^+ (proton) transfer via molecular
        formulas of reactants and products.

4.9   The answer is: (c) a redox reaction. The "water-splitting" reaction is

        $$2\,H_2O(l) \rightarrow 2\,H_2(g) + O_2(g)$$

        In the reaction, the oxidation number of hydrogen decreases from +1 to 0, and the
        oxidation number of oxygen increases from –2 to 0. Hydrogen is reduced and oxygen is
        oxidized.

4.10   *Analyze/Plan.* We are given a total ion concentration for two salt solutions and
        the formulas of the ionic solutes. Decide the total moles of ions produced when one
        mole of the salt dissolves, and the ratio of the moles of the ion in question to the total
        moles of ions. Use this ratio to calculate the concentration of the ion in question.

(a)   When one mole of NaCl dissolves, it produces 1 mole of Cl^- ions and 2 total
        moles of ions. The concentration of Cl^- is: ½(1.2 m*M*) = 0.60 m*M* Cl^-(aq).

(b)   When one mole of $FeCl_3$ dissolves, it produces 3 moles of Cl^- ions and 4 total moles of ions. The concentration of Cl^- is: ¾ (1.2 mM) = 0.90 mM Cl^-(aq).

4.11   *Analyze/Plan.* The plot shows indicator color versus volume of standard solution added. Consider how the indicator changes color in a titration like the one in Figure 4.18, and relate this behavior to the shapes shown in the graph.

Solve. The "green" data set is expected from a titration like the one in Figure 4.18. The indicator color remains constant until the reaction is very near the equivalence point. Within a very small volume of standard solution added, the indicator color changes rapidly. This behavior is shown in green. The red graph shows a constantly changing indicator color.

4.12   *Analyze/Plan.* The purpose of every titration is to determine the equivalence point. Based on the indicator behavior described in the exercise, decide how the amount of reactants and products in the titration beaker relate to the equivalence point. Given this reaction mixture, design an experiment to reach the equivalence point.

Solve. This indicator is colorless in acid and blue in base. When it is added to the beaker, the solution is dark blue, so the solution is quite basic. This means that the amount of base added is already greater than the amount of acid initially present. To reach the equivalence point, more acid must be added to the titration beaker.

First, record the volume of base added. Then, find a standard acid solution (an acid of very well-known concentration) or standardize an acid solution (probably HCl). Rinse and fill a clean buret with the standard acid. Carefully titrate the mixture in the beaker until the blue color fades and finally disappears. Record the volume of standard acid added. Subtract the amount of added acid from the total amount of base added. The remaining amount of base is the volume required to reach equivalence with the original acid sample. This procedure is called "back titration."

General Properties of Aqueous Solutions (Section 4.1)

4.13   (a)   False. Electrolyte solutions conduct electricity because *ions* are moving through the solution.

(b)   True. The conductivity is unchanged as long as the concentration of electrolytes is unchanged. Because ions are mobile in solution, the added presence of uncharged molecules does not inhibit conductivity.

4.14   (a)   False. Methanol is an organic alcohol, and the –OH group is not ionizable. The neutral methanol molecules in solution do not support the movement of charge. The solution does not conduct electricity.

(b)   True. CH_3COOH is a weak electrolyte. When it dissolves in water, a small percentage of the molecules ionize to form H^+ and CH_3COO^- ions. The presence of a small concentration of ions produces a weakly conducting solution, and the specific presence of H^+ makes the solution acidic.

4.15   Statement (b) is most correct. Statements (a) and (c) are incorrect because water is not a strong acid and the hydrogen and oxygen bonds of water are not broken by ionic solids.

4.16　Anions are negatively charged. They will be attracted to and thus physically closer to the partially positive portion of the water molecule, which is the hydrogens.

4.17　*Analyze/Plan.* Given the solute formula, determine the separate ions formed upon dissociation.　*Solve.*

(a)　$FeCl_2(aq) \rightarrow Fe^{2+}(aq) + 2\,Cl^-(aq)$

(b)　$HNO_3(aq) \rightarrow H^+(aq) + NO_3^-(aq)$

(c)　$(NH_4)_2SO_4(aq) \rightarrow 2\,NH_4^+(aq) + SO_4^{2-}(aq)$

(d)　$Ca(OH)_2(aq) \rightarrow Ca^{2+}(aq) + 2\,OH^-(aq)$

4.18　(a)　$MgI_2(aq) \rightarrow Mg^{2+}(aq) + 2\,I^-(aq)$

(b)　$K_2CO_3(aq) \rightarrow 2\,K^+(aq) + CO_3^{2-}(aq)$

(c)　$HClO_4(aq) \rightarrow H^+(aq) + ClO_4^-(aq)$

(d)　$NaCH_3COO(aq) \rightarrow Na^+(aq) + CH_3COO^-(aq)$

4.19　*Analyze/Plan.* Apply the definition of a weak electrolyte to HCOOH.

Solve. When HCOOH dissolves in water, neutral HCOOH molecules, H^+ ions and $HCOO^-$ ions are all present in the solution. $HCOOH(aq) \rightleftharpoons H^+(aq) + HCOO^-(aq)$

4.20　(a)　acetone (nonelectrolyte): $CH_3COCH_3(aq)$ molecules only; hypochlorous acid (weak electrolyte): $HClO(aq)$ molecules, $H^+(aq)$, $ClO^-(aq)$; ammonium chloride (strong electrolyte): $NH_4^+(aq)$, $Cl^-(aq)$

(b)　NH_4Cl, 0.2 mol solute particles; $HClO$, between 0.1 and 0.2 mol particles; CH_3COCH_3, 0.1 mol of solute particles

Precipitation Reactions (Section 4.2)

4.21　*Analyze.* Given: formula of compound. Find: solubility.

Plan. Follow the guidelines in Table 4.1, in light of the anion present in the compound and notable exceptions to the "rules."　*Solve.*

(a)　$Mg\mathbf{Br_2}$: soluble

(b)　$Pb\mathbf{I_2}$: insoluble, Pb^{2+} is an exception to soluble iodides

(c)　$(NH_4)_2\mathbf{CO_3}$: soluble, NH_4^+ is an exception to insoluble carbonates

(d)　$Sr(\mathbf{OH})_2$: soluble, Sr^{2+} is an exception to insoluble hydroxides

(e)　$Zn\mathbf{SO_4}$: soluble

4.22　According to Table 4.1:

(a)　$Ag\mathbf{I}$: insoluble (an exception to the generally soluble iodides)

(b)　$Na_2\mathbf{CO_3}$: soluble, Na^+ is an exception to insoluble carbonates

(c)　$Ba\mathbf{Cl_2}$: soluble

(d)　$Al(\mathbf{OH})_3$: insoluble

(e)　$Zn(\mathbf{CH_3COO})_2$: soluble

4.23   *Analyze.* Given: formulas of reactants. Find: balanced equation including precipitates.

        Plan. Follow the logic in Sample Exercise 4.3.

        Solve. In each reaction, the precipitate is in bold type.

        (a) $Na_2CO_3(aq) + 2\,AgNO_3(aq) \rightarrow \mathbf{Ag_2CO_3(s)} + 2\,NaNO_3(aq)$

        (b)   No precipitate (all nitrates and most sulfates are soluble).

        (c) $FeSO_4(aq) + Pb(NO_3)_2(aq) \rightarrow \mathbf{PbSO_4(s)} + Fe(NO_3)_2(aq)$

4.24   In each reaction, the precipitate is in bold type.

        (a)   No precipitate. [CH_3COO^- gains H^+ to form $CH_3COOH(aq)$, which is soluble.]

        (b) $Cu(NO_3)_2(aq) + 2\,KOH(aq) \rightarrow \mathbf{Cu(OH)_2(s)} + 2\,KNO_3(aq)$

        (c) $Na_2S(aq) + CdSO_4(aq) \rightarrow \mathbf{CdS(s)} + Na_2SO_4(aq)$

4.25   *Analyze/Plan.* Follow the logic in Sample Exercise 4.4. From the complete ionic equation, identify the ions that don't change during the reaction; these are the spectator ions.   *Solve.*

        (a) $2\,K^+(aq) + CO_3^{2-}(aq) + Mg^{2+}(aq) + SO_4^{2-}(aq) \rightarrow MgCO_3(s) + 2\,K^+(aq) + SO_4^{2-}(aq)$
            Spectators: K^+, SO_4^{2-}

        (b) $Pb^{2+}(aq) + 2\,NO_3^-(aq) + 2\,Li^+(aq) + S^{2-}(aq) \rightarrow PbS(s) + 2\,Li^+(aq) + 2\,NO_3^-(aq)$
            Spectators: Li^+, NO_3^-

        (c) $6\,NH_4^+(aq) + 2\,PO_4^{3-}(aq) + 3\,Ca^{2+}(aq) + 6\,Cl^-(aq) \rightarrow Ca_3(PO_4)_2(s) + 6\,NH_4^+(aq)$
            $+ 6\,Cl^-(aq)$ Spectators: NH_4^+, Cl^-

4.26   Spectator ions are those that do not change during reaction.

        (a) $2\,Cr^{3+}(aq) + 3\,CO_3^{2-}(aq) \rightarrow Cr_2(CO_3)_3(s)$; spectators: NH_4^+, SO_4^{2-}

        (b) $Ba^{2+}(aq) + SO_4^{2-}(aq) \rightarrow BaSO_4(s)$; spectators: K^+, NO_3^-

        (c) $Fe^{2+}(aq) + 2\,OH^-(aq) \rightarrow Fe(OH)_2(s)$; spectators: K^+, NO_3^-

4.27   *Analyze.* Given: reactions of unknown salt with HBr, H_2SO_4, $NaOH$. Find: Does the unknown salt contain K^+ or Pb^{2+} or Ba^{2+}?

        Plan. Analyze solubility guidelines for Br^-, SO_4^{2-}, and OH^- and select the cation that produces a precipitate with each of the anions.

        Solve. K^+ forms no precipitates with any of the anions. $BaSO_4$ is insoluble, but $BaCl_2$ and $Ba(OH)_2$ are soluble. Because the unknown salt forms precipitates with all three anions, it must contain Pb^{2+}.

        Check. $PbBr_2$, $PbSO_4$, and $Pb(OH)_2$ are all insoluble according to Table 4.1, so our process of elimination is confirmed by the insolubility of the Pb^{2+} compounds.

4.28 Br^- and NO_3^- can be ruled out because the $BaBr_2$ is soluble and all NO_3^- salts are soluble. CO_3^{2-} forms insoluble salts with the three cations given; it must be the anion in question.

4.29   *Analyze/Plan.* Using Table 4.1, determine the precipitates that could form when each of the unknowns is reacted with $Ba(NO_3)_2$ and $NaCl$.   *Solve.*

    (a)   True. If the unknown is $Al_2(SO_4)_3$, $BaSO_4$ could precipitate.

    (b)   True. If the unknown is $AgNO_3$, $AgCl$ could precipitate.

    (c)   False (two ways). Ag_2SO_4 is a soluble ionic compound, and no combination of the possible unknowns and the two reagents could produce Ag_2SO_4.

    (d)   True. This is the overall correct answer to the question.

    (e)   False, because (c) is false.

4.30   Consider all the possible combinations of $Pb^{2+}(aq)$, $Na^+(aq)$, $Ca^{2+}(aq)$, $CH_3COO^-(aq)$, $S^{2-}(aq)$, and $Cl^-(aq)$. Which of these compounds are insoluble ionic compounds that will precipitate when the solutions are combined?

    $PbS(s)$ and $PbCl_2(s)$ will precipitate.

Acids, Bases, and Neutralization Reactions (Section 4.3)

4.31   *Analyze.* Given: solute and concentration of three solutions. Find: the solution that is most acidic.

    Plan: See Sample Exercise 4.6 and Table 4.2. Determine whether solutes are strong or weak acids or bases or nonelectrolytes. For solutions of equal concentration, strong acids will have greatest concentration of solvated protons. Take varying concentration into consideration when evaluating the same class of solutions.

    Solve. (a) $LiOH$ is a strong base, (b) HI is a strong acid, (c) CH_3OH is a molecular compound and nonelectrolyte. Solution (b), 0.2 M HI, is the most acidic solution.

    Check. The solution concentrations weren't needed to answer the question.

4.32 $NH_3(aq)$ is a weak base, whereas KOH and $Ba(OH)_2$ are strong bases. $NH_3(aq)$ is only slightly ionized, so even (a) 0.6 M NH_3 is less basic than (b) 0.150 M KOH. $Ba(OH)_2$ has twice as many OH^- per mole as KOH, so (c) 0.100 M $Ba(OH)_2$ is more basic than (b) 0.150 M KOH. The most basic solution is (c) 0.100 M $Ba(OH)_2$.

4.33   (a)   False. Sulfuric acid, H_2SO_4, is a diprotic acid; it has two ionizable hydrogen atoms.

    (b)   False. According to Table 4.2, HCl is a strong acid.

    (c)   False. Methanol, CH_3OH, is a molecular nonelectrolyte.

4.34   (a)   True. NH_3 produces OH^- in aqueous solution by reacting with H_2O (hydrolysis): $NH_3(aq) + H_2O(l) \rightleftharpoons NH_4^+(aq) + OH^-(aq)$. The OH^- causes the solution to be basic. $NH_3(aq)$ attracts an H^+ from water, leaving $OH^-(aq)$ in the solution.

    (b)   False. According to Table 4.2, HF is not one of the strong acids.

    (c)   True. H_2SO_4 is a **diprotic** acid; it has two ionizable hydrogens. The first hydrogen completely ionizes to form H^+ and HSO_4^-, but HSO_4^- only **partially** ionizes into H^+ and SO_4^{2-} (HSO_4^- is a weak electrolyte). Thus, an aqueous solution of H_2SO_4 contains a mixture of H^+, HSO_4^-, and SO_4^{2-}, with the concentration of HSO_4^- greater than the concentration of SO_4^{2-}.

4.35   *Analyze.* Given: chemical formulas. Find: classify as acid, base, salt; strong, weak, or nonelectrolyte.

Plan. See Tables 4.2 and 4.3. Ionic or molecular? Ionic, soluble: OH^-, strong base and strong electrolyte; otherwise, salt, strong electrolyte. Molecular: NH_3, weak base and weak electrolyte; H-first, acid; strong acid (Table 4.2), strong electrolyte; otherwise weak acid and weak electrolyte.   *Solve.*

(a)   HF: acid, mixture of ions and molecules (weak electrolyte)

(b) CH_3CN: none of the above, entirely molecules (nonelectrolyte)

(c) $NaClO_4$: salt, entirely ions (strong electrolyte)

(d) $Ba(OH)_2$: base, entirely ions (strong electrolyte)

4.36   Because the solution does conduct some electricity, but less than an equimolar NaCl solution (a strong electrolyte), the unknown solute must be a weak electrolyte. The weak electrolytes in the list of choices are NH_3 and H_3PO_3; because the solution is acidic, the unknown must be H_3PO_3.

4.37   *Analyze.* Given: chemical formulas. Find: electrolyte properties.

Plan. To classify as electrolytes, formulas must be identified as acids, bases, or salts as in Solution 4.35.   *Solve.*

(a) H_2SO_3: H first, so acid; not in Table 4.2, so weak acid; therefore, weak electrolyte

(b) CH_3CH_2OH: not acid, not ionic (no metal cation), contains OH group, but not as anion so not a base; therefore, nonelectrolyte

(c) NH_3: common weak base; therefore, weak electrolyte

(d) $KClO_3$: ionic compound, so strong electrolyte

(e) $Cu(NO_3)_2$: ionic compound, so strong electrolyte

4.38   (a) $LiClO_4$: strong   (b)   HClO: weak   (c) $CH_3CH_2CH_2OH$: non

(d) $HClO_3$: strong   (e) $CuSO_4$: strong   (f) $C_{12}H_{22}O_{11}$: non

4.39   *Plan.* Follow Sample Exercise 4.7.   *Solve.*

(a) $2\ HBr(aq) + Ca(OH)_2(aq) \rightarrow CaBr_2(aq) + 2\ H_2O(l)$

$H^+(aq) + OH^-(aq) \rightarrow H_2O(l)$

(b) $Cu(OH)_2(s) + 2\ HClO_4(aq) \rightarrow Cu(ClO_4)_2(aq) + 2\ H_2O(l)$

$Cu(OH)_2(s) + 2\ H^+(aq) \rightarrow 2\ H_2O(l) + Cu^{2+}(aq)$

(c) $Al(OH)_3(s) + 3\ HNO_3(aq) \rightarrow Al(NO_3)_3(aq) + 3\ H_2O(l)$

$Al(OH)_3(s) + 3\ H^+(aq) \rightarrow 3\ H_2O(l) + Al^{3+}(aq)$

4.40   (a) $2\ CH_3COOH(aq) + Ba(OH)_2(aq) \rightarrow Ba(CH_3COO)_2(aq) + 2\ H_2O(l)$

$CH_3COOH(aq) + OH^-(aq) \rightarrow CH_3COO^-(aq) + H_2O(l)$

(b) $Cr(OH)_3(s) + 3\ HNO_2(aq) \rightarrow Cr(NO_2)_3(aq) + 3\ H_2O(l)$

$Cr(OH)_3(s) + 3\ HNO_2(aq) \rightarrow 3\ H_2O(l) + Cr^{3+}(aq) + 3\ NO_2^-(aq)$

(c) $HNO_3(aq) + NH_3(aq) \rightarrow NH_4NO_3(aq)$

        $H^+(aq) + NH_3(aq) \rightarrow NH_4^+(aq)$

4.41   *Analyze.* Given: names of reactants. Find: gaseous products.

        Plan. Write correct chemical formulas for the reactants, complete and balance the metathesis reaction, and identify either H_2S or CO_2 products as gases.   *Solve.*

        (a) $CdS(s) + H_2SO_4(aq) \rightarrow CdSO_4(aq) + H_2S(g)$

                $CdS(s) + 2 H^+(aq) \rightarrow H_2S(g) + Cd^{2+}(aq)$

        (b) $MgCO_3(s) + 2 HClO_4(aq) \rightarrow Mg(ClO_4)_2(aq) + H_2O(l) + CO_2(g)$

                $MgCO_3(s) + 2 H^+(aq) \rightarrow H_2O(l) + CO_2(g) + Mg^{2+}(aq)$

4.42   (a) $FeO(s) + 2 H^+(aq) \rightarrow H_2O(l) + Fe^{2+}(aq)$

        (b) $NiO(s) + 2 H^+(aq) \rightarrow H_2O(l) + Ni^{2+}(aq)$

4.43   *Analyze.* Given the formulas or names of reactants, write balanced molecular and net ionic equations for the reactions.

        Plan. Write correct chemical formulas for all reactants. Predict products of the neutralization reactions by exchanging ion partners. Balance the complete molecular equation, identify spectator ions by recognizing strong electrolytes, write the corresponding net ionic equation (omitting spectators).   *Solve.*

        (a) $MgCO_3(s) + 2 HCl(aq) \rightarrow MgCl_2(aq) + H_2O(l) + CO_2(g)$

                $MgCO_3(s) + 2 H^+(aq) \rightarrow Mg^{2+}(aq) + H_2O(l) + CO_2(g)$

                $MgO(s) + 2 HCl(aq) \rightarrow MgCl_2(aq) + H_2O(l)$

                $MgO(s) + 2 H^+(aq) \rightarrow Mg^{2+}(aq) + H_2O(l)$

                $Mg(OH)_2(s) + 2 HCl(aq) \rightarrow MgCl_2(aq) + 2 H_2O(l)$

                $Mg(OH)_2(s) + 2 H^+(aq) \rightarrow Mg^{2+}(aq) + 2 H_2O(l)$

        (b)   We can distinguish magnesium carbonate, $MgCO_3(s)$, because its reaction with acid produces $CO_2(g)$, which appears as bubbles. The other two compounds are indistinguishable because the products of the two reactions are exactly the same.

4.44   (a) $K_2O(aq) + H_2O(l) \rightarrow 2 KOH(aq)$, molecular; $O^{2-}(aq) + H_2O(l) \rightarrow 2 OH^-(aq)$, net ionic

        (b)   base (H^+ ion acceptor): $O^{2-}(aq)$;

        (c)   acid (H^+ ion donor): $H_2O(aq)$;

        (d)   spectator: K^+

Oxidation–Reduction Reactions (Section 4.4)

4.45   (a)   False. *Oxidation* is loss of electrons; *reduction* is gain of electrons. (LEO says GER.)

        (b)   True. When a substance is oxidized, its oxidation number increases. When a substance is reduced, its oxidation number decreases.

4.46   (a)   True. Oxidation is loss of electrons; it can occur in the presence of any electron acceptor, not just oxygen.

        (b)   False. Oxidation and reduction can only occur together, not separately. When a substance is oxidized, it loses electrons, but free electrons do not exist under normal conditions. If electrons are lost by one substance they must be gained by another, and vice versa.

4.47   *Analyze.* Given the labeled periodic chart, determine which regions are most and least easily oxidized.

        Plan. Review the definition of oxidation and apply it to the properties of elements in the indicated regions of the chart.   *Solve.*

        (a)   Oxidation is loss of electrons. Elements easily oxidized form positive ions; these are metals. Elements in regions A and B are metals, and their ease of oxidation is shown in Table 4.5.

        (b)   Elements not readily oxidized tend to gain electrons and form negative ions; these are nonmetals. Elements in region D are nonmetals and are least easily oxidized.

4.48   (a) $BaSO_4$; +6   (b) H_2SO_3; +4   (c) SrS; –2   (d) H_2S; –2

        (e)   Sulfur is the third row of group 6A, the third column from the right on the periodic table. That is in region D on the designated chart.

        (f)   Based on these compounds, the range of oxidation numbers for sulfur is +6 to –2. Sulfur and other nonmetals in region D can adopt both positive and negative oxidation numbers. This is also true for the metalloids in region C. These elements have properties of both metals and nonmetals and can thus adopt both positive and negative oxidation numbers.

4.49   *Analyze.* Given the chemical formula of a substance, determine the oxidation number of a particular element in the substance.

        Plan. Follow the logic in Sample Exercise 4.8.   *Solve.*

        (a) +4   (b) +4   (c) +7   (d) +1   (e) +3   (f) –1 (O_2^{2-} is peroxide ion)

4.50   (a) +3   (b) +3   (c) –2   (d) –3   (e) +3   (f) +6

4.51   *Analyze.* Given: chemical reaction. Find: element oxidized or reduced. *Plan.* Assign oxidation numbers to all species. The element whose oxidation number increases (becomes more positive) is oxidized; the one whose oxidation number decreases (becomes more negative) is reduced.   *Solve.*

        (a) $N_2(g)$ [N, 0] → 2 $NH_3(g)$ [N, –3], N is reduced; 3 $H_2(g)$ [H, 0] → 2 $NH_3(g)$ [H, +1], H is oxidized.

        (b) Fe^{2+} → Fe, Fe is reduced; Al → Al^{3+}, Al is oxidized

        (c) Cl_2 → 2 Cl^-, Cl is reduced; 2 I^- → I_2, I is oxidized

        (d) S^{2-} → SO_4^{2-} (S, +6), S is oxidized; H_2O_2 (O, –1) → H_2O (O, –2); O is reduced

4.52   (a)   oxidation–reduction reaction; P is oxidized, Cl is reduced

        (b)   oxidation–reduction reaction; K is oxidized, Br is reduced

(c)   oxidation–reduction reaction; C is oxidized, O is reduced

(d)   precipitation reaction

4.53   *Analyze.* Given: reactants. Find: balanced molecular and net ionic equations.

Plan. Metals oxidized by H^+ form cations. Predict products by exchanging cations and balance. The anions are the spectator ions and do not appear in the net ionic equations.

Solve.

(a) $Mn(s) + H_2SO_4(aq) \rightarrow MnSO_4(aq) + H_2(g)$;

$Mn(s) + 2\,H^+(aq) \rightarrow Mn^{2+}(aq) + H_2(g)$

Products with the metal in a higher oxidation state are possible, depending on reaction conditions and acid concentration.

(b) $2\,Cr(s) + 6\,HBr(aq) \rightarrow 2\,CrBr_3(aq) + 3\,H_2(g)$;

$2\,Cr(s) + 6\,H^+(aq) \rightarrow 2\,Cr^{3+}(aq) + 3\,H_2(g)$

(c) $Sn(s) + 2\,HCl(aq) \rightarrow SnCl_2(aq) + H_2(g)$; $Sn(s) + 2\,H^+(aq) \rightarrow Sn^{2+}(aq) + H_2(g)$

(d) $2\,Al(s) + 6\,HCOOH(aq) \rightarrow 2\,Al(HCOO)_3(aq) + 3\,H_2(g)$;

$2\,Al(s) + 6\,HCOOH(aq) \rightarrow 2\,Al^{3+}(aq) + 6\,HCOO^-(aq) + 3\,H_2(g)$

4.54   (a) $2\,HCl(aq) + Ni(s) \rightarrow NiCl_2(aq) + H_2(g)$; $Ni(s) + 2\,H^+(aq) \rightarrow Ni^{2+}(aq) + H_2(g)$

(b) $H_2SO_4(aq) + Fe(s) \rightarrow FeSO_4(aq) + H_2(g)$; $Fe(s) + 2\,H^+(aq) \rightarrow Fe^{2+}(aq) + H_2(g)$

Products with the metal in a higher oxidation state are possible, depending on reaction conditions and acid concentration.

(c) $2\,HBr(aq) + Mg(s) \rightarrow MgBr_2(aq) + H_2(g)$; $Mg(s) + 2\,H^+(aq) \rightarrow Mg^{2+}(aq) + H_2(g)$

(d) $2\,CH_3COOH(aq) + Zn(s) \rightarrow Zn(CH_3COO)_2(aq) + H_2(g)$;

$Zn(s) + 2\,CH_3COOH(aq) \rightarrow Zn^{2+}(aq) + 2\,CH_3COO^-(aq) + H_2(g)$

4.55   *Analyze.* Given: a metal and an aqueous solution. Find: balanced equation.

Plan. Use Table 4.5. If the metal is above the aqueous solution, reaction will occur; if the aqueous solution is higher, NR. If reaction occurs, predict products by exchanging cations (a metal ion or H^+), then balance the equation.   *Solve.*

(a) $Fe(s) + Cu(NO_3)_2(aq) \rightarrow Fe(NO_3)_2(aq) + Cu(s)$

(b) $Zn(s) + MgSO_4(aq) \rightarrow NR$

(c) $Sn(s) + 2\,HBr(aq) \rightarrow SnBr_2(aq) + H_2(g)$

(d) $H_2(g) + NiCl_2(aq) \rightarrow NR$

(e) $2\,Al(s) + 3\,CoSO_4(aq) \rightarrow Al_2(SO_4)_3(aq) + 3\,Co(s)$

4.56   (a) $Ni(s) + Cu(NO_3)_2(aq) \rightarrow Ni(NO_3)_2(aq) + Cu(s)$

(b) $Zn(NO_3)_2(aq) + MgSO_4(aq) \rightarrow NR$

(c) $Au(s) + HCl(aq) \rightarrow NR$

(d) $2\,Cr(s) + 3\,CoCl_2(aq) \rightarrow 2\,CrCl_3(aq) + 3\,Co(s)$

(e) $H_2(g) + 2\,AgNO_3(aq) \rightarrow 2\,Ag(s) + 2\,HNO_3(aq)$

4.57   (a)   i. $Zn(s) + Cd^{2+}(aq) \rightarrow Cd(s) + Zn^{2+}(aq)$

           ii. $Cd(s) + Ni^{2+}(aq) \rightarrow Ni(s) + Cd^{2+}(aq)$

           Observation (i) indicates that Cd is less active than Zn; observation (ii) indicates that Cd is more active than Ni. Cd is between Zn and Ni on the activity series.

   (b)   Chromium, iron, and cobalt, the three elements between Zn and Ni in Table 4.5, more closely define the position of Cd in the activity series.

   (c)   Place an iron strip in $CdCl_2(aq)$. If $Cd(s)$ is deposited, Cd is less active than Fe; if there is no reaction, Cd is more active than Fe. Do the same test with Co if Cd is less active than Fe or with Cr if Cd is more active than Fe.

4.58 $Br_2 + 2\,NaI \rightarrow 2\,NaBr + I_2$ indicates that Br_2 is more easily reduced than I_2.

        $Cl_2 + 2\,NaBr \rightarrow 2\,NaCl + Br_2$ shows that Cl_2 is more easily reduced than Br_2.

        The order for ease of reduction is $Cl_2 > Br_2 > I_2$. Conversely, the order for ease of oxidation is $I^- > Br^- > Cl^-$.

   (a)   From the information above, the halogen I_2 is most stable (less likely to react) when mixed with other halides, X^-.

   (b) $Cl_2 + 2\,KI \rightarrow 2\,KCl + I_2$

   (c) $Br_2 + LiCl \rightarrow$ no reaction

Concentrations of Solutions (Section 4.5)

4.59   (a)   *Concentration* is an *intensive* property; it is the **ratio** of the amount of solute present in a certain quantity of solvent or solution. This ratio remains constant regardless of how much solution is present.

   (b)   The term *0.50 mol HCl* defines an amount (~18 g) of the pure substance HCl. The term 0.50 *M* HCl is a ratio; it indicates that there are 0.50 mol of HCl solute in 1.0 liter of solution. This same ratio of moles solute to solution volume is present regardless of the volume of solution under consideration.

4.60   *Analyze/Plan.* Follow the logic in Sample Exercise 4.11. *Solve.*

   (a) $M = \dfrac{\text{mol solute}}{\text{L solution}}; \dfrac{35.0 \text{ g } C_{12}H_{22}O_{11}}{1.000 \text{ L}} \times \dfrac{1 \text{ mol } C_{12}H_{22}O_{11}}{342.3 \text{ g } C_{12}H_{22}O_{11}} = 0.102 \, M \, C_{12}H_{22}O_{11}$

   (b)   Add 1.000 L of water to reduce the molarity by a factor of 2. Adding water does not change to amount of solute, but it does change the total volume of solution. A total solution volume of 2.000 L will reduce the molarity by a factor of 2.

4.61   *Analyze/Plan.* Follow the logic in Sample Exercises 4.11 and 4.12. *Solve.*

   (a) $M = \dfrac{\text{mol solute}}{\text{L solution}}; \dfrac{0.175 \text{ mol } ZnCl_2}{150 \text{ mL}} \times \dfrac{1000 \text{ mL}}{1 \text{ L}} = 1.17 \, M \, ZnCl_2$

        Check. $(0.175 / 0.150) > 1.0 \, M$

(b) $\text{mol} = M \times \text{L}; \quad \dfrac{4.50 \text{ mol HNO}_3}{1 \text{ L}} \times \dfrac{1 \text{ mol H}^+}{\text{mol HNO}_3} \times 0.0350 \text{ L} = 0.158 \text{ mol H}^+$

Check. $(4.5 \times .04) \approx 0.16 \text{ mol}$

(c) $\text{L} = \dfrac{\text{mol}}{M}; \quad \dfrac{0.350 \text{ mol NaOH}}{6.00 \text{ mol NaOH/L}} = 0.0583 \text{ L or } 58.3 \text{ mL of } 6.00 \, M \text{ NaOH}$

Check. $(0.325/6.0) > 0.50 \text{ L}.$

4.62   (a) $M = \dfrac{\text{mol solute}}{\text{L solution}}; \quad \dfrac{12.5 \text{ g Na}_2\text{CrO}_4}{0.750 \text{ L}} \times \dfrac{1 \text{ mol Na}_2\text{CrO}_4}{161.97 \text{ g Na}_2\text{CrO}_4} = 0.103 \, M \text{ Na}_2\text{CrO}_4$

(b) $\text{mol} = M \times \text{L}; \quad \dfrac{0.112 \text{ mol KBr}}{1 \text{ L}} \times 0.150 \text{ L} = 1.68 \times 10^{-2} \text{ mol KBr}$

(c) $\text{L} = \dfrac{\text{mol}}{M}; \quad \dfrac{0.150 \text{ mol HCl}}{6.1 \text{ mol HCl/L}} = 2.5 \times 10^{-2} \text{ L or } 25 \text{ mL}$

4.63   *Analyze.* Given molarity, M, and volume, L, find mass of $\text{Na}^+(aq)$ in the blood.

Plan. Calculate moles $\text{Na}^+(aq)$ using the definition of molarity: $M = \dfrac{\text{mol}}{\text{L}}; \text{mol} = M \times \text{L}.$

Calculate mass $\text{Na}^+(aq)$ using the definition moles: $\text{mol} = \text{g/MM}; \text{g} = \text{mol} \times \text{MM}.$ (MM is the symbol for molar mass in this manual.)

Solve. $\dfrac{0.135 \text{ mol}}{\text{L}} \times 5.0 \text{ L} \times \dfrac{23.0 \text{ g Na}^+}{\text{mol Na}^+} = 15.525 = 16 \text{ g Na}^+(aq)$

Check. Because there are more than 0.1 mol/L and we have 5.0 L, there should be more than half a mol (11.5 g) of Na^+. The calculation agrees with this estimate.

4.64   Calculate the mol of Na^+ at the two concentrations; the difference is the mol NaCl required to increase the Na^+ concentration to the desired level.

$\dfrac{0.118 \text{ mol}}{\text{L}} \times 4.6 \text{ L} = 0.5428 = 0.54 \text{ mol Na}^+$

$\dfrac{0.138 \text{ mol}}{\text{L}} \times 4.6 \text{ L} = 0.6348 = 0.63 \text{ mol Na}^+$

$(0.6348 - 0.5428) = 0.092 = 0.09 \text{ mol NaCl (2 decimal places and 1 sig fig)}$

$0.092 \text{ mol NaCl} \times \dfrac{58.5 \text{ g NaCl}}{\text{mol}} = 5.38 = 5 \text{ g NaCl}$

4.65   *Analyze.* Given: g alcohol/100 mL blood; molecular formula of alcohol. Find: molarity (mol/L) of alcohol. *Plan.* Use the molar mass (MM) of alcohol to change (g/100) mL to (mol/100 mL) then mL to L.

Solve. MM of alcohol = $2(12.01) + 6(1.1008) + 1(16.00) = 46.07 \text{ g alcohol/mol}$

$\text{BAC} = \dfrac{0.08 \text{ g alcohol}}{100 \text{ mL blood}} \times \dfrac{1 \text{ mol alcohol}}{46.07 \text{ g alcohol}} \times \dfrac{1000 \text{ mL}}{1 \text{ L}} = 0.0174 = 0.02 \, M \text{ alcohol}$

4.66   *Analyze.* Given: BAC (definition from Exercise 4.65), vol of blood. Find: mass alcohol in bloodstream.

Plan. Change BAC (g/100 mL) to (g/L), then times vol of blood in L.

Solve. BAC = 0.10 g/100 mL

$$\frac{0.10 \text{ g alcohol}}{100 \text{ mL blood}} \times \frac{1000 \text{ mL}}{1 \text{ L}} \times 5.0 \text{ L blood} = 5.0 \text{ g alcohol}$$

4.67   *Plan.* Proceed as in Sample Exercises 4.13.

$$M = \frac{\text{mol}}{\text{L}}; \quad \text{mol} = \frac{\text{g}}{\text{MM}} \quad \text{(MM is the symbol for molar mass in this manual.)}$$

Solve.

(a) $\dfrac{6.86 \text{ mol CH}_3\text{CH}_2\text{OH}}{1 \text{ L}} \times 1.00 \text{ L} \times \dfrac{46.07 \text{ g CH}_3\text{CH}_2\text{OH}}{1 \text{ mol CH}_3\text{CH}_2\text{OH}} = 316.04 = 316 \text{ g CH}_3\text{CH}_2\text{OH}$

Check. $(7 \times 50) \approx 350$ g ethanol (this is an upper limit)

(b) $316.04 \text{ g CH}_3\text{CH}_2\text{OH} \times \dfrac{1 \text{ mL}}{0.789 \text{ g CH}_3\text{CH}_2\text{OH}} = 400.56 = 401 \text{ mL} = 0.401 \text{ L CH}_3\text{CH}_2\text{OH}$

Check. $(320/0.8) \approx 400$ mL ethanol

4.68 $M = \dfrac{\text{mol}}{\text{L}}; \quad \text{mol} = \dfrac{\text{g}}{\text{MM}}$ (MM is the symbol for molar mass in this manual.)

$M = \dfrac{\text{mol solute}}{\text{L solution}}; \quad \dfrac{124 \text{ mg C}_6\text{H}_8\text{O}_6}{0.2366 \text{ L}} \times \dfrac{1 \text{ g}}{1000 \text{ mg}} \times \dfrac{1 \text{ mol C}_6\text{H}_8\text{O}_6}{176.12 \text{ g C}_6\text{H}_8\text{O}_6} = 2.98 \times 10^{-3} \ M \text{ C}_6\text{H}_8\text{O}_{66}$

4.69   *Analyze.* Given: formula and concentration of each solute. Find: concentration of K^+ in each solution. *Plan.* Note mol K^+/mol solute and compare concentrations or total moles.
Solve.

(a) $\text{KCl} \rightarrow K^+ + Cl^-; 0.20 \ M \text{ KCl} = 0.20 \ M \ K^+$

$\text{K}_2\text{CrO}_4 \rightarrow \mathbf{2} \ K^+ + \text{CrO}_4{}^{2-}; 0.15 \ M \text{ K}_2\text{CrO}_4 = 0.30 \ M \ K^+$

$\text{K}_3\text{PO}_4 \rightarrow \mathbf{3} \ K^+ + \text{PO}_4{}^{3-}; 0.080 \ M \text{ K}_3\text{PO}_4 = 0.24 \ M \ K^+$

0.15 M K_2CrO_4 has the highest K^+ concentration.

(b) K_2CrO_4: 0.30 M $K^+ \times 0.0300$ L = 0.0090 mol K^+

K_3PO_4: 0.24 M $K^+ \times 0.0250$ L = 0.0060 mol K^+

30.0 mL of 0.15 M K_2CrO_4 has more K^+ ions.

4.70   (a)   0.10 M $\text{BaI}_2 = 0.2 \ M \ I^-$; 0.25 M KI = 0.25 M I^-

0.25 M KI has the higher I^- concentration.

(b)   0.10 M KI = 0.1 M I^-; 0.040 M $\text{ZnI}_2 = 0.080 \ M \ I^-$; 0.10 M KI has a higher I^- concentration than 0.040 M ZnI_2. Total volume does not affect concentration.

(c) $3.2\ M\ \text{HI} = 3.2\ M\ \text{I}^-$

$$145\ \text{g NaI} \times \frac{1\ \text{mol NaI}}{149.9\ \text{g NaI}} \times \frac{1}{0.150\ \text{L}} = 6.45\ M\ \text{NaI} = 6.45\ M\ \text{I}^-$$

The NaI solution has the higher I^- concentration.

4.71   *Analyze.* Given: molecular formula and solution molarity. Find: concentration (M) of each ion.

Plan. Follow the logic in Sample Exercise 4.12.

Solve.

(a) $\text{NaNO}_3 \rightarrow \text{Na}^+, \text{NO}_3^-$; $0.25\ M\ \text{NaNO}_3 = 0.25\ M\ \text{Na}^+, 0.25\ M\ \text{NO}_3^-$

(b) $\text{MgSO}_4 \rightarrow \text{Mg}^{2+}, \text{SO}_4^{2-}$; $1.3 \times 10^{-2}\ M\ \text{MgSO}_4 = 1.3 \times 10^{-2}\ M\ \text{Mg}^{2+}, 1.3 \times 10^{-2}\ M\ \text{SO}_4^{2-}$

(c) $\text{C}_6\text{H}_{12}\text{O}_6 \rightarrow \text{C}_6\text{H}_{12}\text{O}_6$ (molecular solute); $0.0150\ M\ \text{C}_6\text{H}_{12}\text{O}_6 = 0.0150\ M\ \text{C}_6\text{H}_{12}\text{O}_6$

(d)   *Plan.* There is no reaction between NaCl and $(\text{NH}_4)_2\text{CO}_3$, so this is just a dilution problem, $M_1V_1 = M_2V_2$. Then account for ion stoichiometry.

Solve. $45.0\ \text{mL} + 65.0\ \text{mL} = 110.0\ \text{mL}$ total volume

$$\frac{0.272\ M\ \text{NaCl} \times 45.0\ \text{mL}}{110.0\ \text{mL}} = 0.111\ M\ \text{NaCl}; 0.111\ M\ \text{Na}^+, 0.111\ M\ \text{Cl}^-$$

$$\frac{0.0247\ M\ (\text{NH}_4)_2\text{CO}_3 \times 65.0\ \text{mL}}{110.0\ \text{mL}} = 0.0146\ M\ (\text{NH}_4)_2\text{CO}_3$$

$2 \times (0.0146\ M) = 0.0292\ M\ \text{NH}_4^+, 0.0146\ M\ \text{CO}_3^{2-}$

Check. By adding the two solutions (with no common ions or chemical reaction), we have approximately doubled the solution volume, and reduced the concentration of each ion by approximately a factor of two.

4.72   (a)   *Plan.* These two solutions have common ions. Find the ion concentration resulting from each solution, then add.

Solve. total volume $= 42.0\ \text{mL} + 37.6\ \text{mL} = 79.6\ \text{mL}$

$$\frac{0.170\ M\ \text{NaOH} \times 42.0\ \text{mL}}{79.6\ \text{mL}} = 0.08970 = 0.0897\ M\ \text{NaOH};$$

$0.0897\ M\ \text{Na}^+, 0.0897\ M\ \text{OH}^-$

$$\frac{0.400\ M\ \text{NaOH} \times 37.6\ \text{mL}}{79.6\ \text{mL}} = 0.18894 = 0.189\ M\ \text{NaOH};$$

$0.189\ M\ \text{Na}^+, 0.189\ M\ \text{OH}^-$

$M\ \text{Na}^+ = 0.08970\ M + 0.18894\ M = 0.27864 = 0.279\ M\ \text{Na}^+$

$M\ \text{OH}^- = M\ \text{Na}^+ = 0.279\ M\ \text{OH}^-$

(b)   *Plan.* No common ions; just dilution.

         Solve. 44.0 mL + 25.0 mL = 69.0 mL

$$\frac{0.100 \ M \ Na_2SO_4 \times 44.0 \ mL}{69.0 \ mL} = 0.06377 = 0.0638 \ M \ Na_2SO_4$$

$$2 \times (0.06377 \ M) = 0.1275 = 0.128 \ M \ Na^+; \ 0.0638 \ M \ SO_4^{2-}$$

$$\frac{0.150 \ M \ KCl \times 25.0 \ mL}{69.0 \ mL} = 0.054348 = 0.0543 \ M \ KCl$$

$$0.0543 \ M \ K^+, \ 0.0543 \ M \ Cl^-$$

(c)   *Plan.* Calculate concentration of K^+ and Cl^- due to the added solid. Then sum to get total concentration of Cl^-.

         Solve. $\dfrac{3.60 \ g \ KCl}{75.0 \ mL \ soln} \times \dfrac{1 \ mol \ KCl}{74.55 \ g \ KCl} \times \dfrac{1000 \ mL}{1 \ L} = 0.6439 = 0.644 \ M \ KCl$

         $0.250 \ M \ CaCl_2; \ 2(0.250 \ M) = 0.500 \ M \ Cl^-$

         total $Cl^- = 0.644 \ M + 0.500 \ M = 1.144 \ M \ Cl^-, \ 0.644 \ M \ K^+, \ 0.250 \ M \ Ca^{2+}$

4.73   *Analyze/Plan.* Follow the logic of Sample Exercise 4.14.

        Solve.

      (a) $V_1 = M_2V_2/M_1; \quad \dfrac{0.250 \ M \ NH_3 \times 1000.0 \ mL}{14.8 \ M \ NH_3} = 16.89 = 16.9 \ mL \ 14.8 \ M \ NH_3$

         Check. $250/15 \approx 15 \ M$

      (b) $M_2 = M_1V_1/V_2; \quad \dfrac{14.8 \ M \ NH_3 \times 10.0 \ mL}{500 \ mL} = 0.296 \ M \ NH_3$

         Check. $150/500 \approx 0.30 \ M$

4.74   (a) $V_1 = M_2V_2/M_1; \quad \dfrac{0.500 \ M \ HNO_3 \times 0.110 \ L}{6.0 \ M \ HNO_3} = 0.00917 \ L = 9.2 \ mL \ 6.0 \ M \ HNO_3$

      (b) $M_2 = M_1V_1/V_2; \quad \dfrac{6.0 \ M \ HNO_3 \times 10.0 \ mL}{250 \ mL} = 0.240 \ M \ HNO_3$

4.75   *Analyze/Plan.* Calculate the number of drug molecules in 1.00 mL of the stock solution, using $M \times L$ = moles and Avogadro's number. Then calculate the desired ratio.

        Solve. 1.00 mL = 0.00100 L

$$\frac{1.5 \times 10^{-9} \ mol}{L} \times 0.0010 \ L \times \frac{6.022 \times 10^{23} \ molecules}{mole} = 9.033 \times 10^{11} = 9.0 \times 10^{11} \ molecules$$

$$\frac{9.033 \times 10^{11} \ drug \ molecules}{2.0 \times 10^5 \ cancer \ cells} = 4.517 \times 10^6 = 4.5 \times 10^6$$

4.76   *Analyze/Plan.* The 25.00 mL of antibiotic solution needs to contain a minimum of 1.0×10^8 molecules of the drug. Calculate the moles of drug this represents. The concentration of the stock solution is 5.00×10^{-9} *M*. Then,

L stock solution = mol drug$/M$ solution; L = mol$/5.00 \times 10^{-9}$ *M*.

$$1.0 \times 10^8 \text{ molecules} \times \frac{1 \text{ mol}}{6.022 \times 10^{23} \text{ molecules}} \times \frac{1 \text{ L}}{5.00 \times 10^{-9} \text{ mol}} \times \frac{1000 \text{ mL}}{1 \text{ L}} = 3.3 \times 10^{-5} \text{ mL}$$

A volume of 3.3×10^{-5} mL corresponds to 33×10^{-9} L, or 33 nL (nanoliters).

Two points are of note. First, desired results can be achieved with a very small amount of the drug, which reduces the cost. And, delivering such a small volume of stock solution may be a challenge. A dilution scheme for the stock solution could be employed. If 1.00 mL of the 5.00×10^{-9} *M* antibiotic stock solution is diluted to 1.00 L, and 1.00 mL of this solution is diluted to 0.500 L, the resulting concentration of the diluted stock solution is then 1.00×10^{-14} *M*. Using the relationship L = mol$/M$, we find that 0.017 L or 17 mL of the 1.00×10^{-14} *M* stock solution diluted to 25.00 mL would kill the desired amount of bacteria.

4.77   *Analyze.* Given: density of pure acetic acid, volume pure acetic acid, volume new solution. Find: molarity of new solution. *Plan.* Calculate the mass of acetic acid, CH_3COOH, present in 20.0 mL of the pure liquid.   *Solve.*

$$20.00 \text{ mL acetic acid} \times \frac{1.049 \text{ g acetic acid}}{1 \text{ mL acetic acid}} = 20.98 \text{ g acetic acid}$$

$$20.98 \text{ g } CH_3COOH \times \frac{1 \text{ mol } CH_3COOH}{60.05 \text{ g } CH_3COOH} = 0.349375 = 0.3494 \text{ mol } CH_3COOH$$

$$M = \text{mol/L} = \frac{0.349375 \text{ mol } CH_3COOH}{0.2500 \text{ L solution}} = 1.39750 = 1.398 \text{ } M \text{ } CH_3COOH$$

Check. $(20 \times 1) \approx 20$ g acid; $(20/60) \approx 0.33$ mol acid; $(0.33/0.25 = 0.33 \times 4) \approx 1.33 \text{ } M$

4.78 $$50.000 \text{ mL glycerol} \times \frac{1.2656 \text{ g glycerol}}{1 \text{ mL glycerol}} = 63.280 \text{ g glycerol}$$

$$63.280 \text{ g } C_3H_8O_3 \times \frac{1 \text{ mol } C_3H_8O_3}{92.094 \text{ g } C_3H_8O_3} = 0.687124 = 0.68712 \text{ mol } C_3H_8O_3$$

$$M = \frac{0.687124 \text{ mol } C_3H_8O_3}{0.25000 \text{ L solution}} = 2.7485 \text{ } M \text{ } C_3H_8O_3$$

Solution Stoichiometry and Chemcial Analysis (Section 4.6)

4.79   *Analyze.* Given: volume and molarity $AgNO_3$, molarity HCl. Find: volume HCl or mass of KCl.

(a)   *Plan.* $M \times L = $ mol $AgNO_3 = $ mol Ag^+; balanced equation gives ratio mol HCl/mol $AgNO_3$; mol HCl $\rightarrow$ vol HCl.   *Solve.*

$$\frac{0.200 \text{ mol } AgNO_3}{1 \text{ L}} \times 0.0150 \text{ L} = 3.00 \times 10^{-3} \text{ mol } AgNO_3(aq)$$

$$AgNO_3(aq) + HCl(aq) \rightarrow AgCl(s) + HNO_3(aq)$$

$$\text{mol HCl} = \text{mol AgNO}_3 = 3.00 \times 10^{-3} \text{ mol KCl}$$

$$3.00 \times 10^{-3} \text{ mol HCl} \times \frac{1 \text{ L}}{0.150 \text{ mol HCl}} = 0.0200 \text{ L} = 20.0 \text{ mL } 0.15 \text{ } M \text{ HCl}$$

Check. $(0.2 \times 0.015) = 0.003$ mol; $(0.003/0.15) \approx 0.02$ L HCl

(b) *Plan.* $M \times L = $ mol $AgNO_3 = $ mol Ag^+; balanced equation gives ratio mol KCl/mol $AgNO_3$; mol $KCl \rightarrow$ vol KCl. *Solve.*

$$\frac{0.200 \text{ mol AgNO}_3}{1 \text{ L}} \times 0.0150 \text{ L} = 3.00 \times 10^{-3} \text{ mol AgNO}_3(aq)$$

$$AgNO_3(aq) + KCl(aq) \rightarrow AgCl(s) + KNO_3(aq)$$

$$\text{mol KCl} = \text{mol AgNO}_3 = 3.00 \times 10^{-3} \text{ mol KCl}$$

$$3.00 \times 10^{-3} \text{ mol KCl} \times \frac{74.55 \text{ g KCl}}{1 \text{ mol KCl}} = 0.224 \text{ g KCl}$$

Check. $(0.2 \times 0.015) = 0.003$ mol; $(0.003 \times 75) \approx 0.225$ g KCl

(c) Clearly, the KCl reagent is virtually free relative to the HCl solution. The KCl analysis is more cost-effective.

4.80 *Plan.* $M \times L = $ mol $Cd(NO_3)_2$; balanced equation $\rightarrow$ mol ratio $\rightarrow$ mol $NaOH \rightarrow$ g $NaOH$

Solve. $\dfrac{0.500 \text{ mol Cd(NO}_3)_2}{1 \text{ L}} \times 0.0350 \text{ L} = 0.0175 \text{ mol Cd(NO}_3)_2$

$$Cd(NO_3)_2(aq) + 2 \text{ NaOH}(aq) \rightarrow Cd(OH)_2(s) + 2 \text{ NaNO}_3(aq)$$

$$0.0175 \text{ mol Cd(NO}_3)_2 \times \frac{2 \text{ mol NaOH}}{1 \text{ mol Cd(NO}_3)_2} \times \frac{40.00 \text{ g NaOH}}{1 \text{ mol NaOH}} = 1.40 \text{ g NaOH}$$

4.81 (a) *Analyze.* Given: M and vol base, M acid. Find: vol acid

Plan/Solve. Write the balanced equation for the reaction in question:

$$HClO_4(aq) + NaOH(aq) \rightarrow NaClO_4(aq) + H_2O(l)$$

Calculate the moles of the known substance, in this case NaOH.

$$\text{moles NaOH} = M \times L = \frac{0.0875 \text{ mol NaOH}}{1 \text{ L}} \times 0.0500 \text{ L} = 0.004375 = 0.00438 \text{ mol NaOH}$$

Apply the mole ratio (mol unknown/mol known) from the chemical equation.

$$0.004375 \text{ mol NaOH} \times \frac{1 \text{ mol HClO}_4}{1 \text{ mol NaOH}} = 0.004375 \text{ mol HClO}_4$$

Calculate the desired quantity of unknown, in this case the volume of 0.115 M $HClO_4$ solution.

$$L = \text{mol}/M; \quad L = 0.004375 \text{ mol HClO}_4 \times \frac{1 \text{ L}}{0.115 \text{ mol HClO}_4} = 0.0380 \text{ L} = 38.0 \text{ mL}$$

Check. $(0.09 \times 0.05) = 0.0045$ mol; $(0.0045/0.11) \approx 0.040$ L ≈ 40 mL

(b) Following the logic outlined in part (a):

$$2 \, HCl(aq) + Mg(OH)_2(s) \rightarrow MgCl_2(aq) + 2 \, H_2O(l)$$

$$2.87 \text{ g Mg(OH)}_2 \times \frac{1 \text{ mol Mg(OH)}_2}{58.32 \text{ g Mg(OH)}_2} = 0.049211 = 0.0492 \text{ mol Mg(OH)}_2$$

$$0.0492 \text{ mol Mg(OH)}_2 \times \frac{2 \text{ mol HCl}}{1 \text{ mol Mg(OH)}_2} = 0.0984 \text{ mol HCl}$$

$$L = mol/M = 0.09840 \text{ mol HCl} \times \frac{1 \text{ L HCl}}{0.128 \text{ mol HCl}} = 0.769 \text{ L} = 769 \text{ mL}$$

(c) $AgNO_3(aq) + KCl(aq) \rightarrow AgCl(s) + KNO_3(aq)$

$$785 \text{ mg KCl} \times \frac{1 \times 10^{-3} \text{ g}}{1 \text{ mg}} \times \frac{1 \text{ mol KCl}}{74.55 \text{ g KCl}} \times \frac{1 \text{ mol AgNO}_3}{1 \text{ mol KCl}} = 0.01053 = 0.0105 \text{ mol AgNO}_3$$

$$M = mol/L = \frac{0.01053 \text{ mol AgNO}_3}{0.0258 \text{ L}} = 0.408 \, M \text{ AgNO}_3$$

(d) $HCl(aq) + KOH(aq) \rightarrow KCl(aq) + H_2O(l)$

$$\frac{0.108 \text{ mol HCl}}{1 \text{ L}} \times 0.0453 \text{ L} \times \frac{1 \text{ mol KOH}}{1 \text{ mol HCl}} \times \frac{56.11 \text{ g KOH}}{1 \text{ mol KOH}} = 0.275 \text{ g KOH}$$

4.82 (a) $2 \, HCl(aq) + Ba(OH)_2(aq) \rightarrow BaCl_2(aq) + 2 \, H_2O(l)$

$$\frac{0.101 \text{ mol Ba(OH)}_2}{1 \text{ L Ba(OH)}_2} \times 0.0500 \text{ L Ba(OH)}_2 \times \frac{2 \text{ mol HCl}}{1 \text{ mol Ba(OH)}_2} \times \frac{1 \text{ L HCl}}{0.120 \text{ mol HCl}}$$
$$= 0.0842 \text{ L or } 84.2 \text{ mL HCl soln}$$

(b) $H_2SO_4(aq) + 2 \, NaOH(aq) \rightarrow Na_2SO_4(aq) + 2 \, H_2O(l)$

$$0.200 \text{ g NaOH} \times \frac{1 \text{ mol NaOH}}{40.00 \text{ g NaOH}} \times \frac{1 \text{ mol H}_2SO_4}{2 \text{ mol NaOH}} \times \frac{1 \text{ L H}_2SO_4}{0.125 \text{ mol H}_2SO_4}$$
$$= 0.0200 \text{ L or } 20.0 \text{ mL H}_2SO_4 \text{ soln}$$

(c) $BaCl_2(aq) + Na_2SO_4(aq) \rightarrow BaSO_4(s) + 2 \, NaCl(aq)$

$$752 \text{ mg} = 0.752 \text{ g Na}_2SO_4 \times \frac{1 \text{ mol Na}_2SO_4}{142.1 \text{ g Na}_2SO_4} \times \frac{1 \text{ mol BaCl}_2}{1 \text{ mol Na}_2SO_4} \times \frac{1}{0.0558 \text{ L}}$$
$$= 0.0948 \, M \text{ BaCl}_2$$

(d) $2 \, HCl(aq) + Ca(OH)_2(aq) \rightarrow CaCl_2(aq) + 2 \, H_2O(l)$

$$0.0427 \text{ L HCl} \times \frac{0.208 \text{ mol HCl}}{1 \text{ L HCl}} \times \frac{1 \text{ mol Ca(OH)}_2}{2 \text{ mol HCl}} \times \frac{74.10 \text{ g Ca(OH)}_2}{1 \text{ mol Ca(OH)}_2} = 0.329 \text{ g Ca(OH)}_2$$

4.83 *Analyze/Plan.* See Exercise 4.81(a) for a more detailed approach. *Solve.*

$$\frac{6.0 \text{ mol H}_2SO_4}{1 \text{ L}} \times 0.027 \text{ L} \times \frac{2 \text{ mol NaHCO}_3}{1 \text{ mol H}_2SO_4} \times \frac{84.01 \text{ g NaHCO}_3}{1 \text{ mol NaHCO}_3} = 27 \text{ g NaHCO}_3$$

4.84   See Exercise 4.81(a) for a more detailed approach.

$$\frac{0.115 \text{ mol NaOH}}{1 \text{ L}} \times 0.0425 \text{ L} \times \frac{1 \text{ mol CH}_3\text{COOH}}{1 \text{ mol NaOH}} \times \frac{60.05 \text{ g CH}_3\text{COOH}}{1 \text{ mol CH}_3\text{COOH}}$$

$$= 0.29349 = 0.293 \text{ g CH}_3\text{COOH in 3.45 mL}$$

$$1.00 \text{ qt vinegar} \times \frac{1 \text{ L}}{1.057 \text{ qt}} \times \frac{1000 \text{ mL}}{1 \text{ L}} \times \frac{0.29349 \text{ g CH}_3\text{COOH}}{3.45 \text{ mL vinegar}} = 80.5 \text{ g CH}_3\text{COOH/qt}$$

4.85   *Analyze.* Given: M and vol HCl. Find: MM of base, an alkali metal hydroxide.

Plan. Alkali metal ions have a 1+ charge, so the general formula of an alkali metal hydroxide is MOH. One mol of MOH requires one mol of HCl for neutralization.

(a) $M \times L = \text{mol HCl} = \text{mol MOH.}$ $MM = \dfrac{\text{g MOH}}{1 \text{ mol MOH}}$.   *Solve.*

$$\frac{2.50 \text{ mol HCl}}{1 \text{ L}} \times 0.0170 \text{ L} = 0.0425 \text{ mol HCl} = 0.0425 \text{ mol MOH}$$

$$MM \text{ of MOH} = \frac{4.36 \text{ g MOH}}{0.0425 \text{ mol MOH}} = 102.59 = 103 \text{ g/mol}$$

(b)   MM of alkali metal = MM of MOH – (17.01 g)   *Solve.*

MM of alkali metal = (102.59 g/mol – 17.01 g/mol) = 85.58 = 86 g/mol

The experimental molar mass most closely fits that of Rb^+, 85.47 g/mol

Check. The experimental molar mass matches one of the alkali metals.

4.86   *Analyze/Plan.* Follow the logic in Exercise 4.85. The unknown is a group 2A metal hydroxide, general formula $M(OH)_2$. Two mol HCL are required to neutralize 1 mol $M(OH)_2$. *Solve.*

(a) $$\frac{2.50 \text{ mol HCl}}{1 \text{ L}} \times 0.0569 \text{ L} \times \frac{1 \text{ mol M(OH)}_2}{2 \text{ mol HCl}} = 0.071125 = 0.0711 \text{ mol M(OH)}_2$$

$$MM \text{ of M(OH)}_2 = \frac{8.65 \text{ g M(OH)}_2}{0.071125 \text{ mol M(OH)}_2} = 121.62 = 122 \text{ g/mol}$$

(b)   MM of group 2A metal = MM of $M(OH)_2$ – 2(17.01 g)

MM of group 2A metal = (121.62 g/mol – 34.02 g/mol) = 87.60 = 88 g/mol

The experimental molar mass most closely fits that of Sr^{2+}, 87.62 g/mol

Check. The experimental molar mass matches one of the group 2A metals.

4.87   (a) $NiSO_4(aq) + 2 \text{ KOH}(aq) \rightarrow Ni(OH)_2(s) + K_2SO_4(aq)$

(b)   The precipitate is $Ni(OH)_2$.

(c)   *Plan.* Compare mol of each reactant; mol = $M \times L$

Solve. 0.200 M KOH × 0.1000 L KOH = 0.0200 mol KOH

0.150 M $NiSO_4$ × 0.2000 L $NiSO_4$ = 0.0300 mol $NiSO_4$

1 mol $NiSO_4$ requires 2 mol KOH, so 0.0300 mol $NiSO_4$ requires 0.0600 mol KOH. Because only 0.0200 mol KOH is available, KOH is the limiting reactant.

(d)   *Plan*. The amount of the limiting reactant (KOH) determines amount of product, in this case $Ni(OH)_2$.

        Solve. $0.0200 \text{ mol KOH} \times \dfrac{1 \text{ mol Ni(OH)}_2}{2 \text{ mol KOH}} \times \dfrac{92.71 \text{ g Ni(OH)}_2}{1 \text{ mol Ni(OH)}_2} = 0.927 \text{ g Ni(OH)}_2$

(e)   *Plan/Solve*. Limiting reactant: OH^-: no excess OH^- remains in solution.

        Excess reactant: Ni^{2+}: M Ni^{2+} remaining = mol Ni^{2+} remaining/L solution

        0.0300 mol Ni^{2+} initial $- 0.0100$ mol Ni^{2+} reacted = 0.0200 mol Ni^{2+} remaining

        0.0200 mol $Ni^{2+}/0.3000$ L = 0.0667 M $Ni^{2+}(aq)$

        Spectators: SO_4^{2-}, K^+. These ions do not react, so the only change in their concentration is dilution. The final volume of the solution is 0.3000 L.

        $M_2 = M_1 V_1 / V_2$: 0.200 M $K^+ \times 0.1000$ L/0.3000 L = 0.0667 M $K^+(aq)$

        0.150 M $SO_4^{2-} \times 0.2000$ L/0.3000 L = 0.100 M SO_4^{2-} (aq)

4.88   (a) $2 \text{ HNO}_3(aq) + Sr(OH)_2(s) \rightarrow Sr(NO_3)_2(aq) + 2 \text{ H}_2O(l)$

       (b)   Determine the limiting reactant, then the identity and concentration of ions remaining in solution. Assume that the $H_2O(l)$ produced by the reaction does **not** increase the total solution volume.

          $15.0 \text{ g Sr(OH)}_2 \times \dfrac{1 \text{ mol Sr(OH)}_2}{121.64 \text{ g Sr(OH)}_2} = 0.1233 = 0.123 \text{ mol Sr(OH)}_2$

          mol $OH^- = 2(0.1233)$ mol $Sr(OH)_2 = 0.2466 = 0.247$ mol OH^-

          0.200 M $HNO_3 \times 0.0550$ L $HNO_3 = 0.0110$ mol HNO_3.

        Two mol HNO_3 react with one mol $Sr(OH)_2$, so HNO_3 is the limiting reactant. No excess H^+ remains in solution. The remaining ions are OH^- (excess reactant), Sr^{2+}, and NO_3^- (spectators).

        OH^-: 0.2466 mol OH^- initial $- 0.0110$ mol OH^- react =

                     $0.2356 = 0.236$ mol OH^- remain

                     0.2356 mol $OH^-/0.0550$ L soln = 4.28 M $OH^-(aq)$

        Sr^{2+}: 0.123 mol $Sr^{2+}/0.0550$ L soln = 2.24 M $Sr^{2+}(aq)$

        NO_3^-: 0.0110 mol $NO_3^-/0.0550$ L = 0.200 M $NO_3^-(aq)$

       (c)   The resulting solution is **basic** because of the large excess of $OH^-(aq)$.

4.89   *Analyze*. Given: mass impure $Mg(OH)_2$; M and vol **excess** HCl; M and vol NaOH.

      Find: mass % $Mg(OH)_2$ in sample.   *Plan/Solve*. Write balanced equations.

      $Mg(OH)_2(s) + 2 \text{ HCl}(aq) \rightarrow MgCl_2(aq) + 2 \text{ H}_2O(l)$

      $HCl(aq) + NaOH(aq) \rightarrow NaCl(aq) + H_2O(l)$

      Calculate total moles HCl = M HCl $\times$ L HCl

      $\dfrac{0.2050 \text{ mol HCl}}{1 \text{ L soln}} \times 0.1000 \text{ L} = 0.02050 \text{ mol HCl total}$

mol excess HCl = mol NaOH used = M NaOH × L NaOH

$$\frac{0.1020 \text{ mol NaOH}}{1 \text{ L soln}} \times 0.01985 \text{ L} = 0.0020247 = 0.002025 \text{ mol NaOH}$$

mol HCl reacted with $Mg(OH)_2$ = total mol HCl – excess mol HCl

0.02050 mol total – 0.0020247 mol excess = 0.0184753 = 0.01848 mol HCl reacted

(The result has 5 decimal places and 4 sig. figs.)

Use mol ratio to get mol $Mg(OH)_2$ in sample, then molar mass of $Mg(OH)_2$ to get g pure $Mg(OH)_2$.

$$0.0184753 \text{ mol HCl} \times \frac{1 \text{ mol } Mg(OH)_2}{2 \text{ mol HCl}} \times \frac{58.32 \text{ g } Mg(OH)_2}{1 \text{ mol } Mg(OH)_2} = 0.53874 = 0.5387 \text{ g } Mg(OH)_2$$

$$\text{mass \% } Mg(OH)_2 = \frac{\text{g } Mg(OH)_2}{\text{g sample}} \times 100 = \frac{0.53874 \text{ g } Mg(OH)_2}{0.5895 \text{ g sample}} \times 100 = 91.39 \text{ \% } Mg(OH)_2$$

4.90   *Plan.* $CaCO_3(s) + 2 \text{ HCl}(aq) \rightarrow CaCl_2(aq) + H_2O(l) + CO_2(g)$

   $\text{HCl}(aq) + \text{NaOH}(aq) \rightarrow \text{NaCl}(aq) + H_2O(l)$

total mol HCl – excess mol HCl = mol HCl reacted; mol $CaCO_3$ = (mol HCl)/2;

g $CaCO_3$ = mol $CaCO_3$ × molar mass; mass % = (g $CaCO_3$/g sample) × 100

Solve:

$$\frac{1.035 \text{ mol HCl}}{1 \text{ L soln}} \times 0.03000 \text{ L} = 0.031050 = 0.03105 \text{ mol HCl total}$$

$$\frac{1.010 \text{ mol NaOH}}{1 \text{ L soln}} \times 0.01156 \text{ L} = 0.011676 = 0.01168 \text{ mol HCl excess}$$

0.031050 total – 0.011676 excess = 0.019374 = 0.01937 mol HCl reacted

$$0.019374 \text{ mol HCl} \times \frac{1 \text{ mol } CaCO_3}{2 \text{ mol HCl}} \times \frac{100.09 \text{ g } CaCO_3}{1 \text{ mol } CaCO_3} = 0.96959 = 0.9696 \text{ g } CaCO_3$$

$$\text{mass \% } CaCO_3 = \frac{\text{g } CaCO_3}{\text{g rock}} \times 100 = \frac{0.96959}{1.248} \times 100 = 77.69\% \text{ } CaCO_3$$

Additional Exercises

4.91   (a) $U(s) + 2 \text{ ClF}_3(g) \rightarrow UF_6(g) + Cl_2(g)$

   (b)   This is not a metathesis reaction. The compounds involved are molecular rather than ionic, so the reaction is more complex than ions changing "partners."

   (c)   It is a redox reaction. U is oxidized, Cl is reduced.

4.92   (a)   The precipitate is CdS(s).

   (b) $Na^+(aq)$ and $NO_3^-(aq)$ are spectator ions and remain in solution. Any excess reactant ions also remain in solution.

   (c) $Cd^{2+}(aq) + S^{2-}(aq) \rightarrow CdS(s)$.

   (d)   This is not a redox reaction. (It is a metathesis reaction.)

4.93   The two precipitates formed are $AgCl(s)$ and $SrSO_4(s)$. Because no precipitate forms on addition of hydroxide ion to the remaining solution, the other two possibilities, Ni^{2+} and Mn^{2+}, are absent.

4.94   (a,b)   Expt. 1 No reaction

Expt. 2 $2 Ag^+(aq) + CrO_4^{2-}(aq) \rightarrow Ag_2CrO_4(s)$ red precipitate

Expt. 3 $Ca^{2+}(aq) + CrO_4^{2-}(aq) \rightarrow CaCrO_4(s)$ yellow precipitate

Expt. 4 $2 Ag^+(aq) + C_2O_4^{2-}(aq) \rightarrow Ag_2C_2O_4(s)$ white precipitate

Expt. 5 $Ca^{2+}(aq) + C_2O_4^{2-}(aq) \rightarrow CaC_2O_4(s)$ white precipitate

Expt. 6 $Ag^+(aq) + Cl^-(aq) \rightarrow AgCl(s)$ white precipitate

4.95   (a) $Al(OH)_3(s) + 3 H^+(aq) \rightarrow Al^{3+}(aq) + 3 H_2O(l)$

(b) $Mg(OH)_2(s) + 2 H^+(aq) \rightarrow Mg^{2+}(aq) + 2 H_2O(l)$

(c) $MgCO_3(s) + 2 H^+(aq) \rightarrow Mg^{2+}(aq) + H_2O(l) + CO_2(g)$

(d) $NaAl(CO_3)(OH)_2(s) + 4 H^+(aq) \rightarrow Na^+(aq) + Al^{3+}(aq) + 3 H_2O(l) + CO_2(g)$

(e) $CaCO_3(s) + 2 H^+(aq) \rightarrow Ca^{2+}(aq) + H_2O(l) + CO_2(g)$

[In (c), (d), and (e), one could also write the equation for formation of bicarbonate, e.g., $MgCO_3(s) + H^+(aq) \rightarrow Mg^{2+} + HCO_3^-(aq)$.]

4.96 $4 NH_3(g) + 5 O_2(g) \rightarrow 4 NO(g) + 6 H_2O(g)$.

$N = -3$ $O = 0$ $N = +2$ $O = -2$

(a)   redox reaction   (b)   N is oxidized, O is reduced

$2 NO(g) + O_2(g) \rightarrow 2 NO_2(g)$.

$N = +2$ $O = 0$ $N = +4, O = -2$

(a)   redox reaction   (b)   N is oxidized, O is reduced

$3 NO_2(g) + H_2O(l) \rightarrow 2 HNO_3(aq) + NO(g)$.

$N = +4$ $N = +5$ $N = +2$

(a)   redox reaction

(b)   N is oxidized ($NO_2 \rightarrow HNO_3$), N is reduced ($NO_2 \rightarrow NO$). A reaction where the same element is both oxidized and reduced is called disproportionation.

(c) $0.150 \, M \, HNO_3 \times 1000.0 \, L = 150.00 = 150$ mol HNO_3 required

$$150 \text{ mol } HNO_3 \times \frac{3 \text{ mol } NO_2}{2 \text{ mol } HNO_3} \times \frac{2 \text{ mol } NO}{2 \text{ mol } NO_2} \times \frac{4 \text{ mol } NH_3}{4 \text{ mol } NO} = 225 \text{ mol } NH_3$$

$$225 \text{ mol } HNO_3 \times \frac{17.0305 \text{ g } NH_3}{1 \text{ mol } NH_3} = 3831.86 = 3.83 \times 10^3 \text{ g } NH_3$$

4.97   A metal on Table 4.5 can be oxidized by ions of the elements below it. Or, a metal on the table is able to displace the (metal) cations below it from their compounds.

(a) $Zn(s) + 2\,HNO_3(aq) \rightarrow Zn(NO_3)_2(aq) + H_2(g)$

The substance that inflates the balloon is $H_2(g)$. Of Zn, Cu, and Hg, only Zn is above H on Table 4.5, so only Zn can displace H from HCl.

(b) $35.0\ g\ Zn \times \dfrac{1\ mol\ Zn}{65.39\ g\ Zn} = 0.53525 = 0.535\ mol\ Zn$

$\dfrac{3.00\ mol\ HNO_3}{L} \times 0.150\ L = 0.450\ mol\ HNO_3$

One mol Zn reacts with 2 mol HNO_3, so HNO_3 is the limiting reactant.

$0.450\ mol\ HNO_3 \times \dfrac{1\ mol\ H_2}{2\ mol\ HNO_3} \times \dfrac{2.016\ g\ H_2}{1\ mol\ H_2} = 0.4536 = 0.454\ g\ H_2$

(c)   Both Zn and Cu are above Ag on Table 4.5, so both Zn and Cu can displace Ag from $AgNO_3$. Note that $H_2(g)$ would also displace Ag, but $H^+(aq)$ will not.

$Zn(s) + 2\,AgNO_3(aq) \rightarrow 2\,Ag(s) + Zn(NO_3)_2(aq)$; $Zn^{2+}(aq)$ and $NO_3^-(aq)$ remain

$Cu(s) + 2\,AgNO_3(aq) \rightarrow 2\,Ag(s) + Cu(NO_3)_2(aq)$; $Cu^{2+}(aq)$ and $NO_3^-(aq)$ remain

(d)   0.535 mol Zn [from part (b)]; $42.0\ g\ Zn \times \dfrac{1\ mol\ Zn}{63.546\ g\ Cu} = 0.6609 = 0.661\ mol\ Cu$

$\dfrac{0.750\ mol\ AgNO_3}{L} \times 0.150\ L = 0.1125 = 0.113\ mol\ AgNO_3$

One mol metal reacts with 2 mol $AgNO_3$, so $AgNO_3$ is the limiting reactant for both Zn and Cu.

$0.1125\ mol\ AgNO_3 \times \dfrac{107.87\ g\ Ag}{1\ mol\ Ag} = 12.135 = 12.1\ g\ Ag$ in both reactions

4.98   (a)   Copper is the solvent and tin is the solute. In a solution, the solvent is present in greater amount and the solute is present in lesser amount. There can be more than one solute, but only one solvent.

(b)   The molarity of Sn in the bronze is mol Sn/L bronze. If the bronze contains 10.0% Sn, 100.0 g of bronze contains 10.0 g Sn. Use AW of Sn to calculate mol Sn. Use the density of the bronze, $7.9\ g/cm^3$, to calculate the volume of 100.0 g of this bronze.

$10.0\ g\ Sn \times \dfrac{1\ mol\ Sn}{118.71\ g\ Sn} = 0.08424 = 0.0842\ mol\ Sn$

$100.0\ g\ bronze \times \dfrac{1\ cm^3}{7.9\ g} = 12.658 = 13\ cm^3 = 13\ mL$

$M = \dfrac{mol\ Sn}{L\ bronze} = \dfrac{0.08424\ mol\ Sn}{12.658\ cm^3\ bronze} \times \dfrac{1000\ cm^3}{1\ L} = 6.655 = 6.7\ M$

(c) $Sn(s) + 2\,HCl(aq) \rightarrow SnCl_2(aq) + H_2(g)$

According to the Activity Series of Metals, Table 4.5, Sn is above H and Cu is below it. That is, Sn can be oxidized to $Sn^{2+}(aq)$ by HCl(aq), but Cu(s) cannot. We can treat bronze with HCl(aq) to remove Sn as $Sn^{2+}(aq)$ and leave a pure Cu sample.

[A closer look shows that this method has its problems. In fact, HCl causes some corrosion of the Cu, regardless of its position in the activity series, so we would not be left with a pure Cu sample. And, it would probably remove tin only on the surface of the bronze. A better method, which relies on principles presented in later chapters of the text, follows. Dissolve all metal with nitric acid, precipitate the tin, isolate the remaining solution and electroplate the copper from it.]

4.99   *Plan.* Calculate moles KBr from the two quantities of solution (mol = $M \times L$). Moles $AgNO_3$ required equals total moles KBr present. Change grams $AgNO_3$ to moles $AgNO_3$.   *Solve.*

1.00 *M* KBr $\times$ 0.0350 L = 0.0350 mol KBr;   0.600 *M* KBr $\times$ 0.060 L = 0.0360 mol KBr

0.0350 mol KBr + 0.0360 mol KBr = 0.0710 mol KBr total requires 0.0710 mol $AgNO_3$

$$0.0710 \text{ mol AgNO}_3 \times \frac{169.87 \text{ g AgNO}_3}{\text{mol AgNO}_3} = 12.0607 = 12.1 \text{ g AgNO}_3$$

4.100   (a)   Acid-base. Because of the $-NH_2$ group, dopamine is an organic base similar to NH_3. (The $-OH$ groups are not basic because they are not hydroxide ions. In fact, when they are bonded to carbon atoms in this kind of six-membered ring, they are slightly acidic.)

         (b)   Determine the molecular formula of dopamine and calculate molar mass. Then, moles dopamine = g/molar mass; M = mol/L.

$C_8H_{11}NO_2$, molar mass = 153.18; 400.0 mg dopamine = 0.4000 g dopamine;

250.0 mL solution = 0.2500 L solution

$$0.4000 \text{ g dopamine} \times \frac{1 \text{ mol}}{153.18 \text{ g}} \times \frac{1}{0.2500 \text{ L}} = 0.01045 \, M \text{ dopamine}$$

         (c)   Find the number of dopamine molecules in a 5.00 mm^3 brain that has an increased dopamine concentration of 0.75 μM. M = mol/L; 1 L = 1 dm^3

0.75 μM dopamine solution = $0.75 \times 10^{-6} \, M$ dopamine solution

$$5.00 \text{ mm}^3 \times \frac{1 \text{ L}}{1 \text{ dm}^3} \times \frac{1 \text{ dm}^3}{(100)^3 \text{ mm}^3} = 5.00 \times 10^{-6} \text{ L}$$

$$\frac{0.75 \times 10^{-6} \text{ mol dopamine}}{1 \text{ L soln}} \times 5.00 \times 10^{-6} \text{ L} \times \frac{6.022 \times 10^{23} \text{ molecules}}{\text{mol}} = 2.258 \times 10^{12} =$$

$$2.3 \times 10^{12} \text{ dopamine molecules}$$

4.101   (a) Na^+ must replace the total positive (+) charge due to Ca^{2+} and Mg^{2+}. Think of this as moles of charge rather than moles of particles.

$$\frac{0.020 \text{ mol } Ca^{2+}}{1 \text{ L water}} \times 1.5\times10^3 \text{ L} \times \frac{2 \text{ mol (+) charge}}{1 \text{ mol } Ca^{2+}} = 60 \text{ mol of (+) charge}$$

$$\frac{0.0040 \text{ mol } Mg^{2+}}{1 \text{ L water}} \times 1.5\times10^3 \text{ L} \times \frac{2 \text{ mol (+) charge}}{1 \text{ mol } Mg^{2+}} = 12 \text{ mol of (+) charge}$$

72 moles of (+) charge must be replaced; 72 mol Na^+ are needed.

   (b) $72 \text{ mol } Na^+ \times \dfrac{1 \text{ mol } Na^+}{1 \text{ mol NaCl}} \times \dfrac{58.44 \text{ g NaCl}}{1 \text{ mol NaCl}} = 4208 \text{ g} = 4.2\times10^3 \text{ g NaCl}$

4.102 $H_2C_4H_4O_6 + 2 OH^-(aq) \rightarrow C_4H_4O_6^{2-}(aq) + 2 H_2O(l)$

$$0.02465 \text{ L NaOH soln} \times \frac{0.2500 \text{ mol NaOH}}{1 \text{ L}} \times \frac{1 \text{ mol } H_2C_4H_4O_6}{2 \text{ mol NaOH}} \times \frac{1}{0.0500 \text{ L } H_2C_4H_4O_6}$$

$$= 0.06163 \, M \, H_2C_4H_4O_6 \text{ soln}$$

4.103   (a) $12.50 \text{ g Sr(OH)}_2 \times \dfrac{1 \text{ mol Sr(OH)}_2}{121.64 \text{ g Sr(OH)}_2} \times \dfrac{1}{0.05000 \text{ L}} = 2.0552 = 2.055 \, M \, Sr(OH)_2$

   (b) $2 HNO_3(aq) + Sr(OH)_2(aq) \rightarrow Sr(NO_3)_2(aq) + 2 H_2O(l)$

   (c)   *Plan.* $\text{mol Sr(OH)}_2 = M \times L \rightarrow$ mol ratio $\rightarrow$ mol $HNO_3 \rightarrow M \, HNO_3$.   *Solve.*

$$\frac{2.0552 \text{ mol Sr(OH)}_2}{L} \times 0.0239 \text{ L Sr(OH)}_2 \times \frac{2 \text{ mol } HNO_3}{1 \text{ mol Sr(OH)}_2} \times \frac{1}{0.0375 \text{ L } HNO_3}$$

$$= 2.6197 = 2.62 \, M \, HNO_3$$

4.104   mol OH^- from NaOH(aq) + mol OH^- from $Zn(OH)_2$(s) = mol H^+ from HBr

mol $H^+ = M$ HBr $\times$ L HBr = $0.500 \, M$ HBr $\times 0.350$ L HBr = 0.175 mol H^+

mol OH^- from NaOH = M NaOH $\times$ L NaOH = $0.500 \, M$ NaOH $\times 0.0885$ L NaOH

$$= 0.04425 = 0.0443 \text{ mol } OH^-$$

mol OH^- from $Zn(OH)_2$(s) = 0.175 mol H^+ − 0.04425 mol OH^- from NaOH = 0.13075

$$= 0.131 \text{ mol } OH^- \text{ from } Zn(OH)_2$$

$$0.13075 \text{ mol } OH^- \times \frac{1 \text{ mol } Zn(OH)_2}{2 \text{ mol } OH^-} \times \frac{99.41 \text{ g } Zn(OH)_2}{1 \text{ mol } Zn(OH)_2} = 6.50 \text{ g } Zn(OH)_2$$

Integrative Exercises

4.105   (a)   A metal can be oxidized by ions of the elements below it on Table 4.5. Of the three substances given, K^+(aq) is above Mg(s), but Ag^+(aq) is below it, so $AgNO_3$(aq) will react with Mg(s).

   (b) $Mg(s) + 2 Ag^+(aq) \rightarrow Mg^{2+}(aq) + 2 Ag(s)$

(c) $g\ Mg \rightarrow mol\ Mg \rightarrow$ [via mole ratio] $mol\ Ag^+ \rightarrow$ [via (mol/M)] $vol\ AgNO_3(aq)$

$$5.00\ g\ Mg \times \frac{1\ mol\ Mg}{24.305\ g\ Mg} = 0.2057 = 0.206\ mol\ Mg$$

$$0.2057\ mol\ Mg \times \frac{2\ mol\ AgNO_3}{1\ mol\ Mg} \times \frac{1.00\ L}{2.00\ mol\ AgNO_3} = 0.2057 = 0.206\ L\ AgNO_3(aq)$$

(d) $\dfrac{0.2057\ mol\ Mg^{2+}}{0.2057\ L\ soln} = 1.00\ M\ Mg^{2+}(aq)$

4.106   (a)   At the equivalence point of a titration, $mol\ NaOH\ added = mol\ H^+\ present$

$$M_{NaOH} \times L_{NaOH} = \frac{g\ acid}{MM\ acid} \text{ (for an acid with 1 acidic hydrogen)}$$

$$MM\ acid = \frac{g\ acid}{M_{NaOH} \times L_{NaOH}} = \frac{0.2053\ g}{0.1008\ M \times 0.0150\ L} = 136\ g/mol$$

(b)   Assume 100 g of acid.

$$70.6\ g\ C \times \frac{1\ mol\ C}{12.01\ g\ C} = 5.88\ mol\ C;\ 5.88/1.47 \approx 4$$

$$5.89\ g\ H \times \frac{1\ mol\ H}{1.008\ g\ H} = 5.84\ mol\ H;\ 5.84/1.47 \approx 4$$

$$23.5\ g\ O \times \frac{1\ mol\ O}{16.00\ g\ O} = 1.47\ mol\ O;\ 1.47/1.47 = 1$$

The empirical formula is C_4H_4O.

$$\frac{MM}{FW} = \frac{136}{68.1} = 2; \text{ the molecular formula is } 2 \times \text{ the empirical formula.}$$

The molecular formula is $C_8H_8O_2$.

4.107   (a)   Gold atoms from Au(s) are oxidized. Oxygen atoms from $O_2(g)$ are reduced. The oxidation number of Au changes from 0 to +1 {in $Na[Au(CN)_2]$}. The oxidation number of O changes from 0 to –2 (in NaOH).

(b) $2\ [Au(CN)_2]^-(aq) + Zn(s) \rightarrow 2\ Au(s) + Zn^{2+}(aq) + 4\ CN^-(aq)$

The oxidation of one Zn atom produces 2 electrons, so 2 Au atoms {in the form of $[Au(CN)_2]^-(aq)$} are reduced.

(c)   Use mass % Au in rocks to get mass Au. Calculate moles Au, apply mole ratio of NaCN(aq) to Au(s) from the balanced equation given in the exercise, calculate volume of 0.200 M NaCN(aq) required. 2.00% Au = 2.00 kg Au/100 kg rocks

$$40.0\ kg\ rocks \times \frac{2.00\ kg\ Au}{100\ kg\ rocks} \times \frac{1000\ g}{kg} = 8.00 \times 10^2\ g\ Au$$

$$8.00 \times 10^2\ g\ Au \times \frac{1\ mol\ Au}{196.97\ g\ Au} \times \frac{8\ mol\ NaCN}{4\ mol\ Au} \times \frac{1\ L\ soln}{0.200\ mol\ NaCN} = 40.615 = 40.6\ L$$

4.108   *Plan.* Abbreviate the commercial aqueous ammonia solution as NH_3 soln. Abbreviate citric acid as H_3Cit. Write the balanced equation.

$$\text{vol } NH_3 \text{ soln} \xrightarrow{\text{density}} \text{mass } NH_3 \text{ soln} \xrightarrow{\text{mass \%}} \text{mass } NH_3$$

$$\text{mass } NH_3 \rightarrow \text{mol } NH_3 \rightarrow \text{mol } H_3Cit \rightarrow \text{mass } H_3Cit$$

Solve. $H_3Cit(aq) + 3\ NH_3(aq) \rightarrow 3\ NH_4^+(aq) + Cit^{3-}(aq)$

$$3.43 \times 10^4 \text{ gal } NH_3 \text{ soln} \times \frac{3.785 \text{ L}}{\text{gal}} \times \frac{1000 \text{ mL}}{\text{L}} \times \frac{0.88 \text{ g soln}}{\text{mL soln}} = 1.14246 \times 10^8 = 1.1 \times 10^8 \text{ g } NH_3 \text{ soln}$$

$$1.14246 \times 10^8 \text{ g } NH_3 \text{ soln} \times \frac{30 \text{ g } NH_3}{100 \text{ g } NH_3 \text{ soln}} = 3.4274 \times 10^7 = 3.4 \times 10^7 \text{ g } NH_3$$

$$3.4274 \times 10^7 \text{ g } NH_3 \times \frac{1 \text{ mol } NH_3}{17.0305 \text{ g } NH_3} \times \frac{1 \text{ mol } H_3Cit}{3 \text{ mol } NH_3} \times \frac{192.124 \text{ g } H_3Cit}{1 \text{ mol } H_3Cit} = 1.3 \times 10^8 \text{ g } H_3Cit$$

4.109   (a) $Mg(OH)_2(s) + 2\ HNO_3(aq) \rightarrow Mg(NO_3)_2(aq) + 2\ H_2O(l)$

        (b) $7.75 \text{ g } Mg(OH)_2 \times \dfrac{1 \text{ mol } Mg(OH)_2}{58.32 \text{ g } Mg(OH)_2} = 0.13289 = 0.133 \text{ mol } Mg(OH)_2$

            $0.200\ M\ HNO_3 \times 0.0250 \text{ L} = 0.00500 \text{ mol } HNO_3$

            The 0.00500 mol HNO_3 would neutralize 0.00250 mol $Mg(OH)_2$ and much more $Mg(OH)_2$ is present, so HNO_3 is the limiting reactant.

        (c)   Because HNO_3 limits, 0 mol HNO_3 is present after reaction.

            0.00250 mol $Mg(NO_3)_2$ is produced.

            0.13289 mol $Mg(OH)_2$ initial − 0.00250 mol $Mg(OH)_2$ react

                                        = 0.130 mol $Mg(OH)_2$ remain

4.110   (a)   Calculate the volume of the first part of the river. Get mass of MCHM using density. Then, mass MCHM → mol MCHM → M MCHM

$$7.00 \text{ ft} \times \frac{1 \text{ yd}}{3 \text{ ft}} \times 100.0 \text{ yd} \times 100.0 \text{ yd} \times \frac{(1 \text{ m})^3}{(1.0936)^3 \text{ yd}^3} \times \frac{(10)^3 \text{ dm}^3}{1 \text{ m}^3} = 1.7840 \times 10^7 \text{ dm}^3 =$$

$$1.78 \times 10^7 \text{ L}$$

$$7500 \text{ gal MCHM} \times \frac{3.785 \text{ L}}{\text{gal}} \times \frac{1000 \text{ mL}}{\text{L}} \times \frac{0.9074 \text{ g MCHM}}{\text{mL}} = 2.576 \times 10^7 =$$

$$2.6 \times 10^7 \text{ g MCHM}$$

$$2.576 \times 10^7 \text{ g MCHM} \times \frac{1 \text{ mol MCHM}}{128.21 \text{ g MCHM}} \times \frac{1}{1.784 \times 10^7 \text{ L}} = 0.01126 = 0.011\ M \text{ MCHM}$$

        (b)   A concentration of $1.0 \times 10^{-4}\ M$ is a decrease in concentration of approximately 100 times. It requires a total volume that is 100 times the volume of the first part of the river. If depth and width are constant, the length of the spill must increase by a factor of approximately 100. That is, the spill would need to cover 100(100 yd) = 10,000 yd of the river. For perspective, this is about 5.7 miles, a substantial distance.

We can calculate a more precise distance. Use the dilution formula to calculate the volume of river required to produce a 1.0×10^{-4} M solution, assuming uniform mixing and MCMH concentration throughout the length of the spill.

$0.01126\ M\ \text{MCHM} \times 1.784 \times 10^{7}\ \text{L} = 1.0 \times 10^{-4}\ M \times ?\text{L}$

volume of river $= 2.009 \times 10^{9}\ \text{L} = 2.0 \times 10^{9}\ \text{L}$

$$2.009 \times 10^{9}\,\text{dm}^3 \times \frac{1\,\text{m}^3}{(10)^3\,\text{dm}^3} \times \frac{(1.0936)^3\,\text{yd}^3}{1\,\text{m}^3} \times \frac{1}{100\,\text{yd}} \times \frac{1}{7\,\text{ft}} \times \frac{3\,\text{ft}}{1\,\text{yd}} =$$

$$1.126 \times 10^{4} = 1.1 \times 10^{4}\ \text{yd}$$

The total length of the spill would be 1.1×10^{4} yd. The initial length of 100 yards (or 0.01×10^{4} yd) is barely significant. The spill would need to spread 1.1×10^{4} yards or 6.4 miles farther to achieve a "safe" concentration.

4.111   (a)   By virtue of its –NH group, Ritalin is a weak base like NH_3. Because it is a weak base, it is a weak electrolyte.

         (b)   The molecular formula of Ritalin is $C_{14}H_{19}NO_2$. Molar mass = 233.31 g/mol

$$10.0\ \text{mg ritalin} \times \frac{1 \times 10^{-3}\ \text{g}}{\text{mg}} \times \frac{1\ \text{mol ritalin}}{233.31\ \text{g ritalin}} \times \frac{1\ \text{L}}{5.0\ \text{L}} = 8.572 \times 10^{-6} = 8.6 \times 10^{-6}\ M$$

         (c)   Six hours is two half-lives for Ritalin. One half-life reduces molarity by ½ and two half-lives reduce molarity by ½(½) or ¼. The concentration of Ritalin after six hours will be $(8.572 \times 10^{-6}\ M)/4 = 2.143 \times 10^{-6}\ M = 2.1 \times 10^{-6}\ M$.

4.112 $Ag^{+}(aq) + Cl^{-}(aq) \rightarrow AgCl(s)$

$$\frac{0.2997\ \text{mol Ag}^+}{1\ \text{L}} \times 0.04258\ \text{L} \times \frac{1\ \text{mol Cl}^-}{1\ \text{mol Ag}^+} \times \frac{35.453\ \text{g Cl}^-}{1\ \text{mol Cl}^-} = 0.45242 = 0.4524\ \text{g Cl}^-$$

$$25.00\ \text{mL seawater} \times \frac{1.025\ \text{g}}{\text{mL}} = 25.625 = 25.63\ \text{g seawater}$$

$$\text{mass \% Cl}^- = \frac{0.45242\ \text{g Cl}^-}{25.625\ \text{g seawater}} \times 100 = 1.766\%\ \text{Cl}^-$$

4.113   (a)   **As** O_4^{3-}; +5

         (b) $Ag_3\textbf{PO}_4$ is silver phosphate; $Ag_3\textbf{AsO}_4$ is silver arsenate

         (c) $$0.0250\ \text{L soln} \times \frac{0.102\ \text{mol Ag}^+}{1\ \text{L soln}} \times \frac{1\ \text{mol Ag}_3\text{AsO}_4}{3\ \text{mol Ag}^+} \times \frac{1\ \text{mol As}}{1\ \text{mol Ag}_3\text{AsO}_4} \times \frac{74.92\ \text{g As}}{1\ \text{mol As}}$$

$$= 0.06368 = 0.0637\ \text{g As}$$

$$\text{mass percent} = \frac{0.06368\ \text{g As}}{1.22\ \text{g sample}} \times 100 = 5.22\%\ \text{As}$$

4.114   *Analyze.* Given 10 ppb AsO_4^{3-}, find mass Na_3AsO_4 in 1.00 L of drinking water.

Plan. Use the definition of ppb to calculate g AsO_4^{3-} in 1.0 L of water. Convert g $AsO_4^{3-} \rightarrow$ g Na_3AsO_4 using molar masses. Assume the density of H_2O is 1.00 g/mL.

Solve. 1 billion $= 1 \times 10^9$; 1 ppb $= \dfrac{1 \text{ g solute}}{1 \times 10^9 \text{ g solution}}$

$$\frac{1 \text{ g solute}}{1 \times 10^9 \text{ g solution}} \times \frac{1 \text{ g solution}}{1 \text{ mL solution}} \times \frac{1 \times 10^3 \text{ mL}}{1 \text{ L solution}} = \frac{\text{g } AsO_4^{3-}}{1 \times 10^6 \text{ L } H_2O}$$

$$10 \text{ ppb } AsO_4^{3-} = \frac{10 \text{ g } AsO_4}{1 \times 10^6 \text{ L } H_2O} \times 1 \text{ L } H_2O = 1.0 \times 10^{-5} \text{ g As/L.}$$

$$1.0 \times 10^{-5} \text{ g } AsO_4^{3-} \times \frac{1 \text{ mol } AsO_4^{3-}}{138.92 \text{ g } AsO_4^{3-}} \times \frac{1 \text{ mol } Na_3AsO_4}{1 \text{ mol As}} \times \frac{207.89 \text{ g } Na_3AsO_4^{3-}}{1 \text{ mol } Na_3AsO_4}$$

$$= 1.5 \times 10^{-5} \text{ g } Na_3AsO_4 \text{ in } 1.00 \text{ L } H_2O$$

4.115   (a)   mol HCl initial – mol NH_3 from air = mol HCl remaining

$$= \text{mol NaOH required for titration}$$

$$\text{mol NaOH} = 0.0588 \, M \times 0.0131 \text{ L} = 7.703 \times 10^{-4} = 7.70 \times 10^{-4} \text{ mol NaOH}$$

$$= 7.70 \times 10^{-4} \text{ mol HCl remain}$$

mol HCl initial – mol HCl remaining = mol NH_3 from air

$$(0.0105 \, M \text{ HCl} \times 0.100 \text{ L}) - 7.703 \times 10^{-4} \text{ mol HCl} = \text{mol } NH_3$$

$$10.5 \times 10^{-4} \text{ mol HCl} - 7.703 \times 10^{-4} \text{ mol HCl} = 2.80 \times 10^{-4} = 2.8 \times 10^{-4} \text{ mol } NH_3$$

$$2.8 \times 10^{-4} \text{ mol } NH_3 \times \frac{17.03 \text{ g } NH_3}{1 \text{ mol } NH_3} = 4.77 \times 10^{-3} = 4.8 \times 10^{-3} \text{ g } NH_3$$

(b)   ppm is defined as molecules of $NH_3 / 1 \times 10^6$ molecules in air.

Calculate molecules NH_3 from mol NH_3.

$$2.80 \times 10^{-4} \text{ mol } NH_3 \times \frac{6.022 \times 10^{23} \text{ molecules}}{1 \text{ mol}} = 1.686 \times 10^{20}$$

$$= 1.7 \times 10^{20} \text{ } NH_3 \text{ molecules}$$

Calculate total volume of air processed, then g air using density, then molecules air using molar mass.

$$\frac{10.0 \text{ L}}{1 \text{ min}} \times 10.0 \text{ min} \times \frac{1.20 \text{ g air}}{1 \text{ L air}} \times \frac{1 \text{ mol air}}{29.0 \text{ g air}} \times \frac{6.022 \times 10^{23} \text{ molecules}}{1 \text{ mol}}$$

$$= 2.492 \times 10^{24} = 2.5 \times 10^{24} \text{ air molecules}$$

$$\text{ppm } NH_3 = \frac{1.686 \times 10^{20} \text{ } NH_3 \text{ molecules}}{2.492 \times 10^{24} \text{ air molecules}} \times 1 \times 10^6 = 68 \text{ ppm } NH_3$$

(c)   68 ppm > 50 ppm. The manufacturer is **not** in compliance.

5 Thermochemistry

Visualizing Concepts

5.1 (a) *Analyze/Plan.* The exercise gives the charges and separation of two particles. Use Equation 5.2 to calculate electrostatic potential energy, E_{el}. *Solve.*

$$E_{el} = \frac{\kappa Q_1 Q_2}{d}; \; \kappa = 8.99 \times 10^9 \text{ J-m}/C^2; \; Q_1 = Q_2 = 2.0 \times 10^{-5} \text{ C}; d = 1.0 \text{ cm}$$

$$E_{el} = \frac{8.99 \times 10^9 \text{ J-m}}{C^2} \times \frac{1}{1.0 \text{ cm}} \times \frac{100 \text{ cm}}{m} \times 2.0 \times 10^{-5} \text{ C} \times 2.0 \times 10^{-5} \text{ C} = 359.6 = 3.6 \times 10^2 \text{ J}$$

(b) The spheres are both positively charged, so they will move away from each other. (Like charged particles repel, oppositely charged particles attract.)

(c) As the like charged spheres move apart, the electrostatic potential energy of the system is converted to kinetic energy. As the distance between them approaches infinity, potential energy approaches zero and the kinetic energy of each particle is 1.8×10^2 J, one half of the initial potential energy calculated in part (a).

$$E_k = 1/2 \, mv^2; \; v = (2 \, E_k \, /m)^{1/2}; \; E_k = \frac{1}{2}(3.6 \times 10^2 \text{ J}) = 1.8 \times 10^2 \text{ J}$$

$$v = \left(2 \times \frac{1.8 \times 10^2 \text{ kg-m}^2}{s^2} \times \frac{1}{1.0 \text{ kg}} \right)^{1/2} = 18.97 = 19 \text{ m/s}$$

5.2 (a) The caterpillar uses energy produced by its metabolism of food to climb the twig and increase its potential energy.

(b) Heat, q, is the energy transferred from a hotter to a cooler object. Without knowing the temperature of the caterpillar and its surroundings, we cannot predict the sign of q. It is likely that q is approximately zero, because a small creature like a caterpillar is unlikely to support a body temperature much different from its environmental temperature.

(c) Work, w, is the energy transferred when a force moves an object. When the caterpillar climbs the twig, it does work as its body moves against the force of gravity.

(d) No. The amount of work is independent of time and therefore independent of speed (assuming constant caterpillar speed).

(e) No. Potential energy depends only on the caterpillar's position, so the change in potential energy depends only on the distance climbed, not on the speed of the climb.

5.3   (a)   The internal energy, E, of the products is greater than that of the reactants, so the diagram represents an increase in the internal energy of the system.

        (b) ΔE for this process is positive, (+).

        (c)   If no work is associated with the process, it is endothermic.

5.4   (a)   For an endothermic process, the sign of q is positive; the system gains heat. This is true only for system (iii).

        (b)   In order for ΔE to be less than 0, there is a net transfer of heat or work from the system to the surroundings. The magnitude of the quantity leaving the system is greater than the magnitude of the quantity entering the system. In system (i), the magnitude of the heat leaving the system is less than the magnitude of the work done on the system. In system (iii), the magnitude of the work done by the system is less than the magnitude of the heat entering the system. None of the systems has $\Delta E < 0$.

        (c)   In order for ΔE to be greater than 0, there is a net transfer of work or heat to the system from the surroundings. In system (i), the magnitude of the work done on the system is greater than the magnitude of the heat leaving the system. In system (ii), work is done on the system with no change in heat. In system (iii), the magnitude of the heat gained by the system is greater than the magnitude of the work done on the surroundings. $\Delta E > 0$ for all three systems.

5.5   (a)   No. This distance traveled to the top of a mountain depends on the path taken by the hiker. Distance is a path function, not a state function.

        (b)   Yes. Change in elevation depends only on the location of the base camp and the height of the mountain, not on the path to the top. Change in elevation is a state function, not a path function.

5.6   (a)   State B

        (b) ΔE_{AB} = energy difference between State A and State B.

                  $\Delta E_{AB} = \Delta E_1 + \Delta E_2$ or $\Delta E_{AB} = \Delta E_3 + \Delta E_4$

        (c) ΔE_{CD} = energy difference between State C and State D.

                  $\Delta E_{CD} = \Delta E_2 - \Delta E_4$ or $\Delta E_{CD} = \Delta E_3 - \Delta E_1$

                  (Note that the sign of ΔE depends on the definition of initial and final state, but the magnitude is the absolute value of the difference in energy.)

        (d)   The energy of State E is $\Delta E_1 + \Delta E_4$, whereas the energy of State B is $\Delta E_1 + \Delta E_2$. Because $\Delta E_4 > \Delta E_2$, State E is above State B on the diagram; State E would be the highest energy on the diagram.

5.7   (a)   You, part of the surroundings, do work on the air, part of the system. Energy is transferred to the system via work and the sign of w is (+).

        (b)   The body of the pump (the system) is warmer than the surroundings. Heat is transferred from the warmer system to the cooler surroundings, and the sign of q is (−).

        (c)   The sign of w is positive, and the sign of q is negative, so we cannot absolutely determine the sign of ΔE. It is likely that the heat lost is much smaller than the work done on the system, so the sign of ΔE is probably positive.

5.8   (a)   The temperature of the system and surroundings will equalize, so the temperature of the hotter system will decrease, and the temperature of the colder surroundings will increase. The system loses heat by decreasing its temperature, so the sign of q_{sys} is (–). The surrounding gains heat by increasing its temperature, so the sign of q_{surr} is (+). From the system's perspective, the process is exothermic because it loses heat.

        (b)   If neither volume nor pressure of the system changes, $w = 0$ and $\Delta E = q = \Delta H$. The change in internal energy is equal to the change in enthalpy.

5.9   (a) $w = -P\Delta V$. Because ΔV for the process is (–), the sign of w is (+).

        (b) $\Delta E = q + w$. At constant pressure, $\Delta H = q$. If the reaction is endothermic, the signs of ΔH and q are (+). From (a), the sign of w is (+), so the sign of ΔE is (+). The internal energy of the system increases during the change. (This situation is described by the diagram (ii) in Exercise 5.4.)

5.10   (a) $N_2(g) + O_2(g) \rightarrow 2\,NO(g)$. Because $\Delta V = 0$, $w = 0$.

        (b)   The reaction of two elements to form one mole of a compound fits the definition of a formation reaction. Find the value for enthalpy of formation of $NO(g)$ in Appendix C. $\Delta H = \Delta H_f = 90.37$ kJ for production of 1 mol of $NO(g)$.

5.11   (a) $\Delta H_A = \Delta H_B + \Delta H_C$. Diagram (i) indicates that reaction A can be written as the sum of reactions B and C.

        (b) $\Delta H_Z = \Delta H_X + \Delta H_Y$. Diagram (ii) indicates that reaction Z can be written as the sum of reactions X and Y.

        (c)   Hess's law states that the enthalpy change for a net reaction is the sum of the enthalpy changes of the component steps, regardless of whether the reaction actually occurs via this path.

        (d)   No. The enthalpy relationships are true because enthalpy is a state function, independent of path. Work is not a state function.

5.12   Because mass must be conserved in the reaction A → B, the component elements of A and B must be the same. Further, if $\Delta H_f^\circ > 0$ for both A and B, the energies of both A and B are above the energies of their component elements on the energy diagram.

        (a)   The bold arrow shows the reaction as written; combination of the two thin arrows shows an alternate route from A to B.

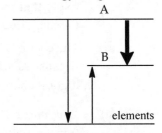

        (b) $\Delta H_{rxn}^\circ = \Delta H_f^\circ \, B - \Delta H_f^\circ \, A$. If the overall reaction is exothermic, the sign of ΔH is (–) and $\Delta H_f^\circ \, A > \Delta H_f^\circ \, B$. This means that the enthalpy of A is the highest energy level on the diagram. This is the situation pictured in the diagram above, but nothing in the given information requires this arrangement. If the reaction is endothermic, $\Delta H_f^\circ \, B > \Delta H_f^\circ \, A$ and the enthalpy of B would be the highest energy level on the diagram.

The Nature of Chemical Energy (Section 5.1)

5.13   *Analyze/Plan.* Use Equation 5.2 to calculate electrostatic potential energy, E_{el}. The distances between the particles are given in the exercise. The charge of an electron and a proton are given in Section 5.1. *Solve.*

(a) $E_{el} = \dfrac{\kappa Q_1 Q_2}{d}$; $d = 53$ pm; $Q_e = -1.60 \times 10^{-19}$ C; $Q_p = 1.60 \times 10^{-19}$ C

$$E_{el} = \frac{8.99 \times 10^9 \text{ J-m}}{C^2 \times 53 \text{ pm}} \times \frac{1 \text{ pm}}{1 \times 10^{-12} \text{ m}} \times -1.60 \times 10^{-19} \text{ C} \times 1.60 \times 10^{-19} \text{ C}$$

$$= -4.342 \times 10^{-18} = -4.3 \times 10^{-18} \text{ J}$$

(b) $E_{el} = \dfrac{8.99 \times 10^9 \text{ J-m}}{C^2 \times 1.0 \text{ nm}} \times \dfrac{1 \text{ nm}}{1 \times 10^{-9} \text{ m}} \times -1.60 \times 10^{-19} \text{ C} \times 1.60 \times 10^{-19} \text{ C}$

$$= -2.301 \times 10^{-19} = -2.3 \times 10^{-19} \text{ J}$$

The change in potential energy is $[-2.3 \times 10^{-19} \text{ J} - (-4.3 \times 10^{-18} \text{ J})] = 4.1 \times 10^{-18} \text{ J}$

(c)   The electrostatic potential energy of the system increases (becomes less negative) as the separation between the oppositely charged particles increases.

5.14   (a) $E_{el} = \dfrac{\kappa Q_1 Q_2}{d}$; $d = 62$ pm; $Q_p = 1.60 \times 10^{-19}$ C

$$E_{el} = \frac{8.99 \times 10^9 \text{ J-m}}{C^2 \times 62 \text{ pm}} \times \frac{1 \text{pm}}{1 \times 10^{-12} \text{ m}} \times (1.60 \times 10^{-19} \text{ C})^2 = 3.712 \times 10^{-18} = 3.7 \times 10^{-18} \text{ J}$$

(b) $E_{el} = \dfrac{8.99 \times 10^9 \text{ J-m}}{C^2 \times 1.0 \text{ nm}} \times \dfrac{1 \text{ nm}}{1 \times 10^{-9} \text{ m}} \times (1.60 \times 10^{-19} \text{ C})^2 = 2.301 \times 10^{-19} = 2.3 \times 10^{-19} \text{ J}$

The change in potential energy is $[2.3 \times 10^{-19} \text{ J} - (3.7 \times 10^{-18} \text{ J})] = -3.5 \times 10^{-18} \text{ J}$

(c)   The electrostatic potential energy of the system decreases as the separation between the like charged particles increases.

5.15   (a)   *Analyze/Plan.* Use the equation for electrostatic attractive force and the distance between particles given in the exercise. Find the charges of a proton and an electron in Section 5.1. *Solve.*

$$F_{el} = \frac{\kappa Q_1 Q_2}{d^2}; \quad d = 1.0 \times 10^2 \text{ pm};$$

$$Q_e = -1.60 \times 10^{-19} \text{ C}; \quad Q_p = 1.60 \times 10^{-19} \text{ C}; \quad 1 \text{ J} = 1 \text{ N-m}$$

$$F_{el} = \frac{8.99 \times 10^9 \text{ J-m}}{C^2 \times (1.0 \times 10^2 \text{ pm})^2} \times \frac{(1 \text{ pm})^2}{(1 \times 10^{-12})^2 \text{ m}^2} \times -1.60 \times 10^{-19} \text{ C} \times 1.60 \times 10^{-19} \text{ C}$$

$$= -2.301 \times 10^{-8} \text{ J/m} = -2.3 \times 10^{-8} \text{ N}$$

(b)   *Analyze/Plan.* Use the formula for gravitational force between two particles and the distance given in the exercise. Find the masses of a proton and an electron on the inside cover of the text. *Solve.*

$$F_g = \frac{G m_1 m_2}{d^2}; \quad G = \frac{6.674 \times 10^{-11} \text{ N-m}^2}{\text{kg}^2};$$

$d = 1.0 \times 10^2$ pm; $m_e = 9.109 \times 10^{-31}$ kg; $m_p = 1.673 \times 10^{-27}$ kg

$$F_g = \frac{6.674 \times 10^{-11} \text{ N-m}^2}{\text{kg}^2 \times (1.0 \times 10^2 \text{ pm})^2} \times \frac{(1 \text{ pm})^2}{(1 \times 10^{-12})^2 \text{ m}^2} \times 9.109 \times 10^{-31} \text{ kg} \times 1.673 \times 10^{-27} \text{ kg}$$

$$= 1.0171 \times 10^{-47} = 1.0 \times 10^{-47} \text{ N}$$

(c) The magnitude of the electrostatic force of attraction is 2.3×10^{-8} N. (The negative sign of the electrostatic force indicates attraction.) The gravitational force is 1.0×10^{-47} N. The electrostatic attraction is 2.3×10^{39} time larger. (This is almost 40 orders of magnitude.)

5.16 (a) $F_{el} = \frac{\kappa Q_1 Q_2}{d^2}$; $d = 75$ pm; $Q_p = 1.60 \times 10^{-19}$ C ; 1 J $= 1$ N-m

$$F_{el} = \frac{8.99 \times 10^9 \text{ J-m}}{C^2 \times (75 \text{ pm})^2} \times \frac{(1 \text{ pm})^2}{(1 \times 10^{-12})^2 \text{ m}^2} \times (1.60 \times 10^{-19} \text{ C})^2 =$$

$$4.091 \times 10^{-8} \text{ J/m} = 4.1 \times 10^{-8} \text{ N}$$

(b) $F_g = \frac{G m_1 m_2}{d^2}$; $G = \frac{6.674 \times 10^{-11} \text{ N-m}^2}{\text{kg}^2}$; $d = 75$ pm; $m_p = 1.673 \times 10^{-27}$ kg

$$F_g = \frac{6.674 \times 10^{-11} \text{ N-m}^2}{\text{kg}^2 \times (75 \text{ pm})^2} \times \frac{(1 \text{ pm})^2}{(1 \times 10^{-12})^2 \text{ m}^2} \times (1.673 \times 10^{-27} \text{ kg})^2 =$$

$$3.321 \times 10^{-44} = 3.3 \times 10^{-44} \text{ N}$$

(c) The magnitude of the electrostatic force of repulsion is 4.1×10^{-8} N. (The positive sign of the electrostatic force indicates repulsion.) The gravitational force is 3.3×10^{-44} N. If the two protons are allowed to move, the electrostatic repulsion is much greater than gravitational attraction and the protons move apart.

5.17 *Analyze/Plan.* We must find the work required to completely separate two oppositely charged particles. Work is the energy required to move an object against a force (Section 1.4). At infinite separation, the electrostatic potential energy of the pair of ions is zero. The magnitude of the work required is equal to the electrostatic potential energy, E_{el}, of the pair of ions. Use Equation 5.2 to calculate E_{el} and work. The charges of the ions and the distance between them are given in the exercise. *Solve.*

$$E_{el} = \frac{\kappa Q_1 Q_2}{d}; \quad d = 0.50 \text{ nm}; Q_{Cl} = -1.6 \times 10^{-19} \text{ C}; Q_{Na} = 1.6 \times 10^{-19} \text{ C}$$

$$E_{el} = \frac{8.99 \times 10^9 \text{ J-m}}{C^2 \times 0.50 \text{ nm}} \times \frac{1 \text{ nm}}{1 \times 10^{-9} \text{ m}} \times -1.6 \times 10^{-19} \text{ C} \times 1.6 \times 10^{-19} \text{ C} =$$

$$-4.603 \times 10^{-19} = -4.6 \times 10^{-19} \text{ J}$$

The sign of E_{el} is negative, so the work required to separate the ions is 4.6×10^{-19} J.

5.18 $E_{el} = \frac{\kappa Q_1 Q_2}{d}$; $d = 0.35$ nm; $Q_O = -3.2 \times 10^{-19}$ C; $Q_{Mg} = 3.2 \times 10^{-19}$ C

$$E_{el} = \frac{8.99 \times 10^9 \text{ J-m}}{C^2 \times 0.35 \text{ nm}} \times \frac{1 \text{ nm}}{1 \times 10^{-9} \text{ m}} \times -3.2 \times 10^{-19} \text{ C} \times 3.2 \times 10^{-19} \text{ C} =$$

$$-2.630 \times 10^{-18} = -2.6 \times 10^{-18} \text{ J}$$

The sign of E_{el} is negative, so the work required to separate the ions is 2.6×10^{-18} J.

5. 19   (a)   Gravity; work is done because the force of gravity is opposed and the pencil is lifted a distance above the desk.

          (b)   Spring force; work is done because the force of the coiled spring is opposed as the spring is compressed over a distance.

5.20   (a)   Electrostatic attraction; no work is done because the particles are held apart at a constant distance.

          (b)   Magnetic attraction; work is done because the nail is moved a distance in opposition to the force of magnetic attraction.

The First Law of Thermodynamics (Section 5.2)

5.21   (a)   Matter cannot leave a closed system. Energy in the form of heat or work can be transferred between a closed system and the surroundings.

          (b)   Neither matter nor energy can leave or enter an isolated system.

          (c)   Any part of the universe not part of the system is called the surroundings.

5.22   (a)   The liquid is an *open* system because it exchanges both matter and energy with the surroundings. Matter exchange occurs when solution flows into and out of the apparatus. The apparatus is not insulated, so energy exchange also occurs. Closed systems exchange energy but not matter, whereas isolated systems exchange neither.

          (b)   If the inlet and outlet are closed, the system can exchange energy but not matter with the surroundings; it becomes a closed system.

5.23   (a)   According to the first law of thermodynamics, energy is conserved.

          (b)   The total *internal energy* (E) of a system is the sum of all the kinetic and potential energies of the system components.

          (c)   The internal energy of a closed system (where no matter exchange with surroundings occurs) increases when work is done on the system by the surroundings and/or when heat is transferred to the system from the surroundings (the system is heated).

5.24   (a) $\Delta E = q + w$

          (b)   The quantities q and w are negative when the system loses heat to the surroundings (it cools) or does work on the surroundings.

5.25   *Analyze.* Given: heat and work. Find: magnitude and sign of ΔE.

       Plan. In each case, evaluate q and w in the expression $\Delta E = q + w$. For an exothermic process, q is negative; for an endothermic process, q is positive.   *Solve.*

          (a) $q = 0.763$ kJ, $w = -840$ J $= -0.840$ kJ. $\Delta E = 0.763$ kJ $- 0.840$ kJ $= -0.077$ kJ. The process is endothermic.

          (b)   q is negative because the system releases heat, and w is positive because work is done on the system. $\Delta E = -66.1$ kJ $+ 44.0$ kJ $= -22.1$ kJ. The process is exothermic.

5.26   In each case, evaluate q and w in the expression ΔE = q + w. For an exothermic process, q is negative; for an endothermic process, q is positive.

   (a)   q is negative and w is positive. ΔE = –0.655 kJ + 0.382 kJ = –0.273 kJ. The process is exothermic.

   (b)   q is positive and w is essentially zero. ΔE = 322 J. The process is endothermic.

5.27   *Analyze.* How do the different physical situations (cases) affect the changes to heat and work of the system upon addition of 100 J of energy?

   Plan. Use the definitions of heat and work and the First Law to answer the questions.

   Solve. If the piston is allowed to move, case (1), the heated gas will expand and push the piston up, doing work on the surroundings. If the piston is fixed, case (2), most of the electrical energy will be manifested as an increase in heat of the system.

   (a)   Because little or no work is done by the system in case (2), the gas will absorb most of the energy as heat; the case (2) gas will have the higher temperature.

   (b)   In Case 1, w is negative because work is done on the surroundings by expansion. Because the transfer of electrical energy is never completely efficient and some energy will be transferred as heat, q is positive. In Case 2, w is zero because no work (expansion) is done. The value of q is positive because all energy is transferred as heat.

   (c)   ΔE is greater for case (2) because the entire 100 J increases the internal energy of the system, rather than a part of the energy doing work on the surroundings.

5.28 $E_{el} = \dfrac{\kappa Q_1 Q_2}{r}$ For two oppositely charged particles, the sign of E_{el} is negative; the closer the particles, the greater the magnitude of E_{el}.

   (a)   The potential energy of oppositely charged spheres increases (becomes less negative) as the particles are separated (r increases).

   (b)   ΔE for the process is positive; the internal energy of the system increases as the oppositely charged particles are separated.

   (c)   Work is done on the system to separate the particles so w is positive. Mechanical separation of macroscopic charged spheres involves no heat transfer. Work alone accounts for the change in energy of the system.

5.29   (a)   A *state function* is a property of a system that depends only on the physical state (pressure, temperature, etc.) of the system, not on the route used by the system to get to the current state.

   (b)   Internal energy and enthalpy **are** state functions; heat **is not** a state function.

   (c)   Volume **is** a state function. The volume of a system depends only on conditions (pressure, temperature, amount of substance), not the route or method used to establish that volume.

5.30   (a)   Independent. Potential energy is a state function.

   (b)   Dependent. Some of the energy released could be employed in performing work, as is done in the body when sugar is metabolized; heat is not a state function.

(c)   Dependent. The work accomplished depends on whether the gasoline is used in an engine, burned in an open flame, or in some other manner. Work is not a state function.

Enthalpy (Sections 5.3 and 5.4)

5.31   *Analyze.* Given, P = 1.0 atm, ΔV = +0.50 L. Find work involved, in J.

Plan. This change is P – V work done at constant P. $w = -P\Delta V$. 1 L-atm = 101.3 J

Solve. $w = -1.0$ atm(0.50 L) = –0.50 L-atm; 0.50 L-atm × 101.3 J/L-atm = –50.65 = –51 J

The negative sign indicates that work is done by the system on the surroundings.

5.32   P = 0.857 atm. ΔV = 1.26 L – 5.00 L = –3.74 L

$w = -0.857$ atm(– 3.74 L) = 3.2052 = 3.21 L-atm;

3.2052 L-atm × 101.3 J/L-atm = 324.69 = 325 J

5.33   (a)   Change in enthalpy (ΔH) is usually easier to measure than change in internal energy (ΔE) because, at constant pressure, $\Delta H = q_p$. The heat flow associated with a process at constant pressure can easily be measured as a change in temperature. Measuring ΔE requires a means to measure both q and w.

(b)   H describes the enthalpy of a system at a certain set of conditions; the value of H depends only on these conditions. q describes energy transferred as heat, an energy *change*, which, in the general case, does depend on how the change occurs. We can equate change in enthalpy, ΔH, with heat, q_p, only for the specific conditions of constant pressure and exclusively P-V work.

(c)   If ΔH is positive, the enthalpy of the system increases, and the process is endothermic.

5.34   (a)   When a process occurs under constant external pressure and only P–V work occurs, the enthalpy change (ΔH) equals the amount of heat transferred. $\Delta H = q_p$.

(b) $\Delta H = q_p$. If the system releases heat, q and ΔH are negative, and the enthalpy of the system decreases.

(c)   If $\Delta H = 0$, $q_p = 0$ and $\Delta E = w$.

5.35   (a)   At constant pressure, $\Delta E = \Delta H - P\Delta V$. To calculate ΔE, more information about the conditions of the reaction must be known. For an ideal gas at constant pressure and temperature, $P\Delta V = RT\Delta n$. We know the value of $\Delta n = -3$ from the chemical reaction. We must know either the temperature, T, or the values of P and ΔV to calculate ΔE from ΔH.

(b) ΔE is larger than ΔH.

(c)   Because the value of Δn is negative, the quantity ($-P\Delta V$) is positive. We add a positive quantity to ΔH to calculate ΔE, so ΔE must be larger.

5.36   (a)   At constant volume ($\Delta V = 0$), $\Delta E = q_v$.

(b) ΔE will be larger than ΔH.

(c)   According to the definition of enthalpy, H = E + PV, so $\Delta H = \Delta E + \Delta(PV)$. For an ideal gas at constant temperature and volume, $\Delta PV = V\Delta P = RT\Delta n$. For this reaction, there are 2 mol of gaseous product and 3 mol of gaseous reactants, so $\Delta n = -1$. Thus $V\Delta P$ or $\Delta(PV)$ is negative. Because $\Delta H = \Delta E + \Delta(PV)$, the negative $\Delta(PV)$ term means that ΔE is larger or less negative than ΔH.

5.37   *Analyze/Plan.* q = 824 J = 0.824 kJ (heat is absorbed by the system), w = 0.65 kJ (work is done on the system).   *Solve.*

$\Delta E = q + w = 0.824$ kJ + 0.65 kJ = 1.47 kJ. $\Delta H = q = 0.824$ kJ (at constant pressure).

Check. The reaction is endothermic.

5.38   The gas is the system. If 0.49 kJ of heat is added, q = +0.49 kJ. Work done by the system decreases the overall energy of the system, so w = −214 J = −0.214 kJ .

$\Delta E = q + w = 0.49$ kJ − 0.214 kJ = 0.276 kJ. $\Delta H = q = 0.49$ kJ (at constant pressure).

5.39   (a) $C_2H_5OH(l) + 3\,O_2(g) \rightarrow 3\,H_2O(g) + 2\,CO_2(g)$ $\Delta H = -1235$ kJ

(b)   *Analyze.* How are reactants and products arranged on an enthalpy diagram?

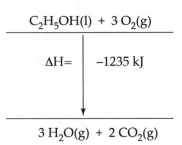

Plan. The substances (reactants or products, collectively) with higher enthalpy are shown on the upper level, and those with lower enthalpy are shown on the lower level.

Solve. For this reaction, ΔH is negative, so the products have lower enthalpy and are shown on the lower level; reactants are on the upper level. The arrow points in the direction of reactants to products and is labeled with the value of ΔH.

5.40   (a) $Ca(OH)_2(s) \rightarrow CaO(s) + H_2O(g)$ (b)

$\Delta H = 109$ kJ

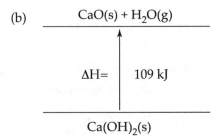

5.41   *Analyze/Plan.* Consider ΔH for the exothermic reaction as written.   *Solve.*

(a) $\Delta H = -284.6$ kJ/2 mol $O_3(g) = -142.3$ kJ/mol $O_3(g)$

(b)   Because ΔH is negative, the reactants, $2\,O_3(g)$ has the higher enthalpy.

5.42   *Plan.* Consider the sign of an enthalpy change that would convert one of the substances into the other.   *Solve.*

(a) $CO_2(s) \rightarrow CO_2(g)$. This change is sublimation, which is endothermic, +ΔH. $CO_2(g)$ has the higher enthalpy.

(b) $H_2 \rightarrow 2\,H$. Breaking the H–H bond requires energy, so the process is endothermic, +ΔH. Two moles of H atoms have higher enthalpy.

120

(c) $H_2O(g) \rightarrow H_2(g) + 1/2\ O_2(g)$. Decomposing H_2O into its elements requires energy and is endothermic, $+\Delta H$. One mole of $H_2(g)$ and 0.5 mol $O_2(g)$ at 25 °C have the higher enthalpy.

(d) $N_2(g)$ at 100 °C $\rightarrow N_2(g)$ at 300 °C. An increase in the temperature of the sample requires that heat is added to the system, $+q$ and $+\Delta H$. $N_2(g)$ at 300 °C has the higher enthalpy.

5.43   *Analyze/Plan.* Follow the strategy in Sample Exercise 5.4.   *Solve.*

(a) Exothermic (ΔH is negative)

(b) $3.55\ \text{g Mg} \times \dfrac{1\ \text{mol Mg}}{24.305\ \text{g Mg}} \times \dfrac{-1204\ \text{kJ}}{2\ \text{mol Mg}} = -87.9\ \text{kJ heat transferred}$

Check. The units of kJ are correct for heat. The negative sign indicates heat is evolved.

(c) $-234\ \text{kJ} \times \dfrac{2\ \text{mol MgO}}{-1204\ \text{kJ}} \times \dfrac{40.30\ \text{g MgO}}{1\ \text{mol Mg}} = 15.7\ \text{g MgO produced}$

Check. Units are correct for mass. $(200 \times 2 \times 40/1200) \approx (16{,}000/1200) > 10\ \text{g}$

(d) $2\ \text{MgO}(s) \rightarrow 2\ \text{Mg}(s) + O_2(g) \qquad \Delta H = +1204\ \text{kJ}$

This is the reverse of the reaction given above, so the sign of ΔH is reversed.

$40.3\ \text{g MgO} \times \dfrac{1\ \text{mol MgO}}{40.30\ \text{g MgO}} \times \dfrac{1204\ \text{kJ}}{2\ \text{mol MgO}} = +602\ \text{kJ heat absorbed}$

Check. 40.3 g MgO is just 1 mol MgO, so the calculated value is the heat absorbed per mol of MgO, 1204 kJ/2 mol MgO = 602 kJ.

5.44   (a) The sign of ΔH is positive, so the reaction is endothermic.

(b) $24.0\ \text{g CH}_3\text{OH} \times \dfrac{1\ \text{mol CH}_3\text{OH}}{32.04\ \text{g CH}_3\text{OH}} \times \dfrac{252.8\ \text{kJ}}{2\ \text{mol CH}_3\text{OH}} = 94.7\ \text{kJ heat absorbed}$

(c) $82.1\ \text{kJ} \times \dfrac{2\ \text{mol CH}_4}{252.8\ \text{kJ}} \times \dfrac{16.04\ \text{g CH}_4}{1\ \text{mol CH}_4} = 10.4\ \text{g CH}_4\ \text{produced}$

(d) The sign of ΔH is reversed for the reverse reaction: $\Delta H = -252.8\ \text{kJ}$

$38.5\ \text{g CH}_4 \times \dfrac{1\ \text{mol CH}_4}{16.04\ \text{g CH}_4} \times \dfrac{-252.8\ \text{kJ}}{2\ \text{mol CH}_4} = -303\ \text{kJ heat released}$

5.45   *Analyze.* Given: balanced thermochemical equation, various quantities of substances and/or enthalpy. *Plan.* Enthalpy is an extensive property; it is "stoichiometric." Use the mole ratios implicit in the balanced thermochemical equation to solve for the desired quantity. Use molar masses to change mass to moles and vice versa where appropriate. *Solve.*

(a) $0.450\ \text{mol AgCl} \times \dfrac{-65.5\ \text{kJ}}{1\ \text{mol AgCl}} = -29.5\ \text{kJ}$

Check. Units are correct; sign indicates heat evolved.

(b) $9.00\ \text{g AgCl} \times \dfrac{1\ \text{mol AgCl}}{143.3\ \text{g AgCl}} \times \dfrac{-65.5\ \text{kJ}}{1\ \text{mol AgCl}} = -4.11\ \text{kJ}$

Check. Units correct; sign indicates heat evolved.

(c) 9.25×10^{-4} mol AgCl $\times \dfrac{+65.5 \text{ kJ}}{1 \text{ mol AgCl}} = 0.0606$ kJ $= 60.6$ J

Check. Units correct; sign of ΔH reversed; sign indicates heat is absorbed during the reverse reaction.

5.46   (a)   1.36 mol $O_2 \times \dfrac{-89.4 \text{ kJ}}{3 \text{ mol } O_2} = -40.53 = -40.5$ kJ

(b)   10.4 g KCl $\times \dfrac{1 \text{ mol KCl}}{74.55 \text{ g KCl}} \times \dfrac{-89.4 \text{ kJ}}{2 \text{ mol KCl}} = -6.2358 = -6.24$ kJ

(c)   Because the sign of ΔH is reversed for the reverse reaction, it seems reasonable that other characteristics would be reversed, as well. If the forward reaction proceeds spontaneously, the reverse reaction is probably not spontaneous. Also, we know from experience that KCl(s) does not spontaneously react with atmospheric O_2(g), even at elevated temperature.

5.47   *Analyze.* Given: balanced thermochemical equation. *Plan.* Follow the guidelines given in Section 5.4 for evaluating thermochemical equations.   *Solve.*

(a)   When a chemical equation is reversed, the sign of ΔH is reversed.

CO_2(g) + 2 H_2O(l) $\rightarrow$ CH_3OH(l) + 3/2 O_2(g) $\Delta H = +726.5$ kJ

(b)   Enthalpy is extensive. If the coefficients in the chemical equation are multiplied by 2 to obtain all integer coefficients, the enthalpy change is also multiplied by 2.

2 CH_3OH(l) + 3 O_2(g) $\rightarrow$ 2 CO_2(g) + 4 H_2O(l) $\Delta H = 2(-726.5)$ kJ $= -1453$ kJ

(c)   The exothermic forward reaction is more likely to be thermodynamically favored.

(d)   Decrease. Vaporization (liquid $\rightarrow$ gas) is endothermic. If the product were H_2O(g), the reaction would be more endothermic and would have a smaller negative ΔH. (Depending on temperature, the enthalpy of vaporization for 2 mol H_2O is about +88 kJ, not large enough to cause the overall reaction to be endothermic.)

5.48   (a)   3 C_2H_2(g) $\rightarrow$ C_6H_6(l) $\Delta H = -630$ kJ

(b) C_6H_6(l) $\rightarrow$ 3 C_2H_2(g) $\Delta H = +630$ kJ

ΔH for the formation of 3 mol of acetylene is 630 kJ. ΔH for the formation of 1 mol of C_2H_2 is then 630 kJ/3 = 210 kJ.

(c)   The exothermic reverse reaction is more likely to be thermodynamically favored.

(d)

If the reactant is in the higher enthalpy gas phase, the overall ΔH for the reaction decreases.

Calorimetry (Section 5.5)

The specific heat of water to four significant figures, **4.184 J/g-K,** will be used in many of the following exercises; temperature units of K and °C will be used interchangeably.

5.49   (a)   J/mol-K or J/mol-°C. Heat capacity is the amount of heat in J required to raise the temperature of an object or a certain amount of substance 1 °C or 1 K. Molar heat capacity is the heat capacity of one mole of substance.

        (b) $\dfrac{J}{g\text{-}°C}$ or $\dfrac{J}{g\text{-}K}$ Specific heat is a particular kind of heat capacity where the amount of substance is 1 g.

        (c)   To calculate heat capacity from specific heat, the **mass** of the particular piece of copper pipe must be known.

5.50   *Analyze.* Both objects are heated to 100 °C. The two hot objects are placed in the same amount of cold water at the same temperature. Object A raises the water temperature more than object B. *Plan.* Apply the definition of heat capacity to heating the water and heating the objects to determine which object has the greater heat capacity.   *Solve.*

        (a)   Both beakers of water contain the same mass of water, so they both have the same heat capacity. Object A raises the temperature of its water more than object B, so more heat was transferred from object A than from object B. Because both objects were heated to the same temperature initially, object A must have absorbed more heat to reach the 100 °C temperature. The greater the heat capacity of an object, the greater the heat required to produce a given rise in temperature. Thus, object A has the greater heat capacity.

        (b)   Because no information about the masses of the objects is given, we cannot compare or determine the specific heats of the objects.

5.51   *Plan.* Manipulate the definition of specific heat to solve for the desired quantity, paying close attention to units. $C_s = q/(m \times \Delta T)$.   *Solve.*

        (a) $\dfrac{4.184\ J}{1\ g\text{-}K}$ or $\dfrac{4.184\ J}{1\ g\text{-}°C}$

        (b) $\dfrac{4.184\ J}{1\ g\text{-}°C} \times \dfrac{18.02\ g\ H_2O}{1\ mol\ H_2O} = \dfrac{75.40\ J}{mol\text{-}°C}$

        (c) $\dfrac{185\ g\ H_2O \times 4.184\ J}{1\ g\text{-}°C} = 774\ J/°C$

        (d) $10.00\ kg\ H_2O \times \dfrac{1000\ g}{1\ kg} \times \dfrac{4.184\ J}{1\ g\text{-}°C} \times \dfrac{1\ kJ}{1000\ J} \times (46.2\ °C - 24.6\ °C) = 904\ kJ$

        Check. $(10 \times 4 \times 20) \approx 800$ kJ; the units are correct. Note that the conversion factors for kg → g and J → kJ cancel. An equally correct form of specific heat would be kJ/kg-°C

5.52   (a)   In Table 5.2, Hg(l) has the smallest specific heat, so it will require the smallest amount of energy to heat 50.0 g of the substance 10 K.

        (b) $50.0\ g\ Hg(l) \times 10\ K \times \dfrac{0.14\ J}{g\text{-}K} = 70\ J$

5.53   *Analyze/Plan.* Follow the logic in Sample Exercise 5.5.   *Solve.*

(a) $80.0 \text{ g } C_8H_{18} \times \dfrac{2.22 \text{ J}}{\text{g-K}} \times (25.0\,°C - 10.0\,°C) = 2.66 \times 10^3 \text{ J (or 2.66 kJ)}$

(b)   *Plan.* Calculate the molar heat capacity of octane and compare it with the molar heat capacity of water, 75.40 J/mol-°C, as calculated in Exercise 5.51(b).   *Solve.*

$$\dfrac{2.22 \text{ J}}{\text{g-K}} \times \dfrac{114.2 \text{ g } C_8H_{18}}{1 \text{ mol } C_8H_{18}} = \dfrac{253.58 \text{ J}}{\text{mol-K}} = \dfrac{254 \text{ J}}{\text{mol-K}}$$

The molar heat capacity of $C_8H_{18}(l)$, 254 J/mol-K, is greater than that of $H_2O(l)$, so it will require more heat to increase the temperature of octane than to increase the temperature of water.

5.54   (a) $\text{specific heat} = \dfrac{\text{J}}{1 \text{ g-}°C} = \dfrac{322 \text{ J}}{100.0 \text{ g} \times (50\,°C - 25\,°C)} = 0.1288 = \dfrac{0.13 \text{ J}}{1 \text{ g-}°C}$

(b)   In general, the greater the heat capacity, the more heat is required to raise the temperature of 1 gram of substance 1 °C. The specific heat of gold is 0.13 J/g-°C, whereas that of iron is 0.45 J/g-°C (Table 5.2). For gold and iron blocks with equal mass, same initial temperature and same amount of heat added, the one with the lower specific heat, gold, will require less heat per °C and have the higher final temperature.

(c) $\dfrac{0.1288 \text{ J}}{1 \text{ g-}°C} \times \dfrac{196.97 \text{ g Au}}{1 \text{ mol Au}} = 25.37 = \dfrac{25 \text{ J}}{\text{mol-}°C}$

5.55   *Analyze.* Because the temperature of the water increases, the dissolving process is exothermic and the sign of ΔH is negative. The heat lost by the NaOH(s) dissolving equals the heat gained by the solution.

Plan/Solve. Calculate the heat gained by the solution. The temperature change is $37.8 - 21.6 = 16.2\,°C$. The total mass of solution is (100.0 g H_2O + 6.50 g NaOH) = 106.5 g.

$$106.5 \text{ g solution} \times \dfrac{4.184 \text{ J}}{1 \text{ g-}°C} \times 16.2\,°C \times \dfrac{1 \text{ kJ}}{1000 \text{ J}} = 7.2187 = 7.22 \text{ kJ}$$

This is the amount of heat lost when 6.50 g of NaOH dissolves.

The heat loss per mole NaOH is

$$\dfrac{-7.2187 \text{ kJ}}{6.50 \text{ g NaOH}} \times \dfrac{40.00 \text{ g NaOH}}{1 \text{ mol NaOH}} = -44.4 \text{ kJ/mol} \quad \Delta H = q_p = -44.4 \text{ kJ/mol NaOH}$$

Check. $(-7/7 \times 40) \approx -40$ kJ; the units and sign are correct.

5.56   (a)   Follow the logic in Solution 5.55. The total mass of the solution is (60.0 g H_2O + 4.25 g NH_4NO_3) = 64.25 = 64.3 g. The temperature change of the solution is $22.0 - 16.9 = -5.1\,°C$. The heat lost by the surroundings is

$$64.25 \text{ g solution} \times \dfrac{4.184 \text{ J}}{1 \text{ g-}°C} \times -5.1\,°C \times \dfrac{1 \text{ kJ}}{1000 \text{ J}} = -1.371 = -1.4 \text{ kJ}$$

That is, 1.4 kJ is absorbed when 4.25 g NH_4NO_3(s) dissolves.

$$\dfrac{+1.371 \text{ kJ}}{4.25 \text{ } NH_4NO_3} \times \dfrac{80.04 \text{ g } NH_4NO_3}{1 \text{ mol } NH_4NO_3} = +25.82 = +26 \text{ kJ/mol } NH_4NO_3$$

(b)   This process is endothermic because the temperature of the surroundings decreases, indicating that heat is absorbed by the system.

5.57 *Analyze/Plan.* Follow the logic in Sample Exercise 5.7. *Solve.*

$q_{bomb} = -q_{rxn}$; $\Delta T = 30.57\ °C - 23.44\ °C = 7.13\ °C$

$$q_{bomb} = \frac{7.854\ kJ}{1\ °C} \times 7.13\ °C = 56.00 = 56.0\ kJ$$

At constant volume, $q_v = \Delta E$. ΔE and ΔH are very similar.

$$\Delta H_{rxn} \approx \Delta E_{rxn} = q_{rxn} = -q_{bomb} = \frac{-56.0\ kJ}{2.200\ g\ C_6H_4O_2} = -25.454 = -25.5\ kJ/g\ C_6H_4O_2$$

$$\Delta H_{rxn} = \frac{-25.454\ kJ}{1\ g\ C_6H_4O_2} \times \frac{108.1\ g\ C_6H_4O_2}{1\ mol\ C_6H_4O_2} = -2.75 \times 10^3\ kJ/mol\ C_6H_4O_2$$

5.58 (a) $C_6H_5OH(s) + 7\ O_2(g) \rightarrow 6\ CO_2(g) + 3\ H_2O(l)$

(b) $q_{bomb} = -q_{rxn}$; $\Delta T = 26.37\ °C - 21.36\ °C = 5.01\ °C$

$$q_{bomb} = \frac{11.66\ kJ}{1\ °C} \times 5.01\ °C = 58.417 = 58.4\ kJ$$

At constant volume, $q_v = \Delta E$. ΔE and ΔH are very similar.

$$\Delta H_{rxn} \approx \Delta E_{rxn} = q_{rxn} = -q_{bomb} = \frac{-58.417\ kJ}{1.800\ g\ C_6H_5OH} = -32.454 = -32.5\ kJ/g\ C_6H_5OH$$

$$\Delta H_{rxn} = \frac{-32.454\ kJ}{1\ g\ C_6H_5OH} \times \frac{94.11\ g\ C_6H_5OH}{1\ mol\ C_6H_5OH} = \frac{-3.054 \times 10^3\ kJ}{mol\ C_6H_5OH}$$

$$= -3.05 \times 10^3\ kJ/mol\ C_6H_5OH$$

5.59 *Analyze.* Given: specific heat and mass of glucose, ΔT for calorimeter. Find: heat capacity, C, of calorimeter. *Plan.* All heat from the combustion raises the temperature of the calorimeter. Calculate heat from combustion of glucose, divide by ΔT for calorimeter to get kJ/°C. $\Delta T = 24.72\ °C - 20.94\ °C = 3.78\ °C$ *Solve.*

(a) $C_{total} = 3.500\ g\ glucose \times \dfrac{15.57\ kJ}{1\ g\ glucose} \times \dfrac{1}{3.78\ °C} = 14.42 = 14.4\ kJ/°C$

(b) Qualitatively, assuming the same exact initial conditions in the calorimeter, twice as much glucose produces twice as much heat, which raises the calorimeter temperature by twice as many °C. Quantitatively,

$$7.000\ g\ glucose \times \frac{15.57\ kJ}{1\ g\ glucose} \times \frac{1\ °C}{14.42\ kJ} = 7.56\ °C$$

Check. Units are correct. ΔT is twice as large as in part (a). The result has 3 sig figs, because the heat capacity of the calorimeter is known to 3 sig figs.

5.60 (a) $C = 2.760\ g\ C_6H_5COOH \times \dfrac{26.38\ kJ}{1\ g\ C_6H_5COOH} \times \dfrac{1}{8.33\ °C} = 8.74055 = 8.74\ kJ/°C$

(b) $\dfrac{8.74055\ kJ}{°C} \times 4.95\ °C \times \dfrac{1}{1.440\ g\ sample} = 30.046 = 30.0\ kJ/g\ sample$

(c) If water is lost from the calorimeter, the heat capacity of the calorimeter decreases.

5 Thermochemistry Solutions to Exercises

Hess's Law (Section 5.6)

5.61   Yes, because internal energy is a state function. Hess's Law works for any state function.

5.62   (a)   *Analyze/Plan.* Arrange the reactions so that in the overall sum, B appears in both reactants and products and can be canceled. This is a general technique for using Hess's Law.   *Solve.*

$$
\begin{array}{ll}
A \rightarrow B & \Delta H = +30 \text{ kJ} \\
B \rightarrow C & \Delta H = +60 \text{ kJ} \\
\hline
A \rightarrow C & \Delta H = +90 \text{ kJ}
\end{array}
$$

(b)

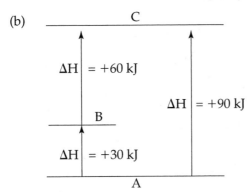

Check. The process of A forming C can be described as A forming B and B forming C.

5.63   *Analyze/Plan.* Follow the logic in Sample Exercise 5.8. Manipulate the equations so that "unwanted" substances can be canceled from reactants and products. Adjust the corresponding sign and magnitude of ΔH.   *Solve.*

$$
\begin{array}{ll}
P_4O_6(s) \rightarrow P_4(s) + 3 O_2(g) & \Delta H = 1640.1 \text{ kJ} \\
P_4(s) + 5 O_2(g) \rightarrow P_4O_{10}(s) & \Delta H = -2940.1 \text{ kJ} \\
\hline
P_4O_6(s) + 2 O_2(g) \rightarrow P_4O_{10}(s) & \Delta H = -1300.0 \text{ kJ}
\end{array}
$$

Check. We have obtained the desired reaction.

5.64
$$
\begin{array}{ll}
2 \text{ C}(s) + O_2(g) + 4 H_2(g) \rightarrow 2 CH_3OH(g) & \Delta H = -402.4 \text{ kJ} \\
2 CO(g) \rightarrow O_2(g) + 2 C(s) & \Delta H = 221.0 \text{ kJ} \\
\hline
2 CO(g) + 4 H_2(g) \rightarrow 2 CH_3OH(g) & \Delta H = -181.4 \text{ kJ} \\
CO(g) + 2 H_2(g) \rightarrow CH_3OH(g) & \Delta H = (-181.4)/2 = -90.7 \text{ kJ}
\end{array}
$$

5.65   *Analyze/Plan.* Follow the logic in Sample Exercise 5.9. Manipulate the equations so that "unwanted" substances can be canceled from reactants and products. Adjust the corresponding sign and magnitude of ΔH.   *Solve.*

$$
\begin{array}{lll}
C_2H_4(g) & \rightarrow 2 H_2(g) + 2 C(s) & \Delta H = -52.3 \text{ kJ} \\
2 C(s) + 4 F_2(g) & \rightarrow 2 CF_4(g) & \Delta H = 2(-680 \text{ kJ}) \\
2 H_2(g) + 2 F_2(g) & \rightarrow 4 HF(g) & \Delta H = 2(-537 \text{ kJ}) \\
\hline
C_2H_4(g) + 6 F_2(g) & \rightarrow 2 CF_4(g) + 4 HF(g) & \Delta H = -2.49 \times 10^3 \text{ kJ}
\end{array}
$$

Check. We have obtained the desired reaction.

126

5.66

$$N_2O(g) \rightarrow N_2(g) + 1/2\,O_2(g) \qquad \Delta H = 1/2\,(-163.2\text{ kJ})$$
$$NO_2(g) \rightarrow NO(g) + 1/2\,O_2(g) \qquad \Delta H = 1/2(113.1\text{ kJ})$$
$$\underline{N_2(g) + O_2(g) \rightarrow 2\,NO(g) \qquad\qquad \Delta H = 180.7\text{ kJ}}$$
$$N_2O(g) + NO_2(g) \rightarrow 3\,NO(g) \qquad \Delta H = 155.7\text{ kJ}$$

Enthalpies of Formation (Section 5.7)

5.67 (a) *Standard conditions* for enthalpy changes are usually P = 1 atm and T = 298 K. For the purpose of comparison, standard enthalpy changes, $\Delta H°$, are tabulated for reactions at these conditions.

(b) *Enthalpy of formation*, ΔH_f, is the enthalpy change that occurs when a compound is formed from its component elements.

(c) *Standard enthalpy of formation*, $\Delta H_f°$, is the enthalpy change that accompanies formation of 1 mole of a substance from elements in their standard states.

5.68 (a) The standard enthalpy of formation for any element in its standard state is zero. Elements in their standard states are the reference point for the enthalpy of formation scale.

(b) $12\,C(s) + 11\,H_2(g) + 11/2\,O_2(g) \rightarrow C_{12}H_{22}O_{11}(s)$

5.69 (a) $1/2\,N_2(g) + O_2(g) \rightarrow NO_2(g)$ $\qquad\qquad \Delta H_f° = 33.84$ kJ

(b) $S(s) + 3/2\,O_2(g) \rightarrow SO_3(g)$ $\qquad\qquad \Delta H_f° = -395.2$ kJ

(c) $Na(s) + 1/2\,Br_2(l) \rightarrow NaBr(s)$ $\qquad\qquad \Delta H_f° = -361.4$ kJ

(d) $Pb(s) + N_2(g) + 3\,O_2(g) \rightarrow Pb(NO_3)_2(s)$ $\qquad \Delta H_f° = -451.9$ kJ

5.70 (a) $H_2(g) + O_2(g) \rightarrow H_2O_2(g)$ $\qquad\qquad \Delta H_f° = -136.10$ kJ

(b) $Ca(s) + C(s) + 3/2\,O_2(g) \rightarrow CaCO_3(s)$ $\qquad \Delta H_f° = -1207.1$ kJ

(c) $1/4\,P_4(s) + 1/2\,O_2(g) + 3/2\,Cl_2(g) \rightarrow POCl_3(l)$ $\qquad \Delta H_f° = -597.0$ kJ

(d) $2\,C(s) + 3\,H_2(g) + 1/2\,O_2(g) \rightarrow C_2H_5OH(l)$ $\qquad \Delta H_f° = -277.7$ kJ

5.71 *Plan.* $\Delta H_{rxn}° = \Sigma n\Delta H_f°\,(\text{products}) - \Sigma n\Delta H_f°\,(\text{reactants})$. Be careful with coefficients, states, and signs. *Solve.*

$$\Delta H_{rxn}° = \Delta H_f°\,Al_2O_3(s) + 2\,\Delta H_f°\,Fe(s) - \Delta H_f°\,Fe_2O_3(s) - 2\,\Delta H_f°\,Al(s)$$
$$\Delta H_{rxn}° = (-1669.8\text{ kJ}) + 2(0) - (-822.16\text{ kJ}) - 2(0) = -847.6\text{ kJ}$$

5.72 Use heats of formation to calculate $\Delta H°$ for the combustion of butane.

$$C_3H_8(g) + 5\,O_2(g) \rightarrow 3\,CO_2(g) + 4\,H_2O(l)$$
$$\Delta H_{rxn}° = 3\,\Delta H_f°\,CO_2(g) + 4\,\Delta H_f°\,H_2O(l) - \Delta H_f°\,C_3H_8(g) - 5\,\Delta H_f°\,O_2(g)$$
$$\Delta H_{rxn}° = 3(-393.5\text{ kJ}) + 4(-285.83\text{ kJ}) - (-103.85\text{ kJ}) - 5(0) = -2219.97 = -2220.0\text{ kJ/mol }C_3H_8$$
$$10.00\text{ g }C_3H_8 \times \frac{1\text{ mol }C_3H_8}{44.096\text{ g }C_3H_8} \times \frac{-2219.97\text{ kJ}}{1\text{ mol }C_3H_8} = -503.4\text{ kJ}$$

5.73 *Plan.* $\Delta H_{rxn}^{\circ} = \Sigma n \Delta H_f^{\circ}$ (products) $- \Sigma n \Delta H_f^{\circ}$ (reactants). Be careful with coefficients, states, and signs. *Solve.*

(a) $\Delta H_{rxn}^{\circ} = 2 \Delta H_f^{\circ} SO_3(g) - 2 \Delta H_f^{\circ} SO_2(g) - \Delta H_f^{\circ} O_2(g)$

$= 2(-395.2 \text{ kJ}) - 2(-296.9 \text{ kJ}) - 0 = -196.6 \text{ kJ}$

(b) $\Delta H_{rxn}^{\circ} = \Delta H_f^{\circ} MgO(s) + \Delta H_f^{\circ} H_2O(l) - \Delta H_f^{\circ} Mg(OH)_2(s)$

$= -601.8 \text{ kJ} + (-285.83 \text{ kJ}) - (-924.7 \text{ kJ}) = 37.1 \text{ kJ}$

(c) $\Delta H_{rxn}^{\circ} = 4 \Delta H_f^{\circ} H_2O(g) + \Delta H_f^{\circ} N_2(g) - \Delta H_f^{\circ} N_2O_4(g) - 4 \Delta H_f^{\circ} H_2(g)$

$= 4(-241.82 \text{ kJ}) + 0 - (9.66 \text{ kJ}) - 4(0) = -976.94 \text{ kJ}$

(d) $\Delta H_{rxn}^{\circ} = \Delta H_f^{\circ} SiO_2(s) + 4 \Delta H_f^{\circ} HCl(g) - \Delta H_f^{\circ} SiCl_4(l) - 2 \Delta H_f^{\circ} H_2O(l)$

$= -910.9 \text{ kJ} + 4(-92.30 \text{ kJ}) - (-640.1 \text{ kJ}) - 2(-285.83 \text{ kJ}) = -68.3 \text{ kJ}$

5.74 (a) $\Delta H_{rxn}^{\circ} = \Delta H_f^{\circ} CaCl_2(s) + \Delta H_f^{\circ} H_2O(g) - \Delta H_f^{\circ} CaO(s) - 2 \Delta H_f^{\circ} HCl(g)$

$= -795.8 \text{ kJ} + (-241.82 \text{ kJ}) - (-635.5 \text{ kJ}) - 2(-92.30 \text{ kJ}) = -217.5 \text{ kJ}$

(b) $\Delta H_{rxn}^{\circ} = 2 \Delta H_f^{\circ} Fe_2O_3(s) - 4 \Delta H_f^{\circ} FeO(s) - \Delta H_f^{\circ} O_2(g)$

$= 2(-822.16 \text{ kJ}) - 4(-271.9 \text{ kJ}) - (0) = -556.7 \text{ kJ}$

(c) $\Delta H_{rxn}^{\circ} = \Delta H_f^{\circ} Cu_2O(s) + \Delta H_f^{\circ} NO_2(g) - 2 \Delta H_f^{\circ} CuO(s) - \Delta H_f^{\circ} NO(g)$

$= -170.7 \text{ kJ} + (33.84 \text{ kJ}) - 2(-156.1 \text{ kJ}) - (90.37 \text{ kJ}) = 85.0 \text{ kJ}$

(d) $\Delta H_{rxn}^{\circ} = 2 \Delta H_f^{\circ} N_2H_4(g) + 2 \Delta H_f^{\circ} H_2O(l) - 4 \Delta H_f^{\circ} NH_3(g) - \Delta H_f^{\circ} O_2(g)$

$= 2(95.40 \text{ kJ}) + 2(-285.83 \text{ kJ}) - 4(-46.19 \text{ kJ}) - (0) = -196.10 \text{ kJ}$

5.75 *Analyze.* Given: combustion reaction, enthalpy of combustion, enthalpies of formation for most reactants and products. Find: enthalpy of formation for acetone.

Plan. Rearrange the expression for enthalpy of reaction to calculate the desired enthalpy of formation. *Solve.*

$\Delta H_{rxn}^{\circ} = 3 \Delta H_f^{\circ} CO_2(g) + 3 \Delta H_f^{\circ} H_2O(l) - \Delta H_f^{\circ} C_3H_6O(l) - 4 \Delta H_f^{\circ} O_2(g)$

$-1790 \text{ kJ} = 3(-393.5 \text{ kJ}) + 3(-285.83 \text{ kJ}) - \Delta H_f^{\circ} C_3H_6O(l) - 4(0)$

$\Delta H_f^{\circ} C_3H_6O(l) = 3(-393.5 \text{ kJ}) + 3(-285.83 \text{ kJ}) + 1790 \text{ kJ} = -248 \text{ kJ}$

5.76 $\Delta H_{rxn}^{\circ} = \Delta H_f^{\circ} Ca(OH)_2(s) + \Delta H_f^{\circ} C_2H_2(g) - 2 \Delta H_f^{\circ} H_2O(l) - \Delta H_f^{\circ} CaC_2(s)$

$-127.2 \text{ kJ} = -986.2 \text{ kJ} + 226.77 \text{ kJ} - 2(-285.83 \text{ kJ}) - \Delta H_f^{\circ} CaC_2(s)$

ΔH_f° for $CaC_2(s) = -60.6 \text{ kJ}$

5.77 (a) $C_8H_{18}(l) + 25/2 O_2(g) \rightarrow 8 CO_2(g) + 9 H_2O(g) \quad \Delta H^{\circ} = -5064.9 \text{ kJ}$

(b) *Plan.* Follow the logic in Solution 5.75 and 5.76. *Solve.*

$\Delta H_{rxn}^{\circ} = 8 \Delta H_f^{\circ} CO_2(g) + 9 \Delta H_f^{\circ} H_2O(g) - \Delta H_f^{\circ} C_8H_{18}(l) - 25/2 \Delta H_f^{\circ} O_2(g)$

$-5064.9 \text{ kJ} = 8(-393.5 \text{ kJ}) + 9(-241.82 \text{ kJ}) - \Delta H_f^{\circ} C_8H_{18}(l) - 25/2(0)$

$\Delta H_f^{\circ} C_8H_{18}(l) = 8(-393.5 \text{ kJ}) + 9(-241.82 \text{ kJ}) + 5064.9 \text{ kJ} = -259.5 \text{ kJ}$

5.78   (a) $C_4H_{10}O(l) + 6\,O_2(g) \rightarrow 4\,CO_2(g) + 5\,H_2O(l)$ $\Delta H° = -2723.7$ kJ

        (b) $\Delta H_{rxn}° = 4\,\Delta H_f°\ CO_2(g) + 5\,\Delta H_f°\ H_2O(l) - \Delta H_f°\ C_4H_{10}O(l) - 6\,\Delta H_f°\ O_2(g)$

               $-2723.7 = 4(-393.5\ \text{kJ}) + 5(-285.83\ \text{kJ}) - \Delta H_f°\ C_4H_{10}O(l) - 6(0)$

               $\Delta H_f°\ C_4H_{10}O(l) = 4(-393.5\ \text{kJ}) + 5(-285.83\ \text{kJ}) + 2723.7\ \text{kJ} = -279.45 = -279.5$ kJ

5.79   (a) $C_2H_5OH(l) + 3\,O_2(g) \rightarrow 2\,CO_2(g) + 3\,H_2O(g)$

        (b) $\Delta H_{rxn}° = 2\,\Delta H_f°\ CO_2(g) + 3\,\Delta H_f°\ H_2O(g) - \Delta H_f°\ C_2H_5OH(l) - 3\,\Delta H_f°\ O_2(g)$

               $= 2(-393.5\ \text{kJ}) + 3(-241.82\ \text{kJ}) - (-277.7\ \text{kJ}) - 3(0) = -1234.76 = -1234.8$ kJ

        (c)   *Plan.* The enthalpy of combustion of ethanol [from part (b)] is −1234.8 kJ/mol. Change mol to mass using molar mass, then mass to volume using density. *Solve.*

               $$\frac{-1234.76\ \text{kJ}}{\text{mol}\ C_2H_5OH} \times \frac{1\ \text{mol}\ C_2H_5OH}{46.06844\ \text{g}} \times \frac{0.789\ \text{g}}{\text{mL}} \times \frac{1000\ \text{mL}}{\text{L}} = -21,147 = -2.11 \times 10^4\ \text{kJ/L}$$

               Check. $(1200/50) \approx 25;\ 25 \times 800 \approx 20,000$

        (d)   *Plan.* The enthalpy of combustion corresponds to any of the molar amounts in the equation as written. Production of −1234.76 kJ also produces 2 mol CO_2. Use this relationship to calculate mass CO_2/kJ.

               $$\frac{2\ \text{mol}\ CO_2}{-1234.76\ \text{kJ}} \times \frac{44.0095\ \text{g}\ CO_2}{\text{mol}} = 0.071284\ \text{g}\ CO_2\ /\ \text{kJ emitted}$$

               Check. The negative sign associated with enthalpy indicates that energy is emitted.

5.80   (a) $CH_3OH(l) + 3/2\,O_2(g) \rightarrow CO_2(g) + 2\,H_2O(g)$

        (b) $\Delta H_{rxn}° = \Delta H_f°\ CO_2(g) + 2\,\Delta H_f°\ H_2O(g) - \Delta H_f°\ CH_3OH(l) - 3/2\,\Delta H_f°\ O_2(g)$

               $= -393.5\ \text{kJ} + 2(-241.82\ \text{kJ}) - (-238.6\ \text{kJ}) - 3/2(0) = -638.54 = -638.5$ kJ

        (c) $$\frac{-638.54\ \text{kJ}}{\text{mol}\ CH_3OH} \times \frac{1\ \text{mol}\ CH_3OH}{32.04\ \text{g}} \times \frac{0.791\ \text{g}}{\text{mL}} \times \frac{1000\ \text{mL}}{\text{L}} = 1.58 \times 10^4\ \text{kJ/L produced}$$

        (d) $$\frac{1\ \text{mol}\ CO_2}{-638.54\ \text{kJ}} \times \frac{44.0095\ \text{g}\ CO_2}{\text{mol}} = 0.06892\ \text{g}\ CO_2/\text{kJ emitted}$$

Bond Enthalpies (Section 5.8)

5.81   (a) $+\Delta H$; energy must be supplied to separate oppositely charged ions.

        (b) $-\Delta H$; energy is released when a chemical bond is formed.

        (c) $+\Delta H$; energy must be supplied to separate a negatively charged electron from a neutral atom.

        (d) $+\Delta H$; energy must be supplied to melt a solid.

5.82   (a) $-\Delta H$; the reactants have four N–O bonds (some of them multiple bonds) while the products have these same N–O bonds plus an N–N bond. Overall the reaction involves formation of a new chemical bond and enthalpy decreases.

        (b) $-\Delta H$; energy is released when a chemical bond is formed.

        (c) $-\Delta H$; energy is released when oppositely charged ions form ionic bonds.

        (d) $+\Delta H$; energy must be supplied to break a chemical bond.

5.83   *Analyze.* Given: structural formulas. Find: enthalpy of reaction.

        Plan. Count the number and kinds of bonds that are broken and formed by the reaction. Use bond enthalpies from Table 5.4 and Equation 5.32 to calculate the overall enthalpy of reaction, ΔH.   *Solve.*

        (a) $\Delta H = D(\text{H–H}) + D(\text{Br–Br}) - 2\, D(\text{H–Br})$

               $= 436\ \text{kJ} + 193\ \text{kJ} - 2(366\ \text{kJ}) = -103\ \text{kJ}$

        (b) $\Delta H = 6\, D(\text{C–H}) + 2\, D(\text{C–O}) + 2\, D(\text{O–H}) + 3\, D(\text{O=O})$

                    $- 4\, D(\text{C=O}) - 8\, D(\text{O–H})$

               $= 6\, D(\text{C–H}) + 2\, D(\text{C–O}) + 3\, D(\text{O=O}) - 4\, D(\text{C=O}) - 6\, D(\text{O–H})$

               $\Delta H = 6(413) + 2(358) + 3(495) - 4(799) - 6(463) = -1295\ \text{kJ}$

5.84   (a) $\Delta H = 3\, D(\text{C–Br}) + D(\text{C–H}) + D(\text{Cl–Cl}) - 3\, D(\text{C–Br}) - D(\text{C–Cl}) - D(\text{H–Cl})$

               $= D(\text{C–H}) + D(\text{Cl–Cl}) - D(\text{C–Cl}) - D(\text{H–Cl})$

               $\Delta H = 413 + 242 - 328 - 431 = -104\ \text{kJ}$

        (b) $\Delta H = 4\, D(\text{C–H}) + 2\, D(\text{O=O}) - 2\, D(\text{C=O}) - 4\, D(\text{O–H})$

               $\Delta H = 4(413) + 2(495) - 2(799) - 4(463) = -808\ \text{kJ}$

5.85   (a)   *Plan.* $\Delta H^{\circ}_{\text{rxn}} = \Sigma n \Delta H^{\circ}_{\text{f}}$ (products) $- \Sigma n \Delta H^{\circ}_{\text{f}}$ (reactants). Be careful with coefficients, states, and signs.   *Solve.*

               $\Delta H^{\circ}_{\text{rxn}} = 2\, \Delta H^{\circ}_{\text{f}}\ \text{Br(g)} - \Delta H^{\circ}_{\text{f}}\ \text{Br}_2(\text{g})$

               $= 2(111.8) - 30.71 = 192.9\ \text{kJ}$

               This reaction is just the breaking of a Br–Br single bond to form Br atoms; reactants and products are all in the gas phase. The enthalpy of reaction represents the bond enthalpy D(Br–Br), 193 kJ.

        (b)   The value of D(Br–Br) in Table 5.4 is 193 kJ, the same as the enthalpy calculated in part (a). The difference between the two values is zero (to three significant figures).

5.86   (a)   The relevant reaction is $\text{N}_2(\text{g}) \rightarrow 2\,\text{N(g)}$.

               $\Delta H^{\circ}_{\text{rxn}} = 2\, \Delta H^{\circ}_{\text{f}}\ \text{N(g)} - \Delta H^{\circ}_{\text{f}}\ \text{N}_2(\text{g})$

               $= 2(472.7) - 0 = 945.4\ \text{kJ}$

               Our estimate for $D(\text{N}\equiv\text{N})$ is 945.4 kJ = 945 kJ.

(b) Calculate the overall enthalpy change for the reaction using standard enthalpies of formation. Use this value for ΔH and bond enthalpies from Table 5.4 to estimate the enthalpy of the nitrogen-nitrogen bond in N_2H_4.

$$\Delta H^o_{rxn} = 2\,\Delta H^o_f\ NH_3(g) - \Delta H^o_f\ N_2H_4(g) - \Delta H^o_f\ H_2(g)$$

$$= 2(-46.19) - 95.40 - 0 = -187.78 = -188\ kJ$$

$$\Delta H = 4\,D(N\text{–}H) + D(N\text{–}N) + D(H\text{–}H) - 6\,D(N\text{–}H)$$

$$\Delta H = D(N\text{–}N) + D(H\text{–}H) - 2\,D(N\text{–}H)$$

$$D(N\text{–}N) = \Delta H - D(H\text{–}H) + 2\,D(N\text{–}H)$$

$$D(N\text{–}N) = (-188) - (436) + 2(391) = 158\ kJ$$

(c) The nitrogen-nitrogen bond in N_2H_4 has an enthalpy of 158 kJ; in N_2 the enthalpy is 945 kJ. We are comparing the same pair of bonded atoms, so it is safe to say that the bond in N_2H_4 is weaker than the bond in N_2.

5.87 (a) $\Delta H = 2\,D(H\text{–}H) + D(O=O) - 4\,D(O\text{–}H) = 2(436) + 495 - 4(463) = -485\ kJ$

(b) The estimate from part (a) is less negative or larger than the true reaction enthalpy. When we use bond enthalpies to estimate reaction enthalpies, we assume all reactants and products are gases. We have estimated the enthalpy change for production of $H_2O(g)$. Because condensation, $[(g) \rightarrow (l)]$, is exothermic, we expect ΔH for production of liquid water to be more negative or smaller than the value we estimated in part (a).

(c) $\Delta H^o_{rxn} = 2\,\Delta H^o_f\ H_2O(l) - 2\,\Delta H^o_f\ H_2(g) - \Delta H^o_f\ O_2(g)$

$$= 2(-285.83) - 2(0) - 0 = -571.66 = -572\ kJ$$

As predicted in part (b), the true enthalpy of reaction is more negative than the result calculated using bond enthalpies.

5.88 (a) $\Delta H = D(H\text{–}H) + D(I\text{–}I) - 2\,D(H\text{–}I) = 436 + 151 - 2(299) = -11\ kJ$

(b) When we use bond enthalpies to estimate reaction enthalpies, we assume all reactants and products are gases. In this case, one of the reactants, I_2, is a solid. Because sublimation, $[(s) \rightarrow (g)]$, is endothermic, the actual starting reactants have a lower enthalpy than gas phase reactants. Energy must be supplied to get to the estimated starting point. The actual enthalpy change for the reactants as written will be more endothermic (more positive, larger) than the value we estimated in part (a).

(c) $\Delta H^o_{rxn} = 2\,\Delta H^o_f\ HI(g) - \Delta H^o_f\ H_2(g) - \Delta H^o_f\ I_2(s)$

$$= 2(25.94) - 0 - 0 = 51.88 = 52\ kJ$$

Foods and Fuels (Section 5.9)

5.89 (a) *Fuel value* is the amount of energy produced when 1 gram of a substance (fuel) is combusted.

(b) The fuel value of fats is 9 kcal/g and of carbohydrates is 4 kcal/g. Therefore, 5 g of fat produce 45 kcal, whereas 9 g of carbohydrates produce 36 kcal; 5 g of fat are a greater energy source.

(c) These products of metabolism are expelled as waste, $H_2O(l)$ primarily in urine and feces, and $CO_2(g)$ as gas when breathing.

5.90   (a)   One gram of fat produces more energy than one gram of carbohydrates when metabolized.

         (b)   For convenience, assume 100 g of chips.

$$12 \text{ g protein} \times \frac{17 \text{ kJ}}{1 \text{ g protein}} \times \frac{1 \text{ Cal}}{4.184 \text{ kJ}} = 48.76 = 49 \text{ Cal}$$

$$14 \text{ g fat} \times \frac{38 \text{ kJ}}{1 \text{ g fat}} \times \frac{1 \text{ Cal}}{4.184 \text{ kJ}} = 127.15 = 130 \text{ Cal}$$

$$74 \text{ g carbohydrates} \times \frac{17 \text{ kJ}}{1 \text{ g carbohydrates}} \times \frac{1 \text{ Cal}}{4.184 \text{ kJ}} = 300.67 = 301 \text{ Cal}$$

total Cal = (48.76 + 127.15 + 300.67) = 476.58 = 480 Cal

$$\% \text{ Cal from fat} = \frac{127.15 \text{ Cal fat}}{476.58 \text{ total Cal}} \times 100 = 26.68 = 27\%$$

(Because the conversion from kJ to Cal was common to all three components, we would have determined the same percentage by using kJ.)

         (c) $25 \text{ g fat} \times \dfrac{38 \text{ kJ}}{\text{g fat}} = x \text{ g protein} \times \dfrac{17 \text{ kJ}}{\text{g protein}}; x = 56 \text{ g protein}$

5.91   (a)   *Plan.* Calculate the Cal (kcal) from each nutritional component of the soup, then sum.   *Solve.*

$$2.5 \text{ g fat} \times \frac{38 \text{ kJ}}{1 \text{ g fat}} = 95.0 \text{ or } 0.95 \times 10^2 \text{ kJ}$$

$$14 \text{ g carbohydrates} \times \frac{17 \text{ kJ}}{1 \text{ g carbohydrate}} = 238 \text{ or } 2.4 \times 10^2 \text{ kJ}$$

$$7 \text{ g protein} \times \frac{17 \text{ kJ}}{1 \text{ g protein}} = 119 \text{ or } 1 \times 10^2 \text{ kJ}$$

total energy = 95.0 kJ + 238 kJ + 119 kJ = 452 or 5×10^2 kJ

$$452 \text{ kJ} \times \frac{1 \text{ kcal}}{4.184 \text{ kJ}} \times \frac{1 \text{ Cal}}{1 \text{kcal}} = 108.03 \text{ or } 1 \times 10^2 \text{ Cal/serving}$$

Check. 100 Cal/serving is a reasonable result; units are correct. The data and the result have 1 sig fig.

         (b)   Sodium does not contribute to the calorie content of the food, because it is not metabolized by the body; it enters and leaves as Na^+.

5.92   Calculate the fuel value in a pound of M&M® candies.

$$96 \text{ g fat} \times \frac{38 \text{ kJ}}{1 \text{ g fat}} = 3648 \text{ kJ} = 3.6 \times 10^3 \text{ kJ}$$

$$320 \text{ g carbohydrate} \times \frac{17 \text{ kJ}}{1 \text{ g carbohydrate}} = 5440 \text{ kJ} = 5.4 \times 10^3 \text{ kJ}$$

$$21 \text{ g protein} \times \frac{17 \text{ kJ}}{1 \text{ g protein}} = 357 \text{ kJ} = 3.6 \times 10^2 \text{ kJ}$$

total fuel value = 3648 kJ + 5440 kJ + 357 kJ = 9445 kJ = 9.4×10^3 kJ/lb

$$\frac{9445 \text{ kJ}}{\text{lb}} \times \frac{1 \text{ lb}}{453.6 \text{ g}} \times \frac{42 \text{ g}}{\text{serving}} = 874.5 \text{ kJ} = 8.7 \times 10^2 \text{kJ/serving}$$

$$\frac{874.5 \text{ kJ}}{\text{serving}} \times \frac{1 \text{ kcal}}{4.184 \text{ kJ}} \times \frac{1 \text{ Cal}}{1 \text{ kcal}} = 209.0 \text{ Cal} = 2.1 \times 10^2 \text{ Cal/serving}$$

Check. 210 Cal is the approximate food value of a candy bar, so the result is reasonable.

5.93   *Plan.* g → mol → kJ → Cal   *Solve.*

$$16.0 \text{ g } C_6H_{12}O_6 \times \frac{1 \text{ mol } C_6H_{12}O_6}{180.2 \text{ g } C_6H_{12}O_6} \times \frac{2812 \text{ kJ}}{\text{mol } C_6H_{12}O_6} \times \frac{1 \text{ Cal}}{4.184 \text{ kJ}} = 59.7 \text{ Cal}$$

Check. 60 Cal is a reasonable result for most of the food value in an apple.

5.94 $$177 \text{ mL} \times \frac{1.0 \text{ g wine}}{1 \text{ mL}} \times \frac{0.106 \text{ g ethanol}}{1 \text{ g wine}} \times \frac{1 \text{ mol ethanol}}{46.1 \text{ g ethanol}} \times \frac{1367 \text{ kJ}}{1 \text{ mol ethanol}} \times \frac{1 \text{ Cal}}{4.184 \text{ kJ}}$$

$$= 133 = 1.3 \times 10^2 \text{ Cal}$$

Check. A "typical" 6 oz. glass of wine has 150–250 Cal, so this is a reasonable result. Note that alcohol is responsible for most of the food value of wine.

5.95   *Plan.* Use enthalpies of formation to calculate molar heat (enthalpy) of combustion using Hess's Law. Use molar mass to calculate heat of combustion per kg of hydrocarbon.   *Solve.*

Propyne: $C_3H_4(g) + 4 O_2(g) \rightarrow 3 CO_2(g) + 2 H_2O(g)$

(a) $\Delta H_{rxn}^\circ = 3(-393.5 \text{ kJ}) + 2(-241.82 \text{ kJ}) - (185.4 \text{ kJ}) - 4(0) = -1849.5$

$$= -1850 \text{ kJ/mol } C_3H_4$$

(b) $$\frac{-1849.5 \text{ kJ}}{1 \text{ mol } C_3H_4} \times \frac{1 \text{ mol } C_3H_4}{40.065 \text{ g } C_3H_4} \times \frac{1000 \text{ g } C_3H_4}{1 \text{ kg } C_3H_4} = -4.616 \times 10^4 \text{ kJ/kg } C_3H_4$$

Propylene: $C_3H_6(g) + 9/2 O_2(g) \rightarrow 3 CO_2(g) + 3 H_2O(g)$

(a) $\Delta H_{rxn}^\circ = 3(-393.5 \text{ kJ}) + 3(-241.82 \text{ kJ}) - (20.4 \text{ kJ}) - 9/2(0) = -1926.4$

$$= -1926 \text{ kJ/mol } C_3H_6$$

(b) $$\frac{-1926.4 \text{ kJ}}{1 \text{ mol } C_3H_6} \times \frac{1 \text{ mol } C_3H_6}{42.080 \text{ g } C_3H_6} \times \frac{1000 \text{ g } C_3H_6}{1 \text{ kg } C_3H_6} = -4.578 \times 10^4 \text{ kJ/kg } C_3H_6$$

Propane: $C_3H_8(g) + 5 O_2(g) \rightarrow 3 CO_2(g) + 4 H_2O(g)$

(a) $\Delta H_{rxn}^\circ = 3(-393.5 \text{ kJ}) + 4(-241.82 \text{ kJ}) - (-103.8 \text{ kJ}) - 5(0) = -2044.0$

$$= -2044 \text{ kJ/mol } C_3H_8$$

(b) $$\frac{-2044.0 \text{ kJ}}{1 \text{ mol } C_3H_8} \times \frac{1 \text{ mol } C_3H_8}{44.096 \text{ g } C_3H_8} \times \frac{1000 \text{ g } C_3H_8}{1 \text{ kg } C_3H_8} = -4.635 \times 10^4 \text{ kJ/kg } C_3H_8$$

(c)   These three substances yield nearly identical quantities of heat per unit mass, but propane is marginally higher than the other two.

5.96 $\Delta H^{\circ}_{rxn} = \Delta H^{\circ}_f\, CO_2(g) + 2\,\Delta H^{\circ}_f\, H_2O(g) - \Delta H^{\circ}_f\, CH_4(g) - 2\,\Delta H^{\circ}_f\, O_2(g)$

= $= -393.5\ kJ + 2(-241.82\ kJ) - (-74.8\ kJ) - 2(0)\ kJ = -802.3\ kJ$

$\Delta H^{\circ}_{rxn} = \Delta H^{\circ}_f\, CF_4(g) + 4\,\Delta H^{\circ}_f\, HF(g) - \Delta H^{\circ}_f\, CH_4(g) - 4\,\Delta H^{\circ}_f\, F_2(g)$

= $= -679.9\ kJ + 4(-268.61\ kJ) - (-74.8\ kJ) - 4(0)\ kJ = -1679.5\ kJ$

The second reaction is twice as exothermic as the first. The "fuel values" of hydrocarbons in a fluorine atmosphere are approximately twice those in an oxygen atmosphere. Note that the difference in ΔH° values for the two reactions is in the ΔH°_f for the products, because the ΔH°_f for the reactants is identical.

5.97   *Analyze/Plan.* Given population, Cal/person/day and kJ/mol glucose, calculate kg glucose/yr. Calculate kJ/yr, then kg/yr. 1 billion = 1×10^9. 365 day = 1 yr. 1 Cal = 1 kcal, 4.184 kJ = 1 kcal = 1 Cal.   *Solve.*

$$7.0 \times 10^9\ persons \times \frac{1500\ Cal}{person\text{-}day} \times \frac{365\ day}{1\ yr} \times \frac{4.184\ kJ}{1\ Cal} = 1.6035 \times 10^{16} = 1.6 \times 10^{16}\ kJ/yr$$

$$\frac{1.6035 \times 10^{16}\ kJ}{yr} \times \frac{1\ mol\ C_6H_{12}O_6}{2803\ kJ} \times \frac{180.2\ g\ C_6H_{12}O_6}{1\ mol\ C_6H_{12}O_6} \times \frac{1\ kg}{1000\ g} = 1.0 \times 10^{12}\ kg\ C_6H_{12}O_6/yr$$

Check. 1×10^{12} kg is 1 *trillion* kg of glucose.

5.98 (a) Use density to change L to g, molar mass to change g to mol, heat of combustion to change mol to kJ. Ethanol is C_2H_5OH, gasoline is C_8H_{18}. From Exercise 5.79 (c), heat of combustion of ethanol is –1234.8 kJ/mol.

$$1.0\ L\ C_2H_5OH \times \frac{1000\ mL}{1\ L} \times \frac{0.79\ g}{1\ mL} \times \frac{1\ mol\ C_2H_5OH}{46.07\ g} \times \frac{1234.8\ kJ}{1\ mol\ C_2H_5OH}$$

$$= 21{,}174 = 2.1 \times 10^4\ kJ/L\ C_2H_5OH$$

$$1.0\ L\ C_8H_{18} \times \frac{1000\ mL}{1\ L} \times \frac{0.70\ g}{1\ mL} \times \frac{1\ mol\ C_8H_{18}}{114.23\ g\ C_8H_{18}} \times \frac{5400\ kJ}{1\ mol\ C_8H_{18}}$$

$$= 33{,}091 = 3.3 \times 10^4\ kJ/L\ C_8H_{18}$$

(b) If density and heat of combustion of E85 are weighted averages of the values for the pure substances, than energy per liter E85 is also a weighted average of energy per liter for the two substances.

kJ/L E85 = 0.15(kJ/L C_8H_{18}) + 0.85(kJ/L C_2H_5OH)

kJ/L E85 = 0.15(33,091 kJ) + 0.85(21,174 kJ) = $22{,}962 = 2.3 \times 10^4$ kJ/L E85

(c) Whether comparing gal or L, all conversion factors for the two fuels cancel, so we can apply the energy ratio directly to the volume under consideration.

The energy ratio for E85 to gasoline is (22,962/33,091) = 0.6939 = 0.69

$$10\ gal\ gas \times \frac{kJ\ from\ E85}{0.6939\ kJ\ from\ gas} = 14.41 = 14\ gal\ E85$$

(d) If the E85/gasoline energy ratio is 0.69, the cost ratio must be 0.69 or less to "break-even" on price. 0.69($3.88) = $2.68/gal E85

Check. 10 gal gas($3.88/gal) = $39; 14.4 gal E85($2.68/gal) = $39.

Additional Exercises

5.99 Like the combustion of $H_2(g)$ and $O_2(g)$ described in Section 5.4, the reaction that inflates airbags is spontaneous after initiation. Spontaneous reactions are usually exothermic, $-\Delta H$. The airbag reaction occurs at constant atmospheric pressure, $\Delta H = q_p$; both are likely to be large and negative. When the bag inflates, work is done by the system on the surroundings, so the sign of w is negative.

5.100 Freezing is an exothermic process (the opposite of melting, which is clearly endothermic). When the system, the soft drink, freezes, it releases energy to the surroundings, the can. Some of this energy does the work of splitting the can. (When water freezes, it expands. It is specifically this expansion that does the work of splitting the can.)

5.101 (a) No work is done when the gas expands.

(b) No work is done because the evacuated flask is truly empty. There is no surrounding substance to be "pushed back."

(c) $\Delta E = q + w$. From part (b), no work is done when the gas expands. The flasks are perfectly insulated, so no heat flows. $\Delta E = 0 + 0 = 0$. The answer is a bit surprising, because a definite change occurred that required no work or heat transfer and consequently involved no energy change.

5.102 (a) $q = 0$, $w > 0$ (work done to system), $\Delta E > 0$

(b) Because the system (the gas) is losing heat, the sign of q is negative.

Two interpretations of the final state in (b) are possible. If the final state in (b) is identical to the final state in (a), $\Delta E(a) = \Delta E(b)$. If the final volumes are identical, case (b) requires either more (non-PV) work or heat input to compress the gas because some heat is lost to the surroundings. (The moral of this story is that the more energy lost by the system as heat, the greater the work on the system required to accomplish the desired change.)

Alternatively, if w is identical in the two cases and q is negative for case (b), then $\Delta E(b) < \Delta E(a)$. Assuming identical final volumes, the final temperature and pressure in (b) are slightly lower than those values in (a).

5.103 $\Delta E = q + w = +38.95 \text{ kJ} - 2.47 \text{ kJ} = +36.48 \text{ kJ}$

$\Delta H = q_p = +38.95 \text{ kJ}$

5.104 If a function sometimes depends on path, then it is simply not a state function. Enthalpy is a state function, so ΔH for the two pathways leading to the same change of state pictured in Figure 5.10 must be the same. However, q is not the same for the both. Our conclusion must be that $\Delta H \neq q$ for these pathways. The condition for $\Delta H = q_p$ (other than constant pressure) is that the only possible work on or by the system is pressure-volume work. Clearly, the work being done in this scenario is not pressure-volume work, so $\Delta H \neq q$, even though the two changes occur at constant pressure.

5.105 Find the heat capacity of 1.7×10^3 gal H_2O.

$$C_{H_2O} = 1.7 \times 10^3 \text{ gal } H_2O \times \frac{4 \text{ qt}}{1 \text{ gal}} \times \frac{1 \text{ L}}{1.057 \text{ qt}} \times \frac{1 \times 10^3 \text{ cm}^3}{1 \text{ L}} \times \frac{1 \text{ g}}{1 \text{ cm}^3} \times \frac{4.184 \text{ J}}{1 \text{ g-}^\circ\text{C}}$$

$$= 2.692 \times 10^7 \text{ J/}^\circ\text{C} = 2.7 \times 10^4 \text{ kJ/}^\circ\text{C; then,}$$

$$\frac{2.692 \times 10^7 \text{ J}}{1\,^\circ\text{C}} \times \frac{1\,\text{g-}^\circ\text{C}}{0.85\text{ J}} \times \frac{1\,\text{kg}}{1 \times 10^3\,\text{g}} \times \frac{1\,\text{brick}}{1.8\,\text{kg}} = 1.8 \times 10^4 \text{ or } 18{,}000 \text{ bricks}$$

Check. $(1.7 \times {\sim}16 \times 10^6)/({\sim}1.6 \times 10^3) \approx 17 \times 10^3$ bricks; the units are correct.

5.106   (a) $q_{Cu} = \dfrac{0.385\text{ J}}{\text{g-K}} \times 121.0 \text{ g Cu} \times (30.1\,^\circ\text{C} - 100.4\,^\circ\text{C}) = -3274.9 = -3.27 \times 10^3 \text{ J}$

The negative sign indicates the 3.27×10^3 J are lost by the Cu block.

(b) $q_{H_2O} = \dfrac{4.184\text{ J}}{\text{g-K}} \times 150.0 \text{ g H}_2\text{O} \times (30.1\,^\circ\text{C} - 25.1\,^\circ\text{C}) = 3138 = 3.1 \times 10^3 \text{ J}$

The positive sign indicates that 3.14×10^3 J are gained by the H_2O.

(c)   The difference in the heat lost by the Cu and the heat gained by the water is 3.275×10^3 J $- 3.138 \times 10^3$ J $= 0.137 \times 10^3$ J $= 1 \times 10^2$ J. The temperature change of the calorimeter is 5.0 °C. The heat capacity of the calorimeter in J/K is

$$0.137 \times 10^3 \text{ J} \times \frac{1}{5.0\,^\circ\text{C}} = 27.4 = 3 \times 10^1 \text{ J/K}.$$

Because q_{H_2O} is known to 1 decimal place, the difference has 1 decimal place and the result has 1 sig fig.

If the rounded results from (a) and (b) are used,

$$C_{calorimeter} = \frac{0.2 \times 10^3 \text{ J}}{5.0\,^\circ\text{C}} = 4 \times 10^1 \text{ J/K}.$$

(d) $q_{H_2O} = 3.275 \times 10^3 \text{ J} = \dfrac{4.184\text{ J}}{\text{g-K}} \times 150.0 \text{ g} \times (\Delta T)$

$\Delta T = 5.22\,^\circ\text{C}$; $T_f = 25.1\,^\circ\text{C} + 5.22\,^\circ\text{C} = 30.3\,^\circ\text{C}$

5.107   (a)   From the mass of benzoic acid that produces a certain temperature change, we can calculate the heat capacity of the calorimeter.

$$\frac{0.235 \text{ g benzoic acid}}{1.642\,^\circ\text{C change observed}} \times \frac{26.38\text{ kJ}}{1 \text{ g benzoic acid}} = 3.7755 = 3.78 \text{ kJ/}^\circ\text{C}$$

Now we can use this experimentally determined heat capacity with the data for caffeine.

$$\frac{1.525\,^\circ\text{C rise}}{0.265 \text{ g caffeine}} \times \frac{3.7755\text{ kJ}}{1\,^\circ\text{C}} \times \frac{194.2 \text{ g caffeine}}{1 \text{ mol caffeine}} = 4.22 \times 10^3 \text{ kJ/mol caffeine}$$

(b)   The overall uncertainty is approximately equal to the sum of the uncertainties due to each effect. The uncertainty in the mass measurement is 0.001/0.235 or 0.001/0.265, about 1 part in 235 or 1 part in 265. The uncertainty in the temperature measurements is 0.002/1.642 or 0.002/1.525, about 1 part in 820 or 1 part in 760. Thus the uncertainty in heat of combustion from each measurement is

$$\frac{4220}{235} = 18 \text{ kJ}; \quad \frac{4220}{265} = 16 \text{ kJ}; \quad \frac{4220}{820} = 5 \text{ kJ}; \quad \frac{4220}{760} = 6 \text{ kJ}$$

The sum of these uncertainties is 45 kJ. In fact, the overall uncertainty is less than this because independent errors in measurement do tend to partially cancel.

5.108   (a) $Mg(s) + 2 H_2O(l) \rightarrow Mg(OH)_2(s) + H_2(g)$

$\Delta H_{rxn}^{\circ} = \Delta H_f^{\circ} \ Mg(OH)_2(s) + \Delta H_f^{\circ} \ H_2(g) - 2 \ \Delta H_f^{\circ} \ H_2O(l) - \Delta H_f^{\circ} \ Mg(s)$

$= -924.7 \ kJ + 0 - 2(-285.83 \ kJ) - 0 = -353.04 = -353.0 \ kJ$

(b)   Use the specific heat of water, 4.184 J/g-°C, to calculate the energy required to heat the water. Use the density of water at 25 °C to calculate the mass of H_2O to be heated. (The change in density of H_2O going from 21 °C to 79 °C does not substantially affect the strategy of the exercise.) Then use the 'heat stoichiometry' in (a) to calculate mass of Mg(s) needed.

$$75 \ mL \times \frac{0.997 \ g \ H_2O}{mL} \times \frac{4.184 \ J}{g\text{-}^{\circ}C} \times 58 \ ^{\circ}C \times \frac{1 \ kJ}{1000 \ J} = 18.146 \ kJ = 18 \ kJ \ required$$

$$18.146 \ kJ \times \frac{1 \ mol \ Mg}{353.04 \ kJ} \times \frac{24.305 \ g \ Mg}{1 \ mol \ Mg} = 1.249 \ g = 1.2 \ g \ Mg \ needed$$

5.109   (a)   For comparison, balance the equations so that 1 mole of CH_4 is burned in each.

$CH_4(g) + O_2(g) \rightarrow C(s) + 2 H_2O(l)$

$CH_4(g) + 3/2 \ O_2(g) \rightarrow CO(g) + 2 H_2O(l)$

$CH_4(g) + 2 O_2(g) \rightarrow CO_2(g) + 2 H_2O(l)$

(b) $\Delta H_{rxn}^{\circ} = \Delta H_f^{\circ} \ C(s) + 2 \ \Delta H_f^{\circ} \ H_2O(l) - \Delta H_f^{\circ} \ CH_4(g) - \Delta H_f^{\circ} \ O_2(g)$

$= 0 + 2(-285.83 \ kJ) - (-74.8) - 0 = -496.9 \ kJ$

$\Delta H_{rxn}^{\circ} = \Delta H_f^{\circ} \ CO(g) + 2 \ \Delta H_f^{\circ} \ H_2O(l) - \Delta H_f^{\circ} \ CH_4(g) - 3/2 \ \Delta H_f^{\circ} \ O_2(g)$

$= (-110.5 \ kJ) + 2(-285.83 \ kJ) - (-74.8 \ kJ) - 3/2(0) = -607.4 \ kJ$

$\Delta H_{rxn}^{\circ} = \Delta H_f^{\circ} \ CO_2(g) + 2 \ \Delta H_f^{\circ} \ H_2O(l) - \Delta H_f^{\circ} \ CH_4(g) - 2 \ \Delta H_f^{\circ} \ O_2(g)$

$= -393.5 \ kJ + 2(-285.83 \ kJ) - (-74.8 \ kJ) - 2(0) = -890.4 \ kJ$

(c)   Assuming that $O_2(g)$ is present in excess, the reaction that produces $CO_2(g)$ represents the most negative ΔH per mole of CH_4 burned. More of the potential energy of the reactants is released as heat during the reaction to give products of lower potential energy. The reaction that produces $CO_2(g)$ is the most "downhill" in enthalpy.

5.110
$$\begin{array}{lll}
2[CH_4(g) + 2 \ O_2(g) \rightarrow CO_2(g) + 2 \ H_2O(l)] & \Delta H^{\circ} = 2(-890.3 \ kJ) \\
C_2H_6(g) \rightarrow C_2H_4(g) + H_2(g) & \Delta H^{\circ} = 136.3 \ kJ \\
1/2[4 \ CO_2(g) + 6 \ H_2O(l) \rightarrow 2 \ C_2H_6(g) + 7 \ O_2(g)] & \Delta H^{\circ} = 1/2(3120.8 \ kJ) \\
1/2[2 \ H_2O(l) \rightarrow 2 \ H_2(g) + O_2(g)] & \Delta H^{\circ} = 1/2(571.6 \ kJ) \\
\hline
2 \ CH_4(g) \rightarrow C_2H_4(g) + 2 \ H_2(g) & \Delta H^{\circ} = 201.9 \ kJ
\end{array}$$

5.111   For nitroethane:

$$\frac{1368 \text{ kJ}}{1 \text{ mol } C_2H_5NO_2} \times \frac{1 \text{ mol } C_2H_5NO_2}{75.072 \text{ g } C_2H_5NO_2} \times \frac{1.052 \text{ g } C_2H_5NO_2}{1 \text{ cm}^3} = 19.17 \text{ kJ/cm}^3$$

For ethanol:

$$\frac{1367 \text{ kJ}}{1 \text{ mol } C_2H_5OH} \times \frac{1 \text{ mol } C_2H_5OH}{46.069 \text{ g } C_2H_5OH} \times \frac{0.789 \text{ g } C_2H_5OH}{1 \text{ cm}^3} = 23.4 \text{ kJ/cm}^3$$

For methylhydrazine:

$$\frac{1307 \text{ kJ}}{1 \text{ mol } CH_6N_2} \times \frac{1 \text{ mol } CH_6N_2}{46.072 \text{ g } CH_6N_2} \times \frac{0.874 \text{ g } CH_6N_2}{1 \text{ cm}^3} = 24.8 \text{ kJ/cm}^3$$

Thus, **methylhydrazine** would provide the most energy per unit volume, with ethanol a close second.

5.112   (a) $3 C_2H_2(g) \rightarrow C_6H_6(l)$

$\Delta H^\circ_{rxn} = \Delta H^\circ_f\ C_6H_6(l) - 3\ \Delta H^\circ_f\ C_2H_2(g) = 49.0 \text{ kJ} - 3(226.77 \text{ kJ}) = -631.31 = -631.3 \text{ kJ}$

(b)   Because the reaction is exothermic (ΔH is negative), the reactant, 3 moles of $C_2H_2(g)$, has more enthalpy than the product, 1 mole of $C_6H_6(l)$.

(c)   The fuel value of a substance is the amount of heat (kJ) produced when 1 gram of the substance is burned. Calculate the molar heat of combustion (kJ/mol) and use this to find kJ/g of fuel.

$C_2H_2(g) + 5/2\ O_2(g) \rightarrow 2 CO_2(g) + H_2O(l)$

$\Delta H^\circ_{rxn} = 2\ \Delta H^\circ_f\ CO_2(g) + \Delta H^\circ_f\ H_2O(l) - \Delta H^\circ_f\ C_2H_2(g) - 5/2\ \Delta H^\circ_f\ O_2(g)$

$\quad = 2(-393.5 \text{ kJ}) + (-285.83 \text{ kJ}) - 226.77 \text{ kJ} - 5/2\ (0) = -1299.6 \text{ kJ/mol } C_2H_2$

$$\frac{-1299.6 \text{ kJ}}{1 \text{ mol } C_2H_2} \times \frac{1 \text{ mol } C_2H_2}{26.036 \text{ g } C_2H_2} = 49.916 = 50 \text{ kJ/g } C_2H_2$$

$C_6H_6(l) + 15/2\ O_2(g) \rightarrow 6 CO_2(g) + 3 H_2O(l)$

$\Delta H^\circ_{rxn} = 6\ \Delta H^\circ_f\ CO_2(g) + 3\ \Delta H^\circ_f\ H_2O(l) - \Delta H^\circ_f\ C_6H_6(l) - 15/2\ \Delta H^\circ_f\ O_2(g)$

$\quad = 6(-393.5 \text{ kJ}) + 3(-285.83 \text{ kJ}) - 49.0 \text{ kJ} - 15/2\ (0) = -3267.5 \text{ kJ/mol } C_6H_6$

$$\frac{-3267.5 \text{ kJ}}{1 \text{ mol } C_6H_6} \times \frac{1 \text{ mol } C_6H_6}{78.114 \text{ g } C_6H_6} = 41.830 = 42 \text{ kJ/g } C_6H_6$$

5.113   The reaction for which we want ΔH is:

$4 NH_3(l) + 3 O_2(g) \rightarrow 2 N_2(g) + 6 H_2O(g)$

Before we can calculate ΔH for this reaction, we must calculate ΔH_f for $NH_3(l)$.

We know that ΔH_f for $NH_3(g)$ is -46.2 kJ/mol, and that for $NH_3(l) \rightarrow NH_3(g)$, $\Delta H = 23.2$ kJ/mol

Thus, $\Delta H_{vap} = \Delta H_f\ NH_3(g) - \Delta H_f\ NH_3(l)$.

$\quad 23.2 \text{ kJ} = -46.2 \text{ kJ} - \Delta H_f\ NH_3(l); \Delta H_f\ NH_3(l) = -69.4 \text{ kJ/mol}$

Then for the overall reaction, the enthalpy change is:

$$\Delta H_{rxn} = 6\,\Delta H_f\,H_2O(g) + 2\,\Delta H_f\,N_2(g) - 4\,\Delta H_f\,NH_3(l) - 3\,\Delta H_f\,O_2$$

$$= 6(-241.82\text{ kJ}) + 2(0) - 4(-69.4\text{ kJ}) - 3(0) = -1173.3\text{ kJ}$$

$$\frac{-1173.3\text{ kJ}}{4\text{ mol NH}_3} \times \frac{1\text{ mol NH}_3}{17.0\text{ g NH}_3} = \frac{0.81\text{ g NH}_3}{1\text{ cm}^3} \times \frac{1000\text{ cm}^3}{1\text{ L}} = \frac{-1.4 \times 10^4\text{ kJ}}{\text{L NH}_3}$$

(This result has 2 significant figures because the density is expressed to 2 figures.)

$$2\,CH_3OH(l) + 3\,O_2(g) \rightarrow 2\,CO_2(g) + 4\,H_2O(g)$$

$$\Delta H = 2(-393.5\text{ kJ}) + 4(-241.82\text{ kJ}) - 2(-239\text{ kJ}) - 3(0) = -1276\text{ kJ}$$

$$\frac{-1276\text{ kJ}}{2\text{ mol CH}_3OH} \times \frac{1\text{ mol CH}_3OH}{32.04\text{ g CH}_3OH} \times \frac{0.792\text{ g CH}_3OH}{1\text{ cm}^3} \times \frac{1000\text{ cm}^3}{1\text{ L}} = \frac{-1.58 \times 10^4\text{ kJ}}{\text{L CH}_3OH}$$

In terms of heat obtained per unit volume of fuel, methanol is a slightly better fuel than liquid ammonia.

5.114 **1,3-butadiene**, C_4H_6, MM = 54.092 g/mol

(a) $C_4H_6(g) + 11/2\,O_2(g) \rightarrow 4\,CO_2(g) + 3\,H_2O(l)$

$$\Delta H_{rxn}^{\circ} = 4\,\Delta H_f^{\circ}\,CO_2(g) + 3\,\Delta H_f^{\circ}\,H_2O(l) - \Delta H_f^{\circ}\,C_4H_6(g) - 11/2\,\Delta H_f^{\circ}\,O_2(g)$$

$$= 4(-393.5\text{ kJ}) + 3(-285.83\text{ kJ}) - 111.9\text{ kJ} + 11/2\,(0) = -2543.4\text{ kJ/mol }C_4H_6$$

(b) $\dfrac{-2543.4\text{ kJ}}{1\text{ mol }C_4H_6} \times \dfrac{1\text{ mol }C_4H_6}{54.092\text{ g}} = 47.020 \rightarrow 47\text{ kJ/g}$

(c) $\%\,H = \dfrac{6(1.008)}{54.092} \times 100 = 11.18\%\ H$

1-butene, C_4H_8, MM = 56.108 g/mol

(a) $C_4H_8(g) + 6\,O_2(g) \rightarrow 4\,CO_2(g) + 4\,H_2O(l)$

$$\Delta H_{rxn}^{\circ} = 4\,\Delta H_f^{\circ}\,CO_2(g) + 4\,\Delta H_f^{\circ}\,H_2O(l) - \Delta H_f^{\circ}\,C_4H_8(g) - 6\,\Delta H_f^{\circ}\,O_2(g)$$

$$= 4(-393.5\text{ kJ}) + 4(-285.83\text{ kJ}) - 1.2\text{ kJ} - 6(0) = -2718.5\text{ kJ/mol }C_4H_8$$

(b) $\dfrac{-2718.5\text{ kJ}}{1\text{ mol }C_4H_8} \times \dfrac{1\text{ mol }C_4H_8}{56.108\text{ g }C_4H_8} = 48.451 \rightarrow 48\text{ kJ/g}$

(c) $\%\,H = \dfrac{8(1.008)}{56.108} \times 100 = 14.37\%\ H$

n-butane, $C_4H_{10}(g)$, MM = 58.124 g/mol

(a) $C_4H_{10}(g) + 13/2\,O_2(g) \rightarrow 4\,CO_2(g) + 5\,H_2O(l)$

$$\Delta H_{rxn}^{\circ} = 4\,\Delta H_f^{\circ}\,CO_2(g) + 5\,\Delta H_f^{\circ}\,H_2O(l) - \Delta H_f^{\circ}\,C_4H_{10}(g) - 13/2\,\Delta H_f^{\circ}\,O_2(g)$$

$$= 4(-393.5\text{ kJ}) + 5(-285.83\text{ kJ}) - (-124.7\text{ kJ}) - 13/2(0)$$

$$= -2878.5\text{ kJ/mol }C_4H_{10}$$

(b) $\dfrac{-2878.5 \text{ kJ}}{1 \text{ mol } C_4H_{10}} \times \dfrac{1 \text{ mol } C_4H_{10}}{58.124 \text{ g } C_4H_{10}} = 49.523 \rightarrow 50 \text{ kJ/g}$

(c) $\% \text{ H} = \dfrac{10(1.008)}{58.124} \times 100 = 17.34\% \text{ H}$

(d)   It is certainly true that as the mass % H increases, the fuel value (kJ/g) of the hydrocarbon increases, given the same number of C atoms. A graph of the data in parts (b) and (c) (see below) suggests that mass % H and fuel value are directly proportional when the number of C atoms is constant.

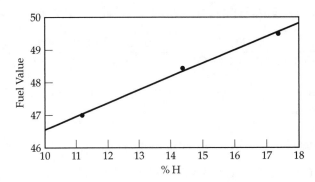

5.115 $\Delta E_p = m \times g \times d$. Be careful with units. $1 \text{ J} = 1 \text{ kg-m}^2/\text{s}^2$

$201 \text{ lb} \times \dfrac{1 \text{ kg}}{2.205 \text{ lb}} \times \dfrac{9.81 \text{ m}}{\text{s}^2} \times \dfrac{45 \text{ ft}}{\text{time}} \times \dfrac{1 \text{ yd}}{3 \text{ ft}} \times \dfrac{1 \text{ m}}{1.0936 \text{ yd}} \times 20 \text{ times}$

$= 2.453 \times 10^5 \text{ kg-m}^2/\text{s}^2 = 2.453 \times 10^5 \text{ J} = 2.5 \times 10^2 \text{ kJ}$

$1 \text{ Cal} = 1 \text{ kcal} = 4.184 \text{ kJ}$

$2.453 \times 10^2 \text{ kJ} \times \dfrac{1 \text{ Cal}}{4.184 \text{ kJ}} = 58.63 = 59 \text{ Cal}$

No, if all work is used to increase the man's potential energy, 20 rounds of stair-climbing will not compensate for one extra order of 245 Cal fries. In fact, more than 59 Cal of work will be required to climb the stairs, because some energy is required to move limbs and some energy will be lost as heat.

5.116   *Plan.* Use dimensional analysis to calculate the amount of solar energy supplied per m^2 in 1 h. Use stoichiometry to calculate the amount of plant energy used to produce sucrose per m^2 in 1 h. Calculate the ratio of energy for sucrose to total solar energy, per m^2 per h.

Solve. $1 \text{ W} = 1 \text{ J/s}$, $1 \text{ kW} = 1 \text{ kJ/s}$

$\dfrac{1.0 \text{ kW}}{\text{m}^2} = \dfrac{1.0 \text{ kJ/s}}{\text{m}^2} = \dfrac{1.0 \text{ kJ}}{\text{m}^2\text{-s}} \times \dfrac{60 \text{ s}}{1 \text{ min}} \times \dfrac{60 \text{ min}}{1 \text{ h}} = \dfrac{3.6 \times 10^3 \text{ kJ}}{\text{m}^2\text{-h}}$

$\dfrac{5645 \text{ kJ}}{\text{mol sucrose}} \times \dfrac{1 \text{ mol sucrose}}{342.3 \text{ g sucrose}} \times \dfrac{0.20 \text{ g sucrose}}{\text{m}^2\text{-h}} = 3.298 = 3.3 \text{ kJ} / \text{m}^2\text{-h for sucrose production}$

$\dfrac{3.298 \text{ kJ for sucrose}}{3.6 \times 10^3 \text{ kJ total solar}} \times 100 = 0.092\% \text{ sunlight used to produce sucrose}$

5.117   (a) $6 CO_2(g) + 6 H_2O(l) \rightarrow C_6H_{12}O_6(s) + 6 O_2(g)$, $\Delta H° = 2803$ kJ

This is the reverse of the combustion of glucose (Section 5.8 and Exercise 5.89), so $\Delta H° = -(-2803)$ kJ $= +2803$ kJ.

$$\frac{5.5 \times 10^{16} \text{ g } CO_2}{\text{yr}} \times \frac{1 \text{ mol } CO_2}{44.01 \text{ g } CO_2} \times \frac{2803 \text{ kJ}}{6 \text{ mol } CO_2} = 5.838 \times 10^{17} = 5.8 \times 10^{17} \text{ kJ}$$

(b) $1 W = 1 J/s$; $1 W\text{-}s = 1 J$

$$\frac{5.838 \times 10^{17} \text{ kJ}}{\text{yr}} \times \frac{1000 \text{ J}}{\text{kJ}} \times \frac{1 \text{ yr}}{365 \text{ d}} \times \frac{1 \text{ d}}{24 \text{ h}} \times \frac{1 \text{ h}}{60 \text{ min}} \times \frac{1 \text{ min}}{60 \text{ s}} \times \frac{1 W\text{-}s}{J}$$

$$\times \frac{1 \text{ MW}}{1 \times 10^6 W} = 1.851 \times 10^7 \text{ MW} = 1.9 \times 10^7 \text{ MW}$$

$$1.9 \times 10^7 \text{ MW} \times \frac{1 \text{ plant}}{10^3 \text{ MW}} = 1.9 \times 10^4 = 19,000 \text{ nuclear power plants}$$

Integrative Exercises

5.118   (a) $mi/h \rightarrow m/s$

$$1050 \frac{mi}{h} \times \frac{1.6093 \text{ km}}{1 \text{ mi}} \times \frac{1000 \text{ m}}{1 \text{ km}} \times \frac{1 \text{ h}}{3600 \text{ s}} = 469.38 = 469.4 \text{ m/s}$$

(b)   Find the mass of one N_2 molecule in kg.

$$\frac{28.0134 \text{ g } N_2}{1 \text{ mol}} \times \frac{1 \text{ mol}}{6.022 \times 10^{23} \text{ molecules}} \times \frac{1 \text{ kg}}{1000 \text{ g}} = 4.6518 \times 10^{-26}$$

$$= 4.652 \times 10^{-26} \text{ kg}$$

$$E_k = 1/2 \text{ mv}^2 = 1/2 \times 4.6518 \times 10^{-26} \text{ kg} \times (469.38 \text{ m/s})^2$$

$$= 5.1244 \times 10^{-21} \frac{\text{kg-m}^2}{\text{s}^2} = 5.124 \times 10^{-21} \text{ J}$$

(c) $$\frac{5.1244 \times 10^{-21} \text{ J}}{\text{molecule}} \times \frac{6.022 \times 10^{23} \text{ molecules}}{1 \text{ mol}} = 3086 \text{ J/mol} = 3.086 \text{ kJ/mol}$$

5.119   (a) $E_p = mgd = 52.0 \text{ kg} \times 9.81 \text{ m/s}^2 \times 10.8 \text{ m} = 5509.3 \text{ J} = 5.51 \text{ kJ}$

(b) $$E_k = 1/2 \text{ mv}^2; \; v = (2E_k/m)^{1/2} = \left(\frac{2 \times 5509.3 \text{ kg-m}^2/\text{s}^2}{52.0 \text{ kg}} \right)^{1/2} = 14.6 \text{ m/s}$$

(c)   Yes, the diver does work on entering (pushing back) the water in the pool.

5.120   (a) $CH_4(g) + 2 O_2(g) \rightarrow CO_2(g) + 2 H_2O(l)$

$\Delta H° = \Delta H_f° \; CO_2(g) + 2 \; \Delta H_f° \; H_2O(l) - \Delta H_f° \; CH_4(g) - 2 \; \Delta H_f° \; O_2(g)$

$= -393.5$ kJ $+ 2(-285.83$ kJ$) - (-74.8$ kJ$) - 2(0) = -890.36 = -890.4$ kJ/mol CH_4

The minus sign indicates that 890.4 kJ are produced per mole of CH_4 burned.

$$\frac{890.36 \text{ kJ}}{\text{mol } CH_4} \times \frac{1000 \text{ J}}{1 \text{ kJ}} \times \frac{1 \text{ mol}}{6.022 \times 10^{23} \text{ molecules } CH_4} = 1.4785 \times 10^{-18}$$

$$= 1.479 \times 10^{-18} \text{ J/molecule}$$

(b) $1 eV = 96.485 kJ/mol$

$$8 keV \times \frac{1000 eV}{1 keV} \times \frac{96.485 kJ}{eV\text{-mol}} \times \frac{1 mol}{6.022 \times 10^{23}} \times \frac{1000 J}{kJ} = 1.282 \times 10^{-15} = 1 \times 10^{-15} J/X\text{-ray}$$

The energy produced by the combustion of 1 molecule of $CH_4(g)$ is much smaller than the energy of the X-ray.

5.121   (a,b) $Ag^+(aq) + Li(s) \rightarrow Ag(s) + Li^+(aq)$

$\Delta H° = \Delta H_f° \ Li^+(aq) - \Delta H_f° \ Ag^+(aq)$

$= -278.5 kJ - 105.90 kJ = -384.4 kJ$

$Fe(s) + 2 Na^+(aq) \rightarrow Fe^{2+}(aq) + 2 Na(s)$

$\Delta H° = \Delta H_f° \ Fe^{2+}(aq) - 2 \ \Delta H_f° \ Na^+(aq)$

$= -87.86 kJ - 2(-240.1 kJ) = +392.3 kJ$

$2 K(s) + 2 H_2O(l) \rightarrow 2 KOH(aq) + H_2(g)$

$\Delta H° = 2 \ \Delta H_f° \ KOH(aq) - 2 \ \Delta H_f° \ H_2O(l)$

$= 2(-482.4 kJ) - 2(-285.83 kJ) = -393.1 kJ$

(c)   Exothermic reactions are more likely to be favored, so we expect the first and third reactions be favored.

(d)   In the activity series of metals, Table 4.5, any metal can be oxidized by the cation of a metal below it on the table.

Ag^+ is below Li, so the first reaction will occur.

Na^+ is above Fe, so the second reaction will not occur.

H^+ (formally in H_2O) is below K, so the third reaction will occur.

These predictions agree with those in part (c).

5.122   (a) $\Delta H° = \Delta H_f° \ NaNO_3(aq) + \Delta H_f° \ H_2O(l) - \Delta H_f° \ HNO_3(aq) - \Delta H_f° \ NaOH(aq)$

$\Delta H° = -446.2 kJ - 285.83 kJ - (-206.6 kJ) - (-469.6 kJ) = -55.8 kJ$

$\Delta H° = \Delta H_f° \ NaCl(aq) + \Delta H_f° \ H_2O(l) - \Delta H_f° \ HCl(aq) - \Delta H_f° \ NaOH(aq)$

$\Delta H° = -407.1 kJ - 285.83 kJ - (-167.2 kJ) - (-469.6 kJ) = -56.1 kJ$

$\Delta H° = \Delta H_f° \ NH_3(aq) + \Delta H_f° \ Na^+(aq) + \Delta H_f° \ H_2O(l) - \Delta H_f° \ NH_4^+(aq) - \Delta H_f° \ NaOH(aq)$

$= -80.29 kJ - 240.1 kJ - 285.83 kJ - (-132.5 kJ) - (-469.6 kJ) = -4.1 kJ$

(b) $H^+(aq) + OH^-(aq) \rightarrow H_2O(l)$

(c)   The $\Delta H°$ values for the first two reactions are nearly identical, -55.8 kJ and -56.1 kJ. The spectator ions by definition do not change during the course of a reaction, so $\Delta H°$ is the enthalpy change for the net ionic equation. Because the first two reactions have the same net ionic equation, it is not surprising that they have the same $\Delta H°$.

(d)   Strong acids are more likely than weak acids to donate H^+. The neutralizations of the two strong acids are energetically favorable, whereas the neutralization of $NH_4^+(aq)$ is significantly less favorable. $NH_4^+(aq)$ is probably a weak acid.

5.123   (a)   mol Cu $= M \times L = 1.00\ M \times 0.0500\ L = 0.0500$ mol

g $=$ mol $\times$ MM $= 0.0500 \times 63.546 = 3.1773 = 3.18$ g Cu

(b)   The precipitate is copper(II) hydroxide, $Cu(OH)_2$.

(c) $CuSO_4(aq) + 2\ KOH(aq) \rightarrow Cu(OH)_2(s) + K_2SO_4(aq)$, complete

$Cu^{2+}(aq) + 2\ OH^-(aq) \rightarrow Cu(OH)_2(s)$, net ionic

(d)   The temperature of the calorimeter rises, so the reaction is exothermic and the sign of q is negative.

$$q = -6.2\ ^\circ C \times 100\ g \times \frac{4.184\ J}{1\ g\text{-}^\circ C} = -2.6 \times 10^3\ J = -2.6\ kJ$$

The reaction as carried out involves only 0.050 mol of $CuSO_4$ and the stoichiometrically equivalent amount of KOH. On a molar basis,

$$\Delta H = \frac{-2.6\ kJ}{0.050\ mol} = -52\ kJ \text{ for the reaction as written in part (c)}$$

5.124   (a) $AgNO_3(aq) + NaCl(aq) \rightarrow NaNO_3(aq) + AgCl(s)$

net ionic equation: $Ag^+(aq) + Cl^-(aq) \rightarrow AgCl(s)$

$\Delta H^o = \Delta H_f^\circ\ AgCl(s) - \Delta H_f^\circ\ Ag^+(aq) - \Delta H_f^\circ Cl^-(aq)$

$\Delta H^\circ = -127.0\ kJ - (105.90\ kJ) - (-167.2\ kJ) = -65.7\ kJ$

(b) ΔH° for the complete molecular equation will be the same as ΔH° for the net ionic equation. $Na^+(aq)$ and $NO_3^-(aq)$ are spectator ions; they appear on both sides of the chemical equation. Because the overall enthalpy change is the enthalpy of the products minus the enthalpy of the reactants, the contributions of the spectator ions cancel.

(c) $\Delta H^\circ = \Delta H_f^\circ\ NaNO_3(aq) + \Delta H_f^\circ\ AgCl(s) - \Delta H_f^\circ\ AgNO_3(aq) - \Delta H_f^\circ\ NaCl(aq)$

$\Delta H_f^\circ\ AgNO_3(aq) = \Delta H_f^\circ\ NaNO_3(aq) + \Delta H_f^\circ\ AgCl(s) - \Delta H_f^\circ\ NaCl(aq) - \Delta H^o$

$\Delta H_f^\circ\ AgNO_3(aq) = -446.2\ kJ + (-127.0\ kJ) - (-407.1\ kJ) - (-65.7\ kJ)$

$\Delta H_f^\circ\ AgNO_3(aq) = -100.4\ kJ/mol$

5.125   (a) 21.83 g $CO_2 \times \dfrac{1\ mol\ CO_2}{44.01\ g\ CO_2} \times \dfrac{1\ mol\ C}{1\ mol\ CO_2} \times \dfrac{12.01\ g\ C}{1\ mol\ C} = 5.9572 = 5.957$ g C

4.47 g $H_2O \times \dfrac{1\ mol\ H_2O}{18.02\ g\ H_2O} \times \dfrac{2\ mol\ H}{1\ mol\ H_2O} \times \dfrac{1.008\ g\ H}{mol\ H} = 0.5001 = 0.500$ g H

The sample mass is $(5.9572 + 0.5001) = 6.457$ g

(b) $5.957 \text{ g C} \times \dfrac{1 \text{ mol C}}{12.01 \text{ g C}} = 0.4960 \text{ mol C}; \quad 0.4960/0.496 = 1$

$0.500 \text{ g H} \times \dfrac{1 \text{ mol H}}{1.008 \text{ g H}} = 0.496 \text{ mol H}; \quad 0.496/0.496 = 1$

The empirical formula of the hydrocarbon is CH.

(c)   Calculate "ΔH_f°" for 6.457 g of the sample.

$6.457 \text{ g sample} + O_2(g) \rightarrow 21.83 \text{ g } CO_2(g) + 4.47 \text{ g } H_2O(g), \; \Delta H_{comb}^{\circ} = -311 \text{ kJ}$

$\Delta H_{comb}^{\circ} = \Delta H_f^{\circ} \, CO_2(g) + \Delta H_f^{\circ} \, H_2O(g) - \Delta H_f^{\circ} \, \text{sample} - \Delta H_f^{\circ} \, O_2(g)$

$\Delta H_f^{\circ} \, \text{sample} = \Delta H_f^{\circ} \, CO_2(g) + \Delta H_f^{\circ} \, H_2O(g) - \Delta H_{comb}^{\circ}$

$\Delta H_f^{\circ} \, CO_2(g) = 21.83 \text{ g } CO_2 \times \dfrac{1 \text{ mol } CO_2}{44.01 \text{ g } CO_2} \times \dfrac{-393.5 \text{ kJ}}{\text{mol } CO_2} = -195.185 = -195.2 \text{ kJ}$

$\Delta H_f^{\circ} \, H_2O(g) = 4.47 \text{ g } H_2O \times \dfrac{1 \text{ mol } H_2O}{18.02 \text{ g } H_2O} \times \dfrac{-241.82 \text{ kJ}}{\text{mol } H_2O} = -59.985 = -60.0 \text{ kJ}$

$\Delta H_f^{\circ} \, \text{sample} = -195.185 \text{ kJ} - 59.985 \text{ kJ} - (-311 \text{ kJ}) = 55.83 = 56 \text{ kJ}$

$H_f^{\circ} = \dfrac{55.83 \text{ kJ}}{6.457 \text{ g sample}} \times \dfrac{13.02 \text{ g}}{\text{CH unit}} = 112.6 = 1.1 \times 10^2 \text{ kJ/CH unit}$

(d)   The hydrocarbons in Appendix C with empirical formula CH are C_2H_2 and C_6H_6.

substance	ΔH_f° / mol	ΔH_f° / CH unit
$C_2H_2(g)$	226.7 kJ	113.4 kJ
$C_6H_6(g)$	82.9 kJ	13.8 kJ
$C_6H_6(l)$	49.0 kJ	8.17 kJ
sample		1.1×10^2 kJ

The calculated value of ΔH_f°/CH unit for the sample is a good match with acetylene, $C_2H_2(g)$.

5.126   (a) $CH_4(g) \rightarrow C(g) + 4 \text{ H}(g)$ (i) reaction given

           $CH_4(g) \rightarrow C(s) + 2 \text{ H}_2(g)$ (ii) reverse of formation

The differences are: the state of C in the products; the chemical form, atoms, or diatomic molecules, of H in the products.

(b)   i. $\Delta H^{\circ} = \Delta H_f^{\circ} \, C(g) + 4 \, \Delta H_f^{\circ} \, H(g) - \Delta H_f^{\circ} \, CH_4(g)$

              $= 718.4 \text{ kJ} + 4(217.94) \text{ kJ} - (-74.8) \text{ kJ} = 1665.0 \text{ kJ}$

     ii. $\Delta H^{\circ} = \Delta H_f^{\circ} \, CH_4 = -(-74.8) \text{ kJ} = 74.8 \text{ kJ}$

The rather large difference in $\Delta H°$ values is due to the enthalpy difference between isolated gaseous C atoms and the orderly, bonded array of C atoms in graphite, C(s), as well as the enthalpy difference between isolated H atoms and H_2 molecules. In other words, it is due to the difference in the enthalpy stored in chemical bonds in C(s) and $H_2(g)$ versus the corresponding isolated atoms.

(c) $CH_4(g) + 4\,F_2(g) \rightarrow CF_4(g) + 4\,HF(g)$ $\Delta H° = -1679.5$ kJ

The $\Delta H°$ value for this reaction was calculated in Solution 5.96.

$$3.45 \text{ g } CH_4 \times \frac{1 \text{ mol } CH_4}{16.04 \text{ g } CH_4} \times 0.21509 = 0.215 \text{ mol } CH_4$$

$$1.22 \text{ g } F_2 \times \frac{1 \text{ mol } F_2}{38.00 \text{ g } F_2} = 0.03211 = 0.0321 \text{ mol } F_2$$

There are fewer mol F_2 than CH_4, but 4 mol F_2 are required for every 1 mol of CH_4 reacted, so clearly F_2 is the limiting reactant.

$$0.03211 \text{ mol } F_2 \times \frac{-1679.5 \text{ kJ}}{4 \text{ mol } F_2} = -13.48 = -13.5 \text{ kJ heat evolved}$$

5.127   (a) $C_2H_5OH(l) + 3\,O_2(g) \rightarrow 2\,CO_2(g) + 3\,H_2O(l)$

       (b) $\Delta H_{rxn}° = 2\,\Delta H_f°\,CO_2(g) + 3\,\Delta H_f°\,H_2O(l) - \Delta H_f°\,C_2H_5OH(l) - 3\,\Delta H_f°\,O_2(g)$

                $= 2(-393.5) \text{ kJ} + 3(-285.83 \text{ kJ}) - (-277.7 \text{ kJ}) - 3(0) = -1366.79 = -1367 \text{ kJ}$

       (c)   oz beer $\rightarrow$ mL beer $\rightarrow$ mL ethanol $\rightarrow$ g ethanol

$$12 \text{ oz beer} \times \frac{29.5735 \text{ mL}}{\text{oz}} \times \frac{0.042 \text{ mL ethanol}}{\text{mL beer}} \times \frac{0.789 \text{ g ethanol}}{\text{mL}} =$$
$$11.76 = 12 \text{ g ethanol}$$

       (d)   The metabolism reaction in part (a) produces 1367 kJ/mol ethanol, part (b).

$$11.76 \text{ g ethanol} \times \frac{1 \text{ mol ethanol}}{46.07 \text{ g ethanol}} \times \frac{1367 \text{ kJ}}{\text{mol ethanol}} \times \frac{1 \text{ Cal}}{4.184 \text{ kJ}} = 83.40 = 83 \text{ Cal}$$

       (e)   Percent Cal from ethanol

$$\frac{83.40 \text{ Cal from ethanol}}{110 \text{ Cal total}} \times 100 = 75.82 = 76\%$$

6 Electronic Structure of Atoms

Visualizing Concepts

6.1 (a) *Analyze/Plan.* We are given the speed of sound in dry air and the frequency of the lowest audible wave. We must find the wavelength of the sound waves. For any wave, speed equals wavelength times frequency. *Solve.*

$$\text{Speed} = \lambda \times \nu; \quad \lambda = \text{speed}/\nu; \quad 20\,\text{Hz} = 20\,\text{s}^{-1} = 20/\text{s}$$

$$\lambda = \text{speed}/\nu; \quad \frac{343\,\text{m}}{\text{s}} \times \frac{\text{s}}{20} = 17.15 = 17\,\text{m} \text{ (to 1 sig fig, 20 m)}$$

(b) *Analyze/Plan.* We are asked to find the frequency of electromagnetic radiation with this wavelength. All electromagnetic radiation in a vacuum has the speed 2.998×10^{8} m/s; the symbol for the speed of light is "c". $\nu = c/\lambda$. *Solve.*

$$\nu = c/\lambda; \quad \frac{2.998 \times 10^{8}\,\text{m}}{\text{s}} \times \frac{1}{17.15\,\text{m}} = 1.7 \times 10^{7}\,\text{s}^{-1} = 1.7 \times 10^{7}\,\text{Hz}$$

$$1.7 \times 10^{7}\,\text{Hz} \times \frac{1\,\text{MHz}}{1 \times 10^{6}\,\text{Hz}} = 17\,\text{MHz} \text{ (20 MHz, to 1 sig fig)}$$

(c) According to Figure 6.4, this would be radio frequency radiation.

6.2 Given: 2450 MHz radiation. Hz = s^{-1}, unit of frequency. M = 1×10^{6}; 2450×10^{6} Hz = 2.45×10^{9} Hz = $2.45 \times 10^{9}\,\text{s}^{-1}$.

(a) Find $2.45 \times 10^{9}\,\text{s}^{-1}$ on the frequency axis of Figure 6.4. The wavelength that corresponds to this frequency is approximately $1 \times 10^{-1} = 0.1$ m or 10 cm.

(b) No, visible radiation has wavelengths of 4×10^{-7} to 7×10^{-7} m, much shorter than 0.1 m.

(c) Energy and wavelength are inversely proportional. Photons of the longer 0.1 m radiation have less energy than visible photons.

(d) Radiation of 0.1 m is in the low energy portion of the microwave region. The appliance is probably a microwave oven. (Appliances with heating elements that glow red or orange give off wavelengths in the visible or the near visible portion of the infrared. The 0.1 m wavelength is too long to belong to these appliances.)

6.3 (a) By inspection, wave (a) has the longer wavelength.

(b) Wave (b) has the higher frequency because it has the shorter wavelength.

(c) Wave (b) has a higher energy because it has a higher frequency (and shorter wavelength).

6.4 (a) (iii) *Betelgeuse* < (i) Sun < (ii) *Rigel*.

(b) Black body radiation

6.5   (a)   Increase. The rainbow has shorter wavelength blue light on the inside and longer wavelength red light on the outside. (See Figure 6.4.)

       (b)   Decrease. Wavelength and frequency are inversely related. Wavelength increases so frequency decreases going from the inside to the outside of the rainbow.

6.6   (a) $n = 1$ to $n = 4$ (b) $n = 1$ to $n = 2$

       (c)   Wavelength and energy are inversely proportional; the smaller the energy, the longer the wavelength. In order of increasing wavelength (and decreasing energy): (iii) $n = 2$ to $n = 4$ < (iv) $n = 1$ to $n = 3$ < (ii) $n = 2$ to $n = 3$ < (i) $n = 1$ to $n = 2$

6.7   (a)   Wave (iii) corresponds to transition C. Transition C represents the smallest energy change, which will emit a photon with the longest wavelength.

       (b)   *Analyze/Plan.* Use Equation 6.6, which describes energy changes in the hydrogen atom, to calculate the energy of the photon emitted for each transition. *Solve.*

$$A: n_i = 2, n_f = 1; \Delta E = -2.18 \times 10^{-18} \text{ J} \left[\frac{1}{n_f^2} - \frac{1}{n_i^2} \right] = -2.18 \times 10^{-18} \text{ J} (1 - 1/4) = -1.635 \times 10^{-18} \text{ J}$$

$$-1.64 \times 10^{-18} \text{ J}$$

$$B: n_i = 3, n_f = 2; \Delta E = -2.18 \times 10^{-18} \text{ J} (1/4 - 1/9) = -3.028 \times 10^{-19} = -3.03 \times 10^{-19} \text{ J}$$

$$C: n_i = 4, n_f = 3; \Delta E = -2.18 \times 10^{-18} \text{ J} (1/9 - 1/16) = -1.0597 \times 10^{-19} = -1.06 \times 10^{-19} \text{ J}$$

The negative signs for ΔE indicate that the photons are emitted.

       (c) $$A: \ \lambda = hc/E = \frac{6.626 \times 10^{-34} \text{ J-s} \times 2.998 \times 10^8 \text{ m/s}}{1.635 \times 10^{-18} \text{ J}} = 1.22 \times 10^{-7} \text{ m}$$

$$B: \ \lambda = hc/E = \frac{6.626 \times 10^{-34} \text{ J-s} \times 2.998 \times 10^8 \text{ m/s}}{3.028 \times 10^{-19} \text{ J}} = 6.56 \times 10^{-7} \text{ m}$$

$$C: \ \lambda = hc/E = \frac{6.626 \times 10^{-34} \text{ J-s} \times 2.998 \times 10^8 \text{ m/s}}{1.0597 \times 10^{-19} \text{ J}} = 1.87 \times 10^{-6} \text{ m}$$

According to Figure 6.4, visible light has wavelengths from 4×10^{-7} m to 7.5×10^{-7} m (400–750 nm). Transition B emits photons of visible light.

6.8   (a) $\psi^2(x)$ will be positive or zero at all values of x and have two maxima with larger magnitudes than the maximum in $\psi(x)$.

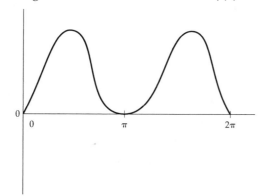

(b)   The greatest probability of finding the electron is at the two maxima in $\psi^2(x)$ at $x = \pi/2$ and $3\pi/2$.

(c)   There is zero probability of finding the electron at $x = \pi$. This value is called a node.

6.9   (a) $l = 1$

      (b) $3p_y$ ($n = 3$ shell, dumbbell shape, node at the nucleus, oriented along the y-axis)

      (c)   (iii) (m_l indicates the orientation of an orbital)

6.10   (a)   The smallest possible n value for a d orbital is 3.

      (b)   For a d orbital, $l = 2$.

      (c)   For a d orbital, the largest possible value of m_l is 2 (possible values are 2, 1, 0, –1, or –2)

      (d)   The probability density goes to zero along the xy and xz planes.

6.11   Configuration B is the correct one for a nitrogen atom in its ground state.

      Configuration C violates the Pauli exclusion principle, that no two electrons can have the same set of quantum numbers. That is, two electrons in the same orbital must have opposite spins.

      Configurations A, C, and D violate the Hund's rule, that electron spin is maximized. That is, electrons will occupy degenerate orbitals singly when possible and they will have the same spin.

6.12   (a)   Group 7A or 17, the halogens, the column second from the right

      (b)   Group 5A or 15

      (c)   Gallium, atomic number 31, at the intersection of row 4 and group 3A or 13

      (d)   All of the B groups, groups 3–12, in the middle of the major part of the table, not including the two rows of f-block elements

The Wave Nature of Light (Section 6.1)

6.13   (a)   Meters (m)   (b) $1/\text{seconds} \, (s^{-1})$ (c)   meters/second ($m\text{-}s^{-1}$ or m/s)

6.14   (a)   Wavelength (λ) and frequency (ν) are inversely proportional; the proportionality constant is the speed of light (c). $\nu = c/\lambda$.

      (b)   Light in the 210–230 nm range is in the ultraviolet region of the spectrum. These wavelengths are slightly shorter than the 400 nm short-wavelength boundary of the visible region.

6.15   (a)   True.

      (b)   False. Ultraviolet light has shorter wavelengths than visible light. [See Solution 6.14(b).]

      (c)   False. X-rays travel at the same speed as microwaves. (X-rays and microwaves are both electromagnetic radiation.)

      (d)   False. Electromagnetic radiation and sound waves travel at different speeds. (Sound is not a form of electromagnetic radiation.)

6.16 (a) False. The frequency of radiation decreases as the wavelength increases.

    (b) True.

    (c) False. Infrared light has lower frequencies than visible light.

    (d) False. The glow from a fireplace and the energy within a microwave oven are both forms of electromagnetic radiation. (A foghorn blast is a form of sound waves, which are not accompanied by oscillating electric and magnetic fields.)

6.17 *Analyze/Plan.* Use the electromagnetic spectrum in Figure 6.4 to determine the wavelength of each type of radiation; put them in order from shortest to longest wavelength. *Solve.*

Wavelength of X-rays < ultraviolet < green light < red light < infrared < radio waves

Check. These types of radiation should read from left to right on Figure 6.4.

6.18 Wavelength of (a) gamma rays < (d) yellow (visible) light < (e) red (visible) light < (b) 93.1 MHz FM (radio) waves < (c) 680 kHz or 0.680 MHz AM (radio) waves

6.19 *Analyze/Plan.* These questions involve relationships between wavelength, frequency, and the speed of light. Manipulate the equation $v = c/\lambda$ to obtain the desired quantities, paying attention to units. *Solve.*

    (a) $v = c/\lambda$; $\dfrac{2.998 \times 10^8 \text{ m}}{\text{s}} \times \dfrac{1}{10 \text{ μm}} \times \dfrac{1 \text{ μm}}{1 \times 10^{-6} \text{ m}} = 3.0 \times 10^{13} \text{ s}^{-1}$

    (b) $\lambda = c/v$; $\dfrac{2.998 \times 10^8 \text{ m}}{s} \times \dfrac{1 \text{ s}}{5.50 \times 10^{14}} = 5.45 \times 10^{-7} \text{ m}$ (545 nm)

    (c) The radiation in (b) is in the visible region and is "visible" to humans. The 10 μm (1×10^{-5} m) radiation in (a) is in the infrared region and is not visible.

    (d) $50.0 \text{ μs} \times \dfrac{1 \text{ s}}{1 \times 10^6 \text{ μs}} \times \dfrac{2.998 \times 10^8 \text{ m}}{\text{s}} = 1.50 \times 10^4 \text{ m}$

Check. Confirm that powers of 10 make sense and units are correct.

6.20 (a) $v = c/\lambda$; $\dfrac{2.998 \times 10^8 \text{ m}}{\text{s}} \times \dfrac{1}{0.86 \text{ nm}} \times \dfrac{1}{1.0 \times 10^{-9} \text{ m}} = 3.5 \times 10^{17} \text{ s}^{-1}$

    (b) $\lambda = c/v$; $\dfrac{2.998 \times 10^8 \text{ m}}{\text{s}} \times \dfrac{1 \text{ s}}{6.4 \times 10^{11}} = 4.7 \times 10^{-4} \text{ m}$

    (c) The radiation in (a) can be observed by an X-ray detector. The 8.6×10^{-10} m wavelength is within the range of 10^{-8} to 10^{-11} m for X-rays.

    (d) $0.38 \text{ ps} \times \dfrac{1 \times 10^{-12} \text{ s}}{1 \text{ ps}} \times \dfrac{2.998 \times 10^8 \text{ m}}{\text{s}} = 1.1 \times 10^{-4} \text{ m}$ (0.11 mm)

6.21 *Analyze/Plan.* $v = c/\lambda$; change nm → m.

Solve. $v = c/\lambda$; $\dfrac{2.998 \times 10^8 \text{ m}}{1 \text{ s}} \times \dfrac{1}{650 \text{ nm}} \times \dfrac{1 \text{ nm}}{1 \times 10^{-9} \text{ m}} = 4.6 \times 10^{14} \text{ s}^{-1}$

The color is red.

Check. $(3000 \times 10^5 / 650 \times 10^{-9}) \approx 4.5 \times 10^{14} \text{ s}^{-1}$; units are correct.

6.22   According to Figure 6.4, ultraviolet radiation has both higher frequency and shorter wavelength than infrared radiation. Looking forward to section 6.2, the energy of a photon is directly proportional to frequency (E = hv), so ultraviolet radiation yields more energy from a photovoltaic device.

Quantized Energy and Photons (Section 6.2)

6.23   (iii) Quantization means that energy changes can only happen in certain allowed increments. If the human growth quantum is one-foot, growth occurs instantaneously in one-foot increments.

6.24   Planck's original hypothesis was that energy could only be gained or lost in discrete amounts (quanta) with a certain minimum size. The size of the minimum energy change is related to the frequency of the radiation absorbed or emitted, $\Delta E = hv$, and energy changes occur only in multiples of hv.

Einstein postulated that light itself is quantized, that the minimum energy of a photon (a quantum of light) is directly proportional to its frequency, $E = hv$. If a photon that strikes a metal surface has less than the threshold energy, no electron is emitted from the surface. If the photon has energy equal to or greater than the threshold energy, an electron is emitted and any excess energy becomes the kinetic energy of the electron.

6.25   *Analyze/Plan.* These questions deal with the relationships between energy, wavelength, and frequency. Use the relationships $E = hv = hc/\lambda$ to calculate the desired quantities. Pay attention to units. *Solve.*

(a) $E = hv = 6.626 \times 10^{-34} \text{ J-s} \times 2.94 \times 10^{14} \text{ s}^{-1} = 1.95 \times 10^{-19} \text{ J}$

(b) $E = hv = \dfrac{hc}{\lambda} = \dfrac{6.626 \times 10^{-34} \text{ J-s} \times 2.998 \times 10^8 \text{ m/s}}{413 \text{ nm}} \times \dfrac{1 \text{ nm}}{1 \times 10^{-9} \text{ m}} = 4.81 \times 10^{-19} \text{ J}$

(c) $\lambda = hc/\Delta E = \dfrac{6.626 \times 10^{-34} \text{ J-s}}{6.06 \times 10^{-19} \text{ J}} \times \dfrac{2.998 \times 10^8 \text{ m}}{1 \text{ s}} = 3.28 \times 10^{-7} \text{ m} = 328 \text{ nm}$

6.26   *Analyze/Plan.* These questions deal with the relationships between energy, wavelength, and frequency. Use the relationships $E = hv = hc/\lambda$ to calculate the desired quantities. Pay attention to units. *Solve.*

(a) $v = c/\lambda; \quad \dfrac{2.998 \times 10^8 \text{ m}}{\text{s}} \times \dfrac{1}{532 \text{ nm}} \times \dfrac{1 \text{ nm}}{1 \times 10^{-9} \text{ m}} = 5.64 \times 10^{14} \text{ s}^{-1}$

(b) $E = hv = 6.626 \times 10^{-34} \text{ J-s} \times 5.64 \times 10^{14} \text{ s}^{-1} = 3.73 \times 10^{-19} \text{ J}$

(c)   The energy gap between the ground and excited states is the energy of a single 532 nm photon emitted when one electron relaxes from the excited to the ground state. $\Delta E = 3.73 \times 10^{-19} \text{ J}$

6.27   *Analyze/Plan.* Use $E = hc/\lambda$; pay close attention to units. *Solve.*

(a) $E = hc/\lambda = 6.626 \times 10^{-34} \text{ J-s} \times \dfrac{2.998 \times 10^8 \text{ m}}{1 \text{ s}} \times \dfrac{1}{3.3 \, \mu\text{m}} \times \dfrac{1 \, \mu\text{m}}{1 \times 10^{-6} \text{ m}}$

$$= 6.0 \times 10^{-20} \text{ J}$$

$$E = hc/\lambda = 6.626 \times 10^{-34} \text{ J-s} \times \frac{2.998 \times 10^8 \text{ m}}{1 \text{ s}} \times \frac{1}{0.154 \text{ nm}} \times \frac{1 \text{ nm}}{1 \times 10^{-9} \text{ m}}$$

$$= 1.29 \times 10^{-15} \text{ J}$$

Check. $(6.6 \times 3/3.3) \times (10^{-34} \times 10^8/10^{-6}) \approx 6 \times 10^{-20} \text{ J}$

$(6.6 \times 3/0.15) \times (10^{-34} \times 10^8/10^{-9}) \approx 120 \times 10^{-17} \approx 1.2 \times 10^{-15} \text{ J}$

The results are reasonable. We expect the longer wavelength 3.3 μm radiation to have the lower energy.

(b)   The 3.3 μm photon is in the infrared and the 0.154 nm (1.54×10^{-10} m) photon is in the X-ray region; the X-ray photon has the greater energy.

6.28 $E = h\nu$

$$\text{AM: } 6.626 \times 10^{-34} \text{ J-s} \times \frac{1010 \times 10^3}{1 \text{ s}} = 6.69 \times 10^{-28} \text{ J}$$

$$\text{FM: } 6.626 \times 10^{-34} \text{ J-s} \times \frac{98.3 \times 10^6}{1 \text{ s}} = 6.51 \times 10^{-26} \text{ J}$$

The FM photon has about 100 times more energy than the AM photon.

6.29   *Analyze/Plan.* Use $E = hc/\lambda$ to calculate J/photon; Avogadro's number to calculate J/mol; photon/J [the result from part (a)] to calculate photons in 1.00 mJ. Pay attention to units. *Solve.*

(a) $E_{photon} = hc/\lambda = \dfrac{6.626 \times 10^{-34} \text{ J-s}}{325 \times 10^{-9} \text{ m}} \times \dfrac{2.998 \times 10^8 \text{ m}}{\text{s}} = 6.1122 \times 10^{-19}$

$$= 6.11 \times 10^{-19} \text{ J/photon}$$

(b) $\dfrac{6.1122 \times 10^{-19} \text{ J}}{1 \text{ photon}} \times \dfrac{6.022 \times 10^{23} \text{ photons}}{1 \text{ mol}} = 3.68 \times 10^5 \text{ J/mol} = 368 \text{ kJ/mol}$

(c) $\dfrac{1 \text{ photon}}{6.1122 \times 10^{-19} \text{ J}} \times 1.00 \text{ mJ} \times \dfrac{1 \times 10^{-3} \text{ J}}{1 \text{ mJ}} = 1.64 \times 10^{15} \text{ photons}$

Check. Powers of 10 (orders of magnitude) and units are correct.

(d)   If the energy of one 325 nm photon breaks exactly one bond, 1 mol of photons break 1 mol of bonds. The average bond energy in kJ/mol is the energy of 1 mol of photons (from part b) 368 kJ/mol.

6.30 $\dfrac{242 \times 10^3 \text{ J}}{\text{mol } Cl_2} \times \dfrac{1 \text{ mol}}{6.022 \times 10^{23} \text{ photons}} = 4.0186 \times 10^{-19} = 4.02 \times 10^{-19} \text{ J/photon}$

$$\lambda = hc/E = \frac{6.626 \times 10^{-34} \text{ J-s}}{4.0186 \times 10^{-19} \text{ J}} \times \frac{2.998 \times 10^8 \text{ m}}{1 \text{ s}} = 4.94 \times 10^{-7} \text{ m} = 494 \text{ nm}$$

According to Figure 6.4, this is visible radiation.

6.31   *Analyze/Plan.* $E = hc/\lambda$ gives J/photon. Use this result with J/s (given) to calculate photons/s. *Solve.*

(a)   The $\sim 1 \times 10^{-6}$ m radiation is infrared but very near the visible edge.

(b) $E_{photon} = hc/\lambda = \dfrac{6.626 \times 10^{-34} \text{ J-s}}{987 \times 10^{-9} \text{ m}} \times \dfrac{2.998 \times 10^{8} \text{ m}}{1 \text{ s}} = 2.0126 \times 10^{-19}$

$$= 2.01 \times 10^{-19} \text{ J/photon}$$

$$\dfrac{0.52 \text{ J}}{32 \text{ s}} \times \dfrac{1 \text{ photon}}{2.0126 \times 10^{-19} \text{ J}} = 8.1 \times 10^{16} \text{ photons/s}$$

Check. $(7 \times 3/1000) \times (10^{-34} \times 10^{8}/10^{-9}) \approx 21 \times 10^{-20} \approx 2.1 \times 10^{-19}$ J/photon

$(0.5/30/2) \times (1/10^{-19}) = 0.008 \times 10^{19} = 8 \times 10^{16}$ photons/s

Units are correct; powers of 10 are reasonable.

6.32   (a) $3.55 \text{ mm} = 3.55 \times 10^{-3}$ m; the radiation is microwave.

     (b) $E_{photon} = hc/\lambda = \dfrac{6.626 \times 10^{-34} \text{ J-s}}{3.55 \times 10^{-3} \text{ m}} \times \dfrac{2.998 \times 10^{8} \text{ m}}{1 \text{ s}} = 5.5957 \times 10^{-23}$

$$= 5.60 \times 10^{-23} \text{ J/photon}$$

$$\dfrac{5.5957 \times 10^{-23} \text{ J}}{1 \text{ photon}} \times \dfrac{3.2 \times 10^{8} \text{ photons}}{1 \text{ s}} \times \dfrac{60 \text{ s}}{1 \text{ min}} \times \dfrac{60 \text{ min}}{1 \text{ hr}} = 6.4463 \times 10^{-11}$$

$$= 6.4 \times 10^{-11} \text{ J/hr}$$

6.33   *Analyze/Plan.* Use $E = h\nu$ and $\nu = c/\lambda$. Calculate the desired characteristics of the photons. Assume 1 photon interacts with 1 electron. Compare E_{min} and E_{120} to calculate maximum kinetic energy of the emitted electron.   *Solve.*

     (a) $E = h\nu = 6.626 \times 10^{-34} \text{ J-s} \times 1.09 \times 10^{15} \text{ s}^{-1} = 7.22 \times 10^{-19}$ J

     (b) $\lambda = c/\nu = \dfrac{2.998 \times 10^{8} \text{ m}}{1 \text{ s}} \times \dfrac{1 \text{ s}}{1.09 \times 10^{15}} = 2.75 \times 10^{-7}$ m $= 275$ nm

     (c) $E_{120} = hc/\lambda = 6.626 \times 10^{-34} \text{ J-s} \times \dfrac{2.998 \times 10^{8} \text{ m}}{1 \text{ s}} \times \dfrac{1}{120 \text{ nm}} \times \dfrac{1 \text{ nm}}{1 \times 10^{-9} \text{ m}}$

$$= 1.655 \times 10^{-18} = 1.66 \times 10^{-18} \text{ J}$$

The excess energy of the 120 nm photon is converted into the kinetic energy of the emitted electron.

$E_k = E_{120} - E_{min} = 16.55 \times 10^{-19}$ J $- 7.22 \times 10^{-19}$ J $= 9.3 \times 10^{-19}$ J/electron

Check. E_{120} must be greater than E_{min} in order for the photon to impart kinetic energy to the emitted electron. Our calculations are consistent with this requirement.

6.34   (a) $\nu = E/h = \dfrac{6.94 \times 10^{-19} \text{ J}}{6.626 \times 10^{-34} \text{ J-s}} = 1.04739 \times 10^{15} = 1.05 \times 10^{15} \text{ s}^{-1}$

     (b) $\lambda = hc/E = \dfrac{6.626 \times 10^{-34} \text{ J-s}}{6.94 \times 10^{-19} \text{ J}} \times \dfrac{2.998 \times 10^{8} \text{ m}}{\text{s}} = 2.86 \times 10^{-7}$ m $= 286$ nm

     (c)   No. The maximum 286 nm wavelength required to eject an electron from titanium is in the ultraviolet region (Figure 6.4).

     (d) $E_{233} = hc/\lambda = \dfrac{6.626 \times 10^{-34} \text{ J-s}}{233 \times 10^{-9} \text{ m}} \times \dfrac{2.998 \times 10^{8} \text{ m}}{\text{s}} = 8.5256 \times 10^{-19} = 8.53 \times 10^{-19} \text{ J}$

$E_K = E_{233} - E_{min} = 8.5256 \times 10^{-19}$ J $- 6.94 \times 10^{-19}$ J $= 1.5856 \times 10^{-19} = 1.59 \times 10^{-19}$ J

Bohr's Model; Matter Waves (Sections 6.3 and 6.4)

6.35   When the electron in a hydrogen atom transitions from $n = 1$ to $n = 3$, the atom "expands." The average distance from the nucleus of an $n = 3$ electron is greater than that of an $n = 1$ electron. The volume where there is a significant probability of finding an electron increases.

6.36   (a)   True.

         (b)   False. As the value of n increases, energy becomes less negative. The energy of the $n = 2$ state is higher (less negative) than that of the $n = 1$ state.

         (c)   True.

6.37   *Analyze/Plan.* An isolated electron is assigned an energy of zero; the closer the electron comes to the nucleus, the more negative its energy. Thus, as an electron moves closer to the nucleus, the energy of the electron decreases and the excess energy is emitted. Conversely, as an electron moves further from the nucleus, the energy of the electron increases and energy must be absorbed.   *Solve.*

         (a)   As the principle quantum number decreases, the electron moves toward the nucleus and energy is emitted.

         (b)   An increase in the radius of the orbit means the electron moves away from the nucleus; energy is absorbed.

         (c)   An isolated electron is assigned an energy of zero. As the electron moves to the $n = 3$ state closer to the H^+ nucleus, its energy becomes more negative (decreases) and energy is emitted.

6.38   (a)   Absorbed.   (b)   Emitted.   (c)   Absorbed.

6.39   *Analyze/Plan.* Equation 6.5: $E = (-2.18 \times 10^{-18} \text{ J})(1/n^2)$.   *Solve.*

         (a) $E_2 = -2.18 \times 10^{-18} \text{ J}/(2)^2 = -5.45 \times 10^{-19} \text{ J}$

              $E_6 = -2.18 \times 10^{-18} \text{ J}/(6)^2 = -6.0556 \times 10^{-20} = -0.606 \times 10^{-19} \text{ J}$

              $\Delta E = E_2 - E_6 = (-5.45 \times 10^{-19} \text{ J}) - (-0.606 \times 10^{-19} \text{ J})$

                  $= -4.844 \times 10^{-19} \text{ J} = -4.84 \times 10^{-19} \text{ J}$

              $\lambda = hc/\Delta E = \dfrac{6.626 \times 10^{-34} \text{ J-s}}{4.844 \times 10^{-19} \text{ J}} \times \dfrac{2.998 \times 10^8 \text{ m}}{\text{s}} = 4.10 \times 10^{-7} \text{ m} = 410 \text{ nm}$

         (b)   The visible range is 400–700 nm, so this line is visible; the observed color is violet.

              Check. We expect E_6 to be a more positive (or less negative) than E_2, and it is. ΔE is negative, which indicates emission. The orders of magnitude make sense and units are correct.

6.40   (a)   The value of n for the electron increases, so ΔE is positive.

       (b)   *Analyze/Plan.* Use Equation 6.6 to calculate ΔE, then $\lambda = hc/\Delta E$.   *Solve.*

$$n_i = 4, n_f = 9; \quad \Delta E = -2.18 \times 10^{-18} \text{ J} \left[\frac{1}{n_f^2} - \frac{1}{n_i^2} \right] = -2.18 \times 10^{-18} \text{ J} \, (1/81 - 1/16)$$

$$\lambda = hc/E = \frac{6.626 \times 10^{-34} \text{ J-s} \times 2.998 \times 10^8 \text{ m/s}}{-2.18 \times 10^{-18} \text{ J} \, (1/81 - 1/16)} = 1.82 \times 10^{-6} \text{ m}$$

This wavelength will be absorbed. Energy is required to move the electron to the higher energy $n = 9$ state. [Note that the denominator of the preceding equation has a positive sign, because $(1/81 - 1/16)$ is negative.]

       (c)   The light in (b) is in the infrared.

6.41   (a)   Statement (ii) is the best explanation. Only lines with $n_f = 2$ represent ΔE values and wavelengths that lie in the visible portion of the spectrum. Lines with $n_f = 3$ have smaller ΔE values and lie in the lower energy, longer wavelength infrared portion of the electromagnetic spectrum.

       (b)   *Analyze/Plan.* Use Equation 6.6 to calculate ΔE, then $\lambda = hc/\Delta E$.   *Solve.*

$$n_i = 3, n_f = 2; \quad \Delta E = -2.18 \times 10^{-18} \text{ J} \left[\frac{1}{n_f^2} - \frac{1}{n_i^2} \right] = -2.18 \times 10^{-18} \text{ J} \, (1/4 - 1/9)$$

$$\lambda = hc/E = \frac{6.626 \times 10^{-34} \text{ J-s} \times 2.998 \times 10^8 \text{ m/s}}{-2.18 \times 10^{-18} \text{ J} \, (1/4 - 1/9)} = 6.56 \times 10^{-7} \text{ m}$$

This is the red line at 656 nm.

$$n_i = 4, n_f = 2; \quad \lambda = hc/E = \frac{6.626 \times 10^{-34} \text{ J-s} \times 2.998 \times 10^8 \text{ m/s}}{-2.18 \times 10^{-18} \text{ J} \, (1/4 - 1/16)} = 4.86 \times 10^{-7} \text{ m}$$

This is the blue-green line at 486 nm.

$$n_i = 5, n_f = 2; \quad \lambda = hc/E = \frac{6.626 \times 10^{-34} \text{ J-s} \times 2.998 \times 10^8 \text{ m/s}}{-2.18 \times 10^{-18} \text{ J} \, (1/4 - 1/25)} = 4.34 \times 10^{-7} \text{ m}$$

This is the blue-violet line at 434 nm.

Check. The calculated wavelengths correspond well to three lines in the H emission spectrum in Figure 6.11, so the results are sensible.

6.42   (a)   From Exercise 6.41, we know that transitions with $n_f = 2$ emit light in the visible region of the electromagnetic spectrum. Transitions with $n_f = 1$ have larger ΔE values and shorter wavelengths than those with $n_f = 2$. These transitions will lie in the ultraviolet region.

       (b) $n_i = 2, n_f = 1; \; \lambda = hc/E = \dfrac{6.626 \times 10^{-34} \text{ J-s} \times 2.998 \times 10^8 \text{ m/s}}{-2.18 \times 10^{-18} \text{ J} \, (1/1 - 1/4)} \times \dfrac{1 \text{ nm}}{1 \times 10^{-9} \text{ m}} = 121 \text{ nm}$

            $n_i = 3, n_f = 1; \; \lambda = hc/E = \dfrac{6.626 \times 10^{-34} \text{ J-s} \times 2.998 \times 10^8 \text{ m/s}}{-2.18 \times 10^{-18} \text{ J} \, (1/1 - 1/9)} \times \dfrac{1 \text{ nm}}{1 \times 10^{-9} \text{ m}} = 103 \text{ nm}$

$$n_i = 4,\ n_f = 1;\ \lambda = hc/E = \frac{6.626\times10^{-34}\ \text{J-s}\times2.998\times10^8\ \text{m/s}}{-2.18\times10^{-18}\ \text{J}\,(1/1-1/16)}\times\frac{1\ \text{nm}}{1\times10^{-9}\ \text{m}} = 97.2\ \text{nm}$$

Note that all three wavelengths are in the ultraviolet region of the electromagnetic spectrum.

6.43　(a)　$93.07\ \text{nm}\times\dfrac{1\times10^{-9}\ \text{m}}{1\ \text{nm}} = 9.307\times10^{-8}\ \text{m}$; this line is in the ultraviolet region.

(b)　*Analyze/Plan.* Only lines with $n_f = 1$ have a large enough ΔE to lie in the ultraviolet region (see Solutions 6.41 and 6.42). Solve Equation 6.6 for n_i, recalling that ΔE is negative for emission. *Solve.*

$$\frac{-hc}{\lambda} = -2.18\times10^{-18}\ \text{J}\left[\frac{1}{n_f^2}-\frac{1}{n_i^2}\right];\quad \frac{hc}{\lambda(2.18\times10^{-18}\ \text{J})} = \left[1-\frac{1}{n_i^2}\right]$$

$$-\frac{1}{n_i^2} = \left[\frac{hc}{\lambda(2.18\times10^{-18}\ \text{J})}-1\right];\quad \frac{1}{n_i^2} = \left[1-\frac{hc}{\lambda(2.18\times10^{-18}\ \text{J})}\right]$$

$$n_i^2 = \left[1-\frac{hc}{\lambda(2.18\times10^{-18}\ \text{J})}\right]^{-1};\quad n_i = \left[1-\frac{hc}{\lambda(2.18\times10^{-18}\ \text{J})}\right]^{-1/2}$$

$$n_i = \left(1-\frac{6.626\times10^{-34}\ \text{J-s}\times2.998\times10^8\ \text{m/s}}{9.307\times10^{-8}\ \text{m}\times2.18\times10^{-18}\ \text{J}}\right)^{-1/2} = 7\ (n\ \text{values must be integers})$$

$$n_i = 7,\ n_f = 1$$

Check. From Solution 6.42, we know that $n_i > 4$ for $\lambda = 93.07$ nm. The calculated result is close to 7, so the answer is reasonable.

6.44　(a)　$1094\ \text{nm}\times\dfrac{1\times10^{-9}\ \text{m}}{1\ \text{nm}} = 1.094\times10^{-6}\ \text{m}$; this line is in the infrared.

(b)　Absorption lines with $n_i = 1$ are in the ultraviolet and with $n_i = 2$ are in the visible. Thus, $n_i \geq 3$, but we do not know the exact value of n_i. Calculate the longest wavelength with $n_i = 3$ ($n_f = 4$). If this is less than 1094 nm, $n_i > 3$.

$$\lambda = hc/E = \frac{6.626\times10^{-34}\ \text{J-s}\times2.998\times10^8\ \text{m/s}}{-2.18\times10^{-18}\ \text{J}\,(1/16-1/9)} = 1.875\times10^{-6}\ \text{m}$$

This wavelength is longer than 1.094×10^{-6} m, but we are in the ballpark. Use $n_i = 3$ and solve for n_f as in Solution 6.43. Note that ΔE is positive because we are dealing with absorption.

$$n_f = \left(\frac{1}{n_i^2}-\frac{hc}{\lambda(2.18\times10^{-18}\ \text{J})}\right)^{-1/2} = \left(1/9-\frac{6.626\times10^{-34}\ \text{J-s}\times2.998\times10^8\ \text{m/s}}{1.094\times10^{-6}\ \text{m}\times2.18\times10^{-18}\ \text{J}}\right)^{-1/2} = 6$$

$$n_f = 6,\ n_i = 3$$

6.45　*Analyze/Plan.* According to Equation 6.6 and several preceding solutions, the greater the energy of light absorbed, the lower the value of n_i. The greater the energy of light, the greater its frequency, $\Delta E = h\nu$.

Solve. The order of increasing frequency (and energy) of light absorbed is:

$n = 4$ to $n = 9 < n = 3$ to $n = 6 < n = 2$ to $n = 3 < n = 1$ to $n = 2$

6.46   According to Equation 6.6 and several preceding solutions, the greater the energy of light emitted, the lower the value of n_f. The greater the energy of light, the shorter its wavelength, $\Delta E = hc/\lambda$. The order of increasing wavelength (and decreasing energy) of light emitted is:

$n = 4$ to $n = 2 < n = 3$ to $n = 2 < n = 5$ to $n = 3 < n = 7$ to $n = 4$

6.47   *Analyze/Plan.* $\lambda = \dfrac{h}{mv}$; $1\,J = \dfrac{1\,\text{kg-m}^2}{s^2}$;

Change mass to kg and velocity to m/s in each case.   *Solve.*

(a) $\dfrac{50\,\text{km}}{1\,\text{hr}} \times \dfrac{1000\,\text{m}}{1\,\text{km}} \times \dfrac{1\,\text{hr}}{60\,\text{min}} \times \dfrac{1\,\text{min}}{60\,\text{s}} = 13.89 = 14\,\text{m/s}$

$\lambda = \dfrac{6.626 \times 10^{-34}\,\text{kg-m}^2\text{-s}}{1\,s^2} \times \dfrac{1}{85\,\text{kg}} \times \dfrac{1\,s}{13.89\,\text{m}} = 5.6 \times 10^{-37}\,\text{m}$

(b) $10.0\,\text{g} \times \dfrac{1\,\text{kg}}{1000\,\text{g}} = 0.0100\,\text{kg}$

$\lambda = \dfrac{6.626 \times 10^{-34}\,\text{kg-m}^2\text{-s}}{1\,s^2} \times \dfrac{1}{0.0100\,\text{kg}} \times \dfrac{1\,s}{250\,\text{m}} = 2.65 \times 10^{-34}\,\text{m}$

(c)   We need to calculate the mass of a single Li atom in kg.

$\dfrac{6.94\,\text{g Li}}{1\,\text{mol Li}} \times \dfrac{1\,\text{kg}}{1000\,\text{g}} \times \dfrac{1\,\text{mol}}{6.022 \times 10^{23}\,\text{Li atoms}} = 1.152 \times 10^{-26} = 1.15 \times 10^{-26}\,\text{kg}$

$\lambda = \dfrac{6.626 \times 10^{-34}\,\text{kg-m}^2\text{-s}}{1\,s^2} \times \dfrac{1}{1.152 \times 10^{-26}\,\text{kg}} \times \dfrac{1\,s}{2.5 \times 10^5\,\text{m}} = 2.3 \times 10^{-13}\,\text{m}$

(d)   Calculate the mass of a single O_3 molecule in kg.

$\dfrac{48.00\,\text{g O}_3}{1\,\text{mol O}_3} \times \dfrac{1\,\text{kg}}{1000\,\text{g}} \times \dfrac{1\,\text{mol}}{6.022 \times 10^{23}\,\text{O}_3\,\text{molecules}} = 7.971 \times 10^{-26}$

$= 7.97 \times 10^{-26}\,\text{kg}$

$\lambda = \dfrac{6.626 \times 10^{-34}\,\text{kg-m}^2\text{-s}}{1\,s^2} \times \dfrac{1}{7.971 \times 10^{-26}\,\text{kg}} \times \dfrac{1\,s}{550\,\text{m}}$

$= 1.51 \times 10^{-11}\,\text{m}$ (15 pm)

6.48   Find the mass of an electron on the inside back cover of the text. $\lambda = h/mv$; change velocity to m/s.

mass of muon $= 206.8 \times 9.1094 \times 10^{-31}\,\text{g} = 1.8838 \times 10^{-28} = 1.88 \times 10^{-28}\,\text{kg}$

$\lambda = \dfrac{6.626 \times 10^{-34}\,\text{kg-m}^2\text{-s}}{1\,s^2} \times \dfrac{1}{1.8838 \times 10^{-28}\,\text{kg}} \times \dfrac{1\,s}{8.85 \times 10^3\,\text{m/s}} = 3.97 \times 10^{-10}\,\text{m}$

$= 3.97\,\text{Å}$

6.49   *Analyze/Plan.* Use v = h/mλ; change wavelength to meters and mass of neutron (back inside cover) to kg.   *Solve.*

$$\lambda = 1.25 \,\text{Å} \times \frac{1 \times 10^{-10} \,\text{m}}{1 \,\text{Å}} = 1.25 \times 10^{-10} \,\text{m}; \ \text{mass} = 1.6749 \times 10^{-27} \,\text{kg}$$

$$v = \frac{6.626 \times 10^{-34} \,\text{kg-m}^2\text{-s}}{1 \,\text{s}^2} \times \frac{1}{1.6749 \times 10^{-27} \,\text{kg}} \times \frac{1}{1.25 \times 10^{-10} \,\text{m}} = 3.16 \times 10^3 \,\text{m/s}$$

Check. $(6.6/1.6/1.25) \times (10^{-34}/10^{-27}/10^{-10}) \approx 3 \times 10^3 \,\text{m/s}$

6.50 $m_e = 9.1094 \times 10^{-31}$ kg (back inside cover of textbook)

$$\lambda = \frac{6.626 \times 10^{-34} \,\text{kg-m}^2\text{-s}}{1 \,\text{s}^2} \times \frac{1}{9.1094 \times 10^{-31} \,\text{kg}} \times \frac{1 \,\text{s}}{9.47 \times 10^6 \,\text{m}} = 7.68 \times 10^{-11} \,\text{m}$$

$$7.68 \times 10^{-11} \,\text{m} \times \frac{1 \,\text{Å}}{1 \times 10^{-10} \,\text{m}} = 0.768 \,\text{Å}$$

Because atomic radii and interatomic distances are on the order of 1–5 Å (Section 2.3), the wavelength of this electron is comparable to the size of atoms.

6.51   *Analyze/Plan.* Use $\Delta x \geq h/4\pi m \Delta v$, paying attention to appropriate units. Note that the uncertainty in speed of the particle (Δv) is important, rather than the speed itself.   *Solve.*

(a) $m = 1.50 \,\text{mg} \times \dfrac{1 \,\text{g}}{1000 \,\text{mg}} \times \dfrac{1 \,\text{kg}}{1000 \,\text{g}} = 1.50 \times 10^{-6} \,\text{kg}; \ \Delta v = 0.01 \,\text{m/s}$

$$\Delta x \geq \frac{6.626 \times 10^{-34} \,\text{J-s}}{4\pi(1.50 \times 10^{-6} \,\text{kg})(0.01 \ \text{m/s})} \geq 3.52 \times 10^{-27} = 4 \times 10^{-27} \,\text{m}$$

(b) $m = 1.673 \times 10^{-24} \,\text{g} = 1.673 \times 10^{-27} \,\text{kg}; \ \Delta v = 0.01 \times 10^4 \,\text{m/s}$

$$\Delta x \geq \frac{6.626 \times 10^{-34} \,\text{J-s}}{4\pi(1.673 \times 10^{-27} \,\text{kg})(0.01 \times 10^4 \,\text{m/s})} \geq 3 \times 10^{-10} \,\text{m}$$

Check. The more massive particle in (a) has a much smaller uncertainty in position.

6.52 $\Delta x \geq = h/4\pi m \Delta v$; use masses in kg, Δv in m/s.

(a) $\dfrac{6.626 \times 10^{-34} \,\text{J-s}}{4\pi(9.109 \times 10^{-31} \,\text{kg})(0.01 \times 10^5 \,\text{m/s})} = 6 \times 10^{-8} \,\text{m}$

(b) $\dfrac{6.626 \times 10^{-34} \,\text{J-s}}{4\pi(1.675 \times 10^{-27} \,\text{kg})(0.01 \times 10^5 \,\text{m/s})} = 3 \times 10^{-11} \,\text{m}$

(c)   For particles moving with the same uncertainty in velocity, the more massive neutron has a much smaller uncertainty in position than the lighter electron. We know the position of the neutron with much greater precision.

Quantum Mechanics and Atomic Orbitals (Sections 6.5 and 6.6)

6.53   (a)   False. The contour representation shows a volume where there is significant electron density. The electron can be moving anywhere within this volume.

(b)   False. The probability of finding an electron at a given distance from the nucleus is the radial probability function. This is the probability density summed over all points that lie a distance r from the nucleus.

6.54   (a)   False. Each electron behaves as a wave and is allowed to spread out over regions that are separated by a node.

       (b)   True. The radial probability function is the sum of the probability density at all points a distance r from the nucleus. The radial probability goes to zero at the distance of the node because the probability density goes to zero at this distance.

       (c)   False. For an s orbital, the number of nodes is $n - 1$.

6.55   (a)   The possible values of l are $(n - 1)$ to 0. $n = 4, l = 3, 2, 1, 0$

       (b)   The possible values of m_l are $-l$ to $+l$. $l = 2, m_l = -2, -1, 0, 1, 2$

       (c)   Because the value of m_l is less than or equal to the value of l, $m_l = 2$ must have an l-value greater than or equal to 2. In terms of elements that have been observed, the possibilities are 2, 3, and 4.

6.56   (a)   For $n = 3$, there are nine uniques combinations of l and m_l (values): ($l = 2$; $m_l = -2, -1, 0, 1, 2; l = 1, m_l = -1, 0, 1; l = 0, m_l = 0$).

       (b)   For $n = 4$, there are sixteen unique combinations of l and m_l (values):

            ($l = 3, m_l = -3$ to $+3; l = 2, m_l = -2$ to $+2; l = 1, m_l = -1$ to $+1; l = 0, = 0$).

            In general, for each principal quantum number n, there are n values for l and n^2 values for m_l. For each shell, there are n kinds of orbitals and n^2 total orbitals.

6.57   (a)   3p: $n = 3, l = 1$ (b)   2s: $n = 2, l = 0$

       (c)   4f: $n = 4, l = 3$ (d)   5d: $n = 5, l = 2$

6.58   (a)   2, 1, 1;   2, 1, 0;   2, 1, -1   (b)   5, 2, 2;   5, 2, 1;   5, 2, 0;   5, 2, -1;   5, 2, -2

6.59   (a)   2, 1 ,0, -1, -2   (b)   ½, -½

6.60   (a)   2, 3   (b)   ½, -½

6.61   Impossible: (a) 1p, only $l = 0$ is possible for $n = 1$; (d) 2d, for $n = 2, l = 1$ or 0, but not 2

6.62

n	l	m_l	orbital
2	1	-1	2p (example)
1	0	0	1s
3	-3	2	not allowed ($l < n$ and + only)
3	2	-2	3d
2	0	-1	not allowed ($m_l = -l$ to $+l$)
0	0	0	not allowed ($n \neq 0$)
4	2	1	4d
5	3	0	5f

6.63

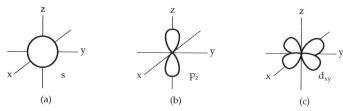

(a)   (b)   (c)

Note that the lobes of the d_{xy} orbital lie in the xy-plane and point between the x-axes and y-axes.

6.64

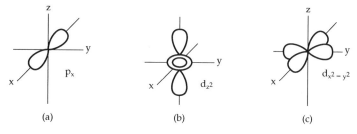

(a)   (b)   (c)

6.65   (a)   The 1s and 2s orbitals of an H atom have the same overall spherical shape. The 2s orbital has a larger radial extension and one node, whereas the 1s orbital has continuous electron density. Because the 2s orbital is "larger," there is greater probability of finding an electron farther from the nucleus in the 2s orbital.

(b)   A single 2p orbital is directional in that its electron density is concentrated along one of the three Cartesian axes of the atom. The $d_{x^2-y^2}$ orbital has electron density along both the x- and y-axes, whereas the p_x orbital has density only along the x-axis.

(c)   The average distance of an electron from the nucleus in a 3s orbital is greater than for an electron in a 2s orbital. In general, for the same kind of orbital, the larger the n value, the greater the average distance of an electron from the nucleus of the atom.

(d)   1s < 2p < 3d < 4f < 6s. In the H atom, orbitals with the same n value are degenerate and energy increases with increasing n value. Thus, the order of increasing energy is given above.

6.66   (a)   In an s orbital, there are $(n-1)$ nodes.

(b)   The $2p_x$ orbital has one node (the yz plane passing through the nucleus of the atom). The 3s orbital has two nodes.

(c)   Probability density, $\psi^2(r)$, is the probability of finding an electron at a single point, r. The radial probability function, $P(r)$, is the probability of finding an electron at any point that is distance r from the nucleus. Figure 6.19 contains plots of $P(r)$ vs. r for 1s, 2s, and 3s orbitals. The most obvious features of these plots are the radii of maximum probability for the three orbitals, and the number and location of nodes for the three orbitals.

By comparing plots for the three orbitals, we see that as n increases, the number of nodes increases and the radius of maximum probability (orbital size) increases.

(d)   2s = 2p < 3s < 4d < 5s. In the hydrogen atom, orbitals with the same n value are degenerate and energy increases with increasing n value.

Many-Electron Atoms and Electron Configurations (Sections 6.7 – 6.9)

6.67   (a)   Yes. An He$^+$ ion has only one electron. In a one-electron particle, the energy of an orbital depends only on the value of n. Therefore, the 2s and 2p orbitals of He$^+$ have the same energy.

        (b)   Yes. In an He atom, the 2s orbital is lower in energy than the 2p orbitals. A helium atom is a two-electron particle. In multi-electron particles, electron-electron repulsions cause orbitals with the same n-value but different l-values to have different energies.

6.68   (a)   The electron with the greater average distance from the nucleus feels a smaller attraction for the nucleus and is higher in energy. Thus the 3p is higher in energy than 3s.

        (b)   Because it has a larger n value, a 3s electron has a greater average distance from the chlorine nucleus than a 2p electron. The 3s electron experiences a smaller attraction for the nucleus and requires less energy to remove from the chlorine atom.

6.69   (a)   No, both configurations follow the Pauli exclusion principle.

        (b)   No, both configurations obey Hund's rule.

        (c)   No. In the absence of a magnetic field, we cannot say which configuration has lower energy. The difference in the two configurations is the m_s value of the electron in the 2s orbital. In the absence of an external magnetic field, electrons that only differ in m_s have the same energy.

6.70   (a)   Ag: [Kr]5s^{1}4d^{10} (1 unpaired electron in the 5s orbital)

        (b)   No. The experiment requires that the atoms in the beam each have an unpaired electron. The electron configuration of cadmium is: [Kr]5s^{2}4d^{10}. There are no unpaired electrons in the configuration of cadmium, so a beam of Cd atoms would not be deflected in a magnetic field.

        (c)   Yes. The electron configuration of fluorine is: [He]2s^{2}2p^5. Each fluorine atom has one unpaired electron in a 2p orbital. A beam of F atoms will be deflected in a magnetic field.

6.71   *Analyze/Plan.* Each subshell has an l-value associated with it. For a particular l-value, permissible m_l-values are $-l$ to $+l$. Each m_l-value represents an orbital, which can hold two electrons.   *Solve.*

        (a) 6   (b) 10   (c) 2   (d) 14

6.72   (a) 2   (b) 14   (c) 2   (d) 2

6.73   (a)   "Valence electrons" are those involved in chemical bonding. They are part (or all) of the outer-shell electrons listed after core electrons in a condensed electron configuration.

        (b)   "Core electrons" are inner shell electrons that have the electron configuration of the nearest noble-gas element.

        (c)   Each box represents an orbital.

        (d)   Each half-arrow in an orbital diagram represents an electron. The direction of the half-arrow represents electron spin.

6.74 | Element | (a) N | (b) Si | (c) Cl |
|---|---|---|---|
| Electron Configuration | $[He]2s^2 2p^3$ | $[Ne]3s^2 3p^2$ | $[Ne]3s^2 3p^5$ |
| Core electrons | 2 | 10 | 10 |
| Valence electrons | 5 | 4 | 7 |
| Unpaired electrons | 3 | 2 | 1 |

6.75   *Analyze/Plan.* Follow the logic in Sample Exercise 6.9.   *Solve.*

     (a)   Cs: $[Xe]6s^1$ (b)   Ni: $[Ar]4s^2 3d^8$

     (c)   Se: $[Ar]4s^2 3d^{10} 4p^4$ (d)   Cd: $[Kr]5s^2 4d^{10}$

     (e)   U: $[Rn]5f^3 6d^1 7s^2$. (Note the U and several other *f*-block elements have irregular *d*- and *f*-electron orders.)

     (f)   Pb: $[Xe]6s^2 4f^{14} 5d^{10} 6p^2$

6.76   (a)   Mg: $[Ne]3s^2$, 0 unpaired electrons

     (b)   Ge: $[Ar]4s^2 3d^{10} 4p^2$, 2 unpaired electrons

     (c)   Br: $[Ar]4s^2 3d^{10} 4p^5$, 1 unpaired electron

     (d)   V: $[Ar]4s^2 3d^3$, 3 unpaired electrons

     (e)   Y: $[Kr]5s^2 4d^1$, 1 unpaired electron

     (f)   Lu: $[Xe]6s^2 4f^{14} 5d^1$, 1 unpaired electron

6.77   (a)   Be, 0 unpaired electrons   (b)   O, 2 unpaired electrons

     (c)   Cr, 6 unpaired electrons   (d)   Te , 2 unpaired electrons

6.78   (a)   7A or 17 (halogens), 1 unpaired electron

     (b)   4B or 4, 2 unpaired electrons

     (c)   3A or 13 (row 4 and below), 1 unpaired electron

     (d)   the f-block elements Sm and Pu, 6 unpaired electrons

6.79   (a)   The orbitals are not filled in order of increasing energy. The fifth electron would fill the 2p subshell (same *n*-value as 2s) before the 3s.

     (b)   The orbitals are not filled in order of increasing energy. After 6s, electrons fill the 4f subshell.

     (c)   The orbitals are not filled in order of increasing energy. The 3p subshell would fill before the 3d because it has the lower *l*-value and the same *n*-value. (If there were more electrons, 4s would also fill before 3d.)

6.80   Count the total number of electrons to assign the element.

     (a)   F: $[He]2s^2 2p^5$ (b)   Ge: $[Ar]4s^2 3d^{10} 4p^2$ (c)   Nb: $[Kr]5s^2 4d^3$

6 Electronic Structure of Atoms   Solutions to Exercises

Additional Exercises

6.81 (a) $\lambda_A = 1.6 \times 10^{-7}$ m $/ 4.5 = 3.56 \times 10^{-8} = 3.6 \times 10^{-8}$ m

$\lambda_B = 1.6 \times 10^{-7}$ m $/ 2 = 8.0 \times 10^{-8}$ m

(b) $\nu = c/\lambda;\ \nu_A = \dfrac{2.998 \times 10^8 \text{ m}}{1 \text{ s}} \times \dfrac{1}{3.56 \times 10^{-8} \text{ m}} = 8.4 \times 10^{15} \text{ s}^{-1}$

$\nu_B = \dfrac{2.998 \times 10^8 \text{ m}}{1 \text{ s}} \times \dfrac{1}{8.0 \times 10^{-8} \text{ m}} = 3.7 \times 10^{15} \text{ s}^{-1}$

(c) A: ultraviolet, B: ultraviolet

6.82 (a) $\nu = c/\lambda = \dfrac{2.998 \times 10^8 \text{ m}}{\text{s}} \times \dfrac{1}{589 \text{ nm}} \times \dfrac{1 \text{ nm}}{1 \times 10^{-9} \text{ m}} = 5.0900 \times 10^{14} = 5.09 \times 10^{14} \text{ s}^{-1}$

(b) $E = h\nu = 6.626 \times 10^{-34} \text{ J-s} \times 5.0900 \times 10^{14} \text{ s}^{-1} \times \dfrac{6.022 \times 10^{23} \text{ photons}}{\text{mol}}$
$\times 0.1 \text{ mol} = 2.03 \times 10^4 \text{ J} = 20.3 \text{ kJ}$

(c) $\Delta E = h\nu = \dfrac{hc}{\lambda} = \dfrac{6.626 \times 10^{-34} \text{ J-s} \times 2.998 \times 10^8 \text{ m/s}}{589 \text{ nm}} \times \dfrac{1 \text{ nm}}{1 \times 10^{-9} \text{ m}} = 3.37 \times 10^{-19} \text{ J}$

(d) The 589 nm light emission is characteristic of Na^+. If the pickle is soaked in a different salt long enough to remove all Na^+, the 589 nm light would not be observed. Emission at a different wavelength, characteristic of the new salt, would be observed.

6.83 (a) Elements that emit in the visible: Ba (blue), Ca (violet-blue), K (violet), Na (yellow/orange). (The other wavelengths are in the ultraviolet.)

(b) Au: shortest wavelength, highest energy

Na: longest wavelength, lowest energy

(c) $\lambda = c/\nu = \dfrac{2.998 \times 10^8 \text{ m/s}}{9.23 \times 10^{14}/\text{s}} \times \dfrac{1 \text{ nm}}{1 \times 10^{-9} \text{ m}} = 325 \text{ nm},\quad \text{Cu}$

6.84 All electromagnetic radiation travels at the same speed, 2.998×10^8 m/s. Change miles to meters and seconds to some appropriate unit of time.

$391 \times 10^6 \text{ mi} \times \dfrac{1.6093 \text{ km}}{1 \text{ mi}} \times \dfrac{1000 \text{ m}}{1 \text{ km}} \times \dfrac{1 \text{ s}}{2.998 \times 10^8 \text{ m}} \times \dfrac{1 \text{ min}}{60 \text{ s}} = 35.0 \text{ min}$

6.85 (a) $\nu = c/\lambda = \dfrac{2.998 \times 10^8 \text{ m/s}}{320 \text{ nm}} \times \dfrac{1 \text{ nm}}{1 \times 10^{-9} \text{ m}} = 9.4 \times 10^{14} \text{ s}^{-1}$

(b) $E = hc/\lambda = \dfrac{6.626 \times 10^{-34} \text{ J-s} \times 2.998 \times 10^8 \text{ m/s}}{3.2 \times 10^{-7} \text{ m}} \times \dfrac{1 \text{ kJ}}{1000 \text{ J}} \times \dfrac{6.022 \times 10^{23} \text{ photons}}{\text{mole}}$
$= 373.8 = 370 \text{ kJ/mol}$

(c) UV-B photons have shorter wavelength and higher energy.

(d) Yes. The higher energy UV-B photons would be more likely to cause sunburn.

6.86   E = hc/λ → J/photon; total energy = power × time; photons = total energy / J / photon

$$E = \frac{6.626 \times 10^{-34} \text{ J-s} \times 2.998 \times 10^8 \text{ m/s}}{780 \times 10^{-9} \text{ m}} = 2.5468 \times 10^{-19} = 2.55 \times 10^{-19} \text{ J/photon}$$

$$0.10 \text{ mW} = \frac{0.10 \times 10^{-3} \text{ J}}{1 \text{ s}} \times 69 \text{ min} \times \frac{60 \text{ s}}{1 \text{ min}} = 0.4140 = 0.41 \text{ J}$$

$$0.4140 \text{ J} \times \frac{1 \text{ photon}}{2.5468 \times 10^{-19} \text{ J}} = 1.626 \times 10^{18} = 1.6 \times 10^{18} \text{ photons}$$

6.87   (a)   If a plant appears orange, it absorbs the complementary (opposite) color on the color wheel. The plant most strongly absorbs blue light in the range 430-490 nm.

       (b) $E = hc/\lambda = \dfrac{6.626 \times 10^{-34} \text{ J-s}}{455 \text{ nm}} \times \dfrac{2.998 \times 10^8 \text{ m}}{1 \text{ s}} \times \dfrac{1 \text{ nm}}{1 \times 10^{-9} \text{ m}} = 4.37 \times 10^{-19} \text{ J}$

6.88   (a) $v = c/\lambda$; $\dfrac{2.998 \times 10^8 \text{ m}}{\text{s}} \times \dfrac{1}{542 \text{ nm}} \times \dfrac{1 \text{ nm}}{1 \times 10^{-9} \text{ m}} = 5.5314 \times 10^{14} = 5.53 \times 10^{14} \text{ s}^{-1}$

       (b)   Calculate J/photon using E = hc/λ; change to kJ/mol.

$$E_{photon} = \frac{6.626 \times 10^{-34} \text{ J-s}}{542 \times 10^{-9} \text{ m}} \times \frac{2.998 \times 10^8 \text{ m}}{\text{s}} = 3.6651 \times 10^{-19} = 3.67 \times 10^{-19} \text{ J/photon}$$

$$\frac{3.6651 \times 10^{-19} \text{ J}}{\text{photon}} \times \frac{6.022 \times 10^{23} \text{ photons}}{\text{mol}} \times \frac{1 \text{ kJ}}{1000 \text{ J}} = 220.71 = 221 \text{ kJ/mol}$$

       (c)   Let E_{total} be the total energy of an incident photon, E_{min} be the minimum energy required to eject an electron, and E_k be the "extra" energy that becomes the kinetic energy of the ejected electron.

$E_{total} = E_{min} + E_k$, $E_k = E_{total} - E_{min} = hv - hv_0$, $E_k = h(v - v_0)$. The slope of the line is the value of h, Planck's constant.

6.89   (a)   When an electron is excited to $n = \infty$, it is completely removed from the atom. The end result of this process is ionization, the production of H^+.

       (b) $n_i = 1$, $n_f = \infty$; $\Delta E = -2.18 \times 10^{-18} \text{ J} \left[\dfrac{1}{n_f^2} - \dfrac{1}{n_i^2} \right] = -2.18 \times 10^{-18} \text{ J} (1/\infty - 1/1) = 2.18 \times 10^{-18} \text{ J}$

$$\lambda = hc/E = \frac{6.626 \times 10^{-34} \text{ J-s}}{2.18 \times 10^{-18} \text{ J}} \times \frac{2.998 \times 10^8 \text{ m}}{1 \text{ s}} = 9.11 \times 10^{-8} \text{ m} = 91.1 \text{ nm}$$

       (c)   If light with a wavelength shorter than 91.1 nm is used to excite the H atom, the excess energy will become the kinetic energy of the ejected electron. (The potential energy of the ejected electron is, by definition, zero. It no longer experiences electrostatic interactions with the H atom.)

       (d)   The frequency associated with the wavelength calculated in part (b) is analogous to v_0 on the plot in Exercise 6.88. Any excess kinetic energy imparted to the ejected electron corresponds to the sloped line to the right of v_0.

6.90   (a)   "blue" cone, $\lambda_{max} = 450$ nm $= 450 \times 10^{-9}$ m

$$E = hc/\lambda = \frac{6.626 \times 10^{-34} \text{ J-s}}{450 \times 10^{-9} \text{ m}} \times \frac{2.998 \times 10^{8} \text{ m}}{1 \text{ s}} = 4.41 \times 10^{-19} \text{ J}$$

"green" cone, $\lambda_{max} = 545$ nm $= 545 \times 10^{-9}$ m

$$E = hc/\lambda = \frac{6.626 \times 10^{-34} \text{ J-s}}{545 \times 10^{-9} \text{ m}} \times \frac{2.998 \times 10^{8} \text{ m}}{1 \text{ s}} = 3.64 \times 10^{-19} \text{ J}$$

"red" cone, $\lambda_{max} = 585$ nm $= 585 \times 10^{-9}$ m

$$E = hc/\lambda = \frac{6.626 \times 10^{-34} \text{ J-s}}{585 \times 10^{-9} \text{ m}} \times \frac{2.998 \times 10^{8} \text{ m}}{1 \text{ s}} = 3.40 \times 10^{-19} \text{ J}$$

(b)   "blue" scattering efficiency $= \left(\dfrac{1}{450}\right)^4$; "green" scattering efficiency $= \left(\dfrac{1}{545}\right)^4$

ratio of "blue" to "green" $= \dfrac{\left(\dfrac{1}{450 \text{ nm}}\right)^4}{\left(\dfrac{1}{545 \text{ nm}}\right)^4} = \left(\dfrac{545}{450}\right)^4 = 2.15$

(c)   Mainly, the shorter wavelengths perceived by the "blue" cone are scattered more efficiently, so there is more of the blue light to see. Also, the amplitude of the absorption curve for the "blue" cone is greater than the amplitudes of the other two curves. This indicates that our eyes are more sensitive to blue light than the other wavelengths. (It is also true that the intensities of the different wavelengths reaching Earth are not the same, but this information is not conveyed in the exercise.)

6.91   (a)   Lines with $n_f = 1$ lie in the ultraviolet (see Solution 6.42) and with $n_f = 2$ lie in the visible (see Solution 6.41). Lines with $n_f = 3$ will have smaller ΔE and longer wavelengths and lie in the infrared.

(b)   Use Equation 6.6 to calculate ΔE, then $\lambda = hc/\Delta E$.

$$n_i = 4, \, n_f = 3; \, \Delta E = -2.18 \times 10^{-18} \text{ J}\left[\frac{1}{n_f^2} - \frac{1}{n_i^2}\right] = -2.18 \times 10^{-18} \text{ J} (1/9 - 1/16)$$

$$\lambda = hc/E = \frac{6.626 \times 10^{-34} \text{ J-s} \times 2.998 \times 10^{8} \text{ m/s}}{-2.18 \times 10^{-18} (1/9 - 1/16)} = 1.87 \times 10^{-6} \text{ m}$$

$$n_i = 5, \, n_f = 3; \, \lambda = hc/E = \frac{6.626 \times 10^{-34} \text{ J-s} \times 2.998 \times 10^{8} \text{ m/s}}{-2.18 \times 10^{-18} (1/9 - 1/25)} = 1.28 \times 10^{-6} \text{ m}$$

$$n_i = 6, \, n_f = 3; \, \lambda = hc/E = \frac{6.626 \times 10^{-34} \text{ J-s} \times 2.998 \times 10^{8} \text{ m/s}}{-2.18 \times 10^{-18} (1/9 - 1/36)} = 1.09 \times 10^{-6} \text{ m}$$

These three wavelengths are all greater than 1 μm or 1×10^{-6} m. They are in the infrared, close to the visible edge (0.7×10^{-6} m).

6.92   (a)   Gaseous atoms of various elements in the sun's atmosphere typically have ground state electron configurations. When these atoms are exposed to radiation from the sun, the electrons change from the ground state to one of several allowed excited states. Atoms absorb the wavelengths of light that correspond to these allowed energy changes. All other wavelengths of solar radiation pass through the atmosphere unchanged. Thus, the dark lines are the wavelengths that correspond to allowed energy changes in atoms of the solar atmosphere. The continuous background is all other wavelengths of solar radiation.

           (b)   The scientist should record the absorption spectrum of pure neon or other elements of interest. The Fraunhofer lines that belong to a particular element will appear at the same wavelength as the lines in the absorption spectrum of that element.

6.93   (a)   not valid, m_l cannot be greater than l.

           (b)   valid

           (c)   valid

           (d)   not valid, the only possible values for m_s are $+ \frac{1}{2}$ and $- \frac{1}{2}$.

           (e)   not valid, the maximum value for l is $(n-1)$.

6.94   (a) He^+ is hydrogen-like because it is a one-electron particle. An He atom has two electrons. The Bohr model is based on the interaction of a single electron with the nucleus but does not accurately account for additional interactions when two or more electrons are present.

           (b)   Divide each energy by the smallest value to find the integer relationship.

                H: $-2.18 \times 10^{-18} / -2.18 \times 10^{-18} = 1; \ Z = 1$

                He^+: $-8.72 \times 10^{-18} / -2.18 \times 10^{-18} = 4; \ Z = 2$

                Li^{2+}: $-1.96 \times 10^{-17} / -2.18 \times 10^{-18} = 9; \ Z = 3$

                The ground-state energies are in the ratio of 1:4:9, which is also the ratio Z^2, the square of the nuclear charge for each particle.

                The ground state energy for hydrogen-like particles is:

                $E = R_H Z^2$. (By definition, $n = 1$ for the ground state of a one-electron particle.)

           (c) $Z = 6$ for C^{5+}. $E = -2.18 \times 10^{-18} \, J \, (6)^2 = -7.85 \times 10^{-17} \, J$

6.95   *Plan.* Calculate v from kinetic energy. $\lambda = h/mv$.   *Solve.*

         $E_k = mv^2/2; \ v^2 = 2E_k/m; \ v = \sqrt{2E_k/m}$

$$ v = \left(\frac{2 \times 2.147 \times 10^{-15} \ \text{kg-m}^2/\text{s}^2}{9.1094 \times 10^{-31} \ \text{kg}} \right)^{1/2} = 6.866 \times 10^7 = 6.87 \times 10^7 \ \text{m/s} $$

$$ \lambda = h/mv = \frac{6.626 \times 10^{-34} \ \text{J-s}}{9.1094 \times 10^{-31} \ \text{kg} \times 6.866 \times 10^7 \ \text{m/s}} \times \frac{1 \ \text{kg-m}^2/\text{s}^2}{1 \ \text{J}} = 1.06 \times 10^{-11} \ \text{m} = 10.6 \ \text{pm} $$

6.96   Heisenberg postulated that the dual nature of matter places a limitation on how precisely we can know both the position and momentum of an object. This limitation is significant at the subatomic particle level. The *Star Trek* transporter (presumably) disassembles humans into their protons, neutrons, and electrons, moves the particles at high speed (possibly the speed of light) to a new location, and reassembles the particles into the human. Heisenberg's uncertainty principle indicates that if we know the momentum (mv) of the moving particles, we can't precisely know their position (x). If a few of the subatomic particles don't arrive in exactly the correct location, the human would not be reassembled in their original form. So, the "Heisenberg compensator" is necessary to make sure that the transported human arrives at the new location intact.

6.97   (a)   A subatomic particle is so small that we must use light to measure its position. The interacting photons will impart some momentum to the subatomic particle, thus disturbing it. The shorter the wavelength of the photon, the more accurate the measurement but the greater the momentum imparted to the particle and the bigger the disturbance.

          (b)   An ongoing discussion in quantum theory is whether we can know the quantum states of a system without observing and thus disturbing the system. One interpretation of quantum theory indicated that a system could have multiple acceptable states before it was observed in a single state. That is, the act of observation defined the state. Schrodinger articulated this question on a macroscopic level with his "cat" paradox. Recently, physicists have devised clever ways to observe "cat states," where the act of observing does not destroy the simultaneous states.

6.98   (a)   Probability density, $[\psi(r)]^2$, is the probability of finding an electron at a single point at distance r from the nucleus. The radial probability function, $4\pi r^2$, is the probability of finding an electron at any point on the sphere defined by radius r. $P(r) = 4\pi r^2 [\psi(r)]^2$.

          (b)   The term $4\pi r^2$ explains the differences in plots of the two functions. Plots of the probability density, $[\psi(r)^2]$, for s orbitals shown in Figure 6.22 each have their maximum value at r = 0, with $(n-1)$ smaller maxima at greater values of r. The plots of radial probability, P(r), for the same s orbitals shown in Figure 6.19 have values of zero at r = 0 and the size of the maxima increases. P(r) is the product of $[\psi(r)]^2$ and $4\pi r^2$. At r = 0, the value of $[\psi(r)]^2$ is finite and large, but the value of $4\pi r^2$ is zero, so the value of P(r) is zero. As r increases, the values of $[\psi(r)]^2$ vary as shown in Figure 6.22, but the values of $4\pi r^2$ increase continuously, leading to the increasing size of P(r) maxima as r increases.

          (c)

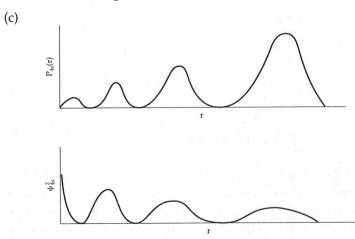

6.99   (a)   The p_z orbital has a nodal plane where z = 0. This is the xy plane.

(b)   The d_{xy} orbital has 4 lobes and 2 nodal planes, the 2 planes where x = 0 and y = 0. These are the yz and xz planes.

(c)   The $d_{x^2-y^2}$ has 4 lobes and 2 nodal planes, the planes where $x^2 - y^2 = 0$. These are the planes that bisect the x and y axes and contain the z axis.

6.100   (a)   This frequency is in the radiowave portion of the electromagnetic spectrum. One megahertz is 1×10^6 Hz; 600 MHz is 6×10^8 Hz. According to Figure 6.4, this frequency is in the radiowave region.

(b) $\Delta E = h\nu = 6.626 \times 10^{-34}$ J-s $\times \dfrac{450 \times 10^6}{s} = 2.98 \times 10^{-25}$ J

(c)   The proton. When radiowaves are absorbed during an NMR or MRI experiment, it is the spin of a nucleus that changes. The nucleus of a hydrogen atom is a single proton.

6.101   If m_s had three allowed values instead of two, each orbital would hold three electrons instead of two. Assuming that the same orbitals are available (that there is no change in the n, l, and m_l values), the number of elements in each of the first four rows would be:

1st row: 1 orbital $\times 3 =$ 3 elements

2nd row: 4 orbitals $\times 3 = 12$ elements

3rd row: 4 orbitals $\times 3 = 12$ elements

4th row: 9 orbitals $\times 3 = 27$ elements

The s-block would be 3 columns wide, the p-block 9 columns wide and the d-block 15 columns wide.

6.102   (a)   Br: [Ar]$4s^2 3d^{10} 4p^5$, 1 unpaired electron

(b)   Ga: [Ar]$4s^2 3d^{10} 4p^1$, 1 unpaired electron

(c)   Hf: [Xe]$6s^2 4f^{14} 5d^2$, 2 unpaired electrons

(d)   Sb: [Kr]$5s^2 4d^{10} 5p^3$, 3 unpaired electrons

(e)   Bi: [Xe]$6s^2 4f^{14} 5d^{10} 6p^3$, 3 unpaired electrons

(f)   Sg: [Rn]$7s^2 5f^{14} 6d^4$, 4 unpaired electrons

6.103   The core would be the electron configuration of element 118. If no new subshell begins to fill, the condensed electron configuration of element 126 would be similar to those of elements vertically above it on the periodic chart, Pu and Sm. The condensed configuration would be [118]$8s^2 6f^6$. On the other hand, the 5g subshell could begin to fill after 8s, resulting in the condensed configuration [118]$8s^2 5g^6$. Exceptions are also possible (likely).

6.104   (a)   Neutral H atoms have a single unpaired electron. For a beam of H atoms to be deflected by a magnetic field, the unpaired electrons must interact with the magnetic field. For a single beam of atoms to be split into two beams that are deflected in opposite directions, the unpaired electron on each atom must have some characteristic that can interact in two different ways with a magnetic field. We call that characteristic "electron spin." The significance of this observation is experimental evidence for electron spin.

(b)   If the strength of the magnetic field were increased, the magnitude of the deflection would increase.

(c)   If H atoms were replaced by He atoms, no deflection would occur. Helium has no unpaired electrons to interact with the magnetic field.

(d)   The electron configuration of Ag is $[Kr]5s^14d^{10}$. Neutral Ag atoms each have a single unpaired electron. Each unpaired electron has two possible m_s values and a beam of Ag atoms will be split in two by the magnetic field.

Integrative Exercises

6.105   (a)   We know the wavelength of microwave radiation, the volume of coffee to be heated, and the desired temperature change. Assume the density and heat capacity of coffee are the same as pure water. We need to calculate: (i) the total energy required to heat the coffee and (ii) the energy of a single photon to find (iii) the number of photons required.

(i)   From Chapter 5, the heat capacity of liquid water is 4.184 J/g-°C.

To find the mass of 200 mL of coffee at 23 °C, use the density of water given in Appendix B.

$$200 \text{ mL} \times \frac{0.997 \text{ g}}{1 \text{ mL}} = 199.4 = 199 \text{ g coffee}$$

$$\frac{4.184 \text{ J}}{1 \text{ g-°C}} \times 199.4 \text{ g} \times (60 \text{ °C} - 23 \text{ °C}) = 3.087 \times 10^4 \text{ J} = 31 \text{ kJ}$$

(ii) $E = hc/\lambda = 6.626 \times 10^{-34} \text{ J-s} \times \dfrac{2.998 \times 10^8 \text{ m}}{1 \text{ s}} \times \dfrac{1}{0.112 \text{ m}}$

$$= \frac{1.77 \times 10^{-24} \text{ J}}{1 \text{ photon}}$$

(iii) $3.087 \times 10^4 \text{ J} \times \dfrac{1 \text{ photon}}{1.774 \times 10^{-24} \text{ J}} = 1.7 \times 10^{28} \text{ photons}$

(The answer has 2 sig figs because the temperature change, 37 °C, has 2 sig figs.)

(b)   1 W = 1 J/s. 900 W = 900 J/s. From part (a), 31 kJ are required to heat the coffee.

$$3.087 \times 10^4 \text{ J} \times \frac{1 \text{ s}}{900 \text{ J}} = 34.30 = 34 \text{ s}$$

6.106 $\Delta H^\circ_{rxn} = \Delta H^\circ_f \ O_2(g) + \Delta H^\circ_f \ O(g) - \Delta H^\circ_f \ O_3(g)$

$\Delta H^\circ_{rxn} = 0 + 247.5 \text{ kJ} - 142.3 \text{ kJ} = +105.2 \text{ kJ}$

$$\frac{105.2 \text{ kJ}}{\text{mol O}_3} \times \frac{1 \text{ mol O}_3}{6.022 \times 10^{23} \text{ molecules}} \times \frac{1000 \text{ J}}{1 \text{ kJ}} = \frac{1.747 \times 10^{-19} \text{ J}}{\text{O}_3 \text{ molecule}}$$

$$\Delta E = hc/\lambda; \lambda = \frac{hc}{\Delta E} = \frac{6.626 \times 10^{-34} \text{ J-s} \times 2.998 \times 10^8 \text{ m/s}}{1.747 \times 10^{-19} \text{ J}} = 1.137 \times 10^{-6} \text{ m}$$

Radiation with this wavelength is in the infrared portion of the spectrum. (Clearly, processes other than simple photodissociation cause O_3 to absorb ultraviolet radiation.)

6.107   (a)   The electron configuration of Zr is $[Kr]5s^2 4d^2$ and that of Hf is $[Xe]6s^2 4f^{14} 5d^2$. Although Hf has electrons in f orbitals as the rare earth elements do, the 4f subshell in Hf is filled, and the 5d electrons primarily determine the chemical properties of the element. Thus, Hf should be chemically similar to Zr rather than the rare earth elements.

         (b) $ZrCl_4(s) + 4\, Na(l) \rightarrow Zr(s) + 4\, NaCl(s)$

                This is an oxidation–reduction reaction; Na is oxidized and Zr is reduced.

         (c) $2\, ZrO_2(s) + 4\, Cl_2(g) + 3\, C(s) \rightarrow 2\, ZrCl_4(s) + CO_2(g) + 2\, CO(g)$

                $$55.4\,g\,ZrO_2 \times \frac{1\,mol\,ZrO_2}{123.2\,g\,ZrO_2} \times \frac{2\,mol\,ZrCl_4}{2\,mol\,ZrO_2} \times \frac{233.0\,g\,ZrCl_4}{1\,mol\,ZrCl_4} = 105\,g\,ZrCl_4$$

         (d)   In ionic compounds of the type MCl_4 and MO_2, the metal ions have a 4+ charge, indicating that the neutral atoms have lost 4 electrons. Zr, $[Kr]5s^2 4d^2$, loses the 4 electrons beyond its Kr core configuration. Hf, $[Xe]6s^2 4f^{14} 5d^2$, similarly loses its four 6s and 5d electrons, but not electrons from the "complete" 4f subshell.

6.108   (a)   Each oxide ion, O^{2-}, carries a 2- charge. Each metal oxide is a neutral compound, so the metal ion or ions must adopt a total positive charge equal to the total negative charge of the oxide ions in the compound. The table below lists the electron configuration of the neutral metal atom, the positive charge of each metal ion in the oxide, and the corresponding electron configuration of the metal ion.

                i.   K: $[Ar]\,4s^1$ 1+ $[Ar]$

                ii.   Ca: $[Ar]\,4s^2$ 2+ $[Ar]$

                iii.   Sc: $[Ar]\,4s^2 3d^1$ 3+ $[Ar]$

                iv.   Ti: $[Ar]\,4s^2 3d^2$ 4+ $[Ar]$

                v.   V: $[Ar]\,4s^2 3d^3$ 5+ $[Ar]$

                vi.   Cr: $[Ar]\,4s^1 3d^5$ 6+ $[Ar]$

                Each metal atom loses all (valence) electrons beyond the Ar core configuration. In K_2O, Sc_2O_3, and V_2O_5, where the metal ions have odd charges, 2 metal ions are required to produce a neutral oxide.

         (b)   i.   potassium oxide
                ii.   calcium oxide
                iii.   scandium (III) oxide
                iv.   titanium (IV) oxide
                v.   vanadium (V) oxide
                vi.   chromium (VI) oxide

                (Roman numerals are required to specify the charges on the transition metal ions, because more than one stable ion may exist.)

(c)   Recall that $\Delta H_f^\circ = 0$ for elements in their standard states. In these reactions, M(s) and $H_2(g)$ are elements in their standard states.

i. $K_2O(s) + H_2(g) \rightarrow 2\,K(s) + H_2O(g)$

$\Delta H^\circ = \Delta H_f^\circ\,H_2O(g) + 2\,\Delta H_f^\circ\,K(s) - \Delta H\,K_2O(s) - \Delta H_f^\circ\,H_2(g)$

$\Delta H^\circ = -241.82\text{ kJ} + 2(0) - (-363.2\text{ kJ}) - 0 = 121.4\text{ kJ}$

ii. $CaO(s) + H_2(g) \rightarrow Ca(s) + H_2O(g)$

$\Delta H^\circ = \Delta H_f^\circ\,H_2O(g) + \Delta H_f^\circ\,Ca(s) - \Delta H_f^\circ\,CaO(s) - \Delta H_f^\circ\,H_2(g)$

$\Delta H^\circ = -241.82\text{ kJ} + 0 - (-635.1\text{ kJ}) - 0 = 393.3\text{ kJ}$

iii. $TiO_2(s) + 2\,H_2(g) \rightarrow Ti(s) + 2\,H_2O(g)$

$\Delta H^\circ = 2\,\Delta H_f^\circ\,H_2O(g) + \Delta H_f^\circ\,Ti(s) - \Delta H_f^\circ\,TiO_2(s) - 2\,\Delta H_f^\circ\,H_2(g)$

$= 2(-241.82) + 0 - (-938.7) - 2(0) = 455.1\text{ kJ}$

iv. $V_2O_5(s) + 5\,H_2(g) \rightarrow 2\,V(s) + 5\,H_2O(g)$

$\Delta H^\circ = 5\,\Delta H_f^\circ\,H_2O(g) + 2\,\Delta H_f^\circ\,V(s) - \Delta H_f^\circ\,V_2O_5(s) - 5\,\Delta H_f^\circ\,H_2(g)$

$= 5(-241.82) + 2(0) - (-1550.6) - 5(0) = 341.5\text{ kJ}$

(d) ΔH_f° becomes more negative moving from left to right across this row of the periodic chart. Because Sc lies between Ca and Ti, the median of the two ΔH_f° values is approximately –785 kJ/mol. However, the trend is clearly not linear. Dividing the ΔH_f° values by the positive charge on the pertinent metal ion produces the values –363, –318, –235, and –310. The value between Ca^{2+} (–318) and Ti^{4+} (–235) is Sc^{3+} (–277). Multiplying (–277) by 3, a value of approximately –830 kJ results. A reasonable range of values for ΔH_f° of $Sc_2O_3(s)$ is then –785 to – 830 kJ/mol.

6.109   (a)   Bohr's theory was based on the Rutherford "nuclear" model of the atom. That is, Bohr theory assumed a dense positive charge at the center of the atom and a diffuse negative charge (electrons) surrounding it. Bohr's theory then specified the nature of the diffuse negative charge. The prevailing theory before the nuclear model was Thomson's plum pudding or watermelon model, with discrete electrons scattered about a diffuse positive charge cloud. Bohr's theory could not have been based on the Thomson model of the atom.

(b)   De Broglie's hypothesis is that electrons exhibit both particle and wave properties. Thomson's conclusion that electrons have mass is a particle property, whereas the nature of cathode rays is a wave property. De Broglie's hypothesis actually rationalizes these two seemingly contradictory observations about the properties of electrons.

6.110   (a) ^{238}U: 92 p, 146 n, 92 e; ^{235}U: 92 p, 143 n, 92 e

In keeping with the definition isotopes, only the number of neutrons is different in the two nuclides. Because the two isotopes have the same number of electrons, they will have the same electron configuration.

(b)   U: $[Rn]7s^2 5f^4$

(c)   From Figure 6.31, the actual electron configuration is $[Rn]7s^2 5f^3 6d^1$. The energies of the 6d and 5f orbitals are very close, and electron configurations of many actinides include 6d electrons.

(d) $^{238}_{92}U \rightarrow {}^{234}_{90}Th + {}^{4}_{2}He$ ^{234}Th has 90 p, 144 n, 90 e. ^{238}U has lost 2 p, 2 n, 2 e.

These are organized into ${}^{4}_{2}He$ shown in the nuclear reaction above.

(e)   From Figure 6.31, the electron configuration of Th is $[Rn]7s^2 6d^2$. This is not really surprising because there are so many rare earth electron configurations that are exceptions to the expected orbital filling order. However, Th is the only rare earth that has two d valence electrons. Furthermore, the configuration of Th is different than that of Ce, the element above it on the periodic chart, so the electron configuration is at least interesting.

7 Periodic Properties of the Elements

Visualizing Concepts

7.1 *Analyze/Plan.* Consider Equation 7.1 in relation to the illustration (and Figure 7.2). The intensity of the bulb represents the nuclear charge, Z. The thickness of the frosting represents the shielding, S. *Solve.*

(a) Moving from boron to carbon, the intensity of the bulb increases because Z increases from 5 to 6. The thickness of the frosting stays the same because the core electron configuration is the same for both atoms (The added electron occupies its own p orbital, so electron repulsion doesn't change much.)

(b) Moving from boron to aluminum, the intensity of the bulb increases because Z increases from 5 to 13. The thickness of the frosting also increases because Al has the core configuration of Ne, while B has the core of He. (We know from the chapter that the increase in Z dominates and Z_{eff} increases slightly going down a column; the 3p valence electron of Al "sees" a brighter light than the 2p valence electron of B.)

7.2 The order of radii is $Br^- > Br > F$, so the largest brown sphere is Br^-, the intermediate blue one is Br, and the smallest red one is F.

7.3 (a) Mg^{2+} is isoelectronic with Ne, K^+, and Cl^- are isoelectronic with Ar, and Se^{2-} is isoelectronic with Kr. The atomic radii of the noble gases increase moving down the column, so this gives the rough order of size for the corresponding isoelectronic ions. The Cl^- ion is larger than K^+ because it has a smaller positive nuclear charge holding the same number and configuration of electrons. The order of ionic radii is then $Mg^{2+} < K^+ < Cl^- < Se^{2-}$; these ions match the spheres moving from left to right.

(b) Ca^{2+} and S^{2-} are both isoelectronic with Ar, as are K^+ and Cl^-. For ions in an isoelectronic series, the larger the nuclear charge, the smaller the ionic radius. Ca^{2+} is smaller than K^+ and fits between the two leftmost spheres. S^{2-} is larger than Cl^- and fits between the two rightmost spheres.

7.4 The red sphere represents a metal and the blue sphere represents a nonmetal. The size of the red sphere decreases on reaction, so it loses one or more electrons and becomes a cation. Metals lose electrons when reacting with nonmetals, so the red sphere represents a metal. The size of the blue sphere increases on reaction, so it gains one or more electrons and becomes an anion. Nonmetals gain electrons when reacting with metals, so the blue sphere represents a nonmetal.

7.5   (a)   The bonding atomic radius of A, r_A, is $d_1/2$. The distance d_2 is the sum of the bonding atomic radii of A and X, $r_A + r_X$. We know that $r_A = d_1/2$, so $d_2 = r_X + d_1/2$, $r_X = d_2 - d_1/2$.

    (b)   The length of the X–X bond is $2r_X$.

          $2r_X = 2(d_2 - d_1/2) = 2d_2 - d_1$.

7.6   (a)   The $3d$ subshell is missing.

    (b)   Statement (ii) is the best description of why the $2s$ and $2p$ subshells have different energies in Na.

    (c)   In a Na atom, the highest energy electron is in the $3s$ subshell.

    (d)   In a Na vapor lamp, the highest energy $3s$ electron is excited into the empty $3p$ subshell.

7.7   The trend for bonding atomic radius (1) is shown in chart (iii).
The trend for first ionization energy (2) is shown in chart (ii).
The trend for effective nuclear charge is shown in chart (i).

7.8   (a) $X + 2F_2 \rightarrow XF_4$

    (b)   If X is a nonmetal, XF_4 is a molecular compound. If X is a metal, XF_4 is ionic. For an ionic compound with this formula, X would have a charge of 4+, and a much smaller bonding atomic radius than F^-. X in the diagram has about the same bonding radius as F, so it is likely to be a nonmetal.

Periodic Table; Effective Nuclear Charge (Sections 7.1 and 7.2)

7.9   (a)   The results are 2, 8, 18, 32.

    (b)   The atomic numbers of the noble gases are 2, 10, 18, 36, 54, and 86. The differences between sequential pairs of these atomic numbers is 8, 8, 18, 18, and 32. These differences correspond to the results in (a). They represent the filling of new subshells when moving across the next row of the periodic chart.

    (c)   The Pauli exclusion principle is the source of the "2" in the expressions in part (a). The Pauli principle states that no two electrons can have the same four quantum numbers. Because m_s has only two possible values, the consequence is that an atomic orbital can hold a maximum of "2" electrons.

7.10   Assuming *eka-* means one place below or under, *eka-manganese* on Figure 7.1 is technetium, Tc.

7.11   (a)   According to Figure 7.1, of the elements listed, only Fe was known before 1700.

    (b)   The seven metals known in ancient times, Fe, Cu, Ag, Sn, Au, Hg, and Pb, are mostly near the bottom of the activity series, Table 4.5. These less active metals are present in nature in elemental form; they can be observed directly and their isolation does not require chemical processing.

7.12   (a)   In order of increasing atomic mass, the element following chlorine is potassium, K.

        (b)   Potassium is a reactive metal that is solid at room temperature and pressure. Elements in group 8A are the noble gases. All are unreactive nonmetals that exist as gases at ambient conditions.

7.13   *Analyze/Plan.* Z_{eff} values for elements 3–18 are shown as the green line in Figure 7.5. Use Equation 7.1 to calculate Z_{eff} values for elements 1 and 2, H and He. Compare the values. *Solve.*

$Z_{eff} = Z - S$, where Z is the atomic number and S is the number of core electrons. For H and He, the valence electrons have $n = 1$, and there are no core electrons; $S = 0$.

H: $1 - 0 = 1$; He: $2 - 0 = 2$.

The minimum value of Z_{eff} from Figure 7.5 is 1 for Li and Na. The maximum value is 8 for Ne and Ar.

For elements 1–18, H, Li and Na have minimum values of Z_{eff}; Ne and Ar have maximum values.

7.14   Statement (iii) is incorrect. Because of the nearly uniform spherical distribution of the core electrons, they screen much more effectively than valence electrons.

7.15   (a)   *Analyze/Plan.* $Z_{eff} = Z - S$. Find the atomic number, Z, of Na and K. Write their electron configurations and count the number of core electrons. Assume S = number of core electrons.

           Solve. Na: $Z = 11$; $[Ne]3s^1$. In the Ne core there are 10 electrons. $Z_{eff} = 11 - 10 = 1$. K: $Z = 19$; $[Ar]4s^1$. In the Ar core there are 18 electrons. $Z_{eff} = 19 - 18 = 1$.

        (b)   *Analyze/Plan.* $Z_{eff} = Z - S$. Write the complete electron configuration for each element to show counting for Slater's rules. $S = 0.35$ [# of electrons with same *n*] + 0.85 [# of electrons with $(n-1)$] + 1[# of electrons with $(n-2)$].

           Solve. Na: $1s^2 2s^2 2p^6 3s^1$. $S = 0.35(0) + 0.85(8) + 1(2) = 8.8$. $Z_{eff} = 11 - 8.8 = 2.2$

           K: $1s^2 2s^2 2p^6 3s^2 3p^6 4s^1$. $S = 0.35(0) + 0.85(8) + 1(10) = 16.8$. $Z_{eff} = 19 - 16.8 = 2.2$

        (c)   For both Na and K, the two values of Z_{eff} are 1.0 and 2.2. The Slater value of 2.2 is closer to the values of 2.51 (Na) and 3.49 (K) obtained from detailed calculations.

        (d)   Both approximations, "core electrons 100% effective" and Slater, yield the same value of Z_{eff} for Na and K. Neither approximation accounts for the gradual increase in Z_{eff} moving down a group.

        (e)   Following the trend from detailed calculations, we predict a Z_{eff} value of approximately 4.5 for Rb.

7.16   Follow the method in the preceding question to calculate Z_{eff} values.

        (a)   Si: $Z = 14$; $[Ne]3s^2 3p^2$. 10 electrons in the Ne core. $Z_{eff} = 14 - 10 = 4$
           Cl: $Z = 17$; $[Ne]3s^2 3p^5$. 10 electrons in the Ne core. $Z_{eff} = 17 - 10 = 7$

        (b)   Si: $1s^2 2s^2 2p^6 3s^2 3p^2$. $S = 0.35(3) + 0.85(8) + 1(2) = 9.85$. $Z_{eff} = 14 - 9.85 = 4.15$
           Cl: $1s^2 2s^2 2p^6 3s^2 3p^5$. $S = 0.35(6) + 0.85(8) + 1(2) = 10.90$. $Z_{eff} = 17 - 10.90 = 6.10$

        (c)   The Slater values of 4.15 (Si) and 6.10 (Cl) are closer to the results of detailed calculations, 4.29 (Si) and 6.12 (Cl).

(d)   The Slater method of approximation more closely approximates the gradual increase in Z_{eff} moving across a row. The "core 100%-effective" approximation underestimates Z_{eff} for Si but overestimates it for Cl. Slater values are closer to detailed calculations and a better indication of the change in Z_{eff} moving from Si to Cl.

(e)   Relative to Si, P has one more proton (Z + 1) and one more 3p electron (S + 0.35). It is reasonable to predict that the difference in Z_{eff} will be +0.65. That is, Z_{eff} for P will be (4.15 + 0.65) = 4.80.

7.17   Krypton has a larger nuclear charge (Z = 36) than argon (Z = 18). The shielding of electrons in the $n = 3$ shell by the 1s, 2s, and 2p core electrons in the two atoms is approximately equal, so the $n = 3$ electrons in Kr experience a greater effective nuclear charge and are thus situated closer to the nucleus.

7.18   Mg < P < K < Ti < Rh. The shielding of electrons in the $n = 3$ shell by 1s, 2s, and 2p core electrons in these elements is approximately equal, so the effective nuclear charge increases as Z increases.

Atomic and Ionic Radii (Section 7.3)

7.19   The quantity described in (b) must be measured experimentally in order to determine the bonding atomic radius of an atom. Bonding atomic radius is a property of a bonded atom. The measurement must be done on an atom participating in a chemical bond.

7.20   (a)   In the figure, the relevant distance is 3.72 Å. This is the distance between two Ar atoms that are touching. One-half this distance is the effective radius of an Ar atom in the close packed structure, 1.86 Å.

(b)   According to Figure 7.7, the bonding atomic radius of Ar is 1.06 Å. This value is significantly smaller than 1.86 Å, the effective radius of Ar in the close packed solid.

(c)   No, the Ar atoms are not held together by chemical bonds in the close packed solid. According to Figure 7.7, an Ar–Ar chemical bond would have a distance of approximately 2.12 Å, a much closer approach than the 3.72 Å in the close packed solid.

7.21   (a)   The atomic (*metallic*) radius of W is the interatomic W–W distance divided by two, 2.74 Å/2= 1.37 Å.

(b)   Under high pressure, we expect atoms in a pure substance to move closer together. That is, the distance between W atoms will decrease.

7.22   Statement (iv) is incorrect. Moving left to right in a particular period, the significant nuclear buildup while adding electrons into the same *d* subshell causes Z to increase and radii to decrease.

7.23   From bonding atomic radii in Figure 7.7, As–I = 1.19 Å + 1.39 Å = 2.58 Å. This is very close to the experimental value of 2.55 Å in AsI_3.

7.24   Bi–I = 2.81 Å = r_{Bi} + r_I. From Figure 7.7, r_I = 1.39 Å.

r_{Bi} = [Bi–I] – r_I = 2.81 Å – 1.39 Å = 1.42 Å.

7.25   *Plan.* Locate each element on the periodic charge and use trends in radii to predict their order.   *Solve.*

    (a)   Cs > K > Li   (b) Pb > Sn > Si   (c) N > O > F

7.26   (a)   Na < Ca < Ba   (b) As < Sn < In

    (c)   Be < Si < Al. This order assumes the increase in radius from the second to the third row is greater than the decrease moving right in the third row. Radii in Figure 7.7 confirm this assumption.

7.27   (a)   False. Cations are smaller than their corresponding neutral atoms. Electrostatic repulsions are reduced by removing an electron from a neutral atom, Z_{eff} increases, and the cation is smaller.

    (b)   True. [See (a) above.]

    (c)   False. I^- is bigger than Cl^-. Going down a column, the n value of the valence electrons increases and they are farther from the nucleus. Valence electrons also experience greater shielding by core electrons. The greater radial extent of the valence electrons outweighs the increase in Z, and the size of particles with like charge increases.

7.28   (a)   As Z stays constant and the number of electrons increases, the electron-electron repulsions increase, the electrons spread apart, and the anion becomes larger. The reverse is true for the cation, which becomes smaller than the neutral atom.

        $I^- > I > I^+$

    (b)   For cations with the same charge, ionic radii increase going down a column because there is an increase in the principle quantum number and the average distance from the nucleus of the outer electrons.

        $Ca^{2+} > Mg^{2+} > Be^{2+}$

    (c)   Fe: $[Ar]4s^2 3d^6$; Fe^{2+}: $[Ar]3d^6$; Fe^{3+}: $[Ar]3d^5$. The 4s valence electrons in Fe are on average farther from the nucleus than the 3d electrons, so Fe is larger than Fe^{2+}. Because there are five 3d orbitals, in Fe^{2+} at least one orbital must contain a pair of electrons. Removing one electron to form Fe^{3+} significantly reduces repulsion, increasing the nuclear charge experienced by each of the other d electrons and decreasing the size of the ion. $Fe > Fe^{2+} > Fe^{3+}$

7.29 Ga^{3+}: none; Zr^{4+}: Kr; Mn^{7+}: Ar; I^-: Xe; Pb^{2+}: Hg

7.30   (a) Cl^-: Ar   (b) Sc^{3+}: Ar

    (c) Fe^{2+}: $[Ar]3d^6$. There is no neutral atom with the same electron configuration. Fe^{2+} has 24 electrons. Neutral Cr has 24 electrons, $[Ar]4s^1 3d^5$. Because transition metals fill the s subshell first but also lose s electrons first when they form ions, many transition metal ions do not have neutral atoms with the same electron configuration.

    (d) Zn^{2+}: $[Ar]3d^{10}$. There is no neutral atom with same electron configuration.

    (e) Sn^{4+}: $[Kr]4d^{10}$; a neutral Pd atom has 46 electrons and an anomalous electron configuration which is the same as the electron configuration of Sn^{4+}.

7.31   (a)   *Analyze/Plan.* Follow the logic in Sample Exercise 7.4.

Solve. Na^+ is smaller. Because F^- and Na^+ are isoelectronic, the ion with the larger nuclear charge, Na^+, has the smaller radius.

(b)   *Analyze/Plan.* The electron configuration of the ions is [Ne] or $[He]2s^2 2p^6$. The ions have either 10 core electrons or 2 core electrons. Apply Equation 7.1 to both cases and check the result.

Solve. F^-: $Z = 9$. For 10 core electrons, $Z_{eff} = 9 - 10 = -1$. Although we might be able to interpret a negative value for Z_{eff}, positive values will be easier to compare; we will assume a He core of 2 electrons.
$F^-, Z = 9$. $Z_{eff} = 9 - 2 = 7$. $Na^+: Z_{eff} = 11 - 2 = 9$

(c)   *Analyze/Plan.* The electron of interest has $n = 2$. There are seven other $n = 2$ electrons, and two $n = 1$ electrons.
Solve. $S = 0.35(7) + 0.85(2) + 1(0) = 4.15$
F^-: $Z_{eff} = 9 - 4.15 = 4.85$. $Na^+: Z_{eff} = 11 - 4.15 = 6.85$

(d)   For isoelectronic ions (without d electrons), the electron configurations and therefore shielding values (S) are the same. Only the nuclear charge changes. So, as nuclear charge (Z) increases, effective nuclear charge (Z_{eff}) increases and ionic radius decreases.

7.32   (a) K^+ (larger Z) is smaller.

(b) Cl^- and K^+: $[Ne]3s^2 3p^6$. 10 core electrons

$Cl^-, Z = 17$. $Z_{eff} = 17 - 10 = 7$
$K^+, Z = 19$. $Z_{eff} = 19 - 10 = 9$

(c)   Valence electron, $n = 3$; seven other $n = 3$ electrons; eight $n = 2$ electrons; two $n = 1$ electrons. $S = 0.35(7) + 0.85(8) + 1(2) = 11.25$
$Cl^-: Z_{eff} = 17 - 11.25 = 5.75$. $K^+: Z_{eff} = 19 - 11.25 = 7.75$

(d)   For isoelectronic ions (without d electrons), the electron configurations and therefore shielding values (S) are the same. Only the nuclear charge changes. So, as nuclear charge (Z) increases, effective nuclear charge (Z_{eff}) increases and ionic radius decreases.

7.33   *Analyze/Plan.* Use relative location on periodic chart and trends in atomic and ionic radii to establish the order.

(a) $Cl < S < K$ (b) $K^+ < Cl^- < S^{2-}$

(c)   Even though K has the largest Z value, the *n*-value of the outer electron is larger than the *n*-value of valence electrons in S and Cl so K atoms are largest. When the 4s electron is removed, K^+ is isoelectronic with Cl^- and S^{2-}. The larger Z value causes the 3p electrons in K^+ to experience the largest effective nuclear charge and K^+ is the smallest ion.

7.34   (a) $Se < Se^{2-} < Te^{2-}$ (b) $Co^{3+} < Fe^{3+} < Fe^{2+}$ (c) $Ti^{4+} < Sc^{3+} < Ca$ (d) $Be^{2+} < Na^+ < Ne$

7.35   (a) O^{2-} is larger than O because the increase in electron repulsions that accompany addition of an electron causes the electron cloud to expand.

   (b) S^{2-} is larger than O^{2-} because for particles with like charges, size increases going down a family.

   (c) S^{2-} is larger than K^+ because the two ions are isoelectronic and K^+ has the larger Z and Z_{eff}.

   (d) K^+ is larger than Ca^{2+} because the two ions are isoelectronic and Ca^{2+} has the larger Z and Z_{eff}.

7.36   Make a table of d(measured), d(ionic radii), d(covalent radii), as well as differences between measured and estimated values. The estimated distances are just the sum of the various ionic radii from Figure 7.8 and covalent radii from Figure 7.7. All distances and differences are given in Å. Use these values to judge accuracy in part (c).

   (a,b)

	d(meas)	(a) d(ion)	(b) Δ(ion – meas)	(c) d(cov)	(c) Δ(cov – meas)
Li–F	2.01	2.09	0.08	1.85	–0.16
Na–Cl	2.82	2.83	0.01	2.68	–0.14
K–Br	3.30	3.34	0.04	3.23	–0.11
Rb–I	3.67	3.72	0.05	3.59	–0.08

   (c)   Distance estimates from bonding atomic radii are not as accurate as those from ionic radii. This indicates that bonding in the series LiF, NaCl, KBr, and RbI is more accurately described as ionic, rather than covalent. The details of these two bonding models will be discussed in Chapter 8.

Ionization Energies; Electron Affinities (Sections 7.4 and 7.5)

7.37 $Al(g) \rightarrow Al^+(g) + e^-$; $Al^+(g) \rightarrow Al^{2+}(g) + e^-$; $Al^{2+}(g) \rightarrow Al^{3+}(g) + e^-$

   The process for the first ionization energy requires the least amount of energy. When an electron is removed from an atom or ion, electrostatic repulsions are reduced, Z_{eff} increases, and the energy required to remove the next electron increases. This rationale is confirmed by the ionization energies listed in Table 7.2.

7.38   (a) $Pb(g) \rightarrow Pb^+(g) + e^-$; $Pb^+(g) \rightarrow Pb^{2+}(g) + e^-$

   (b) $Zr^{3+}(g) \rightarrow Zr^{4+}(g) + e^-$

7.39   *Analyze/Plan.* We are asked about the second ionization energy of three elements. This involves removing an electron from the 1+ ion of each element. Write the electron configurations of each ion and consider the attraction for the nucleus of the valence electron to be lost.   *Solve.*

   The electron configurations are: Li^+, $1s^2$ or [He]; Be^+, $[He]2s^1$; K^+, [Ne] $3s^2 3p^6$ or [Ar]. Be has one more valence electron to lose whereas Li^+ has the stable noble gas configuration of He. It requires much more energy to remove a 1s core electron close to the nucleus of Li^+ than a 2s valence electron farther from the nucleus of Be^+. K^+ also has a stable noble gas configuration. The electron to be lost is a core 2p electron. This electron is farther from the nucleus than the 1s electron in Li^+ and will require less energy to remove. Of these three elements, Li has the highest second ionization energy.

7.40   (a)   False. Ionization energies are always positive quantities.

(b)   False. F has a greater first ionization energy than O, because Z_{eff} for F is greater than Z_{eff} for O.

(c)   True.

(d)   False. Ionization energies are not cumulative. The third ionization energy is the energy needed to remove an electron from a gas phase ion with a 2+ charge.

7.41   (a)   In general, the smaller the atom, the larger its first ionization energy.

(b)   According to Figure 7.10, He has the largest and Cs has the smallest first ionization energy of the nonradioactive elements.

7.42   (a)   Moving from F to I in group 7A, first ionization energies decrease and atomic radii increase. The greater the atomic radius, the smaller the electrostatic attraction of an outer electron for the nucleus and the smaller the ionization energy of the element.

(b)   First ionization energies increase slightly going from K to Kr and atomic sizes decrease. As valence electrons are drawn closer to the nucleus (atom size decreases), it requires more energy to completely remove them from the atom (first ionization energy increases). Each trend has a discontinuity at Ga, owing to the increased shielding of the 4p electrons by the filled 3d subshell.

7.43   *Plan.* Use periodic trends in first ionization energy.   *Solve.*

(a)   Cl   (b)   Ca   (c)   K   (d)   Ge   (e)   Sn

7.44   Greater distance of valence electrons from the nucleus predicts lower first ionization energy in all the pairs of elements below. Z_{eff} decreases moving left along a row but increases slightly moving down column. These trends are not (solely) predictive of first ionization energy for the pairs of elements in this exercise.

(a)   Ba. Recall that transition metals like Ti lose *n*s electrons first when forming ions. The 6s valence electrons in Ba are farther from the nucleus and have a smaller first ionization energy than the 4s valence electrons of Ti.

(b)   Ag. Recall that transition elements lose *n*s electrons first when forming ions. The 5s valence electron of Ag is farther from the nucleus and has a lower first ionization energy than the 4s valence electron of Cu.

(c)   Ge. The 4p valence electrons in Ge have a smaller first ionization energy than the 3p valence electrons in Cl. Going from Cl to Ge, the decrease in Z_{eff} moving four places to the left may more than compensate for the small increase moving one place down. If so, the trends in Z_{eff} and distance of valence electrons from the nucleus cooperate to produce the (significantly) lower first ionization energy for Ge.

(d)   Pb. The 6p valence electrons in Pb are farther from the nucleus and have a smaller first ionization energy than the 5p valence electrons in Sb, despite the buildup in nuclear charge (Z) associated with filling the 4f subshell between Sb and Pb.

7.45   *Plan.* Follow the logic of Sample Exercise 7.7.   *Solve.*

    (a) Co^{2+}: $[Ar]3d^7$ (b) Sn^{2+}: $[Kr]5s^2 4d^{10}$

    (c) Zr^{4+}: $[Kr]$, noble-gas configuration   (d) Ag^+: $[Kr]4d^{10}$

    (e) S^{2-}: $[Ne]3s^2 3p^6$, noble-gas configuration

7.46   (a) Ru^{3+}: $[Kr]3d^5$

    (b) As^{3-}: $[Ar]4s^2 3d^{10}4p^6 = [Kr]$, noble-gas configuration

    (c) Y^{3+}: $[Kr]$, noble-gas configuration   (d) Pd^{2+}: $[Kr]3d^8$

    (e) Pb^{2+}: $[Xe]6s^2 4f^{14}5d^{10}$ (f) Au^{3+}: $[Xe]4f^{14}5d^8$

7.47   *Plan.* Focus on transition metals, which have d electrons in their outer shell. Use Figure 7.15 to find representative oxidations states for transition metals. Note that, by definition, metals lose electrons to form positive ions.

    Solve. Elements in group 10 and beyond have at least eight d electrons. Of these, the group 10 metals all form 2+ ions with the electron configuration nd^8. Of the elements in groups 11 and 12, only Au adopts a sufficiently high positive charge to form an ion with the configuration nd^8. (Other possibilities not listed on Figure 7.15 exist.)

    Ni^{2+}: $[Ar]3d^8$; Pd^{2+}: $[Kr]4d^8$; Pt^{2+}: $[Xe]4f^{14}5d^8$; Au^{3+}: $[Xe]4f^{14}5d^8$

7.48   The 2+ ions of group 8 metals and the 3+ ions of group 9 metals have the electrons configuration nd^6. (Other possibilities not listed on Figure 7.15 exist.)

    Fe^{2+}: $[Ar]3d^6$; Ru^{2+}: $[Kr]4d^6$; Os^{2+}: $[Xe]4f^{14}5d^6$

    Co^{3+}: $[Ar]3d^6$; Rh^{3+}: $[Kr]4d^6$; Ir^{3+}: $[Xe]4f^{14}5d^6$

7.49   *Analyze/Plan.* The second electron affinity of Cl is addition of an electron to Cl^-. Write the chemical equation and electron configurations for the relevant ions. Consider the attraction of the added electron for the Cl nucleus.   *Solve.*

    Second Electron affinity: $Cl^-(g)$ $+ e^-$ $\rightarrow$ $Cl^{2-}(g)$

                                 $[Ne]3s^2 3p^6$ $[Ne]3s^2 3p^6 4s^1$

    The Cl^- ion has the stable noble gas electron configuration of Ar. According to Figure 7.12 the first electron affinity of Ar is positive. Because the Cl^- ion has the electron configuration of Ar and a smaller nuclear charge, we predict that the second electron affinity of Cl will be positive as well. A positive value means that Cl^{2-} ion is unstable and will not form. It is probably not possible to directly measure the second electron affinity of Cl. [Sometimes unstable species have a finite lifetime before decomposing, so a very fast measurement may be possible.]

7.50   No. The process described by electron affinity can be written as: $A + e^- \rightarrow A^-$

    If ΔE for this process is negative, it means that the energy of A^- is lower than the total energy of A plus the energy of a free electron. If electron affinity is negative, the entity that is lower in energy, or more stable, is the added electron. An electron in an atom or ion is stabilized by its attraction for the atomic nucleus and is lower in energy than a free electron.

7.51   *Analyze/Plan.* Write chemical equations for the electron affinity of K^+ and K. Consider the effective nuclear charge, $Z - S$, experienced by the added electron.   *Solve.*

$$K^+(g) \qquad + e^- \rightarrow \qquad K(g) \,; \qquad K(g) \quad + e^- \rightarrow \quad K^-(g)$$

[Ne]$3s^2 3p^6$ or [Ar]   [Ne]$3s^2 3p^6 4s^1$ or [Ar]$4s^1$ [Ar]$4s^1$ [Ar]$4s^2$

The electron affinity of K^+ is more negative. The full nuclear charge, Z, is the same for all K atoms and ions. In both cases, the added electron occupies a 4s subshell; the screening by the Ar core of electrons is the same. The single difference is the electron-electron repulsion experienced by the electron added to neutral K(g). This repulsion raises the energy of K^- relative to K and the added electron. According to Figure 7.12, the electron affinity of K is negative. The electron affinity of K^+ will also be negative and have a greater magnitude.

7.52   Ionization energy of F^-: $F^-(g) \rightarrow F(g) + e^-$

Electron affinity of F: $F(g) + e^- \rightarrow F^-(g)$

The two processes are the reverse of each other. The energies are equal in magnitude but opposite in sign. $I_1 (F^-) = -E (F)$

7.53   *Analyze/Plan.* Consider the definitions of ionization energy and electron affinity, along with the appropriate electron configurations.   *Solve.*

(a)   Ionization energy of Ne: Ne(g) $\rightarrow$ $Ne^+(g)$ + e^-

[He]$2s^2 2p^6$ [He]$2s^2 2p^5$

Electron affinity of F: F(g)   + e^- $\rightarrow$ $F^-(g)$

[He]$2s^2 2p^5$ [He]$2s^2 2p^6$

(b)   The I_1 of Ne is positive, whereas E_1 of F is negative. All ionization energies are positive.

(c)   One process is apparently the reverse of the other, with one important difference. The Z (and Z_{eff}) for Ne is greater than Z (and Z_{eff}) for F^-. We expect the magnitude of I_1(Ne) to be somewhat greater than the magnitude of E_1(F). [Repulsion effects approximately cancel; repulsion decrease upon I_1 causes smaller positive value; repulsion increase upon E_1 causes smaller negative value.]

7.54 $Ca^+(g)$ + e^- $\rightarrow$ Ca(g)

[Ar] $4s^1$ [Ar] $4s^2$

Statements (i) and (ii) are true.

Properties of Metals and Nonmetals (Section 7.6)

7.55   (a)   Decrease

(b)   Increase

(c)   The smaller the first ionization energy of an element, the greater the metallic character of that element. The trends in (a) and (b) are the opposite of the trends in ionization energy.

7.56   Element Y has the greater metallic character. Metallic character increases as ionization energy decreases.

7.57   *Analyze/Plan.* Use Figure 7.13, "Metals, metalloids, and nonmetals," and Figure 7.15, "Representative oxidation states of the elements," to inform our discussion.

  Solve. Agree. An element that commonly forms a cation is a metal. The only exception to this statement shown on Figure 7.15 is antimony, Sb, a metalloid that commonly forms cations. Although Sb is a metalloid, it is far down (in the fifth row) on the chart and likely to have significant metallic character.

7.58   Disagree. According to Figure 7.15, both Sb and Te are metalloids and commonly form ions. Sb forms cations and Te forms anions.

7.59   *Analyze/Plan.* Ionic compounds are formed by combining a metal and a nonmetal; molecular compounds are formed by two or more nonmetals.   *Solve.*

  Ionic: SnO_2, Al_2O_3, Li_2O, Fe_2O_3; molecular: CO_2, H_2O

7.60   Follow the logic in Sample Exercise 7.8. Scandium is a metal, so we expect Sc_2O_3 to be ionic. Metal oxides are usually basic and react with acid to form a salt and water. We choose $HNO_3(aq)$ as the acid for our equation.

  $Sc_2O_3(s) + 6\ HNO_3(aq) \rightarrow 2\ Sc(NO_3)_3(aq) + 3\ H_2O(l)$

  The net ionic equation is:

  $Sc_2O_3(s) + 6\ H^+(aq) \rightarrow 2\ Sc^{3+}(aq) + 3\ H_2O(l)$

7.61   *Analyze/Plan.* Decide whether MnO is an acidic or basic oxide.

  Solve. MnO will react more readily with HCl(aq). Manganese is a metal. Oxides of metals usually act like bases, which react readily with the strong acid HCl(aq).

7.62   The more nonmetallic the central atom, the more acidic the oxide. In order of increasing acidity: $CaO < Al_2O_3 < SiO_2 < CO_2 < P_2O_5 < SO_3$

7.63   *Analyze/Plan.* Cl_2O_7 is a molecular compound formed by two nonmetallic elements. More specifically, it is a nonmetallic oxide and acidic.   *Solve.*

  (a)   Dichlorine heptoxide

  (b)   Elemental chlorine and oxygen are diatomic gases.

    $2\ Cl_2(g) + 7\ O_2(g) \rightarrow 2\ Cl_2O_7(l)$

  (c) Cl_2O_7 is an acidic oxide, so it will be more reactive to base, OH^-.

    $Cl_2O_7(l) + 2\ OH^-(aq) \rightarrow 2\ ClO_4^-(aq) + H_2O(l)$

  (d)   The oxidation state of Cl in Cl_2O_7 is +7. In this oxidation state, the electron configuration of Cl is $[He]2s^22p^6$ or [Ne].

7.64   (a) $XCl_4(l) + 2\ H_2O(l) \rightarrow XO_2(s) + 4\ HCl(g)$

    The second product is HCl(g).

(b) If X were a metal, both the oxide and the chloride would be high melting solids. If X were a nonmetal, XO_2 would be a nonmetallic, molecular oxide and probably gaseous, like CO_2, NO_2, and SO_2. Neither of these statements describes the properties of XO_2 and XCl_4, so X is probably a metalloid.

(c) Use the *Handbook of Chemistry* to find formulas and melting points of oxides, and formulas and boiling points of chlorides of selected metalloids.

metalloid	formula of oxide	m.p. of oxide	formula of chloride	b.p. of chloride
boron	B_2O_3	460 °C	BCl_3	12 °C
silicon	SiO_2	~1700 °C	$SiCl_4$	58 °C
germanium	GeO GeO_2	710 °C ~1100 °C	$GeCl_2$ $GeCl_4$	decomposes 84 °C
arsenic	As_2O_3 As_2O_5	315 °C 315 °C	$AsCl_3$	132 °C

Boron, arsenic, and, by analogy, antimony, do not fit the description of X because the formulas of their oxides and chlorides are wrong. Silicon and germanium, in the same family, have oxides and chlorides with appropriate formulas. Both SiO_2 and GeO_2 melt above 1000 °C, but the boiling point of $SiCl_4$ is much closer to that of XCl_4. Element X is silicon.

7.65 (a) $BaO(s) + H_2O(l) \rightarrow Ba(OH)_2(aq)$

(b) $FeO(s) + 2 HClO_4(aq) \rightarrow Fe(ClO_4)_2(aq) + H_2O(l)$

(c) $SO_3(g) + H_2O(l) \rightarrow H_2SO_4(aq)$

(d) $CO_2(g) + 2 NaOH(aq) \rightarrow Na_2CO_3(aq) + H_2O(l)$

7.66 (a) $K_2O(s) + H_2O(l) \rightarrow 2 KOH(aq)$

(b) $P_2O_3(l) + 3 H_2O(l) \rightarrow 2 H_3PO_3(aq)$

(c) $Cr_2O_3(s) + 6 HCl(aq) \rightarrow 2 CrCl_3(aq) + 3 H_2O(l)$

(d) $SeO_2(s) + 2 KOH(aq) \rightarrow K_2SeO_3(aq) + H_2O(l)$

Group Trends in Metals and Nonmetals (Sections 7.7 and 7.8)

7.67 (a) Ca and Mg are both metals; they tend to lose electrons and form cations when they react. Ca is more reactive because it has a lower ionization energy than Mg. The Ca valence electrons in the 4s orbital are less tightly held because they are farther from the nucleus than the 3s valence electrons of Mg.

(b) K and Ca are both metals; they tend to lose electrons and form cations when they react. K is more reactive because it has the lower first ionization energy. The 4s valence electron in K is less tightly held because it has the same *n* value as the valence electrons of Ca and experiences smaller Z and Z_{eff}.

7.68   Rb: $[Kr]5s^1$, r = 2.11 Å   Ag: $[Kr]5s^1 4d^{10}$, r = 1.53 Å

The electron configurations both have a [Kr] core and a single 5s electron; Ag has a completed 4d subshell as well. The smaller radius of Ag indicates that the 5s electron in Ag experiences a much greater effective nuclear charge than the 5s electron in Rb. Ag has a much larger Z (47 vs. 37), and although the 4d electrons in Ag shield the 5s electron somewhat, the increased shielding does not compensate for the large increase in Z. Ag is much less reactive (less likely to lose an electron) because its 5s electron experiences a much larger effective nuclear charge and is more difficult to remove.

7.69   (a) $2 K(s) + Cl_2(g) \rightarrow 2 KCl(s)$

(b) $SrO(s) + H_2O(l) \rightarrow Sr(OH)_2(aq)$

(c) $4 Li(s) + O_2(g) \rightarrow 2 Li_2O(s)$

(d) $2 Na(s) + S(l) \rightarrow Na_2S(s)$

7.70   (a) $2 Cs(s) + 2 H_2O(l) \rightarrow 2 CsOH(aq) + H_2(g)$

(b) $Sr(s) + 2 H_2O(l) \rightarrow Sr(OH)_2(aq) + H_2(g)$

(c) $2 Na(s) + O_2(g) \rightarrow Na_2O_2(s)$ (See Equation 7.20.)

(d) $Ca(s) + I_2(s) \rightarrow CaI_2(s)$

7.71   (a)   The reactions of the alkali metals with hydrogen and with a halogen are redox reactions. In both classes of reaction, the alkali metal loses electrons and is oxidized. Both hydrogen and the halogen gain electrons and are reduced. Hydrogen or the halogen act as oxidizing agents in these reactions.

$Ca(s) + F_2(g) \rightarrow CaF_2(s)$ $Ca(s) + H_2(g) \rightarrow CaH_2(s)$

(b)   The oxidation number of Ca in both products is +2. The electron configuration is that of Ar, $[Ne]3s^2 3p^6$.

7.72   (a) $2 K(s) + H_2(g) \rightarrow 2 KH(s)$

(b)

$K(g)$	$\rightarrow$ $K^+(g)$	419 kJ	(I_1 of K)
$H(g) + e^- $	$\rightarrow$ $H^-(g)$	−73 kJ	(E_1 of H)
$K(g) + H(g)$	$\rightarrow$ $K^+(g) + H^-(g)$	346 kJ	
$H(g)$	$\rightarrow$ $H^+(g)$	1312 kJ	(I_1 of H)
$K(g) + e^-$	$\rightarrow$ $K^-(g)$	−48 kJ	(E_1 of K)
$K(g) + H(g)$	$\rightarrow$ $K^-(g) + H^+(g)$	1264 kJ	

(c)   Both reactions are endothermic; the first reaction is less unfavorable and therefore more favorable than the second.

(d)   The more energetically favorable reaction in part (c) produces hydride ions (H^-) and potassium ions (K^+), so it is reasonable to describe potassium hydride as containing hydride ions.

7.73 | | **Br** | **Cl** |

(a) $[Ar]4s^2 4p^5$ $[Ne]3s^2 3p^5$

(b) −1   −1

(c) 1140 kJ/mol   1251 kJ/mol

(d) reacts slowly to form HBr+HOBr   reacts slowly to form HCl+HOCl

(e) −325 kJ/mol   −349 kJ/mol

(f) 1.20 Å   1.02 Å

The $n = 4$ valence electrons in Br are farther from the nucleus and less tightly held than the $n = 3$ valence electrons in Cl. Therefore, the ionization energy of Cl is greater, the electron affinity is more negative and the atomic radius is smaller.

7.74 *Plan.* Predict the physical and chemical properties of At based on the trends in properties in the halogen (7A) family.   *Solve.*

(a) F, at the top of the column, is a diatomic gas; I, immediately above At, is a diatomic solid; the melting points of the halogens increase going down the column. At is likely to be a diatomic solid at room temperature.

(b) Like the other halogens, we expect it to be a nonmetal. According to Figure 7.13, there are no metalloids in row 6 of the periodic table, and At is a nonmetal. (Looking forward to Chapter 8, the most likely way for At to satisfy the octet rule is for it to gain an electron to form At⁻, which makes it a nonmetal.)

(c) All halogens form ionic compounds with Na; they have the generic formula NaX. The compound formed by At will have the formula NaAt.

7.75 (a) The term "inert" was dropped because it no longer described all the group 8A elements.

(b) In the 1960s, scientists discovered that Xe would react with substances such as F_2 and PtF_6 that have a strong tendency to remove electrons. Thus, Xe could not be categorized as an "inert" gas.

(c) The group is now called the noble gases.

7.76 (a) Xe has a lower ionization energy than Ne. The valence electrons in Xe are much farther from the nucleus than those of Ne ($n = 5$ vs $n = 2$) and much less tightly held by the nucleus; they are more "willing" to be shared than those in Ne. Also, Xe has empty 5d orbitals that can help to accommodate the bonding pairs of electrons, whereas Ne has all its valence orbitals filled.

(b) In the *CRC Handbook of Chemistry and Physics*, 79th edition, Xe–F bond distances in gas phase molecules are listed as: XeF_2, 1.977 Å; XeF_4, 1.94 Å; XeF_6, 1.89 Å. From Figure 7.7, the sum of atomic radii for Xe and F is (1.40 Å + 0.57 Å) = 1.97 Å. This number represents an "average" or "typical" distance and agrees well with the bond distance in XeF_2. Bond lengths in specific compounds are not exactly equal to the sum of covalent radii. Physical state, electronic, and steric factors affect bond lengths in specific compounds.

7.77   (a) $2 O_3(g) \rightarrow 3 O_2(g)$

    (b) $Xe(g) + F_2(g) \rightarrow XeF_2(g)$

        $Xe(g) + 2 F_2(g) \rightarrow XeF_4(s)$

        $Xe(g) + 3 F_2(g) \rightarrow XeF_6(s)$

    (c) $S(s) + H_2(g) \rightarrow H_2S(g)$

    (d) $2 F_2(g) + 2 H_2O(l) \rightarrow 4 HF(aq) + O_2(g)$

7.78   (a) $Cl_2(g) + H_2O(l) \rightarrow HCl(aq) + HOCl(aq)$

    (b) $Ba(s) + H_2(g) \rightarrow BaH_2(s)$

    (c) $2 Li(s) + S(s) \rightarrow Li_2S(s)$

    (d) $Mg(s) + F_2(g) \rightarrow MgF_2(s)$

Additional Exercises

7.79   Up to $Z = 82$, there are three instances where atomic weights are reversed relative to atomic numbers: Ar and K; Co and Ni; Te and I.

7.80   (a)   2s

    (b)   Slater's rules provide a method for calculating the shielding, S, and Z_{eff} experienced by a particular electron in an atom. Slater assigns a shielding value of 0.35 to electrons with the same n-value, assuming that s and p electrons shield each other to the same extent. However, because s electrons have a finite probability of being very close to the nucleus (Figure 7.4), they shield p electrons more than p electrons shield them. To account for this difference, assign a slightly larger shielding value to s electrons and a slightly smaller shielding value to the p electrons. This will produce a slightly greater S and smaller Z_{eff} for p electrons than for s electrons with the same n-value.

7.81   (a)   P: $[Ne]3s^2 3p^3$. $Z_{eff} = Z - S = 15 - 10 = 5$.

    (b)   Four other $n = 3$ electrons, eight $n = 2$ electrons, two $n = 1$ electron. $S = 0.35(4) + 0.85(8) + 1(2) = 10.2$. $Z_{eff} = Z - S = 15 - 10.2 = 4.8$.

    (c)   The 3s electrons penetrate the [Ne] core electrons (by analogy to Figure 7.4) and experience less shielding than the 3p electrons. That is, S is greater for 3p electrons, owing to the penetration of the 3s electrons, so $Z - S$ (3p) is less than

        $Z - S$ (3s).

    (d)   The 3p electrons are the outermost electrons; they experience a smaller Z_{eff} than 3s electrons and thus a smaller attraction for the nucleus, given equal n-values. The first electron lost is a 3p electron. Each 3p orbital holds one electron, so there is no preference as to which 3p electron will be lost.

7.82   Atomic size (bonding atomic radius) is strongly correlated to Z_{eff}, which is determined by Z and S. Moving across the representative elements, electrons added to ns or np valence orbitals do not effectively screen each other. The increase in Z is not accompanied by a similar increase in S; Z_{eff} increases and atomic size decreases. Moving across the transition elements, electrons are added to $(n–1)d$ orbitals and become part of the core electrons, which do significantly screen the ns valence electrons. The increase in Z is accompanied by a larger increase in S for the ns valence electrons; Z_{eff} increases more slowly and atomic size decreases more slowly.

7.83   (a)   The estimated distances in the table below are the sum of the radii of the group 5A elements and H from Figure 7.7.

bonded atoms	estimated distance	measured distance
P–H	1.38	1.419
As–H	1.50	1.519
Sb–H	1.70	1.707

In general, the estimated distances are very slightly shorter than the measured distances. (Recall that the radii in Figure 7.7 come from measuring many different molecules for each element, not just the bonds listed in this exercise.)

(b)   The principal quantum number of the outer electrons and thus the average distance of these electrons from the nucleus increases from P ($n = 3$) to As ($n = 4$) to Sb ($n = 5$). This causes the systematic increase in M–H distance.

7.84   She is correct. Xenon bonds with certain elements to form compounds, so its bonding atomic radius is an average of experimentally determined values. To date, no compound containing Ne has been observed, so its "atomic radius" is an estimate. Measured values are always more realistic than estimates.

7.85   (a)   Assume that the bonding atomic radius of the element will be one-half of the bond distance in the element.

$r_{As} = (2.48 \text{ Å})/2 = 1.24 \text{ Å};$ $r_{Cl} = (1.99 \text{ Å})/2 = 0.995 \text{ Å}$

The As–Cl distance is the sum of these radii: $1.24 \text{ Å} + 0.995 \text{ Å} = 2.235 = 2.24 \text{ Å}$.

(b)   From Figure 7.7, the predicted As–Cl distance = $1.19 \text{ Å} + 1.02 \text{ Å} = 2.21 \text{ Å}$

7.86   The estimated A–B distance is $(r_A + r_B) = (A–A)/2 + (B–B)/2$. Because the AB_2 molecule is linear, the distance between the two terminal B atoms is twice the A–B distance, $2[(A–A)/2 + (B–B)/2] = (A–A) + (B–B)$. This is just the sum of the bond lengths of the two diatomic molecules. The separation between the two B nuclei in AB_2 is $2.36 \text{ Å} + 1.94 \text{ Å} = 4.30 \text{ Å}$.

7.87   (a)   The most common oxidation state of the chalcogens is –2, whereas that of the halogens is –1.

(b)   The family listed has the larger value of the stated property.
atomic radii, chalcogens
ionic radii of the most common oxidation state, chalcogens
first ionization energy, halogens
second ionization energy, halogens

7.88   Y: $[Kr]5s^2 4d^1$, Z = 39   Zr: $[Kr]5s^2 4d^2$, Z = 40

Y: La: $[Xe]6s^2 5d^1$, Z = 57   Hf: $[Xe]6s^2 4f^{14} 5d^2$, Z = 72

The completed 4f subshell in Hf leads to a much larger change in Z going from Zr to Hf (72 − 40 = 32) than in going from Y to La (57 − 39 = 18). The 4f electrons in Hf do not completely shield the valence electrons, so there is also a larger increase in Z_{eff}. This significant increase in Z_{eff} going from Zr to Hf causes the two elements to have the same radii, even though the valence electrons of Hf have a larger n value than those of Zr. (This phenomenon is called the "lantanide contraction.")

7.89   (a) Co^{4+} is smaller.

(b) Co^{4+}, 0.67 Å < Co^{3+}, 0.75 Å < Li^+, 0.90 Å

Values from WebElements©, CN 6, high spin (for comparing equivalent ion environments)

(c)   As Li^+ ions are inserted, smaller Co^{4+} ions are reduced to larger Co^{3+} ions and the lithium colbalt oxide will expand.

(d)   "Sodium colbalt oxide" will probably not work as an electrode material, because Na^+ ions are much larger than Li^+ ions, which are larger than Co^{4+} and Co^{3+} ions. Na^+ ions would be too large to insert into the electrode without disrupting the structure of the material.

(e)   An alternative metal for a sodium version of the electrode would have redox-active ions with larger ionic radii than the Co^{4+} and Co^{3+} ions. Moving left along the fourth row of the periodic table, Fe^{3+}/Fe^{2+} and Mn^{3+}/Mn^{2+} ion couples are possibilities. Both have radii larger than Co^{4+}/Co^{3+} ions. Mn^{3+} is more redox-active than Fe^{3+} and may be a more effective electrode material.

7.90   (a) $2 Sr(s) + O_2(g) \rightarrow 2 SrO(s)$

(b)   Assume that the corners of the cube are at the centers of the outermost O^{2-} ions, and that the edges each pass through the center of one Sr^{2+} ion. The length of an edge is then $r(O^{2-}) + 2r(Sr^{2+}) + r(O^{2-}) = 2r(O^{2-}) + 2r(Sr^{2+}) = 2(1.32$ Å$) + 2(1.26$ Å$) = 5.16$ Å.

(c)   Density is the ratio of mass to volume.

$$d = \frac{\text{mass SrO in cube}}{\text{vol cube}} = \frac{\text{\# SrO units} \times \text{mass of SrO}}{\text{vol cube}}$$

Calculate the mass of 1 SrO unit in grams and the volume of the cube in cm^3; solve for number of SrO units.

$$\frac{103.62 \text{ g SrO}}{\text{mol}} \times \frac{1 \text{ mol SrO}}{6.022 \times 10^{23} \text{ SrO units}} = 1.7207 \times 10^{-22} = 1.721 \times 10^{-22} \text{ g/SrO unit}$$

$$V = (5.16)^3 \text{ Å}^3 \times \frac{(1 \times 10^{-8})^3 \text{ cm}^3}{\text{Å}^3} = 1.3739 \times 10^{-22} = 1.37 \times 10^{-22} \text{ cm}^3$$

$$d = \frac{\text{number of SrO units} \times 1.7207 \times 10^{-22} \text{ g/SrO unit}}{1.3739 \times 10^{-22} \text{ cm}^3} = 5.10 \text{ g/cm}^3$$

$$\text{number of SrO units} = 5.10 \text{ g/cm}^3 \times \frac{1.3739 \times 10^{-22} \text{ cm}^3}{1.7207 \times 10^{-22} \text{ g/SrO unit}} = 4.07 \text{ units}$$

Because the number of formula units must be an integer, there are four SrO formula units in the cube. Using average values for ionic radii to estimate the edge length probably leads to the small discrepancy.

7.91   C: $1s^2 2s^2 2p^2$. I_1 through I_4 represent loss of the 2p and 2s electrons in the outer shell of the atom. The values of I_1–I_4 increase as expected. The nuclear charge is constant, but removing each electron reduces repulsive interactions between the remaining electrons, so effective nuclear charge increases and ionization energy increases. I_5 and I_6 represent loss of the 1s core electrons. These 1s electrons are much closer to the nucleus and experience the full nuclear charge (they are not shielded), so the values of I_5 and I_6 are significantly greater than I_1–I_4. I_6 is larger than I_5 because all repulsive interactions have been eliminated.

7.92   Only statement (ii) is true.

We expect electron affinities for Group 4A to be more negative than Group 3A, based on increasing effective nuclear charge moving from left to right across a row. Electron-electron repulsion causes the electron affinities of Groups 5A and 6A to be less negative than expected from effective nuclear charge trends.

7.93
$$A(g) \rightarrow A^+(g) + e^- \qquad \text{ionization energy of A}$$
$$A(g) + e^- \rightarrow A^-(g) \qquad \text{electron affinity of A}$$
$$A(g) + A(g) \rightarrow A^+(g) + A^-(g) \qquad \text{ionization energy of A + electron affinity of A}$$

The energy change for the reaction is the ionization energy of A plus the electron affinity of A.

This process is endothermic for both nonmetals and metals. Considering data for Cl and Na from Figures 7.10 and 7.12, the endothermic ionization energy term dominates the exothermic electron affinity term, even for Cl, which has the most exothermic electron affinity listed.

7.94   (a)   O: $[\text{He}]2s^2 2p^4$

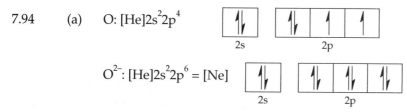

O^{2-}: $[\text{He}]2s^2 2p^6 = [\text{Ne}]$

(b)   O^{3-}: $[\text{Ne}]3s^1$ The third electron would be added to the 3s orbital, which is farther from the nucleus and more strongly shielded by the [Ne] core. The overall attraction of this 3s electron for the O nucleus is not large enough for O^{3-} to be a stable particle.

7.95   (a)   P: $[\text{Ne}]\,3s^2 3p^3$; S: $[\text{Ne}]\,3s^2 3p^4$. In P, each 3p orbital contains a single electron, whereas in S one 3p orbital contains a pair of electrons. Removing an electron from S eliminates the need for electron pairing and reduces electrostatic repulsion, so the overall energy required to remove the electron is smaller than in P, even though Z is greater.

(b)   C: [He] $2s^2 2p^2$; N: [He] $2s^2 2p^3$; O: [He] $2s^2 2p^4$. An electron added to an N atom must be paired in a relatively small 2p orbital, so the additional electron-electron repulsion more than compensates for the increase in Z and the electron affinity is smaller (less exothermic) than that of C. In an O atom, one 2p orbital already contains a pair of electrons, so the additional repulsion from an extra electron is offset by the increase in Z and the electron affinity is greater (more exothermic). Note from Figure 7.12 that the electron affinity of O is only slightly more exothermic than that of C, although the value of Z has increased by 2.

(c) O^+: [He] $2s^2 2p^3$; O^{2+}: [He] $2s^2 2p^2$; F: [He] $2s^2 2p^5$; F^+: [He] $2s^2 2p^4$. Both 'core-only' [Z_{eff} (F) = 7; Z_{eff} (O^+) = 6] and Slater [Z_{eff} (F) = 5.2; Z_{eff} (O^+) = 4.9] predict that F has a greater Z_{eff} than O^+. Variation in Z_{eff} does not offer a satisfactory explanation. The decrease in electron-electron repulsion going from F to F^+ energetically favors ionization and causes it to be less endothermic than the second ionization of O, where there is no significant decrease in repulsion.

(d) Mn^{2+}: [Ar]$3d^5$; Mn^{3+}; [Ar] $3d^4$; Cr^{2+}: [Ar] $3d^4$; Cr^{3+}: [Ar] $3d^3$; Fe^{2+}: [Ar] $3d^6$; Fe^{3+}: [Ar] $3d^5$. The third ionization energy of Mn is expected to be larger than that of Cr because of the larger Z value of Mn. The third ionization energy of Fe is less than that of Mn because going from $3d^6$ to $3d^5$ reduces electron repulsions, making the process less endothermic than predicted by nuclear charge arguments.

7.96   (a) Cl^-, K^+ (b) Mn^{2+}, Fe^{3+} (c) Sn^{2+}, Sb^{3+}

7.97   (a)   (ii)   (b) (v)   (c) (i)

7.98   (a)   The group 2B metals have complete $(n–1)d$ subshells. An additional electron would occupy an np subshell and be substantially shielded by both ns and $(n–1)d$ electrons. Overall this is not a lower energy state than the neutral atom and a free electron.

(b)   Valence electrons in Group 1B elements experience a relatively large effective nuclear charge because of the buildup in Z with the filling of the $(n–1)d$ subshell (and for Au, the 4f subshell.) Thus, the electron affinities are large and negative. Group 1B elements are exceptions to the usual electron filling order and have the generic electron configuration $ns^1(n–1)d^{10}$. The additional electron would complete the ns subshell and experience repulsion with the other ns electron. Going down the group, size of the ns subshell increases and repulsion effects decrease. That is, effective nuclear charge is greater going down the group because it is less diminished by repulsion, and electron affinities become more negative.

7.99   (a)   For both H and the alkali metals, the added electron will complete an ns subshell (1s for H and ns for the alkali metals) so shielding and repulsion effects will be similar. For the halogens, the electron is added to an np subshell, so the energy change is likely to be quite different.

(b) True. Only He has a smaller estimated "bonding" atomic radius, and no known compounds of He exist. The electron configuration of H is $1s^1$. The single 1s electron experiences no repulsion from other electrons and feels the full unshielded nuclear charge. It is held very close to the nucleus. The outer electrons of all other elements that form compounds are shielded by a spherical inner core of electrons and are less strongly attracted to the nucleus, resulting in larger bonding atomic radii.

(c) Ionization is the process of removing an electron from an atom. For the alkali metals, the ns electron being removed is effectively shielded by the core electrons, so ionization energies are low. For the halogens, a significant increase in nuclear charge occurs as the np orbitals fill, and this is not offset by an increase in shielding. The relatively large effective nuclear charge experienced by np electrons of the halogens is similar to the unshielded nuclear charge experienced by the H 1s electron. Both H and the halogens have large ionization energies.

(d) Ionization energy of hydride: $H^-(g) \rightarrow H(g) + e^-$

(e) Electron affinity of hydrogen: $H(g) + e^- \rightarrow H^-(g)$

The two processes in parts (d) and (e) are the exact reverse of one another. The value for the ionization energy of hydride is equal in magnitude but opposite in sign to the electron affinity of hydrogen.

7.100 Because Xe reacts with F_2, and O_2 has approximately the same ionization energy as Xe, O_2 will probably react with F_2. Possible products would be O_2F_2, analogous to XeF_2, or OF_2.

$$O_2(g) + F_2(g) \rightarrow O_2F_2(g)$$
$$O_2(g) + 2 F_2(g) \rightarrow 2 OF_2(g)$$

7.101 The first ionization energies are: Ag, 731 kJ/mol; Mn, 717 kJ/mol. According to Figure 7.13, we define metallic character as showing the opposite trend as ionization energy. That is, the smaller the ionization energy, the greater the metallic character. Because Mn has the smaller ionization energy, it should have the greater metallic character. (It is difficult to predict the relative metallic character of these two elements from trends. Ag is one row lower but four columns further right than Mn; these are opposing trend directions.)

7.102 The most likely product is (i). The ionization energy of K is less than that of H, so K will lose an electron and H will gain one; the (mostly) ionic solid KH is the product. This is one instance when H acts like a nonmetal, even though it appears with the metals on the periodic chart.

7.103 Statement (ii) is the best explanation.

Statements (ii) and (iii) are both true, assuming "smaller" in statement (iii) means more negative. In this reaction, Na or Cs loses an electron and acts like a metal. Statement (ii) about ionization energy is more directly applicable. Electron affinity is about an atom gaining an electron.

7.104 (a) All alkali metals except Li form metal peroxides when they react with oxygen; the formation of a peroxide (or a superoxide) eliminates Li. The lilac-purple flame indicates that the metal is potassium (see Figure 7.22).

(b) $K_2O_2(s)$ + $2 H_2O(l)$ $\rightarrow$ $H_2O_2(aq)$ + $2 KOH(aq)$

    potassium peroxide   hydrogen peroxide

    $2 KO_2(s)$ + $2 H_2O(l) \rightarrow$ $H_2O_2(aq)$ + $2 KOH(aq) + O_2(g)$

    potassium superoxide   hydrogen peroxide

Both potassium peroxide and potassium superoxide react with water to form hydrogen peroxide. The white solid could be either potassium salt.

7.105   (a)   The pros are that Zn and Cd are in the same family, have the same electron configuration and thus similar chemical properties. The same can be said for Zn^{2+} and Cd^{2+} ions. Because of their chemical similarity, we expect Cd^{2+} to easily substitute for Zn^{2+} in flexible molecules. The main difference is that Zn^{2+}, with an ionic radius of 0.88 Å, is much smaller than Cd^{2+}, with an ionic radius of 1.09 Å. Although Zn^{2+} is beneficial in living systems, Cd^{2+} is toxic. This difference in biological function could be related to the size difference and is a definite con.

    (b) Cu^+ is isoelectronic with Zn^{2+}. That is, the two ions have the same number of electrons and the same electron configurations. The ionic radius of Cu^+ is 0.91 Å, very similar to that of Zn^{2+}. We expect Cu^+ to be a reasonable substitute for Zn^{2+} in terms of chemical properties and size. Electrostatic interactions may vary, because of the difference in charges of the two ions. (All ionic radii are taken from WebElements©.)

7.106   (a)   *Plan.* Use qualitative physical (bulk) properties to narrow the range of choices, then match melting point and density to identify the specific element.   *Solve.*

Hardness varies widely in metals and nonmetals, so this information is not too useful. The relatively high density, appearance, and ductility indicate that the element is probably less metallic than copper. Focus on the block of nine main group elements centered around Sn. Pb is not a possibility because it was used as a comparison standard. The melting point of the five elements closest to Pb are:

    Tl, 303.5 °C; In, 156.1 °C; Sn, 232 °C; Sb, 630.5 °C; Bi, 271.3 °C

The best match is In. To confirm this identification, the density of In is 7.3 g/cm^3, also a good match to properties of the unknown element.

    (b)   To write the correct balanced equation, determine the formula of the oxide product from the mass data, assuming the unknown is In.

5.08 g oxide – 4.20 g In = 0.88 g O

4.20 g In/114.82 g/mol = 0.0366 mol In; 0.0366/0.0366 = 1

0.88 g O/16.00 g/mol = 0.0550 mol O; 0.0550/0.0366 = 1.5

Multiplying by two produces an integer ratio of 2 In: 3 O and a formula of In_2O_3. The balanced equation is: $4 In(s) + 3 O_2(g) \rightarrow 2 In_2O_3(s)$

    (c)   According to Figure 7.1, the element In was discovered between 1843 and 1886. The investigator who first recorded this data in 1822 could have been the first to discover In.

7.107   *Plan.* According to the periodic table on the inside cover of your text, element 116 is in group 6A, so element 117 will be in group 7A, the halogens. Write the electron configuration and use information from Figures 7.7, 7.10, 7.12, 7.15, and Table 7.7 along with periodic trends to estimate values for properties. Remember that element 117 is two rows below iodine, and that the increase in Z and Z_{eff} that accompanies filling of the f orbitals will decrease the size of the changes in ionization energy, electron affinity and atomic size.   *Solve.*

Electron configuration: $[Rn]7s^2 5f^{14} 6d^{10} 7p^5$

First ionization energy: 805 kJ/mol

Electron affinity: –235 kJ/mol

Atomic size: 1.65 Å

Common oxidation state: –1

7.108   (a)   Si and Ge are in group 4A and have 4 valence electrons. GaAs and GaP have their first element in group 3A with 3 valence electrons and their second element in group 5A with 5 valence electrons. Cd in CdS and CdSe is in group 2B and has 2 valence electrons, whereas S and Se are in group 6A with 6 valence electrons. In each case, the two elements in the compound semiconductor have an *average* of 4 valence electrons.

(b)   The roman numerals represent the number of valence electrons in the component elements of the compound semiconductor. CdS and CdSe are II-VI materials, whereas GaAs and GaP are III-V materials.

(c)   Replace Ga with In: InP, InAs, InSb; replace Se with Te: CdTe. It is problematic to replace Cd with Hg, because Hg is toxic. ZnS is ionic and an insulator, so Zn may not be a good substitute for Cd.

Integrative Exercises

7.109   (a) $\nu = c/\lambda; 1 \text{ Hz} = 1 \text{ s}^{-1}$

Ne: $\nu = \dfrac{2.998 \times 10^8 \text{ m/s}}{14.610 \text{ Å}} \times \dfrac{1 \text{ Å}}{1 \times 10^{-10} \text{ m}} = 2.052 \times 10^{17} \text{ s}^{-1} = 2.052 \times 10^{17} \text{ Hz}$

Ca: $\nu = \dfrac{2.998 \times 10^8 \text{ m/s}}{3.358 \times 10^{-10} \text{ m}} = 8.928 \times 10^{17} \text{ Hz}$

Zn: $\nu = \dfrac{2.998 \times 10^8 \text{ m/s}}{1.435 \times 10^{-10} \text{ m}} = 20.89 \times 10^{17} \text{ Hz}$

Zr: $\nu = \dfrac{2.998 \times 10^8 \text{ m/s}}{0.786 \times 10^{-10} \text{ m}} = 38.14 \times 10^{17} = 38.1 \times 10^{17} \text{ Hz}$

Sn: $\nu = \dfrac{2.998 \times 10^8 \text{ m/s}}{0.491 \times 10^{-10} \text{ m}} = 61.06 \times 10^{17} = 61.1 \times 10^{17} \text{ Hz}$

(b)

Element	Z	ν	$\nu^{1/2}$
Ne	10	2.052×10^{17}	4.530×10^{8}
Ca	20	8.928×10^{17}	9.449×10^{8}
Zn	30	20.89×10^{17}	14.45×10^{8}
Zr	40	38.14×10^{17}	19.5×10^{8}
Sn	50	61.06×10^{17}	24.7×10^{8}

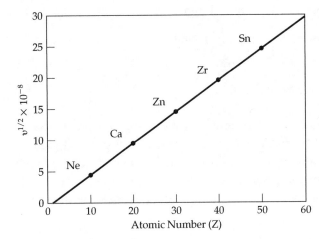

(c)   The plot in part (b) indicates that there is a linear relationship between atomic number and the square root of the frequency of the X-rays emitted by an element. Thus, elements with each integer atomic number should exist. This relationship allowed Moseley to predict the existence of elements that filled "holes" or gaps in the periodic table.

(d)   For Fe, Z = 26. From the graph, $\nu^{1/2} = 12.5 \times 10^{8}$, $\nu = 1.56 \times 10^{18}$ Hz.

$$\lambda = c/\nu = \frac{2.998 \times 10^{8} \text{ m/s}}{1.56 \times 10^{18} \text{ s}^{-1}} \times \frac{1 \text{ Å}}{1 \times 10^{-10} \text{ m}} = 1.92 \text{ Å}$$

(e) $\lambda = 0.980 \text{ Å} = 0.980 \times 10^{-10}$ m

$$\nu = c/\lambda = \frac{2.998 \times 10^{8} \text{ m/s}}{0.980 \times 10^{-10} \text{ m}} = 30.6 \times 10^{17} \text{ Hz}; \quad \nu^{1/2} = 17.5 \times 10^{8}$$

From the graph, $\nu^{1/2} = 17.5 \times 10^{8}$, Z = 36. The element is krypton, Kr.

7.110   (a)   Li: [He]2s^1. Assume that the [He] core is 100% effective at shielding the 2s valence electron $Z_{eff} = Z - S \approx 3 - 2 = 1+$.

(b)   The first ionization energy represents loss of the 2s electron.

ΔE = energy of free electron (n = ∞) – energy of electron in ground state ($n = 2$)

$\Delta E = I_1 = [-2.18 \times 10^{-18} \text{ J } (Z^2/\infty^2)] - [-2/18 \times 10^{-18} \text{ J } (Z^2/2^2)]$

$\Delta E = I_1 = 0 + 2.18 \times 10^{-18} \text{ J } (Z^2/2^2)$

For Li, which is not a one-electron particle, let $Z = Z_{eff}$.

$\Delta E \approx 2.18 \times 10^{-18} \text{ J } (+1^2/4) \approx 5.45 \times 10^{-19} \text{ J/atom}$

(c) Change the result from part (b) to kJ/mol so it can be compared to the value in Table 7.4.

$$5.45 \times 10^{-19} \ \frac{J}{atom} \times \frac{6.022 \times 10^{23} \ atom}{mol} \times \frac{1 \ kJ}{1000 \ J} = 328 \ kJ/mol$$

The value in Table 7.4 is 520 kJ/mol. This means that our estimate for Z_{eff} was a lower limit, that the [He] core electrons do not perfectly shield the 2s electron from the nuclear charge.

(d) From Table 7.4, $I_1 = 520$ kJ/mol.

$$\frac{520 \ kJ}{mol} \times \frac{1000 \ J}{kJ} \times \frac{1 \ mol}{6.022 \times 10^{23} \ atoms} = 8.6350 \times 10^{-19} \ J/atom$$

Use the relationship for I_1 and Z_{eff} developed in part (b).

$$Z_{eff}^2 = \frac{4(8.6350 \times 10^{-19} \ J)}{2.18 \times 10^{-18} \ J} = 1.5844 = 1.58; \ Z_{eff} = 1.26$$

This value, $Z_{eff} = 1.26$, based on the experimental ionization energy, is greater than our estimate from part (a), which is consistent with the explanation in part (c).

7.111 (a) $E = hc/\lambda$; 1 nm = 1×10^{-9} m; 58.4 nm = 58.4×10^{-9} m;

1 eV = 96.485 kJ/mol, 1 eV - mol = 96.485 kJ

$$E = \frac{6.626 \times 10^{-34} \ J\text{-}s \times 2.998 \times 10^{8} \ m/s}{58.4 \times 10^{-9} \ m} = 3.4015 \times 10^{-18} = 3.40 \times 10^{-18} \ J/photon$$

(b) $Hg(g) \rightarrow Hg^+(g) + 1e^-$

(c) $I_1 = E_{58.4} - E_K = 3.4015 \times 10^{-18} J - 1.72 \times 10^{-18} J = 1.6815 \times 10^{-18} =$

$$1.68 \times 10^{-18} \ J/atom$$

$$\frac{1.6815 \times 10^{-18} \ J}{atom} \times \frac{1 \ kJ}{1000 \ J} \times \frac{6.022 \times 10^{23} \ atoms}{mol} = 1012.6 = 1.01 \times 10^3 \ kJ/mol$$

(d) From Figure 7.10, iodine (I) appears to have the ionization energy closest to that of Hg, approximately 1000 kJ/mol.

7.112 (a) The X-ray source had an energy of 1253.6 eV. Change eV to J/photon and use the relationship $\lambda = hc/E$ to find wavelength.

$$1253.6 \ eV \times \frac{1.602 \times 10^{-19} \ J}{eV} = 2.0083 \times 10^{-16} = 2.008 \times 10^{-16} \ J/photon$$

$$\lambda = \frac{6.6261 \times 10^{-34} \ J\text{-}s \times 2.9979 \times 10^8 \ m/s}{2.0083 \times 10^{-16} \ J} = 9.8911 \times 10^{-10} \ m = 0.98911 \ nm = 9.891 \ \text{Å}$$

(b) Express energies of Hg 4f and O 1s electrons in terms of kJ/mol for comparison with data from Figure 7.10 of the text.

$$Hg \ 4f: \ 105 \ eV \times \frac{1.602 \times 10^{-19} \ J}{eV} \times \frac{6.022 \times 10^{23} \ atoms}{mol} = 10,130 = 1.01 \times 10^4 \ kJ/mol$$

$$O \ 1s: \ 531 \ eV \times \frac{1.602 \times 10^{-19} \ J}{eV} \times \frac{6.022 \times 10^{23} \ atoms}{mol} = 51,227 = 5.12 \times 10^4 \ kJ/mol$$

By definition, the first ionization energy is the minimum energy required to remove the first electron from an atom. This first electron is the highest energy valence electron in the neutral atom. We expect the energies of valence electrons to be higher than those of core electrons, and first ionization energies to be less than the energy required to remove a lower energy core electron.

For Hg, the first ionization energy is 1007 kJ/mol, whereas the XPS energy of the 4f electron is 10,100 kJ/mol. The energy required to remove a 4f core electron is 10 times the energy required to remove a 6s valence electron.

For O, the first ionization energy is 1314 kJ/mol, whereas the XPS energy of a 1s electron is 51,200 kJ/mol. The energy required to remove a 1s core electron is 50 times that required to remove a 2p valence electron.

(c) Hg^{2+} : $[Xe]4f^{14}5d^{10}$; valence electrons are 5d

      O^{2-}: $[He]2s^22p^6$ or [Ne]; valence electrons are 2p

7.113   (a) Mg_3N_2

      (b) $Mg_3N_2(s) + 3\,H_2O(l) \rightarrow 3\,MgO(s) + 2\,NH_3(g)$

The driving force is the production of $NH_3(g)$.

      (c)   After the second heating, all the Mg is converted to MgO.

Calculate the initial mass Mg.

$$0.486 \text{ g MgO} \times \frac{24.305 \text{ g Mg}}{40.305 \text{ g MgO}} = 0.293 \text{ g Mg}$$

x = g Mg converted to MgO; y = g Mg converted to Mg_3N_2; x = 0.293 − y

$$\text{g MgO} = x\left(\frac{40.305 \text{ g MgO}}{24.305 \text{ g Mg}}\right); \text{ g } Mg_3N_2 = y\left(\frac{100.929 \text{ g } Mg_3N_2}{72.915 \text{ g Mg}}\right)$$

$$\text{g MgO} + \text{g } Mg_3N_2 = 0.470$$

$$(0.293 - y)\left(\frac{40.305}{24.305}\right) + y\left(\frac{100.929}{72.915}\right) = 0.470$$

$$(0.293 - y)(1.6583) + y(1.3842) = 0.470$$

$$-1.6583\,y + 1.3842\,y = 0.470 - 0.48588$$

$$-0.2741\,y = -0.01588 = -0.016$$

$$y = 0.05794 = 0.058 \text{ g Mg in } Mg_3N_2$$

$$\text{g } Mg_3N_2 = 0.05794 \text{ g Mg} \times \frac{100.929 \text{ g } Mg_3N_2}{72.915 \text{ g Mg}} = 0.0802 = 0.080 \text{ g } Mg_3N_2$$

$$\text{mass \% } Mg_3N_2 = \frac{0.0802 \text{ g } Mg_3N_2}{0.470 \text{ g (MgO} + Mg_3N_2)} \times 100 = 17\%$$

(The final mass % has 2 sig figs because the mass of Mg obtained from solving simultaneous equations has 2 sig figs.)

(d) $3 Mg(s) + 2 NH_3(g) \rightarrow Mg_3N_2(s) + 3 H_2(g)$

$$6.3 \text{ g Mg} \times \frac{1 \text{ mol Mg}}{24.305 \text{ g Mg}} = 0.2592 = 0.26 \text{ mol Mg}$$

$$2.57 \text{ g NH}_3 \times \frac{1 \text{ mol NH}_3}{17.031 \text{ g NH}_3} = 0.1509 = 0.15 \text{ mol NH}_3$$

$$0.2592 \text{ mol Mg} \times \frac{2 \text{ mol NH}_3}{3 \text{ mol Mg}} = 0.1728 = 0.17 \text{ mol NH}_3$$

0.26 mol Mg requires more than the available NH_3 so NH_3 is the limiting reactant.

$$0.1509 \text{ mol NH}_3 \times \frac{3 \text{ mol H}_2}{2 \text{ mol NH}_3} \times \frac{2.016 \text{ g H}_2}{\text{mol H}_2} = 0.4563 = 0.46 \text{ g H}_2$$

(e) $\Delta H^{\circ}_{rxn} = \Delta H^{\circ}_f Mg_3N_2(s) + 3 \Delta H^{\circ}_f H_2(g) - 3 \Delta H^{\circ}_f Mg(s) - 2 \Delta H^{\circ}_f NH_3(g)$

$= -461.08 \text{ kJ} + 3(0) - 3(0) - 2(-46.19) = -368.70 \text{ kJ}$

7.114   (a) $r_{Bi} = r_{BiBr_3} - r_{Br} = 2.63 \text{ Å} - 1.20 \text{ Å} = 1.43 \text{ Å}$

(b) $Bi_2O_3(s) + 6 HBr(aq) \rightarrow 2 BiBr_3(aq) + 3 H_2O(l)$

(c) Bi_2O_3 is soluble in acid solutions because it acts as a base and undergoes acid-base reactions like the one in part (b). It is insoluble in base because it cannot act as an acid. Thus, Bi_2O_3 is a basic oxide, the oxide of a metal. Based on the properties of its oxide, Bi is characterized as a metal.

(d) Bi: $[Xe]6s^2 4f^{14} 5d^{10} 6p^3$. Bi has five outer electrons in the 6p and 6s subshells. If all five electrons participate in bonding, compounds such as BiF_5 are possible. Also, Bi has a large enough atomic radius (1.43 Å) and low-energy orbitals available to accommodate more than four pairs of bonding electrons.

(e) The high ionization energy and relatively large negative electron affinity of F, coupled with its small atomic radius, make it the most electron withdrawing of the halogens. BiF_5 forms because F has the greatest tendency to attract electrons from Bi. Also, the small atomic radius of F reduces repulsions between neighboring bonded F atoms. The strong electron withdrawing properties of F are also the reason that only F compounds of Xe are known.

7.115   (a) $4 KO_2(s) + 2 CO_2(g) \rightarrow 2 K_2CO_3(s) + 3 O_2(g)$

(b) K, +1; O, −1/2 (O_2^- is superoxide ion); C, +4; O, −2 → K, +1; C, +4; O, −2; O, 0

Oxygen (in the form of superoxide) is oxidized (to O_2) and reduced (to O^{2-}).

(c) $18.0 \text{ g CO}_2 \times \dfrac{1 \text{ mol CO}_2}{44.01 \text{ g CO}_2} \times \dfrac{4 \text{ mol KO}_2}{2 \text{ mol CO}_2} \times \dfrac{71.10 \text{ g KO}_2}{1 \text{ mol KO}_2} = 58.2 \text{ g KO}_2$

$18.0 \text{ g CO}_2 \times \dfrac{1 \text{ mol CO}_2}{44.01 \text{ g CO}_2} \times \dfrac{3 \text{ mol O}_2}{2 \text{ mol CO}_2} \times \dfrac{32.00 \text{ g O}_2}{1 \text{ mol O}_2} = 19.6 \text{ g O}_2$

8 Basic Concepts of Chemical Bonding

Visualizing Concepts

8.1   *Analyze/Plan.* Count the number of electrons in the Lewis symbol. This corresponds to the 'A'-group number of the family.   *Solve.*

(a)   Group 4A or 14

(b)   Group 2A or 2

(c)   Group 5A or 15

(These are the appropriate groups in the s and p blocks, where Lewis symbols are most useful.)

8.2   *Analyze.* Given the size and charge of four different ions, determine their ionic bonding characteristics.

Plan. The magnitude of lattice energy is directly proportional to the charges of the two ions and inversely proportional to their separation. $E_{el} = -Q_1Q_2/d$. Apply these concepts to A, B, X, and Y.

(a)   AY and BX have a 1:1 ratio of cations and anions. In an ionic compound, the total positive and negative charges must be equal. To form a 1:1 compound, the magnitude of positive charge on the cation must equal the magnitude of negative charge on the anion. A^{2+} combines with Y^{2-} and B^+ combines with X^- to form 1:1 compounds.

(b)   AY has the larger lattice energy. The A–Y and B–X separations are nearly equal. (A is smaller than B, but X is smaller than Y, so the differences in cation and anion radii approximately cancel.) In AY, $Q_1Q_2 = (2)(2) = 4$, whereas in BX, $Q_1Q_2 = (1)(1) = 1$.

8.3   *Analyze.* Given a schematic "slab" of NaCl(s), answer questions regarding the various ions and the electrostatic interactions among them. *Plan.* $E_{el} = -Q_1Q_2/d$. Use geometry to estimate or calculate distances when needed.   *Solve.*

(a)   The smaller purple balls represent Na^+ cations. Na^+ has a completed $n = 2$ shell, whereas Cl^- has a completed $n = 3$ shell.

(b)   The larger green balls represent Cl^- anions.

(c)   Four. Green-purple interactions are attractive; these are electrostatic attractions between two oppositely charged ions. The sign of E_{el} for these interactions is negative (–).

(d)   Four. Green-green (and purple-purple) interactions are repulsive; these are electrostatic attractions between two ions with the same charge. The sign of E_{el} for these interactions is positive (+).

198

(e)   Larger. Because the anions and cations have the same magnitude of charge (1– and 1+), the magnitude of their interactions depends on the distance between the ions; the shorter the distance, the larger the magnitude of the interaction. The distances between any green and any purple ion are the same, *d*. The magnitude between any two like-colored ions is the hypotenuse of a right triangle with distance $\sqrt{2}\,d$. The shorter attractive interactions have the greater magnitude. Because there are equal numbers of attractive and repulsive interactions, the sum of the attractive interactions is larger.

(f)   Positive. If this pattern of ions was extended indefinitely in two dimensions, the magnitude of the total attractive interactions would be greater than the magnitude of the total repulsive interactions. Lattice energy is the energy required to overcome attractive interactions and separate the particles into gas phase ions. The lattice energy would be positive.

8.4   *Analyze/Plan.* Count the valence electrons in the orbital diagram, take ion charge into account, and find the element with this orbital electron count on the periodic table. Write the complete electron configuration for the ion.   *Solve.*

(a)   This ion has six 4d-electrons. Transition metals, or d-block elements, have valence electrons in d-orbitals. Transition metal ions first lose electrons from the 5s orbital, then from 4d if required by the charge. This 2+ ion has lost two electrons from 5s, none from 4d. The transition metal with six 4d-electrons is ruthenium, Ru.

(b)   The electron configuration of Ru is $[Kr]5s^2 4d^6$. (The configuration of Ru^{2+} is $[Kr]4d^6$).

8.5   *Analyze/Plan.* This question is a "reverse" Lewis structure. Count the valence electrons shown in the Lewis structure. For each atom, assume zero formal charge and determine the number of valence electrons an unbound atom has. Name the element.   *Solve.*

A: 1 shared e^- pair = 1 valence electron + 3 unshared pairs = 7 valence electrons, F

E: 2 shared pairs = 2 valence electrons + 2 unshared pairs = 6 valence electrons, O

D: 4 shared pairs = 4 valence electrons, C

Q: 3 shared pairs = 3 valence electrons + 1 unshared pair = 5 valence electrons, N

X: 1 shared pair = 1 valence electron, no unshared pairs, H

Z: same as X, H

Check. Count the valence electrons in the Lewis structure. Does the number correspond to the molecular formula CH_2ONF? 12 e^- pair in the Lewis structure. $CH_2ONF = 4 + 2 + 6 + 5 + 7 = 24\ e^-$, 12 e^- pair. The molecular formula we derived matches the Lewis structure.

8.6   (a) HNO_2, 18 valence e^-, 9 e^- pairs NO_2^-, 18 valence e^-, 9 e^- pairs

$$H-\ddot{O}-N=\ddot{O} \qquad\qquad \left[\,:\ddot{O}-N=\ddot{O}\,\right]^-$$

(b)   The formal charge on N is zero, in both species.

(c) NO_2^- is expected to exhibit resonance; the double bond can be drawn to either oxygen atom. An alternate resonance structure for HNO_2 can be drawn, but it has nonzero formal charges on the oxygen atoms. This structure is less likely than the one shown above.

(d)   The N=O bond length in HNO_2 will be shorter than the N–O lengths in NO_2^-, assuming that the structure shown above is the main contributor to the structure of HNO_2. This is a reasonable assumption because the Lewis structure in part (a) minimizes formal charges. Because there are two equivalent resonance structures for NO_2^-, the N–O lengths are approximately an average of N–O single and double bond lengths. These are longer than the full N=O double bond in HNO_2.

8.7   *Analyze/Plan.* Because there are no unshared pairs in the molecule, we use single bonds to H to complete the octet of each C atom. For the same pair of bonded atoms, the greater the bond order, the shorter and stronger the bond.   *Solve.*

(a)   Four. Moving from left to right along the molecule, the first C needs two H atoms, the second needs one, the third needs none, and the fourth needs one. The complete molecule is:

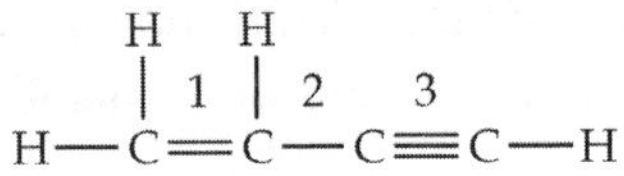

$$H-\underset{\underset{\displaystyle H}{|}}{C}\overset{1}{=}\underset{\underset{\displaystyle H}{|}}{C}-C\overset{2}{\quad}\overset{3}{\equiv}C-H$$

(b)   In order of increasing bond length: 3 < 1 < 2

(c)   Bond 3 is strongest. For the same pair of bonded atoms, the shorter the bond length, the stronger the bond.

8.8   *Analyze/Plan.* Given an oxyanion of the type XO_4^{n-}, find the identity of X from elements in the third period. Use the generic Lewis structure to determine the identity of X, and to draw the ion-specific Lewis structures. Use the definition of formal charge, [# of valence electrons – # of nonbonding electrons – (# of bonding electrons/2)], to draw Lewis structures where X has a formal charge of zero.   *Solve.*

(a)   According to the generic Lewis structure, each anion has 12 nonbonding and 4 bonding electron pairs for a total of 32 electrons. Of these 32 electrons, the 4 O atoms contribute $(4 \times 6) = 24$, and the overall negative charges contribute 1, 2, or 3. # X electrons = 32 – 24 – n.

For n = 1–, X has (32 – 24 –1) = 7 valence electrons. X is Cl, and the ion is ClO_4^-.

For n = 2–, X has (32 – 24 – 2) = 6 valence electrons. X is S, and the ion is SO_4^{2-}.

For n = 3–, X has (32 – 24 – 3) = 5 valence electrons. X is P, and the ion is PO_4^{3-}.

Check. The identity of the ions is confirmed in Table 2.5.

(b)   In the generic Lewis structure, X has 0 nonbonding electrons and (8/2) = 4 bonding electrons. Differences in formal charge are because of different numbers of valence electrons on X.

For PO_4^{3-}, formal charge of P is (5 – 4) = +1.

For SO_4^{2-}, formal charge of S is (6 – 4) = +2.

For ClO_4^-, formal charge of Cl is (7 – 4) = +3.

(c)   To reduce the formal charge of X to zero, X must have more bonding electrons. This is accomplished by changing the appropriate number of lone pairs on O to multiple bonds between X and O.

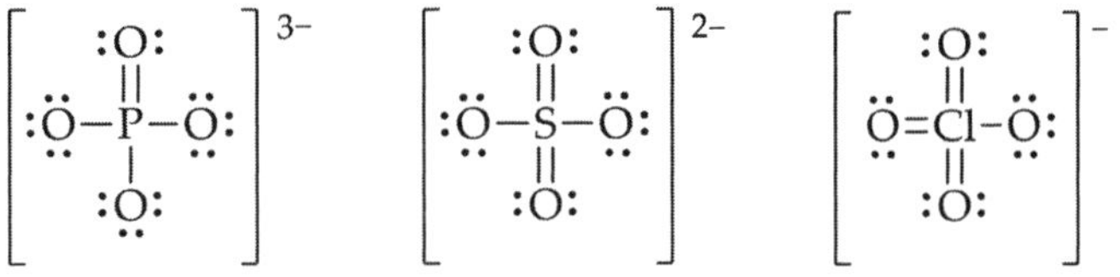

$$
\left[\begin{array}{c} :\!O\!: \\ \| \\ :\ddot{O}\!-\!P\!-\!\ddot{O}\!: \\ | \\ :\underset{\cdot\cdot}{O}\!: \end{array}\right]^{3-}
\qquad
\left[\begin{array}{c} :\!O\!: \\ \| \\ :\ddot{O}\!-\!S\!-\!\ddot{O}\!: \\ \| \\ :\!O\!: \end{array}\right]^{2-}
\qquad
\left[\begin{array}{c} :\!O\!: \\ \| \\ \ddot{O}\!=\!Cl\!-\!\ddot{O}\!: \\ \| \\ :\!O\!: \end{array}\right]^{-}
$$

Lewis Symbols (Section 8.1)

8.9   (a)   False. The total number of electrons in an atom is the same as its atomic number. Valence electrons are those that take part in chemical bonding, those in the outermost shell of the atom.

(b)   N: [He] $2s^2 2p^3$ A nitrogen atom has 5 valence electrons.

              Valence electrons

(c) $1s^2 2s^2 2p^6 \quad 3s^2 3p^2$ The atom (Si) has 4 valence electrons.

      [Ne]   valence electrons

8.10   (a)   False. The valence shell of H is $n = 1$, which holds a maximum of 2 electrons.

(b)   S: [Ne]$3s^2 3p^4$ A sulfur atom has 6 valence electrons, so it must gain 2 electrons to achieve an octet.

(c) $1s^2 2s^2 2p^3$ = [He]$2s^2 2p^3$ The atom (N) has 5 valence electrons and must gain 3 electrons to achieve an octet.

8.11   (a)   Si: $1s^2 2s^2 2p^6 3s^2 3p^2$.

(b)   Four.

(c)   The 3s and 3p electrons are valence electrons.

8.12   (a)   Ti: [Ar]$4s^2 3d^2$. Ti has 4 valence electrons. These valence electrons are available for chemical bonding, whereas core electrons do not participate in chemical bonding.

(b)   Hf: [Xe]$6s^2 4f^{14} 5d^2$

(c)   If Hf and Ti both behave as if they have 4 valence electrons, the 6s and 5d orbitals in Hf behave as valence orbitals and the 4f behaves as a core orbital. This is reasonable because 4f is complete and 4f electrons are, on average, closer to the nucleus than 5d or 6s electrons. The core orbitals for Hf are then [Xe]$4f^{14}$.

8.13   (a) $\cdot\dot{Al}\cdot$ (b) $:\ddot{Br}:$ (c) $:\ddot{Ar}:$ (d) $\dot{Sr}$

8.14   (a) $K\cdot$ (b) $\cdot\dot{As}\cdot$ (c) $\left[:\!Sn\!:\right]^{2+}$ (d) $\left[:\ddot{N}\!:\right]^{3-}$

Ionic Bonding (Section 8.2)

8.15   (a) $\text{Mg}\cdot + :\ddot{\text{O}}: \longrightarrow \text{Mg}^{2+} + \left[:\ddot{\text{O}}:\right]^{2-}$

      (b)   2 electrons are transferred.

      (c)   Mg loses electrons.

8.16   (a) $\text{Ca}\cdot + \cdot\ddot{\text{F}}: + \cdot\ddot{\text{F}}: \longrightarrow \text{Ca}^{2+} + 2\left[:\ddot{\text{F}}:\right]^{-}$

      (b) CaF_2

      (c)   2 electrons are transferred.

      (d)   Ca loses electrons.

8.17   (a) AlF_3 (b) K_2S (c) Y_2O_3 (d) Mg_3N_2

8.18   (a) BaF_2 (b) $CsCl$ (c) Li_3N (d) Al_2O_3

8.19   (a) Sr^{2+}: $[Ar]4s^2 3d^{10} 4p^6 = [Kr]$, noble-gas configuration

      (b) Ti^{2+}: $[Ar]3d^2$

      (c) Se^{2-}: $[Ar]4s^2 3d^{10} 4p^6 = [Kr]$, noble-gas configuration

      (d) Ni^{2+}: $[Ar]3d^8$

      (e) Br^-: $[Ar]4s^2 3d^{10} 4p^6 = [Kr]$, noble-gas configuration

      (f) Mn^{3+}: $[Ar]3d^4$

8.20   (a) Cd^{2+}: $[Kr]4d^{10}$

      (b) P^{3-}: $[Ne]3s^2 3p^6 = [Ar]$, noble-gas configuration

      (c) Zr^{4+}: $[Ar]4s^2 3d^{10} 4p^6 = [Kr]$, noble-gas configuration

      (d) Ru^{3+}: $[Kr]4d^5$

      (e) As^{3-}: $[Ar]4s^2 3d^{10} 4p^6 = [Kr]$, noble-gas configuration

      (f) Ag^+: $[Kr]4d^{10}$

8.21   (a)   Endothermic. *Lattice energy* is the energy required to totally separate 1 mole of solid ionic compound into its gaseous ions. E_{el} for attractive interactions among ions is negative, so the energy required to overcome these attractions and separate the ions is positive.

      (b) $NaCl(s) \rightarrow Na^+(g) + Cl^-(g)$

      (c)   Salts like NaCl that have singly charged ions will have smaller lattice energies compared with salts that have doubly charged ions. The magnitude of lattice energy depends on the magnitudes of the charges of the two ions, their radii, and the arrangement of ions in the lattice. The main factor is the charges because the radii of ions do not vary over a wide range.

8.22   (a)   NaCl, 788 kJ/mol; KF, 808 kJ/mol

Given that crystal structure and ionic charges are the same for the two compounds, the difference in lattice energy is because of the difference in ion separation (d). Lattice energy is inversely proportional to ion separation (d), so we expect the compound with the smaller lattice energy, NaCl, to have the larger ion separation. That is, the Na–Cl distance should be longer than the K–F distance.

     (b)   Na–Cl, 1.16 Å + 1.67 Å = 2.83 Å

K–F, 1.52 Å + 1.19 Å = 2.71 Å

This estimate of the relative ion separations agrees with the estimate from lattice energies. Ionic radii indicate that the Na–Cl distance is longer than the K–F distance.

8.23   *Analyze/Plan.* Assign ion charges by the position of the elements in the periodic table. Lattice energy is directly related to the product of ion charges and inversely related to the ion separation. The dominant factor is ion charges, because the difference between ion separations from one compound to another is not as large as the possible difference between the products of ion charges.   *Solve.*

     (a) Na^+, 1+; Ca^{2+}, 2+

     (b) F^-, 1– ; O^{2-}, 2–

     (c)   CaO will have the larger lattice energy. Lattice energy is directly related to the magnitudes of ion charges. CaO has larger cation and anion charges.

     (d)   Consider the relationship between the lattice energies of CaO and NaF. (Assume the lattice energy of NaF, 910 kJ, has 3 significant figures, similar to other values in Table 8.1.)

$$\frac{CaO}{NaF} = \frac{3414\,kJ}{910\ kJ} = 3.75$$

The ratio of the lattice energies is approximately 4 and the ratio of the two products of cation and anion charges is [(2)(2)/(1)(1)] = 4. If the charges in ScN are 3+ and 3–, respectively, the lattice enthalpy of ScN will be approximately (3)(3)(910) = 8190 kJ.

Since the calculated ratio is less than the integer value of 4, we expect 8190 kJ to be slightly greater than the measured lattice energy of ScN. From Table 8.1, the measured lattice energy of ScN is 7547, slightly less than our estimate. Recall that lattice energy is also inversely related to ion separation, which is probably greater for ScN than NaF. This also predicts that the measured lattice energy of ScN will be less than our estimate, which is based only on the differences in ion charges.

8.24   (a)   According to Equation 8.4, electrostatic attraction increases with increasing charges of the ions and decreases with increasing radius of the ions. Thus, lattice energy (i) increases as the charges of the ions increase and (ii) decreases as the sizes of the ions increase.

     (b)   KI < LiBr < MgS < GaN. Lattice energy increases as the charges on the ions increase. The ions in KI and LiBr all have 1+ and 1– charges. K^+ is larger than Li^+, and I^- is larger than Br^-. The ion separation is larger in KI, so it has the smaller lattice energy.

8.25   (a)   K–F, 1.52 Å + 1.19 Å = 2.71 Å

                   Na–Cl, 1.16 Å + 1.67 Å = 2.83 Å

                   Na–Br, 1.16 Å + 1.82 Å = 2.98 Å

                   Li–Cl, 0.90 Å + 1.67 Å = 2.57 Å

  (b)   The order of decreasing lattice energy should be the order of increasing ion separation: LiCl > KF > NaCl > NaBr

  (c)   From Table 8.1: LiCl, 1030 kJ; KF, 808 kJ; NaCl, 788 kJ; NaBr, 732 kJ The predictions from ionic radii are correct.

8.26   Trend (a) is because of differences in ionic radii. The ions have 1+ and 1– charges in all three compounds. In (b), the ions in the two compounds have different charges, which dominates the lattice energy trend. In (c), ion charge is the main influence, but differences in ionic radii predict the same trend.

8.27   Statement (a) is the best explanation. Equation 8.4 predicts that as the oppositely charged ions approach each other, the energy of interaction will be large and negative. This more than compensates for the energy required to form Ca^{2+} and O^{2-} from the neutral atoms.

8.28 $Ba(s) \rightarrow Ba(g)$; $Ba(g) \rightarrow Ba^+(g) + e^-$; $Ba^+(g) \rightarrow Ba^{2+}(g) + e^-$;

        $I_2(s) \rightarrow 2\,I(g)$; $2\,I(g) + 2\,e^- \rightarrow 2\,I^-(g)$, exothermic;

        $Ba^{2+}(g) + 2\,I^-(g) \rightarrow BaI_2(s)$, exothermic

8.29 $RbCl(s) \rightarrow Rb^+(g) + Cl^-(g)$ ΔH (lattice energy) = ?

        By analogy to NaCl, Figure 8.6, the lattice energy is

$$\Delta H_{latt} = -\Delta H_f^\circ\, RbCl(s) + \Delta H_f^\circ\, Rb(g) + \Delta H_f^\circ\, Cl(g) + I_1\,(Rb) + E\,(Cl)$$
$$= -(-430.5\text{ kJ}) + 85.8\text{ kJ} + 121.7\text{ kJ} + 403\text{ kJ} + (-349\text{ kJ}) = +692\text{ kJ/mol}$$

8.30   (a) $MgCl_2$, 2326 kJ; $SrCl_2$, 2127 kJ. Because the ionic radius of Ca^{2+} is greater than that of Mg^{2+}, but less than that of Sr^{2+}, the ion separation (d) in $CaCl_2$ will be intermediate as well. We expect the lattice energy of $CaCl_2$ to be in the range 2200–2250 kJ.

  (b)   By analogy to Figure 8.6:

$$\Delta H_{latt} = -\Delta H_f^\circ\, CaCl_2 + \Delta H_f^\circ\, Ca(g) + 2\,\Delta H_f^\circ\, Cl(g) + I_1\,(Ca) + I_2\,(Ca) + 2\,E\,(Cl)$$
$$= -(-795.8\text{ kJ}) + 179.3\text{ kJ} + 2(121.7\text{ kJ}) + 590\text{ kJ} + 1145\text{ kJ} + 2(-349\text{ kJ}) = +2256\text{ kJ}$$

        This value is near the range predicted in part (a).

Covalent Bonding, Electronegativity, and Bond Polarity
(Sections 8.3 and 8.4)

8.31   (a)   The bonding in (iii) water and (iv) oxygen is likely to be covalent. (i) iron and (ii) sodium chloride contain metal atoms, which are unlikely to participate in covalent bonding in simple substances. (v) argon is a monatomic gas whose atoms do not strongly interact.

  (b)   Substance XY is likely to be covalent because it is a gas even below room temperature.

8.32   K and Ar. K is an active metal with 1 valence electron. It is most likely to achieve an octet by losing this single electron and to participate in ionic bonding. Ar has a stable octet of valence electrons; it is not likely to form chemical bonds of any type.

8.33   *Analyze/Plan.* Follow the logic in Sample Exercise 8.3.   *Solve.*

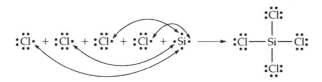

Check. Each pair of shared electrons in $SiCl_4$ is shown as a line; each atom is surrounded by an octet of electrons.

(a)   4   (b) 7   (c) 8   (d) 8   (e) 4

8.34

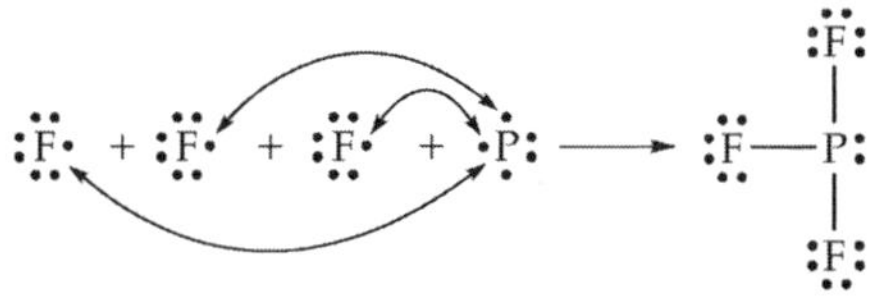

(a)   5   (b) 7   (c) 8   (d) 8   (e) 3

8.35   (a)   :Ö==Ö:

         (b)   There are four bonding electrons (two bonding electron pairs) in the structure of O_2.

         (c)   The greater the number of shared electron pairs between two atoms, the shorter the distance between the atoms. If O_2 has two bonding electron pairs, the O–O distance will be shorter than the O–O single bond distance.

8.36   (a)   The H atoms must be terminal because H can form only one bond.

              14 e⁻, 7 e⁻ pairs

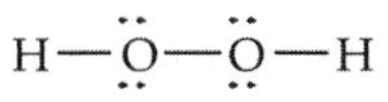

              H—Ö—Ö—H

         (b)   There are two bonding electrons (one bonding electron pair) between the two O atoms.

         (c)   Longer. The oxygen atoms in H_2O_2 share one pair of electrons, whereas those in O_2 share two pairs (Solution 8.35). The fewer the number of shared electron pairs between two atoms, the longer the distance between them.

8.37   Statement (b) is false. Electron affinity is a property of gas phase atoms or ions, whereas electronegativity is a property of bonded atoms in a molecule.

8.38   (a)   The electronegativity of the elements increases going from left to right across a row of the periodic table.

         (b)   Electronegativity generally decreases going down a family of the periodic table.

         (c)   False. Elements with the largest ionization energies are the most electronegative.

8.39   *Plan.* Electronegativity increases going up and to the right in the periodic table.   *Solve.*

(a)   Mg   (b) S   (c) C   (d) As

Check. The electronegativity values in Figure 8.8 confirm these selections.

8.40   Electronegativity increases going up and to the right in the periodic table.

(a)   O   (b) Al   (c) Cl   (d) F

8.41   The bonds in (a), (c), and (d) are polar because the atoms involved differ in electronegativity. The more electronegative element in each polar bond is:

(a)   F   (c) O   (d) I

8.42   The more different the electronegativity values of the two elements, the more polar the bond.

(a)   O–F < C–F < Be–F. This order is clear from the periodic trend.

(b)   S–Br < C–P < O–Cl. Refer to the electronegativity values in Figure 8.8 to confirm the order of bond polarity. The 3 pairs of elements all have the same positional relationship on the periodic table. The more electronegative element is one row above and one column to the left of the less electronegative element. This leads us to conclude that ΔEN is similar for the 3 bonds, which is confirmed by values in Figure 8.8. The most polar bond, O–Cl, involves the most electronegative element, O. Generally, the largest electronegativity differences tend to be between row 2 and row 3 elements. The 2 bonds in this exercise involving elements in row 2 and row 3 do have slightly greater ΔEN than the S–Br bond, between elements in rows 3 and 4.

(c)   C–S < N–O < B–F. You might predict that N–O is least polar because the elements are adjacent on the table. However, the big decrease going from the second row to the third means that the electronegativity of S is not only less than that of O, but essentially the same as that of C. C–S is the least polar.

8.43   (a)   *Analyze/Plan.* Q is the charge at either end of the dipole. Q = μ/r. The values for HBr are μ = 0.82 D and r = 1.41 Å. Change Å to m; use the definition of debyes and the charge of an electron to calculate effective charge in units of *e*.   *Solve.*

$$Q = \frac{\mu}{r} = \frac{0.82\,D}{1.41\,\text{Å}} \times \frac{1\,\text{Å}}{1\times10^{-10}\,m} \times \frac{3.34\times10^{-30}\,C\text{-}m}{1\,D} \times \frac{1\,e}{1.60\times10^{-19}\,C} = 0.12\,e$$

(b)   Decrease. Q = μ/r, μ = Q × r. If r decreases and Q remains the same, μ decreases.

8.44   (a)   The more electronegative element, Br, will have a stronger attraction for the shared electrons and adopt a partial negative charge.

(b)   Q is the charge at either end of the dipole.

$$Q = \frac{\mu}{r} = \frac{1.21\,D}{2.49\,\text{Å}} \times \frac{1\,\text{Å}}{1\times10^{-10}\,m} \times \frac{3.34\times10^{-30}\,C\text{-}m}{1\,D} \times \frac{1\,e}{1.60\times10^{-19}\,C} = 0.1014 = 0.101\,e$$

The charges on I and Br are 0.101 *e*.

8.45   *Analyze/Plan.* Generally, compounds formed by a metal and a nonmetal are described as ionic, whereas compounds formed from two or more nonmetals are covalent. However, substances with metals in a high oxidation state often have properties of molecular compounds. In this exercise we know that one substance in each pair is molecular and one is ionic; we may need to distinguish by comparison.   *Solve.*

(a) SiF_4, metalloid and nonmetal, molecular, silicon tetrafluoride

LaF$_3$, metal and nonmetal, ionic, lanthanum(III) fluoride

(b) $FeCl_2$, metal and nonmetal, ionic, iron(II) chloride

$ReCl_6$, metal in high oxidation state, Re(VI), molecular, rhenium hexachloride

(c) $PbCl_4$, metal and nonmetal, Pb(IV) is relatively high oxidation state, molecular (by contrast with RbCl, which is definitely ionic), lead tetrachloride

RbCl, metal and nonmetal, ionic, rubidium chloride

8.46   Generally, compounds formed by a metal and a nonmetal are described as ionic, whereas compounds formed from two or more nonmetals are covalent. However, substances with metals in a high oxidation states often have properties of molecular compounds.

(a) $TiCl_4$, metal and nonmetal, Ti(IV) is a relatively high oxidation state, molecular (by contrast with CaF_2, which is definitely ionic), titanium tetrachloride

CaF_2, metal and nonmetal, ionic, calcium fluoride

(b) ClF_3, two nonmetals, molecular, chlorine trifluoride

VF_3, metal and nonmetal, ionic, vanadium(III) fluoride

(c) $SbCl_5$, metalloid and nonmetal, molecular, antimony pentachloride

AlF_3, metal and nonmetal, ionic, aluminum fluoride

Lewis Structures; Resonance Structures (Sections 8.5 and 8.6)

8.47   *Analyze.* Counting the **correct number of valence electrons** is the foundation of every Lewis structure.   *Plan/Solve.*

(a)   Count valence electrons: $4 + (4 \times 1) = 8$ e$^-$, 4 e$^-$ pairs. Follow the procedure in Sample Exercise 8.6.

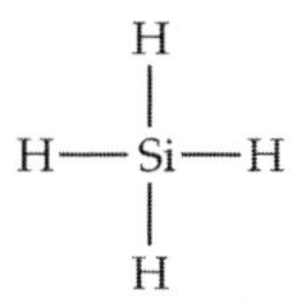

$$\begin{array}{c} H \\ | \\ H - Si - H \\ | \\ H \end{array}$$

(b)   Valence electrons: $4 + 6 = 10$ e$^-$, 5 e$^-$ pairs

:C≡O:

(c)   Valence electrons: $[6 + (2 \times 7)] = 20$ e$^-$, 10 e$^-$ pairs

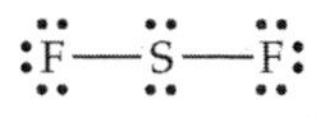

:F̈—S̈—F̈:

i.   Place the S atom in the middle and connect each F atom with a single bond; this requires 2 e$^-$ pairs.

ii.   Complete the octets of the F atoms with nonbonded pairs of electrons; this requires an additional 6 e⁻ pairs.

iii. The remaining 2 e⁻ pairs complete the octet of the central S atom.

(d)   (Draw the structure that obeys the octet rule, for now.) 32 valence e⁻, 16 e⁻ pairs

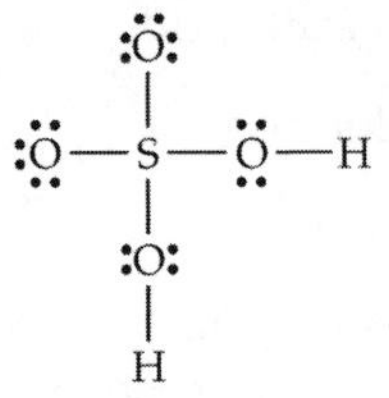

(e)   Follow Sample Exercise 8.8. 20 valence e⁻, 10 e⁻ pairs

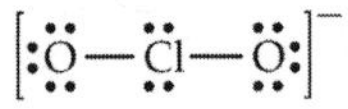

(f)   14 valence e⁻, 7 e⁻ pairs

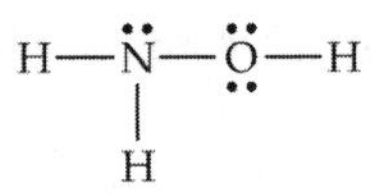

Check. In each molecule, bonding e⁻ pairs are shown as lines, and each atom is surrounded by an octet of electrons (duet for H).

8.48 (a)   12 valence e⁻, 6 e⁻ pairs   (b)   14 valence e⁻, 7 e⁻ pairs

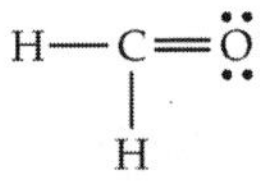

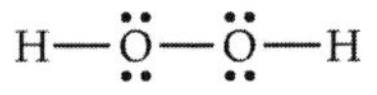

(c)   50 valence e⁻, 25 e⁻ pairs   (d)   26 valence e⁻, 13 e⁻ pairs

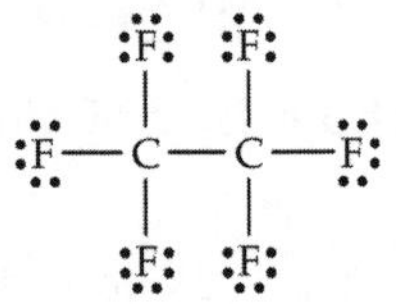

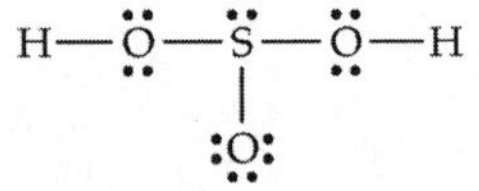

(The Lewis structure that obeys the octet rule)

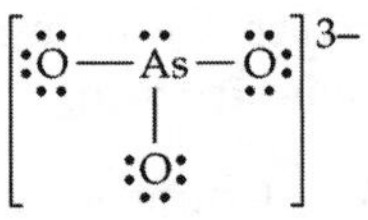

(e)   26 valence e⁻, 13 e⁻ pairs   (f)   NH₂Cl 14 e⁻, 7 e⁻ pairs

(The Lewis structure that obeys the octet rule)

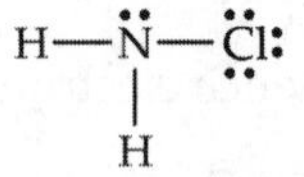

8.49 Statement (b) is most true. (The other four are clearly false.) Keep in mind that when it is necessary to place more than an octet of electrons around an atom to minimize formal charge, there may not be a "best" Lewis structure.

8.50   (a)   26 e⁻, 13 e⁻ pairs

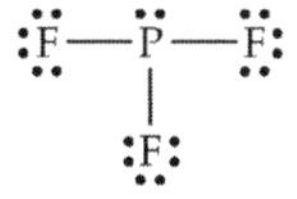

The octet rule is satisfied for all atoms in the structure.

(b)   F is more electronegative than P. Assuming F atoms hold all shared electrons, the oxidation number of each F is –1. The oxidation number of P is +3.

(c)   Assuming perfect sharing, the formal charges on all F and P atoms are 0.

8.51   *Analyze/Plan.* Draw the correct Lewis structure: count valence electrons in each atom, total valence electrons and electron pairs in the molecule or ion; connect bonded atoms with a line, place the remaining e⁻ pairs as needed, in nonbonded pairs or multiple bonds, so that each atom is surrounded by an octet (or duet for H). Calculate formal charges: assign electrons to individual atoms [nonbonding e⁻ + 1/2 (bonding e⁻)]; formal charge = valence electrons – assigned electrons. Assign oxidation numbers, assuming that the more electronegative element holds all electrons in a bond.

Solve. Formal charges are shown near the atoms, oxidation numbers (ox. #) are listed below the structures.

(a)   16 e⁻, 8 e⁻ pairs   (b)   26 valence e⁻, 13 e⁻ pairs

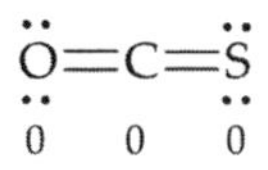

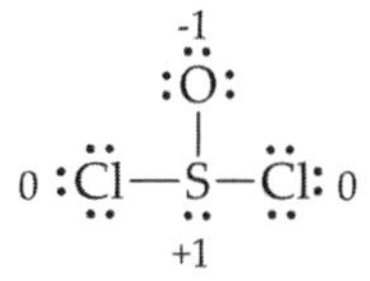

ox. #: O, –2; C, +4; S, –2   ox #: S, +4; Cl, –1; O, –2

(c)   26 valence e⁻, 13 e⁻ pairs   (d)   20 valence e⁻, 10 e⁻ pairs

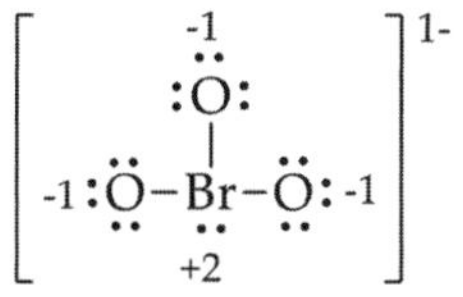

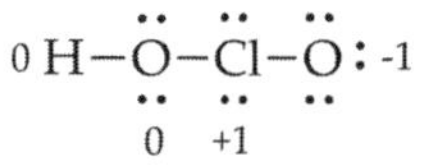

ox. #: Br, +5; O, –2   ox. #: Cl, +3; H, +1; O, –2

Check. Each atom is surrounded by an octet (or duet) and the sum of the formal charges and oxidation numbers is the charge on the particle.

8.52   Formal charges are given near the atoms, oxidation numbers are listed below the structures.

(a)   18 e⁻, 9 e⁻ pairs   (b)   24 e⁻, 12 e⁻ pairs

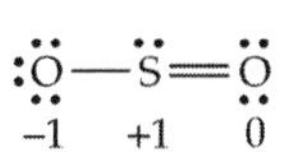

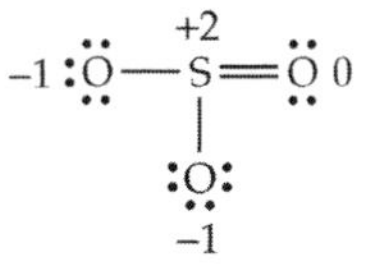

ox. #: S, +4; O, –2   ox. #: S, +6; O, –2

(c)   26 e⁻, 13 e⁻ pairs

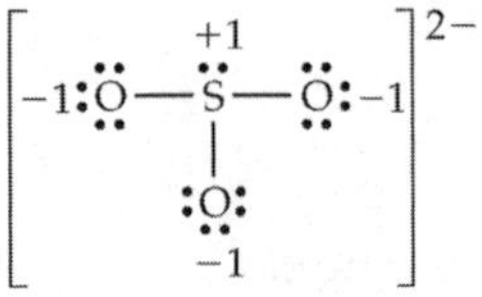

$$\left[-1\,\ddot{\underset{..}{O}} - \overset{+1}{\underset{..}{S}} - \ddot{\underset{..}{O}}{:}\,-1 \right]^{2-}$$

$$\underset{-1}{:\ddot{O}:}$$

ox. #: S, +4; O, –2

(d) $SO_2 < SO_3 < SO_3^{2-}$

Double bonds are shorter than single bonds. SO_2 has two resonance structures with alternating single and double bonds, for an approximate average "one-and-a-half" bond. SO_3 has three resonance structures with one double and two single bonds, for an approximately, "one-and-a-third" bond. SO_3^{2-} has all single bonds. The order of increasing bond length is the order of decreasing bond type.

SO_2 (1.5) $< SO_3$ (1.3) $< SO_3^{2-}$ (1.0).

8.53   (a)   *Plan.* Count valence electrons, draw all possible correct Lewis structures, taking note of alternate placements for multiple bonds.   *Solve.*

18 e⁻, 9 e⁻ pairs

$$\left[\ddot{O} = \ddot{N} - \ddot{O}{:} \right]^{-} \longleftrightarrow \left[{:}\ddot{O} - \ddot{N} = \ddot{O} \right]^{-}$$

Check. The octet rule is satisfied.

(b)   *Plan.* Isoelectronic species have the same number of valence electrons and the same electron configuration.   *Solve.*

A single O atom has 6 valence electrons, so the neutral ozone molecule O_3 is isoelectronic with NO_2^-.

$$\ddot{O} = \ddot{O} - \ddot{O}{:} \longleftrightarrow {:}\ddot{O} - \ddot{O} = \ddot{O}$$

Check. The octet rule is satisfied.

(c)   Because each N–O bond has partial double bond character, the N–O bond length in NO_2^- should be shorter than N–O single bonds but longer than N=O double bonds.

8.54   (a)   18 e⁻, 9 e⁻ pairs

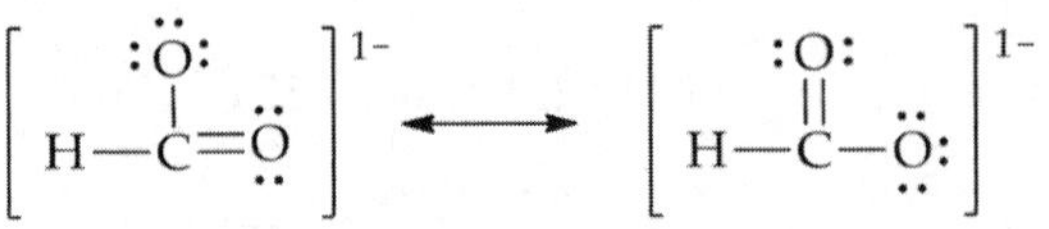

$$\left[\begin{array}{c} :\ddot{O}: \\ | \\ H - C = \ddot{O} \end{array} \right]^{1-} \longleftrightarrow \left[\begin{array}{c} :O: \\ || \\ H - C - \ddot{O}: \end{array} \right]^{1-}$$

(b)   Yes, resonance structures are required to describe the structure.

(c)   The Lewis structure of CO_2 (16 e⁻, 8 e⁻ pairs) is

$$\ddot{O}\!=\!C\!=\!\ddot{O}$$

In CO_2, the C–O bonds are full double bonds with two shared pairs of electrons. In HCO_2^-, the two resonance structures indicate that the C–O bonds have partial, but not full, double bond character. The C–O bond lengths in formate will be longer than those in CO_2.

8.55   *Plan/Solve.* The Lewis structures are as follows:

5 e⁻ pairs   8 e⁻ pairs

$$:C\!\equiv\!O:$$ $$\ddot{O}\!=\!C\!=\!\ddot{O}$$

12 e⁻ pairs

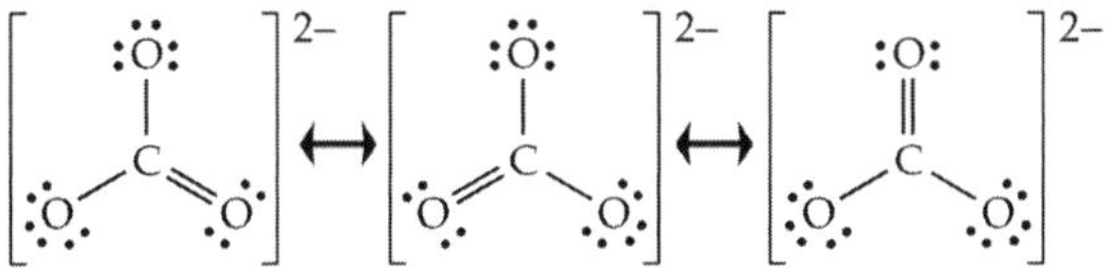

The more pairs of electrons shared by two atoms, the shorter the bond between the atoms. The average number of electron pairs shared by C and O in the three species is 3 for CO, 2 for CO_2, and 1.33 for CO_3^{2-}. The order of bond lengths, from shortest to longest, is: $CO < CO_2 < CO_3^{2-}$.

8.56   The Lewis structures are as follows:

5 e⁻ pairs   9 e⁻ pairs

$$\left[:N\!\equiv\!O:\right]^{+}$$

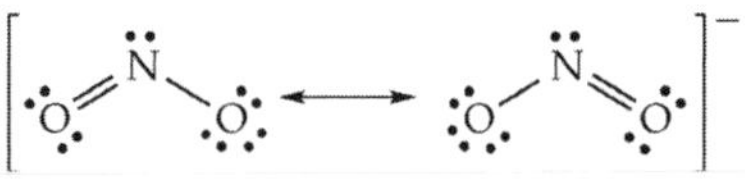

12 e⁻ pairs

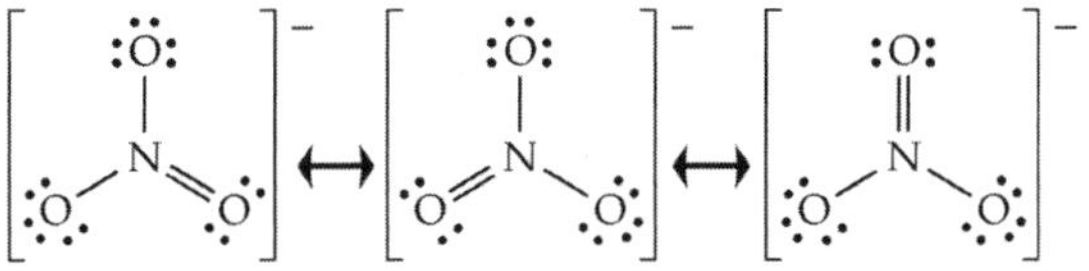

The average number of electron pairs in the N–O bond is 3.0 for NO^+, 1.5 for NO_2^-, and 1.33 for NO_3^-. The more electron pairs shared between two atoms, the shorter the bond. The order of N–O bond lengths from shortest to longest is: $NO^+ < NO_2^- < NO_3^-$.

8.57   (a)   False. Because of resonance (see Figure 8.15), the C–C bonds in benzene are the same length, but they are shorter than a typical single C–C bond and longer than a typical double C–C bond.

        (b)   HCCH, 10 e⁻, 5 e⁻ pr

$$H\!-\!C\!\equiv\!C\!-\!H$$

False. The C–C bond in acetylene is an isolated triple bond; it is shorter than an isolated double bond and therefore shorter than the average C–C bond in benzene.

8.58   (a)

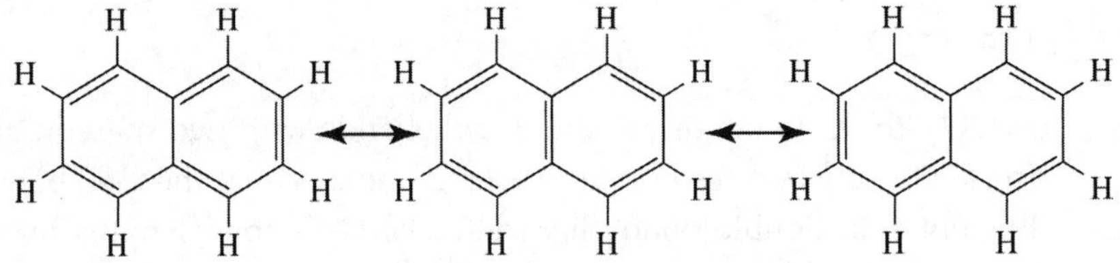

(b)   The resonance model of this molecule has bonds that are neither single nor double, but somewhere in between. This results in bond lengths that are intermediate between C–C single and   double bond lengths.

(c)   Four. Among the three resonance structures, there are four C–C bonds that appear twice as double bonds and once as a single bond. These are shorter than the others. (The other seven C–C bonds appear twice as single bonds and once as a double bond.)

Exceptions to the Octet Rule (Section 8.7)

8.59   *Analyze/Plan.* In order to decide whether a molecule is an exception to the octet rule, examine the Lewis structure. Does the Lewis structure have an odd number of electrons, an atom with less than eight electrons, or an atom with more than eight electrons? *Solve.*

(a) CO_2, 16 e⁻, 8 e⁻pr

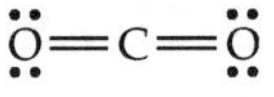

H_2O, 8 e⁻, 4 e⁻pr

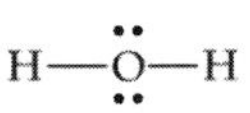

NH_3, 8 e⁻, 4 e⁻pr

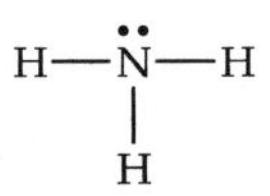

PF_3, 26 e⁻, 13 e⁻pr

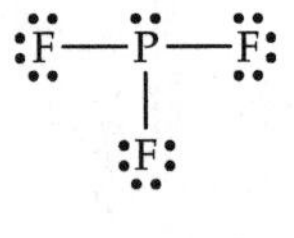

AsF_5, 40 e⁻, 20 e⁻pr

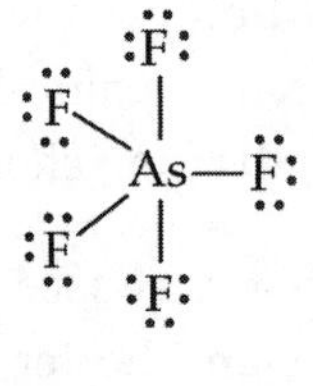

None of the molecules have an odd number of electrons. CO_2, H_2O, NH_3, and PF_3 obey the octet rule. In AsF_5, the central As atom is bound to five F atoms, so it has 10 electrons around it. AsF_5 is an exception to the octet rule.

(b) BH_4^-, 8 e⁻, 4 e⁻ pr

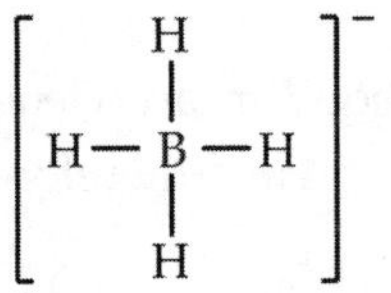

$B_3N_3H_6$, 30 e⁻, 15 e⁻pr

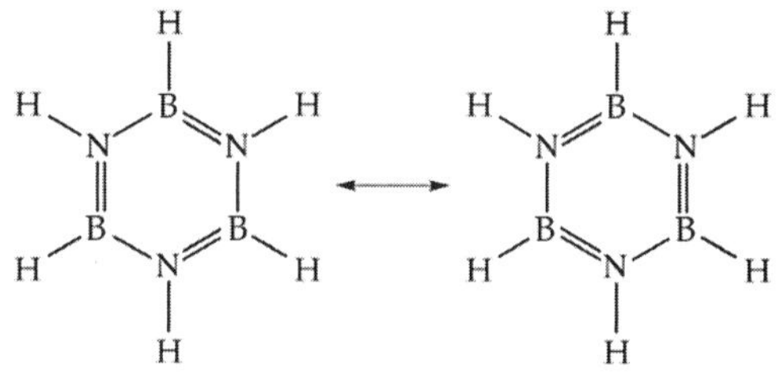

BCl_3, 24 e⁻, 12 e⁻pr

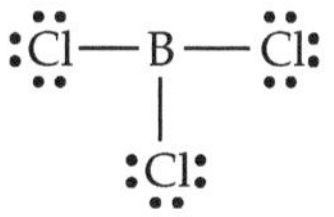

[The structure shown minimizes formal charges and is the dominant form (see Section 8.7)]

BCl_3 is an exception to the octet rule; the B atom has an incomplete octet.

8.60   (a)   7, 1   (b) 6, 2   (c) 5, 3   (d) 4, 4

8.61   *Analyze/Plan.* For each species, count the number of valence electrons and electron pairs. Draw the dominant Lewis structure. For the purpose of this exercise, assume that the dominant Lewis structure is the one that minimizes formal charge.

ClO, 13 e⁻, 6.5 e⁻ pairs

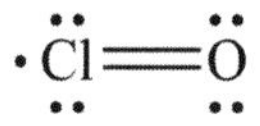

Odd number of electrons
Does not obey the octet rule

ClO⁻, 14 e⁻, 7 e⁻ pairs

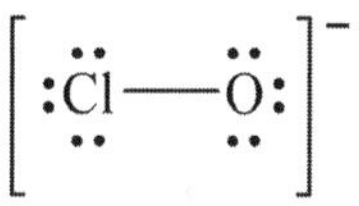

Obeys the octet rule

ClO_2^-, 20 e⁻, 10 e⁻ pairs

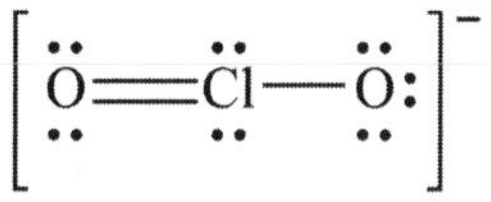

Cl has expanded octet
Does not obey the octet rule

ClO_3^-, 26 e⁻, 13 e⁻ pairs

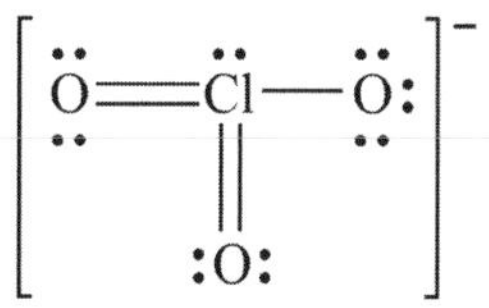

Cl has expanded octet
Does not obey the octet rule

ClO_4^-, 32 e⁻, 16 e⁻ pairs

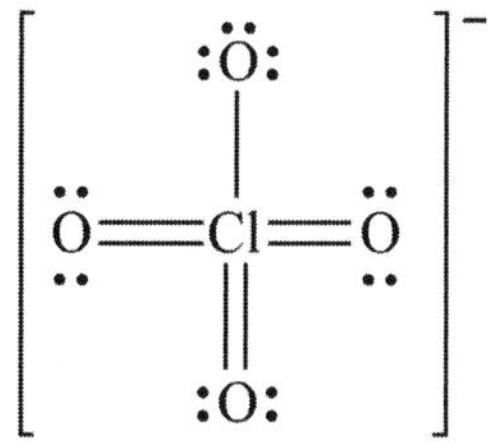

Cl has expanded octet
Does not obey the octet rule

In each species, Cl has a zero formal charge and O atoms that form double bonds have zero formal charge. The O atoms that from single bonds have –1 formal charge. For ClO_2^-, ClO_3^-, and ClO_4^-, structures that do not minimize formal charge but obey the octet rule can be drawn. The octet rule vs minimum formal charge debate is ongoing.

8.62   The second friend is more correct. In the third row and beyond, atoms have the space and available orbitals to accommodate extra electrons. Because atomic radius increases going down a family, elements in the third period and beyond are less subject to destabilization from additional electron-electron repulsions. It is also true, but probably not as important, that elements in the third shell and beyond contain empty d orbitals that are relatively close in energy to valence orbitals (the ones that accommodate the octet).

8.63   (a) $8\,e^-$, $4\,e^-$ pairs $[PH_3]$

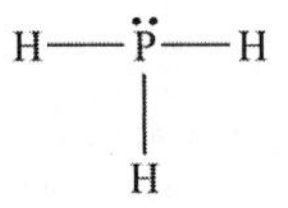

H—P̈—H
│
H

Obeys octet rule.

(b) $6\,e^-$, $3\,e^-$ pairs

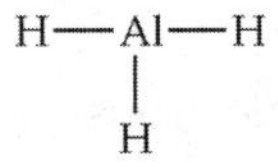

H—Al—H
│
H

Does not obey the octete rule. Central Al has 6 electrons (impossible to satisfy octet rule with only 6 valence electrons).

(c) $16\,e^-$, $8\,e^-$ pairs

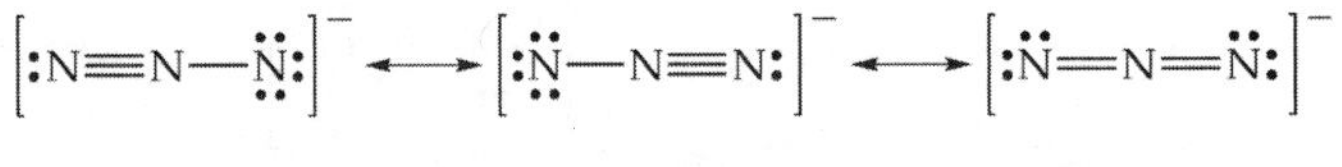

$$\left[:N\equiv N-\ddot{\underset{\cdot\cdot}{N}}:\right]^- \longleftrightarrow \left[:\ddot{\underset{\cdot\cdot}{N}}-N\equiv N:\right]^- \longleftrightarrow \left[:\ddot{N}=N=\ddot{N}:\right]^-$$

3 resonance structures; all obey octet rule.

(d) $20\,e^-$, $10\,e^-$ pairs

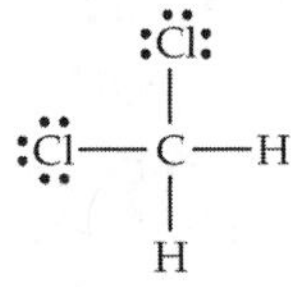

:C̈l:
│
:C̈l—C—H
│
H

Obeys octet rule.

(e) $48\,e^-$, $24\,e^-$ pairs $[SnF_6^{2-}]$

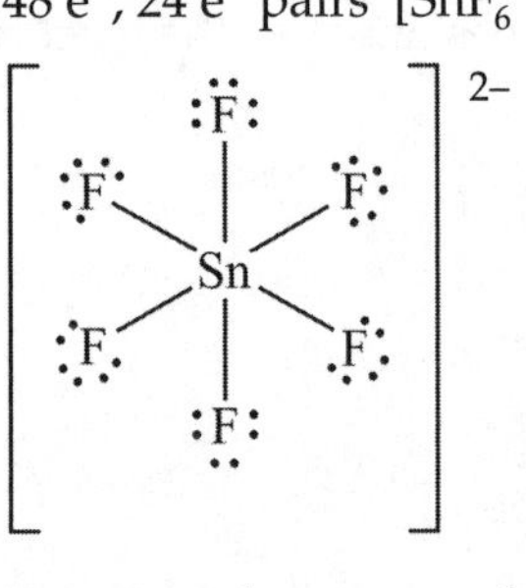

$$\begin{bmatrix} & & :\ddot{F}: & & \\ :\ddot{F}\cdot & & | & & \cdot\ddot{F}: \\ & & Sn & & \\ :\ddot{F}\cdot & & | & & \cdot\ddot{F}: \\ & & :\ddot{F}: & & \end{bmatrix}^{2-}$$

Does not obey octet rule. Central Sn has 12 electrons.

8.64   (a) $11\,e^-$, $5.5\,e^-$ pairs

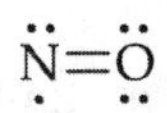

$\dot{\ddot{N}}=\ddot{O}$

Does not obey the octet rule. N has only 7 electrons.

(b)   24 e⁻, 12 e⁻ pairs

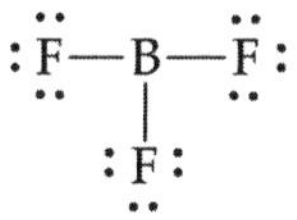

:F̈—B—F̈:
     |
     :F̈:

Does not obey the octet rule. B has only 6 electrons.

(c)   22 e⁻, 11 e⁻ pairs

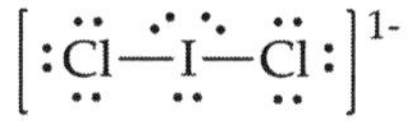

[:C̈l—Ï—C̈l:]¹⁻

Does not obey the octet rule. Central I has 10 electrons.

(d)   32 e⁻, 16 e⁻ pairs

    :B̈r: :B̈r:
      | |
:B̈r—P—Ö: :B̈r—P=Ö
      | |
    :B̈r: :B̈r:

The structure on the left obeys the octet rule, whereas the one on the right minimizes formal charges but does not obey the octet rule. The P in the right structure has 10 electrons.

(e)   36 e⁻, 18 e⁻ pairs

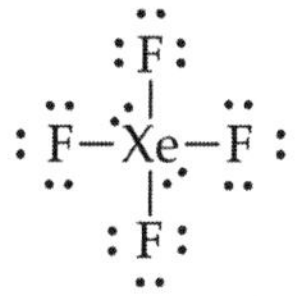

     :F̈:
      ·|
:F̈—Xe—F̈:
     |·
     :F̈:

Does not obey the octet rule. Central Xe has 12 electrons.

8.65   (a)   16 e⁻, 8 e⁻ pairs

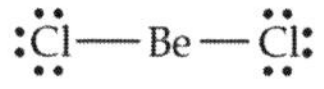

:C̈l—Be—C̈l:

This structure violates the octet rule; Be has only 4 e⁻ around it.

(b) 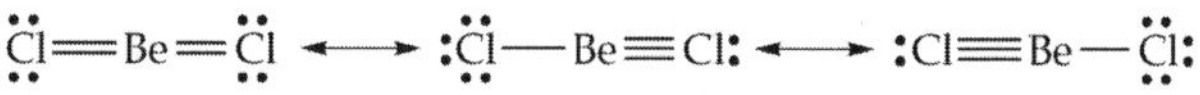   C̈l=Be=C̈l ⟷ :C̈l—Be≡Cl: ⟷ :Cl≡Be—C̈l:

(c)   The formal charges on each of the atoms in the four resonance structures are:

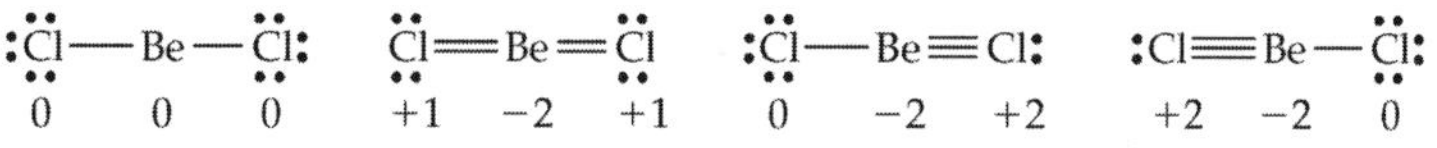

:C̈l—Be—C̈l: C̈l=Be=C̈l   :C̈l—Be≡Cl: :Cl≡Be—C̈l:
 0   0   0   +1   −2   +1   0   −2   +2   +2   −2   0

Formal charges are minimized on the structure that violates the octet rule; this form is probably dominant.

8.66   (a)   26 e⁻, 13 e⁻ pairs

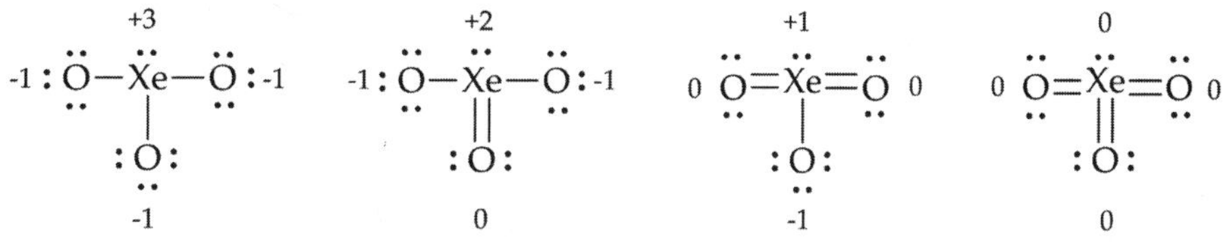

     +3   +2   +1   0
−1:Ö—Xe—Ö:−1   −1:Ö—Xe—Ö:−1   0 Ö=Xe=Ö 0   0 Ö=Xe=Ö 0
     | || | ||
    :Ö: :Ö: :Ö: :Ö:
    −1   0   −1   0

(b)   Yes, the structure with no double bonds obeys the octet rule for all atoms.

(c)   The structure with one double bond has 3 resonance structures (3 possible positions for the double bond), as does the structure with 2 double and 1 single bond (3 possible positions for the single bond). The total number of resonance structures is then 8.

(d)   The structure with 3 double bonds minimizes formal charges on all atoms.

8.67   (a)   *Analyze/Plan.* Given H_2SO_4 with H attached to O, assume S is central and bound to the four O atoms. Draw a Lewis structure where S and O obey the octet rule and H atoms have two electrons and are terminal.   *Solve.*

$32\,e^-$, $16\,e^-$ pr

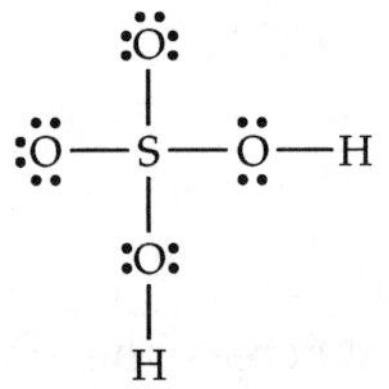

(b)   *Analyze/Plan.* Starting with the Lewis structure from part (a), rearrange electrons to minimize formal charge. Formal charge = valence e^- – assigned e^-. To have a formal charge of zero, both S and O should have 6 assigned electrons.

Assigned e^- = [nonbonding e^- + ½(bonding e^-)]   *Solve.*

For the structure in part (a),

S: assigned electrons = 0 + ½(8) = 4;

                    FC = 6 valence e^- – 4 assigned e^- = +2

O (terminal):   assigned electrons = 6 + ½(2) = 7;

                    FC = 6 valence e^- – 7 assigned e^- = –1

O (bound to H):   assigned electrons = 4 + ½(4) = 6;

                    FC = 6 valence e^- – 6 assigned e^- = 0

To minimize formal charges, change one nonbonding e^- pair on each terminal O atom into a bonding e^- pair between that atom and S. That is, the bonds between the terminal O atoms and S become double bonds.

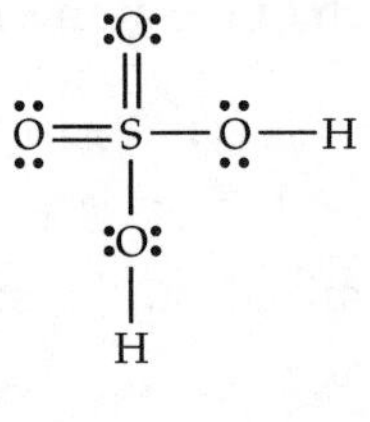

8.68   (a) $32\,e^-$, $16\,e^-$ pairs 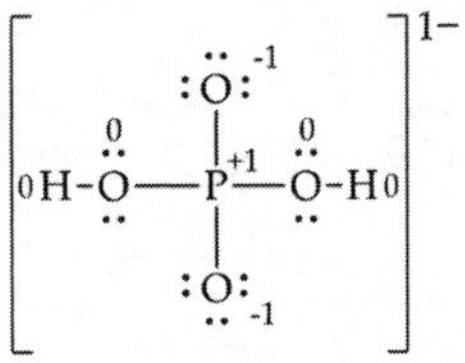   (b)

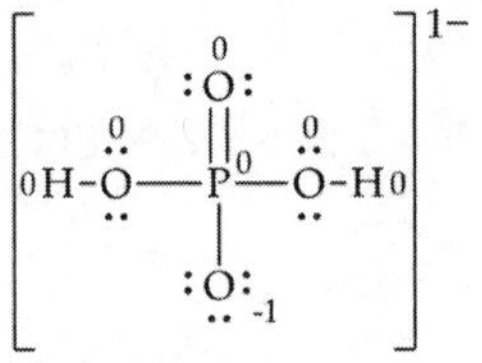

8 Chemical Bonding

Solutions to Exercises

Strengths and Lengths of Covalent Bonds (Section 8.8)

8.69 *Analyze.* Given: structural formulas. Find: enthalpy of reaction.

Plan. Count the number and kinds of bonds that are broken and formed by the reaction. Use bond enthalpies from Table 8.3 and Equation 5.32 to calculate the overall enthalpy of reaction, ΔH. *Solve.*

(a) ΔH = 2 D(O–H) + D(O–O) + 4 D(C–H) + D(C=C)

$\quad$ –2 D(O–H) – 2 D(O–C) – 4 D(C–H) – D(C–C)

ΔH = D(O–O) + D(C=C) – 2 D(O–C) – D(C–C)

$\quad$ = 146 + 614 – 2(358) – 348 = –304 kJ

(b) ΔH = 5 D(C–H) + D(C ≡ N) + D(C=C) – 5 D(C–H) – D(C ≡ N) – 2 D(C–C)

$\quad$ = D(C=C) – 2 D(C–C) = 614 – 2(348) = –82 kJ

(c) ΔH = 6 D(N–Cl) – 3 D(Cl–Cl) – D(N ≡ N)

$\quad$ = 6(200) – 3(242) – 941 = –467 kJ

8.70 (a) ΔH = 3 D(C–Br) + D(C–H) + D(Cl–Cl) – 3 D(C–Br) – D(C–Cl) – D(H–Cl)

$\quad$ = D(C–H) + D(Cl–Cl) – D(C–Cl) – D(H–Cl)

ΔH = 413 + 242 – 328 – 431 = –104 kJ

(b) ΔH = 4 D(C–H) + 2 D(C–S) + 2 D(S–H) + D(C–C) + 2 D(H–Br)

$\quad$ –4 D(S–H) – D(C–C) – 2 D(C–Br) – 4 D(C–H)

$\quad$ = 2 D(C–S) + 2 D(H–Br) – 2 D(S–H) – 2 D(C–Br)

ΔH = 2(259) + 2(366) – 2(339) – 2(276) = 20 kJ

(c) ΔH = 4 D(N–H) + D(N–N) + D(Cl–Cl) – 4 D(N–H) – 2 D(N–Cl)

$\quad$ = D(N–N) + D(Cl–Cl) – 2D(N–Cl)

ΔH = 163 + 242 – 2(200) = 5 kJ

8.71 (a) False. For the same atom pair, the longer the bond the smaller the bond enthalpy.

(b) False. See average bond enthalpies in Table 8.3.

(c) False. The single bond lengths in Table 8.4 are all less than 5 Å.

(d) False. Breaking a chemical bond requires an input of energy.

(e) True.

8.72 (a) True. See Table 8.4. In general for the same atom pair, the greater the number of bonds between atoms, the shorter the bond.

(b) False. The Lewis structure of O_2 has two pairs of bonding electrons and four pairs of nonbonding electrons.

(c)   False. The C–O bond in CO is shorter, because the two atoms share three electron pairs, while in CO_2, each C–O bond involves two shared electron pairs.

(d)   False. The Lewis structure of O_2 has an isolated double bond. Ozone has resonance structures; the average O–O bond length will be shorter than a single but longer than a double bond.

(e)   False. The more electronegative an atom, the more strongly it holds its own electrons. It forms the minimum number of bonds needed to satisfy the octet rule.

8.73   Ionic bond enthalpies depend on the charge and size of the participating ions. The Ca–O bond will be stronger than the Na–Cl bond, because the ion charges are greater.

8.74   Ionic bond enthalpies depend on the charge and size of the participating ions. The Cs–F and Li–F bonds have the same anion and cation charges but Cs^+ has a much larger ionic radius than Li^+. The Cs–F bond, with the larger ion separation, will have the smaller ionic bond enthalpy.

8.75   (a) CH_2, 6 e⁻, 3 e⁻ pr

        H—C̈—H

        The C atom in carbene is extremely electron deficient, which causes carbene to be very reactive.

        (b) $2 CH_2 \rightarrow C_2H_4$

        The product of this reaction contains a C–C double bond, with a typical bond length of 1.34 Å. (See Table 8.4.)

8.76   NO, 11 e⁻, 5.5 e⁻ pr

        N̈=Ö

        NO^+, 10 e⁻, 5 e⁻ pr

        $[:N\equiv O:]^+$

        The bond in NO is longer than the bond in NO^+. In the neutral NO molecule, the N atom has an incomplete octet and the N–O bond is formally a double bond. In NO^+, the odd electron has been lost and the N–O bond is a triple bond. For the same pair of bonded atoms, a double bond is longer than a triple bond.

Additional Exercises

8.77   A triple C–C bond. A 1.15 Å bond length is relatively short for any atom pair. Table 8.4 indicates than an average triple C–C bond length is 1.20 Å.

8.78   A double N–N bond. Table 8.4 indicates than an average double N–N bond length is 1.24 Å. (The bonding atomic radius of N is less than that of C, so we expect N–N bonds of all types to be shorter than C–C bonds.)

8.79   (a)   All the compounds in the series contain Group 2A metal cations with oxidation numbers of +2. The oxidation number of hydrogen is then –1. (H^- is named hydride.)

(b) Be^{2+} is the smallest cation, so BeH_2 has the shortest cation-anion distance. All compounds in the series have the same product of ion charges, so the trend in cation-anion distance dictates the trend in lattice energy. Lattice energy is inversely related to ion separation, so the compound with the largest lattice energy, BeH_2, has the shortest cation-anion distance.

(c)   It costs 3205 kJ of energy to break one mole of BeH_2 into its component gas phase ions. Charge attraction is a stabilizing force; energy is required to overcome it.

(d)   Mg. Lattice energy depends on ionic radius and charge. The charges are the same in the series, so ionic radius is the discriminating factor. The compound with a lattice energy nearest 2870 kJ/mol has the Group 2A cation with the radius most similar to Zn^{2+}.

8.80   (a)

	Compound	Lattice Energy (kJ)			Compound	Lattice Energy (kJ)	
106 kJ	NaCl	788	56 kJ	104 kJ	LiCl	834	55 kJ
	NaBr	732			**LiBr**	**779**	
	Na I	682			Li I	730	

The difference in lattice energy between LiCl and LiI is 104 kJ. The difference between NaCl and NaI is 106 kJ; the difference between NaCl and NaBr is 56 kJ, or 53% of the difference between NaCl and NaI. Applying this relationship to the Li salts, 0.53(104 kJ) = 55 kJ difference between LiCl and LiBr. The approximate lattice energy of LiBr is (834 – 55) kJ = 779 kJ.

(b)

	Compound	Lattice Energy (kJ)			Compound	Lattice Energy (kJ)	
106 kJ	NaCl	788	56 kJ	57 kJ	CsCl	657	30 kJ
	NaBr	732			**CsBr**	**627**	
	Na I	682			Cs I	600	

By analogy to the Na salts, the difference between lattice energies of CsCl and CsBr should be approximately 53% of the difference between CsCl and CsI. The lattice energy of CsBr is approximately 627 kJ.

(c)

	Compound	Lattice Energy (kJ)			Compound	Lattice Energy (kJ)	
578 kJ	MgO	3795	381 kJ	199 kJ	$MgCl_2$	2326	131 kJ
	CaO	3414			**$CaCl_2$**	**2195**	
	SrO	3217			$SrCl_2$	2127	

By analogy to the oxides, the difference between the lattice energies of $MgCl_2$ and $CaCl_2$ should be approximately 66% of the difference between $MgCl_2$ and $SrCl_2$. That is, 0.66(199 kJ) = 131 kJ. The lattice energy of $CaCl_2$ is approximately (2326 – 131) kJ = 2195 kJ.

8.81   The charge on M is likely to be 3+. According to Table 8.1, the lattice energy for an ionic compound with the general formula MX and a charge of 2+ on the metal will be in the range of $3\text{-}4 \times 10^3$ kJ/mol. The charge on M must be greater than 2+. ScN, where the charge on Sc is 3+, has a lattice energy of 7547 kJ/mol. It is reasonable to conclude that the charge on M is 3+, and the M–X distance is greater than the Sc–N distance.

8.82   (a)   Six. Figure 8.4 shows a small part of the CaO (or NaCl) structure. Think of it as a cube with faces and edges. Focus on one purple Ca^{2+} cation in the interior of a face (not on an edge). It is touching 4 O^{2-} anions also in the face. Additionally, this same Ca^{2+} touches an O^{2-} behind and in front of it. The O^{2-} behind it is inside the cube, but the one "in front" of it is in the next layer of ions not shown in the figure. This is a total of six O^{2-} anions touching a single Ca^{2+} cation.

        (b)   Energy would be consumed. Electrostatic attraction holds ion pairs together in a 3-dimensional structure such as the one in Figure 8.4. This energy of electrostatic attraction must be overcome in order to convert a crystal of CaO into a collection of widely separated Ca–O ion pairs.

        (c)   The ionic radii from Figure 7.8 are: Ca^{2+}, 1.14 Å; O^{2-}, 1.26 Å. The lattice energy is 3414 kJ/mol.

$$E_{el} = \frac{\kappa Q_1 Q_2}{d}; d = 2.40 \text{ Å}; Q_1 = 2(1.60 \times 10^{-19} \text{ C}), Q_2 = -2(1.60 \times 10^{-19} \text{ C})$$

$$E = \frac{8.99 \times 10^9 \text{ J-m}}{C^2} \times \frac{-4(1.60 \times 10^{-19} \text{ C})^2}{2.40 \text{ Å}} \times \frac{1 \text{ Å}}{1 \times 10^{-10} \text{ m}} = -3.836 \times 10^{-18} = -3.84 \times 10^{-18} \text{ J}$$

        (d)   On a molar basis: $(-3.836 \times 10^{-18} \text{ J})(6.022 \times 10^{23}) = -2.310 \times 10^6 \text{ J} = -2310 \text{ kJ}$

        Note that the absolute value of this potential energy is less than the lattice energy of CaO, 3414 kJ/mol. The difference represents the additional energy required to separate the individual $Ca^{2+}O^{2-}$ ion pairs from their three-dimensional array similar to the one in Figure 8.4.

        (e)   The electrostatic interactions in a crystal lattice are more complicated than those in a single ion pair.

8.83   By analogy to the Born-Haber cycle for NaCl(s), Figure 8.6, the enthalpy of formation for $NaCl_2$(s) is

$\Delta H_f^\circ NaCl_2(s) = -\Delta H_{latt} NaCl_2 + \Delta H_f^\circ Na(g) + 2\Delta H_f^\circ Cl(g) + I_1(Na) + I_2(Na) + 2 E(Cl)$

        (a) $\Delta H_f^\circ NaCl_2(s) = -\Delta H_{latt} NaCl_2 + 107.7 \text{ kJ} + 2(121.7 \text{ kJ}) + 496 \text{ kJ} + 4562 \text{ kJ} + 2(-349 \text{ kJ})$
                $\Delta H_f^\circ NaCl_2(s) = -\Delta H_{latt} NaCl_2 + 4711 \text{ kJ}$

        The collective energy of the "other" steps in the cycle (vaporization and ionization of Na^{2+}, dissociation of Cl_2 and electron affinity of Cl) is +4711 kJ. In order for the sign of $\Delta H_f^\circ NaCl_2$ to be negative, the lattice energy would have to be greater than 4711 kJ.

        (b) $\Delta H_f^\circ NaCl_2(s) = -(2326 \text{ kJ}) + 4711 \text{ kJ} = 2385 \text{ kJ}$
                This value is large and positive.

8.84   (a)   Yes. If X and Y have different electronegativities, they have different attractions for the electrons in the molecule. The electron density around the more electronegative atom will be greater, producing a charge separation or dipole in the molecule.

        (b)   Yes. $\mu = Qr$. The dipole moment, μ, is the product of the magnitude of the separated charges, Q, and the distance between them, r. The longer the bond between X and Y, the larger the dipole moment.

8.85 (a) B–O. The most polar bond will be formed by the two elements with the greatest difference in electronegativity. Because electronegativity increases moving right and up on the periodic table, the possibilities are B–O and Te–O. These two bonds are likely to have similar electronegativity differences (3 columns apart vs. 3 rows apart). Values from Figure 8.8 confirm the similarity, and show that B–O is slightly more polar.

   (b) Te–I. Both are in the fifth row of the periodic table and have the two largest covalent radii among this group of elements.

   (c) TeI_2. Te needs to participate in two covalent bonds to satisfy the octet rule, and each I atom needs to participate in one bond, so by forming a TeI_2 molecule, the octet rule can be satisfied for all three atoms.

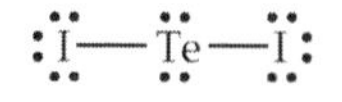

$$:\!\ddot{I}\!—\!\ddot{Te}\!—\!\ddot{I}\!:$$

   (d) B_2O_3. Although this is probably not a purely ionic compound, it can be understood in terms of gaining and losing electrons to achieve a noble-gas configuration. If each B atom were to lose 3 e^- and each O atom were to gain 2 e^-, charge balance and the octet rule would be satisfied.

     P_2O_3. Each P atom needs to share 3 e^- and each O atom 2 e^- to achieve an octet. Although the correct number of electrons seem to be available, a correct Lewis structure is difficult to imagine. In fact, phosphorus (III) oxide exists as P_4O_6 rather than P_2O_3 (Chapter 22).

8.86 (a) $Q = \dfrac{\mu}{r} = \dfrac{1.24\,D}{1.60\,\text{Å}} \times \dfrac{1\,\text{Å}}{1\times10^{-10}\,m} \times \dfrac{3.34\times10^{-30}\,C\text{-}m}{1\,D} \times \dfrac{1\,e}{1.60\times10^{-19}\,C} = 0.1618 = 0.162e$

   (b) From Figure 8.8, the electronegativity of Cl is 3.0 and that of O is 3.5. Because O is the more electronegative element, we expect it to have a partial negative charge in the ClO molecule.

   (c) 13 e^-, 6.5 e^- pairs

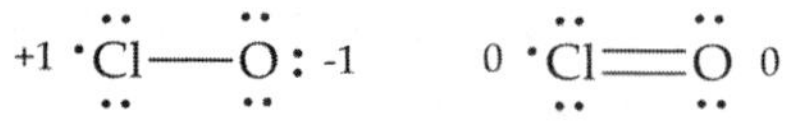

$$+1\ \cdot\ddot{Cl}\!—\!\ddot{O}:\ \text{-}1 \qquad\qquad 0\ \cdot\ddot{Cl}\!=\!\ddot{O}\ 0$$

     According to formal charge arguments, the Lewis structure on the right is dominant. In both structures, the less electronegative Cl atom is electron-deficient. However, the small electronegativity difference and calculated charges both point to a slightly polar covalent molecule. The true bonding situation is a blend of the two extreme Lewis structures, with the right-most structure making the larger contribution.

   (d) Because ClO^- has an overall charge of 1–, the sum of the formal charges in any correct Lewis structure is 1–. We expect the more electronegative O atom to carry the negative formal charge. The best Lewis structure for ClO^- is then

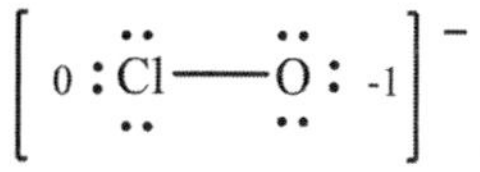

$$\left[\, 0\ :\!\ddot{Cl}\!—\!\ddot{O}:\ \text{-}1 \,\right]^{-}$$

     The formal charge on Cl in this structure is 0.

8.87   (a)   Estimate relative attraction for the bonding electron pair by calculating the relative electronegativity of the two atoms. From Figure 8.8, the electronegativity of Br is 2.8 and of Cl is 3.0.

Br has $2.8/(3.0 + 2.8) = 0.48$ of the charge of the bonding e^- pair.

Cl has $3.0/(3.0 + 2.8) = 0.52$ of the charge of the bonding e^- pair.

This amounts to $0.52 \times 2e = 1.04e$ on Cl or $0.04e$ more than a neutral Cl atom. This implies a -0.04 charge on Cl and $+0.04$ charge on Br.

   (b)   From Figure 7.7, the covalent radius of Br is 1.20 Å and of Cl is 1.02 Å. The Br–Cl separation is 2.22 Å.

$$\mu = Qr = 0.04\,e \times \frac{1.60 \times 10^{-19}\,C}{e} \times 2.22\,\text{Å} \times \frac{1 \times 10^{-10}\,m}{\text{Å}} \times \frac{1\,D}{3.34 \times 10^{-30}\,C\text{-}m} = 0.43\,D$$

   (c) $$Q = \frac{\mu}{r} = \frac{0.57\,D}{2.22\,\text{Å}} \times \frac{1\,\text{Å}}{1 \times 10^{-10}\,m} \times \frac{3.34 \times 10^{-30}\,C\text{-}m}{1\,D} \times \frac{1\,e}{1.60 \times 10^{-19}\,C} = 0.054\,e$$

From this calculation, the partial charge on Br is $+0.054$ and on Cl is -0.054.

8.88   (a) $2\,NaAlH_4(s) \rightarrow 2\,NaH(s) + 2\,Al(s) + 3\,H_2(g)$

   (b)   Hydrogen is the only nonmetal in $NaAlH_4$, so we expect it to be most electronegative. (The position of H on the periodic table is problematic. Its electronegativity does not fit the typical trend for Gp 1A elements.) For the two metals, Na and Al, electronegativity increases moving up and to the right on the periodic table, so Al is more electronegative. The least electronegative element in the compound is Na.

   (c)   Covalent bonds hold polyatomic anions together; elements involved in covalent bonding have smaller electronegativity differences than those that are involved in ionic bonds. Possible covalent bonds in $NaAlH_4$ are Na–H and Al–H. Al and H have a smaller electronegativity difference than Na and H and are more likely to form covalent bonds. The anion has an overall 1– charge, so it can be thought of as four hydride ions and one Al^{3+} ion. The formula is AlH_4^-. For the purpose of counting valence electrons, assume neutral atoms.

$8\,e^-$ $4\,e^-$ pairs

$$\left[\begin{array}{c} H \\ | \\ H{-}Al{-}H \\ | \\ H \end{array} \right]^-$$

   (d)   The formal charge of H in AlH_4^- is 0. (The formal charge of Al is -1. This brings the sum of formal charges to -1, the overall charge of the polyatomic anion.)

8.89   (a) I_3^-, $22\,e^-$, $11\,e^-$ pr

$$\left[\,:\!\ddot{\underset{..}{I}}\!{-}\!\ddot{\underset{..}{I}}\!{-}\!\ddot{\underset{..}{I}}\!: \, \right]^-$$

   (b) F_2, HF, CF_4, SiF_4, etc.

    (c)   No. I_3^- violates the octet rule but is quite stable. Violating the octet rule does not prevent the existence of a substance.

    (d)   This classmate is at least partly correct. The presence of five bonding and nonbonding electron pairs around a central atom as small as F would generate significant electron-electron repulsion. These repulsions would destabilize F_3^-.

8.90   Formal charge (FC) = # valence e^- – (# nonbonding e^- + 1/2 # bonding e^-)

    (a)   18 e^-, 9 e^- pairs

            FC for the central O = 6 – [2 + 1/2 (6)] = +1

    (b)   48 e^-, 24 e^- pairs

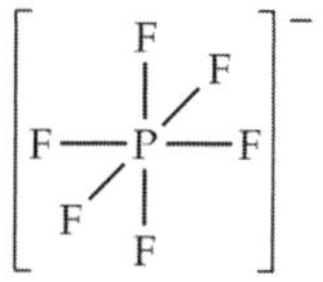

     FC for P = 5 – [0 + 1/2 (12)] = –1

            The three nonbonded pairs on each F have been omitted.

    (c)   17 e^-; 8 e^- pairs, 1 odd e^-

            The odd electron is probably on N because it is less electronegative than O. Assuming the odd electron is on N, FC for N = 5 – [1 + 1/2 (6)] = +1. If the odd electron is on O, FC for N = 5 – [2 + 1/2 (6)] = 0.

    (d)   28 e^-, 14 e^- pairs   (e)   32 e^-, 16 e^- pairs

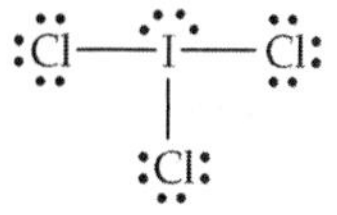

           FC for I = 7 – [4 + 1/2 (6)] = 0   FC for Cl = 7 – [0 + 1/2 (8)] = +3

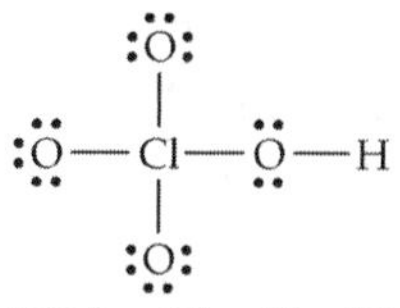

8.91   14e^-, 7 e^- pairs   32 e^-, 16 e^- pairs

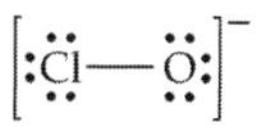

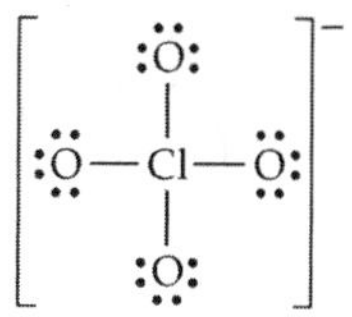

    (a)   FC on Cl in ClO^- = 7 – [6 + 1/2(2)] = 0

    (b)   FC on Cl in ClO_4^- = 7 – [0 + 1/2(8)] = +3

    (c)   The oxidation number of Cl in ClO^- is

            [ON + (–2)] = –1; ON of Cl = +1

    (d)   The oxidation number of Cl in ClO_4^- is

            [ON + 4(–2)] = –1; ON of Cl = +7

(e) The more positive the formal charge and the higher the oxidation number of a bonded atom, the greater the electron deficiency at that atom. The atom with the higher oxidation number is more likely to accept electrons from another compound and be reduced. Perchlorate, ClO_4^-, is much more likely to be reduced.

8.92 (a)

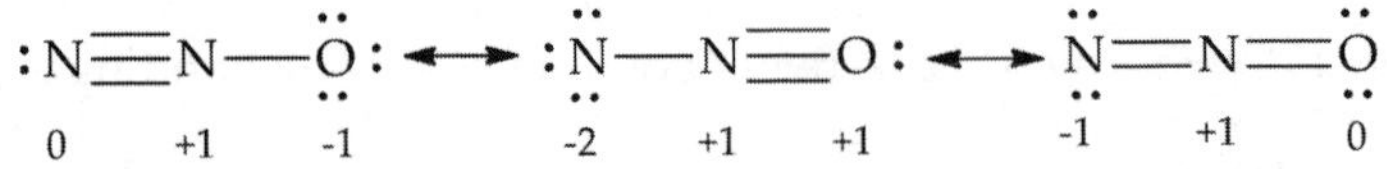

$$:N\!\!\equiv\!\!N\!\!-\!\!\ddot{O}: \longleftrightarrow :\ddot{N}\!\!-\!\!N\!\!\equiv\!\!O: \longleftrightarrow \ddot{N}\!\!=\!\!N\!\!=\!\!\ddot{O}$$
$$\quad 0 \qquad +1 \quad -1 \qquad\qquad -2 \quad +1 \quad +1 \qquad\qquad -1 \quad +1 \quad 0$$

In the leftmost structure, the more electronegative O atom has the negative formal charge, so this structure is likely to be most important.

(b) No single resonance structure rationalizes both observed bond lengths. In general, the more shared pairs of electrons between two atoms, the shorter the bond, and vice versa. That the N–N bond length in N_2O is slightly longer than the typical N≡N indicates that the middle and right resonance structures where the N atoms share less than three electron pairs are contributors to the true structure. That the N–O bond length is slightly shorter than a typical N=O indicates that the middle structure, where N and O share more than two electron pairs, does contribute to the true structure. This physical data indicates that although formal charge can be used to predict which resonance form will be more important to the observed structure, the influence of minor contributors on the true structure cannot be ignored.

8.93 (a) $12 + 3 + 15 = 30$ valence e^-, 15 e^- pairs.

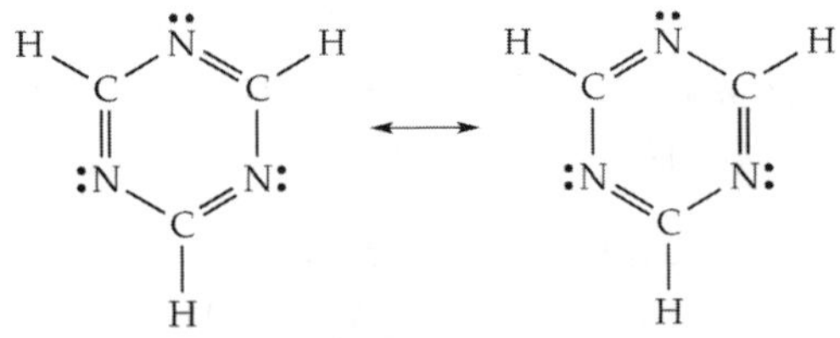

Structures with H bound to N and nonbonded electron pairs on C can be drawn, but the structures above minimize formal charges on the atoms.

(b) The resonance structures indicate that triazine will have six equal C–N bond lengths, intermediate between C–N single and C–N double bond lengths. (See Solutions 8.57 and 8.58.) From Table 8.4, an average C–N length is 1.43 Å, a C=N length is 1.38 Å. The average of these two lengths is 1.405 Å. The C–N bond length in triazine should be in the range 1.40–1.41 Å.

8.94 (a) $24 + 4 + 14 = 42$ valence e^-, 21 e^- pairs.   (b)

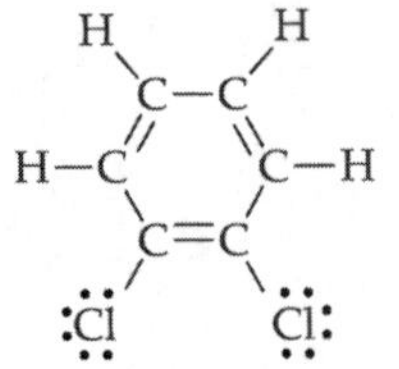

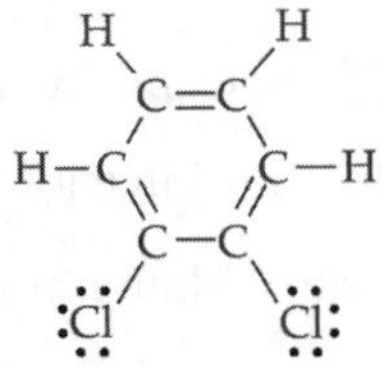

(c) No. In benzene, the six C atoms are equivalent. In ortho-dichlorobenzene, the two C atoms bound to Cl are not equivalent to the four C atoms bound to H. In the two resonance structures above, one has a double bond between the C atoms bound to Cl, and the other has a single bond in this position. The two ortho-dichlorobenzene resonance structures are not equivalent like the resonance structures of benzene.

Integrative Exercises

8.95   (a)   False. The B–A=B structure says nothing about the nonbonding electrons in the molecule. One possible example is NO_2, which has an odd electron.

        (b)   True.

8.96 $\Delta H = 8\ D(C-H) - D(C-C) - 6\ D(C-H) - D(H-H)$

        $= 2\ D(C-H) - D(C-C) - D(H-H)$

        $= 2(413) - 348 - 436 = +42\ kJ$

    $\Delta H = 8\ D(C-H) + 1/2\ D(O=O) - D(C-C) - 6\ D(C-H) - 2\ D(O-H)$

        $= 2\ D(C-H) + 1/2\ D(O=O) - D(C-C) - 2\ D(O-H)$

        $= 2(413) + 1/2\ (495) - 348 - 2(463) = -200\ kJ$

The fundamental difference in the two reactions is the formation of 1 mol of H–H bonds versus the formation of 2 mol of O–H bonds. The latter is much more exothermic, so the reaction involving oxygen is more energetically favorable.

8.97   (a) $\Delta H = 5\ D(C-H) + D(C-C) + D(C-O) + D(O-H) - 6\ D(C-H) - 2\ D(C-O)$

            $= D(C-C) + D(O-H) - D(C-H) - D(C-O)$

            $= 348\ kJ + 463\ kJ - 413\ kJ - 358\ kJ$

        $\Delta H = +40\ kJ$; ethanol has the lower enthalpy

        (b) $\Delta H = 4\ D(C-H) + D(C-C) + 2\ D(C-O) - 4\ D(C-H) - D(C-C) - D(C=O)$

            $= 2\ D(C-O) - D(C=O)$

            $= 2(358\ kJ) - 799\ kJ$

        $\Delta H = -83\ kJ$; acetaldehyde has the lower enthalpy

        (c) $\Delta H = 8\ D(C-H) + 4\ D(C-C) + D(C=C) - 8\ D(C-H) - 2\ D(C-C) - 2\ D(C=C)$

            $= 2\ D(C-C) - D(C=C)$

            $= 2(348\ kJ) - 614\ kJ$

        $\Delta H = +82\ kJ$; cyclopentene has the lower enthalpy

        (d) $\Delta H = 3\ D(C-H) + D(C-N) + D(C \equiv N) - 3\ D(C-H) - D(C-C) - D(C \equiv N)$

            $= D(C-N) - D(C-C)$

            $= 293\ kJ - 348\ kJ$

        $\Delta H = -55\ kJ$; acetonitrile has the lower enthalpy

8.98   (a) $Ti^{2+}: [Ar]3d^2; \; Ca:[Ar]4s^2$.

       (b)   Ca has no unpaired electrons and Ti^{2+} has two. The two valence electrons in Ca are paired in the 4s orbital. Each of the two valence electrons in Ti^{2+} occupies its own 3d orbital (Hund's rule).

       (c)   To be isoelectronic with Ca^{2+}, Ti would have a 4+ charge.

8.99   (a) H_2O_2, 14 e⁻, 7 e⁻ pr $H-\ddot{\underset{..}{O}}-\ddot{\underset{..}{O}}-H$

       (b)   Referring to Table 8.3, the single O–O bond is the weakest bond in H_2O_2.

       (c)   The average bond enthalpy of one mole of single O–O bonds is 146 kJ. The enthalpy of one single O–O bond is

$$\frac{146 \, kJ}{mol} \times \frac{1 \, mol}{6.022 \times 10^{23} \, bonds} \times \frac{1000 \, J}{kJ} = 2.4244 \times 10^{-19} \, J = 2.42 \times 10^{-19} \, J$$

$$\lambda = \frac{hc}{E} = \frac{6.626 \times 10^{-34} \, J\text{-}s}{2.4244 \times 10^{-19} \, J} \times \frac{2.998 \times 10^8 \, m}{s} \times \frac{1 \times 10^9 \, nm}{m} = 819.37 = 819 \, nm$$

[Recall that the wavelength range for visible light is typically 400-750 nm. All visible light has energy sufficient to break a single O–O bond.]

8.100   The pathway to the formation of K_2O can be written:

$2 \, K(s) \rightarrow 2 \, K(g)$	$2 \, \Delta H_f^o \, K(g)$
$2 \, K(g) \rightarrow 2 \, K^+(g) + 2 \, e^-$	$2 \, I_1(K)$
$1/2 \, O_2(g) \rightarrow O(g)$	$\Delta H_f^o \, O(g)$
$O(g) + e^- \rightarrow O^-(g)$	$E_1(O)$
$O^-(g) + e^- \rightarrow O^{2-}(g)$	$E_2(O)$
$2 \, K^+(g) + O^{2-}(g) \rightarrow K_2O(s)$	$-\Delta H_{latt} \, K_2O(s)$
$2 \, K(s) + 1/2 \, O_2(g) \rightarrow K_2O(s)$	$\Delta H_f^o \, K_2O(s)$

$\Delta H_f^o \, K_2O(s) = 2 \, \Delta H_f^o \, K(g) + 2 \, I_1(K) + \Delta H_f^o \, O(g) + E_1(O) + E_2(O) - \Delta H_{latt} \, K_2O(s)$

$E_2(O) = \Delta H_f^o \, K_2O(s) + \Delta H_{latt} \, K_2O(s) - 2 \, \Delta H_f^o \, K(g) - 2 \, I_t(K) - \Delta H_f^o \, O(g) - E_1(O)$

$E_2(O) = -363.2 \, kJ + 2238 \, kJ - 2(89.99) \, kJ - 2(419) \, kJ - 247.5 \, kJ - (-141) \, kJ$
           $= +750 \, kJ$

8.101   To calculate empirical formulas, assume 100 g of sample.

       (a) $\dfrac{76.0 \, g \, Ru}{101.07 \, g/mol} = 0.752 \, mol \, Ru; \; 0.752/0.752 = 1 \, Ru$

              $\dfrac{24.0 \, g \, O}{15.9994 \, g/mol} = 1.50 \, mol \, O; \; 1.50/0.762 = 2 \, O$

              The empirical formula of compound 1 is RuO_2.

       (b) $\dfrac{61.2 \, g \, Ru}{101.07 \, g/mol} = 0.6055 \, mol \, Ru; \; 0.6055/0.6055 = 1 \, Ru$

              $\dfrac{38.8 \, g \, O}{15.9994 \, g/mol} = 2.425 \, mol \, O; \; 2.425/0.6055 = 4 \, O$

              The empirical formula of compound 2 is RuO_4.

(c)   The lower melting yellow compound is molecular. Substances with metals in high oxidation states are often molecular. RuO_4 contains Ru(VIII), whereas RuO_2 contains Ru(IV), so RuO_4 is more likely to be molecular. The yellow compound is RuO_4.

(d)   The very high melting black compound is ionic. The black compound is RuO_2.

(e)   Yellow RuO_4 is molecular.

(f)   Black RuO_2 is ionic.

8.102   (a)   Even though Cl has the greater (more negative) electron affinity, F has a much larger ionization energy, so the electronegativity of F is greater.

F: k(I–EA) = k(1681 – (–328)) = k(2009)

Cl: k(I–EA) = k(1251 – (–349)) = k(1600)

(b)   Electronegativity is the ability of an atom in a molecule to attract electrons to itself. It can be thought of as the ability to hold its own electrons (as measured by ionization energy) and the capacity to attract the electrons of other atoms (as measured by electron affinity). Thus, both properties are relevant to the concept of electronegativity.

(c)   EN = k(I – EA). For F: 4.0 = k(2009), k = 4.0/2009 = 2.0×10^{-3}

(d)   Cl: EN = 2.0×10^{-3} (1600) = 3.2

O: EN = 2.0×10^{-3} (1314 – (–141)) = 2.9

(e)   F: (I+EA)/2 = (1681 – 328)/2 = 676.5 = 677

To scale the value to 4.0 for F, 4.0 = k(677), k = 4.0/677 = 5.9×10^{-3}

Cl: 5.9×10^{-3} (1251 – 349)/2 = 2.7

Br: 5.9×10^{-3} (1140 – 325)/2 = 2.4

I: 5.9×10^{-3} (1008 – 295)/2 = 2.1

On this scale, the electronegativity of Br is 2.4.

8.103   (a)   Assume 100 g.

$$14.52 \text{ g C} \times \frac{1\,\text{mol}}{12.011 \text{ g C}} = 1.209 \text{ mol C}; 1.209/1.209 = 1$$

$$1.83 \text{ g H} \times \frac{1\,\text{mol}}{1.008 \text{ g H}} = 1.816 \text{ mol H}; 1.816/1.209 = 1.5$$

$$64.30 \text{ g Cl} \times \frac{1\,\text{mol}}{35.453 \text{ g Cl}} = 1.814 \text{ mol Cl}; 1.814/1.209 = 1.5$$

$$19.35 \text{ g O} \times \frac{1\,\text{mol}}{15.9994 \text{ g O}} = 1.209 \text{ mol O}; 1.209/1.209 = 1.0$$

Multiplying by 2 to obtain an integer ratio, the empirical formula is $C_2H_3Cl_3O_2$.

(b)   The empirical formula mass is 2(12.0) + 3(1.0) + 3(35.5) + 2(16) = 165.5. The empirical formula is the molecular formula.

(c) $44\ e^-$, $22\ e^-$ pairs

$$
\begin{array}{c}
:\!\ddot{\underset{\cdot\cdot}{Cl}}\!:\quad :\!\ddot{\underset{}{O}}\!-\!H \\
\ |\qquad\quad| \\
:\!\ddot{\underset{\cdot\cdot}{Cl}}\!-\!C\!-\!C\!-\!\ddot{\underset{\cdot\cdot}{O}}\!-\!H \\
|\qquad\quad| \\
:\!\underset{\cdot\cdot}{\overset{\cdot\cdot}{Cl}}\!:\quad\ H
\end{array}
$$

8.104   (a)   Assume 100 g.

$$62.04\ \text{g Ba} \times \frac{1\ \text{mol}}{137.33\ \text{g Ba}} = 0.4518\ \text{mol Ba}; 0.4518/0.4518 = 1.0$$

$$37.96\ \text{g N} \times \frac{1\ \text{mol}}{14.007\ \text{g N}} = 2.710\ \text{mol N}; 2.710/0.4518 = 6.0$$

The empirical formula is BaN_6. Ba has an ionic charge of 2+, so there must be two 1– azide ions to balance the charge. The formula of each azide ion is N_3^-.

(b) $16\ e^-$, $8\ e^-$ pairs

$$
\left[:\!\ddot{N}\!=\!N\!=\!\ddot{N}\!:\right]^- \longleftrightarrow \left[:\!N\!\equiv\!N\!-\!\ddot{\underset{\cdot\cdot}{N}}\!:\right]^- \longleftrightarrow \left[:\!\ddot{\underset{\cdot\cdot}{N}}\!-\!N\!\equiv\!N\!:\right]^-
$$
$$
\quad -1\ \ +1\ \ -1 \qquad\qquad 0\ \ +1\ \ -2 \qquad\qquad -2\ \ +1\ \ \ 0
$$

(c)   The structure with two double bonds minimizes formal charges and is probably the main contributor.

(d)   The two N–N bond lengths will be equal. The two minor contributors would individually cause unequal N–N distances, but collectively they contribute equally to the lengthening and shortening of each bond. The N–N distance will be approximately 1.24 Å, the average N=N distance.

8.105   (a) C_2H_2: $10\ e^-$, $5\ e^-$ pair N_2: $10\ e^-$, $5\ e^-$ pair

                 $H\!-\!C\!\equiv\!C\!-\!H$ $:\!N\!\equiv\!N\!:$

(b)   The enthalpy of formation for N_2 is 0 kJ/mol and for C_2H_2 is 226.77 kJ/mol. N_2 is an extremely stable, unreactive compound. Under appropriate conditions, it can be either oxidized or reduced. C_2H_2 is a reactive gas, used in combination with O_2 for welding and as starting material for organic synthesis.

(c) $2\ N_2(g) + 5\ O_2(g) \rightarrow 2\ N_2O_5(g)$

         $2\ C_2H_2(g) + 5\ O_2(g) \rightarrow 4\ CO_2(g) + 2\ H_2O(g)$

(d) $\Delta H_{rxn}^{\circ}\ (N_2) = 2\ \Delta H_f^{\circ}\ N_2O_5(g) - 2\ \Delta H_f^{\circ}\ N_2(g) - 5\ \Delta H_f^{\circ}\ O_2(g)$

            $= 2(11.30) - 2(0) - 5(0) = 22.60\ \text{kJ}$

     $\Delta H_{ox}^{\circ}\ (N_2) = 11.30\ \text{kJ/mol}\ N_2$

        $\Delta H_{rxn}^{\circ}\ (C_2H_2) = 4\ \Delta H_f^{\circ}\ CO_2(g) + 2\ \Delta H_f^{\circ}\ H_2O(g) - 2\ \Delta H_f^{\circ}\ C_2H_2(g) - 5\ \Delta H_f^{\circ}\ O_2(g)$

              $= 4(-393.5\ \text{kJ}) + 2(-241.82\ \text{kJ}) - 2(226.77\ \text{kJ}) - 5(0) = -2511.18\ \text{kJ}$

     $\Delta H_{ox}^{\circ}\ (C_2H_2) = -1255.6\ \text{kJ/mol}\ C_2H_2$

(e) $N_2(g) + 3\ H_2(g) \rightarrow 2\ NH_3(g)$

        $\Delta H_{rxn}^{\circ}\ (N_2) = 2\ \Delta H_f^{\circ}\ NH_3(g) - \Delta H_f^{\circ}\ N_2(g) - 3\ \Delta H_f^{\circ}\ H_2(g)$

            $= 2(-46.19) - (0) - 3(0) = -92.38\ \text{kJ}$

     $\Delta H_{rxn}^{\circ}\ (N_2) = -46.19\ \text{kJ/mol}\ N_2$

$$C_2H_2(g) + 3\,H_2(g) \rightarrow 2\,CH_4(g)$$

$$\Delta H^o_{rxn}\,(C_2H_2) = 2\,\Delta H^o_f\,CH_4(g) - 2\,\Delta H^o_f\,C_2H_2(g) - 3\,\Delta H^o_f\,H_2(g)$$

$$= 2(-679.9\text{ kJ}) - 226.77\text{ kJ} - 3(0) = -1586.6\text{ kJ}$$

$$\Delta H^o_{rxn}\,(C_2H_2) = -793.3\text{ kJ/mol }C_2H_2$$

8.106   (a)   Assume 100 g of compound

$$69.6\text{ g S} \times \frac{1\text{ mol S}}{32.07\text{ g}} = 2.17\text{ mol S}$$

$$30.4\text{ g N} \times \frac{1\text{ mol N}}{14.01\text{ g}} = 2.17\text{ mol N}$$

S and N are present in a 1:1 mol ratio, so the empirical formula is SN. The empirical formula mass is 46. MM/FW = 184.3/46 = 4 The molecular formula is S_4N_4.

(b)   44 e^-, 22 e^- pairs. Because of its small radius, N is unlikely to have an expanded octet. Begin with alternating S and N atoms in the ring. Try to satisfy the octet rule with single bonds and lone pairs. At least two double bonds somewhere in the ring are required.

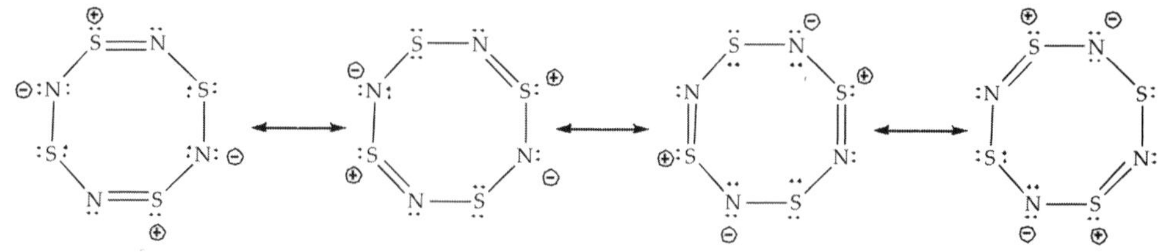

These structures carry formal charges on S and N atoms as shown. Other possibilities include:

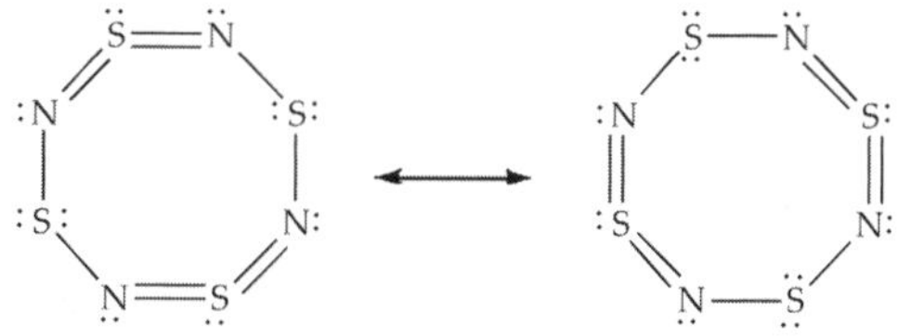

These structures have zero formal charges on all atoms and are likely to contribute to the true structure. Note that the S atoms that are shown with two double bonds are not necessarily linear because S has an expanded octet. Other resonance structures with four double bonds are:

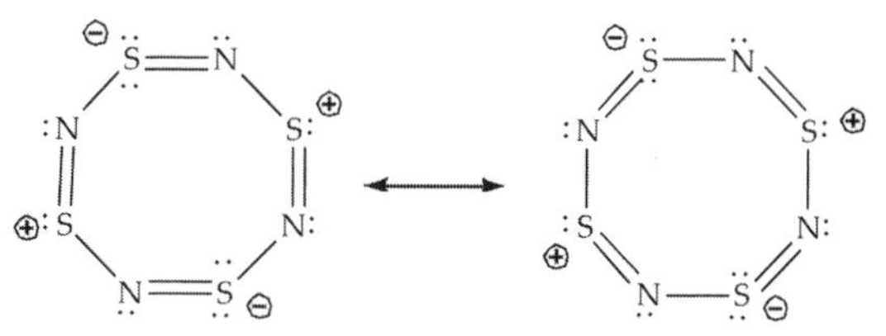

In either resonance structure, the two "extra" electron pairs can be placed on any pair of S atoms in ring, leading to a total of 10 resonance structures. The sulfur atoms alternately carry formal charges of +1 and −1. Without further structural information, it is not possible to eliminate any of the above structures. Clearly, the S_4N_4 molecule stretches the limits of the Lewis model of chemical bonding.

(c) Each resonance structure has 8 total bonds and more than 8 but fewer than 16 bonding e$^-$ pairs, so an "average" bond will be intermediate between a S–N single and double bond. We estimate an average S–N single bond length to be 1.77 Å (sum of bonding atomic radii from Figure 7.7). We do not have a direct value for a S–N double bond length. Comparing double and single bond lengths for C–C (1.34 Å, 1.54 Å), N–N (1.24 Å, 1.47 Å), and O–O (1.21 Å, 1.48 Å) bonds from Table 8.4, we see that, on average, a double bond is approximately 0.23 Å shorter than a single bond. Applying this difference to the S–N single bond length, we estimate the S–N double bond length as 1.54 Å. Finally, the intermediate S–N bond length in S_4N_4 should be between these two values, approximately 1.60–165 Å. (The measured bond length is 1.62 Å.)

(d) $S_4N_4 \rightarrow 4\,S(g) + 4\,N(g)$

$\Delta H = 4\,\Delta H_f^\circ\, S(g) + 4\,\Delta H_f^\circ\, N(g) - \Delta H_f^\circ\, S_4N_4$

$\Delta H = 4(222.8\text{ kJ}) + 4(472.7\text{ kJ}) - 480\text{ kJ} = 2302\text{ kJ}$

This energy, 2302 kJ, represents the dissociation of 8 S–N bonds in the molecule; the average dissociation energy of one S–N bond in S_4N_4 is then 2302 kJ/8 bonds = 287.8 kJ.

8.107 (a) Yes. In the structure shown in the exercise, each P atom needs 1 unshared pair to complete its octet. This is confirmed by noting that only 6 of the 10 valence e$^-$ pairs are bonding pairs.

(b) There are 6 P–P bonds in P_4.

(c) 20 e$^-$, 10 e$^-$ pr

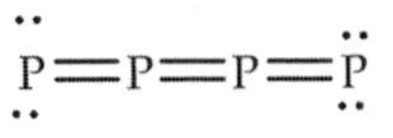

$$\ddot{P}\!\!=\!\!P\!\!=\!\!P\!\!=\!\!\ddot{P}$$

There are no other resonance forms for this structure. The octet rule is satisfied for all atoms. However, it requires P=P, which is uncommon because P has a covalent radius that is too large to accommodate the side-to-side π overlap of parallel p orbitals required for double bond formation.

(d) From left to right, the formal charges are on the P atoms in the linear structure are –1, +1, +1, –1. In the tetrahedral structure, all formal charges are zero. Clearly the linear structure does not minimize formal charge and is probably less stable than the tetrahedral structure, owing to the difficulty of P=P bond formation (see above).

8.108 (a) HCOOH, 18 e$^-$, 9 e$^-$ pr

$$\left[H\!-\!\overset{\displaystyle :\ddot{O}:}{\underset{}{C}}\!=\!\ddot{O}\!-\!H \right]^{1-} \longleftrightarrow \left[H\!-\!\overset{\displaystyle :O:}{\underset{}{C}}\!-\!\ddot{O}\!-\!H \right]^{1-}$$

The resonance structure on the right is the dominant form. The structure on the left can be drawn; no atom violates the octet rule, but formal charges on O are not minimized and it is a very minor form.

(b) $HCOOH(aq) + NaOH(aq) \rightarrow HCOO^-(aq) + H_2O(l) + Na^+(aq)$

(c)

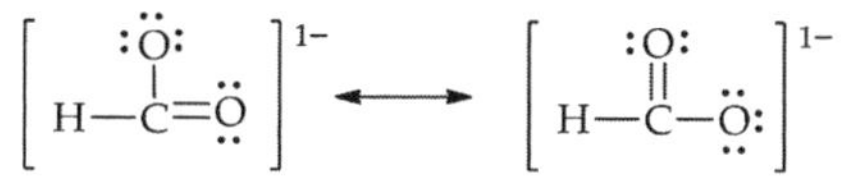

$$\left[H-\overset{\overset{\displaystyle :\ddot{O}:}{|}}{C}=\ddot{O} \right]^{1-} \longleftrightarrow \left[H-\overset{\overset{\displaystyle :O:}{||}}{C}-\ddot{O}: \right]^{1-}$$

(d) $0.785 \text{ mL HCOOH} \times \dfrac{1.220 \text{ g}}{\text{mL}} \times \dfrac{1 \text{ mol HCOOH}}{46.025 \text{ g HCOOH}} = 0.02081 = 0.0208 \text{ mol HCOOH}$

$0.02081 \text{ moL HCOOH} \times \dfrac{1 \text{ mol NaOH}}{1 \text{ mol HCOOH}} \times \dfrac{1 \text{ L NaOH}}{0.100 \text{ } M \text{ NaOH}} \times \dfrac{1000 \text{ mL}}{\text{L}} =$

$208.08 = 208 \text{ mL of } 0.100 \text{ } M \text{ NaOH}$

8.109   (a) NH_3BF_3, 32 e⁻, 16 e⁻ pr

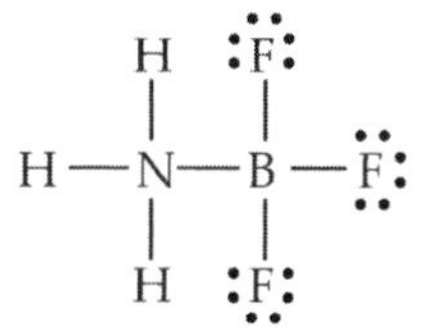

(b)   Moving from left to right across a row of the periodic table, electronegativity increases. The electron density will be greater around the atom with greater electronegativity, in this case N.

$$\overset{\delta+}{B} - \overset{\delta-}{N}$$

(c)   The difference between NH_3BCl_3 and NH_3BF_3 is that Cl has replaced F as the element bound to B. Chlorine is less electronegative and less electron withdrawing than fluorine. This increases the electron density at B and renders the B–N bond less polar in NH_3BCl_3, than NH_3BF_3.

8.110   (a) NH_4^+, 8 e⁻, 4 e⁻ pr; Cl⁻, 8 e⁻, 4 e⁻ pr

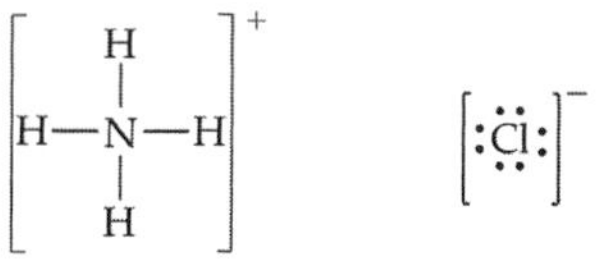

(b)   No. The bond in $NH_4Cl(s)$ is ionic; it is electrostatic attraction among oppositely charged ions.

(c) $14 \text{ g NH}_4\text{Cl} \times \dfrac{1 \text{ mol NH}_4\text{Cl}}{53.495 \text{ g NH}_4\text{Cl}} \times \dfrac{1}{0.5000 \text{ L}} = 0.5234 = 0.52 \text{ } M$

(d) $NH_4Cl(aq) + AgNO_3(aq) \rightarrow AgCl(s) + NH_4NO_3(aq)$

$14 \text{ g NH}_4\text{Cl} \times \dfrac{1 \text{ mol NH}_4\text{Cl}}{53.495 \text{ g NH}_4\text{Cl}} \times \dfrac{1 \text{ mol AgNO}_3}{1 \text{ mol NH}_4\text{Cl}} \times \dfrac{169.88 \text{ g AgNO}_3}{1 \text{ mol AgNO}_3}$

$= 44.459 = 44 \text{ g AgNO}_3$

9 Molecular Geometry and Bonding Theories

Visualizing Concepts

9.1   Removing an atom from the equatorial plane of trigonal bipyramid in Figure 9.3 creates a seesaw shape. It might appear that you could also obtain a seesaw by removing two atoms from the square plane of the octahedron. However, one of the B–A–B angles in the seesaw is 120°, so it must be derived from a trigonal bipyramid.

9.2   (a)   120°

         (b)   If the blue balloon expands, the angle between red and green balloons decreases.

         (c)   (ii)

9.3   *Analyze/Plan.* Visualize the molecular geometry and the electron-domain geometries that could produce it. Confirm your choices with Tables 9.2 and 9.3. In Table 9.3, note that octahedral electron-domain geometry results in only 3 possible molecular geometries: octahedral, square pyramidal, and square planar (not T-shaped, bent, or linear).   *Solve.*

         (a)   2. Molecular geometry: linear. Possible electron-domain geometries: linear, trigonal bipyramidal

         (b)   1. Molecular geometry, T-shaped. Possible electron-domain geometries: trigonal bipyramidal

         (c)   1. Molecular geometry, octahedral. Possible electron-domain geometries: octahedral

         (d)   1. Molecular geometry, square-pyramidal. Possible electron-domain geometries: octahedral

         (e)   1. Molecular geometry, square planar. Possible electron-domain geometries: octahedral

         (f)   1. Molecular geometry, triangular pyramid. Possible electron-domain geometries: trigonal bipyramidal. This is an unusual molecular geometry that is not listed in Table 9.3. It could occur if the equatorial substituents on the trigonal bipyramid were extremely bulky, causing the nonbonding electron pair to occupy an axial position.

9.4   (a)   4 e^- domains

         (b)   The molecule has a nonzero dipole moment, because the C–H and C–F bond dipoles do not cancel each other.

         (c)   (ii)

9.5   (a)   Zero. Moving from left to right along the x-axis of the plot, the distance between the Cl atoms increases. At very large separation, the potential energy of interaction approaches zero.

        (b)   The Cl–Cl bond distance is approximately 2.0 Å. The Cl–Cl bond energy is approximately 240 kJ/mol. The minimum energy for the two atoms represents the stabilization obtained by bringing two Cl atoms together at the optimum (bond) distance. The x-coordinate of the minimum point on the plot is the Cl–Cl bond length; the y-coordinate is the bond strength or enthalpy.

        (c)   Weaker. Under extreme pressure, assume the Cl–Cl separation gets shorter. According to the plot, the potential energy of the atom pair increases and the bond gets weaker as the separation becomes shorter than the optimum bond distance.

9.6   (a)   (iii)

        (b) sp^3

9.7   (a)   Recall that π bonds require p atomic orbitals, so the maximum hybridization of a C atom involved in a double bond is sp^2 and in a triple bond is sp. There are 6 C atoms in the molecule. Starting on the left, the hybridizations are: sp^2, sp^2, sp^3, sp, sp, sp^3.

        (b)   All single bonds are σ bonds. Double and triple bonds each contain 1 σ bond. This molecule has 8 C–H σ bonds and 5 C–C σ bonds, for a total of 13 σ bonds.

        (c)   Double bonds have 1 π bond and triple bonds have 2 π bonds. This molecule has a total of 3 π bonds.

        (d)   Any central atom with sp^2 hybridization will have bond angles of 120° around it. The two left-most C atoms are sp^2 hybridized, so any angle with one of these C atoms central will be 120°. This amounts to 1 H–C–H, 4 H–C–C and 1 C–C–C angle.

9.8   (a)   (i)

        (b)   (iii)

9.9   (a) C_4H_4O (b) 26 valence e^- (c) sp^2 (d) 4 e^- (e) (iii)

9.10   (a)   The lower-energy MO is σ_{1s}, the higher-energy MO is σ^*_{1s}.

        (b) H_2^+ (c) BO = ½   (d) σ_{1s} (the lowest energy available orbital)

9.11   *Analyze/Plan.* σ molecular orbitals (MOs) are symmetric about the internuclear axis, π MOs are not. Bonding MOs have most of their electron density in the area between the nuclei, antibonding MOs have a node between the nuclei.

        (a)   (i)   Two s atomic orbitals (electron density at each nucleus).

                (ii)   Two p atomic orbitals overlapping end to end (node near each nucleus).

                (iii)   Two p atomic orbitals overlapping side to side (node near each nucleus).

        (b)   (i) σ-type (symmetric about the internuclear axis, s orbitals can produce only σ overlap).

                (ii) σ-type (symmetric about internuclear axis)

                (iii) π-type (not symmetric about internuclear axis, side-to-side overlap)

(c)   (i)   antibonding (node between nuclei)

        (ii)   bonding (concentration of electron density between nuclei)

        (iii)   antibonding (node between nuclei)

(d)   (i)   The nodal plane is between the atom centers, perpendicular to the interatomic axis and equidistant from each atom.

        (ii)   There are two nodal planes; both are perpendicular to the interatomic axis. One is left of the left atom and the second is right of the right atom.

        (iii)   There are two nodal planes; one is between the atom centers, perpendicular to the interatomic axis and equidistant from each atom. The second contains the interatomic axis and is perpendicular to the first.

9.12   (a)   The diagram has five electrons in MOs formed by 2p atomic orbitals. C has two 2p electrons, so X must have three 2p electrons. X is N.

          (b)   The molecule has an unpaired electron, so it is paramagnetic.

          (c)   Atom X is N, which is more electronegative than C. The atomic orbitals of the more electronegative N are slightly lower in energy than those of C. The lower-energy π_{2p} bonding molecular orbitals will have a greater contribution from the lower-energy N atomic orbitals. (Higher energy π_{2p}^{*} MOs will have a greater contribution from higher-energy C atomic orbitals.)

Molecular Shapes; the VSEPR Model (Sections 9.1 and 9.2)

9.13   (a)   It is not possible to tell the number of nonbonding electron pairs about the A atom from this information. If AB_2 obeys the octet rule, A would have 0 nonbonding pairs around it, as in CO_2. If AB_2 does not obey the octet rule, there could be 0 or 3 nonbonding pairs around A. Examples are BeH_2 and XeF_2.

          (b)   Three. XeF_2 has 11 electron pairs. Of these, 2 are bonding pairs between Xe and F and 6 are nonbonding pairs around the two F atoms. This leaves three nonbonding electron pairs around Xe.

          (c)   Yes. The electron domain geometry of XeF_2 is trigonal bipyramidal; the 3 nonbonding pairs are equatorial, the 2 bonding pairs are axial, and the molecular geometry is linear.

9.14   (a)   In a symmetrical tetrahedron, the four bond angles are equal to each other, with values of 109.5°. The H–C–H angles in CH_4 and the O–Cl–O angles in ClO_4^{-} will have values close to 109.5°.

          (b)   'Planar' molecules are flat, so trigonal planar BF_3 is flat. In the trigonal pyramidal NH_3 molecule, the central N atom sits out of the plane of the three H atoms; this molecule is not flat.

9.15   A molecule with tetrahedral molecular geometry has an atom at each vertex of the tetrahedron. A trigonal pyramidal molecule has one vertex of the tetrahedron occupied by a nonbonding electron pair rather than an atom. That is, a trigonal pyramid is a tetrahedron with one vacant vertex.

9.16   (a)   three coplanar 120° angles

        (b)   four 109.5° angles

        (c)   90° angles in the equatorial square plane and between axial atoms and those in the square plane, 12 in all; 180° angles between atoms opposite each other, 3 in all

        (d)   one 180° angle

9.17   (a)   Octahedral. There are 6 electron domains around A in an AB_6 molecule. Because none of the 6 electron domains are nonbonding, the electron domain geometry and molecular geometry are octahedral.

        (b)   Octahedral. An AB_4 molecule has 4 bonding electron domains around A. Additionally, this molecules has two nonbonding domains, for a total of 6 electron domains around A. A total of six electron pairs dictates octahedral electron domain geometry.

        (c)   Square planar. For an octahedral electron domain geometry that includes two nonbonding domains, the nonbonding domains are opposite each other to minimize repulsions. The four bonding domains occupy the remaining positions in the octahedron, forming a square plane.

9.18   We expect the nonbonding electron domain in NH_3 to occupy a smaller volume than the one in PH_3. The electronegativity of N, 3.0, is larger than that of P, 2.1. The nonbonding electrons will be more strongly attracted to N than to P, and the volume of the domain will be smaller. This means that the charge density of the nonbonding domain in NH_3 will be greater and it will experience stronger repulsions than the nonbonding domain in PH_3.

9.19   *Analyze/Plan.* Draw the Lewis structure of each molecule and take note of nonbonding (lone) electron pairs about the central atom.   *Solve.*

        (a) SiH_4, 8 valence e⁻, 4 e⁻ pr, 0 nonbonding pairs, no effect on molecular shape

$$\begin{array}{c} H \\ | \\ H\!-\!Si\!-\!H \\ | \\ H \end{array}$$

        (b) PF_3, 26 valence e⁻, 13 e⁻ pr, 1 nonbonding pair on P, influences molecular shape

$$\ddot{\underset{\cdot\cdot}{F}}\!-\!\overset{\cdot\cdot}{P}\!-\!\ddot{\underset{\cdot\cdot}{F}}$$
$$|$$
$$\ddot{\underset{\cdot\cdot}{F}}$$

        (c)   HBr, 8 valence e⁻, 4 e⁻ pr, 3 nonbonding pairs on Br, no effect on molecular shape because Br is not "central"

$$H\!-\!\ddot{\underset{\cdot\cdot}{Br}}\!:$$

        (d)   HCN, 10 valence e⁻, 5 e⁻ pr, 0 nonbonding pairs on C, no effect on molecular shape

$$H\!-\!C\!\equiv\!N\!:$$

        (e) SO_2, 18 valence e⁻, 9 e⁻ pr, 1 nonbonding pair on S, influences molecular shape

$$:\ddot{O}\!-\!\ddot{S}\!=\!\ddot{O} \longleftrightarrow \ddot{O}\!=\!\ddot{S}\!-\!\ddot{O}: \longleftrightarrow \ddot{O}\!=\!\ddot{S}\!=\!\ddot{O}$$

9.20   Draw the Lewis structure of each molecule. If it has nonbonding electron pairs on the central atom, decide whether they will cause bond angles to deviate from ideal values for the particular electron-domain geometry.

(a) H_2S, 8 valence e^-, 4 e^- pr, tetrahedral electron-domain geometry with 2 nonbonding electron pairs on S will cause the bond angle to deviate from ideal 109.5° angles

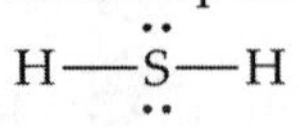

$$H—\overset{\cdot\cdot}{\underset{\cdot\cdot}{S}}—H$$

(b) BCl_3, 24 valence e^-, 12 e^- pr, trigonal planar electron-domain geometry with zero nonbonding pairs on B. We confidently predict 120° angles.

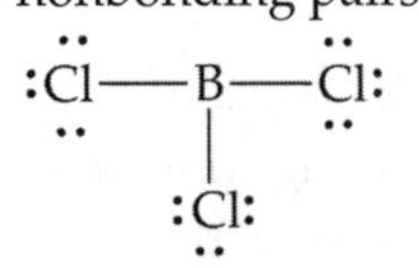

$$:\overset{\cdot\cdot}{\underset{\cdot\cdot}{Cl}}—B—\overset{\cdot\cdot}{\underset{\cdot\cdot}{Cl}}:$$
$$|$$
$$:\overset{}{\underset{\cdot\cdot}{Cl}}:$$

(c) CH_3I, 14 valence e^-, 7 e^- pr, tetrahedral electron-domain geometry with zero nonbonding pairs on C. Because the bonding electron domains are not exactly the same, we predict some deviation from ideal 109.5° angles.

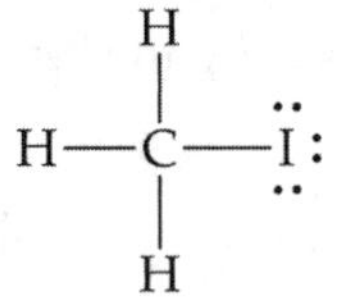

$$\begin{array}{c} H \\ | \\ H—C—\overset{\cdot\cdot}{\underset{\cdot\cdot}{I}}: \\ | \\ H \end{array}$$

(d) CBr_4, 32 valence e^-, 16 e^- pr, tetrahedral electron-domain geometry with zero nonbonding pairs on C. We confidently predict 109.5° angles.

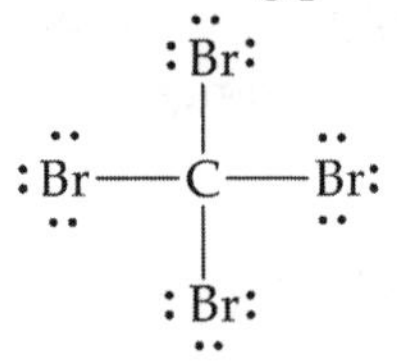

$$\begin{array}{c} :\overset{\cdot\cdot}{Br}: \\ | \\ :\overset{\cdot\cdot}{\underset{\cdot\cdot}{Br}}—C—\overset{\cdot\cdot}{\underset{\cdot\cdot}{Br}}: \\ | \\ :\overset{}{\underset{\cdot\cdot}{Br}}: \end{array}$$

(e) $TeBr_4$, 34 valence e^-, 17 e^- pr, trigonal bipyramidal electron-domain geometry with one nonbonding pair on Te. The structure is similar to SF_4 shown in Sample Exercise 9.2. The bond angles will deviate from ideal values, but perhaps not as much as in SF_4.

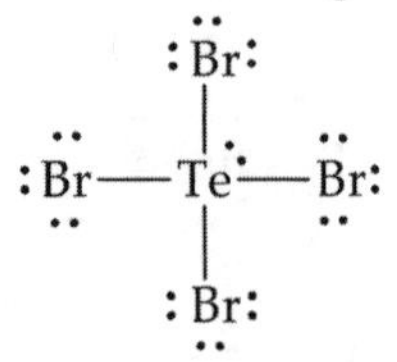

$$\begin{array}{c} :\overset{\cdot\cdot}{Br}: \\ | \\ :\overset{\cdot\cdot}{\underset{\cdot\cdot}{Br}}—Te\overset{\cdot\cdot}{—}\overset{\cdot\cdot}{\underset{\cdot\cdot}{Br}}: \\ | \\ :\overset{}{\underset{\cdot\cdot}{Br}}: \end{array}$$

9.21   *Analyze/Plan.* Draw the Lewis structure of each molecule and count the number of nonbonding (lone) electron pairs. Note that the question asks 'in the molecule' rather than just around the central atom. *Solve.*

(a) $(CH_3)_2S$, 20 valence e^-, 10 e^- pr, 2 nonbonding pairs

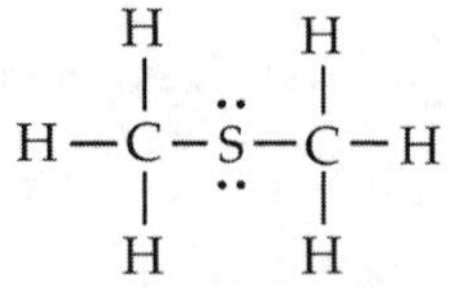

$$\begin{array}{ccc} H & & H \\ | & \overset{\cdot\cdot}{} & | \\ H—C—S—C—H \\ | & \underset{\cdot\cdot}{} & | \\ H & & H \end{array}$$

(b)   HCN, 10 valence e^-, 5 e^- pr, 1 nonbonding pair

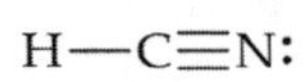

$$H—C\equiv N:$$

(c) H_2C_2, 10 valence e^-, 5 e^- pr, 0 nonbonding pairs

       $H-C\equiv C-H$

(d) CH_3F, 14 valence e^-, 7 e^- pr, 3 nonbonding pairs

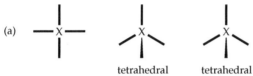

9.22   *Analyze/Plan.* See Table 9.1.   *Solve.*

       (a)   trigonal planar   (b)   tetrahedral

       (c)   trigonal bipyramidal   (d)   octahedral

9.23   *Analyze/Plan.* See Tables 9.2 and 9.3.   *Solve.*

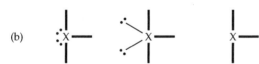

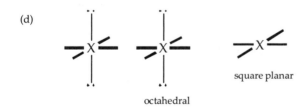

(c)

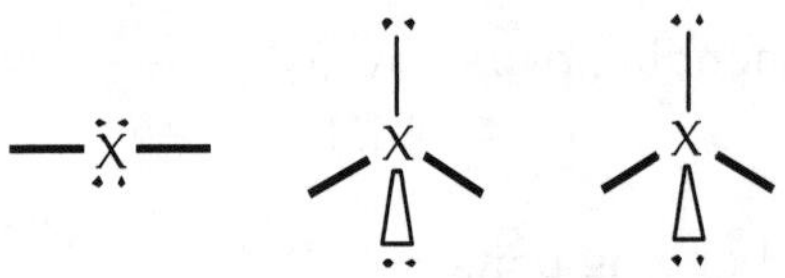

tetrahedral   bent

9.25 *Analyze/Plan.* Follow the logic in Sample Exercises 9.1 and 9.2. *Solve.*

bent (b), linear (l), octahedral (oh), seesaw (ss), square pyramidal (sp), square planar (spl), tetrahedral (td), trigonal bipyramidal (tbp), trigonal planar (tr), trigonal pyramidal (tp), T-shaped (T)

Molecule or ion	Valence electrons	Lewis structure		Electron-domain geometry	Molecular geometry
(a) HCN	10	:N≡C—H	N≡C—H	l	l
(b) SO_3^{2-}	26	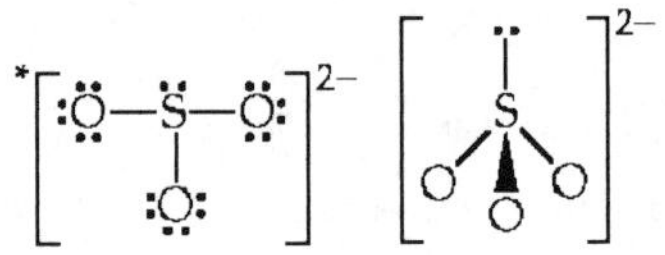		td	tp
(c) SF_4	34			tbp	ss
(d) PF_6^-	48	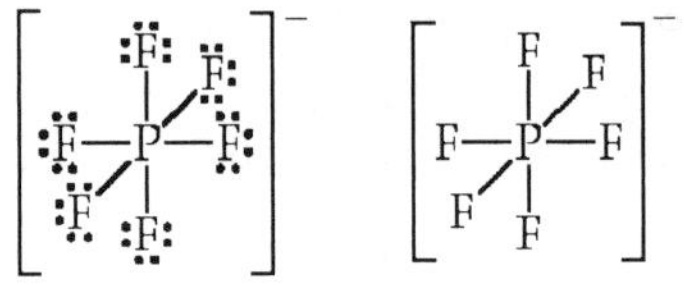		oh	oh
(e) NH_3Cl^+	14	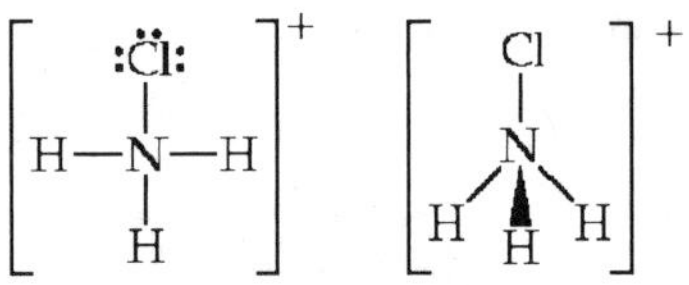		td	td
(f) N_3^-	16			l	l

*More than one resonance structure is possible. All equivalent resonance structures predict the same molecular geometry.

9.26 bent (b), linear (l), octahedral (oh), seesaw (ss), square pyramidal (sp), square planar (spl), tetrahedral (td), trigonal bipyramidal (tbp), trigonal planar (tr), trigonal pyramidal (tp), T-shaped (T)

Molecule or ion	Valence electrons	Lewis structure		Electron-domain Geometry	Molecular geometry
(a) AsF_3	26			td	tp
(b) CH_3^+	6			tr	tr

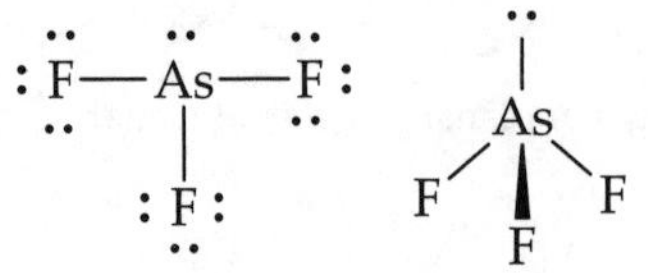

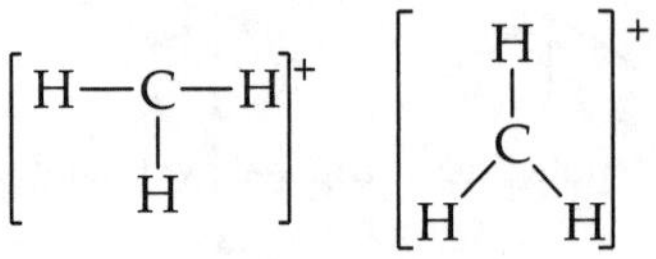

9.26   (Continued). bent (b), linear (l), octahedral (oh), seesaw (ss) square pyramidal (sp), square planar (spl), tetrahedral (td), trigonal bipyramidal (tbp), trigonal planar (tr), trigonal pyramidal (tp), T-shaped (T)

	Molecule or ion	Valence electrons	Lewis structure	Electron-domain geometry	Molecular geometry
(c)	BrF_3	28	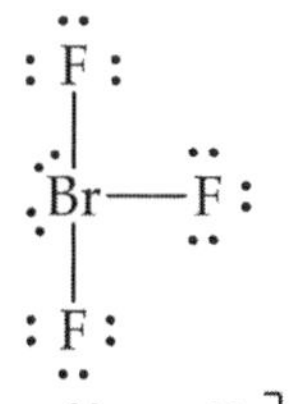 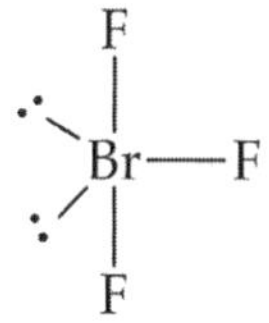	tbp	T
(d)	ClO_3^-	26	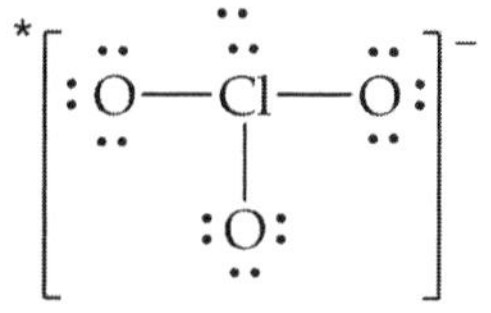 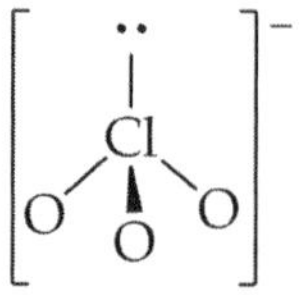	td	tp
(e)	XeF_2	22	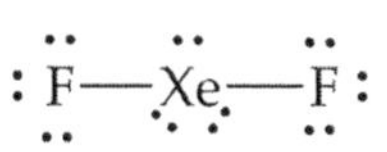 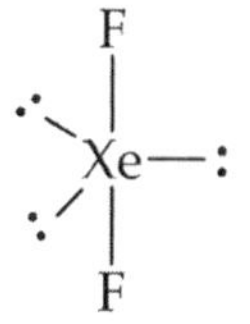	tbp	l
(f)	BrO_2^-	20	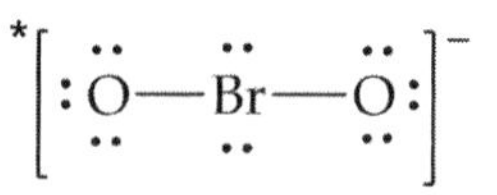 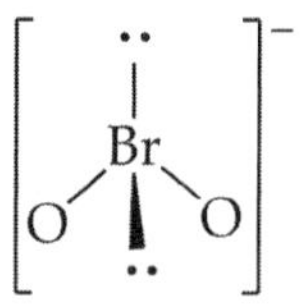	td	b

*More than one resonance structure is possible. All equivalent resonance structures predict the same molecular geometry.

9.27   *Analyze/Plan.* Work backward from molecular geometry, using Tables 9.2 and 9.3. *Solve.*

(a)   Electron-domain geometries: (i), trigonal planar; (ii), tetrahedral; (iii), trigonal bipyramidal

(b)   nonbonding electron domains: (i), 0; (ii), 1; (iii), 2

(c)   N and P. Shape (ii) has three bonding and one nonbonding electron domains. Li and Al would form ionic compounds with F, so there would be no nonbonding electron domains. Assuming that F always has three nonbonding domains, BF_3 and ClF_3 would have the wrong number of nonbonding domains to produce shape ii.

(d)   Cl (also Br and I, because they have seven valence electrons). This T-shaped molecular geometry arises from a trigonal bipyramidal electron-domain geometry with two nonbonding domains (Table 9.3). Assuming each F atom has three nonbonding domains and forms only single bonds with A, A must have seven valence electrons to produce these electron-domain and molecular geometries. It must be in or below the third row of the periodic table, so that it can accommodate more than four electron domains.

9.28   (a)   Electron-domain geometries: (i), octahedral; (ii), tedrahedral; (iii), trigonal bipyramidal

(b)   nonbonding electron domains: (i), 2; (ii), 0; (iii), 1

(c)   S or Se. Shape (iii) has five electron domains, so A must be in or below the third row of the periodic table. This eliminates Be and C. Assuming each F atom has three nonbonding electron domains and forms only single bonds with A, A must have six valence electrons to produce these electron-domain and molecular geometries.

(d)   Xe. (See Table 9.3.) Assuming F behaves typically, A must be in or below the third row and have eight valence electrons. Only Xe fits this description. (Noble-gas elements above Xe have not been shown to form molecules of the type AF_4. See Section 7.8.)

9.29   *Analyze/Plan.* Follow the logic in Sample Exercise 9.3.   *Solve.*

(a)   1, less than 109.5°; 2, less than 109.5°

(b)   3, different than 109.5°; 4, less than 109.5°

(c)   5 – 180°

(d)   6, slightly more than 120°; 7, less than 109.5°; 8, slightly different than 109.5°

9.30   (a)   1, less than 109.5°; 2, less than 120°

(b)   3, close to 109.5°; 4, slightly greater than 120°

(c)   5, less than 109.5°; 6, less than 109.5°

(d)   7, 180°; 8, close to 109.5°

9.31   *Analyze/Plan.* Draw correct Lewis structures for NH_2^-, NH_3, and NH_4^+. The more nonbonding electron domains (lone pairs) around N, the smaller the H–N–H bond angles.   *Solve.*

(a) NH_4^+. There are no lone pairs on N, so this ion has the largest bond angles.

(b) NH_2^-. Amide ion has two bonding and two nonbonding domains around N. The two lone pairs compress the H–N–H bond angle to its smallest value.

9.32   *Analyze/Plan.* Given the formula of each molecule or ion, draw the correct Lewis structure and use principles of VSEPR to answer the question.   *Solve.*

The three nonbonded electron pairs on each F atom have been omitted for clarity.

The two molecules with trigonal bipyramidal electron-domain geometry, PF_5 and SF_4, have more than one F–A–F bond angle.

9.33   *Analyze.* Given: molecular formulas. Find: explain features of molecular geometries.

Plan. Draw the correct Lewis structures for the molecules and use VSEPR to predict and explain observed molecular geometry.   *Solve.*

(a) BrF_4^- 36 e⁻, 18 e⁻ pr BF_4^- 32 e⁻, 16 e⁻ pr

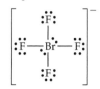   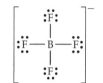

6 e⁻ pairs around Br   4 e⁻ pairs around B,
octahedral e⁻ domain geometry   tetrahedral e⁻ domain geometry
square planar molecular geometry   tetrahedral molecular geometry

The fundamental feature that determines molecular geometry is the number of electron domains around the central atom, and the number of these that are bonding domains. Although BrF_4^- and BF_4^- are both of the form AX_4^-, the central atoms, and thus the number of valence electrons in the two ions, are different. This leads to different numbers of e⁻ domains about the two central atoms. Even though both ions have four bonding electron domains, the six total domains around Br require octahedral domain geometry and square planar molecular geometry, whereas the four total domains about B lead to tetrahedral domain and molecular geometry.

(b) H_2X, 8 e⁻, 4 e⁻ pr

$$
\begin{array}{c}
\ddot{} \\
| \\
X \\
\diagup \;| \;\diagdown \\
H \quad | \quad H \\
\ddot{}
\end{array}
$$

All molecules in the series have tetrahedral electron-domain geometry and bent molecular structure. To a first approximation, the H–X–H angles will be less than 109.5°. Any variation will be because of differences in repulsion among the nonbonding and bonding electron domains. The less electronegative the central atom, the larger the nonbonding electron domain, the greater the effect of repulsive forces on adjacent bonding domains. The less electronegative the central atom, the larger the deviation from ideal tetrahedral angles. The angles will vary as $H_2O > H_2S > H_2Se$.

9.34   (a) ClO_2^-, 20 e⁻, 10 e⁻ pr

$$
\left[\ddot{\underset{..}{O}} - \ddot{\underset{..}{Cl}} - \ddot{\underset{..}{O}} \right]^-
\qquad\qquad
\begin{array}{c}
\ddot{} \\
| \\
Cl \\
\diagup \;| \;\diagdown \\
O \quad | \quad O \\
\ddot{}
\end{array}
$$

2 bonding and 2 nonbonding   bent shape
e⁻ domains around Cl

(More than one resonance structure is possible. All have 2 bonding and two nonbonding domains around Cl and predict bent shape.)

(b) SO_4^{2-}, 32 e⁻, 16 e⁻ pr

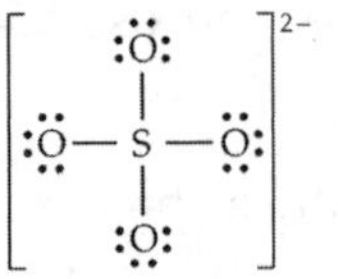

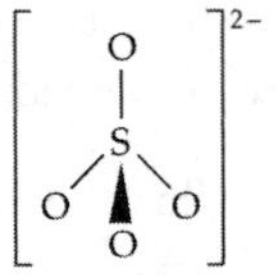

4 bonding e⁻ domains around S   tetrahedral shape

(Many equivalent resonance structures for SO_4^{2-} are possible, but all have 4 bonding e⁻ domains around S and tetrahedral shape.)

(c) NF_3, 26 e⁻, 13 e⁻ pr

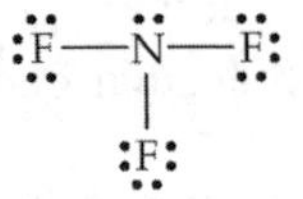

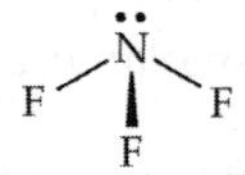

3 bonding and 1 nonbonding   trigonal pyramid shape
e⁻ domains around N

(d) CCl_2Br_2, 32 e⁻, 16 e⁻ pr

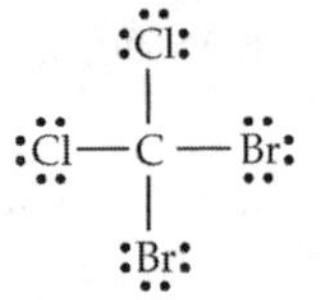

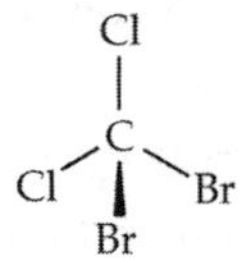

4 bonding e⁻ domains around C   tetrahedral shape

(e) SF_4^{2+}, 32 e⁻, 16 e⁻ pr

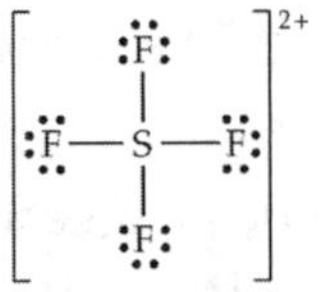

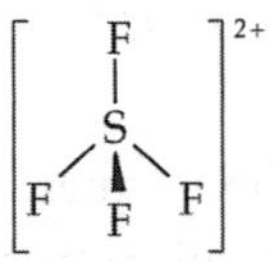

4 bonding e⁻ domains around S   tetrahedral shape

Shapes and Polarity of Polyatomic Molecules (Section 9.3)

9.35   A bond dipole is the asymmetric charge distribution between two bonded atoms with unequal electronegativities. A molecular dipole moment is the three-dimensional sum of all the bond dipoles in a molecule. (A molecular dipole moment is a measurable physical property; a bond dipole is not measurable, unless the molecule is diatomic.)

9.36   For a polar A–X bond in an AX_3 molecule, as the X–A–X bond angle increases from 100° to 120°, the molecular dipole moment decreases. In a symmetrical AX_3 molecule with 120° bond angles, bond dipoles cancel and the molecule is nonpolar. As the bond angle decreases, the resultant of the three bond dipoles becomes larger, and the dipole moment increases.

9.37   *Analyze/plan.* Follow the logic in Sample Exercise 9.4.   *Solve.*

(a) SCl_2, 20 e⁻, 10 e⁻ pr

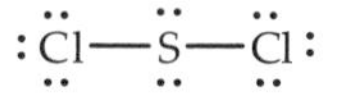

:C̈l—S̈—C̈l:

tetrahedral e⁻ domain geometry
bent molecular geometry

S and Cl have different electronegativities; the S–Cl bonds are polar. The bond dipoles are not opposite each other, so the molecule is polar. The dipole moment vector bisects the Cl–S–Cl bond angle. (A more difficult question is which end of the dipole moment vector is negative. The resultant of the two bond dipoles has its negative end toward the Cl atoms. However, the partial negative charge because of the lone pairs on S points opposite to the negative end of the resultant. A reasonable guess is that the negative end of the dipole moment vector is in the direction of the lone pairs.)

(b) $BeCl_2$, 16 e⁻, 8 e⁻ pr

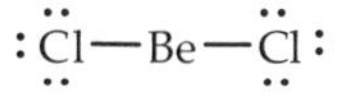

:C̈l—Be—C̈l:

linear electron-domain and molecular geometry

(Resonance structures with Be=Cl can be drawn, but electronegativity arguments predict that most electron density will reside on Cl and that the structure above is the main resonance contributor.) Be and Cl have very different electronegativities, so the Be–Cl bonds are polar. The individual bond dipoles are equal and opposite, so the net molecular dipole moment is zero.

9.38   (a)   If PH_3 is polar, it must have a measurable dipole moment. This means that the three P–H bond dipoles do not cancel. If PH_3 were planar, the P–H bond dipoles would cancel, and the molecule would be nonpolar. The measurable dipole moment of PH_3 is experimental evidence that the molecule cannot be planar.

(b) O_3, 18 e⁻, 9 e⁻ pr; :Ö=Ö—Ö: ⟷ :Ö—Ö=Ö:

trigonal planar e⁻ domain geometry
bent molecular geometry

Because all atoms are the same, the individual bond dipoles are zero. However, the central O atom has a lone pair of electrons that cause an unequal electron (and charge) distribution in the molecule. This lone pair is the source of the dipole moment in O_3.

9.39   (a)   Nonpolar. The BF_3 molecule has polar B–F bonds, but they are arranged in a symmetrical trigonal plane. The individual bond dipoles cancel, leaving the molecule with zero net dipole moment.

(b)   No. The $BF_3{}^{2-}$ ion has 3 bonding and 1 nonbonding electron pairs around B. The added nonbonding electron pair requires that the electron domain geometry is tetrahedral and the shape is a trigonal pyramid.

(c)   Yes. In BF_2Cl the B–F bond dipoles do not exactly cancel with the B–Cl bond dipole, resulting in a net dipole moment.

9.40   (a)   In Exercise 9.27, molecules (ii) and (iii) will have nonzero dipole moments. Molecule (i) has zero nonbonding electron pairs on A, and the 3 A–F dipoles are oriented so that the sum of their vectors is zero (the bond dipoles cancel). Molecules (ii) and (iii) have nonbonding electron pairs on A and their bond dipoles do not cancel. A nonbonding electron pair (or pairs) on a central atom almost guarantees at least a small molecular dipole moment, because no bond dipole exactly cancels a nonbonding pair. (Exceptions are molecular geometries with nonbonding electron domains 180° apart.)

        (b)   In Exercise 9.28, molecules (i) and (ii) have zero dipole moments and are nonpolar. AF_4 molecules will have a zero dipole moment if the 4 A–F bond dipoles are arranged (symmetrically) so that they cancel, and any nonbonding pairs are arranged so that they cancel.

9.41   *Analyze/Plan.* Given molecular formulas, draw correct Lewis structures, determine molecular structure and polarity.   *Solve.*

        (a)   Polar, $\Delta EN > 0$

            I–F

        (b)   Nonpolar, the molecule is linear and the bond dipoles cancel.

            S$=$C$=$S

        (c)   Nonpolar, in a symmetrical trigonal planar structure, the bond dipoles cancel.

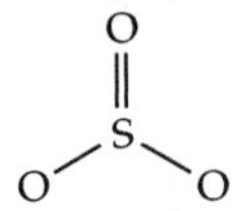

        (d)   Polar, although the bond dipoles are essentially zero, there is an unequal charge distribution because of the nonbonded electron pair on P.

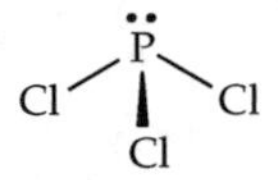

        (e)   Nonpolar, symmetrical octahedron

        (f)   Polar, square pyramidal molecular geometry, bond dipoles do not cancel.

9.42   (a)   Nonpolar, in a symmetrical tetrahedral structure (Figure 9.1) the bond dipoles cancel.

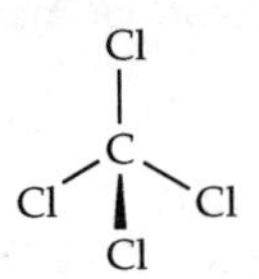

(b)   Polar, there is an unequal charge distribution because of the nonbonded electron pair on N.

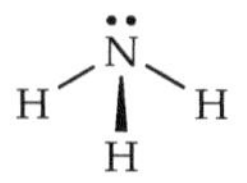

(c)   Polar, there is an unequal charge distribution because of the nonbonded electron pair on S.

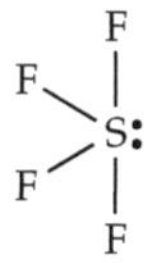

(d)   Nonpolar, the bond dipoles and the nonbonded electron pairs cancel.

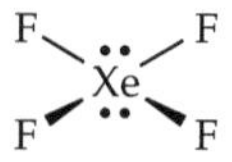

(e)   Polar, the C–H and C–Br bond dipoles are not equal and do not cancel.

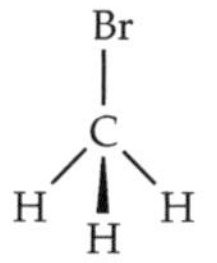

(f)   Nonpolar, in a symmetrical trigonal planar structure, the bond dipoles cancel.

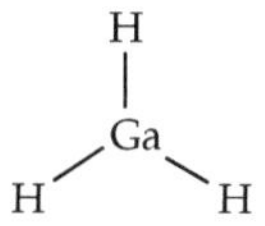

9.43   *Analyze/Plan.* Given molecular formulas, draw correct Lewis structures, analyze molecular structure and determine polarity.   *Solve.*

(a) $C_2H_2Cl_2$, each isomer has 24 e⁻, 12 e⁻ pr. Lewis structures:

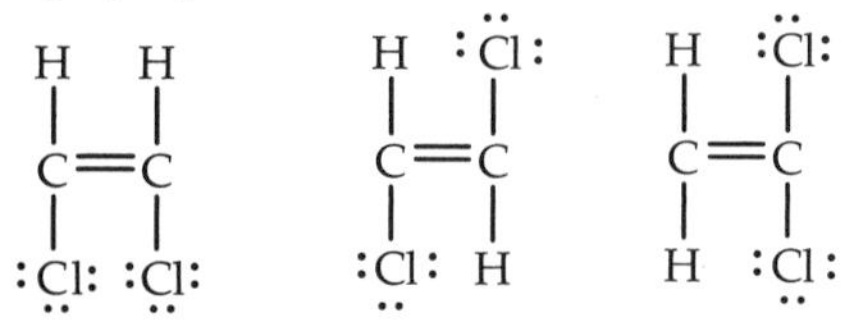

Molecular geometries:

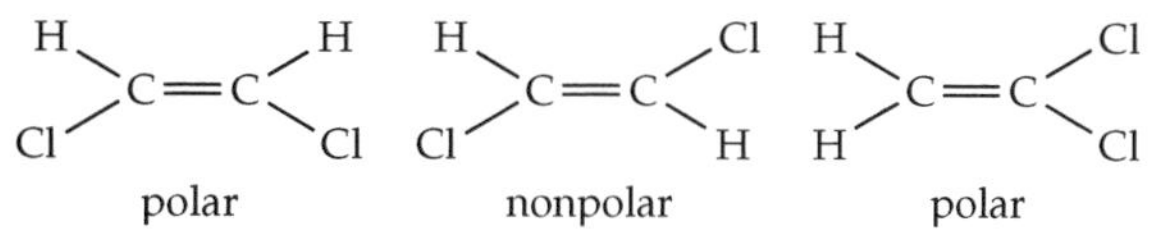

(b)   All three isomers are planar. The molecules on the left and right are polar because the C–Cl bond dipoles do not point in opposite directions. In the middle isomer, the C–Cl bonds and dipoles are pointing in opposite directions (as are the C–H bonds), the molecule is nonpolar and has a measured dipole moment of zero.

(c) C_2H_3Cl (lone pairs on Cl omitted for clarity)

There are four possible placements for Cl:

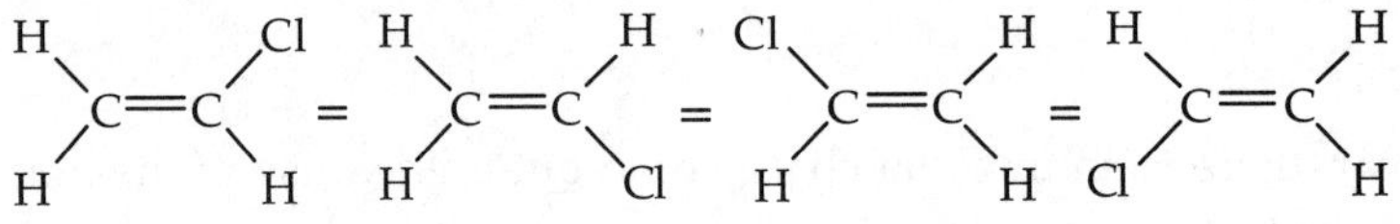

By rotating each of these structures in various directions, it becomes clear that the four structures are equivalent; C_2H_3Cl has only one isomer. Because C_2H_3Cl has only one C–Cl bond, the bond dipoles do not cancel, and the molecule has a dipole moment.

9.44   Each C–Cl bond is polar. The question is whether the vector sum of the C–Cl bond dipoles in each molecule will be nonzero. In the *ortho* and *meta* isomers, the C–Cl vectors are at 60° and 120° angles, respectively, and their resultant dipole moments are nonzero. In the *para* isomer, the C–Cl vectors are opposite, at an angle of 180°, with a resultant dipole moment of zero. The *ortho* and *meta* isomers are polar, the *para* isomer is nonpolar.

Orbital Overlap; Hybrid Orbitals (Sections 9.4 and 9.5)

9.45   (a)   True.

(b)   False. Examples of bonds that could involve an s orbital on one atom and a p orbital on another are H–F, H–Cl, etc.

(c)   True. See Tables 9.2 and 9.3.

(d)   False. A 1s orbital is shaped like a sphere; there are no nodal planes.

(e)   True. All p orbitals have a nodal plane.

9.46   (a)

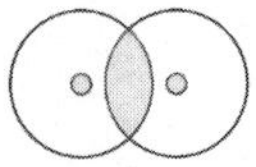

2s   2s

(b)

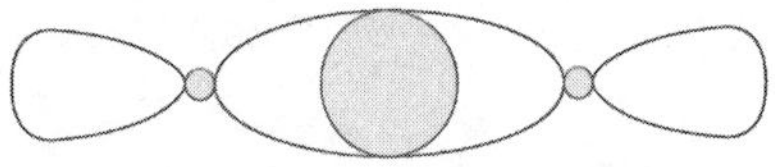

$2p_z$ $2p_z$

(c)

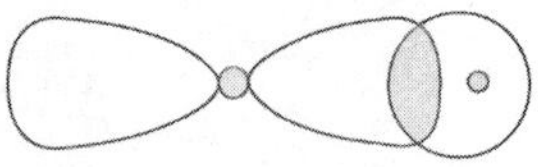

$2p_z$ 2s

9.47   (a)   False. The more orbital overlap in a bond, the stronger the bond.

(b)   True.

(c)   False. Hybrid orbitals are combinations of atomic orbitals on the same atom.

(d)   False. Nonbonding electron pairs (lone pairs) can occupy hybrid orbitals.

9.48   By analogy to the H_2 molecule shown in Figure 9.13, as the distance between the atoms decreases, the overlap between their bonding orbitals increases. According to Figure 7.7, the bonding atomic radius for the halogens is in the order F < Cl < Br < I. The order of bond lengths in the molecules is I–F < I–Cl < I–Br < I–I. If the extent of orbital overlap increases as the distance between atoms decreases, I–F has the greatest overlap and I_2 the least. The order for extent of orbital overlap is I–I < I–Br < I–Cl < I–F.

9.49   (a)   B: $[He]2s^2 2p^1$

         (b)   F, $[He]2s^2 2p^5$

         (c) BF_3, 24 e⁻, 12 e⁻ pairs

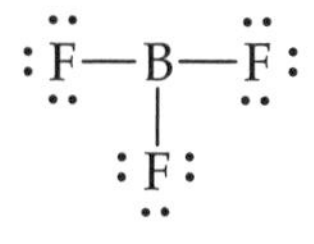

                :F̈—B—F̈:
                        |
                      :F̈:

                sp^2. The three electron domains around B require sp^2 hybrid orbitals.

         (d)   A single 2p orbital is unhybridized. It lies perpendicular to the trigonal plane of the sp^2 hybrid orbitals.

9.50   (a)   S: $[Ne]3s^2 3p^4$

         (b)   Cl, $[Ne]3s^2 2p^5$

         (c) SCl_2, 20 e⁻, 10 e⁻ pr

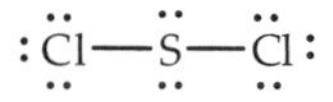

                :C̈l—S̈—C̈l:

                sp^3. The four electron domains (two bonding and two nonbonding) around S require sp^3 hybrid orbitals.

         (d)   No valence atomic orbitals on S remain unhybridized. The sp^3 hybrids use all the 3s and 3p valence orbitals. There are 3d orbitals on S, but these are not considered valence orbitals.

9.51   *Analyze/Plan.* Given the molecular (or ionic) formula, draw the correct Lewis structure and determine the electron-domain geometry, which determines hybridization.   *Solve.*

         (a)   24 e⁻, 12 e⁻ pairs

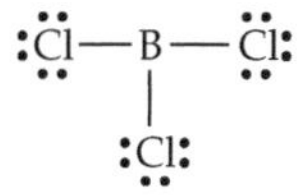

                :C̈l—B—C̈l:
                        |
                      :C̈l:

                3 e⁻ pairs around B, trigonal planar e⁻ domain geometry, sp^2 hybridization

         (b)   32 e⁻, 16 e⁻ pairs

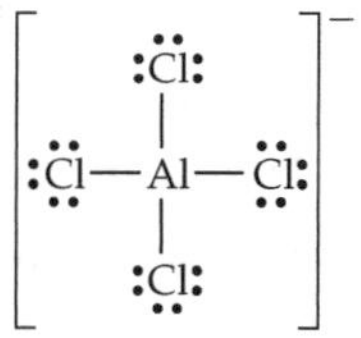

                $$\left[\begin{array}{c} :\ddot{C}l: \\ | \\ :\ddot{C}l—Al—\ddot{C}l: \\ | \\ :\ddot{C}l: \end{array} \right]^{-}$$

                4 e⁻ domains around Al, tetrahedral e⁻ domain geometry, sp^3 hybridization

(c) $16\,e^-$, $8\,e^-$ pairs

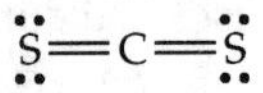

$$\ddot{\underset{..}{S}}=C=\ddot{\underset{..}{S}}$$

$2\,e^-$ domains around C, linear e^- domain geometry, sp hybridization

(d) $8\,e^-$, $4\,e^-$ pairs

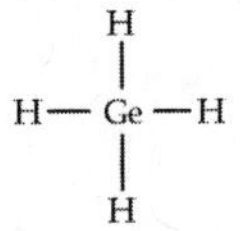

$$H-\underset{\underset{H}{|}}{\overset{\overset{H}{|}}{Ge}}-H$$

$4\,e^-$ pairs around Ge, tetrahedral e^- domain geometry, sp^3 hybridization

9.52   (a) $32\,e^-$, $16\,e^-$ pairs

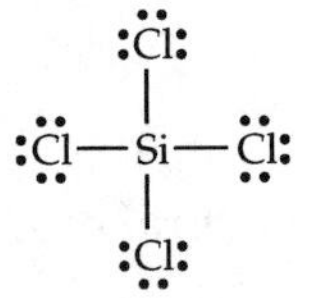

$$:\ddot{\underset{..}{Cl}}-\underset{\underset{:\ddot{\underset{..}{Cl}}:}{|}}{\overset{\overset{:\ddot{Cl}:}{|}}{Si}}-\ddot{\underset{..}{Cl}}:$$

$4\,e^-$ pairs around Si, tetrahedral e^- domain geometry, sp^3 hybridization

(b) $10\,e^-$, $5\,e^-$ pairs

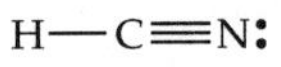

$$H-C\equiv N:$$

$2\,e^-$ domains around C, linear e^- domain geometry, sp hybridization

(c) $24\,e^-$, $12\,e^-$ pairs

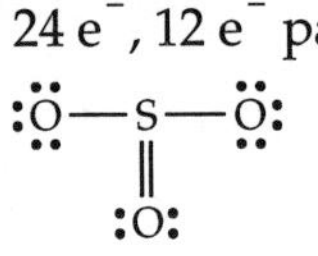

$$:\ddot{\underset{..}{O}}-\underset{\underset{:\ddot{O}:}{\|}}{S}-\ddot{\underset{..}{O}}:$$

(other resonance structures are possible)

$3\,e^-$ domains around S, trigonal planar e^- domain geometry, sp^2 hybridization

(d) $20\,e^-$, $10\,e^-$ pairs

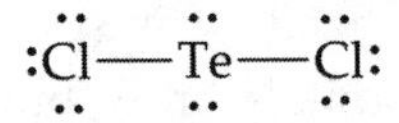

$$:\overset{..}{\underset{..}{Cl}}-\overset{..}{Te}-\overset{..}{\underset{..}{Cl}}:$$

$4\,e^-$ domains around Te, tetrahedral e^- domain geometry, sp^3 hybridization

9.53   Left: No hybrid orbitals discussed in this chapter have angles of 90°; p atomic orbitals are perpendicular to each other.

        Center: Angles of 109.5° are characteristic of sp^3 hybrid orbitals.

        Right: Angles of 120° can be formed by sp^2 hybrids.

9.54   (a)   The three moieties, BH_4^-, CH_4, and NH_4^+, each have 8 valence e^-, $4\,e^-$ pairs, 4 bonding e^- domains, tetrahedral e^- domain and molecular geometry, and sp^3 hybridization at the central atom.

(b)   The electronegativity of the central atoms decreases in the series N > C > B. The question is: where does the electronegativity of H lie in this series? By examination of electronegativity values in Figure 8.8, H is slightly less electronegative than C, and almost the same as B. The magnitude of the bond dipole decreases in the series N–H > C–H > B–H. The negative end of the dipole is toward N, C, and H, respectively.

(c) AlH_4^{-}, SiH_4, and PH_4^{+}. By the same arguments used in part (a), we expect these three moieties to have the same tetrahedral e^{-} domain and molecular geometry and sp^3 hybridization at the central atom as the species in part (a).

Multiple Bonds (Section 9.6)

9.55   (a)      (b)

(c)   A σ bond is generally stronger than a π bond, because there is more extensive orbital overlap.

(d)   Two s orbitals cannot form a π bond. A π bond has no electron density along the internuclear axis. Overlap of s orbitals results in electron density along the internuclear axis. (Another way to say this is that *s* orbitals have the wrong symmetry to form a π bond.)

9.56   (a)   Two unhybridized p orbitals remain, and the atom can form two pi bonds.

(b)   It would be much easier to twist or rotate around a single sigma bond. Sigma bonds are formed by end-to-end overlap of orbitals and the bonding electron density is symmetric about the internuclear axis. Rotating (twisting) around a sigma bond can be done without disrupting either the orbital overlap or bonding electron density, without breaking the bond.

The π part of a double bond is formed by side-to-side overlap of p atomic orbitals perpendicular to the internuclear axis. This π overlap locks the atoms into position and makes twisting difficult. Also, only a small twist (rotation) destroys overlap of the *p* orbitals and breaks the π bond.

9.57   *Analyze/Plan.* Draw the correct Lewis structures, count electron domains and decide hybridization. Molecules with π bonds that require all bonded atoms to be in the same plane are planar. For bond-type counting, single bonds are σ bonds, double bonds consist of one σ and one π bond, triple bonds consist of one σ and two π bonds.   *Solve.*

(a)

$$H-\overset{\overset{\displaystyle H}{|}}{\underset{\underset{\displaystyle H}{|}}{C}}-\overset{\overset{\displaystyle H}{|}}{\underset{\underset{\displaystyle H}{|}}{C}}-H \qquad \overset{H}{\underset{H}{>}}C=C\overset{H}{\underset{H}{<}} \qquad H-C\equiv C-H$$

(b) sp^3 sp^2 sp

(c)   nonplanar   planar   planar

(d)   7 σ, 0 π   5 σ, 1 π   3 σ, 2 π

9.58   (a) 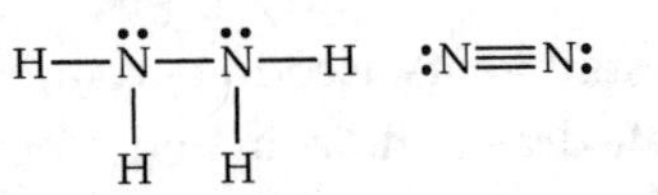   H—N̈—N̈—H   :N≡N:
                | |
                H   H

     (b)   The N atoms in N_2H_4 are sp^3 hybridized; there are no unhybridized p orbitals available for π bonding. In N_2, the N atoms are sp hybridized, with two unhybridized p orbitals on each N atom available to form the two π bonds in the N≡N triple bond.

     (c)   The N–N triple bond in N_2 is significantly stronger than the N–N single bond in N_2H_4, because it consists of one σ and two π bonds, rather than a 'plain' sigma bond. Generally, bond strength increases as the extent of orbital overlap increases. The additional overlap from the two π bonds adds to the strength of the N–N bond in N_2.

9.59   *Analyze/Plan.* Single bonds are σ bonds, double bonds consist of 1 σ and 1 π bond. Each bond is formed by a pair of valence electrons.   *Solve.*

     (a) C_3H_6 has $3(4) + 6(1) = 18$ valence electrons

     (b)   8 pairs or 16 total valence electrons form σ bonds

     (c)   1 pair or 2 total valence electrons form π bonds

     (d)   no valence electrons are nonbonding

     (e)   The left and central C atoms are sp^2 hybridized; the right C atom is sp^3 hybridized.

9.60   (a)   The C with a double bond to O has three electron domains and is sp^2 hybridized; the other three C atoms are sp^3 hybridized.

     (b) $C_4H_8O_2$ has $4(4) + 8(1) + 2(6) = 36$ valence electrons.

     (c)   13 pairs or 26 total valence electrons form σ bonds

     (d)   1 pair or 2 total valence electrons form π bonds

     (e)   4 pairs or 8 total valence electrons are nonbonding

9.61   *Analyze/Plan.* Given the correct Lewis structure, analyze the electron domain geometry at each central atom. This determines the hybridization and bond angles at that atom. *Solve.*

     (a)   ~109° bond angles about the left most C, sp^3; ~120° bond angles about the right-hand C, sp^2

     (b)   The doubly bonded O can be viewed as sp^2, the other as sp^3; the nitrogen is sp^3 with approximately less than 109.5° bond angles.

     (c)   9 σ bonds, 1 π bond

9.62   (a)   1, ~120°; 2, ~120°; 3, less than 109.5°

     (b)   1, sp^2; 2, sp^2; 3, sp^3

     (c)   21 σ bonds

9.63   (a)   In a localized π bond, the electron density is concentrated strictly between the two atoms forming the bond. In a delocalized π bond, parallel p orbitals on more than two adjacent atoms overlap and the electron density is spread over all the atoms that contribute p orbitals to the network. There are still two regions of overlap, above and below the σ framework of the molecule.

        (b)   The existence of more than one resonance form is a good indication that a molecule will have delocalized π bonding.

        (c)

        The existence of more than one resonance form for NO_2 indicates that the π bond is delocalized. From an orbital perspective, the electron-domain geometry around N is trigonal planar, so the hybridization at N is sp^2. This leaves a p orbital on N and one on each O atom perpendicular to the trigonal plane of the molecule, in the correct orientation for delocalized π overlap. Physically, the two N–O bond lengths are equal, indicating that the two N–O bonds are equivalent, rather than one longer single bond and one shorter double bond.

9.64   (a,b) $24\,e^-$, $12\,e^-$ pairs

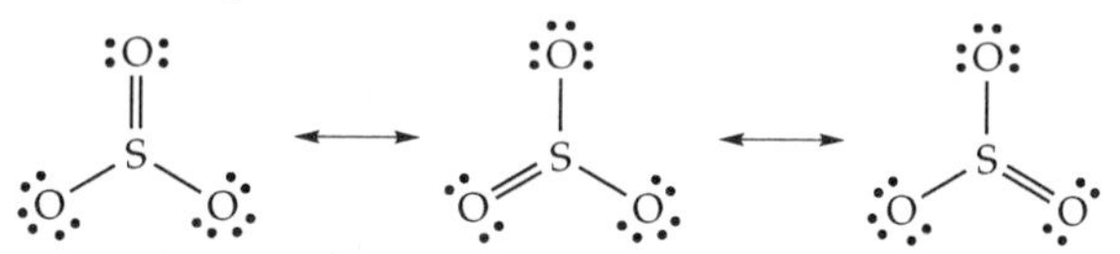

        3 electron domains around S, trigonal planar electron-domain geometry, sp^2 hybrid orbitals

        (c)   The multiple resonance structures indicate delocalized π bonding. All four atoms lie in the trigonal plane of the sp^2 hybrid orbitals. On each atom there is a p atomic orbital perpendicular to this plane in the correct orientation for π overlap. The resulting delocalized π electron cloud is Y-shaped (the shape of the molecule) and has electron density above and below the plane of the molecule.

9.65   *Analzye/Plan.* Follow the logic in Sample Exercise 9.7.

        (a) $18\,e^-$, $9\,e^-$ pairs

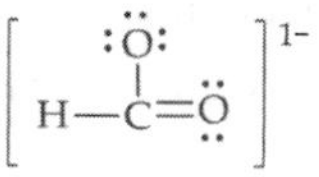

        (b) sp^2

        (c)   Yes, there is one other resonance structure.

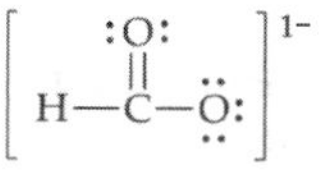

        (d)   There are four electrons in the π system of the molecule. If the C and both O atoms are sp^2 hybridized, there are three bonding electron pairs and four nonbonding electron pairs in the σ system. This leaves two electron pairs or four electrons in the π system.

9.66   (a)   The Lewis structure depicts an anion with a 1– charge. The chemical formula of the given structure is $C_3H_3O_2$. This grouping of atoms has 27 valence electrons, whereas the structure shown has 14 electron pairs or 28 electrons. This means that the structure is an anion with a 1– charge.

      (b) sp^2

      (c)   Yes, there is one other resonance structure.

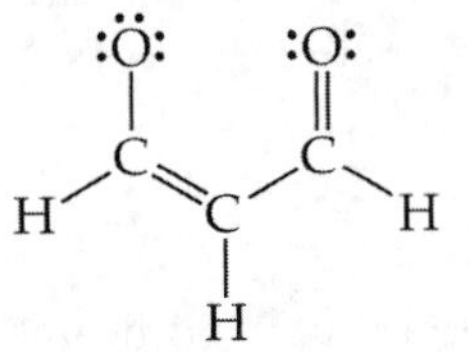

      (d)   There are six electrons in the π system of the molecule. If all the C and O atoms are sp^2 hybridized, there are seven bonding electron pairs and four nonbonding electron pairs in the σ system. This leaves three electron pairs or six electrons in the π system.

9.67   *Analyze/Plan.* Count valence e^- and e^- pairs in each molecule. Complete the Lewis structure by placing nonbonding electron pairs. Analyze the electron-domain geometry at each central atom; visualize and describe the molecular structure.   *Solve.*

      (a) $26 \, e^-, 13 \, e^-$ pairs

            $H-C\equiv C-C\equiv C-C\equiv N:$

            The molecule is linear. Each C atom has 2 bonding e^- domains, linear geometry, and sp hybridization. This requires that all atoms not only lie in the same plane, but in a line.

      (b) $34 \, e^-, 17 \, e^-$ pairs

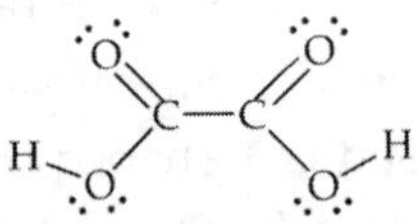

            The two central C atoms each have 3 bonding e^- domains, trigonal planar geometry, and sp^2 hybridization. Each O–C–O group is planar, whereas the terminal H atoms can rotate out of these planes. In principle, there is free rotation about the C–C σ bond, but delocalization of the π electrons is possible if the two planes are coincident. It is possible to put all 8 atoms in the same plane.

      (c) $12 \, e^-, 6 \, e^-$ pairs

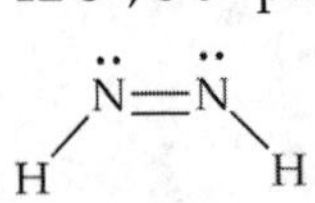

            The molecule is planar. Each N atom has 3 bonding e^- domains, trigonal planar geometry, and sp^2 hybridization. Because the N atoms share a π bond, the planes must be coincident and all 4 atoms are required to lie in this plane. [The structure shown here has H atoms on the same side of the double bond. There is another isomer that has H atoms on opposite sides of the double bond. Both compounds are planar with 120 degree bond angles.]

9.68   (a)   24 e⁻ , 12 e⁻ pairs

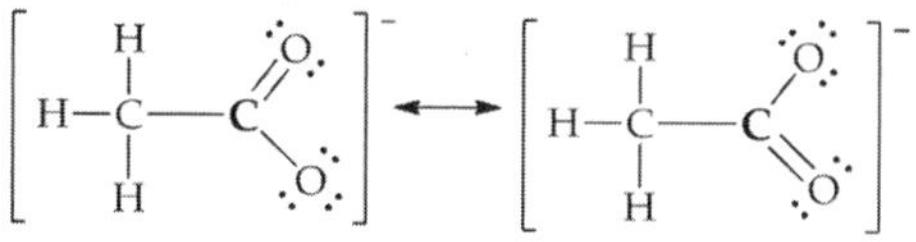

$$\left[\begin{array}{c} H \\ | \\ H-C-C \\ | \\ H \end{array} \begin{array}{c} :\ddot{O}: \\ \diagdown \\ \diagup \\ :\ddot{O}: \end{array} \right]^{-} \longleftrightarrow \left[\begin{array}{c} H \\ | \\ H-C-C \\ | \\ H \end{array} \begin{array}{c} :\ddot{O}: \\ \diagup \\ \diagdown \\ :\ddot{O} \end{array} \right]^{-}$$

The designated C atom has 3 bonding e⁻ domains and sp² hybridization.

     (b)   8 e⁻ , 4 e⁻ pairs

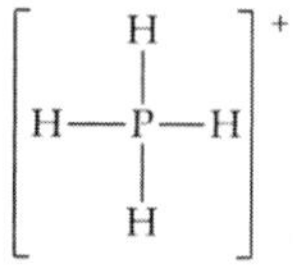

$$\left[\begin{array}{c} H \\ | \\ H-P-H \\ | \\ H \end{array} \right]^{+}$$

The P atom has 4 bonding e⁻ domains and sp³ hybridization

     (c)   24 e⁻ , 12 e⁻ pairs

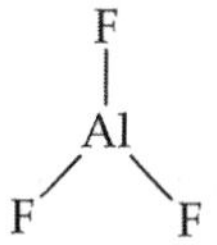

$$\begin{array}{c} F \\ | \\ Al \\ \diagup \quad \diagdown \\ F \qquad F \end{array}$$

The Al atom has 3 bonding e⁻ domains and sp² hybridization

     (d)   16 e⁻ , 8 e⁻ pairs

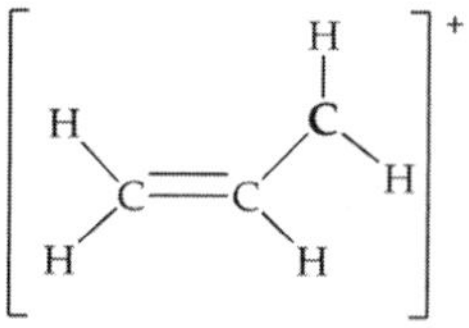

$$\left[\begin{array}{c} \quad\quad H \\ \quad\quad | \\ H \quad\quad C \\ \diagdown \quad\quad \diagdown H \\ C=C \\ \diagup \quad\quad \diagdown \\ H \quad\quad H \end{array} \right]^{+}$$

The designated C atom has 3 bonding e⁻ domains and sp² hybridization.

Molecular Orbitals and Period 2 Diatomic Molecules (Sections 9.7 and 9.8)

9.69   (a)   Hybrid orbitals are mixtures (linear combinations) of atomic orbitals from a single atom; the hybrid orbitals remain localized on that atom. Molecular orbitals are combinations of atomic orbitals from two or more atoms. They are associated with the entire molecule, not a single atom.

     (b)   Each MO, like each AO or hybrid, can hold a maximum of two electrons.

     (c)   Yes, antibonding MOs can have electrons in them.

9.70   (a)   An MO, because the AOs come from two different atoms.

     (b)   A hybrid orbital, because the AOs are on the same atom.

     (c)   Yes. The Pauli exclusion principle, that no two electrons can have the same four quantum numbers, means that an orbital can hold at most two electrons. (Because n, l, and m_l are the same for a particular orbital and m_s has only two possible values, an orbital can hold at most two electrons). This is true for atomic and molecular orbitals.

9.71   (a)

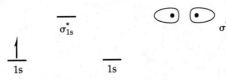

(b)   There is one electron in H_2^+.

(c)

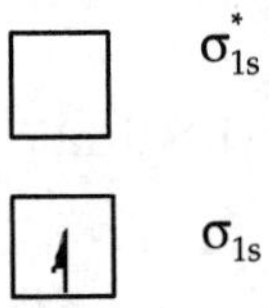

(d)   Bond order $= 1/2\,(1 - 0) = \frac{1}{2}$

(e)   Fall apart. The stability of H_2^+ is because of the lower-energy state of the σ bonding molecular orbital relative to the energy of a H 1s atomic orbital. If the single electron in H_2^+ is excited to the σ^*_{1s} orbital, its energy is higher than the energy of an H 1s atomic orbital and H_2^+ will decompose into a hydrogen atom and a hydrogen ion.

$$H_2^+ \xrightarrow{h\nu} H + H^+.$$

(f)   Statement (i) is true.

9.72   (a)

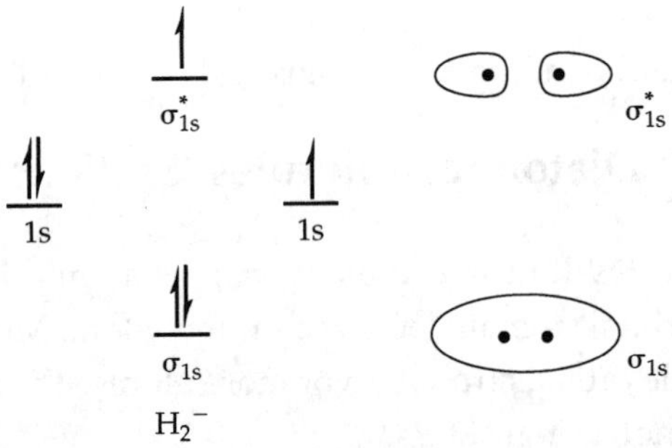

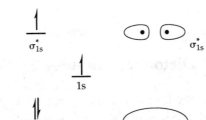

(b)

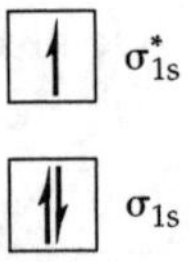

(c)   Bond order $= 1/2\,(2 - 1) = \frac{1}{2}$

(d)   If one electron moves from σ_{1s} to σ^*_{1s}, the bond order becomes $-\frac{1}{2}$. There is a net increase in energy relative to isolated H atoms, so the ion will decompose.

$$H_2^- \xrightarrow{h\nu} H + H^-.$$

(e)   Statement (i) is true.

9.73

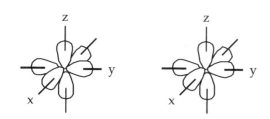

(a) One. With three mutually perpendicular p orbitals on each atom, only one set can be oriented for end-to-end sigma overlap.

(b) Two. The 2p orbitals on each atom not involved in σ bonding can be aligned for side-to-side π overlap.

(c) Three, 1 σ^* and 2 π^*. There are a total of 6 p orbitals on the two atoms. When combining AOs to form MOs, total number of orbitals is conserved. If 3 of the 6 MOs are bonding MOs, as described in (a) and (b), then the remaining 3 MOs must be antibonding. They will have the same symmetry as the bonding MOs, 1 σ^* and 2 π^*.

9.74   (a) True.

(b) False. Pi star molecular orbitals have a nodal plane through the nuclei. This means that there is zero probability of finding an electron in a pi star orbital at the nucleus.

(c) True.

(d) False. Electrons can and do occupy antibonding molecular orbitals.

9.75   (a) When comparing the same two bonded atoms, the greater the bond order, the shorter the bond length and the greater the bond energy. That is, bond order and bond energy are directly related, whereas bond order and bond length are inversely related. When comparing different bonded nuclei, there are no simple relationships (see Solution 8.100).

(b) Be_2, 4 e$^-$ Be_2^+, 3 e$^-$

| ↑↓ | σ_{2s}^* | | ↑ | σ_{2s}^* |

| ↑↓ | σ_{2s} | | ↑↓ | σ_{2s} |

BO = 1/2(2 − 2) = 0   BO = 1/2(2 − 1) = 0.5

Be_2 has a bond order of zero and is not energetically favored over isolated Be atoms; it is not expected to exist. Be_2^+ has a bond order of 0.5 and is slightly lower in energy than isolated Be atoms. It will probably exist under special experimental conditions, but be unstable.

9.76   (a) O_2^{2-} has a bond order of 1.0, whereas O_2^- has a bond order of 1.5. For the same bonded atoms, the greater the bond order the shorter the bond, so O_2^- has the shorter bond.

(b)   The two possible orbital energy level diagrams are:

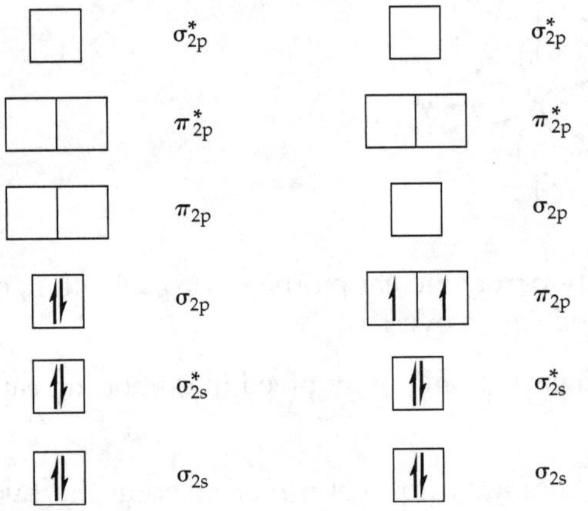

The magnetic properties of a molecule reveal whether it has unpaired electrons. If the σ_{2p} MOs are lower in energy, B_2 has no unpaired electrons. If the π_{2p} MOs are lower in energy than the σ_{2p} MO, there are two unpaired electrons. The magnetic properties of B_2 must indicate that it has unpaired electrons.

(c)   According to Figure 9.43, the two highest-energy electrons of O_2 are in antibonding π_{2p}^{*} MOs and O_2 has a bond order of 2.0. Removing these two electrons to form O_2^{2+} produces an ion with bond order 3.0. O_2^{2+} has a stronger O–O bond than O_2, because O_2^{2+} has a greater bond order.

9.77   (a,b) Substances with no unpaired electrons are weakly repelled by a magnetic field. This property is called *diamagnetism*.

(c) O_2^{2-}, Be_2^{2+} [see Figure 9.43 and Solution 9.75(b)]

9.78   (a)   Substances with unpaired electrons are attracted into a magnetic field. This property is called *paramagnetism*.

(b)   Weigh the substance normally and in a magnetic field, as shown in Figure 9.44. Paramagnetic substances appear to have a larger mass when weighed in a magnetic field.

(c)   See Figures 9.35 and 9.43. O_2^{+}, one unpaired electron; N_2^{2-}, two unpaired electrons; Li_2^{+}, one unpaired electron

9.79

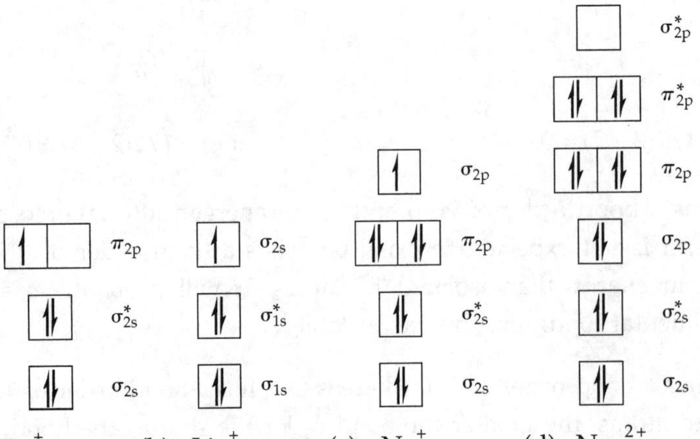

(a) B_2^{+}
       increase

(b) Li_2^{+}
       increase

(c) N_2^{+}
       increase

(d) Ne_2^{2+}
       decrease

Addition of an electron increases bond order if it occupies a bonding MO and decreases stability if it occupies an antibonding MO.

9.80   Determine the number of "valence" (noncore) electrons in each molecule or ion. Use the homonuclear diatomic MO diagram from Figure 9.43 (shown below) to calculate bond order and magnetic properties of each species. The electronegativity difference between heteroatomics increases the energy difference between the 2s AO on one atom and the 2p AO on the other, rendering the "no interaction" MO diagram in Figure 9.43 appropriate.

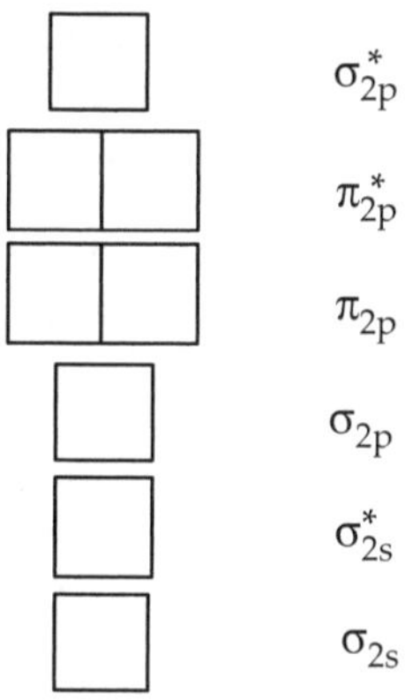

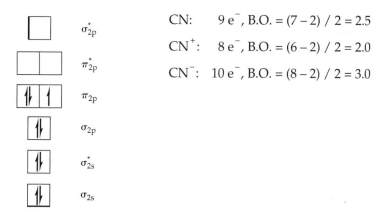

(a) CO^+: $9\,e^-$, B.O. $= (7-2)\,/\,2 = 2.5$, paramagnetic

(b) NO^-: $12\,e^-$, B.O. $= (8-4)\,/\,2 = 2.0$, paramagnetic

(c) OF^+: $12\,e^-$, B.O. $= (8-4)\,/\,2 = 2.0$, paramagnetic

(d) NeF^+: $14\,e^-$, B.O. $= (8-6)\,/\,2 = 1.0$, diamagnetic

9.81   *Analyze/Plan.* Determine the number of "valence" (non-core) electrons in each molecule or ion. Use the homonuclear diatomic MO diagram from Figure 9.43 (shown below) to calculate bond order and magnetic properties of each species. The electronegativity difference between heteroatomics increases the energy difference between the 2s AO on one atom and the 2p AO on the other, rendering the "no interaction" MO diagram in Figure 9.43 appropriate. *Solve.*

CN: $9\,e^-$, B.O. $= (7-2)\,/\,2 = 2.5$

CN^+: $8\,e^-$, B.O. $= (6-2)\,/\,2 = 2.0$

CN^-: $10\,e^-$, B.O. $= (8-2)\,/\,2 = 3.0$

(a) CN^- has the highest bond order and therefore the strongest C–N bond.

(b)   CN and CN^+. CN has an odd number of valence electrons, so it must have an unpaired electron. The electron configuration for CN is shown in the diagram. Removing one electron from the π_{2p} MOs to form CN^+ produces an ion with two unpaired electrons. Adding one electron to the π_{2p} MOs of CN to form CN^- produces an ion with all electrons paired.

9.82 (a) Statement (ii) is the best explanation. The bond order of NO is $[1/2 (8 - 3)] = 2.5$. The electron that is lost is in an antibonding molecular orbital, so the bond order in NO^+ is 3.0. The increase in bond order is the driving force for the formation of NO^+.

(b) To form NO^-, an electron is added to an antibonding orbital, and the new bond order is $[1/2 (8 - 4)] = 2$. The order of increasing bond order and bond strength is: $NO^- < NO < NO^+$. NO^- and NO are paramagnetic with two and one unpaired electrons, respectively. NO^+ is diamagnetic.

(c) NO^+ is isoelectronic with N_2, and NO^- is isoelectronic with O_2.

9.83 (a) $3s, 3p_x, 3p_y, 3p_z$ (b) π_{3p}

(c) Two. Note that there are two degenerate π_{3p} bonding molecular orbitals; each holds two electrons. A total of 4 electrons can be designated as π_{3p}, but no single molecular orbital can hold more than two electrons.

(d) If the MO diagram for P_2 is similar to that of N_2, P_2 will have no unpaired electrons and be diamagnetic.

9.84 (a) I: $5s, 5p_x, 5p_y, 5p_z$; Br: $4s, 4p_x, 4p_y, 4p_z$

(b) By analogy to F_2, the BO of IBr will be 1.

(c) I and Br have valence atomic orbitals with different principal quantum numbers. This means that the radial extensions (sizes) of the valence atomic orbital that contribute to the MO are different. The n = 5 valence AOs on I are larger than the n = 4 valence AOs on Br.

(d) σ_{np}^*

(e) None

Additional Exercises

9.85 (a) The physical basis of VSEPR is the electrostatic repulsion of like-charged particles, in this case groups or domains of electrons. That is, owing to electrostatic repulsion, electron domains will arrange themselves to be as far apart as possible.

(b) The σ-bond electrons are localized in the region along the internuclear axes. The positions of the atoms and geometry of the molecule are thus closely tied to the locations of these electron pairs. Because the π-bond electrons are distributed above and below the plane that contains the σ bonds, these electron pairs do not, in effect, influence the geometry of the molecule. Thus, all σ- and π-bond electrons localized between two atoms are located in the same electron domain.

9.86 (a) Two. If the electron-domain geometry is trigonal bipyramidal, there are five total electron domains around the central atom. An AB_3 molecule has three bonding domains, so there must be two nonbonding domains on A.

(b) (iii) (See Table 9.3.)

9.87   34 e⁻, 17 e⁻ pairs   36 e⁻, 18 e⁻ pairs

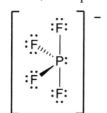   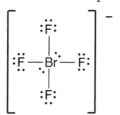

        34 e⁻, 17 e⁻ pairs   32 e⁻, 16 e⁻ pairs

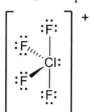   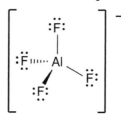

(a)   Three. The P, Br, and Cl central atoms have more than an octet of electrons about them.

(b)   One, AlF_4^-

(c) BrF_4^-

(d) PF_4^- and ClF_4^+

9.88   (a)   40 e⁻, 20 e⁻ pairs

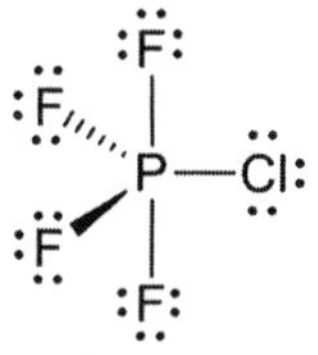

5 e⁻ domains

trigonal-bipyramidal electron-domain geometry

(b)   The greater the electronegativity of the terminal atom, the larger the negative charge centered on the atom, the smaller the effective size of the P–X bonding electron domain. A P–F bond will produce a smaller (and shorter) electron domain than a P–Cl bond.

(c)   The molecular geometry (shape) is also trigonal bipyramidal, because all five electron domains are bonding domains. Because we predicted the P–F electron domain to be smaller, the larger P–Cl bonding domain will occupy the equatorial plane of the molecule, minimizing the number of 90° P–Cl to P–F repulsions. This is the same argument that places a "larger" nonbonding domain in the equatorial position of a molecule like SF_4.

(d)   The molecular geometry is distorted from a perfect trigonal bipyramid because not all electron domains are alike. The 90° P–Cl to P–F repulsions will be greater than the 90° P–F to P–F repulsions, so the F(axial)–P–Cl angles will be greater than 90°. The equatorial F–P–F angles may distort slightly to "make room" for the axial F atoms that are "pushed away" from the equatorial Cl atom.

9.89 For any triangle, the law of cosines gives the length of side c as $c^2 = a^2 + b^2 - 2ab \cos\theta$.

Let the edge length of the cube (uy = vy = vz) = X

The length of the face diagonal (uv) is

$(uv)^2 = (uy)^2 + (vy)^2 - 2(uy)(vy) \cos 90$

$(uv)^2 = X^2 + X^2 - 2(X)(X) \cos 90$

$(uv)^2 = 2X^2; \ uv = \sqrt{2}X$

The length of the body diagonal (uz) is

$(uz)^2 = (vz^2) + (uv)^2 - 2(vz)(uv) \cos 90$

$(uz)^2 = X^2 + (\sqrt{2}X)^2 - 2(X)(\sqrt{2}X) \cos 90$

$(uz)^2 = 3X^2; \ uz = \sqrt{3}X$

For calculating the characteristic tetrahedral angle, the appropriate triangle has vertices u, v, and w. Theta, θ, is the angle formed by sides wu and wv and the hypotenuse is side uv.

$wu = wv = uz/2 = \sqrt{3}/2X; \ uv = \sqrt{2}X$

$(\sqrt{2}X)^2 = (\sqrt{3}/2X)^2 + (\sqrt{3}/2X)^2 - 2(\sqrt{3}/2X)(\sqrt{3}/2) \cos\theta$

$2X^2 = 3/4 X^2 + 3/4 X^2 - 3/2 X^2 \cos\theta$

$2X^2 = 3/2 X^2 - 3/2 X^2 \cos\theta$

$1/2 X^2 = -3/2 X^2 \cos\theta$

$\cos\theta = -(1/2 X^2) / (3/2 X^2) = -1/3 = -0.3333$

$\theta = 109.47°$

9.90 *Analyze/Plan.* For entries where the molecule is listed, follow the logic in Sample Exercises 9.4 and 9.5. For entries where no molecule is listed, decide electron-domain geometry from hybridization (or vice versa). If the molecule is nonpolar, the terminal atoms will be identical. If the molecule is polar, the terminal atoms will be different, or the central atom will have one or more lone pairs, or both. *Solve.*

Molecule	Molecular Structures	Electron Domain Geometry	Hybrdization of Central Atom	Dipole Moment Yes or No
CO_2	O=C=O	linear	sp	no
NH_3		tetrahedral	sp^3	yes
CH_4		tetrahedral	sp^3	no
BH_3		trigonal planar	sp^2	no

SF_4		trigonal bipyramidal	not applicable	yes
SF_6		octahedral	not applicable	no
H_2CO		trigonal planar	sp^2	yes
PF_5		trigonal bipyramidal	not applicable	no
XeF_2	F—Xe—F	trigonal bipyramidal	not applicable	no

9.91   (a) CO_2, 16 valence e^- (b) $(CN)_2$, 18 valence e^-

           $\overset{..}{\underset{..}{O}}{=}C{=}\overset{..}{\underset{..}{O}}$ $:N{\equiv}C{-}C{\equiv}N:$

            2σ 2π $3\sigma, 3\pi$

        (c) H_2CO, 12 valence e^- (d)   HCOOH, 18 valence e^-

           $3\sigma, 1\pi$ $4\sigma, 1\pi$

9.92   (a)

          $3(4) + 3(6) + 6(1) = 36$ e^-, 18 e^- pr

      (b)   There are 11 σ and 1 π bonds.

      (c)   The C=O on the right-hand C atom is shortest. For the same bonded atoms, in this case C and O, the greater the bond order, the shorter the bond.

      (d)   The right-most C has three e^- domains, so the hybridization is sp^2.

      (e)   Bond angles about the right-most C atom are approximately 120°. The middle and left-hand C atoms both have four bonding e^- domains and are sp^3 hybridized. Because the bonding domains about each C atom are not exactly the same, the bond angles will deviate somewhat from 109.5°; we expect larger deviations for the angles around the middle C atom.

9.93   (a)   Square pyramidal

(b)   Yes, there is one nonbonding electron domain on A. If there were only five bonding domains, the shape would be trigonal bipyramidal. With five bonding and one nonbonding electron domains, the molecule has octahedral domain geometry.

(c)   (iii). If the B atoms are halogens, each will have three nonbonding electron pairs; there are five bonding pairs, and A has one nonbonded pair, for a total of $[5(3) + 5 + 1] = 21$ e$^-$ pairs and 42 electrons in the Lewis structure. If the five halogens contribute 35 e$^-$, A must contribute seven valence electrons. A is also a halogen.

9.94   (a)   Cisplatin has a dipole moment. In the square planar trans structure, all equivalent bond dipoles can be oriented opposite each other, for a net dipole moment of zero.

(b)   Cisplatin is the cancer drug. The Cl atoms in cisplatin occupy bonding sites that are in the correct orientation to bind adjacent N atoms in DNA. That is, the Cl–Pt–Cl angle in cisplatin is about 90°, and the N–Pt–N angle in the new DNA complex is also about 90°.

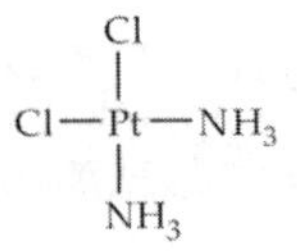

cisplatin 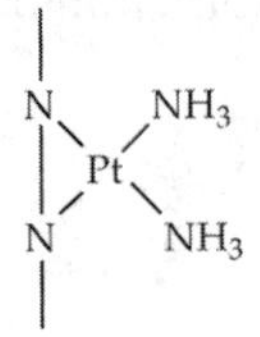   DNA complex

9.95

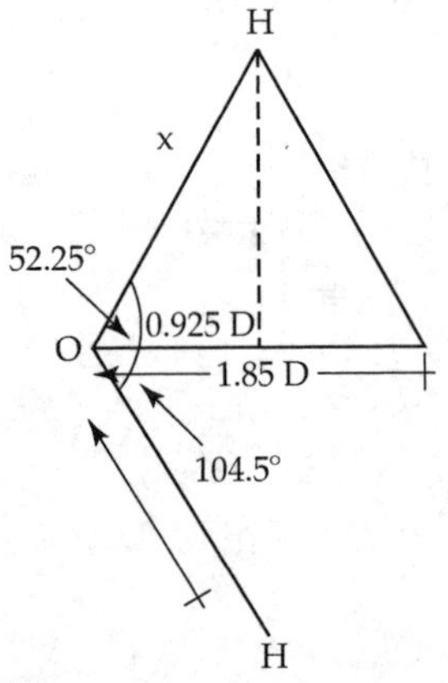

(a)   The bond dipoles in H_2O lie along the O–H bonds with the positive end at H and the negative end at O. The dipole moment vector of the H_2O molecule is the resultant (vector sum) of the two bond dipoles. This vector bisects the H–O–H angle and has a magnitude of 1.85 D with the negative end pointing toward O.

(b)   Because the dipole moment vector bisects the H–O–H bond angle, the angle between one H–O bond and the dipole moment vector is 1/2 the H–O–H bond angle, 52.25°. Dropping a perpendicular line from H to the dipole moment vector creates the right triangle pictured. If x = the magnitude of the O–H bond dipole, x cos (52.25) = 0.925 D. x = 1.51 D.

(c)   The X–H bond dipoles (Table 8.2) and the electronegativity values of X (Figure 8.8) are

	Electronegativity	**Bond dipole**
F	4.0	1.82
O	3.5	1.51
Cl	3.0	1.08

Because the electronegativity of O is midway between the values for F and Cl, the O–H bond dipole should be approximately midway between the bond dipoles of HF and HCl. The value of the O–H bond dipole calculated in part (b) is consistent with this prediction.

9.96   (a) XeF_2, 22 e⁻, 11 e⁻ pr

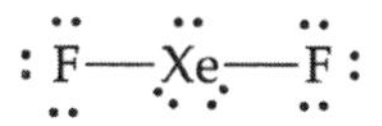

: F̈—Xe̤—F̈ :

5 electron domains around Xe, trigonal bipyramidal electron domain geometry

XeF_4, 36 e⁻, 18 e⁻ pr

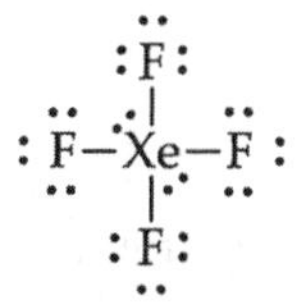

6 electron domains around Xe, octahedral electron domain geometry

XeF_6 50 e⁻, 25 e⁻ pairs

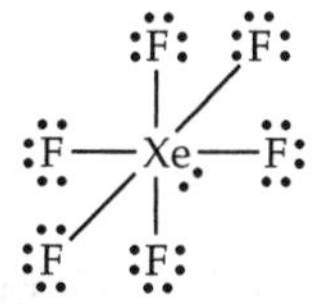

7 electron domains around Xe, pentagonal bipyramidal electron domain geometry

(b) IF_6^-, 50 e⁻, 25 e⁻ pairs

IF_6^- is isoelectronic with XeF_6. The Lewis structure is similar to the one for XeF_6, but with I as the central atom. The electron domain geometry is pentagonal bipyramidal. With respect to molecular geometry, the question is whether the nonbonding domain occupies an axial or equatorial position. The equatorial plane of a pentagonal bipyramid has F–I–F angles of 72°. Placing the nonbonded domain in the equatorial plane would create severe repulsions between it and the adjacent bonded domains. Thus, the nonbonded domain will reside in the axial position. The molecular geometry is a pentagonal pyramid.

9.97 Statements (ii) and (iii) are true.

9.98

(a) The molecule is not planar. The CH_2 planes at each end are twisted 90° from one another.

(b) Allene has no dipole moment.

(c) The bonding in allene would not be described as delocalized. The π electron clouds of the two adjacent C=C are mutually perpendicular. The mechanism for delocalization of π electrons is mutual overlap of parallel p atomic orbitals on adjacent atoms. If adjacent π electron clouds are mutually perpendicular, there is no overlap and no delocalization of π electrons.

9.99 (a) 9σ, 3π

(b) (i) 2, the second and third C atoms from the left

(ii) 2, the rightmost C atom and the N atom

(iii) 1, the leftmost C atom

9.100 (a) $16 \, e^-$, $8 \, e^-$ pairs

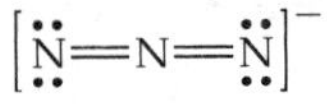

$$\left[\ddot{\ddot{N}} {=\!\!=} N {=\!\!=} \ddot{\ddot{N}} \right]^-$$

There are 2 electron domains around the central nitrogen atom. The N–N–N angle is 180° and the ion is linear.

(b) The central N atom has 2 electron domains and sp hydridization.

(c) The central N atom forms 2 sigma and 2 pi bonds.

9.101 (a) $\ddot{\ddot{O}} {=\!\!=} \ddot{O} {-} \ddot{\ddot{O}} \colon \longleftrightarrow \colon \ddot{\ddot{O}} {-} \ddot{O} {=\!\!=} \ddot{\ddot{O}}$

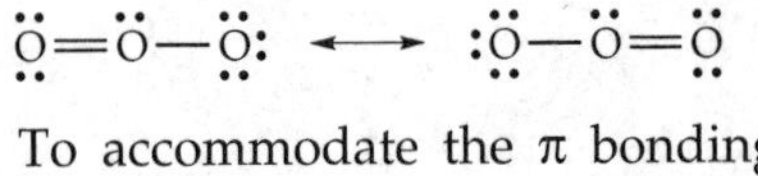

To accommodate the π bonding by all 3 O atoms indicated in the resonance structures above, all O atoms are sp^2 hybridized.

(b) For the first resonance structure, both sigma bonds are formed by overlap of sp^2 hybrid orbitals, the π bond is formed by overlap of atomic p orbitals, one of the nonbonded pairs on the right terminal O atom is in a p atomic orbital, and the remaining five nonbonded pairs are in sp^2 hybrid orbitals.

(c) Only unhybridized p atomic orbitals can be used to form a delocalized π system.

(d) The unhybridized p orbital on each O atom is used to form the delocalized π system, and in both resonance structures one nonbonded electron pair resides in a p atomic orbital. The delocalized π system then contains four electrons, two from the π bond and two from the nonbonded pair in the p orbital.

9.102   (a)   Each C atom is surrounded by three electron domains (two single bonds and one double bond), so bond angles at each C atom will be approximately 120°.

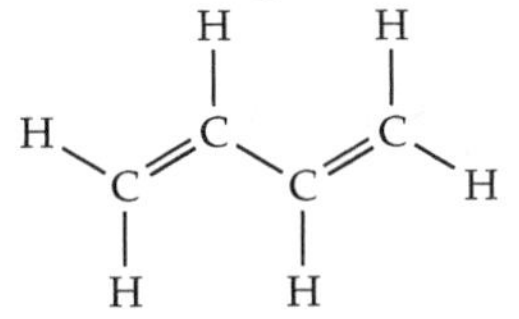

Because rotation around the central C–C single bond is possible, other conformations can be drawn.

     (b)   Each C atom has 3 bonding electron domains and $120°$ bond angles, so all C atoms have sp^2 hybridization.

     (c)   Stronger. When comparing C–C bonds, the shorter the bond, the stronger the bond.

     (d)   Each C atom in butadiene has sp^2 hybridization and one unhybridized 2p orbital. If the C atoms are coplanar, the 4 unhybridized 2p orbitals are parallel and in the correct orientation for overlap. This provides a mechanism for delocalization of the pi electrons across the entire molecule, resulting in the shorter central C–C bond.

9.103   (a) $30\,e^-$, $15\,e^-$ pairs   (b)

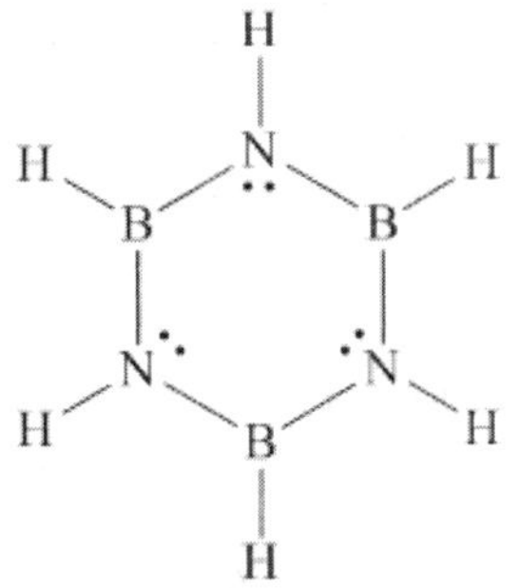

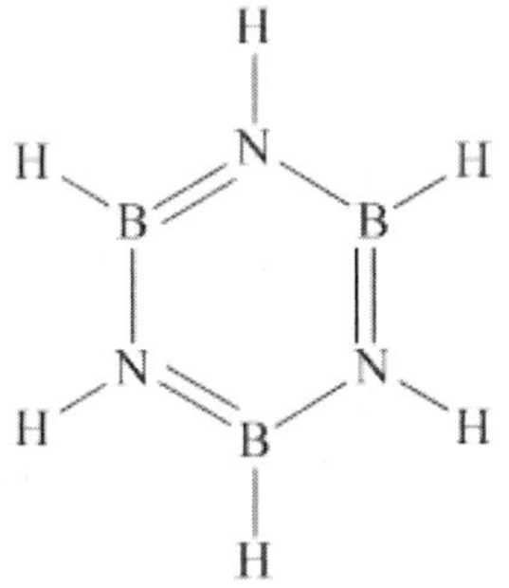

     (c)   In the Lewis structure in part (b), the N atoms have a +1 formal charge and the B atoms have a –1 formal charge. Because N is more electronegative than B, these formal charges do not seem favorable.

     (d)   The Lewis structure in part (b) has two resonance structures.

     (e)   In part (a), the B atoms are sp^2 hybridized and the N atoms are sp^3 hybridized. In part (b), both B and N are sp^2 hybridized. We would not expect the structure in part (a) to lead to a planar molecule, whereas the structure in (b) would be planar.

     (f)   The magnitude of the bond distance is between the values for single and double bonds, which favors multiple resonance structures with alternating single and double bonds, the structure in part (b). That the B–N bond lengths are identical also favors this structure.

     (g)   Six. There are 12 electron pairs in the σ system, which leaves 3 electron pairs in the π system.

9.104   Refer to Figure 9.43 and the *Chemistry Put to Work* box on Orbitals and Energy.

(a)   Ground state, $\sigma_{2s}^2\sigma_{2s}^{*2}\pi_{2p}^4\sigma_{2p}^2$; excited state, $\sigma_{2s}^2\sigma_{2s}^{*2}\pi_{2p}^4\sigma_{2p}^1\pi_{2p}^{*1}$

(b)   Paramagnetic. The first excited state has two unpaired electrons, one in the σ_{2p} and one in the π_{2p}^*.

(c) σ_{2p} to π_{2p}^*

(d) $\Delta E = hc/\lambda;\ \lambda = 170\ nm = 170 \times 10^{-9}\ m$

$$\Delta E_{170} = \frac{hc}{\lambda} = \frac{6.626 \times 10^{-34}\ \text{J-s}}{170 \times 10^{-9}\ m} \times \frac{2.998 \times 10^8\ m}{s} = 1.1685 \times 10^{-18} = 1.2 \times 10^{-18}\ \text{J/photon}$$

$$\frac{1.1685 \times 10^{-18}\ J}{1\ \text{photon}} \times \frac{6.022 \times 10^{23}\ \text{photons}}{1\ \text{mol}} = 7.04 \times 10^5\ \text{J/mol} = 7.0 \times 10^2\ \text{kJ/mol}$$

(e)   Weaker. In the first excited state there is one fewer electron in a bonding MO and one more electron in an antibonding MO. The bond order of the N–N bond in the first excited state is smaller and the bond is weaker than in the ground state.

9.105   (a)   The orbital in the sketch is a σ antibonding MO.

(b)   In H_2^-, there is one electron in the σ^* antibonding MO.

(c)   In H_2^-, BO = ½.

(d)   (iv). For the same two bonded atoms, the smaller the bond order, the weaker and longer the bond.

9.106   Use the MO diagrams in Figure 9.43 to calculate bond order, taking into account the correct number of electrons in each ion.

Ne_2 (BO = 0) < H_2^+ (BO = ½) < B_2 (BO = 1) < F_2^+ (BO = 1.5) < N_2^+ (BO = 2.5)

9.107   (a)   The diagram shows two s atomic orbitals with opposite phases. (See Figures 9.31 and 9.40.) Because they are spherically symmetric, the interaction of s orbitals can only produce a σ molecular orbital. Because the two orbitals in the diagram have opposite phases, the interaction excludes electron density from the region between the nuclei. The resulting MO has a node between the two nuclei and is labeled σ_{2s}^*. The principal quantum number designation is arbitrary, because it defines only the size of the pertinent AOs and MOs. Shapes and phases of MOs depend only on these same characteristics of the interacting AOs.

(b)   The diagram shows two p atomic orbitals with oppositely phased lobes pointing at each other. (See Figure 9.36.) End-to-end overlap produces a σ-type MO; opposite phases mean a node between the nuclei and an antibonding MO. The interaction results in a σ_{2p}^* MO.

(c)   The diagram shows parallel p atomic orbitals with like-phased lobes aligned. (See Figure 9.36.) Side-to-side overlap produces a π-type MO; overlap of like-phased lobes concentrates electron density between the nuclei and a bonding MO. The interaction results in a π_{2p} MO.

9.108   The white solid has the larger HOMO-LUMO gap. The green solid absorbs visible red light and appears green. The white solid does not absorb visible light, but light of higher energy in the ultraviolet region of the spectrum. The white solid has the larger HOMO-LUMO gap.

9.109   We will refer to *azobenzene* (on the left) as A and *hydrazobenzene* (on the right) as H.

     (a)   A: sp^2; H: sp^3

     (b)   A: Fourteen. Each N and C atom has one unhybridized p orbital. H: Twelve. Each C atom has one unhybridized p orbital, but the N atoms have no unhybridized p orbitals.

     (c)   A: ~120°; H: less than 109.5°

     (d)   Because all C and N atoms in A have unhybridized p orbitals, all can participate in delocalized π bonding. The delocalized π system extends over the entire molecule, including both benzene rings and the azo "bridge." In H, the N atoms have no unhybridized p orbitals, so they cannot participate in delocalized π bonding. Each of the benzene rings in H is delocalized, but the network cannot span the N atoms in the bridge.

     (e)   This is consistent with the answer to (d). In order for the unhybridized p orbitals in A to overlap, they must be parallel. This requires a planar σ-bond framework where all atoms in the molecule are coplanar.

     (f)   For a molecule to be useful in a solar energy conversion device, it must absorb visible light. This requires a HOMO-LUMO energy gap in the visible region. For organic molecules, the size of the gap is related to the number of conjugated π bonds; the more conjugated π bonds, the smaller the gap and the more likely the molecule is to be colored. Azobenzene has seven conjugated π bonds (π network delocalized over the entire molecule) and appears red-orange. Hydrazobenzene has only three conjugated π bonds (π network on benzene rings only) and appears white. Thus, the smaller HOMO-LUMO energy gap in A causes it to be both intensely colored and a more useful molecule for solar energy conversion.

9.110   (a)   H: $1s^1$; F: $[He]2s^2 2p^5$

                  When molecular orbitals are formed from atomic orbitals, the total number of orbitals is conserved. Because H and F have a total of five valence AOs ($H_{1s} + F_{2s} + 3F_{2p}$), the MO diagram for HF has five MOs.

     (b)   H and F have a total of eight valence electrons. Because each MO can hold a maximum of two electrons, four of the five MOs would be occupied.

     (c)

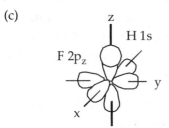

                  If H and F lie on the z-axis, then the $2p_z$ orbital of F will overlap with the 1s orbital of H.

(d)   Because F is more electronegative than H, the valence orbitals on F are at lower energy than those on H.

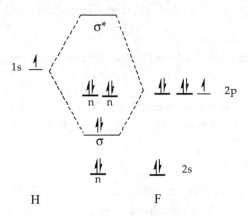

The HF MO diagram has 6 nonbonding, 2 bonding, and 0 antibonding electrons. The BO = [2 – 0]/2 = 1. (Nonbonding electrons do not "count" toward bond order.)

(e)   H—$\ddot{\underset{..}{F}}$:

In the Lewis structure for HF, the nonbonding electrons are on the (more electronegative) F atom, as they are in the MO diagram.

9.111   (a)   CO, 10 e$^-$, 5 e$^-$ pair

:C≡O:

(b)   The bond order for CO, as predicted by the MO diagram in Figure 9.46, is 1/2[8 – 2] = 3.0. A bond order of 3.0 agrees with the triple bond in the Lewis structure.

(c)   Applying the MO diagram in Figure 9.46 to the CO molecule, the highest-energy electrons would occupy the π_{2p} MOs. That is, π_{2p} would be the HOMO, highest occupied molecular orbital. If the true HOMO of CO is a σ-type MO, the order of the π_{2p} and σ_{2p} orbitals must be reversed. Figure 9.42 shows how the interaction of the 2s orbitals on one atom and the 2p orbitals on the other atom can affect the relative energies of the resulting MOs. This 2s–2p interaction in CO is significant enough so that the σ_{2p} MO is higher in energy than the π_{2p} MOs, and the σ_{2p} is the HOMO. This is the same energy order of MOs shown for large 2s–2p interaction in homonuclear diatomic molecules in Figure 9.43.

(d)   We expect the atomic orbitals of the more electronegative element to have lower energy than those of the less electronegative element. When atoms of the two elements combine, the lower-energy atomic orbitals make a greater contribution to the bonding MOs and the higher-energy atomic orbitals make a larger contribution to the antibonding orbitals. Thus, the π_{2p} bonding MOs will have a greater contribution from the more electronegative O atom.

9.112   (a) π_{2p}. The 2p valence atomic orbitals on C form the π molecular orbitals in ethylene. The symbol for the bonding molecular orbital is π_{2p}.

(b) π^*_{2p}. By analogy to diatomic molecular orbital diagrams, the MO closest in energy to π_{2p} is π^*_{2p}. This is the LUMO.

(c)   Weaker. In the excited state, there is one fewer electron in a bonding MO and one more electron in an antibonding MO. This reduces the bond order of the C–C bond and weakens the bond.

(d)   The C–C bond in ethylene is easier to twist when the molecule is in the excited state. In the excited state the C–C bond order is decreased (see above) and the π bond is essentially "broken." There is free rotation around a C–C single bond.

Integrative Exercises

9.113   (a)   Assume 100 g of compound

$$2.1 \text{ g H} \times \frac{1 \text{ mol H}}{1.008 \text{ g H}} = 2.1 \text{ mol H}; \, 2.1/2.1 = 1$$

$$29.8 \text{ g N} \times \frac{1 \text{ mol N}}{14.01 \text{ g N}} = 2.13 \text{ mol N}; \, 2.13/2.1 \approx 1$$

$$68.1 \text{ g O} \times \frac{1 \text{ mol O}}{16.00 \text{ g O}} = 4.26 \text{ mol O}; \, 4.26/2.1 \approx 2$$

The empirical formula is HNO_2; formula weight = 47. Because the approximate molar mass is 50, the molecular formula is HNO_2.

(b)   Assume N is central, because it is unusual for O to be central, and part (d) indicates as much. HNO_2: 18 valence e^-

$$\ddot{O}{=}\ddot{N}{-}\ddot{O}{-}H \longleftrightarrow \, {:}\ddot{O}{-}\ddot{N}{=}\ddot{O}{-}H$$
$$ -1 \quad 0 \quad +1$$

The second resonance form is a minor contributor because of unfavorable formal charges.

(c)   The electron-domain geometry around N is trigonal planar with an O–N–O angle of approximately 120°. If the resonance structure on the right makes a significant contribution to the molecular structure, all four atoms would lie in a plane. If only the left structure contributes, the H could rotate in and out of the molecular plane. The relative contributions of the two resonance structures could be determined by measuring the O–N–O and N–O–H bond angles.

(d)   3 VSEPR e^- domains around N, sp^2 hybridization

(e)   3 σ, 1 π for both structures (or for H bound to N).

9.114   (a) $2SF_4(g) + O_2(g) \rightarrow 2OSF_4(g)$

(b)   40 e^-, 20 e^- pairs

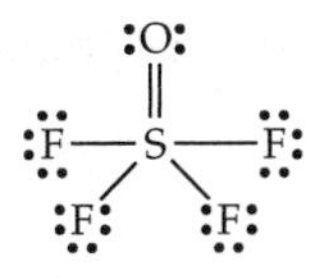

$$\begin{array}{c} {:}\ddot{O}{:} \\ \| \\ {:}\ddot{F}{-}\overset{\displaystyle }{S}{-}\ddot{F}{:} \\ {:}\ddot{F}{:} \quad {:}\ddot{F}{:} \end{array}$$

There must be a double bond drawn between O and S in order for their formal charges to be zero.

(c) $\Delta H = 8D(S-F) + D(O=O) - 8D(S-F) - 2D(S=O)$

$\Delta H = D(O=O) - 2D(S=O) = 495 - 2(523) = -551$ kJ, exothermic

(d)   trigonal-bipyramidal electron-domain geometry

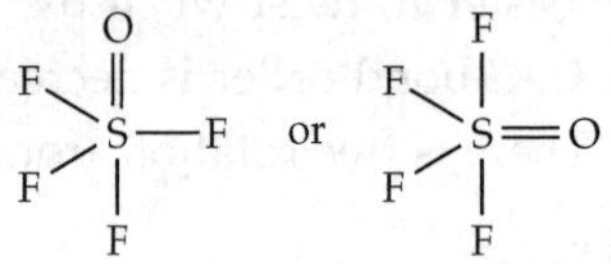

(e)   In the structure on the left, there are 3 equatorial and 1 axial fluorine atoms. In the structure on the right, there are 2 equatorial and 2 axial fluorine atoms.

9.115   (a) PX_3, 26 valence e^-, 13 $^-$ pairs

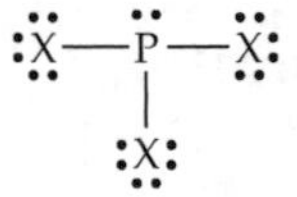

4 electron domains around P, tetrahedral e^- domain geometry,
If all bonding and nonbonding electron domains are the same size, perfect tetrahedral angles are 109.5°. If all bonding electron domains are the same size but the nonbonding domain is larger, bond angles are somewhat less than 109.5°.

(b)   As electronegativity increases (I < Br < Cl < F), the X–P–X angles decreases.

(c)   The greater the electronegativity of X, the larger the magnitude of negative charge centered on X. The more negative charge centered on X, the smaller the P–X bonding domains, the greater the effect of the nonbonding domain and the smaller the bond angle. Also, as the electronegativity of X decreases and the bonding domain size increases, the effect of the large nonbonding domain decreases.

(d) $PBrCl_4$, 40 valence electrons, 20 e^- pairs. The molecule will have trigonal-bipyramidal electron-domain geometry (similar to PCl_5 in Table 9.3). Based on the argument in part (c), the P–Br bond will have greater repulsions with P–Cl bonds than P–Cl bonds have with each other. Therefore, the Br will occupy an equatorial position in the trigonal bipyramid, so that the more unfavorable P–Br to P–Cl repulsions can be situated at larger angles in the equatorial plane.

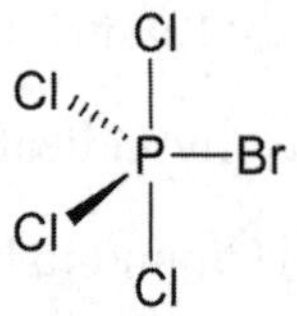

9.116   (a)   Three electron domains around each central C atom, sp^2 hybridization

(b)   A 180° rotation around the C=C double bond is required to convert the trans isomer into the cis isomer. A 90° rotation around the bond eliminates all overlap of the p orbitals that form the π bond and it is broken.

(c)   **average bond enthalpy**

    C=C   614 kJ/mol

    C–C   348 kJ/mol

The difference in these values, 266 kJ/mol, is the average bond enthalpy of a C–C π bond. This is the amount of energy required to break 1 mol of C–C π bonds. The energy per molecule is

$$266 \text{ kJ/mol} \times \frac{1000 \text{ J}}{1 \text{ kJ}} \times \frac{1 \text{ mol}}{6.022 \times 10^{23} \text{ molecules}} = 4.417 \times 10^{-19}$$

$$= 4.42 \times 10^{-19} \text{ J/molecule}$$

(d) $\lambda = hc/E = \dfrac{6.626 \times 10^{-34} \text{ J-s} \times 2.998 \times 10^{8} \text{ m/s}}{4.417 \times 10^{-19} \text{ J}} = 4.50 \times 10^{-7} \text{ m} = 450 \text{ nm}$

(e)   Yes, 450 nm light is in the visible portion of the spectrum. A cis-trans isomerization in the retinal portion of the large molecule rhodopsin is the first step in a sequence of molecular transformations in the eye that leads to vision. The sequence of events enables the eye to detect visible photons, in other words, to see.

9.117   (a)   C $\equiv$ C   839 kJ/mol   (1 σ, 2 π)

             C=C   614 kJ/mol   (1 σ, 1 π)

             C–C   348 kJ/mol   (1 σ)

The contribution from 1 π-bond would be (614–348) 266 kJ/mol. From a second π bond, (839 – 614), 225 kJ/mol. An average π bond contribution would be (266 + 225)/2 = 246 kJ/mol.

This is $\dfrac{246 \text{ kJ/}\pi \text{ bond}}{348 \text{ kJ/}\sigma \text{ bond}} \times 100 = 71\%$ of the average enthalpy of a σ bond.

(b)   N≡N   941 kJ/mol

       N=N   418 kJ/mol

       N–N   163 kJ/mol

first π = (418 – 163) = 255 kJ/mol

second π = (941 – 418) = 523 kJ/mol

average π-bond enthalpy = (255 + 523)/2 = 389 kJ/mol

This is $\dfrac{389 \text{ kJ/}\pi \text{ bond}}{163 \text{ kJ/}\sigma \text{ bond}} \times 100 = 240\%$ of the average enthalpy of a σ bond.

N–N σ bonds are weaker than C–C σ bonds, whereas N–N π bonds are stronger than C–C π bonds. The relative energies of C–C σ and π bonds are similar, whereas N–N π bonds are much stronger than N–N σ bonds.

(c) N_2H_4, 14 valence e^-, 7 e^- pairs

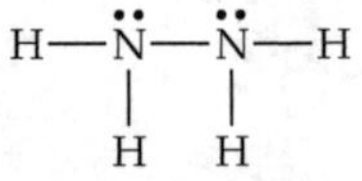

$$H-\overset{..}{N}-\overset{..}{N}-H$$
$$\quad\;\; | \quad\;\; |$$
$$\quad\;\; H \quad\; H$$

4 electron domains around N, sp^3 hybridization

N_2H_2, 12 valence e^-, 6 e^- pairs

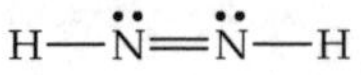

$$H-\overset{..}{N}=\overset{..}{N}-H$$

3 electron domains around N, sp^2 hybridization

N_2, 10 valence e^-, 5 e^- pairs

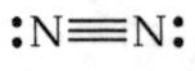

:N≡N:

2 electron domains around N, sp hybridization

(d)   In the three types of N–N bonds, each N atom has a nonbonding or lone pair of electrons. The lone pair to bond pair repulsions are minimized going from less than 109.5° to 120° to 180° bond angles, making the π bonds stronger relative to the σ bond. In the three types of C–C bonds, no lone-pair to bond-pair repulsions exist, and the σ and π bonds have more similar energies.

9.118

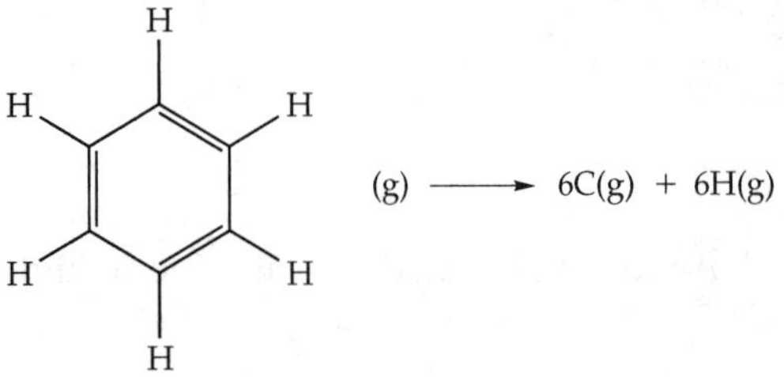

(g) ⟶ 6C(g) + 6H(g)

$\Delta H = 6D(C–H) + 3D(C–C) + 3D(C=C) - 0$

$\quad = 6(413 \text{ kJ}) + 3(348 \text{ kJ}) + 3(614 \text{ kJ})$

$\quad = 5364 \text{ kJ}$

(The products are isolated atoms; there is no bond making.)

According to Hess' law:

$\Delta H° = 6\Delta H_f^° \text{ C(g)} + 6\Delta H_f^° \text{ H(g)} - \Delta H_f^° \text{ C}_6\text{H}_6\text{(g)}$

$\quad = 6(718.4 \text{ kJ}) + 6(217.94 \text{ kJ}) - (82.9 \text{ kJ})$

$\quad = 5535 \text{ kJ}$

The difference in the two results, 171 kJ/mol C_6H_6 is because of the resonance stabilization in benzene. That is, because the π electrons are delocalized, the molecule has a lower overall energy than that predicted for the presence of 3 localized C–C and C=C bonds. Thus, the amount of energy actually required to decompose 1 mole of C_6H_6(g), represented by the Hess' law calculation, is greater than the sum of the localized bond enthalpies (not taking resonance into account) from the first calculation above.

9.119   (a) $3d_{z^2}$

       (b)   Ignoring the donut of the d_{z^2} orbital

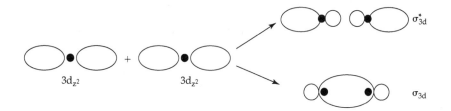

       (c)   A node is generated in σ_{3d}^* because antibonding MOs are formed when AO lobes with opposite phases interact. Electron density is excluded from the internuclear region and a node is formed in the MO.

       (d)   Sc: $[Ar]4s^2 3d^1$ Omitting the core electrons, there are six e^- in the energy-level diagram.

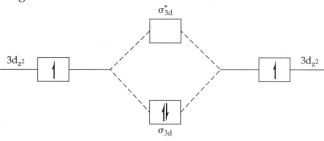

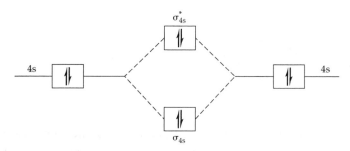

       (e)   The bond order in Sc_2 is $1/2(4-2) = 1.0$.

9.120   (a)   The molecular and empirical formulas of the four molecules are:

            benzene: molecular, C_6H_6; empirical, CH

            napthalene: molecular, $C_{10}H_8$; empirical, C_5H_4

            anthracene: molecular, $C_{14}H_{10}$; empirical, C_7H_5

            tetracene: molecular, $C_{18}H_{12}$, empirical, C_3H_2

       (b)   Yes. Because the compounds all have different empirical formulas, combustion analysis could in principle be used to distinguish them. In practice, the mass % of C in the four compounds is not very different, so the data would have to be precise to at least 3 decimal places and 4 would be better.

       (c) $C_{10}H_8(s) + 12O_2(g) \rightarrow 10CO_2(g) + 4H_2O(g)$

(d) $\Delta H_{comb} = 5D(C=C) + 6D(C-C) + 8D(C-H) + 12D(O=O) - 20D(C=O) - 8D(O-H)$

         $= 5(614) + 6(348) + 8(413) + 12(495) - 20(799) - 8(463)$

         $= -5282 \text{ kJ/mol C}_{10}\text{H}_8$

(e)   Yes. For example, the resonance structures of naphthalene are:

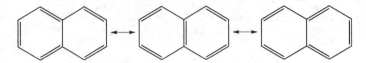

(f)   Colored compounds absorb visible light and appear the color of the visible light that they reflect. Colorless compounds typically absorb shorter wavelength, higher-energy light. The energy of light absorbed corresponds to the energy gap between the HOMO and LUMO of the molecule. That tetracene absorbs longer wavelength, lower-energy visible light indicates that it has the smallest HOMO-LUMO energy gap of the four molecules. Tetracene also has the most conjugated double bonds of the four molecules. We might conclude that the more conjugated double bonds in an organic molecule, the smaller the HOMO-LUMO energy gap. More information about the absorption spectra of anthracene, naphthalene, and benzene is needed to confirm this conclusion.

9.121   (a,b)

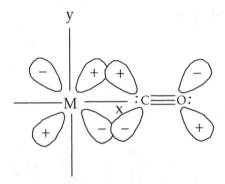

(c)   The two lobes of a p AO have opposite phases. These are shown on the diagram as + and −. An antibonding MO is formed when p AOs with opposite phases interact.

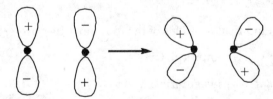

(d)   Note that the d_{xy} AO has lobes that lie between, not on, the x and y axes.

(e)   A π bond forms by overlap of orbitals on M and C. There is electron density above and below, but not along, the M–C axis.

(f)   According to Exercise 9.111, the HOMO of CO is a σ-type MO. So the appropriate MO diagram is shown on the left side of Figure 9.43. A lone CO molecule has 10 valence electrons, the HOMO is σ_{2p} and the bond order is 3.0. The LUMO is π_{2p}^*.

When M and CO interact as shown in the π_{2p}^* diagram, d-π back bonding causes the π_{2p}^* to become partially occupied. Electron density in the π_{2p}^* decreases electron density in the bonding molecular orbitals and decreases the BO of the bound CO. The strength of the C–O bond in a metal–CO complex decreases relative to the strength of the C–O bond in an isolated CO molecule.

9.122   (a)   22 valence e^-, 11 e^- pairs

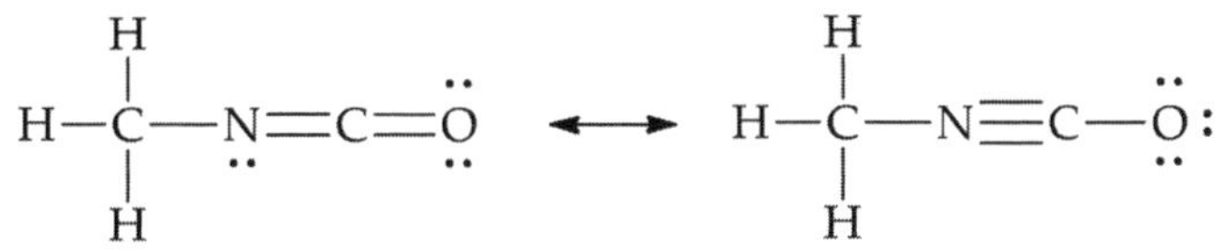

(The structure on the right does not minimize formal charges and will make a minor contribution to the true structure.)

(b)   Both resonance structures predict the same bond angles. We expect H–C–N angles close to 109.5°.

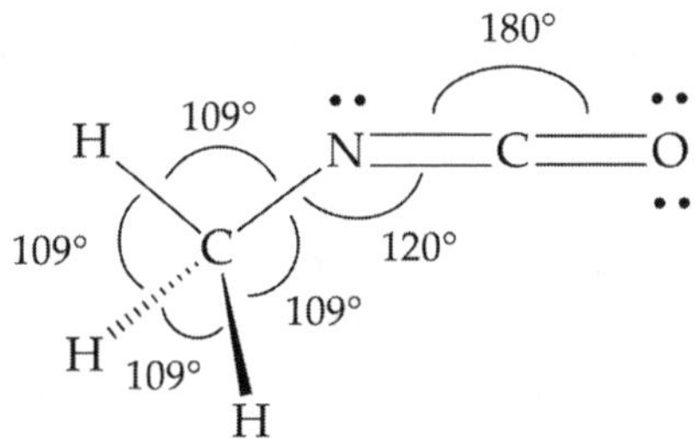

(c)   The two extreme Lewis structures predict different bond lengths. As the true bonding model is some blend of the extreme Lewis structures, the true bond lengths are a blend of the extreme values. Our bond length estimates take into account that the structure minimizing formal charge makes a larger contribution to the true structure.

From Figure 7.7, the sum of the covalent radii for C and H is 1.07Å. We list this value in the table. As corroboration, the value of the O–H bond length given in Exercise 9.95 is 0.96 Å. According to Figure 7.7, the covalent radius of C is 0.10 Å greater than that of O, so we expect the C–H bond length to be approximately 0.10 Å greater than the measured O–H distance, about 1.06 Å. The lengths for isolated C–N, N–C, and C–O bonds are taken from Table 8.4.

Bond	Length (Å) N=C=O	Length (Å) N≡C–O	Length (Å) estimated
C–H	1.07	1.07	1.07
C–N	1.43	1.43	1.43
N–C	1.38	1.16	1.33
C–O	1.23	1.43	1.28

(d)   The molecule will have a dipole moment. The N–C and C–O bond dipoles are opposite each other, but they are not equal. And, there are nonbonding electron pairs that are not directly opposite each other (in either structure) and will not cancel. There will be a resulting dipole.

10 Gases

Visualizing Concepts

10.1   It would be much easier to drink from a straw on Mars. When a straw is placed in a glass of liquid, the liquid level in the straw equals the liquid level in the glass. The atmosphere exerts equal pressure inside and outside the straw. When we drink through a straw, we withdraw air, thereby reducing the pressure on the liquid inside. If only 0.007 atm is exerted on the liquid in the glass, a very small reduction in pressure inside the straw will cause the liquid to rise.

Another approach is to consider the gravitational force on Mars. Because the pull of gravity causes atmospheric pressure, the gravity on Mars must be much smaller than that on Earth. With a very small Martian gravity holding liquid in a glass, it would be very easy to raise the liquid through a straw.

10.2   (a) $V_1/T_1 = V_2/T_2$ (Charles' law)

$V_1/300 \text{ K} = V_2/500 \text{ K}$

$V_2 = 5/3 \, V_1$

(b) $P_1V_1 = P_2V_2$ (Boyle's law)

$1 \text{ atm} \times V_1 = 2 \text{ atm} \times V_2$

$V_2 = 1/2 \, V_1$

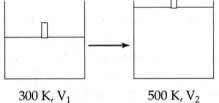 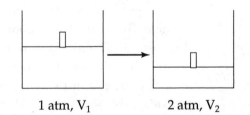

300 K, V_1 500 K, V_2	1 atm, V_1 2 atm, V_2

(c) $V_1/T_1 = V_2/T_2$ (Charles' law)

$V_1/300 \text{ K} = V_2/200 \text{ K}$

$V_2 = 2/3 \, V_1$

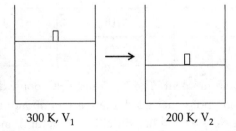

300 K, V_1 200 K, V_2

10.3   Statement (b) is correct. At constant volume and temperature, pressure depends on total number of particles. To reduce the pressure by a factor of 2, the number of particles must be reduced by a factor of 2. At the lower pressure, the container would have half as many particles as at the higher pressure.

Compare the two situations using the ideal-gas law, $PV = nRT$. $P = nRT/V$.

$P_2 = P_1/2$. V, R, and T are the same for the two states.

$n_2RT/V = (n_1RT/V)/2$; $n_2 = n_1/2$

10.4   Statement (d) describes the volume change. At constant pressure and temperature, the container volume is directly proportional to the number of particles present (Avogadro's law). As the reaction proceeds, 3 gas molecules are converted to 2 gas molecules, so the number of particles and the container volume decrease by 33%.

10.5 $PV = nRT$ (ideal-gas equation). In the ideal-gas equation, R is a constant. Given constant V and n (fixed amount of ideal gas), P and T are directly proportional. If P is doubled, T is also doubled. That is, if P is doubled, T increases by a factor of 2.

10.6   Over time, the gases will mix perfectly. Each bulb will contain 4 blue and 3 red atoms. The "blue" gas has the greater partial pressure after mixing, because it has the greater number of particles (at the same T and V as the "red" gas.)

10.7   (a)   Partial pressure depends on the number of particles of each gas present. Red has the fewest particles, then yellow, then blue. $P_{red} < P_{yellow} < P_{blue}$

(b) $P_{gas} = X_{gas} P_t$. Calculate the mole fraction, X_{gas} = [mol gas / total moles] or [particles gas / total particles]. This is true because Avogadro's number is a counting number, and mole ratios are also particle ratios.

X_{red} = 2 red atoms / 10 total atoms = 0.2; P_{red} = 0.2(1.40 atm) = 0.28 atm

X_{yellow} = 3 yellow atoms / 10 total atoms = 0.3; P_{yellow} = 0.3(1.40 atm) = 0.42 atm

X_{blue} = 5 blue atoms / 10 total atoms = 0.5; P_{blue} = 0.5(1.40 atm) = 0.70 atm

Check. (0.28 atm + 0.42 atm + 0.70 atm) = 1.40 atm. The sum of the calculated partial pressures equals the given total pressure.

10.8

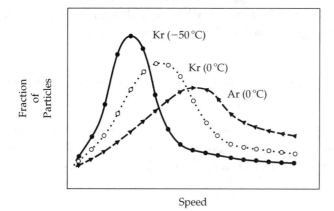

10.9   (a)   Curve B. At constant temperature, the root-mean-square (rms) speed (as well as the average speed) of a collection of gas particles is inversely related to molar mass; the lighter the particle, the faster it moves. Therefore, curve B represents He and curve A represents O_2. Curve B has the higher rms speed and He is the lighter gas. Curve A has the lower rms speed and O_2 is the heavier gas.

         (b)   B. For the same gas, average kinetic energy $(1/2\ mu_{rms}^2)$, and therefore root-mean-square speed (u_{rms}) is directly related to Kelvin temperature. Curve A is the lower temperature and curve B is the higher temperature.

         (c)   The root-mean-square speed. According to Figure 10.12(b), the most probable speed is lowest, then the average speed, and the rms speed is highest. This is true for the distribution of speeds for any gas, including the two in this exercise.

10.10   (a)   Total pressure is directly related to total number of particles (or total mol particles). P(ii) < P(i) = P(iii)

         (b)   Partial pressure of He is directly related to number of He atoms (yellow) or mol He atoms. P_{He}(iii) < P_{He}(ii) < P_{He}(i)

         (c)   Density is total mass of gas per unit volume. We can use the atomic or molar masses of He (4) and N_2 (28), as relative masses of the particles.

$$\text{mass(i)} = 5(4) + 2(28) = 76$$

$$\text{mass(ii)} = 3(4) + 1(28) = 40$$

$$\text{mass(iii)} = 2(4) + 5(28) = 148$$

         Because the container volumes are equal, d(ii) < d(i) < d(iii).

         (d)   At the same temperature, all gases have the same average kinetic energy. The average kinetic energies of the particles in the three containers are equal.

10.11   The NH_4Cl(s) ring will form at location a. The process described in this exercise is diffusion, rather than simple effusion. According to Section 10.8, Graham's law approximates (but does not exactly describe) the diffusion rates of two gases under identical conditions. According to Graham's law, the ratio of rates is inversely related to the ratio of molar masses of the two gases. That is, the lighter gas moves faster than the heavier gas. When introduced into the tube, NH_3, MM = 17, moves faster and therefore farther than HCl, MM = 36. If NH_3 moves farther than HCl, the two gases meet and form NH_4Cl(s) nearer the end where HCl was introduced; this is in the vicinity of location a.

10.12   (a)   The van der Waals constant *a* accounts for the influence of intermolecular forces in lowering the pressure of a real gas.

         (b)   According to the plot, Gas B is closest to ideal behavior, then Gas A and Gas C. Gas B will have the smallest *a* value, then A and C. From the values in Table 10.3, Gas B is N_2, a = 1.39 ; Gas A is CO_2, a = 3.59; Gas C = Cl_2, a = 6.49.

Gas Characteristics; Pressure (Sections 10.1 and 10.2)

10.13   Statement (c) is false. Gaseous molecules are so far apart that there is no barrier to mixing, regardless of the identity of the molecules.

10.14   (a)   g/L. Gas samples are mostly empty space; their volume is large relative to the mass of the molecules present. The unit g/L has a relatively small mass and large volume.

        (b)   Pa and atm. Pressure is defined as force per unit area. The unit N is only a unit of force. The unit kg/m^2 is a unit of mass per unit area, not force per unit area.

        (c) F_2 is most likely to be a gas at room temperature. K_2O is an ionic compound; all simple ionic compounds are solids. Br_2 is a covalent compound with high molar mass. The greater the molar mass, the stronger the dispersion forces and the less likely the compound is to be a gas. (In the series of diatomic halogens, F_2 is a gas, Br_2 is a liquid, and I_2 is a solid at room temperature and ordinary atmospheric pressure.)

10.15   *Analyze.* Given: mass, area. Find: pressure. *Plan.* P = F/A. Data is given in terms of lb and $in.^2$; calculate pressure in $lb/in.^2$ (or psi) for part (c), then use this value to find pressure in and atm for part (b) and kPa for part (a).   *Solve.*

        (a) $P = \dfrac{F}{A} = \dfrac{130\ lb}{0.50\ in.^2} = 260 = 2.6 \times 10^2\ lb/in.^2$

        (b)   1 atm = 101.325 kPa

$$17.687\ atm \times \frac{101.325\ kPa}{1\ atm} = 1792 = 1.8 \times 10^3\ kPa$$

        (c) $14.70\ lb/in.^2 = 1\ atm$

$$260\ lb/in.^2 \times \frac{1\ atm}{14.70\ lb/in.^2} = 17.687 = 18\ atm$$

10.16 $P = m \times a/A$; $1\ Pa = 1\ kg/m\text{-}s^2$; $A = 3.0\ cm \times 4.1\ cm \times 4 = 49.2 = 49\ cm^2$

$$\frac{262\ kg}{49.2\ cm^2} \times \frac{9.81\ m}{s^2} \times \frac{(100)^2\ cm^2}{1\ m^2} = 5.224 \times 10^5\ \frac{kg}{m\text{-}s^2} = 5.2 \times 10^5\ Pa$$

10.17   *Analyze.* Given: 760 mm column of Hg, densities of Hg and glycerol. Find: height of a column of glycerol at the same pressure.

        Plan. Develop a relationship between pressure, height of a column of liquid, and density of the liquid. Relationships that might prove useful: P = F/A; F = m × a; m = d × V(density)(volume); V = A × height   *Solve.*

$$P = \frac{F}{A} = \frac{m \times a}{A} = \frac{d \times V \times a}{A} = \frac{d \times A \times h \times a}{A} = d \times h \times a$$

(a) $P_{Hg} = P_{glycerol}$; Using the relationship derived above: $(d \times h \times a)_{glycerol} = (d \times h \times a)_{Hg}$

Because a, the acceleration due to gravity, is equal in both liquids,
$(d \times h)_{glycerol} = (d \times h)_{Hg}$

$1.26 \text{ g/mL} \times h_{glycerol} = 13.6 \text{ g/mL} \times 760 \text{ mm}$

$h_{glycerol} = \dfrac{13.6 \text{ g/mL} \times 760 \text{ mm}}{1.26 \text{ g/mL}} = 8.203 \times 10^3 = 8.20 \times 10^3 \text{ mm} = 8.20 \text{ m}$

(b) *Plan.* Calculate the height of a column of Hg that exerts the same pressure as 15 ft of water. The density of water is 1.00 g/mL. Then calculate the total pressure, atmospheric plus water, in atmospheres.

$15 \text{ ft H}_2\text{O} \times \dfrac{12 \text{ in.}}{1 \text{ ft}} \times \dfrac{2.54 \text{ cm}}{1 \text{ in.}} \times \dfrac{10 \text{ mm}}{1 \text{ cm}} = 4572 = 4.6 \times 10^3 \text{ mm H}_2\text{O}$

$(d \times h)_{H_2O} = (d \times h)_{Hg}$

$1.00 \text{ g/mL} \times 4572 \text{ mm H}_2\text{O} = 13.6 \text{ g/mL} \times ? \text{ mm Hg}$

$h_{Hg} = \dfrac{1.00 \text{ g/mL} \times 4572 \text{ mm}}{13.6 \text{ g/mL}} = 336.2 = 3.4 \times 10^2 \text{ mm Hg}$

$P_{total} = P_{atm} + P_{H_2O} = 750 \text{ mm Hg} + 336 \text{ mm Hg} = 1086 \text{ mm Hg} = 1.1 \times 10^2 \text{ mm Hg}$

$P_{total} = 1086 \text{ mm Hg} \times \dfrac{1 \text{ atm}}{760 \text{ mm Hg}} = 1.429 = 1.4 \text{ atm}$

10.18 (a) Abbreviate 1-iododecane as 1id. Using the relationship derived in Solution 10.17 for two liquids under the influence of gravity, $(d \times h)_{1id} = (d \times h)_{Hg}$. At 749 torr, the height of an Hg barometer is 749 mm.

$\dfrac{1.20 \text{ g}}{1 \text{ mL}} \times h_{1id} = \dfrac{13.6 \text{ g}}{1 \text{ mL}} \times 749 \text{ mm}; \quad h_{1id} = \dfrac{13.6 \text{ g/mL} \times 749 \text{ mm}}{1.20 \text{ g/mL}}$

$= 8.49 \times 10^3 \text{ mm} = 8.49 \text{ m}$

(b) $21 \text{ ft H}_2\text{O} \times \dfrac{12 \text{ in.}}{1 \text{ ft}} \times \dfrac{2.54 \text{ cm}}{1 \text{ in.}} \times \dfrac{10 \text{ mm}}{1 \text{ cm}} = 6401 = 6.4 \times 10^3 \text{ mm H}_2\text{O}$

$(d \times h)_{H_2O} = (d \times h)_{Hg}$

$1.00 \text{ g/mL} \times 6401 \text{ mm H}_2\text{O} = 13.6 \text{ g/mL} \times ? \text{ mm Hg}$

$h_{Hg} = \dfrac{1.00 \text{ g/mL} \times 6401 \text{ mm}}{13.6 \text{ g/mL}} = 470.7 = 4.7 \times 10^2 \text{ mm Hg}$

$P_{total} = P_{atm} + P_{H_2O} = 742 \text{ mm Hg} + 471 \text{ mm Hg} = 1213 \text{ mm Hg} = 1.2 \times 10^3 \text{ mm Hg}$

$P_{total} = 1213 \text{ mm Hg} \times \dfrac{1 \text{ atm}}{760 \text{ mm Hg}} = 1.596 = 1.6 \text{ atm}$

10.19 *Analyze/Plan.* We are given pressure in one unit and asked to change it to another unit. Select appropriate conversion factors and use dimensional analysis to find the desired quantities. *Solve.*

(a) $265 \text{ torr} \times \dfrac{1 \text{ atm}}{760 \text{ torr}} = 0.349 \text{ atm}$

(b) $265 \text{ torr} \times \dfrac{1 \text{ mm Hg}}{1 \text{ torr}} = 265 \text{ mm Hg}$

(c) $265 \text{ torr} \times \dfrac{1.01325 \times 10^5 \text{ Pa}}{760 \text{ torr}} = 3.53 \times 10^4 \text{ Pa}$

(d) $265 \text{ torr} \times \dfrac{1.01325 \times 10^5 \text{ Pa}}{760 \text{ torr}} \times \dfrac{1 \text{ bar}}{1 \times 10^5 \text{ Pa}} = 0.353 \text{ bar}$

(e) $265 \text{ torr} \times \dfrac{1 \text{ atm}}{760 \text{ torr}} \times \dfrac{14.70 \text{ psi}}{1 \text{ atm}} = 5.13 \text{ psi}$

10.20   (a) $0.912 \text{ atm} \times \dfrac{760 \text{ torr}}{1 \text{ atm}} = 693 \text{ torr}$

      (b) $0.685 \text{ bar} \times \dfrac{1 \times 10^5 \text{ Pa}}{1 \text{ bar}} \times \dfrac{1 \text{ kPa}}{1 \times 10^3 \text{ Pa}} = 68.5 \text{ kPa}$

      (c) $655 \text{ mm Hg} \times \dfrac{1 \text{ atm}}{760 \text{ mm Hg}} = 0.862 \text{ atm}$

      (d) $1.323 \times 10^5 \text{ Pa} \times \dfrac{1 \text{ atm}}{1.01325 \times 10^5 \text{ Pa}} = 1.3057 = 1.306 \text{ atm}$

      (e) $2.50 \text{ atm} \times \dfrac{14.70 \text{ psi}}{1 \text{ atm}} = 36.75 = 36.8 \text{ psi}$

10.21   *Analyze/Plan.* Follow the logic in Solution 10.19.   *Solve.*

      (a) $30.45 \text{ in. Hg} \times \dfrac{25.4 \text{ mm}}{1 \text{ in.}} \times \dfrac{1 \text{ torr}}{1 \text{ mm Hg}} = 773.4 \text{ torr}$

      [The result has 4 sig figs because 25.4 mm/in. is considered to be an exact number.]

      (b) $30.45 \text{ in Hg} = 773.4 \text{ torr}; \quad 773.4 \text{ torr} \times \dfrac{1 \text{ atm}}{760 \text{ torr}} = 1.018 \text{ atm}$

10.22 $882 \text{ mbar} = 0.882 \text{ bar}$

      (a) $0.882 \text{ bar} \times \dfrac{1 \times 10^5 \text{ Pa}}{1 \text{ bar}} \times \dfrac{1 \text{ atm}}{101,325 \text{ Pa}} = 0.8705 = 0.871 \text{ atm}$

      (b) $0.882 \text{ bar} = 0.8705 \text{ atm} \times \dfrac{760 \text{ torr}}{1 \text{ atm}} = 661.55 = 662 \text{ torr}$

      (c) $0.882 \text{ bar} = 661.55 \text{ torr} \times \dfrac{1 \text{ mm Hg}}{1 \text{ torr}} \times \dfrac{1 \text{ cm Hg}}{10 \text{ mm Hg}} \times \dfrac{1 \text{ in. Hg}}{2.54 \text{ cm Hg}} = 26.045 = 26.0 \text{ in. Hg}$

10.23   *Analyze/Plan.* Follow the logic in Sample Exercise 10.2.   *Solve.*

      (i)   The Hg level is lower in the open end than the closed end, so the gas pressure is less than atmospheric pressure.

$$P_{gas} = 0.995 \text{ atm} - \left(52 \text{ cm} \times \dfrac{1 \text{ atm}}{76.0 \text{ cm}} \right) = 0.31 \text{ atm}$$

      (ii)   The Hg level is higher in the open end, so the gas pressure is greater than atmospheric pressure.

$$P_{gas} = 0.995 \text{ atm} + \left(67 \text{ cm Hg} \times \dfrac{1 \text{ atm}}{76.0 \text{ cm Hg}} \right) = 1.8766 = 1.88 \text{ atm}$$

(iii)   This is a closed-end manometer, so $P_{gas} = h$.

$$P_{gas} = 10.3 \text{ cm} \times \frac{1 \text{ atm}}{76.0 \text{ cm}} = 0.136 \text{ atm}$$

10.24   (a)   The atmosphere is exerting 15.4 cm = 154 mm Hg (torr) more pressure than the gas.

$$P_{gas} = P_{atm} - 15.4 \text{ torr} = \left(0.985 \text{ atm} \times \frac{760 \text{ torr}}{1 \text{ atm}} \right) - 15.4 \text{ torr} = 733 \text{ torr}$$

(b)   The gas is exerting 12.3 mm Hg (torr) more pressure than the atmosphere.

$$P_{gas} = P_{atm} + 12.3 \text{ torr} = \left(0.99 \text{ atm} \times \frac{760 \text{ torr}}{1 \text{ atm}} \right) + 12.3 \text{ torr} = 764.7 \text{ torr} = 7.6 \times 10^2 \text{ torr}$$

(Atmospheric pressure of 0.99 atm determines that the result has 2 sig figs.)

The Gas Laws (Section 10.3)

10.25   *Analyze/Plan.* Given certain changes in volume and temperature of a gas contained in a cylinder with a moveable piston, predict which will double the gas pressure. Consider the gas law relationships in Section 10.3.   *Solve.*

(a)   No. P and V are inversely proportional at constant T. If the volume increases by a factor of 2, the pressure *decreases* by a factor of 2.

(b)   No. P and *Kelvin* T are directly proportional at constant V. Doubling °C does increase V, but it does not double it.

(c)   Yes. P and V are inversely proportional at constant T. If the volume decreases by a factor of 2, the pressure increases by a factor of 2.

10.26   *Analyze.* Given: initial P, V, T. Find: final values of P, V, T for certain changes of condition. *Plan.* Select the appropriate gas law relationships from Section 10.3; solve for final conditions, paying attention to units.   *Solve.*

(a) $P_1V_1 = P_2V_2$; the proportionality holds true for any pressure or volume units.

$P_1 = 752 \text{ torr}, V_1 = 5.12 \text{ L}, P_2 = 1.88 \text{ atm}$

$$V_2 = \frac{P_1V_1}{P_2} = \frac{752 \text{ torr} \times 5.12 \text{ L}}{1.88 \text{ atm}} \times \frac{1 \text{ atm}}{760 \text{ torr}} = 2.69 \text{ L}$$

Check. As pressure increases, volume should decrease; our result agrees with this.

(b) $V_1/T_1 = V_2/T_2$; T must be in Kelvins for the relationship to be true.

$V_1 = 5.12 \text{ L}, T_1 = 21 \text{ °C} = 294 \text{ K}, T_2 = 175 \text{ °C} = 448 \text{ K}$

$$V_2 = \frac{V_1T_2}{T_1} = \frac{5.12 \text{ L} \times 448 \text{ K}}{294 \text{ K}} = 7.80 \text{ L}$$

Check. As temperature increases, volume should increase; our result is consistent with this.

10.27   (a)   *Analyze/Plan.* Use Boyle's law, PV = constant or $P_1V_1 = P_2V_2$, and Charles' law, V/T = constant or $V_1/T_1 = V_2/T_2$, to derive Amonton's law for the relationship between P and T at constant V.   *Solve.*

Boyle's law: $P_1V_1 = P_2V_2$ or $\dfrac{P_1}{P_2} = \dfrac{V_2}{V_1}$. At constant V, $\dfrac{V_2}{V_1} = 1$ and $\dfrac{P_1}{P_2} = 1$.

Charles' law: $\dfrac{V_1}{T_1} = \dfrac{V_2}{T_2}$ or $\dfrac{V_1}{V_2} = \dfrac{T_1}{T_2}$. At constant V, $\dfrac{V_1}{V_2} = 1$ and $\dfrac{T_1}{T_2} = 1$.

$\dfrac{P_1}{P_2} = 1$ and $\dfrac{T_1}{T_2} = 1$, then $\dfrac{P_1}{P_2} = \dfrac{T_1}{T_2}$ or $\dfrac{P_1}{T_1} = \dfrac{P_2}{T_2}$ or P/T = constant.

Amonton's law is that pressure and temperature are directly proportional at constant volume.

(b)   *Analyze/Plan.* Recall that the proportional relationships for gases are true for any units of pressure (or volume) but only for Kelvin temperatures. Change °F to K, then use the relationship derived in (a) to calculate the new tire pressure.
°C = 5/9(°F – 32); K = °C + 273.   *Solve.*

T_1: °C = 5/9(75 – 32) = 23.89 = 24 °C.   K = 24 °C + 273 = 297 K. T_1 = 32.0 psi

T_2: °C = 5/9(120 – 32) = 48.89 = 49 °C.   K = 49 °C + 273 = 322 K. P_2 = ?

$\dfrac{P_1}{T_1} = \dfrac{P_2}{T_2}$ or $P_2 = \dfrac{P_1T_2}{T_1} = \dfrac{32.0\ \text{psi} \times 322\ \text{K}}{297\ \text{K}} = 34.694 = 34.7\ \text{psi}$

10.28   According to Avogadro's hypothesis, the mole ratios in the chemical equation will be volume ratios for the gases if they are at the same temperature and pressure.

$$N_2(g) + 3H_2(g) \rightarrow 2NH_3(g)$$

The volumes of H_2 and N_2 are in a stoichiometric $\dfrac{3.6\ \text{L}}{1.2\ \text{L}}$ or $\dfrac{3\ \text{vol}\ H_2}{1\ \text{vol}\ N_2}$ ratio, so either can be used to determine the volume of $NH_3(g)$ produced.

$1.2\ \text{L}\ N_2 \times \dfrac{2\ \text{mol}\ NH_3}{1\ \text{mol}\ N_2} = 2.4\ \text{L}\ NH_3(g)$ produced.

The Ideal-Gas Equation (Section 10.4)

(In *Solutions to Exercises*, the symbol for molar mass is MM.)

10.29   (a)   STP stands for standard temperature, 0 °C (or 273 K), and standard pressure, 1 atm.

(b) $V = \dfrac{nRT}{P}$; $V = 1\ \text{mol} \times \dfrac{0.08206\ \text{L-atm}}{\text{mol-K}} \times \dfrac{273\ \text{K}}{1\ \text{atm}}$

V = 22.4 L for 1 mole of gas at STP

(c)   25 °C + 273 = 298 K

$V = \dfrac{nRT}{P}$; $V = 1\ \text{mol} \times \dfrac{0.08206\ \text{L-atm}}{\text{mol-K}} \times \dfrac{298\ \text{K}}{1\ \text{atm}}$

V = 24.5 L for 1 mol of gas at 1 atm and 25 °C

(d) $\dfrac{0.08206\ \text{L-atm}}{\text{mol-K}} \times \dfrac{1.01325 \times 10^5\ \text{Pa}}{1\ \text{atm}} \times \dfrac{1\ \text{bar}}{10^5\ \text{Pa}} = \dfrac{0.08315\ \text{L-bar}}{\text{mol-K}}$

10.30   Assume 1 mole of Ne at STP. Sample volume = 22.4 L. Calculate the volume occupied by 1 mole of Ne atoms. $1 L = 1 \ dm^3$, $r = 0.69$ Å, $V = 4\pi r^3/3$

$$r = 0.69 \text{ Å} \times \frac{1 \times 10^{-10} \text{ m}}{1 \text{ Å}} \times \frac{10 \text{ dm}}{\text{m}} = 6.9 \times 10^{-10} \text{ dm}$$

$$V = \frac{4 \pi r^3}{3}; \quad \frac{4 \pi (6.9 \times 10^{-10} \text{ dm})^3}{3} \times \frac{6.022 \times 10^{23} \text{ Ne atoms}}{\text{mol}} = 8.287 \times 10^{-4} \text{ dm}^3 = 8.3 \times 10^{-4} \text{ L}$$

$$\frac{8.3 \times 10^{-4} \text{ L occupied by Ne atoms}}{22.4 \text{ L total gas volume}} = 1/27,000 \text{ of the total space occupied by Ne atoms}$$

10.31   *Analyze/Plan.* PV = nRT. At constant volume and temperature, P is directly proportional to n.

Solve. For samples with equal masses of gas, the gas with MM = 30 will have twice as many moles of particles and twice the pressure. Thus, flask A contains the gas with MM = 30 and flask B contains the gas with MM = 60.

10.32   n = g/MM; PV = nRT = gRT/MM; MM = gRT/PV.

2-L flask: MM = 4.8 RT/2.0(x) = 2.4 RT/x

3-L flask: MM = 0.36 RT/3.0 (0.1x) = 1.2 RT/x

The molar masses of the two gases are not equal. The gas in the 2-L flask has a molar mass that is twice as large as the gas in the 3-L flask.

10.33   *Analyze/Plan.* Follow the strategy for calculations involving many variables given in Section 10.4.   *Solve.*

$$T = \frac{PV}{nR} = 2.00 \text{ atm} \times \frac{1.00 \text{ L}}{0.500 \text{ mol}} \times \frac{\text{mol-K}}{0.08206 \text{ L-atm}} = 48.7 \text{ K}$$

K = 27 °C + 273 = 300 K

$$n = \frac{PV}{RT} = 0.300 \text{ atm} \times \frac{0.250 \text{ L}}{300 \text{ K}} \times \frac{\text{mol-K}}{0.08206\text{-atm}} = 3.05 \times 10^{-3} \text{ mol}$$

$$650 \text{ torr} \times \frac{1 \text{ atm}}{760 \text{ torr}} = 0.85526 = 0.855 \text{ atm}$$

$$V = \frac{nRT}{P} = 0.333 \text{ mol} \times \frac{350 \text{ K}}{0.85526 \text{ atm}} \times \frac{0.08206 \text{ L-atm}}{\text{mol-K}} = 11.2 \text{ L}$$

585 mL = 0.585 L

$$P = \frac{nRT}{V} = 0.250 \text{ mol} \times \frac{295 \text{ K}}{0.585 \text{ L}} \times \frac{0.08206 \text{ L-atm}}{\text{mol-K}} = 10.3 \text{ atm}$$

P	V	n	T
2.00 atm	1.00 L	0.500 mol	**48.7 K**
0.300 atm	0.250 L	**3.05×10^{-3} mol**	27 °C
650 torr	**11.2 L**	0.333 mol	350 K
10.3 atm	585 mL	0.250 mol	295 K

10.34 *Analyze/Plan.* Follow the strategy for calculations involving many variables given in Section 10.4. *Solve.*

(a) $n = 1.50$ mol, $P = 1.25$ atm, $T = -6\ °C = 267$ K

$$V = \frac{nRT}{P} = 1.50\ \text{mol} \times \frac{0.08206\ \text{L-atm}}{\text{mol-K}} \times \frac{267\ \text{K}}{1.25\ \text{atm}} = 26.3\ \text{L}$$

(b) $n = 3.33 \times 10^{-3}$ mol, $V = 478$ mL $= 0.478$ L

$$P = 750\ \text{torr} \times \frac{1\ \text{atm}}{760\ \text{torr}} = 0.9868 = 0.987\ \text{atm}$$

$$T = \frac{PV}{nR} = 0.9868\ \text{atm} \times \frac{0.478\ \text{L}}{3.33 \times 10^{-3}\ \text{mol}} \times \frac{1\ \text{mol-K}}{0.08206\ \text{L-atm}} = 1726 = 1.73 \times 10^3\ \text{K}$$

(c) $n = 0.00245$ mol, $V = 413$ mL $= 0.413$ L, $T = 138\ °C = 411$ K

$$P = \frac{nRT}{V}; = 0.00245\ \text{mol} \times \frac{0.08206\ \text{L-atm}}{\text{mol-K}} \times \frac{411\ \text{K}}{0.413\ \text{L}} = 0.200\ \text{atm}$$

(d) $V = 126.5$ L, $T = 54\ °C = 327$ K,

$$P = 11.25\ \text{kPa} \times \frac{1\ \text{atm}}{101.325\ \text{kPa}} = 0.11103 = 0.1110\ \text{atm}$$

$$n = \frac{PV}{RT} = 0.11103\ \text{atm} \times \frac{\text{mol-K}}{0.08206\ \text{L-atm}} \times \frac{126.5\ \text{L}}{327\ \text{K}} = 0.523\ \text{mol}$$

10.35 *Analyze/Plan.* Follow the strategy for calculations involving many variables. *Solve.*

$n = g/MM$; $PV = nRT$; $PV = gRT/MM$; $g = MM \times PV/RT$

$P = 1.0$ atm, $T = 23\ °C = 296$ K, $V = 1.75 \times 10^5\ \text{ft}^3$. Change ft^3 to L, then calculate grams (or kg).

$$1.75 \times 10^5\ \text{ft}^3 \times \frac{(12)^3\ \text{in}^3}{\text{ft}^3} \times \frac{(2.54)^3\ \text{cm}^3}{\text{in}^3} \times \frac{1\ \text{L}}{1 \times 10^3\ \text{cm}^3} = 4.9554 \times 10^6 = 4.96 \times 10^6\ \text{L}$$

$$g = \frac{4.003\ \text{g He}}{1\ \text{mol He}} \times \frac{\text{mol-K}}{0.08206\ \text{L-atm}} \times \frac{1.0\ \text{atm} \times 4.955 \times 10^6\ \text{L}}{296\ \text{K}} = 8.2 \times 10^5\ \text{g} = 820\ \text{kg He}$$

10.36 Find the volume of the tube in cm^3; $1\ \text{cm}^3 = 1$ mL.

$r = d/2 = 2.5\ \text{cm}/2 = 1.25 = 1.3$ cm; $h = 5.5$ m $= 5.5 \times 10^2$ cm

$V = \pi r^2 h = 3.14159 \times (1.25\ \text{cm})^2 \times (5.5 \times 10^2\ \text{cm}) = 2.700 \times 10^3\ \text{cm}^3 = 2.7$ L

$$PV = \frac{g}{MM}RT;\quad g = \frac{MM \times PV}{RT};\quad P = 1.78\ \text{torr} \times \frac{1\ \text{atm}}{760\ \text{torr}} = 2.342 \times 10^{-3} = 2.34 \times 10^{-3}\ \text{atm}$$

$$g = \frac{20.18\ \text{g Ne}}{1\ \text{mol Ne}} \times \frac{\text{mol-K}}{0.08206\ \text{L-atm}} \times \frac{2.342 \times 10^{-3}\ \text{atm} \times 2.700\ \text{L}}{308\ \text{K}} = 5.049 \times 10^{-3} = 5.0 \times 10^{-3}\ \text{g Ne}$$

10.37 *Analyze/Plan.* Follow the strategy for calculations involving many variables. *Solve.*

(a) $V = 2.25$ L; $T = 273 + 37\ °C = 310$ K; $P = 735\ \text{torr} \times \dfrac{1\ \text{atm}}{760\ \text{torr}} = 0.96710 = 0.967$ atm

$PV = nRT$, $n = PV/RT$, number of molecules (#) $= n \times 6.022 \times 10^{23}$

$$\# = \frac{0.9671\ \text{atm} \times 2.25\ \text{L}}{310\ \text{K}} \times \frac{\text{mol-K}}{0.08206\ \text{L-atm}} \times \frac{6.022 \times 10^{23}\ \text{molecules}}{\text{mol}}$$

$$= 5.15 \times 10^{22}\ \text{molecules}$$

(b) $V = 5.0 \times 10^3$ L; $T = 273 + 0\,°C = 273$ K; $P = 1.00$ atm; MM $= 28.98$ g/mol

$$PV = \frac{g}{MM}RT; \quad g = \frac{MM \times PV}{RT}$$

$$g = \frac{28.98 \text{ g air}}{1 \text{ mol air}} \times \frac{\text{mol-K}}{0.08206 \text{ L-atm}} \times \frac{1.00 \text{ atm} \times 5.0 \times 10^3 \text{ L}}{273 \text{ K}}$$

$$= 6468 \text{ g} = 6.5 \times 10^3 \text{ g air} = 6.5 \text{ kg air}$$

10.38   (a) $P_{O_3} = 3.0 \times 10^{-3}$ atm; $T = 250$ K; $V = 1$ L (exact)

$$\text{\# of } O_3 \text{ molecules} = \frac{PV}{RT} \times 6.022 \times 10^{23}$$

$$\text{\#} = \frac{3.0 \times 10^{-3} \text{ atm} \times 1 \text{ L}}{250 \text{ K}} \times \frac{\text{mol-K}}{0.08206 \text{ L-atm}} \times \frac{6.022 \times 10^{23} \text{ molecules}}{\text{mol}}$$

$$= 8.8 \times 10^{19} \; O_3 \text{ molecules}$$

(b) $\text{\# of } CO_2 \text{ molecules} = \dfrac{PV}{RT} \times 6.022 \times 10^{23} \times 0.0004$

$$\text{\#} = \frac{1.0 \text{ atm} \times 2.0 \text{ L}}{300 \text{ K}} \times \frac{\text{mol-K}}{0.08206 \text{ L-atm}} \times \frac{6.022 \times 10^{23} \text{ molecules}}{\text{mol}} \times 0.0004$$

$$= 1.957 \times 10^{19} = 2 \times 10^{19} \; CO_2 \text{ molecules}$$

10.39   *Analyze/Plan.* Follow the strategy for calculations involving many variables.   *Solve.*

(a) $P = \dfrac{nRT}{V}$; $n = 0.29$ kg $O_2 \times \dfrac{1000 \text{ g}}{1 \text{ kg}} \times \dfrac{1 \text{ mol } O_2}{32.00 \text{ g } O_2} = 9.0625 = 9.1$ mol; $V = 2.3$ L;

$T = 273 + 9\,°C = 282$ K

$$P = \frac{9.0625 \text{ mol}}{2.3 \text{ L}} \times \frac{0.08206 \text{ L-atm}}{\text{mol-K}} \times 282 \text{ K} = 91 \text{ atm}$$

(b) $V = \dfrac{nRT}{P}$; $= \dfrac{9.0625 \text{ mol}}{0.95 \text{ atm}} \times \dfrac{0.08206 \text{ L-atm}}{\text{mol-K}} \times 299 \text{ K} = 2.3 \times 10^2$ L

10.40   (a) $V = 0.250$ L, $T = 23\,°C = 296$ K, $n = 2.30$ g $C_3H_8 \times \dfrac{1 \text{ mol } C_3H_8}{44.1 \text{ g } C_3H_8} = 0.052154$

$$= 0.0522 \text{ mol}$$

$$P = \frac{nRT}{V} = 0.052154 \text{ mol} \times \frac{0.08206 \text{ L-atm}}{\text{mol-K}} \times \frac{296 \text{ K}}{0.250 \text{ L}} = 5.07 \text{ atm}$$

(b)   STP $= 1.00$ atm, 273 K

$$V = \frac{nRT}{P} = 0.052154 \text{ mol} \times \frac{0.08206 \text{ L-atm}}{\text{mol-K}} \times \frac{273 \text{ K}}{1.00 \text{ atm}} = 1.1684 = 1.17 \text{ L}$$

(c) $°C = 5/9\,(°F - 32°)$; $K = °C + 273.15 = 5/9\,(130\,°F - 32°) + 273.15 = 327.59 = 328$ K

$$P = \frac{nRT}{V} = 0.052154 \text{ mol} \times \frac{0.08206 \text{ L-atm}}{\text{mol-K}} \times \frac{327.59 \text{ K}}{0.250 \text{ L}} = 5.608 = 5.61 \text{ atm}$$

10.41   *Analyze/Plan.* Follow the strategy for calculations involving many variables.   *Solve.*

$$P = \frac{nRT}{V}; \; n = 35.1 \text{ g } CO_2 \times \frac{1 \text{ mol } CO_2}{44.01 \text{ g } CO_2} = 0.7975 = 0.798 \text{ mol}; V = 4.0 \text{ L}; T = 298 \text{ K}$$

$$P = \frac{0.7975 \text{ mol}}{4.0 \text{ L}} \times \frac{0.08206 \text{ L-atm}}{\text{mol-K}} \times 298 \text{ K} = 4.876 = 4.9 \text{ atm}$$

10.42 Calculate the mass of He that will produce a pressure of 75 atm in the cylinder, then subtract that mass from 5.225 g He to calculate the mass of He that must be released.

$$PV = \frac{g}{MM}RT; \quad g = \frac{MM \times PV}{RT}$$

$$g = \frac{4.0026 \text{ g He}}{1 \text{ mol He}} \times \frac{\text{mol-K}}{0.08206 \text{ L-atm}} \times \frac{75 \text{ atm} \times 0.334 \text{ L}}{296 \text{ K}} = 4.1279 = 4.1 \text{ g He remain}$$

5.225 g He initial – 4.1279 g He remain = 1.0971 = 1.1 g He must be released

10.43 *Analyze/Plan.* Follow the strategy for calculations involving many variables. *Solve.*

$$V = 8.70 \text{ L}, \ T = 24 \text{ }°C = 297 \text{ K}, \ P = 895 \text{ torr} \times \frac{1 \text{ atm}}{760 \text{ torr}} = 1.1776 = 1.18 \text{ atm}$$

(a) $\quad g = \dfrac{MM \times PV}{RT}; \ g = \dfrac{70.91 \text{g Cl}_2}{1 \text{ mol Cl}_2} \times \dfrac{\text{mol-K}}{0.08206 \text{ L-atm}} \times \dfrac{1.1776 \text{ atm}}{297 \text{ K}} \times 8.70 \text{ L}$

$$= 29.8 \text{g Cl}_2$$

(b) $\quad V_2 = \dfrac{P_1 V_1 T_2}{T_1 P_2} = \dfrac{895 \text{ torr} \times 8.70 \text{ L} \times 273 \text{ K}}{297 \text{ K} \times 760 \text{ torr}} = 9.42 \text{ L}$

(c) $\quad T_2 = \dfrac{P_2 V_2 T_1}{P_1 V_1} = \dfrac{876 \text{ torr} \times 15.00 \text{ L} \times 297 \text{ K}}{895 \text{ torr} \times 8.70 \text{ L}} = 501 \text{ K}$

(d) $\quad P_2 = \dfrac{P_1 V_1 T_2}{V_2 T_1} = \dfrac{895 \text{ torr} \times 8.70 \text{ L} \times 331 \text{ K}}{5.00 \text{ L} \times 297 \text{ K}} = 1.73 \times 10^3 \text{ torr} = 2.28 \text{ atm}$

10.44 $\quad T = 23 \text{ }°C = 296 \text{ K}, \ P = 16{,}500 \text{ kPa} \times \dfrac{1 \text{ atm}}{101.325 \text{ kPa}} = 162.84 = 163 \text{ atm}$

$$V = 55.0 \text{ gal} \times \frac{3.7854 \text{ L}}{1 \text{ gal}} = 208.20 = 208 \text{ L}$$

(a) $\quad g = \dfrac{MM \times PV}{RT}; \ g = \dfrac{32.0 \text{ g O}_2}{1 \text{ mol O}_2} \times \dfrac{\text{mol-K}}{0.08206 \text{ L-atm}} \times \dfrac{162.84 \text{ atm}}{296 \text{ K}} \times 208.20 \text{ L}$

$$= 4.4665 \times 10^4 \text{ g O}_2 = 44.7 \text{ kg O}_2$$

(b) $\quad V_2 = \dfrac{P_1 V_1 T_2}{T_1 P_2} = \dfrac{16{,}500 \text{ kPa} \times 208.20 \text{ L} \times 273 \text{ K}}{296 \text{ K} \times 101.325 \text{ kPa}} = 3.13 \times 10^4 \text{ L}$

(c) $\quad T_2 = \dfrac{P_2 T_1}{P_1} = \dfrac{150.0 \text{ atm} \times 296 \text{ K}}{16{,}500 \text{ kPa}} \times \dfrac{101.325 \text{ kPa}}{1 \text{ atm}} = 272.7 = 273 \text{ K}$

(d) $\quad P_2 = \dfrac{P_1 V_1 T_2}{V_2 T_1} = \dfrac{16{,}500 \text{ kPa} \times 208.20 \text{ L} \times 297 \text{ K}}{55.0 \text{ L} \times 296 \text{ K}} = 62{,}671 = 6.27 \times 10^4 \text{ kPa}$

10.45 *Analyze.* Given: mass of cockroach, rate of O_2 consumption, temperature, percent O_2 in air, volume of air. Find: mol O_2 consumed per hour; mol O_2 in 1 quart of air; mol O_2 consumed in 48 hr.

(a) *Plan/Solve.* V of O_2 consumed = rate of consumption × mass × time. n = PV/RT.

$$5.2 \text{ g} \times 1 \text{ hr} \times \frac{0.8 \text{ mL O}_2}{1 \text{ g-hr}} = 4.16 = 4 \text{ mL O}_2 \text{ consumed}$$

$$n = \frac{PV}{RT} = 1 \text{ atm} \times \frac{\text{mol-K}}{0.08206 \text{ L-atm}} \times \frac{0.00416 \text{ L}}{297 \text{ K}} = 1.71 \times 10^{-4} = 2 \times 10^{-4} \text{ mol O}_2$$

(b)   *Plan/Solve.* qt air → L air → L O_2 available. mol O_2 available = PV/RT.

mol O_2/hr (from part (a)) → total mol O_2 consumed. Compare O_2 available and O_2 consumed.

$$1 \text{ qt air} \times \frac{0.946 \text{ L}}{1 \text{ qt}} \times 0.21 \ O_2 \text{ in air} = 0.199 \text{ L } O_2 \text{ available}$$

$$n = 1 \text{ atm} \times \frac{\text{mol-K}}{0.08206 \text{ L-atm}} \times \frac{0.199 \text{ L}}{297 \text{ K}} = 8.17 \times 10^{-3} = 8 \times 10^{-3} \text{ mol } O_2 \text{ available}$$

$$\text{cockroach uses } \frac{1.71 \times 10^{-4} \text{ mol}}{1 \text{ hr}} \times 48 \text{ hr} = 8.21 \times 10^{-3} = 8 \times 10^{-3} \text{ mol } O_2 \text{ consumed}$$

Not only does the cockroach use 20% of the available O_2, it needs all the O_2 in the jar.

10.46   Change mass to kg; 1 hr = 60 min; pay attention to units.

(a) $185 \text{ lb} \times \dfrac{1 \text{ kg}}{2.2046 \text{ lb}} \times \dfrac{47.5 \text{ mL } O_2}{\text{kg-min}} \times 60 \text{ min} = 2.39 \times 10^5 \text{ mL}$

(b) $165 \text{ lb} \times \dfrac{1 \text{ kg}}{2.2046 \text{ lb}} \times \dfrac{65.0 \text{ mL } O_2}{\text{kg-min}} \times 60 \text{ min} = 2.92 \times 10^5 \text{ mL}$

Further Applications of the Ideal-Gas Equation (Section 10.5)

10.47   *Analyze/Plan.* At the same temperature and pressure, the density of a gas increases with increasing molar mass.

HF (20 g/mol) < CO (28 g/mol) < N_2O (44 g/mol) < Cl_2 (71 g/mol)

10.48 CO_2 < SO_2 < HBr. For gases at the same conditions, density is directly proportional to molar mass. The order of increasing molar mass is the order of increasing density. CO_2, 44 g/mol < SO_2, 64 g/mol < HBr, 81 g/mol.

10.49   (c) Because the helium atoms are of lower mass than the average air molecule, the helium gas is less dense than air. The balloon thus weighs less than the air displaced by its volume.

10.50   (b) Xe atoms have a higher mass than N_2 molecules. Because both gases at STP have the same number of molecules per unit volume, the Xe gas must be denser.

10.51   *Analyze/Plan.* Conditions (P, V, T) and amounts of gases are given. Rearrange the relationship PV × MM = gRT to obtain the desired quantity, paying attention (as always!) to units. *Solve.*

(a) $d = \dfrac{MM \times P}{RT}$;   MM = 46.0 g/mol; P = 0.970 atm, T = 35 °C = 308 K

$$d = \frac{46.0 \text{ g } NO_2}{1 \text{ mol}} \times \frac{\text{mol-K}}{0.08206 \text{ L-atm}} \times \frac{0.970 \text{ atm}}{308 \text{ K}} = 1.77 \text{ g/L}$$

(b) $MM = \dfrac{gRT}{PV} = \dfrac{2.50 \text{ g}}{0.875 \text{ L}} \times \dfrac{0.08206 \text{ L-atm}}{\text{mol-K}} \times \dfrac{308 \text{ K}}{685 \text{ torr}} \times \dfrac{760 \text{ torr}}{1 \text{ atm}} = 80.1 \text{ g/mol}$

10.52 (a) $d = \dfrac{MM \times P}{RT}$; MM = 146.1 g/mol, T = 21 °C = 294 K, P = 707 torr

$$d = \frac{146.1 \text{ g}}{1 \text{ mol}} \times \frac{\text{mol-K}}{0.08206 \text{ L-atm}} \times \frac{707 \text{ torr}}{294 \text{ K}} \times \frac{1 \text{ atm}}{760 \text{ torr}} = 5.63 \text{ g/L}$$

(b) $MM = \dfrac{dRT}{P} = \dfrac{7.135 \text{ g}}{1 \text{ L}} \times \dfrac{0.08206 \text{ L-atm}}{\text{mol-K}} \times \dfrac{285 \text{ K}}{743 \text{ torr}} \times \dfrac{760 \text{ torr}}{1 \text{ atm}} = 171 \text{ g/mol}$

10.53 *Analyze/Plan.* Given: mass, conditions (P, V, T) of unknown gas. Find: molar mass. MM = gRT/PV. *Solve.*

$$MM = \frac{gRT}{PV} = \frac{1.012 \text{ g}}{0.354 \text{ L}} \times \frac{0.08206 \text{ L-atm}}{\text{mol-K}} \times \frac{372 \text{ K}}{742 \text{ torr}} \times \frac{760 \text{ torr}}{1 \text{ atm}} = 89.4 \text{ g/mol}$$

10.54 $MM = \dfrac{gRT}{PV} = \dfrac{0.846 \text{ g}}{0.354 \text{ L}} \times \dfrac{0.08206 \text{ L-atm}}{\text{mol-K}} \times \dfrac{373 \text{ K}}{752 \text{ torr}} \times \dfrac{760 \text{ torr}}{1 \text{ atm}} = 73.9 \text{ g/mol}$

10.55 *Analyze/Plan.* Follow the logic in Sample Exercise 10.9. *Solve.*

$$\text{mol O}_2 = \frac{PV}{RT} = 3.5 \times 10^{-6} \text{ torr} \times \frac{1 \text{ atm}}{760 \text{ torr}} \times \frac{\text{mol-K}}{0.08206 \text{ L-atm}} \times \frac{0.452 \text{ L}}{300 \text{ K}} = 8.456 \times 10^{-11}$$

$$= 8.5 \times 10^{-11} \text{ mol O}_2$$

$$8.456 \times 10^{-11} \text{ mol O}_2 \times \frac{2 \text{ mol Mg}}{1 \text{ mol O}_2} \times \frac{24.3 \text{ g Mg}}{1 \text{ mol Mg}} = 4.1 \times 10^{-9} \text{ g Mg (4.1 ng Mg)}$$

10.56 $n_{H_2} = \dfrac{P_{H_2} V}{RT}$; P = 825 torr $\times \dfrac{1 \text{ atm}}{760 \text{ torr}} = 1.0855 = 1.09$ atm; T = 273 + 21 °C = 294 K

$$n_{H_2} = 1.0855 \text{ atm} \times \frac{\text{mol-K}}{0.08206 \text{ L-atm}} \times \frac{145 \text{ L}}{294 \text{ K}} = 6.5242 = 6.52 \text{ mol H}_2$$

$$6.5242 \text{ mol H}_2 \times \frac{1 \text{ mol CaH}_2}{2 \text{ mol H}_2} \times \frac{42.10 \text{ g CaH}_2}{1 \text{ mol CaH}_2} = 137.34 = 137 \text{ g CaH}_2$$

10.57 (a) *Analyze/Plan.* g glucose → mol glucose → mol CO_2 → V CO_2 *Solve.*

$$24.5 \text{ g} \times \frac{1 \text{ mol glucose}}{180.1 \text{ g}} \times \frac{6 \text{ mol CO}_2}{1 \text{ mol glucose}} = 0.8162 = 0.816 \text{ mol CO}_2$$

$$V = \frac{nRT}{P} = 0.8162 \text{ mol} \times \frac{0.08206 \text{ L-atm}}{\text{mol-K}} \times \frac{310 \text{ K}}{0.970 \text{ atm}} = 21.4 \text{ L CO}_2$$

(b) *Analyze/Plan.* g glucose → mol glucose → mol O_2 → V O_2 *Solve.*

$$50.0 \text{ g} \times \frac{1 \text{ mol glucose}}{180.1 \text{ g}} \times \frac{6 \text{ mol O}_2}{1 \text{ mol glucose}} = 1.6657 = 1.67 \text{ mol O}_2$$

$$V = \frac{nRT}{P} = 1.6657 \text{ mol} \times \frac{0.08206 \text{ L-atm}}{\text{mol-K}} \times \frac{298 \text{ K}}{1 \text{ atm}} = 40.7 \text{ L O}_2$$

10.58 Follow the logic in Sample Exercise 10.9. The $H_2(g)$ will be used in a balloon, which operates at atmospheric pressure. Because atmospheric pressure is not explicitly given, assume 1 atm (infinite sig figs).

$$n = \frac{PV}{RT} = 1 \text{ atm} \times \frac{3.1150 \times 10^4 \text{ L}}{295 \text{ K}} \times \frac{\text{mol-K}}{0.08206 \text{ L-atm}} = 1.28678 \times 10^3 = 1.29 \times 10^3 \text{ mol H}_2$$

From the balanced equation, 1 mol of Fe produces 1 mol of H_2, so 1.29×10^3 mol Fe are required.

$$1.28678 \times 10^3 \text{ mol Fe} \times \frac{55.845 \text{ g Fe}}{\text{mol Fe}} \times \frac{1 \text{ kg}}{1000 \text{ g}} = 71.86 = 71.9 \text{ kg Fe}$$

10.59   *Analyze/Plan.* The gas sample is a mixture of $H_2(g)$ and $H_2O(g)$. Find the partial pressure of $H_2(g)$ and then the moles of $H_2(g)$ and $Zn(s)$.   *Solve.*

$P_t = 738 \text{ torr} = P_{H_2} + P_{H_2O}$

From Appendix B, the vapor pressure of H_2O at 24 °C = 22.38 torr

$$P_{H_2} = (738 \text{ torr} - 22.38 \text{ torr}) \times \frac{1 \text{ atm}}{760 \text{ torr}} = 0.9416 = 0.942 \text{ atm}$$

$$n_{H_2} = \frac{P_{H_2} V}{RT} = 0.9416 \text{ atm} \times \frac{\text{mol-K}}{0.08206 \text{ L-atm}} \times \frac{0.159 \text{ L}}{297 \text{ K}} = 0.006143 = 0.00614 \text{ mol } H_2$$

$$0.006143 \text{ mol } H_2 \times \frac{1 \text{ mol Zn}}{1 \text{ mol } H_2} \times \frac{65.39 \text{ g Zn}}{1 \text{ mol Zn}} = 0.402 \text{ g Zn}$$

10.60   The gas sample is a mixture of $C_2H_2(g)$ and $H_2O(g)$. Find the partial pressure of C_2H_2, then moles CaC_2 and C_2H_2.

$P_t = 753 \text{ torr} = P_{C_2H_2} + P_{H_2O}$. P_{H_2O} at 23 °C = 21.07 torr

$$P_{C_2H_2} = (753 \text{ torr} - 21.07 \text{ torr}) \times \frac{1 \text{ atm}}{760 \text{ torr}} = 0.96307 = 0.963 \text{ atm}$$

$$1.524 \text{ g CaC}_2 \times \frac{1 \text{ mol CaC}_2}{64.10 \text{ g}} \times \frac{1 \text{ mol } C_2H_2}{1 \text{ mol CaC}_2} = 0.023775 = 0.02378 \text{ mol } C_2H_2$$

$$V = 0.023775 \text{ mol} \times \frac{0.08206 \text{ L-atm}}{\text{mol-K}} \times \frac{296 \text{ K}}{0.96307 \text{ atm}} = 0.600 \text{ L } C_2H_2$$

Partial Pressures (Section 10.6)

10.61   (a)   When the stopcock is opened, the volume occupied by $N_2(g)$ increases from 2.0 to 5.0 L. At constant T, $P_1V_1 = P_2V_2$. 1.0 atm × 2.0 L = P_2 × 5.0 L; P_2 = 0.40 atm

(b)   When the gases mix, the volume of $O_2(g)$ increases from 3.0 to 5.0 L. At constant T, $P_1V_1 = P_2V_2$. 2.0 atm × 3.0 L = P_2 × 5.0 L; P_2 = 1.2 atm

(c) $P_t = P_{N_2} + P_{O_2} = 0.40 \text{ atm} + 1.2 \text{ atm} = 1.6 \text{ atm}$

10.62   (a)   The partial pressure of gas A is **not affected** by the addition of gas C. The partial pressure of A depends only on moles of A, volume of container, and conditions; none of these factors changes when gas C is added.

(b)   The total pressure in the vessel **increases** when gas C is added, because the total number of moles of gas increases.

(c)   The mole fraction of gas B **decreases** when gas C is added. The moles of gas B stay the same, but the total moles increase, so the mole fraction of B (nB/nt) decreases.

10.63   *Analyze.* Given: amount, V, T of three gases. Find: P of each gas, total P.

Plan. $P = nRT/V$; $P_t = P_1 + P_2 + P_3 +$ *Solve.*

(a) $P_{He} = \dfrac{nRT}{V} = 0.765 \text{ mol} \times \dfrac{0.08206 \text{ L-atm}}{\text{mol-K}} \times \dfrac{298 \text{ K}}{10.00 \text{ L}} = 1.871 = 1.87 \text{ atm}$

$P_{Ne} = \dfrac{nRT}{V} = 0.330 \text{ mol} \times \dfrac{0.08206 \text{ L-atm}}{\text{mol-K}} \times \dfrac{298 \text{ K}}{10.00 \text{ L}} = 0.8070 = 0.807 \text{ atm}$

$P_{Ar} = \dfrac{nRT}{V} = 0.110 \text{ mol} \times \dfrac{0.08206 \text{ L-atm}}{\text{mol-K}} \times \dfrac{298 \text{ K}}{10.00 \text{ L}} = 0.2690 = 0.269 \text{ atm}$

(b) $P_t = 1.871 \text{ atm} + 0.8070 \text{ atm} + 0.2690 \text{ atm} = 2.9470 = 2.95 \text{ atm}$

10.64   Given mass, V and T of O_2 and He, find the partial pressure of each gas. Sum to find the total pressure in the tank.

$V = 10.0 \text{ L}; T = 19 \,°C; 19 + 273 = 292 \text{ K}$

$n_{O_2} = 51.2 \text{ g } O_2 \times \dfrac{1 \text{ mol } O_2}{31.999 \text{ g } O_2} = 1.600 = 1.60 \text{ mol } O_2$

$n_{He} = 32.6 \text{ g He} \times \dfrac{1 \text{ mol He}}{4.0026 \text{ g He}} = 8.1447 = 8.14 \text{ mol He}$

$P_{O_2} = 1.600 \text{ mol} \times \dfrac{0.08206 \text{ L-atm}}{\text{mol-K}} \times \dfrac{292 \text{ K}}{10.0 \text{ L}} = 3.8338 = 3.84 \text{ atm}$

$P_{He} = 8.1447 \text{ mol} \times \dfrac{0.08206 \text{ L-atm}}{\text{mol-K}} \times \dfrac{292 \text{ K}}{10.0 \text{ L}} = 19.5159 = 19.5 \text{ atm}$

$P_t = 3.8338 + 19.5159 = 23.3497 = 23.3 \text{ atm}$

10.65   *Analyze.* Given 407 ppm CO_2 in the atmosphere; 407 L CO_2 in 10^6 total L air. Find: the mole fraction of CO_2 in the atmosphere. *Plan.* Avogadro's law deals with the relationship between volume and moles of a gas.

Solve. Avogadro's law states that volume of a gas at constant temperature and pressure is directly proportional to moles of the gas. Using volume fraction to express concentration assumes that the 407 L CO_2 and 10^6 total L air are at the same temperature and pressure. That is, 407 L is the volume that the number of moles of CO_2 present in 10^6 L air would occupy at atmospheric temperature and pressure. The mole fraction of CO_2 in the atmosphere is then just the volume fraction from the concentration by volume.

$\chi_{CO_2} = \dfrac{407 \text{ L } CO_2}{10^6 \text{ L air}} = 0.000407$

10.66 $\chi_{Xe} = 4/100 = 0.04$; $\chi_{Ne} = \chi_{He} = (1 - 0.04)/2 = 0.48$

$V_t = 0.900 \text{ mm} \times 0.300 \text{ mm} \times 10.0 \text{ mm} \times \dfrac{1 \text{ cm}^3}{10^3 \text{ mm}^3} \times \dfrac{1 \text{ L}}{1000 \text{ cm}^3} = 2.70 \times 10^{-6} \text{ L}$

$P_t = 500 \text{ torr} \times \dfrac{1 \text{ atm}}{760 \text{ torr}} = 0.657895 = 0.658 \text{ atm}$

$n_t = \dfrac{PV}{RT} = 0.657895 \text{ atm} \times \dfrac{\text{mol-K}}{0.08206 \text{ L-atm}} \times \dfrac{2.70 \times 10^{-6} \text{ L}}{298 \text{ K}} \times \dfrac{6.022 \times 10^{23} \text{ atoms}}{\text{mol}}$

$= 4.3743 \times 10^{16} = 4.37 \times 10^{16} \text{ total atoms}$

Xe atoms $= X_{Xe} \times$ total atoms $= 0.04(4.3743 \times 10^{16}) = 1.75 \times 10^{15} = 2 \times 10^{15}$ Xe atoms

Ne atoms $=$ He atoms $= 0.48(4.3743 \times 10^{16}) = 2.10 \times 10^{16} = 2.1 \times 10^{16}$ Ne and He atoms

Assumptions: To calculate total moles of gas and total atoms, we assumed a reasonable room temperature. Because '4% Xe' was not defined, we conveniently assumed mole percent. The 1:1 relationship of Ne to He is assumed to be by volume and not by mass.

10.67   *Analyze.* Given: mass CO_2 at V, T; pressure of air at same V, T. Find: partial pressure of CO_2 at these conditions, total pressure of gases at V, T.

Plan. g $CO_2 \rightarrow$ mol $CO_2 \rightarrow P_{CO_2}$ (via $P = nRT/V$); $P_t = P_{CO_2} + P_{air}$ *Solve.*

$$5.50 \text{ g } CO_2 \times \frac{1 \text{ mol } CO_2}{44.01 \text{ g } CO_2} = 0.12497 = 0.125 \text{ mol } CO_2; \quad T = 273 + 24 \text{ °C} = 297 \text{ K}$$

$$P_{CO_2} = 0.12497 \text{ mol} \times \frac{297 \text{ K}}{10.0 \text{ L}} \times \frac{0.08206 \text{ L-atm}}{\text{mol-K}} = 0.30458 = 0.305 \text{ atm}$$

$$P_{air} = 705 \text{ torr} \times \frac{1 \text{ atm}}{760 \text{ torr}} = 0.92763 = 0.928 \text{ atm}$$

$$P_t = P_{CO_2} + P_{air} = 0.30458 + 0.92763 = 1.23221 = 1.232 \text{ atm}$$

(Result has 3 decimal places and 4 sig figs.)

10.68 $V \ C_2H_5OC_2H_5(l) \rightarrow$ mass $C_2H_5OC_2H_5 \rightarrow$ mol $C_2H_5OC_2H_5 \rightarrow P_{C_2H_5OC_2H_5}$ (at V, T)

$P_t = P_{N_2} + P_{O_2} + P_{C_2H_5OC_2H_5}$; $\quad T = 273.15 + 35.0 \text{ °C} = 308.15 = 308.2 \text{ K}$

(a) $5.00 \text{ mL } C_2H_5OC_2H_5 \times \dfrac{0.7134 \text{ g } C_2H_5OC_2H_5}{\text{mL}} \times \dfrac{1 \text{ mol } C_2H_5OC_2H_5}{74.12 \text{ g } C_2H_5OC_2H_5}$

$$= 0.048125 = 0.0481 \text{ mol } C_2H_5OC_2H_5$$

$$P = \frac{nRT}{V} = 0.048125 \text{ mol} \times \frac{308.15}{6.00 \text{ L}} \times \frac{0.08206 \text{ L-atm}}{\text{mol-K}} = 0.20282 = 0.203 \text{ atm}$$

(b) $P_t = P_{N_2} + P_{O_2} + P_{C_2H_5OC_2H_5} = 0.751 \text{ atm} + 0.208 \text{ atm} + 0.203 \text{ atm} = 1.162 \text{ atm}$

10.69   *Analyze/Plan.* When the sample is cooled, the water vapor condenses and all the gas pressure is because of $CO_2(g)$. The partial pressure CO_2 at 200 °C is equal to the mole fraction of CO_2 times the total pressure of the mixture. Apply Amonton's law (see Solution 10.27) to the CO_2 pressures at the two temperatures.   *Solve.*

For a 3:1 mole ratio of CO_2 to H_2O

$$X_{CO_2} = \frac{3}{4} = 0.75; \quad P_{CO_2} = 0.75 \times 2.00 \text{ atm} = 1.50 \text{ atm}$$

$$T_1 = 200 \text{ °C} + 273 = 473 \text{ K}; \quad T_2 = 10 \text{ °C} + 273 = 283 \text{ K}$$

$$\frac{P_1}{T_1} = \frac{P_2}{T_2} \text{ or } P_2 = \frac{P_1 T_2}{T_1} = \frac{1.50 \text{ atm} \times 283 \text{ K}}{473 \text{ K}} = 0.8975 = 0.9 \text{ atm}$$

[The result has 1 sig fig if you consider the mole ratio to have 1 sig fig. If you think of the mole ratio as exact (experimentally unlikely), the result will have 3 sig figs.]

10.70 $T = 320\ °C + 273 = 593\ K$; $P_t\ (593\ K) = P_{N_2} + P_{O_2}$

       (i)   Use Amonton's law (see Solution 10.27) to calculate P_{N_2} at 593 K.

       (ii)   Use stoichiometry to calculate mol O_2 produced by decomposing 5.15 g Ag_2O.

       (iii)   Use the ideal-gas law to calculate P_{O_2} at 593 K. Sum the pressures.

       (i) $P_1 = 760\ torr = 1\ atm$; $T_1 = 32\ °C + 273 = 305\ K$; $T_2 = 320\ °C + 273 = 593\ K$

$$\frac{P_1}{T_1} = \frac{P_2}{T_2} \text{ or } P_2 = \frac{P_1 T_2}{T_1} = \frac{1.00\ atm \times 593\ K}{305\ K} = 1.9443 = 1.94\ atm$$

       (ii) $2\ Ag_2O(s) \rightarrow 4\ Ag(s) + O_2(g)$

$$5.15\ g\ Ag_2O \times \frac{1\ mol\ Ag_2O}{231.74\ g\ Ag_2O} \times \frac{1\ mol\ O_2}{2\ mol\ Ag_2O} = 0.01111 = 0.0111\ mol\ O_2$$

       (iii) $V = 0.0750\ L$, $T = 593\ K$, $n = 0.0111$

$$P = \frac{nRT}{V};\ P = \frac{0.01111\ mol}{0.0750\ L} \times \frac{0.08206\ L\text{-}atm}{mol\text{-}K} \times 593\ K = 4.876 = 7.209 = 7.21\ atm$$

$$P_t\ (593\ K) = P_{N_2} + P_{O_2} = 1.94\ atm + 7.21\ atm = 9.15\ atm$$

10.71   *Analyze/Plan.* Mole fraction = pressure fraction. Find the desired mole fraction of O_2 and change to mole percent.   *Solve.*

$$X_{O_2} = \frac{P_{O_2}}{P_t} = \frac{0.21\ atm}{8.38\ atm} = 0.025;\ \text{mole \%} = 0.025 \times 100 = 2.5\%$$

10.72   (a) $n_{O_2} = 15.08\ g\ O_2 \times \dfrac{1\ mol}{31.999\ g} = 0.4713\ mol$; $n_{N_2} = 8.17\ g\ N_2 \times \dfrac{1\ mol}{28.02\ g} = 0.292\ mol$

           $n_{H_2} = 2.64\ g\ H_2 \times \dfrac{1\ mol}{2.016\ g} = 1.31\ mol$; $n_t = 0.4713 + 0.292 + 1.31 = 2.07\ mol$

$$X_{O_2} = \frac{n_{O_2}}{n_t} = \frac{0.4713}{2.07} = 0.228;\ X_{N_2} = \frac{n_{N_2}}{n_t} = \frac{0.292}{2.07} = 0.141$$

$$X_{H_2} = \frac{1.31}{2.07} = 0.633$$

       (b) $P_{O_2} = n \times \dfrac{RT}{V};\ P_{O_2} = 0.4713\ mol \times \dfrac{0.08206\ L\text{-}atm}{mol\text{-}K} \times \dfrac{288.15\ K}{15.50\ L} = 0.7190\ atm$

$$P_{N_2} = 0.292\ mol \times \frac{0.08206\ L\text{-}atm}{mol\text{-}K} \times \frac{288.15\ K}{15.50\ L} = 0.445\ atm$$

$$P_{H_2} = 1.31\ mol \times \frac{0.08206\ L\text{-}atm}{mol\text{-}K} \times \frac{288.15\ K}{15.50\ L} = 2.00\ atm$$

10.73   *Analyze/Plan.* $N_2(g)$ and $O_2(g)$ undergo changes of conditions and are mixed. Calculate the new pressure of each gas and add them to obtain the total pressure of the mixture.

$P_2 = P_1 V_1 T_2 / V_2 T_1$; $P_t = P_{N_2} + P_{O_2}$.   *Solve.*

$$P_{N_2} = \frac{P_1 V_1 T_2}{V_2 T_1} = \frac{5.25\ atm \times 1.00\ L \times 293\ K}{12.5\ L \times 299\ K} = 0.41157 = 0.412\ atm$$

$$P_{O_2} = \frac{P_1 V_1 T_2}{V_2 T_1} = \frac{5.25\ atm \times 5.00\ L \times 293\ K}{12.5\ L \times 299\ K} = 2.05786 = 2.06\ atm$$

$$P_t = 0.41157\ atm + 2.05786\ atm = 2.46943 = 2.47\ atm$$

10.74   Calculate the pressure of the gas in the second vessel directly from mass and conditions using the ideal-gas equation.

(a) $P_{SO_2} = \dfrac{gRT}{M\,V} = \dfrac{3.00 \text{ g SO}_2}{64.07 \text{ g SO}_2/\text{mol}} \times \dfrac{0.08206 \text{ L-atm}}{\text{mol-K}} \times \dfrac{299 \text{ K}}{10.0 \text{ L}} = 0.11489 = 0.115 \text{ atm}$

(b) $P_{N_2} = \dfrac{gRT}{M\,V} = \dfrac{2.35 \text{ g N}_2}{28.01 \text{ g N}_2/\text{mol}} \times \dfrac{0.08206 \text{ L-atm}}{\text{mol-K}} \times \dfrac{299 \text{ K}}{10.0 \text{ L}} = 0.20585 = 0.206 \text{ atm}$

(c) $P_t = P_{SO_2} + P_{N_2} = 0.11489 \text{ atm} + 0.20585 \text{ atm} = 0.321 \text{ atm}$

Kinetic-Molecular Theory of Gases; Effusion and Diffusion
(Sections 10.7 and 10.8)

10.75   (a)   Decrease. Increasing the container volume increases the distance between collisions and decreases the number of collisions per unit time.

        (b)   Increase. Increasing temperature increases the rms speed of the gas molecules, which increases the number of collisions per unit time.

        (c)   Decrease. Increasing the molar mass of a gas decreases the rms speed of the molecules, which decreases the number of collisions per unit time.

10.76   (a)   False. The average kinetic energy per molecule in a collection of gas molecules is the same for all gases at the same temperature.

        (b)   True.

        (c)   False. The molecules in a gas sample at a given temperature exhibit a distribution of kinetic energies.

        (d)   True.

        (e)   False. Gas molecules at the same temperature exhibit a distribution of speeds.

10.77   *Analyze/Plan.* Given two gases, compare their rms speeds. Use Equation 10.23.   *Solve.*

$$\frac{u_{rms}(\text{He})}{u_{rms}(\text{WF}_6)} = \sqrt{\frac{\text{MM}(\text{WF}_6)}{\text{MM}(\text{He})}} = \sqrt{\frac{297.9 \text{ g/mol}}{4.003 \text{ g/mol}}} = 8.63$$

The rms speed of WF_6 is approximately 9 times slower than that of He.

10.78   The gas undergoes a chemical reaction that has fewer gas particles in products than in reactants. Mass is conserved when a chemical reaction occurs, so the mass of (flask + contents) remains constant. Pressure is directly proportional to number of particles, so pressure decreases as the number of gaseous particles decreases. One simple example of such a reaction is the dimerization of NO_2: $2\,NO_2(g) \rightarrow N_2O_4$.

10.79   *Analyze/Plan.* Apply the concepts of the Kinetic-Molecular Theory (KMT) to the situation where a gas is heated at constant volume. Determine how the quantities in (a)–(d) are affected by this change.   *Solve.*

        (a)   Average kinetic energy is proportional to temperature (K), so average kinetic energy of the molecules increases.

        (b)   The average kinetic energy of a gas is $1/2\,mu_{rms}^2$. Molecular mass doesn't change as T increases; average kinetic energy increases so rms speed (u) increases. (Also, $u_{rms} = (3RT/MM)^{1/2}$, so u_{rms} is directly related to T.)

(c)   As T and thus rms molecular speed increase, molecular momentum (mu) increases and the strength of an average impact with the container wall increases.

(d)   As T and rms molecular speed increase, the molecules collide more frequently with the container walls, and the total number of collisions per second increases.

10.80   (a)   They have the same number of molecules (equal volumes of gases at the same temperature and pressure contain equal numbers of molecules).

(b) N_2 is more dense because it has the larger molar mass. Because the volumes of the samples and the number of molecules are equal, the gas with the larger molar mass will have the greater density.

(c)   The average kinetic energies are equal (statement 5, section 10.7).

(d) CH_4 will effuse faster. The lighter the gas molecules, the faster they will effuse (Graham's law).

10.81   (a)   *Plan.* The larger the molar mass, the slower the average speed (at constant temperature).

Solve. In order of increasing speed (and decreasing molar mass):
$HBr < NF_3 < SO_2 < CO < Ne$

(b)   *Plan.* Follow the logic of Sample Exercise 10.13.   *Solve.*

$$u_{rms} = \sqrt{\frac{3\,RT}{MM}} = \left(\frac{3 \times 8.314\ \text{kg-m}^2/\text{s}^2\text{-mol-K} \times 298\ \text{K}}{71.0 \times 10^{-3}\ \text{kg/mol}}\right)^{1/2} = 324\ \text{m/s}$$

(c)   *Plan.* Use Equation 10.21 to calculate the most probable speed, u_{mp}. MM of O_3 = 48.0 g/mol; T = 270 K.   *Solve.*

$$u_{mp} = \sqrt{\frac{2\,RT}{MM}} = \left(\frac{2 \times 8.314\ \text{kg-m}^2/\text{s}^2\text{-mol-K} \times 270\ \text{K}}{48.0 \times 10^{-3}\ \text{kg/mol}}\right)^{1/2} = 306\ \text{m/s}$$

10.82   (a)   *Plan.* The greater the molecular (and molar) mass, the smaller the rms and average speeds of the molecules. Calculate the molar mass of each gas, and place them in decreasing order of mass and increasing order of rms and average speed.

Solve. CO = 28 g/mol; SF_6 = 146 g/mol; H_2S = 34 g/mol; Cl_2 = 71 g/mol; HBr = 81 g/mol. In order of increasing speed (and decreasing molar mass):

$SF_6 < HBr < Cl_2 < H_2S < CO$

(b)   *Plan.* Follow the logic of Sample Exercise 10.13.   *Solve.*

$$CO:\ u_{rms} = \sqrt{\frac{3RT}{MM}} = \left(\frac{3 \times 8.314\ \text{kg-m}^2/\text{s}^2\text{-mol-K} \times 300\ \text{K}}{28.0 \times 10^{-3}\ \text{kg/mol}}\right)^{1/2} = 5.17 \times 10^2\ \text{m/s}$$

$$Cl_2:\ u_{rms} = \left(\frac{3 \times 8.314\ \text{kg-m}^2/\text{s}^2\text{-mol-K} \times 300\ \text{K}}{70.9 \times 10^{-3}\ \text{kg/mol}}\right)^{1/2} = 3.25 \times 10^2\ \text{m/s}$$

As expected, the lighter molecule moves at the greater speed.

(c)   *Plan.* From Equations 10.20 and 10.21, we see that the ratio of most probable speed to rms speed is $(2/3)^{1/2}$. Use this ratio and the results from part (b) to calculate most probable speeds.   *Solve.*

$$CO: u_{mp} = (2/3)^{1/2}(5.17 \times 10^2 \text{ m/s}) = 422 \text{ m/s}$$

$$Cl_2: u_{mp} = (2/3)^{1/2}(3.25 \times 10^2 \text{ m/s}) = 265 \text{ m/s}$$

The lighter molecule, CO, has the greater most probable speed. Note that the most probable speed is less than the rms speed, as shown on Figure 10.12(b).

10.83   Statements (a) and (d) are true. Statement (b) is false because effusion is the escape of gas molecules through a tiny hole, while diffusion is the distribution of a gas throughout space or throughout another substance. Statement (c) is false because perfume molecules travel to your nose by the process of diffusion, not effusion.

10.84   Write each proportionality relationship as an equation, then combine them to obtain a formula for mean free path.

The operational symbols and units are: mean free path, λ, meters (m); temperature, T, kelvins (K); pressure, P, atmospheres (atm); diameter of a gas molecule, d, meters (m), constant, R_{mfp}.

$\lambda = \text{constant} \times T; \lambda = \text{constant}/P; \lambda = \text{constant}/d^2$

Combining : $\lambda = \dfrac{R_{mfp} \times T}{P \times d^2}$

The units of R_{mfp} are chosen and arranged so that they cancel the units of measurement, leaving an appropriate length unit for λ.

With the units defined above, R_{mfp} will have units of $\dfrac{\text{atm-m}^3}{K}$.

(Note that $1 \text{ m}^3 = 10^3 \text{ dm}^3 = 1000 \text{ L}$. Substituting, R_{mfp} would have units of $\dfrac{\text{atm-L}}{K}$, with the factor of 1000 incorporated into the value of R_{mfp}.)

10.85   *Plan.* The heavier the molecule, the slower the rate of effusion. Thus, the order for increasing rate of effusion is in the order of decreasing mass.   *Solve.*

rate $^2H^{37}Cl$ < rate $^1H^{37}Cl$ < rate $^2H^{35}Cl$ < rate $^1H^{35}Cl$

10.86 $\dfrac{\text{rate}^{235}U}{\text{rate}^{238}U} = \sqrt{\dfrac{238.05}{235.04}} = \sqrt{1.0128} = 1.0064$

There is a slightly greater rate enhancement for $^{235}U(g)$ atoms than $^{235}UF_6(g)$ molecules (1.0043), because ^{235}U is a greater percentage (100%) of the mass of the diffusing particles than in $^{235}UF_6$ molecules. The masses of the isotopes were taken from *The Handbook of Chemistry and Physics.*

10.87   *Analyze.* Given: relative effusion rates of two gases at same temperature. Find: molecular formula of one of the gases. *Plan.* Use Graham's law to calculate the formula weight of arsenic(III) sulfide, and thus the molecular formula.   *Solve.*

$$\dfrac{\text{rate (sulfide)}}{\text{rate (Ar)}} = \left[\dfrac{39.9}{\text{MM (sulfide)}} \right]^{1/2} = 0.28$$

MM (sulfide) = $39.9 / 0.28^2 = 509$ g/mol (two significant figures)

The empirical formula of arsenic(III) sulfide is As_2S_3, which has a formula mass of 246.1. Twice this is 490 g/mol, close to the value estimated from the effusion experiment. Thus, the formula of the gas phase molecule is As_4S_6.

10.88   The time required is proportional to the reciprocal of the effusion rate.

$$\frac{\text{rate}(X)}{\text{rate}(O_2)} = \frac{105\,s}{31\,s} = \left[\frac{32\,g\,O_2}{MM_x}\right]^{1/2}; \quad MM_x = 32\,g\,O_2 \times \left[\frac{105}{31}\right]^2 = 370\,g/mol\ (\text{two sig figs})$$

Nonideal-Gas Behavior (Section 10.9)

10.89   (a)   Nonideal-gas behavior is observed at very high pressures and/or low temperatures.

                (b)   The real volumes of gas molecules and attractive intermolecular forces between molecules cause gases to behave nonideally.

10.90   Ideal-gas behavior is most likely to occur at high temperature and low pressure, so the atmosphere on Mercury is more likely to obey the ideal-gas law. The higher temperature on Mercury means that the kinetic energies of the molecules will be larger relative to intermolecular attractive forces. Further, the gravitational attractive forces on Mercury are lower because the planet has a much smaller mass. This means that for the same column mass of gas (Figure 10.1), atmospheric pressure on Mercury will be lower.

10.91   Statement (b) is true. The constants a and b are characteristic of a particular gas and are independent of pressure and temperature.

10.92   *Plan.* The constants a and b are part of the correction terms in the van der Waals equation. The smaller the values of a and b, the smaller the corrections and the more ideal the gas.   *Solve.*

             Ar ($a = 1.34$, $b = 0.0322$) will behave more like an ideal gas than CO_2 ($a = 3.59$, $b = 0.0427$) at high pressures.

10.93   *Analyze/Plan.* Follow the logic in Sample Exercise 10.15. Use the ideal-gas equation to calculate pressure in (a), the van der Waals equation in (b). n = 1.00 mol, V = 5.00 L, T = 25 °C = 298 K; $a = 6.49\ L^2$-atm/mol^2, $b = 0.0562$ L/mol.

             (a) $P = \dfrac{nRT}{V} = 1.00\,mol \times \dfrac{298\,K}{5.00\,L} \times \dfrac{0.08206\,\text{L-atm}}{\text{mol-K}} = 4.89$ atm

             (b) $P = \dfrac{nRT}{V - nb} - \dfrac{n^2 a}{V^2};$

                   $P = \dfrac{(1.00\,mol)(298\,K)(0.08206\,\text{L-atm/mol-K})}{5.00\,L - (1.00\,mol)(0.0562\,L/mol)} - \dfrac{(1.00\,mol)^2(6.49\,L^2\text{-atm/mol}^2)}{(5.00\,L)^2}$

                   $P = 4.9463$ atm $- 0.2596$ atm $= 4.6868 = 4.69$ atm

             (c)   From Sample Exercise 10.15, the difference at 22.41 L between the ideal and van der Waals results is (1.00 − 0.990) = 0.010 atm. At 5.00 L, the difference is (4.89 − 4.69) = 0.20 atm. The effects of both molecular attractions, the a correction, and molecular volume, the b correction, increase with decreasing volume. For the a correction, V^2 appears in the denominator, so the correction increases exponentially as V decreases. For the b correction, nb is a larger portion of the total volume as V decreases. That is, 0.0562 L is 1.1% of 5.0 L, but only 0.25% of 22.41 L. Qualitatively, molecular attractions are more important as the amount of free space decreases and the number of molecular collisions increase. Molecular volume is a larger part of the total volume as the container volume decreases.

10.94   *Analyze.* Conditions and amount of $CCl_4(g)$ are given. *Plan.* Use ideal-gas equation and van der Waals equation to calculate pressure of gas at these conditions.   *Solve.*

(a) $P = 1.00 \text{ mol} \times \dfrac{0.08206 \text{ L-atm}}{\text{mol-K}} \times \dfrac{353 \text{ K}}{33.3 \text{ L}} = 0.870 \text{ atm}$

(b) $P = \dfrac{nRT}{V-nb} - \dfrac{an^2}{V^2} = \dfrac{1.00 \times 0.08206 \times 353}{33.3 - (1.00 \times 0.1383)} - \dfrac{20.4(1.00)^2}{(33.3)^2} = 0.855 \text{ atm}$

Check. The van der Waals result indicates that the real pressure will be less than the ideal pressure. That is, intermolecular forces reduce the effective number of particles and the real pressure. This is reasonable for 1 mole of gas at relatively low temperature and pressure.

(c)   According to Table 10.3, CCl_4 has larger *a* and *b* values. That is, CCl_4 experiences stronger intermolecular attractions and has a larger molecular volume than Cl_2 does. CCl_4 will deviate more from ideal behavior at these conditions than Cl_2 will.

10.95   *Analyze.* Given the *b* value of Xe, 0.0510 L/mol, calculate the radius of a Xe atom.

Plan. Use Avogadro's number to change L/mol to L/atom. Use the volume formula, $V = 4/3 \, \pi r^3$ and units conversion to obtain the radius in Å. 1 L = 1 dm^3.   *Solve.*

$\dfrac{0.0510 \text{ L}}{1 \text{ mol Xe}} \times \dfrac{1 \text{ mol Xe}}{6.022 \times 10^{23} \text{ Xe atoms}} \times \dfrac{1 \text{ dm}^3}{1 \text{ L}} = 8.4689 \times 10^{-26} = 8.47 \times 10^{-26} \text{ dm}^3$

$V = 4/3 \, \pi r^3; \quad r^3 = 3 \, V/4 \, \pi; \quad r = \left(3 \, V / 4 \, \pi\right)^{1/3}$

$r = \left(\dfrac{3 \times 8.4689 \times 10^{-26} \text{ dm}^3}{4 \times 3.14159}\right)^{1/3} = 2.7243 \times 10^{-9} = 2.72 \times 10^{-9} \text{ dm}$

$2.72 \times 10^{-9} \text{ dm} \times \dfrac{1 \text{ m}}{10 \text{ dm}} \times \dfrac{1 \text{ Å}}{1 \times 10^{-10} \text{ m}} = 2.72 \text{ Å}$

The calculated value is the nonbonding radius. From Figure 7.7 in Section 7.3, the bonding atomic radius of Xe is 1.40 Å. We expect the nonbonding radius of an atom to be larger than the bonding radius, but our calculated value is nearly twice as large.

10.96   The van der Waals radius we calculate from the *b* parameter in Table 10.3 is more closely associated with the nonbonding atomic radius of an atom. From section 7.3, the nonbonding or *van der Waals* radius is half of the shortest internuclear distance when two nonbonding atoms collide. So, radii calculated from the van der Waals equation are nonbonding radii. According to the kinetic-molecular theory, ideal-gas particles undergo perfectly elastic, billiard-ball collisions, in keeping with the definition of nonbonding radii.

Also, from the results of Exercise 10.95, the atomic radius calculated from the van der Waals *b* value is twice as large as the bonding atomic radius from Figure 7.7. Nonbonding radii are larger than bonding radii because no lasting penetration of electron clouds occurs during a nonbonding collision.

Additional Exercises

10.97   *Analyze.* The height of an Hg column is 760 mm at a pressure of 1.01×10^5 Pa. *Plan.* Develop a relationship between pressure, height of a column of liquid, and density of the liquid. Relationships that might prove useful: $P = F/A$; $F = m \times g$; $m = d \times V$; $V = A \times height$ *Solve.*

$$P = \frac{F}{A} = \frac{m \times g}{A} = \frac{d \times V \times g}{A} = \frac{d \times A \times h \times g}{A} = d \times h \times g; \quad d = \frac{P}{h \times g}$$

$g = 9.81 \text{ m/s}^2; h = 760 \text{ mm} = 0.760 \text{ m}; 1 \text{ Pa} = 1 \text{ kg/m-s}^2$

$$d = \frac{P}{h \times g} = \frac{1.01 \times 10^5 \text{ kg/m-s}^2}{0.760 \text{ m} \times 9.81 \text{ m/s}^2} = \frac{1.3547 \times 10^4 \text{ kg}}{m^3} = \frac{1.35 \times 10^4 \text{ kg}}{m^3}$$

$$d = \frac{1.3547 \times 10^4 \text{ kg}}{m^3} \times \frac{1000 \text{ g}}{\text{kg}} \times \frac{1 \text{ m}^3}{(100)^3 \text{ cm}^3} \times \frac{1 \text{ cm}^3}{1 \text{ mL}} = 13.5 \text{ g/mL}$$

10.98 $P_1V_1 = P_2V_2; V_2 = P_1V_1/P_2$

$$V_2 = \frac{3.0 \text{ atm} \times 1.0 \text{ mm}^3}{730 \text{ torr}} \times \frac{760 \text{ torr}}{1 \text{ atm}} = 3.1 \text{ mm}^3$$

10.99 $PV = nRT$, $n = PV/RT$. Because RT is constant, n is proportional to PV.

Total available $n = (15.0 \text{ L} \times 1.00 \times 10^2 \text{ atm}) - (15.0 \text{ L} \times 1.00 \text{ atm}) = 1485$

$$= 1.49 \times 10^3 \text{ L-atm}$$

Each balloon holds $2.00 \text{ L} \times 1.00 \text{ atm} = 2.00 \text{ L-atm}$

$$\frac{1485 \text{ L-atm available}}{2.00 \text{ L-atm/balloon}} = 742.5 = 742 \text{ balloons}$$

(Only 742 balloons can be filled completely, with a bit of He left over.)

10.100 $P = \frac{nRT}{V}; n = 1.4 \times 10^{-5} \text{ mol}, V = 0.600 \text{ L}, T = 23 \text{ °C} = 296 \text{ K}$

$$P = 1.4 \times 10^{-5} \text{ mol} \times \frac{0.08206 \text{ L-atm}}{\text{mol-K}} \times \frac{296 \text{ K}}{0.600 \text{ L}} = 5.7 \times 10^{-4} \text{ atm} = 0.43 \text{ mm Hg}$$

10.101   (a)   Change mass CO_2 to mol CO_2. $P = 1.00 \text{ atm}, T = 27 \text{ °C} = 300 \text{ K}$.

$$6 \times 10^6 \text{ tons } CO_2 \times \frac{2000 \text{ lb}}{\text{ton}} \times \frac{453.6 \text{ g}}{\text{lb}} \times \frac{1 \text{ mol } CO_2}{44.01 \text{ g } CO_2} = 1.237 \times 10^{11} = 1 \times 10^{11} \text{ mol}$$

$$V = \frac{nRT}{P} = 1.237 \times 10^{11} \text{ mol} \times \frac{300 \text{ K}}{1.00 \text{ atm}} \times \frac{0.08206 \text{ L-atm}}{\text{mol-K}} = 3.045 \times 10^{12} = 3 \times 10^{12} \text{ L}$$

(b) $1.237 \times 10^{11} \text{ mol } CO_2 \times \frac{44.01 \text{ g } CO_2}{\text{mol } CO_2} \times \frac{1 \text{ cm}^3}{1.2 \text{ g}} \times \frac{1 \text{ L}}{1000 \text{ cm}^3} = 4.536 \times 10^9 = 5 \times 10^9 \text{ L}$

(c) $n = 1.237 \times 10^{11} \text{ mol}, P = 70 \text{ atm}, T = 30 \text{ °C} = 303 \text{ K}$

$$V = \frac{nRT}{P} = 1.237 \times 10^{11} \text{ mol} \times \frac{303 \text{ K}}{70 \text{ atm}} \times \frac{0.08206 \text{ L-atm}}{\text{mol-K}} = 4.3939 \times 10^{10} = 4 \times 10^{10} \text{ L}$$

10.102   (a) $n = \dfrac{PV}{RT} = 3.00 \text{ atm} \times \dfrac{\text{mol-K}}{0.08206 \text{ L-atm}} \times \dfrac{110 \text{ L}}{300 \text{ K}} = 13.4 \text{ mol } C_3H_8(g)$

      (b) $\dfrac{0.590 \text{ g } C_3H_8 \text{ (l)}}{1 \text{ mL}} \times 110 \times 10^3 \text{ mL} \times \dfrac{1 \text{ mol } C_3H_8}{44.094 \text{ g}} = 1.47 \times 10^3 \text{ mol } C_3H_8 \text{ (l)}$

      (c)   Using C_3H_8 in a 110-L container as an example, the ratio of moles liquid to moles gas that can be stored in a certain volume is $\dfrac{1.47 \times 10^3 \text{ mol liquid}}{13.4 \text{ mol gas}} = 110$.

             A container with a fixed volume holds many more moles (molecules) of C_3H_8(l) because in the liquid phase the molecules are touching. In the gas phase, the molecules are far apart (statement 2, Section 10.7), and many fewer molecules will fit in the container.

10.103   Vol of room $= 12 \text{ ft} \times 20 \text{ ft} \times 9 \text{ ft} \times \dfrac{12^3 \text{ in.}^3}{1 \text{ft}^3} \times \dfrac{2.54^3 \text{ cm}^3}{1^3 \text{ in.}^3} \times \dfrac{1 \text{ L}}{1000 \text{ cm}^3} = 61,164 = 6 \times 10^4 \text{ L}$

         Calculate the **total** moles of gas in the laboratory at the conditions given.

         $n_t = \dfrac{PV}{RT} = 1.00 \text{ atm} \times \dfrac{\text{mol-K}}{0.08206 \text{ L-atm}} \times \dfrac{61,164 \text{ L}}{297 \text{ K}} = 2510 = 3 \times 10^3 \text{ mol gas}$

         A $Ni(CO)_4$ concentration of 1 part in 10^9 means 1 mol $Ni(CO)_4$ in 1×10^9 total moles of gas.

         $\dfrac{\text{x mol } Ni(CO)_4}{2.510 \times 10^3 \text{ mol gas}} = \dfrac{1}{1 \times 10^9} = 2.510 \times 10^{-6} = 3 \times 10^{-6} \text{ mol } Ni(CO)_4$

         $2.510 \times 10^{-6} \text{ mol } Ni(CO)_4 \times \dfrac{170.74 \text{ g } Ni(CO)_4}{1 \text{ mol } Ni(CO)_4} = 4.286 \times 10^{-3} = 4 \times 10^{-3} \text{ g} = 4 \text{ mg } Ni(CO)_4$

10.104   (a)   mol = g/MM; assume mol Ar = mol X;

             $\dfrac{\text{g Ar}}{39.948 \text{ g/mol}} = \dfrac{\text{g X}}{\text{MM X}}; \dfrac{3.224 \text{ g Ar}}{39.948 \text{ g/mol}} = \dfrac{8.102 \text{ g X}}{\text{MM X}}$

             $\text{MM X} = \dfrac{(8.102 \text{ g X})(39.948 \text{ g/mol})}{3.224 \text{ g Ar}} = 100.39 = 100.4 \text{ g/mol}$

      (b)   Assume mol Ar = mol X. For gases, PV= nRT and n = PV/RT. For moles of the two gases to be equal, the implied assumption is that P, V, and T are constant. Because we use the same container for both gas samples, constant V is a good assumption. The values of P and T are not explicitly stated.

             We also assume that the gases behave ideally. At ambient conditions, this is a reasonable assumption.

10.105   It is simplest to calculate the partial pressure of each gas as it expands into the total volume, then sum the partial pressures.

         $P_2 = P_1V_1/V_2; P_{N_2} = 265 \text{ torr } (1.0 \text{ L}/2.5 \text{ L}) = 106 = 1.1 \times 10^2 \text{ torr}$

         $P_{Ne} = 800 \text{ torr } (1.0 \text{ L}/2.5 \text{ L}) = 320 = 3.2 \times 10^2 \text{ torr}$

         $P_{H_2} = 532 \text{ torr } (0.5 \text{ L}/2.5 \text{ L}) = 106 = 1.1 \times 10^2 \text{ torr}$

         $P_t = P_{N_2} + P_{Ne} + P_{H_2} = (106 + 320 + 106) \text{ torr} = 532 = 5.3 \times 10^2 \text{ torr}$

10.106   **(a)** $n = \dfrac{PV}{RT} = 0.980 \text{ atm} \times \dfrac{\text{mol-K}}{0.08206 \text{ L-atm}} \times \dfrac{0.524 \text{ L}}{347 \text{ K}} = 0.018034 = 0.0180 \text{ mol air}$

$\text{mol O}_2 = 0.018034 \text{ mol air} \times \dfrac{0.2095 \text{ mol O}_2}{1 \text{ mol air}} = 0.003778 = 0.00378 \text{ mol O}_2$

(b) $C_8H_{18}(l) + 25/2 \ O_2(g) \rightarrow 8 \ CO_2(g) + 9 \ H_2O(g)$

(The H_2O produced in an automobile engine is in the gaseous state.)

$0.003778 \text{ mol O}_2 \times \dfrac{1 \text{ mol C}_8 \text{ H}_{18}}{12.5 \text{ mol O}_2} \times \dfrac{114.2 \text{ g C}_8\text{H}_{18}}{1 \text{ mol C}_8\text{H}_{18}} = 0.0345 \text{ g C}_8\text{H}_{18}$

10.107   **(a)**   Pressure percent = mole percent. Change pressure/mole percents to mole fraction. Partial pressure of each gas is mole fraction (X) times total pressure. $P_x = X_x P_t$

$P_{N_2} = 0.748(0.985 \text{ atm}) = 0.737 \text{ atm}; \ P_{O_2} = 0.153(0.985 \text{ atm}) = 0.151 \text{ atm}$

$P_{CO_2} = 0.037(0.985 \text{ atm}) = 0.03645 = 0.036 \text{ atm}$

$P_{H_2O} = 0.062(0.985 \text{ atm}) = 0.06107 = 0.061 \text{ atm}$

(b) $PV = nRT, n = PV/RT; \ P = 0.036 \text{ atm}, V = 0.455 \text{ L}, T = 37 \ ^\circ C = 310 \text{ K}$

$n = 0.03645 \text{ atm} \times \dfrac{0.455 \text{ L}}{310 \text{ K}} \times \dfrac{\text{mol-K}}{0.08206 \text{ L-atm}} = 6.520 \times 10^{-4} = 6.5 \times 10^{-4} \text{ mol}$

(c) $C_6H_{12}O_6 + 6 \ O_2 \rightarrow 6 \ CO_2 + 6 \ H_2O$

$6.520 \times 10^{-4} \text{ mol CO}_2 \times \dfrac{1 \text{ mol C}_6\text{H}_{12}\text{O}_6}{6 \text{ mol CO}_2} \times \dfrac{180.15 \text{ g C}_6\text{H}_{12}\text{O}_6}{1 \text{ mol C}_6\text{H}_{12}\text{O}_6} = 0.01958$

$= 0.020 \text{ g C}_6\text{H}_{12}\text{O}_6$

10.108   V and T are the same for He and O_2.

$P_{He}V = n_{He}RT, \quad P_{He}/n_{He} = RT/V; \quad P_{O_2}/n_{O_2} = RT/V$

$\dfrac{P_{He}}{n_{He}} = \dfrac{P_{O_2}}{n_{O_2}} = n_{O_2} \times \dfrac{P_{O_2} \times n_{He}}{P_{He}}; \ n_{He} = 1.42 \text{ g He} \times \dfrac{1 \text{ mol He}}{4.003 \text{ g He}} = 0.3547 = 0.355 \text{ mol He}$

$n_{O_2} = \dfrac{158 \text{ torr}}{42.5 \text{ torr}} \times 0.355 \text{ mol} = 1.3188 = 1.32 \text{ mol O}_2; 1.3188 \text{ mol O}_2 \times \dfrac{32.00 \text{ g O}_2}{1 \text{ mol O}_2} = 42.2 \text{ g O}_2$

10.109   At constant temperature, an ideal gas at a certain pressure and volume, P_1V_1, expands into a larger volume and lower pressure, P_2V_2. This is a Boyle's law problem.

Let $V_1 = x$ L, $V_2 = 0.800$ L + x. Change 1.50 atm to torr, so the pressure units cancel.

$1.50 \text{ atm} \times 760 \text{ torr/atm} = 1140 \text{ torr}$

$P_1V_1 = P_2V_2$. 1140 torr (x L) = 695 torr [(0.800 + x) L]. Units cancel.

1140 x = 556 + 695 x; 445 x = 556; x = 1.25 L

Check. 1140 (1.25) = 695 (2.05); 1425 = 1425. The algebra is correct.

10.110 (a) The quantity $d/P = MM/RT$ should be a constant at all pressures for an ideal gas. It is not, however, because of nonideal behavior. If we graph d/P versus P, the ratio should approach ideal behavior at low P. At P = 0, $d/P = 2.2525$. Using this value in the formula $MM = d/P \times RT$, MM = 50.46 g/mol.

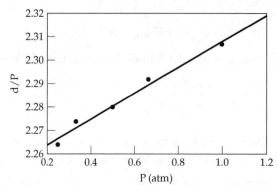

(b) The ratio d/P varies with pressure because of the finite volumes of gas molecules and attractive intermolecular forces.

10.111 Calculate the number of moles of Ar in the vessel:

n = (339.854 – 337.428)/39.948 = 0.060729 = 0.06073 mol

The total number of moles of the mixed gas is the same (Avogadro's law). Thus, the average atomic weight is (339.076 – 337.428)/0.060729 = 27.137 = 27.14. Let the mole fraction of Ne be x. Then,

x (20.183) + (1 – x) (39.948) = 27.137; 12.811 = 19.765 x; x = 0.6482

Neon is thus 64.82 mole percent of the mixture.

10.112 $u_{rms} = (3RT/MM)^{1/2}$; $u_{rms2} = 2 u_{rms1}$; $T_1 = -33 \,°C = 240$ K

$u_{rms1} = (3RT_1/MM)^{1/2}$; $u_{rms1}^2 = 3(240)R/MM = 720R/MM$

$u_{rms1} = (720R/MM)^{1/2}$; $u_{rms2} = 2u_{rms1} = (2)(720R/MM)^{1/2}$

$(2)(720R/MM)^{1/2} = (3RT_2/MM)^{1/2}$

$(2)^2(720R/MM) = 3RT_2/MM$; $(2)^2(720) = 3T_2$

$T_2 = (4)(720)/3 = 960$ K = 687 °C

Increasing the rms speed (u) by a factor of 2 requires heating to 960 K (or 687 °C), increasing the temperature by a factor of 4.

10.113 (a) Assumption 3 states that attractive and repulsive forces between molecules are negligible. All gases in the list are nonpolar. The largest and most structurally complex molecule, SF_6, is most likely to depart from this assumption.

(b) The monatomic gas Ne is smallest and least structurally complex, so it will behave most like an ideal gas.

(c) Root-mean-square speed is inversely related to molecular mass. The lightest gas, CH_4, has the highest rms speed.

(d)   The heaviest and most structurally complex is SF_6. Also, S and F have larger atomic radii than C and H; this means that S–F bonds will be longer than C–H bonds and the volume of SF_6 will be greater than that of CH_4. It is reasonable to assume that SF_6 will occupy the greatest molecular volume relative to total volume. A quantitative measure is the *b* value in Table 10.3, with units of L/mol. Unfortunately, SF_6 does not appear in Table 10.3.

(e)   Average kinetic energy is only related to absolute (K) temperature. At the same temperature, they all have the same average kinetic-molecular energy.

(f)   Rate of effusion is inversely related to molecular mass. The lighter the molecule, the faster it effuses. Ne and CH_4 have smaller molecular masses and effuse faster than N_2.

(g)   If SF_6 occupies the greatest molecular volume [see part (d)], we expect it to have the largest van der Waals *b* parameter.

10.114   (a)   The effect of intermolecular attraction becomes more significant as a gas is compressed to a smaller volume at constant temperature. This compression causes the pressure, and thus the number of intermolecular collisions, to increase. Intermolecular attraction causes some of these collisions to be inelastic, which amplifies the deviation from ideal behavior.

(b)   The effect of intermolecular attraction becomes less significant as the temperature of a gas is increased at constant volume. When the temperature of a gas is increased at constant volume, the pressure of the gas, the number of intermolecular collisions, and the average kinetic energy of the gas particles increase. This higher average kinetic energy means that a larger fraction of the molecules has sufficient kinetic energy to overcome intermolecular attractions, even though there are more total collisions. This increases the fraction of elastic collisions, and the gas more closely obeys the ideal-gas equation.

10.115   The larger and heavier the particle, in this case a single atom, the more likely it is to deviate from ideal behavior. Other than Rn, Xe is the largest (atomic radius = 1.40 Å), heaviest (molar mass = 131.3 g/mol) and most dense (5.90 g/L) noble gas. Its susceptibility to intermolecular interactions is also demonstrated by its ability to form compounds like XeF_4.

10.116   (a)   At STP, the molar volume $= 1\,mol \times \dfrac{0.08206\,L\text{-}atm}{mol\text{-}K} \times \dfrac{273\,K}{1\,atm} = 22.4\,L$

Dividing the value for *b*, 0.0322 L/mol, by 4, we obtain 0.00805 L. Thus, the volume of the Ar atoms is (0.00805/22.4)100 = 0.0359% of the total volume.

(b)   At 200 atm pressure (and 0 °C, standard temperature) the molar volume is 0.112 L, and the volume of the Ar atoms is 7.19% of the total volume.

10.117   (a) $120.00\,kg\,N_2(g) \times \dfrac{1000\,g}{1\,kg} \times \dfrac{1\,mol\,N_2}{28.0135\,g\,N} = 4283.6\,mol\,N_2$

$P = \dfrac{nRT}{V} = 4283.6\,mol \times \dfrac{0.08206\,L\text{-}atm}{mol\text{-}K} \times \dfrac{553\,K}{1100.0\,L} = 176.72 = 177\,atm$

(b)   Rearranging Equation [10.25] to isolate P, $P = \dfrac{nRT}{V - nb} - \dfrac{n^2 a}{V^2}$

$$P = \frac{(4283.6 \text{ mol})(0.08206 \text{ L-atm/mol-K})(553 \text{ K})}{1100.0 \text{ L} - (4283.6 \text{ mol})(0.0391 \text{ L/mol})} - \frac{(4283.6 \text{ mol})^2 (1.39 \text{ L}^2\text{-atm/mol}^2)}{(1100.0 \text{ L})^2}$$

$$P = \frac{194{,}388 \text{ L-atm}}{1100.0 \text{ L} - 167.5 \text{ L}} - 21.1 \text{ atm} = 208.5 \text{ atm} - 21.1 \text{ atm} = 187.4 \text{ atm}$$

(c)   The pressure corrected for the real volume of the N_2 molecules is 208.5 atm, 31.8 atm higher than the ideal pressure of 176.7 atm. The 21.1 atm correction for intermolecular forces reduces the calculated pressure somewhat, but the "real" pressure is still higher than the ideal pressure. The correction for the real volume of molecules dominates. Even though the value of b is small, the number of moles of N_2 is large enough so that the molecular volume correction is larger than the attractive forces correction.

Integrative Exercises

10.118   (a) $MM = \dfrac{gRT}{VP} = \dfrac{1.56 \text{ g}}{1.00 \text{ L}} \times \dfrac{0.08206 \text{ L-atm}}{\text{mol-K}} \times \dfrac{323 \text{ K}}{0.984 \text{ atm}} = 42.0 \text{ g/mol}$

Assume 100 g cyclopropane

$100 \text{ g} \times 0.857 \text{ C} = 85.7 \text{ g C} \times \dfrac{1 \text{ mol C}}{12.01 \text{ g}} = \dfrac{7.136 \text{ mol C}}{7.136} = 1 \text{ mol C}$

$100 \text{ g} \times 0.143 \text{ H} = 14.3 \text{ g H} \times \dfrac{1 \text{ mol H}}{1.008 \text{ g}} = \dfrac{14.19 \text{ mol H}}{7.136} = 2 \text{ mol H}$

The empirical formula of cyclopropane is CH_2 and the empirical formula weight is 12 + 2 = 14 g. The ratio of molar mass to empirical formula weight, 42.0 g/14 g, is 3; therefore, there are three empirical formula units in one cyclopropane molecule. The molecular formula is $3 \times (CH_2) = C_3H_6$.

(b)   Ar is a monoatomic gas. Cyclopropane molecules are larger and more structurally complex, even though the molar masses of Ar and C_3H_6 are similar. If both gases are at the same relatively low temperature, they have approximately the same average kinetic energy, and the same ability to overcome intermolecular attractions. We expect intermolecular attractions to be more significant for the more complex C_3H_6 molecules, and that C_3H_6 will deviate more from ideal behavior at the conditions listed. This conclusion is supported by the a values in Table 10.3. The a values for CH_4 and CO_2, more complex molecules than Ar atoms, are larger than the value for Ar. If the pressure is high enough for the volume correction in the van der Waals equation to dominate behavior, the larger C_3H_6 molecules definitely deviate more than Ar atoms from ideal behavior.

(c)   Cyclopropane, C_3H_6, MM = 42.0 g/mol; methane, CH_4 MM = 16.0. Rate of effusion through a pinhole is inversely related to molar mass. Cyclopropane would effuse through a pinhole slower than methane, because it has the greater molar mass.

10.119   *Plan.* Write the balanced equation for the combustion of methanol. Because amounts of both reactants are given, determine the limiting reactant. Use mole ratios to calculate moles H_2O produced, based on the amount of limiting reactant. Change moles to grams

H_2O, then use density to calculate volume of $H_2O(l)$ produced. Assume the condensed $H_2O(l)$ is at 25 °C, where density = 0.99707 g/mol. *Solve.*

methanol = $CH_3OH(l)$. $2\,CH_3OH(l) + 3\,O_2(g) \rightarrow 2\,CO_2(g) + 4\,H_2O(g)$

$$25.0\text{ mL }CH_3OH \times \frac{0.850\text{ g }CH_3OH}{mL} \times \frac{1\text{ mol }CH_3OH}{32.04\text{ g}} = 0.6632 = 0.663\text{ mol }CH_3OH$$

$$\text{mol }O_2 = n = \frac{PV}{RT} = 1.00\text{ atm} \times \frac{12.5\text{ L}}{273\text{ K}} \times \frac{mol\text{-}K}{0.08206\text{ L-atm}} = 0.5580 = 0.558\text{ mol }O_2$$

$$0.558\text{ mol }O_2 \times \frac{2\text{ mol }CH_3OH}{3\text{ mol }O_2} = 0.372\text{ mol }CH_3OH$$

0.558 mol O_2 can react with only 0.372 mol CH_3OH, so O_2 is the limiting reactant. Note that a large volume of $O_2(g)$ is required to completely react with a relatively small volume of $CH_3OH(l)$.

$$0.558\text{ mol }O_2 \times \frac{4\text{ mol }H_2O}{3\text{ mol }O_2} \times \frac{18.02\text{ g }H_2O}{1\text{ mol }H_2O} \times \frac{1\text{ mL }H_2O}{0.99707\text{ g }H_2O} = 13.446 = 13.4\text{ mL }H_2O$$

10.120   (a)   Get g C from mL CO_2; get g H from mL H_2O. Also calculate mol C and H, to use in part (b). Get g N by subtraction. Calculate % composition.

n = PV/RT. At STP, P = 1 atm, T = 273 K. (STP implies an infinite number of sig figs.)

$$n_{CO_2} = 0.08316\text{ L} \times \frac{1\text{ atm}}{273\text{ K}} \times \frac{mol\text{-}K}{0.08206\text{ L-atm}} = 0.003712\text{ mol }CO_2$$

$$0.003712\text{ mol }CO_2 \times \frac{1\text{ mol C}}{1\text{ mol }CO_2} \times \frac{12.0107\text{ g C}}{mol\text{ C}} = 0.044585 = 0.04458\text{ g }CO_2$$

$$n_{H_2O} = 0.07330\text{ L} \times \frac{1\text{ atm}}{273\text{ K}} \times \frac{mol\text{-}K}{0.08206\text{ L-atm}} = 3.2720 \times 10^{-3}$$

$$= 3.272 \times 10^{-3}\text{ mol }H_2O$$

$$3.2720 \times 10^{-3}\text{ mol }H_2O \times \frac{2\text{ mol H}}{1\text{ mol }H_2O} \times \frac{1.00794\text{ g H}}{mol\text{ H}} = 6.5959 \times 10^{-3}$$

$$= 6.596 \times 10^{-3}\text{ g H}$$

$$\text{mass \% X} = \frac{\text{mass X}}{\text{sample mass}} \times 100; \text{ sample mass} = 100.0\text{ mg} = 0.1000\text{ g}$$

$$\text{\% C} = \frac{0.044585\text{ g}}{0.1000\text{ g}} \times 100 = 44.585 = 44.58\text{\% C}$$

$$\text{\% H} = \frac{6.5959 \times 10^{-3}\text{ g H}}{0.1000\text{ g}} \times 100 = 6.5959 = 6.596\text{\% H}$$

$$\text{\% Cl} = \frac{0.01644\text{ g Cl}}{0.1000\text{ g}} \times 100 = 16.44\text{\% Cl}$$

% N = 100 − 44.58 − 6.596 − 16.44 = 32.38% N

  (b)   0.003712 mol C; $2(3.272 \times 10^{-3}) = 6.544 \times 10^{-3}$ mol H

$$0.01644\text{ g Cl} \times \frac{1\text{ mol C}}{35.453\text{ g Cl}} = 4.637 \times 10^{-4}\text{ mol Cl}$$

0.1000 g sample × 0.3238 mass fraction N = 0.03238 g N

$$0.03238\text{ g N} \times \frac{1\text{ mol N}}{14.0067\text{ g N}} = 0.0023118 = 0.002312\text{ mol N}$$

Divide by the smallest number of mol to find the simplest ratio of moles.

$$\frac{0.003712 \text{ mol C}}{4.637 \times 10^{-4}} = 8.005 \text{ C}$$

$$\frac{6.544 \times 10^{-3} \text{ mol H}}{4.637 \times 10^{-4}} = 14.11 \text{ H}$$

$$\frac{4.637 \times 10^{-4} \text{ mol Cl}}{4.637 \times 10^{-4}} = 1.000 \text{ Cl}$$

$$\frac{0.002312 \text{ mol N}}{4.637 \times 10^{-4}} = 4.985 \text{ N}$$

If we assume 14.11 is "close" to 14 (a reasonable assumption), the empirical formula is $C_8H_{14}N_5Cl$.

(c)   Molar mass of the compound is required to determine molecular formula when the empirical formula is known.

10.121   (a)   *Plan.* Use the ideal-gas law to calculate the moles CO_2 that react.

Solve. P(reacted) = P(initial) – P(final), at constant V, T. Because both CaO and BaO react with CO_2 in a 1:1 mole ratio, mol CaO + mol BaO = mol CO_2. Use molar masses to calculate % CaO in sample.

$$P(\text{reacted}) = 730 \text{ torr} - 150 \text{ torr} = 580 \text{ torr}; 580 \text{ torr} \times \frac{1 \text{ atm}}{760 \text{ torr}} = 0.76316 = 0.763 \text{ atm}$$

$$n = \frac{PV}{RT} = 0.76316 \text{ atm} \times \frac{1.0 \text{ L}}{298 \text{ K}} \times \frac{\text{mol-K}}{0.08206 \text{ L-atm}} = 0.03121 = 0.0312 \text{ mol } CO_2$$

(b)   *Plan.* Use the stoichiometry of the reaction and definition of moles to calculate the mass and Mass % of CaO.

Solve. $CaO(s) + CO_2(s) \rightarrow CaCO_3(s)$. $BaO(s) + CO_2(g) \rightarrow BaCO_3(s)$

mol CO_2 reacted = mol CaO + mol BaO

Let x = g CaO, 4.00 – x = g BaO

$$0.03121 = \frac{x}{56.08} + \frac{4.00 - x}{153.3}$$

0.03121(56.08)(153.3) = 153.3x + 56.08(4.00 – x)

268.3 = (153.3x – 56.08x) + 224.3

43.98 = 97.22x, x = 0.452 = 0.45 g CaO

$$\frac{0.452 \text{ g CaO}}{4.00 \text{ g sample}} \times 100 = 11.3 = 11\% \text{ CaO}$$

(By strict sig fig rules, the result has 2 sig figs, because 268 – 224 = 44 has 0 decimal places and 2 sig figs.)

10.122   (a) $5.00 \text{ g HCl} \times \dfrac{1 \text{ mol HCl}}{36.46 \text{ g HCl}} = 0.1371 = 0.137 \text{ mol HCl}$

$5.00 \text{ g NH}_3 \times \dfrac{1 \text{ mol NH}_3}{17.03 \text{ g NH}_3} = 0.2936 = 0.294 \text{ mol NH}_3$

The gases react in a 1:1 mole ratio, HCl is the limiting reactant and is completely consumed. (0.2936 mol − 0.1371 mol) = 0.1565 = 0.157 mol NH_3 remain in the system. $NH_3(g)$ is the only gas remaining after reaction.

(b) $V_t = 4.00$ L. $P = \dfrac{nRT}{V} = 0.1565 \text{ mol} \times \dfrac{0.08206 \text{ L-atm}}{\text{mol-K}} \times \dfrac{298 \text{ K}}{4.00 \text{ L}} = 0.957$ atm

(c) $0.137 \text{ mol HCl} \times \dfrac{1 \text{ mol NH}_4\text{Cl}}{1 \text{ mol HCl}} \times \dfrac{53.49 \text{ g NH}_4\text{Cl}}{1 \text{ mol NH}_4\text{Cl}} = 7.3284 = 7.33 \text{ g NH}_4\text{Cl}$

10.123 $n = \dfrac{PV}{RT} = 1.00 \text{ atm} \times \dfrac{\text{mol-K}}{0.08206 \text{ L-atm}} \times \dfrac{2.7 \times 10^{12} \text{ L}}{273 \text{ K}} = 1.205 \times 10^{11} = 1.2 \times 10^{11} \text{ mol CH}_4$

$$CH_4(g) + 2\,O_2(g) \rightarrow CO_2(g) + 2\,H_2O(l) \quad \Delta H^\circ = -890.4 \text{ kJ}$$

(At STP, H_2O is in the liquid state.)

$$\Delta H^\circ_{rxn} = \Delta H^\circ_f CO_2(g) + 2\Delta H^\circ_f H_2O(l) - \Delta H^\circ_f CH_4(g) - 2\Delta H^\circ_f O_2(g)$$

$$\Delta H^\circ_{rxn} = -393.5 \text{ kJ} + 2(-285.83 \text{ kJ}) - (-74.8 \text{ kJ}) - 0 = -890.4 \text{ kJ}$$

$$\dfrac{-890.4 \text{ kJ}}{1 \text{ mol CH}_4} \times 1.205 \times 10^{11} \text{ mol CH}_4 = -1.073 \times 10^{14} = -1.1 \times 10^{14} \text{ kJ}$$

The negative sign indicates heat evolved by the combustion reaction.

10.124   (a) 19 e^-, 9.5 e^- pairs

         $\ddot{\text{O}}\!\!\!-\!\!\dot{\text{C}}\text{l}-\ddot{\text{O}}\!:$

         Resonance structures can be drawn with the odd electron on O, but electronegativity considerations predict that it will be on Cl for most of the time.

   (b) ClO_2 is very reactive because it is an odd-electron molecule. Adding an electron (reduction) both pairs the odd electron and completes the octet of Cl. Thus, ClO_2 has a strong tendency to gain an electron and be reduced.

   (c) ClO_2^-, 20 e^-, 10 e^- pairs

         $\left[:\!\ddot{\text{O}}\!\!\!-\!\!\ddot{\text{C}}\text{l}-\ddot{\text{O}}\!:\right]^-$

   (d) 4 e^- domains around Cl, O–Cl–O bond angle ~107° (<109° owing to repulsion by nonbonding domains)

   (e)   Calculate mol Cl_2 from ideal-gas equation; determine limiting reactant; mass ClO_2 via mol ratios.

         $\text{mol Cl}_2 = \dfrac{PV}{RT} = 1.50 \text{ atm} \times \dfrac{2.00 \text{ L}}{294 \text{ K}} \times \dfrac{\text{mol-K}}{0.08206 \text{ L-atm}} = 0.1243 = 0.124 \text{ mol Cl}_2$

         $15.0 \text{ g NaClO}_2 \times \dfrac{1 \text{ mol NaClO}_2}{90.44 \text{ g}} = 0.1659 = 0.166 \text{ mol NaClO}_2$

         2 mol $NaClO_2$ are required for 1 mol Cl_2, so $NaClO_2$ is the limiting reactant. For every 2 mol $NaClO_2$ reacted, 2 mol ClO_2 are produced, so mol ClO_2 = mol $NaClO_2$.

         $0.1659 \text{ mol ClO}_2 \times \dfrac{67.45 \text{ g ClO}_2}{\text{mol}} = 11.2 \text{ g ClO}_2$

10.125   **(a)** $ft^3 \, CH_4 \to L \, CH_4 \to mol \, CH_4 \to mol \, CH_3OH \to g \, CH_3OH \to L \, CH_3OH$

$$10.7 \times 10^9 \, ft^3 \, CH_4 \times \frac{1 \, yd^3}{3^3 \, ft^3} \times \frac{1 \, m^3}{(1.0936)^3 \, yd^3} \times \frac{1 \, L}{1 \times 10^{-3} \, m^3} = 3.03001 \times 10^{11}$$

$$= 3.03 \times 10^{11} \, L \, CH_4$$

$$n = \frac{PV}{RT} = \frac{3.03 \times 10^{11} \, L \times 1.00 \, atm}{298 \, K} \times \frac{mol\text{-}K}{0.08206 \, L\text{-}atm} = 1.2391 \times 10^{10}$$

$$= 1.24 \times 10^{10} \, mol \, CH_4$$

$1 \, mol \, CH_4 = 1 \, mol \, CH_3OH$

$$1.2391 \times 10^{10} \, mol \, CH_3OH \times \frac{32.04 \, g \, CH_3OH}{mol \, CH_3OH} \times \frac{1 \, mL \, CH_3OH}{0.791 \, g} \times \frac{1 \, L}{1000 \, mL}$$

$$= 5.0189 \times 10^8 = 5.02 \times 10^8 \, L \, CH_3OH$$

(b) $CH_4(g) + 2 \, O_2(g) \to CO_2(g) + 2 \, H_2O(l) \quad \Delta H° = -890.4 \, kJ$

$\Delta H° = \Delta H_f° \, CO_2(g) + 2 \, \Delta H_f° \, H_2O(l) - \Delta H_f° \, CH_4(g) - 2 \, \Delta H_f° \, O_2(g)$

$\Delta H° = -393.5 \, kJ + 2(-285.83 \, kJ) - (-74.8 \, kJ) - 0 = -890.4 \, kJ$

$$1.2391 \times 10^{10} \, mol \, CH_4 \times \frac{-890.4 \, kJ}{1 \, mol \, CH_4} = -1.10 \times 10^{13} \, kJ$$

$CH_3OH(l) + 3/2 \, O_2(g) \to CO_2(g) + 2 \, H_2O(l) \quad \Delta H° = -726.6 \, kJ$

$\Delta H° = \Delta H_f° \, CO_2(g) + 2 \, \Delta H_f° \, H_2O(l) - \Delta H_f° \, CH_3OH(l) - 3/2 \, \Delta H_f° \, O_2(g)$

$= -393.5 \, kJ + 2(-285.83 \, kJ) - (-238.6 \, kJ) - 0 = -726.6 \, kJ$

$$1.2391 \times 10^{10} \, mol \, CH_3OH \times \frac{-726.6 \, kJ}{1 \, mol \, CH_3OH} = -9.00 \times 10^{12} \, kJ$$

(c)   Assume a volume of 1.00 L of each liquid.

$$1.00 \, L \, CH_4(l) \times \frac{466 \, g}{1 \, L} \times \frac{1 \, mol}{16.04 \, g} \times \frac{-890.4 \, kJ}{mol \, CH_4} = -2.59 \times 10^4 \, kJ/L \, CH_4$$

$$1.00 \, L \, CH_3OH \times \frac{791 \, g}{1 \, L} \times \frac{1 \, mol}{32.04 \, g} \times \frac{-726.6 \, kJ}{mol \, CH_3OH} = -1.79 \times 10^4 \, kJ/L \, CH_3OH$$

Clearly $CH_4(l)$ has the higher enthalpy of combustion per unit volume.

10.126   After reaction, the flask contains $IF_5(g)$ and whichever reactant is in excess. Determine the limiting reactant, which regulates the moles of IF_5 produced and moles of excess reactant.

$I_2(s) + 5F_2(g) \to 2 \, IF_5(g)$

$$10.0 \, g \, I_2 \times \frac{1 \, mol \, I_2}{253.8 \, g \, I_2} \times \frac{5 \, mol \, F_2}{1 \, mol \, I_2} = 0.1970 = 0.197 \, mol \, F_2$$

$$10.0 \, g \, F_2 \times \frac{1 \, mol \, F_2}{38.00 \, g \, F_2} = 0.2632 = 0.263 \, mol \, F_2 \text{ available}$$

I_2 is the limiting reactant; F_2 is in excess.

$0.263 \, mol \, F_2$ available $- 0.197 \, mol \, F_2$ reacted $= 0.066 \, mol \, F_2$ remain.

$$10.0 \, g \, I_2 \times \frac{1 \, mol \, I_2}{253.8 \, g \, I_2} \times \frac{2 \, mol \, IF_5}{1 \, mol \, I_2} = 0.0788 \, mol \, IF_5 \text{ produced}$$

(a) $P_{IF_5} = \dfrac{nRT}{V} = 0.0788 \text{ mol} \times \dfrac{0.08206 \text{ L-atm}}{\text{mol-K}} \times \dfrac{398 \text{ K}}{5.00 \text{ L}} = 0.515 \text{ atm}$

(b) $\chi_{IF_5} = \dfrac{\text{mol IF}_5}{\text{mol IF}_5 + \text{mol F}_2} = \dfrac{0.0788}{0.0788 + 0.066} = 0.544$

(c)   42 valence e^-, 21 e^- pairs

(d) $0.0788 \text{ mol IF}_5 \times \dfrac{221.90 \text{ g IF}_5}{\text{mol IF}_5} = 17.4857 = 17.5 \text{ g IF}_5 \text{ produced}$

$0.066 \text{ mol F}_2 \times \dfrac{38.00 \text{ g F}_2}{\text{mol F}_2} = 2.508 = 2.5 \text{ g F}_2 \text{ remain}$

Total mass in flask = 17.5 g IF$_5$ + 2.5 g F$_2$ = 20.00 g; mass is conserved.

10.127   (a) $MgCO_3(s) + 2HCl(aq) \rightarrow MgCl_2(aq) + H_2O(l) + CO_2(g)$

$CaCO_3(s) + 2HCl(aq) \rightarrow CaCl_2(aq) + H_2O(l) + CO_2(g)$

(b) $n = \dfrac{PV}{RT} = 743 \text{ torr} \times \dfrac{1 \text{ atm}}{760 \text{ torr}} \times \dfrac{\text{mol-K}}{0.08206 \text{ L-atm}} \times \dfrac{1.72 \text{ L}}{301 \text{ K}}$

$= 0.06808 = 0.0681 \text{ mol CO}_2$

(c)   x = g MgCO$_3$, y = g CaCO$_3$, x + y = 6.53 g

mol MgCO$_3$ + mol CaCO$_3$ = mol CO$_2$ total

$\dfrac{x}{84.32} + \dfrac{y}{100.09} = 0.06808; \quad y = 6.53 - x$

$\dfrac{x}{84.32} + \dfrac{6.53 - x}{100.09} = 0.06808$

100.09x − 84.32x + 84.32(6.53) = 0.06808 (84.32)(100.09)

15.77x + 550.610 = 574.549; x = 1.52 g MgCO$_3$

mass % MgCO$_3$ = $\dfrac{1.52 \text{ g MgCO}_3}{6.53 \text{ g sample}} \times 100 = 23.3\%$

[By strict sig fig rules, the answer has 2 sig figs: 15.77x + 551 (3 digits from 6.53) = 575; 575 − 551 = 24 (no decimal places, 2 sig figs) leads to 1.5 g MgCO$_3$ and 23% MgCO$_3$]

11 Liquids and Intermolecular Forces

Visualizing Concepts

In this chapter we will use the temperature units °C and K interchangeably when designating specific heats and *changes* in temperature.

11.1   (a)   The diagram best describes a liquid.

       (b)   In the diagram, the particles are close together, mostly touching but there is no regular arrangement or order. This rules out a gaseous sample, where the particles are far apart, and a crystalline solid, which has a regular repeating structure in all three directions.

11.2   (a)   (i)   Hydrogen bonding; H–F interactions qualify for this narrowly defined interaction.

            (ii)   London dispersion forces, the only intermolecular forces between nonpolar F_2 molecules.

            (iii)   Ion-dipole forces between Na^+ cation and the negative end of a polar covalent water molecule.

            (iv)   Dipole-dipole forces between oppositely charged portions of two polar covalent SO_2 molecules.

       (b)   London dispersion forces in (ii) are probably the weakest.

11.3   (a)   The viscosity of glycerol will be greater than that of 1-propanol.

       (b)   Viscosity is the resistance of a substance to flow. The stronger the intermolecular forces in a liquid, the greater its viscosity. Hydrogen bonding is the predominant force for both molecules. Glycerol has three times as many –OH groups and many more hydrogen-bonding interactions than 1-propanol, so it experiences stronger intermolecular forces and greater viscosity. (Both molecules have the same carbon-chain length, so dispersion forces are similar.)

11.4   *Analyze.* When heat is added to a liquid, the temperature of the liquid rises. If enough heat is added to reach the boiling point (bp), any excess heat is used to vaporize the liquid. If heat is still available when all the liquid is converted to gas, the temperature of the gas rises.

      Plan. Use the specific heat of $CH_4(l)$ to calculate the amount of heat required to raise the temperature of 32.0 g of $CH_4(l)$ from –170 °C to –161.5 °C. If this is less than 42 kJ, use ΔH_{vap} to calculate the energy required to vaporize the liquid, and so on, until exactly 42.0 kJ has been used to increase the temperature and/or change the state of CH_4.

Solve. Heat the liquid to its boiling point: ΔT = [–161.5 °C – (–170 °C)] = 8.5 °C = 8.5 K
(Note that the magnitude of one degree is the same in Kelvins and Celsius.)

$$\frac{3.48\ J}{g\text{-}K} \times 32.0\ g\ CH_4 \times 8.5\ K = 946.56 = 9.5 \times 10^2\ J = 0.95\ kJ$$

Heating $CH_4(l)$ to its boiling point requires only 0.95 kJ. We have added 42 kJ, so there is definitely enough heat to vaporize the liquid. ΔH_{vap} for $CH_4(l)$ is 8.20 kJ/mol. The 32.0 g sample is 2.00 mol $CH_4(l)$, so the energy required to vaporize the sample at –161.5 °C is (2 × 8.20 kJ/mol =) 16.4 kJ. The energy used to heat the sample to –161.5 °C and vaporize it at this temperature is (0.947 kJ + 16.4 kJ) = 17.347 = 17.3 kJ. We have (42.0 kJ – 17.347 kJ) = 24.653 = 24.7 kJ left to heat the gas.

$$\Delta T = 24.65\ kJ \times \frac{g\text{-}K}{2.22\ J} \times \frac{1000\ J}{kJ} \times \frac{1}{32.0\ g\ CH_4} = 346.99 = 347\ K = 347\ ^\circ C$$

The final temperature of the methane gas, $CH_4(g)$, is then
(–161.5 °C + 346.99 °C) = 185.49 = 185 °C.

11.5   (a)   385 torr. Find 30 °C on the horizontal axis, and follow a vertical line from this point to its intersection with the red vapor pressure curve. Follow a horizontal line from the intersection to the vertical axis and read the vapor pressure.

       (b)   22 °C. Reverse the procedure outlined in part (a). Find 300 torr on the vertical axis, follow it to the curve and down to the value on the horizontal axis.

       (c)   47 °C. The normal boiling point of a liquid is the temperature at which its vapor pressure is 1 atm, or 760 mm Hg. A vapor pressure of 1 atm is very near the top of this diagram, at approximately 47 °C.

11.6   *Analyze/Plan.* We are given the structures of two molecules with the same molecular formula and asked questions about their physical properties. Because the molecules have the same molecular formula, their van der Waals forces are similar. Consider any differences in intermolecular forces experienced by the two molecules.   *Solve.*

       (a)   Propanol, the molecule on the left, has an O–H bond and experiences hydrogen bonding, while ethyl methyl ether does not.

       (b)   We expect propanol to have a larger dipole moment. Both molecules are somewhat polar, but propanol has hydrogen bonding.

       (c)   Propanol boils at 97.2 °C, while ethyl methyl ether boils at 10.8 °C. Molecules in liquid propanol are attracted to each other by hydrogen bonding. More kinetic energy and thus a higher temperature is required to separate the molecules and produce a gas.

11.7   (a)   360 K, the normal boiling point; 260 K, normal freezing point. The left-most line is the freezing/melting curve, the right-most line is the condensation/boiling curve. The normal boiling and freezing points are the temperatures of boiling and freezing at 1 atm pressure.

       (b)   The material is solid in the left-most green (or pale blue) zone, liquid in the blue zone, and gas in the tan zone. (i) gas; (ii) solid; (iii) liquid.

       (c)   The triple point, where all three phases are in equilibrium, is the point where the three lines on the phase diagram meet. For this substance, the triple point is approximately 185 K at 0.45 atm.

11.8   (a)   The substance is in a liquid crystalline state at temperatures T_1 and T_2. At T_1, the molecules are aligned in layers and the long molecular axes are perpendicular to the layer planes; this describes a smectic A phase. At T_2, the long molecular axes are aligned but the ends are not aligned; this describes a nematic phase.

           (b) T_3 is the highest temperature. The molecular arrangement in this phase has the least order, so it represents the highest temperature. (The molecules are closely packed, but not aligned in any way; this describes an ordinary liquid phase.)

Molecular Comparisons of Gases, Liquids, and Solids (Section 11.1)

11.9   (a)   solid < liquid < gas

           (b)   gas < liquid < solid

           (c)   Matter in the gaseous state is most easily compressed, because particles are far apart and there is much empty space.

11.10   (a)   In solids, particles are in essentially fixed positions relative to each other, so the average energy of attraction is stronger than average kinetic energy. In liquids, particles are close together but moving relative to each other. The average attractive energy and average kinetic energy are approximately balanced. In gases, particles are far apart and in constant, random motion. Average kinetic energy is much greater than average energy of attraction.

           (b)   As the temperature of a substance is increased, the average kinetic energy of the particles increases. In a collection of particles (molecules), the state is determined by the strength of interparticle forces relative to the average kinetic energy of the particles. As the average kinetic energy increases, more particles are able to overcome intermolecular attractive forces and move to a less ordered state, from solid to liquid to gas.

           (c)   If a gas is placed under very high pressure, the particles undergo many collisions with the container and with each other. The large number of particle-particle collisions increases the likelihood that intermolecular attractions will cause the molecules to coalesce (liquefy).

11.11   (a)   It increases. Kinetic energy is the energy of motion. As melting occurs, the motion of atoms relative to each other increases, which increases kinetic energy. As the kinetic energy of individual atoms increases, the overall average kinetic energy of the sample increases.

           (b)   It increases. As the atoms move relative to one another, the average distance between them increases. The physical property that corroborates this is density. The density of a solid is usually greater than the density of its liquid, indicating a greater sample volume for the liquid. This greater sample volume is the result of greater distance between atoms in three dimensions.

11.12   (a) $Ar < CCl_4 < Si$. Intermolecular energy of attraction increases from gas to liquid to solid. (See Solution 11.10.)

           (b) $Ar < CCl_4 < Si$. Boiling point increases as intermolecular energy of attraction increases. The greater the intermolecular attractive energy among particles, the greater the kinetic energy and temperature required to overcome this attractive energy and produce the less ordered gaseous state.

11.13   (a)   At standard temperature and pressure, the molar volumes of Cl_2 and NH_3 are nearly the same because they are both gases. In the gas phase, molecules are far apart and most of the volume occupied by the substance is empty space. Differences in molecular characteristics such as weight, shape, and dipole moment have little bearing on the molar volume of a gas.

   The ideal-gas law states that one mole of any gas at STP will occupy a fixed volume. The slight difference in molar volumes of the two gases is predicted by the van der Waals correction, which quantifies deviation from ideal behavior.

   (b)   On cooling to 160 K, both compounds condense from the gas phase to the solid state. Condensation, as the word implies, eliminates most of the empty space between molecules, so we expect a significant decrease in the molar volume.

   (c) $\dfrac{1\,cm^3}{2.02\,g\,Cl_2} \times \dfrac{70.096\,g\,Cl_2}{1\,mol\,Cl_2} = 35.1\,cm^3/mol\,Cl_2 = 0.0351\,L/mol\,Cl_2$

   $\dfrac{1\,cm^3}{0.84\,g\,NH_3} \times \dfrac{17.031\,g\,NH_3}{1\,mol\,NH_3} = 20.3\,cm^3/mol\,NH_3 = 0.0203\,L/mol\,NH_3$

   (d)   Solid state molar volumes are not as similar as those in the gaseous state. In the solid state, most of the empty space is gone, so molecular characteristics do influence molar volumes. Cl_2 is heavier than NH_3 and the Cl–Cl bond distance is almost double the N–H bond distance (Figure 7.7). Intermolecular attractive forces among polar NH_3 molecules bind them more tightly than forces among nonpolar Cl_2 molecules. These factors all contribute to a molar volume for $Cl_2(s)$ that is almost twice that of $NH_3(s)$.

   (e)   Like solids, liquids are condensed phases. That is, there is little empty space between molecules in the liquid state. We expect the molar volumes of the liquids to be closer to those in the solid state than those in the gaseous state.

11.14   (a)   The average distance between molecules is greater in the liquid state. Density is the ratio of the mass of a substance to the volume it occupies. For the same substance in different states, mass will be the same. The smaller the density, the greater the volume occupied, and the greater the distance between molecules. The liquid at 130° has the lower density (1.08 g/cm^3), so the average distance between molecules is greater.

   (b)   Less. For the same mass of compound, the sample with the higher density will occupy the smaller volume.

Intermolecular Forces (Section 11.2)

11.15   (a)   London dispersion forces

   (b)   dipole-dipole forces

   (c)   hydrogen-bonding forces

11.16   (a)   *Intra*molecular interactions are generally stronger than *inter*molecular interactions. That is, interactions within molecules, such as chemical bonds, are stronger than interactions between molecules.

   (b)   Intermolecular interactions are broken when a liquid is converted to a gas.

11.17 (a) SO_2 is a polar covalent molecule, so dipole-dipole and London dispersion forces must be overcome to convert the liquid to a gas.

(b) CH_3COOH is a polar covalent molecule that experiences London dispersion, dipole-dipole, and hydrogen-bonding (O–H bonds) forces. All of these forces must be overcome to convert the liquid to a gas.

(c) H_2S is a polar covalent molecule that experiences London dispersion and dipole-dipole forces, so these must be overcome to change the liquid into a gas. (H–S bonds do not lead to hydrogen-bonding interactions.)

11.18 (a) CH_3OH experiences hydrogen bonding, but CH_3SH does not.

(b) Both gases are influenced by London dispersion forces. The heavier the gas particles, the stronger the London dispersion forces. The heavier Xe is a liquid at the specified conditions, whereas the lighter Ar is a gas.

(c) Both gases are influenced by London dispersion forces. The larger, diatomic Cl_2 molecules are more polarizable, experience stronger dispersion forces, and have the higher boiling point.

(d) Acetone and 2-methylpropane are molecules with similar molar masses and London dispersion forces. Acetone also experiences dipole-dipole forces and has the higher boiling point.

11.19 (a) *Polarizability* increases as molecular size (and thus molecular weight) increases. In order of increasing polarizability: $CH_4 < SiH_4 < SiCl_4 < GeCl_4 < GeBr_4$.

(b) The magnitude of London dispersion forces and thus the boiling points of molecules increase as polarizability increases. The order of increasing boiling points is the order of increasing polarizability:

$$CH_4 < SiH_4 < SiCl_4 < GeCl_4 < GeBr_4$$

11.20 (a) True. A more polarizable molecule can develop a larger transient dipole, increasing the strength of electrostatic attractions and dispersion forces among molecules.

(b) False. The noble gases are all monoatomic. Going down the family, the atomic radius and the size of the electron cloud increase. The larger the electron cloud, the more polarizable the atom, the stronger the London dispersion forces, and the higher the boiling point. (In general, strength of forces and boiling point vary in the same direction, not opposite directions.)

(c) False. Generally, dipole-dipole forces are stronger than dispersion forces for molecules of similar size and mass. The size of the molecule and the magnitude of its dipole moment (if there is one) determine the relative magnitudes of dispersion and dipole-dipole forces.

(d) True. For molecules with similar molecular weights and elemental composition, linear molecules have the possibility for greater contact along and around their surfaces than spherical molecules. Their electron clouds are thus more polarizable, and dispersion forces are greater.

(e) True. The larger the electron cloud, the more easily it can be distorted by transient electronic interactions.

11.21   *Analyze/Plan.* For molecules with similar structures, the strength of dispersion forces increases with molecular size (molecular weight and number of electrons in the molecule).

        Solve: (a) H_2S (b) CO_2 (c) GeH_4

11.22   For molecules with similar structures, the strength of dispersion forces increases with molecular size (molecular weight and number of electrons in the molecule).

(a) Br_2

(b) $CH_3CH_2CH_2CH_2CH_2SH$

(c) $CH_3CH_2CH_2Cl$. These two molecules have the same molecular formula and molecular weight (C_3H_7Cl, molecular weight = 78.5 amu), so the shapes of the molecules determine which has the stronger dispersion forces. According to Figure 11.6, the cylindrical (not branched) molecule will have stronger dispersion forces.

11.23   Both hydrocarbons experience dispersion forces. Rod-like butane molecules can contact each other over the length of the molecule, whereas spherical 2-methylpropane molecules can only touch tangentially. The larger contact surface of butane facilitates stronger forces and produces a higher boiling point.

11.24   Both molecules experience hydrogen bonding through their –OH groups and dispersion forces between their hydrocarbon portions. The position of the –OH group in isopropyl alcohol shields it somewhat from approach by other molecules and slightly decreases the extent of hydrogen bonding. Also, isopropyl alcohol is less rod-like (it has a shorter chain) than propyl alcohol, so dispersion forces are weaker. Because hydrogen bonding and dispersion forces are weaker in isopropyl alcohol, it has the lower boiling point.

11.25   (a)   A molecule must contain H atoms bound to either N, O, or F atoms to participate in hydrogen bonding with like molecules.

(b)   **CH_3NH_2** and **CH_3OH** have N–H and O–H bonds, respectively; they will form hydrogen bonds with other molecules of the same kind. (CH_3F has C–F and C–H bonds, but no H–F bonds.)

11.26   (a)   HF has the higher boiling point because hydrogen bonding is stronger than dipole-dipole forces.

(b) $CHBr_3$ has the higher boiling point because it has the higher molar mass, which leads to greater polarizability and stronger dispersion forces.

(c)   ICl has the higher boiling point because it is a polar molecule. For molecules with similar structures and molar masses, dipole-dipole forces are stronger than dispersion forces.

11.27   (a)   Replacing a hydroxyl hydrogen with a CH_3 group eliminates hydrogen bonding in that part of the molecule. This reduces the strength of intermolecular forces and leads to a (much) lower boiling point.

(b) $CH_3OCH_2CH_2OCH_3$ is a larger, more polarizable molecule with stronger London dispersion forces and thus a higher boiling point.

11.28   (a) C_4H_{10}. Both molecules experience dispersion forces. C_4H_{10} has the higher boiling point due to greater molar mass and similar strength of forces.

        (b) $CH_3CH_2CH_2CH_2OH$ has the higher boiling point due to the influence of hydrogen bonding.

        (c) SO_3. This is a tough call; SO_2 has dipole-dipole forces; SO_3 has dispersion forces but a larger molecular weight. The relative strength of dispersion and dipole-dipole forces depends on the mass and shape of the molecules. SO_3 molecules have greater molecular weight and are planar, so alignment is facile and dispersion forces are strong; SO_3 has the higher boiling point (confirmed by CRC *Handbook of Chemistry and Physics*).

        (d) Cl_2CO has the higher boiling point due to greater molecular weight and stronger dispersion forces. (Note that H_2CO does not have hydrogen bonding, because the H atoms are bound to C, not to O.)

11.29 | Physical Property | H_2O | H_2S |
        | --- | --- | --- |
        | Normal Boiling Point, °C | 100.00 | −60.7 |
        | Normal Melting Point, °C | 0.00 | −85.5 |

Based on its much higher normal melting and boiling point, H_2O has stronger intermolecular forces.

H_2O, with H bound to O, has hydrogen bonding. H_2S, with H bound to S, has dipole-dipole forces. (The electronegativities of H and S, 2.1 and 2.5, respectively, are similar. The H–S bond dipoles in H_2S are not large, but S does have two nonbonded electron pairs. The molecule has medium polarity.) Both molecules have London dispersion forces.

11.30   Statement (c) is the best explanation. Chloroform is somewhat polar and has a larger dipole moment than carbon tetrachloride. The C–H bond in chloroform is not, however, a candidate for hydrogen bonding. The only explanation is that carbon tetrachloride is more polarizable, by virtue of its larger molar mass, than chloroform.

11.31 SO_4^{2-} has a greater negative charge than BF_4^-, so ion-ion electrostatic attractions are greater in sulfate salts. These strong forces limit the ion mobility required for the formation of an ionic liquid. (This is called an electronic effect.)

11.32   The longer the alkyl side chain of the 1-alkyl-3-methylimidazolium cation, the more irregular the shape of the cation. Particles with irregular shapes are more difficult to pack into solids, so the melting point of the salt decreases as the length of the alkyl group and irregularity increases. (This is called a steric effect.)

Select Properties of Liquids (Section 11.3)

11.33   (a)   As temperature increases, surface tension decreases; they are inversely related.

        (b)   As temperature increases, viscosity decreases; they are inversely related.

(c)   Surface tension and viscosity are both directly related to the strength of intermolecular attractive forces. The same attractive forces that cause surface molecules to be difficult to separate cause molecules elsewhere in the sample to resist movement relative to one another. Liquids with high surface tension have intermolecular attractive forces sufficient to produce a high viscosity as well.

11.34   The order of increasing strength of intermolecular forces is also the order of increasing viscosity and surface tension

(a) $CH_3CH_2CH_3 < CH_2Cl_2 < CH_3CH_2OH$

(b) $CH_3CH_2CH_3 < CH_2Cl_2 < CH_3CH_2OH$

(c) $CH_3CH_2CH_3 < CH_2Cl_2 < CH_3CH_2OH$

11.35   (a)   Diagram (ii) shows stronger adhesive forces between the surface and the liquid. These adhesive forces attract the liquid to the surface and flatten the drop.

(b)   Diagram (i) represents water on a nonpolar surface. The stronger hydrogen bonding cohesive forces among water molecules in the liquid prevent the drop from spreading.

(c)   Diagram (ii) represents water on a polar surface. Adhesive dipole-dipole interactions between water molecules and the surface compete successfully with cohesive hydrogen bonding forces in the liquid and the drop spreads.

11.36   (a)   H—N̈—N̈—H   H—Ö—Ö—H   H—Ö—H
                      | |
                      H   H

(b)   All have bonds (N–H or O–H, respectively) capable of forming hydrogen bonds. Hydrogen bonding is the strongest intermolecular interaction between neutral molecules and leads to very strong cohesive forces in liquids. The stronger the cohesive forces in a liquid, the greater the surface tension.

11.37   (a)   The three molecules have similar structures and all experience hydrogen-bonding, dipole-dipole, and dispersion forces. The main difference in the series is the increase in the number of carbon atoms in the alkyl chain, with a corresponding increase in chain length, molecular weight, and strength of dispersion forces. The boiling points, surface tension, and viscosities all increase because the strength of dispersion forces increases.

(b)   Ethylene glycol has an –OH group at both ends of the molecule. This greatly increases the possibilities for hydrogen bonding, so the overall intermolecular attractive forces are greater and the viscosity of ethylene glycol is much greater.

(c)   Water has the highest surface tension but lowest viscosity because it is the smallest molecule in the series. Because water molecules are small, they approach each other closely and form many strong hydrogen bonds. There is no hydrocarbon chain to disrupt hydrogen bond formation or to inhibit their attraction to molecules in the interior of the drop. Water molecules at the surface of a drop are missing a few hydrogen bonds and are strongly pulled into the center of the drop, resulting in high surface tension. The absence of an alkyl chain also means the molecules can move around each other easily, resulting in the low viscosity.

11.38   (a)   For molecules with similar shapes, viscosity usually decreases with decreasing molecular weight. Because n-pentane has one fewer carbon atom and a shorter chain than n-hexane, the molecules are slightly more free to move around each other and n-pentane will have the smaller viscosity.

         (b)   According to Figure 11.6, neopentane is roughly spherical, whereas n-pentane is cylindrical or rod shaped. The spherical neopentane has weaker dispersion forces and the molecules are more free to tumble, so it will have the smaller viscosity.

Phase Changes (Section 11.4)

11.39   (a)   melting, endothermic

         (b)   evaporation or vaporization, endothermic

         (c)   deposition, exothermic

         (d)   condensation, exothermic

11.40   (a)   condensation, exothermic

         (b)   sublimation, endothermic

         (c)   vaporization (evaporation), endothermic

         (d)   freezing, exothermic

11.41   (a)   Melting, (s) → (l)

         (b)   Endothermic. Energy is always required to overcome intermolecular forces that organize molecules into a solid.

         (c)   Heat of vaporization is usually larger than heat of fusion. Vaporization requires enough energy to separate molecules by large distances. Melting or fusion only requires that molecules move relative to each other, not that they are separated.

11.42   (a)   Liquid ethyl chloride at room temperature is far above its boiling point. When the liquid contacts the room temperature surface, heat sufficient to vaporize the liquid is transferred from the surface to the ethyl chloride, and the heat content of the molecules increases. At constant atmospheric pressure, $\Delta H = q$, so the heat content and the enthalpy content of $C_2H_5Cl(g)$ are higher than that of $C_2H_5Cl(l)$. This indicates that the specific heat of the gas is less than that of the liquid, because the heat content of the gas starts at a higher level.

         (b)   Liquid C_2H_5Cl is vaporized (boiled), $C_2H_5Cl(g)$ is warmed to the final temperature, and the solid surface is cooled to the final temperature. The enthalpy of vaporization (ΔH_{vap}) of $C_2H_5Cl(l)$, the specific heat of $C_2H_5Cl(g)$, and the specific heat of the solid surface must be considered.

11.43   *Analyze.* The heat required to vaporize 60 g of H_2O equals the heat lost by the cooled water.

        Plan. Using the enthalpy of vaporization, calculate the heat required to vaporize 60 g of H_2O in this temperature range. Using the specific heat capacity of water, calculate the mass of water than can be cooled 15 °C if this much heat is lost.

Solve. Evaporation of 60 g of water requires:

$$60 \text{ g H}_2\text{O} \times \frac{2.4 \text{ kJ}}{1 \text{ g H}_2\text{O}} = 1.44 \times 10^2 \text{ kJ} = 1.4 \times 10^5 \text{ J}$$

Cooling a certain amount of water by 15 °C:

$$1.44 \times 10^5 \text{ J} \times \frac{1 \text{ g-K}}{4.184 \text{ J}} \times \frac{1}{15 \text{ °C}} = 2294 = 2.3 \times 10^3 \text{ g H}_2\text{O}$$

Check. The units are correct. A surprisingly large mass of water (2300 g ≈ 2.3 L) can be cooled by this method.

11.44 Energy released when 200 g of H_2O is cooled from 15 °C to 0 °C:

$$\frac{4.184 \text{ J}}{\text{g-K}} \times 200 \text{ g H}_2\text{O} \times 15 \text{ °C} = 12.55 \times 10^3 \text{ J} = 13 \text{ kJ}$$

Energy released when 200 g of H_2O is frozen (there is no change in temperature during a change of state):

$$\frac{334 \text{ J}}{\text{g}} \times 200 \text{ g H}_2\text{O} = 6.68 \times 10^4 \text{ J} = 66.8 \text{ kJ}$$

Total energy released = 12.55 kJ + 66.8 kJ = 79.35 = 79.4 kJ

Mass of freon that will absorb 79.4 kJ when vaporized:

$$79.35 \text{ kJ} \times \frac{1 \times 10^3 \text{ J}}{1 \text{ kJ}} \times \frac{1 \text{ g CCl}_2\text{F}_2}{289 \text{ J}} = 275 \text{ g CCl}_2\text{F}_2$$

11.45 *Analyze/Plan.* Follow the logic in Sample Exercise 11.3. *Solve.* Physical data for ethanol, C_2H_5OH, is: mp = –114 °C; ΔH_{fus} = 5.02 kJ/mol; $C_{s(solid)}$ = 0.97 J/g-K; bp = 78 °C; ΔH_{vap} = 38.56 kJ/mol; $C_{s(liquid)}$ = 2.3 J/g-K. *Solve.*

(a) Heat the liquid from 35 °C to 78 °C, ΔT = 43 °C = 43 K.

$$42.0 \text{ g C}_2\text{H}_5\text{OH} \times \frac{2.3 \text{ J}}{\text{g-K}} \times 43 \text{ K} \times \frac{1 \text{ kJ}}{1000 \text{ J}} = 4.1538 = 4.2 \text{ kJ}$$

Vaporize (boil) the liquid at 78 °C, using ΔH_{vap}.

$$42.0 \text{ g C}_2\text{H}_5\text{OH} \times \frac{1 \text{ mol C}_2\text{H}_5\text{OH}}{46.07 \text{ g}} \times \frac{38.56 \text{ kJ}}{\text{mol}} = 35.1535 = 35.2 \text{ kJ}$$

Total energy required is 4.1538 kJ + 35.1535 kJ = 39.3073 = 39.3 kJ.

(b) Heat the solid from –155 °C to –114 °C, ΔT = 41 °C = 41 K.

$$42.0 \text{ g C}_2\text{H}_5\text{OH} \times \frac{0.97 \text{ J}}{\text{g-K}} \times 41 \text{ K} \times \frac{1 \text{ kJ}}{1000 \text{ J}} = 1.6703 = 1.7 \text{ kJ}$$

Melt the solid at –114 °C, using ΔH_{fus}.

$$42.0 \text{ g C}_2\text{H}_5\text{OH} \times \frac{1 \text{ mol C}_2\text{H}_5\text{OH}}{46.07 \text{ g}} \times \frac{5.02 \text{ kJ}}{\text{mol}} = 4.5765 = 4.58 \text{ kJ}$$

Heat the liquid from –114 °C to 78 °C, $\Delta T = 192$ °C $= 192$ K.

$$42.0 \text{ g C}_2\text{H}_5\text{OH} \times \frac{2.3 \text{ J}}{\text{g-K}} \times 192 \text{ K} \times \frac{1 \text{ kJ}}{1000 \text{ J}} = 18.5472 = 19 \text{ kJ}$$

From part (a), vaporizing (boiling) 42.0 g of $\text{C}_2\text{H}_5\text{OH}$ liquid at 78 °C requires

35.1535 kJ = 35.2 kJ.

Total energy required = 1.6703 kJ + 4.5765 kJ + 18.5472 kJ + 35.1535 kJ = 59.9476

= 60 kJ.

Check. The relative energies of the various steps are reasonable; vaporization is the largest. The sum has no decimal places because (19 kJ) has no decimal places.

11.46   Consider the process in steps, using the appropriate thermochemical constant.

Heat the liquid from 10.00 °C to 47.6 °C, $\Delta T = 37.6$ °C $= 37.6$ K, using the specific heat of the liquid.

$$35.0 \text{ g C}_2\text{Cl}_3\text{F}_3 \times \frac{0.91 \text{ J}}{\text{g-K}} \times 37.6 \text{ K} \times \frac{1 \text{ kJ}}{1000 \text{ J}} = 1.1976 = 1.2 \text{ kJ}$$

Boil the liquid at 47.6 °C (320.6 K), using the enthalpy of vaporization.

$$35.0 \text{ g C}_2\text{Cl}_3\text{F}_3 \times \frac{1 \text{ mol C}_2\text{Cl}_3\text{F}_3}{187.4 \text{ g C}_2\text{Cl}_3\text{F}_3} \times \frac{27.49 \text{ kJ}}{\text{mol}} = 5.1342 = 5.13 \text{ kJ}$$

Heat the gas from 47.6 °C to 105.00 °C, $\Delta T = 57.4$ °C $= 57.4$ K, using the specific heat of the gas.

$$35.0 \text{ g C}_2\text{Cl}_3\text{F}_3 \times \frac{0.67 \text{ J}}{\text{g-K}} \times 57.4 \text{ K} \times \frac{1 \text{ kJ}}{1000 \text{ J}} = 1.3460 = 1.3 \text{ kJ}$$

The total energy required is 1.1967 kJ + 5.1342 kJ + 1.3460 kJ = 7.6778 = 7.7 kJ.

11.47   (a)   False. The critical pressure is the pressure required to cause liquefaction at the critical temperature.

(b)   True.

(c)   False. In general, the higher the critical temperature, the higher the critical pressure.

(d)   True. The more intermolecular forces in a substance, the greater the kinetic energy required to overcome them. This greater kinetic energy translates to higher critical temperatures and pressures (as well as melting and boiling points).

11.48   (a) CCl_3F, CCl_2F_2, and CClF_3 are polar molecules that experience dipole-dipole and London dispersion forces with like molecules. CF_4 is a nonpolar compound that experiences only dispersion forces.

(b)   According to Solution 11.47(b), the higher the critical temperature, the stronger the intermolecular attractive forces of a substance. Therefore, the strength of intermolecular attraction increases moving from right to left across the series and as molecular weight increases. $\text{CF}_4 < \text{CClF}_3 < \text{CCl}_2\text{F}_2 < \text{CCl}_3\text{F}$.

(c)   The increasing intermolecular attraction with increasing molecular weight indicates that the critical temperature and pressure of CCl_4 will be greater than that of CCl_3F. Looking at the numerical values in the series, an increase of 88 K in critical temperature and 3.1 atm in critical pressure to the corresponding values for CCl_3F seem reasonable.

Physical Property	CCl_3F	CCl_4 (predicted)	CCl_4 (CRC)
Critical Temperature (K)	471	557	556.6
Critical Pressure (atm)	43.5	46.6	44.6

The predicted values for CCl_4 are in very good agreement with literature values. The key concept is that dispersion, not dipole-dipole, forces dominate the physical properties in the series.

Vapor Pressure (Section 11.5)

11.49   Properties (c) intermolecular attractive forces, (d) temperature, and (e) density of the liquid affect vapor pressure of a liquid.

11.50   A normal boiling point of 56 °C places the liquid-vapor curve for acetone between the curves for diethyl ether and ethanol on Figure 11.25. Following a vertical line of increasing vapor pressure at 25 °C, we first cross the ethanol curve, then the (virtual) acetone curve. This means that, at 25 °C, the vapor pressure of acetone is higher than the vapor pressure of ethanol. (The lower boiling point of acetone is a strong indicator that it will have a higher vapor pressure than ethanol at a given temperature.)

11.51   (a)   *Analyze/Plan.* Given the molecular formulae of several substances, determine the kind of intermolecular forces present, and rank the strength of these forces. The weaker the forces, the more volatile the substance.   *Solve.*

$$CBr_4 < CHBr_3 < CH_2Br_2 < CH_2Cl_2 < CH_3Cl < CH_4$$

(b) $CH_4 < CH_3Cl < CH_2Cl_2 < CH_2Br_2 < CHBr_3 < CBr_4$

(c)   Boiling point increases as the strength of intermolecular forces increases. By analogy to attractive forces in HCl (Section 11.2), the trend will be dominated by dispersion forces, even though four of the molecules ($CHBr_3$, CH_2Br_2, CH_2Cl_2, and CH_3Cl) are polar. Thus, the order of increasing boiling point is the order of increasing molar mass and increasing strength of dispersion forces.

11.52   (a)   False. The heavier (and larger) CBr_4 has stronger dispersion forces, a higher boiling point, lower vapor pressure, and is less volatile.

(b)   True.

(c)   False.

(d)   False.

11.53   (a)   The water in the two pans is at the same temperature, the boiling point of water at the atmospheric pressure of the room. During a phase change, the temperature of a system is constant. All energy gained from the surroundings is used to accomplish the transition, in this case to vaporize the liquid water. The pan of water that is boiling vigorously is gaining more energy and the liquid is being vaporized more quickly than in the other pan, but the temperature of the phase change is the same.

(b) Vapor pressure does not depend on either volume or surface area of the liquid. As long as the containers are at the same temperature, the vapor pressures of water in the two containers are the same.

11.54 Statement (c) is the best explanation for the cool tea. Statements (a) and (b) are somewhat true, but the effects would not be rapid. That the boiling point is lower at lower pressure is more significant. Statement (d) is false.

11.55 *Analyze/Plan.* Follow the logic in Sample Exercise 11.4. The boiling point is the temperature at which the vapor pressure of a liquid equals atmospheric pressure. *Solve.*

(a) The boiling point of ethanol at 200 torr is ~48 °C.

(b) The vapor pressure of ethanol at 60 °C is approximately 340 torr. Thus, at 60 °C ethyl alcohol would boil at an external pressure of 340 torr.

(c) The boiling point of diethyl ether at 400 torr is ~17 °C.

(d) 40 °C is above the normal boiling point of diethyl ether, so the pressure at which 40 °C is the boiling point is greater than 760 torr. According to Figure 11.25, a boiling point of 40 °C requires an external pressure of 1000 torr. (At these conditions, the vapor pressure of diethyl ether is 1000 torr.)

11.56 (a)

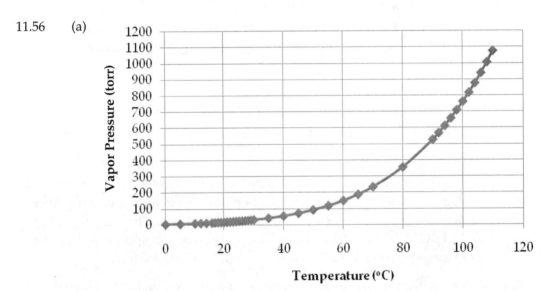

A plot of vapor pressure versus temperature data for H_2O from Appendix B is shown here. The vapor pressure of water at body temperature, 37 °C, is approximately 50 torr.

(b) The data point at 760.0 torr, 100 °C is the normal boiling point of H_2O. This is the temperature at which the vapor pressure of H_2O is equal to a pressure of 1 atm or 760 torr.

(c) At an external (atmospheric) pressure of 633 torr, the boiling point of H_2O is approximately 96 °C.

(d) At an external pressure of 774 torr, the boiling point of water is approximately 100.5 °C.

Phase Diagrams (Section 11.6)

11.57   (a)   The *critical point* is the temperature and pressure beyond which the gas and liquid phases are indistinguishable.

(b)   The gas/liquid line ends at the critical point because at conditions beyond the critical temperature and pressure, there is no distinction between gas and liquid. In experimental terms, a gas cannot be liquefied at temperatures higher than the critical temperature, regardless of pressure.

(At conditions beyond the critical point, a substance is known as a supercritical fluid. Supercritical fluids have many practical applications, such as decaffeination of green coffee beans and dry cleaning.)

11.58   (a)   The *triple point* on a phase diagram represents the temperature and pressure at which the gas, liquid, and solid phases are in equilibrium.

(b)   No. A phase diagram represents a closed system, one where no matter can escape and no substance other than the one under consideration is present; air cannot be present in the system. Even if air is excluded, at 1 atm of external pressure, the triple point of water is inaccessible, regardless of temperature [see Figure 11.28].

11.59   (a)   The water vapor would deposit to form a solid at a pressure of around 4 torr. At higher pressure, perhaps 5 atm or so, the solid would melt to form liquid water. This occurs because the melting point of ice, which is 0 °C at 1 atm, decreases with increasing pressure.

(b)   In thinking about this exercise, keep in mind that the **total** pressure is being maintained at a constant 0.50 atm. That pressure is composed of water vapor pressure and some other pressure, which could come from an inert gas. At 100 °C and 0.50 atm, water is in the vapor phase. As it cools, the water vapor will condense to the liquid at the temperature where the vapor pressure of liquid water is 0.50 atm. From Appendix B, we see that condensation occurs at approximately 82 °C. Further cooling of the liquid water results in freezing to the solid at approximately 0 °C. The freezing point of water increases with decreasing pressure, so at 0.50 atm, the freezing temperature is very slightly above 0 °C.

11.60   (a)   Solid CO_2 sublimes to form $CO_2(g)$ at a temperature of about –60 °C.

(b)   Solid CO_2 melts to form $CO_2(l)$ at a temperature of about –55 °C. The $CO_2(l)$ boils when the temperature reaches approximately –45 °C.

11.61   *Analyze/Plan.* Follow the logic in Sample Exercise 11.5, using the phase diagram for neon.   *Solve.*

(a)   The normal melting point is the temperature where solid becomes liquid at 1 atm pressure. Following a horizontal line at 1 atm to the solid-liquid line, the normal melting point is approximately 24 K.

(b)   Neon sublimes, changes directly from solid to gas, at pressures less than the triple point pressure, approximately 0.5 atm.

(c)   Room temperature is 298 K, in the region where neon is a supercritical fluid. Neon cannot be liquefied at any temperature above the critical temperature, approximately 45 K, regardless of pressure.

11.62   (a)   The normal boiling point is the temperature where liquid becomes gas at 1 atm pressure. Moving vertically down from 1 atm on the liquid-gas line to the temperature axis, the normal boiling point is approximately 27 to 28 K, or –246 °C to –245 °C.

        (b)   The much higher critical temperature and pressure of Ar (150.9 K, 48 atm) compared with those of Ne (25 K, 0.43 atm), indicate that Ar experiences much stronger intermolecular forces than Ne.

11.63   *Analyze/Plan.* Follow the logic in Sample Exercise 11.5, using the phase diagram for methane in Figure 11.30.   *Solve.*

        (a)   According to Sample Exercise 11.5, the triple point of methane (CH_4) is approximately (–180 °C, 0.1 atm). The solid-liquid line in the phase diagram is essentially vertical in the pressure range 0.1-100 atm. This means that conditions at the surface of Titan (–178 °C, 1.6 atm) are very close to the solid-liquid line. Methane on the surface of titan is likely to exist in both solid and liquid forms.

        (b)   Methane is a liquid at –178 °C and 1.6 atm. Moving upward through Titan's atmosphere at a constant temperature of –178 °C, pressure decreases. At a pressure slightly greater than 0.1 atm, we expect to see vaporization to gaseous methane. If we begin with solid methane at 1.6 atm and a temperature slightly below –180 °C, we expect sublimation to gaseous methane at a pressure slightly less than 0.1 atm.

11.64   The density of Ga(s), 5.91 g/cm^3, is less than the density of Ga(l), 6.1 g/cm^3, just above the melting temperature. "Typically" the density of a solid is greater than the density of its liquid. Gallium is then an atypical substance, like water, where the solid state is denser and more compact than the liquid. This results in a backward sloping solid-liquid line on the phase diagram for water, and we also expect to see this unusual feature on the diagram for gallium.

Liquid Crystals (Section 11.7)

11.65   In a nematic liquid crystalline phase, molecules are aligned along their long axes, but the molecular ends are not aligned. In an ordinary liquid, molecules have no orderly arrangement; they are randomly oriented, or amorphous. Both an ordinary liquid and a nematic liquid crystal phase are fluids; molecules are free to move relative to one another. In an ordinary liquid, molecules can move in any direction. In a nematic phase, molecules are free to translate in all dimensions. Molecules cannot tumble or rotate out of the molecular plane, or the order of the nematic phase is lost and the sample becomes an ordinary liquid.

11.66   Reinitzer observed that cholesteryl benzoate has a phase that exhibits properties intermediate between those of the solid and liquid phases. This "liquid-crystalline" phase, formed by melting at 145 °C, is viscous and opaque; its viscosity decreases on heating and it becomes clear at 179 °C.

11.67   (a)   True.

(b)   False. Liquid crystalline molecules are often rod-like with some rigidity in the long direction.

(c)   True. Liquid crystalline is a phase of matter with distinct phase-change temperatures.

(d)   False. If no significant intermolecular forces were present, there would be no driving force for the intermediate ordering typical of liquid crystals.

(e)   False. Molecules containing only carbon and hydrogen do no exhibit the significant intermolecular forces required for liquid crystal formation.

(f)   True.

11.68   (a)   Graph B applies to a liquid crystalline material.

(b)   Melting solid to liquid. Graph A applies to a "regular" material. The constant temperature of the 2–3 segment represents the first phase change, melting solid.

(c)   Melting solid to liquid crystal. Segment 2–3 is the first constant temperature process. For a liquid crystalline material, this is melting solid to liquid crystal.

(d)   Heating liquid. In a regular material, the second heating change is heating the liquid.

(e)   Heating the liquid crystal. In a liquid crystalline material the second heating seqment is heating the liquid crystal.

11.69   Because order is maintained in at least one dimension, the molecules in a liquid-crystalline phase are not totally free to change orientation. This makes the liquid-crystalline phase more resistant to flow, more viscous, than the isotropic liquid.

11.70   A nematic phase is composed of sheets of molecules aligned along their lengths, but with no additional order within the sheet or between sheets. A cholesteric phase also contains this kind of sheet, but with some ordering between sheets. In a cholesteric phase, there is a characteristic angle between molecules in one sheet and those in an adjacent sheet. That is, one sheet of molecules is twisted at some characteristic angle relative to the next, producing a "screw" axis perpendicular to the sheets.

11.71   As the temperature of a substance increases, the average kinetic energy of the molecules increases. More molecules have sufficient kinetic energy to overcome intermolecular attractive forces, so overall ordering of the molecules decreases as temperature increases. Melting provides kinetic energy sufficient to disrupt alignment in one dimension in the solid, producing a smectic phase with ordering in two dimensions. Additional heating of the smectic phase provides kinetic energy sufficient to disrupt alignment in another dimension, producing a nematic phase with one-dimensional order.

11.72   In the nematic phase, molecules are aligned in one dimension, the long dimension of the molecule. In a smectic phase (A or C), molecules are aligned in two dimensions. Not only are the long directions of the molecules aligned, but the ends are also aligned. The molecules are organized into layers; the height of the layer is related to the length of the molecule.

Additional Exercises

11.73 (a) decrease   (b) increase   (c) increase   (d) increase

(e) increase   (f) increase   (g) increase

11.74 (a) 0 K – 54 K. A substance is a solid until its temperature reaches the melting point.

(b) 54 K – 95 K. The density of a substance decreases significantly when it goes from solid to liquid. The data shows that this happens between 90 and 100 K for O_2. It is reasonable to assume the boiling point is in the middle of this range. (A quick Internet search indicates that the boiling point of O_2 is 90.1 K.)

(c) 95 K – 140 K. For temperatures in the table, O_2 is a gas when the density is low.

(d) 95 K. As noted in part (b), O_2 is a liquid at 90 K and a gas at 100 K. The boiling point is somewhere in the range 90 – 100 K (actually 90.1 K)

(e) Dispersion forces only

11.75 (a) Correct.

(b) The lower boiling liquid must experience less total intermolecular forces.

(c) If both liquids are structurally similar nonpolar molecules, the lower boiling liquid has a lower molecular weight than the higher boiling liquid.

(d) Correct.

(e) At their boiling points, both liquids have vapor pressures of 760 mm Hg.

11.76 (a) The cis isomer has stronger dipole-dipole forces; the trans isomer is nonpolar. (Because both molecules have the same molecular weight, we can say that the dipole-dipole and dispersion forces of the cis isomer are stronger than the dispersion-only forces of the trans isomer.)

(b) The molecule with the stronger intermolecular interactions will have the higher boiling point. The cis isomer boils at 60.3 °C and the trans isomer boils at 47.5 °C.

11.77 Statement (b) best explains the data, although statement (d) is also true, because the molecular structures of the compounds are similar. Statements (a) and (c) are false.

11.78 (a) Four, all of them. All covalent compounds exhibit dispersion interactions.

(b) Three. Benzene is nonpolar and exhibits only dispersion forces. When Cl, Br, or OH is substituted for H in benzene, the molecule becomes polar and the molecules exhibit dipole-dipole interactions.

(c) One. The O–H bond in phenol participates in hydrogen bonding.

(d) Bromine is larger and more polarizable than chlorine, so the dispersion forces in bromobenzene are stronger than those in chlorobenzene.

(e) Phenol exhibits hydrogen bonding, which is the strongest intermolecular interaction among covalent molecules.

11.79　　The GC base pair, with more hydrogen bonds, is more stable to heating. To break up a base pair by heating, sufficient thermal energy must be added to break the existing hydrogen bonds. With 50% more hydrogen bonds, the GC pair is definitely more stable (harder to break apart) than the AT pair.

11.80　　The two O–H groups in ethylene glycol are involved in many hydrogen-bonding interactions. Pentane is nonpolar and experiences only dispersion forces.

　　　(a)　　Ethylene glycol is more viscous because of its strong hydrogen bonding.

　　　(b)　　Pentane has the lower boiling point because it experiences weaker intermolecular forces.

　　　(c)　　Ethylene glycol is used as antifreeze because it is viscous and doesn't boil off in the very hot radiator.

　　　(d)　　Pentane is used as the "blowing agent" because it has a low boiling point and is volatile.

11.81　　A plot of the number of carbon atoms versus boiling point follows. For eight C atoms, C_8H_{18}, the boiling point is approximately 130 °C. The more carbon atoms in the hydrocarbon, the longer the chain, the more polarizable the electron cloud, the stronger the London dispersion forces, and the higher the boiling point.

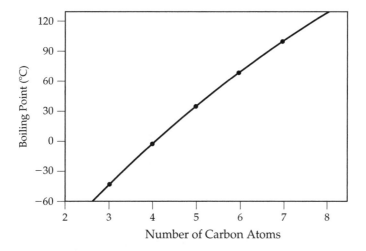

11.82　　Ionic liquids are the liquid phase of ionic compounds. Upon melting, the ions are free to move relative to one another. The ion-ion interparticle attractive forces at work in an ionic liquid are extremely strong relative to dispersion, dipole-dipole, and even hydrogen-bonding forces operating in most molecular solvents. These powerful ion-ion forces must be broken for an ion to escape to the vapor phase. In the distribution of particle energies at room temperature, very few ions have sufficient kinetic energy to escape these interactions and move to the vapor phase. With few particles in the vapor phase, the vapor pressures of ionic liquids are extremely low.

11.83　　(a)　　Sweat, or salt water, on the surface of the body vaporizes to establish its typical vapor pressure at atmospheric pressure. Because the atmosphere is a totally open system, saturated vapor pressure is never reached, and the sweat evaporates continuously. Evaporation is an endothermic process. The heat required to vaporize sweat is absorbed from your body, helping to keep it cool.

(b)   The vacuum pump reduces the pressure of the atmosphere (air + water vapor) above the water. Eventually, atmospheric pressure equals the vapor pressure of water and the water boils. Boiling is an endothermic process, and the temperature drops if the system is not able to absorb heat from the surroundings fast enough. As the temperature of the water decreases, the water freezes. (On a molecular level, the evaporation of water removes the molecules with the highest kinetic energies from the liquid. This decrease in average kinetic energy is what we experience as a temperature decrease.)

11.84   (a)   If the Clausius-Clapeyron equation is obeyed, a graph of ln P versus $1/T$ (K^{-1}) should be linear. Here are the data in a form for graphing.

T (K)	1/T	P (torr)	ln P
280.0	3.571×10^{-3}	32.42	3.479
300.0	3.333×10^{-3}	92.47	4.527
320.0	3.125×10^{-3}	225.1	5.417
330.0	3.030×10^{-3}	334.4	5.812
340.0	2.941×10^{-3}	482.9	6.180

According to the graph, the Clausius-Clapeyron equation is obeyed, to a first approximation.

$$\Delta H_{vap} = -\text{slope} \times R; \quad \text{slope} = \frac{3.479 - 6.180}{(3.571 - 2.941) \times 10^{-3}} = -\frac{2.701}{0.630 \times 10^{-3}} = -4.29 \times 10^3$$

$$\Delta H_{vap} = -(-4.29 \times 10^3) \times 8.314 \text{ J/mol-K} = 35.7 \text{ kJ/mol}$$

(b)   The normal boiling point is the temperature at which the vapor pressure of the liquid equals atmospheric pressure, 760 torr. From the graph,

ln 760 = 6.63, 1/T for this vapor pressure = 2.828×10^{-3}; T = 353.6 K

11.85   (a)   The Clausius-Clapeyron equation is $\ln P = \dfrac{-\Delta H_{vap}}{RT} + C.$

For two vapor pressures, P_1 and P_2, measured at corresponding temperatures T_1 and T_2, the relationship is

$$\ln P_1 - \ln P_2 = \left(\frac{-\Delta H_{vap}}{RT_1} + C \right) - \left(\frac{-\Delta H_{vap}}{RT_2} + C \right)$$

$$\ln P_1 - \ln P_2 = \frac{-\Delta H_{vap}}{R} \left(\frac{1}{T_1} - \frac{1}{T_2} \right) + C - C; \quad \ln \frac{P_1}{P_2} = \frac{-\Delta H_{vap}}{R} \left(\frac{1}{T_1} - \frac{1}{T_2} \right)$$

(b) $P_1 = 13.95$ torr, $T_1 = 298$ K; $P_2 = 144.78$ torr, $T_2 = 348$ K

$$\ln \frac{13.95}{144.78} = \frac{-\Delta H_{vap}}{8.314 \text{ J/mol-K}} \left(\frac{1}{298} - \frac{1}{348} \right)$$

$$-2.33974 \, (8.314 \text{ J/mol-K}) = -\Delta H_{vap} \, (4.821 \times 10^{-4} /K)$$

$$\Delta H_{vap} = 4.035 \times 10^4 = 4.0 \times 10^4 \text{ J/mol} = 40 \text{ kJ/mol}$$

$[(1/T_1) - (1/T_2)]$ has 2 sig figs and so does the result.

(c)   The normal boiling point of a liquid is the temperature at which the vapor pressure of the liquid is 760 torr.

$P_1 = 144.78$ torr, $T_1 = 348$ K; $P_2 = 760$ torr, $T_2 = $ bp of octane

$$\ln\left(\frac{144.78}{760.0}\right) = \frac{-4.035 \times 10^4 \text{ J/mol}}{8.314 \text{ J/mol-K}}\left(\frac{1}{348 \text{ K}} - \frac{1}{T_2}\right)$$

$$\frac{-1.6581}{-4.8533 \times 10^3} = 2.874 \times 10^{-3} - \frac{1}{T_2}; \quad \frac{1}{T_2} = 2.874 \times 10^{-3} - 3.416 \times 10^{-4}$$

$$1/T_2 = 2.532 \times 10^{-3} = 2.53 \times 10^{-3}; \quad T_2 = 395 \text{ K } (122\ ^\circ\text{C})$$

From the plot of boiling point versus number of carbon atoms in Solution 11.81, we read an approximate boiling point for octane of 130 °C. These two temperatures are close, but do differ by more than 5%. Considering experimental uncertainties in the vapor pressure (vp) data, and the empirical nature of the plot, the two values are surprisingly close. The literature boiling point of octane, 126 °C, is exactly midway between our two estimates.

(d) $P_1 = $ vp of octane at –30 °C, $T_1 = 243$ K; $P_2 = 144.78$ torr, $T_2 = 348$ K

$$\ln\frac{P_1}{144.78 \text{ torr}} = \frac{-4.035 \times 10^4 \text{ J/mol}}{8.314 \text{ J/mol-K}}\left(\frac{1}{243} - \frac{1}{348}\right)$$

$$\ln\frac{P_1}{144.78 \text{ torr}} = \frac{-4.035 \times 10^4 \text{ J/mol}}{8.314 \text{ J/mol-K}} \times 1.242 \times 10^{-3} = -6.026 = -6.03$$

$$\frac{P_1}{144.78 \text{ torr}} = e^{-6.026}; \quad P_1 = 0.002415(144.78) = 0.3496 = 0.35 \text{ torr}$$

[This result has 2 sig figs because (ln = –6.03) has 2 decimal places. In a ln or log, the places left of the decimal show order of magnitude, and places right of the decimal show sig figs in the real number.] The result, 0.35 torr at –30 °C, is reasonable, because we expect vapor pressure to decrease as temperature decreases, and we are approaching the freezing point of octane, –57 °C.

11.86   Physical data for the two compounds from the *Handbook of Chemistry and Physics*:

	MM	dipole moment	boiling point
CH_2Cl_2	85 g/mol	1.60 D	40.0 °C
CH_3I	142 g/mol	1.62 D	42.4 °C

(a)   The two substances have very similar molecular structures; each is an unsymmetrical tetrahedron with a single central carbon atom and no hydrogen bonding. Because the structures are very similar, the magnitudes of the dipole-dipole forces should be similar. This is verified by their very similar dipole moments. The heavier compound, CH_3I, will have slightly stronger London dispersion forces. Because the nature and magnitude of the intermolecular forces in the two compounds are nearly the same, it is very difficult to predict which will be more volatile [or which will have the higher boiling point as in part (b)].

(b)   Given the structural similarities discussed in part (a), one would expect the boiling points to be very similar, and they are. Based on its larger molar mass (and dipole-dipole forces being essentially equal) one might predict that CH_3I would have a slightly higher boiling point; this is verified by the known boiling points.

(c)   According to Equation [11.1], $\ln P = \dfrac{-\Delta H_{vap}}{RT} + C$

A plot of $\ln P$ versus $1/T$ for each compound is linear. Because the order of volatility changes with temperature for the two compounds, the two lines must cross at some temperature; the slopes of the two lines, ΔH_{vap} for the two compounds, and the y-intercepts, C, must be different.

(d)

CH₂Cl₂			**CH₃I**		
ln P	**T (K)**	**1/T**	**ln P**	**T (K)**	**1/T**
2.303	229.9	4.351×10^{-3}	2.303	227.4	4.398×10^{-3}
3.689	250.9	3.986×10^{-3}	3.689	249.0	4.016×10^{-3}
4.605	266.9	3.747×10^{-3}	4.605	266.2	3.757×10^{-3}
5.991	297.3	3.364×10^{-3}	5.991	298.5	3.350×10^{-3}

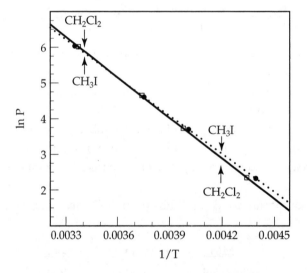

For CH_2Cl_2, $-\Delta H_{vap}/R = \text{slope} = \dfrac{(5.991 - 2.303)}{(3.364 \times 10^{-3} - 4.350 \times 10^{-3})} = \dfrac{-3.688}{0.987 \times 10^{-3}}$

$$= -3.74 \times 10^3 = -\Delta H_{vap}/R$$

$\Delta H_{vap} = 8.314 \, (3.74 \times 10^3) = 3.107 \times 10^4 \, \text{J/mol} = 31.1 \, \text{kJ/mol}$

For CH_3I, $-\Delta H_{vap}/R = \text{slope} = \dfrac{(5.991 - 2.303)}{(3.350 \times 10^{-3} - 4.398 \times 10^{-3})} = \dfrac{-3.688}{1.048 \times 10^{-3}} = -3.519 \times 10^3$

$$= -\Delta H_{vap}/R$$

$\Delta H_{vap} = 8.314 \, (3.519 \times 10^3) = 2.926 \times 10^4 \, \text{J/mol} = 29.3 \, \text{kJ/mol}$

11.87   The normal melting point (nmp) and normal boiling point (nbp) are at a pressure of 1 atm. The triple point (tp) occurs at 1000 Pa. Change this pressure to atm.

$$1000 \times \frac{1 \text{ atm}}{1.01325 \times 10^5 \text{ Pa}} = 0.009869 = 9.87 \times 10^{-3} = 1 \times 10^{-2} \text{ atm}$$

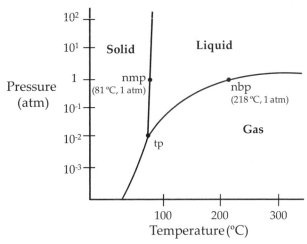

With no information about the critical point of naphthalene, we cannot denote it or the supercritical fluid region.

11.88   When voltage is applied to a liquid crystal display, the molecules align with the voltage and the appearance of the display changes. At low Antarctic temperatures, the liquid crystalline phase is closer to its freezing point. The molecules have less kinetic energy due to temperature and the applied voltage may not be sufficient to overcome orienting forces among the molecules. If some or all of the molecules do not rotate when the voltage is applied, the display will not function properly.

11.89

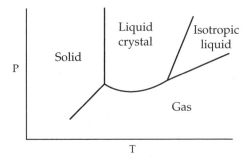

Integrative Exercises

11.90   (a)   Isooctane will have a lower viscosity than octane. Isooctane molecules are more spherical and cannot become as entangled as the flexible chains of octane molecules.

         (b)   In order of increasing boiling point: hexane < heptane < octane < nonane < decane

                Because the molecules have similar structures, the strength of dispersion forces increases with increasing chain length and molar mass. The stronger the dispersion forces the higher the boiling point.

(c)   Statement (i) is the likely explanation. Statement (ii) is false and (iii) is unlikely.

(d)   Statement (ii) is the explanation. *n*-octyl alcohol exhibits hydrogen bonding. Statement (i) is false.

11.91   (a)   24 valence e⁻, 12 e⁻ pairs

$$H \quad :\overset{\|}{O}: \quad H$$

The geometry around the central C atom is trigonal planar, and around the two terminal C atoms, tetrahedral.

(b)   Polar. The C=O bond is quite polar and the dipoles in the trigonal plane around the central C atom do not cancel.

(c)   Dipole-dipole and London dispersion forces

(d)   Because the molecular weights of acetone and 1-propanol are similar, the strength of the London dispersion forces in the two compounds is also similar. The big difference is that 1-propanol has hydrogen bonding, whereas acetone does not. These relatively strong attractive forces lead to the higher boiling point for 1-propanol.

11.92

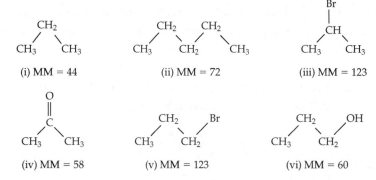

It is useful to draw the structural formulas because intermolecular forces are determined by the size and shape (structure) of molecules.

(a)   *Molar mass*: compounds (i) and (ii) have similar rod-like structures; (ii) has a longer rod. The longer chain leads to greater molar mass, stronger London dispersion forces and higher heat of vaporization.

(b)   *Molecular shape*: compounds (iii) and (v) have the same chemical formula and molar mass but different molecular shapes (they are structural isomers). The more rod-like shape of (v) leads to more contact between molecules, stronger dispersion forces, and higher heat of vaporization.

(c)   *Molecular polarity*: rod-like hydrocarbons (i) and (ii) are essentially nonpolar, owing to free rotation about C–C σ bonds, whereas (iv) is quite polar, owing to the C=O group. (iv) has a smaller molar mass than (ii) but a larger heat of vaporization, which must be due to the presence of dipole-dipole forces in (iv). [Note that (iii) and (iv), with similar shape and molecular polarity, have very similar heats of vaporization.]

(d)   *Hydrogen-bonding interactions*: molecules (v) and (vi) have similar structures, but (vi) has hydrogen bonding and (v) does not. Even though molar mass and thus dispersion forces are larger for (v), (vi) has the higher heat of vaporization. This must be due to hydrogen-bonding interactions.

11.93   Ethanol will evaporate until its vapor fills the flask at a pressure of 40.0 torr. Calculate the mass of 2.00 L of ethanol vapor at 19 °C and a pressure of 40.0 torr. Subtract this mass from the original 1.00 g to find the mass of liquid ethanol remaining.

T = 19 °C + 273.15 = 292 K; MM = 46.07 g/mol;

$$40.0 \text{ torr} \times \frac{1 \text{ atm}}{760 \text{ torr}} = 0.052632 = 0.0526 \text{ atm}$$

$$g = \frac{MM \times PV}{RT}; \; g = \frac{46.07 \text{ g ethanol}}{1 \text{ mol ethanol}} \times \frac{\text{mol-K}}{0.08206 \text{ L-atm}} \times \frac{0.052632 \text{ atm}}{292 \text{ K}} \times 2.0 \text{ L} = 0.202 \text{ g vapor}$$

g ethanol liquid = 1.00 g – 0.202 g vapor = 0.798 = 0.80 g

11.94   (a)   For butane to be stored as a liquid at temperatures above its boiling point (–5 °C), the pressure in the tank must be greater than atmospheric pressure. In terms of the phase diagram of butane, the pressure must be high enough so that, at tank conditions, the butane is "above" the gas-liquid line and in the liquid region of the diagram.

The pressure of a gas is described by the ideal-gas law as P = nRT/V; pressure is directly proportional to moles of gas. The more moles of gas present in the tank the greater the pressure, until sufficient pressure is achieved for the gas to liquefy. At the point where liquid and gas are in equilibrium and temperature is constant, liquid will vaporize or condense to maintain the equilibrium vapor pressure. That is, as long as some liquid is present, the gas pressure in the tank will be constant.

(b)   If butane gas escapes the tank, butane liquid will vaporize (evaporate) to maintain the equilibrium vapor pressure. Vaporization is an endothermic process, so the butane will absorb heat from the surroundings. The temperature of the tank and the liquid butane will decrease.

(c) $$250 \text{ g } C_4H_{10} \times \frac{1 \text{ mol } C_4H_{10}}{58.12 \text{ g } C_4H_{10}} \times \frac{21.3 \text{ kJ}}{\text{mol}} = 91.6 \text{ kJ}$$

$$V = \frac{nRT}{P} = 250 \text{ g} \times \frac{1 \text{ mol}}{58.12 \text{ g}} \times \frac{0.08206 \text{ L-atm}}{\text{mol-K}} \times \frac{308 \text{ K}}{755 \text{ torr}} \times \frac{760 \text{ torr}}{1 \text{ atm}} = 109.44 = 109 \text{ L}$$

11.95   *Plan.*

(i)   Using thermochemical data from Appendix B, calculate the energy (enthalpy) required to melt and heat the H_2O.

(ii)   Using Hess's Law, calculate the enthalpy of combustion, ΔH_{comb}, for C_3H_8.

(iii) Solve the stoichiometry problem.

Solve.

(i)   Heat $H_2O(s)$ from -20 °C to 0.0 °C; 5.50×10^3 g $H_2O \times \dfrac{2.092 \text{ J}}{\text{g-°C}} \times 20 \text{ °C} = 2.301 \times 10^2$

$$= 2.3 \times 10^2 \text{ kJ}$$

Melt $H_2O(s)$; 5.50×10^3 g $H_2O \times \dfrac{6.008 \text{ kJ}}{\text{mol } H_2O} \times \dfrac{1 \text{ mol } H_2O}{18.02 \text{ g } H_2O} = 1834 = 1.83 \times 10^3$ kJ

Heat $H_2O(l)$ from 0 °C to 75 °C; 5.50×10^3 g $H_2O \times \dfrac{4.184 \text{ J}}{\text{g-}^\circ\text{C}} \times 75$ °C $= 1726 = 1.7 \times 10^3$ kJ

Total energy = 230.1 kJ + 1834 kJ + 1726 kJ = 3790 = 3.8×10^3 kJ

(The result is significant to 100 kJ, limited by 1.7×10^3 kJ)

(ii) $C_3H_8(g) + 5\,O_2(g) \rightarrow 3\,CO_2(g) + 4\,H_2O(l)$

Assume that one product is $H_2O(l)$, because this leads to a more negative ΔH_{comb} and fewer grams of $C_3H_8(g)$ required.

$\Delta H_{comb} = 3\,\Delta H_f^\circ\ CO_2(g) + 4\,\Delta H_f^\circ\ H_2O(l) - \Delta H_f^\circ\ C_3H_8(g) - 5\,\Delta H_f^\circ\ O_2(g)$

$= 3(-393.5 \text{ kJ}) + 4\,(-285.83 \text{ kJ}) - (-103.85 \text{ kJ}) - 5(0) = -2219.97 = -2220$ kJ

(iii) 3.790×10^3 kJ required $\times \dfrac{1 \text{ mol } C_3H_8}{2219.97 \text{ kJ}} \times \dfrac{44.096 \text{ g } C_3H_8}{1 \text{ mol } C_3H_8} = 75$ g C_3H_8

(3.8×10^3 kJ required has 2 sig figs and so does the result)

11.96 $P = \dfrac{nRT}{V} = \dfrac{g\ RT}{M\ V}$; $T = 273.15 + 26.0$ °C $= 299.15 = 299.2$ K; $V = 5.00$ L

g $C_6H_6(g) = 7.2146 - 5.1493 = 2.0653$ g $C_6H_6(g)$

$P\,(\text{vapor}) = \dfrac{2.0653 \text{ g}}{78.11 \text{ g/mol}} \times \dfrac{299.15 \text{ K}}{5.00 \text{ L}} \times \dfrac{0.08206 \text{ L-atm}}{\text{mol-K}} \times \dfrac{760 \text{ torr}}{1 \text{ atm}} = 98.660 = 98.7$ torr

11.97   *Plan.* Relative humidity and vp of H_2O at given $T \rightarrow P_{H_2O} \rightarrow$ ideal-gas law $\rightarrow$ mol $H_2O(g) \rightarrow H_2O$ molecules. Change °F $\rightarrow$ °C, volume of room from ft$^3 \rightarrow$ L.

Solve. °C = 5/9 (°F – 32); °C = 5/9 (68 °F – 32) = 20 °C;

$\text{rh} = (P_{H_2O} \text{ in air/vp of } H_2O) \times 100$

From Appendix B, vp of H_2O at 20 °C = 17.54 torr

P_{H_2O} in air = rh × vp of H_2O/100 = 58 × 17.54 torr/100 = 10.173 = 10 torr

$V = 12 \text{ ft} \times 10 \text{ ft} \times 8 \text{ ft} \times \dfrac{12^3 \text{ in}^3}{\text{ft}^3} \times \dfrac{2.54^3 \text{ cm}^3}{\text{in}^3} \times \dfrac{1 \text{ L}}{1000 \text{ cm}^3} = 2.718 \times 10^4 = 3 \times 10^4$ L

(The result has 1 sig fig, as does the measurement 8 ft.)

$PV = nRT; n = PV/RT$

$n = 10.173 \text{ torr} \times \dfrac{1 \text{ atm}}{760 \text{ torr}} \times \dfrac{\text{mol-K}}{0.08206 \text{ L-atm}} \times \dfrac{2.718 \times 10^4 \text{ L}}{293 \text{ K}} = 15.13 = 2 \times 10^1$ mol H_2O

$15.13 \text{ mol } H_2O \times \dfrac{6.022 \times 10^{23} \text{ molecules}}{1 \text{ mol}} = 9.112 \times 10^{24} = 9 \times 10^{24}\ H_2O$ molecules

12 Solids and Modern Materials

Visualizing Concepts

12.1   The red-orange compound is more likely to be a semiconductor and the white one an insulator. The red-orange compound absorbs light in the visible spectrum (red-orange is reflected, so blue-green is absorbed), whereas the white compound does not. This indicates that the red-orange compound has a lower energy electron transition than the white one. Semiconductors have lower energy electron transitions than insulators.

12.2   When choosing a unit cell, remember that the environment of each lattice point must be identical and that unit cells must *tile* to generate the complete two-dimensional lattice. For a given structure, there are often several ways to draw a unit cell. We will select the unit cell with higher symmetry (more 90° or 120° angles) and smaller area ($a \times b$).

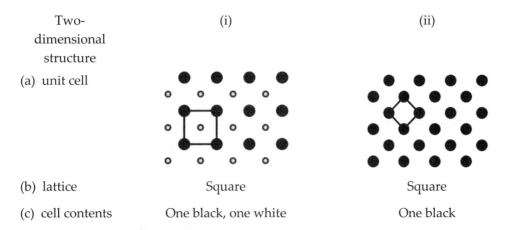

	(i)	(ii)
Two-dimensional structure		
(a) unit cell		
(b) lattice	Square	Square
(c) cell contents	One black, one white	One black

12.3   Sketch (a) represents ductility. Metals can be drawn into wires.

Sketch (b) represents malleability. Metals deform under pressure; they can be pounded into sheets.

12.4   Arrangement (i) represents close packing. Columns of atoms are offset relative to one another to minimize the amount of empty space in the two-dimensional lattice.

12.5   (a)   Clearly, the structure is close packed. The question is: cubic or hexagonal? This is a side view of a close-packed array, like the one in Figure 12.13. The key is the arrangement of the third row relative to the first. Looking at any three rows of cannon balls, there is a ball in the third row directly above (at the same horizontal position as) one in the first row. This is an ABABAB pattern and the structure is hexagonal close packed.

(b)   CN = 12, regardless of whether the structure is hexagonal or cubic close packed.

(c)   CN(1) = 9, CN(2) = 6. The coordination numbers of these two balls are less than 12 because they are on the "surface" of the structure. Of the 12 maximum closest positions, several are unoccupied. Ball 1 is missing two balls that would be in front of it in its own layer, and one ball of the triad in the layer above, for a total of 3 missing and 9 occupied positions. Ball 2 is missing these same three balls, along will all 3 balls in the triad of the layer that would be below it. Ball 2 has six missing and six occupied nearest neighbors.

12.6   Arrangement (b) is more stable. In (a), the cation is so small that its neighboring anions are nearly touching. Very close contacts among like-charged particles produce strong electrostatic repulsions and an unstable arrangement.

12.7   Fragment (b) is more likely to give rise to electrical conductivity. Arrangement (b) has a delocalized system of π electrons, in which electrons are free to move. Mobile electrons are required for electrical conductivity.

12.8   (a)   Band A is the valence band.

          (b)   Band B is the conduction band.

          (c)   Band A (the valence band) consists of bonding molecular orbitals (MOs).

          (d)   This is the electronic structure of a p-type doped semiconductor. The electronic structure shows a few empty MOs or "positive" holes in the valence band. This fits the description of a p-type doped semiconductor.

          (e)   The dopant is Ga. Of the three elements listed, only Ga has fewer valence electrons than Ge, the requirement for a p-type dopant.

12.9   Polymer (a) is more crystalline and has the higher melting point. The polymer chains in cartoon (a) are linear and have two ordered regions shown in the upper left and lower right. The greater the degree of order, the more crystalline the material. The ordered regions of polymer (a) indicate that there are stronger intermolecular forces attracting the chains to each other. Stronger intermolecular forces mean that polymer (a) will have the higher melting point.

12.10   The smaller the nanocrystals, the greater the band gap, E_g, and the shorter the wavelength of emitted light. According to Figure 6.4, the shortest visible wavelengths appear violet or purple and the longest appear red.

          (a)   The 4.0 nm nanocrystals will have the smallest E_g, and emit the longest wavelength. That describes vial 2, the red one.

          (b)   The 2.8 nm nanocrystals will have the largest E_g, and emit the shortest wavelength. That describes vial 1, the green one.

          (c)   The band gap of the 100 nm CdTe crystals is 1.5 eV.

$$\lambda = \frac{hc}{E} = \frac{6.626 \times 10^{-34} \text{ J-s} \times 2.998 \times 10^{8} \text{ m}}{\text{s}} \times \frac{1}{1.5 \text{ eV}} \times \frac{1 \text{ eV}}{1.602 \times 10^{-19} \text{ J}} = 8.27 \times 10^{-7} \text{ m}$$

$$\upsilon = \frac{c}{\lambda} = \frac{2.998 \times 10^{8} \text{ m/s}}{8.27 \times 10^{-7} \text{ m}} = 3.63 \times 10^{14} \text{ s}^{-1}$$

          The visible portion of the electromagnetic spectrum has wavelengths up to 750 nm (7.50×10^{-7} m). The 100 nm CdTe crystals emit a wavelength longer than this, 827 nm (8.27×10^{-7} m). This light is in the IR portion of the spectrum and is not visible to the human eye.

Classification of Solids (Section 12.1)

12.11   Statement (b) best explains the difference. In molecular solids, relatively weak intermolecular forces (hydrogen-bonding, dipole-dipole, dispersion) bind the molecules in the lattice, so relatively little energy is required to disrupt these forces. In covalent-network solids, covalent bonds join atoms into an extended network. Melting or deforming a covalent-network solid means breaking these covalent bonds, which requires a large amount of energy.

12.12   (a)   Covalent-network solid. According to Figure 12.1, diamond and silicon are covalent-network solids. Individual atoms are bound into a three-dimensional network by strong covalent (single) bonds.

          (b)   Covalent-network solid. The physical properties could describe a covalent-network or ionic solid. Silicon and oxygen are both nonmetals, so their bonding is likely to be covalent.

12.13   (a)   hydrogen-bonding forces, dipole-dipole forces, London dispersion forces

          (b)   covalent chemical bonds (mainly)

          (c)   ionic bonds, the Coulombic forces between anions and cations (mainly)

          (d)   metallic bonds

12.14   (a)   metallic

          (b)   molecular or metallic (physical properties of metals vary widely)

          (c)   covalent-network or ionic

          (d)   covalent-network

12.15   (a)   ionic   (b)   metallic

          (c)   covalent-network (Based on melting point, it is probably a network solid. Transition metals in high oxidation states often form bonds with nonmetals that have significant covalent character. It could also be characterized as ionic with some covalent character to the bonds.)

          (d)   molecular   (e)   molecular   (f)   molecular

12.16   (a)   InAs—covalent-network (InAs is a compound semiconductor, with an average of four electrons per atom and somewhat polar covalent bonds.)

          (b)   MgO—ionic crystal (metal and nonmetal)

          (c)   HgS—ionic crystal (metal and nonmetal)

          (d)   In—metallic (metal, Figure 7.13)

          (e)   HBr—molecular (two nonmetals)

12.17   Metallic. The melting point eliminates molecular and covalent-network solids. Because the solid conducts electricity and is insoluble in water, it is metallic. Ionic solids are insulators that are often soluble in water.

12.18   Molecular. The substance is low-melting, which strongly suggests it is molecular. Many molecular solids, such as sucrose, are soluble in water. That an aqueous solution of the white solid does not conduct electricity confirms that this is a molecular solid.

Structures of Solids (Section 12.2)

12.19   *Analyze/Plan.* Crystalline solids have a regular repeat in all three directions. Amorphous solids have no regular repeating structure. Draw diagrams that reflect these definitions. *Solve.*

(a)   (b)

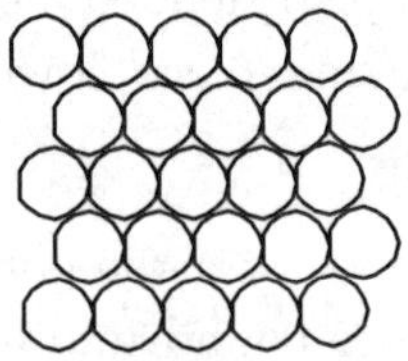

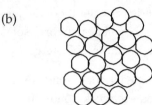

                  crystalline   amorphous

12.20   Statement (e) is the best explanation for the difference in density. In amorphous silica (SiO_2) the regular structure of quartz is disrupted; the loose, disordered structure, has many vacant "pockets" throughout. There are fewer SiO_2 groups per volume in the amorphous solid; the packing is less efficient and less dense.

12.21   *Analyze.* Given two two-dimensional structures, draw and describe the unit cells and lattice vectors. *Plan.* When choosing a unit cell, the environment of each lattice point must be identical and the unit cells must *tile* to generate the complete two-dimensional lattice. For a given structure, there are often several ways to draw a unit cell. The radii of A and B are equal.   *Solve.*

Two-dimensional structure	(i)	(ii)
(a) unit cell		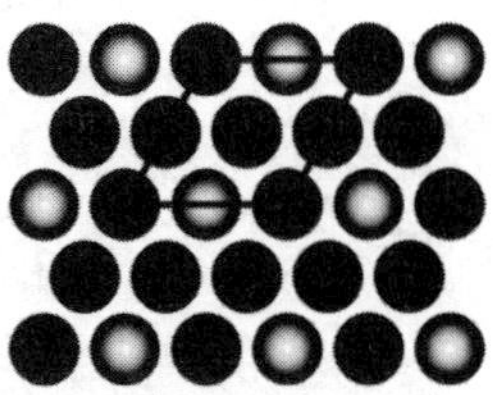
(b) γ, a, b	$\gamma = 90°$, $a = b$	$\gamma = 120°$, $a = b$
(c) lattice type	square	hexagonal

12.22

Two-dimensional structure	(i)	(ii)
(a) unit cell	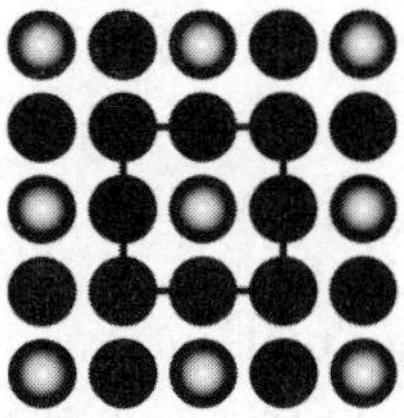	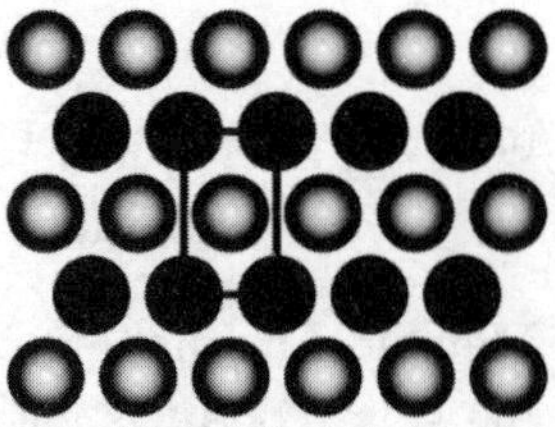
(b) γ, a, b	$\gamma = 90°$, $a = b$	$\gamma = 90°$, $a \neq b$
(c) lattice type	square	rectangular

12.23   *Plan.* Refer to Figure 12.6 to find geometric characteristics of the seven three-dimensional primitive lattices.

Solve. Tetragonal, $a = b \neq c$, $\alpha = \beta = \gamma = 90°$. In the new lattice, one of the edge lengths is longer than (not the same as) the other two, and all angles remain 90°.

12.24   Rhombohedral, $a = b = c$, $\alpha = \beta = \gamma \neq 90°$. In the new lattice, the angles between the unit cell edges remain equal, but have some value less than 90°. The unit cell edge lengths remain equal.

12.25   Choice (e), both rhombohedral and triclinic. According to Figure 12.6, if no lattice vectors are perpendicular to each other, none of the unit cell angles (α, β, γ) are 90°. This is characteristic of two of the three- dimensional primitive lattices: triclinic and rhombohedral.

12.26   Choice (c), rhombohedral. If all three lattice vectors have the same length, $a = b = c$. This is characteristic of the rhombohedral as well as the cubic lattice.

12.27   Choice (b), 2. A body-centered cubic lattice is composed of body-centered cubic unit cells. A unit cell contains the minimum number of atoms when it has atoms only at the lattice points. A body-centered cubic unit cell like this is shown in Figure 12.12(b). There is one atom totally inside the cell (1×1) and one at each corner ($8 \times 1/8$) for a total of 2 atoms in the unit cell. (Only metallic elements have body-centered cubic lattices and unit cells.)

12.28   Choice (d), 4. A face-centered cubic lattice is composed of face-centered cubic unit cells. A unit cell contains the minimum number of atoms when it has atoms only at the lattice points. A face-centered cubic unit cell like this is shown in Figure 12.12(c). There is one atom centered on each face ($6 \times 1/2$) and one at each corner ($8 \times 1/8$) for a total of 4 atoms in the unit cell. (Only metallic elements have face-centered cubic lattices and unit cells.)

12.29   *Analyze.* Given a diagram of the unit cell dimensions and contents of nickel arsenide, determine what kind of lattice this crystal possess, and the empirical formula of the compound. *Plan.* Refer to Figure 12.6 to find geometric characteristics of the seven three-dimensional primitive lattices. Decide where atoms of the two elements are located in the unit cell and use Table 12.1 to help determine the empirical formula.

(a) $a = b = 3.57$ Å. $c = 5.10$ Å $\neq a$ or b. $\alpha = \beta = 90°$, $\gamma = 120°$. This unit cell is hexagonal. There are no atoms in the exact middle of the cell or on the face centers, so it is a primitive hexagonal unit cell. Nickel arsenide has a primitive hexagonal unit cell and crystal lattice.

(b)   There are Ni atoms at each corner of the cell ($8 \times 1/8$) and centered on four of the unit cell edges ($4 \times 1/4$) for a total of 2 Ni atoms. There are 2 As atoms totally inside the cell. The unit cell contains 2 Ni and 2 As atoms; the empirical formula is NiAs.

12.30   (a) $a = b = 3.55$ Å. $c = 6.18$ Å $\neq a$ or b. $\alpha = \beta = \gamma = 90°$. This is a tetragonal unit cell and crystal lattice.

(b)   There is one Al atom totally inside the cell and K atoms at each corner ($8 \times 1/8$). There are F atoms centered on four of the unit cell faces ($4 \times 1/2$) and two F atoms totally inside the cell. The unit cell contains 1 K, 1 Al, and 4 F atoms. The empirical formula is $KAlF_4$.

Metallic Solids (Section 12.3)

12.31   *Analyze/Plan.* Consider body-centered and face-centered structures shown in Figures 12.11 and 12.12 to relate the structures and densities of the metals.

Solve. A body-centered cubic structure has more empty space than a face-centered cubic one. (A faced-centered cubic structure is one of the close-packed structures.) The more empty space, the less dense the solid. We expect potassium, which has the lowest density of the listed elements, to adopt the body-centered cubic structure.

12.32   Metallic: (b) NiCo alloy and (c) W. The lattices of these substances are composed of neutral metal atoms. Delocalization of valence electrons produces metallic properties.

Not metallic: (d) Ge is a metalloid, not a metal. (a) $TiCl_4$ and (e) ScN are ionic compounds; in ionic compounds, electrons are localized on the individual ions, precluding metallic properties.

12.33   *Analyze.* Give diagrams of three structure types, find which is most densely packed and which is least densely packed. *Plan.* Assume that the same element packs in each of the three structures, so that atomic mass and volume are constant. Then we are analyzing the packing efficiency or relative amount of empty space in each structure. *Solve.*

Structure type A has a face-centered cubic unit cell with metal atoms only at the lattice points; this corresponds to a cubic close-packed structure. Structure type B has a body-centered cubic unit cell with metal atoms at the lattice points; this is also a body-centered cubic structure. Structure type C has a hexagonal unit cell with two atoms totally inside the cell. Building up many unit cells into a lattice (Figure 12.14) leads to a hexagonal close-packed structure.

(a)   In both cubic and hexagonal-close packed structures, any individual atom has twelve nearest neighbor atoms. Both structures are close packed and have equal amounts of empty space. Structure types A and C have equally dense packing and are more densely packed than structure type B.

(b)   Structure type B, which is not close packed, has the least dense atom packing.

12.34   (a)   The density of a crystalline solid is (unit cell mass/unit cell volume). Solve for unit cell volume, then use geometry and the properties of a body-centered cubic unit cell to calculate the atomic radius of sodium. There are 2 Na atoms in each body-centered cubic unit cell (Figure 12.12).

$$V = \frac{\text{unit cell mass}}{\rho} = \frac{2 \text{ Na atoms} \times 22.99 \text{ g Na}}{6.022 \times 10^{23} \text{ Na atoms}} \times \frac{cm^3}{0.97 \text{ g}} \times \frac{1 \text{ Å}^3}{(10^{-8} \text{ cm})^3} = 78.71 = 79 \text{ Å}^3$$

For a cubic unit cell, $V = a^3$. We need the relationship between atomic radius and unit cell edge length for a body-centered cubic unit cell. In a body-centered cubic metal structure, the atoms touch along the body diagonal, d_2. Then, $d_2 = 4$ r. From the Pythagorean theorem, $d_2 = \sqrt{3}\ a$. (See Solution 12.36 (c).)

$a = (V)^{1/3} = (78.71)^{1/3} = 4.2857 = 4.3$ Å.

$d_2 = 4\ r_{Na}$; $d_2 = \sqrt{3}\ a$; $r_{Na} = \sqrt{3}\ a/4 = \dfrac{\sqrt{3} \times 4.2857 \text{ Å}}{4} = 1.8557 = 1.9$ Å

(b)   A cubic close-packed metal structure has a face-centered cubic unit cell; there are 4 atoms in each unit cell and atoms touch along the face diagonal, d_1. Then, $d_1 = 4$ r. From the Pythagorean theorem, $d_1 = \sqrt{2}\ a$.

$$d_1 = 4\ r_{Na}\ ;\ \ d_1 = \sqrt{2}\ a;\ \ a = 4\ r_{Na}/\sqrt{2} = \frac{4 \times 1.8557\ \text{Å}}{\sqrt{2}} = 5.2489 = 5.2\ \text{Å} = 5.2 \times 10^{-8}\ \text{cm}$$

$$\rho = \frac{4\ \text{Na atoms}}{(5.2489 \times 10^{-8}\ \text{cm})^3} \times \frac{22.99\ \text{g Na}}{6.022 \times 10^{23}\ \text{Na atoms}} = 1.056 = 1.1\ \text{g/cm}^3$$

Sodium metal with a cubic close-packed structure would not float on water.

12.35   *Analyze.* Given the cubic unit cell edge length and arrangement of Ir atoms, calculate the atomic radius and the density of the metal. *Plan.* In a face-centered cubic metal structure, there is space between the atoms along the unit cell edge, but they touch along the face diagonal. See Figure 12.12. Use the geometry of the right equilateral triangle to calculate the atomic radius. From the definition of density and paying attention to units, calculate the density of Ir(s).   *Solve.*

(a)   The length of the face diagonal of a face-centered cubic unit cell is four times the radius of the atom and $\sqrt{2}$ times the unit cell dimension or edge length, a for cubic unit cells.

$$4\ r = \sqrt{2}\ a;\ \ r = \sqrt{2}\ a/4 = \frac{\sqrt{2} \times 3.833\ \text{Å}}{4} = 1.3552 = 1.355\ \text{Å}$$

(b)   The density of iridium is the mass of the unit cell contents divided by the unit cell volume. There are 4 Ir atoms in a face-centered cubic unit cell.

$$\rho = \frac{4\ \text{Ir atoms}}{(3.833 \times 10^{-8}\ \text{cm})^3} \times \frac{192.22\ \text{g Ir}}{6.022 \times 10^{23}\ \text{Ir atoms}} = 22.67\ \text{g/cm}^3$$

Check. The units of density are correct. Note that Ir is quite dense.

12.36   (a)   In a body-centered cubic unit cell, there is one atom totally inside the unit cell (1×1) and one atom at each of the eight corners ($8 \times 1/8$), for a total of 2 Ca atoms in each unit cell.

(b)   Because atoms are only at lattice points in a body-centered metal structure, each metal atom has an equivalent environment. That is, atoms at the corner of the cell and the interior atom must have equivalent environments. Consider the Ca atom at the middle of the unit cell. It has eight nearest neighbors, the eight Ca atoms at the corners of the cell. Interior atoms in adjacent unit cells are farther from the reference Ca atom and are not "nearest." (For a corner Ca atom, the eight nearest neighbors are the interior atoms in the eight unit cells that include 1/8 of that corner atom.)

(c)   In a body-centered cubic metal structure, the atoms touch along the body diagonal, d_2. Describe the length of the body diagonal in terms of the unit cell dimension a. Define a triangle with sides that are: d_2, the body diagonal of the cube; d_1, the face diagonal of the cube; and a, one edge of the cube. The length of the face diagonal is $\sqrt{2}\ a$ and the edge length is a. By the Pythagorean theorem,

$$d_2 = \sqrt{d_1^2 + a^2} = \sqrt{(\sqrt{2}\ a)^2 + a^2} = \sqrt{3}\ a$$

$$4\ r_{Ca} = \sqrt{3}\ a;\ \ a = 4\ r_{Ca}/\sqrt{3} = \frac{4 \times 1.97\ \text{Å}}{\sqrt{3}} = 4.5495 = 4.55\ \text{Å}$$

(d) $\rho = \dfrac{2 \text{ Ca atoms}}{(4.5495 \times 10^{-8} \text{ cm})^3} \times \dfrac{40.08 \text{ g Ca}}{6.022 \times 10^{23} \text{ Ca atoms}} = 1.4136 = 1.41 \text{ g/cm}^3$

12.37   (a)   *Analyze.* We are given a face-centered cubic unit cell with edge length 5.588 Å.
         Plan. In a face-centered cubic metal structure, the atoms touch along the face
         diagonal of the unit cell. The length of the face diagonal of a face-centered cubic
         unit cell is four times the radius of the atom and $\sqrt{2}$ times the unit cell
         dimension or edge length, *a* for cubic unit cells.   *Solve.*

$$4r = \sqrt{2}\, a; \quad r = \sqrt{2}\, a/4 = \frac{\sqrt{2} \times 5.588 \text{ Å}}{4} = 1.9757 = 1.976 \text{ Å}$$

(b)   *Plan.* The density of calcium is the mass of the unit cell contents divided by the
       unit cell volume. There are 4 Ca atoms in a face-centered cubic unit cell.   *Solve.*

$$\rho = \frac{4 \text{ Ca atoms}}{(5.588 \times 10^{-8} \text{ cm})^3} \times \frac{40.078 \text{ g Ca}}{6.022 \times 10^{23} \text{ Ca atoms}} = 1.526 \text{ g/cm}^3$$

Check.   The units of density are correct.

12.38   *Analyze/Plan.* We are given the radius of atoms in various cubic structures and asked to
         calculate the volume of each structure. Use Figure 12.12 to determine the relationship
         between atomic radius and unit cell edge length *a*. For cubic unit cells, the three edge
         lengths are equal. The unit cell volume is a^3.   *Solve.*

(a)   In a primitive cubic structure, atoms touch along a unit cell edge, *a*.

$a = 2\,r; \; V = 8\,r^3. \; V = 8(1.82)^3 = 48.2285 = 48.2 \text{ Å}^3$

(Primitive cubic metal structures are relatively rare.)

(b)   In a face-centered cubic metal structure, the atoms touch along the face diagonal,
       d_1. Then, $d_1 = 4\,r$. From the Pythagorean theorem, the length of the face diagonal
       equals $\sqrt{2}$ times the unit cell dimension, *a*.

$$4\,r = \sqrt{2}\, a; \; a = \frac{4 \times r}{\sqrt{2}}; \; V = a^3 = \left(\frac{4 \times r}{\sqrt{2}}\right)^3 = 136.411 = 136 \text{ Å}^3$$

12.39   *Analyze.* Given the structure of aluminum metal and the atomic radius of an Al atom,
         find the number of Al atoms in each unit cell and the coordination number of each Al
         atom. Calculate (estimate) the length of the unit cell edge and the density of aluminum
         metal.

Plan. Use Figure 12.12(c) to count the number of Al atoms in one unit cell and use Figure
12.13 to visualize the coordination number of each Al atom. According to Figure 12.12(c),
there is space between the atoms along the unit cell edge, but they touch along the face
diagonal. Use the geometry of the right equilateral triangle and the atomic radius to
calculate the unit cell edge length. From the definition of density and paying attention to
units, calculate the density of aluminum metal.   *Solve.*

(a)   8 corners × 1/8 atom/corner + 6 faces × ½ atom/face = 4 atoms

(b)   Each aluminum atom is in contact with 12 nearest neighbors, 6 in one plane, 3
       above that plane, and 3 below. Its coordination number is thus 12.

(c) The length of the face diagonal of a face-centered cubic unit cell is four times the radius of the metal and $\sqrt{2}$ times the unit cell dimension (usually designated a for cubic cells).

$$4 \times 1.43 \text{ Å} = \sqrt{2} \times a; \quad a = \frac{4 \times 1.43 \text{ Å}}{\sqrt{2}} = 4.0447 = 4.04 \text{ Å} = 4.04 \times 10^{-8} \text{ cm}$$

(d) The density of the metal is the mass of the unit cell contents divided by the volume of the unit cell.

$$\text{density} = \frac{4 \text{ Al atoms}}{(4.0447 \times 10^{-8} \text{ cm})^3} \times \frac{26.98 \text{ g Al}}{6.022 \times 10^{23} \text{ Al atoms}} = 2.71 \text{ g/cm}^3$$

12.40   *Analyze.* Given the atomic arrangement, length of the cubic unit cell edge, and density of the solid, calculate the atomic weight of the element. *Plan.* If we calculate the mass of a single unit cell, and determine the number of atoms in one unit cell, we can calculate the mass of a single atom and of a mole of atoms. *Solve.*

The volume of the unit cell is $(4.078 \times 10^{-8} \text{ cm})^3$. The mass of the unit cell is:

$$\frac{19.30 \text{ g}}{\text{cm}^3} \times \frac{(4.078 \times 10^{-8})^3 \text{ cm}^3}{\text{unit cell}} = 1.3089 \times 10^{-21} = 1.31 \times 10^{-21} \text{ g/unit cell}$$

There are four atoms of the element present in the face-centered cubic unit cell. Thus the atomic weight is:

$$\frac{1.3089 \times 10^{-21} \text{ g}}{\text{unit cell}} \times \frac{1 \text{ unit cell}}{4 \text{ atoms}} \times \frac{6.0221 \times 10^{23} \text{ atoms}}{1 \text{ mol}} = 197.06 = 197 \text{ g/mol}$$

Check. The result is in the range of known atomic weights and the units are correct. The element is most likely gold.

12.41   Statement (b) is false. Alloys are mixtures, not compounds, that vary in composition. One or more of the components of the alloy can be a nonmetal.

12.42   (a)   False. Substitutional and interstitial alloys are both solution alloys.

          (b)   True

          (c)   True

12.43   *Analyze/Plan.* Consider the descriptions of various alloy types in Section 12.3. *Solve.*

          (a) $Fe_{0.97}Si_{0.03}$; interstitial alloy. The radii of Fe and Si are substantially different, so Si could fit in "holes" in the Fe lattice. Also, the small amount of Si relative to Fe is characteristic of an interstitial alloy.

          (b) $Fe_{0.60}Ni_{0.40}$, substitutional alloy. The two metals have very similar atomic radii and are present in similar amounts.

          (c) $SmCo_5$, intermetallic compound. The two elements are present in stoichiometric amounts.

12.44   (a) $Cu_{0.66}Zn_{0.34}$, substitutional alloy; similar atomic radii, substantial amounts of both components

        (b) Ag_3Sn, intermetallic compound; set stoichiometric ratio of components

        (c) $Ti_{0.99}O_{0.01}$, interstitial alloy; very different atomic radii, tiny amount of smaller component

12.45   (a)   True

        (b)   False. Interstitial alloys form between elements with very different bonding atomic radii.

        (c)   False. Nonmetallic elements are typically found in interstitial alloys.

12.46   (a)   True

        (b)   True

        (c)   False. In stainless steel, the chromium atoms replace iron atoms in the structure. (The atomic radii of Fe and Cr are 1.25 Å and 1.27 Å, respectively. Metal atoms with similar radii form substitutional alloys.)

12.47   *Analyze.* Given the color of a gold alloy, find the other element(s) in the alloy and the type of alloy formed. *Plan.* Refer to "Chemistry Put to Work: ALLOYS OF GOLD".   *Solve.*

        (a)   White gold, nickel or palladium, substitutional alloy

        (b)   Rose gold, copper, substitutional alloy

        (c)   Green gold, silver, substitutional alloy

12.48   The term *expansion* implies that unit cell edge lengths will increase. This agrees with the fact that the volume increases. Density is the mass of a unit cell divided by its volume. Since the number of atoms in a unit cell does not change when the metal is heated, but the volume increases, the density will decrease.

Metallic Bonding (Section 12.4)

12.49   (a)   True

        (b)   False. See statement (a).

        (c)   False. Delocalized electrons in metals facilitate the transfer of kinetic energy, which is the basis of thermal conductivity.

        (d)   False. Metals have large thermal conductivities.

12.50   The part of the bar sitting in the dark will feel hot. The structure of the metal and its delocalized electrons extend over the whole bar. The high thermal conductivity of the metal facilitates heat transfer from the part of the bar in the sun to the part of the bar in the shade.

12.51   *Plan.* By analogy to Figure 12.22, the most bonding, lowest energy MOs have the fewest nodes. As energy increases, the number of nodes increases. When constructing an MO diagram from AOs, total number of orbitals is conserved. The MO diagram for a linear chain of six Li atoms will have six MOs, starting with zero nodes and maximum overlap, and ending with five nodes and minimum overlap.   *Solve.*

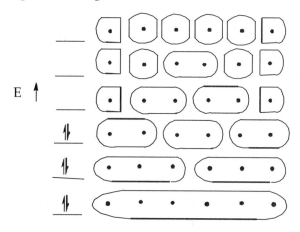

(a)   Six. Six AOs require six MOs.

(b)   Zero nodes in lowest energy orbital

(c)   Five nodes in highest energy orbital

(d)   Two nodes in the HOMO

(e)   Three nodes in the LUMO

(f)   The HOMO-LUMO energy gap for the six-atom diagram is smaller than the one for the four-atom diagram. In general, the more atoms in the chain, the smaller the HOMO-LUMO energy gap.

12.52

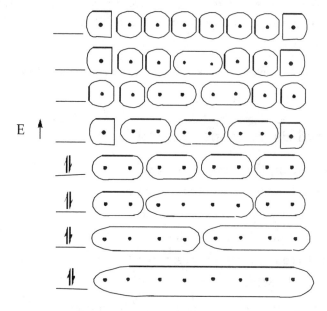

(a) 8   (b) 0   (c) 7   (d) 3   (e) 4   (f) smaller

12.53   *Analyze/Plan.* Consider the definition of ductility, as well as the discussion of metallic bonding in Section 12.4.   *Solve.*

Ductility is the property related to the ease with which a solid can be drawn into a wire. Basically, the softer the solid the more ductile it is. The more rigid the solid, the less ductile it is. For metals, ductility decreases as the strength of metal-metal bonding increases, producing a stiffer lattice less susceptible to distortion.

(a)   Ag is more ductile. Mo, with 6 valence electrons, has a filled bonding band, strong metal-metal interactions, and a rigid lattice. This predicts high hardness and low ductility. Ag, with 11 valence electrons, has a nearly filled antibonding band as well as a filled bonding band. Bonding is weaker than in Mo, and Ag is more ductile.

(b)   Zn is more ductile. Si is a covalent-network solid with all valence electrons localized in bonds between Si atoms. Covalent-network substances are high-melting, hard, and not particularly ductile.

12.54   Statement (d) is false and does not follow from the fact that alkali metals have relatively weak metal-metal bonds. We expect strong metal-metal bonds to be present in metals with high melting points (Figure 12.21).

12.55   The relevant electron configurations are:

Y: $[Kr]5s^2 4d^1$; Zr: $[Kr]5s^2 4d^2$; Nb: $[Kr]5s^2 4d^3$; Mo: $[Kr]5s^1 4d^5$;

The order of increasing melting points is Y < Zr < Nb < Mo.

Moving across the fifth period from Y to Mo, the number of valence electrons increases, from 3 for Y to 6 for Mo. More valence electrons (up to 6) mean increased occupancy of the bonding molecular orbital band, and increased strength of metallic bonding. Melting requires that atoms are moving relative to each other. Stronger metallic bonding requires more energy to break bonds and mobilize atoms, resulting in higher melting points from Y to Mo.

12.56   In each group, choose the metal that has the number of valence electrons closest to six.

(a)   Re   (b)   Mo   (c)   Ru

Ionic and Molecular Solids (Sections 12.5 and 12.6)

12.57   (a)   Sr: Sr atoms occupy the 8 corners of the cube.

           8 corners × 1/8 sphere/corner = 1 Sr atom

   O: O atoms occupy the centers of the 6 faces of the cube.

           6 faces × 1/2 atom/face = 3 O atoms

   Ti: There is 1 Ti atom at the body center of the cube.

   Empirical Formula: $SrTiO_3$

(b)   Six. The Ti atom at the center of the cube is coordinated to the six O atoms at the face centers.

(c)   Twelve. Each Sr atom occupies one corner of 8 unit cells. Sr is coordinated to 3 oxygen positions in each unit cell for a total of 24 oxygen positions. However, each O position is in the center of a cell face, with half-occupancy in each cell.

24 oxygen positions × ½ occupancy = 12 oxygen atoms. Each Ti atom is coordinated to 12 oxygen atoms.

12.58   Co: 8 corners × 1/8 sphere/corner = 1 Co atom

           O: 2 atoms completely inside the unit cell = 2 O atoms

           Formula: CoO_2

           The oxidation number of Co is +4, the oxidation state is Co(IV).

12.59   *Analyze/Plan.* The density of MnS is the mass of the unit cell contents divided by the volume of the unit cell. MnS has the same structure as rock salt, or NaCl. Refer to the NaCl structure in Figures 12.25 and 12.26 to determine the unit cell contents. Use the unit cell edge length to find unit cell volume. Calculate density. *Solve.*

By analogy to the NaCl structure, with Mn replacing Na and S replacing Cl, there are 8 sulfide ions (green spheres) on the corners of the unit cell, and 6 sulfide ions in the middle of the faces. The number of sulfide ions per unit cell is then [8(1/8) + 6(1/2)] = 4. There is 1 manganese ion (purple sphere almost hidden in figure) completely inside the unit cell and 12 manganese ions (purple spheres) along the unit cell edges. The number of manganese ions is then [1 + 12(1/4)] = 4. This result satisfies charge balance requirements.

         $4\ Mn^{2+}$, $4\ S^{2-}$, 4 MnS formula units. The mass of 1 MnS formula unit is 87.003 g/ 6.022×10^{23} NaF units.

$$ d = \frac{4\ MnS\ units}{(5.223\ \text{Å})^3} \times \frac{87.003\ g}{6.0221 \times 10^{23}\ MnS\ units} \times \left(\frac{1\ \text{Å}}{1 \times 10^{-8}\ cm} \right)^3 = 4.056\ g/cm^3 $$

Check. The value for the density of alabandite (MnS) reported in the *CRC Handbook of Chemistry and Physics*, 74th Ed., is 3.99 g/cm^3. The calculated density is within 2% of the reported value.

12.60   Calculate the mass of a single unit cell and then use density to find the volume of a single unit cell. The edge length is the cube root of the volume of a cubic cell. From the previous exercise, there are four PbSe units in a NaCl-type unit cell . The unit cell edge length is designated *a*.

$$ 8.27\ g/cm^3 = \frac{4\ PbSe\ units}{a^3} \times \frac{286.2\ g}{6.022 \times 10^{23}\ PbSe\ units} \times \left(\frac{1\ \text{Å}}{1 \times 10^{-8}\ cm} \right)^3 $$

$a^3 = 229.87\ \text{Å}^3$, $a = 6.13\ \text{Å}$

12.61   *Analyze.* Given the atomic arrangement and length of the unit cell side, calculate the density of HgS and HgSe. Qualitatively and quantitatively compare the densities of the two solids. *Plan.* Calculate the mass and volume of a single unit cell and then use them to find density. The unit cell volume is the cube of edge length. *Solve.*

(a)   According to Figures 12.25 and 12.26, sulfide ions (yellow spheres) occupy the corners and faces of the unit cell (in a face-centered cubic arrangement), for a total of $6(1/2) + 8(1/8) = 4\ S^{2-}$ ions per unit cell. There are 4 mercury ions (gray spheres) totally in the interior of the cell. This means there are 4 HgS units in a unit cell with the zinc blende structure.

$$\text{density} = \frac{4\ \text{HgS units}}{(5.852\ \text{Å})^3} \times \frac{232.655\ \text{g}}{6.022 \times 10^{23}\ \text{HgS units}} \times \left(\frac{1\ \text{Å}}{1 \times 10^{-8}\ \text{cm}}\right)^3 = 7.711\ \text{g/cm}^3$$

(b)   We expect Se^{2-} to have a larger ionic radius than S^{2-}, because Se is below S in the chalcogen family and both ions have the same charge. Thus, HgSe will occupy a larger volume and the unit cell edge will be longer.

(c)   For HgSe, also with the zinc blende structure:

$$\text{density} = \frac{4\ \text{HgSe units}}{(6.085\ \text{Å})^3} \times \frac{279.55\ \text{g HgSe}}{6.022 \times 10^{23}\ \text{HgSe units}} \times \left(\frac{1\ \text{Å}}{1 \times 10^{-8}\ \text{cm}}\right)^3 = 8.241\ \text{g/cm}^3$$

Even though HgSe has a larger unit cell volume than HgS, it also has a larger molar mass. The mass of Se is more than twice that of S, whereas the radius of Se^{2-} is only slightly larger than that of S^{2-} (Figure 7.8). The greater mass of Se accounts for the greater density of HgSe.

12.62   (a)   Each cell edge goes through two half Rb^+ ions (at the corners) and one full I^- ion (centered on the edge). The length of an edge, a, is then

$a = 2r_{Rb} + 2r_I = 2(1.66\ \text{Å}) + 2(2.06\ \text{Å}) = 7.44\ \text{Å}$

(b) $$\text{density} = \frac{4\ \text{RbI units}}{(7.44\ \text{Å})^3} \times \frac{212.37\ \text{g}}{6.022 \times 10^{23}\ \text{RbI units}} \times \left(\frac{1\ \text{Å}}{1 \times 10^{-8}\ \text{cm}}\right)^3 = 3.43\ \text{g/cm}^3$$

(c)   In the CsCl-type structure, I^- anions sit at the corners of the cube and an Rb^+ cation sits completely inside the unit cell (at or near the body center). Assume that the anions and cations touch along the body diagonal d_2, so that

$d_2 = 2r_{Rb} + 2r_I = 2(1.66\ \text{Å}) + 2(2.06\ \text{Å}) = 7.44\ \text{Å}$. (This is the edge length of the NaCl-type structure.)

From Solution 12.36, the relationship between the body diagonal and edge of a cube is: $d_2 = \sqrt{3} \times a$; $a = d_2/\sqrt{3}$. $a = 7.44\ \text{Å}/\sqrt{3} = 4.2955 = 4.30\ \text{Å}$

(d)   There is one RbI unit ($8 \times 1/8\ I^-$ anions and 1 Rb^+ cation) in the CsCl-type unit cell.

$$\text{density} = \frac{1\ \text{RbI unit}}{(4.2955\ \text{Å})^3} \times \frac{212.37\ \text{g}}{6.022 \times 10^{23}\ \text{RbI units}} \times \left(\frac{1\ \text{Å}}{1 \times 10^{-8}\ \text{cm}}\right)^3 = 4.45\ \text{g/cm}^3$$

The density of the CsCl-type structure is greater than the density of the NaCl-type structure, owing to the much smaller unit cell volume.

12.63   *Analyze.* Given that CuI, CsI, and NaI uniquely adopt one of the structure types pictured in Figure 12.26, match the ionic compound with its structure. Use ionic radii to inform your decision. *Plan.* Note the relative cation and anion radii in the three structures in Figure 12.26. Match these ratios with those of CuI, CsI, and NaI. *Solve.*

(a) In the CsCl structure, the anion and cation have about the same radius; in the NaCl structure, the anion is somewhat larger than the cation; in the ZnS structure the anion is much larger than the cation.

In the three compounds given, Cs^+ (r = 1.81 Å) and I^- (r = 2.06 Å) have the most similar radii; CsI will adopt the CsCl-type structure. The radii of Na^+ (r = 1.16 Å) and I^- (r = 2.06 Å) are somewhat different; NaI will adopt the NaCl-type structure. The radii of Cu^+ (r = 0.74 Å) and I^- (r = 2.06 Å) are very different; CuI has the ZnS-type structure.

(b) As cation size decreases, coordination number of the anion decreases. In CsI, I^- has a coordination number of eight; in NaI, I^- has a coordination number of six; in CuI, I^- has a coordination number of four.

12.64 (a) In CaF_2 the ionic radii are very similar, Ca^{2+} (r = 1.14 Å) and F^- (r = 1.19 Å). In ZnF_2 the cation radius is smaller and the ionic radii are more different, Zn^{2+} (r = 0.88 Å) and F^- (r = 1.19 Å). Cations in both structures in the exercise are shown with equal radii, so direct inspections does not answer the question. We can, however, refer to the trend that, for compounds with the same cation/anion ratio, as cation size decreases, coordination number of the anion decreases. The anion coordination number (CN) for the rutile structure is 3 and for the fluorite structure is 4. The compound with the smaller cation, ZnF_2, will adopt the rutile structure and the compound with the larger cation, CaF_2, will adopt the fluorite structure. (Detailed analysis of coordination numbers follows.)

(b) Rutile (top) structure, cation (blue) coordination number (CN) = 6, anion (green) CN = 3. [The blue cation in the interior or the cell is coordinated to the six (green) anions associated with the same cell; either of the green interior anions is associated with the triangle of blue cations located inside and at the two nearest corners of the cell.]

Fluorite (bottom) structure, cation (blue) CN = 8, anion (green) CN = 4. [A blue cation at one of the face centers is coordinated to the four nearest green anions inside the cell and to four identical anions in an adjacent cell that also contains the cation. Any of the green interior anions is coordinated to a tetrahedron of blue cations located at one corner and the three nearest face centers.]

12.65 *Analyze.* Given three magnesium compounds in which the coordination number (CN) of Mg^{2+} is six, determine the coordination number of the anion. *Plan.* Use Equation 12.1 with the cation/anion ratio of each compound and the Mg^{2+} coordination number to calculate the anion coordination number in each compound. *Solve.*

(a) MgS: 1 cation, 1 anion, cation CN = 6

$$\frac{\text{number of cations per formula unit}}{\text{number of anions per formula unit}} = \frac{\text{anion coordination number}}{\text{cation coordination number}}$$

$$\text{anion CN} = \frac{\text{cation CN} \times \text{\# of cations per formula unit}}{\text{\# of anions per formula unit}} = \frac{6 \times 1}{1} = 6$$

(b) MgF_2: 1 cation, 2 anions, cation CN = 6

$$\text{anion CN} = \frac{\text{cation CN} \times \text{\# of cations per formula unit}}{\text{\# of anions per formula unit}} = \frac{6 \times 1}{2} = 3$$

(c)   MgO: 1 cation, 1 anion, cation CN = 6, anion CN = 6. The cation/anion ratio and cation CN are the same as in part (a), so the anion CN is the same (6).

12.66 $$\frac{\text{\# of cations per formula unit}}{\text{\# of anions per formula unit}} = \frac{\text{anion coordination number}}{\text{cation coordination number}}$$

$$\text{cation CN} = \frac{\text{anion CN} \times \text{\# of anions per formula unit}}{\text{\# of cations per formula unit}}$$

(a) AlF_3:   1 cation, 3 anions, anion CN = 2; cation CN = $(2 \times 3)/1 = 6$

(b) Al_2O_3:   2 cations, 3 anions, anion CN = 6; cation CN = $(6 \times 3)/2 = 9$

(c)   AlN: 1 cation, 1 anion, anion CN = 4; cation CN = $(4 \times 1)/1 = 4$

12.67   (a)   False. Although both molecular solids and covalent-network solids have covalent bonds, the melting points of molecular solids are much lower because intermolecular forces among their molecules are much weaker than covalent bonds among atoms in a covalent-network solid.

(b)   True. The statement is true if there are no significant differences in polarity or molar mass of the molecules being compared.

12.68   (a)   False. Melting point is a bulk property, not a molecular property.

(b)   True. Strengths of intermolecular forces determine properties of the bulk material like melting point.

Covalent-Network Solids (Section 12.7)

12.69   (a)   Ionic solids are much more likely to dissolve in water. Polar water molecules can disrupt ionic bonds to surround and separate ions and form a solution, but they cannot break the covalent bonds of a covalent-network solid.

(b)   Covalent-network solids can become better conductors of electricity via chemical substitution. Most semiconductors are covalent-network solids, and doping or chemical substitution changes their electrical properties.

12.70   The extended network of localized covalent bonds in a covalent-network solid produces solids that are inflexible, hard, and high-melting. The delocalized nature of metallic bonding leads to flexibility, because atoms can move relative to one another without "breaking" bonds. Metals have a wide range of hardness and melting point, depending on the occupancy of the bands.

(a)   Ductility, metallic solid

(b)   Hardness, covalent-network solid and metallic solid, depending on the strength of metallic bonding

(c)   High melting point, covalent-network solid and metallic solid, depending on the strength of metallic bonding

12.71   *Analyze/Plan.* Follow the logic in Sample Exercise 12.3.   *Solve.*

    (a)   CdS. Both semiconductors contain Cd; S and Te are in the same family, and S is higher.

    (b)   GaN. Ga is in the same family and higher than In; N is in the same family and higher than P.

    (c)   GaAs. Both semiconductors contain As; Ga and In are in the same family and Ga is higher.

12.72   (a)   InP. P is in the same family and higher than As.

    (b)   AlP. Al and P are horizontally separated, resulting in greater bond polarity, and Al and P are in the row above Ge. Band gap values in Table 12.4 confirm this order.

    (c)   AgI. The four elements are in the same row; Ag and I are farther apart than Cd and Te.

12.73   *Analyze.* Given: GaAs. Find: dopant to make n-type semiconductor.

    Plan. An n-type semiconductor has extra negative charges. If the dopant replaces a few Ga atoms, it should have more valence electrons than Ga, Group 3A.

    Solve. The obvious choice is a Group 4A element, either Ge or Si. Ge would be closer to Ga in bonding atomic radius (Figure 7.7).

12.74   p-type semiconductors have a slight electron deficit. If the dopant replaces As, Group 5A, it should have fewer than five valence electrons. The dopant will be a 4A element, probably Si or Ge. Si would be closer to As in bonding atomic radius.

12.75   (a)   *Analyze.* Given: 1.1 eV. Find: wavelength in meters that corresponds to the energy 1.1 eV. *Plan.* Use dimensional analysis to find wavelength.

    Solve. $1 \text{ eV} = 1.602 \times 10^{-19}$ J (inside back cover of text); $\lambda = hc/E$

$$\lambda = 6.626 \times 10^{-34} \text{ J-s} \times \frac{3.00 \times 10^8 \text{ m}}{\text{s}} \times \frac{1}{1.1 \text{ eV}} \times \frac{1 \text{ eV}}{1.602 \times 10^{-19} \text{ J}} = 1.128 \times 10^{-6} = 1.1 \times 10^{-6} \text{ m}$$

    (b)   Si can absorb energies ≥ 1.1 eV, or wavelengths $\leq 1.1 \times 10^{-6}$ m. A wavelength of 1.1×10^{-6} m corresponds to 1100 nm. The range of wavelengths in the visible portion of the spectrum is 300 to 750 nm. Silicon absorbs all wavelengths in the visible portion of the solar spectrum.

    (c)   Si absorbs wavelengths less than 1100 nm. This corresponds to approximately 80–90% of the total area under the curve.

12.76   (a)   From Table 12.4, the band gap of CdTe is 1.50 eV (or 145 kJ/mol).

    (b) $\lambda = \dfrac{hc}{E} = \dfrac{6.626 \times 10^{-34} \text{ J-s} \times 2.998 \times 10^8 \text{ m}}{\text{s}} \times \dfrac{1}{1.50 \text{ eV}} \times \dfrac{1 \text{ eV}}{1.602 \times 10^{-19} \text{ J}}$

$$= 8.267 \times 10^{-7} = 8.27 \times 10^{-7} \text{ m}$$

    (c, d)   CdTe can absorb energies ≥ 1.50 eV or wavelengths $\leq 8.27 \times 10^{-7}$ m. A wavelength of 8.27×10^{-7} m corresponds to 827 nm. CdTe absorbs a smaller range of wavelengths and thus a smaller portion of the solar spectrum than Si.

12.77   *Plan/Solve.* Follow the logic in Solution 12.75.

$$\lambda = hc/E = 6.626 \times 10^{-34} \text{ J-s} \times \frac{3.00 \times 10^{8} \text{ m}}{\text{s}} \times \frac{1}{1.74 \text{ eV}} \times \frac{1 \text{ eV}}{1.602 \times 10^{-19} \text{ J}} = 7.131 \times 10^{-7}$$

$$= 7.13 \times 10^{-7} \text{ m} = 713 \text{ nm}$$

The emitted light with a wavelength of 713 nm is red light in the visible region of the electromagnetic spectrum.

12.78   The band gap, ΔE, of GaAs is 1.43 eV.

$$\lambda = \frac{hc}{E} = \frac{6.626 \times 10^{-34} \text{ J-s} \times 2.998 \times 10^{8} \text{ m}}{\text{s}} \times \frac{1}{1.43 \text{ eV}} \times \frac{1 \text{ eV}}{1.602 \times 10^{-19} \text{ J}} = 8.67 \times 10^{-7} \text{ m (867 nm)}$$

The visible portion of the electromagnetic spectrum has wavelengths up to 750 nm (7.50×10^{-7} m). GaAs emits longer, 867 nm, light in the IR portion of the spectrum.

12.79   *Analyze/Plan.* From Table 12.4, E_g for GaAs (x = 0) is 1.43 eV and for GaP (x = 1) is 2.26 eV. If E_g varies linearly with x, the band gap for x = 0.5 should be approximately the average of the two extreme values.

Solve. (1.43 + 2.26)/2 = 1.845 = 1.85 eV.

$$\lambda = hc/E = \frac{6.626 \times 10^{-34} \text{ J-s} \times 2.998 \times 10^{8} \text{ m}}{\text{s}} \times \frac{1}{1.845 \text{ eV}} \times \frac{1 \text{ eV}}{1.602 \times 10^{-19} \text{ J}}$$

$$= 6.721 \times 10^{-7} \text{ m} = 672 \text{ nm}$$

12.80   Reverse the logic in Solution 12.79. Given λ, calculate E_g. Then solve for x assuming the value of E_g is a linear combination of the stoichiometric contributions of GaP and GaAs. That is, $2.26 x + 1.43(1 - x) = E_g$.

$$E_g = \frac{hc}{\lambda} = \frac{6.626 \times 10^{-34} \text{ J-s}}{6.60 \times 10^{-7} \text{ m}} \times \frac{2.998 \times 10^{8} \text{ m}}{\text{s}} \times \frac{1 \text{ eV}}{1.602 \times 10^{-19} \text{ J}} = 1.8788 = 1.88 \text{ eV}$$

$2.26 x + 1.43(1 - x) = 1.8788; 2.26 x - 1.43 x + 1.43 = 1.8788$

$0.83 x = 0.4488; x = 0.5407 = 0.54$

Check. From Solution 12.79, an E_g of 1.85 eV corresponds to x = 0.5. E_g =1.88 eV should have a very similar composition, and x = 0.54 is very close to x = 0.5. The P/As composition is very sensitive to wavelength, and provides a useful mechanism to precisely tune the wavelength of the diode.

Polymers (Section 12.8)

12.81   (a)   A monomer is a small molecule with low molecular mass that can be joined with other monomers to form a polymer. It is the repeating unit of a polymer.

(b)   Ethene (ethylene) is commonly used as a monomer, while ethanol and methane are not. Methane does not have reactive groups that can be chemically linked into polymers. Ethanol is reactive because of its one alcohol functional group, but it cannot form long polymer chains. Alcohol monomers used to form polyesters have two alcohol groups.

12.82   *n*-decane does not have a sufficiently high chain length or molecular mass to be considered a polymer.

12.83   A value of 100 amu is too low to be the molecular weight of a polymer; it would not represent enough monomer units. Often a single monomer has a molecular weight greater than 100 amu. Reasonable values for a polymer's molecular weight are 10,000 amu (typical for low density polyethylene), 100,000 amu, and 1,000,000 amu (typical for high density polyethylene).

12.84   True. In an addition polymerization, all atoms present in the monomer are present in the polymer. (We are assuming 100% yield. However, there is usually some unreacted monomer as well as new polymer.)

12.85   *Analyze.* Given two types of reactant molecules, we are asked to write a condensation reaction with an ester product. *Plan.* A condensation reaction occurs when two smaller molecules combine to form a larger molecule and a small molecule, often water. Consider the structures of the two reactants and how they could combine to join the larger fragments and split water.   *Solve.*

A carboxylic acid contains the $-\overset{\overset{\text{O}}{\|}}{\text{C}}-\text{OH}$ functional group; an alcohol contains the –OH functional group. These can be arranged to form the $-\overset{\overset{\text{O}}{\|}}{\text{C}}-\text{O}-\text{C}$ ester functional group and H_2O. Condensation reaction to form an ester:

$$CH_3-\overset{\overset{\text{O}}{\|}}{\text{C}}\boxed{-\text{O}-\text{H} + \text{H}}-\text{O}-CH_2-CH_3 \longrightarrow CH_3-\overset{\overset{\text{O}}{\|}}{\text{C}}-\text{O}-CH_2CH_3 + H_2O$$

| acetic acid | ethanol | | ethyl acetate |

If a dicarboxylic acid (two –COOH groups, usually at opposite ends of the molecule) and a dialcohol (two –OH groups, usually at opposite ends of the molecule) are combined, there is the potential for propagation of the polymer chain at both ends of both monomers. Polyethylene terephthalate (Table 12.6) is an example of a polyester formed from the monomers ethylene glycol and terephthalic acid.

12.86
$$n\ HO-\overset{\overset{\text{O}}{\|}}{\text{C}}-CH_2-CH_2-\overset{\overset{\text{O}}{\|}}{\text{C}}\boxed{-\text{O}-\text{H}\ +\ \text{H}}-\overset{\overset{\text{H}}{|}}{\text{N}}-CH_2-CH_2-NH_2 \quad \right]^n$$

$$\left[-\overset{\overset{\text{O}}{\|}}{\text{C}}-CH_2-CH_2-\overset{\overset{\text{O}}{\|}}{\text{C}}-NH-CH_2-CH_2-NH-\right] + 2n\ H_2O$$

12.87   *Analyze/Plan.* In an addition polymerization, the monomer contains a multiple bond. We expect the monomer associated with this polymer to have either a double or triple bond.

Solve. This polymer is similar to polyethylene (Figure 12.35), with two hydrogen atoms replaced by chlorine atoms. The monomer in polyethylene is ethylene or ethene. The monomer for this polymer is dichloroethylene or dichloroethene.

$$\underset{\text{Cl}}{\overset{\text{Cl}}{>}}\text{C}=\text{C}\underset{\text{H}}{\overset{\text{H}}{<}}$$

12.88   (a)   By analogy to polyisoprene, Figure 12.42,

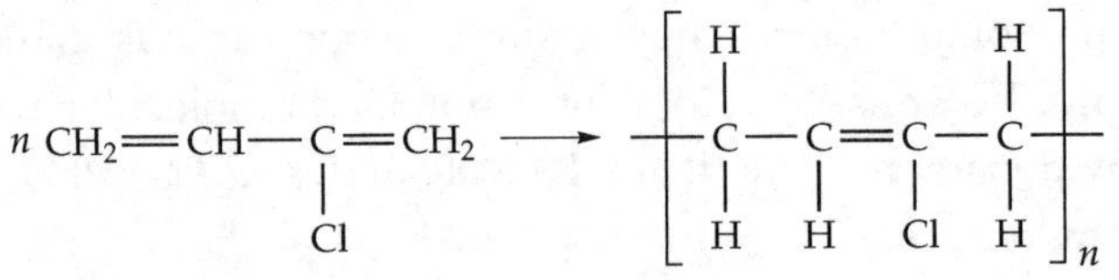

$$n\ CH_2{=}CH{-}C{=}CH_2 \longrightarrow$$

(b) 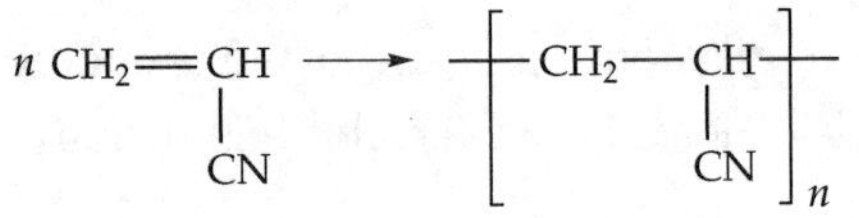 $n\ CH_2{=}CH \longrightarrow$

12.89   *Analyze/Plan.* Given the structure of a polymer, identify the monomers that react to form the polymer. Compare the polymer structure to other condensation polymers shown in Table 12.6.   *Solve.*

Kevlar is a polymer similar to nylon. The connecting unit is

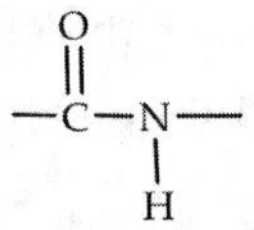

According to Figure 12.37, the monomers involved in this type polymer are a diacid, a molecule with two carboxylic acid groups, and a diamine, a molecule with two –NH$_2$ groups. The spacers in Kevlar are both benzene rings. The monomers used to produce Kevlar are:

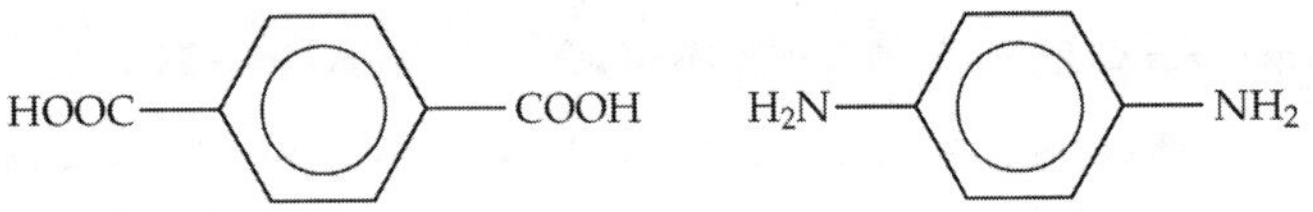

12.90   (a)

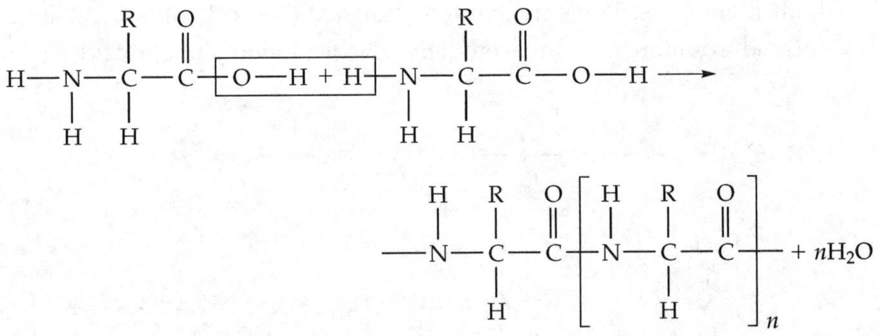

(b)

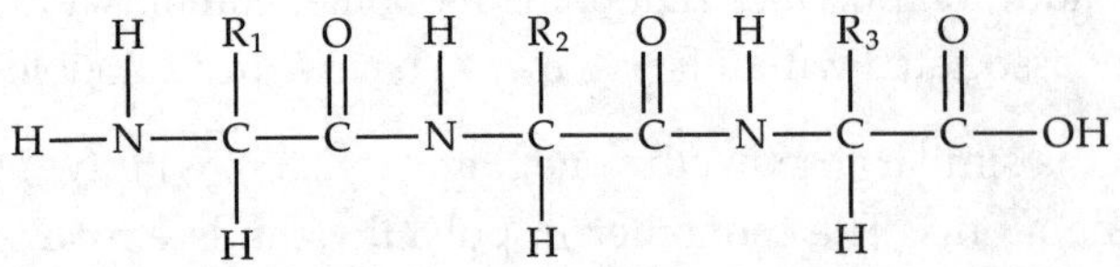

In the peptide shown here, the N terminus amino acid contains R1 and the C terminus amino acid contains R3.

(c)   Six (R1–R2–R3, R2–R3–R1, R3–R1–R2, R1–R3–R2, R3–R2–R1, R2–R1–R3)

12.91 (a) Most of a polymer backbone is composed of σ bonds. The geometry around individual atoms is tetrahedral with bond angles of 109°, so the polymer is not flat, and there is relatively free rotation around the σ bonds. The flexibility of the molecular chains causes flexibility of the bulk material. Flexibility is enhanced by molecular features that inhibit order, such as branching, and diminished by features that encourage order, such as cross-linking or delocalized π electron density.

(b) Less flexible. Cross-linking is the formation of chemical bonds between polymer chains. It reduces flexibility of the molecular chains and increases the hardness of the material. Cross-linked polymers are less chemically reactive because of the links.

12.92 At the molecular level, the longer, unbranched chains of HDPE fit closer together and have more crystalline (ordered, aligned) regions than the shorter, branched chains of LDPE. Closer packing leads to higher density.

12.93 Low degree of crystallinity. A good plastic wrap is extremely flexible; the lower the degree of crystallinity, the less rigid and more flexible the polymer.

12.94 (a) True

(b) True

(c) True

Nanomaterials (Section 12.9)

12.95 Continuous energy bands of molecular orbitals require a large number of atoms contributing a large number of atomic orbitals to the molecular orbital scheme. If a solid has dimensions 1-10 nm, nanoscale dimensions, there may not be enough contributing atomic orbitals to produce continuous energy bands of molecular orbitals.

12.96 (a) Calculate the wavelength of light that corresponds to 2.4 eV, then look at a visible spectrum such as Figure 6.4 to find the color that corresponds to this wavelength.

$$\lambda = \frac{hc}{E} = \frac{6.626 \times 10^{-34} \text{ J-s} \times 3.00 \times 10^8 \text{ m}}{s} \times \frac{1}{2.4 \text{ eV}} \times \frac{1 \text{ eV}}{1.602 \times 10^{-19} \text{ J}}$$
$$= 5.167 \times 10^{-7} = 5.2 \times 10^{-7} \text{ m (520 nm)}$$

The emitted 520 nm light is green.

(b) Yes. The wavelength of blue light is shorter than the wavelength of green light. Emitting a shorter wavelength requires a larger band gap. As particle size decreases, band gap increases. Appropriately sized CdS quantum dots, far smaller than a large CdS crystal, would have a band gap greater than 2.4 eV and could emit shorter wavelength blue light.

(c) No. The wavelength of red light is greater than the wavelength of green light. The large CdS crystal has plenty of AOs contributing to the MO scheme to ensure maximum delocalization. A bigger crystal will not reduce the size of the band gap. The 2.4 eV band gap of the large CdS crystal represents the minimum energy and maximum wavelength of light that can be emitted by this material.

12.97   (a)   False. As particle size decreases, the band gap increases. The smaller the particle, the fewer AOs that contribute to the MO scheme, the more localized the bonding and the larger the band gap.

       (b)   False. The wavelength of emitted light corresponds to the energy of the band gap. As particle size decreases, band gap increases and wavelength decreases ($E = hc/\lambda$).

12.98   True. Blue light has short wavelengths, corresponding to a relatively large band gap. As particle size decreases, band gap increases and wavelength decreases. We could begin with a semiconductor with a smaller band gap and make it a nanoparticle to increase E_g and decrease wavelength. (Nanoparticle size becomes one more way to tune the properties of semiconductors.)

12.99   *Analyze.*   Given: Au, 4 atoms per unit cell, 4.08 Å cell edge, volume of sphere = $4/3 \, \pi \, r^3$. Find: Au atoms in 20 nm diameter sphere.

     Plan. Relate the number of Au atoms in the volume of 1 cubic unit cell to the number of Au atoms in a 20 nm diameter sphere. Change units to Å (you could just as well have chosen nm as the common unit), calculate the volumes of the unit cell and sphere, and use a ratio to calculate atoms in the sphere.

     Solve.   vol. of unit cell = $(4.08 \text{ Å})^3 = 67.9173 = 67.9 \text{ Å}^3$

$$20 \text{ nm diameter} = 10 \text{ nm radius}; 10 \text{ nm} \times \frac{1 \times 10^{-9} \text{ m}}{1 \text{ nm}} \times \frac{1 \text{ Å}}{1 \times 10^{-10} \text{ m}} = 100 \text{ Å radius}$$

     (Note that 1 nm = 10 Å.)

     vol. of sphere = $4/3 \times 3.14159 \times (100 \text{ Å})^3 = 4.18879 \times 10^6 = 4.19 \times 10^6 \text{ Å}^3$

$$\frac{4 \text{ Au atoms}}{67.9173 \text{ Å}^3} = \frac{x \text{ Au atoms}}{4.18879 \times 10^6 \text{ Å}^3}; x = 2.46699 \times 10^5 = 2.47 \times 10^5 \text{ Au atoms}$$

12.100   (a)   There are four InP formula units in each cubic unit cell. Calculate the number of unit cells contained in a 3.00 nm cubic crystal and a 5.00 nm crystal. The volume of one unit cell is

$$(5.869)^3 \text{ Å}^3 \times \frac{1 \text{ nm}^3}{10^3 \text{ Å}^3} = 0.20216 = 0.2022 \text{ nm}^3$$

        The volume of the 3.00 nm crystal is $(3.00)^3 \text{ nm}^3 = 27.0 \text{ nm}^3$

$$\frac{27.0 \text{ nm}^3/\text{crystal}}{0.20216 \text{ nm}^3/\text{unit cell}} \times \frac{4 \text{ InP units}}{\text{unit cell}} = 534.23 = 534 \text{ InP units}$$

        That is, 534 In atoms and 534 P atoms.

        The volume of the 5.00 nm crystal is $(5.00)^3 \text{ nm}^3 = 125 \text{ nm}^3$

$$\frac{125 \text{ nm}^3/\text{crystal}}{0.20216 \text{ nm}^3/\text{unit cell}} \times \frac{4 \text{ InP units}}{\text{unit cell}} = 2473.3 = 2.473 \times 10^3 \text{ InP units}$$

        That is, 2473 In atoms and 2473 P atoms.

      (b)   As a quantum dot gets smaller, its band gap gets larger. Larger band gaps correspond to a higher energy and shorter wavelength of emitted light. Blue light has a shorter wavelength than orange light (Figure 6.4). Therefore, the smaller 3.00 nm cube emits blue light and the 5.00 nm cube emits orange light.

12.101   Statement (b) is correct.

      (a)   Neither graphite nor graphene is molecular. Both are covalent-network solids with covalent-network bonding.

      (c)   Neither graphite nor graphene is an insulator or a metal.

      (d)   Both are pure carbon.

      (e)   The carbon atoms in both are sp^2 hybridized.

12.102   Statement (e) is correct. Statements (b) and (c) both support the notion that buckyballs are discrete molecules and not extended materials.

Additional Exercises

12.103   (a)   NaCl, ionic; $MgCl_2$, ionic; PCl_3, molecular; SCl_2, molecular. The melting points of the solids are definitive. PCl_3 and SCl_2 are molecular solids because of their low melting points. NaCl and $MgCl_2$ are ionic solids because of their relatively high melting points. A covalent-network solid has a much higher melting point than either of these compounds.

      (b) $CaCl_2$ has a much higher melting point than $SiCl_4$. The ionic solid $CaCl_2$ is composed of a metal cation and a nonmetal anion. The molecular solid $SiCl_4$ is composed of a metalloid, Si, and a nonmetal, Cl.

12.104   According to Figure 12.6, a tetragonal unit cell has a square base, with the third lattice vector perpendicular to the base but with a different length. In other words, $a = b \neq c$, $\alpha = \beta = \gamma = 90°$. To create a face-centered tetragonal unit cell, add a lattice point in the center of each face (both square and rectangular faces). Draw a second face-centered tetragonal unit cell above or below the first one.

      The square base of a new tetragonal unit cell can be drawn by connecting the face centers of the four rectangular faces. Connecting these four face centers with those of the adjacent (old) tetragonal cell creates the new tetragonal unit cell. The lattice point at the center of the old square face becomes the body center of the new unit cell.

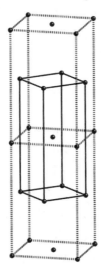

12.105   Orthorhombic, $a \neq b \neq c$, $\alpha = \beta = \gamma = 90°$. In the new lattice, the edge lengths are all different, and all angles remain $90°$.

12.106   Qualitatively from Figure 12.12, a face-centered cubic structure has a greater portion of its volume occupied by metal atoms and less "empty" space than a body-centered structure. The face-centered structure will have the greater density.

Quantitatively, use the atomic radius of iron, 1.32 Å from Figure 7.7, to estimate the unit cell edge length in each of the structures, then calculate the estimated density of both structures. Recall that a face-centered cubic structure has 4 atoms per unit cell, and a body-centered structure has 2.

Face-centered cubic: $4\, r_{Fe} = \sqrt{2}\, a;\quad a = 4\, r_{Fe}/\sqrt{2} = \dfrac{4 \times 1.32\ \text{Å}}{\sqrt{2}} = 3.7335 = 3.73\ \text{Å}$

$\rho = \dfrac{4\ \text{Fe atoms}}{(3.7335 \times 10^{-8}\ \text{cm})^3} \times \dfrac{55.845\ \text{g Fe}}{6.022 \times 10^{23}\ \text{Fe atoms}} = 7.1277 = 7.13\ \text{g/cm}^3$

Body-centered cubic: $4\, r_{Fe} = \sqrt{3}\, a;\quad a = 4\, r_{Fe}/\sqrt{3} = \dfrac{4 \times 1.32\ \text{Å}}{\sqrt{3}} = 3.0484 = 3.05\ \text{Å}$

$\rho = \dfrac{2\ \text{Fe atoms}}{(3.0484 \times 10^{-8}\ \text{cm})^3} \times \dfrac{55.845\ \text{g Fe}}{6.022 \times 10^{23}\ \text{Fe atoms}} = 6.5472 = 6.55\ \text{g/cm}^3$

The face-centered cubic structure has the greater density.

12.107   The metallic properties of malleability, ductility, and high electrical and thermal conductivity are results of the delocalization of valence electrons throughout the lattice. Delocalization occurs because metal atom valence orbitals of nearest-neighbor atoms interact to produce nearly continuous molecular orbital energy bands. When C atoms are introduced into the metal lattice, their valence orbitals do not have the same energies as metal orbitals, and the interaction is different. This causes a discontinuity in the band structure and limits delocalization of electrons. The properties of the carbon-infused metal begin to resemble those of a covalent-network lattice with localized electrons. The substance is harder and less conductive than the pure metal.

12.108   Density is the mass of the unit cell contents divided by the unit cell volume [(edge length)^3]. Refer to Figure 12.12(c) to determining the number of Ni atoms and Figure 12.17 for the number of Ni_3Al units in each unit cell.

Ni: There are 4 Ni atoms in each face-centered cubic unit cell ($8 \times 1/8$ at the corners, $6 \times 1/2$ on the face centers).

$\text{density} = \dfrac{4\ \text{Ni atoms}}{(3.53\ \text{Å})^3} \times \dfrac{58.6934\ \text{g Ni}}{6.022 \times 10^{23}\ \text{Ni atoms}} \times \left(\dfrac{\text{Å}}{1 \times 10^{-8}\ \text{cm}}\right)^3 = 8.86\ \text{g/cm}^3$

Ni_3Al: There is 1 Ni_3Al unit in each cubic unit cell. According to Figure 12.17, Ni is at the face centers ($6 \times 1/2 = 3$ Ni atoms) and Al is at the corners ($8 \times 1/8 = 1$ Al atom); the stoichiometry is correct.

$\text{density} = \dfrac{1\ Ni_3Al\ \text{unit}}{(3.56\ \text{Å})^3} \times \dfrac{203.062\ \text{g}\ Ni_3Al}{6.022 \times 10^{23}\ \text{Ni atoms}} \times \left(\dfrac{\text{Å}}{1 \times 10^{-8}\ \text{cm}}\right)^3 = 7.47\ \text{g/cm}^3$

The density of the Ni_3Al alloy (intermetallic compound) is ~85% of the density of pure Ni. The sizes of the two unit cells are very similar. In Ni_3Al, one out of every four Ni atoms is replaced with an Al atom. The mass of an Al atom is (~27 amu) is about half that of a Ni atom (~59 amu); the mass of the unit cell contents of Ni_3Al is ~7/8 (87.5%) that of Ni, and the densities show the same relationship.

12.109   Ni$_3$Al: Ni is at the face centers ($6 \times 1/2 = 3$ Ni atoms) and Al is at the corners ($8 \times 1/8 = 1$ Al atom). The atom ratio in the structure matches the empirical formula.

Nb$_3$Sn: The top, front, and side faces of the unit cell are clearly visible; each has 2 Nb atoms centered on it, as opposed to totally inside the cell. The three opposite faces are not completely visible in the diagram, but must be the same by translational symmetry. Two Nb atoms centered on each of 6 faces ($12 \times 1/2 = 6$ Nb atoms). One Sn atom completely inside the unit cell and one at each corner ($1 + 8 \times 1/8 = 2$ Sn atoms). The atom ratio is 6 Nb : 2 Sn or 3 Nb : 1 Sn; the atom ratio in the structure matches the empirical formula.

SmCo$_5$: One Sm atom at each corner ($8 \times 1/8 = 1$ Sm atom). One Co atom totally inside the unit cell (1 Co atom), two Co atoms centered on the top and bottom faces ($4 \times 1/2 = 2$ Co atoms), four Co atoms centered on 4 side faces ($4 \times 1/2 = 2$ Co atoms), for a total of ($1 + 2 + 2 = 5$ Co atoms). The ratio is 1 Sm : 5 Co, which matches the empirical formula.

12.110   (a)   CsCl, primitive cubic lattice (Figure 12.25)

(b)   Au, face-centered cubic lattice (Figure 12.18)

(c)   NaCl, face-centered cubic lattice (Figure 12.25)

(d)   Po, primitive cubic lattice, rare for metals (Section 12.3)

(e)   ZnS, face-centered cubic lattice (Figure 12.25)

12.111   Calculate the wavelength of light that corresponds to an energy of 2.20 eV.

$$\lambda = \frac{hc}{E} = \frac{6.626 \times 10^{-34} \text{ J-s} \times 2.998 \times 10^{8} \text{ m}}{s} \times \frac{1}{2.20 \text{ eV}} \times \frac{1 \text{ eV}}{1.602 \times 10^{-19} \text{ J}}$$

$$= 5.636 \times 10^{-7} = 5.64 \times 10^{-7} \text{ m (564 nm)}$$

[The absorbed 564-nm light is yellow-green and cinnabar appears the complementary color, red-violet or vermillion.]

12.112   Aluminum, density = 2.70 g/mL, is indeed more dense than silicon, density = 2.33 g/mL. This difference is significant, but not as large as one might expect relative to the difference in electrical conductivity. It is true that under very high pressure, about 12 GPa or 120,000 atm, the structure of silicon changes from the diamond structure to a close-packed structure. The conductivity also changes from that of a semiconductor to that of a metal. The structure change can be detected by monitoring the conductivity (or resistivity) of the material.

In terms of atomic properties, aluminum adopts a metallic (close-packed) structure because it is electron deficient. With only three valence electrons, it is difficult for an aluminum atom to acquire a complete octet by covalent bonding. A close-packed structure (face-centered cubic) provides each aluminum atom with twelve nearest neighbors and the possibility of electron delocalization to satisfy its bonding needs. On the other hand, each silicon atom has four valence electrons and can complete its octet by forming four bonds in a covalent-network structure with localized bonding.

12.113   (a)   Zinc sulfide, ZnS (Figure 12.26).

(b)   Covalent. Silicon and carbon are both nonmetals, and their electronegativities are similar.

(c)   Silicon carbide is hard and high-melting because it is a covalent-network solid. In SiC, the C atoms form a face-centered cubic array with Si atoms occupying alternate tetrahedral holes in the lattice. This means that the coordination numbers of both Si and C are 4; each Si is bound to four C atoms in a tetrahedral arrangement, and each C is bound to four Si atoms in a tetrahedral arrangement, producing an extended three-dimensional network. SiC is high-melting because a great deal of chemical energy is stored in the covalent Si–C bonds. Melting requires breaking covalent Si–C bonds, which takes a huge amount of thermal energy. It is hard because the three-dimensional lattice resists any change that would weaken the Si–C bonding network.

12.114   (a) $V = (\text{edge length})^3 = (0.5\,\text{mm})^3 = 0.125 = 0.1\,\text{mm}^3$

$$d = \frac{m}{V}; \; m = d \times V = \frac{8.96\,\text{g}}{\text{cm}^3} \times 0.125\,\text{mm}^3 \times \frac{1\,\text{cm}^3}{(10)^3\,\text{mm}^3} = 1.12 \times 10^{-3} = 1 \times 10^{-3}\,\text{g Cu}$$

$$1.12 \times 10^{-3}\,\text{g Cu} \times \frac{1\,\text{mol Cu}}{63.546\,\text{g Cu}} \times \frac{6.022 \times 10^{23}\,\text{Cu atoms}}{\text{mol Cu}} = 1.06 \times 10^{19} = 1 \times 10^{19}\,\text{Cu atoms}$$

(b) $$\frac{1 \times 10^{-19}\,\text{J range of energy levels}}{1.06 \times 10^{19}\,\text{Cu atoms}} = 9 \times 10^{-39}\,\text{J/Cu atoms}$$

This spacing is substantially smaller than the 1×10^{-18} J separation between energy levels in the hydrogen atom.

12.115   Semiconductors have a filled valence band and an empty conduction bond, separated by a characteristic difference in energy, the band gap, E_g. When a semiconductor is heated, more electrons have sufficient energy to jump the band gap, and conductivity increases. Metals have a partially filled continuous energy band. Heating a metal increases the average kinetic energy of the metal atoms, usually through increased vibrations within the lattice. The greater vibrational energy of the atoms leads to imperfections in the lattice and discontinuities in the energy band. Thermal vibrations create barriers to electron delocalization and reduce the conductivity of the metal.

12.116   (a)   There are 8 Na atoms and 4 O atoms in the unit cell. The red O atoms adopt a face-centered cubic arrangement. There are O atoms at each corner of the cube, and in the center of each face.

8 corners × 1/8 atom/corner + 6 faces × ½ atom/face = 4 O atoms

The 8 green Na atoms are completely in the interior of the unit cell.

Check. These numbers of atoms agree with the empirical formula of Na_2O.

(b)   *Analyze/Plan.* Visualize the coordination number and environment of the sodium ion.

Solve. In this structure, Na atoms sit completely inside the unit cell (but not at the body center). Like Zn atoms in the ZnS structure (Figure 12.26), they have 4 nearest-neighbor anions in a tetrahedral coordination environment.

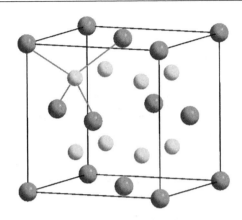

[In Na_2O, the 4 O atoms (dark gray) around a particular Na atom (light gray) are the O atom on the nearest corner and 3 oxygen atoms on adjacent face centers that form a triangular face of the coordination tetrahedron. For the upper, left, front Na atom, this is the O atom on the top, left, front corner and the 3 oxygen atoms on the top, left, and front faces.]

(c)   There are 8 Na atoms and 4 O atoms or 4 Na_2O formula units in the cubic unit cell. Density is unit cell contents divided by unit cell volume.

$$\text{density} = \frac{4\,Na_2O\ \text{units}}{(5.550\ \text{Å})^3} \times \frac{61.979\ \text{g}\ Na_2O}{6.0221 \times 10^{23}\ Na_2O\ \text{units}} \times \left(\frac{\text{Å}}{1 \times 10^{-8}\ \text{cm}}\right)^3 = 2.408\ \text{g/cm}^3$$

12.117   (a)

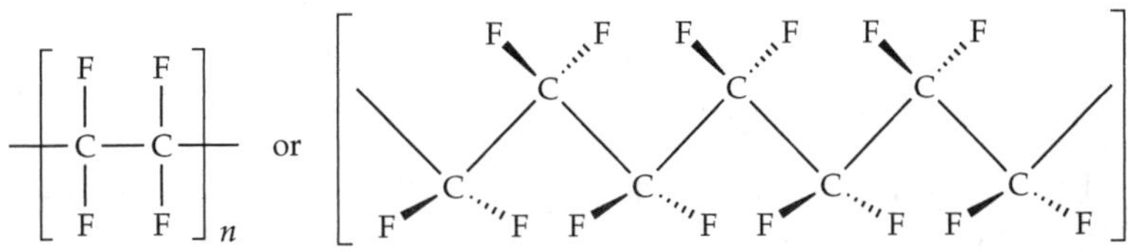

$$\left[\begin{array}{c} \text{F} \quad \text{F} \\ | \quad\ | \\ -\text{C}-\text{C}- \\ | \quad\ | \\ \text{F} \quad \text{F} \end{array}\right]_n \quad \text{or}$$

(b)   Teflon^{TM} is formed by addition polymerization.

12.118

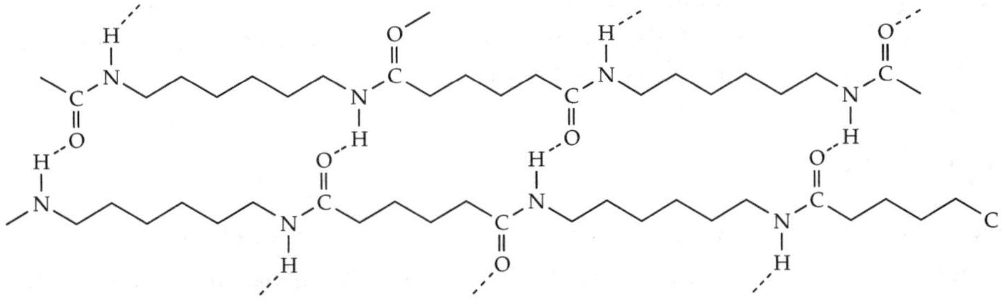

Hydrogen bonding occurs between $-\overset{\overset{\text{O}}{\|}}{\underset{\underset{\text{H}}{|}}{\text{C}}}-\text{N}-$ amide groups of adjacent chains.

12.119   X-ray _diffraction_ is the phenomenon that enables us to measure interatomic distances in crystals. Diffraction is most efficient when the wavelength of light is similar to the size of the object (e.g., the slit) doing the diffracting. Interatomic distances are on the order of 1-10 Å, and the wavelengths of X-rays are also in this range. Visible light has wavelengths of 400-700 nm, or 4000-7000 Å, too long to be diffracted efficiently by atoms (electrons) in crystals.

12.120 $n\lambda = 2d \sin \theta$; $n = 1$, $\lambda = 1.54$ Å, $\theta = 14.22$ °; calculate d.

$$d = \frac{n\lambda}{2 \sin \theta} = \frac{1 \times 1.54 \text{ Å}}{2 \sin(14.22)} = 3.1346 = 3.13 \text{ Å}$$

12.121   Germanium is in the same family but below Si on the periodic chart. This means that Ge will probably have bonding characteristics and crystal structure similar to those of Si. Because Ge has a larger bonding atomic radius than Si, we expect a larger unit cell and *d*-spacing for Ge. In Bragg's law, $n\lambda = 2d \sin \theta$, *d* and $\sin \theta$ are inversely proportional. That is, the larger the *d*-spacing, the smaller the value of $\sin \theta$ and θ. In a diffraction experiment, we expect a Ge crystal to diffract X-rays at a smaller θ-angle than a Si crystal, assuming the X-rays have the same wavelength.

12.122   (a)   Both diamond (d = 3.5 g/cm^3) and graphite (d = 2.3 g/cm^3) are covalent-network solids with efficient packing arrangements in the solid state; there is relatively little empty space in their respective crystal lattices. Diamond, with bonded C–C distances of 1.54 Å in all directions, is more dense than graphite, with shorter C–C distances within carbon sheets but longer 3.35 Å separations between sheets (Figure 12.29). Buckminsterfullerene has much more empty space, both inside each C_{60} "ball" and between balls, than either diamond or graphite, so its density will be considerably less than 2.3 g/cm^3.

         (b)   In a face-centered cubic unit cell, there are 4 complete C_{60} units.

$$\frac{4 \text{ } C_{60} \text{ units}}{(14.2 \text{ Å})^3} \times \frac{720.66 \text{ g}}{6.022 \times 10^{23} \text{ } C_{60} \text{ units}} \times \left(\frac{1 \text{ Å}}{1 \times 10^{-8} \text{ cm}}\right)^3 = 1.67 \text{ g/cm}^3$$

         (1.67 g/cm^3 is the smallest density of the three allotropes: diamond, graphite, and buckminsterfullerene.)

12.123   (b)   Increase. Promotion of an electron from the valence band to the conduction band generates a mobile electron in the conduction band and a mobile hole in the valence band. Both phenomena promote delocalization and increase the conductivity of the semiconductor.

Integrative Exercises

12.124   The karat scale is based on mass%, not mol%. In each case, change mol% to mass % Au. Then, karat = mass fraction Au $\times$ 24. Determine color using mass% and Figure 12.18.

         (a)   Assume 0.50 mol Ag and 0.50 mol Au.

$$0.50 \text{ mol Ag} \times \frac{107.87 \text{ g Ag}}{\text{mol Ag}} = 53.935 = 54 \text{ g Ag}$$

$$0.50 \text{ mol Au} \times \frac{196.97 \text{ g Au}}{\text{mol Au}} = 98.485 = 98 \text{ g Au}$$

$$\text{mass \% Au} = \frac{98.485 \text{ g Au}}{(98.485 + 53.935) \text{ g total}} \times 100 = 64.61 = 65 \text{ \% Au}$$

         karat = 0.6461 $\times$ 24 = 15.5064 = 16 karat Au

         On Figure 12.18, the Au/Ag alloy line is labeled in terms of mass% Ag. For an alloy that is 65% Au and 35% Ag, the color is greenish yellow.

(b)   Assume 0.50 mol Cu and 0.50 mol Au.

$$0.50 \text{ mol Cu} \times \frac{63.546 \text{ g Cu}}{\text{mol Cu}} = 31.773 = 32 \text{ g Cu}$$

$$0.50 \text{ mol Au} = 98.485 = 98 \text{ g Au}$$

$$\text{mass\% Au} = \frac{98.485 \text{ g Au}}{(98.485 + 31.773) \text{ g total}} \times 100 = 75.61 = 76 \text{ \% Au}$$

$$\text{karat} = 0.7561 \times 24 = 18.15 = 18 \text{ karat Au}$$

On Figure 12.18, for an alloy that is 76% Au and 24% Cu, the color is reddish gold.

12.125   *Analyze.* Given: mass % of Al, Mg, O; density, unit cell edge length. Find: number of each type of atom. *Plan.* We are not given the type of cubic unit cell, primitive, body centered, face centered. So we must calculate the number of formula units in the unit cell, using density, cell volume, and formula weight. Begin by determining the empirical formula and formula weight from mass % data. *Solve.* Assume 100 g spinel.

$$37.9 \text{ g Al} \times \frac{1 \text{ mol Al}}{26.98 \text{ g Al}} = 1.405 \text{ mol Al}; \ 1.405/0.7036 \approx 2$$

$$17.1 \text{ g Mg} \times \frac{1 \text{ mol Mg}}{24.305 \text{ g Mg}} = 0.7036 \text{ mol Mg}; \ 0.7036/0.7036 = 1$$

$$45.0 \text{ g O} \times \frac{1 \text{ mol O}}{16.00 \text{ g Al}} = 2.813 \text{ mol O}; \ 2.813/0.7036 \approx 4$$

The empirical formula is Al_2MgO_4; formula weight = 142.3 g/mol

Calculate the number of formula units per unit cell.

$$8.09 \text{ Å} = 8.09 \times 10^{-10} \text{ m} = 8.09 \times 10^{-8} \text{ cm}; \ V = (8.09 \times 10^{-8})^3 \text{ cm}^3$$

$$\frac{3.57 \text{ g}}{\text{cm}^3} \times (8.09 \times 10^{-8})^3 \text{ cm}^3 \times \frac{1 \text{ mol}}{142.3 \text{ g}} \times \frac{6.022 \times 10^{23} \text{ units}}{\text{mol}} = 7.999 = 8$$

There are 8 formula units per unit cell, for a total of 16 Al atoms, 8 Mg atoms, and 32 O atoms.

[The relationship between density (d), unit cell volume (V), number of formula units (Z), formula weight (FW), and Avogadro's number (N) is a useful one. It can be rearranged to calculate any single variable, knowing values for the others. For densities in g/cm^3 and unit cell volumes in cm^3 the relationship is Z = (N × d × V)/FW.]

12.126   Refer to Section 12.7 and Figure 12.29.

(a)   In diamond, each C atom is bound to 4 other C atoms. According to VSEPR, the geometry around a central atom with 4 bonding electron pairs is tetrahedral and the C–C–C bond angles are 109°.

(b)   Within a graphite sheet, each C atom is bound to 3 other C atoms in a trigonal planar arrangement. The C–C–C bond angles are 120°.)

(c)   Within a sheet, sp^2 hybrid orbitals are involved in the sigma bonding network framework. This leaves atomic p orbitals to be involved in interactions between sheets. These interactions are responsible for stacking.

12.127 (a)

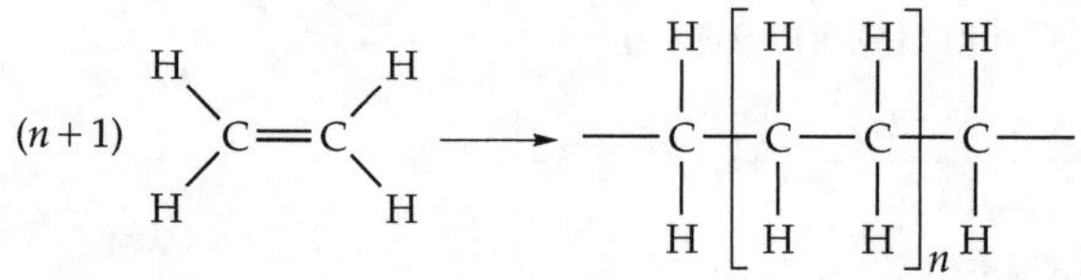

HDPE

$$\Delta H = D(C=C) - 2D(C-C) = 614 - 2(348) = -82 \text{ kJ/mol } C_2H_4$$

(b) $(n+1)$ HOOC—$(CH_2)_6$—COOH + $(n+1)$ H_2N—$(CH_2)_6$—NH_2 $\longrightarrow$

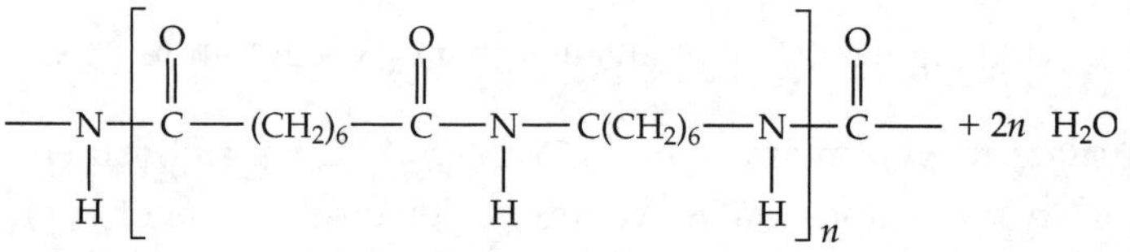

Nylon 6,6

$$\Delta H = 2D(C-O) + 2D(N-H) - 2D(C-N) - 2D(H-O)$$

$$\Delta H = 2(358) + 2(391) - 2(293) - 2(463) = -14 \text{ kJ/mol}$$

(This is –14 kJ/mol of either reactant.)

(c) $(n+1)$ HOOC—⬡—COOH + $(n+1)$ HO—CH_2—CH_2—OH $\longrightarrow$

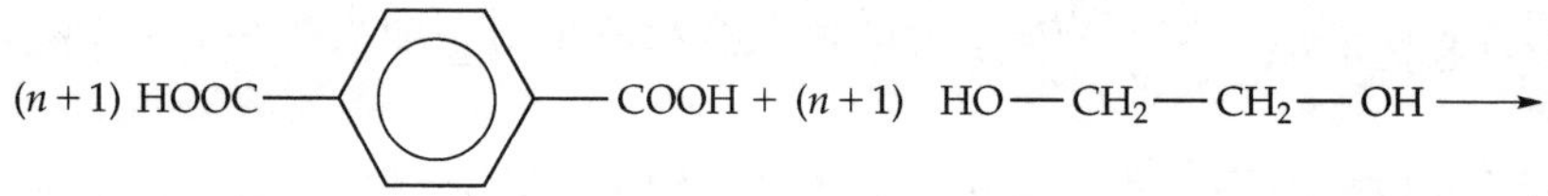

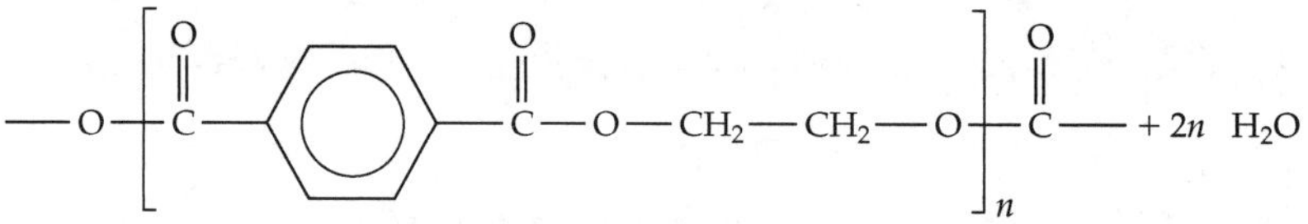

PET

$$\Delta H = 2D(C-O) + 2D(O-H) - 2D(C-O) - 2D(O-H) = 0 \text{ kJ}$$

12.128 (a) sp^3 hybrid orbitals at C, 109° bond angles around C

(b)

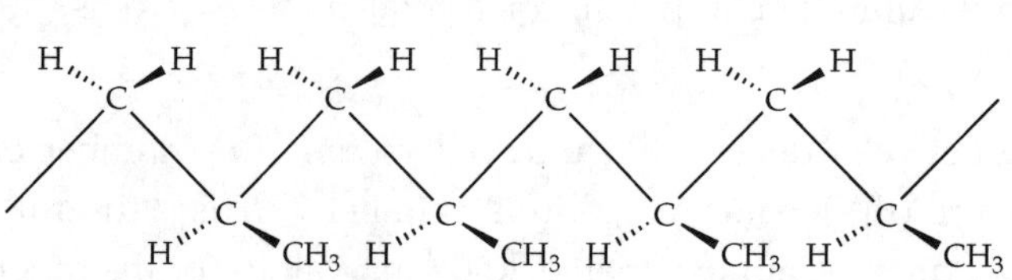

isotactic

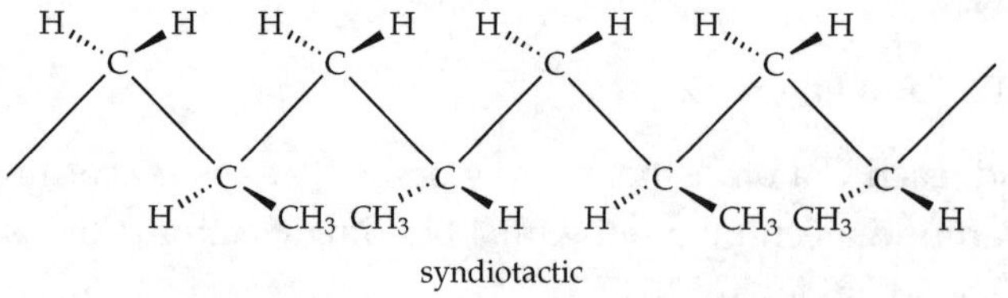

syndiotactic

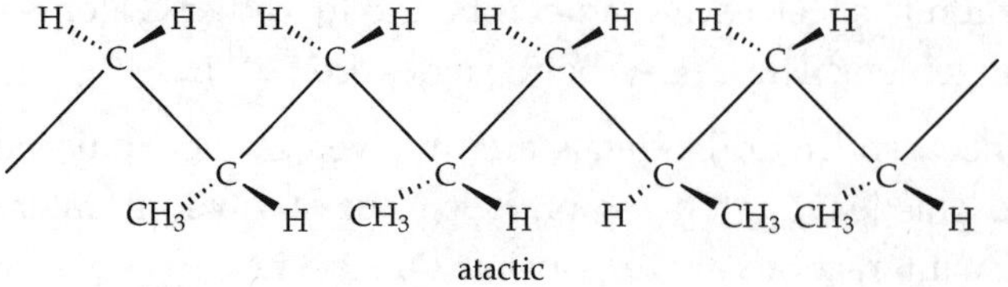

atactic

Isotactic polypropylene has the highest degree of crystallinity and highest melting point. The regular shape of the polymer backbone allows for close, orderly (almost zipper-like) contact between chains. This maximizes dispersion forces between chains and produces higher order (crystallinity) and melting point. Atactic polypropylene has the least order and the lowest melting point.

(c)   Cotton, with $-C-$ groups and polyester, with $-\overset{\overset{\text{O}}{\|}}{C}-O-C$ groups
with OH below the first C

Both participate in hydrogen-bonding interactions with H_2O molecules. These are strong intermolecular forces that hold the "moisture" at the surface of the fabric next to the skin. Polypropylene has no strong interactions with water, and capillary action "wicks" the moisture away from the skin.

12.129   (a)

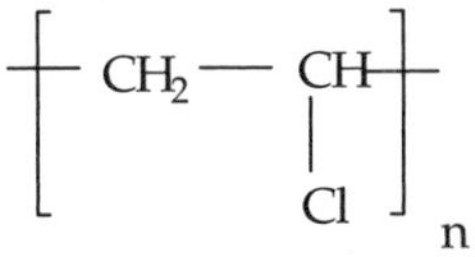

$\left[CH_2 - \underset{\underset{Cl}{|}}{CH} \right]_n$

C—Cl	328 kJ/mol	← lowest
C—C	348 kJ/mol	
C—H	413 kJ/mol	

(b)   C–Cl bonds are weakest, so they are most likely to break upon heating.

(c)   The repeating unit in polyvinyl chloride consists of two C atoms, each in a different environment. Consider the net changes in these two C atoms when the polymer is converted to diamond a high pressure.

Diamond is a covalent-network structure where each C atom is tetrahedrally bound to four other C atoms [Figure 12.29(a)].

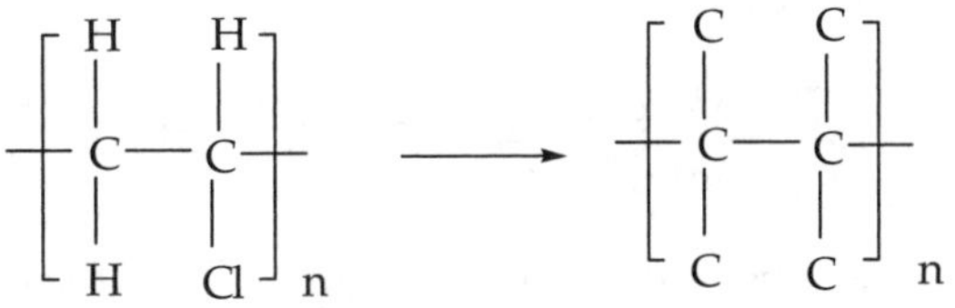

$$\left[\begin{matrix} H & H \\ | & | \\ C & -C \\ | & | \\ H & Cl \end{matrix} \right]_n \longrightarrow \left[\begin{matrix} C & C \\ | & | \\ C & -C \\ | & | \\ C & C \end{matrix} \right]_n$$

Assume that there is no net change to the C–C bonds in the structure, even though they may be broken and reformed. The net change to the 2-C vinyl chloride unit is then breaking three C–H bonds and one C–Cl bond, and making four C–C bonds.

$\Delta H = 3D(C–H) – D(C–Cl) – 4D(C–C) = 3(413) + 328 – 4(348)$

$= 523$ kJ/vinyl chloride unit

12.130   (a)   Follow the logic outlined in Solution 12.99.

vol. of unit cell $= (5.43 \text{ Å})^3 = 160.1030 = 1.60 \times 10^2 \text{ Å}^3$

$1 \text{ cm}^3 \times \dfrac{(1)^3 \text{ Å}^3}{(1 \times 10^{-8})^3 \text{ cm}^3} = 1 \times 10^{24} \text{ Å}^3$ (volume of material)

$\dfrac{4 \text{ Si atoms}}{160.103 \text{ Å}^3} = \dfrac{x \text{ Si atoms}}{1 \times 10^{24} \text{ Å}^3}$; $x = 2.4984 \times 10^{22} = 2.50 \times 10^{22}$ Si atoms

(To 1 sig fig, the result is 2×10^{22} Si atoms.)

(b)   1 ppm phosphorus = 1 P atom per 1×10^6 Si atoms

$$\frac{1 \text{ P atom}}{1 \times 10^6 \text{ Si atoms}} = \frac{x \text{ P atoms}}{2.4984 \times 10^{22} \text{ Si Atoms}}; x = 2.4984 \times 10^{22}$$

$$= 2.50 \times 10^{16} \text{ P atoms}$$

$$2.4984 \times 10^{16} \text{ P atom} \times \frac{1 \text{ mol}}{6.022 \times 10^{23} \text{ atoms}} \times \frac{30.97376 \text{ g P}}{\text{mol}} \times \frac{1 \text{ mg}}{1 \times 10^{-3} \text{ g}}$$

$$= 1.29 \times 10^{-3} \text{ mg P } (1.29 \, \mu\text{g})$$

12.131   *Analyze/Plan.* Write a balanced chemical equation for the reaction. Calculate moles $FeTiO_3$ to be produced, then moles and grams of the two reactants required to produce this amount of product.   *Solve.*

(a) $FeO(s) + TiO_2(s) \rightarrow FeTiO_3(s)$

(b)   The formula weight of $FeTiO_3$ is 151.71 g/mol.

(c) $2.500 \text{ g FeTiO}_3 \times \dfrac{1 \text{ mol}}{151.710 \text{ g FeTiO}_3} = 0.0164788 = 0.01648 \text{ mol FeTiO}_3$

(d)   The coefficients in the chemical equation are all 1, so we need 0.01648 moles of each of the reactants to produce 0.01648 mol $FeTiO_3(s)$ (assuming the reaction goes to completion). The molar mass of FeO is 71.844 g/mol. The molar mass of TiO_2 is 79.8558 g/mol.

$$0.0164788 \text{ mol FeO} \times \frac{71.844 \text{ g FeO}}{\text{mol FeO}} = 1.18391 = 1.184 \text{ g FeO}$$

(e) $0.0164788 \text{ mol TiO}_2 \times \dfrac{79.8558 \text{ g TiO}_2}{\text{mol}} = 1.31593 = 1.316 \text{ g TiO}_2$

12.132   The bonding atomic radius of a Si atom is 1.11 Å (Figure 7.7), so the diameter is 2.22 Å.

$$14 \text{ nm} \times \frac{1 \times 10^{-9} \text{ m}}{1 \text{ nm}} \times \frac{1 \text{ Å}}{1 \times 10^{-10} \text{ m}} \times \frac{1 \text{ Si atom}}{2.22 \text{ Å}} = 63 \text{ Si atoms.}$$

13 Properties of Solutions

Visualizing Concepts

13.1 <u>(a) < (b) < (c).</u> In Section 13.1, *entropy* is qualitatively defined as randomness or dispersal in space. In container (a), the two kinds of particles are not mixed and the particles are close together, so (a) has the least entropy. In container (b), the particles occupy approximately the same volume as container (a) but the two kinds of particles are homogeneously mixed, so the degree of dispersal and randomness is greater than in (a). In container (c), the two kinds of particles are homogeneously mixed and they occupy a larger volume than in (b), so (c) has the greatest entropy.

13.2 (a) The oxygen atom of the water molecule is associated with the cation. In the polar water molecule, the partial negative charge is localized on the oxygen atom. The electrostatic attractive interaction is between the positive charge of the cation and the partially negative charge of the oxygen atom.

(b) Statement (c). The smaller ionic radius of lithium ion means that the positive charge is localized over a smaller volume; it is essentially a point charge. This increases the strength of the interaction between lithium cation and each individual water molecule.

13.3 (a) No.

(b) The ionic solid with the smaller lattice energy will be more soluble in water. Lattice energy is the main component of ΔH_{solute}, the enthalpy required to separate solute particles. The solid with the smaller lattice energy will have the less endothermic ΔH_{solute}. Assuming $\Delta H_{solvent}$ and ΔH_{mix} are the same for the two solids, the one with the smaller lattice energy will have a less endothermic ΔH_{soln} and be more soluble in water.

13.4 Statements (a) and (d) are true. Statement (b) is false because of the average distance between particles. Statement (c) is false because gases, like all pure substances, have characteristic boiling points.

13.5 Diagram (b) is the best representation of a saturated solution. There is some undissolved solid with particles that are close together and ordered, in contact with a solution containing mobile, separated solute particles. As much solute has dissolved as can dissolve, leaving some undissolved solid in contact with the saturated solution.

13.6 Statement (b) is the best explanation. Statement (a) is false because gas molecules are in constant random motion. Statement (c) is false because only N, O, and F atoms can formally act as hydrogen bond acceptors. Statement (d) is false because lighter gases are more likely to form saturated solutions with water. The term saturated means that all solute that can dissolve has dissolved. Less soluble solutes, in this case lighter gases, are more likely to form saturated solutions.

13.7   Vitamin B_6 is more water soluble and vitamin E is more fat soluble. The B_6 molecule contains three –OH groups and the —N̈— that can enter into many hydrogen-bonding interactions with water. Its relatively small molecular size indicates that dispersion forces will not play a large role in intermolecular interactions and that hydrogen bonding will dominate. On the other hand, the long, rod-like hydrocarbon chain of vitamin E will lead to stronger dispersion forces among it and the mostly nonpolar fats. Although vitamin E has one –OH and one —Ö— group, the long hydrocarbon chain prevents water from surrounding and separating the vitamin E molecules, reducing its water solubility.

13.8   The second friend is correct. As pressure decreases, gas solubility decreases. The first bubbles were from air gases dissolved at atmospheric pressure but not dissolved at lower pressure. The second batch of bubbles is from the water boiling. As the pressure in the vessel decreases, eventually the vapor pressure of water equals external pressure, and the water boils.

13.9   (a)   Yes, the *molarity* changes with a change in temperature. Molarity is defined as moles solute per unit volume of solution. If solution volume is different, molarity is different.

         (b)   No, *molality* does not change with change in temperature. Molality is defined as moles solute per kilogram of solvent. Even though the volume of solution has changed because of increased kinetic energy, the mass of solute and solvent have not changed, and the molality stays the same.

13.10   (a)   The blue line represents the solution. According to Raoult's law, the presence of a nonvolatile solute lowers the vapor pressure of a volatile solvent. At any given temperature, the blue line has a lower vapor pressure and represents the solution.

         (b)   The boiling point of a liquid is the temperature at which the vapor pressure of the liquid is equal to atmospheric pressure. Assuming atmospheric pressure of 1.0 atm, the boiling point of the solvent (red line) is approximately 64 °C. The boiling point of the solution is approximately 70 °C.

13.11   Ideally, 0.50 L. If the volume outside the balloon is very large compared to 0.25 L, solvent will flow across the semipermeable membrane until the molarities of the inner and outer solutions are equal, 0.1 *M*. This requires an "inner" solution volume twice as large as the initial volume, or 0.50 L. (In reality, osmosis across the balloon membrane is not perfect. The solution concentration inside the balloon will be slightly greater than 0.1 *M* and the volume of the balloon will be slightly less than 0.50 L.)

13.12   Beaker (b) represents a liquid-liquid emulsion such as milk. Beaker (a) represents a true solution. Beaker (c) has some alignment of molecules and represents a liquid crystal.

The Solution Process (Section 13.1)

13.13   (a)   False. It is generally true that solute-solvent interactions must balance the sum of solute-solute and solvent-solvent interactions. If solute-solute interactions are significantly stronger than solute-solvent interactions, dissolving will probably not occur. Entropy of mixing can compensate for a slightly endothermic enthalpy of solution, but not a large one.

         (b)   False. If a solution forms, enthalpy of mixing is exothermic, a negative number.

         (c)   True.

13.14 (a) False. NaCl is more soluble in water than it is in benzene because the enthalpy of mixing for NaCl and water is much more exothermic. Ion-dipole interactions between NaCl and water are much stronger than the interactions between ionic NaCl and nonpolar benzene. (And, water is more dense than benzene.)

(b) True

(c) True

13.15 *Analyze/Plan*. Decide whether the solute and solvent in question are ionic, polar covalent, or nonpolar covalent. Draw Lewis structures as needed. Then state the appropriate type of solute-solvent interaction. *Solve*.

(a) CCl_4, nonpolar; benzene, nonpolar; dispersion forces

(b) methanol, polar with hydrogen bonding; water, polar with hydrogen bonding; hydrogen bonding

(c) KBr, ionic; water, polar; ion-dipole forces

(d) HCl, polar; CH_3CN, polar; dipole-dipole forces

13.16 From weakest to strongest solvent-solute interactions:

(b), dispersion forces < (c), hydrogen bonding < (a), ion-dipole

13.17 (a) Very soluble.

(b) ΔH_{mix} will be the largest negative number. ΔH_{solute} and $\Delta H_{solvent}$ are both positive (endothermic); they represent the energy required to overcome attractive interactions in the solute and solvent, respectively. ΔH_{mix} represents the attractive interactions between solute and solvent particles. In order for ΔH_{soln} to be negative (exothermic), ΔH_{mix} must have a greater magnitude than (ΔH_{solute} + $\Delta H_{solvent}$).

13.18 (a) This solution process is endothermic. The enthalpy of the solution is greater than the enthalpy of the unmixed solute plus solvent.

(b) The solution forms because the favorable entropy of mixing outweighs the increase in enthalpy by the solution.

13.19 (a) Lattice energy is the amount of energy required to completely separate a mole of solid ionic compound into its gaseous ions (Section 8.2). For ionic solutes, this corresponds to ΔH_{solute} (solute-solute interactions) in Equation 13.1.

(b) In Equation 13.1, ΔH_{mix} is always exothermic. Formation of attractive interactions, no matter how weak, always lowers the energy of the system, relative to the energy of the isolated particles.

13.20 ΔH_{mix} is much more negative (exothermic) than $\Delta H_{solvent}$ or ΔH_{solute}. Both $\Delta H_{solvent}$ and ΔH_{solute} will be endothermic, because separating solvent molecules or solute ions requires energy. Because ΔH_{soln} is exothermic, ΔH_{mix} must be exothermic, and not just more negative than the other two, but more negative than the sum of the other two. This is not surprising, because ΔH_{mix} involves formation of many ion-dipole interactions, strong interparticle forces.

13.21   (a) ΔH_{soln} is nearly zero. ΔH_{soln} is determined by the relative magnitudes of the "old" solute-solute (ΔH_{solute}) and solvent-solvent ($\Delta H_{solvent}$) interactions and the new solute-solvent interactions (ΔH_{mix}); $\Delta H_{soln} = \Delta H_{solute} + \Delta H_{solvent} + \Delta H_{mix}$. Because the solute and solvent in this case experience very similar London dispersion forces, the energy required to separate them individually and the energy released when they are mixed are approximately equal. $\Delta H_{solute} + \Delta H_{solvent} \approx -\Delta H_{mix}$ and ΔH_{soln} is nearly zero.

       (b)   The entropy of the system increases when heptane and hexane form a solution. The change involves taking two pure liquids and forming a (homogeneous) mixture. The result is an increase in randomness or disorder. From part (a), the enthalpy of mixing is nearly zero, so the increase in entropy is the driving force for mixing in all proportions.

13.22   Statement (c) is the best explanation. Statement (a) is true, but it doesn't explain the enthalpy of solution. Statement (b) is false. With a positive enthalpy of solution, entropy of mixing must be favorable. Statement (d) involves molar mass; molar mass influences dispersion forces that are not important in the solubility of ionic solids.

Saturated Solutions; Factors Affecting Solubility (Sections 13.2 and 13.3)

13.23   (a)   Supersaturated, because the solution contains more solute than a saturated solution at this temperature.

       (b)   The bits of glass scraped from the vessel act as a seed crystal, a place where solute molecules can align to form a crystal. The excess chromium nitrate is crystallizing out.

       (c)   324 g dissolved at 35 °C – 208 g soluble at 15 °C = 116 g crystals form

13.24   (a) $$\frac{1.22 \text{ mol } MnSO_4 \cdot H_2O}{1 \text{ L soln}} \times \frac{169.0 \text{ g } MnSO_4 \cdot H_2O}{1 \text{ mol}} \times 0.100 \text{ L}$$

                                                   $= 20.6 \text{ g } MnSO_4 \cdot H_2O/100 \text{ mL}$

             The 1.22 *M* solution is unsaturated.

       (b)   Add a known mass, say 5.0 g, of $MnSO_4 \cdot H_2O$, to the unknown solution. If the solid dissolves, the solution is unsaturated. If there is undissolved $MnSO_4 \cdot H_2O$, filter the solution and weigh the solid. If there is less than 5.0 g of solid, some of the added $MnSO_4 \cdot H_2O$, dissolved and the unknown solution is unsaturated. If there is exactly 5.0 g, no additional solid dissolved and the unknown is saturated. If there is more than 5.0 g, excess solute has precipitated and the solution is supersaturated.

13.25   *Analyze/Plan.* On Figure 13.15, find the solubility curve for the appropriate solute. Find the intersection of 40 °C and 40 g solute on the graph. If this point is below the solubility curve, more solute can dissolve and the solution is unsaturated. If the intersection is on or above the curve, the solution is saturated.   *Solve.*

       (a)   unsaturated   (b)   saturated   (c)   saturated   (d)   unsaturated

13.26   (a)   at 30 °C, $\dfrac{10 \text{ g KClO}_3}{100 \text{ g H}_2\text{O}} \times 250 \text{ g H}_2\text{O} = 25 \text{ g KClO}_3$

      (b) $\dfrac{66 \text{ g Pb(NO}_3)_2}{100 \text{ g H}_2\text{O}} \times 250 \text{ g H}_2\text{O} = 165 = 1.7 \times 10^2 \text{ g Pb(NO}_3)_2$

      (c) $\dfrac{3 \text{ g Ce}_2(\text{SO}_4)_3}{100 \text{ g H}_2\text{O}} \times 250 \text{ g H}_2\text{O} = 7.5 = 8 \text{ g Ce}_2(\text{SO}_4)_3$

13.27   (a)   We expect the liquids water and glycerol to be miscible in all proportions. Glycerol has an –OH group on each C atom in the molecule. This structure facilitates strong hydrogen bonding similar to that in water. When two liquids have very similar intermolecular interactions, entropy of mixing drives the solution process and the two liquids are usually miscible in all proportions (Solution 13.21). Like dissolves like.

      (b)   Hydrogen bonding, dipole-dipole forces, London dispersion forces

13.28   The most likely reason is statement (b), although statement (c) also contributes.

Many substances are called "oil," but they typically contain molecules composed mostly of carbon and hydrogen with fairly high molecular weights. Both statements contribute to the fact that oil molecules are nonpolar and experience strong dispersion forces. The properties of water are dominated by its strong hydrogen bonding. Oil and water have very dissimilar intermolecular interactions and the liquids are not miscible.

There are examples of high molecular weight compounds (e.g., sugars and proteins) with many hydrogen-bonding interactions that are soluble in water, so statement (b) is the better answer.

13.29   *Analyze/Plan.* Evaluate molecules in the four common laboratory solvents for strength of intermolecular interactions with nonpolar solutes. *Solve.* Toluene, $C_6H_5CH_3$, is the best solvent for nonpolar solutes. Without polar groups or nonbonding electron pairs, it forms only dispersion interactions with itself and other molecules. The enthalpy of solution, ΔH_{soln}, is essentially zero (as in Solution 13.21) and solution occurs because of the favorable entropy of mixing.

13.30   We expect alanine to be more soluble in water than hexane. Alanine has a –COOH and a –NH₂ group available to form hydrogen bonds with water molecules. Although there are some potential dispersion forces between the terminal –CH₃ group of alanine and hexane molecules, we expect the hydrogen bonding between alanine and water to be stronger. Stronger intermolecular attractive forces between alanine and water lead to a more negative ΔH_{mix} and more negative (smaller positive) ΔH_{soln} for water than for hexane.

13.31   (a)   Despite the presence of the –COOH group, stearic acid is more soluble in nonpolar CCl_4 than in polar (hydrogen bonding) water. Dispersion interactions among nonpolar $CH_3(CH_2)_{16}$– chains dominate the properties of stearic acid.

      (b)   Dioxane will be more soluble in water than cyclohexane will, because dioxane can act as a hydrogen bond acceptor.

13.32   The red part of the molecule, a carboxyl group able to form hydrogen bonds, contributes to its water solubility. The gray and white parts, a phenyl ring and several –CH₃ groups, form a large nonpolar area that contributes to its water insolubility.

13.33   *Analyze/Plan.* Hexane is a nonpolar hydrocarbon that experiences dispersion forces with other nonpolar molecules. Solutes that primarily experience dispersion forces will be more soluble in hexane.   *Solve.*

     (a) CCl_4 is more soluble because dispersion forces among nonpolar CCl_4 molecules are similar to dispersion forces in hexane. Ionic bonds in $CaCl_2$ are unlikely to be broken by weak solute-solvent interactions. For $CaCl_2$, ΔH_{solute} is large, relative to ΔH_{mix}.

     (b)   Benzene, C_6H_6, is also a nonpolar hydrocarbon and will be more soluble in hexane. Glycerol experiences hydrogen bonding with itself; these solute-solute interactions are less likely to be overcome by weak solute-solvent interactions.

     (c)   Octanoic acid, $CH_3(CH_2)_6COOH$, will be more soluble than acetic acid CH_3COOH. Both solutes experience hydrogen bonding by –COOH groups, but octanoic acid has a long, rod-like hydrocarbon chain with dispersion forces similar to those in hexane, facilitating solubility in hexane.

13.34   *Analyze/Plan.* Water, H_2O, is a polar solvent that forms hydrogen bonds with other H_2O molecules. The more soluble solute in each case will have intermolecular interactions that are most similar to the hydrogen bonding in H_2O.   *Solve.*

     (a)   Glucose, $C_6H_{12}O_6$, is more soluble because it is capable of hydrogen bonding (Figure 13.10). Nonpolar C_6H_{12} is capable only of dispersion interactions and does not have strong intermolecular interactions with polar (hydrogen bonding) H_2O.

     (b)   Ionic sodium propionate, CH_3CH_2COONa, is more soluble. Sodium propionate is a crystalline solid, whereas propionic acid is a liquid. The increase in disorder or entropy when an ionic solid dissolves leads to significant water solubility, despite the strong ion-ion forces (large ΔH_{solute}) present in the solute (see Solution 13.22).

     (c)   HCl is more soluble because it is a strong electrolyte and completely ionized in water. Ionization leads to ion-dipole solute-solvent interactions, and an increase in disorder. CH_3CH_2Cl is a molecular solute capable of relatively weak dipole-dipole solute-solvent interactions and is much less soluble in water.

13.35   (a)   False. The lower the temperature, the more soluble most gases are in water.

     (b)   True.

     (c)   False. In a supersaturated solution all solute remains dissolved.

     (d)   True. Solubility of liquids and solids usually increases with increasing temperature.

13.36   (a)   True.

     (b)   False. The solubility of most ionic solids increases as the temperature of the solution increases.

     (c)   False. The solubility of gases in water decreases as temperature increases because the kinetic energy of the gas particles increases and more solute particles have sufficient energy to escape the solution. Also, interactions between gaseous solutes and water typically do not involve hydrogen bonding.

     (d)   True. See Figure 13.15.

13.37   *Analyze/Plan.* Follow the logic in Sample Exercise 13.2.   *Solve.*

$$S_{He} = 3.7 \times 10^{-4} \, M/atm \times 1.5 \, atm = 5.6 \times 10^{-4} \, M$$

$$S_{N_2} = 6.0 \times 10^{-4} \, M/atm \times 1.5 \, atm = 9.0 \times 10^{-4} \, M$$

13.38 $650 \, torr \times \dfrac{1 \, atm}{760 \, torr} = 0.855 \, atm$; $P_{O_2} = X_{O_2}(P_t) = 0.21(0.855 \, atm) = 0.1796 = 0.18 \, atm$

$$S_{O_2} = kP_{O_2} = \frac{1.38 \times 10^{-3} \, mol}{L\text{-}atm} \times 0.1796 \, atm = 2.5 \times 10^{-4} \, M$$

Concentrations of Solutions (Section 13.4)

13.39   *Analyze/Plan.* Follow the logic in Sample Exercise 13.3.   *Solve.*

(a) $\text{mass \%} = \dfrac{\text{mass solute}}{\text{total mass solution}} \times 100 = \dfrac{10.6 \, g \, Na_2SO_4}{10.6 \, g \, Na_2SO_4 + 483 \, g \, H_2O} \times 100 = 2.15\%$

(b) $\text{ppm} = \dfrac{\text{mass solute}}{\text{total mass solution}} \times 10^6$; $\dfrac{2.86 \, g \, Ag}{1 \, ton \, ore} \times \dfrac{1 \, ton}{2000 \, lb} \times \dfrac{1 \, lb}{453.6 \, g} \times 10^6 = 3.15 \, ppm$

13.40   (a) $\text{mass \%} = \dfrac{\text{mass solute}}{\text{total mass solution}} \times 100$

$$\text{mass solute} = 0.035 \, mol \, I_2 \times \frac{253.8 \, g \, I_2}{1 \, mol \, I_2} = 8.883 = 8.9 \, g \, I_2$$

$$\text{mass \% } I_2 = \frac{8.883 \, g \, I_2}{8.883 \, g \, I_2 + 125 \, g \, CCl_4} \times 100 = 6.635 = 6.6\% \, I_2$$

(b) $\text{ppm} = \dfrac{\text{mass solute}}{\text{total mass solution}} \times 10^6 = \dfrac{0.0079 \, g \, Sr^{2+}}{1 \times 10^3 \, g \, H_2O} \times 10^6 = 7.9 \, ppm \, Sr^{2+}$

13.41   *Analyze/Plan.* Given masses of CH_3OH and H_2O, calculate moles of each component.

(a)   Mole fraction CH_3OH = (mol CH_3OH)/(total mol)

(b)   mass % CH_3OH = [(g CH_3OH)/(total mass)] × 100

(c)   molality CH_3OH = (mol CH_3OH)/(kg H_2O).   *Solve.*

(a) $14.6 \, g \, CH_3OH \times \dfrac{1 \, mol \, CH_3OH}{32.04 \, g \, CH_3OH} = 0.4557 = 0.456 \, mol \, CH_3OH$

$$184 \, g \, H_2O \times \frac{1 \, mol \, H_2O}{18.02 \, g \, H_2O} = 10.211 = 10.2 \, mol \, H_2O$$

$$X_{CH_3OH} = \frac{0.4557}{0.4557 + 10.211} = 0.04272 = 0.0427$$

(b) $\text{mass \% } CH_3OH = \dfrac{14.6 \, g \, CH_3OH}{14.6 \, g \, CH_3OH + 184 \, g \, H_2O} \times 100 = 7.35\% \, CH_3OH$

(c) $m = \dfrac{0.4557 \, mol \, CH_3OH}{0.184 \, kg \, H_2O} = 2.477 = 2.48 \, m \, CH_3OH$

13.42　(a)　$\dfrac{20.8 \text{ g C}_6\text{H}_5\text{OH}}{94.11 \text{ g/mol}} = 0.2210 = 0.221 \text{ mol C}_6\text{H}_5\text{OH}$

$\dfrac{425 \text{ g CH}_3\text{CH}_2\text{OH}}{46.07 \text{ g/mol}} = 9.2251 = 9.23 \text{ mol CH}_3\text{CH}_2\text{OH}$

$X_{\text{C}_6\text{H}_5\text{OH}} = \dfrac{0.2210}{0.2210 + 9.2251} = 0.02340 = 0.0234$

(b)　$\text{mass \%} = \dfrac{20.8 \text{ g C}_6\text{H}_5\text{OH}}{20.8 \text{ g C}_6\text{H}_5\text{OH} + 425 \text{ g CH}_3\text{CH}_2\text{OH}} \times 100 = 4.67\% \text{ C}_6\text{H}_5\text{OH}$

(c)　$m = \dfrac{0.2210 \text{ mol C}_6\text{H}_5\text{OH}}{0.425 \text{ kg CH}_3\text{CH}_2\text{OH}} = 0.5200 = 0.520 \; m \text{ C}_6\text{H}_5\text{OH}$

13.43　*Analyze/Plan.* Given mass solute and volume solution, calculate mol solute, then molarity = mol solute/L solution. Or, for dilution, $M_c \times L_c = M_d \times L_d$. *Solve.*

(a)　$M = \dfrac{\text{mol solute}}{\text{L soln}}; \; \dfrac{0.540 \text{ g Mg(NO}_3)_2}{0.2500 \text{ L soln}} \times \dfrac{1 \text{ mol Mg(NO}_3)_2}{148.3 \text{ g Mg(NO}_3)_2} = 1.46 \times 10^{-2} \; M \text{ Mg(NO}_3)_2$

(b)　$\dfrac{22.4 \text{ g LiClO}_4 \cdot 3\text{H}_2\text{O}}{0.125 \text{ L soln}} \times \dfrac{1 \text{ mol LiClO}_4 \cdot 3\text{H}_2\text{O}}{160.4 \text{ g LiClO}_4 \cdot 3\text{H}_2\text{O}} = 1.12 \; M \text{ LiClO}_4 \cdot 3\text{H}_2\text{O}$

(c)　$M_c \times L_c = M_d \times L_d; \; 3.50 \; M \text{ HNO}_3 \times 0.0250 \text{ L} = ?M \text{ HNO}_3 \times 0.250 \text{ L}$
250 mL of 0.350 M HNO$_3$

13.44　(a)　$M = \dfrac{\text{mol solute}}{\text{L soln}}; \; \dfrac{15.0 \text{ g Al}_2(\text{SO}_4)_3}{0.250 \text{ L soln}} \times \dfrac{1 \text{ mol Al}_2(\text{SO}_4)_3}{342.2 \text{ g Al}_2(\text{SO}_4)_3} = 0.175 \; M \text{ Al}_2(\text{SO}_4)_3$

(b)　$\dfrac{5.25 \text{ g Mn(NO}_3)_2 \cdot 2\text{H}_2\text{O}}{0.175 \text{ L soln}} \times \dfrac{1 \text{ mol Mn(NO}_3)_2 \cdot 2\text{H}_2\text{O}}{215.0 \text{ g Mn(NO}_3)_2 \cdot 2\text{H}_2\text{O}} = 0.140 \; M \text{ Mn(NO}_3)_2$

(c)　$M_c \times L_c = M_d \times L_d; \; 9.00 \; M \text{ H}_2\text{SO}_4 \times 0.0350 \text{ L} = ?M \text{ H}_2\text{SO}_4 \times 0.500 \text{ L}$
500 mL of 0.630 M H$_2$SO$_4$

13.45　*Analyze/Plan.* Follow the logic in Sample Exercise 13.4.　*Solve.*

(a)　$m = \dfrac{\text{mol solute}}{\text{kg solvent}}; \; \dfrac{8.66 \text{ g C}_6\text{H}_6}{23.6 \text{ g CCl}_4} \times \dfrac{1 \text{ mol C}_6\text{H}_6}{78.11 \text{ g C}_6\text{H}_6} \times \dfrac{1000 \text{ g CCl}_4}{1 \text{ kg CCl}_4} = 4.70 \; m \text{ C}_6\text{H}_6$

(b)　The density of H$_2$O = 0.997 g/mL = 0.997 kg/L.

$\dfrac{4.80 \text{ g NaCl}}{0.350 \text{ L H}_2\text{O}} \times \dfrac{1 \text{ mol NaCl}}{58.44 \text{ g NaCl}} \times \dfrac{1 \text{ L H}_2\text{O}}{0.997 \text{ kg H}_2\text{O}} = 0.235 \; m \text{ NaCl}$

13.46　(a)　$16.0 \text{ mol H}_2\text{O} \times \dfrac{18.02 \text{ g H}_2\text{O}}{1 \text{ mol H}_2\text{O}} = 288.3 \text{ g H}_2\text{O} = 0.288 \text{ kg H}_2\text{O}$

$m = \dfrac{1.12 \text{ mol KCl}}{0.2883 \text{ kg H}_2\text{O}} = 3.8846 = 3.88 \; m \text{ KCl}$

(b)　$m = \dfrac{\text{mol solute}}{\text{kg solute}}; \; \text{mol S}_8 = m \times \text{kg C}_{10}\text{H}_8 = 0.12 \; m \times 0.1000 \text{ kg C}_{10}\text{H}_8 = 0.012 \text{ mol}$

$0.012 \text{ mol S}_8 \times \dfrac{256.5 \text{ g S}_8}{1 \text{ mol S}_8} = 3.078 = 3.1 \text{ g S}_8$

13.47   *Analyze/Plan.* Assume 1 L of solution. Density gives the total mass of 1 L of solution. The g H_2SO_4/L are also given in the problem. Mass % = (mass solute/total mass solution) × 100. Calculate mass solvent from mass solution and mass solute. Calculate moles solute and solvent and use the appropriate definitions to calculate mole fraction, molality, and molarity.   *Solve.*

(a) $\dfrac{571.6 \text{ g } H_2SO_4}{1 \text{ L soln}} \times \dfrac{1 \text{ L soln}}{1329 \text{ g soln}} = 0.430098 \text{ g } H_2SO_4/\text{g soln}$

      mass % is thus $0.4301 \times 100 = 43.01\% \ H_2SO_4$

(b)   In a liter of solution there are $1329 - 571.6 = 757.4 = 757 \text{ g } H_2O$.

      $\dfrac{571.6 \text{ g } H_2SO_4}{98.09 \text{ g/mol}} = 5.827 \text{ mol } H_2SO_4$; $\dfrac{757.4 \text{ g } H_2O}{18.02 \text{ g/mol}} = 42.03 = 42.0 \text{ mol } H_2O$

      $X_{H_2SO_4} = \dfrac{5.827}{42.03 + 5.827} = 0.122$

      (The result has 3 sig figs because (g H_2O) resulting from subtraction is limited to 3 sig figs.)

(c)   molality $= \dfrac{5.827 \text{ mol } H_2SO_4}{0.7574 \text{ kg } H_2O} = 7.693 = 7.69 \ m \ H_2SO_4$

(d)   molarity $= \dfrac{5.827 \text{ mol } H_2SO_4}{1 \text{ L soln}} = 5.827 \ M \ H_2SO_4$

13.48   (a)   mass % $= \dfrac{\text{mass } C_6H_8O_6}{\text{total mass solution}} \times 100$;

         $\dfrac{80.5 \text{ g } C_6H_8O_6}{80.5 \text{ g } C_6H_8O_6 + 210 \text{ g } H_2O} \times 100 = 27.71 = 27.7\% \ C_6H_8O_6$

   (b)   mol $C_6H_8O_6 = \dfrac{80.5 \text{ g } C_6H_8O_6}{176.1 \text{ g/mol}} = 0.4571 = 0.457 \text{ mol } C_6H_8O_6$

         mol $H_2O = \dfrac{210 \text{ g } H_2O}{18.02 \text{ g/mol}} = 11.654 = 11.7 \text{ mol } H_2O$

         $X_{C_6H_8O_6} = \dfrac{0.4571 \text{ mol } C_6H_8O_6}{0.4571 \text{ mol } C_6H_8O_6 + 11.654 \text{ mol } H_2O} = 0.0377$

   (c) $m = \dfrac{0.4571 \text{ mol } C_6H_8O_6}{0.210 \text{ kg } H_2O} = 2.18 \ m \ C_6H_8O_6$

   (d) $M = \dfrac{\text{mol } C_6H_8O_6}{\text{L solution}}$; $290.5 \text{ g soln} \times \dfrac{1 \text{ mL}}{1.22 \text{ g}} \times \dfrac{1 \text{ L}}{1000 \text{ mL}} = 0.2381 = 0.238 \text{ L}$

         $M = \dfrac{0.4571 \text{ mol } C_6H_8O_6}{0.2381 \text{ L soln}} = 1.92 \ M \ C_6H_8O_6$

13.49   *Analyze/Plan.* Given: 98.7 mL of $CH_3CN(l)$, 0.786 g/mL; 22.5 mL CH_3OH, 0.791 g/mL. Use the density and volume of each component to calculate mass and then moles of each component. Use the definitions to calculate mole fraction, molality, and molarity. *Solve.*

(a)　　$\text{mol CH}_3\text{CN} = \dfrac{0.786 \text{ g}}{1 \text{ mL}} \times 98.7 \text{ mL} \times \dfrac{1 \text{ mol CH}_3\text{CN}}{41.05 \text{ g CH}_3\text{CN}} = 1.8898 = 1.89 \text{ mol}$

$\text{mol CH}_3\text{OH} = \dfrac{0.791 \text{ g}}{1 \text{ mL}} \times 22.5 \text{ mL} \times \dfrac{1 \text{ mol CH}_3\text{OH}}{32.04 \text{ g CH}_3\text{OH}} = 0.5555 = 0.556 \text{ mol}$

$X_{\text{CH}_3\text{OH}} = \dfrac{0.5555 \text{ mol CH}_3\text{OH}}{1.8898 \text{ mol CH}_3\text{CN} + 0.5555 \text{ mol CH}_3\text{OH}} = 0.227$

(b)　　Assuming CH_3OH is the solute and CH_3CN is the solvent,

$98.7 \text{ mL CH}_3\text{CN} \times \dfrac{0.786 \text{ g}}{1 \text{ mL}} \times \dfrac{1 \text{ kg}}{1000 \text{ g}} = 0.07758 = 0.0776 \text{ kg CH}_3\text{CN}$

$m_{\text{CH}_3\text{OH}} = \dfrac{0.5555 \text{ mol CH}_3\text{OH}}{0.07758 \text{ kg CH}_3\text{CN}} = 7.1604 = 7.16 \ m \ \text{CH}_3\text{OH}$

(c)　　The total volume of the solution is 121.2 mL, assuming volumes are additive.

$M = \dfrac{0.5555 \text{ mol CH}_3\text{OH}}{0.1212 \text{ L solution}} = 4.58 \ M \ \text{CH}_3\text{OH}$

13.50　　Given: 8.10 g C_4H_4S, 1.065 g/mL; 250.0 mL C_7H_8, 0.867 g/mL

(a)　　$\text{mol C}_4\text{H}_4\text{S} = 8.10 \text{ g C}_4\text{H}_4\text{S} \times \dfrac{1 \text{ mol C}_4\text{H}_4\text{S}}{84.15 \text{ g C}_4\text{H}_4\text{S}} = 0.09626 = 0.0963 \text{ mol C}_4\text{H}_4\text{S}$

$\text{mol C}_7\text{H}_8 = \dfrac{0.867 \text{ g}}{1 \text{ mL}} \times 250.0 \text{ mL} \times \dfrac{1 \text{ mol C}_7\text{H}_8}{92.14 \text{ g C}_7\text{H}_8} = 2.352 = 2.35 \text{ mol}$

$X_{\text{C}_4\text{H}_4\text{S}} = \dfrac{0.09626 \text{ mol C}_4\text{H}_4\text{S}}{0.09626 \text{ mol C}_4\text{H}_4\text{S} + 2.352 \text{ mol C}_7\text{H}_8} = 0.03932 = 0.0393$

(b)　　$m_{\text{C}_4\text{H}_4\text{S}} = \dfrac{\text{mol C}_4\text{H}_4\text{S}}{\text{kg C}_7\text{H}_8}; \ 250.0 \text{ mL} \times \dfrac{0.867 \text{ g}}{1 \text{ mL}} \times \dfrac{1 \text{ kg}}{1000 \text{ g}} = 0.2168 = 0.217 \text{ kg C}_7\text{H}_8$

$m_{\text{C}_4\text{H}_4\text{S}} = \dfrac{0.09626 \text{ mol C}_4\text{H}_4\text{S}}{0.2168 \text{ kg C}_7\text{H}_8} = 0.444 \ m \ \text{C}_4\text{H}_4\text{S}$

(c)　　$8.10 \text{ g C}_4\text{H}_4\text{S} \times \dfrac{1 \text{ mL}}{1.065 \text{ g}} = 7.606 = 7.61 \text{ mL C}_4\text{H}_4\text{S};$

$V_{\text{soln}} = 7.61 \text{ mL C}_4\text{H}_4\text{S} + 250.0 \text{ mL C}_7\text{H}_8 = 257.6 \text{ mL}$

$M_{\text{C}_4\text{H}_4\text{S}} = \dfrac{0.09626 \text{ mol C}_4\text{H}_4\text{S}}{0.2576 \text{ L soln}} = 0.374 \ M \ \text{C}_4\text{H}_4\text{S}$

13.51　　_Analyze/Plan._ Given concentration and volume of solution use definitions of the appropriate concentration units to calculate amount of solute; change amount to moles if needed.　_Solve._

(a)　　$\text{mol} = M \times \text{L}; \ \dfrac{0.250 \text{ mol SrBr}_2}{1 \text{ L soln}} \times 0.600 \text{ L} = 0.150 \text{ mol SrBr}_2$

(b)　　Assume that for dilute aqueous solutions, the mass of the solvent is the mass of solution. Use proportions to get mol KCl.

$\dfrac{0.180 \text{ mol KCl}}{1 \text{ kg H}_2\text{O}} = \dfrac{x \text{ mol KCl}}{0.0864 \text{ kg H}_2\text{O}}; \ x = 1.56 \times 10^{-2} \text{ mol KCl}$

(c)   Use proportions to get mass of glucose, then change to mol glucose.

$$\frac{6.45 \text{ g C}_6\text{H}_{12}\text{O}_6}{100 \text{ g soln}} = \frac{\text{x g C}_6\text{H}_{12}\text{O}_6}{124.0 \text{ g soln}}; \; x = 8.00 \text{ g C}_6\text{H}_{12}\text{O}_6$$

$$8.00 \text{ g C}_6\text{H}_{12}\text{O}_6 \times \frac{1 \text{ mol C}_6\text{H}_{12}\text{O}_6}{180.2 \text{ g C}_6\text{H}_{12}\text{O}_6} = 4.44 \times 10^{-2} \text{ mol C}_6\text{H}_{12}\text{O}_6$$

13.52   (a) $\dfrac{1.50 \text{ mol HNO}_3}{1 \text{ L soln}} \times 0.255 \text{ L} = 0.3825 = 0.383 \text{ mol HNO}_3$

(b)   Assume that for dilute aqueous solutions, the mass of the solvent is the mass of solution.

$$\frac{1.50 \text{ mol NaCl}}{1 \text{ kg H}_2\text{O}} = \frac{\text{x mol}}{50.0 \times 10^{-6} \text{ kg}}; \; x = 7.50 \times 10^{-5} \text{ mol NaCl}$$

(c) $\dfrac{1.50 \text{ g C}_{12}\text{H}_{22}\text{O}_{11}}{100 \text{ g soln}} = \dfrac{\text{x g C}_{12}\text{H}_{22}\text{O}_{11}}{75.0 \text{ g soln}}; \; x = 1.125 = 1.13 \text{ g C}_{12}\text{H}_{22}\text{O}_{11}$

$$1.125 \text{ g C}_{12}\text{H}_{22}\text{O}_{11} \times \frac{1 \text{ mol C}_{12}\text{H}_{22}\text{O}_{11}}{342.3 \text{ g C}_{12}\text{H}_{22}\text{O}_{11}} = 3.287 \times 10^{-3} = 3.29 \times 10^{-3} \text{mol C}_{12}\text{H}_{22}\text{O}_{11}$$

13.53   *Analyze/Plan.* When preparing solution, we must know amount of solute and solvent. Use the appropriate concentration definition to calculate amount of solute. If this amount is in moles, use molar mass to get grams; use mass in grams directly. Amount of solvent can be expressed as total volume or mass of solution. Combine mass solute and solvent to produce the required amount (mass or volume) of solution.   *Solve.*

(a) $\text{mol} = M \times L; \; \dfrac{1.50 \times 10^{-2} \text{ mol KBr}}{1 \text{ L soln}} \times 0.75 \text{ L} \times \dfrac{119.0 \text{ g KBr}}{1 \text{ mol KBr}} = 1.3 \text{ g KBr}$

Weigh out 1.3 g KBr, dissolve in water, dilute with stirring to 0.75 L (750 mL).

(b)   Mass of solution is required, but density is not specified. Use molality to calculate mass fraction, and then the masses of solute and solvent needed for 125 g of solution.

$$\frac{0.180 \text{ mol KBr}}{1000 \text{ g H}_2\text{O}} \times \frac{119.0 \text{ g KBr}}{1 \text{ mol KBr}} = 21.42 = 21.4 \text{ g KBr/kg H}_2\text{O}. \text{ Thus,}$$

$$\text{mass fraction} = \frac{21.42 \text{ g KBr}}{1000 + 21.42} = 0.02097 = 0.0210$$

In 125 g of the 0.180 *m* solution, there are

$$(125 \text{ g soln}) \times \frac{0.02097 \text{ g KBr}}{1 \text{ g soln}} = 2.621 = 2.62 \text{ g KBr}$$

Weigh out 2.62 g KBr, dissolve it in 125 − 2.62 = 122.38 = 122 g H$_2$O to make exactly 125 g of 0.180 *m* solution.

(c)   Using solution density, calculate the total mass of 1.85 L of solution, and from the mass % of KBr, the mass of KBr required.

$$1.85 \text{ L soln} \times \frac{1000 \text{ mL}}{1 \text{ L}} \times \frac{1.10 \text{ g soln}}{1 \text{ mL}} = 2035 = 2.04 \times 10^3 \text{ g soln}$$

0.120 (2035 g soln) = 244.2 = 244 g KBr

Dissolve 244 g KBr in water, dilute with stirring to 1.85 L.

(d)   Calculate moles KBr needed to precipitate 16.0 g AgBr. $AgNO_3$ is present in excess.

$$16.0 \text{ g AgBr} \times \frac{1 \text{ mol AgBr}}{187.8 \text{ g AgBr}} \times \frac{1 \text{ mol KBr}}{1 \text{ mol AgBr}} = 0.08520 = 0.0852 \text{ mol KBr}$$

$$0.0852 \text{ mol KBr} \times \frac{1 \text{ L soln}}{0.150 \text{ mol KBr}} = 0.568 \text{ L soln}$$

Weigh out 0.0852 mol KBr (10.1 g KBr), dissolve it in a small amount of water, and dilute to 0.568 L.

13.54   (a) $\dfrac{0.110 \text{ mol } (NH_4)_2SO_4}{1 \text{ L soln}} \times 1.50 \text{ L} \times \dfrac{132.2 \text{ g } (NH_4)_2SO_4}{1 \text{ mol } (NH_4)_2SO_4} = 21.81 = 21.8 \text{ g } (NH_4)_2SO_4$

Weigh 21.8 g $(NH_4)_2SO_4$, dissolve in a small amount of water, continue adding water with thorough mixing up to a total solution volume of 1.50 L.

(b)   Determine the mass fraction of Na_2CO_3 in the solution:

$$\frac{0.65 \text{ mol } Na_2CO_3}{1000 \text{ g } H_2O} \times \frac{106.0 \text{ g } Na_2CO_3}{1 \text{ mol } Na_2CO_3} = 68.9 \text{ g} = \frac{69 \text{ g } Na_2CO_3}{1000 \text{ g } H_2O}$$

$$\text{mass fraction} = \frac{68.9 \text{ g } Na_2CO_3}{1000 \text{ g } H_2O + 68.9 \text{ g } Na_2CO_3} = 0.06446 = 0.064$$

In 225 g of solution, there are 0.06446(225) = 14.503 = 15 g Na_2CO_3.

Weigh out 15 g Na_2CO_3 and dissolve it in 225 – 15 = 210 g H_2O to make exactly 225 g of solution. (210 g H_2O/0.997 g H_2O/mL @ 25 °C = 211 mL H_2O)

[Carrying 3 sig figs, weigh 14.5 g Na_2CO_3 and dissolve it in 225 – 14.5 = 210.5 g H_2O. This produces a solution that is much closer to 0.65 m.]

(c) $1.20 \text{ L} \times \dfrac{1000 \text{ mL}}{1 \text{ L}} \times \dfrac{1.16 \text{ g}}{1 \text{ mL}} = 1392 \text{ g solution}$; 0.150(1392 g soln) = 209 g $Pb(NO_3)_2$

Weigh 209 g $Pb(NO_3)$ and add (1392 – 209) = 1183 g H_2O to make exactly (1392 = 1.39×10^3) g or 1.20 L of solution.

(1183 g H_2O/0.997 g/mL @ 25 °C = 1187 mL H_2O)

(d)   Calculate the mol HCl necessary to neutralize 5.5 g $Ba(OH)_2$.

$Ba(OH)_2(s) + 2 \text{ HCl}(aq) \rightarrow BaCl_2(aq) + 2 H_2O(l)$

$$5.5 \text{ g } Ba(OH)_2 + \frac{1 \text{ mol } Ba(OH)_2}{171 \text{ g } Ba(OH)_2} \times \frac{2 \text{ mol HCl}}{1 \text{ mol } Ba(OH)_2} = 0.0643 = 0.064 \text{ mol HCl}$$

$$M = \frac{\text{mol}}{L}; \ L = \frac{\text{mol}}{M} = \frac{0.0643 \text{ mol HCl}}{0.50 \text{ M HCl}} = 0.1287 = 0.13 \text{ L} = 130 \text{ mL}$$

130 mL of 0.50 M HCl are needed.

$M_c \times L_c = M_d \times L_d$; 6.0 $M \times L_c$ = 0.50 $M \times$ 0.1287 L; L_c = 0.01072 L = 11 mL

Using a pipette, measure exactly 11 mL of 6.0 M HCl and dilute with water to a total volume of 130 mL.

13.55 *Analyze/Plan.* Assume a solution volume of 1.00 L. Calculate the mass of 1.00 L of solution and the mass of HNO_3 in 1.00 L of solution. Mass % = (mass solute/mass solution) × 100. *Solve.*

$$1.00 \text{ L} \times \frac{1000 \text{ mL}}{1 \text{ L}} \times \frac{1.42 \text{ g soln}}{\text{mL soln}} = 1.42 \times 10^3 \text{ g soln}$$

$$16 \, M = \frac{16 \text{ mol HNO}_3}{1 \text{ L soln}} \times \frac{63.02 \text{ g HNO}_3}{1 \text{ mol HNO}_3} = 1008 = 1.0 \times 10^3 \text{ g HNO}_3$$

$$\text{mass \%} = \frac{1008 \text{ g HNO}_3}{1.42 \times 10^3 \text{ g soln}} \times 100 = 71\% \text{ HNO}_3$$

13.56 *Analyze/Plan.* Assume 1.00 L of solution. Calculate mass of 1 L of solution using density. Calculate mass of NH_3 using mass %, then mol NH_3 in 1.00 L. *Solve.*

$$1.00 \text{ L soln} \times \frac{1000 \text{ mL}}{1 \text{ L}} \times \frac{0.90 \text{ g soln}}{1 \text{ mL soln}} = 9.0 \times 10^2 \text{ g soln/L}$$

$$\frac{900 \text{ g soln}}{1.00 \text{ L soln}} \times \frac{28 \text{ g NH}_3}{100 \text{ g soln}} \times \frac{1 \text{ mol NH}_3}{17.03 \text{ g NH}_3} = 14.80 = 15 \text{ mol NH}_3/\text{L soln} = 15 \, M \text{ NH}_3$$

13.57 *Analyze.* Given: 80.0% Cu, 20.0% Zn by mass; density = 8750 kg/m^3. Find: (a) *m* of Zn (b) *M* of Zn

(a) *Plan.* In the brass alloy, Zn is the solute (lesser component) and Cu is the solvent (greater component). *m* = mol Zn/kg Cu. 1 m^3 brass alloy weighs 8750 kg. 80.0% is Cu, 20.0% is Zn. Change g Zn → mol Zn and solve for *m*. *Solve.*

$$8750 \text{ kg brass} \times \frac{80 \text{ g Cu}}{100 \text{ g brass}} = 7.00 \times 10^3 \text{ kg Cu}$$

$$8750 \text{ kg brass} - 7000 \text{ kg Cu} = 1750 \text{ kg Zn}$$

$$1750 \text{ kg Zn} \times \frac{1000 \text{ g}}{\text{kg}} \times \frac{1 \text{ mol Zn}}{65.39 \text{ g Zn}} = 26,762.5 = 2.68 \times 10^4 \text{ mol Zn}$$

$$m = \frac{2.676 \times 10^4 \text{ mol Zn}}{7000 \text{ kg Cu}} = 3.82 \, m \text{ Zn}$$

(b) *Plan.* *M* = mol Zn/L brass. Use mol Zn from part (a). Change 1 m^3 → L brass and calculate *M*. *Solve.*

$$1 \text{ m}^3 \times \frac{(10)^3 \text{ dm}^3}{\text{m}^3} \times \frac{1 \text{ L}}{1 \text{ dm}^3} = 1000 \text{ L}$$

$$M = \frac{2.676 \times 10^4 \text{ mol Zn}}{1000 \text{ L brass}} = 26.76 = 26.8 \, M \text{ Zn}$$

13.58 (a)

$$\frac{0.0500 \text{ mol C}_8\text{H}_{10}\text{N}_4\text{O}_2}{1 \text{ kg CHCl}_3} \times \frac{194.2 \text{ g C}_8\text{H}_{10}\text{N}_4\text{O}_2}{1 \text{ mol C}_8\text{H}_{10}\text{N}_4\text{O}_2} = 9.7100$$

$$= 9.71 \text{ g C}_8\text{H}_{10}\text{N}_4\text{O}_2/\text{kg CHCl}_3$$

$$\frac{9.710 \text{ g C}_8\text{H}_{10}\text{N}_4\text{O}_2}{9.710 \text{ g C}_8\text{H}_{10}\text{N}_4\text{O}_2 + 1000.00 \text{ g CHCl}_3} \times 100 = 0.9617 = 0.962\% \text{ C}_8\text{H}_{10}\text{N}_4\text{O}_2 \text{ by mass}$$

(b) $1000 \text{ g CHCl}_3 \times \dfrac{1 \text{ mol CHCl}_3}{119.4 \text{ CHCl}_3} = 8.375 = 8.38 \text{ mol CHCl}_3$

$$X_{C_8H_{10}N_4O_2} = \dfrac{0.0500}{0.0500 + 8.375} = 0.00593$$

13.59   *Analyze.* Given: 4.6% CO_2 by volume (in air), 1 atm total pressure. Find: partial pressure and molarity of CO_2 in air.

Plan. 4.6% CO_2 by volume means 4.6 mL of CO_2 could be isolated from 100 mL of air, at the same temperature and pressure. According to Avogadro's law, equal volumes of gases at the same temperature and pressure contain equal numbers of moles. By inference, the volume ratio of CO_2 to air, 4.6/100 or 0.046, is also the mole ratio.   *Solve.*

(a) $P_{CO_2} = X_{CO_2} \times P_t = 0.046 \,(1 \text{ atm}) = 0.046 \text{ atm}$

(b) $M = \text{mol } CO_2/\text{L air} = n/V.$ $PV = nRT, \; M = n/V = P/RT$

$$M_{CO_2} = \dfrac{P_{CO_2}}{RT} = \dfrac{0.046 \text{ atm}}{310 \text{ K}} \times \dfrac{\text{mol-K}}{0.08206 \text{ L-atm}} = 1.8 \times 10^{-3} \; M$$

13.60   (a)   For gases at the same temperature and pressure, volume % = mol %. The volume and mol % of CO_2 in this breathing air is 4.0%.

(b) $P_{CO_2} = X_{CO_2} \times P_t = 0.040 \,(1 \text{ atm}) = 0.040 \text{ atm}$

$$M_{CO_2} = \dfrac{P_{CO_2}}{RT} = \dfrac{0.040 \text{ atm}}{310 \text{ K}} \times \dfrac{\text{mol-K}}{0.08206 \text{ L-atm}} = 1.6 \times 10^{-3} \; M$$

Colligative Properties (Section 13.5)

13.61   (a)   False   (b) True   (c) True   (d) False

13.62   (a)   False. Adding solvent decreases the molality of the solution and elevates the freezing point of the solution.

(b)   False. The solid that forms is nearly pure solvent.

(c)   False. The more concentrated the solution, the lower the freezing point.

(d)   True

(e)   True

13.63   *Analyze/Plan.* H_2O vapor pressure will be determined by the mole fraction of H_2O in the solution. The vapor pressure of pure H_2O at 20 °C = 17.5 torr.   *Solve.*

The density of water at 20 °C is not exactly 1 g/mL. From Appendix B, the density of water at 25 °C is 0.99707 g/mL. For the purpose of this calculation, assume this is the density at 20 °C.

$$\dfrac{10.0 \text{ g } C_6H_{12}O_6}{180.15 \text{ g/mol}} = 0.055509 = 0.0555 \text{ mol}; \quad \dfrac{997 \text{ g } H_2O}{18.02 \text{ g/mol}} = 55.327 = 55.3 \text{ mol}$$

$$P_{H_2O} = X_{H_2O} \, P^{\circ}_{H_2O} = \dfrac{55.3 \text{ mol } H_2O}{55.3 + 0.0555} \times 17.5 \text{ torr} = 17.48 = 17.5 \text{ torr}$$

$$\dfrac{10.0 \text{ g } C_{12}H_{22}O_{11}}{342.3 \text{ g/mol}} = 0.029214 = 0.0292 \text{ mol}; \quad \dfrac{997 \text{ g } H_2O}{18.02 \text{ g/mol}} = 55.327 = 55.3 \text{ mol}$$

$$P_{H_2O} = X_{H_2O} \, P_{H_2O}^{\circ} = \dfrac{55.3 \text{ mol } H_2O}{55.3 + 0.0292} \times 17.5 \text{ torr} = 17.49 = 17.5 \text{ torr}$$

Because these two solutions are so dilute, they have essentially the same vapor pressure. Generally, the less concentrated solution, the one with fewer moles of solute per kilogram of solvent, will have the higher vapor pressure.

13.64   *Analyze/Plan.* Calculate the vapor pressure predicted by Raoult's law and compare it to the experimental vapor pressure. Assume ethylene glycol (eg) is the solute.   *Solve.*

$$X_{H_2O} = X_{eg} = 0.500; \quad P_A = X_A P_A^{\circ} = 0.500(149) \text{ torr} = 74.5 \text{ torr}$$

The experimental vapor pressure (P_A), 67 torr, is less than the value predicted by Raoult's law for an ideal solution. The solution is not ideal.

Check. An ethylene glycol-water solution has extensive hydrogen bonding, which causes deviation from ideal behavior. We expect the experimental vapor pressure to be less than the ideal value and it is.

13.65   (a)   *Analyze/Plan.* H_2O vapor pressure will be determined by the mole fraction of H_2O in the solution. The vapor pressure of pure H_2O at 338 K (65 °C) = 187.5 torr.

Solve.

$$\dfrac{22.5 \text{ g } C_{12}H_{22}O_{11}}{342.3 \text{ g/mol}} = 0.06573 = 0.0657 \text{ mol}; \quad \dfrac{200.0 \text{ g } H_2O}{18.02 \text{ g/mol}} = 11.09878 = 11.10 \text{ mol}$$

$$P_{H_2O} = X_{H_2O} \, P_{H_2O}^{\circ} = \dfrac{11.09878 \text{ mol } H_2O}{11.09878 + 0.06573} \times 187.5 \text{ torr} = 186.4 \text{ torr}$$

(b)   *Analyze/Plan.* For this problem, it will be convenient to express Raoult's law in terms of the lowering of the vapor pressure of the solvent, ΔP_A.

$\Delta P_A = P_A^{\circ} - X_A P_A^{\circ} = P_A^{\circ} (1 - X_A)$. $1 - X_A = X_B$, the mole fraction of the *solute* particles

$\Delta P_A = X_B P_A^{\circ}$; the vapor pressure of the solvent (A) is lowered according to the mole fraction of solute (B) particles present.   *Solve.*

$$P_{H_2O} \text{ at } 40 \text{ °C} = 55.3 \text{ torr}; \quad \dfrac{340 \text{ g } H_2O}{18.02 \text{ g/mol}} = 18.868 = 18.9 \text{ mol } H_2O$$

$$X_{C_3H_8O_2} = \dfrac{2.88 \text{ torr}}{55.3 \text{ torr}} = \dfrac{y \text{ mol } C_3H_8O_2}{y \text{ mol } C_3H_8O_2 + 18.868 \text{ mol } H_2O} = 0.05208 = 0.0521$$

$$0.05208 = \dfrac{y}{y + 18.868}; \quad 0.05208 \, y + 0.98263 = y; \quad 0.94792 \, y = 0.98263,$$

$$y = 1.0366 = 1.04 \text{ mol } C_3H_8O_2$$

This result has 3 sig figs because (0.340 kg water) has 3 sig figs.

$$1.0366 \text{ mol } C_3H_8O_2 \times \dfrac{76.09 \text{ g } C_3H_8O_2}{\text{mol } C_3H_8O_2} = 78.88 = 78.9 \text{ g } C_3H_8O_2$$

13.66   (a) H_2O vapor pressure will be determined by the mole fraction of H_2O in the solution. The vapor pressure of pure H_2O at 343 K (70 °C) = 233.7 torr.

$$\frac{28.5 \text{ g C}_3\text{H}_8\text{O}_3}{92.10 \text{ g/mol}} = 0.3094 = 0.309 \text{ mol}; \quad \frac{125 \text{ g H}_2\text{O}}{18.02 \text{ g/mol}} = 6.937 = 6.94 \text{ mol}$$

$$P_{\text{H}_2\text{O}} = \frac{6.937 \text{ mol H}_2\text{O}}{6.937 + 0.309} \times 233.7 \text{ torr} = 223.7 = 224 \text{ torr}$$

(b)   Calculate X_B by vapor pressure lowering; $X_B = \Delta P_A / P_A°$. [See Solution 13.65(b).] Given moles solvent, calculate moles solute from the definition of mole fraction.

$$X_{\text{C}_2\text{H}_6\text{O}_2} = \frac{10.0 \text{ torr}}{100 \text{ torr}} = 0.100$$

$$\frac{1.00 \times 10^3 \text{ g C}_2\text{H}_5\text{OH}}{46.07 \text{ g/mol}} = 21.71 = 21.7 \text{ mol C}_2\text{H}_5\text{OH; let } y = \text{mol C}_2\text{H}_6\text{O}_2$$

$$X_{\text{C}_2\text{H}_6\text{O}_2} = \frac{y \text{ mol C}_2\text{H}_6\text{O}_2}{y \text{ mol C}_2\text{H}_6\text{O}_2 + 21.71 \text{ mol C}_2\text{H}_5\text{OH}} = 0.100 = \frac{y}{y + 21.71}$$

$0.100 \, y + 2.171 = y; \, 0.900 \, y = 2.171; \, y = 2.412 = 2.41 \text{ mol C}_2\text{H}_6\text{O}_2$

$$2.412 \text{ mol C}_2\text{H}_6\text{O}_2 \times \frac{62.07 \text{ g}}{1 \text{ mol}} = 150 \text{ g C}_2\text{H}_6\text{O}_2$$

13.67   *Analyze/Plan.* At 63.5 °C, $P°_{\text{H}_2\text{O}} = 175$ torr, $P°_{\text{Eth}} = 400$ torr. Let G = the mass of H_2O and/or C_2H_5OH. *Solve.*

(a) $$X_{\text{Eth}} = \frac{\dfrac{G}{46.07 \text{ g/mol C}_2\text{H}_5\text{OH}}}{\dfrac{G}{46.07 \text{ g/mol C}_2\text{H}_5\text{OH}} + \dfrac{G}{18.02 \text{ g/mol H}_2\text{O}}}$$

Multiplying top and bottom of the right side of the equation by 1/G gives:

$$X_{\text{Eth}} = \frac{1/46.07}{1/46.07 + 1/18.02} = \frac{0.02171}{0.02171 + 0.05549} = 0.2812$$

(b) $P_t = P_{\text{Eth}} + P_{\text{H}_2\text{O}}; \, P_{\text{Eth}} = X_{\text{Eth}} \times P°_{\text{Eth}}; \, P_{\text{H}_2\text{O}} = X_{\text{H}_2\text{O}} P°_{\text{H}_2\text{O}}$

$X_{\text{Eth}} = 0.2812, \, P_{\text{Eth}} = 0.2812 \, (400 \text{ torr}) = 112.48 = 112 \text{ torr}$

$X_{\text{H}_2\text{O}} = 1 - 0.2812 = 0.7188; \quad P_{\text{H}_2\text{O}} = 0.7188(175 \text{ torr}) = 125.8 = 126 \text{ torr}$

$P_t = 112.5 \text{ torr} + 125.8 \text{ torr} = 238.3 = 238 \text{ torr}$

(c) $X_{\text{Eth}} \text{ in vapor} = \dfrac{P_{\text{Eth}}}{P_{\text{total}}} = \dfrac{112.5 \text{ torr}}{238.3 \text{ torr}} = 0.4721 = 0.472$

13.68   (a)   Because C_6H_6 and C_7H_8 form an ideal solution, we can use Raoult's law. Because both components are volatile, both contribute to the total vapor pressure of 35 torr.

$P_t = P_{\text{C}_6\text{H}_6} + P_{\text{C}_7\text{H}_8}; \, P_{\text{C}_6\text{H}_6} = X_{\text{C}_6\text{H}_6} P°_{\text{C}_6\text{H}_6}; \, P_{\text{C}_7\text{H}_8} = X_{\text{C}_7\text{H}_8} P°_{\text{C}_7\text{H}_8}$

$X_{\text{C}_7\text{H}_8} = 1 - X_{\text{C}_6\text{H}_6}; \, P_T = X_{\text{C}_6\text{H}_6} P°_{\text{C}_6\text{H}_6} + (1 - X_{\text{C}_6\text{H}_6}) P°_{\text{C}_7\text{H}_8}$

$35 \text{ torr} = X_{\text{C}_6\text{H}_6} (75 \text{ torr}) + (1 - X_{\text{C}_6\text{H}_6}) 22 \text{ torr}$

$13 \text{ torr} = 53 \text{ torr} \, (X_{\text{C}_6\text{H}_6}); \, X_{\text{C}_6\text{H}_6} = \dfrac{13 \text{ torr}}{53 \text{ torr}} = 0.2453 = 0.25; \, X_{\text{C}_7\text{H}_8} = 0.7547 = 0.75$

(b) $P_{C_6H_6} = 0.2453\,(75\text{ torr}) = 18.4\text{ torr};\quad P_{C_7H_8} = 0.7547\,(22\text{ torr}) = 16.6\text{ torr}$

In the vapor, $X_{C_6H_6} = \dfrac{P_{C_6H_6}}{P_t} = \dfrac{18.4\text{ torr}}{18.4\text{ torr} + 16.6\text{ torr}} = 0.53;\quad X_{C_7H_8} = 0.47$

13.69   (a)   Because NaCl is a soluble ionic compound and a strong electrolyte, there are 2 mol of dissolved particles for every 1 mol of NaCl solute. $C_6H_{12}O_6$ is a molecular solute, so there is 1 mol of dissolved particles per mol solute. Boiling point elevation is directly related to total moles of dissolved particles; 0.10 *m* NaCl has more dissolved particles so its boiling point is higher than 0.10 *m* $C_6H_{12}O_6$.

(b)   In solutions of strong electrolytes like NaCl, electrostatic attractions between ions lead to ion pairing. Ion pairing reduces the effective number of particles in solution, decreasing the **change** in boiling point. The actual boiling point is then lower than the calculated boiling point for a 0.10 *m* solution.

13.70   *Analyze/Plan.* ΔT_b depends on mol dissolved particles. Assume 100 g of each solution, calculate mol solute and mol dissolved particles. Glucose and sucrose are molecular solutes, but $NaNO_3$ dissociates into 2 mol particles per mol solute.   *Solve.*

10% by mass means 10 g solute in 100 g solution. If we have 10 g of each solute, the one with the smallest molar mass will have the largest mol solute. The molar masses are: glucose, 180.2 g/mol; sucrose, 342.3 g/mol; $NaNO_3$, 85.0 g/mol. $NaNO_3$ has most mol solute, and twice as many dissolved particles, so it will have the highest boiling point. Sucrose has least mol solute and lowest boiling point. Glucose is intermediate.

In order of increasing boiling point: 10% sucrose < 10% glucose < 10% $NaNO_3$.

13.71   *Analyze/Plan.* Rank the solutions in order of increasing boiling point. All solutes are nonvolatile. The more nonvolatile solute particles, the higher the boiling point of the solution.   *Solve.*

Solve. Because LiBr and $Zn(NO_3)_2$ are electrolytes, the particle concentrations in these solutions are 0.10 *m* and 0.15 *m*, respectively (although ion-ion attractive forces may decrease the effective concentrations somewhat). Thus, the order of increasing particle concentration and boiling point is:

0.050 *m* LiBr < 0.120 *m* glucose < 0.050 *m* $Zn(NO_3)_2$

13.72   0.030 *m* phenol > 0.040 *m* glycerin = 0.020 *m* KBr. Phenol is very slightly ionized in water, but not enough to match the number of particles in a 0.040 *m* glycerin solution. Assuming the ideal van't Hoff factor of 2.00, the KBr solution is 0.040 *m* in particles, so it has the same freezing point as 0.040 *m* glycerin, which is a nonelectrolyte. (The measured van't Hoff factor for 0.040 *m* KBr will be slightly less than 2. We expect the measured order of freezing points to be 0.030 *m* phenol > 0.020 *m* KBr > 0.040 *m* glycerin.)

13.73   *Analyze/Plan.* $\Delta T = K(m)$; first, calculate the **molality** of each solution.   *Solve.*

(a)   0.22 *m*

(b) $2.45\text{ mol }CHCl_3 \times \dfrac{119.4\text{ g }CHCl_3}{\text{mol }CHCl_3} = 292.53\text{ g} = 0.293\text{ kg};$

$\dfrac{0.240\text{ mol }C_{10}H_8}{0.29253\text{ kg }CHCl_3} = 0.8204 = 0.820\ m$

(c) $1.50 \text{ g NaCl} \times \dfrac{1 \text{ mol NaCl}}{58.44 \text{ g NaCl}} \times \dfrac{2 \text{ mol particles}}{1 \text{ mol NaCl}} = 0.05133 = 0.0513 \text{ mol particles}$

$m = \dfrac{0.05133 \text{ mol NaCl}}{0.250 \text{ kg H}_2\text{O}} = 0.20534 = 0.205 \ m$

(d) $2.04 \text{ g KBr} \times \dfrac{1 \text{ mol KBr}}{119.0 \text{ g KBr}} \times \dfrac{2 \text{ mol particles}}{1 \text{ mol KBr}} = 0.03429 = 0.0343 \text{ mol particles}$

$4.82 \text{ g C}_6\text{H}_{12}\text{O}_6 \times \dfrac{1 \text{ mol C}_6\text{H}_{12}\text{O}_6}{180.2 \text{ g C}_6\text{H}_{12}\text{O}_6} = 0.02675 = 0.0268 \text{ mol particles}$

$m = \dfrac{(0.03429 + 0.02675) \text{ mol particles}}{0.188 \text{ kg H}_2\text{O}} = 0.32465 = 0.325 \ m$

Solve. Then, $\text{fp} = T_f - K_f(m)$; $\text{bp} = T_b + K_b(m)$; T in °C

	m	T_f	$-K_f(m)$	fp	T_b	$+K_b(m)$	bp
(a)	0.22	−114.6	−1.99(0.22) = −0.44	−115.0	78.4	1.22(0.22) = 0.27	78.7
(b)	0.820	−63.5	−4.68(0.820) = −3.84	−67.3	61.2	3.63(0.820) = 2.98	64.2
(c)	0.205	0.0	−1.86(0.205) = −0.381	−0.4	100.0	0.51(0.205) = 0.10	100.1
(d)	0.325	0.0	−1.86(0.325) = −0.605	−0.6	100.0	0.51(0.325) = 0.17	100.2

13.74 $\Delta T = K(m)$; first calculate the **molality** of the solute particles.

(a) $0.25 \ m$

(b) $\dfrac{20.0 \text{ g C}_{10}\text{H}_{22}}{0.0500 \text{ kg CHCl}_3} \times \dfrac{1 \text{ mol C}_{10}\text{H}_{22}}{142.3 \text{ g C}_{10}\text{H}_{22}} = 2.811 = 2.81 \ m$

(c) $3.50 \text{ g NaOH} \times \dfrac{1 \text{ mol NaOH}}{40.00 \text{ g NaOH}} \times \dfrac{2 \text{ mol particles}}{1 \text{ mol NaOH}} = 0.1750 = 0.175 \text{ mol particles}$

$m = \dfrac{0.1750 \text{ mol NaCl}}{0.175 \text{ kg H}_2\text{O}} = 1.000 = 1.00 \ m$

(d) $m = \dfrac{0.45 \text{ mol eg} + 2(0.15) \text{ mol KBr}}{0.150 \text{ kg H}_2\text{O}} = \dfrac{0.75 \text{ mol particles}}{0.150 \text{ kg H}_2\text{O}} = 5.0 \ m$

Then, $\text{fp} = T_f - K_f(m)$; $\text{bp} = T_b + K_b(m)$; T in °C

	m	T_f	$-K_f(m)$	fp	T_b	$+K_b(m)$	bp
(a)	0.25	−114.6	−1.99(0.25) = −0.50	−115.1	78.4	1.22(0.25) = 0.31	78.7
(b)	2.81	−63.5	−4.68(2.81) = −13.2	−76.7	61.2	3.63(2.81) = 10.2	71.4
(c)	1.00	0.0	−1.86(1.00) = −1.86	−1.9	100.0	0.51(1.00) = 0.51	100.5
(d)	5.0	0.0	−1.86(5.0) = −9.3	−9.3	100.0	0.51(5.0) = 2.6	102.6

13.75   *Analyze.* Given freezing point of solution and mass of solvent, calculate mass of solute.

Plan. Use $\Delta T_f = K_f(m)$ to calculate the required molality, and then apply the definition of molality to calculate moles and grams of $C_2H_6O_2$.

Solve. fp of solution = −5.00 °C; fp of solvent (H_2O) = 0.0 °C

$\Delta T_f = 5.00\,^\circ C = K_f(m); \ 5.00\,^\circ C = 1.86\,^\circ C / m(m)$

$$m = \frac{5.00\,^\circ C}{1.86\,^\circ C / m} = 2.688 = 2.69 \ m \ C_2H_6O_2$$

$$m = \frac{\text{mol } C_2H_6O_2}{\text{kg } H_2O} = C_2H_6O_2 = m \times \text{kg } H_2O$$

$2.688 \ m \ C_2H_6O_2 \times 1.00 \ \text{kg } H_2O = 2.688 = 2.69 \ \text{mol } C_2H_6O_2$

$2.688 \ m \ C_2H_6O_2 \times \dfrac{62.07 \ \text{g } C_2H_6O_2}{1 \ \text{mol}} = 166.84 = 167 \ \text{g } C_2H_6O_2$

13.76   Use ΔT_b = find m of aqueous solution, and then use m to calculate ΔT_f and freezing point. $K_b = 0.51$, $K_f = 1.86$.

bp = 105.0 °C; $\Delta T_b = 105.0\,^\circ C - 100.0\,^\circ C = 5.0\,^\circ C$

$$\Delta T_b = K_b(m); \quad m = \frac{\Delta T_b}{K_b} = \frac{5.0\,^\circ C}{0.51} = 9.804 = 9.8 \ m$$

$\Delta T_f = 1.86\,^\circ C / m \times 9.804 \ m = 18.24 = 18\,^\circ C$; freezing point = $0.0\,^\circ C - 18.24\,^\circ C = -18\,^\circ C$

13.77   *Analyze/Plan.* $\Pi = MRT$; $T = 25\,^\circ C + 273 = 298$ K; $M = \text{mol } C_9H_8O_4/L$ soln   *Solve.*

$$M = \frac{44.2 \ \text{mg } C_9H_8O_4}{0.358 \ L} \times \frac{1 \ g}{1000 \ \text{mg}} \times \frac{1 \ \text{mol } C_9H_8O_4}{180.2 \ \text{g } C_9H_8O_4} = 6.851 \times 10^{-4} = 6.85 \times 10^{-4} \ M$$

$$\Pi = \frac{6.851 \times 10^{-4} \ \text{mol}}{L} \times \frac{0.08206 \ \text{L-atm}}{\text{mol-K}} \times 298 \ K = 0.01675 = 0.0168 \ \text{atm} = 12.7 \ \text{torr}$$

13.78 $\Pi = MRT$; $T = 20\,^\circ C + 273 = 293$ K

$$M \ (\text{of ions}) = \frac{\text{mol NaCl} \times 2}{L \ \text{soln}} = \frac{3.4 \ \text{g NaCl}}{1 \ L \ \text{soln}} \times \frac{1 \ \text{mol NaCl}}{58.4 \ \text{g NaCl}} \times \frac{2 \ \text{mol ions}}{1 \ \text{mol NaCl}} = 0.116 = 0.12 \ M$$

$$\Pi = \frac{0.116 \ \text{mol}}{L} \times \frac{0.08206 \ \text{L-atm}}{\text{mol-K}} \times 293 \ K = 2.8 \ \text{atm}$$

13.79   *Analyze/Plan.* Follow the logic in Sample Exercise 13.10 to calculate the molar mass of adrenaline based on the boiling point data. Use the structure to obtain the molecular formula and molar mass. Compare the two values.   *Solve.*

$\Delta T_b = K_b \ m$; $m = \dfrac{\Delta T_b}{K_b} = \dfrac{+0.49}{5.02} = 0.0976 = 0.098 \ m$ adrenaline

$$m = \frac{\text{mol adrenaline}}{\text{kg } CCl_4} = \frac{\text{g adrenaline}}{\text{MM adrenaline} \times \text{kg } CCl_4}$$

$$\text{MM adrenaline} = \frac{\text{g adrenaline}}{m \times \text{kg } CCl_4} = \frac{0.64 \ \text{g adrenaline}}{0.0976 \ m \times 0.0360 \ \text{kg } CCl_4} = 1.8 \times 10^2 \ \text{g/mol adrenaline}$$

Check. The molecular formula is $C_9H_{13}NO_3$, MM = 183 g/mol. The values agree to 2 sig figs, the precision of the experimental value.

13.80 $\Delta T_f = 5.5 - 4.1 = 1.4$; $m = \dfrac{\Delta T_f}{K_f} = \dfrac{1.4}{5.12} = 0.273 = 0.27\ m$

MM lauryl alcohol $= \dfrac{g\ \text{lauryl alcohol}}{m \times \text{kg}\ C_6H_6} = \dfrac{5.00\ g\ \text{lauryl alcohol}}{0.273 \times 0.100\ \text{kg}\ C_6H_6}$

$= 1.8 \times 10^2\ g/\text{mol lauryl alcohol}$

13.81   *Analyze/Plan.* Follow the logic in Sample Exercise 13.11.   *Solve.*

$\Pi = MRT$; $M = \dfrac{\Pi}{RT}$; $T = 25\ ^{\circ}C + 273 = 298\ K$

$M = 0.953\ \text{torr} \times \dfrac{1\ \text{atm}}{760\ \text{torr}} \times \dfrac{\text{mol-K}}{0.08206\ \text{L-atm}} \times \dfrac{1}{298\ K} = 5.128 \times 10^{-5} = 5.13 \times 10^{-5}\ M$

$\text{mol} = M \times L = 5.128 \times 10^{-5} \times 0.210\ L = 1.077 \times 10^{-5} = 1.08 \times 10^{-5}\ \text{mol lysozyme}$

$MM = \dfrac{g}{\text{mol}} = \dfrac{0.150\ g}{1.077 \times 10^{-5}\ \text{mol}} = 1.39 \times 10^4\ g/\text{mol lysozyme}$

13.82 $M = P/RT = \dfrac{0.605\ \text{atm}}{298\ K} \times \dfrac{\text{mol-K}}{0.08206\ \text{L-atm}} = 0.02474 = 0.0247\ M$

$MM = \dfrac{g}{M \times L} = \dfrac{2.35\ g}{0.02474\ M \times 0.250\ L} = 380\ g/\text{mol}$

13.83   *Analyze/Plan.* $i = \Pi$ (measured) $/ \Pi$ (calculated for a nonelectrolyte);

Π (calculated) $= MRT$.   *Solve.*

Π (calculated) $= \dfrac{0.010\ \text{mol}}{L} \times \dfrac{0.08206\ \text{L-atm}}{\text{mol-K}} \times 298\ K = 0.2445 = 0.24\ \text{atm}$

$i = 0.674\ \text{atm}/0.2445\ \text{atm} = 2.756 = 2.8$

13.84   If these were ideal solutions, they would have equal ion concentrations and equal ΔT_f values. Data in Table 13.4 indicates that the van't Hoff factors (i) for both salts are less than the ideal values. For 0.030 m NaCl, i is between 1.87 and 1.94, about 1.92. For 0.020 m K_2SO_4, i is between 2.32 and 2.70, about 2.62. From Equation 13.15,

ΔT_f (measured) $= i \times \Delta T_f$ (calculated for nonelectrolyte)

NaCl: ΔT_f (measured) $= 1.92 \times 0.030\ m \times 1.86\ ^{\circ}C/m = 0.11\ ^{\circ}C$

K_2SO_4: ΔT_f (measured) $= 2.62 \times 0.020\ m \times 1.86\ ^{\circ}C/m = 0.097\ ^{\circ}C$

0.030 m NaCl would have the larger ΔT_f.

(The deviations from ideal behavior are due to ion pairing in the two electrolyte solutions. K_2SO_4 has more extensive ionpairing and a larger deviation from ideality because of the higher charge on SO_4^{2-} relative to Cl^-.)

Colloids (Section 13.6)

13.85   (a)   No. In the gaseous state, the particles are far apart and intermolecular attractive forces are small. When two gases combine, all terms in Equation 13.1 are essentially zero and the mixture is always homogeneous.

(b)   The outline of a light beam passing through a colloid is visible, whereas light passing through a true solution is invisible unless collected on a screen. This is the Tyndall effect. To determine whether Faraday's (or anyone's) apparently homogeneous dispersion is a true solution or a colloid, shine a beam of light on it and see if the light is scattered.

13.86   The best answer is (b) emulsion (Table 13.5).

13.87   The best emulsifying agent is (d) $CH_3(CH_2)_{11}COONa$. A good emulsifying agent has a polar (or ionic) end to interact with hydrophilic substances, and an nonpolar end to interact with hydrophobic substances. Choices (c) and (d) fit this description, but the ionic end of (d) will help stabilize the colloid.

13.88   The presence of aerosols in the atmosphere decreases the amount of sunlight that arrives at Earth's surface, compared to an "aerosol-free" atmosphere. All colloids scatter light (the Tyndall effect). Aerosols in the atmosphere scatter the incoming sunlight. Although some of this scattered light will eventually reach Earth, some will not.

13.89   (a)   No. Adsorbed ions stabilize hydrophobic colloids in water. The hydrophobic/hydrophilic nature of the protein will determine which electrolyte at which concentration will be the most effective precipitating salt.

        (b)   Stronger. If a protein has been "salted out," protein-protein interactions are sufficiently strong so that the protein molecules "stick together" and form a solid. Before the electrolyte is added, protein-protein interactions are weaker than protein-water interactions and the protein molecules remained suspended in solution.

        (c)   The first hypothesis seems plausible, because ion-dipole interactions among electrolytes and water molecules are stronger than dipole-dipole and hydrogen-bonding interactions between water and protein molecules. However, this ignores the strength of ion-dipole interactions between the electrolyte and protein molecules. And, we know from Figure 13.27 that ions are adsorbed on the surface of hydrophobic colloids. With the right protein and electrolyte, the second hypothesis also seems plausible.

                The van't Hoff effect is a result of ion pairing. We know from Table 13.4 that the effect of ion pairing increases with concentration and the charge on the ions of the electrolyte. If we could measure the charge and adsorbed water content of protein molecules as a function of salt concentration, then we could distinguish between these two hypotheses.

13.90   (a)   Head.

        (b)   Tail.

        (c)   The charged $-COO^-$ head will experience ion-dipole, dipole-dipole, and hydrogen bonding interactions with water. The hydrocarbon tail of sodium stearate will experience dispersion forces with hydrophobic grease.

Additional Exercises

13.91 (a) Hydrochloride. The hydrochloride is a salt, an ionic compound, with the possibility of ion-dipole interactions in addition to hydrogen bonding and dipole-dipole interactions. Of the two forms, it will be more soluble in water.

(b) Free base. Both forms have several hydrogen bond receptors (O and N atoms with nonbonded electron pairs), but the free base is less soluble because it does not have the possibility of ion-dipole interactions.

(c) *Analyze/Plan.* Calculate the molar mass of the free base, then moles and molarity. mol = g/molar mass; M = mol/L. *Solve.*

The molar mass of the free base is 303.353 g/mol. 6.70 mL = 0.00670 L ethanol

$$1.00 \text{ g free base} \times \frac{1 \text{ mol free base}}{303.353 \text{ g free base}} \times \frac{1}{0.00670 \text{ L ethanol}} = 0.492 \, M \text{ free base}$$

(d) Use the method from part (c) to calculate molarity of the hydrochloride.

The molar mass of the hydrochloride is 339.814 g/mol. 0.400 mL = 0.000400 L

$$1.00 \text{ g hydrochloride} \times \frac{1 \text{ mol hydrochloride}}{339.814 \text{ g free base}} \times \frac{1}{0.000400 \text{ L water}} = 7.36 \, M \text{ hydrochloride}$$

(e) *Analyze/Plan.* According to the chemical reaction given in the exercise, the free base reacts in a 1:1 mole ratio with HCl(aq). Calculate moles of free base in 1.00 kg and then liters of 12.0 M HCl(aq) required. 1.00 kg = 1.00×10^3 g. *Solve.*

$$1.00 \times 10^3 \text{ g free base} \times \frac{1 \text{ mol free base}}{303.353 \text{ g free base}} \times \frac{1 \text{ L}}{12.0 \text{ mol HCl}} = 0.275 \text{ L} = 275 \text{ mL}$$

13.92 (a) True

(b) False. A saturated solution in contact with undissolved solute exists in a state of dynamic equilibrium. Both dissolving and crystallization occur simultaneously and at the same rates.

(c) True

13.93 Assume that the density of the solution is 1.00 g/mL.

(a) $$4 \text{ ppm O}_2 = \frac{4 \text{ mg O}_2}{1 \text{ kg soln}} = \frac{4 \times 10^{-3} \text{ g O}_2}{1 \text{ L soln}} \times \frac{1 \text{ mol O}_2}{32.0 \text{ g O}_2} = 1.25 \times 10^{-4} = 1 \times 10^{-4} \, M$$

(b) $$S_{O_2} = kP_{O_2}; \; P_{O_2} = S_{O_2}/k = \frac{1.25 \times 10^{-4} \text{ mol}}{L} \times \frac{\text{L-atm}}{1.71 \times 10^{-3} \text{ mol}} = 0.0731 = 0.07 \text{ atm}$$

$$0.0731 \text{ atm} \times \frac{760 \text{ torr}}{1 \text{ atm}} = 55.6 = 60 \text{ torr}$$

13.94 (a) $$S_{Rn} = kP_{Rn}; \; k = S_{Rn}/P_{Rn} = 7.27 \times 10^{-3} \, M/1 \text{ atm} = 7.27 \times 10^{-3} \text{ mol/L-atm}$$

(b) $$P_{Rn} = \chi_{Rn}P_{total}; \; P_{Rn} = 3.5 \times 10^{-6}(32 \text{ atm}) = 1.12 \times 10^{-4} = 1.1 \times 10^{-4} \text{ atm}$$

$$S_{Rn} = k \, P_{Rn}; \; S_{Rn} = \frac{7.27 \times 10^{-3} \text{ mol}}{\text{L-atm}} \times 1.12 \times 10^{-4} \text{ atm} = 8.1 \times 10^{-7} \, M$$

13.95   0.10% by mass means 0.10 g glucose/100 g blood.

      (a)   ppm glucose $= \dfrac{\text{g glucose}}{\text{g solution}} \times 10^6 = \dfrac{0.10 \text{ g glucose}}{100 \text{ g blood}} \times 10^6 = 1000$ ppm glucose

      (b) m = mol glucose/kg solvent. Assume that the mixture of nonglucose components is the "solvent."

            mass solvent = 100 g blood – 0.10 g glucose = 99.9 g solvent = 0.0999 kg solvent

            mol glucose $= 0.10 \text{ g} \times \dfrac{1 \text{ mol}}{180.2 \text{ g } C_6H_{12}O_6} = 5.55 \times 10^{-4} = 5.6 \times 10^{-4}$ mol glucose

            $m = \dfrac{5.55 \times 10^{-4} \text{ mol glucose}}{0.0999 \text{ kg solvent}} = 5.6 \times 10^{-3}$ m glucose

      (c)   To calculate molarity, solution volume must be known. The density of blood is needed to relate mass and volume.

13.96   *Analyze.* Given 13 ppt Au in seawater, find grams of Au in 1.0×10^3 gal seawater. The definition of ppt is (mass solute/mass solution) $\times 10^{12}$. *Plan.* Assume seawater is a dilute aqueous solution with a density of 1.00 g/mL. Use the definition of ppt to calculate g Au. *Solve.*

            $\dfrac{13 \text{ g Au}}{1 \times 10^{12} \text{ g soln}} \times \dfrac{1.0 \text{ g soln}}{\text{mL soln}} \times \dfrac{1000 \text{ mL}}{1 \text{ L}} \times \dfrac{3.7854 \text{ L}}{\text{gal}} \times 1.0 \times 10^3 \text{ gal soln} = 4.9 \times 10^{-5}$ g Au

13.97   *Analyze.* The definition of ppb is (mass solute/mass solution) $\times 10^9$. *Plan.* Use the definition to get g Pb and g solution. Change g Pb to mol Pb, g solution to L solution, calculate molarity. *Solve.*

      (a) $9.0 \text{ ppb} = \dfrac{9.0 \text{ g Pb}}{1 \times 10^9 \text{ g soln}} \times 10^9$

            For dilute aqueous solutions (drinking water) assume that the density of the solution is the density of H_2O.

            $\dfrac{9.0 \text{ g Pb}}{1 \times 10^9 \text{ g soln}} \times \dfrac{1.0 \text{ g soln}}{\text{mL soln}} \times \dfrac{1000 \text{ mL}}{1 \text{ L}} \times \dfrac{1 \text{ mol Pb}}{207.2 \text{ g Pb}} = 4.34 \times 10^{-8}$ $M = 4.3 \times 10^{-8}$ M

      (b)   Change 60 m^3 H_2O to cm^3 (mL) H_2O to g H_2O (or g soln).

            $60 \text{ m}^3 \times \dfrac{100^3 \text{ cm}^3}{\text{m}^3} \times \dfrac{1 \text{ g } H_2O}{\text{cm}^3 H_2O} = 6.0 \times 10^7$ g H_2O or soln

            $\dfrac{9.0 \text{ g Pb}}{1 \times 10^9 \text{ g soln}} \times 6.0 \times 10^7 \text{ g soln} = 0.54$ g Pb

13.98   (a) $\dfrac{1.80 \text{ mol LiBr}}{1 \text{ L soln}} \times \dfrac{86.85 \text{ g LiBr}}{1 \text{ mol LiBr}} = 156.3 = 156$ g LiBr

            1 L soln = 826 g soln; g CH_3CN = 826 – 156.3 = 669.7 = 670 g CH_3CN

            m LiBr $= \dfrac{1.80 \text{ mol LiBr}}{0.6697 \text{ kg } CH_3CN} = 2.69$ m

      (b) $\dfrac{669.7 \text{ g } CH_3CN}{41.05 \text{ g/mol}} = 16.31 = 16.3$ mol CH_3CN; $X_{\text{LiBr}} = \dfrac{1.80}{1.80 + 16.31} = 0.0994$

      (c)   mass % $= \dfrac{669.7 \text{ g } CH_3CN}{826 \text{ g soln}} \times 100 = 81.1\%$ CH_3CN

13.99   Mole fraction ethyl alcohol, $X_{C_2H_5OH} = \dfrac{P_{C_2H_5OH}}{P^{\circ}_{C_2H_5OH}} = \dfrac{8 \text{ torr}}{100 \text{ torr}} = 0.08$

$\dfrac{620 \times 10^3 \text{ g } C_{24}H_{50}}{338.6 \text{ g/mol}} = 1.83 \times 10^3 \text{ mol } C_{24}H_{50}; \quad \text{let } y = \text{mol } C_2H_5OH$

$X_{C_2H_5OH} = 0.08 = \dfrac{y}{y + 1.83 \times 10^3}; \quad 0.92 \, y = 146.4; \quad y = 1.6 \times 10^2 \text{ mol } C_2H_5OH$

(Strictly speaking, y should have 1 sig fig because 0.08 has 1 sig fig, but this severely limits the calculation.)

$1.6 \times 10^2 \text{ mol } C_2H_5OH \times \dfrac{46 \text{ g } C_2H_5OH}{1 \text{ mol}} = 7.4 \times 10^3 \text{ g or } 7.4 \text{ kg } C_2H_5OH$

13.100   *Analyze.* Given vapor pressure of both pure water and the aqueous solution and moles H_2O find moles of solute in the solution.

Plan. Use vapor pressure lowering, $P_A = X_A P^{\circ}_A$, to calculate X_A, mole fraction solvent, and then use the definition of mole fraction to calculate moles solute particles. Because NaCl is a strong electrolyte, there is 1 mol NaCl for every 2 mol solute particles.

Solve.

$X_{H_2O} = P_{soln}/P_{H_2O} = 25.7/31.8 = 0.80818 = 0.808$

$X_{H_2O} = \dfrac{\text{mol } H_2O}{\text{mol ions} + \text{mol } H_2O}; \quad 0.80818 = \dfrac{0.115}{(\text{mol ions} + 0.115)}$

$0.80818 \, (0.115 + \text{mol ions}) = 0.115; \, 0.80818 \, (\text{mol ions}) = 0.115 - 0.092940$

mol ions $= 0.02206/0.80818 = 0.02730 = 0.0273;$

mol NaCl $=$ mol ions$/2 = 0.02730/2 = 0.01365 = 0.0137$ mol NaCl

$0.01365 \text{ mol NaCl} \times \dfrac{58.443 \text{ g } C_2H_5OH}{1 \text{ mol}} = 0.7977 = 0.798 \text{ g NaCl}$

13.101   (a)   The solvent vapor pressure over each solution is determined by the total particle concentrations present in the solutions. When the particle concentrations are equal, the vapor pressures will be equal and equilibrium established. The particle concentration of the nonelectrolyte is just 0.050 *M*, the ion concentration of the NaCl is 2 × 0.035 *M* = 0.070 *M*. Solvent will diffuse from the less concentrated nonelectrolyte solution. The level of the NaCl solution will rise, and the level of the nonelectrolyte solution will fall.

        (b)   Let x = volume of solvent transferred

$\dfrac{0.050 \, M \times 30.0 \text{ mL}}{(30.0 - x) \text{ mL}} = \dfrac{0.070 \, M \times 30.0 \text{ mL}}{(30.0 + x) \text{ mL}}; \quad 1.5(30.0 + x) = 2.1(30.0 - x)$

$45 + 1.5 \, x = 63 - 2.1 \, x; \; 3.6 \, x = 18; \; x = 5.0 = 5 \text{ mL transferred}$

The volume in the nonelectrolyte beaker is (30.0 − 5.0) = 25.0 mL; in the NaCl beaker (30.0 + 5.0) = 35.0 mL.

13.102   *Analyze/Plan.*   A nonelectrolyte is dissolved in ethanol, C_2H_5OH, at its normal boiling point, 78.4 °C. At the normal boiling point, the vapor pressure of a liquid equals 1 atm or 760 torr. The resulting solution is not boiling because the solute has raised the boiling point above 78.4 °C. The vapor pressure of the solution at this temperature and pressure is 7.40×10^2 or 740 torr. Use Raoult's law, Equation 13.10, to calculate the mole fraction of ethanol in the solution. Use the definition of mole fraction to calculate the molar mass of the solute.   *Solve.*

$$P_{solution} = X_{solvent}P^{\circ}_{solvent}; \quad X_{ethanol} = P_{solution}/P^{\circ}_{solvent}$$

$$X_{ethanol} = 740 \text{ torr}/760 \text{ torr} = 0.97368 = 0.974$$

mol ethanol = 100.0 g ethanol / 46.068 g/mol = 2.1707 = 2.171 mol

$$X_{Ethanol} = \frac{\text{mol ethanol}}{\text{mol ethanol} + \text{mol solute}} = \frac{2.1707 \text{ mol ethanol}}{2.1707 \text{ mol ethanol} + \dfrac{9.15 \text{ g solute}}{\text{molar mass solute}}}$$

Multiply top and bottom by molar mass solute (MM), then multiply both sides by (2.1707MM + 9.15).

$$0.97368(2.1707\text{MM} + 9.15) = 2.1707\text{MM}$$

$$8.9092 = (2.1707 - 2.1136)\text{MM}; \quad \text{MM} = 156.03 = 156 \text{ g/mol}$$

13.103   (a)   0.100 *m* K_2SO_4 is 0.300 *m* in particles. H_2O is the solvent.

$$\Delta T_f = K_f m = -1.86(0.300) = -0.558; \quad T_f = 0.0 - 0.558 = -0.558 \text{ °C} = -0.6 \text{ °C}$$

(b) ΔT_f (nonelectrolyte) $= -1.86(0.100) = -0.186; \quad T_f = 0.0 - 0.186 = -0.186 \text{ °C} = -0.2 \text{ °C}$

T_f (measured) $= i \times T_f$ (nonelectrolyte)

From Table 13.4, *i* for 0.100 *m* K_2SO_4 = 2.32

T_f (measured) $= 2.32(-0.186 \text{ °C}) = -0.432 \text{ °C} = -0.4 \text{ °C}$

13.104   (a) $K_b = \dfrac{\Delta T_b}{m}; \quad \Delta T_b = 47.46 \text{ °C} - 46.30 \text{ °C} = 1.16 \text{ °C}$

$$m = \frac{\text{mol solute}}{\text{kg CS}_2} = \frac{0.250 \text{ mol}}{400.0 \text{ mL CS}_2} \times \frac{1 \text{ mL CS}_2}{1.261 \text{ g CS}_2} \times \frac{1000 \text{ g}}{1 \text{ kg}} = 0.4956 = 0.496 \text{ } m$$

$$K_b = \frac{1.16 \text{ °C}}{0.4956 \text{ } m} = 2.34 \text{ °C}/m$$

(b) $m = \dfrac{\Delta T_b}{K_b} = \dfrac{(47.08 - 46.30) \text{ °C}}{2.34 \text{ °C}/m} = 0.333 = 0.33 \text{ } m$

$$m = \frac{\text{mol unknown}}{\text{kg CS}_2}; \quad m \times \text{kg CS}_2 = \frac{\text{g unknown}}{\text{MM unknown}}; \quad \text{MM} = \frac{\text{g unknown}}{m \times \text{kg CS}_2}$$

$$50.0 \text{ mL CS}_2 \times \frac{1.261 \text{ g CS}_2}{1 \text{ mL}} \times \frac{1 \text{ kg}}{1000 \text{ g}} = 0.06305 = 0.0631 \text{ kg CS}_2$$

$$\text{MM} = \frac{5.39 \text{ g unknown}}{0.333 \text{ m} \times 0.06305 \text{ kg CS}_2} = 257 = 2.6 \times 10^2 \text{ g/mol}$$

13.105 $M = \dfrac{\Pi}{RT} = \dfrac{57.1 \text{ torr}}{298 \text{ K}} \times \dfrac{1 \text{ atm}}{760 \text{ torr}} \times \dfrac{\text{mol-K}}{0.08206 \text{ L-atm}} = 3.072 \times 10^{-3} = 3.07 \times 10^{-3} \ M$

$\dfrac{0.036 \text{ g solute}}{100 \text{ g H}_2\text{O}} \times \dfrac{1000 \text{ g H}_2\text{O}}{1 \text{ kg H}_2\text{O}} = 0.36 \text{ g solute/kg H}_2\text{O}$

Assuming molarity and molality are the same in this dilute solution, we can then say 0.36 g solute = 3.072×10^{-3} mol; MM = 117 g/mol. Because the salt is completely ionized, the formula weight of the lithium salt is **twice** this calculated value, or **234 g/mol**. The organic portion, $C_nH_{2n+1}O_2^-$, has a formula weight of 234 – 7 = 227 g. Subtracting 32 for the oxygens, and 1 to make the formula C_nH_{2n}, we have C_nH_{2n}, MM = 194 g/mol. Because each CH_2 unit has a mass of 14, n ≈ 194/14 ≈ 14. The formula for our salt is $LiC_{14}H_{29}O_2$.

Integrative Exercises

13.106   Because these are very dilute solutions, assume that the density of the solution ≈ the density of H_2O ≈ 1.0 g/mL at 25 °C. Then, 100 g solution = 100 g H_2O = 0.100 kg H_2O.

     (a) CF_4 : $\dfrac{0.0015 \text{ g CF}_4}{0.100 \text{ kg H}_2\text{O}} \times \dfrac{1 \text{ mol CF}_4}{88.00 \text{ g CF}_4} = 1.7 \times 10^{-4} \ m$

            $CClF_3$: $\dfrac{0.009 \text{ g CClF}_3}{0.100 \text{ kg H}_2\text{O}} \times \dfrac{1 \text{ mol CClF}_3}{104.46 \text{ g CClF}_3} = 8.6 \times 10^{-4} \ m = 9 \times 10^{-4} \ m$

            CCl_2F_2 : $\dfrac{0.028 \text{ g CCl}_2\text{F}_2}{0.100 \text{ kg H}_2\text{O}} \times \dfrac{1 \text{ mol CCl}_2\text{F}_2}{120.9 \text{ g CCl}_2\text{F}_2} = 2.3 \times 10^{-3} \ m$

            $CHClF_2$: $\dfrac{0.30 \text{ g CHClF}_2}{0.100 \text{ kg H}_2\text{O}} \times \dfrac{1 \text{ mol CHClF}_2}{86.47 \text{ g CHClF}_2} = 3.5 \times 10^{-2} \ m$

     (b)   Dipole moment. CCl_2F_2 has the largest molar mass, but it is not the most soluble. None of the molecules are capable of hydrogen bonding with water; F atoms bound to C are not hydrogen bond acceptors and H bound to C is not a hydrogen bond donor.

     (c)   Air is 21% O_2 by volume. The volume of $O_2(g)$ in a baby's lungs is

            0.21(15 mL) = 3.15 = 3.2 mL = 0.0032 L

            Assume air pressure in the lungs is 1 atm and body temperature is 37 °C or 310 K.

            $n = \dfrac{PV}{RT} = 1 \text{ atm} \times \dfrac{\text{mol-K}}{0.08206 \text{ L-atm}} \times \dfrac{0.00315 \text{ L}}{310 \text{ K}} = 1.238 \times 10^{-4} = 1.2 \times 10^{-4} \text{ mol O}_2$

            A volume of 66 mL of $O_2(g)$ dissolves in 100 mL of the fluorinated liquid. That is 66% O_2 by volume. [(66/100)100 = 66]

            In a 15 mL volume of the liquid, the volume of $O_2(g)$ is

            0.66(15 mL) = 9.90 = 9.9 mL = 0.0099 L

            $n = \dfrac{PV}{RT} = 1 \text{ atm} \times \dfrac{\text{mol-K}}{0.08206 \text{ L-atm}} \times \dfrac{0.00990 \text{ L}}{310 \text{ K}} = 3.892 \times 10^{-4} = 3.9 \times 10^{-4} \text{ mol O}_2$

13.107   (a) $\dfrac{0.015 \text{ g N}_2}{1 \text{ L blood}} \times \dfrac{1 \text{ mol N}_2}{28.01 \text{ g N}_2} = 5.355 \times 10^{-4} = 5.4 \times 10^{-4} \text{ mol N}_2/\text{L blood}$

       (b)   At 100 ft, the partial pressure of N_2 in air is 0.78 (4.0 atm) = 3.12 atm. This is just four times the partial pressure of N_2 at 1.0 atm air pressure. According to Henry's law, $S_g = kP_g$, a fourfold increase in P_g results in a fourfold increase in S_g, the solubility of the gas. Thus, the solubility of N_2 at 100 ft is $4(5.355 \times 10^{-4} M) = 2.142 \times 10^{-3} = 2.1 \times 10^{-3} M$.

       (c)   If the diver suddenly surfaces, the amount of N_2/L blood released is the difference in the solubilities at the two depths:

           $(2.142 \times 10^{-3} \text{ mol/L} - 5.355 \times 10^{-4} \text{ mol/L}) = 1.607 \times 10^{-3} = 1.6 \times 10^{-3} \text{ mol N}_2/\text{L blood.}$

           At surface conditions of 1.0 atm external pressure and 37 °C = 310 K,

           $V = \dfrac{nRT}{P} = 1.607 \times 10^{-3} \text{ mol} \times \dfrac{310 \text{ K}}{1.0 \text{ atm}} \times \dfrac{0.08206 \text{ L-atm}}{\text{mol-K}} = 0.041 \text{ L}$

           That is, 41 mL of tiny N_2 bubbles are released from each liter of blood.

13.108   The stronger the intermolecular forces, the higher the heat (enthalpy) of vaporization.

       (a)   None of the substances are capable of hydrogen bonding in the pure liquid, and they have similar molar masses. All intermolecular forces are van der Waals forces, dipole-dipole, and dispersion forces. In decreasing order of strength of forces: acetone > acetaldehyde > ethylene oxide > cyclopropane

           The first three compounds have dipole-dipole and dispersion forces, the last only dispersion forces.

       (b)   The order of solubility in hexane should be the reverse of the order above. The least polar substance, cyclopropane, will be most soluble in hexane. Ethanol, CH_3CH_2OH, is capable of hydrogen bonding with the three polar compounds. Thus, acetaldehyde, acetone, and ethylene oxide should be more soluble than cyclopropane, but without further information we cannot distinguish among the polar molecules.

13.109   (a)   The central atom and the number of electron-pair domains about it are: (i) Cl, 4; (ii) B, 4; (iii) P, 6; (iv) Al, 4; (v) B, 4

       (b)   The electron-domain geometry around B in BARF is tetrahedral.

       (c)   The central P atom in anion (iii) has an expanded octet. As drawn, the central Cl atom in anion (i) also has an expanded octet. Note that multiple resonance structures for ClO_4^- can be drawn, including one where Cl obeys the octet rule. The structure shown in this exercise is the one that minimizes formal charge.

       (d)   BARF is the largest anion; it will have the strongest dispersion forces that promote solubility in nonpolar solvents.

13.110   (a) $Zn(s) + H_2SO_4(aq) \rightarrow ZnSO_4(aq) + H_2(g)$

$$2.050 \text{ g Zn} \times \frac{1 \text{ mol Zn}}{65.39 \text{ g Zn}} = 0.03135 \text{ mol Zn}$$

$$1.00 \text{ } M \text{ H}_2\text{SO}_4 \times 0.0150 \text{ L} = 0.0150 \text{ mol H}_2\text{SO}_4$$

Because Zn and H_2SO_4 react in a 1:1 mole ratio, H_2SO_4 is the limiting reactant; 0.0150 mol of $H_2(g)$ are produced.

   (b) $P = \dfrac{nRT}{V} = \dfrac{0.0150 \text{ mol}}{0.122 \text{ L}} \times \dfrac{0.08206 \text{ L-atm}}{\text{mol-K}} \times 298 \text{ K} = 3.0066 = 3.01 \text{ atm}$

   (c) $S_{H_2} = kP_{H_2} = \dfrac{7.8 \times 10^{-4} \text{ mol}}{\text{L-atm}} \times 3.0066 \text{ atm} = 0.002345 = 2.3 \times 10^{-3} \text{ } M$

$$\frac{0.002345 \text{ mol H}_2}{\text{L soln}} \times 0.0150 \text{ L} = 3.518 \times 10^{-5} = 3.5 \times 10^{-5} \text{ mol dissolved H}_2$$

$$\frac{3.5 \times 10^{-5} \text{ mol dissolved H}_2}{0.0150 \text{ mol H}_2 \text{ produced}} \times 100 = 0.23\% \text{ dissolved H}_2$$

This is approximately 2.3 parts per thousand; for every 10,000 H_2 molecules, 23 are dissolved. It was reasonable to ignore dissolved $H_2(g)$ in part (b).

13.111   (a) $\dfrac{1.3 \times 10^{-3} \text{ mol CH}_4}{\text{L soln}} \times 4.0 \text{ L} = 5.2 \times 10^{-3} \text{ mol CH}_4$

$$V = \frac{nRT}{P} = \frac{5.2 \times 10^{-3} \text{ mol} \times 298 \text{ K}}{1.0 \text{ atm}} \times \frac{0.08206 \text{ L-atm}}{\text{mol-K}} = 0.13 \text{ L}$$

   (b)   No. The three hydrocarbons are all nonpolar. The order of increasing water solubility is the order of increasing polarizability.

   (c)   Hydrocarbons can experience only dispersion forces with water.

   (d)

methane   ethane   ethylene

Ethylene is the only one of the hydrocarbons that possesses a π bond. That ethylene is more soluble in water than ethane means that is has stronger dispersion interactions with water and that the π cloud is more polarizable. The π cloud is outside the molecular framework and the σ cloud is inside. The π cloud is more easily deformed by approaching molecules, making it more polarizable than the σ cloud. In common vernacular, the π cloud is mushy and sticks out, exposing it to the attack of surrounding water molecules.

   (e)   NO is most soluble because it is polar. It is more soluble in water than O_2 or N_2 because it has a dipole moment and has dipole-dipole interactions with water. The three molecules have similar molar masses and their dispersion forces with water are similar.

   (f) H_2S has dipole-dipole and dispersion forces with water. While H–S bonds technically do not qualify for hydrogen bonding, dipole-dipole interactions with water are strong enough so that a small amount of H–S bond dissociation does occur. H_2S is weakly acidic in water. This encourages the water solubility of H_2S.

(g) SO_2 has dispersion and dipole-dipole interactions with water. In fact, the dipole-dipole forces are strong enough so that SO_2 reacts with water to form H_2SO_3, a weak acid. The large solubility of SO_2 is a sure sign that a chemical process has occurred.

13.112   The resulting solution is very dilute, so assume ideal behavior. Assume the amount of water consumed in the reaction is negligible. Ignore the solubility of $H_2(g)$ in the solution (see Solution 3.110).

$$1.0 \text{ mm}^3 \times \frac{0.535 \text{ g}}{\text{cm}^3} \times \frac{1^3 \text{ cm}^3}{10^3 \text{ mm}^3} = 5.35 \times 10^{-4} = 5.4 \times 10^{-4} \text{ g Li}$$

$$5.35 \times 10^{-4} \text{ g Li} \times \frac{1 \text{ mol Li}}{6.941 \text{ g Li}} = 7.708 \times 10^{-5} = 7.7 \times 10^{-5} \text{ mol Li}$$

mol Li = mol LiOH; 2 mol ions per mol LiOH

7.708×10^{-5} mol Li = 7.708×10^{-5} mol LiOH = 1.542×10^{-4} mol ions = 1.5×10^{-4} mol ions

$$m = \frac{1.542 \times 10^{-4} \text{ mol ions}}{0.500 \text{ L } H_2O} \times \frac{1 \text{ L}}{1000 \text{ mL}} \times \frac{1 \text{ mL } H_2O}{0.997 \text{ g } H_2O} \times \frac{1000 \text{ g}}{1 \text{ kg}} = 3.092 \times 10^{-4} = 3.1 \times 10^{-4} \, m$$

$\Delta T_f = K_f\, m = -1.86(3.092 \times 10^{-4}) = -5.8 \times 10^{-4} \text{ °C}$; $T_f = 0.00000 - 0.00058 = -0.00058$ °C

The freezing point of the LiOH(aq) solution is essentially zero.

13.113 $X_{CHCl_3} = X_{C_3H_6O} = 0.500$

(a)   For an ideal solution, Raoult's law is obeyed.

$P_t = P_{CHCl_3} + P_{C_3H_6O}$; $P_{CHCl_3} = 0.5(300 \text{ torr}) = 150$ torr

$P_{C_3H_6O} = 0.5(360 \text{ torr}) = 180$ torr; $P_t = 150$ torr + 180 torr = 330 torr

(b)   The mixing of the two liquids is exothermic. According to Coulomb's law, electrostatic attractive forces lead to an overall lowering of the energy of the system. Thus, when the two liquids mix and hydrogen bonds are formed, the energy of the system is decreased and $\Delta H_{soln} < 0$.

13.114   (a)   True solutions do not scatter light, colloids do. Below the critical micelle concentration, cmc, the mixture of solvent and surfactant is a true solution. Above the cmc, the mixture is a colloid. The micelles are too large to be perfectly mixed in the solvent. They are suspended in the solvent, resulting in a colloid that scatters light.

(b)   Surfactant monomers are anions; the "head" carries a negative charge. Below the cmc, each monomer is an independent particle. Above the cmc, many monomers aggregate into one micelle, drastically reducing the effective number of particles "in solution." (A micelle does have a greater negative charge than a monomer.) This dramatically changes the ionic conductivity.

(c)   The interior of a micelle is a hydrophobic environment. If a dye molecule becomes entrapped in a micelle, it will fluoresce. In the absence of micelles, the dye will not fluoresce in an aqueous solution.

At low sodium stearate concentrations, the fluorescent intensity will be low. As the surfactant concentration increases, fluorescent intensity increases gradually until the cmc is reached (assuming a few micelles form at concentrations below cmc). At the cmc, there is a large increase in fluorescent intensity. At concentrations greater than the cmc, fluorescent intensity remains high and probably increases gradually.

14 Chemical Kinetics

Visualizing Concepts

14.1   *Analyze/Plan.* Consider the chemical reaction that occurs in the cylinders of an automobile engine. How are the droplets related to the reaction, and how does droplet size affect the rate of the reaction?

Solve. The reaction occurring in the cylinder is the combustion of gasoline. Gasoline is injected into the cylinders in the form of a spray, as shown in the photos. This is a heterogeneous reaction, because gasoline is a liquid and oxygen (from air) is a gas. The rate of a heterogeneous reaction depends on the surface area of the liquid or solid reactant, in this case, the surface area of the droplets in the spray. The smaller the droplets, the greater the surface area exposed to oxygen, the faster the combustion reaction.

In the case of a clogged injector, larger droplets lead to slower combustion. Uneven combustion in the various cylinders can cause the engine to run roughly and decrease fuel economy.

14.2   *Analyze/Plan.* Given the plot of [X] vs time, answer questions about reaction speed and rate. Consider the definitions of average reaction rate and instantaneous rate.   *Solve.*

(a)   True. X is a product, because its concentration increases with time.

(b)   False. The speed of a reaction is its rate, or how quickly the concentration of a reactant or product changes over time. This graph shows how [X] increases over time. The rate at any particular time, the instantaneous rate, is the slope of the tangent to the curve at that time. Visualizing the tangents at points 0, 1, 2, and 3, we see that the slopes of these lines are decreasing with time. That is, the rate of reaction is decreasing; the reaction is slowing down as time progresses.

(c)   True. The average rate of reaction between any two points on the graph is the slope of the line connecting the two points. Points 1 and 2 are earlier in the reaction when more reactants are available, so the average rate of formation of products is greater. As reactants are used up, the rate of X production decreases, and the average rate between points 2 and 3 is smaller.

(d)   False. The graph shows the build-up of [X] as the reaction progresses. [X] will not decrease once it reaches its maximum. (In the case of a chemical equilibrium, [X] will increase until reaching its equilibrium concentration, but it will not decrease after equilibrium is established.)

14.3   (a)   Chemical equation (iv), B → 2A, is consistent with the data. The concentration of A increases with time, and the concentration of B decreases with time, so B must be a reactant and A must be a product. The ending concentration of A is approximately twice as large as the starting concentration of B, so mole ratio of A:B is 2:1. The reaction is B → 2A.

        (b)   Rate = $-\Delta[B]/\Delta t = \frac{1}{2}\,\Delta[A]/\Delta t$

14.4   *Analyze/Plan.* Given a plot of increase in [M] over time, answer questions about reaction rate and progress. Consider the definition of reaction rate.   *Solve.*

        (a)   Yes. The plot of [M] versus time from t = 0 to t = 15 is a straight line, so [M] increases at a constant rate and the reaction occurs at a constant rate. The rate is zero after t = 15 min.

        (b)   Yes. [M] does not change after 15 min. This means that no more M is being produced and the reaction is no longer occurring.

        (c)   Statement (ii) is correct. After t = 15 min, 0.00 mol K and 0.20 mol L remain in solution. When an additional 0.20 mol K is added at 30 min, the second half of the reaction occurs. The plot of [M] versus t looks like the first half of the reaction, from t = 0 to t = 15 min.

14.5   *Analyze.* Given three mixtures and the order of reaction in each reactant, determine which mixture will have the fastest initial rate.

        Plan. Write the rate law. Count the number of reactant molecules in each container. The three containers have equal volumes and total numbers of molecules. Use the molecule count as a measure of concentration of NO and O_2. Calculate the initial rate for each container and compare.

        Solve. Rate = $k[NO]^2[O_2]$; rate is proportional to $[NO]^2[O_2]$

Container	[NO]	[O_2]	$[NO]^2[O_2] \propto$ rate
(1)	5	4	100
(2)	7	2	98
(3)	3	6	54

        The relative rates in containers (1) and (2) are very similar, with (1) having the slightly faster initial rate.

14.6   *Plan.* For a first-order reaction, a plot of ln[A] versus time is linear, as shown in the diagram. The slope is –k, and the intercept is $[A]_0$. According to the Arrhenius equation, 14.21, k increases with increasing temperature.   *Solve.*

        (a)   Graphs 1 and 2 have the same slope and thus the same rate constant, k. These experiments are done at the same temperature. The y-intercepts of the two graphs are different; the experiments had different initial concentrations of A.

        (b)   Graphs 2 and 3 have the same y-intercept and thus the same starting concentration of A. The slopes of the two graphs are different, so their rate constants are different and they occur at different temperatures. Graph 3, with the smaller slope and k value will occur at the lower temperature.

14.7   *Analyze.* Given concentrations of reactants and products at two times, as represented in the diagram, find $t_{1/2}$ for this first-order reaction.

Plan. For a first-order reaction, $t_{1/2} = 0.693/k$; $t_{1/2}$ depends only on k. Use Equation 14.12 to solve for k.   *Solve.*

(a)   Because reactants and products are in the same container, use number of particles as a measure of concentration. The red dots are reactant A, and the blue are product B. $[A]_0 = 16$, $[A]_{30} = 4$, $t = 30$ min.

$$\ln\frac{[A]_t}{[A]_0} = -kt. \quad \ln(4/16) = -k(30 \text{ min}); \quad \frac{-1.3863}{-30 \text{ min}} = k;$$

$k = 0.046210 = 0.0462 \text{ min}^{-1}$

$t_{1/2} = 0.693/k = 0.693/0.046210 = 15 \text{ min}$

By examination, $[A]_0 = 16$, $[A]_{30} = 4$. After 1 half-life, $[A] = 8$; after a second half-life, $[A] = 4$. Thirty minutes represents exactly 2 half-lives, so $t_{1/2} = 15$ min. [This is more straightforward than the calculation, but a less general method.]

(b)   After 4 half-lives, $[A]_t = [A]_0 \times 1/2 \times 1/2 \times 1/2 \times 1/2 = [A]_0/16$. In general, after n half-lives, $[A] = [A]_0/2^n$.

14.8   (a)   Plot (v) is zero order.

(b)   Plot (i) is first order.

(c)   Plot (iii) is second order.

14.9   *Analyze/Plan.* The reaction profile has a single high point (peak), so the reaction occurs in a single step. This step is necessarily the rate-determining step.   *Solve.*

(1)   Total potential energy of the reactants.

(2) E_a, activation energy of the reaction. This is the difference in energy between the potential energy of the activated complex (transition state) and the potential energy of the reactants.

(3) ΔE, net energy change for the reaction. This is the difference in energy between the products and reactants. (Under appropriate conditions, this could also be ΔH.) For this reaction, the energy of products is lower than the energy of reactants, and the reaction releases energy to the surroundings.

(4)   Total potential energy of the products.

14.10   *Analyze/Plan.* On a plot of ln k versus $1/T$, the slope is $-E_a/R$ and the y-intercept is ln A, where E_a is activation energy and A is the frequency factor.   *Solve.*

(a)   Blue. The magnitude of the slope of the blue line is greater than that of the red line.

(b)   Red. The y-intercept of the red line is greater than that of the blue line.

14.11   (a)   False. The red pathway is slower, because it has the greater activation energy, E_a.

(b)   True. For both reactions, the difference in potential energy between the products and the activated complex is greater than the difference between the reactants and the activated complex.

(c)   True. ΔE is the difference between the energy of the reactants and the energy of the products.

14.12   (a) $NO_2 + F_2 \rightarrow NO_2F + F$

               $NO_2 + F \rightarrow NO_2F$

       (b) $2NO_2 + F_2 \rightarrow 2NO_2F$

       (c)   F is the intermediate, because it is produced and then consumed during the reaction.

       (d) $Rate = k[NO_2][F_2]$

14.13   This is the profile of a two-step mechanism, $A \rightarrow B$ and $B \rightarrow C$. There is one intermediate, B. Because there are two energy maxima, there are two transition states. The $B \rightarrow C$ step is faster, because its activation energy is smaller. For the overall reaction $A \rightarrow C$, ΔE is negative, because the potential energy of the products is lower than the potential energy of the reactants.

14.14   The most likely transition state shows the relative geometry of both reactants and products. It is reasonable to assume that multiple bonds, with greater total bond energy, remain intact at the expense of single bonds. In the black-and-white diagram below, open circles represent the red balls and closed circles represent the blue.

14.15   (a) $A_2 + AB + AC \rightarrow BA_2 + A + AC$

               $BA_2 + A + AC \rightarrow A_2 + BA_2 + C$

               _____

               net: $AB + AC \rightarrow BA_2 + C$

       (b)   A is the intermediate; it is produced and consumed.

       (c) A_2 is the catalyst; it is consumed and reproduced.

14.16

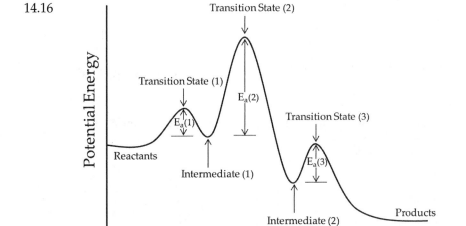

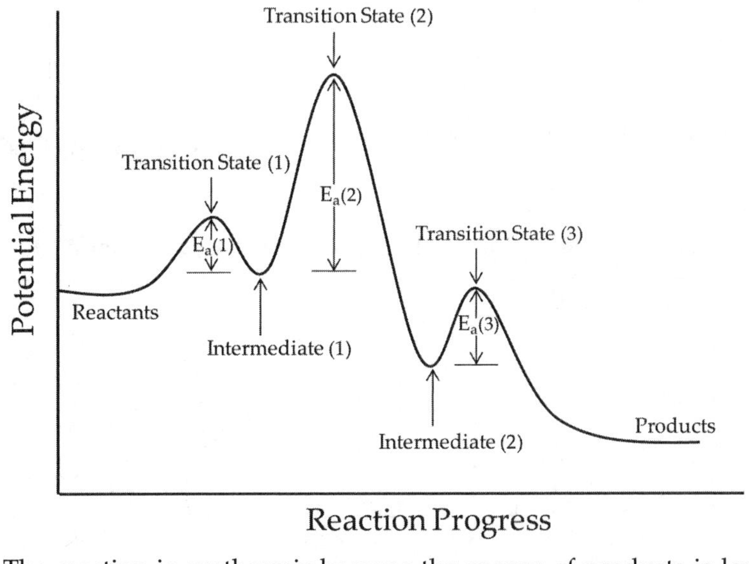

The reaction is exothermic because the energy of products is lower than the energy of reactants. The two intermediates are formed at different rates because $E_a(1) \neq E_a(2)$. To have two intermediates, the mechanism must have at least three steps.

Reaction Rates (Sections 14.1 and 14.2)

14.17   (a)   *Reaction rate* is the change in the amount of products or reactants in a given amount of time; it is the speed of a chemical reaction.

            (b)   Rates depend on concentration of reactants, physical state (or surface area) of reactants, temperature, and reaction activation energy/presence of catalyst.

            (c)   No, the rate of disappearance of reactants is not necessarily the same as the rate of appearance of products. The stoichiometry of the reaction (mole ratios of reactants and products) must be known to relate rate of disappearance of reactants to rate of appearance of products.

14.18   (a) M/s

            (b)   As temperature increases, reaction rate increases.

            (c)   As a reaction proceeds, the instantaneous reaction rate decreases.

14.19   *Analyze/Plan.* Given mol A at a series of times in minutes, calculate mol B produced, molarity of A at each time, change in M of A at each 10 min interval, and ΔM A/s. For this reaction, mol B produced equals mol A consumed. M of A or [A] = mol A/0.100 L. The average rate of disappearance of A for each 10 minute interval is

$$-\frac{\Delta[A]}{s} = -\frac{[A]_1 - [A]_0}{10 \text{ min}} \times \frac{1 \text{ min}}{60 \text{ s}}$$

Solve.

Time (min)	Mol A	(a) Mol B	[A]	Δ[A]	(b) Rate −(Δ[A]/s)
0	0.065	0.000	0.65		
10	0.051	0.014	0.51	−0.14	2.3×10^{-4}
20	0.042	0.023	0.42	−0.09	2×10^{-4}
30	0.036	0.029	0.36	−0.06	1×10^{-4}
40	0.031	0.034	0.31	−0.05	0.8×10^{-4}

     (c) $\dfrac{\Delta M_B}{\Delta t} = \dfrac{(0.029 - 0.014) \text{ mol}/0.100 \text{ L}}{(30 - 10) \text{ min}} \times \dfrac{1 \text{ min}}{60 \text{ s}} = 1.25 \times 10^{-4} = 1.3 \times 10^{-4} \; M/s$

14.20

Time (s)	Mol A	(a) Mol B	Δ Mol A	(b) Rate −(Δ mol A/s)
0	0.100	0.000		
40	0.067	0.033	−0.033	8.3×10^{-4}
80	0.045	0.055	−0.022	5.5×10^{-4}
120	0.030	0.070	−0.015	3.8×10^{-4}
160	0.020	0.080	−0.010	2.5×10^{-4}

     (c)   (ii) The volume of the container must be known to report the rate in units of concentration (mol/L) per time.

14.21   (a)   *Analyze/Plan.* Follow the logic in Sample Exercises 14.1 and 14.2.   *Solve.*

Time (s)	Time Interval (s)	Concentration (M)	ΔM	Rate (M/s)
0		0.0165		
2000	2000	0.0110	−0.0055	28×10^{-7}
5000	3000	0.00591	−0.0051	17×10^{-7}
8000	3000	0.00314	−0.00277	9.23×10^{-7}
12,000	4000	0.00137	−0.00177	4.43×10^{-7}
15,000	3000	0.00074	−0.00063	2.1×10^{-7}

(b) $-\dfrac{\Delta M}{\Delta t} = -\dfrac{(0.00074 - 0.0165)\,M}{(15,000 - 0)\,s} = 1.0507 \times 10^{-6} = 1.05 \times 10^{-6}\ M/s$

(c) $-\dfrac{\Delta M}{\Delta t} = -\dfrac{(0.00137 - 0.0110)\,M}{(12,000 - 2000)\,s} = 9.63 \times 10^{-7}\ M/s$

$-\dfrac{\Delta M}{\Delta t} = -\dfrac{(0.00074 - 0.00314)\,M}{(15,000 - 8000)\,s} = 3.43 \times 10^{-7}\ M/s$

The average rate between t = 2000 and t = 12,000 s is greater. In general, the rate of a reaction decreases over time.

(d)   From the slopes of the lines in the figure at right, the rates are: at 5000 s, $12 \times 10^{-7}\ M/s$; at 8000 s, $5.8 \times 10^{-7}\ M/s$.

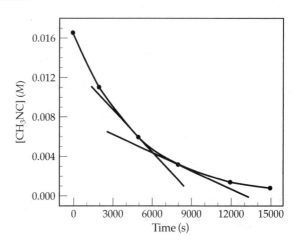

14.22   (a)

Time (min)	Time Interval (min)	Concentration (M)	ΔM	Rate (M/s)
0.0		1.85		
54.0	54.0	1.58	−0.27	8.3×10^{-5}
107.0	53.0	1.36	−0.22	6.9×10^{-5}
215.0	108	1.02	−0.34	5.2×10^{-5}
430.0	215	0.580	−0.44	3.4×10^{-5}

(b) $-\dfrac{\Delta M}{\Delta t} = -\dfrac{(1.85 - 0.580)\,M}{(430 - 0)\,min} \times \dfrac{1\,min}{60\,s} = 4.9225 \times 10^{-5} = 4.92 \times 10^{-5}\ M/s$

401

(c) $-\dfrac{\Delta M}{\Delta t} = -\dfrac{(1.02-1.58)\,M}{(215-54)\,\text{min}} \times \dfrac{1\,\text{min}}{60\,\text{s}} = 5.8 \times 10^{-5}\,M/s$

$-\dfrac{\Delta M}{\Delta t} = -\dfrac{(0.580-1.36)\,M}{(430-107)\,\text{min}} \times \dfrac{1\,\text{min}}{60\,\text{s}} = 4.0 \times 10^{-5}\,M/s$

The average rate between t = 54.0 and t = 215.0 min is greater. In general, the rate of a reaction decreases over time.

(d) From the slopes of the lines in the figure at the right, the rates are: at 75.0 min, $4.2 \times 10^{-3}\,M/\text{min}$, or $7.0 \times 10^{-5}\,M/s$; at 250 min, $2.1 \times 10^{-3}\,M/\text{min}$ or $3.5 \times 10^{-5}\,M/s$.

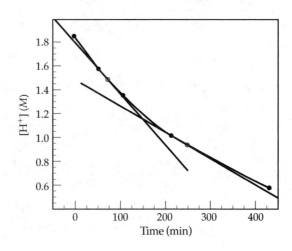

14.23   *Analyze/Plan.* Follow the logic in Sample Exercise 14.3.   *Solve.*

(a) $-\Delta[H_2O_2]/\Delta t = \Delta[H_2]/\Delta t = \Delta[O_2]/\Delta t$

(b) $-\Delta[N_2O]/2\Delta t = \Delta[N_2]/2\Delta t = \Delta[O_2]/\Delta t$

$-\Delta[N_2O]/\Delta t = \Delta[N_2]/\Delta t = 2\Delta[O_2]/\Delta t$

(c) $-\Delta[N_2]/\Delta t = \Delta[NH_3]/2\Delta t;\ -\Delta[H_2]/3\Delta t = \Delta[NH_3]/2\Delta t$

$-2\Delta[N_2]/\Delta t = \Delta[NH_3]/\Delta t;\ -\Delta[H_2]/\Delta t = 3\Delta[NH_3]/2\Delta t$

(d) $-\Delta[\,C_2H_5NH_2]/\Delta t = \Delta[\,C_2H_4]/\Delta t = \Delta[NH_3]/\Delta t$

14.24   (a)   Rate $= -\Delta[H_2O]/2\Delta t = \Delta[H_2]/2\Delta t = \Delta[O_2]/\Delta t$

(b)   Rate $= -\Delta[SO_2]/2\Delta t = -\Delta[O_2]/\Delta t = \Delta[SO_3]/2\Delta t$

(c)   Rate $= -\Delta[NO]/2\Delta t = -\Delta[H_2]/2\Delta t = \Delta[N_2]/\Delta t = \Delta[H_2O]/2\Delta t$

(d)   Rate $= -\Delta[N_2]/\Delta t = -\Delta[H_2]/2\Delta t = \Delta[N_2H_4]/\Delta t$

14.25   *Analyze/Plan.* Use Equation 14.4 to relate the rate of disappearance of reactants to the rate of appearance of products. Use this relationship to calculate desired quantities.   *Solve.*

(a) $\Delta[H_2O]/2\Delta t = -\Delta[H_2]/2\Delta t = -\Delta[O_2]/\Delta t$

H_2 is burning, $-\Delta[H_2]/\Delta t = 0.48$ mol/s

O_2 is consumed, $-\Delta[O_2]/\Delta t = -\Delta[H_2]/2\Delta t = 0.48$ mol/s/2 $= 0.24$ mol/s

H_2O is produced, $+\Delta[H_2O]/\Delta t = -\Delta[H_2]/\Delta t = 0.48$ mol/s

(b)   The change in total pressure is the sum of the changes of each partial pressure. NO and Cl_2 are disappearing and NOCl is appearing.

$-\Delta P_{NO}/\Delta t = -56$ torr/min

$-\Delta P_{Cl_2}/\Delta t = \Delta P_{NO}/2\Delta t = -28$ torr/min

$+\Delta P_{NOCl}/\Delta t = -\Delta P_{NO}/\Delta t = +56$ torr/min

$\Delta P_T/\Delta t = -56$ torr/min $- 28$ torr/min $+ 56$ torr/min $= -28$ torr/min

14.26   (a) $-\Delta[C_2H_4]/\Delta t = \Delta[CO_2]/2\Delta t = \Delta[H_2O]/2\Delta t$

$-2\Delta[C_2H_4]/\Delta t = \Delta[CO_2]/\Delta t = \Delta[H_2O]/\Delta t$

C_2H_4 is burning, $-\Delta[C_2H_4]/\Delta t = 0.036$ M/s

CO_2 and H_2O are produced, at twice the rate that C_2H_4 is consumed.

$\Delta[CO_2]/\Delta t = \Delta[H_2O]/\Delta t = 2(0.036)$ $M/s = 0.072$ M/s

(b)   In this reaction, pressure is a measure of concentration.

$-\Delta[N_2H_4]/\Delta t = -\Delta[H_2]/\Delta t = \Delta[NH_3]/2\Delta t$

N_2H_4 is consumed, $-\Delta[N_2H_4]/\Delta t = 74$ torr/h

H_2 is consumed, $-\Delta[H_2]/\Delta t = 74$ torr/h

NH_3 is produced at twice the rate that N_2H_4 and H_2 are consumed,

$\Delta[NH_3]/\Delta t = -2\Delta[N_2H_4]/\Delta t = 2(74)$ torr/h $= 148$ torr/h

$\Delta P_T/\Delta t = (+148$ torr/h $- 74$ torr/h $- 74$ torr/h$) = 0$ torr/h

Rate Laws (Section 14.3)

14.27   *Analyze/Plan.* Follow the logic in Sample Exercise 14.5.   *Solve.*

(a)   If [A] is doubled, there will be no change in the rate or the rate constant. The overall rate is unchanged because [A] does not appear in the rate law; the rate constant changes only with a change in temperature.

(b)   The reaction is zero order in A, second order in B, and second order overall.

(c)   Units of $k = \dfrac{M/s}{M^2} = M^{-1}s^{-1}$

14.28   (a)   Rate $= k[A][C]^2$

(b)   Rate is proportional to [A], rate doubles

(c)   Rate is not affected by [B], no change

(d)   Rate changes as $[C]^2$, rate increases by a factor of 3^2 or 9

(e)   Rate increases by a factor of $(3)(3)^2 = 27$

(f)   Rate decreases by a factor of $(1/2)(1/2)^2 = 1/8$

14.29 *Analyze/Plan.* Follow the logic in Sample Exercise 14.5. *Solve.*

(a) Rate $= k[N_2O_5] = 4.82 \times 10^{-3}\ s^{-1}\ [N_2O_5]$

(b) Rate $= 4.82 \times 10^{-3}\ s^{-1}\ (0.0240\ M) = 1.16 \times 10^{-4}\ M/s$

(c) Rate $= 4.82 \times 10^{-3}\ s^{-1}\ (0.0480\ M) = 2.31 \times 10^{-4}\ M/s$

When the concentration of N_2O_5 doubles, the rate of the reaction doubles.

(d) Rate $= 4.82 \times 10^{-3}\ s^{-1}\ (0.0120\ M) = 5.78 \times 10^{-5}\ M/s$

When the concentration of N_2O_5 is halved, the rate of the reaction is halved.

14.30 (a) Rate $= k[H_2][NO]^2$

(b) Rate $= (6.0 \times 10^4\ M^{-2}\ s^{-1})\ (0.035\ M)^2\ (0.015\ M) = 1.1\ M/s$

(c) Rate $= (6.0 \times 10^4\ M^{-2}\ s^{-1})\ (0.10\ M)^2\ (0.010\ M) = 6.0\ M/s$

(d) Rate $= (6.0 \times 10^4\ M^{-2}\ s^{-1})\ (0.010\ M)^2\ (0.030\ M) = 0.18\ M/s$

14.31 *Analyze/Plan.* Write the rate law and rearrange to solve for k. Use the given data to calculate k, including units. *Solve.*

(a, b) Rate $= k[CH_3Br][OH^-]$; $k = \dfrac{\text{rate}}{[CH_3Br][OH^-]}$

at 298 K, $k = \dfrac{0.0432\ M/s}{(5.0 \times 10^{-3}\ M)(0.050\ M)} = 1.7 \times 10^2\ M^{-1}s^{-1}$

(c) Because the rate law is first order in $[OH^-]$, if $[OH^-]$ is tripled, the rate triples.

(d) If $[OH^-]$ and $[CH_3Br]$ both triple, the rate increases by a factor of $(3)(3) = 9$.

14.32 (a, b) Rate $= k[C_2H_5Br][OH^-]$; $k = \dfrac{\text{rate}}{[C_2H_5Br][OH^-]}$

at 298 K, $k = \dfrac{1.7 \times 10^{-7}\ M/s}{[0.0477\ M][0.100\ M]} = 3.6 \times 10^{-5}\ M^{-1}s^{-1}$

(c) Adding an equal volume of ethyl alcohol reduces both $[C_2H_5Br]$ and $[OH^-]$ by a factor of two. New rate $= (1/2)(1/2) = 1/4$ of old rate.

14.33 *Analyze/Plan.* Follow the logic in Sample Exercise 14.6. *Solve.*

(a) From the data given, when $[OCl^-]$ doubles, rate doubles. When $[I^-]$ doubles, rate doubles. The reaction is first order in both $[OCl^-]$ and $[I^-]$. Rate $= k[OCl^-][I^-]$.

(b) Using the first set of data:

$k = \dfrac{\text{rate}}{[OCl^-][I^-]} = \dfrac{1.36 \times 10^{-4}\ M/s}{(1.5 \times 10^{-3}\ M)(1.5 \times 10^{-3}\ M)} = 60.444 = 60\ M^{-1}\ s^{-1}$

(c) Rate $= \dfrac{60.444}{M\text{-s}}(2.0 \times 10^{-3}\ M)(5.0 \times 10^{-4}\ M) = 6.0444 \times 10^{-5} = 6.0 \times 10^{-5}\ M/s$

14.34 (a) From the data given, when $[ClO_2]$ increases by a factor of 3 (experiment 2 to experiment 1), the rate increases by a factor of 9. When $[OH^-]$ increases by a factor of 3 (experiment 2 to experiment 3), the rate increases by a factor of 3. The reaction is second order in $[ClO_2]$ and first order in $[OH^-]$. Rate $= k[ClO_2]^2[OH^-]$.

(b)   Using data from experiment 2:

$$k = \frac{rate}{[ClO_2]^2 [OH^-]} = \frac{0.00276 \, M/s}{(0.020 \, M)^2 (0.030 \, M)} = 2.3 \times 10^2 \, M^{-2}s^{-1}$$

(c)   Rate $= 2.3 \times 10^2 \, M^{-2} \, s^{-1} \, (0.100 \, M)^2 (0.050 \, M) = 0.115 = 0.12 \, M/s$

14.35   *Analyze/Plan.*   Follow the logic in Sample Exercise 14.6 to deduce the rate law. Rearrange the rate law to solve for k and deduce units. Calculate a k value for each set of concentrations and then average the three values.   *Solve.*

(a)   Doubling $[NH_3]$ while holding $[BF_3]$ constant doubles the rate (experiments 1 and 2). Doubling $[BF_3]$ while holding $[NH_3]$ constant doubles the rate (experiments 4 and 5).

Thus, the reaction is first order in both BF_3 and NH_3; rate $= k[BF_3][NH_3]$.

(b)   The reaction is second order overall.

(c)   From experiment 1: $k = \dfrac{0.2130 \, M/s}{(0.250 \, M)(0.250 \, M)} = 3.41 \, M^{-1} \, s^{-1}$

(Any of the five sets of initial concentrations and rates could be used to calculate the rate constant k. The average of these 5 values is $k_{avg} = 3.408 = 3.41 \, M^{-1}s^{-1}$.)

(d)   Rate $= 3.408 \, M^{-1}s^{-1}(0.100 \, M)(0.500 \, M) = 0.1704 = 0.170 \, M/s$.

14.36   *Analyze/Plan.*   Follow the logic in Sample Exercise 14.6 to deduce the rate law. Rearrange the rate law to solve for k and deduce units. Calculate a k value for each set of concentrations and then average the three values.   *Solve.*

(a)   Doubling [NO] while holding $[O_2]$ constant increases the rate by a factor of 4 (experiments 1 and 2). Doubling $[O_2]$ while holding [NO] constant doubles the rate (experiments 2 and 3). The reaction is second order in [NO] and first order in $[O_2]$. Rate $= k[NO]^2[O_2]$.

(b, c) From experiment 1: $k_1 = \dfrac{1.41 \times 10^{-2} \, M/s}{(0.0126 \, M)^2 (0.0125 \, M)} = 7105 = 7.11 \times 10^3 \, M^{-2}s^{-1}$

$k_2 = 0.113/(0.0252)^2(0.0250) = 7118 = 7.12 \times 10^3 \, M^{-2} \, s^{-1}$

$k_3 = 5.64 \times 10^{-2}/(0.0252)^2(0.125) = 7105 = 7.11 \times 10^3 \, M^{-2} \, s^{-1}$

$k_{avg} = (7105 + 7118 + 7105)/3 = 7109 = 7.11 \times 10^3 \, M^{-2} \, s^{-1}$

(d)   Rate $= 7.109 \times 10^3 \, M^{-2}s^{-1} \, (0.0750 \, M)^2(0.0100 \, M) = 0.3999 = 0.400 \, M/s$.

(e)   The data are given in terms of the disappearance of NO. Use Equation 14.4 to relate the disappearance of NO to the disappearance of O_2.
       $-\Delta[NO]/2\Delta t = -[O_2]/\Delta t$

       For the concentrations given in part (d), $\Delta[NO]/\Delta t = 0.400 \, M/s$.

       $\Delta [O_2]/\Delta t = \Delta[NO]/2\Delta t = 0.400 \, M/s/2 = 0.200 \, M/s$

14.37   *Analyze/Plan.* Follow the logic in Sample Exercise 4.6 to deduce the rate law. Rearrange the rate law to solve for k and deduce units. Calculate a k value for each set of concentrations and then average the three values.   *Solve.*

(a) Increasing [NO] by a factor of 2.5 while holding [Br_2] constant (experiments 1 and 2) increases the rate by a factor 6.25 or $(2.5)^2$. Increasing [Br_2] by a factor of 2.5 while holding [NO] constant increases the rate by a factor of 2.5. The rate law for the appearance of NOBr is: Rate = $\Delta[NOBr]/\Delta t = k[NO]^2[Br_2]$.

(b) From experiment 1: $k_1 = \dfrac{24\ M/s}{(0.10\ M)^2\,(0.20\ M)} = 1.20 \times 10^4 = 1.2 \times 10^4\ M^{-2}\ s^{-1}$

$k_2 = 150/(0.25)^2(0.20) = 1.20 \times 10^4 = 1.2 \times 10^4\ M^{-2}\ s^{-1}$

$k_3 = 60/(0.10)^2(0.50) = 1.20 \times 10^4 = 1.2 \times 10^4\ M^{-2}\ s^{-1}$

$k_4 = 735/(0.35)^2(0.50) = 1.2 \times 10^4 = 1.2 \times 10^4\ M^{-2}\ s^{-1}$

$k_{avg} = (1.2 \times 10^4 + 1.2 \times 10^4 + 1.2 \times 10^4 + 1.2 \times 10^4)/4 = 1.2 \times 10^4\ M^{-2}\ s^{-1}$

(c) Use the reaction stoichiometry and Equation 14.4 to relate the designated rates. $\Delta[NOBr]/2\Delta t = -\Delta[Br_2]/\Delta t$; the rate of disappearance of Br_2 is half the rate of appearance of NOBr.

(d) Note that the data are given in terms of appearance of NOBr.

$$\dfrac{-\Delta[Br_2]}{\Delta t} = \dfrac{k[NO]^2[Br_2]}{2} = \dfrac{1.2 \times 10^4}{2\ M^2\ s} \times (0.075\ M)^2 \times (0.250\ M) = 8.4\ M/s$$

14.38   (a) Increasing [$S_2O_8^{2-}$] by a factor of 1.5 while holding [I^-] constant increases the rate by a factor of 1.5 (Experiments 1 and 2). Doubling [$S_2O_8^{2-}$] and increasing [I^-] by a factor of 1.5 triples the rate ($2 \times 1.5 = 3$, experiments 1 and 3). Thus the reaction is first order in both [$S_2O_8^{2-}$] and [I^-]; rate = $k\,[S_2O_8^{2-}]\,[I^-]$.

(b) $k = \text{rate}/[S_2O_8^{2-}]\,[I^-]$

$k_1 = 2.6 \times 10^{-6}\ M/s/(0.018\ M)(0.036\ M) = 4.01 \times 10^{-3} = 4.0 \times 10^{-3}\ M^{-1}s^{-1}$

$k_2 = 3.9 \times 10^{-6}/(0.027)(0.036) = 4.01 \times 10^{-3} = 4.01 \times 10^{-3} = 4.0 \times 10^{-3}\ M^{-1}s^{-1}$

$k_3 = 7.8 \times 10^{-6}/(0.036)(0.054) = 4.01 \times 10^{-3} = 4.01 \times 10^{-3} = 4.0 \times 10^{-3}\ M^{-1}s^{-1}$

$k_4 = 1.4 \times 10^{-5}/(0.050)(0.072) = 3.89 \times 10^{-3} = 3.9 \times 10^{-3}\ M^{-1}s^{-1}$

$k_{avg} = 3.98 \times 10^{-3} = 4.0 \times 10^{-3}\ M^{-1}s^{-1}$

(c) $-\Delta[S_2O_8^{2-}]/\Delta t = -\Delta[I^-]/3\Delta t$; the rate of disappearance of $S_2O_8^{2-}$ is one-third the rate of disappearance of I^-.

(d) Note that the data are given in terms of disappearance of $S_2O_8^{2-}$.

$$\dfrac{-\Delta[I^-]}{\Delta t} = \dfrac{-3\Delta[S_2O_8^{2-}]}{\Delta t} = 3(3.98 \times 10^{-3}\ M^{-1}s^{-1})(0.025\ M)(0.050\ M) = 1.5 \times 10^{-5}\ M/s$$

Change of Concentration with Time (Section 14.4)

14.39   (a) A graph of ln[A] versus time yields a straight line for a first-order reaction.

(b) On graph of ln[A] versus time, the rate constant, k, is the (–slope) of the straight line.

14.40   (a) A graph of 1/[A] versus time yields a straight line for a second-order reaction.

(b) On a graph of 1/[A] versus time, the slope of the straight line is the rate constant, k.

(c) The half-life of a second-order reaction remains the same as the reaction proceeds. According to Equation 14.19, the half-life of a second-order reaction does depend on $[A]_0$, $t_{1/2} = 1/k[A]_0$, but this quantity is constant over the course of the reaction.

14.41   *Analyze/Plan.* The half-life of a first-order reaction depends only on the rate constant, $t_{1/2} = 0.693/k$. Use this relationship to calculate k for a given $t_{1/2}$, and, at a different temperature, $t_{1/2}$ given k.   *Solve.*

     (a) $t_{1/2} = 2.3 \times 10^5$ s; $t_{1/2} = 0.693/k$, $k = 0.693/t_{1/2}$

          $k = 0.693/2.3 \times 10^5$ s $= 3.0 \times 10^{-6}$ s^{-1}

     (b) $k = 2.2 \times 10^{-5}$ s^{-1}. $t_{1/2} = 0.693/2.2 \times 10^{-5}$ s$^{-1} = 3.15 \times 10^4 = 3.2 \times 10^4$ s

14.42   (a)   For a first-order reaction, $t_{1/2} = 0.693/k$.

          $t_{1/2} = 0.693/0.271$ s$^{-1} = 2.5572 = 2.56$ s

     (b)   For a first-order reaction, $\ln[A]_t - \ln[A]_0 = -kt$. $\ln[A]_t = -kt + \ln[A]_0$

          $[A]_0 = 0.050$ *M* I_2, $t = 5.12$ s, $k = 0.271$ s^{-1}

          $\ln[I_2] = -0.271$ s^{-1} (5.12 s) + ln(0.050)

          $\ln[I_2] = -4.3833$, $[I_2] = 0.0125$ *M*

          Check. 5.12 s is 2 half-lives. $[I_2]$ should be reduced by a factor of 4, and it is.

14.43   *Analyze/Plan.* Follow the logic in Sample Exercise 14.7. In this reaction, pressure is a measure of concentration. In (a) we are given k, $[A]_0$, t and asked to find $[A]_t$, using Equation 14.13, the integrated form of the first-order rate law. In (b), $[A_t] = 0.1[A_0]$, find t. *Solve.*

     (a) $\ln P_t = -kt + \ln P_0$; $P_0 = 450$ torr; $t = 60$ s

          $\ln P_{60} = -4.5 \times 10^{-2}$ s^{-1}(60) + ln (450) $= -2.70 + 6.109 = 3.409$

          $P_{60} = 30.24 = 30$ torr

     (b) $P_t = 0.10 \ P_0$; $\ln (P_t/P_0) = -kt$

          $\ln (0.10 \ P_0/P_0) = -kt$, $\ln (0.10) = -kt$; $-\ln (0.10)/k = t$

          $t = -(-2.303)/4.5 \times 10^{-2}$ s$^{-1} = 51.2 = 51$ s

   Check. From part (a), the pressure at 60 s is 30 torr, $P_t \sim 0.07 \ P_0$. In part (b) we calculate the time where $P_t = 0.10 \ P_0$ to be 51 s. This time should be smaller than 60 s, and it is. Data and results in the two parts are consistent.

14.44   (a)   Using Equation 14.13 for a first-order reaction: $\ln [A]_t = -kt + \ln [A]_0$

          5.0 min = 300 s; $[N_2O_5]_0 = (0.0250$ mol/2.0 L) $= 0.0125 = 0.013$ *M*

          $\ln [N_2O_5]_{300} = -(6.82 \times 10^{-3}$ s$^{-1})(300$ s) + ln (0.0125)

          $\ln [N_2O_5]_{300} = -2.0460 + (-4.3820) = -6.4280 = -6.43$

          $[N_2O_5]_{300} = 1.616 \times 10^{-3} = 1.6 \times 10^{-3}$ *M*

          mol $N_2O_5 = 1.616 \times 10^{-3}$ *M* $\times 2.0$ L $= 3.2 \times 10^{-3}$ mol

     (b) $[N_2O_5]_t = 0.010$ mol/2.0 L $= 0.0050$ *M*; $[N_2O_5]_0 = 0.0125$ *M*

          $\ln (0.0050) = -(6.82 \times 10^{-3}$ s$^{-1})$ (t) + ln (0.0125)

          $t = \dfrac{-[\ln(0.0050) - \ln(0.0125)]}{(6.82 \times 10^{-3} \text{s}^{-1})} = 134.35 = 1.3 \times 10^2$ s $\times \dfrac{1\,\text{min}}{60\,\text{s}} = 2.24 = 2.2$ min

     (c) $t_{1/2} = 0.693/k = 0.693/6.82 \times 10^{-3}$ s$^{-1} = 101.6 = 102$ s or 1.69 min

14.45   *Analyze/Plan.* Given reaction order, various values for t and P_t, find the rate constant for the reaction at this temperature. For a first-order reaction, a graph of ln P versus t is linear with as slope of –k.   *Solve.*

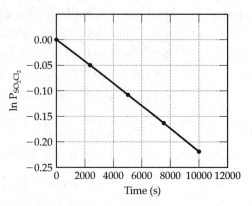

Time (s)	$P_{SO_2Cl_2}$	$\ln P_{SO_2Cl_2}$
0	1.000	0
2500	0.947	–0.0545
5000	0.895	–0.111
7500	0.848	–0.165
10,000	0.803	–0.219

Graph ln $P_{SO_2Cl_2}$ versus time. (Pressure is a satisfactory unit for a gas, because the concentration in moles/liter is proportional to P.) The graph is linear with slope -2.19×10^{-5} s^{-1} as shown on the figure. The rate constant k = –slope = 2.19×10^{-5} s^{-1}.

14.46

Time (s)	P_{CH_3NC}	$\ln P_{CH_3NC}$
0	502	6.219
2,000	335	5.814
5,000	180	5.193
8,000	95.5	4.559
12,000	41.7	3.731
15,000	22.4	3.109

A graph of ln P vs t is linear with a slope of -2.08×10^{-4} s^{-1}. The rate constant, k, = –slope = 2.08×10^{-4} s^{-1}. Half-life = $t_{1/2}$ = $0.693/k = 3.33 \times 10^3$ s.

14.47   *Analyze/Plan.* Given: mol A, t. Change mol to M at various times. Make both first- and second-order plots to see which is linear.   *Solve.*

(a)

Time (min)	mol A	[A] (M)	ln[A]	1/mol A
0	0.065	0.65	–0.43	1.5
10	0.051	0.51	–0.67	2.0
20	0.042	0.42	–0.87	2.4
30	0.036	0.36	–1.02	2.8
40	0.031	0.31	–1.17	3.2

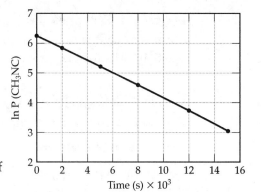

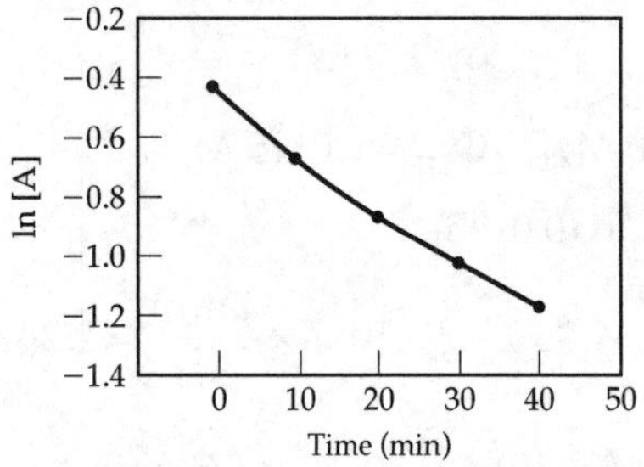

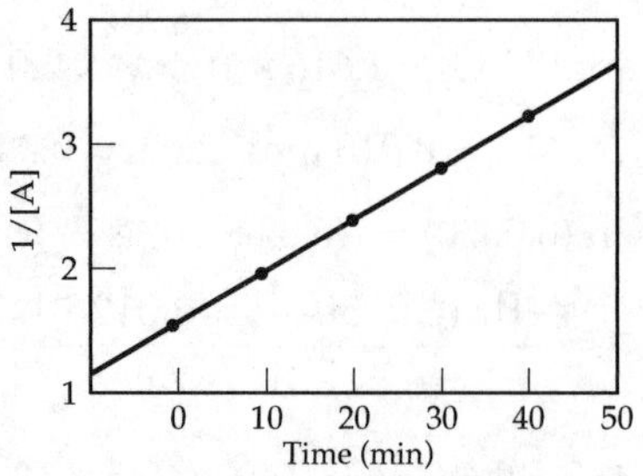

The plot of 1/[A] vs time is linear, so the reaction is second order in [A].

(b)   For a second-order reaction, a plot of 1/[A] versus t is linear with slope k.

k = slope = $(3.2 - 2.0) \, M^{-1}$ / 30 min = $0.040 \, M^{-1} \, min^{-1}$

(The best fit to the line yields slope = $0.042 \, M^{-1} \, min^{-1}$.)

(c) $t_{1/2} = 1/k[A]_0 = 1/(0.040 \, M^{-1} \, min^{-1})(0.65 \, M) = 38.46 = 38$ min

(Using the "best-fit" slope, $t_{1/2} = 37$ min.)

14.48   (a)   Make both first- and second-order plots to see which is linear. Moles is a satisfactory concentration unit, because volume is constant.

Time (s)	mol A	ln (mol A)	1/mol A
0	0.1000	−2.303	10.00
40	0.067	−2.70	14.9
80	0.045	−3.10	22.2
120	0.030	−3.51	33.3
160	0.020	−3.91	50.0

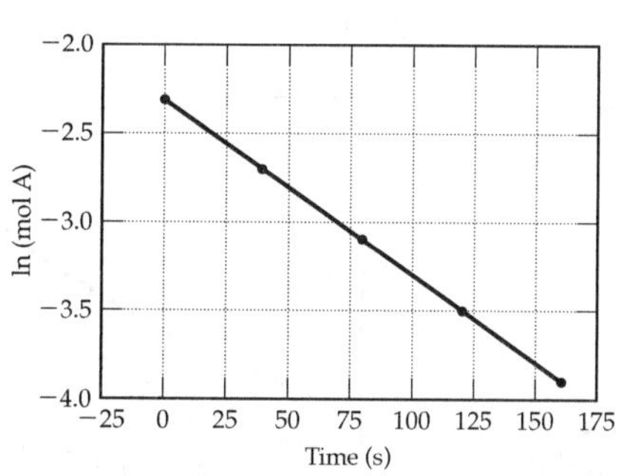

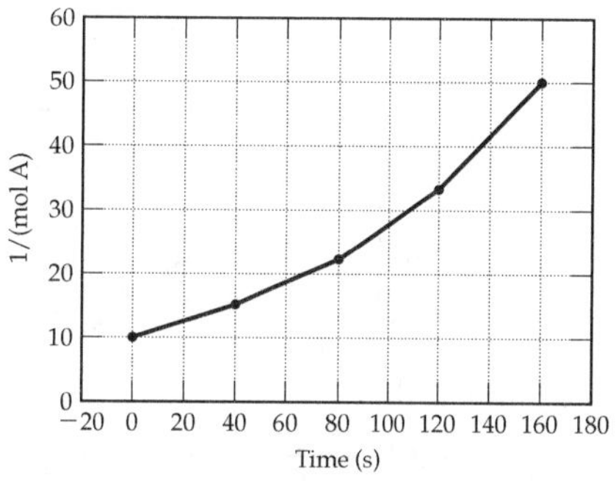

The plot of ln (mol A) vs time is linear, so the reaction is first order in A.

(b)   k = −slope = − [−3.91 − (−2.70)]/120 = 0.010083 = $0.0101 \, s^{-1}$

(The best fit to this line yields the same value for the slope, 0.01006 = $0.0101 \, s^{-1}$)

(c) $t_{1/2} = 0.693/k = 0.693/0.010083 \, s^{-1} = 68.7$ s

14.49   *Analyze/Plan.* Make both first- and second-order plots to see which is linear.   *Solve.*

(a)

Time (s)	[NO₂](M)	ln [NO₂]	1/[NO₂]
0.0	0.100	−2.303	10.0
5.0	0.017	−4.08	59
10.0	0.0090	−4.71	110
15.0	0.0062	−5.08	160
20.0	0.0047	−5.36	210

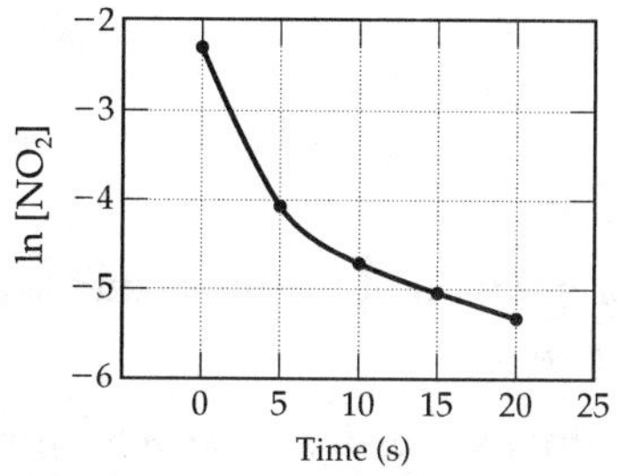

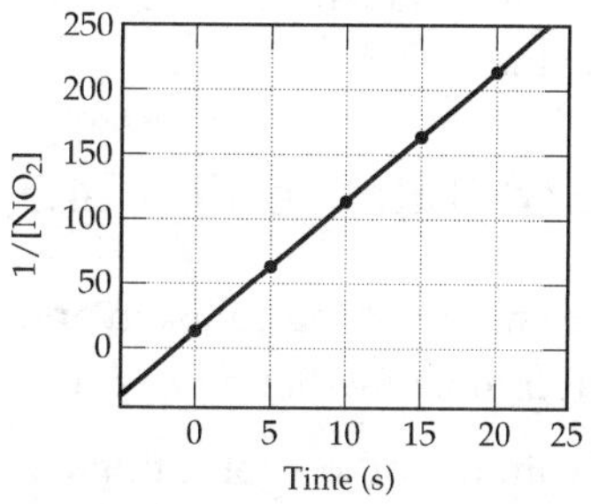

$[NO_2](M)$, $\ln [NO_2]$, $1/[NO_2]$

The plot of 1/[NO₂] versus time is linear, so the reaction is second order in NO₂.

(b)   The slope of the line is $(210 - 59) \, M^{-1} / 15.0 \, s = 10.07 = 10 \, M^{-1}s^{-1} = k$. (The slope of the best-fit line is $10.02 = 10 \, M^{-1}s^{-1}$.)

(c)   From the results above, the rate law is: rate $= k[NO_2]^2 = 10 \, M^{-1}s^{-1}[NO_2]^2$
Using the rate law, calculate the rate at each of the given initial concentrations.
Rate @ $0.200 \, M = 10 \, M^{-1}s^{-1}[NO_2]^2 = 10 \, M^{-1}s^{-1}[0.200 \, M]^2 = 0.400 \, M/s$
Rate @ $0.100 \, M = 10 \, M^{-1}s^{-1}[NO_2]^2 = 10 \, M^{-1}s^{-1}[0.100 \, M]^2 = 0.100 \, M/s$
Rate @ $0.050 \, M = 10 \, M^{-1}s^{-1}[NO_2]^2 = 10 \, M^{-1}s^{-1}[0.050 \, M]^2 = 0.025 \, M/s$

14.50   (a)   Make both first- and second-order plots to see which is linear.

Time (min)	$[C_{12}H_{22}O_{11}](M)$	$\ln [C_{12}H_{22}O_{11}]$	$1/[C_{12}H_{22}O_{11}]$
0	0.316	–1.152	3.16
39	0.274	–1.295	3.65
80	0.238	–1.435	4.20
140	0.190	–1.661	5.26
210	0.146	–1.924	6.85

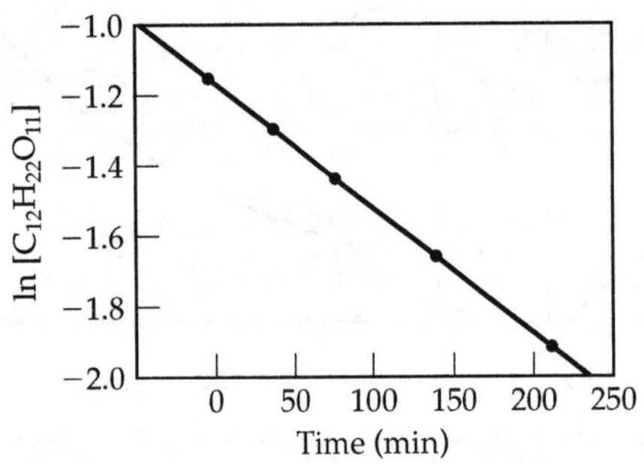

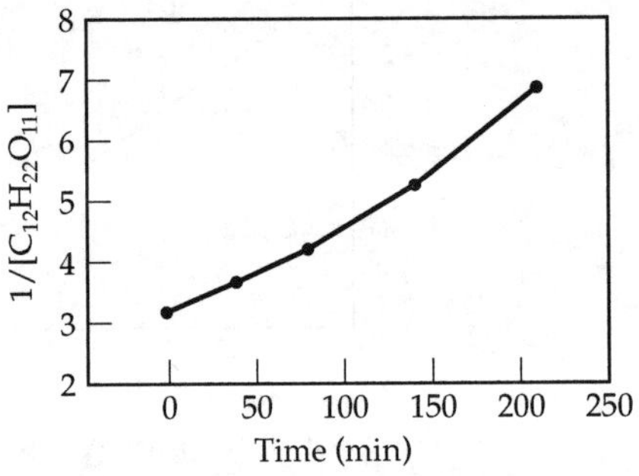

The plot of $\ln [C_{12}H_{22}O_{11}]$ is linear, so the reaction is first order in $C_{12}H_{22}O_{11}$.

(b) $k = -\text{slope} = -[-1.924 - (-1.295)]/171 \text{ min} = 3.68 \times 10^{-3} \text{ min}^{-1}$

(The slope of the best-fit line is $-3.67 \times 10^{-3} \text{ min}^{-1}$.)

(c)   For a reaction zero order in sucrose, the rate does not change as [sucrose] changes. A plot of [sucrose] versus time is linear with negative slope, until all reactant is consumed. $[\text{sucrose}]_t = -kt + [\text{sucrose}]_0$

@39 min, [sucrose] $= -3.68 \times 10^{-3} \text{ min}^{-1}(39 \text{ min}) + 0.316 \, M = 0.17 \, M$

@80 min, [sucrose] $= -3.68 \times 10^{-3} \text{ min}^{-1}(80 \text{ min}) + 0.316 \, M = 0.022 \, M$

@140 min, [sucrose] $= -3.68 \times 10^{-3} \text{ min}^{-1}(140 \text{ min}) + 0.316 \, M = 0 \, M$

@210 min, [sucrose] $= 0 \, M$. All sucrose is consumed at $(0.316/3.68 \times 10^{-3} \text{ min}^{-1}) = 85.9$ min.

Temperature and Rate (Section 14.5)

14.51   (a)   The energy of the collision and the orientation of the molecules when they collide determine whether a reaction will occur.

(b)   Assuming other conditions remain the same, the rate and therefore the rate constant usually increase with an increase in reaction temperature.

(c) The fraction of molecules with energy greater than the activation energy changes most dramatically with temperature. Frequency of collision and the orientation factor are lumped into the frequency factor, A, which is considered to be constant with temperature.

14.52 (a) The orientation factor is less important in $H + Cl \rightarrow HCl$, because the reactants are monatomic and spherical (nondirectional); all collision orientations are equally effective.

(b) The orientation factor does not depend on temperature. It is combined with frequency of collision into the frequency factor, A, which is essentially constant with temperature.

14.53 *Analyze/Plan.* Given the temperature and energy, use Equation 14.20 to calculate the fraction of Ar atoms that have at least this energy. *Solve.*

$f = e^{-E_a/RT}$ $E_a = 10.0 \text{ kJ/mol} = 1.00 \times 10^4 \text{ J/mol}; \; T = 400 \text{ K } (127 \,^{\circ}C)$

$$-E_a/RT = -\frac{1.00 \times 10^4 \text{ J/mol}}{400 \text{ K}} \times \frac{\text{mol-K}}{8.314 \text{ J}} = -3.0070 = -3.01$$

$f = e^{-3.0070} = 4.9 \times 10^{-2}$

At 400 K, approximately 1 out of 20 molecules has this kinetic energy.

14.54 (a) $f = e^{-E_a/RT}$ $E_a = 160 \text{ kJ/mol} = 1.60 \times 10^5 \text{ J/mol}, \; T = 500 \text{ K}$

$$-E_a/RT = -\frac{1.60 \times 10^5 \text{ J/mol}}{500 \text{ K}} \times \frac{\text{mol-K}}{8.314 \text{ J}} = -38.489 = -38.5$$

$f = e^{-38.489} = 1.924 \times 10^{-17} = 2 \times 10^{-17}$

(b) $-E_a/RT = -\dfrac{1.60 \times 10^5 \text{ J/mol}}{520 \text{ K}} \times \dfrac{\text{mol-K}}{8.314 \text{ J}} = -37.009 = -37.0$

$f = e^{-37.009} = 8.45712 \times 10^{-17} = 8.46 \times 10^{-17}$

$$\frac{f \text{ at } 520 \text{ K}}{f \text{ at } 500 \text{ K}} = \frac{8.46 \times 10^{-17}}{1.92 \times 10^{-17}} = 4.41$$

An increase of 20 K means that 4.41 times more molecules have this energy.

14.55 *Analyze/Plan.* Use the definitions of activation energy $(E_{max} - E_{react})$ and ΔE $(E_{prod} - E_{react})$ to sketch the graph and calculate E_a for the reverse reaction. *Solve.*

(a)   (b) $E_a(\text{reverse}) = 73 \text{ kJ}$

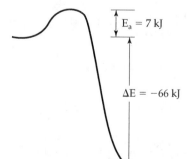

$E_a = 7 \text{ kJ}$

$\Delta E = -66 \text{ kJ}$

14.56   *Analyze/Plan.* Use the definitions of activation energy ($E_{max} - E_{react}$) and ΔE ($E_{prod} - E_{react}$) to sketch the graph and calculate E_a for the reverse reaction.   *Solve.*

(a)   (b) E_a(reverse) = 18 kJ

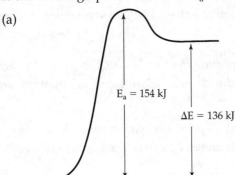

$E_a = 154$ kJ

$\Delta E = 136$ kJ

14.57   (a)   False. If you compare two reactions with similar collision factors, the one with the larger activation energy will be *slower*.

(b)   False. A reaction that has a small rate constant will have either a small frequency factor (A), a large activation energy (E_a), or both.

(c)   True.

14.58   (a)   False. If you measure the rate constant for a reaction at different temperatures, you can calculate the overall activation energy, E_a, for the reaction.

(b)   False. Exothermic reactions are not necessarily faster than endothermic reactions. (The rate of a reaction is not determined by the overall enthalpy change going from reactants to products.)

(c)   False. If you double the temperature for a reaction, there is no change to the activation energy, E_a.

14.59   The order of slowest reaction to fastest reaction is: rate (c) < rate (a) < rate (b). Assuming all collision factors (A) to be the same, reaction rate depends only on E_a; it is independent of ΔE.

14.60 E_a for the reverse reaction is:

(a) 45 – (–25) = 70 kJ   (b) 35 – (–10) = 45 kJ   (c) 55 – 10 = 45 kJ

Based on the magnitude of E_a, the reverse of reactions (b) and (c) occur at the same rate, which is faster than the reverse of reaction (a).

14.61   *Analyze/Plan.* Given k_1, at T_1, calculate k_2 at T_2. Change T to Kelvins, then use the Equation 14.23 to calculate k_2.   *Solve.*

$T_1 = 20\ °C + 273 = 293$ K; $T_2 = 60\ °C + 273 = 333$ K; $k_1 = 2.75 \times 10^{-2}\,s^{-1}$

(a) $\ln\left(\dfrac{k_1}{k_2}\right) = \dfrac{E_a}{R}\left(\dfrac{1}{333} - \dfrac{1}{293}\right) = \dfrac{75.5 \times 10^3\ J/mol}{8.314\ J/mol}(-4.1 \times 10^{-4})$

$\ln(k_1/k_2) = -3.7229 = -3.7;\ \ k_1/k_2 = 0.0242 = 0.02;\ \ k_2 = \dfrac{0.0275\ s^{-1}}{0.0242} = 1.14 = 1\,s^{-1}$

(b) $\ln\left(\dfrac{k_1}{k_2}\right) = \dfrac{125 \times 10^3\ J/mol}{8.314\ J/mol}\left(\dfrac{1}{333} - \dfrac{1}{293}\right) = -6.1638 = -6.2$

$k_1/k_2 = 2.104 \times 10^{-3} = 2 \times 10^{-3};\ \ k_2 = \dfrac{0.0275\ s^{-1}}{2.104 \times 10^{-3}} = 13.07 = 1 \times 10\,s^{-1}$

(c)   The method in parts (a) and (b) assumes that the collision model and thus the Arrhenious equation describe the kinetics of the reactions. That is, activation energy is constant over the temperature range under consideration. There is no assumption about temperature dependence of the frequency factor, because it drops out of the difference equation by subtraction.

14.62 $T_1 = 737\,^\circ C + 273 = 1010\ K,\ k_1 = 0.0796\ M^{-1}s^{-1}$;

$T_2 = 947\,^\circ C + 273 = 1220\ K,\ k_2 = 0.0815\ M^{-1}s^{-1}$

$$\ln\left(\frac{k_1}{k_2}\right) = \frac{E_a}{R}\left(\frac{1}{T_2} - \frac{1}{T_1}\right)$$

$$\ln\left(\frac{0.0796}{0.0815}\right) = \frac{E_a}{8.314\ J/mol}\left(\frac{1}{1220} - \frac{1}{1010}\right)$$

$$-0.023589 = \frac{E_a\,(-1.704\times10^{-4})}{8.314\ J/mol}$$

$$E_a = \frac{8.314\,(-0.023589)\ J/mol}{(-1.704\times10^{-4})} = 1.151\times10^3\ J/mol = 1.15\ kJ/mol$$

14.63   *Analyze/Plan.* Follow the logic in Sample Exercise 14.11.   *Solve.*

k	ln k	T(K)	1/T(× 10³)
0.0521	−2.955	288	3.47
0.101	−2.293	298	3.36
0.184	−1.693	308	3.25
0.332	−1.103	318	3.14

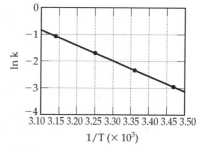

The slope, -5.64×10^3, equals $-E_a/R$. Thus, $E_a = 5.64\times10^3\times8.314\ J/mol = 46.9\ kJ/mol$.

14.64

k	ln k	T(K)	1/T(× 10³)
0.028	−3.58	600	1.67
0.22	−1.51	650	1.54
1.3	0.26	700	1.43
6.0	1.79	750	1.33
23	3.14	800	1.25

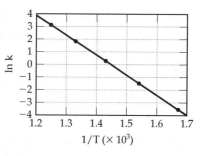

Using the relationship $\ln k = \ln A - E_a/RT$, the slope, $-15.94\times10^3 = -16\times10^3$, is $-E_a/R$. $E_a = 15.94\times10^3\times8.314\ J/mol = 1.3\times10^2\ kJ/mol$. To calculate A, we will use the rate data at 700 K. From the equation given above, $0.262 = \ln A - 15.94\times10^3/700$; $\ln A = 0.262 + 22.771$. $A = 1.0\times10^{10}$.

Reaction Mechanisms (Section 14.6)

14.65   (a)   An *elementary reaction* is a process that occurs in a single event; the order is given by the coefficients in the balanced equation for the reaction.

   (b)   A *unimolecular* elementary reaction involves only one reactant molecule; the activated complex is derived from a single molecule. A *bimolecular* elementary reaction involves two reactant molecules in the activated complex and the overall process.

   (c)   A *reaction mechanism* is a series of elementary reactions that describe how an overall reaction occurs and explain the experimentally determined rate law.

   (d)   A *rate-determining step* is the slowest step in a reaction mechanism. It limits the overall reaction rate.

14.66   (a)   No. An intermediate is a substance that is produced and then consumed during a chemical reaction. It could be a product in the first step of a reaction mechanism, but not a reactant.

   (b)   On a reaction profile, an intermediate is a valley. It is a product of one step in the mechanism and a reactant in the next.

   (c)   Unimolecular. An elementary reaction occurs in a single step. The order of an elementary reaction is given by the coefficients in the chemical reaction. For the decomposition of Cl_2, the coefficient of Cl_2 is one.

   (d)   A *transition state* is a high energy complex formed when one or more reactants collide and distort in a way that can lead to formation of product(s). An *intermediate* is the product of an early elementary reaction in a multistep reaction mechanism. A transition state occurs at an energy maximum or peak of a reaction profile as in Figure 14.19. An intermediate exists at an energy minimum or trough of a reaction profile. Every reaction, single- or multi-step, has a transition state. Only multistep reactions have intermediates.

14.67   *Analyze/Plan.* Elementary reactions occur as a single step, so the molecularity is determined by the number of reactant molecules; the rate law reflects reactant stoichiometry.   *Solve.*

   (a)   unimolecular, rate = $k[Cl_2]$

   (b)   bimolecular, rate = $k[OCl^-][H_2O]$

   (c)   bimolecular, rate = $k[NO][Cl_2]$

14.68   (a)   bimolecular, rate = $k[NO]^2$

   (b)   unimolecular, rate = $k[C_3H_6]$

   (c)   unimolecular, rate = $k[SO_3]$

14.69   *Analyze/Plan.* Use the definitions of the term *intermediate*, along with the characteristics of reaction profiles, to answer the questions.   *Solve.*

   This is a three-step mechanism, A $\rightarrow$ B, B $\rightarrow$ C, and C $\rightarrow$ D.

   (a)   There are 2 intermediates, B and C.

   (b)   There are 3 energy maxima in the reaction profile, so there are 3 transition states.

(c)   Step C → D has the lowest activation energy, so it is fastest.

(d)   The energy of D is slightly greater than the energy of A, so ΔE for the overall reaction is positive.

14.70   (a)   Two elementary reactions; two energy maxima

        (b)   One intermediate; one energy minimum between reactants and products

        (c)   The second step is rate-limiting; second energy maximum and E_a is larger.

        (d)   For the overall reaction, ΔE is negative, because the potential energy of the products is lower than the potential energy of the reactants.

14.71   *Analyze/Plan.* Follow the logic in Sample Exercise 14.14.   *Solve.*

        (a) $H_2(g) + ICl(g) \rightarrow HI(g) + HCl(g)$

                $\underline{HI(g) + ICl(g) \rightarrow I_2(g) + HCl(g)}$

                $H_2(g) + 2\,ICl(g) \rightarrow I_2(g) + 2\,HCl(g)$

        (b)   Intermediates are produced and consumed during reaction. HI is the intermediate.

        (c)   The slow step determines the rate law for the overall reaction. If the first step is slow, the observed rate law is: Rate = $k[H_2][ICl]$.

14.72   (a) $2H_2O_2(aq) \rightarrow 2H_2O(l) + O_2(g)$

        (b) $IO^-(aq)$ is the intermediate.

        (c)   Rate = $k[H_2O_2]\,[I^-]$

14.73   (a)   *Analyze.* Given data on concentration of a reactant versus time, determine whether the proposed reaction mechanism is consistent with the data.

                Plan. Based on the graph, decide the order of reaction with respect to [NO]. Write the two possible rate laws, depending on which step is rate-determining. Decide if one of the rate laws, and thus the mechanism, is consistent with the rate data.

                Solve. The graph of 1/[NO] versus time is linear with positive slope, indicating that the reaction is second order in [NO]. The rate law will include $[NO]^2$.

                If the first step is slow, the observed rate law is the rate law for this step: rate = $k[NO][Cl_2]$. Because the observed rate law is second order in [NO], the second step must be slow relative to the first step. Follow the logic in Sample Exercise 14.15 for determining the rate law of a mechanism with a fast initial step.

                From the rate-determining second step, rate = $k[NOCl_2][NO]$.

                Assuming the first step is a fast equilibrium, $k_1[NO][Cl_2] = k_{-1}[NOCl_2]$.

                Solving for $[NOCl_2]$ in terms of $[NO][Cl_2]$, $[NOCl_2] = \dfrac{k_1}{k_{-1}}[NO][Cl_2]$

                $Rate = \dfrac{k_2 k_1}{k_{-1}}[NO][Cl_2][NO] = [NO]^2[Cl_2]$

                This rate law is second order in [NO]. It is consistent with the observed data.

        (b)   The linear plot guarantees that the overall rate law will include $[NO]^2$. Because the data were obtained at constant $[Cl_2]$, we have no information about reaction order with respect to $[Cl_2]$.

415

14.74   (a)   i. $HBr + O_2 \rightarrow HOOBr$

        ii. $HOOBr + HBr \rightarrow 2\,HOBr$

        iii. $2\,HOBr + 2\,HBr \rightarrow 2\,H_2O + 2\,Br_2$
        _____

                    $4\,HBr + O_2 \rightarrow 2\,H_2O + 2\,Br_2$

        (b)   The observed rate law is: rate = k[HBr][O₂], the rate law for the first elementary
              step. The first step must be rate-determining.

        (c)   HOOBr and HOBr are both intermediates; HOOBr is produced in i and
              consumed in ii and HOBr is produced in ii and consumed in iii.

        (d)   Because the first step is rate-determining, it is possible that neither of the
              intermediates accumulates enough to be detected. This does not disprove the
              mechanism, but indicates that steps ii and iii are very fast, relative to step i.

Catalysis (Section 14.7)

14.75   (a)   A catalyst is a substance that changes (usually increases) the speed of a chemical
              reaction without undergoing a permanent chemical change itself.

        (b)   A homogeneous catalyst is in the same phase as the reactants; a heterogeneous
              catalyst is in a different phase and is usually a solid.

        (c)   A catalyst has no effect on the overall enthalpy change for a reaction. A catalyst
              does affect activation energy, E_a, which is one way that it changes reaction rate. It
              can also affect the frequency factor, A.

14.76   (a)   The smaller the particle size of a solid catalyst, the greater the surface area. The
              greater the surface area, the more active sites and the greater the increase in reaction
              rate.

        (b)   Adsorption is the binding of reactants onto the surface of the heterogeneous
              catalyst. It is usually the first step in the catalyzed reaction.

14.77   KBr(s) is added to H_2O_2(aq) at t = 0. Assume the KBr(s) dissolves instantly. As the
        reaction proceeds, the Br⁻ catalyst is consumed and then regenerated.

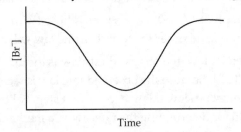

14.78   For an acid-catalyzed reaction in solution, H^+ is a homogeneous catalyst. It is consumed
        and then regenerated during the reaction. (This assumes that H^+ is present in excess and
        that H^+ is not a reactant, that the reactants are neither acids nor bases.) The [H^+] is a
        maximum at t = 0 and when the reaction is complete.

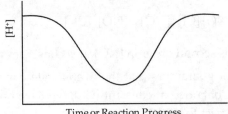

14.79   (a) $2[NO_2(g) + SO_2(g) \rightarrow NO(g) + SO_3(g)]$

$\underline{2\,NO(g) + O_2(g) \rightarrow 2\,NO_2(g)}$

$2\,SO_2(g) + O_2(g) \;\rightarrow 2\,SO_3(g)$

(b) $NO_2(g)$ is a catalyst because it is consumed and then reproduced in the reaction sequence.

(c) $NO(g)$ is an intermediate, because it is produced and then consumed during the reaction.

(d)   Because NO_2 is in the same state as the other reactants, this is homogeneous catalysis.

14.80   (a) $2[NO(g) + N_2O(g) \rightarrow N_2(g) + NO_2(g)]$

$\underline{2\,NO_2(g) \;\rightarrow 2\,NO(g) + O_2(g)}$

$2\,N_2O(g) \;\rightarrow 2\,N_2(g) + O_2(g)$

(b)   NO serves as a catalyst in this reaction. It is present when the reaction sequence begins and after the last step is completed.

(c)   No. The proposed mechanism cannot be ruled out if there is no build-up of NO_2. In this reaction, NO_2 functions as an intermediate; it is produced and then consumed during the reaction. If there is no measurable build-up of NO_2, the first step is slow relative to the second. As soon as NO_2 is produced by the slow first step, it is consumed by the faster second step.

14.81   (a)   When using a powdered metal catalyst, only a small percentage of the metal atoms are at the surface of the bulk material and catalytically active. Use of chemically stable supports such as alumina and silica makes it possible to obtain very large surface areas per unit mass of the precious metal catalyst. This is so because the metal can be deposited in a very thin, even monomolecular, layer on the surface of the support.

(b)   The greater the surface area of the catalyst, the more reaction sites, and the greater the rate of the catalyzed reaction.

14.82   (a)   Catalytic converters are heterogeneous catalysts that adsorb gaseous CO and hydrocarbons and speed up their oxidation to $CO_2(g)$ and $H_2O(g)$. They also adsorb nitrogen oxides, NO_x, and speed up their reduction to $N_2(g)$ and $O_2(g)$. If a catalytic converter is working effectively, the exhaust gas should have very small amounts of the undesirable gases CO, $(NO)_x$, and hydrocarbons.

(b)   The high temperatures could increase the rate of the desired catalytic reactions given in part (a). It could also increase the rate of undesirable reactions such as corrosion, which decrease the lifetime of the catalytic converter.

(c)   The rate of flow of exhaust gases over the converter will determine the rate of adsorption of CO, $(NO)_x$, and hydrocarbons onto the catalyst and thus the rate of conversion to desired products. Too fast an exhaust flow leads to less than maximum adsorption. A very slow flow leads to back pressure and potential damage to the exhaust system. Clearly the flow rate must be adjusted to balance chemical and mechanical efficiency of the catalytic converter.

14.83   As illustrated in Figure 14.23, the two C–H bonds that exist on each carbon of the ethylene molecule before adsorption are retained in the process in which a D atom is added to each C (assuming we use D_2 rather than H_2). To put two deuteriums on a single carbon, it is necessary that one of the already existing C–H bonds in ethylene be broken while the molecule is adsorbed, so the H atom moves off as an adsorbed atom, and is replaced by a D. This requires a larger activation energy than simply adsorbing C_2H_4 and adding one D atom to each carbon.

14.84   Just as the π electrons in C_2H_4 are attracted to the surface of a hydrogenation catalyst, the nonbonding electron density on S causes compounds of S to be attracted to these same surfaces. Strong interactions could cause the sulfur compounds to be permanently attached to the surface, blocking active sites and reducing adsorption of alkenes for hydrogenation.

14.85   Let k and E_a equal the rate constant and activation energy without the enzyme (uncatalyzed). Let k_c and E_{ac} equal the rate constant and activation energy with the enzyme (catalyzed). A is the same for the uncatalyzed and catalyzed reactions. The difference in activation energies is $E_{ac} - E_a$, $k_c = 1.0 \times 10^6 \text{ s}^{-1}$, $k = 0.039 \text{ s}^{-1}$, T = 25 °C = 298 K.

According to Equation 14.22, $\ln k = E_a/RT + \ln A$. Subtracting $\ln k$ from $\ln k_c$

$$\ln k_c - \ln k = \left[\frac{-E_{ac}}{RT}\right] + \ln A - \left[\frac{-E_a}{RT}\right] - \ln A$$

$$\ln (k_c/k) = \frac{E_a - E_{ac}}{RT}; \quad E_a - E_{ac} = RT \ln (k_c/k)$$

$$E_a - E_{ac} = \frac{8.314 \text{ J}}{\text{mol-K}} \times 298 \text{ K} \times \ln \frac{1.0 \times 10^6}{0.039} = 42,267 \text{ J} = 42.267 \text{ kJ} = 42 \text{ kJ}$$

Carbonic anyhdrase lowers the activation energy of the reaction by 42 kJ.

14.86   (a) $(NH_2)_2C{=}O(aq) + H_2O(l) \rightarrow CO_2(g) + 2 NH_3(aq)$

       (b)   From Solution 14.85, $E_a - E_{ac} = RT \ln (k_c/k)$. $k_c = 3.4 \times 10^4 \text{ s}^{-1}$, $k = 4.15 \times 10^{-5} \text{ s}^{-1}$, T = 100 °C = 373 K.

$$E_a - E_{ac} = \frac{8.314 \text{ J}}{\text{mol-K}} \times 373 \text{ K} \times \ln \frac{3.4 \times 10^4}{4.15 \times 10^{-5}} = 63,647 \text{ J} = 63.647 \text{ kJ} = 64 \text{ kJ}$$

       (c)   Because reaction rate always increases with increasing temperature, we expect the rate of the catalyzed reaction to be significantly greater at 100 °C than at 21 °C.

       (d)   The 64 kJ difference between activation energies for the catalyzed and uncatalyzed reaction calculated in part (b) is a minimum difference. Because we expect the value of k_c to be significantly greater than $3.4 \times 10^4 \text{ s}^{-1}$, the difference in activation energies for the catalyzed and uncatalyzed reactions will be greater than 64 kJ.

14.87   *Analyze/Plan.* Let k = the rate constant for the uncatalyzed reaction,
                k_c = the rate constant for the catalyzed reaction

According to Equation 14.22, $\ln k = -E_a/RT + \ln A$

Subtracting $\ln k$ from $\ln k_c$,

$$\ln k_c - \ln k = -\left[\frac{55 \text{ kJ/mol}}{RT} + \ln A\right] - \left[-\frac{95 \text{ kJ/mol}}{RT} + \ln A\right]. \quad \textit{Solve.}$$

(a) RT = 8.314 J/mol-K × 298 K × 1 kJ/1000 J = 2.478 kJ/mol; ln A is the same for both reactions.

$$\ln (k_c / k) = \frac{95\ kJ/mol - 55\ kJ/mol}{2.478\ kJ/mol}; \quad k_c / k = 1.024 \times 10^7 = 1 \times 10^7$$

The catalyzed reaction is approximately 10,000,000 (ten million) times faster at 25 °C.

(b) RT = 8.314 J/mol-K × 398 K × 1 kJ/1000 J = 3.309 kJ/mol

$$\ln (k_c / k) = \frac{40\ kJ/mol}{3.309\ kJ/mol}; \quad k_c / k = 1.778 \times 10^5 = 2 \times 10^5$$

The catalyzed reaction is 200,000 times faster at 125 °C.

14.88 Let k and E_a equal the rate constant and activation energy for the uncatalyzed reaction. Let k_c and E_{ac} equal the rate constant and activation energy of the catalyzed reaction. A is the same for the uncatalyzed and catalyzed reactions. $k_c / k = 1 \times 10^5$, T = 37 °C = 310 K.

According to Equation 14.22, $\ln k = -E_a / RT + \ln A$. Subtracting ln k from ln k_c

$$\ln k_c - \ln k = \left[\frac{-E_{ac}}{RT} \right] + \ln A - \left[\frac{-E_a}{RT} \right] - \ln A$$

$$\ln (k_c / k) = \frac{E_a - E_{ac}}{RT}; \quad E_a - E_{ac} = RT \ln (k_c / k)$$

$$E_a - E_{ac} = \frac{8.314\ J}{mol\text{-}K} \times 310\ K \times \ln (1 \times 10^5) = 2.966 \times 10^4\ J = 29.66\ kJ = 3 \times 10^1 kJ$$

The enzyme must lower the activation energy by 30 kJ to achieve a 1×10^5-fold increase in reaction rate.

Additional Exercises

14.89 (a) False. If this is not an elementary reaction, we need more information to write the correct rate law.

(b) True. If the reaction is elementary, the rate law in part (a) is correct, and the reaction is second order overall.

(c) False. If the reaction is elementary, the reverse reaction is second order.

(d) False. The relationship between the forward and reverse activation energies depends on whether ΔE for the reaction is positive or negative.

14.90 $$Rate = \frac{-\Delta[H_2S]}{\Delta t} = \frac{\Delta[Cl^-]}{2\Delta t} = k[H_2S][Cl_2]$$

$$\frac{-\Delta[H_2S]}{\Delta t} = -(3.5 \times 10^{-2}\ M^{-1}s^{-1})(2.0 \times 10^{-4}\ M)(0.025\ M) = -1.75 \times 10^{-7} = -1.8 \times 10^{-7}\ M/s$$

$$\frac{\Delta[Cl^-]}{\Delta t} = \frac{-2\Delta[H_2S]}{\Delta t} = -2(-1.75 \times 10^{-7}\ M/s) = 3.5 \times 10^{-7}\ M/s$$

14.91   (a) $\text{Rate} = \dfrac{-\Delta[\text{NO}]}{2\Delta t} = \dfrac{-\Delta[\text{O}_2]}{\Delta t} = \dfrac{9.3 \times 10^{-5} \ M/s}{2} = 4.7 \times 10^{-5} \ M/s$

(b, c) $\text{Rate} = k[\text{NO}]^2[\text{O}_2]; \ k = \text{rate}/[\text{NO}]^2[\text{O}_2]$

$$k = \dfrac{4.7 \times 10^{-5} \ M/s}{(0.040 \ M)^2 (0.035 \ M)} = 0.8393 = 0.84 \ M^{-2} s^{-1}$$

(d)   Because the reaction is second order in NO, if the [NO] is increased by a factor of 1.8, the rate would increase by a factor of 1.8^2, or $(3.24) = 3.2$.

14.92   *Analyze/Plan.* Using the relationship rate $= k[\text{A}]^x$, determine the value of x that produces a rate law to match the described situation.   *Solve.*

(a) $x = 0$. The rate of reaction does not depend on $[\text{A}]_0$, so the reaction is zero-order in A.

(b) $x = 2$. When $[\text{A}]_0$ increases by a factor of 3, rate increases by a factor of $(3)^2 = 9$.

(c) $x = 3$. When $[\text{A}]_0$ increases by a factor of 2, rate increases by a factor of $(2)^3 = 8$.

14.93   (a)   The rate increases by a factor of nine when $[\text{C}_2\text{O}_4^{2-}]$ triples (compare experiments 1 and 2). The rate doubles when $[\text{HgCl}_2]$ doubles (compare experiments 2 and 3). The apparent rate law is: $\text{Rate} = k[\text{HgCl}_2][\text{C}_2\text{O}_4^{2-}]^2$

(b) $k = \dfrac{\text{rate}}{[\text{HgCl}_2] \, [\text{C}_2\text{O}_4^{2-}]^2}$ Using the data for Experiment 1,

$$k = \dfrac{(3.2 \times 10^{-5} \ M/s)}{[0.164 \ M][0.15 \ M]^2} = 8.672 \times 10^{-3} = 8.7 \times 10^{-3} \ M^{-2}s^{-1}$$

(c) $\text{Rate} = (8.672 \times 10^{-3} \ M^{-2}s^{-1})(0.100 \ M)(0.25 \ M)^2 = 5.4 \times 10^{-5} \ M/s$

14.94   (a)   Compare experiments 2 and 3, where the $[\text{X}]_0$ is the same and $[\text{Z}]_0$ increases by a factor of 1.5. The rate increases by a factor or 2.25, or $(1.5)^2$. The reaction is second order in [Z]. Next compare experiments 1 and 2, where both $[\text{X}]_0$ and $[\text{Z}]_0$ increase by a factor of two. The rate increases by a factor of eight. We know that doubling $[\text{Z}]_0$ increases rate by a factor of four; doubling $[\text{X}]_0$ then increases rate by a factor of two. The reaction is first order in $[\text{X}]_0$. The apparent rate law is: $\text{Rate} = k[\text{X}][\text{Z}]^2$

Check. Compare experiments 1 and 3. Here $[\text{X}]_0$ doubles and $[\text{Z}]_0$ triples. If our rate law is correct, the rate should increase by a factor of $(2)(3)^2 = 18$. This is indeed the relationship between the rates of Experiments 1 and 3.

(b) $k = \dfrac{\text{rate}}{[\text{X}][\text{Z}]^2}$. Using the data for Experiment 2,

$$k = \dfrac{(3.2 \times 10^2 \ M/s)}{[0.50 \ M][0.50 \ M]^2} = 2.560 \times 10^3 = 2.6 \times 10^3 \ M^{-2}s^{-1}$$

The same value is obtained using data from the other two experiments.

(c) $\text{Rate} = (2.56 \times 10^3 \ M^{-2}s^{-1})(0.75 \ M)(1.25 \ M)^2 = 3.0 \times 10^3 \ M/s$

14.95   (a)   Because the units of the rate constant are $M^{-1}s^{-1}$, the reaction is second order overall and second order in NO_2.

(b)   If $[NO_2]_0 = 0.100 \ M$ and $[NO_2]_t = 0.025 \ M$, use the integrated form of the second order rate equation, $\dfrac{1}{[A]_t} = kt + \dfrac{1}{[A]_0}$, Equation 14.14, to solve for t.

$$\frac{1}{0.025 \ M} = 0.63 \ M^{-1}s^{-1}(t) + \frac{1}{0.100 \ M} ; \ \frac{(40-10) \ M^{-1}}{0.63 \ M^{-1}s^{-1}} = t = 47.62 = 48 \ s.$$

14.96   For a first-order reaction, $t_{1/2} = 0.633/k$. For a second-order reaction, $t_{1/2} = 1/k[A]_0$. Half-life is constant over the course of the reaction for first-order reactions. Although second-order half-life does not appear to depend on t, the value of "$[A]_0$" does change over the course of the reaction, and $t_{1/2}$ increases with time. (For a zero reaction, $t_{1/2}$ decreases with time.)

The rate law for reaction (1) must be first order, because that is the only reaction type that has a constant half-life. Reaction (2), where $t_{1/2}$ increases with time, is second order.

14.97   *Analyze/Plan.* Given k and $[A]_0$, use the integrated form of the first-order rate law to calculate [A] at t = 660 s. Then, Rate = k[A].

For a first-order reaction, $\ln[A]_t - \ln[A]_0 = -kt$. $\ln[A]_t = -kt + \ln[A]_0$

$[A]_0 = 2.5 \times 10^{-2} \ M$, t = 660 s, $k = 3.2 \times 10^{-3} \ s^{-1}$

$\ln [A] = -3.2 \times 10^{-3} \ s^{-1} (660 \ s) + \ln(0.025)$

$\ln [A] = -5.8009, \ [A] = 3.025 \times 10^{-3} \ M = 3.0 \times 10^{-3} \ M$

Rate $= k[A] = (3.2 \times 10^{-3} \ s^{-1})(3.025 \times 10^{-3} \ M) = 9.7 \times 10^{-6} \ M/s$

14.98   (a) $t_{1/2} = 0.693/k = 0.693/7.0 \times 10^{-4} \ s^{-1} = 990 = 9.9 \times 10^{2} \ s$

(b)   Use Equation 14.23 to calculate the desired temperature, T_2. Assuming the same initial concentration of H_2O_2, if the rate of reaction doubles, the rate constant, k, doubles. Then,

$k_2 = 2 \ k_1$, $k_1/k_2 = 1/2$; $E_a = 75 \ kJ/mol = 7.5 \times 10^4 \ J/mol$; $T_1 = 300 \ K$, $T_2 = ?$

$$\ln\left(\frac{k_1}{k_2}\right) = \frac{E_a}{R}\left[\frac{1}{T_2} - \frac{1}{T_1}\right]; \ \ln(1/2) = \frac{7.5 \times 10^4 \ J/mol}{8.314 \ J/mol}\left[\frac{1}{T_2} - \frac{1}{300}\right]$$

$$-0.69315 = 9.0209 \times 10^3 \left[\frac{1}{T_2} - 3.3333 \times 10^{-3}\right]; \ \frac{1}{T_2} = \frac{-0.69315}{9.0209 \times 10^3} + 3.3333 \times 10^{-3}$$

$T_2 = 307 \ K$

Note that a relatively small change in temperature is required to double the reaction rate.

14.99   *Analyze.* Given rate constants for the decay of two radioisotopes, determine half-lives, decay rates, and amount remaining after three half-lives. *Plan.* Determine reaction order. Based on reaction-order, select the appropriate relationships for (a) rate constant and half-life and (c) rate-constant, time and concentration. In this example, mass is a measure of concentration.

Solve. Decay of radioisotopes is a first-order process, because only one species is involved and the decay is not initiated by collision.

(a)   For a first-order process, $t_{1/2} = 0.693/k$.

^{241}Am: $t_{1/2} = 0.693/1.6 \times 10^{-3}$ yr$^{-1} = 433.1 = 4.3 \times 10^2$ yr

^{125}I: $t_{1/2} = 0.693/0.011$ day$^{-1} = 63.00 = 63$ days

(b)   For a given sample size, half of the ^{241}Am sample decays in 433 years, whereas half of the ^{125}I sample decays in 63 days. ^{125}I decays at a much faster rate.

(c)   For a first-order process, $\ln[A]_t - \ln[A]_0 = -kt$. $\ln[A]_t = -kt + \ln[A]_0$.

$[A]_0 = 1.0$ mg; $t = 3\, t_{1/2}$.

^{241}Am: $t = 3\, t_{1/2} = 3(433.1 \text{ yr}) = 1.299 \times 10^3 = 1.3 \times 10^3$ yr

$\ln[Am]_t = -1.6 \times 10^{-3}$ yr$^{-1} (1.299 \times 10^3 \text{ yr}) - \ln(1.0) = -2.079 - 0 = -2.08$

$[Am]_t = 0.125 = 0.13$ mg

or, mass ^{241}Am remaining $= 1.0$ mg$/2^3 = 0.125 = 0.13$ mg

^{125}I: For the same size starting sample and number of elapsed half-lives, the same mass, 0.13 mg ^{125}I, will remain. (The difference is that the elapsed time of 3 half-lives (for ^{125}I is $3(63) = 189$ days $= 0.52$ yr, versus 433 yr for ^{241}Am.)

(d)   Again, for a first-order process, $\ln[A]_t - \ln[A]_0 = -kt$. $\ln[A]_t = -kt + \ln[A]_0$.

$[A]_0 = 1.0$ mg; $t = 4$ days.

$k_{Am} = 1.6 \times 10^{-3}$ yr^{-1} (1 yr/365 days) $= 4.3836 \times 10^{-6} = 4.4 \times 10^{-6}$ day^{-1}

^{241}Am: $\ln[Am]_t = -4.4 \times 10^{-6}$ day^{-1} (4 days) $- \ln(1.0) = -1.7 \times 10^{-5} - 0 = -1.7 \times 10^{-5}$

The amount of ^{241}Am remaining after 4 days is 0.99998 mg, to three significant figures, 1.00 mg.

^{125}I: $\ln[I]_t = -0.011$ day^{-1} (4 days) $- \ln(1.0) = -0.044 - 0 = -0.044$

The amount of ^{125}I remaining after 4 days is 0.956 mg.

14.100 (a) $k = (8.56 \times 10^{-5} \, M/\text{s})/(0.200\, M) = 4.28 \times 10^{-4}$ s^{-1}

(b) $\ln[\text{urea}] = -(4.28 \times 10^{-4}\text{s}^{-1} \times 4.00 \times 10^3 \text{ s}) + \ln(0.500)$

$\ln[\text{urea}] = -1.712 - 0.693 = -2.405 = -2.41$; $[\text{urea}] = 0.0903 = 0.090\, M$

(c) $t_{1/2} = 0.693/k = 0.693/4.28 \times 10^{-4}$ s$^{-1} = 1.62 \times 10^3$ s

14.101 (a) $A = \varepsilon bc$, Equation 14.5. $A = 0.605$, $\varepsilon = 5.60 \times 10^3$ cm^{-1} M^{-1}, $b = 1.00$ cm

$c = \dfrac{A}{\varepsilon b} = \dfrac{0.605}{(5.60 \times 10^3 \text{ cm}^{-1}\, M^{-1})(1.00 \text{ cm})} = 1.080 \times 10^{-4} = 1.08 \times 10^{-4}\, M$

(b)   Calculate $[c]_t$ using Beer's law. We calculated $[c]_0$ in part (a). Use Equation 14.13 to calculate k.

$A_{30} = \varepsilon bc_{30}$; $c_{30} = \dfrac{A_{30}}{\varepsilon b} = \dfrac{0.250}{(5.60 \times 10^3 \text{ cm}^{-1}\, M^{-1})(1.00 \text{ cm})} = 4.464 \times 10^{-5}\, M$

$\ln[c]_t = -kt + \ln[c]_0$; $\dfrac{\ln[c]_0 - \ln[c]_t}{t} = k$; $t = 30 \text{ min} \times \dfrac{60 \text{ s}}{\text{min}} = 1800$ s

$k = (\ln(1.080 \times 10^{-4}) - \ln(4.464 \times 10^{-5}))/1800 \text{ s} = 4.910 \times 10^{-4} = 4.91 \times 10^{-4}$ s^{-1}

(c)   For a first-order reaction, $t_{1/2} = 0.693/k$.

$t_{1/2} = 0.693/4.910 \times 10^{-4} \, s^{-1} = 1.411 \times 10^{3} = 1.41 \times 10^{3} \, s = 23.5$ min

(d) $A_t = 0.100$; calculate c_t using Beer's law, then t from the first-order integrated rate equation.

$$c_t = \frac{A}{\varepsilon b} = \frac{0.100}{(5.60 \times 10^{3} \, cm^{-1} \, M^{-1})(1.00 \, cm)} = 1.786 \times 10^{-5} = 1.79 \times 10^{-5} \, M$$

$$t = \frac{\ln[c]_0 - \ln[c]_t}{k} = \frac{\ln(1.080 \times 10^{-4}) - \ln(1.786 \times 10^{-5})}{4.910 \times 10^{-4} \, s^{-1}}$$

$t = 3.666 \times 10^{3} = 3.67 \times 10^{3} \, s = 61.1$ min

14.102   Calculate [dye] at each time, using Beer's law, $A = \varepsilon bc$; calculate ln[dye] and 1/[dye] and plot these quantities vs time in two separate graphs. The straight-line plot indicates the order of reaction with respect to [dye].

$$A_0 = \varepsilon b c_0; \, c_0 = \frac{A_0}{\varepsilon b} = \frac{1.254}{(4.7 \times 10^{4} \, cm^{-1} \, M^{-1})(1.00 \, cm)} = 2.668 \times 10^{-5} = 2.7 \times 10^{-5} \, M$$

Time (min)	A at 608 nm	[dye]	ln [dye]	1/[dye]
0	1.254	2.7×10^{-5}	−10.53	3.7×10^{4}
30	0.941	2.0×10^{-5}	−10.82	5.0×10^{4}
60	0.752	1.6×10^{-5}	−11.04	6.3×10^{4}
90	0.672	1.4×10^{-5}	−11.16	7.0×10^{4}
120	0.545	1.2×10^{-5}	−11.36	8.6×10^{4}

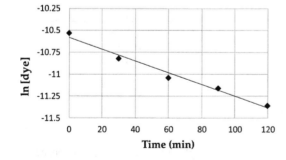
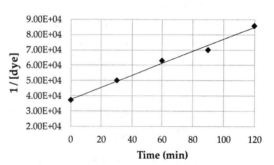

Although the graphs are not absolutely definitive, the plot of 1/[dye] vs time appears to be more linear. (The data point at t = 90 min us "out of line" in both plots and is suspect. More precision and accuracy in the experimental data would be helpful.) Assuming the reaction is second order with respect to the dye, the rate law is: Rate = $k[dye]^2$.

$k = $ slope $ = (8.6 \times 10^{4} - 3.7 \times 10^{4}) \, M^{-1}/(120{-}0)$min $ = 4.1 \times 10^{2} \, M^{-1} \, min^{-1}$

(The best-fit slope and k value is $3.9 \times 10^{2} \, M^{-1} \, min^{-1}$.)

14.103

Time (s)	$[C_5H_6]$ (M)	$\ln[C_5H_6]$	$1/[C_5H_6]$
0	0.0400	–3.219	25.0
50	0.0300	–3.507	33.3
100	0.0240	–3.730	41.7
150	0.0200	–3.912	50.0
200	0.0174	–4.051	57.5

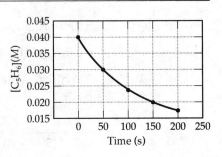

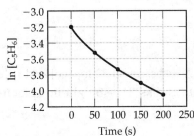

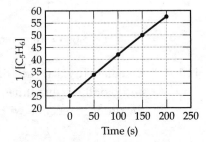

(a)   The plot of $1/[C_5H_6]$ vs time is linear and the reaction is second order.

(b)   The slope of the line in the plot of $1/[C_5H_6]$ vs time is the value of k.

   $k = \text{slope} = (50.0 - 25.0)\ M^{-1}/(150-0)s = 0.167\ M^{-1}\,s^{-1}$

   (The best-fit slope and k value is $0.163\ M^{-1}\,s^{-1}$.)

14.104

$\ln k$	$1/T$
–24.17	3.33×10^{-3}
–20.72	3.13×10^{-3}
–17.32	2.94×10^{-3}
–15.24	2.82×10^{-3}

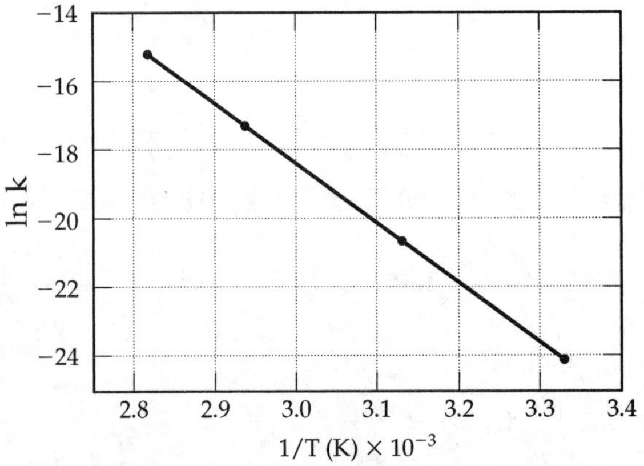

The calculated slope is -1.751×10^4.
The activation energy E_a, equals
$-$ (slope) $\times$ (8.314 J/mol). Thus,
$E_a = 1.8 \times 10^4 (8.314) = 1.5 \times 10^5$
J/mol $= 1.5 \times 10^2$ kJ/mol.
(The best-fit slope is $-1.76 \times 10^4 =$
-1.8×10^4 and the value of E_a is
1.5×10^2 kJ/mol.)

14.105   *Analyze.* Given the time required to reach the "sour" point, at two different temperatures, estimate the activation energy for the reaction.

Plan/Solve. The warmer reaction at 28 °C is faster than the cooler reaction at 5 °C. The ratio of the rates of reaction is the inverse ratio of times.

$$\frac{\text{rate}_{28}}{\text{rate}_5} = \frac{k_{28}[\text{sour milk}]_{28}}{k_5[\text{sour milk}]_5} = \frac{t_5}{t_{28}} = \frac{48\ \text{hr}}{4\ \text{hr}} = 12$$

Assume the concentration of sour milk at 28 °C is the same as the concentration of sour milk at 5 °C. That is, $[\text{sour milk}]_{28} = [\text{sour milk}]_5$. Then, the ratio of rate constants, k_{28}/k_5, equals 12.

$T_1 = 28\ °C = 301\ K;\ T_2 = 5\ °C = 278\ K;\ k_1/k_2 = 12$

$$\ln\left(\frac{k_1}{k_2}\right) = \frac{E_a}{R}\left[\frac{1}{T_2} - \frac{1}{T_1}\right];\ \ln(12) = \frac{E_a}{8.314\ J/mol}\left[\frac{1}{278} - \frac{1}{301}\right]$$

$$E_a = \frac{\ln(12)(8.314\ J/mol)}{2.749 \times 10^{-4}} = 7.516 \times 10^4\ J = 75\ kJ/mol$$

14.106   (a) $T_1 = 77\ °F;\ °C = 5/9\ (°F - 32) = 5/9\ (77 - 32) = 25\ °C = 298\ K$

                $T_2 = 59\ °F;\ °C = 5/9\ (59-32) = 15\ °C = 288\ K;\ k_1/k_2 = 6$

$$\ln\left(\frac{k_1}{k_2}\right) = \frac{E_a}{R}\left[\frac{1}{T_2} - \frac{1}{T_1}\right];\ \ln(6) = \frac{E_a}{8.314\ J/mol}\left[\frac{1}{288} - \frac{1}{298}\right]$$

$$E_a = \frac{\ln(6)(8.314\ J/mol)}{1.165 \times 10^{-4}} = 1.28 \times 10^5\ J = 1.3 \times 10^2\ kJ/mol$$

                $T_1 = 77\ °F = 25\ °C = 298\ K;\ T_2 = 41\ °F = 5\ °C = 278\ K,\ k_1/k_2 = 40$

$$\ln(40) = \frac{E_a}{8.314\ J/mol}\left[\frac{1}{278} - \frac{1}{298}\right];\ E_a = \frac{\ln(40)(8.314\ J/mol)}{2.414 \times 10^{-4}}$$

                $E_a = 1.27 \times 10^5\ J = 1.3 \times 10^2\ kJ/mol$

                The values are amazingly consistent, considering the precision of the data.

      (b)   For a first-order reaction, $t_{1/2} = 0.693/k$, $k = 0.693/t_{1/2}$

                k_1 at $298\ K = 0.693/2.7\ yr = 0.257 = 0.26\ yr^{-1}$

                $T_1 = 298\ K,\ T_2 = 273 - 15\ °C = 258\ K$

$$\ln\left(\frac{0.257}{k_2}\right) = \frac{1.27 \times 10^5\ J}{8.314\ J/mol}\left[\frac{1}{258} - \frac{1}{298}\right] = 7.94727 = 7.95$$

                $0.257/k_2 = e^{7.94727} = 2.828 \times 10^3;\ k_2 = 0.257/2.828 \times 10^3 = 9.088 \times 10^{-5} = 9.1 \times 10^{-5}\ yr^{-1}$

                $t_{1/2} = 0.693/k = 0.693/9.088 \times 10^{-5} = 7.625 \times 10^3\ yr = 7.6 \times 10^3\ yr$

14.107   (a)

$$NO(g) + NO(g) \rightarrow N_2O_2(g)$$
$$\underline{N_2O_2(g) + H_2(g) \rightarrow N_2O(g) + H_2O(g)}$$
$$2\ NO(g) + N_2O_2(g) + H_2(g) \rightarrow N_2O_2(g) + N_2O(g) + H_2O(g)$$
$$2\ NO(g) + H_2(g) \rightarrow N_2O(g) + H_2O(g)$$

      (b)   First reaction: $-\Delta[NO]/\Delta t = k[NO][NO] = k[NO]^2$

                Second reaction: $-\Delta[H_2]/\Delta t = k[H_2][N_2O_2]$

      (c) N_2O_2 is the intermediate: it is produced in the first step and consumed in the second.

      (d)   Because $[H_2]$ appears in the rate law, the second step must be slow relative to the first.

14.108   (a)

$$Cl(g) + O_3(g) \rightarrow ClO(g) + O_2(g)$$
$$\underline{ClO(g) + O(g) \rightarrow Cl(g) + O_2(g)}$$
$$Cl(g) + O_3(g) + ClO(g) + O(g) \rightarrow ClO(g) + O_2(g) + Cl(g) + O_2(g)$$
$$O_3(g) + O(g) \rightarrow 2O_2(g)$$

      (b) $Cl(g)$ is the catalyst. It is consumed in the first step and reproduced in the second.

      (c) $ClO(g)$ is the intermediate. It is produced in the first step and consumed in the second.

14.109 (a)
$$O_3(g) \rightarrow O_2(g) + O(g)$$
$$O(g) + O_3(g) \rightarrow 2\,O_2(g)$$
$$\overline{}$$
$$2\,O_3(g) \rightarrow 3\,O_2(g)$$

(b) Follow the logic in Sample Exercise 14.15 for determining the rate law of a mechanism with a fast initial step.

From the rate determining second step, rate $= k_2\,[O][O_3]$

If the first step is a fast equilibrium, $k_1\,[O_3] = k_{-1}\,[O_2][O]$

Solving for $[O]$ in terms of $[O_3]$ and $[O_2]$, $[O] = \dfrac{k_1[O_3]}{k_{-1}[O_2]}$

$$\text{Rate} = \frac{k_2 k_1[O_3][O_3]}{k_{-1}[O_2]} = k\frac{[O_3]^2}{[O_2]}$$

(c) $O(g)$ is an intermediate; it is produced and then consumed during the reaction.

(d) If the reaction occurred in a single step, the rate law would change. It would be rate $= k[O_3]^2$.

14.110 (a)
$$Cl_2(g) \rightleftharpoons 2Cl(g)$$
$$Cl(g) + CHCl_3(g) \rightarrow HCl(g) + CCl_3(g)$$
$$Cl(g) + CCl_3(g) \rightarrow CCl_4(g)$$
$$\overline{Cl_2(g) + 2\,Cl(g) + CHCl_3(g) + CCl_3(g) \rightarrow 2\,Cl(g) + HCl(g) + CCl_3(g) + CCl_4(g)}$$
$$Cl_2(g) + CHCl_3(g) \rightarrow HCl(g) + CCl_4(g)$$

(b) $Cl(g), CCl_3(g)$

(c) Reaction 1 - unimolecular, Reaction 2 - bimolecular, Reaction 3 - bimolecular

(d) Reaction 2, the slow step, is rate determining.

(e) If Reaction 2 is rate determining, rate $= k_2[CHCl_3][Cl]$. Cl is an intermediate formed in reaction 1, an equilibrium. By definition, the rates of the forward and reverse processes are equal; $k_1\,[Cl_2] = k_{-1}\,[Cl]^2$. Solving for $[Cl]$ in terms of $[Cl_2]$,

$$[Cl]^2 = \frac{k_1}{k_{-1}}[Cl_2]; \quad [Cl] = \left(\frac{k_1}{k_{-1}}[Cl_2]\right)^{1/2}$$

Substituting into the overall rate law

$$\text{rate} = k_2\left(\frac{k_1}{k_{-1}}\right)^{1/2}[CHCl_3][Cl_2]^{1/2} = k[CHCl_3][Cl_2]^{1/2} \text{ (The overall order is 3/2.)}$$

14.111 (a) Rate $= k_1[A][B]$

(b) Follow the logic in Sample Exercise 14.15.

Rate $= k_2[A][X]$; $k_1[A][B] = k_{-1}[C][X]$; $[X] = \dfrac{k_1}{k_{-1}}\dfrac{[A][B]}{[C]}$

Substituting for $[X]$ in the rate expression,

$$\text{Rate} = k_2[A]\frac{k_1}{k_{-1}}\frac{[A][B]}{[C]} = k\frac{[A]^2[B]}{[C]}$$

(c) (iii) The result of part (b) might be surprising because $[C]$ appears in the rate law, and $[C]$ has a negative reaction order.

14.112   (a) $(CH_3)_3AuPH_3 \rightarrow C_2H_6 + (CH_3)AuPH_3$

(b) $(CH_3)_3Au$, $(CH_3)Au$, and PH_3 are intermediates.

(c)   Reaction 1 is unimolecular, Reaction 2 is unimolecular, Reaction 3 is bimolecular.

(d)   Reaction 2, the slow one, is rate determining.

(e)   If Reaction 2 is rate determining, Rate = $k_2[(CH_3)_3Au]$.

$(CH_3)_3Au$ is an intermediate formed in Reaction 1, an equilibrium. By definition, the rates of the forward and reverse processes in Reaction 1 are equal:

$k_1[(CH_3)_3 AuPH_3] = k_{-1}[(CH_3)_3Au][PH_3]$; solving for $[(CH_3)_3Au]$,

$$[(CH_3)_3\ Au] = \frac{k_1[(CH_3)_3\ AuPH_3]}{k_{-1}[PH_3]}$$

Substituting into the rate law

$$Rate = \left(\frac{k_2\ k_1}{k_{-1}}\right)\frac{[(CH_3)_3\ AuPH_3]}{[PH_3]} = \frac{k[(CH_3)_3\ AuPH_3]}{[PH_3]}$$

(f)   The rate is inversely proportional to $[PH_3]$, so adding PH_3 to the $(CH_3)_3AuPH_3$ solution would decrease the rate of the reaction.

14.113   *Analyze/Plan.* Use the structure and unit cell edge of Pt, along with the formulas for volume and surface area of a sphere, to calculate the number of Pt atoms in a 2-nm sphere and on the surface of a 2-nm sphere.

(a)   For a Pt sphere with a 2.0 nm diameter, radius = 1.0 nm.

$$V = 4/3\ \pi r^3 = \frac{4\ \pi(1.0\ nm)^3}{3} \times \frac{10^3\ \text{Å}^3}{1^3\ nm^3} = 4.188879 \times 10^3 = 4.2 \times 10^3\ \text{Å}^3$$

In a face-centered cubic metal structure, there are 4 metal atoms per unit cell. The volume of the unit cell is $(3.924\ \text{Å})^3 = 60.42\ \text{Å}^3$

$$\frac{4\ \text{Pt atoms}}{60.42\ \text{Å}^3} \times 4.1889 \times 10^3\ \text{Å}^3 = 277.3 = 2.8 \times 10^2\ \text{Pt atoms in a 2.0-nm sphere}$$

(b)   Assume that the "footprint" of an atom is its cross-sectional area, the area of a circle with the radius of the atom. The area of this circle is πr^2. The diameter, d, of a Pt atom is 2.8 Å, so r = d/2 = 1.4 Å. The footprint of the Pt atom is then

$\pi (1.4\ \text{Å})^2 = 6.1575 = 6.2\ \text{Å}^2$

The surface area of the 2.0-nm sphere is

$$4\ \pi r^2 = 4\ \pi (1.0\ nm)^2 \times \frac{10^2\ \text{Å}^2}{1^2\ nm^2} = 12.56637 \times 10^2 = 1.3 \times 10^3\ \text{Å}^2$$

$$\frac{1\ \text{Pt atoms}}{6.1575\ \text{Å}^2} \times 1.2566 \times 10^3\ \text{Å}^2 = 204.1 = 2.0 \times 10^2\ \text{surface Pt atoms on a 2.0-nm sphere.}$$

(c) $$\frac{204\ \text{surface Pt atoms}}{277\ \text{total Pt atoms}} \times 100 = 74\%\ \text{Pt atoms on the surface}$$

(d)   For a 5.0-nm Pt sphere, radius = 2.5 nm

$$V = 4/3 \, \pi r^3 = \frac{4 \pi (2.50 \text{ nm})^3}{3} \times \frac{10^3 \, \text{Å}^3}{1^3 \, \text{nm}^3} = 65.4498 \times 10^3 = 6.5 \times 10^4 \, \text{Å}^3$$

$$\frac{4 \text{ Pt atoms}}{60.42 \, \text{Å}^3} \times 65.4498 \times 10^3 \, \text{Å}^3 = 4333 = 4.3 \times 10^3 \text{ Pt atoms in a 5.0-nm sphere}$$

The surface area of the 5.0-nm sphere is

$$4 \pi r^2 = 4 \pi (2.5 \text{ nm})^2 \times \frac{10^2 \, \text{Å}^2}{1^2 \, \text{nm}^2} = 7853.98 = 7.9 \times 10^3 \, \text{Å}^2$$

$$\frac{1 \text{ Pt atoms}}{6.1575 \, \text{Å}^2} \times 7.854 \times 10^3 \, \text{Å}^2 = 1275.5 = 1.3 \times 10^3 \text{ surface Pt atoms on a 5.0-nm sphere}$$

$$\frac{1276 \text{ surface Pt atoms}}{4333 \text{ total Pt atoms}} \times 100 = 29\% \text{ Pt atoms on the surface}$$

The calculations in parts (b) and (d) overestimate the number of Pt atoms on the surface of the sphere, because they do not account for empty space between atoms. For the purpose of comparison, it is most important that we use the same method for both spheres.

[Alternatively, use one face of a face-entered cubic unit cell as a model for the surface area that Pt atoms will occupy. On a face, there is the cross-section of 1 Pt atoms in the center and ¼ Pt atom at each corner. This amounts to the cross-sections of two Pt atoms in $(3.924 \text{ Å})^2 = 15.398 = 15.4 \, \text{Å}^2$.]

$$\frac{2 \text{ Pt atoms}}{15.40 \, \text{Å}^3} \times 1.2566 \times 10^3 \, \text{Å}^2 = 163.2 = 1.6 \times 10^2 \text{ surface Pt atoms on a 2.0-nm sphere.}$$

$$\frac{163 \text{ surface Pt atoms}}{277 \text{ total Pt atoms}} \times 100 = 59\% \text{ Pt atoms on the surface}$$

[Similarly, in a 5.0-nm sphere, there are 1.0×10^3 Pt atoms on the surface, 24% of the total Pt atoms.]

(e)   Both surface models predict that the 2.0-nm sphere will be more catalytically active, because it has a much greater percentage of its atoms on the surface, where they can participate in the chemical reaction.

14.114   *Enzyme*: carbonic anhydrase; *substrate*: carbonic acid (H_2CO_3);
turnover number: 1×10^7 molecules/s.

14.115   Let k and E_a equal the rate constant and activation energy for the uncatalyzed reaction. Let k_c and E_{ac} equal the rate constant and activation energy of the enzyme-catalyzed reaction. Assume A is the same for the uncatalyzed and catalyzed reactions.

$k_c/k = 5000$, T = 37 °C = 310 K.

According to Equation 14.22, $\ln k = -E_a/RT + \ln A$. Subtracting ln k from ln k_c

$$\ln k_c - \ln k = \left[\frac{-E_{ac}}{RT} \right] + \ln A - \left[\frac{-E_a}{RT} \right] - \ln A$$

$$\ln (k_c/k) = \frac{E_a - E_{ac}}{RT}; \quad E_a - E_{ac} = RT \ln (k_c/k)$$

$$E_a - E_{ac} = \frac{8.314 \text{ J}}{\text{mol-K}} \times 310 \text{ K} \times \ln (5000) = 2.195 \times 10^4 \text{ J/mol} = 21.95 \text{ kJ/mol} = 22 \text{ kJ/mol}$$

The enzyme must lower the activation energy by 22 kJ/mol to increase the reaction rate by a factor of 5000.

14.116   (a)   The rate law for the slow step is Rate = k_2[ES], where ES is an intermediate. Use relationships from the fast equilibrium step to substitute for [ES].

rate of the forward reaction = k_1[E][S]

rate of the reverse reaction = k_{-1}[ES]

For an equilibrium, rate forward = rate reverse, k_1[E][S] = k_{-1}[ES];

$[ES] = \dfrac{k_1}{k_{-1}}[E][S]$; Rate = k_2[ES] = $\dfrac{k_2 k_1}{k_{-1}}[E][S] = k[E][S]$

(b)   E + I $\rightleftharpoons$ EI

Integrative Exercises

14.117   *Analyze/Plan.* $2\,N_2O_5 \rightarrow 4\,NO_2 + O_2$ Rate = $k[N_2O_5] = 1.0 \times 10^{-5}\,s^{-1}\,[N_2O_5]$

Use the integrated rate law for a first-order reaction, Equation 14.13, to calculate $k[N_2O_5]$ at 20.0 h. Build a stoichiometry table to determine mol O_2 produced in 20.0 h. Assuming that $O_2(g)$ is insoluble in chloroform, calculate the pressure of O_2 in the 10.0 L container.   *Solve.*

$20.0\,h \times \dfrac{60\,min}{1\,h} \times \dfrac{60\,s}{1\,min} = 7.20 \times 10^4\,s$; $[N_2O_5]_0 = 0.600\,M$

$\ln [A]_t - \ln [A]_0 = -kt$; $\ln [N_2O_5]_t = -kt + \ln [N_2O_5]_0$

$\ln [N_2O_5]_t = -1.0 \times 10^{-5}\,s^{-1}\,(7.20 \times 10^4\,s) + \ln(0.600) = -0.720 - 0.511 = -1.231$

$[N_2O_5]_t = e^{-1.231} = 0.292\,M$

N_2O_5 was present initially as 1.00 L of 0.600 *M* solution.

mol $N_2O_5 = M \times L = 0.600$ mol N_2O_5 initial, 0.292 mol N_2O_5 at 20.0 h

	$2\,N_2O_5$	$\rightarrow$	$4\,NO_2$	+	O_2
t = 0	0.600 mol		0		0
change	–0.308 mol		0.616 mol		0.154 mol
t = 20 h	0.292 mol		0.616 mol		0.154 mol

[Note that the reaction stoichiometry is applied to the "change" line.]

PV = nRT; P = nRT/V; V = 10.0 L, T = 45 °C = 318 K, n = 0.154 mol

$P = 0.154\,mol \times \dfrac{318\,K}{10.0\,L} \times \dfrac{0.08206\,L\text{-}atm}{mol\text{-}K} = 0.402\,atm$

14.118   (a) $\ln k = -E_a/RT + \ln A$; $E_a = 86.8$ kJ/mol = 8.68×10^4 J/mol; T = 35 °C + 273 = 308 K; A = $2.10 \times 10^{11}\,M^{-1}\,s^{-1}$

$\ln k = \dfrac{-8.68 \times 10^4\,J/mol}{308\,K} \times \dfrac{mol\text{-}K}{8.314\,J} + \ln(2.10 \times 10^{11}\,M^{-1}\,s^{-1})$

$\ln k = -33.8968 + 26.0704 = -7.8264$; k = $3.99 \times 10^{-4}\,M^{-1}\,s^{-1}$

(b) $\dfrac{0.335 \text{ g KOH}}{0.250 \text{ L soln}} \times \dfrac{1 \text{ mol KOH}}{56.1 \text{ g KOH}} = 0.02389 = 0.0239 \ M \text{ KOH}$

$\dfrac{1.453 \text{ g C}_2\text{H}_5\text{I}}{0.250 \text{ L soln}} \times \dfrac{1 \text{ mol C}_2\text{H}_5\text{I}}{156.0 \text{ g C}_2\text{H}_5\text{I}} = 0.03726 = 0.0373 \ M \text{ C}_2\text{H}_5\text{I}$

If equal volumes of the two solutions are mixed, the initial concentrations in the reaction mixture are 0.01194 M KOH and 0.01863 M C$_2$H$_5$I. Assuming the reaction is first order in each reactant:

Rate = k[C$_2$H$_5$I][OH$^-$] = $3.99 \times 10^{-4} \ M^{-1}\text{s}^{-1}$ (0.01194 M)(0.01863 M) = $8.88 \times 10^{-8} \ M/\text{s}$

(c) Because C$_2$H$_5$I and OH$^-$ react in a 1 : 1 mole ratio and equal volumes of the solutions are mixed, the reactant with the smaller concentration, KOH, is the limiting reactant.

(d) T = 50 °C + 273 = 323 K

$\ln k = \dfrac{-8.68 \times 10^4 \text{ J/mol}}{323 \text{ K}} \times \dfrac{\text{mol-K}}{8.314 \text{ J}} + \ln (2.10 \times 10^{11} M^{-1}\text{s}^{-1})$

$\ln k = -32.3227 + 26.0704 = -6.2523; \ k = 1.93 \times 10^{-3} \ M^{-1}\text{s}^{-1}$

14.119 Obtaining data for an "Arrhenius" plot like this requires running the reaction several times, each at a different temperature. Different rates and rate constants (k) are obtained at each temperature. We expect a straight line, according to the relationship:
$\ln k = -E_a/RT + \ln A$. The slope of the graph is $-E_a$ and the y-intercept is the orientation factor, A. The graph in the exercise demonstrates these characteristics, times two! The graph indicates that reaction requires two different activation energies, depending on temperature.

Assuming that reactants and products are the same at all temperatures, the reaction proceeds through different pathways, depending on temperature. This could mean two totally different reaction mechanisms, or a multi-step mechanism where different steps are rate-determining at different temperatures.

14.120 (a) $\ln k = -E_a/RT + \ln A$, Equation 14.22. $E_a = 6.3$ kJ/mol = 6.3×10^3 J/mol

T = 100 °C + 273 = 373 K

$\ln k = \dfrac{-6.3 \times 10^3 \text{ J/mol}}{8.314 \text{ J/K-mol} \times 373 \text{ K}} + \ln (6.0 \times 10^8 M^{-1}\text{s}^{-1})$

$\ln k = -2.032 + 20.212 = 18.181 = 18.2; k = 7.87 \times 10^7 = 8 \times 10^7 \ M^{-1}\text{s}^{-1}$

(b) NO, 11 valence e$^-$, 5.5 e$^-$ pair (Assume the less electronegative N atom will be electron deficient.)

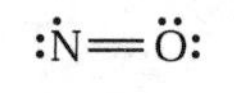

ONF, 18 valence e$^-$, 9 e$^-$ pr

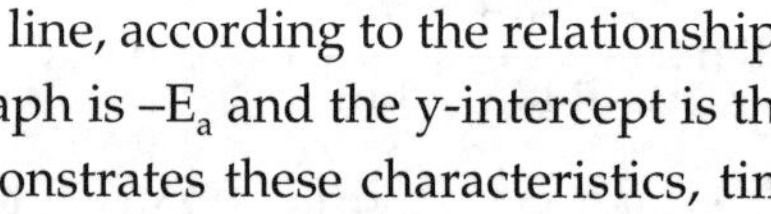

:Ö═N̈—F̈: ⟷ (:Ö—N̈═F̈:)

:N̈═Ö:

The resonance form on the right is a very minor contributor to the true bonding picture, because of high formal charges and the unlikely double bond involving F.

(c)   ONF has trigonal planar electron domain geometry, which leads to a "bent" structure with a bond angle of approximately 120°.

(d)

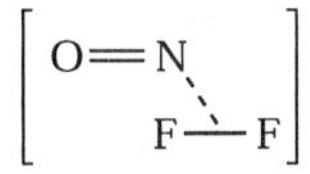

$$\left[\begin{array}{c} O = N \\ \quad\quad\ F - F \end{array} \right]$$

(e)   The electron deficient NO molecule is attracted to electron-rich F_2, so the driving force for formation of the transition state is greater than simple random collisions.

14.121 (a) $\Delta H_{rxn}^{o} = 2\,\Delta H_f^{o}\ H_2O(g) + 2\,\Delta H_f^{o}\ Br_2(g) - 4\,\Delta H_f^{o}\ HBr(g) - \Delta H_f^{o}\ O_2(g)$

$\Delta H_{rxn}^{\circ} = 2(-241.82) + 2(30.71) - 4(-36.23) - (0) = -277.30\ kJ$

(b)   Because the rate of the uncatalyzed reaction is very slow at room temperature, the magnitude of the activation energy for the rate-determining first step must be quite large. At room temperature, the reactant molecules have a distribution of kinetic energies (Chapter 10), but very few molecules even at the high end of the distribution have sufficient energy to form an activated complex. E_a for this step must be much greater than 3/2 RT, the average kinetic energy of the sample.

(c) $20\ e^-$, $10\ e^-$ pr

$$H - \overset{..}{\underset{..}{O}} - \overset{..}{\underset{..}{O}} - \overset{..}{\underset{..}{Br}}:$$

The intermediate resembles hydrogen peroxide, H_2O_2.

14.122 (a)   D(Cl–Cl) = 242 kJ/mol Cl_2

$$\frac{242\ kJ}{mol\ Cl_2} \times \frac{1000\ J}{kJ} \times \frac{1\ mol}{6.022 \times 10^{23}\ molecules} = 4.019 \times 10^{-19} = 4.02 \times 10^{-19}\ J$$

$$\lambda = hc/E = \frac{6.626 \times 10^{-34}\ J\text{-}s \times 2.998 \times 10^8\ m/s}{4.019 \times 10^{-19}\ J} = 4.94 \times 10^{-7}\ m$$

This wavelength, 494 nm, is in the visible portion of the spectrum.

(b)

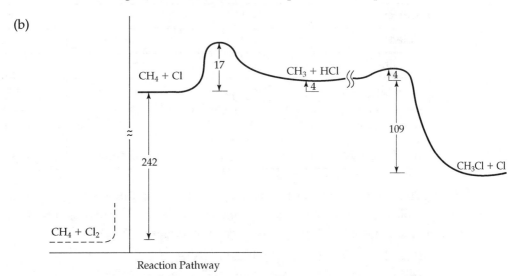

Reaction Pathway

(c)   Because D(Cl–Cl) is 242 kJ/mol, $CH_4(g) + Cl_2(g)$ should be about 242 kJ below the starting point on the diagram. For the reaction

$CH_4(g) + Cl_2(g) \rightarrow CH_3(g) + HCl(g) + Cl(g)$, E_a is 242 + 17 = 259 kJ.

(From bond dissociation enthalpies, ΔH for the overall reaction

$CH_4(g) + Cl_2(g) \rightarrow CH_3Cl(g) + Cl(g)$ is –104 kJ,

so the graph above is simply a sketch of the relative energies of some of the steps in the process.)

(d) CH_3, 7 valence e⁻, odd electron species

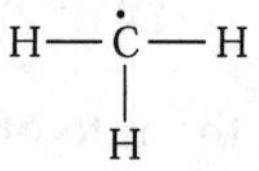

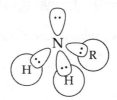

(e)   This sequence is called a chain reaction because Cl· radicals are regenerated in Reaction 4, perpetuating the reaction. Absence of Cl· terminates the reaction, so Cl· + Cl· → Cl_2 is a termination step.

14.123   (a)   A generic Lewis structure for a primary amine is shown below. There are four electron domains about nitrogen, so the hybridization is sp^3. The hybrid orbital picture is shown on the right.

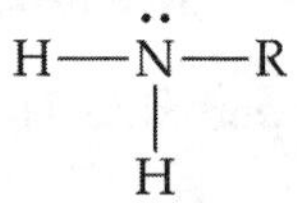

(b)   A reactant that is attracted to the lone pair of electrons on nitrogen will produce a tetrahedral intermediate. This can be a moiety with a full, partial, or even transient positive charge. Steric hindrance will not be large, because two of the atoms bound to nitrogen are small hydrogens.

14.124   (a)

Molecule	NO	NO_2	N_2
Valence e⁻	11	17	10
e⁻ pairs	5.5	8.5	5
Lewis structure	:N̈=Ö:	:Ö—N̈=Ö:	:N≡N:
Bond order	2	1.5	3
Bond energy	607 kJ/mol	404 kJ/mol	941 kJ/mol

(b)   The bond energies in the table above are from Table 8.3 of the text. The NO_2 molecule has two resonance forms and the bond order is 1.5. To obtain an approximate bond energy, average the energies for N=O and N–O:

(607 kJ + 201 kJ)/2 = –404 kJ/bond in NO_2

We know that resonance stabilizes a molecule, so the actual bond energy is probably somewhat greater than this value.

Use Avogadro's number to calculate energy in J/bond, then $\lambda = hc/E$ to calculate wavelength and region of the electromagnetic spectrum. These are the maximum wavelengths that would cause complete bond dissociation.

$$\frac{941\,kJ}{mol\,N_2} \times \frac{1000\,J}{kJ} \times \frac{1\,mol}{6.022 \times 10^{23}\,molecules} = 1.5626 \times 10^{-18} = 1.56 \times 10^{-18}\,J$$

$$\lambda = hc/E = \frac{6.626 \times 10^{-34}\,J\text{-}s \times 2.998 \times 10^8\,m/s}{1.5626 \times 10^{-18}\,J} = 1.27 \times 10^{-7}\,m$$

For NO, the energy per bond is 1.01×10^{-18} J and the wavelength is 1.97×10^{-7} m.

For NO_2, the energy per bond is 6.71×10^{-19} J and the wavelength is 2.96×10^{-7} m.

The three "bond dissociation" wavelengths, 127 nm, 197 nm, and 296 nm, are all in the ultraviolet region, near but not in the visible range of 400-700 nm. We expect longer wavelength electronic excitations for the three gases to be in the visible and near UV, in the same relative order as the bond dissociation wavelengths.

(c)   The experiment requires a UV-VIS spectrometer and a gas flow cell that can be attached to the exhaust stream. According to Beer's law, as the concentration of absorbing species decreases, so does the absorbance. By monitoring a different wavelength of maximum absorption (longer than 297 nm) for each gas, we can measure the concentration of each gas at some point in time. We would monitor the stream before the catalytic converter to establish starting concentrations, then after the converter to observe changes. If the catalyst is working, we expect the two longer wavelength peaks for NO and NO_2 to decrease in size, and the shorter wavelength peak for N_2 to increase.

14.125   (a)   No. Ethane, C_2H_6, has only sigma bonds. The electron density is all localized along the C–H and C–C bonds. The molecule has no pi electrons to interact with the metal surface.

        (b)   Yes. Ammonia, NH_3, has a nonbonding electron pair on nitrogen. This electron domain extends away from the sigma framework of the N–H bonds and is available to interact with the metal surface of the catalyst.

15 Chemical Equilibrium

Visualizing Concepts

15.1 (a) $k_f > k_r$. According to the Arrhenius equation [14.21], $k = Ae^{-E_a/RT}$. As the magnitude of E_a increases, k decreases. On the energy profile, E_a is the difference in energy between the starting point and the energy at the top of the barrier. Clearly this difference is smaller for the forward reaction, so $k_f > k_r$.

(b) From the Equation [15.5], the equilibrium constant $= k_f / k_r$. Because $k_f > k_r$, the equilibrium constant for the process shown in the energy profile is greater than 1.

15.2 Yes, the system is in equilibrium in boxes 4 and 5. The first box is pure reactant A. As the reaction proceeds, some A changes to B. In the fourth and fifth boxes, the relative amounts (concentrations) of A and B are constant. Although the reaction is ongoing the rates of $A \rightarrow B$ and $B \rightarrow A$ are equal, and the relative amounts of A and B are constant.

15.3 *Analyze.* Given box diagram and reaction type, determine whether K greater or smaller than one for the equilibrium mixture depicted in the box.

Plan. Assign species in the box to reactants and products. Write an equilibrium expression in terms of concentrations. Find the relationship between numbers of moles (molecules in the diagram) and concentration. Calculate K.

Solve. Let red = A, blue = X, red and blue pairs = AX. (The colors of A and X are arbitrary.) There are 3A, 2B, and 8AX in the box.

M = mol/L. Because each particle represents one mole, we can use numbers of particles in place of moles in the molarity formula. (The mole is a counting unit for particles, so mol ratios and particle ratios are equivalent.) V = 1 L, so in this case, [A] = number of A particles.

$$K = \frac{[AX]}{[A][X]}; \quad [AX] = 8/V = 8; [A] = 3/V = 3; [X] = 2/V = 2. \ K = \frac{8}{[3][2]} = \frac{8}{6} = 1.33$$

K is greater than one.

15.4 *Analyze/Plan.* Given that element A = red and element B = blue, evaluate the species in the reactant and product boxes, and write the reaction. Answer the remaining questions based on the balanced equation. *Solve.*

(a) reactants: $4A_2 + 4B$; products: $4A_2B$

balanced equation: $A_2 + B \rightarrow A_2B$

(b) $K_c = \frac{[A_2B]}{[A_2][B]}$

(c)   Evaluate concentrations and the value of K for the box in equilibrium, the one on the right of the diagram. $[A_2B] = 0.4\ M$; $[A_2] = 0.1\ M$; $[B] = 0.1\ M$

$$K_c = \frac{[A_2B]}{[A_2][B]} = \frac{[0.4]}{[0.1][0.1]} = 40$$

(d) $\Delta n = \Sigma n(\text{prod}) - \Sigma n(\text{react}) = 1 - 2 = -1$.

(e) $K_p = K_c(RT)^{\Delta n}$, Equation 15.15. Assume a temperature of $25\ ^\circ C = 298\ K$.

$$K_p = 40[(0.0821)(298)]^{-1} = \frac{40}{(0.0821)(298)} = 1.635 = 2$$

15.5   The reaction $A(g) + B(g) \rightleftharpoons AB(g)$ has the larger equilibrium constant. At 40s and 50s, there are more diatomic products and fewer monoatomic reactants than in the other reaction. An equilibrium constant is the ratio of concentrations of products to concentrations of reactants at equilibrium. Another way to say this is that the $A(g) + B(g)$ reaction favors products more than the $X(g) + Y(g)$ reaction. (The $X(g) + Y(g)$ reaction reaches equilibrium more quickly than the $A(g) + B(g)$ reaction, but this is a matter of kinetics, not equilibrium constants.)

15.6   *Analyze/Plan.* The reaction with the largest equilibrium constant has the largest ratio of products to reactants. Count product and reactant molecules. Calculate ratios and compare. *Solve.*

$K = \dfrac{[C_2H_4X_2]}{[C_2H_4][X_2]}$. Use numbers of molecules as an adequate measure of concentration.

(Although the volume terms don't cancel, they are the same for all parts. For the purpose of comparison, we can ignore volume.)   *Solve.*

(a)   8 $C_2H_4Cl_2$, 2 Cl_2, 2 C_2H_4. $K = \dfrac{8}{(2)(2)} = 2$

(b)   6 $C_2H_4Br_2$, 4 Br_2, 4 C_2H_4. $K = \dfrac{6}{(4)(4)} = 0.375 = 0.4$

(c)   3 $C_2H_4I_2$, 7 I_2, 7 C_2H_4. $K = \dfrac{3}{(7)(7)} = 0.0612 = 0.06$

From the smallest to the largest equilibrium constant, (c) < (b) < (a).

Check. By inspection, there are the fewest product molecules and the most reactant molecules in (c); most product and least reactant in (a).

15.7   Statement (b) is definitely true. The reaction is a heterogeneous equilibrium for which $K_p = P_{O_2}$. The larger volume of vessel B requires that more $PbO_2(s)$ must decompose to produce the equilibrium pressure of $O_2(g)$. (After heating, both vessels will have less than the 5.0 g $PbO_2(s)$ that was present initially. If the "less" refers to the amount of solid present initially, statement (a) is also true.)

15.8   *Analyze.* Given box diagrams, reaction type, and value of K_c, determine whether each reaction mixture is at equilibrium.

Plan. Analyze the contents of each box, express them as concentrations (see Solution 15.3). Write the equilibrium expression, calculate Q for each mixture, and compare it to K_c. If Q = K, the mixture is at equilibrium. If Q < K, the reaction shifts right (more product). If Q > K, the reaction shifts left (more reactant).

Solve. $K_c = \dfrac{[AB]^2}{[A_2][B_2]}$.

For this reaction, $\Delta n = 0$, so the volume terms cancel in the equilibrium expression. In this case, the number of each kind of particle can be used as a representation of moles (see Solution 5.3) and molarity.

(a)   Mixture (i): $1A_2, 1B_2, 6AB$; $Q = \dfrac{6^2}{(1)(1)} = 36$

   $Q > K_c$, the mixture is not at equilibrium.

   Mixture (ii): $3A_2, 2B_2, 3AB$; $Q = \dfrac{3^2}{(3)(2)} = 1.5$

   $Q = K_c$, the mixture is at equilibrium.

   Mixture (iii): $3A_2, 3B_2, 2AB$; $Q = \dfrac{2^2}{(3)(3)} = 0.44$

   $Q < K_c$, the mixture is not at equilibrium.

(b)   Mixture (i) proceeds toward reactants.

   Mixture (iii) proceeds toward products.

15.9   For the reaction $A_2(g) + B(g) \rightleftharpoons A(g) + AB(g)$, $\Delta n = 0$ and $K_p = K_c$. We can evaluate the equilibrium expression in terms of concentration. Also, because $\Delta n = 0$, the volume terms in the expression cancel and we can use number of particles as a measure of moles and molarity. The mixture contains $2A$, $4AB$, and $2A_2$.

   $K_c = \dfrac{[A][AB]}{[A_2][B]} = \dfrac{(2)(4)}{(2)(B)} = 2; B = 2$

   2 B atoms should be added to the diagram.

15.10   *Analyze.* Given the diagram and reaction type, calculate the equilibrium constant K_c.

   Plan. Analyze the contents of the cylinder. Express them as concentrations, using number of particles as a measure of moles, and $V = 2$ L. Write the equilibrium expression in terms of concentration and calculate K_c.   *Solve.*

(a)   The mixture contains $2A_2, 2B, 4AB$. $[A_2] = 2/2 = 1$, $[B] = 2/2 = 1$, $[AB] = 4/2 = 2$.

   $K_c = \dfrac{[AB]^2}{[A_2][B]^2} = \dfrac{(2)^2}{(1)(1)^2} = 4$

(b)   A decrease in volume favors the reaction with fewer moles of gas. This reaction has two moles of gas in products and three in reactants, so a decrease in volume favors products. The number of AB (product) molecules will increase.

   Note that a change in volume does not change the value of K_c. If V decreases, the number of AB molecules must increase to maintain the equilibrium value of K_c.

15.11   If temperature increases, K of an endothermic reaction increases and K of an exothermic reaction decreases. Calculate the value of K for the two temperatures and compare. For this reaction, $\Delta n = 0$ and $K_p = K_c$. We can ignore volume and use number of particles as a measure of moles and molarity. $K_c = [A][AB]/[A_2][B]$

(1)   300 K, 3A, 5AB, $1A_2$, 1B; $K_c = (3)(5)/(1)(1) = 15$

(2)   500 K, 1A, 3AB, $3A_2$, 3B; $K_c = (1)(3)/(3)(3) = 0.33$

K_c decreases as T increases, so the reaction is exothermic.

15.12   (a)   Exothermic. In both reaction mixtures (orange and blue), [AB] decreases as T increases.

       (b)   In the reaction, there are fewer moles of gas in products than reactants, so greater pressure favors production of products. At any single temperature, [AB] is greater at P = y than at P = x. Because the concentration of the product, AB, is greater at P = y, P = y is the greater pressure.

Equilibrium; The Equilibrium Constant (Sections 15.1–15.4)

15.13   *Analyze/Plan.* Given the forward and reverse rate constants, calculate the equilibrium constant using Equation 15.5. At equilibrium, the rates of the forward and reverse reactions are equal. Write the rate laws for the forward and reverse reactions and use their equality to answer part (b).   *Solve.*

       (a) $K_c = \dfrac{k_f}{k_r}$, Equation 15.5; $K_c = \dfrac{4.7 \times 10^{-3}\ s^{-1}}{5.8 \times 10^{-1}\ s^{-1}} = 8.1 \times 10^{-3}$

           [For this reaction, $K_p = K_c = 8.1 \times 10^{-3}$]

       (b)   At equilibrium, the partial pressure of A is greater than the partial pressure of B. $rate_f = rate_r; k_f[A] = k_r[B]$

           At equilibrium, the two rates are equal. Because $k_f < k_r$, [A] must be greater than [B] and the partial pressure of A is greater than the partial pressure of B.

15.14   (a)   The reactant $I_2(g)$ predominates at equilibrium. The value of K_c is much less than one, which means that the denominator of the K expression is much larger than the numerator.

       (b)   The reverse reaction has the greater rate constant. $K_c = k_f/k_r$; if K_c is small, k_r is larger than k_f and the reverse reaction has the greater rate constant.

15.15   *Analyze/Plan.* Follow the logic in Sample Exercises 15.1 and 15.5.   *Solve.*

       (a) $K_c = \dfrac{[N_2O][NO_2]}{[NO]^3}$ (b) $K_c = \dfrac{[CS_2][H_2]^4}{[CH_4][H_2S]^2}$

       (c) $K_c = \dfrac{[CO]^4}{[Ni(CO)_4]}$ (d) $K_c = \dfrac{[H^+][F^-]}{[HF]}$

       (e) $K_c = \dfrac{[Ag^+]^2}{[Zn^{2+}]}$ (f) $K_c = [H^+][OH^-]$

       (g) $K_c = [H^+]^2[OH^-]^2$

           homogeneous: (a), (b), (d), (f), (g); heterogeneous: (c), (e)

15.16 (a) $K_c = \dfrac{[O_2]^3}{[O_3]^2}$ (b) $K_c = \dfrac{1}{[Cl_2]^2}$

(c) $K_c = \dfrac{[C_2H_6]^2[O_2]}{[C_2H_4]^2[H_2O]^2}$ (d) $K_c = \dfrac{[CH_4]}{[H_2]^2}$

(e) $K_c = \dfrac{[Cl_2]^2}{[HCl]^4[O_2]}$ (f) $K_c = \dfrac{[CO_2]^{16}[H_2O]^{18}}{[O_2]^{25}}$

(g) $K_c = \dfrac{[CO_2]^{16}}{[O_2]^{25}}$

homogeneous: (a), (c); heterogeneous: (b), (d), (e), (f), (g)

15.17 *Analyze.* Given the value of K_c or K_p, predict the contents of the equilibrium mixture.

Plan. If K_c or $K_p \gg 1$, products dominate; if K_c or $K_p \ll 1$, reactants dominate. *Solve.*

(a) mostly reactants ($K_c \ll 1$)

(b) mostly products ($K_p \gg 1$)

15.18 (a) equilibrium lies to right, favoring products ($K_p \gg 1$)

(b) equilibrium lies to left, favoring reactants ($K_c \ll 1$)

15.19 (a) True.

(b) False. A single-headed arrow indicates that the reaction "goes to completion," that the equilibrium constant is extremely large.

(c) False. The value of the equilibrium constant gives no information about the speed of a reaction.

15.20 (a) False. The value of Δn is not zero.

(b) False. Equilibrium constants are not expressed with units.

(c) True. For a gas phase equilibrium, increasing pressure (by decreasing volume) favors the reaction that produces fewer moles of gas. In this equilibrium, the forward reaction has fewer moles of gas. When the forward reaction is favored, the value of K increases.

15.21 *Analyze/Plan.* Follow the logic in Sample Exercise 15.2. *Solve.*

$PCl_3(g) + Cl_2(g) \rightleftharpoons PCl_5(g)$, $K_c = 0.042$. $\Delta n = 1 - 2 = -1$

$K_p = K_c(RT)^{\Delta n} = 0.042(RT)^{-1} = 0.042/RT$

$K_p = \dfrac{0.042}{(0.08206)(500)} = 0.001024 = 1.0 \times 10^{-3}$

15.22 $SO_2(g) + Cl_2(g) \rightleftharpoons SO_2Cl_2(g)$, $K_p = 34.5$. $\Delta n = 1 - 2 = -1$

$K_p = K_c(RT)^{\Delta n}$; $34.5 = K_c(RT)^{-1} = K_c/RT$;

$K_c = 34.5\,RT = 34.5(0.08206)(303) = 857.81 = 858$

15.23   *Analyze.* Given K_c for a chemical reaction, calculate K_c for the reverse reaction.

   Plan. Evaluate which species are favored by examining the magnitude of K_c. The equilibrium expressions for the reaction and its reverse are the reciprocals of each other, and the values of K_c are also reciprocal.   *Solve.*

   (a)   For the reaction as written, $K_c < 1$, which means that reactants are favored. At this temperature, the equilibrium favors NO and Br_2.

   (b) $K_c(\text{forward}) = \dfrac{[NOBr]^2}{[NO]^2[Br_2]} = 1.3 \times 10^{-2}$

   $K_c(\text{reverse}) = \dfrac{[NO]^2[Br_2]}{[NOBr]^2} = \dfrac{1}{1.3 \times 10^{-2}} = 76.92 = 77$

   (c) $K_{c2}(\text{reverse}) = \dfrac{[NO][Br_2]^{1/2}}{[NOBr]} = (K_c(\text{reverse}))^{1/2} = (76.92)^{1/2} = 8.8$

15.24 $2\,H_2(g) + S_2(g) \rightleftharpoons 2\,H_2S(g)$, $K_c = 1.08 \times 10^7$ at 700 °C. $\Delta n = 2 - 3 = -1$.

   (a) $K_p = K_c(RT)^{\Delta n}$; $T = 700\ °C + 273 = 973\ K$.

   $K_p = 1.08 \times 10^7\,(RT)^{-1} = \dfrac{1.08 \times 10^7}{(0.08206)(973)} = 1.35 \times 10^5$

   (b)   Mostly H_2S. Both K_p and K_c are much greater than one, so the product, H_2S, is favored at equilibrium.

   (c) $H_2(g) + \tfrac{1}{2}\,S_2(g) \rightleftharpoons H_2S(g)$; $K_{c2} = \dfrac{[H_2S]}{[H_2][S_2]^{1/2}}$

   $K_{c2} = (K_c)^{1/2} = (1.08 \times 10^7)^{1/2} = 3.29 \times 10^3$

   $K_{p2} = (K_p)^{1/2} = (1.35 \times 10^5)^{1/2} = 368$

15.25   *Analyze.* Given K_p for a reaction, calculate K_p for a related reaction.

   Plan. The algebraic relationship between the K_p values is the same as the algebraic relationship between equilibrium expressions.

   Solve. $K_p = \dfrac{P_{SO_3}}{P_{SO_2} \times P_{O_2}^{1/2}} = 1.85$

   (a) $K_p = \dfrac{P_{SO_2} \times P_{O_2}^{1/2}}{P_{SO_3}} = \dfrac{1}{1.85} = 0.541$

   (b) $K_p = \dfrac{P_{SO_3}^2}{P_{SO_2}^2 \times P_{O_2}} = (1.85)^2 = 3.4225 = 3.42$

   (c) $K_p = K_c(RT)^{\Delta n}$; $\Delta n = 2 - 3 = -1$; $T = 1000\ K$

   $K_p = K_c(RT)^{-1} = K_c/RT$; $K_c = K_p(RT)$

   $K_c = 3.4225(0.08206)(1000) = 280.85 = 281$

15.26 $K_p = \dfrac{P_{HCl}^4 \times P_{O_2}}{P_{Cl_2}^2 \times P_{H_2O}^2} = 0.0752$

     (a) $K_p = \dfrac{P_{Cl_2}^2 \times P_{H_2O}^2}{P_{HCl}^4 \times P_{O_2}} = \dfrac{1}{0.0752} = 13.298 = 13.3$

     (b) $K_p = \dfrac{P_{HCl}^2 \times P_{O_2}^{1/2}}{P_{Cl_2} \times P_{H_2O}} = (0.0752)^{1/2} = 0.2742 = 0.274$

     (c) $K_p = K_c(RT)^{\Delta n}$; $\Delta n = 2.5 - 2 = 0.5$; $T = 480\ ^\circ C + 273 = 753\ K$

         $K_p = K_c(RT)^{1/2}$, $K_c = K_p/(RT)^{1/2} = 0.2742/[0.08206 \times 753]^{1/2} = 0.03488 = 0.0349$

15.27   *Analyze/Plan.* Follow the logic in Sample Exercise 15.4.   *Solve.*

$$CoO(s) + H_2(g) \rightleftharpoons Co(s) + H_2O(g) \qquad\qquad K_1 = 67$$
$$Co(s) + CO_2(g) \rightleftharpoons CoO(s) + CO(g) \qquad\qquad K_2 = 1/490$$

$$CoO(s) + H_2(g) + Co(s) + CO_2(g) \rightleftharpoons Co(s) + H_2O\ (g) + CoO(s) + CO(g)$$

$$H_2(g) + CO_2(g) \rightleftharpoons H_2O(g) + CO(g)$$

$$K_c = K_1 \times K_2 = 67 \times \frac{1}{490} = 0.1367 = 0.14$$

15.28 $2NO(g) + Br_2(g) \rightleftharpoons 2NOBr(g) \qquad\qquad K_1 = 2.0$

            $N_2(g) + O_2(g) \rightleftharpoons 2NO(g) \qquad\qquad K_2 = \dfrac{1}{2.1 \times 10^{30}}$

$$2NO(g) + Br_2(g) + N_2(g) + O_2(g) \rightleftharpoons 2NOBr(g) + 2NO(g)$$

$$N_2(g) + O_2(g) + Br_2(g) \rightleftharpoons 2NOBr(g)$$

$$K_c = K_1 \times K_2 = 2.0 \times \frac{1}{2.1 \times 10^{30}} = 9.524 \times 10^{-31} = 9.5 \times 10^{-31}$$

15.29   *Analyze/Plan.* Follow the logic in Sample Exercise 15.5.   *Solve.*

     (a) $K_p = P_{O_2}$

     (b) $K_c = [Hg(solv)]^4 [O_2(solv)]$

15.30   (a) $K_p = 1/P_{SO_2}$

     (b) $K_c = \dfrac{[Na_2SO_3]}{[Na_2O][SO_2]}$

Calculating Equilibrium Constants (Section 15.5)

15.31   *Analyze/Plan.* Calculate molarity of reactants and products. Follow the logic in Sample Exercise 15.7 using concentrations rather than pressures.   *Solve.*

$$[CH_3OH] = \frac{0.0406\ mol}{2.00\ L} = 0.0203\ M$$

$$[CO] = \frac{0.170\ mol\ CO}{2.00\ L} = 0.0850\ M; \quad [H_2] = \frac{0.302\ mol\ H_2}{2.00\ L} = 0.151\ M$$

$$K_c = \frac{[CH_3OH]}{[CO][H_2]^2} = \frac{0.0203}{(0.0850)(0.151)^2} = 10.4743 = 10.5$$

15.32 $K_c = \dfrac{[H_2][I_2]}{[HI]^2} = \dfrac{(4.79\times10^{-4})(4.79\times10^{-4})}{(3.53\times10^{-3})^2} = 0.018413 = 0.0184$

15.33   *Analyze/Plan.* Follow the logic in Sample Exercise 15.7.   *Solve.*

     (a) $2NO(g) + Cl_2(g) \rightleftharpoons 2NOCl(g)$

          $K_p = \dfrac{P_{NOCl}^2}{P_{NO}^2 \times P_{Cl_2}} = \dfrac{(0.28)^2}{(0.095)^2(0.171)} = 50.80 = 51$

     (b) $K_p = K_c(RT)^{\Delta n}$; $\Delta n = 2 - 3 = -1$; $K_p = K_c(RT)^{-1} = K_c/(RT)$

          $K_c = K_p(RT) = 50.80(0.08206 \times 500) = 2.1 \times 10^3$

15.34   (a) $K_p = \dfrac{P_{PCl_5}}{P_{PCl_3} \times P_{Cl_2}} = \dfrac{1.30\ atm}{0.124\ atm \times 0.157\ atm} = 66.8$

     (b)   Because $K_p > 1$, products (the numerator of the K_p expression) are favored over reactants (the denominator of the K_p expression).

     (c) $K_p = K_c(RT)^{\Delta n}$; $\Delta n = 1 - 2 = -1$; $K_p = K_c(RT)^{-1} = K_c/(RT)$

          $K_c = K_p(RT) = 66.8(0.08206 \times 450) = 2.5 \times 10^3$

15.35   *Analyze/Plan.* Follow the logic in Sample Exercise 15.8. Because the container volume is 1.0 L, mol = *M*.   *Solve.*

     (a)   First calculate the change in [NO], 0.062 − 0.10 = −0.038 = −0.04 *M*. From the stoichiometry of the reaction, calculate the changes in the other pressures. Finally, calculate the equilibrium pressures.

	$2NO(g)$	$+$	$2H_2(g)$	$\rightleftharpoons$	$N_2(g)$	$+$	$2H_2O(g)$
initial	0.10 *M*		0.050 *M*		0 *M*		0.10 *M*
change	−0.038 *M*		−0.038 *M*		+0.019 *M*		+0.038 *M*
equil.	0.062 *M*		0.012 *M*		0.019 *M*		0.138 *M*

          Strictly speaking, the change in [NO] has two decimal places and thus one sig fig. This limits equilibrium pressures to one sig fig for all but H_2O, and K_c to one sig fig. We compute the extra figures and then round.

     (b) $K_c = \dfrac{[N_2][H_2O]^2}{[NO]^2[H_2]^2} = \dfrac{(0.019)(0.138)^2}{(0.062)^2(0.012)^2} = \dfrac{(0.02)(0.14)^2}{(0.06)^2(0.01)^2} = 653.7 = 7 \times 10^2$

15.36   (a)   Calculate the initial concentrations of $H_2(g)$ and $Br_2(g)$ and the equilibrium concentration of $H_2(g)$. *M* = mol/L.

          $[H_2]_{init} = 1.374\ g\ H_2 \times \dfrac{1\ mol\ H_2}{2.0159\ g\ H_2} \times \dfrac{1}{2.00\ L} = 0.34079 = 0.341\ M$

          $[Br_2] = 70.31\ g\ Br_2 \times \dfrac{1\ mol\ Br_2}{159.81\ g\ Br_2} \times \dfrac{1}{2.00\ L} = 0.21998 = 0.220\ M$

$$[H_2]_{equil} = 0.566 \text{ g } H_2 \times \frac{1 \text{ mol } H_2}{2.0159 \text{ g } H_2} \times \frac{1}{2.00 \text{ L}} = 0.14038 = 0.140 \ M$$

	$H_2(g)$	+	$Br_2(g)$	$\rightleftharpoons$	$2HBr(g)$
initial	0.34079 M		0.21998 M		0
change	−0.20041 M		−0.20041 M		+2(0.20041) M
equil.	0.14038 M		0.01957 M		0.40082 M

The change in H_2 is (0.34079 − 0.14038 = 0.20041 = 0.200). The changes in $[Br_2]$ and [HBr] are set by stoichiometry, resulting in the equilibrium concentrations shown in the table.

(b) $K_c = \dfrac{[HBr]^2}{[H_2][Br_2]} = \dfrac{(0.40082)^2}{(0.14038)(0.01957)} = \dfrac{(0.401)^2}{(0.140)(0.020)} = 58.48 = 58$

The equilibrium concentration of Br_2 has 3 decimal places and 2 sig figs, so the value of K_c has 2 sig figs.

15.37   *Analyze/Plan.* Follow the logic in Sample Exercise 15.8, using partial pressures, rather than concentrations.   *Solve.*

(a) $P = nRT/V; P_{CO_2} = 0.2000 \text{ mol} \times \dfrac{500 \text{ K}}{2.000\text{L}} \times \dfrac{0.08206 \text{ L-atm}}{\text{mol-K}} = 4.1030 = 4.10 \text{ atm}$

$P_{H_2} = 0.1000 \text{ mol} \times \dfrac{500 \text{ K}}{2.000\text{L}} \times \dfrac{0.08206 \text{ L-atm}}{\text{mol-K}} = 2.0515 = 2.05 \text{ atm}$

$P_{H_2O} = 0.1600 \times \dfrac{500 \text{ K}}{2.000\text{L}} \times \dfrac{0.08206 \text{ L-atm}}{\text{mol-K}} = 3.2824 = 3.28 \text{ atm}$

(b)   The change in P_{H_2O} is 3.51 − 3.28 = 0.2276 = 0.23 atm. From the reaction stoichiometry, calculate the change in the other pressures and the equilibrium pressures.

	$CO_2(g)$	+	$H_2(g)$	$\rightleftharpoons$	$CO(g)$	+	$H_2O(g)$
initial	4.10 atm		2.05 atm		0 atm		3.28 atm
change	−0.23 atm		−0.23 atm		+0.23		+0.23 atm
equil.	3.87 atm		1.82 atm		0.23 atm		3.51 atm

(c) $K_p = \dfrac{P_{CO} \times P_{H_2O}}{P_{CO_2} \times P_{H_2}} = \dfrac{(0.23)(3.51)}{(3.87)(1.82)} = 0.1146 = 0.11$

Without intermediate rounding, equilibrium pressures are $P_{H_2O} = 3.51$, $P_{CO} = 0.2276, P_{H_2} = 1.8239, P_{CO_2} = 3.8754$ and $K_p = 0.1130 = 0.11$, in good agreement with the value above.

(d) $K_p = K_c(RT)^{\Delta n}; \ \Delta n = 2 - 2 = 0; \ K_p = K_c(RT)^0; \ K_c = K_p = 0.11$

15.38 (a)

	$N_2O_4(g)$	$\rightleftharpoons$	$2NO_2(g)$
initial	1.500 atm		1.000 atm
change	+0.244 atm		−0.488 atm
equil.	1.744 atm		0.512 atm

The change in P_{NO_2} is $(1.000 - 0.512) = -0.488$ atm, so the change in $P_{N_2O_4}$ is $+(0.488/2) = +0.244$ atm.

(b) $K_p = \dfrac{P_{NO_2}^2}{P_{N_2O_4}} = \dfrac{(0.512)^2}{(1.744)} = 0.1503 = 0.150$

(c) $K_p = K_c(RT)^{\Delta n}$; $\Delta n = 2 - 1 = 1$; $K_p = K_c(RT)^1 = K_c(RT)$

$K_c = K_p/(RT) = 0.1503/(0.08206 \times 298) = 6.15 \times 10^{-3}$

15.39   *Analyze/Plan.* Follow the logic in Sample Exercise 15.8. $mM = 10^{-3}\,M$

	$X(aq)$	+	$Y(aq)$	$\rightleftharpoons$	$XY(aq)$
initial	1.0 mM		1.0 mM		0
change	−0.80 mM		−0.80 mM		+0.80 mM
equil.	0.20 mM		0.20 mM		0.80 mM

$K_c = \dfrac{[XY]}{[X][Y]} = \dfrac{(0.80 \times 10^{-3})}{(0.20 \times 10^{-3})(0.20 \times 10^{-3})} = 2.0 \times 10^4$

15.40   The initial concentrations of drug candidate and protein are the same in the two experiments, and the two reactions have the same stoichiometry. At equilibrium, the concentration of B-protein complex is greater than the concentration of A-protein complex, so drug B is the better choice for further research. Calculation of equilibrium constants for the two reactions confirms this conclusion.

	$A(aq)$	+	$protein(aq)$	$\rightleftharpoons$	$A\text{-}protein(aq)$
initial	$2.00 \times 10^{-6}\,mM$		$1.50 \times 10^{-6}\,mM$		0
change	$-1.00 \times 10^{-6}\,mM$		$-1.00 \times 10^{-6}\,mM$		$+1.00 \times 10^{-6}\,mM$
equil.	$1.00 \times 10^{-6}\,mM$		$0.50 \times 10^{-6}\,mM$		$1.00 \times 10^{-6}\,mM$

$K_c = \dfrac{[A\text{-protein}]}{[A][protein]} = \dfrac{(1.00 \times 10^{-6})}{(1.00 \times 10^{-6})(0.50 \times 10^{-6})} = 2.0 \times 10^6$

	$B(aq)$	+	$protein(aq)$	$\rightleftharpoons$	$B\text{-}protein(aq)$
initial	$2.00 \times 10^{-6}\,mM$		$1.50 \times 10^{-6}\,mM$		0
change	$-1.40 \times 10^{-6}\,mM$		$-1.40 \times 10^{-6}\,mM$		$+1.40 \times 10^{-6}\,mM$
equil.	$0.60 \times 10^{-6}\,mM$		$0.10 \times 10^{-6}\,mM$		$1.40 \times 10^{-6}\,mM$

$K_c = \dfrac{[B\text{-protein}]}{[B][protein]} = \dfrac{(1.40 \times 10^{-6})}{(0.60 \times 10^{-6})(0.10 \times 10^{-6})} = 2.3 \times 10^7$

Applications of Equilibrium Constants (Section 15.6)

15.41   (a)   If $Q_c < K_c$, the reaction will proceed in the direction of more products, to the right.

       (b)   If $Q_c = K_c$, the system is in equilibrium; the concentrations used to calculate Q must be equilibrium concentrations.

15.42   (a)   If $Q_c > K_c$, the reaction will proceed in the direction of more reactants, to the left.

       (b) $Q_c = 0$ if the concentration of any product is zero.

15.43   *Analyze/Plan.* Follow the logic in Sample Exercise 15.9. We are given molarities, so we calculate Q directly and decide on the direction to equilibrium.   *Solve.*

$$K_c = \frac{[CO][Cl_2]}{[COCl_2]} = 2.19 \times 10^{-10} \text{ at } 100 \,°C$$

       (a) $Q = \dfrac{(3.3 \times 10^{-6})(6.62 \times 10^{-6})}{(2.00 \times 10^{-3})} = 1.1 \times 10^{-8}; Q > K$

       The reaction will proceed left to attain equilibrium.

       (b) $Q = \dfrac{(1.1 \times 10^{-7})(2.25 \times 10^{-6})}{(4.50 \times 10^{-2})} = 5.5 \times 10^{-12}; Q < K$

       The reaction will proceed right to attain equilibrium.

       (c) $Q = \dfrac{(1.48 \times 10^{-6})^2}{(0.0100)} = 2.19 \times 10^{-10}; Q = K$

       The reaction is at equilibrium.

15.44   Calculate the reaction quotient in each case, compare with

$$K_p = \frac{P_{NH_3}^2}{P_{N_2} \times P_{H_2}^3} = 4.51 \times 10^{-5}$$

       (a) $Q = \dfrac{(98)^2}{(45)(55)^3} = 1.3 \times 10^{-3}$

       Because $Q > K_p$, the reaction will shift toward reactants to achieve equilibrium.

       (b) $Q = \dfrac{(57)^2}{(143)(0)^3} = \infty$

       Because $Q > K_p$, reaction must shift toward reactants to achieve equilibrium. There must be **some** H_2 present to achieve equilibrium. In this example, the only source of H_2 is the decomposition of NH_3.

       (c) $Q = \dfrac{(13)^2}{(27)(82)^3} = 1.1 \times 10^{-5}$

       Q is only slightly less than K_p, so the reaction will shift slightly toward products to achieve equilibrium.

15.45 *Analyze/Plan.* We are given concentrations, so write the K_c expression and solve for $[Cl_2]$. Change molarity to partial pressure using the ideal gas equation and the definition of molarity. *Solve.*

$$K_c = \frac{[SO_2][Cl_2]}{[SO_2Cl_2]}; \quad [Cl_2] = \frac{K_c[SO_2Cl_2]}{[SO_2]} = \frac{(0.078)(0.108)}{0.052} = 0.16200 = 0.16\ M$$

$$PV = nRT, \quad P = \frac{n}{V}RT; \quad \frac{n}{V} = M; \quad P = M\,RT; \quad T = 100\ °C + 273 = 373\ K$$

$$P_{Cl_2} = \frac{0.16200\ \text{mol}}{L} \times \frac{0.08206\ \text{L-atm}}{\text{mol-K}} \times 373\ K = 4.959 = 5.0\ \text{atm}$$

Check. $K_c = \dfrac{(0.052)(0.162)}{(0.108)} = 0.078.$ Our values are self-consistent.

15.46 $K_p = \dfrac{P_{SO_3}^2}{P_{SO_2}^2 \times P_{O_2}}; \quad P_{SO_3} = \left(K_p \times P_{SO_2}^2 \times P_{O_2}\right)^{1/2} = [(0.345)(0.135)^2(0.455)]^{1/2} = 0.0535\ \text{atm}$

15.47 *Analyze/Plan.* Write the equilibrium constant expression. In each case, change masses to molarities, solve for the equilibrium molarity of the desired component, and calculate mass of that substance present at equilibrium. *Solve.*

$$K_c = \frac{[Br]^2}{[Br_2]} = 1.04 \times 10^{-3}$$

$$[Br_2] = \frac{0.245\ \text{g Br}_2}{0.200\ L} \times \frac{1\ \text{mol Br}_2}{159.8\ \text{g Br}_2} = 0.007666 = 0.00767\ M$$

$$[Br] = (K_c[Br_2])^{1/2} = [(1.04 \times 10^{-3})(0.007666)]^{1/2} = 0.002824 = 0.00282\ M$$

$$\frac{0.002824\ \text{mol Br}}{L} \times 0.200\ L \times \frac{79.90\ \text{g Br}}{\text{mol}} = 0.0451\ \text{g Br(g)}$$

Check. $K_c = (0.002824)^2/(0.007666) = 1.04 \times 10^{-3}$

15.48 $K_c = \dfrac{[HI]^2}{[H_2][I_2]} = 55.3; \quad [HI] = (K_c[H_2][I_2])^{1/2}$

$$[H_2] = \frac{0.056\ \text{g H}_2}{2.00\ L} \times \frac{1\ \text{mol H}_2}{2.016\ \text{g H}_2} = 0.01389 = 0.014\ M$$

$$[I_2] = \frac{4.36\ \text{g I}_2}{2.00\ L} \times \frac{1\ \text{mol I}_2}{253.8\ \text{g I}_2} = 0.008589 = 0.00859\ M$$

$$[HI] = [(55.3)\,(0.01389)\,(0.008589)]^{1/2} = 0.08122 = 0.081\ M$$

$$0.08122\ M\ HI \times 2.00\ L \times \frac{127.9\ \text{g HI}}{\text{mol HI}} = 20.78 = 21\ \text{g HI}$$

Check. $K_c = \dfrac{(0.08122)^2}{(0.01389)(0.008589)} = 55.3$

15.49   *Analyze/Plan.* Write the equilibrium constant expression. In each case, change masses to molarities, solve for the equilibrium molarity of the desired component, and calculate mass of that substance present at equilibrium.   *Solve.*

$$K_c = \frac{[I]^2}{[I_2]} = 3.1 \times 10^{-5}$$

$$[I] = \frac{2.67 \times 10^{-2}\ g\,I}{10.0\ L} \times \frac{1\,mol\,I}{126.9\ g\,I} = 2.1040 \times 10^{-5} = 2.10 \times 10^{-5}\ M$$

$$[I_2] = \frac{[I]^2}{K_c} = \frac{(2.104 \times 10^{-5})^2}{3.1 \times 10^{-5}} = 1.428 \times 10^{-5} = 1.43 \times 10^{-5}\ M$$

$$\frac{1.428 \times 10^{-5}\ mol\,I_2}{L} \times 10.0\ L \times \frac{253.8\ g\,I_2}{mol\,I_2} = 0.0362\ g\,I_2$$

Check. $K_c = \dfrac{(2.104 \times 10^{-5})^2}{1.428 \times 10^{-5}} = 3.1 \times 10^{-5}$

15.50 $PV = nRT; P = \dfrac{gRT}{MM\,V}$

$$P_{SO_3} = \frac{1.17\ g\,SO_3}{80.06\ g/mol} \times \frac{0.08206\ L\text{-}atm}{mol\text{-}K} \times \frac{700\ K}{2.00\ L} = 0.4197 = 0.420\ atm$$

$$P_{O_2} = \frac{0.105\ g\,O_2}{32.00\ g/mol} \times \frac{0.08206\ L\text{-}atm}{mol\text{-}K} \times \frac{700\ K}{2.00\ L} = 0.09424 = 0.0942\ atm$$

$$K_p = 3.0 \times 10^4 = \frac{P_{SO_3}^2}{P_{SO_2}^2 \times P_{O_2}}; P_{SO_2} = \left[P_{SO_3}^2 / (K_p)(P_{O_2}) \right]^{1/2}$$

$$P_{SO_2} = [(0.4197)^2 / (3.0 \times 10^4)(0.09424)]^{1/2} = 7.894 \times 10^{-3} = 7.9 \times 10^{-3}\ atm$$

$$g\,SO_2 = \frac{MM\,PV}{RT} = \frac{64.06\ g\,SO_2}{mol\,SO_2} \times \frac{mol\text{-}K}{0.08206\ L\text{-}atm} \times \frac{7.894 \times 10^{-3}\ atm \times 2.00\ L}{700\ K}$$

$$= 0.01761 = 0.018\ g\,SO_2$$

Check. $K_p = [(0.4197)^2 / (7.894 \times 10^{-3})^2 (0.09424)] = 3.0 \times 10^4$

15.51   *Analyze/Plan.* Follow the logic in Sample Exercise 15.11. Because molarity of NO is given directly, we can construct the equilibrium table straight away.   *Solve.*

	2NO(g) ⇌	N$_2$(g) +	O$_2$(g)	$K_c = \dfrac{[N_2][O_2]}{[NO]^2} = 2.4 \times 10^3$
initial	0.175 *M*	0	0	
change	−2x	+x	+x	
equil.	0.175 − 2x	+x	+x	

$$2.4 \times 10^3 = \frac{x^2}{(0.175 - 2x)^2}; (2.4 \times 10^3)^{1/2} = \frac{x}{0.175 - 2x}$$

$x = (2.4 \times 10^3)^{1/2} (0.175 - 2x); \; x = 8.573 - 97.98x; \; 98.98x = 8.573, \; x = 0.08662 = 0.087 \; M$

$[N_2] = [O_2] = 0.087 \; M; \; [NO] = 0.175 - 2(0.08662) = 0.00177 = 0.002 \; M$

Check. $K_c = (0.08662)^2 / (0.00177)^2 = 2.4 \times 10^3$

15.52 $[Br_2] = 0.25 \; mol/3.0 \; L = 0.08333 = 0.083 \; M; \; [Cl_2] = 0.55 \; mol/3.0 \; L = 0.1833 = 0.18 \; M$

$$Br_2(g) \;+\; Cl_2(g) \;\rightleftharpoons\; 2BrCl(g) \qquad K_c = \frac{[BrCl]^2}{[Br_2][Cl_2]} = 7.0$$

initial	0.083 *M*	0.18 *M*	0
change	–x	–x	+2x
equil.	(0.083 – x)	(0.18 – x)	+2x

$7.0 = \dfrac{(2x)^2}{(0.08333 - x)(0.1833 - x)}; \quad 4x^2 = 7.0(0.0153 - 0.2666x + x^2); \quad 0 = 0.1069 - 1.8662x + 3x^2$

$x = \dfrac{1.8662 \pm \sqrt{(-1.8662)^2 - 4(3)(0.1069)}}{2(3)} = 0.06387 = 0.064 \; M$

(The 0.56 *M* quadratic solution is not chemically meaningful.)

$[BrCl] = 2x = 0.1277 = 0.13 \; M; \qquad [Br_2] = 0.08333 - 0.06387 = 0.01946 = 0.019 \; M$

$[Cl_2] = 0.1833 - 0.06387 = 0.1195 = 0.12 \; M$

Check. $K_c = (0.1277)^2 / (0.01946)(0.1195) = 7.0125 = 7.0$

15.53   *Analyze/Plan.* Write the K_p expression, substitute the stated pressure relationship, and solve for P_{Br_2}.   *Solve.*

$$K_p = \frac{P_{NO}^2 \times P_{Br_2}}{P_{NOBr}^2}$$

When $P_{NOBr} = P_{NO}$, these terms cancel and $P_{Br_2} = K_p = 0.416$ atm. This is true for all cases where $P_{NOBr} = P_{NO}$.

15.54 $K_c = [NH_3][H_2S] = 1.2 \times 10^{-4}$. Because of the stoichiometry, equilibrium concentrations of H_2S and NH_3 will be equal; call this quantity y. Then, $y^2 = 1.2 \times 10^{-4}$, $y = 0.010954 = 0.011 \; M$.

15.55   (a) $CaSO_4(s) \rightleftharpoons Ca^{2+}(aq) + SO_4^{2-}(aq) \qquad K_c = [Ca^{2+}][SO_4^{2-}] = 2.4 \times 10^{-5}$

       At equilibrium, $[Ca^{2+}] = [SO_4^{2-}] = x$

       $K_c = 2.4 \times 10^{-5} = x^2; \; x = 4.9 \times 10^{-3} \; M \; Ca^{2+} \text{ and } SO_4^{2-}$

   (b)   A saturated solution of $CaSO_4(aq)$ is $4.9 \times 10^{-3} \; M$.

       1.4 L of this solution contain:

       $\dfrac{4.9 \times 10^{-3} \; mol}{L} \times 1.4 \; L \times \dfrac{136.14 \; g \, CaSO_4}{mol} = 0.9337 = 0.94 \; g \, CaSO_4$

       A bit more than 1.0 g $CaSO_4$ is needed to have some undissolved $CaSO_4(s)$ in equilibrium with 1.4 L of saturated solution.

15.56   (a)   *Analyze/Plan.* If only $PH_3BCl_3(s)$ is present initially, the equation requires that the equilibrium concentrations of $PH_3(g)$ and $BCl_3(g)$ are equal. Write the K_c expression and solve for $x = [PH_3] = [BCl_3]$.   *Solve.*

$K_c = [PH_3][BCl_3]$; $1.87 \times 10^{-3} = x^2$; $x = 0.043243 = 0.0432$ M PH_3 and BCl_3

(b)   Because the mole ratios are 1:1:1, mol $PH_3BCl_3(s)$ required = mol PH_3 or BCl_3 produced.

$$\frac{0.043243 \text{ mol } PH_3}{L} \times 0.250 \text{ L} = 0.01081 = 0.0108 \text{ mol } PH_3 = 0.0108 \text{ mol } PH_3BCl_3$$

$$0.01081 \text{ mol } PH_3BCl_3 \times \frac{151.2 \text{ g } PH_3BCl_3}{1 \text{ mol } PH_3BCl_3} = 1.6346 = 1.63 \text{ g } PH_3BCl_3$$

In fact, some $PH_3BCl_3(s)$ must remain for the system to be in equilibrium, so a bit more than 1.63 g PH_3BCl_3 is needed.

15.57   *Analyze/Plan.* Follow the approach in Solution 15.51. Calculate [IBr] from mol IBr and construct the equilibrium table.

Solve. [IBr] = 0.500 mol/2.00 L = 0.250 M

Because no I_2 or Br_2 was present initially, the amounts present at equilibrium are produced by the reverse reaction and stoichiometrically equal. Let these amounts equal x. The amount of HBr that reacts is then 2x. Substitute the equilibrium molarities (in terms of x) into the equilibrium expression and solve for x.

	I_2	+	Br_2	$\rightleftharpoons$	2IBr	$K_c = \dfrac{[IBr]^2}{[I_2][Br_2]} = 280$
initial	0 M		0 M		0.250 M	
change	+x M		+x M		–2x M	
equil.	x M		x M		(0.250 – 2x) M	

$K_c = 280 = \dfrac{(0.250 - 2x)^2}{x^2}$; taking the square root of both sides

$16.733 = \dfrac{0.250 - 2x}{x}$; $16.733x + 2x = 0.250$; $18.733x = 0.250$

$x = 0.013345 = 0.0133$ M; $[I_2] = [Br_2] = 0.0133$ M

$[IBr] = 0.250 - 2(0.013345) = 0.2233 = 0.223$ M

Check. $\dfrac{(0.2233)^2}{(0.013345)^2} = 280$. Our values are self-consistent.

15.58 $CaCrO_4(s) \rightleftharpoons Ca^{2+}(aq) + CrO_4^{2-}(aq)$ $K_c = [Ca^{2+}][CrO_4^{2-}] = 7.1 \times 10^{-4}$

At equilibrium, $[Ca^{2+}] = [CrO_4^{2-}] = x$

$K_c = 7.1 \times 10^{-4} = x^2$, $x = 0.0266 = 0.027$ M Ca^{2+} and CrO_4^{2-}

15.59   *Analyze/Plan.* Follow the logic in sample Exercise 15.11, using torr in place of *M*. We are given K_p, so we use pressure in torr directly in the equilibrium expression.

$$CH_4(g) \quad + \quad I_2(g) \rightleftharpoons CH_3I(g) + \quad HI(g)$$

	$CH_4(g)$	$I_2(g)$	$CH_3I(g)$	$HI(g)$
initial	105.1 torr	7.96 torr	0 torr	0 torr
change	−x torr	−x torr	+x torr	+x torr
equil.	105.1−x torr	7.96−x torr	+x torr	+x torr

$$K_p = 2.26 \times 10^{-4} = \frac{x^2}{(105.1-x)(7.96-x)}; \; x^2 = 2.26 \times 10^{-4}(836.6 - 113.1x + x^2)$$

$$0.999774 \, x^2 + 0.02555 \, x - 0.18907 = 0; \; x = \frac{-0.02555 \pm \sqrt{(0.02555)^2 - 4(0.999774)(-0.18907)}}{2(0.999774)}$$

x = 0.422 torr (The negative solution is not chemically meaningful.)

at equilibirum: $P_{CH_3I} = P_{HI} = 0.422$ torr; $P_{CH_4} = 104.7$ torr; $P_{I_2} = 7.54$ torr

15.60

$$CH_3COOH(solv) + \quad CH_3CH_2OH(solv) \rightleftharpoons CH_3COOCH_2CH_3(g) + \quad H_2O(solv)$$

	$CH_3COOH(solv)$	$CH_3CH_2OH(solv)$	$CH_3COOCH_2CH_3(g)$	$H_2O(solv)$
initial	0.275 *M*	3.85 *M*	0 *M*	0 *M*
change	−x *M*	−x *M*	+x *M*	+x *M*
equil.	0.275−x *M*	3.85−x *M*	+x *M*	+x *M*

$$K_c = 6.68 = \frac{x^2}{(0.275-x)(3.85-x)}; \; x^2 = 6.68(1.059 - 4.125 \, x + x^2)$$

$$0 = 5.68 \, x^2 - 27.56 \, x + 7.072; \; x = \frac{27.56 \pm \sqrt{(-27.56)^2 - 4(5.68)(7.072)}}{2(5.68)} = 0.27185 = 0.272 \, M$$

(The 4.58 *M* quadratic solution is not chemically meaningful.)

$$\frac{0.27185 \text{ mol ethyl acetate}}{L} \times 15.0 \, L \times \frac{88.10 \text{ g ethyl acetate}}{mol} = 359.25 = 359 \text{ g ethyl acetate}$$

LeChâtelier's Principle (Section 15.7)

15.61   *Analyze/Plan.* Follow the logic in Sample Exercise 15.12. *Solve.*

(a)   Shift equilibrium to the right; more $SO_3(g)$ is formed, the amount of $SO_2(g)$ decreases.

(b)   Heating an exothermic reaction decreases the value of K. More SO_2 and O_2 will form, the amount of SO_3 will decrease. This is fundamentally different than shifting the relative amounts of reactants and products to maintain K; here, the equilibrium position itself changes.

(c)   Because Δn = −1, a change in volume will affect the equilibrium position and favor the side with more moles of gas. The amounts of SO_2 and O_2 increase and the amount of SO_3 decreases; equilibrium shifts to the left.

(d)   No effect. Speeds up the forward and reverse reactions equally.

(e)   No effect. The noble gas does not appear in the equilibrium expression; the partial pressures of reactants and products do not change upon addition of a noble gas.

(f)   Shift equilibrium to the right; amounts of SO_2 and O_2 decrease.

15.62 $4\,NH_3(g) + 5\,O_2(g) \rightleftharpoons 4\,NO(g) + 6\,H_2O(g)$

        (a)   increase $[NH_3]$, increase yield NO

        (b)   increase $[H_2O]$, decrease yield NO

        (c)   decrease $[O_2]$, decrease yield NO

        (d)   decrease container volume, decrease yield NO (fewer moles gas in reactants)

        (e)   add catalyst, no change

        (f)   increase temperature, decrease yield NO (reaction is exothermic)

15.63   *Analyze/Plan.* Given certain changes to a reaction system, determine the effect on K_p, if any. Only changes in temperature cause changes to the value of K_p.   *Solve.*

        (a)   no effect   (b)   no effect   (c)   no effect

        (d)   increase equilibrium constant   (e)   no effect

15.64   (a)   The reaction must be endothermic ($+\Delta H$) if heating increases the fraction of products.

        (b)   There must be more moles of gas in the products if increasing the volume of the vessel increases the fraction of products.

15.65   *Analyze/Plan.* Use Hess's law, $\Delta H° = \Sigma\Delta H_f°$ products $- \Sigma\Delta H_f°$ reactants, to calculate $\Delta H°$. According to the sign of $\Delta H°$, describe the effect of temperature on the value of K. According to the value of Δn, describe the effect of changes to container volume.   *Solve.*

        (a) $\Delta H° = \Delta H_f°\ NO_2(g) + \Delta H_f°\ N_2O(g) - 3\Delta H_f°\ NO(g)$

              $\Delta H° = 33.84\ kJ + 81.6\ kJ - 3(90.37\ kJ) = -155.7\ kJ$

        (b)   Because the reaction is exothermic, the equilibrium constant will decrease with increasing temperature.

        (c)   A change in volume at constant temperature will affect the fraction of products in the equilibrium mixture because Δn does not equal zero. An increase in container volume would favor reactants, whereas a decrease in volume would favor products.

15.66   (a) $\Delta H° = \Delta H_f°\ CH_3OH(g) - \Delta H_f°\ CO(g) - 2\Delta H_f°\ H_2(g)$

                  $= -201.2\ kJ - (-110.5\ kJ) - 0\ kJ$

                  $= -90.7\ kJ$

        (b)   The reaction is exothermic; an increase in temperature would decrease the value of K and decrease the yield. A low temperature is needed to maximize yield.

        (c)   Increasing total pressure would increase the partial pressure of each gas, shifting the equilibrium toward products. The extent of conversion to CH_3OH increases as the total pressure increases.

15.67   For this reaction, there are more moles of product gas than moles of reactant gas. An increase in total pressure increases the partial pressure of each gas, shifting the equilibrium toward reactants. An increase in pressure favors formation of ozone.

15.68   (a)   Low temperature. For an exothermic reaction such as this, decreasing temperature increases the value of K and the amount of products at equilibrium.

            (b)   No. Because there are equal numbers of moles of gas in the products and reactants, the equilibrium yield of products cannot be changed by changing pressure.

15.69   (a)   Endothermic. Bond breaking is always an endothermic process.

            (b)   The equilibrium constant increases. For an endothermic reaction, heat is a "reactant." An increase in temperature and heat favors the forward reaction and the value of K_c increases.

            (c)   The forward rate constant increases by a larger amount than the reverse rate constant. If $K_c = k_f/k_r$ and the value of K_c increases, the value of k_f must increase by a greater amount than the value of k_r.

15.70   False. When the temperature of an exothermic reaction increases, the rate constants of both the forward and the reverse reactions increase, but the value of the reverse rate constant increases by a greater amount.

Additional Exercises

15.71   (a)   Because both the forward and reverse processes are elementary steps, we can write the rate laws directly from the chemical equation.

$$\text{rate}_f = k_f\,[CO][Cl_2] = \text{rate}_r = k_r\,[COCl][Cl]$$

$$\frac{k_f}{k_r} = \frac{[COCl]\,[Cl]}{[CO][Cl_2]} = K$$

$$K_c = \frac{k_f}{k_r} = \frac{1.4 \times 10^{-28}\,M^{-1}\,s^{-1}}{9.3 \times 10^{10}\,M^{-1}\,s^{-1}} = 1.5 \times 10^{-39}$$

            For a homogeneous equilibrium in the gas phase, we usually write K in terms of partial pressures. In this exercise, concentrations are more convenient because the rate constants are expressed in terms of molarity. For this reaction, the value of K is the same regardless of how it is expressed, because there is no change in the moles of gas in going from reactants to products.

            (b)   Because the K is quite small, reactants are much more plentiful than products at equilibrium.

15.72 $2\,A(g) \rightleftharpoons B(g)$, $K_c = 1$

$$\frac{[B]}{[A]^2} = 1, \quad [B] = [A]^2 \text{ and } [A] = [B]^{1/2}$$

15.73 $CH_4(g) + H_2O(g) \rightarrow CO(g) + 3\,H_2(g)$

$$K_p = \frac{P_{CO} \times P_{H_2}^3}{P_{CH_4} \times P_{H_2O}}\,;\; P = \frac{g\,RT}{MM\,V}\,;\; T = 1000\,K$$

$$P_{CO} = \frac{8.62\,g}{28.01\,g/mol} \times \frac{0.08206\,L\text{-atm}}{mol\text{-}K} \times \frac{1000\,K}{5.00\,L} = 5.0507 = 5.05\,atm$$

$$P_{H_2} = \frac{2.60\,g}{2.016\,g/mol} \times \frac{0.08206\,L\text{-atm}}{mol\text{-}K} \times \frac{1000\,K}{5.00\,L} = 21.1663 = 21.2\,atm$$

$$P_{CH_4} = \frac{43.0\,g}{16.04\,g/mol} \times \frac{0.08206\,L\text{-atm}}{mol\text{-}K} \times \frac{1000\,K}{5.00\,L} = 43.9973 = 44.0\,atm$$

$$P_{H_2O} = \frac{48.4\,g}{18.02\,g/mol} \times \frac{0.08206\,L\text{-atm}}{mol\text{-}K} \times \frac{1000\,K}{5.00\,L} = 44.0811 = 44.1\,atm$$

$$K_p = \frac{(5.0507)(21.1663)^3}{(43.9973)(44.0811)} = 24.6949 = 24.7$$

$$K_p = K_c(RT)^{\Delta n}, \; K_c = K_p/(RT)^{\Delta n}; \; \Delta n = 4 - 2 = 2$$

$$K_c = (24.6949)/[(0.08206)(1000)]^2 = 3.6673 \times 10^{-3} = 3.67 \times 10^{-3}$$

15.74 $[SO_2Cl_2] = \dfrac{2.00\,mol}{2.00\,L} = 1.00\,M$

The change in $[SO_2Cl_2] = 0.56(1.00\,M) = 0.56\,M$

	$SO_2Cl_2(g)$	$\rightleftharpoons$	$SO_2(g)$	$+$	$Cl_2(g)$	$K_c = \dfrac{[SO_2][Cl_2]}{[SO_2Cl_2]}$
initial	1.00 M		0		0	
change	−0.56 M		+0.56 M		+0.56 M	
equil.	0.44 M		+0.56 M		+0.56 M	

(a) $K_c = \dfrac{(0.56)^2}{0.44} = 0.7127 = 0.71$

(b) $K_p = K_c(RT)^{\Delta n}; \; \Delta n = 2 - 1 = 1; \quad K_p = (0.7127)(0.08206)(303) = 17.7214 = 18$

(c)   Increase. There are more moles of gas in the products, so increasing the container volume will shift equilibrium toward products.

(d) $[SO_2Cl_2] = \dfrac{2.00\,mol}{15.00\,L} = 0.13333 = 0.133\,M$. Let x equal the change in $[SO_2Cl_2]$.

The equilibrium concentrations are: $[SO_2Cl_2] = (0.13333 - x)$; $[SO_2] = [Cl_2] = x$

$K_c = \dfrac{(x)^2}{(0.13333 - x)} = 0.7127;$ solving the quadratic, $x = 0.1148 = 0.11\,M$

(We expect the decomposition to be greater than 56%, so we must use the quadratic formula to solve for x.)

% decomposition = $(0.1148/0.1333) \times 100 = 86\%$; the increase in volume does shift the equilibrium toward products.

15.75   (a)   Exothermic. The values of K_c in the table decrease as temperature increases; the reverse reaction is favored. This is the case if heat is a "product" of the reaction.

        (b) $\Delta H_{rxn}^{\circ} = \Sigma n \Delta H_f^{\circ}$ (products) $- \Sigma n \Delta H_f^{\circ}$ (reactants). Be careful with coefficients, states, and signs.

              $\Delta H_{rxn}^{\circ} = 2 \Delta H_f^{\circ} \ NH_3(g) - 3 \Delta H_f^{\circ} \ H_2(g) - \Delta H_f^{\circ} \ N_2(g)$

              $\qquad = 2(-46.19 \text{ kJ}) - 3(0 \text{ kJ}) - 0 \text{ kJ} = -92.38 \text{ kJ}$

              Yes, the calculated value of ΔH is negative, which agrees with the prediction from part (a) that the reaction is exothermic.

        (c) $[NH_3] = 0.025 \text{ mol}/1.00 \text{ L} = 0.025 \ M$

	N_2	$+$	$3 H_2$	$\rightleftharpoons$	$2 NH_3$		$K_c = \dfrac{[NH_3]^2}{[N_2][H_2]^3} = 0.058$
initial	$0 \ M$		$0 \ M$		$0.0250 \ M$		
change	$+x \ M$		$+3x \ M$		$-2x \ M$		
equil.	$x \ M$		$3x \ M$		$(0.0250 - 2x) \ M$		

              $K_c = 0.058 = \dfrac{(0.0250 - 2x)^2}{(x)(3x)^3} = \dfrac{(0.0250 - 2x)^2}{9x^4}$. Take the square root of both sides.

              $0.24083 = \dfrac{0.0250 - 2x}{3x^2}; \quad 0.72249x^2 + 2x - 0.0250 = 0; \quad x = 0.01244 = 0.012 \ M$

              $[NH_3]$ at equilibrium $= 0.025 - 2(0.01244) = 0.000120 = 1.2 \times 10^{-4} \ M$

              or, $[NH_3]$ at equilibrium $= 0.025 - 2(0.012) = 0.001 \ M$

              In this case, intermediate rounding changes the result by an order of magnitude. However, by either method, a very small concentration of $NH_3(g)$ is present at equilibrium.

15.76   (a) $K_p = \dfrac{P_{Br_2} \times P_{NO}^2}{P_{NOBr}^2}; P = \dfrac{gRT}{MM \times V}; T = 100 \ {}^{\circ}C + 273 = 373 \text{ K}$

              $P_{Br_2} = \dfrac{4.19 g}{159.8 \text{ g}/\text{mol}} \times \dfrac{0.08206 \text{ L-atm}}{\text{mol-K}} \times \dfrac{373}{5.00 \text{ L}} = 0.16051 = 0.161 \text{ atm}$

              $P_{NO} = \dfrac{3.08 g}{30.01 \text{ g}/\text{mol}} \times \dfrac{0.08206 \text{ L-atm}}{\text{mol-K}} \times \dfrac{373}{5.00 \text{ L}} = 0.62828 = 0.628 \text{ atm}$

              $P_{NOBr} = \dfrac{3.22 \text{ g NOBr}}{109.9 \text{ g}/\text{mol}} \times \dfrac{0.08206 \text{ L-atm}}{\text{mol-K}} \times \dfrac{373}{5.00 \text{ L}} = 0.17936 = 0.179 \text{ atm}$

              $K_p = \dfrac{(0.16051)(0.62828)^2}{(0.17936)^2} = 1.9695 = 1.97 \qquad K_p = K_c(RT)^{\Delta n}, \Delta n = 3 - 2 = 1$

              $K_c = K_p/RT = 1.9695/(0.08206)(373) = 0.064345 = 0.0643$

        (b) $P_t = P_{Br_2} + P_{NO} + P_{NOBr} = 0.16051 + 0.62828 + 0.17936 = 0.96815 = 0.968 \text{ atm}$

(c)   All NO and Br_2 present at equilibrium came from the decomposition of the original NOBr. The mass of original NOBr is the sum of the masses of all compounds at equilibrium.

Original g NOBr = 4.19 gBr_2 + 3.08 g NO + 3.22 g NOBr = 10.49 g

15.77   (a)

	A(g)	$\rightleftharpoons$	2B(g)
initial	0.75 atm		0
change	–0.39 atm		+0.78 atm
equil.	0.36 atm		0.78 atm

$P_t = P_A + P_B = 0.36 \text{ atm} + 0.78 \text{ atm} = 1.14 \text{ atm}$

(b) $K_p = \dfrac{(P_B)^2}{P_A} = \dfrac{(0.78)^2}{0.36} = 1.690 = 1.7$

(c)   Increasing the volume of the flask favors the reaction with more moles of gas. Doing the reaction in a larger flask maximizes the yield of B.

15.78   (a) $K_p = \dfrac{P_{NH_3}^2}{P_{N_2} \times P_{H_2}^3} = 4.34 \times 10^{-3}; T = 300\ °C + 273 = 573\,K$

$P_{NH_3} = \dfrac{gRT}{MM \times V} = \dfrac{1.05\,g}{17.03\,g/mol} \times \dfrac{0.08206\ L\text{-}atm}{mol\text{-}K} \times \dfrac{573\,K}{1.00\,L} = 2.899 = 2.90\ atm$

	$N_2(g)$ +	$3\,H_2(g)$	$\rightleftharpoons$	$2\,NH_3(g)$
initial	0 atm	0 atm		?
change	x	3x		–2x
equil.	x atm	3x atm		2.899 atm

(Remember, only the change line reflects the stoichiometry of the reaction.)

$K_p = \dfrac{(2.899)^2}{(x)(3x)^3} = 4.34 \times 10^{-3}; 27\,x^4 = \dfrac{(2.899)^2}{4.34 \times 10^{-3}}; x^4 = 71.725$

$x = 2.910 = 2.91\ atm = P_{N_2}; P_{H_2} = 3x = 8.730 = 8.73\ atm$

$g_{N_2} = \dfrac{MM \times PV}{RT} = \dfrac{28.02g\ N_2}{mol\ N_2} \times \dfrac{mol\text{-}K}{0.08206\ L\text{-}atm} \times \dfrac{2.910\ atm \times 1.00\,L}{573\,K} = 1.73\ g\ N_2$

$g_{H_2} = \dfrac{2.016\,g\ H_2}{mol\ H_2} \times \dfrac{mol\text{-}K}{0.08206\ L\text{-}atm} \times \dfrac{8.730\ atm \times 1.00\,L}{573\,K} = 0.374\ g\ H_2$

(b)   The initial $P_{NH_3} = 2.899\ atm + 2(2.910\ atm) = 8.719 = 8.72\ atm$

$g_{NH_3} = \dfrac{17.03\,g\ NH_3}{mol\ NH_3} \times \dfrac{mol\text{-}K}{0.08206\ L\text{-}atm} \times \dfrac{8.719\ atm \times 1.00\,L}{573\,K} = 3.16\ g\ NH_3$

(c) $P_t = P_{N_2} + P_{H_2} + P_{NH_3} = 2.910\ atm + 8.730\ atm + 2.899\ atm = 14.54\ atm$

15.79

$$\text{2IBr} \quad \rightleftharpoons \quad \text{I}_2 \quad + \quad \text{Br}_2$$

initial	0.025 atm	0	0
change	–2x	x	x
equil.	(0.025 – 2 x) atm	x	x

$$K_p = 8.5\times10^{-3} = \frac{P_{I_2} \times P_{Br_2}}{P_{IBr}^2} = \frac{x^2}{(0.025-2x)^2}; \quad \text{taking the square root of both sides}$$

$$\frac{x}{0.025-2x} = (8.5\times10^{-3})^{1/2} = 0.0922; \; x = 0.0922(0.025-2x)$$

$$x + 0.184\,x = 0.002305; \; 1.184\,x = 0.002305; \; x = 0.001947 = 1.9 \times 10^{-3}$$

At equilibrium, $P_{I_2} = P_{Br_2} = x = 1.9\times10^{-3}$ atm

P_{IBr} at equilibrium = $0.025 - 2(1.947 \times 10^{-3}) = 0.02111 = 0.021$ atm

Check. $K_p = (0.001947)^2/(0.02111)^2 = 8.5 \times 10^{-3}$; the calculated concentrations are self-consistent.

15.80   (a) $K_p = 0.052; \; K_p = K_c(RT)^{\Delta n}; \; \Delta n = 2 - 0 = 2; \; K_c = K_p/(RT)^2$

           $K_c = 0.052/[0.08206)(333)]^2 = 6.964 \times 10^{-5} = 7.0 \times 10^{-5}$

   (b) PH_3BCl_3 is a solid and its concentration is taken as a constant, C.

$$[BCl_3] = \frac{0.0500 \text{ g } BCl_3}{1.500 \text{ L}} \times \frac{\text{mol } BCl_3}{117.17 \text{ g } BCl_3} = 2.8449 \times 10^{-4} = 2.84 \times 10^{-4}\, M \text{ BCl}_3$$

$$PH_3BCl_3 \quad \rightleftharpoons \quad PH_3 \quad + \quad BCl_3$$

initial	C	0 M	$2.84 \times 10^{-4}\, M$
change		+x M	+x M
equil.	C	+x M	$(2.84 \times 10^{-4} + x)\, M$

$$K_c = [PH_3][BCl_3]; \; 6.964\times10^{-5} = x(2.84\times10^{-4} + x); \; x^2 + 2.84\times10^{-4}x - 6.964\times10^{-5} = 0$$

$$x = \frac{-2.84\times10^{-4} \pm [(2.84\times10^{-4})^2 - 4(1)(-6.964\times10^{-5})]^{1/2}}{2(1)} = 0.008204 = 8.2\times10^{-3}\, M \text{ PH}_3$$

Check. $K_c = (8.2 \times 10^{-3})(2.84 \times 10^{-4} + 8.2 \times 10^{-3}) = 7.0 \times 10^{-5}$.

15.81 $K_p = P_{NH_3} \times P_{H_2S}; \; P_t = 0.614$ atm

If the equilibrium amounts of NH_3 and H_2S are solely due to the decomposition of $NH_4HS(s)$, the equilibrium pressures of the two gases are equal, and each is 1/2 of the total pressure.

$P_{NH_3} = P_{H_2S} = 0.614 \text{ atm}/2 = 0.307$ atm

$K_p = (0.307)^2 = 0.0943$

15.82   Initial $P_{SO_3} = \dfrac{gRT}{MM\ V} = \dfrac{0.831\,g}{80.07\,g/mol} \times \dfrac{0.08206\ L\text{-}atm}{mol\text{-}K} \times \dfrac{1100\ K}{1.00\ L} = 0.9368 = 0.937\ atm$

	2 SO$_3$	⇌	2 SO$_2$	+	O$_2$
initial	0.9368 atm		0		0
change	–2 x		+2x		+x
equil.	0.9368–2 x		2x		x
[equil.]	0.2104 atm		0.7264 atm		0.3632 atm

$P_t = (0.9368 - 2x) + 2x + x;\ 0.9368 + x = 1.300\ atm;\ x = 1.300 - 0.9368 = 0.3632 = 0.363\ atm$

$K_p = \dfrac{P_{SO_2}^2 \times P_{O_2}}{P_{SO_3}^2} = \dfrac{(0.7264)^2\,(0.3632)}{(0.2104)^2} = 4.3292 = 4.33$

$K_p = K_c(RT)^{\Delta n};\ \Delta n = 3 - 2 = 1;\ K_p = K_c(RT)$

$K_c = K_p/RT = 4.3292/[(0.08206)(1100)] = 0.04796 = 0.0480$

15.83   In general, the reaction quotient is of the form $Q = \dfrac{P_{NOCl}^2}{P_{NO}^2 \times P_{Cl_2}}$.

(a) $Q = \dfrac{(0.11)^2}{(0.15)^2\,(0.31)} = 1.7$

$Q > K_p$. The mixture is not at equilibrium. It will shift to the left and produce more reactants as it moves toward equilibrium.

(b) $Q = \dfrac{(0.050)^2}{(0.12)^2\,(0.10)} = 1.7$

$Q > K_p$. The mixture is not at equilibrium. It will shift to the left and produce more reactants as it moves toward equilibrium.

(c) $Q = \dfrac{(5.10\times10^{-3})^2}{(0.15)^2\,(0.20)} = 5.8\times10^{-3}$

$Q < K_p$. The mixture is not at equilibrium. It will shift to the right and produce more products as it moves toward equilibrium.

15.84 $K_c = [CO_2] = 0.0108;\ [CO_2] = \dfrac{g\ CO_2}{44.01\,g/mol} \times \dfrac{1}{10.0\ L}$

In each case, calculate $[CO_2]$ and determine the position of the equilibrium.

(a) $[CO_2] = \dfrac{4.25\,g}{44.01\,g/mol} \times \dfrac{1}{10.0\ L} = 9.657\times10^{-3} = 9.66\times10^{-3}\ M$

$Q = 9.66\times10^{-3} < K_c$. The reaction proceeds to the right to achieve equilibrium and the amount of $CaCO_3(s)$ decreases.

(b) $[CO_2] = \dfrac{5.66\,g\ CO_2}{44.01\,g/mol} \times \dfrac{1}{10.0\ L} = 0.0129\ M$

$Q = 0.0129 > K_c$. The reaction proceeds to the left to achieve equilibrium and the amount of $CaCO_3(s)$ increases.

(c)   6.48 g CO_2 means $[CO_2] > 0.0129\ M;\ Q > 0.0129 > K_c$, the amount of $CaCO_3$ increases.

15.85

	$CO_2(g)$	+	$H_2(g)$	$\rightleftharpoons$	$CO(g)$	+	$H_2O(g)$
initial	1.50 mol		1.50 mol		0		0
change	$-x$		$-x$		$+x$		$+x$
equil.	$(1.50-x)$mol		$(1.50-x)$mol		x		x

Since $\Delta n = 0$, the volume terms cancel and we can use moles in place of molarity in the K expression.

$$K_c = 0.802 = \frac{[CO][H_2O]}{[CO_2][H_2O]} = \frac{x^2}{(1.50-x)^2}$$

Take the square root of both sides.

$(0.802)^{1/2} = x/(1.50-x);\ \ 0.8955(1.50-x) = x$

$1.3433 = 1.8955x,\ \ x = 0.7087 = 0.709$ mol

$[CO] = [H_2O] = 0.7087\ \text{mol}/3.00\ \text{L} = 0.236\ M$

$[CO_2] = [H_2] = (1.50 - 0.709)\text{mol}/3.00\ \text{L} = 0.264\ M$

15.86   (a) $[CO_2] = \dfrac{25.0\ \text{g}\ CO_2}{44.01\ \text{g/mol}} \times \dfrac{1}{3.00\ \text{L}} = 0.18935 = 0.189\ M$

		$C(s)$	+	$CO_2(g)$	$\rightleftharpoons$	$2CO(g)$
initial		excess		0.189		0
change		$-x$		$-x$		$+2x$
equil.				$0.189-x$		$+2x$

$K_c = 1.9 = \dfrac{[CO]^2}{[CO_2]} = \dfrac{(2x)^2}{0.189-x};\ \ 4x^2 = 1.9(0.189-x);\ \ 4x^2 + 1.9x - 0.36 = 0.$

Solve the quadratic for x.

$x = \dfrac{-1.9 \pm \sqrt{(1.9)^2 - 4(4)(-0.36)}}{2(4)} = 0.14505 = 0.15\ M$

$[CO] = 2x = 2(0.14505) = 0.2901 = 0.29\ M$

$\dfrac{0.2901\ \text{mol}\ CO}{L} \times 3.00\ \text{L} \times \dfrac{28.01\ \text{g}\ CO}{\text{mol}} = 24.38 = 24\ \text{g}\ CO$

(b)   The amount of C(s) consumed is related to x. Change M to mol to g C.

$\dfrac{0.14505\ \text{mol}}{L} \times 3.00\ \text{L} \times 12.01\ \text{g} = 5.226 = 5.2\ \text{g}\ \text{C consumed}$

(c)   A smaller vessel at the same temperature increases the total pressure of the mixture. The equilibrium shifts to form fewer total moles of gas, which favors reactants. The yield of CO product will be smaller in a smaller vessel.

(d)   The two K_c values are 0.133 at 298 K and 1.9 at 1000 K. The reaction is endothermic, because K is larger at higher temperature.

15.87 $K_p = \dfrac{P_{CO_2}}{P_{CO}} = 6.0 \times 10^2$

If P_{CO} is 150 torr, P_{CO_2} can never exceed $760 - 150 = 610$ torr. Then $Q = 610/150 = 4.1$. Because this is far less than K, the reaction will shift in the direction of more product. Reduction will therefore occur.

15.88   The anecdote tells us that increasing the volume of the reaction container, the furnace, had no effect on the amount of unreacted CO(g), the amount of CO(g) expelled. This is true for reactions that have the same number of moles of gaseous products and reactants, as this one does. It also means that $K_p = K_c$.

15.89   (a) $CCl_4(g) \rightleftharpoons C(s) + 2\,Cl_2(g)$

initial	2.00 atm	0 atm
change	$-x$ atm	$+2x$ atm
equil.	$(2.00 - x)$ atm	$2x$ atm

$K_p = 0.76 = \dfrac{P_{Cl_2}^2}{P_{CCl_4}} = \dfrac{(2x)^2}{(2.00 - x)}$

$1.52 - 0.76x = 4x^2; \quad 4x^2 + 0.76x - 1.52 = 0$

Using the quadratic formula, $a = 4$, $b = 0.76$, $c = -1.52$

$x = \dfrac{-0.76 \pm \sqrt{(0.76)^2 - 4(4)(-1.52)}}{2(4)} = \dfrac{-0.76 + 4.99}{8} = 0.5287 = 0.53$ atm

Fraction CCl_4 reacted $= \dfrac{x \text{ atm}}{2.00 \text{ atm}} = \dfrac{0.5287}{2.00} = 0.264 = 26\%$

(b) $P_{Cl_2} = 2x = 2(0.5287) = 1.06$ atm

$P_{CCl_4} = 2.00 - x = 2.00 - 0.5287 = 1.47$ atm

15.90   (a) $Q = \dfrac{P_{PCl_5}}{P_{PCl_3} \times P_{Cl_2}} = \dfrac{(0.20)}{(0.50)(0.50)} = 0.80$

0.80 (Q) > 0.0870 (K), the reaction proceeds to the left.

(b) $PCl_3(g)$ $+$ $Cl_2(g)$ $\rightleftharpoons$ $PCl_5(g)$

initial	0.50 atm	0.50 atm	0.20 atm
change	$+x$ atm	$+x$ atm	$-x$ atm
equil.	$(0.50 + x)$ atm	$(0.50 + x)$ atm	$(0.20 - x)$ atm

(Because the reaction proceeds to the left, P_{PCl_5} must decrease and P_{PCl_3} and P_{Cl_2} must increase.)

$K_p = 0.0870 = \dfrac{(0.20 - x)}{(0.50 + x)(0.50 + x)}; \quad 0.0870 = \dfrac{(0.20 - x)}{(0.250 + 1.00\,x + x^2)}$

$0.0870(0.250 + 1.00\,x + x^2) = 0.20 - x; \quad -0.17825 + 1.0870\,x + 0.0870\,x^2 = 0$

$$x = \frac{-1.0870 \pm \sqrt{(1.0870)^2 - 4(0.0870)(-0.17825)}}{2(0.0870)} = \frac{-1.0870 + 1.1152}{0.174} = 0.162$$

$$P_{PCl_3} = (0.50 + 0.162)\,\text{atm} = 0.662 \qquad P_{Cl_2} = (0.50 + 0.162)\,\text{atm} = 0.662\,\text{atm}$$

$$P_{PCl_5} = (0.20 - 0.162)\,\text{atm} = 0.038\,\text{atm}$$

To two decimal places, the pressures are 0.66, 0.66, and 0.04 atm, respectively. When substituting into the K_p expression, pressures to three decimal places yield a result much closer to 0.0870.

(c)   Increasing the volume of the container favors the process where more moles of gas are produced, so the reverse reaction is favored and the equilibrium shifts to the left; the mole fraction of Cl_2 increases.

(d)   For an exothermic reaction, increasing the temperature decreases the value of K; more reactants and fewer products are present at equilibrium and the mole fraction of Cl_2 increases.

15.91   *Analyze/Plan.* Calculate the equilibrium pressures of H_2, I_2, and HI; use them to calculate K_p. Set up a new equilibrium table and calculate new equilibrium pressures. *Solve.*

$$\frac{P}{n} = \frac{RT}{V} = \frac{0.08206\,\text{L-atm}}{\text{mol-K}} \times \frac{731\,\text{K}}{5.00} = 11.997\,\frac{\text{atm}}{\text{mol}}$$

$$P_{H_2} = P_{I_2} = 0.112\,\text{mol} \times 11.997\,\frac{\text{atm}}{\text{mol}} = 1.344 = 1.34\,\text{atm}$$

$$P_{HI} = 0.775\,\text{mol} \times 11.997\,\frac{\text{atm}}{\text{mol}} = 9.298 = 9.30\,\text{atm}$$

$$H_2(g) + I_2(g) \rightleftharpoons 2HI(g); \qquad K_p = \frac{P_{HI}^2}{P_{H_2} \times P_{I_2}} = \frac{(9.298)^2}{(1.344)^2} = 47.861 = 47.9$$

$$P_{HI}\,(\text{added}) = 0.200\,\text{mol} \times \frac{11.997\,\text{atm}}{\text{mol}} = 2.3994 = 2.40\,\text{atm}$$

	$H_2(g)$	+	$I_2(g)$	$\rightleftharpoons$	$2HI(g)$
initial	1.34 atm		1.34 atm		9.30 atm + 2.40 atm
change	+x atm		+x atm		–2x atm
equil.	(1.34 + x) atm		(1.34 + x) atm		(11.70 – 2x) atm

$$K_p = 47.86 = \frac{(11.70 - 2x)^2}{(1.34 + x)^2}. \quad \text{Take the square root of both sides:}$$

$$6.918 = \frac{11.70 - 2x}{1.34 + x}; \; 9.270 + 6.918\,x = 11.70 - 2x; \; 8.918\,x = 2.430; \; x = 0.27248 = 0.272$$

$$P_{H_2} = P_{I_2} = 1.34 + 0.272 = 1.612 = 1.61\,\text{atm}; \; P_{HI} = 11.70 - 2(0.272) = 11.156 = 11.16\,\text{atm}$$

$$\text{Check. } \frac{(11.156)^2}{(1.612)^2} = 47.89 = 47.9$$

15.92   (a)   Because the volume of the vessel = 1.00 L, mol = M. The reaction will proceed to the left to establish equilibrium.

$$A(g) + \quad 2B(g) \quad \rightleftharpoons \quad 2\,C(g)$$

initial	0 M	0 M	1.00 M
change	+x M	+2 x M	–2 x M
equil.	x M	2 x M	(1.00 – 2 x) M

At equilibrium, [C] = (1.00 – 2x) M, [B] = 2x M.

(b)   x must be less than 0.50 M (so that [C], 1.00 –2 x, is not less than zero).

(c) $K_c = \dfrac{[C]^2}{[A][B]^2}$; $\dfrac{(1.00-2x)^2}{(x)(2x)^2} = 0.25$

$1.00 - 4x + 4x^2 = 0.25(4x)^3$; $x^3 - 4x^2 + 4x - 1 = 0$

(d)

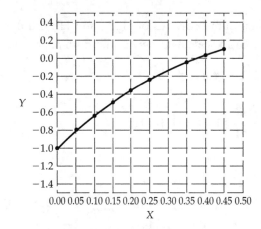

X	Y
0.0	−1.000
0.05	−0.810
0.10	−0.639
0.15	−0.487
0.20	−0.352
0.25	−0.234
0.35	−0.047
0.40	+0.024
0.45	+0.081
~0.383	0.00

(e)   From the plot, x ≈ 0.383 M

[A] = x = 0.383 M; [B] = 2 x = 0.766 M

[C] = 1.00 – 2x = 0.234 M

Using the K_c expression as a check:

$K_c = 0.25$; $\dfrac{(0.234)^2}{(0.383)(0.766)^2} = 0.24$;   the estimated values are reasonable.

15.93 $K_p = \dfrac{P_{O_2} \times P_{CO}^2}{P_{CO_2}^2} \approx 1 \times 10^{-13}$; $P_{O_2} = (0.03)(1\,atm) = 0.03\,atm$

$P_{CO} = (0.002)(1\,atm) = 0.002\,atm$; $P_{CO_2} = (0.12)(1\,atm) = 0.12\,atm$

$Q = \dfrac{(0.03)(0.002)^2}{(0.12)^2} = 8.3 \times 10^{-6} = 8 \times 10^{-6}$

Because Q > K_p, the system will shift to the left to attain equilibrium. Thus, a catalyst that promoted the attainment of equilibrium would result in a lower CO content in the exhaust.

15.94 (a) At 700 K: $K_c = \dfrac{k_f}{k_r} = \dfrac{1.8 \times 10^{-3}\ M^{-1}s^{-1}}{6.3 \times 10^{-2}\ M^{-1}s^{-1}} = 0.02857 = 0.029$

(b) At 800 K: $K_c = \dfrac{k_f}{k_r} = \dfrac{0.097\ M^{-1}s^{-1}}{2.6\ M^{-1}s^{-1}} = 0.03731 = 0.037$

The value of K_c increases when the temperature increases. The reaction is endothermic.

Integrative Exercises

15.95 Calculate the initial $[IO_4^-]$, and then construct an equilibrium table to determine $[H_4IO_6^-]$ at equilibrium.

$$M_c \times V_c = M_d \times L_d;\quad \frac{0.905\ M \times 25.0\ mL}{500.0\ mL} = M_d = 0.04525 = 0.0453\ M\ IO_4^-$$

$$IO_4^-(aq) \;+\; 2H_2O(l) \;\rightleftharpoons\; H_4IO_6^-(aq)$$

	IO_4^-	$H_4IO_6^-$
initial	0.0453 M	0
change	−x	+x
equil.	0.0453 − x	+x

$$K_c = 3.5 \times 10^{-2} = \frac{[H_4IO_6^-]}{[IO_4^-]} = \frac{x}{(0.0453 - x)}$$

Because K_c is relatively large and $[IO_4^-]$ is relatively small, we cannot assume x is small relative to 0.0453.

$0.035(0.04525 - x) = x;\ 0.001584 - 0.035\,x = x;\ 0.001584 = 1.035\,x$

$x = 0.001584/1.035 = 0.001530 = 0.0015\ M\ H_4IO_6^-$ at equilibrium

15.96 (a)

$CoO(s) + H_2(g) \rightleftharpoons Co(s) + H_2O(g)$	$K_1 = 67$
$CO(g) + H_2O(g) \rightleftharpoons H_2(g) + CO_2(g)$	$1/K_2 = 1/0.14$

$CoO(s)+H_2(g)+CO(g)+H_2O(g) \rightleftharpoons Co(s)+H_2O(g)+H_2(g)+CO_2(g)$ $K_c = K_1/K_2$

$CoO(s) + CO(g) \rightleftharpoons Co(s) + CO_2(g)$ $K_c = 4.8 \times 10^2$

(b) Based on the results of part (a), CO(g) is a stronger reducing agent than H_2(g) at 823 K. In the first reaction above, H_2(g) reduces CoO(s) and the value of K is 67. In the fourth reaction, CO(g) reduces CoO(s) and the value of K is 480. The much larger equilibrium constant for the fourth reaction indicates that products are more favored than the products in the first reaction.

(c) $5.00\ g\ CoO(s) \times \dfrac{1\ mol\ CoO(s)}{74.932\ g\ CoO(s)} = 0.066727 = 0.0667\ mol\ CoO(s)$

$n = \dfrac{PV}{RT} = \dfrac{1.00\ atm}{0.08206\ L\text{-}atm/mol\text{-}K} \times \dfrac{2.5 \times 10^{-1}\ L}{298\ K} = 0.010223 = 0.010\ mol\ CO$

(d)

	CoO(s) +	CO(g) ⇌	Co(s) +	CO₂(g)
initial	0.0667 mol	0.010 mol	0 mol	0 mol
change	–x mol	–x mol	x mol	x mol
equil.	0.0667 – x mol	0.0102 – x mol	x mol	x mol

$$K_c = \frac{67}{0.14} = \frac{[CO_2]}{[CO(g)]} = \frac{x}{0.0102-x}; \quad 478.57(0.010-x) = x$$

$4.7857 - 478.57\, x = x; \quad 4.7857 = 479.57\, x; \quad x = 0.009979 = 0.010$ mol

At equilibrium, mol CoO(s) = 0.0667 – 0.009979 = 0.05672 = 0.057 mol CoO(s)

g CoO(s) = 0.05672 mol CoO × 74.932 g/mol = 4.250 = 4.3 g CoO(s)

[Note that the equilibrium concentration of CO is essentially zero. With an equilibrium constant of 4.8×10^2, the reaction "goes to completion" and CO is the limiting reactant.]

15.97   Consider the energy profile for an exothermic reaction.

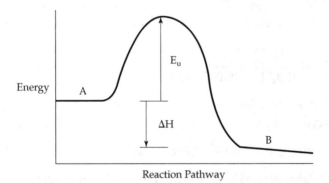

The activation energy in the forward direction, E_{af}, equals E_u and the activation energy in the reverse reaction, E_{ar}, equals $E_u - \Delta H$. (The same is true for an endothermic reaction because the sign of ΔH is the positive and $E_{ar} < E_{af}$.) For the reaction in question,

$$K = \frac{k_f}{k_r} = \frac{A_f e^{-E_{af}/RT}}{A_r e^{-E_{ar}/RT}}$$

Because the ln form of the Arrhenius equation is easier to manipulate, we will consider ln K.

$$\ln K = \ln\left(\frac{k_f}{k_r}\right) = \ln k_f - \ln k_r = \frac{-E_{af}}{RT} + \ln A_f - \left[\frac{-E_{ar}}{RT} + \ln A_r\right]$$

Substituting E_u for E_{af} and $(E_u - \Delta H)$ for E_{ar}

$$\ln K = \frac{-E_u}{RT} + \ln A_f - \left[\frac{-(E_u - \Delta H)}{RT} + \ln A_r\right]; \quad \ln K = \frac{-E_u + (E_u - \Delta H)}{RT} + \ln A_f - \ln A_r$$

$$\ln K = \frac{-\Delta H}{RT} + \ln \frac{A_f}{A_r}$$

For the catalyzed reaction, $E_{cat} < E_u$ and $E_{af} = E_{cat}$, $E_{ar} = E_{cat} - \Delta H$. The catalyst does not change the value of ΔH.

$$\ln K_{cat} = \frac{-E_{cat} + (E_{cat} - \Delta H)}{RT} + \ln A_f - \ln A_r$$

$$\ln K_{cat} = \frac{-\Delta H}{RT} + \frac{\ln A_f}{A_r}$$

Thus, assuming A_f and A_r are not changed by the catalyst, $\ln K = \ln K_{cat}$ and $K = K_{cat}$.

15.98   (a)   For the reaction $SO_2(l) \rightleftharpoons SO_2(g)$, $K_p = P_{SO_2}$. From the phase diagram, as T increases P_{SO_2} and K_p increase. For an endothermic reaction, K increases as T increases. The phase diagram tells us that the vaporization of $SO_2(l)$ is an endothermic process.

       (b)   Read P_{SO_2} from the liquid-gas line on the phase diagram at $0\,^{\circ}$C and $100\,^{\circ}$C. Note that the pressure axis of the phase diagram is logarithmic with respect to pressure, but linear with respect to logP. In terms of logP, the axis labels would be –1, 0, 1 and 2. The logP values at $0\,^{\circ}$C and $100\,^{\circ}$C are approximately 0.25 and 1.4. The values of P_{SO_2} and K_p are $10^{0.25}$ and $10^{1.4}$, 1.8, and 25, respectively.

       (c)   It is not possible to calculate an equilibrium constant between the gas and liquid phases in the supercritical region, because they do not exist separately in this region. That is, the gas and liquid phases are indistinguishable in the supercritical region.

       (d)   Gases are most ideal at high temperature and low pressure. The red dot at slightly greater than $240\,^{\circ}$C is the point where $SO_2(g)$ most closely approaches ideal behavior.

       (e)   The point near $15\,^{\circ}$C is the one at the lowest temperature, but it is also at low pressure. In general, the closer the pressure and temperature conditions are to the point of phase transition, the less ideal the behavior of the gas (because it is nearly a liquid). This describes the point near $115\,^{\circ}$C and 20 atm, which is at relatively high pressure and near the liquid-gas line.

15.99   (a)   The larger rate constant is k_r. $K_c = k_f/k_r$ and the value of K_c is much less than one (3.1×10^{-4}). The value of k_r must be larger than the value of k_f.

       (b) $K_c = \dfrac{k_f}{k_r}$; $k_r = \dfrac{k_f}{K_c} = \dfrac{0.27 s^{-1}}{3.1 \times 10^{-4}} = 870.97 = 8.7 \times 10^2$

       (c)   The forward reaction is bond breaking, which is always endothermic.

       (d)   The reverse rate constant k_r will increase. An increase in temperature causes both k_r and k_f to increase. For an endothermic reaction, we expect the value of K_c to increase with an increase in temperature, so the increase in k_f will be greater than the increase in k_r.

15.100   (a) $H_2O(l) \rightleftharpoons H_2O(g)$; $K_p = P_{H_2O}$

       (b)   At 30°C, the vapor pressure of $H_2O(l)$ is 31.82 torr. $K_p = P_{H_2O} = 31.82$ torr

                $K_p = 31.82$ torr $\times$ 1 atm$/760$ torr $= 0.041868 = 0.04187$ atm

(c)   From part (b), the value of K_p is the vapor pressure of the liquid at that temperature. By definition, vapor pressure = atmospheric pressure = 1 atm at the normal boiling point. $K_p = 1$ atm

15.101 (a)   VSEPR indicates that each O atom has four electron domains about it and thus adopts tetrahedral geometry. One O atom has two covalent bonds to H and two hydrogen bonds to H atoms on the second water molecule. The O atom on the second water molecule has two nonbonding electron pairs. The water dimer is not symmetrical.

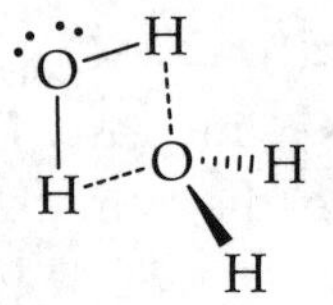

        (b)   Hydrogen-bonding is the intermolecular interaction involved in water dimer formation.

        (c)   Water dimer formation is exothermic, because the value of K decreases as temperature increases.

15.102 The O_2-binding reaction occurs in aqueous solution, so we will write a K_c expression. The amount of $O_2(g)$ will appear as a pressure. By convention, reactions which involve gaseous and aqueous substances have mixed equilibrium expressions written in terms of both pressures and molar concentrations.

$$K_c = \frac{[\text{Hb-}(O_2)_4]}{P_{O_2}^4 \times [\text{Hb}]}$$

The *P50 value* is the partial pressure at which 50% of the hemoglobin is saturated with $O_2(g)$. At this partial pressure, the concentrations of O_2-bound hemoglobin and free hemoglobin are equal, $[\text{Hb} - (O_2)_4] = [\text{Hb}]$. Substitute the two P50 values into the K_c expression and compare the values for fetal and adult hemoglobin.

$$\text{at P50, } K_{cF} = \frac{1}{P_{O_2}^4} = \frac{1}{19^4} = 7.7 \times 10^{-6}; \quad K_{cA} = \frac{1}{P_{O_2}^4} = \frac{1}{26.8^4} = 1.94 \times 10^{-6}$$

Comparing the two values, $K_{cF} / K_{cA} = 7.7 \times 10^{-6}/1.94 \times 10^{-6} \approx 4$. The equilibrium constant for O_2-binding by fetal hemoglobin is approximately four times that by adult hemoglobin.

16 Acid–Base Equilibria

Visualizing Concepts

16.1 *Analyze.* From the structures decide which reactant fits the description of a Brønsted–Lowry (B-L) acid, a B-L base, a Lewis acid, and a Lewis base. *Plan.* A B-L acid is an H^+ donor, and a B-L base is an H^+ acceptor. A Lewis acid is an electron pair acceptor, and a Lewis base is an electron pair donor. *Solve.*

(a) HCl is a B-L acid, because it donates H^+ during reaction. NH_3 is a B-L base, because it accepts H^+ during reaction.

(b) By virtue of its unshared electron pair, NH_3 is the electron pair donor and a Lewis base. HCl is the electron pair acceptor and a Lewis acid.

16.2 *Plan.* The stronger the acid, the greater the extent of ionization. The stronger the acid, the weaker its conjugate base. In an acid–base reaction, equilibrium will favor the side with the weaker acid and base. *Solve.*

(a) HY is stronger than HX. Starting with six HY molecules, four are dissociated; of six HX molecules, only two are dissociated. Because it is dissociated to a greater extent, HY is the stronger acid.

(b) If HY is the stronger acid, Y^- is the weaker base and X^- is the stronger base.

(c) HX and Y^-, the reactants, are the weaker acid and base. Equilibrium lies to the left, and $K_c < 1$.

16.3 (a) True. Solution A is the color of methyl orange in an acidic solution.

(b) False. Methyl orange turns yellow at a pH slightly greater than 4, so solution B could be at any pH greater than 4.

(c) True. The basic color of any indicator occurs at higher pH than the acidic color does.

16.4 *Analyze/Plan.* The pH reading on the meter will identify the solution. Use the definition of pH or pOH to calculate the concentration of the solution. *Solve.*

(a) KOH(aq). The pH meter reads 12.08, indicating that the solution is strongly basic. KOH(aq) is the only choice that is a base.

(b) pOH = 14.00 – pH; pOH = 14.00 – 12.08 = 1.92

pOH = –log $[OH^-]$; $[OH^-] = 10^{-pOH}$; $[OH^-] = 10^{-1.92} = 0.01202 = 0.012\ M$

(c) The pH scale for aqueous solutions is based on the value of K_w for water. K_w is an equilibrium constant whose value depends on temperature. If temperature is significantly different from 25 °C, the value of K_w is different from 1×10^{-14} and relationships like {pH + pOH = 14} do not hold true.

16.5   *Plan.* Strong acids are completely ionized. The acid that is least ionized is the weakest and has the smallest K_a value. At equal concentrations, the weakest acid has the smallest $[H^+]$ and highest pH.   *Solve.*

(a)   HY is a strong acid. There are no neutral HY molecules in the solution, only H^+ cations and Y^- anions.

(b)   HX has the smallest K_a value. It has the most neutral acid molecules and fewest ions.

(c)   HX has the fewest H^+ ions and, therefore, the highest pH.

16.6   *Analyze/Plan.* We are shown a plot of concentration solution versus $[H^+]$. Use the definition of acids and bases along with relationships between total solute concentration and $[H^+]$ for strong or weak acids and bases to answer the questions.   *Solve.*

(a)   The substance is a strong acid. It is an acid because $[H^+]$ increases as solute concentration increases. It is a strong acid because $[H^+]$ is directly proportional to solute concentration. This is not true for weak acids.

(b)   Yes. A strong acid is completely ionized in aqueous solution, so $[H^+]$ = solute concentration = 0.18 M. pH = $-\log[H^+]$ = $-\log(0.18)$ = 0.74.

(c)   No. In pure water, there is a finite concentration of $H^+(aq)$ and $[OH^-]$, owing to the autoionization of water.

16.7   Statement (e) is true. Only line C shows the trend in percent ionization of a weak acid when initial acid concentration increases. Statement (f) is false because the value of K_a, and thus the strength of a weak acid, does not change with acid concentration.

16.8   *Analyze/Plan.* Write the formula of each molecule and compare them to the entries in Tables 16.2 and 16.4. Select the molecule that fits the definition of an acid and the one that fits the definition of a base.   *Solve.*

(a)   Molecule A acts as a base. It is hydroxyl amine, NH_2OH, an entry in Table 16.4. Molecule A is an H^+ acceptor because of the nonbonded electron pair on the N atom of the amine ($-NH_2$) group, not because it contains an $-OH$ group. The presence of an $-OH$ group in an organic molecule does not mean that the molecule is a base.

(b)   Molecule B acts as an acid; it is formic acid, $HCOOH$. It is similar to CH_3COOH, an entry in Table 16.2. The H atom bonded to O is ionizable and $HCOOH$ is an H^+ donor. In general, organic molecules that contain a carboxyl ($-COOH$) group are acids.

(c)   Molecule C acts as neither an acid nor a base; it is methanol, CH_3OH. In organic molecules, the $-OH$ functional group is an alcohol. The H atom bonded to O is not ionizable, and the $-OH$ group does not dissociate in aqueous solution. An alcohol is neither an acid nor a base.

16.9   (a)   Basic. Because of the amine group (N, with a lone electron pair, bound to two C atoms and an H atom), we expect phenylephrine solution to be basic. [Note that $-OH$ bound to a benzene ring is very weakly acidic (K_a for phenol is 1.3×10^{-10}), but probably not acidic enough to dominate the acid/base properties of the molecule.]

(b)   Phenylephrine is a neutral molecule, whereas phenylephrine hydrochloride is a salt. It is the product when HCl reacts with phenylephrine. The cation in the salt has an additional H atom bound to the N of phenylephrine. The anion is chloride.

(c)   Acidic. The cation of phenylephrine hydrochloride is the conjugate acid of phenylephrine, so a solution of the salt is acidic. (Chloride anion is a negligible base.)

16.10   Diagram C best represents an aqueous solution of NaF; it contains mostly Na^+ and F^-, along with a few HF molecules and OH^- ions. The HF and OH^- are present because F^- is a weak Brønsted–Lowry base; it accepts H^+ from a water molecule, producing HF and OH^-. The solution is basic because it contains OH^-.

16.11   *Plan.* Evaluate the structures to determine if the molecules are binary acids or oxyacids. Consider the trends in acid strength for both classes of acids.   *Solve.*

(a)   If X is the same atom on both molecules, the molecule (b) is more acidic. The carboxylate anion, the conjugate base of this carboxylic acid, is stabilized by resonance, whereas the conjugate base of (a) is not resonance-stabilized. Stabilization of the conjugate base causes the ionization equilibrium to favor products, and (b) is the stronger acid.

(b)   Increasing the electronegativity of X increases the strength of both acids. As X becomes more electronegative and attracts more electron density, the O–H bond becomes weaker and more polar. This increases the likelihood of ionization, which increases acid strength. An electronegative X group also stabilizes the anionic conjugate bases by delocalizing the negative charge. This causes the ionization equilibrium to favor products, and the values of K_a to increase.

16.12   *Analyze/Plan.* Consider the definitions of Lewis, Arrhenius, and Brønsted-Lowry acids and bases.   *Solve.*

(a) PCl_4^+, Lewis acid; Cl^-, Lewis base. There are no hydrogen atoms or ions in any of the reactants or products. Only the Lewis definition applies.

(b) NH_3, Lewis base; BF_3, Lewis acid. No hydrogen ions are transferred in the reaction. Only the Lewis definition applies. Also, from Lewis structures, we know that NH_3 has a nonbonded electron pair on N, and BF_3 is electron deficient.

(c) $[Al(H_2O)_6]^{3+}$, Brønsted-Lowry acid; H_2O, Brønsted-Lowry base. This is an H^+ transfer reaction and fits the Brønsted-Lowry definition. Because the reaction occurs in aqueous solution, $[Al(H_2O)_6]^{3+}$ also fits the definition of an Arrhenius acid.

Arrhenius and Brønsted–Lowry Acids and Bases (Sections 16.1 and 16.2)

16.13 $HCl(g) + NH_3(g) \rightarrow NH_4^+Cl^-(s)$. HCl is the B-L (Brønsted–Lowry) acid; it donates an H^+ to NH_3 to form NH_4^+. NH_3 is the B-L base; it accepts the H^+ from HCl.

16.14   Statement (e) is false. One type of compound that contains an –OH group and is not (usually) a Brønsted-Lowry base is an alcohol.

16.15   *Analyze/Plan.* Follow the logic in Sample Exercise 16.1. A conjugate base has one less H^+ than its conjugate acid. A conjugate acid has one more H^+ than its conjugate base. *Solve.*

    (a)   (i) IO_3^- (ii) NH_3

    (b)   (i) OH^- (ii) H_3PO_4

16.16   A conjugate base has one less H^+ than its conjugate acid. A conjugate acid has one more H^+ than its conjugate base.

    (a)   (i) $HCOO^-$ (ii) PO_4^{3-}

    (b)   (i) HSO_4^- (ii) $CH_3NH_3^+$

16.17   *Analyze/Plan.* Use the definitions of B-L acids and bases, and conjugate acids and bases to make the designations. Evaluate the changes going from reactant to product to inform your choices.   *Solve.*

	B-L Acid	+	**B-L Base**	$\rightleftharpoons$	**Conjugate Acid**	+	**Conjugate Base**
(a)	$NH_4^+(aq)$		$CN^-(aq)$		$HCN(aq)$		$NH_3(aq)$
(b)	$H_2O(l)$		$(CH_3)_3N(aq)$		$(CH_3)_3NH^+(aq)$		$OH^-(aq)$
(c)	$HCOOH(aq)$		$PO_4^{3-}(aq)$		$HPO_4^{2-}(aq)$		$HCOO^-(aq)$

16.18

	B-L Acid	+	**B-L Base**	$\rightleftharpoons$	**Conjugate Acid**	+	**Conjugate Base**
(a)	$HBrO(aq)$		$H_2O(l)$		$H_3O^+(aq)$		$BrO^-(aq)$
(b)	$HSO_4^-(aq)$		$HCO_3^-(aq)$		$H_2CO_3(aq)$		$SO_4^{2-}(aq)$
(c)	$H_3O^+(aq)$		$HSO_3^-(aq)$		$H_2SO_3(aq)$		$H_2O(l)$

16.19   *Analyze/Plan.* Follow the logic in Sample Exercise 16.2.   *Solve.*

    (a) Acid: $HSO_3^-(aq)$ + $H_2O(l)$ $\rightleftharpoons$ $SO_3^{2-}(aq) + H_3O^+(aq)$

                B-L acid   B-L base   conj. base   conj. acid

        Base: $HSO_3^-(aq)$ + $H_2O(l)$ $\rightleftharpoons$ $H_2SO_3(aq) + OH^-(aq)$

                B-L base   B-L acid   conj. acid   conj. base

    (b) H_2SO_3 is the conjugate acid of HSO_3^-.

        SO_3^{2-} is the conjugate base of HSO_3^-.

16.20   (a) $H_2C_6H_7O_5^-(aq) + H_2O(l)$ $\rightleftharpoons$ $H_3C_6H_7O_5(aq) + OH^-(aq)$

    (b) $H_2C_6H_7O_5^-(aq) + H_2O(l)$ $\rightleftharpoons$ $HC_6H_7O_5^{2-}(aq) + H_3O^+(aq)$

    (c) $H_3C_6H_7O_5$ is the conjugate acid of $H_2C_6H_7O_5^-$.

        $HC_6H_7O_5^{2-}$ is the conjugate base of $H_2C_6H_7O_5^-$.

16.21   *Analyze/Plan.* Based on the chemical formula, decide whether the base is strong, weak, or negligible. Is it the conjugate of a strong acid (negligible base), weak acid (weak base), or negligible acid (strong base)? Also check Figure 16.4. To write the formula of the conjugate acid, add a single H and increase the particle charge by one.   *Solve.*

    (a) CH_3COO^-, weak base; CH_3COOH, weak acid

    (b) HCO_3^-, weak base; H_2CO_3, weak acid

    (c) O^{2-}, strong base; OH^-, negligible acid

(d) Cl^-, negligible base; HCl, strong acid

(e) NH_3, weak base; NH_4^+, weak acid

16.22   *Analyze/Plan.*   Based on the chemical formula, decide whether the acid is strong, weak, or negligible. Is it one of the known seven strong acids (Section 16.5)? Also check Figure 16.4. To write the formula of the conjugate base, remove a single H and decrease the particle charge by one.   *Solve.*

(a)   HCOOH, weak acid; $HCOO^-$, weak base

(b) H_2, negligible acid; H^-, strong base

(c) CH_4, negligible acid; CH_3^-, strong base

(d)   HF, weak acid; F^-, weak base

(e) NH_4^+, weak acid; NH_3, weak base

16.23   *Analyze/Plan.* Based on the chemical formula, determine the strength of acids and bases by checking the known strong acids (Section 16.5). Recall the paradigm "The stronger the acid, the weaker its conjugate base, and vice versa."   *Solve.*

(a)   HBr. It is one of the seven strong acids (Section 16.5).

(b) F^-. HCl is a stronger acid than HF, so F^- is the stronger conjugate base.

16.24   (a) $HClO_3$. It is one of the seven strong acids (Section 16.5). Also, in a series of oxyacids with the same central atom (Cl), the acid with more O atoms is stronger (Section 16.10).

       (b) HS^-. H_2SO_4 is a stronger acid than H_2S, so HS^- is the stronger conjugate base. In fact, because H_2SO_4 is one of the seven strong acids, HSO_4^- is a negligible base.

16.25   *Analyze/Plan.* Acid–base equilibria favor formation of the weaker acid and base. Compare the relative strengths of the substances acting as acids on opposite sides of the reaction arrow. (Bases can also be compared; the conclusion should be the same.)   *Solve.*

	Base	+	**Acid**	$\rightleftharpoons$	**Conjugate Acid**	+	**Conjugate Base**
(a)	$O^{2-}(aq)$	+	$H_2O(l)$	$\rightleftharpoons$	$OH^-(aq)$	+	$OH^-(aq)$

H_2O is a stronger acid than OH^-, so the equilibrium lies to the right.

(b)	$HS^-(aq)$	+	$CH_3COOH(aq)$	$\rightleftharpoons$	$H_2S(aq)$	+	$CH_3COO^-(aq)$

CH_3COOH is a stronger acid than H_2S, so the equilibrium lies to the right.

(c)	$NO_2^-(aq)$	+	$H_2O(l)$	$\rightleftharpoons$	$HNO_2(aq)$	+	$OH^-(aq)$

HNO_2 is a stronger acid than H_2O, so the equilibrium lies to the left.

16.26

	Base	+	**Acid**	$\rightleftharpoons$	**Conjugate Acid**	+	**Conjugate Base**
(a)	$OH^-(aq)$	+	$NH_4^+(aq)$	$\rightleftharpoons$	$H_2O(l)$	+	$NH_3(aq)$

OH^- is a stronger base than NH_3 (Figure 16.4), so the equilibrium lies to the right.

(b)	$CH_3COO^-(aq)$	+	$H_3O^+(aq)$	$\rightleftharpoons$	$CH_3COOH(aq)$	+	$H_2O(l)$

H_3O^+ is a stronger acid than CH_3COOH (Figure 16.4), so the equilibrium lies to the right.

(c)	$F^-(aq)$	+	$HCO_3^-(aq)$	$\rightleftharpoons$	$CO_3^{2-}(aq)$	+	$HF(aq)$

CO_3^{2-} is a stronger base than F^-, so the equilibrium lies to the left.

Autoionization of Water (Section 16.3)

16.27   Statement (ii) is correct. In pure water, the only source of H^+ is the autoionization reaction, which produces equal concentrations of H^+ and OH^-. As the temperature of water changes, the value of K_w changes, and the pH at which $[H^+] = [OH^-]$ changes.

16.28   (a) $H_2O(l) \rightleftharpoons H^+(aq) + OH^-(aq)$

   (b) $K_w = [H^+][OH^-]$

   (c)   Statement (iii) is true. If a solution is basic, it contains more OH^- than H^+.

16.29   *Analyze/Plan.* Follow the logic in Sample Exercise 16.5. In pure water at 25 °C, $[H^+] = [OH^-] = 1 \times 10^{-7}$ M. If $[H^+] > 1 \times 10^{-7}$ M, the solution is acidic; if $[H^+] < 1 \times 10^{-7}$ M, the solution is basic.   *Solve.*

   (a) $[H^+] = \dfrac{K_w}{[OH^-]} = \dfrac{1.0 \times 10^{-14}}{4.5 \times 10^{-4}\ M} = 2.2 \times 10^{-11}\ M < 1 \times 10^{-7}\ M;\ \text{basic}$

   (b) $[H^+] = \dfrac{K_w}{[OH^-]} = \dfrac{1.0 \times 10^{-14}}{8.8 \times 10^{-9}\ M} = 1.1 \times 10^{-6}\ M > 1 \times 10^{-7}\ M;\ \text{acidic}$

   (c) $[OH^-] = 100[H^+];\ K_w = [H^+] \times 100[H^+] = 100[H^+]^2;$

   $[H^+] = (K_w/100)^{1/2} = 1.0 \times 10^{-8}\ M < 1 \times 10^{-7}\ M;\ \text{basic}$

16.30   In pure water at 25 °C, $[H^+] = [OH^-] = 1 \times 10^{-7}$ M. If $[OH^-] > 1 \times 10^{-7}$ M, the solution is basic; if $[OH^-] < 1 \times 10^{-7}$ M, the solution is acidic.

   (a) $[OH^-] = \dfrac{K_w}{[H^+]} = \dfrac{1.0 \times 10^{-14}}{0.0505\ M} = 1.98 \times 10^{-13}\ M < 1 \times 10^{-7}\ M;\ \text{acidic}$

   (b) $[OH^-] = \dfrac{K_w}{[H^+]} = \dfrac{1.0 \times 10^{-14}}{2.5 \times 10^{-10}\ M} = 4.0 \times 10^{-5}\ M > 1 \times 10^{-7}\ M;\ \text{basic}$

   (c) $[H^+] = 1000[OH^-];\ K_w = 1000[OH^-][OH^-] = 1000[OH^-]^2$

   $[OH^-] = (K_w/1000)^{1/2} = 3.2 \times 10^{-9}\ M < 1 \times 10^{-7}\ M;\ \text{acidic}$

16.31   *Analyze/Plan.* Follow the logic in Sample Exercise 16.4. Note that the value of the equilibrium constant (in this case, K_w) changes with temperature.   *Solve.*

   At 0 °C, $K_w = 1.2 \times 10^{-15} = [H^+][OH^-]$

   In pure water, $[H^+] = [OH^-];\ 1.2 \times 10^{-15} = [H^+]^2;\ [H^+] = (1.2 \times 10^{-15})^{1/2}$

   $[H^+] = [OH^-] = 3.5 \times 10^{-8}\ M$

16.32 $K_w = [D^+][OD^-];$ for pure D_2O, $[D^+] = [OD^-];\ 8.9 \times 10^{-16} = [D^+]^2;$

   $[D^+] = [OD^-] = 3.0 \times 10^{-8}\ M$

The pH Scale (Section 16.4)

16.33   *Analyze/Plan.* A change of one pH unit (in either direction) is:

$$\Delta pH = pH_2 - pH_1 = -(\log[H^+]_2 - \log[H^+]_1) = -\log\frac{[H^+]_2}{[H^+]_1} = \pm1. \text{ The antilog of } +1 \text{ is } 10;$$

the antilog of -1 is 1×10^{-1}. Thus, a ΔpH of one unit represents an increase or decrease in $[H^+]$ by a factor of 10.   *Solve.*

(a) $\Delta pH = \pm2.00$ is a change of $10^{2.00}$; $[H^+]$ changes by a factor of 100.

(b) $\Delta pH = \pm0.5$ is a change of $10^{0.50}$; $[H^+]$ changes by a factor of 3.2.

16.34 $[H^+]_A = 250\,[H^+]_B$. From Solution 16.33, $\Delta pH = -\log\dfrac{[H^+]_B}{[H^+]_A}$

$$\Delta pH = -\log\frac{[H^+]_B}{250\,[H^+]_B} = -\log\left(\frac{1}{250}\right) = 2.40$$

The pH of solution A is 2.40 pH units lower than the pH of solution B, because $[H^+]_A$ is 250 times greater than $[H^+]_B$. The greater the $[H^+]$, the lower the pH of the solution.

16.35   *Analyze/Plan.* At 25 °C, $[H^+][OH^-] = 1 \times 10^{-14}$; pH + pOH = 14. Use these relationships to complete the table. If pH < 7, the solution is acidic; if pH > 7, the solution is basic. *Solve.*

$[H^+]$	$[OH^-]$	pH	pOH	acidic or basic
$7.5 \times 10^{-3}\,M$	$1.3 \times 10^{-12}\,M$	2.12	11.88	acidic
$2.8 \times 10^{-5}\,M$	$3.6 \times 10^{-10}\,M$	4.56	9.44	acidic
$5.6 \times 10^{-9}\,M$	$1.8 \times 10^{-6}\,M$	8.25	5.75	basic
$5.0 \times 10^{-9}\,M$	$2.0 \times 10^{-6}\,M$	8.30	5.70	basic

Check. pH + pOH = 14; $[H^+][OH^-] = 1 \times 10^{-14}$

16.36

pH	pOH	$[H^+]$	$[OH^-]$	acidic or basic
5.25	8.75	$5.6 \times 10^{-6}\,M$	$1.8 \times 10^{-9}\,M$	acidic
11.98	2.02	$1.1 \times 10^{-12}\,M$	$9.6 \times 10^{-3}\,M$	basic
9.36	4.64	$4.4 \times 10^{-10}\,M$	$2.3 \times 10^{-5}\,M$	basic
12.93	1.07	$1.2 \times 10^{-13}\,M$	$8.5 \times 10^{-2}\,M$	basic

16.37   *Analyze/Plan.* Based on the pH and a new value of the equilibrium constant K_w, calculate equilibrium concentrations of $H^+(aq)$ and $OH^-(aq)$. The definition of pH remains pH = $-\log[H^+]$.   *Solve.*

pH = 7.40; $[H^+] = 10^{-pH} = 10^{-7.40} = 4.0 \times 10^{-8}\,M$

$K_w = 2.4 \times 10^{-14} = [H^+][OH^-]$; $[OH^-] = 2.4 \times 10^{-14} / [H^+]$

$[OH^-] = 2.4 \times 10^{-14} / 4.0 \times 10^{-8} = 6.0 \times 10^{-7}\,M$; pOH = $-\log(6.0 \times 10^{-7}) = 6.22$

Alternately, pH + pOH = pK_w. At 37 °C, pH + pOH = $-\log(2.4 \times 10^{-14})$

pH + pOH = 13.62; pOH = 13.62 − 7.40 = 6.22

$[OH^-] = 10^{-pOH} = 10^{-6.22} = 6.0 \times 10^{-7}\,M$

16.38   The pH ranges from 5.2 to 5.6; pOH ranges from (14.0–5.2 =) 8.8 to (14.0–5.6 =) 8.4.

$[H^+] = 10^{-pH}$, $[OH^-] = 10^{-pOH}$

$[H^+] = 10^{-5.2} = 6.31 \times 10^{-6} = 6 \times 10^{-6} \, M$; $[H^+] = 10^{-5.6} = 2.51 \times 10^{-6} = 3 \times 10^{-6} \, M$

The range of $[H^+]$ is $6 \times 10^{-6} \, M$ to $3 \times 10^{-6} \, M$.

$[OH^-] = 10^{-8.8} = 1.58 \times 10^{-9} = 2 \times 10^{-9} \, M$; $[OH^-] = 10^{-8.4} = 3.98 \times 10^{-9} = 4 \times 10^{-9} \, M$

The range of $[OH^-]$ is $2 \times 10^{-9} \, M$ to $4 \times 10^{-9} \, M$.

(The pH has one decimal place, so concentrations are reported to 1 sig fig.)

16.39   *Analyze/Plan.* We are given the behavior of an unknown solution in two indicators. Use information from Figure 16.8 to answer the questions about the solution.   *Solve.*

     (a)   Acidic. Both indicators change color at pH less than 7.

     (b)   The range of possible integer pH values for the solution is 4 to 6. Methyl orange changes color near pH 4 and bromthymol (or bromothymol) blue changes near pH 6.

     (c)   Methyl red changes color slightly above pH 5. It would help narrow the range of possible pH values.

16.40   (a)   Acidic. Bromthymol (or bromothymol) blue, which changes color at a lower pH than phenolphthalein, is yellow below pH 6, so the solution is acidic.

     (b)   (ii), a maximum pH. The solution is the lower-pH color for both indicators, so we know only that the maximum pH is 6.

     (c)   From Figure 16.8, methyl violet, thymol blue, methyl orange, and methyl red would help determine the pH of the solution more precisely. These indicators change colors at pH values from approximately 1 to 5. [One strategy would be to start at the low end of the pH range with methyl violet and work up.]

Strong Acids and Bases (Section 16.5)

16.41   (a)   True.

     (b)   True.

     (c)   False. A 1.0 *M* strong acid solution has a pH of 0.

16.42   (a)   True.

     (b)   True.

     (c)   False. Base strength should not be confused with solubility. Base strength describes the tendency of a dissolved molecule [formula unit for ionic compounds such as $Mg(OH)_2$] to dissociate into cations and hydroxide ions. $Mg(OH)_2$ is a strong base because each $Mg(OH)_2$ unit that dissolves also dissociates into $Mg^{2+}(aq)$ and $OH^-(aq)$. $Mg(OH)_2$ is not very soluble, so relatively few $Mg(OH)_2$ units dissolve when the solid compound is added to water.

16.43   *Analyze/Plan.* Follow the logic in Sample Exercise 16.8. Strong acids are completely ionized, so $[H^+]$ = original acid concentration and pH = $-\log[H^+]$. For the solutions obtained by dilution, use the "dilution" formula, $M_1V_1 = M_2V_2$, to calculate molarity of the acid.   *Solve.*

(a) $8.5 \times 10^{-3}\ M\ HBr = 8.5 \times 10^{-3}\ M\ H^+$; pH = $-\log(8.5 \times 10^{-3}) = 2.07$

(b) $\dfrac{1.52\ g\ HNO_3}{0.575\ L\ soln} \times \dfrac{1\ mol\ HNO_3}{63.02\ g\ HNO_3} = 0.041947 = 0.0419\ M\ HNO_3$

$[H^+] = 0.0419\ M$; pH = $-\log(0.041947) = 1.377$

(c) $M_c \times V_c = M_d \times V_d$; $0.250\ M \times 0.00500\ L = ?\ M \times 0.0500\ L$

$M_d = \dfrac{0.250\ M \times 0.00500\ L}{0.0500\ L} = 0.0250\ M\ HCl$

$[H^+] = 0.0250\ M$; pH = $-\log(0.0250) = 1.602$

(d) $[H^+]_{total} = \dfrac{mol\ H^+\ from\ HBr + mol\ H^+\ from\ HCl}{total\ L\ solution}$

$[H^+]_{total} = \dfrac{(0.100\ M\ HBr \times 0.0100\ L) + (0.200\ M \times 0.0200\ L)}{0.0300\ L}$

$[H^+]_{total} = \dfrac{1.00 \times 10^{-3}\ mol\ H^+ + 4.00 \times 10^{-3}\ mol\ H^+}{0.0300\ L} = 0.1667 = 0.167\ M$

pH = $-\log(0.1667\ M) = 0.778$

16.44   For a strong acid, which is completely ionized, $[H^+]$ = the initial acid concentration.

(a) $0.0167\ M\ HNO_3 = 0.0167\ M\ H^+$; pH = $-\log(0.0167) = 1.777$

(b) $\dfrac{0.225\ g\ HClO_3}{2.00\ L\ soln} \times \dfrac{1\ mol\ HClO_3}{84.46\ g\ HClO_3} = 1.332 \times 10^{-3} = 1.33 \times 10^{-3}\ M\ HClO_3$

$[H^+] = 1.33 \times 10^{-3}\ M$; pH = $-\log(1.332 \times 10^{-3}) = 2.875$

(c) $M_c \times V_c = M_d \times V_d$; $0.500\ L = 500\ mL$

$1.00\ M\ HCl \times 15.00\ mL\ HCl = M_d\ HCl \times 500\ mL\ HCl$

$M_d\ HCl = \dfrac{1.00\ M \times 15.00\ mL}{500\ mL} = 3.00 \times 10^{-2}\ M\ HCl = 3.00 \times 10^{-2}\ M\ H^+$

pH = $-\log(3.00 \times 10^{-2}) = 1.523$

(d) $[H^+]_{total} = \dfrac{mol\ H^+\ from\ HCl + mol\ H^+\ from\ HI}{total\ L\ solution}$;   mol = $M \times L$

$[H^+]_{total} = \dfrac{(0.020\ M\ HCl \times 0.0500\ L) + (0.010\ M\ HI \times 0.125\ L)}{0.175\ L}$

$[H^+]_{total} = \dfrac{1.0 \times 10^{-3}\ mol\ H^+ + 1.25 \times 10^{-3}\ mol\ H^+}{0.175\ L} = 0.01286 = 0.013\ M$

pH = $-\log(0.01286) = 1.89$

16.45   *Analyze/Plan.* Follow the logic in Sample Exercise 16.9. Strong bases dissociate completely upon dissolving. $pOH = -\log[OH^-]$; $pH = 14 - pOH$.   *Solve.*

(a)   Pay attention to the formula of the base to get $[OH^-]$.

$$[OH^-] = 2[Sr(OH)_2] = 2(1.5 \times 10^{-3}\ M) = 3.0 \times 10^{-3}\ M\ OH^-$$

$$pOH = -\log(3.0 \times 10^{-3}) = 2.52;\ pH = 14 - pOH = 11.48$$

(b) $mol/LiOH = g\ LiOH/molar\ mass\ LiOH.$ $[OH^-] = [LiOH].$

$$\frac{2.250\ g\ LiOH}{0.2500\ L\ soln} \times \frac{1\ mol\ LiOH}{23.948\ g\ LiOH} = 0.37581 = 0.3758\ M\ LiOH = [OH^-]$$

$$pOH = -\log(0.37581) = 0.4250;\ pH = 14 - pOH = 13.5750$$

(c)   Use the dilution formula to get the $[NaOH] = [OH^-]$.

$$M_c \times V_c = M_d \times V_d;\ 0.175\ M \times 0.00100\ L = ?\ M \times 2.00\ L$$

$$M_d = \frac{0.175\ M \times 0.00100\ L}{2.00\ L} = 8.75 \times 10^{-5}\ M\ NaOH = [OH^-]$$

$$pOH = -\log(8.75 \times 10^{-5}) = 4.058;\ pH = 14 - pOH = 9.942$$

(d)   Consider total mol OH^- from KOH and $Ca(OH)_2$ as well as total solution volume.

$$[OH^-]_{total} = \frac{mol\ OH^-\ from\ KOH + mol\ OH^-\ from\ Ca(OH)_2}{total\ L\ soln}$$

$$[OH^-]_{total} = \frac{(0.105\ M \times 0.00500\ L) + 2(9.5 \times 10^{-2}\ M \times 0.0150\ L)}{0.0200\ L}$$

$$[OH^-]_{total} = \frac{0.525 \times 10^{-3}\ mol\ OH^- + 2.85 \times 10^{-3}\ mol\ OH^-}{0.0200\ L} = 0.16875 = 0.17\ M$$

$$pOH = -\log(0.16875) = 0.77;\ pH = 14 - pOH = 13.23$$

$(9.5 \times 10^{-2}\ M$ has 2 sig figs, so the $[OH^-]$ has 2 sig figs and pH and pOH have 2 decimal places.)

16.46   For a strong base, which is completely dissociated, $[OH^-]$ = the initial base concentration. Then, $pOH = -\log[OH^-]$ and $pH = 14 - pOH$.

(a) $0.182\ M\ KOH = 0.182\ M\ OH^-$; $pOH = -\log(0.182) = 0.740$; $pH = 14 - 740 = 13.260$

(b) $\dfrac{3.165\ g\ KOH}{0.5000\ L} \times \dfrac{1\ mol\ KOH}{56.106\ g\ KOH} = 0.112822 = 0.1128\ M = [OH^-]$

$$pOH = -\log(0.112822) = 0.9476;\ pH = 14 - pOH = 13.0524$$

(c) $M_c \times V_c = M_d \times V_d$

$0.0105\ M\ Ca(OH)_2 \times 10.0\ mL = M_d\ Ca(OH)_2 \times 500\ mL$

$$M_d\ Ca(OH)_2 = \frac{0.0105\ M\ Ca(OH)_2 \times 10.0\ mL}{500.0\ mL} = 2.10 \times 10^{-4}\ M\ Ca(OH)_2$$

$$Ca(OH)_2(aq) \rightarrow Ca^{2+}(aq) + 2OH^-(aq)$$

$$[OH^-] = 2[Ca(OH)_2] = 2(2.10 \times 10^{-4}\ M) = 4.20 \times 10^{-4}\ M$$

$$pOH = -\log(4.20 \times 10^{-4}) = 3.377;\ pH = 14 - pOH = 10.623$$

(d) $[OH^-]_{total} = \dfrac{\text{mol } OH^- \text{ from NaOH} + \text{mol } OH^- \text{ from Ba(OH)}_2}{\text{total L solution}}$

$$\dfrac{(8.2 \times 10^{-3} \, M \times 0.0400 \, L) + 2(0.015 \, M \times 0.0200 \, L)}{0.0600 \, L}$$

$[OH^-]_{total} = \dfrac{3.28 \times 10^{-4} \, \text{mol } OH^- + 6.0 \times 10^{-4} \, \text{mol } OH^-}{0.0600 \, L} = 0.01547 = 0.015 \, M \, OH^-$

$pOH = -\log(0.01547) = 1.81; \, pH = 14 - 1.81 = 12.19$

16.47 *Analyze/Plan.* pH $\rightarrow$ pOH $\rightarrow$ $[OH^-] = [NaOH]$. *Solve.*

$pOH = 14 - pH = 14.00 - 11.50 = 2.50$

$pOH = 2.50 = -\log[OH^-]; \, [OH^-] = 10^{-2.50} = 3.2 \times 10^{-3} \, M$

$[OH^-] = [NaOH] = 3.2 \times 10^{-3} \, M$

16.48 $pOH = 14 - pH = 14.00 - 10.05 = 3.95$

$pOH = 3.95 = -\log[OH^-]; \, [OH^-] = 10^{-3.95} = 1.122 \times 10^{-4} \, M = 1.1 \times 10^{-4} \, M$

$[OH^-] = 2[Ca(OH)_2]; \, [Ca(OH)_2] = [OH^-] \, / \, 2 = 1.122 \times 10^{-4} \, M \, /2 = 5.6 \times 10^{-5} \, M$

Weak Acids (Section 16.6)

16.49 *Analyze/Plan.* Remember that K_a = [products]/[reactants]. If $H_2O(l)$ appears in the equilibrium reaction, it will **not** appear in the K_a expression, because it is a pure liquid. *Solve.*

(a) $HBrO_2(aq) \rightleftharpoons H^+(aq) + BrO_2^-(aq); \, K_a = \dfrac{[H^+][BrO_2^-]}{[HBrO_2]}$

$HBrO_2(aq) + H_2O(l) \rightleftharpoons H_3O^+(aq) + BrO_2^-(aq); \, K_a = \dfrac{[H_3O^+][BrO_2^-]}{[HBrO_2]}$

(b) $C_2H_5COOH(aq) \rightleftharpoons H^+(aq) + C_2H_5COO^-(aq); \, K_a = \dfrac{[H^+][C_2H_5COO^-]}{[C_2H_5COOH]}$

$C_2H_5COOH(aq) + H_2O(l) \rightleftharpoons H_3O^+(aq) + C_2H_5COO^-(aq);$

$$K_a = \dfrac{[H_3O^+][C_2H_5COO^-]}{[C_2H_5COOH]}$$

16.50 (a) $C_6H_5COOH(aq) \rightleftharpoons H^+(aq) + C_6H_5COO^-(aq); \, K_a = \dfrac{[H^+][C_6H_5COO^-]}{[HC_6H_5COOH]}$

$C_6H_5COOH(aq) + H_2O(l) \rightleftharpoons H_3O^+(aq) + C_6H_5COO^-(aq);$

$$K_a = \dfrac{[H_3O^+][C_6H_5COO^-]}{[HC_6H_5COOH]}$$

(b) $HCO_3^-(aq) \rightleftharpoons H^+(aq) + CO_3^{2-}(aq); \, K_a = \dfrac{[H^+][CO_3^{2-}]}{[HCO_3^-]}$

$HCO_3^-(aq) + H_2O(l) \rightleftharpoons H_3O^+(aq) + CO_3^{2-}(aq); \, K_a = \dfrac{[H_3O^+][CO_3^{2-}]}{[HCO_3^-]}$

16.51 *Analyze/Plan.* Follow the logic in Sample Exercise 16.10. *Solve.*

$$CH_3CH(OH)COH \rightleftharpoons H^+(aq) + CH_3CH(OH)COO^-(aq); \quad K_a = \frac{[H^+][CH_3CH(OH)COO^-]}{[CH_3CH(OH)COOH]}$$

$$[H^+] = [CH_3CH(OH)COO^-] = 10^{-2.44} = 3.63 \times 10^{-3} = 3.6 \times 10^{-3} \ M$$

$$[CH_3CH(OH)COOH] = 0.10 - 3.63 \times 10^{-3} = 0.0964 = 0.096 \ M$$

$$K_a = \frac{(3.63 \times 10^{-3})^2}{(0.0964)} = 1.4 \times 10^{-4}$$

16.52 $$C_6H_5CH_2COOH(aq) \rightleftharpoons H^+(aq) + C_6H_5CH_2COO^-(aq); \quad K_a = \frac{[H^+][C_6H_5CH_2COO^-]}{[C_6H_5CH_2COOH]}$$

$$[H^+] = [C_6H_5CH_2COO^-] = 10^{-2.68} = 2.09 \times 10^{-3} = 2.1 \times 10^{-3} \ M$$

$$[C_6H_5CH_2COOH] = 0.085 - 2.09 \times 10^{-3} = 0.0829 = 0.083 \ M$$

$$K_a = \frac{(2.09 \times 10^{-3})^2}{0.0829} = 5.3 \times 10^{-5}$$

16.53 *Analyze/Plan.* Write the equilibrium reaction and the K_a expression. Use percent ionization to get equilibrium concentration of $[H^+]$, and by stoichiometry, $[X^-]$ and $[HX]$. Calculate K_a *Solve.*

$$[H^+] = 0.110 \times [CH_2ClCOOH]_{initial} = 0.0110 \ M$$

	$CH_2ClCOOH(aq)$	$\rightleftharpoons$	$H^+(aq)$	+	$CH_2ClCOO^-(aq)$
initial	0.100 M		0		0
equil.	0.089 M		0.0110 M		0.0110 M

$$K_a = \frac{[H^+][CH_2ClCOO^-]}{[CH_2ClCOOH]} = \frac{(0.0110)^2}{0.089} = 1.4 \times 10^{-3}$$

16.54 $$[H^+] = 0.132 \times [BrCH_2COOH]_{initial} = 0.0132 \ M$$

	$BrCH_2COOH(aq)$	$\rightleftharpoons$	$H^+(aq)$	+	$BrCH_2COO^-(aq)$
initial	0.100 M		0		0
equil.	0.087		0.0132 M		0.0132 M

$$K_a = \frac{[H^+][BrCH_2COO^-]}{[BrCH_2COOH]} = \frac{(0.0132)^2}{0.087} = 2.0 \times 10^{-3}$$

16.55 *Analyze/Plan.* Write the equilibrium reaction and the K_a expression.

$$[H^+] = 10^{-pH} = [CH_3COO^-]; \ [CH_3COOH] = x - [H^+].$$

Substitute into the K_a expression and solve for x. *Solve.*

$$[H^+] = 10^{-pH} = 10^{-2.90} = 1.26 \times 10^{-3} = 1.3 \times 10^{-3} \ M$$

$$K_a = 1.8 \times 10^{-5} = \frac{[H^+][CH_3COO^-]}{[CH_3COOH]} = \frac{(1.26 \times 10^{-3})^2}{(x - 1.26 \times 10^{-3})}$$

$$1.8 \times 10^{-5} (x - 1.26 \times 10^{-3}) = (1.26 \times 10^{-3})^2;$$

$$1.8 \times 10^{-5} x = 1.585 \times 10^{-6} + 2.266 \times 10^{-8} = 1.608 \times 10^{-6};$$

$$x = 0.08931 = 0.089 \ M \ CH_3COOH$$

16.56 $[H^+] = 10^{-pH} = 10^{-3.65} = 2.239 \times 10^{-4} = 2.2 \times 10^{-4} \, M$

$$K_a = 6.8 \times 10^{-4} = \frac{[H^+][F^-]}{[HF]} = \frac{(2.239 \times 10^{-4})^2}{x - 2.239 \times 10^{-4}}$$

$6.8 \times 10^{-4}(x - 2.239 \times 10^{-4}) = (2.239 \times 10^{-4})^2;$

$6.8 \times 10^{-4} \, x = 1.522 \times 10^{-7} + 0.501 \times 10^{-7} = 2.024 \times 10^{-7}$

$x = 2.976 \times 10^{-4} = 3.0 \times 10^{-4} \, M \, HF$

16.57   *Analyze/Plan.* Follow the logic in Sample Exercise 16.12. Write K_a, construct the equilibrium table, solve for $x = [H^+]$, and then get equilibrium $[C_6H_5COO^-]$ and $[C_6H_5COOH]$ by substituting $[H^+]$ for x.   *Solve.*

	$C_6H_5COOH(aq)$	$\rightleftharpoons$	$H^+(aq)$	$+$	$C_6H_5COO^-(aq)$
initial	0.050 M		0		0
equil.	(0.050 – x) M		x M		x M

$$K_a = \frac{[H^+][C_6H_5COO^-]}{[C_6H_5COOH]} = \frac{x^2}{(0.050-x)} \approx \frac{x^2}{0.050} = 6.3 \times 10^{-5}$$

$x^2 = 0.050 \, (6.3 \times 10^{-5}); \, x = 1.8 \times 10^{-3} \, M = [H^+] = [H_3O^+] = [C_6H_5COO^-]$

$[C_6H_5COOH] = 0.050 - 0.0018 = 0.048 \, M$

Check. $\dfrac{1.8 \times 10^{-3} \, M \, H^+}{0.050 \, M \, C_6H_5COOH} \times 100 = 3.6\%$ ionization; the approximation is valid

16.58

	$HClO_2(aq)$	$\rightleftharpoons$	$H^+(aq)$	$+$	$ClO_2^-(aq)$
initial	0.0125 M		0		0
equil.	(0.0125 – x) M		x M		x M

$$K_a = \frac{[H^+][ClO_2^-]}{[HClO_2]} = \frac{x^2}{(0.0125-x)} \approx \frac{x^2}{0.0125} = 1.1 \times 10^{-2}$$

Assuming x is small relative to 0.0125, $x^2 = 0.0125(0.011); \, x = 1.2 \times 10^{-2} \, M.$

Clearly, x is not small relative to 0.0125, so we must solve the quadratic formula for $[H^+]$.

$x^2 = 0.011 \, (0.0125 - x); \, x^2 + 0.011x - 1.38 \times 10^{-4} = 0$

$$x = \frac{-0.011 \pm \sqrt{(0.011)^2 - 4(1)(-1.38 \times 10^{-4})}}{2(1)} = 0.007452 = 0.0075 \, M;$$

$[H^+] = [H_3O^+] = [ClO_2^-] = 0.0075 \, M; \, [HClO_2] = 0.0125 - 0.0075 = 0.005045 = 5.0 \times 10^{-3} \, M$

Check. $K_a = \dfrac{(7.5 \times 10^{-3})^2}{5.0 \times 10^{-3}} = 0.011;$ our results agree

16.59   *Analyze/Plan.* Follow the logic in Sample Exercise 16.12.   *Solve.*

(a)

	$C_2H_5COOH(aq)$	$\rightleftharpoons$	$H^+(aq)$	$+$	$C_2H_5COO^-(aq)$
initial	0.095 M		0		0
equil.	(0.095 – x) M		x M		x M

$$K_a = \frac{[H^+][C_2H_5COO^-]}{[C_2H_5COOH]} = \frac{x^2}{(0.095-x)} \approx \frac{x^2}{0.095} = 1.3 \times 10^{-5}$$

$x^2 = 0.095(1.3 \times 10^{-5}); \, x = 1.111 \times 10^{-3} = 1.1 \times 10^{-3} \, M \, H^+; \, pH = 2.95$

Check. $\dfrac{1.1 \times 10^{-3}\ M\ \mathrm{H^+}}{0.095\ M\ \mathrm{C_2H_5COO}H} \times 100 = 1.2\%$ ionization; the approximation is valid

(b) $K_a = \dfrac{[\mathrm{H^+}][\mathrm{CrO_4^{2-}}]}{[\mathrm{HCrO_4^-}]} = \dfrac{x^2}{(0.100 - x)} \approx \dfrac{x^2}{0.100} = 3.0 \times 10^{-7}$

$x^2 = 0.100(3.0 \times 10^{-7}); x = 1.732 \times 10^{-4} = 1.7 \times 10^{-4}\ M\ \mathrm{H^+}$

$\mathrm{pH} = -\log(1.732 \times 10^{-4}) = 3.7614 = 3.76$

Check. $\dfrac{1.7 \times 10^{-4}\ M\ \mathrm{H^+}}{0.100\ M\ \mathrm{HCrO_4^-}} \times 100 = 0.17\%$ ionization; the approximation is valid

(c)   Follow the logic in Sample Exercise 16.15. $\mathrm{pOH} = -\log[\mathrm{OH^-}]; \mathrm{pH} = 14 - \mathrm{pOH}$

	$\mathrm{C_5H_5N(aq)} + \mathrm{H_2O(l)}$	$\rightleftharpoons$	$\mathrm{C_5H_5NH^+(aq)}$	+	$\mathrm{OH^-(aq)}$
initial	0.120 M		0		0
equil.	(0.120 − x) M		x M		x M

$K_b = \dfrac{[\mathrm{C_5H_5NH^+}][\mathrm{OH^-}]}{[\mathrm{C_5H_5N}]} = \dfrac{x^2}{(0.120 - x)} \approx \dfrac{x^2}{0.120} = 1.7 \times 10^{-9}$

$x^2 = 0.120(1.7 \times 10^{-9}); x = 1.428 \times 10^{-5} = 1.4 \times 10^{-5}\ M\ \mathrm{OH^-}; \mathrm{pH} = 9.15$

Check. $\dfrac{1.4 \times 10^{-5}\ M\ \mathrm{OH^-}}{0.120\ M\ \mathrm{C_5H_5N}} \times 100 = 0.012\%$ ionization; the approximation is valid

16.60   (a)

	$\mathrm{HOCl(aq)}$	$\rightleftharpoons$	$\mathrm{H^+(aq)}$	+	$\mathrm{OCl^-(aq)}$
initial	0.095 M		0		0
equil.	(0.095 − x) M		x M		x M

$K_a = \dfrac{[\mathrm{H^+}][\mathrm{OCl^-}]}{[\mathrm{HOCl}]} = \dfrac{x^2}{(0.095 - x)} \approx \dfrac{x^2}{0.095} = 3.0 \times 10^{-8}$

$x^2 = 0.095(3.0 \times 10^{-8}); x = [\mathrm{H^+}] = 5.3 \times 10^{-5}\ M, \mathrm{pH} = 4.27$

Check. $\dfrac{5.3 \times 10^{-5}\ M\ \mathrm{H^+}}{0.095\ M\ \mathrm{HOCl}} \times 100 \approx 0.056\%$ ionization

The approximation is nearly valid. To 2 sig figs, the quadratic formula gives the same $[\mathrm{H^+}]$.

(b)

	$\mathrm{H_2NNH_2(aq)} + \mathrm{H_2O(l)}$	$\rightleftharpoons$	$\mathrm{H_2NNH_3^+(aq)}$	+	$\mathrm{OH^-(aq)}$
initial	0.0085 M		0		0
equil.	(0.0085 − x) M		x M		x M

$K_b = \dfrac{[\mathrm{H_2NNH_3^+}][\mathrm{OH^-}]}{[\mathrm{H_2NNH_2}]} = \dfrac{x^2}{(0.0085 - x)} \approx \dfrac{x^2}{0.0085} = 1.3 \times 10^{-6}$

$x^2 = 0.0085(1.3 \times 10^{-6}); x = [\mathrm{OH^-}] = 1.051 \times 10^{-4} = 1.1 \times 10^{-4}\ M$

Clearly, 1.1×10^{-4} M OH$^-$ is not small compared to 8.5×10^{-3} M H$_2$NNH$_2$, and we must solve the quadratic.

$$x^2 = 1.3 \times 10^{-6}(0.0085 - x); \ x^2 + 1.3 \times 10^{-6}x - 1.105 \times 10^{-8} = 0$$

$$x = \frac{-1.3 \times 10^{-6} \pm \sqrt{(1.3 \times 10^{-6})^2 - 4(1)(-1.105 \times 10^{-8})}}{2(1)} = 1.0447 \times 10^{-4}$$

$$= 1.0 \times 10^{-4} \text{ M OH}^-$$

pOH = 3.981 = 3.98; pH = 14 − pOH = 14 − 3.981 = 10.019 = 10.02

Check. Although this solution has more than 12% ionization, the difference in [OH$^-$] between the estimate and the quadratic is not great.

(c)　　　　　　　HONH$_2$(aq) + H$_2$O(l) $\rightleftharpoons$ HONH$_3^+$(aq) + OH$^-$(aq)

initial　　　　　　　0.165 M　　　　　　　0　　　　0

equil.　　　　　　(0.165 − x) M　　　　　x M　　　x M

$$K_b = \frac{[\text{HONH}_3^+][\text{OH}^-]}{[\text{HONH}_2]} = \frac{x^2}{(0.165-x)} \approx \frac{x^2}{0.165} = 1.1 \times 10^{-8}$$

$$x^2 = 0.165(1.1 \times 10^{-8}); \ x = [\text{OH}^-] = 4.3 \times 10^{-5} \text{ M, pH} = 9.63$$

Check. $\dfrac{4.3 \times 10^{-5} \text{ M OH}^-}{0.165 \text{ M HONH}_2} \times 100 = 0.026\%$ ionization; the approximation is valid

16.61　*Analyze/Plan.* $K_a = 10^{-pK_a}$. Follow the logic in Sample Exercise 16.13. *Solve.*

Let $[\text{H}^+] = [\text{NC}_7\text{H}_4\text{SO}_3^-] = z$. $K_a = $ antilog $(-2.32) = 4.79 \times 10^{-3} = 4.8 \times 10^{-3}$.

$$\frac{z^2}{0.10 - z} = 4.79 \times 10^{-3}. \quad \text{As } K_a \text{ is relatively large, solve the quadratic.}$$

$$z^2 + 4.79 \times 10^{-3}z - 4.79 \times 10^{-4} = 0$$

$$z = \frac{-4.79 \times 10^{-3} \pm \sqrt{(4.79 \times 10^{-3})^2 - 4(1)(-4.79 \times 10^{-4})}}{2(1)} = \frac{-4.79 \times 10^{-3} \pm \sqrt{1.937 \times 10^{-3}}}{2}$$

$$z = 1.96 \times 10^{-2} = 2.0 \times 10^{-2} \text{ M H}^+; \text{ pH} = -\log(1.96 \times 10^{-2}) = 1.71$$

16.62　Calculate the initial concentration of HC$_9$H$_7$O$_4$.

$$2 \text{ tablets} \times \frac{500 \text{ mg}}{\text{tablet}} \times \frac{1 \text{ g}}{1000 \text{ mg}} \times \frac{1 \text{ mol HC}_9\text{H}_7\text{O}_4}{180.2 \text{ g HC}_9\text{H}_7\text{O}_4} = 0.005549 = 0.00555 \text{ mol HC}_9\text{H}_7\text{O}_4$$

$$\frac{0.005549 \text{ mol HC}_9\text{H}_7\text{O}_4}{0.250 \text{ L}} = 0.02220 = 0.0222 \text{ M HC}_9\text{H}_7\text{O}_4$$

　　　　　　　HC$_9$H$_7$O$_4$(aq) $\rightleftharpoons$ C$_9$H$_7$O$_4^-$ + H$^+$(aq)

initial　　　　　0.0222 M　　　　0 M　　　0 M

equil.　　　　(0.0222 − x)　　　x M　　　x M

$$K_a = 3.3 \times 10^{-4} = \frac{[\text{H}^+][\text{C}_9\text{H}_7\text{O}_4^-]}{[\text{HC}_9\text{H}_7\text{O}_4]} = \frac{x^2}{(0.0222 - x)}$$

Assuming x is small compared to 0.0222,

$$x^2 = 0.0222\,(3.3 \times 10^{-4}); x = [H^+] = 2.7 \times 10^{-3}\ M$$

$$\frac{2.7 \times 10^{-3}\ M\ H^+}{0.0222\ M\ HC_9H_7O_4} \times 100 = 12\%\ \text{ionization; the approximation is not valid}$$

Using the quadratic formula, $x^2 + 3.3 \times 10^{-4}\,x - 7.325 \times 10^{-6} = 0$

$$x = \frac{-3.3 \times 10^{-4} \pm \sqrt{(3.3 \times 10^{-4})^2 - 4(1)(-7.325 \times 10^{-6})}}{2(1)} = \frac{-3.3 \times 10^{-4} \pm \sqrt{2.941 \times 10^{-5}}}{2}$$

$$x = 2.547 \times 10^{-3} = 2.5 \times 10^{-3}\ M\ H^+; pH = -\log(2.547 \times 10^{-3}) = 2.594 = 2.59$$

16.63 *Analyze/Plan.* Follow the logic in Sample Exercise 16.12 and 16.13. *Solve.*

(a)

	$HN_3(aq)$	$\rightleftharpoons$	$H^+(aq)$	$+$	$N_3^-(aq)$
initial	$0.400\ M$		0		0
equil.	$(0.400 - x)\ M$		$x\ M$		$x\ M$

$$K_a = \frac{[H^+][N_3^-]}{[HN_3]} = 1.9 \times 10^{-5}; \frac{x^2}{(0.400-x)} \approx \frac{x^2}{0.400} = 1.9 \times 10^{-5}$$

$$x = 0.00276 = 2.8 \times 10^{-3}\ M = [H^+];\ \%\ \text{ionization} = \frac{2.76 \times 10^{-3}}{0.400} \times 100 = 0.69\%$$

(b) $1.9 \times 10^{-5} \approx \dfrac{x^2}{0.100}; x = 0.00138 = 1.4 \times 10^{-3}\ M\ H^+$

$$\%\ \text{ionization} = \frac{1.38 \times 10^{-3}\ M\ H^+}{0.100\ M\ HN_3} \times 100 = 1.4\%$$

(c) $1.9 \times 10^{-5} \approx \dfrac{x^2}{0.0400}; x = 8.72 \times 10^{-4} = 8.7 \times 10^{-4}\ M\ H^+$

$$\%\ \text{ionization} = \frac{8.72 \times 10^{-4}\ M\ H^+}{0.0400\ M\ HN_3} \times 100 = 2.2\%$$

Check. Notice that a tenfold dilution [part (a) versus part (c)] leads to a slightly more than threefold increase in percent ionization.

16.64 (a) $C_2H_5COOH(aq) \rightleftharpoons H^+(aq) + C_2H_5COO^-(aq)$

$$K_a = 1.3 \times 10^{-5} = \frac{[H^+][C_2H_5COO^-]}{[C_2H_5COOH]} = \frac{x^2}{0.250 - x}$$

$$x^2 \approx 0.250\,(1.3 \times 10^{-5}); x = 1.803 \times 10^{-3} = 1.8 \times 10^{-3}\ M\ H^+$$

$$\%\ \text{ionization} = \frac{1.803 \times 10^{-3}\ M\ H^+}{0.250\ M\ C_2H_5COOH} \times 100 = 0.721\%$$

(b) $\dfrac{x^2}{0.0800} \approx 1.3 \times 10^{-5};\ x = 1.020 \times 10^{-3} = 1.0 \times 10^{-3}\ M\ H^+$

$$\%\ \text{ionization} = \frac{1.020 \times 10^{-3}\ M\ H^+}{0.0800\ M\ C_2H_5COOH} \times 100 = 1.27\%$$

(c) $\dfrac{x^2}{0.0200} \approx 1.3 \times 10^{-5}$; $x = 5.099 \times 10^{-4} = 5.1 \times 10^{-4} \ M\,H^+$

% ionization $= \dfrac{5.099 \times 10^{-4} \ M\,H^+}{0.0200 \ M\,C_2H_5COOH} \times 100 = 2.55\%$

16.65   *Analyze/Plan.* Follow the logic in Sample Exercise 16.14. Citric acid is a triprotic acid with three K_a values that do not differ by more than 10^3. We must consider all three steps. Also, $C_6H_5O_7{}^{3-}$ is only produced in step 3.   *Solve.*

$H_3C_6H_5O_7(aq) \rightleftharpoons H^+(aq) + H_2C_6H_5O_7{}^-(aq)$ $K_{a1} = 7.4 \times 10^{-4}$

$H_2C_6H_5O_7{}^-(aq) \rightleftharpoons H^+(aq) + HC_6H_5O_7{}^{2-}(aq)$ $K_{a2} = 1.7 \times 10^{-5}$

$HC_6H_5O_7{}^{2-}(aq) \rightleftharpoons H^+(aq) + C_6H_5O_7{}^{3-}(aq)$ $K_{a3} = 4.0 \times 10^{-7}$

(a)   To calculate the pH of a 0.040 M solution, assume initially that only the first ionization is important:

	$H_3C_6H_5O_7(aq)$	$\rightleftharpoons$	$H^+(aq)$	$+$	$H_2C_6H_5O_7{}^-(aq)$
initial	0.040 M		0		0
equil.	$(0.040 - x)\ M$		$x\ M$		$x\ M$

$K_{a1} = \dfrac{[H^+][H_2C_6H_5O_7{}^-]}{[H_3C_6H_5O_7]} = \dfrac{x^2}{(0.040 - x)} = 7.4 \times 10^{-4}$

$x^2 = (0.040 - x)(7.4 \times 10^{-4}); \ \ x^2 \approx (0.040)(7.4 \times 10^{-4}); \ \ x = 0.00544 = 5.4 \times 10^{-3} \ M$

As this value for x is rather large in relation to 0.040, a better approximation for x can be obtained by substituting this first estimate into the expression for x^2, and then solving again for x:

$x^2 = (0.040 - x)\,(7.4 \times 10^{-4}) = (0.040 - 5.44 \times 10^{-3})\,(7.4 \times 10^{-4})$

$x^2 = 2.557 \times 10^{-5}; \ x = 5.057 \times 10^{-3} = 5.1 \times 10^{-3} \ M$

(This is the same result obtained from the quadratic formula.)

The correction to the value of x, though not large, is significant. Does the second ionization produce a significant additional concentration of H^+?

	$H_2C_6H_5O_7{}^-(aq)$	$\rightleftharpoons$	$H^+(aq)$	$+$	$HC_6H_5O_7{}^{2-}(aq)$
initial	$5.1 \times 10^{-3} \ M$		$5.1 \times 10^{-3} \ M$		0
equil.	$(5.1 \times 10^{-3} - y)$		$(5.1 \times 10^{-3} + y)$		y

$K_{a2} = \dfrac{[H^+][HC_6H_5O_7{}^{2-}]}{[H_2C_6H_5O_7{}^-]} = 1.7 \times 10^{-5}; \ \ \dfrac{(5.1 \times 10^{-3} + y)(y)}{(5.1 \times 10^{-3} - y)} = 1.7 \times 10^{-5}$

Assume that y is small relative to 5.1×10^{-3}; that is, that additional ionization of $H_2C_6H_5O_7{}^-$ is small, then

$\dfrac{(5.1 \times 10^{-3})y}{(5.1 \times 10^{-3})} = 1.7 \times 10^{-5} \ M; \ y = 1.7 \times 10^{-5} \ M$

This value is indeed small compared to $5.1 \times 10^{-3} \ M$; $[H^+]$ and pH are determined by the first ionization step. pH $= -\log(5.057 \times 10^{-3}) = 2.30$.

(b)   Yes. We started the calculation by assuming that only the first step made a significant contribution to $[H^+]$ and pH. Calculation proved this assumption to be true. Next, we assumed $[H^+]$ from the first ionization was small relative to 0.040 M citric acid; this assumption was not valid. Finally, we assumed that additional ionization of $H_2C_6H_5O_7^-$ was small, which was true.

(c)   The concentration of citrate ion, $[C_6H_5O_7^{3-}]$, is much less than $[H^+]$. Because the second ionization does not contribute significantly to $[H^+]$, we know that $[HC_6H_5O_7^{2-}]$ is less than $[H^+]$. The third ionization is even less extensive, so $[C_6H_5O_7^{3-}]$ is much less than $[H^+]$.

16.66 $H_2C_4H_4O_6(aq) \rightleftharpoons H^+(aq) + HC_4H_4O_6^-(aq) \qquad K_{a1} = 1.0 \times 10^{-3}$

$HC_4H_4O_6^-(aq) \rightleftharpoons H^+(aq) + C_4H_4O_6^{2-}(aq) \qquad K_{a2} = 4.6 \times 10^{-5}$

Begin by calculating the $[H^+]$ from the first ionization. The equilibrium concentrations are $[H^+] = [HC_4H_4O_6^-] = x$, $[H_2C_4H_4O_6] = 0.25 - x$.

$$K_{a1} = \frac{[H^+][HC_4H_4O_6^-]}{[H_2C_4H_4O_6]} = \frac{x^2}{0.25 - x}; \; x^2 + 1.0 \times 10^{-3}\,x - 2.5 \times 10^{-4} = 0$$

Using the quadratic formula, $x = 1.532 \times 10^{-2} = 0.015\ M\ H^+$ from the first ionization. Next, calculate the H^+ contribution from the second ionization.

	$HC_4H_4O_6^-(aq)$	$\rightleftharpoons$	$H^+(aq)$	$+$	$C_4H_4O_6^{2-}(aq)$
initial	0.015		0.015		0
equil.	$(0.015 - y)$		$(0.015 + y)$		y

$$K_{a2} = \frac{(0.015 + y)(y)}{(0.015 - y)} = 4.6 \times 10^{-5}; \text{assuming y is small compared to 0.015,}$$

$y = 4.6 \times 10^{-5}\ M\ C_4H_4O_6^{2-}(aq)$

This approximation is reasonable, because 4.6×10^{-5} is only 0.3% of 0.015.

$[H^+] = 0.015\ M$ (first ionization) $+ 4.6 \times 10^{-5}$ (second ionization)

Because 4.6×10^{-5} is 0.3% of 0.015 M, it can be safely ignored when calculating total $[H^+]$.

$pH = -\log(0.01532) = 1.18148 = 1.181$

Assumptions:

(1)   The ionization can be treated as a series of steps (valid by Hess's law).

(2)   The extent of ionization in the second step (y) is small relative to that from the first step (valid for this acid and initial concentration). This assumption was used twice, to calculate the value of y from K_{a2} and to calculate total $[H^+]$ and pH.

Weak Bases (Section 16.7)

16.67   (a) $HONH_3^+$

(b)   When hydroxylamine acts as a base, the nitrogen atom accepts a proton.

(c)   14 e^-, 7 e^- pairs

FC on N is +1   FC on O is +1

In neutral hydroxylamine, both O and N have zero formal charges. Nitrogen is less electronegative than oxygen and more likely to share a lone pair of electrons with an incoming (and electron-deficient) H^+. The resulting cation with the +1 formal charge on N is more stable than the one with the +1 formal charge on O.

16.68   (a)   *Analyze/Plan.* To determine relative strength, compare the K_b values of the two bases. *Solve.*

$$K_b \text{ for OCl}^- = \frac{K_w}{K_a \text{ for HClO}} = \frac{1.0 \times 10^{-14}}{3.0 \times 10^{-8}} = 3.3 \times 10^{-7}$$

K_b for hydroxylamine is 1.1×10^{-8}. OCl^- is a stronger base than hydroxylamine.

(b)   When OCl^- acts as a base, the O atom is the proton acceptor.

(c) $14\,e^-, 7\,e^-$ pairs $\left[:\ddot{C}l - \ddot{O}: \right]^-$

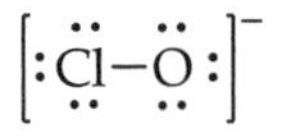

In OCl^-, the –1 formal charge is on O. H^+ attaches to the atom with the negative formal charge.

16.69   *Analyze/Plan.* Remember that K_b = [products]/[reactants]. If $H_2O(l)$ appears in the equilibrium reaction, it will not appear in the K_b expression, because it is a pure liquid. *Solve.*

(a) $(CH_3)_2NH(aq) + H_2O(l) \rightleftharpoons (CH_3)_2NH_2^+(aq) + OH^-(aq)$; $K_b = \dfrac{[(CH_3)_2NH_2^+][OH^-]}{[(CH_3)_2NH]}$

(b) $CO_3^{2-}(aq) + H_2O(l) \rightleftharpoons HCO_3^-(aq) + OH^-(aq)$; $K_b = \dfrac{[HCO_3^-][OH^-]}{[CO_3^{2-}]}$

(c) $HCOO^-(aq) + H_2O(l) \rightleftharpoons HCOOH(aq) + OH^-(aq)$; $K_b = \dfrac{[HCOOH][OH^-]}{[HCOO^-]}$

16.70   (a) $C_3H_7NH_2(aq) + H_2O(l) \rightleftharpoons C_3H_7NH_3^+(aq) + OH^-(aq)$; $K_b = \dfrac{[C_3H_7NH_3^+][OH^-]}{[C_3H_7NH_2]}$

(b) $HPO_4^{2-}(aq) + H_2O(l) \rightleftharpoons H_2PO_4^-(aq) + OH^-(aq)$; $K_b = \dfrac{[H_2PO_4^-][OH^-]}{[HPO_4^{2-}]}$

(c) $C_6H_5CO_2^-(aq) + H_2O(l) \rightleftharpoons C_6H_5CO_2H(aq) + OH^-(aq)$; $K_b = \dfrac{[C_6H_5CO_2H][OH^-]}{[C_6H_5CO_2^-]}$

16.71   *Analyze/Plan.* Follow the logic in Sample Exercise 16.15.   *Solve.*

$$C_2H_5NH_2(aq) + H_2O(l) \rightleftharpoons C_2H_5NH_3^+(aq) + OH^-(aq)$$

initial	0.075 M	0	0
equil.	(0.075 – x) M	x M	x M

$$K_b = \frac{[C_2H_5NH_3^+][OH^-]}{[C_2H_5NH_2]} = \frac{(x)(x)}{(0.075-x)} \approx \frac{x^2}{0.075} = 6.4 \times 10^{-4}$$

$x^2 = 0.075\,(6.4 \times 10^{-4})$; $x = [OH^-] = 6.9 \times 10^{-3}\,M$; $pH = 11.84$

Check. $\dfrac{6.9 \times 10^{-3} \, M \, OH^-}{0.075 \, M \, C_2H_5NH_2} \times 100 = 9.2\%$ ionization; the assumption is not valid

To obtain a more precise result, the K_b expression is rewritten in standard quadratic form and solved via the quadratic formula.

$$\dfrac{x^2}{0.075 - x} = 6.4 \times 10^{-4}; \, x^2 + 6.4 \times 10^{-4} \, x - 4.8 \times 10^{-5} = 0$$

$$x = \dfrac{-b \pm \sqrt{b^2 - 4ac}}{2a} = \dfrac{-6.4 \times 10^{-4} \pm \sqrt{(6.4 \times 10^{-4})^2 - 4(1)(-4.8 \times 10^{-5})}}{2}$$

$x = 6.62 \times 10^{-3} = 6.6 \times 10^{-3} \, M \, OH^-$; pOH = 2.18; pH = 14.00 − pOH = 11.82

Note that the pH values obtained using the two algebraic techniques are very similar.

16.72
$$BrO^-(aq) + H_2O(l) \rightleftharpoons HOBr(aq) + OH^-(aq)$$

initial	0.724 M	0	0
equil.	(0.724 – x) M	x M	x M

$$K_b = \dfrac{[HOBr][OH^-]}{[BrO^-]} = \dfrac{x^2}{0.724 - x} \approx \dfrac{x^2}{0.724} = 4.0 \times 10^{-6}$$

$$x^2 = 0.724 \, (4.0 \times 10^{-6}); \, x = [OH^-] = 1.70 \times 10^{-3} = 1.7 \times 10^{-3} \, M; \, pH = 11.23$$

Check. $\dfrac{1.7 \times 10^{-3} \, M \, OH^-}{0.724 \, M \, BrO^-} \times 100 = 0.24\%$ hydrolysis; the approximation is valid

16.73 *Analyze/Plan.* Based on the pH and initial concentration of base, calculate all equilibrium concentrations. pH → pOH → [OH⁻] at equilibrium. Construct the equilibrium table and calculate other equilibrium concentrations. Substitute into the K_b expression and calculate K_b. *Solve.*

(a) $[OH^-] = 10^{-pOH}$; pOH = 14 − pH = 14.00 − 11.33 = 2.67

$[OH^-] = 10^{-2.67} = 2.138 \times 10^{-3} = 2.1 \times 10^{-3} \, M$

$$C_{10}H_{15}ON(aq) + H_2O(l) \rightleftharpoons C_{10}H_{15}ONH^+(aq) + OH^-(aq)$$

initial	0.035 M	0	0
equil.	0.033 M	$2.1 \times 10^{-3} \, M$	$2.1 \times 10^{-3} \, M$

(b) $K_b = \dfrac{[C_{10}H_{15}ONH^+][OH^-]}{[C_{10}H_{15}ON]} = \dfrac{(2.138 \times 10^{-3})^2}{(0.03286)} = 1.4 \times 10^{-4}$

16.74 (a) pOH = 14.00 − 9.95 = 4.05; $[OH^-] = 10^{-4.05} = 8.91 \times 10^{-5} = 8.9 \times 10^{-5} \, M$

$$C_{18}H_{21}NO_3(aq) + H_2O(l) \rightleftharpoons C_{18}H_{21}NO_3H^+(aq) + OH^-(aq)$$

initial	0.0050 M	0	0
equil.	$(0.0050 - 8.9 \times 10^{-5})$	$8.9 \times 10^{-5} \, M$	$8.9 \times 10^{-5} \, M$

$K_b = \dfrac{[C_{18}H_{21}NO_3H^+][OH^-]}{[C_{18}H_{21}NO_3]} = \dfrac{(8.91 \times 10^{-5})^2}{(0.0050 - 8.91 \times 10^{-5})} = 1.62 \times 10^{-6} = 1.6 \times 10^{-6}$

(b) $pK_b = -\log(K_b) = -\log(1.62 \times 10^{-6}) = 5.79$

The K_a – K_b Relationship; Acid–Base Properties of Salts
(Sections 16.8 and 16.9)

16.75   *Analyze/Plan.* Refer to Equation 16.6 and Sample Exercise 16.17.   *Solve.*

     (a) $C_6H_5OH(aq) + H_2O(l) \rightleftharpoons H_3O^+(aq) + C_6H_5O^-(aq)$

     (b) $K_b = K_w/K_a = 1.0 \times 10^{-14} / 1.3 \times 10^{-10} = 7.7 \times 10^{-5}$

     (c)   Phenol is a stronger acid than water. The benchmark for acid strength in water is 1.0×10^{-14}. All acids listed in Table D.1 of Appendix D are stronger acids than water. (The one notable exception is K_{a2} for H_2S. HS^-, the product of the first ionization of H_2S, has a K_a value of 1×10^{-19}.)

16.76   The stronger a base, the weaker its conjugate acid. From the K_a values in Table 16.3, place the conjugate acids of these oxyanions in order of increasing K_a value, increasing acid strength, and decreasing conjugate base strength. Use K_{a2} for H_2SO_4, H_2CO_3, and H_2SO_3 and K_{a3} for H_3PO_4.

     In order of increasing K_a value and acid strength: $HPO_4^{2-} < HCO_3^- < HSO_3^- < HSO_4^-$

     In order of decreasing base strength: $PO_4^{3-} > CO_3^{2-} > SO_3^{2-} > SO_4^{2-}$

16.77   *Analyze/Plan.* Based on K_a, determine relative strengths of the acids and their conjugate bases. The greater the magnitude of K_a, the stronger the acid and the weaker the conjugate base. K_b (conjugate base) $= K_w/K_a$.   *Solve.*

     (a)   Acetic acid is stronger, because it has the larger K_a value.

     (b)   Hypochlorite ion is the stronger base because the weaker acid, hypochlorous acid, has the stronger conjugate base.

     (c) K_b for $CH_3COO^- = K_w/K_a$ for $CH_3COOH = 1.0 \times 10^{-14}/1.8 \times 10^{-5} = 5.6 \times 10^{-10}$

            K_b for $ClO^- = K_w/K_a$ for $HClO = 1 \times 10^{-14}/3.0 \times 10^{-8} = 3.3 \times 10^{-7}$

            Note that K_b for ClO^- is greater than K_b for CH_3COO^-.

16.78   (a)   Ammonia is the stronger base because it has the larger K_b value.

     (b)   Hydroxylammonium is the stronger acid because the weaker base, hydroxylamine, has the stronger conjugate acid.

     (c) K_a for $NH_4^+ = K_w/K_b$ for $NH_3 = 1.0 \times 10^{-14}/1.8 \times 10^{-5} = 5.6 \times 10^{-10}$

            K_a for $HONH_3^+ = K_w/K_b$ for $HONH_2 = 1.0 \times 10^{-14}/1.1 \times 10^{-8} = 9.1 \times 10^{-7}$

            Note that K_a for $HONH_3^+$ is larger than K_a for NH_4^+.

16.79   *Analyze.* When the solute in an aqueous solution is a salt, evaluate the acid/base properties of the component ions.

     (a)   *Plan.* NaBrO is a soluble salt and, thus, a strong electrolyte. When it is dissolved in H_2O, it dissociates completely into Na^+ and BrO^-. [NaBrO] = [Na^+] = [BrO^-] = 0.10 *M*. Na^+ is the conjugate acid of the strong base NaOH and, thus, does not influence the pH of the solution. BrO^-, on the other hand, is the conjugate base of the weak acid HBrO and *does* influence the pH of the solution. Like any other weak base, it hydrolyzes water to produce $OH^-(aq)$. Solve the equilibrium problem to determine [OH^-].   *Solve.*

$$BrO^-(aq) + H_2O(l) \rightleftharpoons HBrO(aq) + OH^-(aq)$$

initial	0.10 M	0	0
equil.	$(0.10 - x)\,M$	$x\,M$	$x\,M$

$$K_b \text{ for } BrO^- = \frac{[HBrO][OH^-]}{[BrO^-]} = \frac{K_w}{K_a \text{ for } HBrO} = \frac{1 \times 10^{-14}}{2.5 \times 10^{-9}} = 4.00 \times 10^{-6} = 4.0 \times 10^{-6}$$

$$4.00 \times 10^{-6} = \frac{(x)(x)}{(0.10 - x)}; \text{ assume the percent of } BrO^- \text{ that hydrolyzes is small}$$

$$x^2 = 0.10\,(4.00 \times 10^{-6}); \; x = [OH^-] = 6.32 \times 10^{-4} = 6.3 \times 10^{-4}\,M$$

$$pOH = 3.20; \; pH = 14 - 3.20 = 10.80$$

(b) *Plan.* $NaHS(aq) \rightarrow Na^+(aq) + HS^-(aq)$

HS^- is the conjugate base of H_2S and its hydrolysis reaction will determine the $[OH^-]$ and pH of the solution [see similar explanation for NaBrO in part (a)]. We will assume the process $HS^-(aq) \rightleftharpoons H^+(aq) + S^-(aq)$ will not significantly affect the $[OH^-]$ in solution because K_{a2} for H_2S is so small. Solve the equilibrium problem for $[OH^-]$. *Solve.*

$$HS^-(aq) + H_2O(l) \rightleftharpoons H_2S(aq) + OH^-(aq)$$

initial	0.080 M	0	0
equil.	$(0.080 - x)\,M$	x	x

$$K_b = \frac{[H_2S][OH^-]}{[HS^-]} = \frac{K_w}{K_a \text{ for } H_2S} = \frac{1.0 \times 10^{-14}}{9.5 \times 10^{-8}} = 1.053 \times 10^{-7} = 1.1 \times 10^{-7}$$

$$1.053 \times 10^{-7} = \frac{x^2}{(0.080 - x)}; \; x^2 = 0.080\,(1.053 \times 10^{-7}); \; x = 9.177 \times 10^{-5} = 9.2 \times 10^{-5}\,M\,OH^-$$

(Assume x is small compared to 0.080); $pOH = 4.04; \; pH = 14 - 4.04 = 9.96$

Check. $\dfrac{9.2 \times 10^{-5}\,M\,OH^-}{0.080\,M\,HS^-} \times 100 = 0.12\%$ hydrolysis; the approximation is valid

(c) *Plan.* For the two salts present, Na^+ and Ca^{2+} are negligible acids. NO_2^- is the conjugate base of HNO_2 and will determine the pH of the solution. *Solve.*

Calculate total $[NO_2^-]$ present initially.

$[NO_2^-]_{total} = [NO_2^-]$ from $NaNO_2 + [NO_2^-]$ from $Ca(NO_2)_2$

$[NO_2^-]_{total} = 0.10\,M + 2(0.20\,M) = 0.50\,M$

The hydrolysis equilibrium is:

$$NO_2^-(aq) + H_2O(l) \rightleftharpoons HNO_2(aq) + OH^-(aq)$$

initial	0.50 M	0	0
equil.	$(0.50 - x)\,M$	$x\,M$	$x\,M$

$$K_b = \frac{[HNO_2][OH^-]}{[NO_2^-]} = \frac{K_w}{K_a \text{ for } HNO_2} = \frac{1.0 \times 10^{-14}}{4.5 \times 10^{-4}} = 2.22 \times 10^{-11} = 2.2 \times 10^{-11}$$

$$2.2 \times 10^{-11} = \frac{x^2}{(0.50 - x)} \approx \frac{x^2}{0.50}; \; x^2 = 0.50\,(2.22 \times 10^{-11})$$

$$x = 3.33 \times 10^{-6} = 3.3 \times 10^{-6}\,M\,OH^-; \; pOH = 5.48; \; pH = 14 - 5.48 = 8.52$$

16.80   (a)   Proceeding as in Solution 16.79(a):

$$F^-(aq) + H_2O(l) \rightleftharpoons HF(aq) + OH^-(aq)$$

initial	0.105 M	0 M	0 M
equil.	(0.105 – x) M	x M	x M

$$K_b \text{ for } F^- = \frac{[HF][OH^-]}{[F^-]} = \frac{K_w}{K_a \text{ for HF}} = \frac{1.0 \times 10^{-14}}{6.8 \times 10^{-4}} = 1.47 \times 10^{-11} = 1.5 \times 10^{-11}$$

$$1.5 \times 10^{-11} = \frac{(x)(x)}{(0.105 - x)}; \text{ assume the amount of } F^- \text{ that hydrolyzes is small}$$

$$x^2 = 0.105(1.47 \times 10^{-11}); x = [OH^-] = 1.243 \times 10^{-6} = 1.2 \times 10^{-6} \, M$$

pOH = 5.91; pH = 14 – 5.91 = 8.09

(b) $Na_2S(aq) \rightarrow S^{2-}(aq) + 2Na^+(aq)$

$S^{2-}(aq) + H_2O(l) \rightleftharpoons HS^-(aq) + OH^-(aq)$

As in part (a), $[OH^-] = [HS^-] = x$; $[S^{2-}] = 0.035 \, M$

$$K_b = \frac{[HS^-][OH^-]}{[S^{2-}]} = \frac{K_w}{K_a \text{ for HS}^-} = \frac{1.0 \times 10^{-14}}{1 \times 10^{-19}} = 1 \times 10^5$$

Because $K_b \gg 1$, this equilibrium lies far to the right and $[OH^-] = [HS^-] = 0.035 \, M$. K_b for $HS^- = 1.05 \times 10^{-7}$; $[OH^-]$ produced by further hydrolysis of HS^- amounts to $6.1 \times 10^{-5} \, M$. The second hydrolysis step does not make a significant contribution to the total $[OH^-]$ and pH.

$[OH^-] = 0.035 \, M$; pOH = 1.46, pH = 12.54

(c)   As in Solution 16.79(c), calculate $[CH_3COO^-]$.

$[CH_3COO^-]_t = [CH_3COO^-]$ from $NaCH_3COO + [CH_3COO^-]$ from $Ba(CH_3COO)_2$

$[CH_3COO^-]_t = 0.045 \, M + 2(0.055 \, M) = 0.155 \, M$

The hydrolysis equilibrium is:

$$CH_3COO^-(aq) + H_2O(l) \rightleftharpoons CH_3COOH(aq) + OH^-(aq)$$

$$K_b = \frac{[CH_3COOH][OH^-]}{[CH_3COO^-]} = \frac{K_w}{K_a \text{ for CH}_3\text{COOH}} = \frac{1.0 \times 10^{-14}}{1.8 \times 10^{-5}} = 5.56 \times 10^{-10}$$

$$= 5.6 \times 10^{-10}$$

$[OH^-] = [CH_3COOH] = x$; $[CH_3COO^-] = 0.155 - x$

$$K_b = 5.56 \times 10^{-10} = \frac{x^2}{(0.155 - x)}; \text{ assume x is small compared to 0.155 } M$$

$$x^2 = 0.155 \, (5.56 \times 10^{-10}); x = [OH^-] = 9.280 \times 10^{-6} = 9.3 \times 10^{-6}$$

pH = 14 + log (9.280×10^{-6}) = 8.97

16.81   *Analyze/Plan.* The salt dissociates to form Na^+ and CH_3COO^-. Na^+ is a negligible base; the hydrolysis equilibrium of CH_3COO^- determines the pH of the solution.

Solve. The hydrolysis equilibrium is:

$$CH_3COO^-(aq) + H_2O(l) \rightleftharpoons CH_3COOH(aq) + OH^-(aq)$$

$$K_b = \frac{[CH_3COOH][OH^-]}{[CH_3COO^-]} = \frac{K_w}{K_a \text{ for CH}_3\text{COOH}} = \frac{1.0 \times 10^{-14}}{1.8 \times 10^{-5}} = 5.56 \times 10^{-10} = 5.6 \times 10^{-10}$$

$[CH_3COO^-] = x; [OH^-] = [CH_3COOH] = 10^{-pOH}$

$pOH = 14.00 - pH = 14.00 - 9.70 = 4.30; [OH^-] = 10^{-4.30} = 5.012 \times 10^{-5} = 5.0 \times 10^{-5}$

$K_b = 5.56 \times 10^{-10} = \dfrac{(5.012 \times 10^{-5})^2}{x}; \quad x = 4.518 = 4.5\ M\ NaCH_3COO$

Note that no assumption was required in this calculation.

16.82   (a) $C_5H_5NH^+(aq) + H_2O(l) \rightleftharpoons C_5H_5N(aq) + H_3O^+(aq)$

       (b) $K_a = \dfrac{[C_5H_5N][H^+]}{[C_5H_5NH^+]} = \dfrac{K_w}{K_b\ for\ C_5H_5N} = \dfrac{1.0 \times 10^{-14}}{1.7 \times 10^{-9}} = 5.882 \times 10^{-6} = 5.9 \times 10^{-6}$

       (c) $[C_5H_5NH^+] = x; [C_5H_5N] = [H^+] = 10^{-pH} = 10^{-2.95} = 1.122 \times 10^{-3} = 1.1 \times 10^{-3}\ M$

           $K_a = 5.882 \times 10^{-6} = \dfrac{(1.122 \times 10^{-3})^2}{x}; \quad x = 0.2140 = 0.21\ M\ C_5H_5NH^+$

           Note that no assumption was required in this calculation.

16.83   *Analyze/Plan.* Based on the formula of a salt, predict whether an aqueous solution will be acidic, basic, or neutral. Evaluate the acid–base properties of both ions and determine the overall effect on solution pH.   *Solve.*

       (a)   acidic; NH_4^+ is a weak acid, Br^- is negligible.

       (b)   acidic; Fe^{3+} is a highly charged metal cation and a Lewis acid; Cl^- is negligible.

       (c)   basic; CO_3^{2-} is the conjugate base of HCO_3^-; Na^+ is negligible.

       (d)   neutral;   both K^+ and ClO_4^- are negligible.

       (e)   acidic; $HC_2O_4^-$ is amphoteric, but K_a for the acid dissociation (6.4×10^{-5}) is much greater than K_b for the base hydrolysis ($1.0 \times 10^{-14}\ /\ 5.9 \times 10^{-2} = 1.7 \times 10^{-13}$).

16.84   (a)   acidic; Al^{3+} is a highly charged metal cation and a Lewis acid; Cl^- is negligible.

       (b)   neutral;   both Na^+ and Br^- are negligible.

       (c)   basic; ClO^- is the conjugate base of $HClO$; Na^+ is negligible.

       (d)   acidic; $CH_3NH_3^+$ is the conjugate acid of CH_3NH_2; NO_3^- is negligible.

       (e)   basic; SO_3^{2-} is the conjugate base of H_2SO_3; Na^+ is negligible.

16.85   *Plan.* Estimate pH using relative base strength and then calculate to confirm prediction. NaCl is a neutral salt, so it is not the unknown. The unknown is a relatively weak base, because a pH of 8.08 is not very basic. Because F^- is a weaker base than OCl^-, the unknown is probably NaF. Calculate K_b for the unknown from the data provided.   *Solve.*

$[OH^-] = 10^{-pOH}; pOH = 14.00 - pH = 14.00 - 8.08 = 5.92$

$[OH^-] = 10^{-5.92} = 1.202 \times 10^{-6} = 1.2 \times 10^{-6}\ M = [HX]$

$[NaX] = [X^-] = 0.050\ mol\ salt/0.500\ L = 0.10\ M$

$K_b = \dfrac{[OH^-][HX]}{[X^-]} = \dfrac{(1.202 \times 10^{-6})^2}{(0.10 - 1.2 \times 10^{-6})} = \dfrac{(1.202 \times 10^{-6})^2}{0.10} = 1.4 \times 10^{-11}$

$K_b\ for\ F^- = K_w/K_a\ for\ HF = 1.0 \times 10^{-14}/6.8 \times 10^{-4} = 1.5 \times 10^{-11}$

The unknown is NaF.

16.86   *Plan.* Estimate pH of salt solution by evaluating the ions in the salts. Calculate to confirm if necessary.   *Solve.*

KBr: salt of strong acid and strong base, neutral solution. The unknown is probably KBr. Check the others to be sure.

NH_4Cl: salt of a weak base and a strong acid, acidic solution

KCN: salt of a strong base and a weak acid, basic solution

K_2CO_3:   salt of a strong base and a weak acid (HCO_3^-), basic solution

Only KBr fits the acid–base properties of the unknown.

Acid–Base Character and Chemical Structure (Section 16.10)

16.87   (a) HNO_3 is a stronger acid than HNO_2 because it has one more nonprotonated oxygen atom and, thus, a higher oxidation number on N.

(b) H_2S is a stronger acid than H_2O. For binary hydrides, acid strength increases going down a family.

(c) H_2SO_4 is a stronger acid than H_2SeO_4. For oxyacids, the greater the electronegativity of the central atom, the stronger the acid.

(d) CCl_3COOH is stronger than CH_3COOH because the electronegative Cl atoms withdraw electron density from other parts of the molecule, which weakens the O–H bond and makes H^+ easier to remove. Also, the electronegative Cl delocalizes negative charge on the carboxylate anion. This stabilizes the conjugate base, favoring products in the ionization equilibrium and increasing K_a.

16.88   (a)   HCl is a stronger acid than HF. For binary hydrides, acid strength increases going down a column.

(b) H_3PO_4 is a stronger acid than H_3AsO_4. For oxyacids, the more electronegative the central atom, the stronger the acid.

(c) $HBrO_3$ is a stronger acid than $HBrO_2$ because it has one more nonprotonated oxygen and a higher oxidation number on Br.

(d) $H_2C_2O_4$ is a stronger acid than $HC_2O_4^-$. The first ionization of a polyprotic acid is always stronger because H^+ is more tightly held by an anion.

(e) C_6H_5COOH is stronger than C_6H_5OH. The conjugate base of benzoic acid, $C_6H_5COO^-$, is stabilized by resonance, whereas the conjugate base of phenol, $C_6H_5O^-$, is not. C_6H_5COOH has greater tendency to form its conjugate base and is the stronger acid.

16.89   (a) BrO^- (HClO is the stronger acid because Cl is more electronegative than Br, so BrO^- is the stronger base.)

(b) BrO^- ($HBrO_2$ has more nonprotonated O atoms and is the stronger acid, so BrO^- is the stronger base.)

(c) HPO_4^{2-} (larger negative charge, greater attraction for H^+)

16.90　(a)　NO_2^-　　(HNO_3 is the stronger acid because it has more nonprotonated O atoms, so NO_2^- is the stronger base.)

　　　　(b)　PO_4^{3-}　　(K_a for $HAsO_4^{2-}$ is greater than K_a for HPO_4^{2-}, so K_b for PO_4^{3-} is greater and PO_4^{3-} is the stronger base. Note that P is more electronegative than As and H_3PO_4 is a stronger acid than H_3AsO_4, which could lead to the conclusion that AsO_4^{3-} is the stronger base. As in all cases, the measurement of base strength, K_b, supercedes the prediction. Chemistry is an experimental science.

　　　　(c)　CO_3^{2-}　　(The more negative the anion, the stronger the attraction for H^+.)

16.91　(a)　True.

　　　　(b)　False. In a series of acids that have the same central atom, acid strength increases with the number of nonprotonated oxygen atoms bonded to the central atom.

　　　　(c)　False. H_2Te is a stronger acid than H_2S because the H–Te bond is longer, weaker, and more easily ionized than the H–S bond. Binary hydride acid strength increase going down a family.

16.92　(a)　True.

　　　　(b)　False. For oxyacids with the same structure but different central atom, the acid strength *increases* as the electronegativity of the central atom increases.

　　　　(c)　False. HF is a weak acid, weaker than the other hydrogen halides, primarily because the H–F bond energy is exceptionally high.

Lewis Acids and Bases (Section 16.11)

16.93　$NH_3(aq) + H_2O(l) \rightleftharpoons NH_4^+(aq) + OH^-(aq)$

Ammonia, NH_3, acts as an Arrhenius base because it increases the concentration of hydroxide ion, OH^-, in aqueous solution. It acts like a Brønsted-Lowry base because it is a proton, H^+, acceptor. It acts like a Lewis base because it is an electron pair donor. If a substance is an Arrhenius base, it must also be a Brønsted–Lowry base and a Lewis base.

16.94　(a)　$F^-(aq) + H_2O(l) \rightleftharpoons HF(aq) + OH^-(aq)$

　　　　(b)　Basic. The F^- ion is the conjugate base of the weak acid HF.

　　　　(c)　The F^- ion acts as a Lewis base, donating an electron pair to water.

16.95　*Analyze/Plan.* Identify each reactant as an electron pair donor (Lewis base) or electron pair acceptor (Lewis acid). Remember that a Brønsted–Lowry acid is necessarily a Lewis acid, and a Brønsted–Lowry base is necessarily a Lewis base (Solution 16.93).　*Solve.*

	Lewis Acid	**Lewis Base**
(a)	$Fe(ClO_4)_3$ or Fe^{3+}	H_2O
(b)	H_2O	CN^-
(c)	BF_3	$(CH_3)_3N$
(d)	HIO	NH_2^-

16.96

	Lewis Acid	**Lewis Base**
(a)	HNO_2 (or H^+)	OH^-
(b)	$FeBr_3$ (Fe^{3+})	Br^-
(c)	Zn^{2+}	NH_3
(d)	SO_2	H_2O

16.97 (a) Cu^{2+}, higher cation charge

(b) Fe^{3+}, higher cation charge

(c) Al^{3+}, smaller cation radius, same charge

16.98 (a) $ZnBr_2$, smaller cation radius, same charge

(b) $Cu(NO_3)_2$, higher cation charge

(c) $NiBr_2$, smaller cation radius, same charge

Additional Exercises

16.99 (a) Correct.

(b) Incorrect. A Brønsted–Lowry acid must have ionizable hydrogen. Lewis acids are electron pair acceptors, but need not have ionizable hydrogen.

(c) Correct.

(d) Incorrect. K^+ is a negligible Lewis acid because it is the conjugate of strong base KOH. Its relatively large ionic radius and low positive charge render it a poor attractor of electron pairs.

(e) Correct.

16.100 Calculate moles OH^-, calculate moles H^+, determine which is in excess after neutralization, calculate pH

$$0.300 \text{ g Ca(OH)}_2 \times \frac{1 \text{ mol Ca(OH)}_2}{74.093 \text{ g Ca(OH)}_2} \times \frac{2 \text{ mol OH}^-}{1 \text{ mol Ca(OH)}_2} = 8.0979 \times 10^{-3} = 8.10 \times 10^{-3} \ M \ OH^-$$

$1.40 \ M \ HNO_3 \times 0.0500 \text{ L} = 0.0700 \text{ mol } H^+$; H^+ is in excess

$0.0700 \text{ mol } H^+ - 0.00810 \text{ mol } OH^- = 0.0619 \text{ mol } H^+$ remain

$0.0619 \text{ mol } H^+ / 0.0750 \text{ L} = 0.82536 = 0.825 \ M \ H^+$; pH = $-\log (0.82536) = 0.03356 = 0.0834$

16.101 $H_3C_6H_5O_7 \ + \ CH_3NH_2 \ \rightarrow \ CH_3NH_3^+ + H_2C_6H_5O_7^-$

citric acid   methylamine   odorless salt

$$H_3C_6H_5O_7 \ \rightleftharpoons \ H^+ + H_2C_6H_5O_7^- \qquad K_{a1} = 7.4 \times 10^{-4}$$

$$CH_3NH_2 + H_2O \rightleftharpoons CH_3NH_3^+ + OH^- \qquad K_b = 4.4 \times 10^{-4}$$

$$H^+ + OH^- \rightleftharpoons H_2O \qquad 1/K_w = 1/1.0 \times 10^{-14}$$

$$H_3C_6H_5O_7 + CH_3NH_2 + H_2O + H^+ + OH^- \ \rightleftharpoons \ H_2C_6H_5O_7^- + CH_3NH_3^+ + H^+ + OH^- + H_2O$$

$$H_3C_6H_5O_7 + CH_3NH_2 \ \rightleftharpoons \ H_2C_6H_5O_7^- + CH_3NH_3^+$$

$$K = \frac{K_{a_1} \times K_b}{K_w} = \frac{(7.4 \times 10^{-4})(4.4 \times 10^{-4})}{1.0 \times 10^{-14}} = 3.256 \times 10^7 = 3.3 \times 10^7$$

16.102   Statements (a), (d), and (f) are true. Statements (b), (c), and (e) are the opposites of the true statements.

16.103   *Analyze/Plan.* Brønsted-Lowry acids are H^+ donors, Brønsted-Lowry bases are H^+ acceptors. Examine the structures of the molecules and ions for acidic or basic functional groups.   *Solve.*

(a)   Bicarbonate ion, HCO_3^-, <u>both</u> acid and base. Bicarbonate ion is amphiprotic; it can act like an H^+ donor or acceptor.

(b)   Prozac, <u>base</u>. Prozac contains an –NH– (amine) group. The nonbonded electron pair on N causes it to be a hydrogen ion acceptor.

(c)   PABA, <u>both</u> acid and base. The molecule contains both a –COOH (carboxylic acid) group and an $-NH_2$ (amine) group. It can act like an H^+ donor and acceptor.

(d)   TNT, <u>neither</u> acid nor base. TNT contains $-NO_2$ (nitro) groups. The N atoms in these groups do not have a nonbonded electron pair and are neither acidic nor basic.

(e)   N-Methylpyridinium ion, <u>neither</u> acid nor base. The N atom in this ion forms four covalent bonds and does not have a nonbonded electron pair. It cannot act as an H^+ acceptor.

16.104   Upon dissolving, Li_2O dissociates to form Li^+ and O^{2-}. According to Equation 16.22, O^{2-} is completely protonated in aqueous solution.

$$Li_2O(s) + H_2O(l) \rightarrow 2Li^+(aq) + 2OH^-(aq)$$

Thus, initial $[Li_2O] = [O_2^-]$; $[OH^-] = 2[O^{2-}] = 2[Li_2O]$

$$[Li_2O] = \frac{mol\ Li_2O}{L\ solution} = 2.50\ g\ Li_2O \times \frac{1\ mol\ Li_2O}{29.88\ g\ Li_2O} \times \frac{1}{1.500\ L} = 0.0558 = 0.0558\ M$$

$[OH^-] = 0.11156 = 0.112\ M$; pOH $= 0.9525 = 0.953$; pH $= 14.00 - $ pOH $= 13.0475 = 13.048$

16.105   (a)   The conjugate base of benzoic acid is benzoate anion, $C_6H_5COO^-$. The conjugate acid of aniline is anilinium cation, $C_6H_5NH_3^+$.

(b)   To compare relative acidity, compare the K_a values for benzoic acid and anilinium ion.

$$K_a\ for\ C_6H_5NH_3^+ = \frac{K_w}{K_b\ for\ C_6H_5NH_2} = \frac{1.0 \times 10^{-14}}{4.3 \times 10^{-10}} = 2.3256 \times 10^{-5} = 2.3 \times 10^{-5}$$

K_a for C_6H_5COOH, 6.3×10^{-5}, is greater than K_a for $C_6H_5NH_3^+$, 2.3×10^{-5}. The 0.10 M solution of benzoic acid will be somewhat more acidic.

(c)
$$C_6H_5COOH \rightleftharpoons H^+ + C_6H_5COO^- \qquad K_a = 6.3 \times 10^{-5}$$
$$C_6H_5NH_2 + H_2O \rightleftharpoons C_6H_5NH_3^+ + OH^- \qquad K_b = 4.3 \times 10^{-10}$$
$$H^+ + OH^- \rightleftharpoons H_2O \qquad 1/K_w = 1/1.0 \times 10^{-14}$$

$$\overline{C_6H_5COOH + C_6H_5NH_2 + H_2O + H^+ + OH^- \rightleftharpoons C_6H_5COO^- + C_6H_5NH_3^+ + H^+ + OH^- + H_2O}$$
$$C_6H_5COOH + C_6H_5NH_2 \rightleftharpoons C_6H_5COO^- + C_6H_5NH_3^+$$

$$K = \frac{K_a \times K_b}{K_w} = \frac{(6.3 \times 10^{-5})(4.3 \times 10^{-10})}{1.0 \times 10^{-14}} = 2.7090 = 2.7$$

492

16.106 Assume T = 25 °C. If $[OH^-] = 2.5 \times 10^{-9}$ M, pOH = 8.60 and pH = 5.40. This does not make sense (!) because NaOH is a strong base. Usually, we assume that $[H^+]$ and $[OH^-]$ from the autoionization of water do not contribute to the overall $[H^+]$ and $[OH^-]$. However, for acid or base solute concentrations less than 1×10^{-6} M, the autoionization of water produces significant $[H^+]$ and $[OH^-]$ and we must consider it when calculating pH.

	$H_2O(l)$ $\rightleftharpoons$	$[H^+]$	+	$[OH^-]$
initial	C	0		2.5×10^{-9} M
equil.	C	x		$(x + 2.5 \times 10^{-9})$ M

$K_w = 1.0 \times 10^{-14} = [H^+][OH^-] = (x)(x + 2.5 \times 10^{-9}); \ x^2 + 2.5 \times 10^{-9}\,x - 1.0 \times 10^{-14} = 0$

From the quadratic formula, $x = \dfrac{-2.5 \times 10^{-9} \pm \sqrt{(2.5 \times 10^{-9})^2 - 4(1)(-1 \times 10^{-14})}}{2(1)}$

$$= 9.876 \times 10^{-8} = 9.9 \times 10^{-8} \ M \ H^+$$

$[H^+] = 9.9 \times 10^{-8}$ M; $[OH^-] = (9.876 \times 10^{-8} + 2.5 \times 10^{-9}) = 1.013 \times 10^{-7} = 1.0 \times 10^{-7}$ M

pH = 7.0054 = 7.01

Check: $[9.876 \times 10^{-8}][1.013 \times 10^{-7}] = 1.0 \times 10^{-14}$. Now our answer makes sense. The very small concentration of OH^- from the solute raises the solution pH to slightly more than 7.

16.107 (a) False. $H_2C_2O_4$ has no capacity to accept H^+.

(b) True.

(c) True. $HC_2O_4^-$ can act like either a Brønsted–Lowry acid or base. It is a stronger acid than base ($K_{a2} > K_{b1}$), so a solution of the salt will be acidic.

16.108 $H_2Suc(aq) \rightleftharpoons H^+(aq) + HSuc^-(aq) \qquad K_{a1} = 6.9 \times 10^{-5}$

$HSuc^-(aq) \rightleftharpoons H^+(aq) + Suc^{2-}(aq) \qquad K_{a2} = 2.5 \times 10^{-6}$

(a) Calculating the $[H^+]$ from the first ionization. The equilibrium concentrations are $[H^+] = [HSuc^-] = x$; $[H_2Suc] = 0.32 - x$.

$K_{a1} = \dfrac{[H^+][HSuc^-]}{[H_2Suc]} = \dfrac{x^2}{0.32 - x}$; assume x is small relative to 0.32

$x^2 = (0.32)(6.9 \times 10^{-5}); \ x = 0.0046989 = 0.0047 \ M \ H^+; \ pH = 2.328 = 2.33$

(b) To calculate $[Suc^{2-}]$, consider the second ionization equilibrium. Initial $[HSuc^-]$ and $[H^+]$ are 0.0047 M, from the first ionization. Then,

$[Suc^{2-}] = y$; $[H^+] = 0.0047 + y$; $[HSuc^-] = 0.0047 - y$.

$K_{a2} = \dfrac{(0.0047 + y)(y)}{(0.0047 - y)} = 2.5 \times 10^{-6}$; assuming y is small compared to 0.0047,

$y = 2.5 \times 10^{-6} \ M \ Suc^{2-}$. This assumption is reasonable, because 2.5×10^{-6} is only 0.05% of 0.0047.

(c)   The assumption in part (a) is reasonable. From part (b), $[Suc^{2-}]$ and $[H^+]$ from the second ionization are 2.5×10^{-6} M. This is only 0.05% of 0.0047 M, $[H^+]$ from the first ionization. Only the first dissociation is relevant for calculating pH.

(d)   Acidic. $HSuc^-$ can act like either a Brønsted–Lowry acid or base. It is a stronger acid than base ($K_{a2} > K_{b1}$), so a solution of the salt will be acidic.

16.109 (a) $K_b = K_w/K_a$; $pK_b = 14 - pK_a$; $pK_b = 14 - 4.84 = 9.16$

(b) K_a for butyric acid (buCOOH) is $10^{-4.84} = 1.4454 \times 10^{-5} = 1.4 \times 10^{-5}$

$$K_a = \frac{[H^+][buCOO^-]}{[buCOOH]}; \quad [H^+] = [buCOO^-] = x; \quad [buCOOH] = 0.050 - x$$

$$1.4454 \times 10^{-5} = \frac{x^2}{0.050 - x}; \quad \text{assume } x \text{ is small relative to } 0.050$$

$$x^2 = 7.227 \times 10^{-7}; x = [H^+] = 8.501 \times 10^{-4} = 8.5 \times 10^{-4} \, M \, H^+; pH = 3.07$$

(This represents 1.7% ionization, so the approximation is valid.)

(c) K_b for butyrate anion (buCOO⁻) is $10^{-9.16} = 6.918 = 6.918 \times 10^{-10} = 6.9 \times 10^{-10}$

$$K_b = \frac{[OH^-][buCOOH]}{[buCOO^-]}; \quad [OH^-] = [buCOOH] = x; \quad [buCOO^-] = 0.050 - x$$

$$6.918 \times 10^{-10} = \frac{x^2}{0.050 - x}; \quad \text{assume } x \text{ is small relative to } 0.050$$

$$x^2 = 3.459 \times 10^{-11}; x = [OH^-] = 5.881 \times 10^{-6} = 5.9 \times 10^{-6} \, M \, OH^-$$

$$pOH = 5.23; pH = 8.77$$

16.110 *Analyze/Plan.* Evaluate the acid–base properties of the cation and anion to determine whether a solution of the salt will be acidic, basic, or neutral.   *Solve.*

(i) NH_4NO_3: NH_4^+, weak conjugate acid of NH_3; NO_3^-, negligible conjugate base of HNO_3; acidic solution

(ii) $NaNO_3$: Na^+, negligible conjugate acid of NaOH; NO_3^-, negligible conjugate base of HNO_3; neutral solution

(iii) CH_3COONH_4: NH_4^+, weak conjugate acid of NH_3, $K_a = K_w/1.8 \times 10^{-5} = 5.6 \times 10^{-10}$; CH_3COO^-, weak conjugate base of CH_3COOH, $K_b = K_w/1.8 \times 10^{-5} = 5.6 \times 10^{-10}$; neutral solution ($K_a$ for the cation and K_b for the anion are accidentally equal, producing a neutral solution)

(iv)   NaF: Na^+, negligible conjugate acid of NaOH; F^-, weak conjugate base of HF, $K_b = K_w/6.8 \times 10^{-4} = 1.5 \times 10^{-11}$; basic solution

(v) CH_3COONa: Na^+, negligible; CH_3COO^-, weak base, $K_b = 5.6 \times 10^{-10}$; basic solution

In order of increasing acidity (and decreasing pH):
$0.1 \, M \, CH_3COONa > 0.1 \, M \, NaF > 0.1 \, M \, CH_3COONH_4 = 0.1 \, M \, NaNO_3 > 0.1 \, M \, NH_4NO_3$;
(v) > (iv) > (iii) ~ (ii) > (i)
(iv) and (v) are both bases, and (v) has the greater K_b value and higher pH. (ii) and (iii) are both neutral and (i) is acidic.

16.111 Calculate K_b for A^- and then K_a for HA.

$A^- = [NaA] = [A^-] = 0.25\ M;\ [OH^-] = 10^{-pOH};\ pOH = 14.00 - pH = 14.00 - 9.29 = 4.71$

$[OH^-] = 10^{-4.71} = 1.950 \times 10^{-5} = 2.0 \times 10^{-5}\ M = [HX]$

$$K_b = \frac{[OH^-][HA]}{[A^-]} = \frac{(1.950 \times 10^{-5})^2}{(0.25 - 1.950 \times 10^{-5})} = \frac{(1.950 \times 10^{-5})^2}{0.25} = 1.521 \times 10^{-9} = 1.5 \times 10^{-9}$$

K_a for HA $= K_w/K_b$ for $A^- = 1.0 \times 10^{-14}/1.521 \times 10^{-9} = 6.576 \times 10^{-6} = 6.6 \times 10^{-6}$

16.112 The value of pK_{a2} can only be choice (iii).

Calculate K_{a1}, assuming that only the first ionization determines pH.

$[H_2A] = 0.10\ M;\ [H^+] = [HA^-] = 10^{-pH} = 10^{-3.30} = 5.0119 \times 10^{-4} = 5.0 \times 10^{-4}$

$$K_{a1} = \frac{[H^+][A^-]}{[HA]} = \frac{(5.0119 \times 10^{-4})^2}{(0.10 - 5.0119 \times 10^{-4})} = \frac{(5.0119 \times 10^{-4})^2}{0.0995} = 2.524 \times 10^{-6} = 2.5 \times 10^{-6}$$

$pK_{a1} = 5.60$. pK_{a2} must be greater than pK_{a1}, which eliminates choices (i) and (ii).

If a solution of the salt NaHA is acidic, then $K_{a2} > K_{b1}$ and $pK_{a2} < pK_{b1}$.

$pK_{b1} = 14.00 - pK_{a1} = 14.00 - 5.60 = 8.84$. This eliminates choice (iv).

16.113 Call each compound in the neutral form Q.

Then, $Q(aq) + H_2O(l) \rightleftharpoons QH^+(aq) + OH^-$. $K_b = [QH^+][OH^-]/[Q]$

The ratio in question is $[QH^+]/[Q]$, which equals $K_b/[OH^-]$ for each compound. At pH = 2.5, pOH = 11.5, $[OH^-]$ = antilog (–11.5) = $3.16 \times 10^{-12} = 3 \times 10^{-12}\ M$. Now calculate $K_b/[OH^-]$ for each compound:

Nicotine $\qquad \dfrac{[QH^+]}{[Q]} = 7 \times 10^{-7}/3.16 \times 10^{-12} = 2 \times 10^5$

Caffeine $\qquad \dfrac{[QH^+]}{[Q]} = 4 \times 10^{-14}/3.16 \times 10^{-12} = 1 \times 10^{-2}$

Strychnine $\qquad \dfrac{[QH^+]}{[Q]} = 1 \times 10^{-6}/3.16 \times 10^{-12} = 3 \times 10^5$

Quinine $\qquad \dfrac{[QH^+]}{[Q]} = 1.1 \times 10^{-6}/3.16 \times 10^{-12} = 3.5 \times 10^5$

For all the compounds except caffeine, the protonated form has a much higher concentration than the neutral form. However, for caffeine, a very weak base, the neutral form dominates.

16.114 (a) Consider the formation of the zwitterion as a series of steps (Hess's law).

$$NH_2-CH_2-COOH + H_2O \rightleftharpoons NH_2-CH_2-COO^- + H_3O^+ \qquad K_a$$
$$NH_2-CH_2-COOH + H_2O \rightleftharpoons {}^+NH_3-CH_2-COOH + OH^- \qquad K_b$$
$$H_3O^+ + OH^- \rightleftharpoons 2H_2O \qquad 1/K_w$$

$$\overline{}$$

$$NH_2-CH_2-COOH \rightleftharpoons {}^+NH_3-CH_2-COO^- \qquad \frac{K_a \times K_b}{K_w}$$

$$K = \frac{K_a \times K_b}{K_w} = \frac{(4.3 \times 10^{-3})(6.0 \times 10^{-5})}{1.0 \times 10^{-14}} = 2.6 \times 10^7$$

(b) As glycine exists as the zwitterion in aqueous solution, the pH is determined by the following equilibrium.

$$^+NH_3-CH_2-COO^- + H_2O \rightleftharpoons NH_2-CH_2-COO^- + H_3O^+$$

$$K_a = \frac{[NH_2-CH_2-COO^-][H_3O^+]}{[^+NH_3-CH_2-COO^-]} = \frac{K_w}{K_b} = \frac{1.0 \times 10^{-14}}{6.0 \times 10^{-5}} = 1.67 \times 10^{-10} = 1.7 \times 10^{-10}$$

$$x = [H_3O^+] = [NH_2-CH_2-COO^-]; \quad K_a = 1.67 \times 10^{-10} = \frac{(x)(x)}{(0.050-x)} \approx \frac{x^2}{0.050}$$

$$x = [H_3O^+] = 2.89 \times 10^{-6} = 2.9 \times 10^{-6} \, M; \, pH = 5.54$$

(c) In strongly basic solution (pH 13), the $-NH_3^+$ group would be deprotonated, so glycine would be in the form $H_2NCH_2CO_2^-$. In a strongly acidic (pH 1) solution, the $-CO_2^-$ function would be protonated, so glycine would exist as $^+H_3NCH_2COOH$.

16.115 Answer (c) is correct. Water itself is amphiprotic; it can act like an H^+ donor or acceptor. The autoionization equilibrium, Equation 16.12, is also unique in that it describes water acting as an acid and a base simultaneously. The value of K_w is the value of K_a for water acting like an acid and K_b for water acting like a base. Thus, the pK_b of water is 14 (and the pK_a of water is 14.)

Integrative Exercises

16.116 At 25 °C, $[H^+] = [OH^-] = 1.0 \times 10^{-7} \, M$

$$\frac{1.0 \times 10^{-7} \, mol \, H^+}{1 L \, H_2O} \times 0.0010 \, L \times \frac{6.022 \times 10^{23} \, H^+ \, ions}{mol \, H^+} = 6.0 \times 10^{13} \, H^+ \, ions$$

16.117 *Analyze.* Based on the mass % and density of concentrated HCl, calculate volume of concentrated solution required to produce 10.0 L of HCl with pH = 2.05. *Plan.* Calculate molarity of concentrated solution from density and mass %. Calculate molarity of dilute solution from pH. Use the dilution formula to calculate volume (mL) of concentrated solution required.   *Solve.*

$$\frac{1.18 \, g \, conc. \, soln.}{mL \, conc. \, soln.} \times \frac{36.0 \, g \, HCl}{100 \, g \, conc. \, soln.} \times \frac{1000 \, mL}{1 \, L} \times \frac{1 \, mol \, HCl}{36.46 \, g \, HCl} = 11.651 \, mol \, HCl/L$$

$$= 11.7 \, M \, HCl/L$$

For the dilute HCl solution, $[H^+] = 10^{-pH} = 10^{-2.05} = 8.913 \times 10^{-3} = 8.9 \times 10^{-3} \, M \, HCl$

$$M_c \times L_c = M_d \times M_d; \, 11.651 \times L_c = 8.913 \times 10^{-3} \, M \times 10.0 \, L;$$

$$L_c = 7.650 \times 10^{-3}; \, 7.650 \times 10^{-3} \, L \times \frac{1000 \, mL}{1 \, L} = 7.65 = 7.7 \, mL \, conc. \, HCl$$

16.118 $[H^+] = 10^{-pH} = 10^{-2} = 1 \times 10^{-2} \, M \, H^+; \, 1 \times 10^{-2} \, M \times 0.400 \, L = 4.0 \times 10^{-3} = 4 \times 10^{-3} \, mol \, H^+$

$$HCl(aq) + HCO_3^-(aq) \rightarrow Cl^-(aq) + H_2O(l) + CO_2(g)$$

$$4 \times 10^{-3} \, mol \, H^+ = 4 \times 10^{-3} \, mol \, HCO_3^- \times \frac{84.01 \, g \, NaHCO_3}{1 \, mol \, HCO_3^-} = 0.336 = 0.3 \, g \, NaHCO_3$$

16.119   *Analyze.* If pH were directly related to CO_2 concentration, this exercise would be simple. Unfortunately, we must solve the equilibrium problem for the diprotic acid H_2CO_3 to calculate $[H^+]$ and pH. We are given ppm CO_2 in the atmosphere at two different times and the pH that corresponds to one of these CO_2 levels. We are asked to find pH at the other atmospheric CO_2 level.

Plan. Assume all dissolved CO_2 is present as H_2CO_3 (aq) (Sample Exercise 16.14).

pH $\rightarrow [H^+] \rightarrow [H_2CO_3]$. Although H_2CO_3 is a diprotic acid, the two K_a values differ by more than 10^3, so we can ignore the second ionization when calculating $[H_2CO_3]$. Change 380 ppm CO_2 to pressure and calculate the Henry's law constant for CO_2. Calculate the dissolved $[CO_2] = [H_2CO_3]$ at 315 ppm and then solve the K_{a1} expression for $[H^+]$ and pH. *Solve.*

(a) $H_2CO_3(aq) \rightleftharpoons H^+(aq) + HCO_3^-(aq)$

$$K_{a1} = 4.3 \times 10^{-7} = \frac{[H^+][HCO_3^-]}{[H_2CO_3]}; \quad [H^+] = 10^{-5.4} = 3.98 \times 10^{-6} = 4 \times 10^{-6}\ M$$

$$[H^+] = [HCO_3^-]; \quad [H_2CO_3] = x - 4 \times 10^{-6}$$

$$4.3 \times 10^{-7} = \frac{(3.98 \times 10^{-6})^2}{(x - 3.98 \times 10^{-6})}; \quad 4.3 \times 10^{-7}\ x = 1.585 \times 10^{-11} + 1.711 \times 10^{-12}$$

$$x = 1.756 \times 10^{-11}/4.3 \times 10^{-7} = 4.084 \times 10^{-5} = 4 \times 10^{-5}\ M\ H_2CO_3$$

380 ppm = 380 mol $CO_2/1 \times 10^6$ mol air = 0.000380 mol % CO_2

Because of the properties of gases, mol % = pressure %. $P_{CO_2} = 0.000380$ atm. According to Equation 13.4, $S_{CO_2} = kP_{CO_2}$;

$$4.084 \times 10^{-5}\ \text{mol/L} = k(3.80 \times 10^{-4}\ \text{atm}); \quad k = 0.1075 = 0.1\ \text{mol/L-atm}.$$

Forty years ago, $S_{CO_2} = \dfrac{0.1075\ \text{mol}}{\text{L-atm}} \times 3.15 \times 10^{-4}\ \text{atm} = 3.385 \times 10^{-5}$

$$= 3 \times 10^{-5}\ M$$

Now solve K_{a1} for $[H^+]$ at this $[H_2CO_3]$. $[H^+] = x$.

We cannot assume x is small, because $[H_2CO_3]$ is so low.

$$4.3 \times 10^{-7} = x^2/(3.385 \times 10^{-5} - x); \quad x^2 + 4.3 \times 10^{-7}\ x - 1.456 \times 10^{-11} = 0$$

$$x = \frac{-4.3 \times 10^{-7} \pm \sqrt{(4.3 \times 10^{-7})^2 - 4(-1.456 \times 10^{-11})}}{2} = \frac{-4.3 \times 10^{-7} + 7.644 \times 10^{-6}}{2}$$

$$= 3.607 \times 10^{-6} = 4 \times 10^{-6}\ M\ H^+; \quad [H^+] = 4 \times 10^{-6}\ M; \quad \text{pH} = 5.443 = 5.4$$

(Note that, to the precision that the pH data is reported, the change in atmospheric CO_2 leads to no change in pH.)

(b)   From part (a), $[H_2CO_3]$ today $= 4.084 \times 10^{-5}\ M$

$$\frac{4.084 \times 10^{-5}\ \text{mol}\ H_2CO_3}{1\ L} \times 20.0\ L = 8.168 \times 10^{-4} = 8 \times 10^{-4}\ \text{mol}\ CO_2$$

$$V = \frac{nRT}{P} = 8.168 \times 10^{-4}\ \text{mol} \times \frac{298\ K}{1.0\ \text{atm}} \times \frac{0.08206\ \text{L-atm}}{\text{mol-K}} = 0.01997 = 0.02\ L = 20\ \text{mL}$$

16.120 (a) $K_w = [H^+][OH^-] = 5.48 \times 10^{-14}$. In pure water, $[H^+] = [OH^-]$.

$[H^+]^2 = 5.48 \times 10^{-14}$ $[H^+] = 2.34 \times 10^{-7}$ M; pH = 6.63

(b) The value of K_w increases with increasing temperature, so the sign of ΔH is positive. The autoionization of water is endothermic.

16.121 (a) 24 valence e^-, 12 e^- pairs

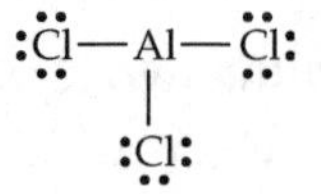

The formal charges on all atoms are zero. Structures with multiple bonds lead to nonzero formal charges. There are three electron domains about Al. The electron-domain geometry and molecular structure are trigonal planar.

(b) The Al atom in $AlCl_3$ has an incomplete octet and is electron deficient. It "needs" to accept another electron pair, to act like a Lewis acid.

(c)

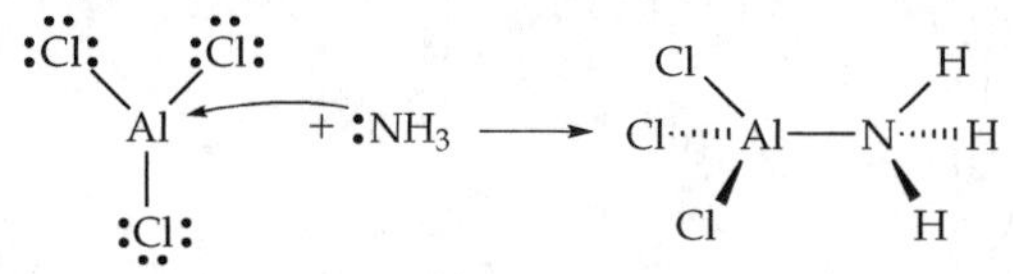

Both the Al and N atoms in the product have tetrahedral geometry.

(d) The Lewis theory is most appropriate. H^+ and $AlCl_3$ are both electron pair acceptors, Lewis acids.

16.122 *Plan.* Use acid ionization equilibrium to calculate the total moles of particles in solution. Use density to calculate kg solvent. From the molality (m) of the solution, calculate ΔT_b and T_b.   *Solve.*

$$HSO_4^-(aq) \quad \rightleftharpoons \quad H^+(aq) \quad + \quad SO_4^{2-}(aq)$$

initial	0.10 M	0	0
equil.	0.10 – x M	x M	x M
	0.071 M	0.029 M	0.029 M

$K_a = 1.2 \times 10^{-2} = \dfrac{[H^+][SO_4^{2-}]}{[HSO_4^-]} = \dfrac{x^2}{0.10-x}$; K_a is relatively large, so use the quadratic.

$x^2 + 0.012\,x - 0.0012 = 0$; $x = \dfrac{-0.012 \pm \sqrt{(0.012)^2 - 4(1)(-0.0012)}}{2}$; $x = 0.029$ M H^+, SO_4^{2-}

Total ion concentration = 0.10 M Na^+ + 0.071 M HSO_4^- + 0.029 M H^+ + 0.029 M SO_4^{2-}

$= 0.229 = 0.23$ M

Assume 100.0 mL of solution. 1.002 g/mL × 100.0 mL = 100.2 g solution.

0.10 M $NaHSO_4$ × 0.1000 L = 0.010 mol $NaHSO_4$ × $\dfrac{120.1\,g\ NaHSO_4}{mol\ NaHSO_4}$

$= 1.201 = 1.2$ g $NaHSO_4$

100.2 g soln – 1.201 g $NaHSO_4$ = 99.0 g = 0.099 kg H_2O

$m = \dfrac{mol\ ions}{kg\ H_2O} = \dfrac{0.229\ M \times 0.1000\ L}{0.0990\ kg} = 0.231 = 0.23$ m ions

$\Delta T_b = K_b(m) = 0.52\ °C/m \times (0.23\ m) = +0.12\ °C$; $T_b = 100.0 + 0.12 = 100.1\ °C$

16.123 Rx 1: $\Delta H = D(H\text{–}F) + 2D(H\text{–}O) - 3D(H\text{–}O) = D(H\text{–}F) - D(H\text{–}O)$

$\Delta H = 567\ kJ - 463\ kJ = 104\ kJ$

Rx 2: $\Delta H = D(H\text{–}Cl) + 2D(H\text{–}O) - 3D(H\text{–}O) = D(H\text{–}Cl) - D(H\text{–}O)$

$\Delta H = 431\ kJ - 463\ kJ = -32\ kJ$

The reaction involving HCl is exothermic, whereas the reaction involving HF is endothermic, owing to the smaller bond dissociation enthalpy of H–Cl. HCl is a stronger acid than HF, and the enthalpy of ionization for HCl is exothermic, whereas that of HF is endothermic. This is consistent with the trend in acid strength for binary acids with heavy atoms (X) in the same family. That is, the longer and weaker the H–X bond, the stronger the acid (and the more exothermic the ionization reaction).

16.124 Calculate M of the solution from osmotic pressure and K_b using the equilibrium expression for the hydrolysis of cocaine. Let Coc = cocaine and $CocH^+$ be the conjugate acid of cocaine.

$$\Pi = MRT;\ M = \Pi/RT = \frac{52.7\ torr}{288\ K} \times \frac{1\ atm}{760\ torr} \times \frac{mol\text{-}K}{0.08206\ L\text{-}atm}$$

$$= 0.002934 = 2.93 \times 10^{-3}\ M\ Coc$$

$pH = 8.53;\ pOH = 14 - pH = 5.47;\ [OH^-] = 10^{-5.47} = 3.39 \times 10^{-6} = 3.4 \times 10^{-6}\ M$

	$Coc(aq) + H_2O(l)$	$\rightleftharpoons$	$CocH^+(aq)$	$+$	$OH^-(aq)$
initial	$2.93 \times 10^{-3}\ M$		0		0
equil.	$(2.93 \times 10^{-3} - 3.4 \times 10^{-6})\ M$		$3.4 \times 10^{-6}\ M$		$3.4 \times 10^{-6}\ M$

$$K_b = \frac{[CocH^+][OH^-]}{[Coc]} = \frac{(3.39 \times 10^{-6})^2}{(2.934 \times 10^{-3} - 3.39 \times 10^{-6})} = 3.9 \times 10^{-9}$$

Note that % hydrolysis is small in this solution, so $3.39 \times 10^{-6}\ M$ is small compared to $2.93 \times 10^{-3}\ M$ and can be ignored in the denominator of the calculation.

16.125 (a) $rate = k[IO_3^-][SO_3^{2-}][H^+]$

(b) $\Delta pH = pH_2 - pH_1 = 3.50 - 5.00 = -1.50$

$\Delta pH = -\log[H^+]_2 - (-\log[H^+]_1);\ -\Delta pH = \log[H^+]_2 - \log[H^+]_1$

$-\Delta pH = \log[H^+]_2 / [H^+]_1;\ [H^+]_2 / [H^+]_1 = 10^{-\Delta pH}$

$[H^+]_2/[H^+]_1 = 10^{1.50} = 31.6 = 32$. The rate will increase by a factor of 32 if $[H^+]$ increases by a factor of 32. The reaction goes faster at lower pH.

(c) As H^+ does not appear in the overall reaction, it is either a catalyst or an intermediate. An intermediate is produced and then consumed during a reaction, so its contribution to the rate law can usually be written in terms of concentrations of other reactants (Sample Exercise 14.15). A catalyst is present at the beginning and end of a reaction and can appear in the rate law if it participates in the rate-determining step (Solution 14.78). This reaction is pH dependent because H^+ is a homogeneous catalyst that participates in the rate-determining step.

16.126   (a)   (i) $HCO_3^-(aq) \rightleftharpoons H^+(aq) + CO_3^{2-}(aq)$ $K_1 = K_{a2}$ for $H_2CO_3 = 5.6 \times 10^{-11}$

                   $H^+(aq) + OH^-(aq) \rightleftharpoons H_2O(l)$ $K_2 = 1/K_w = 1 \times 10^{14}$

                $\overline{HCO_3^-(aq) + OH^-(aq) \rightleftharpoons CO_3^{2-}(aq) + H_2O(l)}$ $K = K_1 \times K_2 = 5.6 \times 10^3$

        (ii) $NH_4^+(aq) \rightleftharpoons H^+(aq) + NH_3(aq)$ $K_1 = K_a$ for $NH_4^+ = 5.6 \times 10^{-10}$

                   $CO_3^{2-}(aq) + H^+(aq) \rightleftharpoons HCO_3^-(aq)$ $K_2 = 1/K_{a2}$ for $H_2CO_3 = 1.8 \times 10^{10}$

                $\overline{NH_4^+(aq) + CO_3^{2-}(aq) \rightleftharpoons HCO_3^-(aq) + NH_3(aq)}$ $K = K_1 \times K_2 = 10$

   (b)   Both (i) and (ii) have K > 1, although K = 10 is not *much* greater than 1. Both could be written with a single arrow. (This is true in general when a strong acid or strong base, $H^+(aq)$ or $OH^-(aq)$, is a reactant.)

17 Additional Aspects of Aqueous Equilibria

Visualizing Concepts

17.1 *Analyze.* Given diagrams showing equilibrium mixtures of HX and X^- with different compositions, decide which has the highest pH. HX is a weak acid and X^- is its conjugate base. *Plan.* Evaluate the contents of the boxes. Use acid–base equilibrium principles to relate $[H^+]$ to box composition. *Solve.*

Use the following acid ionization equilibrium to describe the mixtures: $HX(aq) \rightleftharpoons H^+(aq) + X^-(aq)$. Each box has 4 HX molecules, but differing amounts of X^- ions. The greater the amount of X^- (conjugate base), for the same amount of HX (weak acid), the lower the amount of H^+ and the higher the pH. The middle box, with most X^-, has least H^+ and highest pH.

17.2 (a) The yellow solution has the higher pH. According to Figure 16.8, methyl orange is yellow above pH 4.5 and red (really pink) below pH 3.5. The beaker on the left has a pH greater than 4.5, and the one on the right has pH less than 3.5. (By calculation, pH of left beaker = 4.7, pH of right beaker = 2.9.) The right beaker, with lower pH and greater $[H^+]$, is pure acetic acid. The left beaker contains equal amounts of the weak acid and its conjugate base, acetic acid and acetate ion. Adding the "common-ion" acetate (in the form of sodium acetate) shifts the acid ionization equilibrium to the left, decreases $[H^+]$, and raises pH.

(b) When small amounts of NaOH are added, the left beaker is better able to maintain its pH. For solutions of the same weak acid, pH depends on the *ratio* of conjugate base to conjugate acid. Small additions of base (or acid) have the least effect when this ratio is close to one. The left beaker is a buffer because it contains a weak conjugate acid/conjugate base pair and resists rapid pH change upon addition of small amounts of strong base or acid.

17.3 Statement (b) is correct, $[HA] > [A^-]$. Buffers prepared from weak acids (HA) and their conjugate bases (A^-, usually in the form of a salt) have pH values in a range of approximately 2 pH units, centered around pK_a for the weak acid. If concentration of the weak acid is greater than concentration of the conjugate base, $pH < pK_a$. If concentration of the conjugate base is greater than concentration of the weak acid, $pH > pK_a$. This is generally true for buffers containing a weak conjugate acid/conjugate base (CA/CB) pair.

$[CA] > [CB]$, pH of buffer $< pK_a$ of CA

$[CA] < [CB]$, pH of buffer $> pK_a$ of CA

17.4   *Analyze/Plan.* When strong acid is added to a buffer, it reacts with conjugate base (CB) to produce conjugate acid (CA). [CA] increases and [CB] decreases. The opposite happens when strong base is added to a buffer, [CB] increases and [CA] decreases. Match these situations to the drawings.   *Solve.*

The buffer begins with equal concentrations of HX and X^-.

(a)   After addition of strong acid, [HX] will increase and $[X^-]$ will decrease. Drawing (3) fits this description.

(b)   Adding of strong base causes [HX] to decrease and $[X^-]$ to increase. Drawing (1) matches the description.

(c)   Drawing (2) shows both [HX] and $[X^-]$ to be smaller than the initial concentrations shown on the left. This situation cannot be achieved by adding strong acid or strong base to the original buffer.

17.5   *Analyze/Plan.* Consider the reaction $HA + OH^- \rightarrow A^- + H_2O$. What are the major species present in solution at the listed stages of the titration? Which diagram represents these species?   *Solve.*

(a)   *Before addition of NaOH*, the solution is mostly HA. The only A^- is produced by the ionization equilibrium of HA and is too small to appear in the diagram. This situation is shown in diagram (iii), which contains only HA.

(b)   *After addition of NaOH but before the equivalence point*, some, but not all, HA has been converted to A^-. The solution contains a mixture of HA and A^-; this is shown in diagram (i).

(c)   *At the equivalence point*, all HA has been converted to A^-, with no excess HA or OH^- present. This is shown in diagram (iv).

(d)   *After the equivalence point*, the same amount of A^- as at the equivalence point is present, plus some excess OH^-. This is diagram (ii).

17.6   *Analyze/Plan.* In each case, the first substance is in the buret, and the second is in the flask. If acid is in the flask, the initial pH is low; with base in the flask, the pH starts high. Strong acids have lower pH than weak acids; strong bases have higher pH than weak bases. Polyprotic acids and bases have more than one "jump" in pH.   *Solve.*

(a)   Strong base in flask, pH starts high, ends low as acid is added. Only diagram (ii) fits this description.

(b)   Weak acid in flask, pH starts low, but not extremely low. Diagrams (i), (iii), and (iv) all start at low pH and get higher. Diagram (i) has very low initial pH, and likely has strong acid in the flask. Diagram (iv) has two pH jumps, so it has a polyprotic acid in the flask. Diagram (iii) best fits the profile of adding a strong base to a weak acid.

(c)   Strong acid in the flask, pH starts very low, diagram (i).

(d)   Polyprotic acid, more than one pH jump, diagram (iv).

17.7   *Analyze.* Given two titration curves where 0.10 M NaOH is the titrant, decide which represents the more concentrated acid, and which the stronger acid.

Plan. For equal volumes of acid, concentration is related to volume of titrant (0.10 M NaOH) at the equivalence points. To determine K_a, pH = pK_a half-way to the equivalence point.

Solve.

(a)   Both acids have one ionizable hydrogen, because there is one "jump" in each titration curve. For equal volumes of acid, and the same titrant, the more concentrated acid requires a greater volume of titrant to reach equivalence. The equivalence point of the blue curve is at 25 mL NaOH and of the red curve is at 35 mL NaOH. The red acid is more concentrated.

(b)   According to the Henderson–Hasselbach equation, $pH = pK_a + \log\dfrac{[\text{conj. base}]}{[\text{conj. acid}]}$.

At half-way to the equivalence point, [conj. acid] = [conj. base] and pH = K_a of the conjugate acid. For the blue curve, half-way is 12.5 mL NaOH. The pH at this volume is approximately 7.0. For the red curve, half-way is 17.5 mL NaOH. The pH at this volume is approximately 4.2. A pK_a of 7 corresponds to K_a of 1×10^{-7}, whereas pK_a of 4.2 corresponds to K_a of 6×10^{-5}. The red acid has the larger K_a value.

Note that the stronger acid, the one with the larger K_a value, has a larger change in pH (jump) at the equivalence point. Also note that initial acid pH was not a definitive measure of acid strength, because the acids have different starting concentrations. Both K_a values and concentration contribute to solution pH.

17.8   *Analyze/Plan.* The beaker of saturated $Cd(OH)_2(aq)$ contains undissolved $Cd(OH)_2(s)$, $Cd^{2+}(aq)$, and $OH^-(aq)$. Decide how amounts of each of these three components change when HCl(aq) is added.   *Solve.*

When HCl(aq) is added, it reacts with $OH^-(aq)$ to form $H_2O(l)$ and $Cl^-(aq)$. (Both have been omitted from the figure.) When $OH^-(aq)$ is removed from solution, more $Cd(OH)_2(s)$ dissolves to replace it; $[Cd^{2+}(aq)]$ increases, $[OH^-(aq)]$ decreases and the amount of undissolved $Cd(OH)_2(s)$ decreases. In the resulting solution, $[Cd^{2+}(aq)]$ is greater than $[OH^-(aq)]$ and there is less undissolved solid on the bottom of the beaker. Beaker A accurately represents the solution after equilibrium is reestablished.

17.9   *Analyze/Plan.* Common anions or cations decrease the solubility of salts. Ions that participate in acid–base or complex ion equilibria increase solubility.   *Solve.*

(a) CO_2^{3-} from $BaCO_3$ reacts with H^+ from HNO_3, causing solubility of $BaCO_3$ to increase with increasing HNO_3 concentration. This behavior matches the right diagram.

(b)   Extra CO_2^{3-} from Na_2CO_3 decreases the solubility of $BaCO_3$. Solubility of $BaCO_3$ decreases as $[Na_2CO_3]$ increases. This behavior matches the left diagram.

(c) $NaNO_3$ has no common ions, nor does it enter into acid–base or complex ion equilibria with Ba^{2+} or CO_3^{2-}; it does not affect the solubility of $BaCO_3$. This behavior is shown in the center diagram.

17.10   *Analyze/Plan.* Calculate the molarity of the solution assuming all $Ca(OH)_2(s)$ dissolves. Use this concentration along with the K_{sp} expression for $Ca(OH)_2$ to answer the questions.   *Solve.*

(a) $[Ca(OH)_2] = \dfrac{0.370 \text{ g } Ca(OH)_2}{0.500 \text{ L soln}} \times \dfrac{1 \text{ mol } Ca(OH)_2}{74.093 \text{ g } Ca(OH)_2} = 0.00998745 = 0.00999 \, M$

$[Ca^{2+}] = 0.00999 \, M$; $[OH^-] = 2(0.00998745) = 0.0199749 = 0.0200 \, M$;

$K_{sp} = [Ca^{2+}][OH^-]^2$. Calculate the reaction quotient using the calculated molarities. If it is equal to or greater than K_{sp}, the resulting solution is saturated. $Q = (0.0098745)(0.0199749)^2 = 3.99 \times 10^{-6}$. $Q < K_{sp}$ (6.5×10^{-6}) and the solution is not saturated.

(b)   Consider the beakers individually.

(i)   The 50 mL of 1.0 M HCl is more than enough to neutralize 50 mL of 0.0200 M OH^-(aq). No precipitate forms.

(ii)   NaCl does not react with $Ca(OH)_2$ and the two compounds contain no common ions. No precipitate forms.

(iii) $CaCl_2$ does contain a common ion. Calculate Q for the resulting solution to see if $Ca(OH)_2$ precipitates. $[OH^-]$ in the new solution is 0.00999 M, because it is diluted by a factor of 2. $[Ca^{2+}] = (1.0 + 0.00999)/2 = 0.5050 \, M$. $Q = (0.5050)(0.00999)^2 = 5.04 \times 10^{-5}$. $Q > K_{sp}$ (6.5×10^{-6}) and $Ca(OH)_2$ precipitates.

(iv)   A common ion with a different concentration; $[Ca^{2+}] = (0.10 + 0.00999)/2 = 0.0550 = 0.055 \, M$. $Q = (0.0550)(0.00999)^2 = 5.49 \times 10^{-6}$. $Q \approx K_{sp}$ (6.5×10^{-6}); the solution is very nearly saturated, but no precipitate forms.

17.11   Statement (c) explains the shape of the graph. Solubility is high initially, at low pH, and then decreases to a minimum as pH increases. This depicts formation of an insoluble hydroxide as $[H^+]$ decreases and $[OH^-]$ increases. Additional base then reacts with the insoluble hydroxide to dissolve it. This curve depicts the behavior of an amphoteric insoluble hydroxide; it dissolves upon addition of either acid or base.

17.12   According to Figure 17.23, the two precipitating agents are 6 M HCl (first) and H_2S in 0.2 M HCl (second).

Cation A = Ag^+ (precipitates as AgCl)

Cation B = Cu^+ (precipitates as CuS, acid insoluble)

Cation C = Ni^{2+} (remains in acidic solution)

The Common-Ion Effect (Section 17.1)

17.13   Statement (a) is most correct. The common ion can be either the cation or anion of a salt. The common-ion effect applies to all ions if they are "common" to the salt in question, but not to noncommon ions. Common ions do no affect the equilibrium constant.

17.14 The added salt is soluble and increases $[HB^+]$ in the solution. For a generic weak base B, $K_b = \dfrac{[HB^+][OH^-]}{[B]}$.

(a) Stay the same. Addition of a common ion such as HB^+ does not change the equilibrium constant.

(b) Increase. To maintain the value of the equilibrium constant, addition of HB^+ requires that [B] also increases.

(c) Decrease. Additional HB^+ reacts with OH^-, lowering the pH of the solution.

17.15 *Analyze/Plan.* Follow the logic in Sample Exercise 17.1. *Solve.*

(a)

	$C_2H_5COOH(aq)$	$\rightleftharpoons$	$H^+(aq)$	$+$	$C_2H_5COO^-(aq)$
i	0.085 *M*				0.060 *M*
c	$-x$		$+x$		$+x$
e	(0.085 – x) *M*		$+x\ M$		(0.060 + x) *M*

$$K_a = 1.3 \times 10^{-5} = \frac{[H^+][C_2H_5COO^-]}{[C_2H_5COOH]} = \frac{(x)(0.060+x)}{(0.085-x)}$$

Assume x is small compared to 0.060 and 0.085.

$$1.3 \times 10^{-5} = \frac{0.060\,x}{0.085}; x = 1.8 \times 10^{-5} = [H^+], pH = 4.73$$

Check. Because the extent of ionization of a weak acid or base is suppressed by the presence of a conjugate salt, the 5% rule usually holds true in buffer solutions.

(b)

	$(CH_3)_3N(aq) + H_2O(l)$	$\rightleftharpoons$	$(CH_3)_3NH^+(aq)$	$+$	$OH^-(aq)$
i	0.075 *M*		0.10 *M*		
c	$-x$		$+x$		$+x$
e	(0.075 – x) *M*		(0.10 + x) *M*		$+x\ M$

$$K_b = 6.4 \times 10^{-5} = \frac{[OH^-][(CH_3)_3NH^+]}{[(CH_3)_3N]} = \frac{(x)(0.10+x)}{(0.075-x)} \approx \frac{0.10\,x}{0.075}$$

$$x = 4.8 \times 10^{-5} = [OH^-], pOH = 4.32, pH = 14.00 - 4.32 = 9.68$$

Check. In a buffer, if [conj. acid] > [conj. base], pH < pK_a of the conj. acid. If [conj. acid] < [conj. base], pH > pK_a of the conj. acid. In this buffer, pK_a of $(CH_3)_3NH^+$ is 9.81. $[(CH_3)_3NH^+] > [(CH_3)_3N]$ and pH = 9.68, less than 9.81.

(c) mol = $M \times L$; mol $CH_3COOH = 0.15\ M \times 0.0500\ L = 7.5 \times 10^{-3}$ mol

mol $CH_3COO^- = 0.20\ M \times 0.0500\ L = 0.010$ mol

	$CH_3COOH(aq)$	$\rightleftharpoons$	$H^+(aq)$	$+$	$CH_3COO^-(aq)$
i	7.5×10^{-3} mol		0		0.010 mol
c	$-x$		$+x$		$+x$
e	$(7.5 \times 10^{-3} - x)$ mol		$+x$ mol		(0.010 + x) mol

$[CH_3COOH(aq)] = (7.5 \times 10^{-3} - x)$ mol/0.1000 L;

$[CH_3COO^-(aq)] = (0.010 + x)$ mol/0.1000 L

$$K_a = 1.8 \times 10^{-5} = \frac{[H^+][CH_3COO^-]}{[CH_3COOH]} = \frac{(x)(0.010+x)/0.1000\,L}{(0.0075-x)/0.1000\,L} \approx \frac{x(0.010)}{0.0075}$$

$x = 1.35 \times 10^{-5}\,M = 1.4 \times 10^{-5}\,M\,H^+; pH = 4.87$

Check. pK_a for $CH_3COOH = 4.74$. $[CH_3COO^-] > [CH_3COOH]$, pH of buffer = 4.87, greater than 4.74.

17.16   *Analyze/Plan.* Follow the logic in Sample Exercise 17.1.   *Solve.*

(a)   HCOOH is a weak acid, and HCOONa contains the common ion $HCOO^-$, the conjugate base of HCOOH. Solve the common-ion equilibrium problem.

	HCOOH(aq)	$\rightleftharpoons$	H^+(aq)	+	$HCOO^-$(aq)
i	0.100 M				0.250 M
c	$-x$		$+x$		$+x$
e	$(0.100 - x)\,M$		$+x\,M$		$(0.250 + x)\,M$

$$K_a = 1.8 \times 10^{-4} = \frac{[H^+][HCOO^-]}{[HCOOH]} = \frac{(x)(0.250+x)}{(0.100-x)} \approx \frac{0.250\,x}{0.100}$$

$x = 7.20 \times 10^{-5} = 7.2 \times 10^{-5}\,M = [H^+], pH = 4.14$

Check. Because the extent of ionization of a weak acid or base is suppressed by the presence of a conjugate salt, the 5% rule usually holds true in buffer solutions.

(b) C_5H_5N is a weak base, and C_5H_5NHCl contains the common ion $C_5H_5NH^+$, which is the conjugate acid of C_5H_5N. Solve the common-ion equilibrium problem.

	C_5H_5N(aq) + H_2O(l)	$\rightleftharpoons$	$C_5H_5NH^+$(aq)	+	OH^-(aq)
i	0.510 M		0.450 M		
c	$-x$		$+x$		$+x$
e	$(0.510 - x)\,M$		$(0.450 + x)\,M$		$+x\,M$

$$K_b = 1.7 \times 10^{-9} = \frac{[C_5H_5NH^+][OH^-]}{[C_5H_5N]} = \frac{(0.450+x)(x)}{(0.510-x)} \approx \frac{0.450\,x}{0.510}$$

$x = 1.927 \times 10^{-9} = 1.9 \times 10^{-9}\,M = [OH^-], pOH = 8.715, pH = 14.00 - 8.715 = 5.29$

Check. In a buffer, if [conj. acid] > [conj. base], pH < pK_a of the conj. acid. If [conj. acid] < [conj. base], pH > pK_a of the conj. acid. In this buffer, pK_a of $C_5H_5NH^+$ is 5.23. $[C_5H_5NH^+] < [C_5H_5N]$ and pH = 5.29, greater than 5.23.

(c)   mol = $M \times L$; mol HF = 0.050 $M \times 0.055$ L = $2.75 \times 10^{-3} = 2.8 \times 10^{-3}$ mol;

mol $F^- = 0.10\,M \times 0.125$ L = 0.0125 = 0.013 mol

	HF(aq)	$\rightleftharpoons$	H^+(aq)	+	F^-(aq)
i	2.75×10^{-3} mol		0		0.0125 mol
c	$-x$		$+x$		$+x$
e	$(2.75 \times 10^{-3} - x)$ mol		$+x$		$(0.0125 + x)$ mol

$[HF] = (2.75 \times 10^{-3} + x)/0.180$ L; $[F^-] = (0.0125 + x)/0.180$ L

Note that the volumes will cancel when substituted into the K_a expression.

$$K_a = 6.8 \times 10^{-4} = \frac{[H^+][F^-]}{[HF]} = \frac{x(0.0125+x)/0.180}{(2.75\times 10^{-3}-x)/0.180} \approx \frac{x(0.0125)}{0.00275}$$

$$x \doteq 1.50 \times 10^{-4} = 1.5 \times 10^{-4}\ M\ H^+; pH = 3.83$$

Check. K_a for HF = 3.17. [HF] < [F$^-$], pH of buffer = 3.83, greater than 3.17.

17.17   *Analyze/Plan.* We are asked to calculate % ionization of (a) a weak acid and (b) a weak acid in a solution containing a common ion, its conjugate base. Calculate % ionization as in Sample Exercise 16.13. In part (b), the concentration of the common ion is 0.085 M, not x, as in part (a). *Solve.*

$$buCOOH(aq) \rightleftharpoons H^+(aq) + buCOO^-(aq)\quad K_a = \frac{[H^+][buCOO^-]}{[buCOOH]} = 1.5 \times 10^{-5}$$

equil (a)   0.0075 – x M x M x M

equil (b)   0.0075 – x M x M 0.085 + x M

(a) $K_a = 1.5 \times 10^{-5} = \dfrac{x^2}{0.0075-x} \approx \dfrac{x^2}{0.0075}$; x = [H$^+$] = 3.354×10^{-4} = 3.4×10^{-4} M H$^+$

$$\% \text{ ionization} = \frac{3.4 \times 10^{-4}\ M\ H^+}{0.0075\ M\ buCOOH} \times 100 = 4.5\% \text{ ionization}$$

(b) $K_a = 1.5 \times 10^{-5} = \dfrac{(x)(0.085+x)}{0.0075-x} \approx \dfrac{0.085\,x}{0.0075}$; x = 1.3×10^{-6} M H$^+$

$$\% \text{ ionization} = \frac{1.3 \times 10^{-6}\ M\ H^+}{0.0075\ M\ buCOOH} \times 100 = 0.018\% \text{ ionization}$$

Check. Percent ionization is much smaller when the "common ion" is present.

17.18 $CH_3CH(OH)COOH \rightleftharpoons H^+(aq) + CH_3CH(OH)COO^-$

equil (a)   0.125 – x M x M x M

equil (b)   0.125 – x M x M 0.0075 + x M

$$K_a = \frac{[H^+][CH_3CH(OH)\,COO^-]}{[CH_3CH(OH)COOH]} = 1.4 \times 10^{-4}$$

(a) $K_a = 1.4 \times 10^{-4} = \dfrac{x^2}{0.125-x} \approx \dfrac{x^2}{0.125}$; x = [H$^+$] = 4.18×10^{-3} M = 4.2×10^{-3} M H$^+$

$$\% \text{ ionization} = \frac{4.2 \times 10^{-3}\ M\ H^+}{0.125\ M\ CH_3CH(OH)COOH} \times 100 = 3.4\% \text{ ionization}$$

(b) $K_a = 1.4 \times 10^{-4} = \dfrac{(x)(0.0075+x)}{0.125-x} \approx \dfrac{0.0075\,x}{0.125}$; x = 2.3×10^{-3} M H$^+$

$$\% \text{ ionization} = \frac{2.3 \times 10^{-3}\ M\ H^+}{0.125\ M\ CH_3CH(OH)COOH} \times 100 = 1.9\% \text{ ionization}$$

Buffered Solutions (Section 17.2)

17.19   Only solution (a) is a buffer. CH_3COOH and CH_3COONa are a weak conjugate acid/conjugate base pair that acts as a buffer; CH_3COOH reacts with added base and CH_3COO^- reacts with added acid, leaving $[H^+]$ relatively unchanged. Solution (b) contains only a weak acid, which has no capacity to react with added acid. Although solution (c) contains a conjugate acid/conjugate base pair, Cl^- is a negligible base. In general, the conjugate bases of strong acids are negligible and mixtures of strong acids and their conjugate salts do not act as buffers.

17.20   Only solution (a) is a buffer. NaOH is a strong base and will react with CH_3COOH to form CH_3COONa. As long as CH_3COOH is present in excess, the resulting solution will contain both the conjugate acid $CH_3COOH(aq)$ and the conjugate base $CH_3COO^-(aq)$, the requirements for a buffer.

          Solution (b) contains a large excess and is essentially a strong base. Solution (c) is essentially a strong acid; HCl determines $[H^+]$ and pH of the solution. Solution (d) contains two salts; one is the conjugate base of a weak acid. It can react with added acid, but no added base.

17.21   *Analyze/Plan.* Follow the logic in Sample Exercise 17.3. Assume that % ionization is small in these buffers (Solutions 17.17 and 17.18).   *Solve.*

(a) $K_a = \dfrac{[H^+][CH_3CH(OH)COO^-]}{[CH_3CH(OH)COOH]}$; $[H^+] = \dfrac{[K_a][CH_3CH(OH)COOH]}{[CH_3CH(OH)COO^-]}$

        $[H^+] = \dfrac{1.4 \times 10^{-4}\,(0.12)}{(0.11)}$; $[H^+] = 1.53 \times 10^{-4} = 1.5 \times 10^{-4}\,M$; pH = 3.82

(b)   mol = $M \times L$; total volume = 85 mL + 95 mL = 180 mL

        $[H^+] = \dfrac{K_a[CH_3CH(OH)COOH]}{[CH_3CH(OH)COO^-]} = \dfrac{1.4 \times 10^{-4}(0.13\,M \times 0.085\,L)/0.180\,L}{(0.15\,M \times 0.095\,L)/0.180\,L}$

        $[H^+] = \dfrac{1.4 \times 10^{-4}\,(0.13 \times 0.085)}{(0.15 \times 0.095)}$; $[H^+] = 1.086 \times 10^{-4} = 1.1 \times 10^{-4}\,M$; pH = 3.96

17.22   Assume that % ionization is small in these buffers (Solutions 17.17 and 17.18).

(a)   The conjugate acid in this buffer is HCO_3^-, so use K_{a2} for H_2CO_3, 5.6×10^{-11}

        $K_a = \dfrac{[H^+][CO_3^{2-}]}{[HCO_3^-]}$; $[H^+] = \dfrac{K_a[HCO_3^-]}{[CO_3^{2-}]} = \dfrac{5.6 \times 10^{-11}\,(0.105)}{(0.125)}$

        $[H^+] = 4.70 \times 10^{-11} = 4.7 \times 10^{-11}\,M$; pH = 10.33

(b)   mol = $M \times L$; total volume = 140 mL = 0.140 L

        $[H^+] = \dfrac{K_a(0.20\,M \times 0.065\,L)/0.140\,L}{(0.15\,M \times 0.075\,L)/0.140\,L} = \dfrac{5.6 \times 10^{-11}(0.20 \times 0.065)}{(0.15 \times 0.075)}$

        $[H^+] = 6.47 \times 10^{-11} = 6.5 \times 10^{-11}\,M$; pH = 10.19

17.23 (a) *Analyze/Plan.* Follow the logic in Sample Exercises 17.1 and 17.3. As in Sample Exercise 17.1, start by calculating concentrations of the components. *Solve.*

$$CH_3COOH(aq) \rightleftharpoons H^+(aq) + CH_3COO^-(aq); \ K_a = 1.8 \times 10^{-5} = \frac{[H^+][CH_3COO^-]}{[CH_3COOH]}$$

$$[CH_3COOH] = 0.150 \ M$$

$$[CH_3COO^-] = \frac{20.0 \ g \ CH_3COONa}{0.500 \ L \ soln} \times \frac{1 \ mol \ CH_3COONa}{82.04 \ g \ CH_3COONa} = 0.488 \ M$$

$$[H^+] = \frac{K_a[CH_3COOH]}{[CH_3COO^-]} = \frac{1.8 \times 10^{-5}(0.150 - x)}{(0.488 + x)} \approx \frac{1.8 \times 10^{-5}(0.150)}{(0.488)}$$

$$[H^+] = 5.533 \times 10^{-6} = 5.5 \times 10^{-6} \ M, \ pH = 5.26$$

(b) *Plan.* On the left side of the equation, write all ions present in solution after HCl or NaOH is added to the buffer. Using acid–base properties and relative strengths, decide which ions will combine to form new products. *Solve.*

$$Na^+(aq) + CH_3COO^-(aq) + H^+(aq) + Cl^-(aq) \rightarrow CH_3COOH(aq) + Na^+(aq) + Cl^-(aq)$$

(c) $$CH_3COOH(aq) + Na^+(aq) + OH^-(aq) \rightarrow CH_3COO^-(aq) + H_2O(l) + Na^+(aq)$$

17.24 NH_4^+/NH_3 is a basic buffer. Either the hydrolysis of NH_3 or the dissociation of NH_4^+ can be used to determine the pH of the buffer. Using the dissociation of NH_4^+ leads directly to $[H^+]$ and facilitates use of the Henderson–Hasselbach relationship.

(a) $$NH_4^+(aq) \rightleftharpoons H^+(aq) + NH_3(aq)$$

$$K_a = \frac{K_w}{K_b} = \frac{1.0 \times 10^{-14}}{1.8 \times 10^{-5}} = 5.56 \times 10^{-10} = 5.6 \times 10^{-10}$$

$$[NH_3] = 1.00 \ M \ NH_3$$

$$[NH_4^+] = \frac{10.0 \ g \ NH_4Cl}{0.250 \ L} \times \frac{1 \ mol \ NH_4Cl}{53.50 \ g \ NH_4Cl} = 0.74766 = 0.748 \ M \ NH_4^+$$

$$K_a = \frac{[H^+][NH_3]}{[NH_4^+]}; [H^+] = \frac{K_a[NH_4^+]}{[NH_3]} = \frac{5.56 \times 10^{-10}(0.74766 - x)}{(1.00 + x)} \approx \frac{5.56 \times 10^{-10}(0.74766)}{(1.00)}$$

$$[H^+] = 4.1537 \times 10^{-10} = 4.15 \times 10^{-10} \ M, \ pH = 9.382$$

(b) $$NH_3(aq) + H^+(aq) + NO_3^-(aq) \rightarrow NH_4^+(aq) + NO_3^-(aq)$$

(c) $$NH_4^+(aq) + Cl^-(aq) + K^+(aq) + OH^-(aq) \rightarrow NH_3(aq) + H_2O(l) + Cl^-(aq) + K^+(aq)$$

17.25 *Analyze/Plan.* Follow the logic in Sample Exercises 16.12 and 17.5. *Solve.*

(a) $$K_a = 6.8 \times 10^{-4} = \frac{x^2}{1.00 - x} = \frac{x^2}{1.00}; \ x = [H^+] = 0.02608 = 0.026 \ M; \ pH = 1.58$$

There is 2.6% ionization, so the approximation is valid.

(b) In this problem, $[F^-]$ is the unknown.

$$pH = 3.00, [H^+] = 10^{-3.00} = 1.0 \times 10^{-3}; [HF] = 1.00 - 0.0010 = 0.999 \ M$$

$$K_a = 6.8 \times 10^{-4} = \frac{1.0 \times 10^{-3} [F^-]}{0.999}; \ [F^-] = 0.6793 = 0.68 \ M$$

$$\frac{0.6793 \ mol \ NaF}{1 \ L} \times \frac{41.990 \ g \ NaF}{1 \ mol \ NaF} \times 1.25 \ L = 35.654 = 36 \ g \ NaF$$

17.26   (a) $C_6H_5COOH(aq) \rightleftharpoons H^+(aq) + C_6H_5COO^-(aq)$

$$K_a = 6.3 \times 10^{-5} = \frac{[H^+][C_6H_5COO^-]}{[C_6H_5COOH]}; \; [C_6H_5COOH] = 0.0200 \, M;$$

$$[H^+] = [C_6H_5COO^-] = x$$

$$K_a = 6.3 \times 10^{-5} \approx \frac{x^2}{0.02000}; \; x = [H^+] = 1.123 \times 10^{-3} = 1.1 \times 10^{-3} M; \; pH = 2.95$$

Note that C_6H_5COOH is 5.6% ionized. Solving the quadratic for $[H^+]$ yields $(1.0914 \times 10^{-3} =) 1.1 \times 10^{-3} M \, H^+$, pH = 2.96; this is not a significant difference.

(b) $[H^+] = \dfrac{K_a[C_6H_5COOH]}{[C_6H_5COO^-]}; \quad [H^+] = 10^{-4.00} = 1.0 \times 10^{-4} \, M$

$[C_6H_5COOH] = 0.0200 \, M$; calculate $[C_6H_5COO^-]$. Because the common ion $C_6H_5COO^-$ reduces % ionization, we assume the 5% approximation is valid.

$$[C_6H_5COO^-] = \frac{K_a[C_6H_5COOH]}{[H^+]} = \frac{6.3 \times 10^{-5} \, (0.0200)}{1.0 \times 10^{-4}} = 0.01260 = 0.013 \, M$$

$$\frac{0.0126 \, mol \, C_6H_5COONa}{L} \times 1.50 \, L \times \frac{144.11 \, g \, C_6H_5COONa}{1 \, mol \, C_6H_5COONa}$$

$$= 2.724 = 2.7 \, g \, C_6H_5COONa$$

17.27   *Analyze/Plan.* Follow the logic in Sample Exercises 17.3 and 17.6.   *Solve.*

(a) $K_a = \dfrac{[H^+][CH_3COO^-]}{[CH_3COOH]}; [H^+] = \dfrac{K_a[CH_3COOH]}{[CH_3COO^-]}$

$$[H^+] \approx \frac{1.8 \times 10^{-5} \, (0.10)}{(0.13)} = 1.385 \times 10^{-5} = 1.4 \times 10^{-5} \, M; \, pH = 4.86$$

(b)

$CH_3COOH(aq)$	+	$KOH(aq)$	$\rightarrow$	$CH_3COO^-(aq) + H_2O(l) + K^+(aq)$
0.10 mol		0.02 mol		0.13 mol
–0.02 mol		–0.02 mol		+0.02 mol
0.08 mol		0 mol		0.15 mol

$$[H^+] = \frac{1.8 \times 10^{-5} \, (0.08 \, mol/1.00 \, L)}{(0.15 \, mol/1.00 \, L)} = 9.60 \times 10^{-6} = 1 \times 10^{-5} \, M; \, pH = 5.02 = 5.0$$

(c)

$CH_3COO^-(aq)$	+	$HNO_3(aq)$	$\rightarrow$	$CH_3COOH(aq) + NO_3^-(aq)$
0.13 mol		0.02 mol		0.10 mol
–0.02 mol		–0.02 mol		+0.02 mol
0.11 mol		0 mol		0.12 mol

$$[H^+] = \frac{1.8 \times 10^{-5} \, (0.12 \, mol/1.00 \, L)}{(0.11 \, mol/1.00 \, L)} = 1.96 \times 10^{-5} = 2.0 \times 10^{-5} \, M; \, pH = 4.71$$

17.28   (a) $K_a = \dfrac{[H^+][C_2H_5COO^-]}{[C_2H_5COOH]}$; $[H^+] = \dfrac{K_a[C_2H_5COOH]}{[C_2H_5COO^-]}$

Because this expression contains a ratio of concentrations, we can ignore total volume and work directly with moles.

$$[H^+] = \frac{1.3 \times 10^{-5}(0.15-x)}{(0.10+x)} \approx \frac{1.3 \times 10^{-5}(0.15)}{0.10} = 1.950 \times 10^{-5} = 2.0 \times 10^{-5} \ M, \ pH = 4.71$$

     (b)

$C_2H_5COOH(aq)$	+	$OH^-(aq)$	$\rightarrow$	$C_2H_5COO^-(aq) + H_2O(l)$
0.15 mol		0.01 mol		0.10 mol
−0.01 mol		−0.01 mol		+0.01 mol
0.14 mol		0 mol		0.11 mol

$$[H^+] \approx \frac{1.3 \times 10^{-5}\ (0.14)}{(0.11)} = 1.6545 \times 10^{-5} = 1.7 \times 10^{-5} \ M; \ pH = 4.78$$

     (c)

$C_2H_5COO^-(aq)$	+	$HI(aq)$	$\rightarrow$	$C_2H_5COOH(aq) + I^-(aq)$
0.10 mol		0.01 mol		0.15 mol
−0.01 mol		−0.01 mol		+0.01 mol
0.09 mol		0 mol		0.16 mol

$$[H^+] \approx \frac{1.3 \times 10^{-5}\ (0.16)}{(0.09)} = 2.3111 \times 10^{-5} = 2 \times 10^{-5} \ M; \ pH = 4.6$$

17.29   *Analyze/Plan.* Calculate the [conj. base]/[conj. acid] ratio in the H_2CO_3/HCO_3^- blood buffer. Write the acid dissociation equilibrium and K expression. Find K_a for H_2CO_3 in Appendix D.1. Calculate $[H^+]$ from the pH and solve for the ratio.   *Solve.*

$$H_2CO_3(aq) \ \rightleftharpoons \ H^+(aq) + HCO_3^- \ (aq) \quad K_a = \frac{[H^+][HCO_3^-]}{[H_2CO_3]}; \ \frac{[HCO_3^-]}{[H_2CO_3]} = \frac{K_a}{[H^+]}$$

     (a)   at pH = 7.4, $[H^+] = 10^{-7.4} = 4.0 \times 10^{-8} M$; $\dfrac{[HCO_3^-]}{[H_2CO_3]} = \dfrac{4.3 \times 10^{-7}}{4.0 \times 10^{-8}} = 11$

     (b)   at pH = 7.1, $[H^+] = 7.9 \times 10^{-8} \ M$; $\dfrac{[HCO_3^-]}{[H_2CO_3]} = 5.4$

17.30 $\dfrac{6.5 \text{ g NaH}_2\text{PO}_4}{0.355 \text{ L soln}} \times \dfrac{1 \text{ mol NaH}_2\text{PO}_4}{120 \text{ g NaH}_2\text{PO}_4} = 0.153 = 0.15 \ M$

$\dfrac{8.0 \text{ g Na}_2\text{HPO}_4}{0.355 \text{ L soln}} \times \dfrac{1 \text{ mol Na}_2\text{HPO}_4}{142 \text{ g Na}_2\text{HPO}_4} = 0.159 = 0.16 \ M$

Use Equation 17.9 to find the pH of the buffer. K_a for $H_2PO_4^-$ is K_{a2} for H_3PO_4, 6.2×10^{-8}

$$pH = -\log(6.2 \times 10^{-8}) + \log \frac{0.159}{0.153} = 7.2076 + 0.0167 = 7.22$$

17.31   *Analyze.* Given six solutions, decide which two should be used to prepare a pH 3.50 buffer. Calculate the volumes of the two 0.10 M solutions needed to make approximately 1 L of buffer.

Plan. A buffer must contain a conjugate acid/conjugate base (CA/CB) pair. By examining the chemical formulas, decide which pairs of solutions could be used to make a buffer. If there is more than one possible pair, calculate pK_a for the acids. A buffer is most effective when its pH is within 1 pH unit of pK_a for the conjugate acid component. Select the pair with pK_a nearest to 3.50. Use Equation 17.9 to calculate the [CB]/[CA] ratio and the volumes of 0.10 M solutions needed to prepare 1 L of buffer.   *Solve.*

There are three CA/CB pairs:

HCOOH/HCOONa, $pK_a = 3.74$

CH_3COOH/CH_3COONa, $pK_a = 4.74$

H_3PO_4/NaH_2PO_4, $pK_a = 2.12$

The most appropriate solutions are HCOOH/HCOONa, because pK_a for HCOOH is nearest to 3.50.

$$pH = pK_a + \log\frac{[CB]}{[CA]}; \quad 3.50 = 3.7447 + \log\frac{[HCOONa]}{[HCOOH]}$$

$$\log\frac{[HCOONa]}{[HCOOH]} = -0.2447; \quad \frac{[HCOONa]}{[HCOOH]} = 0.5692 = 0.57$$

Because we are making a total of 1 L of buffer,

let y = vol HCOONa and (1 − y) = vol HCOOH.

$$0.5692 = \frac{[HCOONa]}{[HCOOH]} = \frac{(0.10\,M \times y)/1\,L}{[0.10\,M \times (1-y)]/1\,L}; \quad 0.5692[0.10(1-y)] = 0.10\,y;$$

$0.05692 = 0.15692\,y; \quad y = 0.3627 = 0.36$ L

360 mL of 0.10 M HCOONa, 640 mL of 0.10 M HCOOH

Check. The pH of the buffer is less than pK_a for the conjugate acid, indicating that the amount of CA in the buffer is greater than the amount of CB. This agrees with our result.

17.32   The solutes listed contain three possible conjugate acid/conjugate base (CA/CB) pairs. These are:

HCOOH/HCOONa, $pK_a = 3.74$

CH_3COOH/CH_3COONa, $pK_a = 4.74$

HCN/NaCN, $pK_a = 9.31$

For maximum buffer capacity, pK_a should be within 1 pH unit of the buffer. The acetic acid/acetate pair is most appropriate for a buffer with pH 5.00.

$$pH = pK_a + \log\frac{[CB]}{[CA]}; 5.00 = 4.745 + \log\frac{[CH_3COONa]}{[CH_3COOH]}$$

$$\log\frac{[CH_3COONa]}{[CH_3COOH]} = 0.2553; \quad \frac{[CH_3COONa]}{[CH_3COOH]} = 1.800 = 1.8$$

Because we are making a total of 1 L of buffer,
let y = vol CH_3COONa and $(1 - y)$ = vol CH_3COOH.

$$1.800 = \frac{[CH_3COONa]}{[CH_3COOH]} = \frac{(0.10\,M \times y)/1.0\,L}{[0.10\,M \times (1-y)]/1.0\,L} = \frac{0.10\,y}{0.10 - 0.10\,y}$$

$1.800(0.10 - 0.10\,y) = 0.10\,y; \quad 0.1800 = 0.2800\,y; \quad y = 0.6429 = 0.64\,L$

640 mL of 0.10 M CH_3COONa, 360 mL of CH_3COOH

Check. pH (buffer) > pK_a (CA) and the calculated amount of CB in the buffer is greater than the amount of CA.

Acid–Base Titrations (Section 17.3)

17.33   (a)   Curve B. The initial pH is lower and the equivalence point region is steeper.

         (b)   pH at the approximate equivalence point of curve A = 8.0

                 pH at the approximate equivalence point of curve B = 7.0

         (c)   Volume of base required to reach the equivalence point depends only on moles of acid present; it is independent of acid strength. Because acid B requires 40 mL and acid A requires only 30 mL, more moles of acid B are being titrated. For equal volumes of A and B, the concentration of acid B is greater.

         (d) pK_a of the weak acid is approximately 4.5. In the titration of a weak acid, pH equals pK_a of the weak acid at the volume half-way to the equivalence point. On curve A, the equivalence point is at 30 mL, half-way is 15 mL, and the pH there is 4.5.

17.34   (a)   False. The quantity of base required to reach the equivalence point is the same in the two titrations, assuming both acids have the same initial concentrations.

         (b)   False. The pH is higher initially in the titration of a weak acid.

         (c)   False. The pH is higher at the equivalence point in the titration of a weak acid.

17.35   (a)   False. The same volume of NaOH(aq) is required to reach the equivalence point of both titrations, because moles of acid to be titrated are the same in both flasks.

         (b)   True. CH_3COONa, the salt formed in the titration of CH_3COOH, produces a basic solution, whereas $NaNO_3$, formed in the titration of HNO_3, produces a neutral solution.

         (c)   True. Even though the pH values at the equivalence points of the two titrations are different, phenolphthalein changes color over a wide range of pH values and is appropriate for both titrations.

17.36   (a)   False. The pH at the beginning of the titration of the weaker acid, CH_3COOH, will be higher.

         (b)   True. Past the equivalence point, the titration curves are very similar (but not identical.

         (c)   False. According to Figures 17.12 and 17.13, methyl red is suitable for the titration of the strong acid HNO_3, but not for the titration of the weak acid CH_3COOH.

17.37   *Analyze.* Given reactants, predict whether pH at the equivalence point of a titration is less than, equal to, or greater than 7.

Plan. At the equivalence point of a titration, only product is present in solution; there is no excess of either reactant. Determine the product of each reaction and whether a solution of it is acidic, basic, or neutral.   *Solve.*

(a) $NaHCO_3(aq) + NaOH(aq) \rightarrow Na_2CO_3(aq) + H_2O(l)$

At the equivalence point, the major species in solution are Na^+ and CO_3^{2-}. Na^+ is negligible and CO_3^{2-} is the CB of HCO_3^-. The solution is basic, above pH 7.

(b) $NH_3(aq) + HCl(aq) \rightarrow NH_4Cl(aq)$

At the equivalence point, the major species are NH_4^+ and Cl^-. Cl^- is negligible and NH_4^+ is the CA of NH_3. The solution is acidic, below pH 7.

(c) $KOH(aq) + HBr(aq) \rightarrow KBr(aq) + H_2O(l)$

At the equivalence point, the major species are K^+ and Br^-; both are negligible. The solution is at pH 7.

17.38   (a) $HCOOH(aq) + NaOH(aq) \rightarrow HCOONa(aq) + H_2O(l)$

At the equivalence point, the major species are Na^+ and $HCOO^-$. Na^+ is negligible and $HCOO^-$ is the CB of $HCOOH$. The solution is basic, above pH 7.

(b) $Ca(OH)_2(aq) + 2\,HClO_4(aq) \rightarrow Ca(ClO_4)_2(aq) + 2\,H_2O(l)$

At the equivalence point, the major species are Ca^{2+} and ClO_4^-; both are negligible. The solution is at pH 7.

(c) $C_5H_5N(aq) + HNO_3(aq) \rightarrow C_5H_5NH^+NO_3^-(aq)$

At the equivalence point, the major species are $C_5H_5NH^+$ and NO_3^-. NO_3^- is negligible and $C_5H_5NH^+$ is the CA of C_5H_5N. The solution is acidic, below pH 7.

17.39   The second color change, from yellow to blue near pH = 8.5, is more suitable for the titration of a weak acid with a strong base. The salt present at the equivalence point of this type of titration produces a slightly basic solution. The second color change of thymol blue is in the correct pH range to show (indicate) the equivalence point.

17.40   (a)   At the equivalence point, moles HA added = moles B initially present = $0.10\,M \times 0.0300\,L = 0.0030$ moles HA added.

(b) $BH^+(aq)$.

(c)   Less than 7. The predominant form at equivalence, BH^+, is a weak acid.

(d)   Because the pH at the equivalence point will be less than 7, methyl red would be more appropriate.

17.41   *Analyze/Plan.* We are asked to calculate the volume of 0.0850 *M* NaOH required to titrate various acid solutions to their equivalence point. At the equivalence point, moles base added equals moles acid initially present. Solve the stoichiometry problem, recalling that mol = *M* × L. In part (c), calculate molarity of HCl from g/L and proceed as outlined earlier.   *Solve.*

(a) $40.0 \text{ mL HNO}_3 \times \dfrac{0.0900 \text{ mol HNO}_3}{1000 \text{ mL soln}} \times \dfrac{1 \text{ mol NaOH}}{1 \text{ mol HNO}_3} \times \dfrac{1000 \text{ mL soln}}{0.0850 \text{ mol NaOH}}$

$= 42.353 = 42.4 \text{ mL NaOH soln}$

(b) $35.0 \text{ mL CH}_3\text{COOH} \times \dfrac{0.0850 \, M \text{ CH}_3\text{COOH}}{1000 \text{ mL soln}} \times \dfrac{1 \text{ mol NaOH}}{1 \text{ mol CH}_3\text{COOH}} \times \dfrac{1000 \text{ mL soln}}{0.0850 \text{ mol NaOH}}$

$= 35.0 \text{ mL NaOH soln}$

(c) $\dfrac{1.85 \text{ g HCl}}{1 \text{ L soln}} \times \dfrac{1 \text{ mol HCl}}{36.46 \text{ g HCl}} = 0.05074 = 0.0507 \, M \text{ HCl}$

$50.0 \text{ mL HCl} \times \dfrac{0.05074 \text{ mol HCl}}{1000 \text{ mL}} \times \dfrac{1 \text{ mol NaOH}}{1 \text{ mol HCl}} \times \dfrac{1000 \text{ mL soln}}{0.0850 \text{ mol NaOH}}$

$= 29.847 = 29.8 \text{ mL NaOH soln}$

17.42 (a) $45.0 \text{ mL NaOH} \times \dfrac{0.0950 \text{ mol NaOH}}{1000 \text{ mL soln}} \times \dfrac{1 \text{ mol HCl}}{1 \text{ mol NaOH}} \times \dfrac{1000 \text{ mL soln}}{0.105 \text{ mol HCl}}$

$= 40.7 \text{ mL HCl soln}$

(b) $22.5 \text{ mL NH}_3 \times \dfrac{0.118 \text{ mol NH}_3}{1000 \text{ mL soln}} \times \dfrac{1 \text{ mol HCl}}{1 \text{ mol NH}_3} \times \dfrac{1000 \text{ mL soln}}{0.105 \text{ mol HCl}}$

$= 25.3 \text{ mL HCl soln}$

(c) $125.0 \text{ mL} \times \dfrac{1.35 \text{ g NaOH}}{1000 \text{ mL}} \times \dfrac{1 \text{ mol NaOH}}{40.00 \text{ g NaOH}} \times \dfrac{1 \text{ mol HCl}}{1 \text{ mol NaOH}} \times \dfrac{1000 \text{ mL soln}}{0.105 \text{ mol HCl}}$

$= 40.2 \text{ mL HCl soln}$

17.43 *Analyze/Plan.* Follow the logic in Sample Exercise 17.7 for the titration of a strong acid with a strong base. *Solve.*

moles $H^+ = M_{HBr} \times L_{HBr} = 0.200 \, M \times 0.0200 \text{ L} = 4.00 \times 10^{-3} \text{ mol}$

moles $OH^- = M_{NaOH} \times L_{NaOH} = 0.200 \, M \times L_{NaOH}$

	mL_{HBr}	mL_{NaOH}	Total Volume	Moles H^+	Moles OH^-	Molarity Excess Ion	pH
(a)	20.0	15.0	35.0	4.00×10^{-3}	3.00×10^{-3}	$0.0286(H^+)$	1.544
(b)	20.0	19.9	39.9	4.00×10^{-3}	3.98×10^{-3}	$5 \times 10^{-4}(H^+)$	3.3
(c)	20.0	20.0	40.0	4.00×10^{-3}	4.00×10^{-3}	$1 \times 10^{-7}(H^+)$	7.0
(d)	20.0	20.1	40.1	4.00×10^{-3}	4.02×10^{-3}	$5 \times 10^{-4}(OH^-)$	10.7
(e)	20.0	35.0	55.0	4.00×10^{-3}	7.00×10^{-3}	$0.0545(OH^-)$	12.737

molarity of excess ion = moles ion/total vol in L

(a) $\dfrac{4.00 \times 10^{-3} \text{ mol H}^+ - 3.00 \times 10^{-3} \text{ mol OH}^-}{0.0350 \text{ L}} = 0.0286 \, M \text{ H}^+$

(b) $\dfrac{4.00 \times 10^{-3} \text{ mol H}^+ - 3.98 \times 10^{-3} \text{ mol OH}^-}{0.0339 \text{ L}} = 5.01 \times 10^{-4} = 5 \times 10^{-4} \, M \text{ H}^+$

(c)   equivalence point, mol H^+ = mol OH^-

NaBr does not hydrolyze, so $[H^+] = [OH^-] = 1 \times 10^{-7}\ M$

(d) $\dfrac{4.02 \times 10^{-3}\ \text{mol OH}^- - 4.00 \times 10^{-3}\ \text{mol H}^+}{0.0401\ \text{L}} = 4.99 \times 10^{-4} = 5 \times 10^{-4}\ M\ \text{OH}^-$

(e) $\dfrac{7.00 \times 10^{-3}\ \text{mol OH}^- - 4.00 \times 10^{-3}\ \text{mol H}^+}{0.0550\ \text{L}} = 0.054545 = 0.0545\ M\ \text{OH}^-$

17.44   moles $OH^- = M_{\text{KOH}} \times L_{\text{KOH}} = 0.150\ M \times 0.0200\ L = 3.00 \times 10^{-3}\ \text{mol}$

moles $H^+ = M_{\text{HClO}_4} \times L_{\text{HClO}_4} = 0.125\ M \times L_{\text{HClO}_4}$

	mL_{KOH}	mL_{HClO4}	Total Volume	Moles OH^-	Moles H^+	Molarity Excess Ion	pH
(a)	20.0	20.0	40.0	3.00×10^{-3}	2.50×10^{-3}	$0.013(\text{OH}^-)$	12.10
(b)	20.0	23.0	43.0	3.00×10^{-3}	2.88×10^{-3}	$2.9 \times 10^{-3}(\text{OH}^-)$	11.46
(c)	20.0	24.0	44.0	3.00×10^{-3}	3.00×10^{-3}	$1.0 \times 10^{-7}(\text{OH}^-)$	7.00
(d)	20.0	25.0	45.0	3.00×10^{-3}	3.13×10^{-3}	$2.8 \times 10^{-3}(\text{H}^+)$	2.56
(e)	20.0	30.0	50.0	3.00×10^{-3}	3.75×10^{-3}	$0.015(\text{H}^+)$	1.82

molarity of excess ion $= \dfrac{\text{moles ion}}{\text{total vol in L}}$

(a) $\dfrac{3.00 \times 10^{-3}\ \text{mol OH}^- - 2.50 \times 10^{-3}\ \text{mol H}^+}{0.0400\ \text{L}} = 0.0125 = 0.013\ M\ \text{OH}^-$

(b) $\dfrac{3.00 \times 10^{-3}\ \text{mol OH}^- - 2.875 \times 10^{-3}\ \text{mol H}^+}{0.0430\ \text{L}} = 2.91 \times 10^{-3} = 2.9 \times 10^{-3}\ M\ \text{OH}^-$

(c)   equivalence point, mol H^+ = mol OH^-

$KClO_4$ does not hydrolyze, so $[H^+] = [OH^-] = 1 \times 10^{-7}\ M$

(d) $\dfrac{3.125 \times 10^{-3}\ \text{mol H}^+ - 3.00 \times 10^{-3}\ \text{mol OH}^-}{0.0450\ \text{L}} = 2.78 \times 10^{-3} = 2.8 \times 10^{-3}\ M\ \text{H}^+$

(e) $\dfrac{3.75 \times 10^{-3}\ \text{mol H}^+ - 3.00 \times 10^{-3}\ \text{mol OH}^-}{0.0500\ \text{L}} = 0.0150 = 0.015\ M\ \text{H}^+$

17.45   *Analyze/Plan.* Follow the logic in Sample Exercise 17.8 for the titration of a weak acid with a strong base.   *Solve.*

(a)   At 0 mL, only weak acid, CH_3COOH, is present in solution. Using the acid ionization equilibrium

$$CH_3COOH(aq) \rightleftharpoons H^+(aq) + CH_3COO^-(aq)$$

initial	0.150 M	0	0
equil.	0.150 − x M	x M	x M

$$K_a = \frac{[H^+][CH_3COO^-]}{[CH_3COOH]} = 1.8 \times 10^{-5} \text{ (Appendix D)}$$

$$1.8 \times 10^{-5} = \frac{x^2}{(0.150 - x)} \approx \frac{x^2}{0.150}; \ x^2 = 2.7 \times 10^{-6}; \ x = [H^+] = 0.001643$$

$$= 1.6 \times 10^{-3} \ M; \ pH = 2.78$$

(b–f) Calculate the moles of each component after the acid–base reaction takes place.

Moles CH_3COOH originally present $= M \times L = 0.150 \ M \times 0.0350 \ L = 5.25 \times 10^{-3}$ mol.

Moles NaOH added $= M \times L = 0.150 \ M \times y$ mL.

		NaOH(aq)	+	CH_3COOH (aq) →	$CH_3COONa(aq) + H_2O(l)$
		(0.150 $M \times$ 0.0175 L) =			
(b)	before rx	2.625×10^{-3} mol		5.25×10^{-3} mol	
	after rx	**0**		2.625×10^{-3} mol	2.63×10^{-3} mol
		(0.150 $M \times$ 0.0345 L) =			
(c)	before rx	5.175×10^{-3} mol		5.25×10^{-3} mol	
	after rx	**0**		0.075×10^{-3} mol	5.18×10^{-3} mol
		(0.150 $M \times$ 0.0350 L) =			
(d)	before rx	5.25×10^{-3} mol		5.25×10^{-3} mol	
	after rx	**0**		**0**	5.25×10^{-3} mol
		(0.150 $M \times$ 0.0355 L) =			
(e)	before rx	5.325×10^{-3} mol		5.25×10^{-3} mol	
	after rx	0.075×10^{-3} mol		**0**	5.25×10^{-3} mol
		(0.150 $M \times$ 0.0500 L) =			
(f)	before rx	7.50×10^{-3} mol		5.25×10^{-3} mol	
	after rx	2.25×10^{-3} mol		**0**	5.25×10^{-3} mol

Calculate the molarity of each species (M = mol/L) and solve the appropriate equilibrium problem in each part.

(b) V_T = 35.0 mL CH_3COOH + 17.5 mL NaOH = 52.5 mL = 0.0525 L

$$[CH_3COOH] = \frac{2.625 \times 10^{-3} \text{ mol}}{0.0525} = 0.0500 \ M$$

$$[CH_3COO^-] = \frac{2.625 \times 10^{-3} \text{ mol}}{0.0525} = 0.0500 \ M$$

$$CH_3COOH(aq) \rightleftharpoons H^+(aq) + CH_3COO^-(aq)$$

equil. $0.0500 - x \ M$ $x \ M$ $0.0500 + x \ M$

$$K_a = \frac{[H^+][CH_3COO^-]}{[CH_3COOH]}; \ [H^+] = \frac{K_a[CH_3COOH]}{[CH_3COO^-]}$$

$$[H^+] = \frac{1.8 \times 10^{-5}(0.0500 - x)}{(0.0500 + x)} = 1.8 \times 10^{-5} \ M \ H^+; \ pH = 4.74$$

(c) $[CH_3COOH] = \dfrac{7.5 \times 10^{-5}\,mol}{0.0695\,L} = 0.001079 = 1.1 \times 10^{-3}\ M$

$[CH_3COO^-] = \dfrac{5.175 \times 10^{-3}\,mol}{0.0695\,L} = 0.07446 = 0.074\ M$

$[H^+] = \dfrac{1.8 \times 10^{-5}\,(1.079 \times 10^{-3} - x)}{(0.07446 + x)} \approx 2.6 \times 10^{-7}\ M\ H^+;\ \ pH = 6.58$

(d) At the equivalence point, only CH_3COO^- is present.

$[CH_3COO^-] = \dfrac{5.25 \times 10^{-3}\,mol}{0.0700\,L} = 0.0750\ M$

The pertinent equilibrium is the base hydrolysis of CH_3COO^-.

$$CH_3COO^-(aq) + H_2O(l) \ \rightleftharpoons\ CH_3COOH(aq)\ +\ OH^-(aq)$$

initial	0.0750 M	0	0
equil.	0.0750 – x M	x	x

$K_b = \dfrac{K_w}{K_a\ for\ CH_3COOH} = \dfrac{1.0 \times 10^{-14}}{1.8 \times 10^{-5}} = 5.56 \times 10^{-10} = 5.6 \times 10^{-10} = \dfrac{[CH_3COOH][OH^-]}{[CH_3COO^-]}$

$5.56 \times 10^{-10} = \dfrac{x^2}{0.0750 - x};\ \ x^2 \approx 5.56 \times 10^{-10}(0.0750);\ \ x = 6.458 \times 10^{-6}$

$= 6.5 \times 10^{-6}\ M\ OH^-$

$pOH = -\log(6.458 \times 10^{-6}) = 5.19;\ pH = 14.00 - pOH = 8.81$

(e) After the equivalence point, the excess strong base determines the pOH and pH. The $[OH^-]$ from the hydrolysis of CH_3COO^- is small and can be ignored.

$[OH^-] = \dfrac{0.075 \times 10^{-3}\,mol}{0.0705\,L} = 1.064 \times 10^{-3} = 1.1 \times 10^{-3}\ M;\ \ pOH = 2.97$

$$pH = 14.00 - 2.97 = 11.03$$

(f) $[OH^-] = \dfrac{2.25 \times 10^{-3}\,mol}{0.0850\,L} = 0.0265\ M\ OH^-;\ \ pOH = 1.577;\ \ pH = 14.00 - 1.577 = 12.423$

17.46 (a) Weak base problem: $K_b = 1.8 \times 10^{-5} = \dfrac{[NH_4^+][OH^-]}{[NH_3]}$

At equilibrium, $[OH^-] = x,\ [NH_3] = (0.030 - x);\ [NH_4^+] = x$

$1.8 \times 10^{-5} = \dfrac{x^2}{(0.050 - x)} \approx \dfrac{x^2}{0.050};\ \ x = [OH^-] = 9.487 \times 10^{-4} = 9.5 \times 10^{-4}\ M$

$pH = 14.00 - 3.02 = 10.98$

(b–f) Calculate mol NH_3 and mol NH_4^+ after the acid–base reaction takes place.
$0.050\ M\ NH_3 \times 0.0300\ L = 1.5 \times 10^{-3}$ mol NH_3 present initially.

$$NH_3(aq) \quad + \quad HCl(aq) \quad \rightarrow \quad NH_4^+(aq) + Cl^-(aq)$$

$(0.025\ M \times 0.0200\ L) =$

(b) before rx 1.5×10^{-3} mol 0.50×10^{-3} mol   0 mol

after rx   **1.0×10^{-3} mol**   **0 mol**   **5.0×10^{-4} mol**

$(0.025\ M \times 0.0590\ L) =$

(c) before rx 1.5×10^{-3} mol 1.475×10^{-3} mol   0 mol

after rx   **2.5×10^{-5} mol**   **0 mol**   **1.475×10^{-3} mol**

$(0.025\ M \times 0.0600\ L) =$

(d) before rx 1.5×10^{-3} mol 1.5×10^{-3} mol   0 mol

after rx   **0 mol**   **0 mol**   **1.5×10^{-3} mol**

$(0.025\ M \times 0.0610\ L) =$

(e) before rx 1.5×10^{-3} mol 1.525×10^{-3} mol   0 mol

after rx   **0 mol**   **2.5×10^{-5} mol**   **1.5×10^{-3} mol**

$(0.025\ M \times 0.0650\ L) =$

(f) before rx 1.5×10^{-3} mol 1.625×10^{-3} mol   0 mol

after rx   **0 mol**   **1.25×10^{-4} mol**   **1.5×10^{-3} mol**

(b) Using the acid dissociation equilibrium for NH_4^+ (so that we calculate $[H^+]$ directly), $NH_4^+(aq) \rightleftharpoons H^+(aq) + NH_3(aq)$

$$K_a = \frac{[H^+][NH_3]}{[NH_4^+]} = \frac{K_w}{K_b\ \text{for}\ NH_3} = \frac{1.0 \times 10^{-14}}{1.8 \times 10^{-5}} = 5.56 \times 10^{-10} = 5.6 \times 10^{-10}$$

$$[NH_3] = \frac{1.0 \times 10^{-3}\ \text{mol}}{0.0500\ L} = 0.020\ M; [NH_4^+] = \frac{5.0 \times 10^{-4}\ \text{mol}}{0.0500\ L} = 0.010\ M$$

$$[H^+] = \frac{5.56 \times 10^{-10}\ [NH_4^+]}{[NH_3]} \approx \frac{5.56 \times 10^{-10}\ (0.010)}{(0.020)} = 2.78 \times 10^{-10};\ \ pH = 9.56$$

(We will assume $[H^+]$ is small compared to $[NH_3]$ and $[NH_4^+]$.)

(c) $[NH_3] = \dfrac{2.5 \times 10^{-5}\ \text{mol}}{0.0890\ L} = 2.8 \times 10^{-4}\ M; [NH_4^+] = \dfrac{1.475 \times 10^{-3}\ \text{mol}}{0.0890\ L} = 0.017\ M$

$$[H^+] = \frac{5.56 \times 10^{-10}\ (0.017)}{(2.8 \times 10^{-4})} = 3.38 \times 10^{-8} = 3.4 \times 10^{-8}\ M;\ pH = 7.47$$

(d) At the equivalence point, $[H^+] = [NH_3] = x$

$$[NH_4^+] = \frac{1.5 \times 10^{-3} \, M}{0.0900 \, L} = 0.01667 = 0.017 \, M$$

$$5.56 \times 10^{-10} = \frac{x^2}{0.01667}; \quad x = [H^+] = 3.043 \times 10^{-6} = 3.0 \times 10^{-6} \, M; \quad pH = 5.52$$

(e) Past the equivalence point, $[H^+]$ from the excess HCl determines the pH.

$$[H^+] = \frac{2.5 \times 10^{-5} \, mol}{0.0910 \, L} = 2.747 \times 10^{-4} = 2.7 \times 10^{-4} \, M; \quad pH = 3.56$$

(f) Past the equivalence point, $[H^+]$ from the excess HCl determines the pH.

$$[H^+] = \frac{1.25 \times 10^{-4} \, mol}{0.0950 \, L} = 1.316 \times 10^{-3} = 1.3 \times 10^{-3} \, M; \quad pH = 2.88$$

17.47 *Analyze/Plan.* Calculate the pH at the equivalence point for the titration of several bases with 0.200 M HBr. The volume of 0.200 M HBr required in all cases equals the volume of base and the final volume = $2V_{base}$. The concentration of the salt produced at the

equivalence point is $\dfrac{0.200 \, M \times V_{base}}{2 \, V_{base}} = 0.100 \, M.$

In each case, identify the salt present at the equivalence point, determine its acid–base properties (Section 16.9), and solve the pH problem. *Solve.*

(a) NaOH is a strong base; the salt present at the equivalence point, NaBr, does not affect the pH of the solution. 0.100 M NaBr, pH = 7.00.

(b) $HONH_2$ is a weak base, so the salt present at the equivalence point is $HONH_3^+Br^-$. This is the salt of a strong acid and a weak base, so it produces an acidic solution.

0.100 M $HONH_3^+Br^-$; $\qquad\qquad HONH_3^+(aq) \rightleftharpoons H^+(aq) + HONH_2$

$\qquad\qquad$ [equil.] $\quad$ 0.100 – x $\qquad\qquad$ x $\qquad\qquad$ x

$$K_a = \frac{[H^+][HONH_2]}{[HONH_3^+]} = \frac{K_w}{K_b} = \frac{1.0 \times 10^{-14}}{1.1 \times 10^{-8}} = 9.09 \times 10^{-7} = 9.1 \times 10^{-7}$$

Assume x is small with respect to [salt].

$K_a = x^2/0.100; \; x = [H^+] = 3.02 \times 10^{-4} = 3.0 \times 10^{-4} \, M, \, pH = 3.52$

(c) $C_6H_5NH_2$ is a weak base and $C_6H_5NH_3^+Br^-$ is an acidic salt.

0.100 M $C_6H_5NH_3^+Br^-$. $\quad$ Proceeding as in (b):

$$K_a = \frac{[H^+][C_6H_5NH_2]}{[C_6H_5NH_3^+]} = \frac{K_w}{K_b} = 2.33 \times 10^{-5} = 2.3 \times 10^{-5}$$

$[H^+]^2 = 0.100(2.33 \times 10^{-5}); [H^+] = 1.52 \times 10^{-3} = 1.5 \times 10^{-3} \, M, \, pH = 2.82$

17.48   The volume of NaOH solution required in all cases is

$$V_{base} = \frac{V_{acid} \times M_{acid}}{M_{base}} = \frac{(0.100)\,V_{acid}}{(0.080)} = 1.25\,V_{acid}$$

The total volume at the equivalence point is $V_{base} + V_{acid} = 2.25\,V_{acid}$

The concentration of the salt at the equivalence point is $\dfrac{M_{acid}\,V_{acid}}{2.25\,V_{acid}} = \dfrac{0.100}{2.25} = 0.0444\,M$

(a) $0.0444\,M$ NaBr, pH $= 7.00$

(b) $0.0444\,M$ NaClO$_2$; $ClO_2^-(aq) + H_2O(l) \;\rightleftharpoons\; HClO_2(aq) + OH^-(aq)$

$$K_b = \frac{[HClO_2][OH^-]}{[ClO_2^-]} = \frac{K_w}{K_a} = \frac{1.0 \times 10^{-14}}{1.1 \times 10^{-2}} = 9.09 \times 10^{-13} = 9.1 \times 10^{-13}$$

$[HClO_2] = [OH^-]$; $[ClO_2^-] \approx 0.0444\,M$

$[OH^-]^2 \approx 0.0444(9.09 \times 10^{-13})$; $[OH^-] = 2.01 \times 10^{-7} = 2.0 \times 10^{-7}\,M$, pOH $= 6.70$;

pH $= 7.30$

Note that HClO$_2$ is a relatively strong acid (large K_a value), so the pH at the equivalence point is not much greater than 7.0. Because [OH$^-$] from the hydrolysis of ClO$_2^-$ is very small, the autoionization equilibrium should be considered for a more accurate value of the equivalence point pH.

Let $[H^+] = x$, $[OH^-] = (2.0 \times 10^{-7}\,M + x)$; $1.0 \times 10^{-14} = (x)(2.0 \times 10^{-7}\,M + x)$

Solving the quadratic equation gives a pH of 7.38.

(c) $C_6H_5COO^-(aq) + H_2O(l) \;\rightleftharpoons\; C_6H_5COOH(aq) + OH^-(aq)$

$$K_b = \frac{[C_6H_5COO^-][OH^-]}{[C_6H_5COOH]} = \frac{K_w}{K_a} = \frac{1.0 \times 10^{-14}}{6.3 \times 10^{-5}} = 1.59 \times 10^{-10} = 1.6 \times 10^{-10}$$

$[OH^-]^2 \approx 0.0444(1.59 \times 10^{-8})$; $[OH^-] = 2.655 \times 10^{-6} = 2.7 \times 10^{-6}\,M$, pH $= 8.42$

Solubility Equilibria and Factors Affecting Solubility
(Sections 17.4 and 17.5)

17.49   (a)   True.

(b)   False. The solubility product of a slightly soluble salt is the square of the solubility if the salt contains one cation and one anion.

(c)   False. The common-ion effect is in play for solubility equilibria as well as acid–base equilibria.

(d)   True. The common-ion effect does not change the equilibrium constant.

17.50   (a)   MZ$_2$ has the larger numerical value for the solubility product constant, $4s^2$ versus s^2.

(b)   The $[M^{2+}]$ is the same in the two saturated solutions, because the molar solubilities are the same and there is one mole of $[M^{2+}]$ in one mole of either salt.

(c) $[M^{2+}] = 4 \times 10^{-4}\,M$. After the addition, mol M^{2+} increase, but so does the total volume. The $[M^{2+}]$ remains the same.

17.51   *Analyze/Plan.* Follow the example in Sample Exercise 17.10.   *Solve.*

$K_{sp} = [Ag^+][I^-]$; $K_{sp} = [Sr^{2+}][SO_4^{2-}]$; $K_{sp} = [Fe^{2+}][OH^-]^2$; $K_{sp} = [Hg_2^{2+}][Br^-]^2$

17.52   (a)   False. Solubility is the amount (grams, moles) of solute that will dissolve in a certain volume of solution. Solubility-product constant is an equilibrium constant, the product of the molar concentrations of all the dissolved ions in solution.

(b) $K_{sp} = [Mn^{2+}][CO_3^{2-}]$; $K_{sp} = [Hg^{2+}][OH^-]^2$; $K_{sp} = [Cu^{2+}]^3[PO_4^{3-}]^2$

17.53   *Analyze/Plan.* Follow the logic in Sample Exercise 17.11.   *Solve.*

(a) $CaF_2(s) \rightleftharpoons Ca^{2+}(aq) + 2F^-(aq)$; $K_{sp} = [Ca^{2+}][F^-]^2$

The molar solubility is the moles of CaF_2 that dissolve per liter of solution. Each mole of CaF_2 produces **1** mol Ca^{2+}(aq) and **2** mol F^-(aq).

$[Ca^{2+}] = 1.24 \times 10^{-3}$ M; $[F^-] = 2 \times 1.24 \times 10^{-3}$ $M = 2.48 \times 10^{-3}$ M

$K_{sp} = (1.24 \times 10^{-3})(2.48 \times 10^{-3})^2 = 7.63 \times 10^{-9}$

(b) $SrF_2(s) \rightleftharpoons Sr^{2+}(aq) + 2F^-(aq)$; $K_{sp} = [Sr^{2+}][F^-]^2$

Transform the gram solubility to molar solubility.

$$\frac{1.1 \times 10^{-2} \text{g SrF}_2}{0.100 \text{ L}} \times \frac{1 \text{ mol SrF}_2}{125.6 \text{ g SrF}_2} = 8.76 \times 10^{-4} = 8.8 \times 10^{-4} \text{ mol SrF}_2 / \text{L}$$

$[Sr^{2+}] = 8.76 \times 10^{-4}$ M; $[F^-] = 2(8.76 \times 10^{-4}$ $M)$

$K_{sp} = (8.76 \times 10^{-4})(2(8.76 \times 10^{-4}))^2 = 2.7 \times 10^{-9}$

(c) $Ba(IO_3)_2(s) \rightleftharpoons Ba^{2+}(aq) + 2IO_3^-(aq)$; $K_{sp} = [Ba^{2+}][IO_3^-]^2$

Because 1 mole of dissolved $Ba(IO_3)_2$ produces 1 mole of Ba^{2+}, the molar solubility of $Ba(IO_3)_2 = [Ba^{2+}]$. Let $x = [Ba^{2+}]$; $[IO_3^-] = 2x$.

$K_{sp} = 6.0 \times 10^{-10} = (x)(2x)^2$; $4x^3 = 6.0 \times 10^{-10}$; $x^3 = 1.5 \times 10^{-10}$; $x = 5.3 \times 10^{-4}$ M

The molar solubility of $Ba(IO_3)_2$ is 5.3×10^{-4} mol/L.

17.54   (a) $PbBr_2(s) \rightleftharpoons Pb^{2+}(aq) + 2Br^-(aq)$

$K_{sp} = [Pb^{2+}][Br^-]^2$; $[Pb^{2+}] = 1.0 \times 10^{-2}$ M, $[Br^-] = 2.0 \times 10^{-2}$ M

$K_{sp} = (1.0 \times 10^{-2} M)(2.0 \times 10^{-2} M)^2 = 4.0 \times 10^{-6}$

(b) $AgIO_3(s) \rightleftharpoons Ag^+(aq) + IO_3^-(aq)$; $K_{sp} = [Ag^+][IO_3^-]$

$$[Ag^+] = [IO_3^-] = \frac{0.0490 \text{ g AgIO}_3}{1.00 \text{ L soln}} \times \frac{1 \text{ mol AgIO}_3}{282.8 \text{ g AgIO}_3} = 1.733 \times 10^{-4} = 1.73 \times 10^{-4} \text{ M}$$

$K_{sp} = (1.733 \times 10^{-4} M)(1.733 \times 10^{-4} M) = 3.00 \times 10^{-8}$

(c) $Ca(OH)_2(s) \rightleftharpoons Ca^{2+}(aq) + 2OH^-(aq)$; $K_{sp} = [Ca^{2+}][OH^-]^2$

$[Ca^{2+}] = x$, $[OH^-] = 2x$; $K_{sp} = 6.5 \times 10^{-6} = (x)(2x)^2$

$6.5 \times 10^{-6} = 4x^3$; $x = [Ca^{2+}] = 0.01176 = 0.012$ M; $[OH^-] = 0.02351 = 0.024$ M

$pH = 14 - pOH = 14 - 1.629 = 12.37$

17.55   *Analyze/Plan.* Given gram solubility of a compound, calculate K_{sp}. Write the dissociation equilibrium and K_{sp} expression. Change gram solubility to molarity of the individual ions, taking the stoichiometry of the compound into account. Calculate K_{sp}. *Solve.*

$$CaC_2O_4(s) \rightleftharpoons Ca^{2+}(aq) + C_2O_4^{2-}(aq); \quad K_{sp} = [Ca^{2+}][C_2O_4^{2-}]$$

$$[Ca^{2+}] = [C_2O_4^{2-}] = \frac{0.0061 \text{ g } CaC_2O_4}{1.00 \text{ L soln}} \times \frac{1 \text{ mol } CaC_2O_4}{128.1 \text{ g } CaC_2O_4} = 4.76 \times 10^{-5} = 4.8 \times 10^{-5} \, M$$

$$K_{sp} = (4.76 \times 10^{-5} \, M)(4.76 \times 10^{-5} \, M) = 2.3 \times 10^{-9}$$

17.56 $$PbI_2(s) \rightleftharpoons Pb^{2+}(aq) + 2\,I^-(aq); \quad K_{sp} = [Pb^{2+}][I^-]^2$$

$$[Pb^{2+}] = \frac{0.54 \text{ g } PbI_2}{1.00 \text{ L soln}} \times \frac{1 \text{ mol } PbI_2}{461.0 \text{ g } PbI_2} = 1.17 \times 10^{-3} = 1.2 \times 10^{-3} \, M$$

$$[I^-] = 2[Pb^{2+}]; \; K_{sp} = [Pb^{2+}](2[Pb^{2+}])^2 = 4[Pb^{2+}]^3 = 4(1.17 \times 10^{-3})^3 = 6.4 \times 10^{-9}$$

17.57   *Analyze/Plan.* Follow the logic in Sample Exercises 17.12 and 17.13.   *Solve.*

(a) $$AgBr(s) \rightleftharpoons Ag^+(aq) + Br^-(aq); \quad K_{sp} = [Ag^+][Br^-] = 5.0 \times 10^{-13}$$

molar solubility $= x = [Ag^+] = [Br^-]; K_{sp} = x^2$

$x = (5.0 \times 10^{-13})^{1/2}; x = 7.1 \times 10^{-7}$ mol AgBr/L

(b)   Molar solubility $= x = [Br^-]; [Ag^+] = 0.030 \, M + x$

$K_{sp} = (0.030 + x)(x) \approx 0.030(x)$

$5.0 \times 10^{-13} = 0.030(x); x = 1.7 \times 10^{-11}$ mol AgBr/L

(c)   Molar solubility $= x = [Ag^+]$

There are two sources of Br^-: NaBr(0.10 M) and AgBr(x M)

$K_{sp} = (x)(0.10 + x)$; assume x is small compared to 0.10 M.

$5.0 \times 10^{-13} = 0.10 \, (x); x \approx 5.0 \times 10^{-12}$ mol AgBr/L

17.58 $$LaF_3(s) \rightleftharpoons La^{3+}(aq) + 3\,F^-(aq); \quad K_{sp} = [La^{3+}][F^-]^3$$

(a)   molar solubility $= x = [La^{3+}]; [F^-] = 3x$

$K_{sp} = 2 \times 10^{-19} = (x)(3x)^3; 2 \times 10^{-19} = 27\,x^4; x = (7.41 \times 10^{-21})^{1/4}, x = 9.28 \times 10^{-6}$

$$= 9 \times 10^{-6} \, M \, La^{3+}$$

$$\frac{9.28 \times 10^{-6} \text{ mol } LaF_3}{1 \text{ L}} \times \frac{195.9 \text{ g } LaF_3}{1 \text{ mol}} = 1.82 \times 10^{-3} = 2 \times 10^{-3} \text{ g } LaF_3/L$$

(b)   molar solubility $= x = [La^{3+}]$

There are two sources of F^-: KF (0.010 M) and LaF_3 (3x M)

$K_{sp} = (x)(0.010 + 3x)^3$; assume x is small compared to 0.010 M.

$2 \times 10^{-19} = (0.010)^3 \, x; x = 2 \times 10^{-19}/1.0 \times 10^{-6} = 2 \times 10^{-13} \, M \, La^{3+}$

$$\frac{2 \times 10^{-13} \text{ mol } LaF_3}{1 \text{ L}} \times \frac{195.9 \text{ g } LaF_3}{1 \text{ mol}} = 3.92 \times 10^{-11} = 4 \times 10^{-11} \text{ g } LaF_3/L$$

(c) molar solubility = x, $[F^-] = 3x$, $[La^{3+}] = 0.050\ M + x$

$K_{sp} = (0.050 + x)(3x)^3$; assume x is small compared to $0.050\ M$.

$2 \times 10^{-19} = (0.050)(27\ x^3) = 1.35\ x^3$; $x = (1.48 \times 10^{-19})^{1/3} = 5.29 \times 10^{-7} = 5 \times 10^{-7}\ M$

$$\frac{5.29 \times 10^{-7}\ \text{mol}\ LaF_3}{1\ L} \times \frac{195.9\ \text{g}\ LaF_3}{1\ \text{mol}} = 1.04 \times 10^{-4} = 1 \times 10^{-4}\ \text{g}\ LaF_3/L$$

17.59 *Analyze/Plan.* Given a saturated solution of CaF_2 in contact with undissolved $CaF_2(s)$, consider the effect of adding $CaCl_2(s)$. The two salts have the Ca^{2+} ion in common.

Solve. As $CaCl_2$ is added, $[Ca^{2+}]$ increases, K_{sp} is exceeded, and additional CaF_2 precipitates until equilibrium is reestablished. At the new equilibrium position:

(a) Increase. The additional Ca^{2+} from $CaCl_2$ decreases the solubility of CaF_2.

(b) Increase. We have added $CaCl_2$, which contains Ca^{2+}.

(c) Decrease. After $CaCl_2$ is added, $CaF_2(s)$ precipitates, which decreases $[F^-]$.

17.60 As KI is added, $[I^-]$ increases, K_{sp} is exceeded, and additional PbI_2 precipitates until equilibrium is reestablished. At the new equilibrium position:

(a) Increase. The additional I^- from KI decreases the solubility of PbI_2.

(b) Decrease. After KI is added, additional $PbI_2(s)$ precipitates.

(c) Increase. A small amount of the additional I^- precipitates as $PbI_2(s)$, but most of it increases the concentration of I^- ions in solution.

17.61 *Analyze/Plan.* We are asked to calculate the solubility of a slightly soluble hydroxide salt at various pH values. This is a common ion problem; pH tells us not only $[H^+]$ but also $[OH^-]$, which is an ion common to the salt. Use pH to calculate $[OH^-]$ and then proceed as in Sample Exercise 17.13. *Solve.*

$Mn(OH)_2(s) \rightleftharpoons Mn^{2+}(aq) + 2\ OH^-(aq)$; $K_{sp} = 1.6 \times 10^{-13}$

Because $[OH^-]$ is set by the pH of the solution, the solubility of $Mn(OH)_2$ is just $[Mn^{2+}]$.

(a) pH = 7.0, pOH = 14 − pH = 7.0, $[OH^-] = 10^{-pOH} = 1.0 \times 10^{-7}\ M$

$K_{sp} = 1.6 \times 10^{-13} = [Mn^{2+}](1.0 \times 10^{-7})^2$; $[Mn^{2+}] = \dfrac{1.6 \times 10^{-13}}{1.0 \times 10^{-14}} = 16\ M$

$$\frac{16\ \text{mol}\ Mn(OH)_2}{1\ L} \times \frac{88.95\ \text{g}\ Mn(OH)_2}{1\ \text{mol}\ Mn(OH)_2} = 1423 = 1.4 \times 10^3\ \text{g}\ Mn(OH)_2/L$$

Check. Note that the solubility of $Mn(OH)_2$ in pure water is $3.6 \times 10^{-5}\ M$, and the pH of the resulting solution is 9.0. The relatively low pH of a solution buffered to pH 7.0 actually increases the solubility of $Mn(OH)_2$.

(b) pH = 9.5, pOH = 4.5, $[OH^-] = 3.16 \times 10^{-5} = 3.2 \times 10^{-5}\ M$

$K_{sp} = 1.6 \times 10^{-13} = [Mn^{2+}](3.16 \times 10^{-5})^2$; $[Mn^{2+}] = \dfrac{1.6 \times 10^{-13}}{1.0 \times 10^{-9}} = 1.6 \times 10^{-4}\ M$

$1.6 \times 10^{-4}\ M\ Mn(OH)_2 \times 88.95\ \text{g/mol} = 0.0142 = 0.014\ \text{g/L}$

(c)   pH = 11.8, pOH = 2.2, $[OH^-] = 6.31 \times 10^{-3} = 6.3 \times 10^{-3} \, M$

$$K_{sp} = 1.6 \times 10^{-13} = [Mn^{2+}](6.31 \times 10^{-3})^2; \quad [Mn^{2+}] = \frac{1.6 \times 10^{-13}}{3.98 \times 10^{-5}} = 4.0 \times 10^{-9} \, M$$

$$4.02 \times 10^{-9} \, M \, Mn(OH)_2 \times 88.95 \, g/mol = 3.575 \times 10^{-7} = 3.6 \times 10^{-7} \, g/L$$

17.62 $Ni(OH)_2(s) \rightleftharpoons Ni^{2+}(aq) + 2\,OH^-(aq); \quad K_{sp} = 6.0 \times 10^{-16}$

Because the $[OH^-]$ is set by the pH of the solution, the solubility of $Ni(OH)_2$ is just $[Ni^{2+}]$.

(a)   pH = 8.0, pOH = 14 − pH = 6.0, $[OH^-] = 10^{-pOH} = 1 \times 10^{-6} \, M$

$$K_{sp} = 6.0 \times 10^{-16} = [Ni^{2+}](1.0 \times 10^{-6})^2; [Ni^{2+}] = \frac{6.0 \times 10^{-16}}{1.0 \times 10^{-12}} = 6.0 \times 10^{-4} = 6 \times 10^{-4} \, M$$

(b)   pH = 10.0, pOH = 4.0, $[OH^-] = 1.0 \times 10^{-4} = 1 \times 10^{-4} \, M$

$$K_{sp} = 6.0 \times 10^{-16} = [Ni^{2+}][1.0 \times 10^{-4}]^2; [Ni^{2+}] = \frac{6.0 \times 10^{-16}}{1.0 \times 10^{-8}} = 6.0 \times 10^{-8} = 6 \times 10^{-8} \, M$$

(c)   pH = 12.0, pOH = 2.0, $[OH^-] = 1.0 \times 10^{-2} = 1 \times 10^{-2} \, M$

$$K_{sp} = 6.0 \times 10^{-16} = [Ni^{2+}][1.0 \times 10^{-2}]^2; [Ni^{2+}] = \frac{6.0 \times 10^{-16}}{1.0 \times 10^{-4}} = 6.0 \times 10^{-12} = 6 \times 10^{-12} \, M$$

17.63   *Analyze/Plan.* Follow the logic in Sample Exercise 17.14. *Solve.*

If the anion of the salt is the conjugate base of a weak acid, it will combine with H^+, reducing the concentration of the free anion in solution, thereby causing more salt to dissolve. More soluble in acid: (a) $ZnCO_3$, (b) ZnS, (d) AgCN, (e) $Ba_3(PO_4)_2$.

17.64   If the anion in the slightly soluble salt is the conjugate base of a strong acid, there will be no reaction.

(a) $MnS(s) + 2\,H^+(aq) \rightarrow H_2S(aq) + Mn^{2+}(aq)$

(b) $PbF_2(s) + 2\,H^+(aq) \rightarrow 2\,HF(aq) + Pb^{2+}(aq)$

(c) $AuCl_3(s) + H^+(aq) \rightarrow$ no reaction

(d) $Hg_2C_2O_4(s) + 2\,H^+(aq) \rightarrow H_2C_2O_4(aq) + Hg_2^{2+}(aq)$

(e) $CuBr(s) + H^+(aq) \rightarrow$ no reaction

17.65   *Analyze/Plan.* Follow the logic in Sample Exercise 17.15. *Solve.*

The formation equilibrium is

$$Ni^{2+}(aq) + 6NH_3(aq) \rightleftharpoons Ni(NH_3)_6^{2+}(aq) \quad K_f = \frac{[Ni(NH_3)_6^{2+}]}{[Ni^{2+}][NH_3]^6} = 1.2 \times 10^9$$

$$1.25 \, g \, NiCl_2 \times \frac{1 \, mol \, NiCl_2}{129.62 \, g \, NiCl_2} \times \frac{1}{0.1000 \, L} = 0.096436 = 0.0964 \, M \, NiCl_2$$

Assume that nearly all the Ni^{2+} is in the form $Ni(NH_3)_6^{2+}$.

$[Ni(NH_3)_6^{2+}] = 0.0964\ M$; $[Ni^{2+}] = x$; $[NH_3] = 0.20\ M$

$$1.2 \times 10^9 = \frac{(0.0964)}{x(0.20)^6};\ x = 1.26 \times 10^{-6} = 1.3 \times 10^{-6}\ M = [Ni^{2+}]$$

$[Ni(NH_3)_6^{2+}] = 0.0964\ M - 1.26 \times 10^{-6}\ M = 0.0964\ M$

Note that our assumption that most of the Ni^{2+} is present as $Ni(NH_3)_6^{2+}$ is true.

17.66 $NiC_2O_4(s) \rightleftharpoons Ni^{2+}(aq) + C_2O_4^{2-}(aq);\quad K_{sp} = [Ni^{2+}][C_2O_4^{2-}] = 4 \times 10^{-10}$

When the salt has just dissolved, $[C_2O_4^{2-}]$ will be 0.020 M. Thus, $[Ni^{2+}]$ must be less than 4×10^{-10} / 0.020 = 2×10^{-8} M. To achieve this low $[Ni^{2+}]$, we must complex the Ni^{2+} ion with NH_3: $Ni^{2+}(aq) + 6\ NH_3(aq) \rightleftharpoons Ni(NH_3)_6^{2+}(aq)$. Essentially all Ni(II) is in the form of the complex, so $[Ni(NH_3)_6^{2+}] = 0.020$. Find K_f for $Ni(NH_3)_6^{2+}$ in Table 17.1.

$$K_f = \frac{[Ni(NH_3)_6^{2+}]}{[Ni^{2+}][NH_3]^6} = \frac{(0.020)}{(2 \times 10^{-8})[NH_3]^6} = 1.2 \times 10^9;\ [NH_3]^6 = 8.33 \times 10^{-4};$$

$$[NH_3] = 0.307 = 0.3\ M$$

17.67   *Analyze/Plan.* Calculate the solubility of AgI in pure water according to the method in Sample Exercise 17.12. Obtain K_{eq} for the complexation reaction, making use of pertinent K_{sp} and K_f values from Appendix D.3 and Table 17.1. Write the dissociation equilibrium for AgI and the formation reaction for $Ag(CN)_2^-$. Use algebra to manipulate these equations and their associated equilibrium constants to obtain the desired reaction and its equilibrium constant. Finally, use this K_{eq} value to calculate the solubility of AgI in 0.100 M NaCN solution.   *Solve.*

(a) $AgI(s) \rightleftharpoons Ag^+(aq) + I^-(aq);\quad K_{sp} = [Ag^+][I^-] = 8.3 \times 10^{-17}$

molar solubility = $x = [Ag^+] = [I^-]$; $K_{sp} = x^2$

$x = (8.3 \times 10^{-17})^{1/2}$; $x = 9.1 \times 10^{-9}$ mol AgI/L

(b) $AgI(s) \rightleftharpoons Ag^+(aq) + I^-(aq)$

$Ag^+(aq) + 2\ CN^-(aq) \rightleftharpoons Ag(CN)_2^-(aq)$

$AgI(s) + 2\ CN^-(aq) \rightleftharpoons Ag(CN)_2^-(aq) + I^-(aq)$

$$K = K_{sp} \times K_f = [Ag^+][I^-] \times \frac{[Ag(CN)_2^-]}{[Ag^+][CN^-]^2} = (8.3 \times 10^{-17})(1 \times 10^{21}) = 8 \times 10^4$$

(c)   K is much greater than one for the reaction of AgI(s) with CN^-. This means that the reaction goes to completion. For a AgI(s) in 0.100 M NaCN solution, CN^- is the limiting reactant. Two moles of CN^- react with one mole of AgI, so the solublility of AgI in 0.100 M NaCN is (0.100/2) = 0.0500 M.

17.68 According to Appendix D.3, K_{sp} for $Ag_2S(s)$ is of the type

$$Ag_2S(s) + H_2O(l) \rightleftharpoons 2\,Ag^+(aq) + HS^-(aq) + OH^-(aq) \qquad K_{sp}$$

$$HS^-(aq) + H^+(aq) \rightleftharpoons H_2S(aq) \qquad 1/K_{a1}$$

$$2[Ag^+(aq) + 2\,Cl^-(aq) \rightleftharpoons AgCl_2^-(aq)] \qquad K_f^2$$

$$Ag_2S(s) + H_2O(l) + H^+(aq) + 4\,Cl^-(aq) \rightleftharpoons 2\,AgCl_2^-(aq) + H_2S(aq)$$

Add $H^+(aq)$ to each side to obtain the overall reacation

$$Ag_2S(s) + 2\,H^+(aq) + 4\,Cl^-(aq) \rightleftharpoons 2\,AgCl_2^-(aq) + H_2S(aq)$$

$$K = \frac{K_{sp} \times K_f^2}{K_{a1}} = \frac{(6 \times 10^{-51})(1.1 \times 10^5)^2}{(9.5 \times 10^{-8})} = 7.64 \times 10^{-34} = 8 \times 10^{-34}$$

Precipitation and Separation of Ions (Section 17.6)

17.69 *Analyze/Plan.* Follow the logic in Sample Exercise 17.16. Precipitation conditions: will Q (see Chapter 15) exceed K_{sp} for the compound? *Solve.*

    (a) In base, Ca^{2+} can form $Ca(OH)_2(s)$.

        $Ca(OH)_2(s) \rightleftharpoons Ca^{2+}(aq) + 2\,OH^-(aq)$; $K_{sp} = [Ca^{2+}][OH^-]^2$

        $Q = [Ca^{2+}][OH^-]^2$; $[Ca^{2+}] = 0.050\,M$; $pOH = 14 - 8.0 = 6.0$; $[OH^-] = 1.0 \times 10^{-6}\,M$

        $Q = (0.050)(1.0 \times 10^{-6})^2 = 5.0 \times 10^{-14}$; $K_{sp} = 6.5 \times 10^{-6}$ (Appendix D.3)

        $Q < K_{sp}$, no $Ca(OH)_2$ precipitates.

    (b) $Ag_2SO_4(s) \rightleftharpoons 2\,Ag^+(aq) + SO_4^{2-}(aq)$; $K_{sp} = [Ag+]^2[SO_4^{2-}]$

        $[Ag^+] = \dfrac{0.050\,M \times 100\,mL}{110\,mL} = 4.545 \times 10^{-2} = 4.5 \times 10^{-2}\,M$

        $[SO_4^{2-}] = \dfrac{0.050\,M \times 10\,mL}{110\,mL} = 4.545 \times 10^{-3} = 4.5 \times 10^{-3}\,M$

        $Q = (4.545 \times 10^{-2})^2 (4.545 \times 10^{-3}) = 9.4 \times 10^{-6}$; $K_{sp} = 1.5 \times 10^{-5}$

        $Q < K_{sp}$, no Ag_2SO_4 precipitates.

17.70 (a) $Co(OH)_2(s) \rightleftharpoons Co^{2+}(aq) + 2\,OH^-(aq)$; $K_{sp} = [Co^{2+}][OH^-]^2 = 1.3 \times 10^{-15}$

        $pH = 8.5$; $pOH = 14 - 8.5 = 5.5$; $[OH^-] = 10^{-5.5} = 3.16 \times 10^{-6} = 3 \times 10^{-6}\,M$

        $Q = (0.020)(3.16 \times 10^{-6})^2 = 2 \times 10^{-13}$; $Q > K_{sp}$, $Co(OH)_2$ will precipitate.

    (b) $AgIO_3(s) \rightleftharpoons Ag^+(aq) + IO_3^-(aq)$; $K_{sp} = [Ag+][IO_3^-] = 3.1 \times 10^{-8}$

        $[Ag^+] = \dfrac{0.010\,M\,Ag^+ \times 0.020\,L}{0.030\,L} = 6.667 \times 10^{-3} = 6.7 \times 10^{-3}\,M$

        $[IO_3^-] = \dfrac{0.015\,M\,IO_3^- \times 0.010\,L}{0.030\,L} = 5.000 \times 10^{-3} = 5.0 \times 10^{-3}\,M$

        $Q = (6.667 \times 10^{-3})(5.00 \times 10^{-3}) = 3.3 \times 10^{-5}$; $Q > K_{sp}$, $AgIO_3$ will precipitate.

17.71   *Analyze/Plan.* We are asked to calculate pH necessary to precipitate $Mn(OH)_2(s)$ if the resulting Mn^{2+} concentration is $\leq 1\ \mu g/L$.

$Mn(OH)_2(s) \rightleftharpoons Mn^{2+}(aq) + 2\ OH^-(aq); K_{sp} = [Mn^{2+}][OH^-]^2 = 1.6 \times 10^{-13}$

At equilibrium, $[Mn^{2+}][OH^-]^2 = 1.6 \times 10^{-13}$. Change concentration $Mn^{2+}(aq)$ to mol/L and solve for $[OH^-]$.   *Solve.*

$$\frac{1\ \mu g\ Mn^{2+}}{1\ L} \times \frac{1 \times 10^{-6}\ g}{1\ \mu g} \times \frac{1\ mol\ Mn^{2+}}{54.94\ g\ Mn^{2+}} = 1.82 \times 10^{-8} = 2 \times 10^{-8}\ M\ Mn^{2+}$$

$1.6 \times 10^{-13} = (1.82 \times 10^{-8})[OH^-]^2; [OH^-]^2 = 8.79 \times 10^{-6}; [OH^-] = 2.96 \times 10^{-3} = 3 \times 10^{-3}\ M$

$pOH = 2.53; pH = 14 - 2.53 = 11.47 = 11.5$

17.72 $PbI_2(s) \rightleftharpoons Pb^{2+}(aq) + 2\ I^-(aq);\ K_{sp} = [Pb^{2+}][I^-]^2 = 8.49 \times 10^{-9}$

(This K_{sp} value is taken from *CRC Handbook of Chemistry and Physics*, 74th edition.)

$$[Pb^{2+}] = \frac{0.10\ M \times 0.2\ mL}{10.2\ mL} = 1.96 \times 10^{-3} = 2 \times 10^{-3}\ M;$$

$$[I^-] = \left(\frac{8.49 \times 10^{-9}}{1.96 \times 10^{-3}\ M}\right)^{1/2} = 2.08 \times 10^{-3} = 2 \times 10^{-3}\ M$$

$$\frac{2.08 \times 10^{-3}\ mol\ I^-}{1\ L} \times \frac{126.90\ g\ I^-}{1\ mol\ I^-} \times 0.0102\ L = 2.69 \times 10^{-3}\ g\ I^- = 3 \times 10^{-3}\ g\ I^-$$

17.73   *Analyze/Plan.* We are asked which ion will precipitate first from a solution containing $Pb^{2+}(aq)$ and $Ag^+(aq)$ when $I^-(aq)$ is added. Follow the logic in Sample Exercise 17.17. Calculate $[I^-]$ needed to initiate precipitation of each ion. The cation that requires lower $[I^-]$ will precipitate first.   *Solve.*

$$Ag^+: K_{sp} = [Ag^+][I^-]; 8.3 \times 10^{-17} = (2.0 \times 10^{-4})[I^-]; [I^-] = \frac{8.3 \times 10^{-17}}{2.0 \times 10^{-4}} = 4.2 \times 10^{-13}\ M\ I^-$$

$$Pb^{2+}: K_{sp} = [Pb^{2+}][I^-]^2; 7.9 \times 10^{-9} = (1.5 \times 10^{-3})[I^-]^2;\ [I^-] = \left(\frac{7.9 \times 10^{-9}}{1.5 \times 10^{-3}}\right)^{1/2} = 2.3 \times 10^{-3}\ M\ I^-$$

AgI will precipitate first, at $[I^-] = 4.2 \times 10^{-13}\ M$.

17.74   (a)   Precipitation will begin when $Q = K_{sp}$.

$BaSO_4: K_{sp} = [Ba^{2+}][SO_4^{2-}] = 1.1 \times 10^{-10}$

$1.1 \times 10^{-10} = (0.010)[SO_4^{2-}]; [SO_4^{2-}] = 1.1 \times 10^{-8}\ M$

$SrSO_4: K_{sp} = [Sr^{2+}][SO_4^{2-}] = 3.2 \times 10^{-7}$

$3.2 \times 10^{-7} = (0.010)[SO_4^{2-}]; [SO_4^{2-}] = 3.2 \times 10^{-5}\ M$

The $[SO_4^{2-}]$ necessary to begin precipitation is the smaller of the two values, $1.1 \times 10^{-8}\ M\ SO_4^{2-}$.

(b) Ba^{2+} precipitates first, because it requires the smaller $[SO_4^{2-}]$.

(c) Sr^{2+} will begin to precipitate when $[SO_4^{2-}]$ in solution (not bound in $BaSO_4$) reaches $3.2 \times 10^{-5}\ M$.

17.75   *Analyze/Plan.* We are asked which ion will precipitate first when dilute $Ag^+(aq)$ is added to a solution containing $0.20\ M\ CrO_4^{2-}$, $0.10\ M\ CO_3^{2-}$, and $0.10\ M\ Cl^-$. The anions are present at different concentrations and their silver compounds have different stoichiometry, so we cannot directly compare K_{sp} values. Follow the logic in Sample Exercise 17.17. Calculate $[Ag^+]$ needed to initiate precipitation of each ion. The anion that requires lowest $[Ag^+]$ will precipitate first, and so on.   *Solve.*

Ag_2CrO_4: $K_{sp} = [Ag^+]^2[CrO_4^{2-}] = 1.2 \times 10^{-12}$

$1.2 \times 10^{-12} = [Ag^+]^2(0.20)$; $[Ag^+]^2 = 6.0 \times 10^{-12}$; $[Ag^+] = 2.4 \times 10^{-6}\ M$

Ag_2CO_3: $K_{sp} = [Ag^+]^2[CO_3^{2-}] = 8.1 \times 10^{-12}$

$8.1 \times 10^{-12} = [Ag^+]^2(0.10)$; $[Ag^+]^2 = 8.1 \times 10^{-11}$; $[Ag^+] = 9.0 \times 10^{-6}\ M$

$AgCl$: $K_{sp} = [Ag^+][Cl^-] = 1.8 \times 10^{-10}$

$1.8 \times 10^{-10} = [Ag^+](0.010)$; $[Ag^+] = 1.8 \times 10^{-8}$

AgCl requires the smallest $[Ag^+]$ for precipitation and it will precipitate first. The other two will precipitate almost simultaneously.

17.76   It is not appropriate to compare K_{sp} values directly, because the stoichiometries of the the two precipitates are different.

(a)   Precipitation will begin when $Q = K_{sp}$.

$CaSO_4$: $K_{sp} = [Ca^{2+}][SO_4^{2-}] = 2.4 \times 10^{-5}$

$2.4 \times 10^{-5} = (0.20)[SO_4^{2-}]$; $[SO_4^{2-}] = 1.2 \times 10^{-4}\ M$

Ag_2SO_4: $K_{sp} = [Ag^+]^2[SO_4^{2-}] = 1.5 \times 10^{-5}$

$1.5 \times 10^{-5} = (0.30)^2[SO_4^{2-}]$; $[SO_4^{2-}] = 1.7 \times 10^{-4}\ M$

$CaSO_4$ requires the smaller $[SO_4^{2-}]$ for precipitation and it will precipitate first.

(b)   The $[SO_4^{2-}]$ necessary to begin precipitation is the smaller of the two values, $1.2 \times 10^{-4}\ M\ SO_4^{2-}$.

$1.2 \times 10^{-4}\ M = \dfrac{1.0\ M\ SO_4^{2-} \times x\ L}{(0.010 + x\ L)}$; $x = (0.010)1.2 \times 10^{-4} = 1.2 \times 10^{-6}\ L$.

We assume x is small compared to 0.010 L. The required volume is then $1.2 \times 10^{-6}\ L$ or 0.0012 mL or 1.2 μL. If one drop is approximately 0.2 mL, precipitation will begin as the first drop of $1.0\ M\ Na_2SO_4$ solution is added.

Qualitative Analysis for Metallic Elements (Section 17.7)

17.77   *Analyze/Plan.* Use Figure 17.23 and the description of the five qualitative analysis "groups" in Section 17.7 to analyze the given data. Ag^+ is in Group 1, Al^{3+} is in Group 3, Mg^{2+} is in Group 4, and Na^+ is in Group 5.   *Solve.*

The first two experiments eliminate Group 1 and 2 ions (Figure 17.23). The presence of a precipitate after the third experiment means that a Group 3 cation is present, in this case Al^{3+}. The fact that no insoluble phosphates form in the filtrate from the third experiment rules out Group 4 ions. Ag^+ (Group 1) and Mg^{2+} (Group 4) are definitely absent. Al^{3+} (Group 3) is definitely present and Na^+ (Group 5) is possibly present.

17.78   Initial solubility in water rules out CdS and HgO. Formation of a precipitate on addition of HCl indicates the presence of $Pb(NO_3)_2$ (formation of $PbCl_2$). Formation of a precipitate on addition of H_2S at pH 1 probably indicates $Cd(NO_3)_2$ (formation of CdS). (This test can be misleading because enough Pb^{2+} can remain in solution after filtering $PbCl_2$ to lead to visible precipitation of PbS.) Absence of a precipitate on addition of H_2S at pH 8 indicates that $ZnSO_4$ is not present. The yellow flame test indicates presence of Na^+. In summary, $Pb(NO_3)_2$ and Na_2SO_4 are definitely present, $Cd(NO_3)_2$ is probably present, and CdS, HgO, and $ZnSO_4$ are definitely absent.

17.79   *Analyze/Plan.* We are asked to devise a procedure to separate various pairs of ions in aqueous solutions. In each case, refer to Figure 17.23 to find a set of conditions where the solubility of the two ions differs. Construct a procedure to generate these conditions.   *Solve.*

(a) Cd^{2+} is in Gp. 2, but Zn^{2+} is not. Make the solution acidic using 0.2 *M* HCl; saturate with H_2S. CdS will precipitate, but ZnS will not.

(b) $Cr(OH)_3$ is amphoteric, but $Fe(OH)_3$ is not. Add excess base; $Fe(OH)_3(s)$ precipitates, but Cr^{3+} forms the soluble complex $Cr(OH)_4^-$.

(c) Mg^{2+} is a member of Gp. 4, but K^+ is not. Add $(NH_4)_2HPO_4$ to a basic solution; Mg^{2+} precipitates as $MgNH_4PO_4$, but K^+ remains in solution.

(d) Ag^+ is a member of Gp. 1, but Mn^{2+} is not. Add 6 *M* HCl; precipitate Ag^+ as AgCl(s); Mn^{2+} remains soluble.

17.80   (a)   Make the solution slightly acidic and saturate with H_2S; CdS will precipitate, but Na^+ remains in solution.

(b)   Make the solution acidic and saturate with H_2S; CuS will precipitate, but Mg^{2+} remains in solution.

(c)   Add HCl; $PbCl_2$ precipitates. (It is best to carry out the reaction in an ice-water bath to reduce the solubility of $PbCl_2$.)

(d)   Add dilute HCl; AgCl precipitates, but Hg^{2+} remains in solution.

17.81   (a)   Because phosphoric acid is a weak acid, the concentration of free $PO_4^{3-}(aq)$ in an aqueous phosphate solution is low except in strongly basic media. In less basic media, the solubility product of the phosphates of interest is not exceeded.

(b) K_{sp} for those cations in Group 3 is much larger. Thus, to exceed K_{sp}, a higher $[S^{2-}]$ is required. This is achieved by making the solution more basic.

(c)   They should all redissolve in strongly acidic solution; for example, in 12 *M* HCl (the chlorides of all Group 3 metals are soluble).

17.82   The addition of $(NH_4)_2HPO_4$ could result in precipitation of salts from metal ions of the other groups. The $(NH_4)_2HPO_4$ will render the solution basic, so metal hydroxides as well as insoluble phosphates could form. It is essential to separate the metal ions of a group from other metal ions before carrying out the specific tests for that group.

Additional Exercises

17.83   *Analyze/Plan.*   Follow the approach for deriving the Henderson–Hasselbach (H–H) equation from the K_a expression shown in Section 17.2. Begin with a general K_b expression. *Solve.*

$$B(aq) + H_2O(l) \rightleftharpoons BH^+(aq) + OH^-(aq); \quad K_b = \frac{[BH^+][OH^-]}{[B]}$$

$pOH = -\log[OH^-]$; rearrange K_a to solve for $[OH^-]$

$$[OH^-] = \frac{K_b[B]}{[BH^+]}; \text{ take the } -\log \text{ of both sides}$$

$$-\log[OH^-] = -\log K_b + (-\log[B] - (-\log[BH^+]))$$

$$pOH = pK_b + \log[BH^+] - \log[B]$$

$$pOH = pK_b + \log\frac{[BH^+]}{[B]}$$

17.84 $H_2CO_3(aq) \quad \rightleftharpoons \quad H^+(aq) \quad + \quad HCO_3^-(aq) \qquad K_{a1} = 4.3 \times 10^{-7} \qquad pK_{a1} = 6.37$

   $HCO_3^-(aq) \quad \rightleftharpoons \quad H^+(aq) \quad + \quad CO_3^{2-}(aq) \qquad K_{a2} = 5.6 \times 10^{-11} \qquad pK_{a2} = 10.25$

Use the two equilibrium constant expressions and the total carbonate concentration to solve for the three concentrations.

$[H^+] = 10^{-5.60} = 2.512 \times 10^{-6} = 2.5 \times 10^{-6}\ M.$

$$K_{a1} = \frac{[H^+][HCO_3^-]}{[H_2CO_3]}, \ [H_2CO_3] = \frac{[H^+][HCO_3^-]}{K_{a1}}; \quad K_{a2} = \frac{[H^+][CO_3^{2-}]}{[HCO_3^-]}, \ [CO_3^{2-}] = \frac{K_{a2}[HCO_3^-]}{[H^+]}$$

$$[H_2CO_3] + [HCO_3^-] + [CO_3^{2-}] = 1.0 \times 10^{-5}$$

$$\frac{[H^+][HCO_3^-]}{K_{a1}} + [HCO_3^-] + \frac{K_{a2}[HCO_3^-]}{[H^+]} = 1.0 \times 10^{-5}$$

Multiply by $K_{a1}[H^+]$.

$$[H^+]^2[HCO_3^-] + K_{a1}[H^+][HCO_3^-] + K_{a1}K_{a2}[HCO_3^-] = (1.0 \times 10^{-5})K_{a1}[H^+]$$

$$[HCO_3^-]([H^+]^2 + K_{a1}[H^+] + K_{a1}K_{a2}) = (1.0 \times 10^{-5})K_{a1}[H^+]$$

$$[HCO_3^-] = \frac{(1.0 \times 10^{-5})K_{a1}[H^+]}{[H^+]^2 + K_{a1}[H^+] + K_{a1}K_{a2}} = \frac{(1.0 \times 10^{-5})(4.3 \times 10^{-7})(2.5 \times 10^{-6})}{(2.5 \times 10^{-6})^2 + (4.3 \times 10^{-7})(2.5 \times 10^{-6}) + (4.3 \times 10^{-7})(5.6 \times 10^{-11})}$$

$$[HCO_3^-] = 1.468 \times 10^{-6} = 1.5 \times 10^{-6}\ M$$

$$[H_2CO_3] = \frac{[H^+][HCO_3^-]}{K_{a1}} = \frac{2.5 \times 10^{-6}(1.5 \times 10^{-6})}{4.3 \times 10^{-7}} = 8.7 \times 10^{-6}\ M$$

$$[CO_3^{2-}] = \frac{K_{a2}[HCO_3^-]}{[H^+]} = \frac{5.6 \times 10^{-11}(1.5 \times 10^{-6})}{2.5 \times 10^{-6}} = 3.4 \times 10^{-11}\ M$$

Check.   First, pH of the raindrop (5.6) is less than pK_{a1} (6.37). We expect $[H_2CO_3]$ to be greater than $[HCO_3^-]$, and it is. Second, the calculated total carbon species concentration is $[H_2CO_3] + [HCO_3^-] + [CO_3^{2-}] = 8.7 \times 10^{-6} + 1.5 \times 10^{-6} + 3.4 \times 10^{-11} = 1.0 \times 10^{-5}\ M.$
The calculated results are self-consistent.

17.85 The equilibrium of interest is

$$HC_5H_3O_3(aq) \rightleftharpoons H^+(aq) + C_5H_3O_3^-(aq); K_a = 6.76 \times 10^{-4} = \frac{[H^+][C_5H_3O_3^-]}{[HC_5H_3O_3]}$$

Begin by calculating $[HC_5H_3O_3]$ and $[C_5H_3O_3^-]$ for each case.

(a) $\dfrac{25.0 \text{ g } HC_5H_3O_3}{0.250 \text{ L soln}} \times \dfrac{1 \text{ mol } HC_5H_3O_3}{112.1 \text{ g } HC_5H_3O_3} = 0.8921 = 0.892 \, M \, HC_5H_3O_3$

$\dfrac{30.0 \text{ g } NaC_5H_3O_3}{0.250 \text{ L soln}} \times \dfrac{1 \text{ mol } NaC_5H_3O_3}{134.1 \text{ g } NaC_5H_3O_3} = 0.8949 = 0.895 \, M \, C_5H_3O_3^-$

$[H^+] = \dfrac{K_a[HC_5H_3O_3]}{[C_5H_3O_3^-]} = \dfrac{6.76 \times 10^{-4}(0.8921-x)}{(0.8949+x)} \approx \dfrac{6.76 \times 10^{-4}(0.8921)}{(0.8949)}$

$[H^+] = 6.74 \times 10^{-4} \, M, \text{ pH} = 3.171$

(b) For dilution, $M_1V_1 = M_2V_2$

$[HC_5H_3O_3] = \dfrac{0.250 \, M \times 30.0 \text{ mL}}{125 \text{ mL}} = 0.0600 \, M$

$[C_5H_3O_3^-] = \dfrac{0.220 \, M \times 20.0 \text{ mL}}{125 \text{ mL}} = 0.0352 \, M$

$[H^+] \approx \dfrac{6.76 \times 10^{-4}(0.0600)}{0.0352} = 1.15 \times 10^{-3} \, M, \text{ pH} = 2.938$

(yes, $[H^+]$ is < 5% of 0.0352 M)

(c) $0.0850 \, M \times 0.500 \text{ L} = 0.0425 \text{ mol } HC_5H_3O_3$

$1.65 \, M \times 0.0500 \text{ L} = 0.0825 \text{ mol NaOH}$

	$HC_5H_3O_3(aq)$	+	NaOH(aq)	→	$NaC_5H_3O_3(aq) + H_2O(l)$
initial	0.0425 mol		0.0825 mol		
reaction	–0.0425 mol		–0.0425 mol		+0.0425 mol
after	0 mol		0.0400 mol		0.0425 mol

The strong base NaOH dominates the pH; the contribution of $C_5H_3O_3^-$ is negligible. This combination would be "after the equivalence point" of a titration. The total volume is 0.550 L.

$[OH^-] = \dfrac{0.0400 \text{ mol}}{0.550 \text{ L}} = 0.0727 \, M; \text{ pOH} = 1.138, \text{ pH} = 12.862$

17.86 $K_a = \dfrac{[H^+][In^-]}{[HIn]}$; at pH = 4.68, $[HIn] = [In^-]$; $[H^+] = K_a$; pH = pK_a = 4.68

17.87 (a) $HA(aq) + B(aq) \rightleftharpoons HB^+(aq) + A^-(aq)$ $K_{eq} = \dfrac{[HB^+][A^-]}{[HA][B]}$

(b) Note that the solution is slightly basic because B is a stronger base than HA is an acid. (Or, equivalently, that A^- is a stronger base than HB^+ is an acid.) Thus, a little of the A^- is used up in reaction: $A^-(aq) + H_2O(l) \rightleftharpoons HA(aq) + OH^-(aq)$. Because pH is not very far from neutral, it is reasonable to assume that the reaction in part (a) has gone far to the right, and that $[A^-] \approx [HB^+]$ and $[HA] \approx [B]$. Then,

$$K_a = \frac{[A^-][H^+]}{[HA]} = 8.0 \times 10^{-5}; \text{ when } pH = 9.2, [H^+] = 6.31 \times 10^{-10} = 6 \times 10^{-10} M$$

$$\frac{[A^-]}{[HA]} = 8.0 \times 10^{-5}/6.31 \times 10^{-10} = 1.268 \times 10^5 = 1 \times 10^5$$

From the earlier assumptions, $\dfrac{[A^-]}{[HA]} = \dfrac{[HB^+]}{[B]}$, so $K_{eq} \approx \dfrac{[A^-]^2}{[HA]^2} = 1.608 \times 10^{10} = 2 \times 10^{10}$

(c) K_b for the reaction $B(aq) + H_2O(l) \rightleftharpoons BH^+(aq) + OH^-(aq)$ can be calculated by noting that the equilibrium constant for the reaction in part (a) can be written as $K = K_a (HA) \times K_b (B) / K_w$. (You should prove this to yourself.) Then,

$$K_b(B) = \frac{K \times K_w}{K_a (HA)} = \frac{(1.608 \times 10^{10})(1.0 \times 10^{-14})}{8.0 \times 10^{-5}} = 2.010 = 2$$

K_b (B) is larger than K_a (HA), as it must be if the solution is basic.

17.88 (a) $K_a = \dfrac{[H^+][HCOO^-]}{[HCOOH]}; \quad [H^+] = \dfrac{K_a[HCOOH]}{[HCOO^-]}$

Buffer A: $[HCOOH] = [HCOO^-] = \dfrac{1.00 \text{ mol}}{1.00 \text{ L}} = 1.00 \, M$

$$[H^+] = \frac{1.8 \times 10^{-4} (1.00 \, M)}{(1.00 \, M)} = 1.8 \times 10^{-4} M, pH = 3.74$$

Buffer B: $[HCOOH] = [HCOO^-] = \dfrac{0.010 \text{ mol}}{1.00 \text{ L}} = 0.010 \, M$

$$[H^+] = \frac{1.8 \times 10^{-4} (0.010 \, M)}{(0.010 \, M)} = 1.8 \times 10^{-4} M, pH = 3.74$$

The pH values of the two buffers are equal because they both contain HCOOH and HCOONa and the $[HCOOH] / [HCOO^-]$ *ratio* is the same in both solutions.

(b) Buffer A has the greater capacity because it contains the greater absolute concentrations of HCOOH and HCOO$^-$.

(c) Buffer A:

	HCOO$^-$	+	HCl		$\rightarrow$	HCOOH	+	Cl$^-$
	1.00 mol		0.001 mol			1.00 mol		
	0.999 mol		0			1.001 mol		

$$[H^+] = \frac{1.8 \times 10^{-4}(1.001)}{(0.999)} = 1.8 \times 10^{-4} M, pH = 3.74$$

(In a buffer calculation, volumes cancel and we can substitute moles directly into the K_a expression.)

Buffer B:

	HCOO$^-$	+	HCl		$\rightarrow$	HCOOH	+	Cl$^-$
	0.010 mol		0.001 mol			0.010 mol		
	0.009 mol		0			0.011 mol		

$$[H^+] = \frac{1.8 \times 10^{-4} (0.011)}{(0.009)} = 2.2 \times 10^{-4} M, pH = 3.66$$

(d) Buffer A: $1.00\,M\,HCl \times 0.010\,L = 0.010\,mol\,H^+$ added

mol HCOOH $= 1.00 + 0.010 = 1.01$ mol

mol HCOO$^- = 1.00 - 0.010 = 0.99$ mol

$$[H^+] = \frac{1.8 \times 10^{-4}\,(1.01)}{(0.99)} = 1.8 \times 10^{-4}\,M,\ pH = 3.74$$

Buffer B: mol HCOOH $= 0.010 + 0.010 = 0.020$ mol $= 0.020\,M$

mol HCOO$^- = 0.010 - 0.010 = 0.000$ mol

The solution is no longer a buffer; the only source of HCOO$^-$ is the dissociation of HCOOH. Adding 10 mL of 1.00 M HCl exceeds the buffer capacity of this buffer.

$$K_a = \frac{[H^+][COO^-]}{[HCOOH]} = \frac{x^2}{(0.020 - x)\,M}$$

The extent of ionization is greater than 5%; from the quadratic formula, $x = [H^+] = 1.8 \times 10^{-3}$, pH = 2.74.

17.89 $\dfrac{0.15\,mol\,CH_3COOH}{1\,L\,soln} \times 0.750\,L = 0.1125 = 0.11\,mol\,CH_3COOH$

$0.1125\,mol\,CH_3COOH \times \dfrac{60.05\,g\,CH_3COOH}{1\,mol\,CH_3COOH} \times \dfrac{1\,g\,gl\,acetic\,acid}{0.99\,g\,CH_3COOH} \times \dfrac{1.00\,mL\,gl\,acetic\,acid}{1.05\,g\,gl\,acetic\,acid}$

$= 6.5\,mL$ glacial acetic acid

At pH 4.50, $[H^+] = 10^{-4.50} = 3.16 \times 10^{-5} = 3.2 \times 10^{-5}\,M$; this is small compared to $0.15\,M\,CH_3COOH$.

$K_a = \dfrac{(3.16 \times 10^{-5})\,[CH_3COO^-]}{0.15} = 1.8 \times 10^{-5}; [CH_3COO^-] = 0.0854 = 0.085\,M$

$\dfrac{0.0854\,mol\,CH_3COONa}{1\,L\,soln} \times 0.750\,L \times \dfrac{82.03\,g\,CH_3COONa}{1\,mol\,CH_3COONa} = 5.253 = 5.25\,g\,CH_3COONa$

17.90 (a) For a monoprotic acid (one H$^+$ per mole of acid), at the equivalence point, moles OH$^-$ added = moles H$^+$ originally present

$M_B \times V_B = g\,acid/molar\,mass$

$$MM = \frac{g\,acid}{M_B \times V_B} = \frac{0.2140\,g}{0.0950\,M \times 0.0300\,L} = 75.09 = 75.1\,g/mol$$

(b) Addition of 15.0 mL of 0.0950 M NaOH is half way to the equivalence point of the titration. At this point, the solution in the flask is a buffer where [HA] = [A$^-$] and pH of the solution equals pK$_a$ of HA. At 15.0 mL, pH = pK$_a$ = 6.50

$K_a = 10^{-6.50} = 3.16 \times 10^{-7} = 3.2 \times 10^{-7}$

17.91 (a) For a monoprotic acid (one H$^+$ per mole of acid), at the equivalence point moles OH$^-$ added = moles H$^+$ originally present

$M_B \times V_B = g\,acid/molar\,mass$

$$MM = \frac{g\,acid}{M_B \times V_B} = \frac{0.1687\,g}{0.1150\,M \times 0.0155\,L} = 94.642 = 94.6\,g/mol$$

(b)   initial mol HA $= \dfrac{0.1687 \, g}{94.642 \, g/mol} = 1.783 \times 10^{-3} = 1.78 \times 10^{-3}$ mol HA

mol OH$^-$ added to pH 2.85 $= 0.1150 \, M \times 0.00725 \, L = 8.338 \times 10^{-4}$

$= 8.34 \times 10^{-4}$ mol OH$^-$

	HA(aq)	+	NaOH(aq)	$\rightarrow$	NaA(aq) + H$_2$O
before rx	1.783×10^{-3} mol		0.834×10^{-3} mol		0
change	-0.834×10^{-3} mol		-0.834×10^{-3} mol		0.834×10^{-3} mol
after rx	0.949×10^{-3} mol		0		0.834×10^{-3} mol

$[HA] = \dfrac{9.49 \times 10^{-4} \, \text{mol}}{0.0325 \, L} = 0.02919 = 0.0292 \, M$

$[A^-] = \dfrac{8.34 \times 10^{-4} \, \text{mol}}{0.0325 \, L} = 0.02565 = 0.0257 \, M$

$[H^+] = 10^{-2.85} = 1.413 \times 10^{-3} = 1.4 \times 10^{-3}$

The mixture after reaction (a buffer) can be described by the acid dissociation equilibrium.

	HA(aq)	$\rightleftharpoons$	H$^+$(aq)	+	A$^-$(aq)
initial	$0.0292 \, M$		0		$0.0257 \, M$
equil.	$(0.0292 - 1.4 \times 10^{-3} \, M)$		$1.4 \times 10^{-3} \, M$		$(0.0257 + 1.4 \times 10^{-3}) \, M$

$K_a = \dfrac{[H^+][A^-]}{[HA]} \approx \dfrac{(1.413 \times 10^{-3})(0.02707)}{(0.02778)} = 1.4 \times 10^{-3}$

(Although we have carried extra figures through the calculation to avoid rounding errors, the data dictate an answer with 2 sig figs.)

17.92   At the equivalence point of a titration, moles strong base added equals moles weak acid initially present. $M_B \times V_B = $ mol base added = mol acid initial.

At the half-way point, the volume of base is one-half of the volume required to reach the equivalence point, and the moles base delivered equals one-half of the moles acid initially present. This means that one-half of the weak acid HA is converted to the conjugate base A$^-$. If exactly half of the acid reacts, mol HA = mol A$^-$ and [HA] = [A$^-$] at the half-way point.

From Equation 17.9, $pH = pK_a + \log \dfrac{[\text{conj. base}]}{[\text{conj. acid}]} = pK_a + \log \dfrac{[A^-]}{[HA^-]}$.

If $[A^-]/[HA] = 1$, $\log(1) = 0$ and $pH = pK_a$ of the weak acid being titrated.

17.93   If 50.0 mL base is required to reach the equivalence point, addition of 25.0 mL of base is half-way to the equivalence point. At this point, $[A^-] = [HA]$ and $[A^-]/[HA] = 1$. The pK_a of the weak acid being titrated is equal to the pH of the solution, 3.62.

17.94   Assume that H_3PO_4 will react with NaOH in a stepwise fashion: (This is not unreasonable, because the three K_a values for H_3PO_4 are significantly different.)

$$H_3PO_4(aq) \quad + \quad NaOH(aq) \quad \rightarrow \quad H_2PO_4^-(aq) + Na^+(aq) + H_2O(l)$$

before	0.20 mol	0.30 mol	0 mol
after	0 mol	0.10 mol	0.20 mol

$$H_2PO_4^-(aq) \quad + \quad NaOH(aq) \quad \rightarrow \quad HPO_4^-(aq) + Na^+(aq) + H_2O(l)$$

before	0.20 mol	0.10 mol	0.25 mol
after	0.10 mol	0	0.35 mol

Thus, after all NaOH has reacted, the resulting 1.00 L solution is a buffer containing 0.10 mol $H_2PO_4^-$ and 0.35 mol HPO_4^{2-}. $H_2PO_4^-(aq) \rightleftharpoons H^+(aq) + HPO_4^{2-}(aq)$

$$K_a = 6.2 \times 10^{-8} = \frac{[HPO_4^{2-}][H^+]}{[H_2PO_4^-]}; [H^+] = \frac{6.2 \times 10^{-8} \, (0.10 \, M)}{0.35 \, M} = 1.77 \times 10^{-8} = 1.8 \times 10^{-8} \, M;$$

$$pH = 7.75$$

17.95   The pH of a buffer system is centered around pK_a for the conjugate acid component. For a diprotic acid, two conjugate acid/conjugate base pairs are possible.

$$H_2X(aq) \rightleftharpoons H^+(aq) + HX^-(aq); \quad K_{a1} = 2 \times 10^{-2}; \quad pK_{a1} = 1.70$$

$$HX^-(aq) \rightleftharpoons H^+(aq) + X^{2-}(aq); \quad K_{a2} = 5.0 \times 10^{-7}; \quad pK_{a2} = 6.30$$

Clearly, HX^- / X^{2-} is the more appropriate combination for preparing a buffer with pH = 6.50. The $[H^+]$ in this buffer = $10^{-6.50} = 3.16 \times 10^{-7} = 3.2 \times 10^{-7} \, M$. Using the K_{a2} expression to calculate the $[X^{2-}] / [HX^-]$ ratio:

$$K_{a2} = \frac{[H^+][X^{2-}]}{[HX^-]}; \frac{K_{a2}}{[H^+]} = \frac{[X^{2-}]}{[HX^-]} = \frac{5.0 \times 10^{-7}}{3.16 \times 10^{-7}} = 1.58 = 1.6$$

Because X^{2-} and HX^- are present in the same solution, the ratio of concentrations is also a ratio of moles.

$$\frac{[X^{2-}]}{[HX^-]} = \left(\frac{mol \, X^{2-}/L \, soln}{mol \, HX^-/L \, soln} \right) = \frac{mol \, X^{2-}}{mol \, HX^-} = 1.58; mol \, X^{2-} = (1.58) \, mol \, HX^-$$

In the 1.0 L of 1.0 M H_2X, there is 1.0 mol of material containing X^{2-}.

Thus, mol HX^- + 1.58 (mol HX^-) = 1.0 mol. 2.58 (mol HX^-) = 1.0;

mol HX^- = 1.0 / 2.58 = 0.39 mol HX^-; mol X^{2-} = 1.0 – 0.39 = 0.61 mol X^{2-}.

Thus, enough 1.0 M NaOH must be added to produce 0.39 mol HX^- and 0.61 mol X^{2-}.

Considering the neutralization in a stepwise fashion (see discussion of titrations of polyprotic acids in Section 17.3).

$$H_2X(aq) \quad + \quad NaOH(aq) \quad \rightarrow \quad HX^-(aq) + H_2O(l)$$

before	1.0 mol	1 mol	0
after	0	0	1.0 mol

$$\text{HX}^-(\text{aq}) \quad + \quad \text{NaOH(aq)} \quad \rightarrow \quad \text{X}^{2-}(\text{aq}) + \text{H}_2\text{O(l)}$$

before	1.0		0.61
change	−0.61	−0.61	+0.61
after	0.39	0	0.61

Starting with 1.0 mol of H_2X, 1.0 mol of NaOH is added to completely convert it to 1.0 mol of HX^-. Of that 1.0 mol of HX^-, 0.61 mol must be converted to 0.61 mol X^{2-}. The total moles of NaOH added is $(1.00 + 0.61) = 1.61$ mol NaOH.

$$\text{L NaOH} = \frac{\text{mol NaOH}}{M \text{ NaOH}} = \frac{1.61 \text{ mol}}{1.0 \text{ } M} = 1.6 \text{ L of } 1.0 \text{ } M \text{ NaOH}$$

17.96 $CH_3CH(OH)COO^-$ will be formed by reaction $CH_3CH(OH)COOH$ with NaOH.

$0.1000 \text{ } M \times 0.02500 \text{ L} = 2.500 \times 10^{-3}$ mol $CH_3CH(OH)COOH$; b = mol NaOH needed

$$\text{CH}_3\text{CH(OH)COOH} \quad + \quad \text{NaOH} \quad \rightarrow \quad \text{CH}_3\text{CH(OH)COO}^- + \text{H}_2\text{O} + \text{Na}^+$$

initial	2.500×10^{-3} mol	b mol	
rx	−b mol	−b mol	+b mol
after rx	$(2.500 \times 10^{-3} - b)$ mol	0	b mol

$$K_a = \frac{[\text{H}^+][\text{CH}_3\text{CH(OH)COO}^-]}{[\text{CH}_3\text{CH(OH)COOH}]}; K_a = 1.4 \times 10^{-4}; [\text{H}^+] = 10^{-\text{pH}} = 10^{-3.75} = 1.778 \times 10^{-4} = 1.8 \times 10^{-4} M$$

Because solution volume is the same for reaction $CH_3CH(OH)COOH$ and $CH_3CH(OH)COO^-$, we can use moles in the equation for $[H^+]$.

$$K_a = 1.4 \times 10^{-4} = \frac{1.778 \times 10^{-4} \text{ (b)}}{(2.500 \times 10^{-3} - b)}; 0.7874 \,(2.500 \times 10^{-3} - b) = b, 1.969 \times 10^{-3} = 1.7874 \text{ b},$$

$b = 1.10 \times 10^{-3} = 1.1 \times 10^{-3}$ mol OH^-

(The precision of K_a dictates that the result has 2 sig figs.)

Substituting this result into the K_a expression gives $[H^+] = 1.8 \times 10^{-4}$. This checks and confirms our result. Calculate volume NaOH required from $M = \text{mol/L}$.

$$1.10 \times 10^{-3} \text{ mol OH}^- \times \frac{1 \text{ L}}{1.000 \text{ mol}} \times \frac{1 \text{ } \mu\text{L}}{1 \times 10^{-6} \text{ L}} = 1.1 \times 10^3 \text{ } \mu\text{L} \,(1.1 \text{ mL})$$

17.97 (a) $PbCO_3(s) \rightleftharpoons Pb^{2+}(\text{aq}) + CO_3^{2-}(\text{aq})$

$K_{sp} = [Pb^{2+}][CO_3^{2-}] = 7.4 \times 10^{-14}$. molar solubility $= s = [Pb^{2+}] = [CO_3^{2-}]$

$K_{sp} = s^2 = 7.4 \times 10^{-14}$. $s = [Pb^{2+}] = 2.7203 \times 10^{-7} = 2.7 \times 10^{-7} M$

(b) For very dilute aqueous solutions, assume the solution density is 1.0 g/mL.

$$\text{ppb} = \frac{\text{g solute}}{10^9 \text{ g solution}} = \frac{1 \times 10^{-6} \text{ g solute}}{1 \times 10^3 \text{ g solution}} = \frac{\mu\text{g solute}}{\text{L solution}}$$

$$\frac{2.7203 \times 10^{-7} \text{ mol Pb}^{2+}}{\text{L}} \times \frac{207.2 \text{ g Pb}^{2+}}{1 \text{ mol Pb}^{2+}} \times \frac{1 \mu\text{g}}{1 \times 10^{-6} \text{ g}} = \frac{56.365 \,\mu\text{g Pb}^{2+}}{\text{L}} = 56 \text{ ppb}$$

(c) The solubility of $PbCO_3$ increases as pH is lowered. When pH is lowered, $[H^+]$ increases. The $H^+(\text{aq})$ reacts with $CO_3^{2-}(\text{aq})$ to form HCO_3^- and $H_2CO_3(\text{aq})$. This shifts the solubility equilibrium to the right and increases the solubility of $PbCO_3$.

(d) A saturated solution of lead carbonate, with a lead concentration of 56 ppb, exceeds the EPA acceptable lead level of 15 ppb.

17.98 (a) CdS: 8.0×10^{-28}; CuS: 6×10^{-37}. CdS has greater molar solubility.

(b) PbCO$_3$: 7.4×10^{-14}; BaCrO$_4$: 2.1×10^{-10}. BaCrO$_4$ has greater molar solubility.

(c) Because the stoichiometry of the two complexes is not the same, K_{sp} values can not be compared directly; molar solubilities must be calculated from K_{sp} values.

$Ni(OH)_2$: $K_{sp} = 6.0 \times 10^{-16} = [Ni^{2+}][OH^-]^2$; $[Ni^{2+}] = x$, $[OH^-] = 2x$

$6.0 \times 10^{-16} = (x)(2x)^2 = 4x^3$; $x = 5.3 \times 10^{-6}$ M Ni^{2+}

Note that [OH$^-$] from the autoionization of water is less than 1% of [OH$^-$] from Ni(OH)$_2$ and can be neglected.

$NiCO_3$: $K_{sp} = 1.3 \times 10^{-7} = [Ni^{2+}][CO_3^{2-}]$; $[Ni^{2+}] = [CO_3^{2-}] = x$

$1.3 \times 10^{-7} = x^2$; $x = 3.6 \times 10^{-4}$ M Ni^{2+}

NiCO$_3$ has greater molar solubility than Ni(OH)$_2$, but the values are much closer than expected from inspection of K_{sp} values alone.

(d) Again, molar solubilities must be calculated for comparison.

Ag_2SO_4: $K_{sp} = 1.5 \times 10^{-5} = [Ag^+]^2[SO_4^{2-}]$; $[SO_4^{2-}] = x$, $[Ag^+] = 2x$

$1.5 \times 10^{-5} = (2x)^2(x) = 4x^3$; $x = 1.6 \times 10^{-2}$ M SO$_4^{2-}$

AgI: $K_{sp} = 8.3 \times 10^{-17} = [Ag^+][I^-]$; $[Ag^+] = [I^-] = x$

$8.3 \times 10^{-17} = x^2$; $x = 9.1 \times 10^{-9}$ M Ag$^+$

Ag$_2$SO$_4$ has greater molar solubility than AgI.

17.99 (a) $K_{sp} = 4.5 \times 10^{-9} = [Ca^{2+}][CO_3^{2-}]$; $s = [Ca^{2+}] = [CO_3^{2-}]$

$s^2 = 4.5 \times 10^{-9}$, $s = 6.708 \times 10^{-5} = 6.7 \times 10^{-5}$

(b)

$$CaCO_3(s) \rightleftharpoons Ca^{2+}(aq) + CO_3^{2-}(aq) \qquad\qquad K_{sp}$$

$$CO_3^{2-}(aq) + H_2O(l) \rightleftharpoons HCO_3^-(aq) + OH^-(aq) \qquad\qquad K_b$$

$$CaCO_3(s) + H_2O(l) \rightleftharpoons Ca^{2+}(aq) + HCO_3^-(aq) + OH^-(aq) \qquad K$$

$K_b = K_w / K_a$ for HCO$_3^-$

$$K = K_{sp} \times K_b = \frac{K_{sp} \times K_w}{K_a \text{ for } HCO_3^-} = \frac{4.5 \times 10^{-9} \times 1 \times 10^{-14}}{5.6 \times 10^{-11}} = 8.036 \times 10^{-13} = 8.0 \times 10^{-13}$$

(c) $K = 8.036 \times 10^{-13} = [Ca^{2+}][HCO_3^-][OH^-] = s^3$; $s = 9.297 \times 10^{-5} = 9.3 \times 10^{-5}$ M

(d) pH = 8.3, pOH = 14 − 8.3 = 5.7. $[OH^-] = 10^{-5.7} = 1.995 \times 10^{-6} = 2 \times 10^{-6}$ M

$8.036 \times 10^{-13} = s^2(1.995 \times 10^{-6})$, $s = 6.346 \times 10^{-4} = 6 \times 10^{-4}$ M

(e) pH = 7.5, pOH = 14 − 7.5 = 6.5. $[OH^-] = 10^{-6.5} = 3.162 \times 10^{-7} = 3 \times 10^{-7}$ M

$8.036 \times 10^{-13} = s^2(3.162 \times 10^{-7})$, $s = 1.549 \times 10^{-3} = 2 \times 10^{-3}$ M

The drop in pH from 8.3 to 7.5 significantly increases (from 6.7×10^{-5} M to 1.5×10^{-3} M) the molar solubility of CaCO$_3$(s).

17.100   (a)   Hydroxyapatite: $K_{sp} = [Ca^{2+}]^5[PO_4^{3-}]^3[OH^-]$

            Fluoroapatite: $K_{sp} = [Ca^{2+}]^5[PO_4^{3-}]^3[F^-]$

      (b)   For each mole of apatite dissolved, one mole of OH^- or F^- is formed. Express molar solubility, s, in terms of $[OH^-]$ and $[F^-]$.

            Hydroxyapatite: $[OH^-] = s$, $[Ca^{2+}] = 5s$, $[PO_4^{3-}] = 3s$

            $K_{sp} = 6.8 \times 10^{-27} = (5s)^5(3s)^3 (s) = 84{,}375 \, s^9$

            $s^9 = 8.059 \times 10^{-32} = 8.1 \times 10^{-32}$.

            Use logs to find s. $s = 3.509 \times 10^{-4} = 3.5 \times 10^{-4} \, M \, Ca_5(PO_4)_3OH$.

            Fluoroapatite: $[F^-] = s$, $[Ca^{2+}] = 5s$, $[PO_4^{3-}] = 3s$

            $K_{sp} = 1.0 \times 10^{-60} = (5s)^5(3s)^3 (s) = 84{,}375 \, s^9$

            $s^9 = 1.185 \times 10^{-65} = 1.2 \times 10^{-65}$; $s = 6.109 \times 10^{-8} = 6.1 \times 10^{-8} \, M \, Ca_5(PO_4)_3F$

17.101   (a) K_b for $PO_4^{3-} = K_w / K_{a3} = (1.0 \times 10^{-14}) / (4.2 \times 10^{-13}) = 0.02381 = 2.4 \times 10^{-2}$

      (b) $PO_4^{3-}(aq) + H_2O(l) \rightleftharpoons HPO_4^{2-}(aq) + OH^-(aq)$

            $[HPO_4^{2-}] = [OH^-] = x$, $[PO_4^{3-}] = (1 \times 10^{-3}) - x$

            $K_b = 2.4 \times 10^{-2} = \dfrac{[HPO_4^{2-}][OH^-]}{[PO_4^{3-}]} = \dfrac{x^2}{(1 \times 10^{-3}) - x}$

            Because K_b is relatively large and $[PO_4^{3-}]$ is small, x may be significant relative to $(1 \times 10^{-3}) \, M$. Use the quadratic formula to solve for x.

            $x^2 + 2.4 \times 10^{-2} \, x - 2.4 \times 10^{-5} = 0$; $x = 9.615 \times 10^{-4} = 1 \times 10^{-3} \, M \, OH^-$

            $pOH = 3.017 = 3.0$, $pH = 10.983 = 11.0$

17.102   *Analyze/Plan.*   Calculate the solubility of $Mg(OH)_2$ in 0.50 M NH_4Cl. Find K_{sp} for $Mg(OH)_2$ in Appendix D.3. NH_4^+ is a weak acid, which will increase the solubility of $Mg(OH)_2$. Combine the various interacting equilibria to obtain an overall reaction. Calculate K for this reaction and use it to calculate solubility (s) for $Mg(OH)_2$ in 0.50 M NH_4Cl.   *Solve.*

$$Mg(OH)_2(s) \rightleftharpoons Mg^{2+}(aq) + 2\,OH^-(aq) \qquad\qquad K_{sp}$$

$$2\,NH_4^+(aq) \rightleftharpoons 2\,NH_3(aq) + 2\,H^+(aq) \qquad\qquad K_a$$

$$2\,H^+(aq) + 2\,OH^-(aq) \rightleftharpoons 2\,H_2O(l) \qquad\qquad 1/K_w$$

$$Mg(OH)_2(s) + 2\,NH_4^+(aq) + 2\,H^+(aq) + 2\,OH^-(aq) \rightleftharpoons Mg^{2+}(aq) + 2\,NH_3(aq)$$
$$+ 2\,OH^-(aq) + 2\,H^+(aq) + 2\,H_2O(l)$$

$$Mg(OH)_2(s) + 2\,NH_4^+(aq) \rightleftharpoons Mg^{2+}(aq) + 2\,NH_3(aq) + 2\,H_2O(l)$$

$$K = \frac{[Mg^{2+}][NH_3]^2}{[NH_4^+]^2} = \frac{K_{sp} \times K_a^2}{K_w^2}; \quad K_a \text{ for } NH_4^+ = \frac{K_w}{K_b \text{ for } NH_3}; \quad \frac{K_a}{K_w} = \frac{1}{K_b}$$

$$K = \frac{K_{sp} \times K_a^2}{K_w^2} = \frac{K_{sp}}{K_b^2} = \frac{1.8 \times 10^{-11}}{(1.8 \times 10^{-5})^2} = 5.556 \times 10^{-2} = 5.6 \times 10^{-2}$$

Let $[Mg^{2+}] = s$, $[NH_3] = 2s$, $[NH_4^+] = 0.50 - 2s$

$$K = 5.6 \times 10^{-2} = \frac{[Mg^{2+}][NH_3]^2}{[NH_4^+]^2} = \frac{s(2s)^2}{(0.5 - 2s)^2} = \frac{4s^3}{0.25 - 2s + 4s^2}$$

$5.6 \times 10^{-2}(0.25 - 2s + 4s^2) = 4s^3$; $4s^3 - 0.222s^2 + 0.111s - 1.39 \times 10^{-2} = 0$

Clearly, 2s is not small relative to 0.50. Solving the third-order equation, s = 0.1054 = 0.11 M. The solubility of $Mg(OH)_2$ in 0.50 M NH_4Cl is 0.11 mol/L.

Check. Substitute s = 0.1054 into the K expression.

$$K = \frac{4(0.1054)^3}{[0.50 - 2(0.1054)]^2} = 5.6 \times 10^{-2}.$$

The solubility and K value are consistent, to the precision of the K_{sp} and K_b values.

17.103 $K_{sp} = [Ba^{2+}][MnO_4^-]^2 = 2.5 \times 10^{-10}$

$[MnO_4^-]^2 = 2.5 \times 10^{-10}/2.0 \times 10^{-8} = 0.0125$; $[MnO_4^-] = \sqrt{0.0125} = 0.11\ M$

17.104 $[Ca^{2+}][CO_3^{2-}] = 4.5 \times 10^{-9}$; $[Fe^{2+}][CO_3^{2-}] = 2.1 \times 10^{-11}$

Because $[CO_3^{2-}]$ is the same for both equilibria:

$$[CO_3^{2-}] = \frac{4.5 \times 10^{-9}}{[Ca^{2+}]} = \frac{2.1 \times 10^{-11}}{[Fe^{2+}]}; \text{ rearranging } \frac{[Ca^{2+}]}{[Fe^{2+}]} = \frac{4.5 \times 10^{-9}}{2.1 \times 10^{-11}} = 214 = 2.1 \times 10^2$$

17.105 $PbSO_4(s) \rightleftharpoons Pb^{2+}(aq) + SO_4^{2-}(aq)$; $K_{sp} = 6.3 \times 10^{-7} = [Pb^{2+}][SO_4^{2-}]$

$SrSO_4(s) \rightleftharpoons Sr^{2+}(aq) + SO_4^{2-}(aq)$; $K_{sp} = 3.2 \times 10^{-7} = [Sr^{2+}][SO_4^{2-}]$

Let $x = [Pb^{2+}]$, $y = [Sr^{2+}]$, $x + y = [SO_4^{2-}]$

$$\frac{x(x+y)}{y(x+y)} = \frac{6.3 \times 10^{-7}}{3.2 \times 10^{-7}}; \frac{x}{y} = 1.9688 = 2.0; x = 1.969\ y = 2.0\ y$$

$y(1.969\ y + y) = 3.2 \times 10^{-7}$; $2.969\ y^2 = 3.2 \times 10^{-7}$; $y = 3.283 \times 10^{-4} = 3.3 \times 10^{-4}$

$x = 1.969\ y$; $x = 1.969(3.283 \times 10^{-4}) = 6.464 \times 10^{-4} = 6.5 \times 10^{-4}$

$[Pb^{2+}] = 6.5 \times 10^{-4}\ M$, $[Sr^{2+}] = 3.3 \times 10^{-4}\ M$, $[SO_4^{2-}] = (3.283 + 6.464) \times 10^{-4} = 9.7 \times 10^{-4}\ M$

17.106 $MgC_2O_4(s) \rightleftharpoons Mg^{2+}(aq) + C_2O_4^{2-}(aq)$

$K_{sp} = [Mg^{2+}][C_2O_4^{2-}] = 8.6 \times 10^{-5}$

If $[Mg^{2+}]$ is to be $3.0 \times 10^{-2}\ M$, $[C_2O_4^{2-}] = 8.6 \times 10^{-5}/3.0 \times 10^{-2} = 2.87 \times 10^{-3} = 2.9 \times 10^{-3}\ M$

The oxalate ion undergoes hydrolysis:

$C_2O_4^{2-}(aq) + H_2O(l) \rightleftharpoons HC_2O_4^-(aq) + OH^-(aq)$

$$K_b = \frac{[HC_2O_4^-][OH^-]}{[C_2O_4^{2-}]} = 1.0 \times 10^{-14}/6.4 \times 10^{-5} = 1.56 \times 10^{-10} = 1.6 \times 10^{-10}$$

$$[Mg^{2+}] = 3.0 \times 10^{-2}\, M, [C_2O_4^{2-}] = 2.87 \times 10^{-3} = 2.9 \times 10^{-3}\, M$$

$$[HC_2O_4^-] = (3.0 \times 10^{-2} - 2.87 \times 10^{-3})\, M = 2.71 \times 10^{-2} = 2.7 \times 10^{-2}\, M$$

$$[OH^-] = 1.56 \times 10^{-10} \times \frac{[C_2O_4^{2-}]}{[HC_2O_4^-]} = 1.56 \times 10^{-10} \times \frac{(2.87 \times 10^{-3})}{(2.71 \times 10^{-2})} = 1.652 \times 10^{-11}$$

$$[OH^-] = 1.7 \times 10^{-11}\, M; \text{ pOH} = 10.78, \text{ pH} = 3.22$$

17.107   (a)   Express the molar solubility of $Mg_3(AsO_4)_2$ as s. For each mole of $Mg_3(AsO_4)_2$
that dissolves, 3 moles of Mg^{2+} and 2 moles of AsO_4^{3-} are formed. In terms of s,
$[Mg^{2+}] = 3s$ and $[AsO_4^{3-}] = 2s$.

$$K_{sp} = [Mg^{2+}]^3[AsO_4^{3-}]^2 = [3s]^3[2s]^2 = 108s^5 = 2.1 \times 10^{-20}.\ s^5 = 1.9444 \times 10^{-22}$$

Use logs to find s. $s = 4.547 \times 10^{-5} = 4.5 \times 10^{-5}\, M\ Mg_3(AsO_4)_2.$

(b)   A saturated solution of $Mg_3(AsO_4)_2$ contains AsO_4^{3-} anion, the conjugate base of
$HAsO_4^{2-}$. Base hydrolysis of AsO_4^{3-} changes the pH of the solution. Because
H_3AsO_4 is a polyprotic acid, base hydrolysis can occur in three steps. For
H_3AsO_4, the pK_a and corresponding pK_b values are significantly different from
each other and we can consider each step separately.

$$AsO_4^{3-}\,(aq) + H_2O(l) \rightleftharpoons HAsO_4^{2-}\,(aq) + OH^-\,(aq) \qquad pK_{b3} = 14 - pK_{a3}$$

$$HAsO_4^{2-}\,(aq) + H_2O(l) \rightleftharpoons H_2AsO_4^-\,(aq) + OH^-\,(aq) \qquad pK_{b2} = 14 - pK_{a2}$$

$$H_2AsO_4^-\,(aq) + H_2O(l) \rightleftharpoons H_3AsO_4\,(aq) + OH^-\,(aq) \qquad pK_{b1} = 14 - pK_{a1}$$

$$pK_{b3} = 14 - pK_{a3} = 14 - 11.50 = 2.5;\ K_{b3} = 3.1623 \times 10^{-3} = 3.2 \times 10^{-3}$$

$$[AsO_4^{3-}] = 2s = 9.09 \times 10^{-5}\, M;\ [OH^-] = [HAsO_4^{2-}] = x$$

$$K_{b3} = \frac{[HAsO_4^{2-}][OH^-]}{[AsO_4^{3-}]} = \frac{x^2}{(9.09 \times 10^{-5} - x)} = 3.2 \times 10^{-3}$$

Because K_{b3} is relatively large and $[AsO_4^{3-}]$ is small, we must use the quadratic
equation to solve for x. $x^2 + 3.16 \times 10^{-3} - 2.88 \times 10^{-7} = 0$

$x = 8.847 \times 10^{-5} = 8.8 \times 10^{-5}\, M\ OH^-\,(aq);$ pOH = 4.05, pH = 9.95

Now consider the second hydrolysis step.

$$pK_{b2} = 14 - pK_{a2} = 14 - 6.98 = 7.02;\ K_{b2} = 9.550 \times 10^{-8} = 9.6 \times 10^{-8}$$

$$[H_2AsO_4^-] = y;\ [OH^-] = 8.8 \times 10^{-8}\, M + y;\ [HAsO_4^{2-}] = 8.8 \times 10^{-5}\, M - y$$

$$K_{b2} = \frac{[H_2AsO_4^{2-}][OH^-]}{[HAsO_4^{2-}]} = \frac{y(8.8 \times 10^{-5} + y)}{(8.8 \times 10^{-5} - y)} = 9.6 \times 10^{-8}$$

With these values of K_{b2} and $[HAsO_4^{2-}]$, it is reasonable to assume y is small
relative to 8.8×10^{-5}.

$y = 9.6 \times 10^{-8}\, M;\ [OH^-] = 8.847 \times 10^{-5} + 9.550 \times 10^{-8} = 8.857 \times 10^{-5} = 8.9 \times 10^{-5}\, M.$

pOH = 4.05, pH = 9.95

Because the second hydrolysis step did not produce a significant amount of
OH^-(aq), and K_{b1} is even smaller than K_{b2}, we need not consider the third step.
The pH of a saturated solution of $Mg_3(AsO_4)_2$ is 9.95.

17.108
$$Zn(OH)_2(s) \rightleftharpoons Zn^{2+}(aq) + 2\,OH^-(aq) \qquad K_{sp} = 3.0 \times 10^{-16}$$

$$Zn^{2+}(aq) + 4\,OH^-(aq) \rightleftharpoons Zn(OH)_4^{2-}(aq) \qquad\qquad K_f = 4.6 \times 10^{17}$$

$$Zn(OH)_2(s) + 2\,OH^-(aq) \rightleftharpoons Zn(OH)_4^{2-}(aq) \qquad K = K_{sp} \times K_f = 138 = 1.4 \times 10^2$$

$$K = 138 = 1.4 \times 10^2 = \frac{[Zn(OH)_4^{2-}]}{[OH^-]^2}$$

If 0.015 mol $Zn(OH)_2$ dissolves, 0.015 mol $Zn(OH)_4^{2-}$ should be present at equilibrium.

$$[OH^-]^2 = \frac{(0.015)}{138}; [OH^-] = 1.043 \times 10^{-2}\ M;\ [OH^-] \geq 1.0 \times 10^{-2}\ M \text{ or } pH \geq 12.02$$

17.109 (a) $Cd(OH)_2(s) \rightleftharpoons Cd^{2+}(aq) + 2\,OH^-(aq); K_{sp} = 2.5 \times 10^{-14} = [Cd^{2+}][OH^-]^2.$

$[Cd^{2+}] = s;\ [OH^-] = 2s;\ K_{sp} = 2.5 \times 10^{-14} = 4s^3.\qquad s = 1.8 \times 10^{-5}\ M.$

(b)
$$Cd(OH)_2(s) \rightleftharpoons Cd^{2+}(aq) + 2\,OH^-(aq) \qquad\qquad K_{sp} = 2.5 \times 10^{-14}$$

$$Cd^{2+}(aq) + 4\,Br^-(aq) \rightleftharpoons CdBr_4^{2-}(aq) \qquad\qquad K_f = 5 \times 10^3$$

$$Cd(OH)_2(s) + 4\,Br^-(aq) \rightleftharpoons CdBr_4^{2-}(aq) + 2\,OH^-(aq) \qquad K$$

$$K = K_{sp} \times K_f = (2.5 \times 10^{-14})(5 \times 10^3) = 1.25 \times 10^{-10} = 1 \times 10^{-10}$$

The desired molar solubility of $Cd(OH)_2$ is 1.0×10^{-3}. Assume all soluble Cd^{2+} is present as $CdBr_4^{2-}$. $[CdBr_4^{2-}] = 1.0 \times 10^{-3};\ [OH^-] = 2(1.0 \times 10^{-3}) = 2.0 \times 10^{-3}\ M.$

Let c = initial $[NaBr]$ = initial $[Br^-]$; $[Br^-]$ at equilibrium = $c - 4(1.0 \times 10^{-3}) = (c - 4.0 \times 10^{-3}).$

$$K = 1.25 \times 10^{-10} = \frac{[CdBr_4^{2-}][OH^-]^2}{[Br^-]^4} = \frac{(1.0 \times 10^{-3})(2.0 \times 10^{-3})^2}{(c - 4.0 \times 10^{-3})^4}$$

Assume c is large relative to 4.0×10^{-3}.

$(1.25 \times 10^{-10})c^4 = 4.0 \times 10^{-9}; c = (32)^{1/4} = 2.378 = 2\ M.$ The approximation is valid. 4.0×10^{-3} is about 0.2% of 2 M. Check this result in the equilibrium expression.

$$K = \frac{(1.0 \times 10^{-3})(2.0 \times 10^{-3})^2}{(2.378 - 4.0 \times 10^{-3})^4} = 1.26 \times 10^{-10}.\ \text{Our calculations are consistent.}$$

Integrative Exercises

17.110 (a) Complete ionic ($CHO_2^- = HCOO^-$)

$H^+(aq) + Cl^-(aq) + Na^+(aq) + HCOO^-(aq) \rightarrow HCOOH(aq) + Na^+(aq) + Cl^-(aq)$

Na^+ and Cl^- are spectator ions.

Net ionic: $H^+(aq) + HCOO^-(aq) \rightleftharpoons HCOOH(aq)$

(b) The net ionic equation in part (a) is the reverse of the dissociation of HCOOH.

$$K = \frac{1}{K_a} = \frac{1}{1.8 \times 10^{-4}} = 5.55 \times 10^3 = 5.6 \times 10^3$$

(c)   For Na^+ and Cl^-, this is just a dilution problem.

$M_1V_1 = M_2V_2$; V_2 is 50.0 mL + 50.0 mL = 100.0 mL

Cl^-: $\dfrac{0.15\,M \times 50.0\,mL}{100.0\,mL} = 0.075\,M$; Na^+: $\dfrac{0.15\,M \times 50.0\,mL}{100.0\,mL} = 0.075\,M$

H^+ and $HCOO^-$ react to form HCOOH. Because K >> 1, the reaction essentially goes to completion.

$0.15\,M \times 0.0500\,mL = 7.5 \times 10^{-3}\,mol\,H^+$

$\underline{0.15\,M \times 0.0500\,mL = 7.5 \times 10^{-3}\,mol\,HCOO^-}$

$ = 7.5 \times 10^{-3}\,mol\,HCOOH$

Solve the weak acid problem to determine $[H^+]$, $[HCOO^-]$, and $[HCOOH]$ at equilibrium.

$K_a = \dfrac{[H^+][HCOO^-]}{[HCOOH]}$; $[H^+] = [HCOO^-] = x\,M$; $[HCOOH] = \dfrac{(7.5 \times 10^{-3} - x)\,mol}{0.100\,L}$

$\phantom{K_a = \dfrac{[H^+][HCOO^-]}{[HCOOH]}; [H^+]} = (0.075 - x)\,M$

$1.8 \times 10^{-4} = \dfrac{x^2}{(0.075 - x)} \approx \dfrac{x^2}{0.075}$; $x = 3.7 \times 10^{-3}\,M\,H^+$ and $HCOO^-$

$[HCOOH] = (0.075 - 0.0037) = 0.071\,M$

$\dfrac{[H^+]}{[HCOOH]} \times 100 = \dfrac{3.7 \times 10^{-3}}{0.075} \times 100 = 4.9\%\,dissociation$

In summary:

$[Na^+] = [Cl^-] = 0.075\,M$, $[HCOOH] = 0.071\,M$, $[H^+] = [HCOO^-] = 0.0037\,M$

17.111   (a)   For a monoprotic acid (one H^+ per mole of acid), at the equivalence point moles OH^- added = moles H^+ originally present

$M_B \times V_B$ = g acid/molar mass

$MM = \dfrac{g\,acid}{M_B \times V_B} = \dfrac{0.1044\,g}{0.0500\,M \times 0.02210\,L} = 94.48 = 94.5\,g/mol$

(b)   11.05 mL is exactly half-way to the equivalence point (22.10 mL). When half of the unknown acid is neutralized, $[HA] = [A^-]$, $[H^+] = K_a$ and $pH = pK_a$.

$K_a = 10^{-4.89} = 1.3 \times 10^{-5}$

(c)   From Appendix D, Table D.1, acids with K_a values close to 1.3×10^{-5} are

Name	K_a	Formula	Molar Mass
propionic	1.3×10^{-5}	C_2H_5COOH	74.1
butanoic	1.5×10^{-5}	C_3H_7COOH	88.1
acetic	1.8×10^{-5}	CH_3COOH	60.1
hydroazoic	1.9×10^{-5}	HN_3	43.0

Of these, butanoic has the closest match for K_a and molar mass, but the agreement is not good.

17.112 $n = \dfrac{PV}{RT} = 735 \text{ torr} \times \dfrac{1 \text{ atm}}{760 \text{ torr}} \times \dfrac{7.5 \text{ L}}{295 \text{ K}} \times \dfrac{\text{mol-K}}{0.08206 \text{ L-atm}} = 0.300 = 0.30 \text{ mol } NH_3$

$0.40\,M \times 0.50\,L = 0.20 \text{ mol } HCl$

	$HCl(aq)$	+	$NH_3(g)$	$\rightarrow$	$NH_4^+(aq)$	+	$Cl^-(aq)$
before	0.20 mol		0.30 mol				
after	0		0.10 mol		0.20 mol		0.20 mol

The solution will be a buffer because of the substantial concentrations of NH_3 and NH_4^+ present. Use K_a for NH_4^+ to describe the equilibrium.

$$NH_4^+(aq) \rightleftharpoons NH_3(aq) + H^+(aq)$$

equil. $0.20 - x$ $0.10 + x$ x

$K_a = \dfrac{1.0 \times 10^{-14}}{1.8 \times 10^{-5}} = 5.56 \times 10^{-10} = 5.6 \times 10^{-10}$; $K_a = \dfrac{[NH_3][H^+]}{[NH_4^+]}$; $[H^+] = \dfrac{K_a[NH_4^+]}{[NH_3]}$

Because this expression contains a ratio of concentrations, volume will cancel and we can substitute moles directly. Assume x is small compared to 0.10 and 0.20.

$[H^+] = \dfrac{5.56 \times 10^{-10}\,(0.20)}{(0.10)} = 1.111 \times 10^{-9} = 1.1 \times 10^{-9}\,M$, $pH = 8.95$

17.113 Calculate the initial M of aspirin in the stomach and solve the equilibrium problem to find equilibrium concentrations of $C_8H_7O_2COOH$ and $C_8H_7O_2COO^-$. At $pH = 2$, $[H^+] = 1 \times 10^{-2}$.

$\dfrac{325 \text{ mg}}{\text{tablet}} \times 2 \text{ tablets} \times \dfrac{1\,g}{1000 \text{ mg}} \times \dfrac{1 \text{ mol } C_8H_7O_2COOH}{180.2\,g\ C_8H_7O_2COOH} \times \dfrac{1}{1\,L} = 3.61 \times 10^{-3} = 4 \times 10^{-3}\,M$

	$C_8H_7O_2COOH(aq)$	$\rightleftharpoons$	$C_8H_7O_2COO^-(aq)$	+	$H^+(aq)$
initial	$3.61 \times 10^{-3}\,M$		0		$1 \times 10^{-2}\,M$
equil.	$(3.61 \times 10^{-3} - x)\,M$		$x\,M$		$(1 \times 10^{-2} + x)\,M$

$K_a = 3 \times 10^{-5} = \dfrac{[H^+][C_8H_7O_2COO^-]}{[C_8H_7O_2COOH]} = \dfrac{(0.01+x)(x)}{(3.61 \times 10^{-3} - x)} \approx \dfrac{0.01\,x}{3.61 \times 10^{-3}}$

$x = [C_8H_7O_2COO^-] = 1.08 \times 10^{-5} = 1 \times 10^{-5}\,M$

$\% \text{ ionization} = \dfrac{1.08 \times 10^{-5}\,M\ C_8H_7O_2COO^-}{3.61 \times 10^{-3}\,M\ C_8H_7O_2COOH} \times 100 = 0.3\%$

(% ionization is small, so the approximation was valid.)

% aspirin molecules = $100.0\% - 0.3\% = 99.7\%$ molecules

17.114 According to Equation 13.4, $S_g = kP_g$

$S_{CO_2} = 3.1 \times 10^{-2}\,\dfrac{\text{mol}}{\text{L-atm}} \times 1.10 \text{ atm} = 0.0341 = \dfrac{0.034 \text{ mol}}{L} = 0.034\,M\ CO_2$

$CO_2(g) + H_2O(l) \rightarrow H_2CO_3(aq)$; $0.0341\,M\ CO_2 = 0.0341\,M\ H_2CO_3$

Consider the stepwise dissociation of $H_2CO_3(aq)$.

	$H_2CO_3(aq)$	$\rightleftharpoons$	$H^+(aq)$	$+$	$HCO_3^-(aq)$
initial	0.0341 M		0		0
equil.	$(0.0341 - x)$ M		x		x

$$K_{a1} = \frac{[H^+][HCO_3^-]}{[H_2CO_3]} = \frac{x^2}{(0.0341-x)} \approx \frac{x^2}{0.0341} \approx 4.3 \times 10^{-7}$$

$x^2 = 1.47 \times 10^{-8}$; $x = 1.2 \times 10^{-4}$ M H^+; pH = 3.92

$K_{a2} = 5.6 \times 10^{-11}$; assume the second ionization does not contribute significantly to $[H^+]$.

17.115 $Ca(OH)_2(aq) + 2\,HCl(aq) \rightarrow CaCl_2(aq) + 2\,H_2O$

mmol HCl = $M \times$ mL = 0.0983 $M \times$ 11.23 mL = 1.1039 = 1.10 mmol HCl

mmol $Ca(OH)_2$ = mmol HCl/2 = 1.1039/2 = 0.55195 = 0.552 mmol $Ca(OH)_2$

$$[Ca^{2+}] = \frac{0.55195 \text{ mmol}}{50.00 \text{ mL}} = 0.01104 = 0.0110 \text{ } M$$

$[OH^-] = 2[Ca^{2+}] = 0.02208 = 0.0221$ M

$K_{sp} = [Ca^{2+}][OH^-]^2 = (0.01104)(0.02208)^2 = 5.38 \times 10^{-6}$

The value in Appendix D.3 is 6.5×10^{-6}, a difference of 17%. Because a change in temperature does change the value of an equilibrium constant, the solution may not have been kept at 25 $^\circ$C. It is also possible that experimental errors led to the difference in K_{sp} values.

17.116 $\Pi = MRT$, $M = \dfrac{\Pi}{RT} = \dfrac{21 \text{ torr}}{298 \text{ K}} \times \dfrac{1 \text{ atm}}{760 \text{ torr}} \times \dfrac{\text{mol-K}}{0.08206 \text{ L-atm}} = 1.13 \times 10^{-3} = 1.1 \times 10^{-3}$ M

$SrSO_4(s) \rightleftharpoons Sr^{2+}(aq) + SO_4^{2-}(aq)$; $K_{sp} = [Sr^{2+}][SO_4^{2-}]$

The total particle concentration is 1.13×10^{-3} M. Each mole of $SrSO_4$ that dissolves produces 2 mol of ions, so $[Sr^{2+}] = [SO_4^{2-}] = 1.13 \times 10^{-3}$ $M/2 = 5.65 \times 10^{-4} = 5.7 \times 10^{-4}$ M.

$K_{sp} = (5.65 \times 10^{-4})^2 = 3.2 \times 10^{-7}$

17.117 For very dilute aqueous solutions, assume the solution density is 1 g/mL.

$$ppb = \frac{g \text{ solute}}{10^9 \text{ g solution}} = \frac{1 \times 10^{-6} \text{ g solute}}{1 \times 10^3 \text{ g solution}} = \frac{\mu g \text{ solute}}{L \text{ solution}}$$

(a) $K_{sp} = [Ag^+][Cl^-] = 1.8 \times 10^{-10}$; $[Ag^+] = (1.8 \times 10^{-10})^{1/2} = 1.34 \times 10^{-5} = 1.3 \times 10^{-5}$ M

$$\frac{1.34 \times 10^{-5} \text{ mol Ag}^+}{L} \times \frac{107.9 \text{ g Ag}^+}{1 \text{ mol Ag}^+} \times \frac{1 \mu g}{1 \times 10^{-6} \text{ g}} = \frac{1.4 \times 10^3 \text{ } \mu g \text{ Ag}^+}{L}$$

$$= 1.4 \times 10^3 \text{ ppb} = 1.4 \text{ ppm}$$

(b) $K_{sp} = [Ag^+][Br^-] = 5.0 \times 10^{-13}$; $[Ag^+] = (5.0 \times 10^{-13})^{1/2} = 7.07 \times 10^{-7} = 7.1 \times 10^{-7}$ M

$$\frac{7.07 \times 10^{-7} \text{ mol Ag}^+}{L} \times \frac{107.9 \text{ g Ag}^+}{1 \text{ mol Ag}^+} \times \frac{1 \mu g}{1 \times 10^{-6} \text{ g}} = 76 \text{ ppb}$$

(c) $K_{sp} = [Ag^+][I^-] = 8.3 \times 10^{-17}; [Ag^+] = (8.3 \times 10^{-17})^{1/2} = 9.11 \times 10^{-9} = 9.1 \times 10^{-9} \, M$

$$\frac{9.11 \times 10^{-9} \, \text{mol Ag}^+}{L} \times \frac{107.9 \, \text{g Ag}^+}{1 \, \text{mol Ag}^+} \times \frac{1 \, \mu g}{1 \times 10^{-6} \, g} = 0.98 \, \text{ppb}$$

AgBr(s) would maintain $[Ag^+]$ in the correct range.

17.118   To determine precipitation conditions, we must know K_{sp} for CaF_2(s) and calculate Q under the specified conditions. $K_{sp} = 3.9 \times 10^{-11} = [Ca^{2+}][F^-]^2$

$[Ca^{2+}]$ and $[F^-]$. The term 1 ppb means 1 part per billion or 1 g solute per billion g solution. Assume that the density of this very dilute solution is the density of water.

$$1 \, \text{ppb} = \frac{1 \, \text{g solute}}{1 \times 10^9 \, \text{g solution}} \times \frac{1 \, \text{g solution}}{1 \, \text{mL solution}} \times \frac{1 \times 10^3 \, \text{mL}}{1 \, L} = \frac{1 \times 10^{-6} \, \text{g solute}}{1 \, L \, \text{solution}}$$

$$\frac{1 \times 10^{-6} \, \text{g solute}}{1 \, L \, \text{solution}} \times \frac{1 \, \mu g}{1 \times 10^{-6} \, g} = 1 \, \mu g / 1 \, L$$

$$8 \, \text{ppb Ca}^{2+} \times \frac{1 \, \mu g}{1 \, L} = \frac{8 \, \mu g \, \text{Ca}^{2+}}{1 \, L} = \frac{8 \times 10^{-6} \, \text{g Ca}^{2+}}{1 \, L} \times \frac{1 \, \text{mol Ca}^{2+}}{40 \, g} = 2 \times 10^{-7} \, M \, \text{Ca}^{2+}$$

$$1 \, \text{ppb F}^- \times \frac{1 \, \mu g}{1 \, L} = \frac{1 \, \mu g \, F^-}{1 \, L} = \frac{1 \times 10^{-6} \, \text{g F}^-}{1 \, L} \times \frac{1 \, \text{mol F}^-}{19.0 \, g} = 5 \times 10^{-8} \, M \, F^-$$

$Q = [Ca^{2+}][F^-]^2 = (2 \times 10^{-7})(5 \times 10^{-8})^2 = 5 \times 10^{-22}$

$5 \times 10^{-22} < 3.9 \times 10^{-11}$, $Q < K_{sp}$, no CaF_2 will precipitate

17.119   (a) $CH_3CH(OH)COOH(aq) + HCO_3^-(aq) \rightarrow H_2CO_3(aq) + CH_3CH(OH)COO^-(aq)$

$H_2CO_3(aq) \rightleftharpoons H_2O(l) + CO_2(g)$

(b) $$\frac{2.16 \, \text{g NaHCO}_3}{mL} \times \frac{236.6 \, mL}{48 \, \text{tsp}} \times \frac{1 \, \text{mol NaHCO}_3}{84.01 \, \text{g NaHCO}_3} \times 0.5 \, \text{tsp} = 0.06337$$

$$= 0.0634 \, \text{mol NaHCO}_3$$

mol $NaHCO_3$ = mol HCO_3^-;
at neutralization, mol HCO_3^- = mol $CH_3CH(OH)COOH$

$$\frac{0.06337 \, \text{mol CH}_3CH(OH)COOH}{1 \, \text{cup milk}} \times \frac{1 \, \text{cup}}{236.6 \, mL} \times \frac{1000 \, mL}{L} = 0.2678 =$$

$$0.268 \, M \, CH_3CH(OH)COOH$$

(c) $5/9(^\circ F - 32) = {}^\circ C$; $5/9(350 - 32) = 176.67 = 177 \, {}^\circ C$; $177 \, {}^\circ C + 273 = 450 \, K$

mol CO_2 = mol HCO_3^-

$$V = \frac{nRT}{P} = 0.06337 \, \text{mol} \times \frac{450 \, K}{1 \, \text{atm}} \times \frac{0.08206 \, \text{L-atm}}{\text{mol-K}} = 2.34 \, L \, CO_2$$

17.120   Statements (b) and (d) follow from this observation.

When HF behaves like this, it is acting like a base, rather than an acid. This agrees with statements (b) and (d) and renders (a) incorrect. As for (c), if HF were thermodynamically unstable, it would decompose rather than gain H^+.

18 Chemistry of the Environment

Visualizing Concepts

18.1   *Analyze.* Given that one mole of an ideal gas at 1 atm and 298 K occupies 22.4 L, is the volume of one mole of ideal gas in the middle of the stratosphere greater or less than 22.4 L?

Plan. Consider the relationship between pressure, temperature, and volume of an ideal gas. Use Figure 18.1 to estimate the pressure and temperature in the middle of the stratosphere, and compare the two sets of temperature and pressure.

Solve. According to the ideal-gas law, PV = nRT, so V = nRT/P. Because n and R are constant for this exercise, V is proportional to T/P.

(a)   Greater. The stratosphere ranges from 10 to 50 km, so the middle is at approximately 30 km. At this altitude, T ≈ 230 K, P ≈ 40 torr (from Figure 18.1). Because we are comparing T/P ratios, either atm or torr can be used as pressure units; we will use torr.

At sea level: T/P = 298 K/760 torr = 0.39

At 30 km: T/P = 230 K/40 torr = 5.75

The proportionality constant (T/P) is much greater at 30 km than sea level, so the volume of 1 mol of an ideal gas is greater at this altitude. The decrease in temperature at 30 km is more than offset by the substantial decrease in pressure.

(b)   No. Volume is proportional to T/P, not simply T. The relative volumes of one mole of an ideal gas at 50 km and 85 km depend on the temperature and pressure at the two altitudes. From Figure 18.1,

50 km: T ≈ 270 K, P ≈ 20 torr, T/P = 270 K/20 torr = 13.5

85 km: T ≈ 190 K, P < 0.01 torr, T/P = 190 K/0.01 torr = 19,000

Again, the slightly lower temperature at 85 km is more than offset by a much lower pressure. One mole of an ideal gas will occupy a much larger volume at 85 km than 50 km.

(c)   The thermosphere, stratopause, and low-altitude troposphere. Gases behave most ideally at high temperature and low pressure. Pressure is minimum and temperature is high in the thermosphere. The stratopause (the boundary between the stratosphere and mesosphere) and the troposphere at low altitude are other regions with temperature maxima and relatively low pressures.

18.2   Molecules in the upper atmosphere tend to have multiple bonds because they have sufficiently high bond dissociation enthalpies (Table 8.3) to survive the incoming high-energy radiation from the Sun. According to Table 8.3, for the same two bonded atoms, multiple bonds have higher bond dissociation enthalpies than single bonds. Molecules with single bonds are likely to undergo photodissociation in the presence of the high-energy, short-wavelength solar radiation present in the upper atmosphere.

18.3   (a)   A = troposphere, 0–10 km; B = stratosphere, 12–50 km; C = mesosphere, 50–85 km.

(b)   Ozone is a pollutant in the troposphere and filters UV radiation in the stratosphere.

(c)   Infrared radiation from Earth is most strongly reflected back in the troposphere.

(d)   Assuming the "boundary" between the stratosphere and mesosphere is at 50 km, only region C in the diagram is involved in an aurora borealis.

(e)   The concentration of water vapor is greatest near Earth's surface in region A and decreases with altitude. Water's single bonds are susceptible to photodissociation in regions B and C, so its concentration is likely to be very low in these regions. The relative concentration of CO_2, with strong double bonds, increases in regions B and C, because it is less susceptible to photodissociation.

18.4   *Analyze.* Given granite, marble, bronze, and other solid materials, what observations and measurements indicate whether the material is appropriate for an outdoor sculpture? If the material changes (erodes) over time, what chemical processes are responsible?

Plan. An appropriate material resists chemical and physical changes when exposed to environmental conditions. An inappropriate material undergoes chemical reactions with substances in the troposphere, degrading the structural strength of the material and the sculpture.   *Solve.*

(a)   The appearance and mass of the material upon environmental exposure are indicators of both chemical and physical changes. If the appearance and mass of the material are unchanged after a period of time, the material is well suited for the sculpture because it is inert to chemical and physical changes. Changes in the color or texture of the material's surface indicate that a chemical reaction has occurred, because a different substance with different properties has formed. A decrease in mass indicates that some of the material has been lost, by either chemical reaction or physical change. An increase in mass indicates corrosion. If the mass of the material is unchanged, it is probably inert to chemical and physical environmental changes and suitable for sculpture.

(b)   The two main chemical processes that lead to erosion are reaction with acid rain and corrosion or air oxidation, which is encouraged by acid conditions (see Section 20.8).

Acid rain is primarily H_2SO_3 and/or H_2SO_4, which reacts directly with carbonate minerals such as marble and limestone. Acidic conditions created by acid rain encourage corrosion of metals such as iron, steel, and bronze. Corrosion produces metal oxides, which may or may not cling to the surface of the material. If the oxides are washed away, the material will lose mass after corrosion. Physical erosion due to the effects of wind and rain on soft materials such as sandstone also causes mass to decrease.

18.5   The Sun.

18.6   *Analyze/Plan.* Given salinity, calculate salt concentration in ppm. A salinity of 35 denotes that there are 35 g of dry salt per kg of seawater.

ppm = g salt/1×10^6 g seawater.   *Solve.*

$$\frac{35 \text{ g salt}}{1 \text{ kg seawater}} \times \frac{1 \text{ kg seawater}}{1000 \text{ g seawater}} \times 1 \times 10^6 = 3.5 \times 10^4 \text{ ppm salt}$$

35,000 ppm total salt – 500 ppm salt remain = 34,500 ppm salt must be removed

$$\frac{34,500 \text{ ppm salt removed}}{35,000 \text{ ppm salt total}} \times 100 = 98.57 = 99\% \text{ salt removed}$$

Of the total dissolved salts in seawater, 99% must be removed in order for the water to be considered freshwater.

18.7 $CO_2(g)$ dissolves in seawater to form $H_2CO_3(aq)$. The basic pH of the ocean encourages ionization of $H_2CO_3(aq)$ to form $HCO_3^-(aq)$ and $CO_3^{2-}(aq)$. Under the correct conditions, carbon is removed from the ocean as $CaCO_3(s)$ (sea shells, coral, chalk cliffs). As carbon is removed, more $CO_2(g)$ dissolves to maintain the balance of complex and interacting acid–base and precipitation equilibria.

18.8   The first stage at the Carlsbad plant is a physical filtration using anthracite coal (carbon), sand, and gravel as the porous medium. This filtration will remove particles larger than the smallest filter pore size from the remaining solution. It will not remove dissolved salts; ions are much smaller than the pore size of any part of the filter.

18.9   *Plan.* Follow the yellow arrows on the diagram to find potential routes for environmental contamination at a fracking well site.   *Solve.*

Above ground, the two main avenues for escape of contaminants are leakage and evaporation. Methane, hydrogen sulfide, and other petroleum compounds that are gases at atmospheric conditions can leak from the well head. Heavier petroleum compounds (those containing from three to six carbon atoms) are also released by fracking, mix with the resulting aqueous solution, and end up in the wastewater ponds. Volatile organic compounds produced by fracking and from the fracking liquid can evaporate from the ponds. Wastewwater ponds, if unlined, can seep into nearby water sources or over flow due to rainfall.

Below ground, petroleum gases and fracking liquid can migrate into groundwater, both deep and shallow aquifers. (Fracking mobilizes petroleum and gases, which enables these underground "leaks.")

The extent of underground leaking depends on the integrity of the well plumbing and casings, as well as the real permeability of the "impermeable" layers. Although there are several avenues for contamination, they can be minimized by careful attention to geology, engineering, and oversight.

18.10   Some of the missing CO_2 is absorbed by "land plants" (vegetation other than trees). These plants are directly or indirectly used as food by land animals, or they complete an uninterrupted life cycle. Either way, natural decomposition occurs and the remains are incorporated into the soil. (Fossil fuels were formed millions of years ago by natural decomposition of buried dead organisms.) Soil is the largest land-based carbon reservoir. The amount of carbon-storing capacity of soil is affected by erosion, soil fertility, and other complex factors. For more details, search the Internet for "carbon budget."

Earth's Atmosphere (Section 18.1)

18.11   (a)   The temperature profile of the atmosphere (Figure 18.1) is the basis of its division into regions. The center of each peak or trough in the temperature profile corresponds to a new region.

        (b)   Troposphere, 0–12 km; stratosphere, 12–50 km; mesosphere, 50–85 km; thermosphere, 85–110 km.

18.12   (a)   Boundaries between regions of the atmosphere are at maxima and minima (peaks and valleys) in the atmospheric temperature profile. For example, in the troposphere, temperature decreases with altitude, whereas in the stratosphere, it increases with altitude. The temperature minimum is the tropopause boundary.

        (b)   From Figure 18.1, atmospheric pressure in the troposphere ranges from 760 torr to 200 torr, whereas pressure in the stratosphere ranges from 200 torr to 20 torr. Gas density (g/L) is directly proportional to pressure. The much lower density of the stratosphere means it has the smaller mass, despite having a larger volume than the troposphere.

18.13   *Analyze/Plan.* Given O_3 concentration in ppm, calculate partial pressure. Use the definition of ppm to get mol fraction O_3. For gases, mole fraction = pressure fraction. Use the ideal-gas law to find mol O_3/L air and Avogadro's number to get molecules.

$$P_{O_3} = X_{O_3} \times P_{atm}; \quad 0.441 \text{ ppm } O_3 = \frac{0.441 \text{ mol } O_3}{1 \times 10^6 \text{ mol air}} = 4.41 \times 10^{-7} = X_{O_3} \quad \text{Solve.}$$

        (a) $P_{O_3} = X_{O_3} \times P_{atm} = 4.41 \times 10^{-7}(0.67 \text{ atm}) = 2.955 \times 10^{-7} = 3.0 \times 10^{-7} \text{ atm}$

        (b) $n = \dfrac{PV}{RT} = \dfrac{2.955 \times 10^{-7} \text{ atm} \times 1.0 \text{ L}}{298 \text{ K}} \times \dfrac{\text{mol-K}}{0.08206 \text{ L-atm}} = 1.208 \times 10^{-8} = 1.2 \times 10^{-8} \text{ mol } O_3$

$$1.208 \times 10^{-8} \text{ mol } O_3 \times \frac{6.022 \times 10^{23} \text{ molecules}}{\text{mol}} = 7.277 \times 10^{15} = 7.3 \times 10^{15} \text{ } O_3 \text{ molecules}$$

18.14 $P_{Ar} = X_{Ar} \times P_{atm}; \quad P_{Ar} = 0.00934 (1.05 \text{ bar}) = 0.009807 = 9.81 \times 10^{-3} \text{ bar}$

$$P_{Ar} = 0.009807 \text{ bar} \times \frac{10^5 \text{ Pa}}{\text{bar}} \times \frac{760 \text{ torr}}{101,325 \text{ Pa}} = 7.3559 = 7.36 \text{ torr}$$

$P_{CO_2} = X_{CO_2} \times P_{atm}; \quad P_{CO_2} = 0.000400 (1.05 \text{ bar}) = 0.0004200 = 4.20 \times 10^{-4} \text{ bar}$

$$P_{CO_2} = 0.0004200 \text{ bar} \times \frac{10^5 \text{ Pa}}{\text{bar}} \times \frac{760 \text{ torr}}{101,325 \text{ Pa}} = 0.3150 = 0.315 \text{ torr}$$

18.15   *Analyze/Plan.* Given CO concentration in ppm, calculate number of CO molecules in 1.0 L air at given conditions. ppm CO $\rightarrow X_{O_3} \rightarrow$ atm CO $\rightarrow$ mol CO $\rightarrow$ molecules CO. Use the ideal-gas law to change atm CO to mol CO and Avogadro's number to get molecules. *Solve.*

$$3.5 \text{ ppm CO} = \frac{3.5 \text{ mol CO}}{1 \times 10^6 \text{ mol air}} = 3.5 \times 10^{-6} = X_{CO}$$

$$P_{CO} = X_{CO} \times P_{atm} = 3.5 \times 10^{-6} \times 759 \text{ torr} \times \frac{1 \text{ atm}}{760 \text{ torr}} = 3.495 \times 10^{-6} = 3.5 \times 10^{-6} \text{ atm}$$

$$n_{CO} = \frac{P_{CO}V}{RT} = \frac{3.495 \times 10^{-6} \text{ atm} \times 1.0 \text{ L}}{295 \text{ K}} \times \frac{\text{mol-K}}{0.08206 \text{ L-atm}} = 1.444 \times 10^{-7} = 1.4 \times 10^{-7} \text{ mol CO}$$

$$1.444 \times 10^{-7} \text{ mol CO} \times \frac{6.022 \times 10^{23} \text{ molecules}}{\text{mol}} = 8.695 \times 10^{16} = 8.7 \times 10^{16} \text{ CO molecules}$$

18.16   (a)   ppm Ne = mol Ne/1×10^6 mol air; $X_{Ne} = 1.818 \times 10^{-5}$ mol Ne/mol air

$$\frac{1.818 \times 10^{-5} \text{ mol Ne}}{1 \text{ mol air}} = \frac{x \text{ mol Ne}}{1 \times 10^6 \text{ mol air}}; \ x = 18.18 \text{ ppm Ne}$$

   (b) $P_{Ne} = X_{Ne} \times P_{atm} = 1.818 \times 10^{-5} \times 730 \text{ torr} \times \dfrac{1 \text{ atm}}{760 \text{ torr}} = 1.7462 \times 10^{-5} = 1.75 \times 10^{-5} \text{ atm}$

       T = 296 K

$$\frac{n_{Ne}}{V} = \frac{P_{Ne}}{RT} = \frac{1.7462 \times 10^{-5} \text{ atm}}{296 \text{ K}} \times \frac{\text{mol-K}}{0.08206 \text{ L-atm}} = 7.1892 \times 10^{-7} = 7.19 \times 10^{-7} \text{ mol/L}$$

$$\frac{7.1892 \times 10^{-7} \text{ mol Ne}}{L} \times \frac{6.022 \times 10^{23} \text{ atoms}}{\text{mol}} = 4.3293 \times 10^{17}$$

$$= 4.33 \times 10^{17} \text{ Ne atoms/L}$$

18.17   *Analyze/Plan.* Given bond dissociation energy in kJ/mol, calculate the wavelength of a single photon that will rupture a C–Br bond. kJ/mol → J/molecule. $\lambda = hc/E$. ($\lambda = hc/E$ describes the energy/wavelength relationship of a single photon.)   *Solve.*

   (a) $\dfrac{276 \times 10^3 \text{ J}}{1 \text{ mol}} \times \dfrac{1 \text{ mol}}{6.022 \times 10^{23} \text{ molecules}} = 4.583 \times 10^{-19} = 4.58 \times 10^{-19}$ J/molecule

$$\lambda = \frac{hc}{E} = \frac{(6.626 \times 10^{-34} \text{ J-sec})(3.00 \times 10^8 \text{ m/sec})}{4.583 \times 10^{-19} \text{ J}} = 4.337 \times 10^{-7} \text{ m} = 434 \text{ nm}$$

   (b)   This 434 nm wavelength is visible electromagnetic radiation.

18.18 $\dfrac{339 \times 10^3 \text{ J}}{1 \text{ mol}} \times \dfrac{1 \text{ mol}}{6.022 \times 10^{23} \text{ molecules}} = 5.6294 \times 10^{-19} = 5.63 \times 10^{-19}$ J/molecule

$$\lambda = \frac{hc}{E} = \frac{(6.626 \times 10^{-34} \text{ J-sec})(3.00 \times 10^8 \text{ m/sec})}{5.6294 \times 10^{-19} \text{ J}} = 3.53 \times 10^{-7} \text{ m} = 353 \text{ nm}$$

$$\frac{293 \times 10^3 \text{ J}}{1 \text{ mol}} \times \frac{1 \text{ mol}}{6.022 \times 10^{23} \text{ molecules}} = 4.8655 \times 10^{-19} = 4.87 \times 10^{-19} \text{ J/molecule}$$

$$\lambda = \frac{(6.626 \times 10^{-34} \text{ J-sec})(3.00 \times 10^8 \text{ m/sec})}{4.8655 \times 10^{-19} \text{ J}} = 4.09 \times 10^{-7} \text{ m} = 409 \text{ nm}$$

Photons of wavelengths longer than 409 nm cannot cause rupture of the C–Cl bond in either CF_3Cl or CCl_4. Photons with wavelengths between 409 and 353 nm can cause C–Cl bond rupture in CCl_4, but not in CF_3Cl.

18 Chemistry of the Environment   Solutions to Exercises

18.19 (a) *Photodissociation* is cleavage of the O=O bond such that two neutral O atoms are produced: $O_2(g) \rightarrow 2O(g)$.

Photoionization is absorption of a photon with sufficient energy to eject an electron from an O_2 molecule: $O_2(g) + h\nu \rightarrow O_2^+ + e^-$.

(b) Photoionization of O_2 requires 1205 kJ/mol. Photodissociation requires only 495 kJ/mol. At lower elevations, solar radiation with wavelengths corresponding to 1205 kJ/mol or shorter has already been absorbed, whereas the longer wavelength radiation has passed through relatively well. Below 90 km, the increased concentration of O_2 and the availability of longer wavelength radiation cause the photodissociation process to dominate.

18.20 Photodissociation of N_2 is relatively unimportant compared to photodissociation of O_2 for two reasons. The bond dissociation energy of N_2, 941 kJ/mol, is much higher than that of O_2, 495 kJ/mol. Photons with a wavelength short enough to photodissociate N_2 are not as abundant as the ultraviolet photons that lead to photodissociation of O_2. Also, N_2 does not absorb these photons as readily as O_2 so even if a short-wavelength photon is available, it may not be absorbed by an N_2 molecule.

18.21 (a) A wavelength of 145 nm is in the ultraviolet portion of the electromagnetic spectrum. (See Figure 6.4.)

(b) *Analyze/Plan.* $E = hc/\lambda$. 145 nm = 1.45×10^{-7} m. Change J/photon to kJ/mol. Compare to the bond energy of O_2, 495 kJ/mol. *Solve.*

$$E = \frac{hc}{\lambda} = \frac{(6.626 \times 10^{-34} \text{ J-sec})(3.00 \times 10^8 \text{ m/sec})}{1.45 \times 10^{-7} \text{ m}} = 1.37 \times 10^{-18} \text{ J/photon}$$

$$\frac{1.37 \times 10^{-18} \text{ J}}{\text{photon}} \times \frac{1 \text{ kJ}}{1000 \text{ J}} \times \frac{6.022 \times 10^{23} \text{ photons}}{1 \text{ mol}} = 825.55 = 826 \text{ kJ/mol}$$

The 145-nm photon has more than enough energy to photodissociate O_2.

According to Table 18.3, the photoionization energy of O_2 is 1205 kJ/mol. The 145-nm photon does not have enough energy to photoionize O_2.

18.22 (a) Energy is inversely related to wavelengths, $E = hc/\lambda$. Therefore, photons from UV-C, with the shortest wavelengths, have the highest energy and are most harmful to living tissue.

(b) Atmospheric N_2, O_2, and atomic oxygen absorb wavelengths shorter than 240 nm, a portion of UV-C. In the absence of ozone, wavelengths in UV-A and UV-B are not absorbed.

(c) No, even when appropriate concentrations of ozone are present in the stratosphere, not all UV light is absorbed before reaching Earth's surface. Nearly all of UV-C is absorbed by N_2, O_2, and atomic oxygen, and the remainder is absorbed by stratospheric ozone. Some, but not all of UV-B is absorbed by ozone. The UV-A region is not filtered by ozone, N_2, O_2, or atomic oxygen.

Human Activities and Earth's Atmosphere (Section 18.2)

18.23   The oxidation state of oxygen in O_3 , O_2, and O is zero (0). Reactions in which oxygen changes only from one of these species to another do not involve changes in oxidation state. Examples: $O(g) + O(g) \rightarrow O_2(g)$; $2\,O_3(g) \rightarrow 3\,O_2(g)$.

Ozone depletion reactions that involve a halogen oxide such as ClO do involve a change in oxidation state for oxygen. In ClO, the oxidation state of oxygen is either +1 or +2, but it is not zero. A reaction involving ClO and one of the oxygen species with a zero oxidation state does involve a change in the oxidation state of oxygen atoms.

18.24   Reactions (a) and (b).

It is those star* reactions, Equations 18.3–18.5. For example, the reaction of $O_2(g) + O(g)$ is exothermic and produces a high-energy $O_3^*(g)$ molecule with 105 kJ of energy to disperse. This energy is transferred through collisions, primarily to $N_2(g)$ and $O_2(g)$ molecules. The overall kinetic energy (translational, vibrational, and rotational energy) of these molecules (M*) increases and the temperature of the stratosphere is kept relatively high. (Recall that the temperature of a gas is directly proportional to its average kinetic energy.)

18.25   (a)   A *chlorofluorocarbon* is a compound that contains chlorine, fluorine, and carbon. A *hydrofluorocarbon* contains hydrogen, fluorine, and carbon; it contains hydrogen in place of chlorine.

(b)   CFCs are harmful because they undergo photodissociation to produce Cl atoms that catalyze the destructions of ozone. HFCs are potentially less harmful to the ozone layer because they contain no C–Cl bonds. Their relatively stronger C–F bonds require more energy to undergo photodissociation, energy that is unlikely to be available in the stratosphere. (HFCs are still a powerful greenhouse gas. Montreal Protocal members have recently agreed to limit HFCs as well as CFCs.)

18.26 $32\,e^-$, $16\,e^-$ pr

$$
\begin{array}{c}
\ddot{\mathrm{F}}: \\
| \\
:\ddot{\underset{\cdot\cdot}{\mathrm{Cl}}}\!-\!\mathrm{C}\!-\!\ddot{\underset{\cdot\cdot}{\mathrm{Cl}}}: \\
| \\
:\ddot{\underset{\cdot\cdot}{\mathrm{Cl}}}:
\end{array}
$$

CFC–11, $CFCl_3$, contains C–Cl bonds that can be cleaved by UV light in the stratosphere to produce Cl atoms. It is chlorine in atomic form that catalyzes the destruction of stratospheric ozone. CFC–11 is chemically inert and resists decomposition in the troposphere, so that it eventually reaches the stratosphere in molecular form.

18.27   (a)   *Analyze/Plan.* Given bond enthalpies in kJ/mol, calculate the maximum wavelength of a single photon that will rupture a C–F and a C–Cl bond, respectively. kJ/mol $\rightarrow$ J/molecule. λ = hc/E. (λ = hc/E describes the energy/wavelength relationship of a single photon.)   *Solve.*

$$\frac{485 \times 10^3\,\mathrm{J}}{1\,\mathrm{mol}} \times \frac{1\,\mathrm{mol}}{6.022 \times 10^{23}\,\mathrm{C-F\ bonds}} = 8.054 \times 10^{-19} = 8.05 \times 10^{-19}\,\mathrm{J/C-F\ bond}$$

$$\lambda = \frac{hc}{E} = \frac{(6.626 \times 10^{-34}\,\mathrm{J\text{-}sec})(3.00 \times 10^8\,\mathrm{m/sec})}{8.054 \times 10^{-19}\,\mathrm{J}} = 2.47 \times 10^{-7}\,\mathrm{m} = 247\,\mathrm{nm}$$

$$\frac{328 \times 10^3 \text{ J}}{1 \text{ mol}} \times \frac{1 \text{ mol}}{6.022 \times 10^{23} \text{ C–Cl bonds}} = 5.447 \times 10^{-19} = 5.45 \times 10^{-19} \text{ J/C–Cl bond}$$

$$\lambda = \frac{hc}{E} = \frac{(6.626 \times 10^{-34} \text{ J-sec})(3.00 \times 10^8 \text{ m/sec})}{5.447 \times 10^{-19} \text{ J}} = 3.65 \times 10^{-7} \text{ m} = 365 \text{ nm}$$

(b)   The maximum wavelength that can dissociate a C–F bond is 247 nm; shorter wavelengths are also effective. Since most wavelengths shorter than 240 nm are absorbed in the upper atmosphere, few effective photons will reach the lower atmosphere. We don't expect the photodissociation of C–F bonds to be significant in the lower atmosphere. (The dissociation of C–Cl bonds will be significant.)

18.28   (a)   The products of the reaction are $ClO(g)$ and $O_2(g)$.

        (b)   The average bond dissociation enthalpy of C–Cl is 328 kJ/mol, while that of C–Br is 276 kJ/mol (Table 8.3). Since a C–Cl bond has a higher dissociation enthalpy, the maximum wavelength required to dissociate it will be shorter and more energetic than the wavelength required to dissociate a C–Br bond. A photon capable of dissociating a C–Cl bond will also have energy sufficient to dissociate a C–Br bond.

        (c)   Yes, we expect the substance $CFBr_3$ to accelerate depletion of the ozone layer. Photons capable of dissociating C–Cl bonds are also capable of dissociating C–Br bonds. And, longer effective wavelengths are also available in the lower atmosphere. Bromine atoms will react with ozone to form $BrO(g)$ and $O_2(g)$. Ozone will be depleted.

18.29   *Analyze/Plan.* Write and balance equations for the reaction of $NO(g)$ and $NO_2(g)$ with water. First assume a simple acid–base reaction with a single product, $HNO_3(aq)$; these equations can't be balanced. Reaction of $NO(g)$ and $NO_2(g)$ to produce $HNO_3(aq)$ [or $HNO_2(aq)$] are redox reactions. An Internet search reveals that the reactions are as shown below. *Solve.*

$2 NO_2(g) + H_2O(l) \rightleftharpoons HNO_2(aq) + HNO_3(aq)$

$2 NO(g) + O_2(aq) + H_2O(l) \rightleftharpoons HNO_2(aq) + HNO_3(aq)$ or

$4 NO_2(g) + O_2(aq) + 2 H_2O(l) \rightleftharpoons 4 HNO_3(aq)$

$4 NO(g) + 3 O_2(aq) + 2 H_2O(l) \rightleftharpoons 4 HNO_3(aq)$

18.30   Rainwater is naturally acidic because of the presence of $CO_2(g)$ in the atmosphere. All oxides of nonmetals produce acidic solutions when dissolved in water. Even in the absence of polluting gases such as SO_2, SO_3, NO, and NO_2, CO_2 causes rainwater to be acidic. The important equilibria are:

$$CO_2(g) + H_2O(l) \rightleftharpoons H_2CO_3(aq) \rightleftharpoons H^+(aq) + HCO_3^-(aq)$$

18.31   (a)   Acid rain is primarily $H_2SO_4(aq)$.

             $H_2SO_4(aq) + CaCO_3(s) \rightarrow CaSO_4(s) + H_2O(l) + CO_2(g)$

        (b)   The $CaSO_4(s)$ would be much less reactive with acidic solution, because it would require a strongly acidic solution to shift the relevant equilibrium to the right.

             $CaSO_4(s) + 2 H^+(aq) \rightleftharpoons Ca^{2+}(aq) + 2 HSO_4^-(aq)$

             Note, however, that $CaSO_4(s)$ is brittle and easily dislodged; it provides none of the structural strength of limestone.

18.32   (a) $Fe(s) + O_2(g) + 4\,H_3O^+(aq) \rightarrow Fe^{2+}(aq) + 6\,H_2O(l)$

       (b)   No. Silver is a "noble" metal. It is relatively resistant to oxidation, and much more resistant than iron. In Table 4.5, The Activity Series of Metals in Aqueous Solution, Ag is much, much lower than Fe and it is below hydrogen, whereas Fe is above hydrogen. This means that Fe is susceptible to oxidation by acid, but Ag is not.

18.33   *Analyze/Plan.* Given wavelength of a photon, place it in the electromagnetic spectrum, calculate its energy in kJ/mol, and compare it to an average bond dissociation energy. Use Figure 6.4; $E(J/photon) = hc/\lambda$. J/photon → kJ/mol.   *Solve.*

       (a)   Ultraviolet (Figure 6.4).

       (b) $E_{photon} = hc/\lambda = \dfrac{6.626 \times 10^{-34}\ \text{J-s} \times 3.00 \times 10^8\ \text{m/s}}{335 \times 10^{-9}\ \text{m}} = 5.934 \times 10^{-19}$

$$= 5.93 \times 10^{-19}\ \text{J/photon}$$

$$\dfrac{5.934 \times 10^{-19}\ \text{J}}{1\ \text{photon}} \times \dfrac{6.022 \times 10^{23}\ \text{photons}}{1\ \text{mol}} \times \dfrac{1\ \text{kJ}}{1000\ \text{J}} = 357\ \text{kJ/mol}$$

       (c)   The average C–H bond energy from Table 8.3 is 413 kJ/mol. The energy calculated in part (b), 357 kJ/mol, is the energy required to break 1 mol of C–H bonds in formaldehyde, CH_2O. The C–H bond energy in CH_2O must be less than the "average" C–H bond energy.

       (d) $H-\overset{\displaystyle :\!O\!:}{\underset{\displaystyle \|}{C}}-H + h\nu \longrightarrow H-\overset{\displaystyle :\!O\!:}{\underset{\displaystyle \|}{C}}\cdot + H\cdot$

18.34   (a)   Visible (Figure 6.4).

       (b) $E_{photon} = hc/\lambda = \dfrac{6.626 \times 10^{-34}\ \text{J-s} \times 3.00 \times 10^8\ \text{m/s}}{420 \times 10^{-9}\ \text{m}} = 4.733 \times 10^{-19}$

$$= 4.73 \times 10^{-19}\ \text{J/photon}$$

$$\dfrac{4.733 \times 10^{-19}\ \text{J}}{1\ \text{photon}} \times \dfrac{6.022 \times 10^{23}\ \text{photons}}{1\ \text{mol}} \times \dfrac{1\ \text{kJ}}{1000\ \text{J}} = 285\ \text{kJ/mol}$$

       (c) $\ddot{O}{=}\dot{N}{-}\ddot{\underset{..}{O}}{:} + h\nu \longrightarrow \ddot{O}{=}\dot{N}\cdot + :\ddot{\underset{.}{O}}\cdot$

18.35   (a)   Four sources transfer energy to the atmosphere, surface radiation, evapotranspiration, incoming solar radiation, and convective heating. Surface radiation makes the largest contribution. The total energy absorbed by the atmosphere is $[350 + 78 + 67 + 24] = 519\ \text{W/m}^2$.

       (b)   Of the $519\ \text{W/m}^2$ transferred to the atmosphere, $324\ \text{W/m}^2$ are radiated back to the surface. The percentage is $(324/519) \times 100 = 62.4\%$.

18.36   There is a much larger temperature variation from day to night on Mars than on Earth. This indicates that the greenhouse gases in Mars' atmosphere do not absorb infrared radiation as effectively as those on Earth. Even though atmospheric pressure at the surface of Mars is only 6×10^{-3} atm, 96% of that is due to CO_2. The partial pressure of CO_2 in Mars' atmosphere is then 0.0058 atm (0.006 atm to 1 sig fig). The partial pressure of CO_2 in Earth's atmosphere is only 0.0004 atm, because the amount of CO_2 in Earth's atmosphere is so much smaller. If CO_2 were the main greenhouse gas contributing to daily temperature variation, Mars, with more CO_2, would have a smaller variation than Earth, not larger. The presence of potent greenhouse gas $H_2O(g)$ in Earth's atmosphere is the main reason for our small overnight temperature variation. Thus, the composition of the Mars atmosphere, the absence of water vapor, plays the largest role in the wide daily temperature variation.

Earth's Water (Section 18.3)

18.37   *Analyze/Plan.* Given salinity and density, calculate molarity. A salinity of 5.6 denotes that there are 5.6 g of dry salt per kg of water. 1.03 g/mL = 1.03 kg/L   *Solve.*

$$\frac{5.6 \text{ g NaCl}}{1 \text{ kg soln}} \times \frac{1.03 \text{ kg soln}}{1 \text{ L soln}} \times \frac{1 \text{ mol NaCl}}{58.44 \text{ g NaCl}} \times \frac{1 \text{ mol Na}^+}{1 \text{ mol NaCl}} = 0.0987 = 0.099 \ M \ \text{Na}^+$$

18.38   If the phosphorous is present as $H_2PO_4^-$, there is a 1:1 ratio between the molarity of phosphorus and molarity of phosphate. Thus, we can calculate the molarity based on the given mass of P.

$$\frac{0.07 \text{ g P}}{1 \times 10^6 \text{ g H}_2\text{O}} \times \frac{1 \text{ mol P}}{31 \text{ g P}} \times \frac{1 \text{ mol PO}_4^{3-}}{1 \text{ mol P}} \times \frac{1 \times 10^3 \text{ g H}_2\text{O}}{1 \text{ L H}_2\text{O}} = 2.26 \times 10^{-6} = 2 \times 10^{-6} \ M \ \text{PO}_4^{3-}$$

18.39   *Analyze/Plan.* Given the power of sunlight per square meter striking Earth's surface, the enthalpy of evaporation of water, and specific heat capacity of water, calculate the amount of energy delivered by the Sun over a 12-hour day. Use this amount of energy to calculate: (a) how many grams of water can be evaporated and (b) the temperature of a 10.0 cm by 1 square meter volume of water after 12 hours in the sunlight, assuming no evaporation. Calculate the mass of this volume of water using density at 25 $^\circ$C.   *Solve.*

(a) $\dfrac{168 \text{ W}}{\text{m}^2} \times \dfrac{1 \text{ J/s}}{1 \text{ W}} = \dfrac{168 \text{ J}}{\text{m}^2\text{-s}}$

$\dfrac{168 \text{ J}}{\text{m}^2\text{-s}} \times 1.00 \text{ m}^2 \times 12 \text{ h} \times \dfrac{60 \text{ min}}{1 \text{ h}} \times \dfrac{60 \text{ s}}{1 \text{ min}} \times \dfrac{1 \text{ kJ}}{1000 \text{ J}} = 7257.6 = 7.26 \times 10^3 \text{ kJ}$

$7257.6 \text{ kJ} \times \dfrac{1 \text{ mol H}_2\text{O}}{40.67 \text{ kJ}} \times \dfrac{18.02 \text{ g H}_2\text{O}}{1 \text{ mol H}_2\text{O}} = 3215.7 = 3.22 \times 10^3 \text{ g H}_2\text{O}$

(b) $1.00 \text{ m}^2 \times 10.0 \text{ cm} \times \dfrac{(100)^2 \text{ cm}^2}{1 \text{ m}^2} \times \dfrac{0.99707 \text{ g}}{1 \text{ cm}^3} = 99{,}707 = 9.97 \times 10^4 \text{ gH}_2\text{O}$

$7257.6 \text{ kJ} \times \dfrac{1000 \text{ J}}{1 \text{ kJ}} \times \dfrac{1 \text{ g-}^\circ\text{C}}{4.184 \text{ J}} \times \dfrac{1}{99{,}707 \text{ g H}_2\text{O}} = 17.397 = 17.4 \ ^\circ\text{C}$

The final temperature is 26 $^\circ$C + 17.4 $^\circ$C = 43.4 $^\circ$C.

18.40   (a) $\dfrac{168\ W}{m^2} \times \dfrac{1\ J/s}{1\ W} = \dfrac{168\ J}{m^2\text{-}s}$

$\dfrac{168\ J}{m^2\text{-}s} \times 1.00\ m^2 \times 12\ h \times \dfrac{60\ min}{1\ h} \times \dfrac{60\ s}{1\ min} \times \dfrac{1\ kJ}{1000\ J} = 7257.6 = 7.26 \times 10^3\ kJ$

$7257.6\ kJ \times \dfrac{1\ mol\ H_2O}{6.01\ kJ} \times \dfrac{18.02\ g\ H_2O}{1\ mol\ H_2O} = 21{,}761 = 2.18 \times 10^4\ g\ H_2O\ (ice)$

(b) $1.00\ m^2 \times 1.00\ cm \times \dfrac{(100)^2\ cm^2}{1\ m^2} \times \dfrac{0.99987\ g}{1\ cm^3} = 9998.7 = 1.00 \times 10^4\ g\ H_2O\ (ice\ at\ 0\ ^\circ C)$

$7257.6\ kJ \times \dfrac{1000\ J}{1\ kJ} \times \dfrac{1\ g\text{-}^\circ C}{2.032\ J} \times \dfrac{1}{9998.7\ g\ H_2O} = 357.21 = 357.2\ ^\circ C$

Assuming no phase changes, the final temperature is $-5\ ^\circ C + 357\ ^\circ C = 352\ ^\circ C$. Clearly the ice melts. This agrees with the result from part (a), which shows that sunlight striking 1.00 square meter of ice for 12 hours provides enough energy to melt 2.18×10^4 g ice, twice the mass in the first centimeter of a square meter of ice.

18.41   *Analyze/Plan.* g $Mg(OH)_2 \rightarrow$ mol $Mg(OH)_2 \rightarrow$ mol ratio $\rightarrow$ mol CaO $\rightarrow$ g CaO.   *Solve.*

$1000\ lb\ Mg(OH)_2 \times \dfrac{453.6\ g}{lb} \times \dfrac{1\ mol\ Mg(OH)_2}{58.33\ g\ Mg(OH)_2} \times \dfrac{1\ mol\ CaO}{1\ mol\ Mg(OH)_2} \times \dfrac{56.08\ g\ CaO}{1\ mol\ CaO}$

$$= 4.361 \times 10^5\ g\ CaO$$

18.42   0.05 ppb Au $= 0.05$ g Au$/1 \times 10^9$ g seawater

$\$1{,}000{,}000 \times \dfrac{1\ troy\ oz\ Au}{\$1300} \times \dfrac{31.1035\ g}{troy\ oz} = 2.3926 \times 10^4\ g = 2.39 \times 10^4\ g\ Au\ needed$

$2.3926 \times 10^4\ g\ Au \times \dfrac{1 \times 10^9\ g\ seawater}{0.05\ g\ Au} \times \dfrac{1\ mL\ seawater}{1.03\ g\ seawater} \times \dfrac{1\ L}{1000\ mL} = 4.6458 \times 10^{11}$

$$= 5 \times 10^{11}\ L\ seawater$$

5×10^{11} L seawater is needed if the process is 100% efficient; because it is only 50% efficient, twice as much seawater is needed.

$4.6458 \times 10^{11} \times 2 = 9.2916 \times 10^{12} = 9 \times 10^{12}$ L seawater

Note that the 1 sig fig in 0.05 ppb Au limits the precision of the calculation.

18.43   *Analyze/Plan.* Use molar concentrations of the six major ions in seawater from Table 18.5. Calculate charge in coulombs by multiplying molarity $\times$ integer charge $\times$ Faraday's constant (coulombs/mol). Sum the coulombic charges of the anions, the cations, and compare.   *Solve.*

$\dfrac{0.55\ mol\ Cl^-}{L\ seawater} \times \dfrac{1\ mol\ charge}{mol\ Cl^-} \times \dfrac{9.64853365\ C}{mol} = 5.30669351 = 5.3\ C$

$\dfrac{0.028\ mol\ SO_4^{2-}}{L\ seawater} \times \dfrac{2\ mol\ charge}{mol\ SO_4^{2-}} \times \dfrac{9.64853365\ C}{mol} = 0.54031788 = 0.54\ C$

$\dfrac{0.47\ mol\ Na^+}{L\ seawater} \times \dfrac{1\ mol\ charge}{mol\ K^+} \times \dfrac{9.64853365\ C}{mol} = 4.53481082 = 4.5\ C$

$$\frac{0.054 \text{ mol Mg}^{2+}}{\text{L seawater}} \times \frac{2 \text{ mol charge}}{\text{mol Mg}^{2+}} \times \frac{9.64853365 \text{ C}}{\text{mol}} = 1.04204163 = 1.0 \text{ C}$$

$$\frac{0.010 \text{ mol Ca}^{2+}}{\text{L seawater}} \times \frac{2 \text{ mol charge}}{\text{mol Mg}^{2+}} \times \frac{9.64853365 \text{ C}}{\text{mol}} = 0.19297067 = 0.19 \text{ C}$$

$$\frac{0.010 \text{ mol K}^+}{\text{L seawater}} \times \frac{1 \text{ mol charge}}{\text{mol K}^+} \times \frac{9.64853365 \text{ C}}{\text{mol}} = 0.0964853365 = 0.096 \text{ C}$$

anion charge: $[5.30669351 + 0.54031788] = 5.84701139 = 5.8 \text{ C}$

cation charge: $[4.53481082 + 1.04204164 + 0.19297067 + 0.0964853365] = 5.86630846 = 5.9 \text{ C}$

The two numbers vary in the third significant figure. This is not surprising, because the molarities of the various ions are given to two significant figures.

18.44   (a)   We need to replace the 18 billion gal of water per day used for irrigation. One billion is 1×10^9.

$$\frac{18 \times 10^9 \text{ gal}}{\text{d}} \times \frac{365 \text{ d}}{\text{yr}} \times \frac{3.7854 \text{ L}}{\text{gal}} \times \frac{1 \text{ dm}^3}{\text{L}} \times \frac{1 \text{ m}^3}{(10)^3 \text{ dm}^3} \times \frac{1 \text{ km}^3}{(1000)^3 \text{ m}^3} \times \frac{1}{6 \times 10^5 \text{ km}^2}$$

$$= 4.145 \times 10^{-5} = 4 \times 10^{-5} \text{ km/yr}$$

$$\frac{4.145 \times 10^{-5} \text{ km}}{\text{yr}} \times \frac{1000 \text{ m}}{\text{km}} \times \frac{100 \text{ cm}}{\text{m}} \times \frac{1 \text{ in}}{2.54 \text{ cm}} = 1.632 = 2 \text{ in/yr is needed}$$

However, only 2% of rainfall actually recharges to aquifer, so $(1.632/0.02) = 81.59$ = 80 in/year annual rainfall is required to replace water removed for irrigation. (Data limits the calculated result to 1 sig fig.)

          (b)   The process of dissolving accounts for the presence of arsenic in well water. If minerals in the rock feeding or holding the aquifer are somewhat soluble, ions can dissolve in the water. If arsenic, usually in the form of arsenic oxide anions, is present in the somewhat soluble minerals, it can leach into the aquifer and find its way into wells.

Human Activities and Water Quality (Section 18.4)

18.45   *Analyze/Plan.* Given temperature and the concentration difference between the two solutions, ($\Delta M = 0.22 - 0.01 = 0.21$ *M*), calculate the minimum pressure for reverse osmosis. Use the relationship $\Pi = MRT$ from Section 13.5. This is the pressure required to halt osmosis from the more dilute (0.01 *M*) to the more concentrated (0.22 *M*) solution. Slightly more pressure will initiate reverse osmosis.   *Solve.*

$$\Pi = \Delta MRT = \frac{0.21 \text{ mol}}{\text{L}} \times \frac{0.08206 \text{ L-atm}}{\text{mol-K}} \times 298 \text{ K} = 5.135 = 5.1 \text{ atm}$$

The minimum pressure required to initiate reverse osmosis is greater than 5.1 atm.

18.46   Calculate the total ion concentration of seawater by summing the molarities given in Table 18.5. Then use $\Pi = \Delta MRT$ to calculate pressure.

$$M_{\text{total}} = 0.55 + 0.47 + 0.028 + 0.054 + 0.010 + 0.010 + 2.3 \times 10^{-3} + 8.3 \times 10^{-4}$$

$$+ 4.3 \times 10^{-4} + 9.1 \times 10^{-5} + 7.0 \times 10^{-5} = 1.1257 = 1.13 \text{ } M$$

$$\Pi = \frac{(1.1257 - 0.02)\,\text{mol}}{\text{L}} \times \frac{0.08206\,\text{L} \times \text{atm}}{\text{mol-K}} \times 297\,\text{K} = 26.948 = 26.9\,\text{atm}$$

Check. The largest numbers in the molarity sum have 2 decimal places, so M_{total} has 2 decimal places and 3 sig figs. ΔM also has 2 decimal places and 3 sig figs so the calculated pressure has 3 sig figs. Units are correct.

18.47   *Analyze/Plan.* Under aerobic conditions, excess oxygen is present and decomposition leads to oxidized products, the element in its maximum oxidation state combined with oxygen. Under anaerobic conditions, little or no oxygen is present so decomposition leads to reduced products, the element in its minimum oxidation state combined with hydrogen. *Solve.*

   (a) CO_2, HCO_3^-, H_2O, SO_4^{2-}, NO_3^-.

   (b) $CH_4(g)$, $H_2S(g)$, $NH_3(g)$.

18.48   (a)   Decomposition of organic matter by aerobic bacteria depletes dissolved O_2. A low dissolved oxygen concentration indicates the presence of organic pollutants.

   (b)   In general, gas solubility decreases with increasing temperature. According to Figure 13.16, the solubility of $O_2(g)$ at 20 $^\circ$C is approximately 1.4 mM, and at 30 $^\circ$C is 1.2 mM. This is a 14.3% decrease in solubility over a typical atmospheric temperature range.

   Colder natural water has a greater maximum possible $O_2(g)$ solubility. A general increase in global average temperature accompanied by an increase in water temperature decreases water quality by decreasing the amount of dissolved oxygen.

18.49   *Analyze/Plan.* Given the balanced equation, calculate the amount of one reactant required to react exactly with a certain amount of the other reactants. Solve the stoichiometry problem. g $C_{18}H_{29}SO_3^- \rightarrow$ mol $\rightarrow$ mol ratio $\rightarrow$ mol $O_2 \rightarrow$ g O_2.   *Solve.*

$$10.0\,\text{g}\,C_{18}H_{29}SO_3^- \times \frac{1\,\text{mol}\,C_{18}H_{29}SO_3^-}{325\,\text{g}\,C_{18}H_{29}SO_3^-} \times \frac{51\,\text{mol}\,O_2}{2\,\text{mol}\,C_{18}H_{29}SO_3^-} \times \frac{32.0\,\text{g}\,O_2}{1\,\text{mol}\,O_2} = 25.1\,\text{g}\,O_2$$

Notice that the mass of O_2 required is 2.5 times greater than the mass of biodegradable material.

18.50   Water at 9 ppm O_2 is 50% depleted when the concentration drops by 4.5 ppm.

$$1,200,000\,\text{persons} \times \frac{59\,\text{g}\,O_2}{1\,\text{person}} \times \frac{1 \times 10^6\,\text{g}\,H_2O}{4.5\,\text{g}\,O_2} \times \frac{1\,\text{L}\,H_2O}{1 \times 10^3\,\text{g}\,H_2O} = 1.57 \times 10^{10} = 2 \times 10^{10}\,\text{L}\,H_2O$$

18.51   *Analyze/Plan.* The reaction is metathesis.   *Solve.*

   $Mg^{2+}(aq) + Ca(OH)_2(s) \rightarrow Mg(OH)_2(s) + Ca^{2+}(aq)$

   [The excess $Ca^{2+}(aq)$ is removed as $CaCO_3$ by naturally occurring bicarbonate or added Na_2CO_3.]

18.52   *Analyze/Plan.* Given $[Ca^{2+}]$ and $[Mg^{2+}]$, calculate mol $Ca(OH)_2$ and Na_2CO_3 needed to remove the cations. Consider the chemical equations and reaction stoichiometry for each ion.   *Solve.*

$Ca(OH)_2$ is added to remove Mg^{2+} as $Mg(OH)_2(s)$; Na_2CO_3 removes the original and added Ca^{2+}.

$$Mg^{2+}(aq) + Ca(OH)_2(aq) \rightarrow Mg(OH)_2(s) + Ca^{2+}(aq)$$
$$Ca^{2+}(aq) + Na_2CO_3(aq) \rightarrow CaCO_3(s) + Na^{2+}(aq)$$

One mol $Ca(OH)_2$ is needed for each mol of $Mg^{2+}(aq)$ present.

$$1.200 \times 10^3 \text{ L } H_2O \times \frac{7.0 \times 10^{-4} \text{ mol } Mg^{2+}}{L} \times \frac{1 \text{ mol } Ca(OH)_2}{1 \text{ mol } Mg(OH)_2^-} \times = 0.84 \text{ mol } Ca(OH)_2$$

Total mol Ca^{2+} = mol Ca^{2+} originally present + mol Ca^{2+} from added $Ca(OH)_2$

$$1.200 \times 10^3 \text{ L } H_2O \times \frac{5.0 \times 10^{-4} \text{ mol } Ca^{2+}}{L} = 0.60 \text{ mol } Ca^{2+}(aq)\text{original}$$

0.84 mol Ca^{2+} added + 0.60 mol $Ca^{2+}(aq)$ original = 1.44 mol Ca^{2+} total = 1.44 mol Na_2CO_3
Softening requires 0.84 mol $Ca(OH)_2$ and 1.44 mol Na_2CO_3.

18.53   (a)   *Trihalomethanes* are a class of molecules with one central carbon atom bound to one hydrogen and three halogen atoms. They are produced by the reaction of dissolved chlorine with organic matter naturally present in water, and are by-products of water disinfection via chlorination.

        (b)

```
        Cl                  Cl
        |                   |
   H — C — Cl          H — C — Br
        |                   |
        Cl                  Cl
```

18.54   (a)   The most likely origin of bromate ion, BrO_3^-, in municipal water supplies is oxidation of dissolved bromide ion, Br^-. Bromide can react with ozone in a two-step process to form bromate. The ozone might be produced photochemically, or be part of the water disinfection process.

        (b) BrO_3^- is an oxidizing agent. Hyponitrite ion, NO^-, has one less O atom than nitrite ion, NO_2^-.

        $$BrO_3^-(aq) + 2 NO^-(aq) \rightarrow BrO^-(aq) + 2 NO_2^-(g) \text{ or}$$
        $$BrO_3^-(aq) + NO^-(aq) \rightarrow BrO^-(aq) + NO_3^-(g)$$

Green Chemistry (Section 18.5)

18.55   The fewer steps in a process, the less waste (solvents as well as unusable by-products) is generated. It is probably true that a process with fewer steps requires less energy at the site of the process, and it is certainly true that the less waste the process generates, the less energy is required to clean or dispose of the waste.

18.56   Catalysts increase the rate of a reaction by lowering activation energy, E_a. For an uncatalyzed reaction that requires extreme temperatures and pressures to generate product at a viable rate, finding a suitable catalyst reduces the required temperature and/or pressure, which reduces the amount of energy used to run the process. A catalyst can also increase rate of production, which would reduce the net time and thus energy required to generate a certain amount of product.

18.57   (a)

(b)   •   **Prevention (1).** The alternative process eliminates production of 3-chlorobenzoic acid by-product, chlorine-containing waste that must be treated.

   •   **Atom economy (2).** Most of the starting atoms are in the final product.

   •   **Less hazardous chemical synthesis (3)** and **Inherently safer for accident prevention (12).** The starting material of the alternative process, if it is less concentrated than 30% by mass, is not shock sensitive, and the by-product is nontoxic water. The low molar mass of water means that a small amount of "waste" is generated.

   •   **Catalysis (9)** and **Design for energy efficiency (6).** The alternative process is catalyzed, which could mean that the process will be more energy efficient than the Baeyer–Villiger reaction (see Solution 18.58).

   •   **Raw materials should be renewable (7).** The catalyst can be recovered from the reaction mixture and reused. We don't have information about solvents or other auxiliary substances.

18.58 $scCO_2$ achieves maximum conversion much faster than CH_2Cl_2 solvent. This reduces processing time, temperature, and energy requirements. It also results in fewer unwanted by-products to be separated and processed. Although use of $scCO_2$ can increase the amount of a greenhouse gas released to the environment, it eliminates use of CH_2Cl_2, which is implicated in stratospheric ozone depletion. Use of $scCO_2$ rather than CH_2Cl_2 is a good green trade-off. (If the CO_2 used to form $scCO_2$ can be captured from some other industrial process, the net release of CO_2 is the same, and the use of CH_2Cl_2 is avoided.)

(In either solvent, the reaction is catalyzed, which usually leads to decreased processing temperatures and times, and greater energy efficiency.)

18.59   (a)   Water as a solvent is much "greener" than benzene, which is a known carcinogen. Water fits criteria: (5) safer solvent, (7) renewable feedstock, and (12) inherently safer for accident prevention.

   (b)   Reaction temperature of 500 K rather than 1000 K is "greener," according to criteria (6) design for energy efficiency and (12) inherently safer chemistry for accident prevention. Also, low temperature is less likely to produce undesirable by-products that have to be separated and treated as waste, which fits criterion (1).

   (c)   Sodium chloride as a by-product rather than chloroform ($CHCl_3$) is "greener," according to criteria: (1) prevention, (3) less hazardous chemical systems, and (12) inherently safer.

18.60   (a)   The catalyzed reaction that can be run close to room temperature and for a shorter time is definitely greener, according to criteria (6) design for energy efficiency and (9) catalysis.

(b)   The reagent obtained from corn husks is greener, by criteria (7) use of renewable feedstocks.

(c)   Neither process is totally "ungreen," because recycling of unavoidable by-products is always desirable. However, by criterion (2) atom economy, the process that produces no by-products is greener.

Additional Exercises

18.61   (a)   *Acid rain* is rain with an elevated $[H^+]$ and thus a low pH. The additional H^+ is produced by the dissolution of sulfur and nitrogen oxides such as $SO_3(g)$ and $NO_2(g)$ in rain droplets to form sulfuric and nitric acid, $H_2SO_4(aq)$ and $HNO_3(aq)$.

High levels of CO_2, mostly the product of burning fossil fuels, can also increase the pH of rain. The effect is not as great as for dissolved sulfur and nitrogen oxides, because H_2CO_3 is a much weaker acid than sulfuric or nitric acid.

(b)   A *greenhouse gas* absorbs infrared or "heat" radiation emitted from Earth's surface and serves to maintain a relatively constant temperature on the surface. These include $H_2O(g)$, CH_4, and CO_2. A significant buildup of greenhouse gases in the atmosphere could cause a corresponding increase in the average surface temperature and stimulate global climate.

(c)   *Photochemical smog* is an unpleasant collection of atmospheric pollutants initiated by photochemical dissociation of NO_2 to form NO and O atoms. The major components are $NO(g)$, $NO_2(g)$, $CO(g)$, and unburned hydrocarbons, all produced by automobile engines, and $O_3(g)$, ozone.

(d)   *Ozone depletion* is the reduction of O_3 concentration in the stratosphere, most notably over Antarctica. It is caused by reactions between O_3 and Cl atoms originating from CFCs, as well as other chlorine-containing organic compounds. Depletion of the ozone layer allows damaging ultraviolet radiation disruptive to the ecosystem to reach Earth's surface.

18.62 MM_{avg} at the surface = 83.8(0.17) + 16.0(0.38) + 32.0(0.45) = 34.73 = 35 g/mol
Next, calculate the percentage composition at 200 km. The fractions can be "normalized" by saying that the 0.45 fraction of O_2 is converted into *two* 0.45 fractions of O atoms, then dividing by the total fractions, 0.17 + 0.38 + 0.45 + 0.45 = 1.45:

$$MM_{avg} = \frac{83.8(0.17) + 16.0(0.38) + 16.0(0.90)}{1.45} = 23.95 = 24 \text{ g/mol}$$

18.63   Stratospheric ozone is formed and destroyed in a cycle of chemical reactions. The decomposition of O_3 to O_2 and O produces oxygen atoms, an essential ingredient for the production of ozone. Although single O_3 molecules exist for only a few seconds, new O_3 molecules are constantly reformed. This cyclic process ensures a finite concentration of O_3 in the stratosphere available to absorb ultraviolet radiation. (This explanation assumes that the cycle is not disrupted by outside agents such as CFCs.)

18.64
$$2[Cl(g)+O_3(g)\rightarrow ClO(g)+O_2(g)] \qquad [18.7]$$
$$2[ClO(g) + h\nu \rightarrow O(g)+Cl(g)] \qquad [18.9]$$
$$O(g) + O(g) \rightarrow O_2(g)$$

$$2\,Cl(g)+2\,O_3(g)+2\,ClO(g) + 2\,O(g) \rightarrow 2\,ClO(g)+3\,O_2(g)+2\,Cl(g)$$

$$2\,O_3(g) \xrightarrow{\;Cl\;} 3\,O_2(g) \qquad\qquad [18.10]$$

Note that Cl(g) fits the definition of a catalyst in this reaction.

18.65   CFCs, primarily $CFCl_3$ and CF_2Cl_2, are chemically inert and water insoluble. These properties make them valuable as propellants, refrigerants, and foaming agents because they are virtually unreactive in the *troposphere* (lower atmosphere) and do not initiate or propagate undesirable reactions. Further, they are water insoluble and not removed from the atmosphere by rain; they do not end up in the fresh water supply.

These properties render CFCs a long-term problem in the *stratosphere*. Because CFCs are inert and water insoluble, they are not removed from the troposphere by reaction or dissolution and have very long lifetimes. Virtually the entire mass of released CFCs eventually diffuses into the stratosphere where conditions are right for photo-dissociation and the production of Cl atoms. Cl atoms catalyze the destruction of ozone, O_3.

18.66   (a)   The production of Cl atoms in the stratosphere is the result of the photodissociation of a C–Cl bond in the CFC molecule.

$$CF_2Cl_2(g) \xrightarrow{\;h\nu\;} CF_2Cl(g) + Cl(g)$$

According to Table 8.3, the bond dissociation energy of a C–Br bond is

276 kJ/mol, whereas the value for a C–Cl bond is 328 kJ/mol. Photodissociation of $CBrF_3$ to form Br atoms requires less energy than the production of Cl atoms and should occur readily in the stratosphere.

(b) $CBrF_3(g) \xrightarrow{h\nu} CF_3(g)+Br(g)$

$Br(g) + O_3(g) \rightarrow BrO(g) + O_2(g)$

Also, under certain conditions

$BrO(g) + BrO(g) \rightarrow Br_2O_2(g)$

$Br_2O_2(g) + h\nu \rightarrow O_2(g) + 2Br(g)$

18.67   (a)   A CFC has C–Cl bonds and C–F bonds. In an HFC, the C–Cl bonds are replaced by C–H bonds.

(b)   The longer a halogen-containing molecule exists in the stratosphere, the greater the likelihood that it will encounter light with energy sufficient to dissociate a carbon–halogen bond. Free halogen atoms catalyze the destruction of ozone.

(c)   The bond dissociation enthalpy of a C–F bond is 485 kJ/mol, much more than for a C–Cl bond, 328 kJ/mol (Table 8.3). Although HFCs have long lifetimes in the stratosphere, it is infrequent that light with energy sufficient to dissociate a C–F bond will reach an HFC molecule. F atoms are much less likely than Cl atoms to be produced by photodissociation in the stratosphere.

(d)   The main disadvantage of HFCs as replacements for CFCs is that they are potent greenhouse gases. Although HFCs are far less threatening to stratospheric ozone, they may contribute to global climate change.

18.68   From Section 18.2:

$$N_2(g) + O_2(g) \rightleftharpoons 2\,NO(g) \quad \Delta H = +180.8 \text{ kJ} \quad [18.11]$$

$$2\,NO(g) + O_2(g) \rightleftharpoons 2\,NO_2(g) \quad \Delta H = -113.1 \text{ kJ} \quad [18.12]$$

In an endothermic reaction, heat is a reactant. As the temperature of the reaction increases, the addition of heat favors formation of products and the value of K increases. The reverse is true for exothermic reactions; as temperature increases, the value of K decreases. Thus, K for reaction [18.11], which is endothermic, increases with increasing temperature and K for reaction [18.12], which is exothermic, decreases with increasing temperature.

18.69   (a) $CH_4(g) + 2\,O_2(g) \rightarrow CO_2(g) + 2\,H_2O(g)$

(b) $2\,CH_4(g) + 3\,O_2(g) \rightarrow 2\,CO(g) + 4\,H_2O(g)$

(c)   vol $CH_4 \rightarrow$ vol $O_2 \rightarrow$ volume air ($X_{O_2} = 0.20948$)

Equal volumes of gases at the same temperature and pressure contain equal numbers of moles (Avogadro's law). If 2 moles of O_2 are required for 1 mole of CH_4, 2.0 L of pure O_2 are needed to burn 1.0 L of CH_4.

$$\text{vol } O_2 = X_{O_2} \times \text{vol}_{air} = \frac{\text{vol } O_2}{X_{O_2}} = \frac{2.0 \text{ L}}{0.20948} = 9.5 \text{ L air}$$

18.70   (a) $2\,SO_2(g) + O_2(g) \rightarrow 2\,SO_3(g)$

$SO_2(g) + O_3(g) \rightarrow SO_3(g) + O_2(g)$

$SO_2(g) + H_2O(l) \rightarrow H_2SO_3(l, \text{ aerosol})$

$SO_3(g) + H_2O(l) \rightarrow H_2SO_4(l, \text{ aerosol})$

(b)   The finely dispersed liquid droplets in the aerosol reflect sunlight into space. Less warming solar radiation reaches Earth's surface.

(c)   In the stratosphere, aerosol particles act as a heterogeneous catalyst for ozone destruction by halogens. That is, they provide a platform to attract and orient the reactants in ozone depletion processes. The depletion reactions occur at a greater rate than in the absence of the aerosol catalyst.

18.71   (a)   According to Section 13.3, the solubility of gases in water decreases with increasing temperature. From the graph, this is also true for $CO_2(g)$. Comparing the graph in this exercise with Figure 13.16, the general shape of the solubility versus temperature curve for $CO_2(g)$ is similar to that for other gases. [Although the solubility units are different on the two graphs, it seems that $CO_2(g)$ is significantly more soluble than the gases in Figure 13.16.]

(b)   If the solubility of $CO_2(g)$ in the ocean decreased because of climate change, more $CO_2(g)$ would be released into the atmosphere, perpetuating a cycle of increasing temperature and concomitant release of $CO_2(g)$ from the ocean.

18.72   Most of the 390 watts/m^2 radiated from Earth's surface is in the infrared region of the spectrum. Tropospheric gases, particularly $H_2O(g)$, $CH_4(g)$, and $CO_2(g)$, absorb much of this radiation and prevent it from escaping into space (Figures 18.12 and 18.13). The energy absorbed by these so-called greenhouse gases warms the atmosphere close to Earth's surface and makes the planet livable.

18.73   Given 168 watts/m^2 at 10% efficiency, find the land area needed to produce 12,000 megawatts. 13,200 megawatts = $13,200 \times 10^6 = 1.32 \times 10^{10}$ watts.

168 watts/m^2 (0.10) = 16.8 watts/m^2 solar energy possible with current technology.

$$1.32 \times 10^{10} \text{ watts} \times \frac{1 \text{ m}^2}{16.8 \text{ watts}} = 7.857 \times 10^8 = 7.9 \times 10^8 \text{ m}^2$$

The land area of New York City is 830 km^2, which is 830×10^6 m^2. The area needed for solar energy harvesting to provide peak power would then be $\dfrac{7.857 \times 10^8 \text{ m}^2}{830 \times 10^6 \text{ m}^2} = 0.95$

times the land area of New York City.

18.74   (a) $NO(g) + h\nu \rightarrow N(g) + O(g)$

         (b) $NO(g) + h\nu \rightarrow NO^+(g) + e^-$

         (c) $NO(g) + O_3(g) \rightarrow NO_2(g) + O_2(g)$

         (d) $3 \, NO_2(g) + H_2O(l) \rightarrow 2 \, HNO_3(aq) + NO(g)$

18.75   (a) $CO_3{}^{2-}$ is a relatively strong Brønsted–Lowry base and produces OH^- in aqueous solution according to the hydrolysis reaction:

$$CO_3{}^{2-}(aq) + H_2O(l) \; \rightleftharpoons \; HCO_3{}^-(aq) + OH^-(aq), \quad K_b = 1.8 \times 10^{-4}$$

If $[OH^-(aq)]$ is sufficient for the reaction quotient, Q, to exceed K_{sp} for $Mg(OH)_2$, the solid will precipitate.

         (b) $\dfrac{125 \text{ mg Mg}^{2+}}{1 \text{ kg soln}} \times \dfrac{1 \text{ g Mg}^{2+}}{1000 \text{ mg Mg}^{2+}} \times \dfrac{1.00 \text{ kg soln}}{1.00 \text{ L soln}} \times \dfrac{1 \text{ mol Mg}^{2+}}{24.305 \text{ g Mg}^{2+}} = 5.143 \times 10^{-3}$

$$= 5.14 \times 10^{-3} \, M \text{ Mg}^{2+}$$

$$\frac{4.0 \text{ g Na}_2CO_3}{1.0 \text{ L soln}} \times \frac{1 \text{ mol CO}_3{}^{2-}}{106.0 \text{ g Na}_2CO_3} = 0.03774 = 0.038 \, M \text{ CO}_3{}^{2-}$$

$$K_b = 1.8 \times 10^{-4} = \frac{[HCO_3{}^-][OH^-]}{[CO_3{}^{2-}]} \approx \frac{x^2}{0.03774}; \quad x = [OH^-] = 2.606 \times 10^{-3}$$

$$= 2.6 \times 10^{-3} \, M$$

(This represents 6.9% hydrolysis, but the result will not be significantly different using the quadratic formula.)

$Q = [Mg^{2+}][OH^-]^2 = (5.143 \times 10^{-3})(2.606 \times 10^{-3})^2 = 3.5 \times 10^{-8}$

K_{sp} for $Mg(OH)_2 = 1.6 \times 10^{-12}$; $Q > K_{sp}$, so $Mg(OH)_2$ will precipitate.

18.76   (a)   15 ppb = 15 g Pb in 1×10^9 g solution. For very dilute solutions, assume the density of the solution is 1.0 g/mL. 1.0×10^9 g solution. = 1.0×10^9 mL solution.

$$\frac{15 \text{ g Pb}}{1.0 \times 10^9 \text{ g solution}} \times \frac{1 \text{ mol Pb}}{106.42 \text{ g Pb}} \times \frac{1000 \text{ mL}}{1 \text{ L}} = 1.4 \times 10^{-7} \, M$$

(b)   Change μg/dL to g/1.0×10^9 mL.

$$\frac{1.6 \text{ μg}}{1.0 \times 10^9 \text{ g solution}} \times \frac{1 \times 10^{-6} \text{ g}}{\text{μg}} \times \frac{1 \text{ dL}}{100 \text{ mL}} \times 1 \times 10^9 \text{ mL} = 16 \text{ ppb Pb}$$

18.77   *Plan.* Calculate the volume of air above Los Angeles and the volume of pure O_3 that would be present at the 84 ppb level. For gases at the same temperature and pressure, volume fractions equal mole fractions.   *Solve.*

$$V_{air} = 4000 \text{ mi}^2 \times \frac{(1.6093)^2 \text{ km}^2}{\text{mi}^2} \times \frac{(1000)^2 \text{ m}^2}{1 \text{ km}^2} \times 100 \text{ m} \times \frac{1 \text{ L}}{1 \times 10^{-3} \text{ m}^3} = 1.036 \times 10^{15}$$

$$= 1.0 \times 10^{15} \text{ L air}$$

$$84 \text{ ppb O}_3 = \frac{84 \text{ mol O}_3}{1 \times 10^9 \text{ mol air}} = 8.4 \times 10^{-8} = X_{O_3}$$

$$V \text{ (pure O}_3) = 8.4 \times 10^{-8} (1.036 \times 10^{15} \text{ L air}) = 8.702 \times 10^7 = 8.7 \times 10^7 \text{ L O}_3$$

Values for P and T are required to calculate mol O_3 from volume O_3, using the ideal-gas law. Because these are not specified in the exercise, we will make a reasonable assumption for a sunny April day in Los Angeles. The city is near sea level and temperatures are moderate throughout the year, so P = 1 atm and T = 25 °C (78 °F) are reasonable values. PV = nRT, n = PV/RT.

$$n = 1.000 \text{ atm} \times \frac{8.702 \times 10^7 \text{ L}}{298 \text{ K}} \times \frac{\text{mol-K}}{0.08206 \text{ L-atm}} = 3.558 \times 10^6 = 3.6 \times 10^6 \text{ mol O}_3$$

Check. Using known conditions to make reasonable estimates and assumptions is a valuable skill for problem solving. Knowing when assumptions are required is an important step in the learning process.

Integrative Exercises

18.78   (a) $$0.016 \text{ ppm NO}_2 = \frac{0.016 \text{ mol NO}_2}{1 \times 10^6 \text{ mol air}} = 1.6 \times 10^{-8} = X_{NO_2}$$

$$P_{NO_2} = X_{NO_2} \times P_{atm} = 1.6 \times 10^{-8} (755 \text{ torr}) = 1.208 \times 10^{-5} = 1.2 \times 10^{-5} \text{ torr}$$

(b) $$n = \frac{PV}{RT}; \text{ molecules} = n \times \frac{6.022 \times 10^{23} \text{ molecules}}{\text{mol}} = \frac{PV}{RT} \times \frac{6.022 \times 10^{23} \text{ molecules}}{\text{mol}}$$

$$V = 15 \text{ ft} \times 14 \text{ ft} \times 8 \text{ ft} \times \frac{12^3 \text{ in}^3}{\text{ft}^3} \times \frac{2.54^3 \text{ cm}^3}{\text{in}^3} \times \frac{1 \text{ L}}{1000 \text{ cm}^3} = 4.757 \times 10^4 = 5 \times 10^4 \text{ L}$$

$$1.208 \times 10^{-5} \text{ torr} \times \frac{1 \text{ atm}}{760 \text{ torr}} \times \frac{4.757 \times 10^4 \text{ L}}{293 \text{ K}} \times \frac{\text{mol-K}}{0.08206 \text{ L-atm}}$$

$$\times \frac{6.022 \times 10^{23} \text{ molecules}}{\text{mol}} = 1.894 \times 10^{19} = 2 \times 10^{19} \text{ molecules}$$

18.79   (a) $$8,376,726 \text{ tons coal} \times \frac{83 \text{ ton C}}{100 \text{ ton coal}} \times \frac{44.01 \text{ ton CO}_2}{12.01 \text{ ton C}} = 2.5 \times 10^7 \text{ ton CO}_2$$

$$8,376,726 \text{ tons coal} \times \frac{2.5 \text{ ton S}}{100 \text{ ton coal}} \times \frac{64.07 \text{ ton SO}_2}{32.07 \text{ ton S}} = 4.2 \times 10^5 \text{ ton SO}_2$$

(b) $CaO(s) + SO_2(g) \rightarrow CaSO_3(s)$

$$4.18 \times 10^5 \text{ ton } SO_2 \times \frac{55 \text{ ton } SO_2 \text{ removed}}{100 \text{ ton } SO_2 \text{ produced}} \times \frac{120.15 \text{ ton } CaSO_3}{64.07 \text{ ton } SO_2}$$

$$= 4.3 \times 10^5 \text{ ton } CaSO_3$$

18.80   *Coarse sand* is removed by coarse sand filtration. *Finely divided particles* and some *bacteria* are removed by precipitation with aluminum hydroxide. Remaining *harmful bacteria* are removed by ozonation. *Trihalomethanes* are removed by either aeration or activated carbon filtration; use of activated carbon might be preferred because it does not involve release of TCMs into the atmosphere. *Dissolved organic substances* are oxidized (and rendered less harmful, but not removed) by both aeration and ozonation. Dissolved *nitrates* and *phosphates* are not removed by any of these processes, but are rendered less harmful by adequate aeration.

18.81   Calculate the molar concentration of impurity that would have an absorbance of 0.0001. This is the minimum concentration of the impurity detectable by absorption spectroscopy.

$A = \varepsilon bc$; A = absorbance, ε = extinction coefficient, b = path length, c = molarity. The common path length is 1 cm.

$$c = \frac{A}{\varepsilon b} = 0.0001 \times \frac{M \times cm}{3.45 \times 10^3} \times \frac{1}{1 \, cm} = 2.8986 \times 10^{-8} = 3 \times 10^{-8} \, M$$

Because we do not have the identity of the impurity, we cannot calculate the corresponding concentration in ppb. We can calculate a maximum molar mass for the impurity, such that a 3×10^{-8} M solution is 50 ppb. A concentration of 50 ppb corresponds to 50 g impurity per 10^9 L solution.

$$\frac{50 \text{ g impurity}}{10^9 \text{g solution}} \times \frac{1000 \text{ g solution}}{L \text{ solution}} \times \frac{1 L \text{ solution}}{2.8986 \times 10^{-8} \text{ mol impurity}} = 1725 \text{ g impurity/mol}$$

In this calculation, molar mass is directly proportional to ppm concentration. This means that a 50 ppm solution or any impurity with a molar mass less than or equal to 1725 g/mol will be observable by absorption spectroscopy. Concentrations less than 50 ppm are probably observable, because 1725 is a large molar mass. (The calculated molar mass is more correctly represented with one sig fig as 2×10^3 g/mol.)

18.82   (a) $H - \ddot{O} - H \longrightarrow H\cdot + \cdot\ddot{O} - H$

(b) $\Delta H = 2D(O-H) - D(O-H) = D(O-H) = 463$ kJ/mol

$$\frac{463 \text{ kJ}}{\text{mol } H_2O} \times \frac{1 \text{ mol } H_2O}{6.022 \times 10^{23} \text{ molecules}} \times \frac{1000 \text{ J}}{\text{kJ}} = 7.688 \times 10^{-19}$$

$$= 7.69 \times 10^{-19} \text{ J/} H_2O \text{ molecule}$$

$$\lambda = \frac{hc}{\Delta E} = \frac{6.626 \times 10^{-34} \text{ J-sec} \times 2.998 \times 10^8 \text{ m/s}}{7.688 \times 10^{-19} \text{ J}} = 2.58 \times 10^{-7} \text{ m} = 258 \text{ nm}$$

This wavelength is in the UV region of the spectrum, close to the visible.

(c)
$$OH(g) + O_3(g) \rightarrow HO_2(g) + O_2(g)$$
$$HO_2(g) + O(g) \rightarrow OH(g) + O_2(g)$$
$$\overline{OH(g) + O_3(g) + HO_2(g) + O(g) \rightarrow HO_2(g) + 2\,O_2(g) + OH(g)}$$
$$O_3(g) + O(g) \rightarrow 2\,O_2(g)$$

OH(g) is the catalyst in this overall reaction, another pathway for the destruction of ozone.

18.83   According to Equation 14.12, $\ln([A]_t / [A]_o) = -kt$. $[A]_t = 0.10\,[A]_o$.

$$\ln(0.10\,[A]_o / [A]_o) = \ln(0.10) = -(2 \times 10^{-6}\,s^{-1})\,t$$

$$t = -\ln(0.10) / 2 \times 10^{-6}\,s^{-1} = 1.151 \times 10^6\,s$$

$$1.151 \times 10^6\,s \times \frac{1\,min}{60\,s} \times \frac{1\,h}{60\,min} \times \frac{1\,day}{24\,h} = 13.3\,days\,(1 \times 10\,days)$$

The value of the rate constant limits the result to 1 sig fig. This implies that there is minimum uncertainty of ±1 in the tens place of our answer. Realistically, the remediation could take anywhere from 1 to 20 days.

18.84   (i) $ClO(g) + O_3(g) \rightarrow ClO_2(g) + O_2(g)$

$$\Delta H_i = \Delta H_f^{\circ}\,ClO_2(g) + \Delta H_f^{\circ}\,O_2(g) - \Delta H_f^{\circ}\,ClO(g) - \Delta H_f^{\circ}\,O_3(g)$$

$$\Delta H_i = 102 + 0 - 101 - (142.3) = -141\,kJ$$

(ii) $ClO_2(g) + O(g) \rightarrow ClO(g) + O_2(g)$

$$\Delta H_{ii} = \Delta H_f^{\circ}\,ClO(g) + \Delta H_f^{\circ}\,O_2(g) - \Delta H_f^{\circ}\,ClO_2(g) - \Delta H_f^{\circ}\,O(g)$$

$$\Delta H_{ii} = 101 + 0 - 102 - (247.5) = -249\,kJ$$

(overall) $ClO(g) + O_3(g) + ClO_2(g) + O(g) \rightarrow ClO_2(g) + O_2(g) + ClO(g) + O_2(g)$

$$O_3(g) + O(g) \rightarrow 2O_2(g)$$

$$\Delta H = \Delta H_i + \Delta H_{ii} = -141\,kJ + (-249)\,kJ = -390\,kJ$$

Because the enthalpies of both (i) and (ii) are distinctly exothermic, it is possible that the $ClO - ClO_2$ pair could be a catalyst for the destruction of ozone.

18.85   (a)   Assume the density of water at 20 °C is the same as at 25 °C.

$$1.00\,gal \times \frac{4\,qt}{1\,gal} \times \frac{1\,L}{1.057\,qt} \times \frac{1000\,mL}{1\,L} \times \frac{0.99707\,g\,H_2O}{1\,mL} = 3773$$

$$= 3.77 \times 10^3\,g\,H_2O$$

The $H_2O(l)$ must be heated from 20 °C to 100 °C and then vaporized at 100 °C.

$$3.773 \times 10^3\,g\,H_2O \times \frac{4.184\,J}{g\,°C} \times 80\,°C \times \frac{1\,kJ}{1000\,J} = 1263 = 1.3 \times 10^3\,kJ$$

$$3.773 \times 10^3\,g\,H_2O \times \frac{1\,mol\,H_2O}{18.02\,g\,H_2O} \times \frac{40.67\,kJ}{mol\,H_2O} = 8516 = 8.52 \times 10^3\,kJ$$

$$energy = 1263\,kJ + 8516\,kJ = 9779 = 9.8 \times 10^3\,kJ/gal\,H_2O$$

(b)   According to Solution 5.18, 1 kwh = 3.6×10^6 J.

$$\frac{9779 \text{ kJ}}{\text{gal H}_2\text{O}} \times \frac{1000 \text{ J}}{\text{kJ}} \times \frac{1 \text{ kwh}}{3.6 \times 10^6 \text{ J}} \times \frac{\$0.085}{\text{kwh}} = \$0.23/\text{gal}$$

(c) $\dfrac{\$0.23}{\$1.26} \times 100 = 18\%$ of the total cost is energy

18.86   (a)   A rate constant of $M^{-1}\text{s}^{-1}$ is indicative of a reaction that is second order overall. For the reaction given, the rate law is probably rate = $k[\text{O}][\text{O}_3]$. (Although rate = $k[\text{O}]^2$ or $k[\text{O}_3]^2$ are possibilities, it is difficult to envision a mechanism consistent with either one that would result in two molecules of O_2 being produced.)

(b)   Yes. Most atmospheric processes are initiated by collision. One could imagine an activated complex of four O atoms collapsing to form two O_2 molecules. Also, the rate constant is large, which is less likely for a multistep process. The reaction is analogous to the destruction of O_3 by Cl atoms (Equation 18.7), which is also second order with a large rate constant.

(c) $\Delta H^\circ = 2 \Delta H_f^\circ \text{ O}_2(g) - \Delta H_f^\circ \text{ O}(g) - \Delta H_f^\circ \text{ O}_3(g)$

$\Delta H^\circ = 0 - 247.5 \text{ kJ} - 142.3 \text{ kJ} = -389.8 \text{ kJ}$

The reaction is exothermic, so energy is released; the reaction would raise the temperature of the stratosphere.

18.87   (a)   Holding one reactant concentration constant and changing the other, evaluate the effect this has on the initial rate. Use these observations to write the rate law.

Compare Experiments 1 and 3. $[\text{O}_3]$ is constant, $[\text{H}]$ doubles, initial rate doubles. The reaction is first order in $[\text{H}]$.

Compare Experiments 2 and 1. $[\text{H}]$ is constant, $[\text{O}_3]$ doubles, initial rate doubles. The reaction is first order in $[\text{O}_3]$.

rate = $k[\text{O}_3][\text{H}]$

(b)   Calculate a value for the rate constant for each experiment and average them to obtain a single representative value.

rate = $k[\text{O}_3][\text{H}]$; k = rate/$[\text{O}_3][\text{H}]$

$$k_1 = \frac{1.88 \times 10^{-14} \text{ M/s}}{(5.17 \times 10^{-33} \text{ M})(3.22 \times 10^{-26} \text{ M})} = 1.1293 \times 10^{44} = 1.13 \times 10^{44}$$

$$k_2 = \frac{9.44 \times 10^{-15} \text{ M/s}}{(2.59 \times 10^{-33} \text{ M})(3.25 \times 10^{-26} \text{ M})} = 1.1215 \times 10^{44} = 1.12 \times 10^{44}$$

$$k_3 = \frac{3.77 \times 10^{-14} \text{ M/s}}{(5.19 \times 10^{-33} \text{ M})(6.46 \times 10^{-26} \text{ M})} = 1.1245 \times 10^{44} = 1.12 \times 10^{44}$$

$$k_{avg} = (1.1293 \times 10^{44} + 1.1215 \times 10^{44} + 1.1245 \times 10^{44})/3 = 1.1251 \times 10^{44} =$$
$$1.13 \times 10^{44} \text{ } M^{-1} \text{s}^{-1}$$

18.88   rate = $k[CF_3CH_2F][OH]$. $k = 1.6 \times 10^8\ M^{-1}\ s^{-1}$ at 4 °C.

$[CF_3CH_2F] = 6.3 \times 10^8$ molecules/cm^3, $[OH] = 8.1 \times 10^5$ molecules/cm^3

Change molecules/cm^3 to mol/L (M) and substitute into the rate law.

$$\frac{6.3 \times 10^8\ \text{molecules}}{\text{cm}^3} \times \frac{1\ \text{mol}}{6.022 \times 10^{23}\ \text{molecules}} \times \frac{1000\ \text{cm}^3}{1\ \text{L}} =$$

$$1.0462 \times 10^{-12} = 1.0 \times 10^{-12}\ M\ CF_3CH_2F$$

$$\frac{8.1 \times 10^5\ \text{molecules}}{\text{cm}^3} \times \frac{1\ \text{mol}}{6.022 \times 10^{23}\ \text{molecules}} \times \frac{1000\ \text{cm}^3}{1\ \text{L}} =$$

$$1.3451 \times 10^{-15} = 1.3 \times 10^{-15}\ M\ OH$$

$$\text{rate} = \frac{1.6 \times 10^8}{M\text{-s}} \times 1.0462 \times 10^{-12}\ M \times 1.3451 \times 10^{-15}\ M = 2.2515 \times 10^{-19} = 2.3 \times 10^{-19}\ M/s$$

18.89   (a)   According to Table 18.1, the mole fraction of CO_2 in air is 0.000375.

$P_{CO_2} = X_{CO_2} \times P_{atm} = 0.000400\ (1.00\ \text{atm}) = 4.00 \times 10^{-4}$ atm

$C_{CO_2} = kP_{CO_2} = 3.1 \times 10^{-2}\ M/\text{atm} \times 4.00 \times 10^{-4}\ \text{atm} = 1.24 \times 10^{-5} = 1.2 \times 10^{-5}\ M$

(b) H_2CO_3 is a weak acid, so the $[H^+]$ is regulated by the equilibria:

$H_2CO_3(aq) \rightleftharpoons H^+(aq) + HCO_3^-(aq)$ $K_{a1} = 4.3 \times 10^{-7}$

$HCO_3^-(aq) \rightleftharpoons H^+(aq) + CO_3^{2-}(aq)$ $K_{a2} = 5.6 \times 10^{-11}$

Because the value of K_{a2} is small compared to K_{a1}, we will assume that most of the $H^+(aq)$ is produced by the first dissociation.

$$K_{a1} = 4.3 \times 10^{-7} = \frac{[H^+][HCO_3^-]}{[H_2CO_3]};\ [H^+] = [HCO_3^-] = x,\ [H_2CO_3] = 1.24 \times 10^{-5} - x$$

Because K_{a1} and $[H_2CO_3]$ have similar values, we cannot assume x is small compared to 1.2×10^{-5}.

$$4.3 \times 10^{-7} = \frac{x^2}{(1.24 \times 10^{-5} - x)};\ 5.332 \times 10^{-12} - 4.3 \times 10^{-7}\ x = x^2$$

$$0 = x^2 + 4.3 \times 10^{-7}\ x - 5.332 \times 10^{-12}$$

$$x = \frac{-4.3 \times 10^{-7} \pm \sqrt{(4.3 \times 10^{-7})^2 - 4(1)(-5.332 \times 10^{-12})}}{2(1)}$$

$$x = \frac{-4.3 \times 10^{-7} \pm \sqrt{1.85 \times 10^{-13} + 2.133 \times 10^{-11}}}{2} = \frac{-4.3 \times 10^{-7} \pm 4.638 \times 10^{-6}}{2}$$

The negative result is meaningless; $x = 2.104 \times 10^{-6} = 2.1 \times 10^{-6}\ M\ H^+$; pH = 5.68

Because this $[H^+]$ is quite small, the $[H^+]$ from the autoionization of water might be significant. Calculation shows that for $[H^+] = 2.1 \times 10^{-6}\ M$ from H_2CO_3, $[H^+]$ from $H_2O = 5.2 \times 10^{-9}\ M$, which we can ignore.

18.90 (a) $Al(OH)_3(s) \rightleftharpoons Al^{3+}(aq) + 3\,OH^-(aq)$ $K_{sp} = 1.3 \times 10^{-33} = [Al^{3+}][OH^-]^3$

This is a precipitation conditions problem. At what $[OH^-]$ (we can get pH from $[OH^-]$) will $Q = 1.3 \times 10^{-33}$, the requirement for the onset of precipitation?

$Q = 1.3 \times 10^{-33} = [Al^{3+}][OH^-]^3$. Find the molar concentration of $Al_2(SO_4)_3$ and thus $[Al^{3+}]$.

$$\frac{5.0\,lb\,Al_2(SO_4)_3}{2000\,gal\,H_2O} \times \frac{453.6\,g}{1\,lb} \times \frac{1\,mol\,Al_2\,(SO_4)_3}{342.2\,g\,Al_2(SO_4)_3} \times \frac{1\,gal}{4\,qt} \times \frac{1\,qt}{0.946\,L}$$

$$= 8.758 \times 10^{-4}\,M\,Al_2(SO_4)_3 = 1.752 \times 10^{-3} = 1.8 \times 10^{-3}\,M\,Al^{3+}$$

$Q = 1.3 \times 10^{-33} = (1.752 \times 10^{-3})[OH^-]^3;\ [OH^-]^3 = 7.42 \times 10^{-31}$

$[OH^-] = 9.054 \times 10^{-11} = 9.1 \times 10^{-11}\,M$; pOH = 10.04; pH = 14 − 10.04 = 3.96

(b) $CaO(s) + H_2O(l) \rightarrow Ca^{2+}(aq) + 2\,OH^-(aq);\ [OH^-] = 9.054 \times 10^{-11}\,mol/L$

$$mol\,OH^- = \frac{9.054 \times 10^{-11}\,mol}{1\,L} \times 2000\,gal \times \frac{4\,qt}{1\,gal} \times \frac{0.946\,L}{1\,qt} = 6.852 \times 10^{-7}$$

$$= 6.9 \times 10^{-7}\,mol\,OH^-$$

$$6.852 \times 10^{-7}\,mol\,OH^- \times \frac{1\,mol\,CaO}{2\,mol\,OH^-} \times \frac{56.1\,g\,CaO}{1\,mol\,CaO} \times \frac{1\,lb}{453.6\,g} = 4.2 \times 10^{-8}\,lb\,CaO$$

This is a *very* small amount of CaO, about 20 μg.

18.91 (a) Process (i) is greener, because it does not involve the toxic reactant phosgene ($COCl_2$) and the by-product is water, not HCl.

(b) Reaction (i): C in CO_2 is linear with sp hybridization; C in R–N=C=O is linear with sp hybridization; C in the urethane monomer is trigonal planar with sp^2 hybridization. Reaction (ii): C in $COCl_2$ is trigonal planar with sp^2 hybridization; C in R–N=C=O is linear with sp hybridization; C in the urethane monomer is trigonal planar with sp^2 hybridization.

(c) Traditionally, industrial processes are conducted at higher temperatures to speed up reactions and encourage formation of product. However, this is not a green solution, because it requires additional energy. Using Le Châtelier's principle, we could either "push" or "pull" the reaction toward products. The "push" requires that we increase the amount of reactants, again not a green approach. The greenest way to promote formation of the isocyanate is to "pull" the reaction forward by removing by-product from the reaction mixture. In reaction (i), remove water; in reaction (ii), remove HCl.

18.92 (a) The various forms of carbonate in water are related by the following equilibria:

$H_2CO_3(aq) \rightleftharpoons H^+(aq) + HCO_3^-(aq)$ $K_{a1} = 4.3 \times 10^{-7}$

$HCO_3^-(aq) \rightleftharpoons H^+(aq) + CO_3^{2-}(aq)$ $K_{a2} = 5.6 \times 10^{-11}$

$K_{a1} = 4.3 \times 10^{-7} = \dfrac{[H^+][HCO_3^-]}{[H_2CO_3]};\quad K_{a2} = 5.6 \times 10^{-11} = \dfrac{[H^+][CO_3^{2-}]}{[HCO_3^-]}$

$[H^+] = 10^{-pH} = 10^{-5.6} = 2.5119 \times 10^{-6} = 3 \times 10^{-6}\,M$

Also, $[H_2CO_3] + [HCO_3^-] + [CO_3^{2-}] = 1.0 \times 10^{-5}\,M$

We now have 3 equations in 3 unknowns, so we can solve explicitly for one. Solve for $[HCO_3^-]$ (because it appears in both K_a expressions) and then substitute to find $[H_2CO_3]$ and $[CO_3^{2-}]$.

$$1.0 \times 10^{-5} = \frac{[H^+][HCO_3^-]}{K_{a1}} + [HCO_3^-] + \frac{K_{a2}[HCO_3^-]}{[H^+]}$$

$$1.0 \times 10^{-5} = \frac{2.5119 \times 10^{-6}[HCO_3^-]}{4.3 \times 10^{-7}} + [HCO_3^-] + \frac{5.6 \times 10^{-11}[HCO_3^-]}{2.5119 \times 10^{-6}}$$

$$1.0 \times 10^{-5} = 5.8416[HCO_3^-] + [HCO_3^-] + 2.2294 \times 10^{-5}[HCO_3^-]$$

$$[HCO_3^-] = \frac{1.0 \times 10^{-5}}{6.8416} = 1.4616 \times 10^{-6} = 1.5 \times 10^{-6} M$$

Note that $[CO_3^{2-}]$ is very small compared to $[H_2CO_3]$ and $[HCO_3^-]$.

$$[H_2CO_3] = \frac{(2.5119 \times 10^{-6})(1.4616 \times 10^{-6})}{4.3 \times 10^{-7}} = 8.5383 \times 10^{-6} = 8.5 \times 10^{-6} M$$

$$[CO_3^{2-}] = \frac{(5.6 \times 10^{-11})(1.4616 \times 10^{-6})}{2.5119 \times 10^{-6}} = 3.2586 \times 10^{-11} = 3.3 \times 10^{-11} M$$

Check. $1.5 \times 10^{-6} M + 8.5 \times 10^{-6} M + 3.3 \times 10^{-11} M = 1.0 \times 10^{-5} M$

(b) To test for sulfur-containing species, we must first remove the various forms of carbonate. One method is to exploit the solubility differences between carbonate and sulfate salts. Most sulfates are soluble, whereas most carbonates are not. However, K_{sp} values for carbonates are relatively large, and $[CO_3^{2-}]$ in the raindrop is very small. Precipitating insoluble carbonates will shift the acid dissociation equilibria to the right, but precipitation may not be the best method for effectively removing carbonates.

A different method involves removing carbonates as $CO_2(g)$. Heating the rainwater will decrease the solubility of $CO_2(g)$, which will bubble off as a gas. Slightly acidifying the solution will encourage this process, by shifting the acid dissociation equilibria toward H_2CO_3 and $CO_2(g)$.

After removal of carbonates, sulfates are precipitated with $Ba^{2+}(aq)$. The amount of precipitate is small, but it does cause turbidity in the solution. Turbidity is detected by instrumental methods that measure light scattering by colloids.

19 Chemical Thermodynamics

Visualizing Concepts

19.1   (a)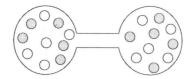

(b)   ΔS is positive, because the disorder of the system increases. Each gas has greater motional freedom as it expands into the second bulb, and there are many more possible arrangements for the mixed gases.

By definition, ideal gases experience no attractive or repulsive intermolecular interactions, so ΔH for the mixing of ideal gases is zero, assuming heat exchange only between the two bulbs.

(c)   The process is spontaneous and, therefore, irreversible. It is inconceivable that the gases would reseparate.

(d)   The entropy change of the surroundings is related to ΔH for the system. Because we are mixing ideal gases and ΔH = 0, ΔH_{surr} is also zero, assuming heat exchange only between the two bulbs.

19.2   (a)   Based on experience, the process is spontaneous. We know that 1,1-difluoro-ethane is a gas at atmospheric pressure, so the pressure inside the can must be much greater than atmospheric in order for the substance to be liquefied. When the nozzle is pressed and the system is open to the lower pressure of the atmosphere, the liquid vaporizes spontaneously. The 1,1-difluoroethane gas escapes the nozzle without external assistance.

(b)   We expect q_{sys} to be positive. We know that ΔH is positive for the vaporization of a gas. Because the change does not occur at constant pressure, q_{sys} and ΔH are not equal, but the sign of q_{sys} is still positive.

(c)   ΔS is definitely positive for this process. Because the process is spontaneous and ΔH is positive, ΔS must be positive and large so that ΔG is negative. It is also true that the system, the 1,1-difluoroethane molecules, occupy a larger volume and have greater motional freedom after vaporization.

(d)   The operation of the keyboard cleaner definitely depends more on entropy change than heat flow.

19.3   (a)   The process depicted is a change of state from a solid to a gas. ΔS is positive because of the greater motional freedom of the particles. ΔH is positive because both melting and boiling are endothermic processes.

(b)   The sign of ΔS_{surr} is negative, and the magnitude is less than or equal to ΔS_{sys}. If the process is spontaneous, the second law states that $\Delta S_{univ} \geq 0$. Because ΔS_{sys} is positive, ΔS_{surr} must be negative. If the change occurs via a reversible pathway, $\Delta S_{univ} = 0$ and $\Delta S_{surr} = -\Delta S_{sys}$. If the pathway is irreversible, the magnitude of ΔS_{sys} is greater than the magnitude of ΔS_{surr}, but the sign of ΔS_{surr} is still negative.

19.4   Both ΔH and ΔS for this reaction are positive.

The reaction involves breaking five blue–blue and twenty blue–red bonds and then forming twenty blue–red bonds. The net change is breaking five blue–blue bonds. Enthalpies for bond breaking (Section 8.8) are always positive.

In the depicted reaction, both reactants and products are in the gas phase (they are far apart and randomly placed). There are twice as many molecules (or moles) of gas in the products, so ΔS is positive for this reaction.

19.5   *Analyze/Plan.* Consider the physical changes that occur when a substance is heated. How do these changes affect the entropy of the substance?   *Solve.*

(a)   Both 1 and 2 represent changes in entropy at constant temperature; these are phase changes. Because 1 happens at a lower temperature, it represents melting (fusion), and 2 represents vaporization.

(b)   The substance changes from solid to liquid in 1, from liquid to gas in 2. The larger volume and greater motional freedom of the gas phase causes ΔS for vaporization to (always) be larger than ΔS for fusion.

(c)   For a perfect crystal at $T = 0$ K, the value of S is zero. This is the third law of thermodynamics.

19.6   (a)   We expect the enthalpy of combustion of the two isomers to be very similar. The molecular formulas of the two molecules are the same, so the balanced chemical equations for the two combustion reactions are identical. In the calculation of combustion enthalpy from standard enthalpies of formation of products and reactants, the only difference will be in the standard enthalpies of formation of the two isomers.

(b)   We expect *n*-pentane to have the higher standard molar entropy. The rod-shaped *n*-pentane has more possible vibrational and rotational motions than the almost-spherical neopentane. That is, *n*-pentane has greater motional energy, which results in a higher standard molar entropy than that of neopentane.

19.7   (a)   At 300 K, $\Delta H = T\Delta S$. Because $\Delta G = \Delta H - T\Delta S$, $\Delta G = 0$ at this point. When $\Delta G = 0$, the system is at equilibrium.

(b)   The reaction is spontaneous when ΔG is negative. This condition is met when $T\Delta S > \Delta H$. From the diagram, $T\Delta S > \Delta H$ when $T > 300$ K. The reaction is spontaneous at temperatures above 300 K.

19.8   (a)   At equilibrium, $\Delta G = 0$. On the diagram, $\Delta G = 0$ at 250 K. The system is at equilibrium at 250 K.

(b)   A reaction is spontaneous when ΔG is negative. The reaction is spontaneous at temperatures greater than 250 K.

    (c) $\Delta G = \Delta H - T\Delta S$, in the form of $y = b + mx$. ΔH is the y intercept of the graph (where $T = 0$) and is positive.

    (d)   The slope of the graph is $-\Delta S$. The slope is negative, so ΔS is positive. [Also, ΔG decreases as T increases, so the $T\Delta S$ term must become more negative and ΔS is positive.]

19.9   (a)   *Analyze.* The boxes depict three different mixtures of reactants and products for the reaction $A_2 + B_2 \rightleftharpoons 2AB$.

            Plan. $K_c = 1 = \dfrac{[AB]^2}{[A][B]}$. Calculate Q for each box, using number of molecules as a measure of concentration. If $Q = 1$, the system is at equilibrium.   *Solve.*

            Box 1: $K = \dfrac{(3)^2}{(3)(3)} = 1$

            Box 2: $Q = \dfrac{(1)^2}{(4)(4)} = \dfrac{1}{16} = 0.0625 = 0.06$

            Box 3: $Q = \dfrac{(7)^2}{(1)(1)} = \dfrac{49}{1} = 49$

            Box 1 is at equilibrium.

    (b)   Box 2.

    (c)   Qualitatively, Box 3 is farthest from equilibrium, so it has the largest magnitude of ΔG (driving force to reach equilibrium), then Box 2, and then Box 1, where $\Delta G = 0$.

            Box 1 < Box 2 < Box 3

            Quantitatively, $\Delta G = \Delta G° - RT\ln Q$. For Box 1, $\Delta G = 0$ and $K = 1$, so $\Delta G° = 0$.

            Box 2: $\Delta G = 0 - RT\ln(0.0625) = 2.77\ RT$

            Box 3: $\Delta G = 0 - RT\ln(49) = -3.89\ RT$

            Quantitative treatment confirms the order for magnitude of ΔG as Box 1 < Box 2 < Box 3.

19.10   (a)   True. When $\Delta G = 0$, the reaction is at equilibrium.

    (b)   False. At equilibrium, there is a mixture of reactants and products.

    (c)   False. There are fewer moles of gas in the products than the reactants, so ΔS is negative.

    (d)   False. At the left and right extremes of the graph, reactants and products are gases at 1 atm pressure; they are in their standard states. The quantity "x" is the difference in free energy between reactants and products in their standard states, $\Delta G°$.

    (e)   True. ΔG is a measure of the driving force for a reaction to reach equilibrium.

Spontaneous Processes (Section 19.1)

19.11   *Analyze/Plan.* Follow the logic in Sample Exercise 19.1.   *Solve.*

    (a)   Spontaneous; at ambient temperature, ripening happens without intervention.

    (b)   Spontaneous; sugar is soluble in water, and even more soluble in hot coffee.

(c)   Spontaneous; N_2 molecules are stable relative to isolated N atoms.

(d)   Spontaneous; under certain atmospheric conditions, lightning occurs.

(e)   Nonspontaneous; CO_2 and H_2O are in contact continuously at atmospheric conditions in nature and do not form CH_4 and O_2.

19.12   (a)   Nonspontaneous; at 1 atm, ice does not melt spontaneously at temperatures below its normal melting point.

(b)   Nonspontaneous; a mixture cannot be separated without outside intervention.

(c)   Spontaneous.

(d)   Spontaneous. The reaction is spontaneous but slow unless encouraged by a catalyst or spark.

(e)   Spontaneous; the very polar HCl molecules readily dissolve in water to form concentrated HCl(aq).

19.13   (a)   True, assuming the conditions are the same for the forward and reverse reactions.

(b)   False. A spontaneous process occurs without outside intervention. This definition says nothing about how quickly the process occurs. Spontaneity is a thermodynamic property, while rate is a kinetic property.

(c)   False. All spontaneous processes are real processes and real processes are irreversible.

(d)   True.

(e)   False. The maximum amount of work can be accomplished by a reversible process.

19.14   (a)   Yes. While most spontaneous processes are exothermic, some, such the melting of ice at room temperature, are endothermic.

(b)   Yes. Melting is an example of a process that is spontaneous at one temperature, the melting point, but nonspontaneous at other temperatures. Other phase changes are also examples of this behavior.

(c)   No. While the forward and reverse processes can be induced, the processes cannot be reversed with an infinitesimally small change in some property of the system.

(d)   Yes. The reversible pathway can accomplish the maximum amount of work on its surroundings because no entropy is lost to the universe. $\Delta S_{universe} = 0$.

19.15   *Analyze/Plan.* Define the system and surroundings. Use the appropriate definition to answer the specific questions.   *Solve.*

(a)   Water is the system. Heat must be added to the system to evaporate the water. The process is endothermic.

(b)   At 1 atm, the reaction is spontaneous at temperatures above 100 °C.

(c)   At 1 atm, the reaction is nonspontaneous at temperatures below 100 °C.

(d)   The two phases are in equilibrium at 100 °C.

19.16   (a)   Exothermic. If melting requires heat and is endothermic, freezing must be exothermic.

         (b)   At 1 atm (indicated by the term *normal* freezing point), the freezing of *n*-octane is spontaneous at temperatures below –57 °C.

         (c)   At 1 atm, the freezing of *n*-octane is nonspontaneous at temperatures above –57 °C.

         (d)   At 1 atm and –57 °C, the normal freezing point of *n*-octane, the solid and liquid phases are in equilibrium. That is, at the freezing point, *n*-octane molecules escape to the liquid phase at the same rate as liquid *n*-octane solidifies, assuming no heat is exchanged between *n*-octane and the surroundings.

19.17   *Analyze/Plan.* Consider the definitions of the terms *reversible, isothermal,* and *state function* to answer the questions.   *Solve.*

         (a)   No. Temperature is a state function, so a change in temperature does not depend on pathway. (As we are talking about an ideal gas, a reversible pathway may be possible for this change in state. An irreversible pathway is always possible.)

         (b)   No. An isothermal process occurs at constant temperature.

         (c)   No. ΔE is a state function. $\Delta E = q + w$; q and w are not state functions. Their values do depend on path, but their sum, ΔE, does not.

19.18   (a)   Yes, because ΔE is a state function. $(1 \rightarrow 2) = -\Delta E\ (2 \rightarrow 1)$

         (b)   No. We can say nothing about the values of q and w because we have no information about the paths.

         (c)   The magnitudes of the work are equal, but the signs are opposite. If the changes of state are reversible, the two paths are the same and $w\ (1 \rightarrow 2) = -w\ (2 \rightarrow 1)$. This is the maximum realizable work from this system.

19.19   *Analyze/Plan.* Define the system and surroundings. Use the appropriate definition to answer the specific questions.   *Solve.*

         (a)   An ice cube can melt reversibly at the conditions of temperature and pressure where the solid and liquid are in equilibrium. At 1 atm external pressure, the normal melting point of water is 0 °C.

         (b)   No. We know that melting is endothermic, so ΔH for melting ice is a positive, nonzero value. Also, since ΔH is a state function, the nonzero value is independent of path. Whether the ice cube melts reversibly or irreversibly, ΔH for the process is not zero.

19.20   (a)   The detonation of an explosive is definitely not reversible. The farflung debris from the explosion (system and surroundings) cannot be perfectly reconstructed, even with a large input of energy.

         (b)   The quantity q is related to ΔH. As the detonation is highly exothermic, q is large and negative.

               If only P–V work is done and P is constant, $\Delta H = q$. Although these conditions probably do not apply to a detonation, we can still predict the sign of q, based on ΔH, if not its exact magnitude.

(c)   The sign (and magnitude) of w depend on the path of the process, the exact details of how the detonation is carried out. It seems clear, however, that work will be done by the system on the surroundings in almost all circumstances (buildings collapse, earth and air are moved), so the sign of w is probably negative.

Entropy and the Second Law of Thermodynamics (Section 19.2)

19.21   (a)   True.

         (b)   False. For a reversible process, the entropy change of the universe is zero.

         (c)   True.

         (d)   False. For a reversible process, the entropy change to the system need not be zero, but it must be matched by the entropy change to the surroundings.

19.22   (a)   True.

         (b)   False. For an irreversible process, the net entropy change of the system and surroundings must be positive.

         (c)   False. For a spontaneous (and thus irreversible) process, the net entropy change of the system and surroundings must be positive. The entropy change of the system could be negative if the entropy change of the surroundings was large and positive.

         (d)   True. For an isothermal process, $\Delta S = q_{rev}/T$.

19.23   (a) $Br_2(l) \rightarrow Br_2(g)$, entropy increases, more mol gas in products, greater motional freedom.

         (b) $\Delta S = \dfrac{\Delta H}{T} = \dfrac{29.6\ kJ}{mol\ Br_2(l)} \times 1.00\ mol\ Br_2(l) \times \dfrac{1}{(273.15+58.8)K} \times \dfrac{1000\ J}{1\ kJ} = 89.2\ J/K$

19.24   (a) $Ga(l) \rightarrow Ga(s)$, ΔS is negative, less motional freedom

         (b) $\Delta H = 60.0\ g\ Ga \times \dfrac{1\ mol\ Ga}{69.723\ g\ Ga} \times \dfrac{-5.59\ kJ}{mol\ Ga} = -4.81046 = -4.81\ kJ$

                 $\Delta S = \dfrac{\Delta H}{T} = -4.81046\ kJ \times \dfrac{1000\ J}{1\ kJ} \times \dfrac{1}{(273.15+29.8)K} = -15.9\ J/K$

19.25   (a)   False. The second law states that entropy is conserved for a reversible process.

         (b)   True. In a reversible process, $\Delta S_{sys} + \Delta S_{surr} = 0$. If ΔS_{sys} is positive, ΔS_{surr} must be negative.

         (c)   False. Because ΔS_{univ} must be positive for a spontaneous process, ΔS_{surr} must be greater than –4.2 J/K. Entropy is not conserved for a spontaneous process.

19.26   (a)   Not necessarily. The only thing we know for sure is that the entropy of the universe increases for a spontaneous process.

       (b) ΔS_{surr} is positive and greater than the magnitude of the decrease in ΔS_{sys}.

       (c) $\Delta S_{sys} = 78$ J/K.

19.27   *Analyze.* Consider ΔS for the isothermal expansion of 0.200 mol of an ideal gas at 27 °C and an initial volume of 10.0 L.

       (a)   Whenever an ideal gas expands isothermally, we expect an increase in entropy, or positive ΔS, owing to the greater volume available for motion of the particles.

       (b)   *Plan.* Use the relationship $\Delta S_{sys} = nR \ln(V_2/V_1)$, Equation 19.3.

             Solve. $\Delta S_{sys} = 0.200 \, (8.314 \text{ J/mol-K})(\ln \, [18.5 \text{ L}/10.0 \text{ L}]) = 1.02$ J/K.

             Check. We expect ΔS to be positive when the motional freedom of a gas increases, and our calculation agrees with this prediction.

       (c)   No. The temperature at which the expansion occurs is not needed to calculate the entropy change, as long as the process is isothermal.

19.28   (a)   According to Boyle's law, pressure and volume are inversely proportional at constant amount and temperature. If the pressure of an ideal gas increases, volume decreases. We expect a decrease in entropy, or negative ΔS, for the isothermal compression of an ideal gas, owing to the smaller volume available for motion of the particles.

       (b)   According to Boyle's law, $P_1V_1 = P_2V_2$ at constant n and T.

             $0.750 \text{ atm} \times V_1 = 1.20 \text{ atm} \times V_2; \; V_2/V_1 = 0.750 \text{ atm}/1.20 \text{ atm} = 0.62500 = 0.625$

             $\Delta S_{sys} = nR \ln \, (V_2/V_1) = 0.600 \text{ mol} \, (8.314 \text{ J/mol-K})(\ln 0.625) = -2.34$ J/K

             Check. An increase in pressure results in a decrease in volume at constant T, so we expect ΔS to be negative, and it is.

       (c)   No. The temperature at which the compression (increase in pressure, decrease in volume) occurs is not needed to calculate the entropy change, as long as the process is isothermal.

The Molecular Interpretation of Entropy and the Third Law of Thermodynamics (Section 19.3)

19.29   (a)   Yes, the expansion is spontaneous.

       (b)   The ideal gas is the system, and everything else, including the vessel containing the vacuum, is the surroundings. There is literally nothing inside the vessel containing the vacuum, no gas molecules and no physical barriers. As the ideal gas expands into the vacuum, there is nothing for it to "push back," so no work is done. Mathematically, $w = -P_{ext}\Delta V$. Because the gas expands into a vacuum, $P_{ext} = 0$ and $w = 0$.

       (c)   Entropy. The "driving force" for the expansion of the gas is the increase in entropy associated with greater volume, more motional freedom, and more possible positions for the gas particles.

19.30 (a) A thermodynamic *state* is a set of conditions, usually temperature and pressure, that defines the properties of a bulk material. A *microstate* is a single possibility for all the positions and kinetic energies of all the molecules in a sample; it is a snapshot of positions and speeds at a particular instant.

(b) According to Equation 19.5 (Boltzmann law), the more possible microstates for a macroscopic state, the greater the entropy of the state. If S decreases going from A to B, then A has more microstates than B. Or, if ΔS is negative, the number of microstates decreases.

(c) According to part (b), if the number of microstates available to a system decreases, ΔS_{sys} is negative. For a spontaneous process, ΔS_{univ} is positive, so ΔS_{surr} is positive (and the magnitude is greater than that of ΔS_{sys}).

19.31 (a) The higher the temperature, the broader the distribution of molecular speeds and kinetic energies available to the particles. At higher temperature, the wider range of accessible kinetic energies leads to more microstates for the system.

(b) A decrease in volume reduces the number of possible positions for the particles and leads to fewer microstates for the system.

(c) Going from liquid to gas, particles have greater translational motion, which increases the number of positions available to the particles and the number of microstates for the system.

19.32 (a) ΔH_{vap} for H_2O at 25 °C = 44.02 kJ/mol; at 100 °C = 40.67 kJ/mol

$$\Delta S = \frac{q_{rev}}{T} = \frac{44.02 \text{ kJ}}{\text{mol}} \times \frac{1000 \text{ J}}{\text{kJ}} \times \frac{1}{298 \text{ K}} = 148 \text{ J/mol-K}$$

$$\Delta S = \frac{q_{rev}}{T} = \frac{40.67 \text{ kJ}}{\text{mol}} \times \frac{1000 \text{ J}}{\text{kJ}} \times \frac{1}{373 \text{ K}} = 109 \text{ J/mol-K}$$

(b) At both temperatures, the liquid → gas phase transition is accompanied by an increase in entropy, as expected. That the magnitude of the increase is greater at the lower temperature requires some explanation.

In the liquid state, there are significant hydrogen bonding interactions between H_2O molecules. This reduces the number of possible molecular positions and the number of microstates. Liquid water at 100° has sufficient kinetic energy to have broken many hydrogen bonds, so the number of microstates for $H_2O(l)$ at 100° is greater than the number of microstates for $H_2O(l)$ at 25 °C. The difference in the number of microstates upon vaporization at 100 °C is smaller, and the magnitude of ΔS is smaller.

19.33 *Analyze/Plan.* Consider the conditions that lead to an increase in entropy: more mol gas in products than reactants, increase in volume of sample and, therefore, number of possible arrangements, more motional freedom of molecules, and so on. *Solve.*

(a) More gaseous particles means more possible arrangements and greater disorder; ΔS is positive.

(b) S_{sys} increases when a banana ripens. Starches are polysaccharides, large molecules that break down into sugars, smaller monosaccharides and disaccharides, when ripening occurs. Formation of the sugars increases the number of molecules and the entropy.

   S_{sys} clearly increases in 19.11 (b), where there is an increase in volume and possible arrangements for the sample.

   In 19.11 (c), the system goes from two moles of gaseous reactants to one mole of gaseous products, and S_{sys} decreases.

   In 19.11 (d), the entropy of the universe clearly increases, but the definition of the system in a lightning strike is more problematic.

   In 19.11 (e), the specified state is room temperature and 1 atm pressure. This means that H_2O is present as a liquid; there is then 1 mol of gaseous reactants (CO_2) and 3 mol of gaseous products (CH_4 and 2 O_2), so S_{sys} increases. (The reaction is not spontaneous because of the very large positive ΔH_{sys} for the reaction as written.)

19.34   (a)   Solids are much more ordered than gases, so ΔS is negative.

   (b)   The entropy of the system increases in Exercise 19.12 (a) and (e). There is more motional freedom for the system in both cases. In (b), (c), and (d), there is less motional freedom after the change and the entropy of the system decreases.

19.35   *Analyze/Plan.* Consider the conditions that lead to an increase in entropy: more mol gas in products than reactants, increase in volume of sample and, therefore, number of possible arrangements, more motional freedom of molecules, and so on.   *Solve.*

   (a)   S increases; translational motion is greater in the liquid than the solid.

   (b)   S decreases; volume and translational motion decrease going from the gas to the liquid.

   (c)   S increases; volume and translational motion are greater in the gas than the solid.

19.36   (a)   When temperature increases, the range of accessible molecular speeds and kinetic energies increases. This produces more microstates and an increase in entropy.

   (b)   When the volume of a gas increases (even at constant T), there are more possible positions for the particles, more microstates, and greater entropy.

   (c)   When equal volumes of two miscible liquids are mixed, the volume of the sample and, therefore, the number of possible arrangements increases. This produces more microstates and an increase in entropy.

19.37   (a)   False. The entropy of a pure crystalline substance at absolute zero is zero.

   (b)   True.

   (c)   False. Monoatomic gases have no rotational or vibrational states.

   (d)   True.

19.38   (a)   True. (From the Boltzmann relationship, $S = k \ln W$.)

       (b)   False. The translational, rotational, and vibrational degrees of freedom of CO_2 (or any molecule) are determined by its structure. As long as heating does not change the basic molecular structure, it does not change the degrees of freedom. The additional kinetic energy due to heating is distributed as more translational, vibrational, and rotational motion. That is, number of microstates [see Solution 19.30 (a)] increases, but the degrees of freedom do not.

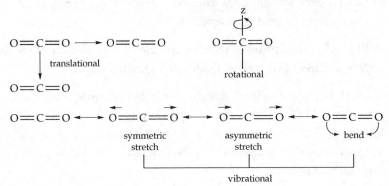

       (c)   False. At a given temperature, $CO_2(g)$ has more microstates and, thus, greater entropy than $Ar(g)$. Because $CO_2(g)$ is a triatomic molecule, it has multiple rotational and vibrational microstates not available to monatomic $Ar(g)$.

19.39   *Analyze/Plan.* Consider the factors that lead to higher entropy: more mol gas in products than reactants, increase in volume of sample and, therefore, number of possible arrangements, more motional freedom of molecules, and so on.   *Solve.*

       (a) $Ar(g)$ (gases have higher entropy due primarily to much larger volume)

       (b) $He(g)$ at 1.5 atm (larger volume and more motional freedom)

       (c)   1 mol of $Ne(g)$ in 15.0 L (larger volume provides more motional freedom)

       (d) $CO_2(g)$ (more motional freedom)

19.40   (a)   One mole of $O_3(g)$ at 300 °C, 0.01 atm ($O_3(g)$ is a more complex molecule and has more vibrational degrees of freedom.)

       (b)   1 mol $H_2O(g)$ at 100 °C, 1 atm [larger volume occupied by $H_2O(g)$]

       (c)   0.5 mol $CH_4(g)$ at 298 K, 20-L volume (more complex molecule, more rotational and vibrational degrees of freedom)

       (d)   100 g of $Na_2SO_4(aq)$ at 30 °C (more motional freedom in aqueous solution)

19.41   *Analyze/Plan.* Consider the markers of an increase in entropy for a chemical reaction: liquids or solutions formed from solids, gases formed from either solids or liquids, increase in mol gas during reaction.   *Solve.*

       (a)   ΔS negative (moles of gas decrease)

       (b)   ΔS positive (gas produced, increased disorder)

       (c)   ΔS negative (moles of gas decrease)

       (d)   ΔS is small and probably positive [moles of gas same in reactants and products, $H_2O(g)$ is more structurally complex than $H_2(g)$]

19.42   (a)   Au(l) → Au(s); negative ΔS, less motional freedom in the solid

        (b) $Cl_2(g)$ → 2 Cl(g); positive ΔS, moles of gas increase

        (c)   CO(g) + 2 H_2(g) → CH_3OH(l); negative ΔS, moles of gas decrease

        (d)   3 $Ca(NO_3)_2$(aq) + 2 $(NH_4)_3PO_4$(aq) → $Ca_3(PO_4)_2$(s) + 6 NH_4NO_3(aq); ΔS is negative, less motional freedom, fewer moles of ions in aqueous solution

Entropy Changes in Chemical Reactions (Section 19.4)

19.43   (a)

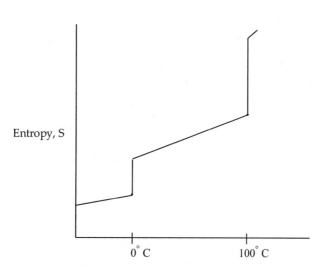

        (b)   Boiling water, at 100 °C, has a much larger entropy change than melting ice at 0 °C. Before and after melting, H_2O molecules are touching. And there is actually a small decrease in volume going from solid to liquid water. Boiling drastically increases the distance between molecules and the volume of the sample. The increase in available molecular positions is much greater for boiling than melting, so the entropy change is also greater.

19.44   Melting = –126.5 °C; boiling = 97.4 °C

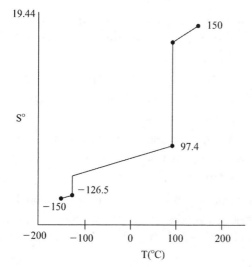

19.45   *Analyze/Plan.* Given two molecules in the same state, predict which will have the higher molar entropy. In general, for molecules in the same state, the more atoms in the molecule, the more degrees of freedom, the greater the number of microstates, and the higher the standard entropy, $S°$.

     (a) $C_2H_6(g)$ has more degrees of freedom and larger $S°$.

     (b) $CO_2(g)$ has more degrees of freedom and larger $S°$.

19.46   Propylene will have a higher $S°$ at 25 °C. At this temperature, both are gases, so there are no lattice effects. Because they have the same molecular formula, only the details of their structures are different. In propylene, there is free rotation around the C–C single bond, whereas in cyclopropane the 3-membered ring severely limits rotation. The greater motional freedom of the propylene molecule leads to a higher absolute entropy.

19.47   *Analyze/Plan.* Consider the conditions that lead to an increase in entropy: more mol gas in products than reactants, increase in volume of sample and, therefore, number of possible arrangements, more motional freedom of molecules, and so on.   *Solve.*

     (a) $Sc(g)$ will have the higher standard entropy at 25 °C. In general, the gas phase of a substance has a larger $S°$ than the solid phase because of the greater volume and motional freedom of the molecules. $Sc(s)$, 34.6 J/mol-K; $Sc(g)$, 174.7 J/mol-K.

     (b) $NH_3(g)$ will have the higher standard entropy at 25 °C. Molecules in the gas phase have more motional freedom than molecules in solution. $NH_3(g)$, 192.5 J/mol-K; $NH_3(aq)$, 111.3 J/mol-K.

     (c) $O_3(g)$ will have the higher standard entropy at 25 °C. The triatomic molecule will have more vibrational degrees of freedom than the diatomic molecule. $O_2(g)$, 205.0 J/K; $O_3(g)$, 237.6 J/K.

     (d)   C(graphite) will have the higher standard entropy at 25 °C. Diamond is a network covalent solid with each C atom tetrahedrally bound to four other C atoms. Graphite consists of sheets of fused planar 6-membered rings with each C atom bound in a trigonal planar arrangement to three other C atoms. The internal entropy in graphite is greater because there is translational freedom among the planar sheets of C atoms while there is very little vibrational freedom within the network covalent diamond lattice. C(diamond), 2.43 J/mol-K; C(graphite) 5.69 J/mol-K.

19.48   (a) $C_6H_6(g)$ will have the higher standard entropy at 25 °C. Molecules in the gas phase have larger volume and greater motional freedom than those in the liquid state. $C_6H_6(l)$, 172.8 J/K; $C_6H_6(g)$, 269.2 J/K.

     (b) $CO_2(g)$ will have the higher standard entropy at 25 °C. The more complex CO_2 molecule has more vibrational degrees of freedom and a slightly higher entropy. $CO(g)$, 197.9 J/mol-K; $CO_2(g)$, 213.6 J/mol-K.

     (c)   Two moles of $NO_2(g)$ will have the higher standard entropy at 25 °C. More particles have a greater number of arrangements or microstates. 1 mol $N_2O_4(g)$, 304.3 J/K; 2 mol $NO_2(g)$, 2(240.45) = 480.90 J/K.

     (d) $HCl(g)$ will have a greater standard entropy at 25 °C. The greater motional freedom of HCl molecules in the gas phase outweighs the greater number of particles in $HCl(aq)$. [$HCl(aq)$ is ionized into $H^+(aq)$ and $Cl^-(aq)$.] $HCl(g)$, 186.69 J/K; $HCl(aq)$, 56.5 J/K.

19.49   For elements with similar structures, the heavier the atoms, the lower the vibrational frequencies at a given temperature. This means that more vibrations can be accessed at a particular temperature resulting in a greater absolute entropy for the heavier elements.

19.50   (a)   C(diamond), $S° = 2.43$ J/mol-K; C(graphite), $S° = 5.69$ J/mol-K. Diamond is a network covalent solid with each C atom tetrahedrally bound to four other C atoms. Graphite consists of sheets of fused planar 6-membered rings with each C atom bound in a trigonal planar arrangement to three other C atoms. The internal entropy in graphite is greater because there is translational freedom among the planar sheets of C atoms whereas there is very little translational or vibrational freedom within the covalent-network diamond lattice.

         (b) $S°$ for buckminsterfullerene will be ≥ 10 J/mol-K. $S°$ for graphite is twice $S°$ for diamond, and $S°$ for the fullerene should be higher than that of graphite. The 60-atom "bucky" balls have more flexibility than graphite sheets. Also, the balls have translational freedom in three dimensions, whereas graphite sheets have it in only two directions. Because of the ball structure, there is more empty space in the fullerene lattice than in graphite or diamond; essentially, 60 C-atoms in fullerene occupy a larger volume than 60 C-atoms in graphite or diamond. Thus, the fullerene has additional "molecular" complexity, more degrees of translational freedom, and occupies a larger volume, all features that point to a higher absolute entropy.

19.51   *Analyze/Plan.* Follow the logic in Sample Exercise 19.5.   *Solve.*

         (a) $\Delta S° = S°\ C_2H_6(g) - S°\ C_2H_4(g) - S°\ H_2(g)$

                $= 229.5 - 219.4 - 130.58 = -120.5$ J/K

                $\Delta S°$ is negative because there are fewer moles of gas in the products.

         (b) $\Delta S° = 2S°\ NO_2(g) - \Delta S°\ N_2O_4(g) = 2(240.45) - 304.3 = +176.6$ J/K

                $\Delta S°$ is positive because there are more moles of gas in the products.

         (c) $\Delta S° = \Delta S°\ BeO(s) + \Delta S°\ H_2O(g) - \Delta S°\ Be(OH)_2(s)$

                $= 13.77 + 188.83 - 50.21 = +152.39$ J/K

                $\Delta S°$ is positive because the product contains more total particles and more moles of gas.

         (d) $\Delta S° = 2S°\ CO_2(g) + 4S°\ H_2O(g) - 2S°\ CH_3OH(g) - 3S°\ O_2(g)$

                $= 2(213.6) + 4(188.83) - 2(237.6) - 3(205.0) = +92.3$ J/K

                $\Delta S°$ is positive because the product contains more total particles and more moles of gas.

19.52   (a) $\Delta S° = S°\ NH_4NO_3(s) - S°\ HNO_3(g) - S°\ NH_3(g)$

                $= 151 - 266.4 - 192.5 = -307.9 = -308$ J/K

                $\Delta S°$ is large and negative because all reactants are gases (2 moles) and the product is a solid.

(b) $\Delta S° = 4S° \, Fe(s) + 3S° \, O_2(g) - 2S° \, Fe_2O_3(s)$

  $= 4(27.15) + 3(205.0) - 2(89.96) = 543.68 = 543.7 \, J/K$

$\Delta S°$ is large and positive because the reaction produces 3 moles of gas and the reactant is a solid.

(c) $\Delta S° = S° \, CaCl_2(s) + S° \, CO_2(g) + S° \, H_2O(l) - S° \, CaCO_3(s) - 2S° \, HCl(g)$

  $= 104.6 + 213.6 + 69.91 - 92.88 - 2(186.69) = -78.15 \, J/K$

$\Delta S°$ is small and negative because the products contain one fewer mole of gas, but one more mole of liquid. Note the very small standard entropy for $H_2O(l)$, owing to its strength of hydrogen bonding. If the products included one mole of a different liquid, the magnitude of the entropy change would be even smaller.

(d) $\Delta S° = S° \, C_6H_6(l) + 6S° \, H_2(g) - 3S° \, C_2H_6(g)$

  $= 172.8 + 6(130.58) - 3(229.5) = 267.78 = 267.8 \, J/K$

$\Delta S°$ is positive because there are more moles of gas in the products.

Gibbs Free Energy (Sections 19.5 and 19.6)

19.53   (a)   Yes. $\Delta G = \Delta H - T\Delta S$

(b)   No. If ΔG is positive, the process is nonspontaneous.

(c)   No. There is no relationship between ΔG and rate of reaction. A spontaneous reaction, one with a $-\Delta G$, may occur at a very slow rate. For example, $2 \, H_2(g) + O_2(g) \rightarrow 2 \, H_2O(g)$, $\Delta G = -457 \, kJ$ is very slow if not initiated by a spark.

19.54   (a)   No. $\Delta G = \Delta G° + RT \ln Q$. The relative magnitudes of ΔG and $\Delta G°$ depend on the value of Q.

(b)   For a process that occurs at constant temperature and pressure, the system is at equilibrium when $\Delta G = 0$.

(c)   No. Activation energy is related to the rate constant, k. The sign and magnitude of ΔG give no information about rate.

19.55   *Analyze/Plan.* Consider the definitions of $\Delta H°$, $\Delta S°$, and $\Delta G°$, along with sign conventions. $\Delta G° = \Delta H° - T\Delta S°$.   *Solve.*

(a) $\Delta H°$ is negative; the reaction is exothermic.

(b) $\Delta S°$ is negative; the reaction leads to decrease in disorder (increase in order) of the system.

(c) $\Delta G° = \Delta H° - T\Delta S° = -35.4 \, kJ - 298 \, K \, (-0.0855 \, kJ/K) = -9.921 = -9.9 \, kJ$

(d)   At 298 K, $\Delta G°$ is negative. If all reactants and products are present in their standard states, the reaction is spontaneous (in the forward direction) at this temperature.

19.56   (a) $\Delta H°$ is positive; the reaction is endothermic.

(b) $\Delta S°$ is positive; the reaction leads to an increase in disorder.

(c) $\Delta G° = \Delta H° - T\Delta S° = 23.7 \, kJ - 298 \, K \, (0.0524 \, kJ/K) = 8.0848 = 8.08 \, kJ$

(d)   At 298 K, $\Delta G°$ is positive. If all reactants and products are present in their standard states, the reaction is spontaneous in the reverse direction at this temperature; it is nonspontaneous in the forward direction.

19.57   *Analyze/Plan.* Follow the logic in Sample Exercises 19.6 and 19.7. Calculate $\Delta H°$ according to Equation 5.31, $\Delta S°$ by Equation 19.8, and $\Delta G°$ by Equation 19.14. Then, use $\Delta H°$ and $\Delta S°$ to calculate $\Delta G°$ using Equation 19.12, $\Delta G° = \Delta H° - T\Delta S°$.   *Solve.*

(a) $\Delta H° = 2(-268.61) - [0 + 0] = -537.22$ kJ

$\Delta S° = 2(173.51) - [130.58 + 202.7] = 13.74 = 13.7$ J/K

$\Delta G° = 2(-270.70) - [0 + 0] = -541.40$ kJ

$\Delta G° = -537.22$ kJ $- 298(0.01374)$ kJ $= -541.31$ kJ

(b) $\Delta H° = -106.7 - [0 + 2(0)] = -106.7$ kJ

$\Delta S° = 309.4 - [5.69 + 2(222.96)] = -142.21 = -142.2$ J/K

$\Delta G° = -64.0 - [0 + 2(0)] = -64.0$ kJ

$\Delta G° = -106.7$ kJ $- 298(-0.14221)$ kJ $= -64.3$ kJ

(c) $\Delta H° = 2(-542.2) - [2(-288.07) + 0] = -508.26 = -508.3$ kJ

$\Delta S° = 2(325) - [2(311.7) + 205.0] = -178.4 = -178$ J/K

$\Delta G° = 2(-502.5) - [2(-269.6) + 0] = -465.8$ kJ

$\Delta G° = -508.26$ kJ $- 298(-0.1784)$ kJ $= -455.097 = -455.1$ kJ

(The discrepancy in $\Delta G°$ values is because of experimental uncertainties in the tabulated thermodynamic data.)

(d) $\Delta H° = -84.68 + 2(-241.82) - [2(-201.2) + 0] = -165.92 = -165.9$ kJ

$\Delta S° = 229.5 + 2(188.83) - [2(237.6) + 130.58] = 1.38 = 1.4$ J/K

$\Delta G° = -32.89 + 2(-228.57) - [2(-161.9) + 0] = -166.23 = -166.2$ kJ

$\Delta G° = -165.92$ kJ $- 298(0.00138)$ kJ $= -166.33 = -166.3$ kJ

19.58   (a) $\Delta H° = 2(-1139.7) - 4(0) + 3(0) = -2279.4$ kJ

$\Delta S° = 2(81.2) - 4(23.6) - 3(205.0) = -547.0$ J/K

$\Delta G° = 2(-1058.1) - 4(0) - 3(0) = -2116.2$ kJ

$\Delta G° = -2279.4$ kJ $- 298$ K$(-0.5470$ kJ/K$) = -2116.4$ kJ

(b) $\Delta H° = -553.5 - 393.5 - (-1216.3) = 269.3$ kJ

$\Delta S° = 70.42 + 213.6 - 112.1 = 171.92 = 171.9$ J/K

$\Delta G° = -525.1 - 394.4 - (-1137.6) = 218.1$ kJ

$\Delta G° = 269.3$ kJ $- 298$ K $(0.1719$ kJ/K$) = 218.1$ kJ

(c)   Assume the reactant is P(g), not P(s).

$\Delta H° = 2(-1594.4) + 5(0) - 2(316.4) - 10(-268.61) = -1135.5$ kJ

$\Delta S° = 2(300.8) + 5(130.58) - 2(163.2) - 10(173.51) = -807.0$ J/K

$\Delta G° = 2(-1520.7) + 5(0) - 2(280.0) - 10(-270.70) = -894.4$ kJ

$\Delta G° = -1135.5$ kJ $- 298$ K$(-0.8070$ kJ/K$) = -895.014 = -895.0$ kJ

(The small discrepancy in $\Delta G°$ values is because of experimental uncertainties in tabulated thermodynamic data.)

(d) $\Delta H° = -284.5 - (0) - (0) = -284.5$ kJ

$\Delta S° = 122.5 - 64.67 - 205.0 = -147.2$ J/K

$\Delta G° = -240.6 - (0) - (0) = -240.6$ kJ

$\Delta G° = -284.5$ kJ $- 298$ K $(-0.1472$ kJ/K$) = -240.634 = -240.6$ kJ

19.59   *Analyze/Plan.* Follow the logic in Sample Exercise 19.7.   *Solve.*

(a) $\Delta G° = 2\Delta G° SO_3(g) - [2\Delta G° SO_2(g) + \Delta G° O_2(g)]$

$= 2(-370.4) - [2(-300.4) + 0] = -140.0$ kJ, spontaneous

(b) $\Delta G° = 3\Delta G° NO(g) - [\Delta G° NO_2(g) + \Delta G° N_2O(g)]$

$= 3(86.71) - [51.84 + 103.59] = +104.70$ kJ, nonspontaneous

(c) $\Delta G° = 4\Delta G° FeCl_3(s) + 3\Delta G° O_2(g) - [6\Delta G° Cl_2(g) + 2\Delta G° Fe_2O_3(s)]$

$= 4(-334) + 3(0) - [6(0) + 2(-740.98)] = +146$ kJ, nonspontaneous

(d) $\Delta G° = \Delta G° S(s) + 2\Delta G° H_2O(g) - [\Delta G° SO_2(g) + 2\Delta G° H_2(g)]$

$= 0 + 2(-228.57) - [(-300.4) + 2(0)] = -156.7$ kJ, spontaneous

19.60   (a) $\Delta G° = 2\Delta G° AgCl(s) - [2\Delta G° Ag(s) + \Delta G° Cl_2(g)]$

$= 2(-109.7) - 2(0) - 0 = -219.4$ kJ, spontaneous

(b) $P_4O_{10}(s) + 16 H_2(g) \rightarrow 4 PH_3(g) + 10 H_2O(g)$

$\Delta G° = 4\Delta G° PH_3(g) + 10\Delta G° H_2O(g) - [\Delta G° P_4O_{10}(s) + 16\Delta G° H_2(g)]$

$= 4(13.4) + 10(-228.57) - [-2675.2] - 16(0) = 443.1$ kJ, nonspontaneous

(c) $\Delta G° = \Delta G° CF_4(g) + 4\Delta G° HF(g) - [\Delta G° CH_4(g) + 4\Delta G° F_2(g)]$

$= -635.1 + 4(-270.70) - (-50.8) - 4(0) = -1667.1$ kJ, spontaneous

(d) $\Delta G° = 2\Delta G° H_2O(l) + \Delta G° O_2(g) - 2\Delta G° H_2O_2(l)$

$= 2(-237.13) + 0 - 2(-120.4) = -233.5$ kJ, spontaneous

19.61   *Analyze/Plan.* Follow the logic in Sample Exercise 19.8(a).   *Solve.*

(a) $2 C_8H_{18}(l) + 25 O_2(g) \rightarrow 16 CO_2(g) + 18 H_2O(l)$

(b)   Because there are more moles of gas in the reactants, $\Delta S°$ is negative, which makes $-T\Delta S$ positive. $\Delta G°$ is less negative than $\Delta H°$. (This argument is true for the reaction as written. If the products are all in the gas phase, there are more moles of gas in the products and $\Delta G°$ is more negative than $\Delta H°$.)

19.62　(a)　$\Delta G°$ should be less negative than $\Delta H°$. Products contain fewer moles of gas, so $\Delta S°$ is negative. $\Delta G° = \Delta H° - T\Delta S°$; $-T\Delta S°$ is positive, so $\Delta G°$ is less negative than $\Delta H°$.

(b)　We can estimate $\Delta S°$ using a similar reaction and then use $\Delta G° = \Delta H° - T\Delta S°$ (estimate) to get a ballpark figure. There are no sulfite salts listed in Appendix C, so use a reaction such as $CO_2(g) + CaO(s) \rightarrow CaCO_3(s)$ or $CO_2(g) + BaO(s) \rightarrow BaCO_3(s)$. Or, calculate both $\Delta S°$ values and use the average as your estimate.

19.63　*Analyze/Plan.* Based on the signs of ΔH and ΔS for a particular reaction, assign a category from Table 19.3 to each reaction.　*Solve.*

(a)　(iii) ΔG is negative at low temperatures, positive at high temperatures. That is, the forward reaction is spontaneous at lower temperatures but not spontaneous at higher temperatures.

(b)　(ii) ΔG is positive at all temperatures. The forward reaction is nonspontaneous at all temperatures.

(c)　(iv) ΔG is positive at low temperatures, negative at high temperatures. That is, the forward reaction will proceed spontaneously at high temperature, but not at low temperature.

19.64　$\Delta G° = \Delta H° - T\Delta S°$

(a)　$\Delta G° = -844 \text{ kJ} - 298 \text{ K}(-0.165 \text{ kJ/K}) = -795 \text{ kJ}$, spontaneous

(b)　$\Delta G° = +572 \text{ kJ} - 298 \text{ K}(0.179 \text{ kJ/K}) = +519 \text{ kJ}$, nonspontaneous

To be spontaneous, ΔG must be negative ($\Delta G < 0$).

Thus, $\Delta H° - T\Delta S° < 0$; $\Delta H° < T\Delta S°$; $T > \Delta H°/\Delta S°$; $T > \dfrac{572 \text{ kJ}}{0.179 \text{ kJ/K}} = 3.20 \times 10^3 \text{ K}$

19.65　*Analyze/Plan.* We are told that the reaction is barely spontaneous and endothermic, and asked to estimate the sign and magnitude of ΔS. If a reaction is spontaneous, $\Delta G < 0$. Use this information with Equation 19.11 to solve the problem.　*Solve.*

At 390 K, $\Delta G < 0$; $\Delta G = \Delta H - T\Delta S < 0$

$23.7 \text{ kJ} - 390 \text{ K} (\Delta S) < 0$; $23.7 \text{ kJ} < 390 \text{ K} (\Delta S)$; $\Delta S > 23.7 \text{ kJ}/390 \text{ K}$

$\Delta S > 0.06077 \text{ kJ/K}$ or $\Delta S > 60.8 \text{ J/K}$

The reaction is spontaneous and endothermic, so the sign of ΔS must be positive. Because the reaction is "barely" spontaneous, the magnitude will not be much greater than 61 J/K.

19.66　At 45 °C or 318 K, $\Delta G > 0$. $\Delta G = \Delta H - T\Delta S > 0$.

$\Delta H - 318 \text{ K} (72 \text{ J/K}) > 0$; $\Delta H > +2.3 \times 10^4 \text{ J}$; $\Delta H > +23 \text{ kJ}$

The reaction is nonspontaneous and has a positive ΔS, so it must be endothermic. Because it is "barely" nonspontaneous, the magnitude will not be much greater than 23 kJ.

19.67　*Analyze/Plan.* Use Equation 19.11 to calculate T when $\Delta G = 0$. This is similar to calculating the temperature of a phase transition in Sample Exercise 19.10. Use Table 19.3 to determine whether the reaction is spontaneous or nonspontaneous above this temperature.　*Solve.*

(a) $\Delta G = \Delta H - T\Delta S$; $0 = -32$ kJ $- T(-98$ J/K$)$; 32×10^3 J $= T(98$ J/K$)$

       $T = 32 \times 10^3$ J$/(98$ J/K$) = 326.5 = 330$ K

(b)   Nonspontaneous. The sign of ΔS is negative, so as T increases, ΔG becomes more positive.

19.68 ΔG is negative when $T\Delta S > \Delta H$ or $T > \Delta H/\Delta S$.

       $\Delta H° = \Delta H°$ CH$_3$OH $+ \Delta H°$ CO(g) $- \Delta H°$ CH$_3$COOH(l)

          $= -201.2 - 110.5 - (-487.0) = 175.3$ kJ

       $\Delta S° = S°$ CH$_3$OH $+ S°$ CO(g) $- S°$ CH$_3$COOH(l) $= 237.6 + 197.9 - 159.8 = 275.7$ J/K

       $T > \dfrac{175.3 \text{ kJ}}{0.2757 \text{ kJ/K}} = 635.8$ K

       The reaction is spontaneous above 635.8 K (363 °C).

19.69   *Analyze/Plan.* Given a chemical equation and thermodynamic data (values of $\Delta H°_f$, $\Delta G°_f$, and $S°$) for reactants and products, predict the variation of $\Delta G°$ with temperature and calculate $\Delta G°$ at 800 K and 1000 K. Use Equations 5.31 and 19.8 to calculate $\Delta H°$ and $\Delta S°$, respectively; use these values to calculate $\Delta G°$ at various temperatures, using Equation 19.12. The signs of $\Delta H°$ and $\Delta S°$ determine the variation of $\Delta G°$ with temperature.   *Solve.*

(a)   Calculate $\Delta H°$ and $\Delta S°$ to determine the sign of $T\Delta S°$.

       $\Delta H° = 3\Delta H°$ NO(g) $- \Delta H°$ NO$_2$(g) $- \Delta H°$ N$_2$O(g)

          $= 3(90.37) - 33.84 - 81.6 = 155.7$ kJ

       $\Delta S° = 3S°$ NO(g) $- S°$ NO$_2$(g) $- S°$ N$_2$O(g)

          $= 3(210.62) - 240.45 - 220.0 = 171.4$ J/K

       $\Delta G° = \Delta H° - T\Delta S°$. Because $\Delta S°$ is positive, $-T\Delta S°$ becomes more negative as T increases and $\Delta G°$ becomes more negative.

(b) $\Delta G° = \Delta H° - T\Delta S° = 155.7$ kJ $- (800$ K$)(0.1714$ kJ/K$)$

       $\Delta G° = 155.7$ kJ $- 137$ kJ $= 19$ kJ

       Because $\Delta G°$ is positive at 800 K, the reaction is not spontaneous at this temperature.

(c) $\Delta G° = 155.7$ kJ $- (1000$ K$)(0.1714$ kJ/K$) = 155.7$ kJ $- 171.4$ kJ $= -15.7$ kJ

       $\Delta G°$ is negative at 1000 K and the reaction is spontaneous at this temperature.

19.70   (a) $\Delta H° = \Delta H°_f$ CH$_3$OH(g) $- \Delta H°_f$ CH$_4$(g) $- 1/2\, \Delta H°_f$ O$_2$(g)

          $= -201.2 - (-74.8) - (1/2)(0) = -126.4$ kJ

       $\Delta S° = S°$ CH$_3$OH(g) $- S°$ CH$_4$(g) $- 1/2\, S°$ O$_2$(g)

          $= 237.6 - 186.3 - 1/2(205.0) = -51.2$ J/K $= -0.0512$ kJ/K

(b) $\Delta G° = \Delta H° - T\Delta S°$. $-T\Delta S°$ is positive, so $\Delta G°$ will increase (becomes more positive) as temperature increases.

(c) $\Delta G° = \Delta H° - T\Delta S° = -126.4 \text{ kJ} - 298 \text{ K}(-0.0512 \text{ kJ/K}) = -111.1 \text{ kJ}$

The reaction is spontaneous at 298 K because $\Delta G°$ is negative at this temperature. In this case, $\Delta G°$ could have been calculated from ΔG_f^o values in Appendix C, because these values are tabulated at 298 K.

(d) No. The reaction is at equilibrium when $\Delta G° = 0$.

$\Delta G° = \Delta H° - T\Delta S° = 0$. $\Delta H° = T\Delta S°$, $T = \Delta H°/\Delta S°$

$T = -126.4 \text{ kJ}/-0.0512 \text{ kJ/K} = 2469 = 2470 \text{ K}$

This temperature is so high that the reactants and products are likely to decompose. At standard conditions, equilibrium is functionally unattainable for this reaction.

19.71   *Analyze/Plan.* Follow the logic in Sample Exercise 19.10.   *Solve.*

(a) $\Delta S_{vap}^o = \Delta H_{vap}^o/T_b$; $T_b = \Delta H_{vap}^o/\Delta S_{vap}^o$

$\Delta H_{vap}^o = \Delta H° \; C_6H_6(g) - \Delta H° \; C_6H_6(l) = 82.9 - 49.0 = 33.9 \text{ kJ}$

$\Delta S_{vap}^o = S° \; C_6H_6(g) - S° \; C_6H_6(l) = 269.2 - 172.8 = 96.4 \text{ J/K}$

$T_b = 33.9 \times 10^3 \text{ J}/96.4 \text{ J/K} = 351.66 = 352 \text{ K} = 79 \text{ °C}$

(b) From the *Handbook of Chemistry and Physics*, 74th edition, $T_b = 80.1$ °C. The values are remarkably close; the small difference is due to deviation from ideal behavior by $C_6H_6(g)$ and experimental uncertainty in both the boiling point measurement and the thermodynamic data.

19.72   (a) The transformation of $I_2(s)$ to $I_2(g)$ is sublimation. The temperature at which the free-energy change for this process is zero is the sublimation temperature. According to Sample Exercise 19.10, $T_{sub} = \Delta H_{sub}^o/\Delta S_{sub}^o$.

Use data from Appendix C to calculate ΔH_{sub}^o and ΔS_{sub}^o for $I_2(s)$.

$I_2(s) \rightarrow I_2(l)$ melting

$\underline{I_2(l) \rightarrow I_2(g) \text{ boiling}}$

$I_2(s) \rightarrow I_2(g)$ sublimation

$\Delta H_{sub}^o = \Delta H_f^o I_2(g) - \Delta H_f^o \; I_2(s) = 62.25 - 0 = 62.25 \text{ kJ}$

$\Delta S_{sub}^o = S° \; I_2(g) - S° \; I_2(s) = 260.57 - 116.73 = 143.84 \text{ J/K} = 0.14384 \text{ kJ/K}$

$T_{sub} = \dfrac{\Delta H_{sub}^o}{\Delta S_{sub}} = \dfrac{62.25 \text{ kJ}}{0.14384 \text{ kJ/K}} = 432.8 \text{ K} = 159.6 \text{ °C}$

[Note that some assumptions were necessary to make this estimate, that both $I_2(s)$ and $I_2(g)$ are present in their standard states and, consequently, $\Delta G_{sub} = \Delta G_{sub}^o = 0$. We also assume that the values of ΔH_{sub}^o and ΔS_{sub}^o are the same at 298 K and at the sublimation temperature.]

(b) T_m for $I_2(s) = 386.85 \text{ K} = 113.7 \text{ °C}$; $T_b = 457.4 \text{ K} = 184.3 \text{ °C}$
(from WebElements™ 2013)

(c)   The boiling point of I_2 is closer to the sublimation temperature. Both boiling and sublimation begin with molecules in a condensed phase (little space between molecules) and end in the gas phase (large intermolecular distances). Separation of the molecules is the main phenomenon that determines both ΔH and ΔS, so it is not surprising that the ratio of $\Delta H/\Delta S$ is similar for sublimation and boiling.

19.73   *Analyze/Plan.* We are asked to write a balanced equation for the combustion of acetylene, calculate $\Delta H°$ for this reaction, and calculate maximum useful work possible by the system. Combustion is combination with O_2 to produce CO_2 and H_2O. Calculate $\Delta H°$ using data from Appendix C and Equation 5.31. The maximum obtainable work is ΔG (Equation 19.18), which can be calculated from data in Appendix C and Equation 19.14.   *Solve.*

(a) $C_2H_2(g) + 5/2\, O_2(g) \rightarrow 2\, CO_2(g) + H_2O(l)$

(b) $\Delta H° = 2\Delta H°\, CO_2(g) + \Delta H°\, H_2O(l) - \Delta H°\, C_2H_2(g) - 5/2\Delta H°\, O_2(g)$

        $= 2(-393.5) - 285.83 - 226.77 - 5/2(0)$

        $= -1299.6$ kJ produced/mol C_2H_2 burned

(c) $w_{max} = \Delta G° = 2\Delta G°\, CO_2(g) + \Delta G°\, H_2O(l) - \Delta G°\, C_2H_2(g) - 5/2\, \Delta G°\, O_2(g)$

        $= 2(-394.4) - 237.13 - 209.2 - 5/2(0) = -1235.1$ kJ

The negative sign indicates that the system does work on the surroundings; the system can accomplish a maximum of 1235.1 kJ of work on its surroundings.

19.74   (a) $CH_4(g) + 2\, O_2(g) \rightarrow CO_2(g) + 2\, H_2O(l)$

        $\Delta H° = \Delta H_f°\, CO_2(g) + 2\Delta H_f°\, H_2O(l) - \Delta H_f°\, CH_4(g) - 2\Delta H_f°\, O_2(g)$

           $= (-393.5) + 2(-285.83) - (-74.8) - 2(0) = -890.4$ kJ/mol CH_4 burned

(b) $w_{max} = \Delta G° = \Delta G_f°\, CO_2(g) + 2\Delta G_f°\, H_2O(l) - \Delta G_f°\, CH_4(g) - 2\Delta G_f°\, O_2(g)$

           $= (-394.4) + 2(-237.13) - (-50.8) - 2(0) = -817.9$ kJ

The system can accomplish at most 817.86 kJ of work per mole of CH_4 on the surroundings.

Free Energy and Equilibrium (Section 19.6)

19.75   *Analyze/Plan.* We are given a chemical reaction and asked to predict the effect of the partial pressure of $O_2(g)$ on the value of ΔG for the system. Consider the relationship $\Delta G = \Delta G° + RT \ln Q$, where Q is the reaction quotient.   *Solve.*

(a) ΔG decreases. $O_2(g)$ appears in the denominator of Q for this reaction. An increase in pressure of O_2 decreases Q and ΔG gets smaller or becomes more negative. Increasing the concentration or partial pressure of a reactant increases the tendency for a reaction to occur.

(b) ΔG increases. $O_2(g)$ appears in the numerator of Q for this reaction. Increasing the pressure of O_2 increases Q and ΔG becomes more positive. Increasing the concentration or partial pressure of a product decreases the tendency for the reaction to occur.

(c) ΔG increases. $O_2(g)$ appears in the numerator of Q for this reaction. An increase in pressure of O_2 increases Q and ΔG becomes more positive. Because pressure of O_2 is raised to the third power in Q, an increase in pressure of O_2 will have the largest effect on ΔG for this reaction. Increasing the concentration or partial pressure of a product decreases the tendency for the reaction to occur.

19.76   Consider the relationship $\Delta G = \Delta G° + RT \ln Q$, where Q is the reaction quotient.

(a) ΔG decreases. $H_2(g)$ appears in the denominator of Q for this reaction. An increase in pressure of H_2 decreases Q and ΔG becomes smaller or more negative. Increasing the concentration or partial pressure of a reactant increases the tendency for a reaction to occur.

(b) ΔG increases. $H_2(g)$ appears in the numerator of Q for this reaction. Increasing the pressure of H_2 increases Q and ΔG becomes more positive. Increasing the concentration or partial pressure of a product decreases the tendency for the reaction to occur.

(c) ΔG decreases. $H_2(g)$ appears in the denominator of Q for this reaction. An increase in pressure of H_2 decreases Q and ΔG becomes smaller or more negative. Increasing the concentration or partial pressure of a reactant increases the tendency for a reaction to occur.

19.77   *Analyze/Plan.* Given a chemical reaction, we are asked to calculate $\Delta G°$ from Appendix C data and ΔG for a given set of initial conditions. Use Equation 19.14 to calculate $\Delta G°$ and Equation 19.19 to calculate ΔG. Follow the logic in Sample Exercise 19.11 when calculating ΔG.   *Solve.*

(a) $\Delta G° = \Delta G° \, N_2O_4(g) - 2\Delta G° \, NO_2(g) = 98.28 - 2(51.84) = -5.40 \text{ kJ}$

(b) $\Delta G = \Delta G° + RT \ln P_{N_2O_4}/P_{NO_2}^2$

$$= -5.40 \text{ kJ} + \frac{8.314 \times 10^{-3} \text{ kJ}}{\text{mol-K}} \times 298 \text{ K} \times \ln[1.60/(0.40)^2] = 0.3048 = 0.30 \text{ kJ}$$

19.78   (a) $\Delta G° = \Delta G° \, C_3H_8(g) + 2\Delta G° \, H_2(g) \rightarrow 3\Delta G° \, CH_4(g)$

$$= -23.47 + 2(0) - 3(-50.8) = 128.9 \text{ kJ}$$

(b) $\Delta G = \Delta G° + RT \ln[P_{C_3H_8} \times P_{H_2}^2 / P_{CH_4}^3]$

$$= 128.9 + \frac{8.314 \times 10^{-3} \text{ kJ}}{\text{mol-K}} \times 298 \text{ K} \times \ln[(0.0100) \times (0.0180)^2 / (40.0)^3]$$

$$= 128.9 - 58.735 = 70.165 = 70.2 \text{ kJ}$$

19.79   *Analyze/Plan.* Given a chemical reaction, we are asked to calculate K using $\Delta G_f°$ data from Appendix C. Calculate $\Delta G°$ using Equation 19.14. Then, $\Delta G° = -RT \ln K$, Equation 19.20; $\ln K = -\Delta G°/RT$ *Solve.*

(a) $\Delta G° = 2\Delta G° \, HI(g) - \Delta G° \, H_2(g) - \Delta G° \, I_2(g)$

$$= 2(1.30) - 0 - 19.37 = -16.77 \text{ kJ}$$

$$\ln K = \frac{-(-16.77 \text{ kJ}) \times 10^3 \text{ J/kJ}}{8.314 \text{ J/K} \times 298 \text{ K}} = 6.76876 = 6.769; \quad K = 870$$

(b) $\Delta G° = \Delta G° \, C_2H_4(g) + \Delta G° \, H_2O(g) - \Delta G° \, C_2H_5OH(g)$

         $= 68.11 - 228.57 - (-168.5) = 8.04 = 8.0 \text{ kJ}$

         $\ln K = \dfrac{-(8.04 \text{ kJ}) \times 10^3 \text{ J/kJ}}{8.314 \text{ J/K} \times 298 \text{ K}} = -3.24511 = -3.25; \ K = 0.039$

(c) $\Delta G° = \Delta G° \, C_6H_6(g) - 3\Delta G° \, C_2H_2(g) = 129.7 - 3(209.2) = -497.9 \text{ kJ}$

         $\ln K = \dfrac{-\Delta G°}{RT} = \dfrac{-(-497.9 \text{ kJ}) \times 10^3 \text{ J/KJ}}{8.314 \text{ J/K} \times 298 \text{ K}} = 200.963 = 201.0; \ K = 2 \times 10^{87}$

19.80 $\Delta G° = -RT \ln K; \ \ln K = -\Delta G° / RT;$ at 298 K, RT = 2.4776 = 2.478 kJ

(a) $\Delta G° = \Delta G° \, NaOH(s) + \Delta G° \, CO_2(g) - \Delta G° \, NaHCO_3(s)$

         $= -379.5 + (-394.4) - (-851.8) = +77.9 \text{ kJ}$

         $\ln K = \dfrac{-\Delta G°}{RT} = \dfrac{-77.9 \text{ kJ}}{2.478 \text{ kJ}} = -31.442 = -31.4; \quad K = 2 \times 10^{-14}$

         $K = P_{CO_2} = 2 \times 10^{-14}$

(b) $\Delta G° = 2\Delta G° \, HCl(g) + \Delta G° \, Br_2(g) - 2\Delta G° \, HBr(g) - \Delta G° \, Cl_2(g)$

         $= 2(-95.27) + 3.14 - 2(-53.22) - 0 = -80.96 \text{ kJ}$

         $\ln K = \dfrac{-(-80.96)}{2.4776} = +32.68; \quad K = 1.6 \times 10^{14}$

         $K = \dfrac{P_{HCl}^2 \times P_{Br_2}}{P_{HBr}^2 \times P_{Cl_2}} = 1.6 \times 10^{14}$

(c)   From Solution 19.59(a), $\Delta G°$ at 298 K = -140.0 kJ.

         $\ln K = \dfrac{-\Delta G°}{RT} = \dfrac{-(-140.0)}{2.4776} = 56.51; \quad K = 3.5 \times 10^{24}$

         $K = \dfrac{P_{SO_3}^2}{P_{SO_2}^2 \times P_{O_2}} = 3.5 \times 10^{24}$

19.81   *Analyze/Plan.* Given a chemical reaction and thermodynamic data in Appendix C, calculate the equilibrium pressure of $CO_2(g)$ at two temperatures. $K = P_{CO_2}$. Calculate $\Delta G°$ at the two temperatures using $\Delta G° = \Delta H° - T\Delta S°$ and then calculate K and P_{CO_2}. *Solve.*

   $\Delta H° = \Delta H° \, BaO(s) + \Delta H° \, CO_2(g) - \Delta H° \, BaCO_3(s)$

     $= -553.5 + -393.5 - (-1216.3) = +269.3 \text{ kJ}$

   $\Delta S° = S° \, BaO(s) + S° \, CO_2(g) - S° \, BaCO_3(s)$

     $= 70.42 + 213.6 - 112.1 = 171.92 \text{ J/K} = 0.1719 \text{ kJ/K}$

(a) ΔG at 298 K = 269.3 kJ $-$ 298 K (0.17192 kJ/K) = 218.07 = 218.1 kJ

         $\ln K = \dfrac{-\Delta G°}{RT} = \dfrac{-218.07 \times 10^3 \text{ J}}{8.314 \text{ J/K} \times 298 \text{ K}} = -88.017 = -88.02$

         $K = 6.0 \times 10^{-39}; \quad P_{CO_2} = 6.0 \times 10^{-39} \text{ atm}$

(b) ΔG at 1100 K = 269.3 kJ – 1100 K (0.17192 kJ) = 80.19 = +80.2 kJ

$$\ln K = \frac{-\Delta G^\circ}{RT} = \frac{-80.19 \times 10^3 J}{8.314 \text{ J/K} \times 1100 \text{ K}} = -8.768 = -8.77$$

$K = 1.6 \times 10^{-4}; \quad P_{CO_2} = 1.6 \times 10^{-4} \text{ atm}$

19.82 $K = P_{CO_2}$. Calculate ΔG° at the two temperatures using $\Delta G^\circ = \Delta H^\circ - T\Delta S^\circ$ and then calculate K and P_{CO_2}.

$\Delta H^\circ = \Delta H^\circ \text{ PbO(s)} + \Delta H^\circ \text{ CO}_2\text{(g)} - \Delta H^\circ \text{ PbCO}_3\text{(s)}$

   $= -217.3 - 393.5 + 699.1 = 88.3 \text{ kJ}$

$\Delta S^\circ = S^\circ \text{ PbO(s)} + S^\circ \text{ CO}_2\text{(g)} - S^\circ \text{ PbCO}_3\text{(s)}$

   $= 68.70 + 213.6 - 131.0 = 151.3 \text{ J/K or } 0.1513 \text{ kJ/K}$

(a) $\Delta G^\circ = \Delta H^\circ - T\Delta S^\circ$. At 673 K, $\Delta G^\circ = 88.3 \text{ kJ} - 673 \text{ K}(0.1513 \text{ kJ/K}) = -13.525$

$= -13.5 \text{ kJ}$

$$\ln K = \frac{-\Delta G^\circ}{RT} = \frac{-(-13.525 \times 10^3)J}{8.314 \text{ J/K} \times 673 \text{ K}} = 2.4172 = 2.42$$

$K = P_{CO_2} = 11.214 = 11 \text{ atm}$

(b) $\Delta G^\circ = \Delta H^\circ - T\Delta S^\circ$. At 453 K, $\Delta G^\circ = 88.3 \text{ kJ} - 453 \text{ K} (0.1513 \text{ kJ}) = 19.7611$

$= 19.8 \text{ kJ}$

$$\ln K = \frac{-(19.7611 \times 10^3 \text{ J})}{8.314 \text{ J/K} \times 453 \text{ K}} = -5.2469 = -5.25; \quad K = P_{CO_2} = 5.3 \times 10^{-3} \text{ atm}$$

19.83   *Analyze/Plan.*   Given an acid dissociation equilibrium and the corresponding K_a value, calculate ΔG° and ΔG for a given set of concentrations. Use Equation 19.20 to calculate ΔG° and Equation 19.19 to calculate ΔG.   *Solve.*

(a) $HNO_2(aq) \rightleftharpoons H^+(aq) + NO_2^-(aq)$

(b) $\Delta G^\circ = -RT \ln K_a = -(8.314 \times 10^{-3})(298) \ln (4.5 \times 10^{-4}) = 19.0928 = 19.1 \text{ kJ}$

(c) $\Delta G = 0$ at equilibrium

(d) $\Delta G = \Delta G^\circ + RT \ln Q$

$$= 19.09 \text{ kJ} + (8.314 \times 10^{-3})(298) \ln \frac{(5.0 \times 10^{-2})(6.0 \times 10^{-4})}{0.20} = -2.725 = -2.7 \text{ kJ}$$

19.84   (a) $CH_3NH_2(aq) + H_2O(l) \rightleftharpoons CH_3NH_3^+(aq) + OH^-(aq)$

(b) $\Delta G^\circ = -RT \ln K_b = -(8.314 \times 10^{-3})(298) \ln (4.4 \times 10^{-4}) = 19.148 = 19.1 \text{ kJ}$

(c) $\Delta G = 0$ at equilibrium

(d) $\Delta G = \Delta G^\circ + RT \ln Q; \; [OH^-] = 1 \times 10^{-14}/6.7 \times 10^{-9} = 1.4925 \times 10^{-6} = 1.5 \times 10^{-6}$

$$= 19.148 + (8.314 \times 10^{-3})(298) \ln \frac{(2.4 \times 10^{-3})(1.4925 \times 10^{-6})}{0.098} = -23.28 = -23.3 \text{ kJ}$$

Additional Exercises

19.85   (a)   The thermodynamic quantities T, E, and S are state functions. T is directly related to the distribution of molecular speeds, which does not depend on the path from one state to another.

        (b)   The quantities q and w do depend on the path taken from one state to another.

        (c)   There is only one *reversible* path between states.

        (d)   Isothermal processes occur at constant T. As the process is reversible, q is q_{rev} and w is w_{max}.

$$\Delta E = q_{rev} + w_{max}; \quad \Delta S = \frac{q_{rev}}{T}$$

19.86   (a)   The sign of $\Delta S°$ at room temperature is positive, because there are more moles of gas in the products than the reactants. (Also, because the process is spontaneous and $\Delta H°$ is positive at room temperature, $\Delta S°$ must be positive so that ΔG is negative.)

        (b) $\Delta S°$ is defined as the entropy change for the reaction with all reactants and products in their standard states, calculated at 298 K. The value of $\Delta S°$ does not change with a change in reaction conditions. The value of ΔG will change, depending on the partial pressure of water vapor in the container.

19.87   (a)   False. The essential question is whether the reaction proceeds far to the right before arriving at equilibrium. The position of equilibrium, which is the essential aspect, is dependent not only on ΔH but on the entropy change as well.

        (b)   True.

        (c)   True.

        (d)   False. *Non*spontaneous processes in general require that work be done to force them to proceed. Spontaneous processes occur without application of work.

        (e)   False. Such a process *might* be spontaneous, but would not necessarily be so. Spontaneous processes are those that are exothermic and/or that lead to increased disorder in the system.

19.88

Process	ΔH	ΔS
(a)	+	+
(b)	−	−
(c)	+	+
(d)	+	+
(e)	−	+

19.89   There is no mistake in the calculation. The second law states that in any spontaneous process there is an increase in the entropy of the universe. Although there may be a decrease in entropy of the system, as in the present case, this decrease is more than offset by an increase in entropy of the surroundings.

19.90   (a) Microstates are possible arrangements for the system. For each die, there are six possibilities for the top face, resulting in (6)(6) = 36 possible arrangements or microstates. (The face that appears on top of one die is not related to or determined by the face on top of the other die.) The two arrangements of top faces shown in the exercise are two of the 36 possible microstates.

   (b) The left pair of dice belongs to state III; the right pair belongs to state VII.

   (c) There are eleven possible states (II through XII; I is not a possibility).

   (d) The state with the most microstates has the highest entropy. State VII has six microstates and the highest entropy. The microstates are (1+6), (2+5), (3+4), (4+3), (5+2), and (6+1). States VI and VIII, on either side of VII, have five microstates. Moving farther away from VII, the number of microstates decreases until we reach the two extremes, II and XII, which each have one microstate.

   (e) States II and XII, with one microstate each, have the lowest entropy.

   (f) The Boltzmann relationship is $S = k \ln W$, where S is the absolute entropy, k is the Botzmann constant, and W is the total number of microstates of the system.

$$S = 1.38 \times 10^{-23} \text{ J/K } (\ln 36) = 4.9453 \times 10^{-23} = 4.95 \times 10^{-23} \text{ J/K}$$

19.91   If $NH_4NO_3(s)$ dissolves spontaneously in water, $\Delta G = \Delta H - T\Delta S$. If ΔG is negative and ΔH is positive, the sign of ΔS must be positive. Furthermore, $T\Delta S > \Delta H$ at room temperature.

19.92   (a) The sign of q for expansion is (+). Vaporization is an endothermic process; the enthaphy of the system increases and q is positive. Our system is the refrigerant. Because the expansion does not occur at constant pressure, q is not exactly equal to ΔH, but its sign is positive.

   (b) The sign of q for compression is (−). Compression is the reverse of expansion, and it has the opposite sign.

   (c) The expansion chamber is inside the house and the compression chamber is outside. During expansion, q_{sys} increases and q_{sur} decreases. The air surrounding the expansion chamber is cooled and then distributed throughout the house to cool it. If expansion occurred outside, the cool air would be wasted. Compression releases heat to the surroundings; it occurs outside so that the released heat can be dissipated by the outside air.

   (d) No. Heat can flow reversibly between a system and its surroundings only if the two have an infinitesimally small difference in temperature and the amount of heat transferred is infinitesimally small. There is no mechanism in our system to regulate the amount of heat transferred. When the liquid flows into the low pressure chamber, all the liquid vaporizes, not an infinitesimally small amount.

   (e) A spontaneous process occurs without outside intervention. In an air conditioner, expansion (vaporization) of the refrigerant is spontaneous, but compression (condensation) to the liquid state is nonspontaneous. Cooling the house from 31 $^\circ$C to 24 $^\circ$C is nonspontaneous. [Note that all spontaneous processes are irreversible, but not all irreversible processes are spontaneous.]

19.93 At the normal boiling point of a liquid, $\Delta G = 0$ and $\Delta H_{vap} = T\Delta S_{vap}$; $T = \Delta H_{vap}/\Delta S_{vap}$. By Trouton's rule, $\Delta S_{vap} = 88$ J/mol-K. The process of vaporization is:

(a) $Br_2(l) \rightleftharpoons Br_2(g)$

$\Delta H_{vap} = \Delta H_f^\circ \; Br_2(g) - \Delta H_f^\circ \; Br_2(l) = 30.71 \text{ kJ} - 0 = 30.71 \text{ kJ}$

$T_b = \dfrac{\Delta H_{vap}}{\Delta S_{vap}} = \dfrac{30.71 \text{ kJ}}{88 \text{ J/mol-K}} \times \dfrac{1000 \text{ J}}{\text{kJ}} = 349 = 3.5 \times 10^2 \text{ K}$

(b) According to WebElements™ 2013, the normal boiling point of $Br_2(l)$ is 332 K. Trouton's rule provides a good "ballpark" estimate.

Trouton's rule assumes that the entropy of vaporization of most molecules is due to the much greater motional freedom of the gaseous state relative to the liquid state. This assumption is incorrect for liquids with strong intermolecular forces (usually hydrogen bonding) in either the liquid or the gaseous state.

There are also the usual experimental uncertainties in the measurement of ΔH_f° for $Br_2(g)$ and the normal boiling point of $Br_2(l)$.

19.94 (a) $1/2 \, N_2(g) + 3/2 \, H_2(g) \rightarrow NH_3(g)$

$C(s) + 1/2 \, O_2(g) \rightarrow CO(g)$

(b) In the first reaction, there are fewer moles of gas in the products than reactants, so ΔS is negative. If $\Delta G_f^\circ = \Delta H_f^\circ - T\Delta S_f^\circ$ and ΔS_f° is negative, $-T\Delta S_f^\circ$ is positive and ΔG_f° is more positive than ΔH_f°.

(c) Condition (iii), when ΔS_f° is negative.

19.95 (a) (i) $Ti(s) + 2 \, Cl_2(g) \rightarrow TiCl_4(g)$

$\Delta H^\circ = \Delta H^\circ \; TiCl_4(g) - \Delta H^\circ \; Ti(s) - 2\Delta H^\circ \; Cl_2(g)$

$= -763.2 - 0 - 2(0) = -763.2 \text{ kJ}$

$\Delta S^\circ = 354.9 - 30.76 - 2(222.96) = -121.78 = -121.8 \text{ J/K}$

$\Delta G^\circ = -726.8 - 0 - 2(0) = -726.8 \text{ kJ}$

$\ln K = \dfrac{-\Delta G^\circ}{RT} = \dfrac{-(-726.8) \text{ kJ}}{2.4777 \text{ kJ}} = 293.337 = 293.3; \quad K = 2 \times 10^{127}$

(ii) $C_2H_6(g) + 7 \, Cl_2(g) \rightarrow 2 \, CCl_4(g) + 6 \, HCl(g)$

$\Delta H^\circ = 2\Delta H^\circ \; CCl_4(g) + 6\Delta H^\circ \; HCl(g) - \Delta H^\circ \; C_2H_6(g) - 7\Delta H^\circ \; Cl_2(g)$

$= 2(-106.7) + 6(-92.30) - (-84.68) - 7(0) = -682.52 = -682.5 \text{ kJ}$

$\Delta S^\circ = 2(309.4) + 6(186.69) - 229.5 - 7(222.96) = -51.28 = -51.4 \text{ J/K}$

$\Delta G^\circ = 2(-64.0) + 6(-95.27) - (-32.89) - 7(0) = -666.73 = -666.7 \text{ kJ}$

$\ln K = \dfrac{-\Delta G^\circ}{RT} = \dfrac{-(-666.73) \text{ kJ}}{2.4777 \text{ kJ}} = 269.0923 = 269.1; \quad K = 7 \times 10^{116}$

     (iii) $BaO(s) + CO_2(g) \rightarrow BaCO_3(s)$

$$\Delta H° = \Delta H° \, BaCO_3(s) - \Delta H° \, BaO(s) - \Delta H° \, CO_2(g)$$

$$= -1216.3 - (-553.5) - (-393.5) = -269.3 \text{ kJ}$$

$$\Delta S° = 112.1 - 70.42 - 213.6 = -171.9 \text{ J/K}$$

$$\Delta G° = -1137.6 - (-525.1) - (-394.4) = -218.1 \text{ kJ}$$

$$\ln K = \frac{-\Delta G°}{RT} = \frac{-(-218.1) \text{ kJ}}{2.4777 \text{ kJ}} = 88.0252 = 88.03; \quad K = 1.7 \times 10^{38}$$

(b)   (i), (ii), and (iii) all have negative $\Delta G°$ values and are spontaneous at standard conditions and 25 °C. Essentially, these reactions all go to completion; they have unimaginably large K values.

(c) $\Delta G° = \Delta H° - T\Delta S°$. All three reactions have $-\Delta H°$ and $-\Delta S°$. They all have $-\Delta G°$ at 25 °C, and $\Delta G°$ becomes more positive as T increases.

19.96 $\Delta G = \Delta G° + RT \ln Q; \ln K = -\Delta G°/RT$

(a) $Q = \dfrac{P_{NH_3}^2}{P_{N_2} \times P_{H_2}^3} = \dfrac{(1.2)^2}{(2.6)(5.9)^3} = 2.697 \times 10^{-3} = 2.7 \times 10^{-3}$

$$\Delta G° = 2\Delta G° \, NH_3(g) - \Delta G° \, N_2(g) - 3\Delta G° \, H_2(g)$$

$$= 2(-16.66) - 0 - 3(0) = -33.32 \text{ kJ}$$

$$\ln K = \frac{-\Delta G°}{RT} = \frac{-(-33.32) \text{ kJ}}{2.4777 \text{ kJ}} = 13.448 = 13.45; \quad K = 6.9 \times 10^5$$

$$\Delta G = -33.32 \text{ kJ} + \frac{8.314 \times 10^{-3} \text{ kJ}}{\text{mol-K}} \times 298 \text{ K} \times \ln(2.69 \times 10^{-3})$$

$$\Delta G = -33.32 - 14.66 = -47.98 = -48.0 \text{ kJ}$$

(b) $Q = \dfrac{P_{N_2}^3 \times P_{H_2O}^4}{P_{N_2H_4}^2 \times P_{NO_2}^2} = \dfrac{(0.5)^3(0.3)^4}{(5.0 \times 10^{-2})^2(5.0 \times 10^{-2})^2} = 162 = 2 \times 10^2$

$$\Delta G° = 3\Delta G° \, N_2(g) + 4\Delta G° \, H_2O(g) - 2\Delta G° \, N_2H_4(g) - 2\Delta G° \, NO_2(g)$$

$$= 3(0) + 4(-228.57) - 2(159.4) - 2(51.84) = -1336.8 \text{ kJ}$$

$$\ln K = \frac{-\Delta G°}{RT} = \frac{-(-1336.8) \text{ kJ}}{2.4777 \text{ kJ}} = 539.533 = 539.53; \quad K = 2.1 \times 10^{234}$$

$$\Delta G = -1336.8 \text{ kJ} + 2.478 \ln 162 = -1324.2 = -1.32 \times 10^3 \text{ kJ}$$

(c) $Q = \dfrac{P_{N_2} \times P_{H_2}^2}{P_{N_2H_4}} = \dfrac{(1.5)(2.5)^2}{0.5} = 18.75 = 2 \times 10^1$

$$\Delta G° = \Delta G° \, N_2(g) + 2\Delta G° \, H_2(g) - \Delta G° \, N_2H_4(g)$$

$$= 0 + 2(0) - 159.4 = -159.4 \text{ kJ}$$

$$\ln K = \frac{-\Delta G°}{RT} = \frac{-(-159.4) \text{ kJ}}{2.4777 \text{ kJ}} = 64.334 = 64.33; \quad K = 8.7 \times 10^{27}$$

$$\Delta G = -159.4 \text{ kJ} + 2.478 \ln 18.75 = -152.1 = -152 \text{ kJ}$$

19.97

Reaction	(a) Sign of $\Delta H°$	(a) Sign of $\Delta S°$	(b) K > 1?	(c) Variation in K as Temp. Increases
(i)	–	–	yes	decrease
(ii)	+	+	no	increase
(iii)	+	+	no	increase
(iv)	+	+	no	increase

(a)   Note that at a particular temperature, positive $\Delta H°$ leads to a smaller value of K, whereas positive $\Delta S°$ increases the value of K.

19.98   (a) $K = \dfrac{X_{CH_3COOH}}{X_{CH_3OH}P_{CO}}$

$\Delta G° = -RT \ln K; \ln K = -\Delta G/RT$

$\Delta G° = \Delta G° \ CH_3COOH(l) - \Delta G° \ CH_3OH(l) - \Delta G° \ CO(g)$

$\quad = -392.4 - (-166.23) - (-137.2) = -89.0 \text{ kJ}$

$\ln K = \dfrac{-(-89.0 \text{ kJ})}{(8.314 \times 10^{-3} \text{ kJ/K})(298 \text{ K})} = 35.922 = 35.9; \quad K = 4 \times 10^{15}$

(b) $\Delta H° = \Delta H° \ CH_3COOH(l) - \Delta H° \ CH_3OH(l) - \Delta H° \ CO(g)$

$\quad = -487.0 - (-238.6) - (-110.5) = -137.9 \text{ kJ}$

The reaction is exothermic, so the value of K will decrease with increasing temperature, and the mole fraction of CH_3COOH will also decrease. Elevated temperatures must be used to increase the speed of the reaction. Thermodynamics cannot predict the rate at which a reaction reaches equilibrium.

(c) $\Delta G° = -RT \ln K; K = 1, \ln K = 0, \Delta G° = 0$

$\Delta G° = \Delta H° - T\Delta S°;$ when $\Delta G° = 0, \Delta H° = T\Delta S°$

$\Delta S° = S° \ CH_3COOH(l) - S° \ CH_3OH(l) - S° \ CO(g)$

$\quad = 159.8 - 126.8 - 197.9 = -164.9 \text{ J/K} = -0.1649 \text{ kJ/K}$

$-137.9 \text{ kJ} = T(-0.1649 \text{ kJ/K}), T = 836.3 \text{ K}$

The equilibrium favors products up to 836 K or 563 °C, so the elevated temperatures to increase the rate of reaction can be safely employed.

19.99   (a)   First, calculate $\Delta G°$ for each reaction:

For $C_6H_{12}O_6(s) + 6 O_2(g) \rightleftharpoons 6 CO_2(g) + 6 H_2O(l)$ (A)

$\quad \Delta G° = 6(-237.13) + 6(-394.4) - (-910.4) + 6(0) = -2878.8 \text{ kJ}$

For $C_6H_{12}O_6(s) \rightleftharpoons 2 C_2H_5OH(l) + 2 CO_2(g)$ (B)

$\quad \Delta G° = 2(-394.4) + 2(-174.8) - (-910.4) = -228.0 \text{ kJ}$

For (A), $\ln K = 2879 \times 10^3/(8.314)(298) = 1162; K = 5 \times 10^{504}$

For (B), $\ln K = 228 \times 10^3/(8.314)(298) = 92.026 = 92.0; K = 9 \times 10^{39}$

(b)　Both these values for K are unimaginably large. However, K for reaction (A) is larger, because $\Delta G°$ is more negative. The magnitude of the work that can be accomplished by coupling a reaction to its surroundings is measured by ΔG. According to these calculations, considerably more work can in principle be obtained from reaction (A), because $\Delta G°$ is more negative.

19.100　(a)　$\Delta G° = -RT \ln K$ (Equation 19.20); $\ln K = -\Delta G°/RT$

Use $\Delta G° = \Delta H° - T\Delta S°$ to get $\Delta G°$ at the two temperatures. Calculate $\Delta H°$ and $\Delta S°$ using data in Appendix C.

$2 CH_4(g) \rightarrow C_2H_6(g) + H_2(g)$

$\Delta H° = \Delta H° \, C_2H_6(g) + \Delta H° \, H_2(g) - 2\Delta H° \, CH_4(g) = -84.68 + 0 - 2(-74.8) = 64.92$

$= 64.9 \text{ kJ}$

$\Delta S° = S° \, C_2H_6(g) + S° \, H_2(g) - 2S° \, CH_4(g) = 229.5 + 130.58 - 2(186.3) = -12.52$

$= -12.5 \text{ J/K}$

at 298 K, $\Delta G = 64.92 \text{ kJ} - 298 \text{ K}(-12.52 \times 10^{-3} \text{ kJ/K}) = 68.65 = 68.7 \text{ kJ}$

$\ln K = \dfrac{-68.65 \text{ kJ}}{(8.314 \times 10^{-3} \text{ kJ/K})(298 \text{ K})} = -27.709 = -27.7, \ K = 9.25 \times 10^{-13} = 9 \times 10^{-13}$

at 773 K, $\Delta G = 64.9 \text{ kJ} - 773 \text{ K}(-12.52 \times 10^{-3} \text{ J/K}) = 74.598 = 74.6 \text{ kJ}$

$\ln K = \dfrac{-74.598 \text{ kJ}}{(8.314 \times 10^{-3} \text{ kJ/K})(773 \text{ K})} = -11.607 = -11.6, \ K = 9.1 \times 10^{-6}$

Because the reaction is endothermic, the value of K increases with an increase in temperature.

$2 CH_4(g) + 1/2 \, O_2(g) \rightarrow C_2H_6(g) + H_2O(g)$

$\Delta H° = \Delta H° \, C_2H_6(g) + \Delta H° \, H_2O(g) - 2\Delta H° \, CH_4(g) - 1/2 \, \Delta H° \, O_2(g)$

$= -84.68 + (-241.82) - 2(-74.8) - 1/2 \, (0) = -176.9 \text{ kJ}$

$\Delta S° = S° \, C_2H_6(g) + S° \, H_2O(g) - 2S° \, CH_4(g) - 1/2 \, S° \, O_2(g)$

$= 229.5 + 188.83 - 2(186.3) - 1/2 \, (205.0) = -56.77 = -56.8 \text{ J/K}$

at 298 K, $\Delta G = -176.9 \text{ kJ} - 298 \text{ K}(-56.77 \times 10^{-3} \text{ kJ/K}) = -159.98 = -160.0 \text{ kJ}$

$\ln K = \dfrac{-(-159.98 \text{ kJ})}{(8.314 \times 10^{-3} \text{ kJ/K})(298 \text{ K})} = 64.571 = 64.57; \ K = 1.1 \times 10^{28}$

at 773 K, $\Delta G = -176.9 \text{ kJ} - 773 \text{ K} \, (-56.77 \times 10^{-3} \text{ kJ/K}) = -133.02 = -133.0 \text{ kJ}$

$\ln K = \dfrac{-(-133.02 \text{ kJ})}{(8.314 \times 10^{-3} \text{ kJ/K})(773 \text{ K})} = 20.698 = 20.70; \ K = 9.750 \times 10^8 = 9.8 \times 10^8$

Because this reaction is exothermic, the value of K decreases with increasing temperature.

(b)　The difference in $\Delta G°$ for the two reactions is primarily enthalpic; the first reaction is endothermic and the second exothermic. Both reactions have $-\Delta S°$, which inhibits spontaneity.

(c)   This is an example of coupling a useful but nonspontaneous reaction with a spontaneous one to spontaneously produce a desired product.

$$2\,CH_4(g) \quad \rightarrow \quad C_2H_6(g) + H_2(g) \qquad \Delta G^o_{298} = +68.7\ kJ,\ nonspontaneous$$

$$H_2(g) + 1/2\,O_2(g) \quad \rightarrow \quad H_2O(g) \qquad \Delta G^o_{298} = -228.57\ kJ,\ spontaneous$$

$$2\,CH_4(g) + 1/2\,O_2(g) \quad \rightarrow \quad C_2H_6(g) + H_2O(g) \qquad \Delta G^o_{298} = -159.9\ kJ,\ spontaneous$$

(d) $CH_4(g) + 2\,O_2(g) \rightarrow CO_2(g) + 2\,H_2O(g)$

19.101 ΔG° for the metabolism of glucose is:

$$6\Delta G^{\circ}\ CO_2(g) + 6\Delta G^{\circ}\ H_2O(l) - \Delta G^{\circ}\ C_6H_{12}O_6(s) - 6\Delta G^{\circ}\ O_2(g)$$

$$\Delta G^{\circ} = 6(-394.4) + 6(-237.13) - (-910.4) + 6(0) = -2878.8\ kJ$$

mol ATP $= -2878.8\ kJ \times 1\ mol\ ATP/(-30.5\ kJ) = 94.4\ mol\ ATP/mol\ glucose$

Note that this calculation is done at standard conditions, not metabolic conditions. A more accurate answer would be obtained using ΔG values that reflect actual concentration, partial pressure, and pH in a cell.

19.102   (a)   The equilibrium of interest here can be written as:

$$K^+\,(plasma) \;\rightleftharpoons\; K^+\,(muscle)$$

Because the fluid is aqueous on both sides of the cell membrane, assume that the equilibrium constant for this process is exactly 1; that is, $\Delta G^{\circ} = 0$. However, ΔG is not zero because the concentrations are not the same on both sides of the membrane. Use Equation 19.19 to calculate ΔG:

$$\Delta G = \Delta G^{\circ} + RT\ \ln\ \frac{[K^+\,(muscle)]}{[K^+\,(plasma)]}$$

$$= 0 + (8.314)(310)\ \ln\ \frac{(0.15)}{(5.0 \times 10^{-3})} = 8766\ J = 8.8\ kJ$$

(b)   Note that ΔG is positive. This means that work must be done on the system (blood plasma plus muscle cells) to move the K^+ ions "uphill," as it were. The minimum amount of work possible is given by the value for ΔG. This value represents the minimum amount of work (8.8 kJ) required to transfer one mole of K^+ ions from the blood plasma at 5×10^{-3} M to muscle cell fluids at 0.15 M, assuming constancy of concentrations. In practice, a larger than minimum amount of work is required.

Integrative Exercises

19.103   (a)   At the boiling point, vaporization is a reversible process, so $\Delta S^{\circ}_{vap} = \Delta H^{\circ}_{vap}/T$.

acetone: $\Delta S^{\circ}_{vap} = \Delta H^{\circ}_{vap}/T = (29.1\,kJ/mol)/329.25\ K = 88.4\ J/mol\text{-}K$

dimethyl ether: $\Delta S^{\circ}_{vap} = (21.5\,kJ/mol)/248.35\ K = 86.6\ J/mol\text{-}K$

ethanol: $\Delta S^{\circ}_{vap} = (38.6\,kJ/mol)/351.6\ K = 110\ J/mol\text{-}K$

octane: $\Delta S^{\circ}_{vap} = (34.4\,kJ/mol)/398.75\ K = 86.3\ J/mol\text{-}K$

pyridine: $\Delta S^{\circ}_{vap} = (35.1\,kJ/mol)/388.45\ K = 90.4\ J/mol\text{-}K$

(b)   Ethanol is the only liquid listed that does not follow *Trouton's rule* and it is also the only substance that exhibits hydrogen bonding in the pure liquid. Hydrogen bonding leads to more ordering in the liquid state and a greater than usual increase in entropy upon vaporization. The rule appears to hold for liquids with London dispersion forces (octane) and ordinary dipole-dipole forces (acetone, dimethyl ether, pyridine), but not for those with hydrogen bonding.

(c)   Owing to strong hydrogen bonding interactions, water probably does not obey Trouton's rule.

From Appendix B, ΔH_{vap}° at 100 °C = 40.67 kJ/mol.

$\Delta S_{vap}^{\circ} = (40.67 \text{ kJ/mol})/373.15 \text{ K} = 109.0 \text{ J/mol-K}$

(d)   Use $\Delta S_{vap}^{\circ} = 88 \text{ J/mol-K}$, the middle of the range for Trouton's rule, to estimate ΔH_{vap}° for chlorobenzene.

$\Delta H_{vap}^{\circ} = \Delta S_{vap}^{\circ} \times T = 88 \text{ J/mol-K} \times 404.95 \text{ K} = 36 \text{ kJ/mol}$

19.104   The entropy of activation for a bimolecular process is usually negative. By definition, the activated complex of a bimolecular process is composed of two molecules. Forming the activated complex reduces the number of particles in the system from two to one and the entropy of this process will usually be negative.

19.105   Calculate $\Delta G°$, $\Delta H°$, and $\Delta S°$ using data from Appendix C. Use $\Delta G° = \Delta H° - T\Delta S°$ to calculate the temperature at which $\Delta G°$ changes sign. Couple this temperature with the sign of $\Delta G°$ at 298 K to state a temperature range over which the reaction is spontaneous.

$\Delta G° = 2\Delta G° \text{ CO}_2(g) + 3\Delta G° \text{ Fe}(s) - 2\Delta G° \text{ Fe}_3\text{O}_4(s) - 2\Delta G° \text{ C}(s, \text{graphite})$

$= 2(-394.4) + 3(0) - (-1014.2) - 2(0) = 225.4 \text{ kJ}$

$\Delta H° = 2\Delta H° \text{ CO}_2(g) + 3\Delta H° \text{ Fe}(s) - 2\Delta H° \text{ Fe}_3\text{O}_4(s) - 2\Delta H° \text{ C}(s, \text{graphite})$

$= 2(-393.5) + 3(0) - (-1117.1) - 2(0) = 330.1 \text{ kJ}$

$\Delta S° = 2\Delta S° \text{ CO}_2(g) + 3\Delta S° \text{ Fe}(s) - 2\Delta S° \text{ Fe}_3\text{O}_4(s) - 2\Delta S° \text{ C}(s, \text{graphite})$

$= 2(213.6) + 3(27.15) - 146.4 - 2(5.69) = 350.87 = 350.9 \text{ J/K} = 0.3509 \text{ kJ/K}$

$\Delta H° = T\Delta S°; T = \Delta H°/\Delta S° = 330.1 \text{ kJ}/0.3509 \text{ J/K} = 940.7 \text{ K} = 667.6 \text{ °C}$

The reaction will be spontaneous at temperatures above 940.7 K or 667.6 °C. Even though $\Delta S°$ is large and positive, a very high temperature is required to overcome the large positive enthalpy of reaction.

19.106   (a) $O_2(g) \xrightarrow{h\nu} 2O(g)$; S increases because there are more moles of gas in the products.

(b) $O_2(g) + O(g) \rightarrow O_3(g)$; S decreases because there are fewer moles of gas in the products.

(c)   S increases as the gas molecules diffuse into the larger volume of the stratosphere; there are more possible positions and, therefore, more motional freedom.

(d) $NaCl(aq) \rightarrow NaCl(s) + H_2O(l)$; ΔS decreases as the mixture (seawater, greater disorder) is separated into pure substances (fewer possible arrangements, more order).

19.107   The heat lost by the hot water (iv) in the cup equals the heat gained by the ice cube. The ice cube is heated from –20 °C to 0 °C (i), the ice melts at 0 °C (ii), the new liquid heats to the final temperature (iii). (iv) = (i) + (ii) + (iii) . The 500 mL of hot water weights 500 g; the ice cube weighs 20 g.

(i) $20\,g \times \dfrac{2.03\,J}{1\,g\text{-}°C} \times (20\,°C) = 812 = 8.1 \times 10^2\,J$

(ii) $20\,g \times \dfrac{1\,mol}{18.02\,g} \times \dfrac{6.01\,kJ}{mol} \times \dfrac{1000\,J}{kJ} = 6670.4 = 6.7 \times 10^3\,J$

(iii) $20\,g \times \dfrac{4.184\,J}{1\,g\text{-}°C} \times (T_f - 0\,°C) = 83.60\,(T_f - 0\,°C) = 84\,(T_f)$

(iv) $500\,g \times \dfrac{4.184\,J}{1\,g\text{-}°C} \times (83\,°C - T_f) = 2{,}092\,(83\,°C - T_f) = (2.09 \times 10^3)\,(83\,°C - T_f)$

$2092\,(83\,°C - T_f) = 812\,J + 6670.4\,J + 83.60\,(T_f)$;

$173{,}636 - [2092\,(T_f)] = 7482.4 + 83.60\,(T_f)$

$166{,}153.6 = 2175.6\,(T_f)$; $T_f = 76.37 = 76\,°C$

19.108   (a)   16 e⁻, 8 e⁻ pairs. The C–S bond order is approximately 2.

$\ddot{\underset{..}{S}} = C = \ddot{\underset{..}{S}}$

(b)   2 e⁻ domains around C, linear e⁻ domain geometry, linear molecular structure.

(c) $CS_2(l) + 3\,O_2(g) \rightarrow CO_2(g) + 2\,SO_2(g)$

(d)   ΔH° = ΔH° $CO_2(g)$ + 2 ΔH° $SO_2(g)$ – ΔH° $CS_2(l)$ – 3 ΔH° $O_2(g)$

   = –393.5 + 2(–296.9) – (89.7) – 3(0) = –1077.0 kJ

ΔG° = ΔG° $CO_2(g)$ + 2 ΔG° $SO_2(g)$ – ΔG° $CS_2(l)$ – 3 ΔG° $O_2(g)$

   = –394.4 + 2(–300.4) – (65.3) – 3(0) = –1060.5 kJ

The reaction is exothermic (–ΔH°) and spontaneous (–ΔG°) at 298 K.

(e)   vaporization: $CS_2(l) \rightarrow CS_2(g)$

$\Delta G^\circ_{vap} = \Delta H^\circ_{vap} - T\Delta S^\circ_{vap}$; $\Delta S^\circ_{vap} = (\Delta H^\circ_{vap} - \Delta G^\circ_{vap})/T$

$\Delta G^\circ_{vap} = \Delta G^\circ\,CS_2(g) - \Delta G^\circ\,CS_2(l) = 67.2 - 65.3 = 1.9\,kJ$

$\Delta H^\circ_{vap} = \Delta H^\circ\,CS_2(g) - \Delta H^\circ\,CS_2(l) = 117.4 - 89.7 = 27.7\,kJ$

$\Delta S^\circ_{vap} = (27.7 - 1.9)\,kJ/298\,K = 0.086577 = 0.0866\,kJ/K = 86.6\,J/K$

ΔS_{vap} is always positive, because the gas phase occupies a greater volume and has more motional freedom and a larger absolute entropy than the liquid.

(f)   At the boiling point, $\Delta G = 0$ and $\Delta H_{vap} = T_b \Delta S_{vap}$.

$T_b = \Delta H_{vap}/\Delta S_{vap} = 27.7 \text{ kJ}/0.086577 \text{ kJ/K} = 319.9 = 320 \text{ K}$

$T_b = 320 \text{ K} = 47 \text{ °C. } CS_2$ is a liquid at 298 K, 1 atm

19.109   (a) $Ag(s) + 1/2 \, N_2(g) + 3/2 \, O_2(g) \rightarrow AgNO_3(s)$; S decreases because there are fewer moles of gas in the product.

(b) $\Delta G_f^o = \Delta H_f^o - T\Delta S^o$; $\Delta S^o = (\Delta G_f^o - \Delta H_f^o)/(-T) = (\Delta H_f^o - \Delta G_f^o)/T$

$\Delta S^o = -124.4 \text{ kJ} - (-33.4 \text{ kJ})/298 \text{ K} = -0.305 \text{ kJ/K} = -305 \text{ J/K}$

ΔS^o is relatively large and negative, as anticipated from part (a).

(c)   Dissolving of $AgNO_3$ can be expressed as

$AgNO_3(s) \rightarrow AgNO_3 \text{ (aq, 1 } M)$

$\Delta H^o = \Delta H^o \, AgNO_3(aq) - \Delta H^o \, AgNO_3(s) = -101.7 - (-124.4) = +22.7 \text{ kJ}$

$\Delta H^o = \Delta H^o \, MgSO_4(aq) - \Delta H^o \, MgSO_4(s) = -1374.8 - (-1283.7) = -91.1 \text{ kJ}$

Dissolving $AgNO_3(s)$ is endothermic $(+\Delta H^o)$, but dissolving $MgSO_4(s)$ is exothermic $(-\Delta H^o)$.

(d) $AgNO_3$: $\Delta G^o = \Delta G_f^o \, AgNO_3(aq) - \Delta G_f^o \, AgNO_3(s) = -34.2 - (-33.4) = -0.8 \text{ kJ}$

$\Delta S^o = (\Delta H^o - \Delta G^o) / T = [22.7 \text{ kJ} - (-0.8 \text{ kJ})] / 298 \text{ K} = 0.0789 \text{ kJ/K} = 78.9 \text{ J/K}$

$MgSO_4$: $\Delta G^o = \Delta G_f^o \, MgSO_4(aq) - \Delta G_f^o \, MgSO_4(s) = -1198.4 - (-1169.6) = -28.8 \text{ kJ}$

$\Delta S^o = (\Delta H^o - \Delta G^o) / T = [-91.1 \text{ kJ} - (-28.8 \text{ kJ})] / 298 \text{ K} = -0.209 \text{ kJ/K} = -209 \text{ J/K}$

(e)   In general, we expect dissolving a crystalline solid to be accompanied by an increase in positional disorder and an increase in entropy; this is the case for $AgNO_3$ ($\Delta S° = +78.9$ J/K). However, for dissolving $MgSO_4(s)$, there is a substantial decrease in entropy ($\Delta S = -209$ J/K). According to Section 13.5, ion-pairing is a significant phenomenon in electrolyte solutions, particularly in concentrated solutions where the charges of the ions are greater than 1. According to Table 13.4, a 0.1 m $MgSO_4$ solution has a van't Hoff factor of 1.21. That is, for each mole of $MgSO_4$ that dissolves, there are only 1.21 moles of "particles" in solution instead of 2 moles of particles. For a 1 m solution, the factor is even smaller. Also, the exothermic enthalpy of mixing indicates substantial interactions between solute and solvent. Substantial ion-pairing coupled with ion–dipole interactions with H_2O molecules lead to a decrease in entropy for $MgSO_4(aq)$ relative to $MgSO_4(s)$.

19.110   (a) $K = P_{NO_2}^2 / P_{N_2O_4}$

Assume equal amounts means equal number of moles. For gases, $P = n(RT/V)$. In an equilibrium mixture, RT/V is a constant, so moles of gas are directly proportional to partial pressure. Gases with equal partial pressures will have equal moles of gas present. The condition $P_{NO_2} = P_{N_2O_4}$ leads to the expression $K = P_{NO_2}$. The value of K then depends on P_t for the mixture. For any particular value of P_t, the condition of equal moles of the two gases can be achieved at some temperature. For example, $P_{NO_2} = P_{N_2O_4} = 1.0$ atm, $P_t = 2.0$ atm.

$$K = \frac{(1.0)^2}{1.0} = 1.0; \quad \ln K = 0; \quad \Delta G^\circ = 0 = \Delta H^\circ - T\Delta S^\circ; \quad T = \Delta H^\circ / \Delta S^\circ$$

$$\Delta H^\circ = 2\Delta H^\circ \ NO_2(g) - \Delta H^\circ \ N_2O_4(g) = 2(33.84) - 9.66 = +58.02 \ kJ$$

$$\Delta S^\circ = 2S^\circ \ NO_2(g) - S^\circ \ N_2O_4(g) = 2(240.45) - 304.3 = 0.1766 \ kJ/K$$

$$T = \frac{58.02 \ kJ}{0.1766 \ kJ/K} = 328.5 \ K \ \text{or} \ 55.5 \ ^\circ C$$

(b) $P_t = 1.00 \ atm; \ P_{N_2O_4} = x, \ P_{NO_2} = 2x; \ x + 2x = 1.00 \ atm$

      $x = P_{N_2O_4} = 0.3333 = 0.333 \ atm; \quad P_{NO_2} = 0.6667 = 0.667 \ atm$

      $K = \dfrac{(0.6667)^2}{0.3333} = 1.334 = 1.33; \qquad \Delta G^\circ = -RT \ln K = \Delta H^\circ - T\Delta S^\circ$

      $-(8.314 \times 10^{-3} \ kJ/K)(\ln 1.334) \ T = 58.02 \ kJ - (0.1766 \ kJ/K) \ T$

      $(-0.00239 \ kJ/K) \ T + (0.1766 \ kJ/K) \ T = 58.02 \ kJ$

      $(0.1742 \ kJ/K) \ T = 58.02 \ kJ; \ T = 333.0 \ K$

(c) $P_t = 10.00 \ atm; \ x + 2x = 10.00 \ atm$

      $x = P_{N_2O_4} = 3.3333 = 3.333 \ atm; \quad P_{NO_2} = 6.6667 = 6.667 \ atm$

      $K = \dfrac{(6.6667)^2}{3.3333} = 13.334 = 13.33; \quad -RT \ln K = \Delta H^\circ - T\Delta S^\circ$

      $-(8.314 \times 10^{-3} \ kJ/K)(\ln 13.334) \ T = 58.02 \ kJ - (0.1766 \ kJ/K) \ T$

      $(-0.02154 \ kJ/K) \ T + (0.1766 \ kJ/K) \ T = 58.02 \ kJ$

      $(0.15506 \ kJ/K) \ T = 58.02 \ kJ; \ T = 374.2 \ K$

19.111   (a) $\Delta G^\circ = 3\Delta G_f^\circ \ S(s) + 2\Delta G_f^\circ \ H_2O(g) - \Delta G_f^\circ \ SO_2(g) - 2\Delta G_f^\circ \ H_2S(g)$

            $= 3(0) + 2(-228.57) - (-300.4) - 2(-33.01) = -90.72 = -90.7 \ kJ$

            $\ln K = \dfrac{-\Delta G^\circ}{RT} = \dfrac{-(-90.72 \ kJ)}{(8.314 \times 10^{-3} \ kJ/K)(298 \ K)} = 36.6165 = 36.6; \ K = 7.99 \times 10^{15}$

                                                       $= 8 \times 10^{15}$

  (b)   The reaction is highly spontaneous at 298 K and feasible in principle. However, use of $H_2S(g)$ produces a severe safety hazard for workers and the surrounding community.

  (c) $P_{H_2O} = \dfrac{25 \ torr}{760 \ torr/atm} = 0.033 \ atm$

      $K = \dfrac{P_{H_2O}^2}{P_{SO_2} \times P_{H_2S}^2}; \quad P_{SO_2} = P_{H_2S} = x \ atm$

      $K = 7.99 \times 10^{15} = \dfrac{(0.033)^2}{x(x)^2}; \quad x^3 = \dfrac{(0.033)^2}{7.99 \times 10^{15}}$

      $x = 5 \times 10^{-7} \ atm$

(d) $\Delta H° = 3\Delta H_f° \ S(s) + 2\Delta H_f° \ H_2O(g) - \Delta H_f° \ SO_2(g) - 2\Delta H_f° \ H_2S(g)$

   $= 3(0) + 2(-241.82) - (-296.9) - 2(-20.17) = -146.4 \ kJ$

$\Delta S° = 3S° \ S(s) + 2S° \ H_2O(g) - S° \ SO_2(g) - 2S° \ H_2S(g)$

   $= 3(31.88) + 2(188.83) - 248.5 - 2(205.6) = -186.4 \ J/K$

The reaction is exothermic ($-\Delta H$), so the value of K_{eq} will decrease with increasing temperature. The negative $\Delta S°$ value means that the reaction will become nonspontaneous at some higher temperature. The process will be less effective at elevated temperatures.

19.112   (a)   When the rubber band is stretched, the molecules become more ordered, so the entropy of the system decreases, ΔS_{sys} is negative.

(b) $\Delta S_{sys} = q_{rev}/T$. Because ΔS_{sys} is negative, q_{rev} is negative and heat is emitted by the system.

(c)   The unstretched rubber band feels cooler. This confirms our answer to (b). If heat is emitted by the system when it is stretched, the surroundings feel warmer. Upon return to the initial state, heat is absorbed by the system (the rubber band) and the surroundings (your lip) feel cooler.

20 Electrochemistry

Visualizing Concepts

20.1   In this analogy, the electron is analogous to the proton (H^+). Just as acid–base reactions can be viewed as proton-transfer reactions, redox reactions can be viewed as electron-transfer reactions.

Oxidizing agents are reduced; they gain electrons. A strong oxidizing agent would be analogous to a strong base.

20.2   An antioxidant is a reducing agent that reacts with certain oxidizing agents in the body. The objective is that the antioxidant will be oxidized in preference to human cells or DNA.

20.3   *Analyze/Plan.* Apply the definitions of oxidation, reduction, anode and cathode to the diagram. Recall relationship between atomic and ionic size from Chapter 7.   *Solve.*

(a)   Oxidation. The gray spheres are uniformly sized and closely aligned; they represent an elemental solid. The diagram shows atoms from the surface of the solid going into solution. In a voltaic cell, this happens when metal atoms on an electrode surface are oxidized. They lose electrons, form cations, and move into solution.

(b)   Anode. Oxidation occurs at the anode.

(c)   When a neutral atom loses a valence electron, Z_{eff} for the remaining electrons increases, and the radius of the resulting cation is smaller than the radius of the neutral atom. The neutral atoms in the electrode are represented by larger spheres than the cations moving into solution.

20.4   *Analyze/Plan.* Consider the voltaic cell pictured in Figure 20.5 as a model. The reaction in a voltaic cell is spontaneous. To generate a *standard* emf, substances must be present in their standard states.

(a)   A concentration of 1 *M* is the standard state for ions in solution. Ions, but not solution, must be able to flow between compartments to complete the circuit, so that the cell can develop an emf. Add 1 *M* A^{2+}(aq) to the beaker with the A(s) electrode. Add 1 *M* B^{2+}(aq) to the beaker with the B(s) electrode. Add a salt bridge to enable the flow of ions from one compartment to the other.

(b)   Reduction occurs at the cathode. For the reaction to occur spontaneously (and thus generate an emf), the half-reaction with the greater E_{red}° will be the reduction half-reaction. In this cell, it is the half-reaction involving A(s) and A^{2+}(aq). The A electrode functions as the cathode.

(c) According to Figure 20.5, electrons flow through the external circuit from the anode to the cathode. In this example, B is the anode and A is the cathode, so electrons flow from B to A through the external circuit.

(d) $E_{cell}^o = E_{red}^o(\text{cathode}) - E_{red}^o(\text{anode})$

$E_{cell}^o = -0.10\ V - (-1.10\ V) = 1.00\ V$

20.5 $A(aq) + B(aq) \rightarrow A^-(aq) + B^+(aq)$

(a) Reduction occurs at the cathode; oxidation occurs at the anode.

$A(aq) + 1e^- \rightarrow A^-(aq)$ occurs at the cathode.

$B(aq) \rightarrow B^+(aq) + 1e^-$ occurs at the anode.

(b) In a voltaic cell, the anode is at higher potential energy than the cathode. The anode reaction, $B(aq) \rightarrow B^+(aq) + 1e^-$, is higher in potential energy.

(c) $\Delta G^\circ = -nFE^\circ$; the signs of ΔG° and E° (or ΔG and E) are opposite. Because this is a spontaneous reaction, ΔG° is negative and E° is positive.

20.6 *Analyze.* Given a series of reduction half-reactions and their standard electrode potentials (E_{red}^o), draw conclusions about their relative strengths as oxidizing and reducing agents. *Plan.* The reactant with the largest E_{red}^o is the easiest to reduce and the strongest oxidizing agent. The reduced form of this substance, the product of the reduction half-reaction, is the most difficult to oxidize and the weakest reducing agent. Conversely, the reactant with the smallest E_{red}^o is the hardest to reduce and the weakest oxidizing agent. The reduced form of this substance, the product of the reduction half-reaction, is the easiest to oxidize and the strongest reducing agent. *Solve.*

(a) $A^+(aq)$ is the strongest oxidizing agent, and D^{3+} is the weakest oxidizing agent.

(b) $D(s)$ is the strongest reducing agent, and $A(s)$ is the weakest reducing agent.

(c) Reactants with more positive E_{red}^o than $C^{3+}(aq)$ will oxidize $C^{2+}(aq)$. Both $A^+(aq)$ and $B^{2+}(aq)$ will oxidize $C^{2+}(aq)$.

20.7 *Analyze.* Given a redox reaction with a negative E°, answer questions regarding ΔG°, the equilibrium constant (K), and work (w). *Plan.* $\Delta G^\circ = -nFE^\circ$; $\Delta G^\circ = -RT \ln K$; $w_{max} = -nFE$. *Solve.*

(a) The signs of ΔG° and E° are opposite. If E° is negative, ΔG° is positive. (The reaction is not spontaneous in the forward direction.)

(b) If ΔG° is positive, $\ln K$ is negative and $K < 1$. Also, K is less than one for a nonspontaneous reaction.

(c) No. If E° is negative, the sign of w is positive. A positive value for w means that work is done on the system by the surroundings. An electrochemical cell based on this reaction cannot accomplish work on its surroundings.

20.8 *Analyze.* Given the voltaic cell shown in the diagram, answer questions about the cell and the effect of solution concentration on cell potential, E.

Plan. Use the definition of a voltaic cell and standard emf, along with the Nernst equation, $E = \Delta E^\circ - (0.0592\ V/n)\log Q$, to answer the questions. *Solve.*

(a)   A voltaic cell involves a spontaneous redox reaction, one with positive E^o_{cell}. To achieve a positive E^o_{cell}, the half-reaction with the more positive E^o_{red} occurs at the cathode. For this cell the two half reactions are

$$Ag^+(aq) + e^- \rightarrow Ag(s), \quad E^o_{red} = 0.80 \text{ V}$$

$$Fe^{2+}(aq) + 2e^- \rightarrow Fe(s), \quad E^o_{red} = -0.44 \text{ V}$$

The Ag(s) electrode is the cathode.

(b)   The standard emf is just E^o_{cell}.

$$E^o_{cell} = E^o_{red}(\text{cathode}) - E^o_{red}(\text{anode}) = 0.80 \text{ V} - (-0.44 \text{ V}) = 1.24 \text{ V}$$

The cell in the diagram is at standard conditions, with solid metal electrodes and 1 M aqueous solutions, so the potential on the meter in the circuit is the standard emf.

(c) $2 \, Ag^+(aq) + Fe(s) \rightarrow 2 \, Ag(s) + Fe^{2+}(aq); n = 2; \quad E = E^o - \dfrac{0.0592}{2} \log \dfrac{[Fe^{2+}]}{[Ag^+]^2}$

The solution in the cathode half-cell is $Ag^+(aq)$. If $[Ag^+(aq)]$ increases by a factor of 10, the change in cell voltage is $E - E^o = -\dfrac{0.0592}{2} \log \dfrac{[1]}{[10]^2} = 0.059 \text{ V}$.

(d)   The solution in the anode half-cell is $Fe^{2+}(aq)$. If $[Fe^{2+}(aq)]$ increases by a factor of 10, the change in cell voltage is $E - E^o = -\dfrac{0.0592}{2} \log \dfrac{[10]}{[1]^2} = -0.030 \text{ V}$.

20.9   *Analyze/Plan.* Consider the Nernst equation, which describes the variation of potential (emf) with respect to changes in concentration.

Solve. $E = E^o - \dfrac{0.0592}{n} \log Q$.

(a)   For this half-reaction, $E^o_{red} = 0.80 \text{ V}; Q = 1/[Ag^+]$

$$E = 0.80 - \frac{0.0592}{1} \log \frac{1}{[Ag^+]}; \quad E = 0.80 + \frac{0.0592}{1} \log [Ag^+]$$

The y-intercept of the graph is E^o. The slope of the line is $+0.0592$. So, as $[Ag^+]$ and $\log[Ag^+]$ increase, E increases. The line that describes this behavior is line 1.

(b)   When $\log[Ag^+] = 0$, $[Ag^+] = 1 \, M$; this is the standard state for $Ag^+(aq)$, so $E_{red} = E^o_{red} = 0.80 \text{ V}$.

20.10   (a)   Zinc is the anode. Metallic zinc cannot be reduced; it must be oxidized. Oxidation occurs at the anode.

         (b)   The energy density of the silver oxide battery is most similar to the nickel-cadmium battery. Energy density (see Figure 20.23) is related to the molar masses of the electrode materials and voltage per cell of the battery. The molar mass of $(Ag_2O + Zn)$ is most like that of $(Ni + Cd)$. The cell potential for a Ni-Cd battery is 1.3 V; for the reaction in the silver oxide battery, the cell potential is about 1.1 V. The Li-ion (3.7 V) and lead-acid (2.0 V) batteries have larger cell potentials. (The standard reduction potential of Ag_2O is 0.34 V. From *Handbook of Chemistry and Physics*, 74th edition)

20.11   Beaker B. The process of iron corrosion includes Fe being oxidized and O_2 being reduced. The reduction of O_2 requires H^+. Beaker B, which contains dilute HCl(aq), provides the H^+ required to encourage corrosion. Iron in contact with a solution above pH 9 does not corrode.

20.12   (a)   +2

        (b) $MgCl_2(l)$

        (c)   Oxidation occurs at the anode. The anode half-reaction is:

              $$2\ Cl^-(aq) \rightarrow Cl_2(g) + 2e^-$$

        (d)   Reduction occurs at the cathode. The cathode half-reaction is:

              $$Mg^{2+}(aq) + 2e^- \rightarrow Mg(l)$$

Oxidation–Reduction Reactions (Section 20.1)

20.13   (a)   *Oxidation* is the loss of electrons.

        (b)   The electrons appear on the products side (right side) of an oxidation half-reaction.

        (c)   The *oxidant* is the reactant that is reduced; it gains the electrons that are lost by the substance being oxidized.

        (d)   An *oxidizing agent* is the substance that promotes oxidation. That is, it gains electrons that are lost by the substance being oxidized. It is the same as the oxidant.

20.14   (a)   *Reduction* is the gain of electrons.

        (b)   The electrons appear on the reactants side (left side) of a reduction half-reaction.

        (c)   The *reductant* is the reactant that is oxidized; it provides the electrons that are gained by the substance being reduced.

        (d)   A *reducing agent* is the substance that promotes reduction. It donates the electrons gained by the substance that is reduced. It is the same as the reductant.

20.15   (a)   True.

        (b)   False. Fe^{3+} is reduced to Fe^{2+}, so it is the oxidizing agent, and Co^{2+} is the reducing agent.

        (c)   True.

20.16   (a)   False. If something is reduced, it gains electrons.

        (b)   True.

        (c)   True. Oxidation can be thought of as a gain of oxygen atoms. Looking forward, this view will be useful for organic reactions, Chapter 24.

20.17   *Analyze/Plan.* Given a chemical equation, we are asked to determine the oxidation number of each reactant and product, and the total number of electrons transferred in the overall reaction. Assign oxidation numbers according to the rules given in Section 4.4. To find electrons transferred, count the electrons gained by an atom that is reduced and multiply by the number of atoms reduced. Check by counting electrons for the atom oxidized.   *Solve.*

        (a)   (i)   Reactants: I, +5; O, –2; C +2; O, –2. Products: I, 0; C, +4; O, –2

              (ii) The total number of electrons transferred is 10. (Each I atom gains 5 electrons, for a total of 10 electrons gained; each C atom loses 2 electrons for a total of 10 electrons lost; electrons balance.)

(b) (i) Reactants: Hg, +2; N, –2; H, +1. Products: Hg, 0; N, 0; H, +1.

(ii) The total number of electrons transferred is 4. (Each Hg gains 2 electrons for a total of 4 electrons gained; each N loses 2 electrons for a total of 4 electrons lost; electrons balance.)

(c) (i) Reactants: H, +1; S, –2; N, +5; O, –2. Products: S, 0; N, +2; O, –2; H, +1; O, –2.

(ii) The total electrons transferred is 6. (Each N atom gains 3 electrons for a total of 6 electrons gained; each S atom loses 2 electrons for a total of 6 electrons lost; electrons balance.)

20.18 (a) (i) Reactants: Mn, +7; O, –2; S, –2; H, +1; O, –2. Products: S, 0; Mn, +4; O, –2; O, –2; H, +1.

(ii) The total number of electrons transfered is 6.

(b) (i) Reactants: H, +1; O, –1; Cl, +7; O, –2; H, +1; O, –2; Products: Cl, +4; O, –2; H, +1; O, –2; O, 0.

(ii) The total number of electrons transferred is 8. (Two Cl atoms each gain 4 electron, eight O atoms each lose 1 electron by changing oxidation number from –1 to 0.)

(c) (i) Reactants: Ba, +2; O, –2; H, +1; H, +1; O, –1; Cl, +4; O, –2. Products: Ba, +2; Cl, +3; O, –2; H, +1; O, –2; O, 0.

(ii) The total number of electrons transferred is 2. (Two Cl atoms each gain 1 electron, two O atoms each lose 1 electron by changing oxidation number from –1 to 0.)

20.19 (a) No oxidation–reduction.

(b) I is oxidized from –1 to +5; Cl is reduced from +1 to –1.

(c) S is oxidized from +4 to +6; N is reduced from +5 to +2.

20.20 (a) No oxidation–reduction.

(b) Pb is reduced from +4 to +2; O is oxidized from –2 to 0.

(c) S is reduced from +6 to +4; Br is oxidized from –1 to 0.

Balancing Oxidation–Reduction Reactions (Section 20.2)

20.21 *Analyze/Plan.* Write the balanced chemical equation and assign oxidation numbers. The substance oxidized is the reductant and the substance reduced is the oxidant. *Solve.*

(a) $TiCl_4(g) + 2 Mg(l) \rightarrow Ti(s) + 2 MgCl_2(l)$

(b) $Mg(l)$ is oxidized; $TiCl_4(g)$ is reduced.

(c) $Mg(l)$ is the reductant; $TiCl_4(g)$ is the oxidant.

20.22 (a) $2 N_2H_4(g) + N_2O_4(g) \rightarrow 3 N_2(g) + 4 H_2O(g)$

(b) $N_2H_4(g)$ is oxidized; $N_2O_4(g)$ is reduced.

(c) $N_2H_4(g)$ serves as the reducing agent; it is itself oxidized. $N_2O_4(g)$ serves as the oxidizing agent; it is itself reduced.

20.23 *Analyze/Plan.* Follow the logic in Sample Exercises 20.2 and 20.3. If the half-reaction occurs in basic solution, balance as in acid, then add OH⁻ to each side. *Solve.*

(a) $Sn^{2+}(aq) \rightarrow Sn^{4+}(aq) + 2 e^-$, oxidation

(b) $TiO_2(s) + 4 H^+(aq) + 2 e^- \rightarrow Ti^{2+}(aq) + 2 H_2O(l)$, reduction

(c) $ClO_3^-(aq) + 6 H^+(aq) + 6 e^- \rightarrow Cl^-(aq) + 3 H_2O(l)$, reduction

(d) $N_2(g) + 8 H^+(aq) + 6 e^- \rightarrow 2 NH_4^+(aq)$, reduction

(e) $4 OH^-(aq) \rightarrow O_2(g) + 2 H_2O(l) + 4 e^-$, oxidation

(f) $SO_3^{2-}(aq) + 2 OH^-(aq) \rightarrow SO_4^{2-}(aq) + H_2O(l) + 2 e^-$, oxidation

(g) $N_2(g) + 6 H_2O + 6 e^- \rightarrow 2 NH_3(g) + 6 OH^-(aq)$, reduction

20.24 (a) $Mo^{3+}(aq) + 3 e^- \rightarrow Mo(s)$, reduction

(b) $H_2SO_3(aq) + H_2O(l) \rightarrow SO_4^{2-}(aq) + 4 H^+(aq) + 2 e^-$, oxidation

(c) $NO_3^-(aq) + 4 H^+(aq) + 3 e^- \rightarrow NO(g) + 2 H_2O(l)$, reduction

(d) $O_2(g) + 4 H^+(aq) + 4 e^- \rightarrow 2 H_2O(l)$, reduction

(e) $O_2(g) + 2 H_2O(l) + 4 e^- \rightarrow 4 OH^-(aq)$, reduction

 (O_2 is reduced to OH^-, not H_2O, in basic solution)

(f) $Mn^{2+}(aq) + 4 OH^-(aq) \rightarrow MnO_2(s) + 2 H_2O(l) + 2 e^-$, oxidation

(g) $Cr(OH)_3(s) + 5 OH^-(aq) \rightarrow CrO_4^{2-}(aq) + 4 H_2O(l) + 3 e^-$, oxidation

20.25 *Analyze/Plan.* Follow the logic in Sample Exercises 20.2 and 20.3 to balance the given equations. Use the method in Sample Exercise 20.1 to identify oxidizing and reducing agents. *Solve.*

(a) $Cr_2O_7^{2-}(aq) + I^-(aq) + 8 H^+ \rightarrow 2 Cr^{3+}(aq) + IO_3^-(aq) + 4 H_2O(l)$

 oxidizing agent, $Cr_2O_7^{2-}$; reducing agent, I^-

(b) The half-reactions are:

$$4 [MnO_4^-(aq) + 8 H^+(aq) + 5 e^- \rightarrow Mn^{2+}(aq) + 4 H_2O(l)]$$
$$\underline{5 [CH_3OH(aq) + H_2O(l) \rightarrow HCOOH(aq) + 4 H^+(aq) + 4 e^-]}$$
$$4 MnO_4^-(aq) + 5 CH_3OH(aq) + 12 H^+(aq) \rightarrow 4 Mn^{2+}(aq) + 5 HCOOH(aq) + 11 H_2O(l)$$

 oxidizing agent, MnO_4^-; reducing agent, CH_3OH

(c)
$$I_2(s) + 6 H_2O(l) \rightarrow 2 IO_3^-(aq) + 12 H^+(aq) + 10 e^-$$
$$\underline{5 [OCl^-(aq) + 2 H^+(aq) + 2 e^- \rightarrow Cl^-(aq) + H_2O(l)]}$$
$$I_2(s) + 5 OCl^-(aq) + H_2O(l) \rightarrow 2 IO_3^-(aq) + 5 Cl^-(aq) + 2 H^+(aq)]$$

 oxidizing agent, OCl^-; reducing agent, I_2

(d)
$$As_2O_3(s) + 5 H_2O(l) \rightarrow 2 H_3AsO_4(aq) + 4 H^+(aq) + 4 e^-$$
$$\underline{2 NO_3^-(aq) + 6 H^+(aq) + 4 e^- \rightarrow N_2O_3(aq) + 3 H_2O(l)}$$
$$As_2O_3(s) + 2 NO_3^-(aq) + 2 H_2O(l) + 2 H^+(aq) \rightarrow 2 H_3AsO_4(aq) + N_2O_3(aq)$$

 oxidizing agent, NO_3^-; reducing agent, As_2O_3

613

(e)
$$2\,[MnO_4^-(aq)+2\,H_2O(l)+3\,e^- \rightarrow MnO_2(s)+4\,OH^-]$$
$$\underline{\hspace{0.5cm} Br^-(aq)+6\,OH^-(aq) \rightarrow BrO_3^-(aq)+3\,H_2O(l)+6\,e^- \hspace{0.5cm}}$$
$$2\,MnO_4^-(aq)+Br^-(aq)+H_2O(l) \rightarrow 2\,MnO_2(s)+BrO_3^-(aq)+2\,OH^-(aq)$$

oxidizing agent, MnO_4^-; reducing agent, Br^-

(f) $Pb(OH)_4^{2-}(aq) + ClO^-(aq) \rightarrow PbO_2(s) + Cl^-(aq) + 2\,OH^-(aq) + H_2O(l)$

oxidizing agent, ClO^-; reducing agent, $Pb(OH)_4^{2-}$

20.26   (a)
$$3\,[NO_2^-(aq)+H_2O(l) \rightarrow NO_3^-(aq)+2\,H^+(aq)+2\,e^-]$$
$$\underline{Cr_2O_7^{2-}(aq)+14\,H^+(aq)+6\,e^- \rightarrow 2\,Cr^{3+}(aq)+7\,H_2O(l)}$$

Net: $3\,NO_2^-(aq) + Cr_2O_7^{2-}(aq) + 8\,H^+(aq) \rightarrow 3\,NO_3^-(aq)+2\,Cr^{3+}(aq)+4\,H_2O(l)$

oxidizing agent, $Cr_2O_7^{2-}$; reducing agent, NO_2^-

(b)   The oxidation half-reaction involves S, and is listed in Appendix E. The reduction half-reaction involves N, and must be written and balanced, according to the procedure in Section 20.2.

$$HNO_3(aq) \rightarrow N_2O(g)$$
$$2\,HNO_3(aq) \rightarrow N_2O(g)$$
$$2\,HNO_3(aq) \rightarrow N_2O(g) + 5\,H_2O(l)$$
$$2\,HNO_3(aq)+8\,H^+(aq) \rightarrow N_2O(g) + 5\,H_2O(l)$$
$$2\,HNO_3(aq)+8\,H^+ +8\,e^- \rightarrow N_2O(g) + 5\,H_2O(l)$$

$$2\,[S(s)+3\,H_2O(l) \rightarrow H_2SO_3(aq)+4\,H^+ +4\,e^-]$$
$$\underline{2\,HNO_3(aq)+8\,H^+ +8\,e^- \rightarrow N_2O(g) + 5\,H_2O(l)}$$
$$2\,HNO_3(aq)+2\,S(s)+H_2O(l) \rightarrow 2\,H_2SO_3(aq)+N_2O(g)$$

oxidizing agent, HNO_3; reducing agent, S.

(c)
$$2\,[Cr_2O_7^{2-}(aq)+14\,H^+(aq)+6\,e^- \rightarrow 2\,Cr^{3+}(aq)+7\,H_2O(l)]$$
$$\underline{3\,[CH_3OH(aq)+H_2O(l) \rightarrow HCOOH(aq)+4\,H^+(aq)+4\,e^-]}$$

Net: $2\,Cr_2O_7^{2-}(aq)+3\,CH_3OH(aq)+16\,H^+(aq) \rightarrow 4\,Cr^{3+}(aq)+3\,HCOOH(aq)+11\,H_2O(l)$

oxidizing agent, $Cr_2O_7^{2-}$; reducing agent, CH_3OH

(d)   The half-reaction involving N_2H_4 is given in Appendix E in base. We add $4\,H^+(aq)$ to each side and reverse the reaction to obtain the oxidation half-reaction shown below.

$$2\,[2\,BrO_3^-(aq)+12\,H^+(aq)+10\,e^- \rightarrow Br_2(l)+6\,H_2O(l)]$$
$$\underline{5\,[N_2H_4(aq) \rightarrow N_2(g)+4\,H^+(aq)+4\,e^-]}$$

Net: $4\,BrO_3^-(aq)+5\,N_2H_4(aq)+4\,H^+(aq) \rightarrow 2\,Br_2(l)+5\,N_2(g)+12\,H_2O(l)$

oxidizing agent, BrO_3^-; reducing agent, N_2H_4

(e) Write and balance each half-reaction, and then sum to get the overall reaction. Follow the procedure in Sample Exercise 20.3 for reactions in basic solution.

oxidation: $Al(s) \rightarrow AlO_2^-(aq)$

$$Al(s) + 2\,H_2O(l) \rightarrow AlO_2^-(aq) + 4\,H^+(aq) + 3\,e^-$$

$$Al(s) + 4\,OH^-(aq) \rightarrow AlO_2^-(aq) + 2\,H_2O(l) + 3\,e^-$$

reduction: $NO_2^-(aq) \rightarrow NH_4^+(aq)$

$$NO_2^-(aq) + 8\,H^+(aq) + 6\,e^- \rightarrow NH_4^+(aq)\,) + 2\,H_2O(l)$$

$$NO_2^-(aq) + 6\,H_2O(l) + 6\,e^- \rightarrow NH_4^+(aq) + 8\,OH^-(aq)$$

$$NO_2^-(aq) + 6\,H_2O(l) + 6\,e^- \rightarrow NH_4^+(aq) + 8\,OH^-(aq)$$

$$2[Al(s) + 4\,OH^-(aq) \rightarrow AlO_2^-(aq) + 2\,H_2O(l) + 3\,e^-]$$

$$NO_2^-(aq) + 6\,H_2O(l) + 6\,e^- \rightarrow NH_4^+(aq) + 8\,OH^-(aq)$$

$$NO_2^-(aq) + 2\,Al(s) + 2\,H_2O(l) \rightarrow NH_4^+(aq) + 2\,AlO_2^-(aq)$$

oxidizing agent, NO_2^-; reducing agent, Al

(f) $H_2O_2(aq) + 2\,e^- \rightarrow O_2(g) + 2\,H^+(aq)$

Because the reaction is in base, the H^+ can be "neutralized" by adding $2\,OH^-$ to each side of the equation: $H_2O_2(aq) + 2\,OH^-(aq) + 2\,e^- \rightarrow O_2(g) + 2\,H_2O(l)$. The other half reaction is: $2\,[ClO_2(aq) + e^- \rightarrow ClO_2^-(aq)]$.

Net: $H_2O_2(aq) + 2\,ClO_2(aq) + 2\,OH^-(aq) \rightarrow O_2(g) + 2\,ClO_2^-(aq) + 2\,H_2O(l)$

oxidizing agent, ClO_2; reducing agent, H_2O_2

Voltaic Cells (Section 20.3)

20.27   (a)   False. Oxidation occurs at the anode; reduction occurs at the cathode.

       (b)   True. The terms voltaic and galvanic are interchangeable.

       (c)   True. The oxidation half-reaction at the anode generates free electrons which flow from the anode to the cathode.

20.28   (a)   True. Oxidation occurs at the anode.

       (b)   True. A voltaic cell utilizes a spontaneous redox reaction and a spontaneous reaction has a positive emf.

       (c)   True. Ions must be able to migrate between the anode and cathode compartments in order to maintain charge balance and complete the electrical circuit.

20.29   *Analyze/Plan.* Follow the logic in Sample Exercise 20.4.   *Solve.*

       (a) $Fe(s)$ is oxidized, $Ag^+(aq)$ is reduced.

       (b) $Ag^+(aq) + 1e^- \rightarrow Ag(s)$; $Fe(s) \rightarrow Fe^{2+}(aq) + 2e^-$

       (c) $Fe(s)$ is the anode, $Ag(s)$ is the cathode.

       (d) $Fe(s)$ is negative; $Ag(s)$ is positive.

       (e)   Electrons flow from the $Fe(-)$ electrode toward the $Ag(+)$ electrode.

       (f)   Cations migrate toward the $Ag(s)$ cathode; anions migrate toward the $Fe(s)$ anode.

20.30   (a)   Al(s) is oxidized, Ni^{2+}(aq) is reduced.

       (b) $Al(s) \rightarrow Al^{3+}(aq) + 3e^-$; $Ni^{2+}(aq) + 2e^- \rightarrow Ni(s)$

       (c)   Al(s) is the anode; Ni(s) is the cathode.

       (d)   Al(s) is negative (–); Ni(s) is positive (+).

       (e)   Electrons flow from the Al(–) electrode toward the Ni(+) electrode.

       (f)   Cations migrate toward the Ni(s) cathode; anions migrate toward the Al(s) anode.

Cell Potentials under Standard Conditions (Section 20.4)

20.31   (a)   One *volt* is the potential energy difference required to impart 1 J of energy to a charge of 1 coulomb. 1 V = 1 J/C.

       (b)   Yes. All voltaic cells involve a spontaneous redox reaction that produces a positive cell potential or emf.

20.32   (a)   In a voltaic cell, the anode has the higher potential energy for electrons. To achieve a lower potential energy, electrons flow from the anode to the cathode.

       (b)   The units of electrical potential are volts. A potential of one volt imparts one joule of energy to one coulomb of charge.

20.33   (a) $2\,H^+(aq) + 2\,e^- \rightarrow H_2(g)$

       (b) $H_2(g) \rightarrow 2\,H^+(aq) + 2\,e^-$

       (c)   A *standard* hydrogen electrode is a hydrogen electrode where the components are at standard conditions, $1\,M\,H^+$(aq) and H_2(g) at 1 atm, such that $E^o = 0$ V.

20.34   (a)   For a reduction potential to be a *standard reduction potential*, the substances in the reaction or half-reaction must be at standard conditions, $1\,M$ aqueous solutions and 1 atm gas pressures.

       (b) $E^o_{red} = 0$ V for a standard hydrogen electrode.

       (c)   It is not possible to measure the standard reduction potential of a single half-reaction because each voltaic cell consists of two half-reactions and only the potential of a complete cell can be measured.

20.35   *Analyze/Plan.* Follow the logic in Sample Exercise 20.5.   *Solve.*

       (a)   The two half-reactions are:

$$Tl^{3+}(aq) + 2\,e^- \rightarrow Tl^+(aq) \qquad\qquad \text{cathode } E^o_{red} = ?$$
$$2\,[Cr^{2+}(aq) \rightarrow Cr^{3+}(aq) + e^-] \qquad\qquad \text{anode} \quad E^o_{red} = -0.41 \text{ V}$$

       (b) $E^o_{cell} = E^o_{red} \text{ (cathode)} - E^o_{red} \text{ (anode)}$

            $1.19 \text{ V} = E^o_{red} - (-0.41 \text{ V})$

              $E^o_{red} = 1.19 \text{ V} - 0.41 \text{ V} = 0.78 \text{ V}$

(c)

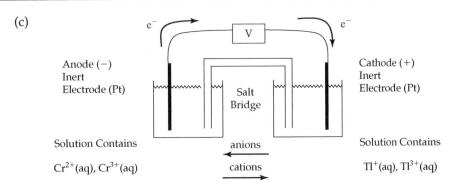

Anode (−)
Inert
Electrode (Pt)

Salt
Bridge

Cathode (+)
Inert
Electrode (Pt)

Solution Contains

Cr^{2+}(aq), Cr^{3+}(aq)

anions

cations

Solution Contains

Tl^{+}(aq), Tl^{3+}(aq)

Note that because Cr^{2+}(aq) is readily oxidized, it would be necessary to keep oxygen out of the left-hand cell compartment.

20.36   (a) $PdCl_4^{2-}(aq) + 2\ e^- \rightarrow Pd(s) + 4\ Cl^-$ cathode $E^o_{red} = ?$

$Cd(s) \rightarrow Cd^{2+}(aq) + 2\ e^-$ anode $E^o_{red} = -0.40$ V

(b) $E^o_{cell} = E^o_{red}$ (cathode) $- E^o_{red}$ (anode); 1.03 V $= E^o_{red} - (-0.40$ V);

$E^o_{red} = 1.03$ V $- 0.40 = 0.63$ V

(c)

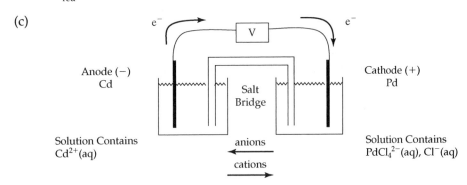

Anode (−)
Cd

Salt
Bridge

Cathode (+)
Pd

Solution Contains
Cd^{2+}(aq)

anions

cations

Solution Contains
$PdCl_4^{2-}$(aq), Cl^-(aq)

20.37   *Analyze/Plan.* Follow the logic in Sample Exercise 20.6.   *Solve.*

(a) $Cl_2(g) \rightarrow 2\ Cl^-(aq) + 2\ e^-$ $E^o_{red} = 1.36$ V

$I_2(s) + 2\ e^- \rightarrow 2\ I^-(aq)$ $E^o_{red} = 0.54$ V

$E° = 1.36$ V $- 0.54$ V $= 0.82$ V

(b) $Ni(s) \rightarrow Ni^{2+}(aq) + 2\ e^-$ $E^o_{red} = -0.28$ V

$2\ [Ce^{4+}(aq) + 1\ e^- \rightarrow Ce^{3+}(aq)]$ $E^o_{red} = 1.61$ V

$E° = 1.61$ V $- (-0.28$ V$) = 1.89$ V

(c) $Fe(s) \rightarrow Fe^{2+}(aq) + 2\ e^-$ $E^o_{red} = -0.44$ V

$2\ [Fe^{3+}(aq) + 1\ e^- \rightarrow Fe^{2+}(aq)]$ $E^o_{red} = 0.77$ V

$E° = 0.77$ V $- (-0.44$ V$) = 1.21$ V

(d) $2\ [NO_3^-(aq) + 4\ H^+ + 3\ e^- \rightarrow NO(g) + 2\ H_2O(l)]$ $E^o_{red} = 0.96$ V

$3\ [Cu(s) \rightarrow Cu^{2+}(aq) + 2\ e^-]$ $E^o_{red} = 0.34$ V

$E° = 0.96$ V $- (0.34$ V$) = 0.62$ V

20.38   (a) $F_2(g) + 2\ e^- \rightarrow 2\ F^-(aq)$ $E^o_{red} = 2.87\ V$

                $H_2(g) \rightarrow 2\ H^+(aq) + 2\ e^-$ $E^o_{red} = 0.00\ V$

            $E° = 2.87\ V - 0.00\ V = 2.87\ V$

  (b) $Cu^{2+}(aq) + 2\ e^- \rightarrow Cu(s)$ $E^o_{red} = 0.34\ V$

                 $Ca(s) \rightarrow Ca^{2+}(aq) + 2\ e^-$ $E^o_{red} = -2.87\ V$

            $E° = 0.34\ V - (-2.87\ V) = 3.21\ V$

  (c) $Fe^{2+}(aq) + 2\ e^- \rightarrow Fe(s)$ $E^o_{red} = -0.44\ V$

              $2[Fe^{2+}(aq) \rightarrow Fe^{3+}(aq) + 1\ e^-]$ $E^o_{red} = 0.77\ V$

            $E° = -0.44\ V - 0.77\ V = -1.21\ V$

  (d) $2\ ClO_3^-(aq) + 12\ H^+ + 10\ e^- \rightarrow Cl_2(g) + 6\ H_2O(l)$ $E^o_{red} = 1.47\ V$

              $5\ [2\ Br^-(aq) \rightarrow Br_2(l) + 2\ e^-]$ $E^o_{red} = 1.07\ V$

            $E° = 1.47\ V - 1.07\ V = 0.40\ V$

20.39   *Analyze/Plan.* Given four half-reactions, find E^o_{red} from Appendix E and combine them to obtain a desired E_{cell}. (a) The largest E_{cell} will combine the half-reaction with the most positive E^o_{red} as the cathode reaction and the one with the most negative E^o_{red} as the anode reaction. (b) The smallest positive E^o_{cell} will combine two half-reactions whose E^o_{red} values are closest in magnitude **and** sign.   *Solve.*

  (a) $3\ [Ag^+(aq) + 1\ e^- \rightarrow Ag(s)]$ $E^o_{red} = 0.80$

               $\underline{Cr(s) \rightarrow Cr^{3+}(aq) + 3\ e^-}$ $E^o_{red} = -0.74$

          $3\ Ag^+(aq) + Cr(s) \rightarrow 3\ Ag(s) + Cr^{3+}(aq)$ $E° = 0.80 - (-0.74) = 1.54\ V$

  (b)   Two of the combinations have equal E° values.

         $2\ [Ag^+(aq) + 1\ e^- \rightarrow Ag(s)]$ $E^o_{red} = 0.80\ V$

            $\underline{Cu(s) \rightarrow Cu^{2+}(aq) + 2\ e^-}$ $E^o_{red} = 0.34\ V$

        $2\ Ag^+(aq) + Cu(s) \rightarrow 2\ Ag(s) + Cu^{2+}(aq)$ $E° = 0.80\ V - 0.34\ V = 0.46\ V$

         $3\ [Ni^{2+}(aq) + 2\ e^- \rightarrow Ni(s)]$ $E^o_{red} = -0.28\ V$

           $\underline{2\ [Cr(s) \rightarrow Cr^{3+}(aq) + 3\ e^-]}$ $E^o_{red} = -0.74\ V$

       $3\ Ni^{2+}(aq) + 2\ Cr(s) \rightarrow 3\ Ni(s) + 2\ Cr^{3+}(aq)$ $E° = -0.28\ V - (-0.74\ V) = 0.46\ V$

20.40   (a) $2\ [Au(s) + 4\ Br^-(aq) \rightarrow AuBr_4^-(aq) + 3\ e^-]$ $E^o_{red} = -0.86\ V$

              $\underline{3\ [2e^- + IO^-(aq) + H_2O(l) \rightarrow I^-(aq) + 2\ OH^-(aq)]}$ $E^o_{red} = 0.49\ V$

    $2\ Au(s) + 8\ Br^-(aq) + 3\ IO^-(aq) + 3\ H_2O(l) \rightarrow 2\ AuBr_4^-(aq) + 3\ I^-(aq) + 6\ OH^-(aq)$

                                         $E° = 0.49 - (-0.86) = 1.35\ V$

  (b) $2\ [Eu^{2+}(aq) \rightarrow Eu^{3+}(aq) + 1\ e^-]$ $E^o_{red} = -0.43\ V$

          $\underline{Sn^{2+}(aq) + 2\ e^- \rightarrow Sn(s)}$ $E^o_{red} = -0.14\ V$

      $2\ Eu^{2+}(aq) + Sn^{2+}(aq) \rightarrow 2\ Eu^{3+}(aq) + Sn(s)$ $E° = -0.14 - (-0.43) = 0.29\ V$

20.41   *Analyze/Plan.* Given the description of a voltaic cell, answer questions about this cell. Combine ideas in Sample Exercises 20.4 and 20.7. The reduction half-reactions are:

$$Cu^{2+}(aq) + 2\,e^- \rightarrow Cu(s) \qquad E° = 0.34\ V$$
$$Sn^{2+}(aq) + 2\,e^- \rightarrow Sn(s) \qquad E° = -0.14\ V$$

Solve.

(a)   It is evident that Cu^{2+} is more readily reduced. Therefore, Cu serves as the cathode, Sn as the anode.

(b)   The copper electrode gains mass as Cu is plated out, the Sn electrode loses mass as Sn is oxidized.

(c)   The overall cell reaction is $Cu^{2+}(aq) + Sn(s) \rightarrow Cu(s) + Sn^{2+}(aq)$

(d) $E° = 0.34\ V - (-0.14\ V) = 0.48\ V$

20.42   (a)   The two half-reactions are:

$$Cd^{2+}(aq) + 2\,e^- \rightarrow Cd(s) \qquad\qquad E° = -0.40\ V$$
$$Cl_2(g) + 2\,e^- \rightarrow 2\,Cl^-(aq) \qquad\qquad E° = 1.36\ V$$

Because $E°$ for the reduction of Cl_2 is greater, Cl_2 is reduced at the cathode, the Pt electrode. Cd(s) is oxidized at the anode, the Cd electrode.

(b)   The Cd anode loses mass as $Cd^{2+}(aq)$ is produced.

(c) $Cl_2(g) + Cd(s) \rightarrow Cd^{2+}(aq) + 2Cl^-(aq)$

(d) $E° = 1.36\ V - (-0.40\ V) = 1.76\ V$

Strengths of Oxidizing and Reducing Agents (Section 20.4)

20.43   *Analyze/Plan.* The more readily a substance is oxidized, the stronger it is as a reducing agent. In each case choose the half-reaction with the more negative reduction potential and the given substance on the right.   *Solve.*

(a)   Mg(s) (–2.37 V vs. –0.44 V)

(b)   Ca(s) (–2.87 V vs. –1.66 V)

(c) H_2(g, acidic) (0.00 V vs. 0.14 V)

(d) IO_3^- (aq) [Both IO_3^- (aq) and BrO_3^-(aq) are good oxidizing agents, but IO_3^-(aq) has the smaller positive reduction potential. (1.20 V vs. 1.52 V)]

20.44   Follow the logic in Sample Exercise 20.8. In each case, choose the half-reaction with the more positive reduction potential and with the given substance on the left.

(a) Cl_2(g) (1.36 V vs. 1.07 V)

(b) Cd^{2+}(aq) (–0.40 V vs. –0.76 V)

(c) ClO_3^-(aq) (Cl^-(aq) is in its minimum oxidation state and cannot act as an oxidizing agent)

(d) O_3(g) (2.07 V vs. 1.78 V)

20.45 *Analyze/Plan.* If the substance is on the left of a reduction half-reaction, it will be an oxidant; if it is on the right, it will be a reductant. *Solve.*

(a) $Cl_2(aq)$: oxidant (on the left, $E^o_{red} = 1.36$ V)

(b) $MnO_4^-(aq, acidic)$: oxidant (on the left, $E^o_{red} = 1.51$ V)

(c) $Ba(s)$: reductant (on the right, $E^o_{red} = -2.90$ V)

(d) $Zn(s)$: reductant (on the right, $E^o_{red} = -0.76$ V)

20.46 If the substance is on the left of a reduction half-reaction, it will be an oxidant; if it is on the right, it will be a reductant.

(a) $Ce^{3+}(aq)$: reductant (on the right, $E^o_{red} = 1.61$ V)

(b) $Ca(s)$: reductant (on the right, $E^o_{red} = -2.87$ V)

(c) $ClO_3^-(aq)$: oxidant (on the left, $E^o_{red} = 1.47$ V)

(d) $N_2O_5(g)$: oxidant (N has maximum oxidation number, +5; can only be reduced and act as oxidant.)

20.47 *Analyze/Plan.* Follow the logic in Sample Exercise 20.8. *Solve.*

(a) Arranged in order of increasing strength as oxidizing agents (and increasing reduction potential):

$$Cu^{2+}(aq) < O_2(g) < Cr_2O_7^{2-}(aq) < Cl_2(g) < H_2O_2(aq)$$

(b) Arranged in order of increasing strength as reducing agents (and decreasing reduction potential):

$$H_2O_2(aq) < I^-(aq) < Sn^{2+}(aq) < Zn(s) < Al(s)$$

20.48 (a) The strongest oxidizing agent is the species most readily reduced, as evidenced by a large, positive reduction potential. That species is H_2O_2. The weakest oxidizing agent is the species that least readily accepts an electron. We expect that it will be very difficult to reduce $Zn(s)$; indeed, $Zn(s)$ acts as a comparatively strong **reducing** agent. No potential is listed for reduction of $Zn(s)$, but we can safely assume that it is less readily reduced than any of the other species present.

(b) The strongest reducing agent is the species most easily oxidized (the largest negative reduction potential). Zn, $E^o_{red} = -0.76$ V, is the strongest reducing agent and F^-, $E^o_{red} = 2.87$ V, is the weakest.

20.49 *Analyze/Plan.* To reduce Eu^{3+} to Eu^{2+}, we need an oxidizing agent, one of the reduced species from Appendix E. It must have a greater tendency to be oxidized than Eu^{3+} has to be reduced. That is, E^o_{red} must be more negative than –0.43 V. *Solve.*

Any of the **reduced** species in Appendix E from a half-reaction with a reduction potential more negative than –0.43 V will reduce Eu^{3+} to Eu^{2+}. From the list of possible reductants in the exercise, Al and $H_2C_2O_4$ will reduce Eu^{3+} to Eu^{2+}.

20.50 Any oxidized species from Appendix E with a reduction potential greater than 0.59 V will oxidize RuO_4^{2-} to RuO_4^-. From the list of possible oxidants in the exercise, $Br_2(l)$ and $BrO_3^-(aq)$ will definitely oxidize RuO_4^{2-} to RuO_4^-. $Sn^{2+}(aq)$ will not, and $O_2(g)$ depends on conditions. In base, it will not, but in strongly acidic solution, it will.

Free Energy and Redox Reactions (Section 20.5)

20.51 *Analyze/Plan.* In each reaction, $Fe^{2+} \rightarrow Fe^{3+}$ will be the oxidation half-reaction and one of the other given half-reactions will be the reduction half-reaction. Follow the logic in Sample Exercise 20.10 to calculate $E°$, $\Delta G°$ and K for each reaction. *Solve.*

(a) $2\,Fe^{2+}(aq) + S_2O_6^{2-}(aq) + 4\,H^+(aq) \rightarrow 2\,Fe^{3+}(aq) + 2\,H_2SO_3(aq)$

$E° = 0.60\,V - 0.77\,V = -0.17\,V$

$2\,Fe^{2+}(aq) + N_2O(aq) + 2\,H^+(aq) \rightarrow 2\,Fe^{3+}(aq) + N_2(g) + H_2O(l)$

$E° = -1.77\,V - 0.77\,V = -2.54\,V$

$Fe^{2+}(aq) + VO_2^+(aq) + 2\,H^+(aq) \rightarrow Fe^{3+}(aq) + VO^{2+}(aq) + H_2O(l)$

$E° = 1.00\,V - 0.77\,V = +0.23\,V$

(b) $\Delta G° = -nFE°$ For the first reaction,

$\Delta G° = -2\,mol \times \dfrac{96{,}485\,J}{1\,V\text{-mol}} \times (-0.17\,V) = 3.280 \times 10^4 = 3.3 \times 10^4\,J$ or 33 kJ

For the second reaction, $\Delta G° = -2(96{,}485)(-2.54) = 4.901 \times 10^5 = 4.90 \times 10^2\,kJ$

For the third reaction, $\Delta G° = -1(96{,}485)(0.23) = -2.22 \times 10^4\,J = -22\,kJ$

(c) $\Delta G° = -RT \ln K$; $\ln K = -\Delta G°/RT$

For the first reaction,

$\ln K = \dfrac{-3.281 \times 10^4\,J}{(8.314\,J/mol\text{-}K)(298\,K)} = -13.243 = -13$; $K = e^{-13.2428} = 1.78 \times 10^{-6} = 2 \times 10^{-6}$

[Convert ln to log; the number of decimal places in the log is the number of sig figs in the result.]

For the second reaction,

$\ln K = \dfrac{-4.902 \times 10^5\,J}{8.314\,J/mol\text{-}K \times 298\,K} = -197.86 = -198$; $K = e^{-198} = 1.23 \times 10^{-86} = 10^{-86}$

For the third reaction,

$\ln K = \dfrac{-(-2.22 \times 10^4\,J)}{8.314\,J/mol\text{-}K \times 298\,K} = 8.958 = 9.0$; $K = e^{9.0} = 7.77 \times 10^3 = 8 \times 10^3$

Check. The equilibrium constants calculated here are indicators of equilibrium position, but are not particularly precise numerical values.

20.52 (a)

$$2\,I^-(aq) \rightarrow I_2(s) + 2\,e^- \qquad E^o_{red} = 0.54\,V$$

$$\dfrac{Hg_2^{2+}(aq) + 2\,e^- \rightarrow 2\,Hg(l) \qquad E^o_{red} = 0.79\,V}{2\,I^-(aq) + Hg_2^{2+}(aq) \rightarrow I_2(s) + 2\,Hg(l) \quad E° = 0.79 - 0.54 = 0.25\,V}$$

$\Delta G° = -nFE° = -2\,mol\,e^- \times \dfrac{96.5\,kJ}{V\text{-mol}\,e^-} \times 0.25\,V = -48.829 = -49\,kJ$

$\ln K = \dfrac{-(-4.8829 \times 10^4\,J)}{(8.314\,J/mol\text{-}K)(298\,K)} = 19.708 = 20$; $K = e^{19.708} = 3.62 \times 10^8 = 10^8$

(b)
$$3\,[Cu^+(aq) \rightarrow Cu^{2+}(aq) + 1\ e^-] \qquad E^o_{red} = 0.15\ V$$
$$\underline{NO_3^-(aq) + 4\ H^+(aq) + 3\ e^- \rightarrow NO(g) + H_2O(l) \qquad E^o_{red} = 0.96\ V}$$
$$3\ Cu^+(aq) + NO_3^-(aq) + 4\ H^+(aq) \rightarrow 3\ Cu^{2+}(aq) + NO(g) + 2\ H_2O(l)$$

$$E^o = 0.96 - 0.15 = 0.81\ V;\ \Delta G^o = -3(96.5)(0.81) = -2.345 \times 10^2\ kJ = -2.3 \times 10^5\ J$$

$$\ln K = \frac{-(-2.345 \times 10^5\ J)}{(8.314\ J/mol\text{-}K)(298\ K)} = 94.65 = 95;\quad K = e^{95} = 1.3 \times 10^{41} = 10^{41}$$

(c)
$$2\,[Cr(OH)_3(s) + 5\ OH^-(aq) \rightarrow CrO_4^{2-}(aq) + 4\ H_2O(l) + 3\ e^-] \qquad E^o_{red} = -0.13\ V$$
$$\underline{3\,[ClO^-(aq) + H_2O(l) + 2\ e^- \rightarrow Cl^-(aq) + 2\ OH^-(aq)] \qquad E^o_{red} = 0.89\ V}$$
$$2\ Cr(OH)_3(s) + 3\ ClO^-(aq) + 4\ OH^-(aq) \rightarrow 2\ CrO_4^{2-}(aq) + 3\ Cl^-(aq) + 5\ H_2O(l)$$

$$E^o = 0.89 - (-0.13) = 1.02\ V;\ \Delta G^o = -6(96.5)(1.02) = -590.58\ kJ = -5.91 \times 10^5\ J$$

$$\ln K = \frac{-(-5.9058 \times 10^5\ J)}{(8.314\ J/mol\text{-}K)(298\ K)} = 238.37 = 238;\quad K = 3.3 \times 10^{103} = 10^{103}$$

This is an unimaginably large number.

20.53   *Analyze/Plan.* Given K, calculate ΔG^o and E^o. Reverse the logic in Sample Exercise 20.10. According to Equation 19.20, $\Delta G^o = -RT \ln K$. According to Equation 20.12, $\Delta G^o = -nFE^o$, $E^o = -\Delta G^o/nF$.   *Solve.*

$$K = 1.5 \times 10^{-4}$$

$$\Delta G^o = -RT \ln K = -(8.314\ J/mol\text{-}K)(298)\ln(1.5 \times 10^{-4}) = 2.181 \times 10^4\ J = 21.8\ kJ$$

$$E^o = -\Delta G^o/nF;\ n = 2;\ F = 96.5\ kJ/mol\ e^-$$

$$E^o = \frac{-21.81\ kJ}{2\ mol\ e^- \times 96.5\ kJ/V\text{-}mol\ e^-} = -0.113\ V$$

Check. The unit of ΔG^o is actually kJ/mol, which means kJ per "mole of reaction," or for the reaction as written. Because we do not have a specific reaction, we interpret the unit as referring to the overall reaction.

20.54 $K = 8.7 \times 10^4;\ \Delta G^o = -RT \ln K;\ E^o = -\Delta G^o/nF;\ n = 1;\ T = 298\ K$

$$\Delta G^o = -8.314\ J/mol\text{-}K \times 298\ K \times \ln(8.7 \times 10^4) = -2.818 \times 10^4\ J = -28.2\ kJ$$

$$E^o = -\Delta G^o/nF = \frac{-(-28.18\ kJ)}{1e^- \times 96.5\ kJ/V\text{-}mol\ e^-} = 0.292\ V$$

20.55   *Analyze/Plan.* Given E^o_{red} values for half reactions, calculate the value of K for a given redox reaction. Follow the logic in Sample Exercise 20.10. For each reaction, calculate E^o from E^o_{red}, then use Equation 20.13 to calculate K.

Solve. $E^o = \dfrac{RT}{nF} \ln K;\ \ln K = \dfrac{nF}{RT}E^o;\ \dfrac{F}{RT} = \dfrac{96{,}485\ J/V\text{-}mol}{8.314\ J/K\text{-}mol \times 298.15\ K} = 38.924/V$

(a) $E^o = -0.28 - (-0.44) = 0.16\ V,\ n = 2\ (Ni^{2+} + 2e^- \rightarrow Ni)$

$$\ln K = 2(38.924/V)(0.16\ V) = 12.456 = 12;\ K = 2.57 \times 10^5 = 3 \times 10^5$$

(b) $E^o = 0 - (-0.28) = 0.28\ V;\ n = 2\ (2H^+ + 2e^- \rightarrow H_2)$

$$\ln K = 2(38.924/V)(0.28\ V) = 21.797 = 22;\ K = 2.93 \times 10^9 = 3 \times 10^9$$

(c) $E° = 1.51 - 1.07 = 0.44$ V; $n = 10$ ($2 MnO_4^- + 10 e^- \rightarrow 2 Mn^{+2}$)

      $\ln K = 10(38.924/V)(0.44 V) = 171.265 = 170$; $K = 2.40 \times 10^{74} = 10^{74}$

Check. Note that small differences in E° values lead to large changes in the magnitude of K. Sig fig rules limit the precision of K values.

20.56   At 298 K, $\ln K = n(38.924/V)E°$. See Solution 20.55 for a more complete development.

(a) $E° = 0.80 V - 0.34 V = 0.46$ V; $n = 2$ ($2Ag^+ + 2e^- \rightarrow 2Ag$)

      $\ln K = 2(38.924/V)0.46 V = 35.810 = 35$; $K = 3.57 \times 10^{15} = 4 \times 10^{15}$

(b) $E° = 1.61 V - 0.32 V = 1.29$ V; $n = 3$ ($3 Ce^{4+} + 3e^- \rightarrow 3 Ce^{3+}$)

      $\ln K = 3(38.924/V)1.29 V = 150.635 = 150$; $K = 2.63 \times 10^{65} = 10^{65}$

(c) $E° = 0.36 V - (-0.23 V) = 0.59$ V; $n = 4$ ($4 Fe(CN)_6^{3-} + 4e^- \rightarrow 4 Fe(CN)_6^{4-}$)

      $\ln K = 4(38.924/V)0.59 V = 91.860 = 92$; $K = 7.84 \times 10^{39} = 10^{40}$

20.57   *Analyze/Plan.* At 298 K, $\ln K = n(38.924/V)E°$. See Solution 20.55 for a more complete development. *Solve.*

(a) $\ln K = 1(38.924/V)0.177 V = 6.890 = 6.89$; $K = 982 = 9.8 \times 10^2$

(b) $\ln K = 2(38.924/V)0.177 V = 13.779 = 13.8$; $K = 9.64 \times 10^5 = 1 \times 10^6$

(c) $\ln K = 3(38.924/V)0.177 V = 20.669 = 20.7$; $K = 9.47 \times 10^8 = 1 \times 10^9$

20.58   At 298 K, $\ln K = n(38.924/V)E°$. See Solution 20.55 for a more complete development.

$$\ln K = n (38.924/V) E°; \quad n = \frac{\ln K}{(38.924/V)E°}; \quad n = \frac{\ln 5.5 \times 10^5}{(38.924/V) 0.17 V} = 1.9975 = 2$$

20.59   *Analyze/Plan.* Given a spontaneous chemical reaction, calculate the maximum possible work for a given amount of reactant at standard conditions. Separate the equation into half-reactions and calculate cell emf. Use Equation 20.14, $w_{max} = -nFE$, to calculate maximum work. At standard conditions, $E = E°$. *Solve.*

$$I_2(s) + 2e^- \rightarrow 2 I^-(aq) \qquad\qquad E°_{red} = 0.54 V$$
$$\underline{Sn(s) \rightarrow Sn^{2+}(aq) + 2 e^- \qquad\qquad E°_{red} = -0.14 V}$$
$$I_2(s) + Sn(s) \rightarrow 2 I^-(aq) + Sn^{2+}(aq) \qquad E° = 0.54 V - (-0.14 V) = 0.68 V$$

$w_{max} = -2(96.5)(0.68) = -131.24 = 1.3 \times 10^2$ kJ/mol Sn

$$\frac{-131.24 \text{ kJ}}{\text{mol Sn}(s)} \times \frac{1 \text{ mol Sn}}{118.71 \text{ g Sn}} \times 75.0 \text{ g Sn} \times \frac{1000 \text{ J}}{\text{kJ}} = -8.3 \times 10^4 \text{ J}$$

Check. The (−) sign indicates that work is done by the cell.

20.60   For this cell at standard conditions, $E° = 1.10$ V.

$w_{max} = \Delta G° = -nFE° = -2(96.5)(1.10) = -212.3 = -212$ kJ/mol Cu

$$50.0 \text{ g Cu} \times \frac{1 \text{ mol Cu}}{63.55 \text{ g Cu}} \times \frac{-212.3 \text{ kJ}}{\text{mol Cu}} = -167 \text{ kJ} = -1.67 \times 10^5 \text{ J}$$

Cell EMF under Nonstandard Conditions (Section 20.6)

20.61 (a) In the Nernst equation, $Q = 1$ if all reactants and products are at standard conditions.

(b) Yes. The Nernst equation is applicable to cell emf at nonstandard conditions, so it must be applicable at temperatures other than 298 K. There are two terms in the Nernst equation. First, values of $E°$ at temperatures other than 298 K are required. Then, in the form of Equation 20.16, there is a variable for T in the second term. In the short-hand form of Equation 20.18, the value 0.0592 assumes 298 K. A different coefficient would apply to cells at temperatures other than 298 K.

20.62 Decrease. As the spontaneous chemical reaction of the voltaic cell proceeds, the concentrations of products increase and the concentrations of reactants decrease.

20.63 *Analyze/Plan.* Given a circumstance, determine its effect on cell emf. Each circumstance changes the value of Q. An increase in Q reduces emf; a decrease in Q increases emf. *Solve.*

$$Zn(s) + 2H^+(aq) \rightarrow Zn^{2+}(aq) + H_2(g); \quad E = E° - \frac{0.0592}{n} \log Q; \quad Q = \frac{[Zn^{2+}] \, P_{H_2}}{[H^+]^2}$$

(a) P_{H_2} increases, Q increases, E decreases

(b) $[Zn^{2+}]$ increases, Q increases, E decreases

(c) $[H^+]$ decreases, Q increases, E decreases

(d) No effect; does not appear in the Nernst equation

20.64 $$Al(s) + 3Ag^+(aq) \rightarrow Al^{3+}(aq) + 3Ag(s); \quad E = E° - \frac{0.0592}{n} \log Q; \quad Q = \frac{[Al^{3+}]}{[Ag^+]^3}$$

Any change that causes the reaction to be less spontaneous (that causes Q to increase and ultimately shifts the equilibrium to the left) will result in a less positive value for E.

(a) Increases E by decreasing $[Al^{3+}]$ on the right side of the equation, which decreases Q.

(b) No effect; the "concentrations" of pure solids and liquids do not influence the value of K for a heterogeneous equilibrium.

(c) No effect; the concentration of Ag^+ and the value of Q are unchanged.

(d) Decreases E; forming AgCl(s) decreases the concentration of Ag^+, which increases Q.

20.65 *Analyze/Plan.* Follow the logic in Sample Exercise 20.11. *Solve.*

(a)
$$Ni^{2+}(aq) + 2\,e^- \rightarrow Ni(s) \qquad E°_{red} = -0.28 \text{ V}$$
$$\underline{Zn(s) \rightarrow Zn^{2+}(aq) + 2\,e^- \qquad E°_{red} = -0.76 \text{ V}}$$
$$Ni^{2+}(aq) + Zn(s) \rightarrow Ni(s) + Zn^{2+}(aq) \qquad E° = -0.28 - (-0.76) = 0.48 \text{ V}$$

(b) $$E = E° - \frac{0.0592}{n} \log \frac{[Zn^{2+}]}{[Ni^{2+}]}; \quad n = 2$$

$$E = 0.48 - \frac{0.0592}{2} \log \frac{(0.100)}{(3.00)} = 0.48 - \frac{0.0592}{2} \log (0.0333)$$

$$E = 0.48 - \frac{0.0592(-1.477)}{2} = 0.48 + 0.0437 = 0.5237 = 0.52 \text{ V}$$

(c) $$E = 0.48 - \frac{0.0592}{2} \log \frac{(0.900)}{(0.200)} = 0.48 - 0.0193 = 0.4607 = 0.46 \text{ V}$$

20.66 (a)

$$3\,[Ce^{4+}(aq)+1\,e^{-}\rightarrow Ce^{3+}(aq)] \qquad E^{o}_{red}=1.61\ V$$

$$\underline{Cr(s)\rightarrow Cr^{3+}(aq)+3\,e^{-} \qquad E^{o}_{red}=-0.74\ V}$$

$$3\,Ce^{4+}(aq)+Cr(s)\rightarrow 3\,Ce^{3+}(aq)+Cr^{3+}(aq) \qquad E^{o}=1.61-(-0.74)=2.35\ V$$

(b) $E = E^{o} - \dfrac{0.0592}{n}\,\log \dfrac{[Ce^{3+}]^3\,[Cr^{3+}]}{[Ce^{4+}]^3};\quad n=3$

$$E = 2.35 - \frac{0.0592}{3}\,\log \frac{(0.10)^3\,(0.010)}{(3.0)^3} = 2.35 - \frac{0.0592}{3}\,\log\,(3.704\times10^{-7})$$

$$E = 2.35 - \frac{0.0592\,(-6.431)}{3} = 2.35 + 0.127 = 2.48\ V$$

(c) $E = 2.35 - \dfrac{0.0592}{3}\,\log\,\dfrac{(2.0)^3\,(1.5)}{(0.010)^3} = 2.35 - 0.1397 = 2.21\ V$

20.67 *Analyze/Plan.* Follow the logic in Sample Exercise 20.11. *Solve.*

(a)

$$4\,[Fe^{2+}(aq)\rightarrow Fe^{3+}(aq)+1\,e^{-}] \qquad E^{o}_{red}=0.77\ V$$

$$\underline{O_2(g)+4\,H^+(aq)+4\,e^{-}\rightarrow 2\,H_2O(l) \qquad E^{o}_{red}=1.23\ V}$$

$$4\,Fe^{2+}(aq)+O_2(g)+4\,H^+(aq)\rightarrow 4\,Fe^{3+}(aq)+2\,H_2O(l) \qquad E^{o}=1.23-0.77=0.46\ V$$

(b) $E = E^{o} - \dfrac{0.0592}{n}\,\log \dfrac{[Fe^{3+}]^4}{[Fe^{2+}]^4[H^+]^4 P_{O_2}};\quad n=4,\ [H^+]=10^{-3.50}=3.2\times10^{-4}\ M$

$$E = 0.46\ V - \frac{0.0592}{4}\,\log \frac{(0.010)^4}{(1.3)^4\,(3.2\times10^{-4})^4\,(0.50)} = 0.46 - \frac{0.0592}{4}\,\log\,(7.0\times10^{5})$$

$$E = 0.46 - \frac{0.0592}{4}\,(5.845) = 0.46 - 0.0865 = 0.3735 = 0.37\ V$$

20.68 (a)

$$2\,[Fe^{3+}(aq)+1\,e^{-}\rightarrow Fe^{2+}(aq)] \qquad E^{o}_{red}=0.77\ V$$

$$\underline{H_2(g)\rightarrow 2H^+(aq)+2\,e^{-} \qquad E^{o}_{red}=0.00\ V}$$

$$2\,Fe^{3+}(aq)+H_2(g)\rightarrow 2\,Fe^{2+}(aq)+2\,H^+(aq) \qquad E^{o}=0.77-0.00=0.77\ V$$

(b) $E = E^{o} - \dfrac{0.0592}{n}\,\log \dfrac{[Fe^{2+}]^2[H^+]^2}{[Fe^{3+}]^2 P_{H_2}};\quad [H^+]=10^{-pH}=1.0\times10^{-4},\ n=2$

$$E = 0.77 - \frac{0.0592}{2}\,\log \frac{(0.0010)^2\,(1.0\times10^{-4})^2}{(3.50)^2\,(0.95)} = 0.77 - \frac{0.0592}{2}\,\log\,(8.6\times10^{-16})$$

$$E = 0.77 - \frac{0.0592(-15.066)}{2} = 0.77 + 0.446 = 1.216 = 1.22\ V$$

20.69 *Analyze/Plan.* We are given a concentration cell with Zn electrodes. Use the definition of a concentration cell in Section 20.6 to answer the stated questions. Use Equation 20.18 to calculate the cell emf. For a concentration cell, Q = [dilute]/[concentrated]. *Solve.*

(a) The compartment with the more dilute solution will be the anode. That is, the compartment with $[Zn^{2+}] = 1.00\times10^{-2}\ M$ is the anode.

(b) Because the oxidation half-reaction is the opposite of the reduction half-reaction, E° is zero.

(c) $E = E° - \dfrac{0.0592}{n} \log Q; \; Q = [Zn^{2+}, \text{dilute}]/[Zn^{2+}, \text{conc.}]$

        $E = 0 - \dfrac{0.0592}{2} \log \dfrac{(1.00 \times 10^{-2})}{(1.8)} = 0.0668 \; V$

(d)   In the anode compartment, $Zn(s) \rightarrow Zn^{2+}(aq)$, so $[Zn^{2+}]$ increases from $1.00 \times 10^{-2} \; M$. In the cathode compartment, $Zn^{2+}(aq) \rightarrow Zn(s)$, so $[Zn^{2+}]$ decreases from $1.8 \; M$.

20.70   (a)   The compartment with $0.0150 \; M \; Cl^-(aq)$ is the cathode.

         (b) $E° = 0 \; V$

         (c) $E = E° - \dfrac{0.0592}{n} \log Q; \; Q = [Cl^-, \text{dilute}]/[Cl^-, \text{conc.}]$

              $E = 0 - \dfrac{0.0592}{1} \log \dfrac{(0.0150)}{(2.55)} = -0.13204 = -0.1320 \; V$

         (d)   In the anode compartment, $[Cl^-]$ will decrease from $2.55 \; M$. In the cathode, $[Cl^-]$ will increase from $0.0150 \; M$.

20.71   *Analyze/Plan.* Follow the logic in Sample Exercise 20.12.   *Solve.*

$E = E° - \dfrac{0.0592}{2} \log \dfrac{[P_{H_2}][Zn^{2+}]}{[H^+]^2}; \; E° = 0.0 \; V - (-0.76 \; V) = 0.76 \; V$

$0.684 = 0.76 - \dfrac{0.0592}{2} \times (\log [P_{H_2}][Zn^{2+}] - 2 \log [H^+])$

$\quad\quad = 0.76 - \dfrac{0.0592}{2} \times (-0.5686 - 2 \log [H^+])$

$0.684 = 0.76 + 0.0168 + 0.0592 \log [H^+]; \; \log [H^+] = \dfrac{0.684 - 0.0168 - 0.76}{0.0592}$

$\log [H^+] = -1.5676 = -1.6; \; [H^+] = 0.0271 = 0.03 \; M; \; pH = 1.6$

20.72   (a) $E° = -0.14 \; V - (-0.13 \; V) = -0.01 \; V; \; n = 2$

              $0.22 = -0.01 - \dfrac{0.0592}{2} \log \dfrac{[Pb^{2+}]}{[Sn^{2+}]} = -0.01 - \dfrac{0.0592}{2} \log \dfrac{[Pb^{2+}]}{1.00}$

              $\log [Pb^{2+}] = \dfrac{-0.23 \, (2)}{0.0592} = -7.770 = -7.8; \; [Pb^{2+}] = 1.7 \times 10^{-8} = 2 \times 10^{-8} \; M$

         (b)   For $PbSO_4(s)$, $K_{sp} = [Pb^{2+}][SO_4^{2-}] = (1.0)(1.7 \times 10^{-8}) = 1.7 \times 10^{-8}$

Batteries and Fuel Cells (Section 20.7)

20.73   *Analyze/Plan.* Given mass of a reactant (Pb), calculate mass of product (PbO_2), and coulombs of charge transferred. This is a stoichiometry problem; we need the balanced equation for the chemical reaction that occurs in the lead-acid battery.

The overall cell reaction is:

$Pb(s) + PbO_2(s) + 2 H^+(aq) + 2 HSO_4^-(aq) \rightarrow 2 PbSO_4(s) + 2 H_2O(l)$ *Solve.*

         (a) $g \; Pb \rightarrow mol \; Pb \rightarrow mol \; PbO_2 \rightarrow g \; PbO_2$

              $402 \; g \; Pb \times \dfrac{1 \; mol \; Pb}{207.2 \; g \; Pb} \times \dfrac{1 \; mol \; PbO_2}{1 \; mol \; Pb} \times \dfrac{239.2 \; g \; PbO_2}{1 \; mol \; PbO_2} = 464 \; g \; PbO_2$

(b)   From the half-reactions for the lead-acid battery, 2 mol electrons are transferred for each mol of Pb reacted. From Section 20.5, 96,485 C/mol e$^-$.

$$402 \text{ g Pb} \times \frac{1 \text{ mol Pb}}{207.2 \text{ g Pb}} \times \frac{2 \text{ mol e}^-}{1 \text{ mol Pb}} \times \frac{96,485 \text{ C}}{1 \text{ mol e}^-} = 374,392 = 3.74 \times 10^5 \text{ C}$$

20.74   (a)   The overall cell reaction is:

$$2 \text{ MnO}_2(s) + \text{Zn}(s) + 2 \text{ H}_2\text{O}(l) \rightarrow 2 \text{ MnO(OH)}(s) + \text{Zn(OH)}_2(s)$$

$$4.50 \text{ g Zn} \times \frac{1 \text{ mol Zn}}{65.39 \text{ g Zn}} \times \frac{2 \text{ mol MnO}_2}{1 \text{ mol Zn}} \times \frac{86.94 \text{ g MnO}_2}{1 \text{ mol MnO}_2} = 12.0 \text{ g MnO}_2$$

(b)   Two mol e$^-$ are transferred for every mol of Zn reacted. 96,485 C/mol e$^-$

$$4.50 \text{ g Zn} \times \frac{1 \text{ mol Zn}}{65.39 \text{ g Zn}} \times \frac{2 \text{ mol e}^-}{1 \text{ mol Zn}} \times \frac{96,485 \text{ C}}{1 \text{ mol e}^-} = 13,280 = 1.32 \times 10^4 \text{ C}$$

20.75   *Analyze/Plan.* We are given a redox reaction and asked to write half-reactions, calculate E°, and indicate whether Li(s) is the anode or cathode. Determine which reactant is oxidized and which is reduced. Separate into half-reactions, find E_{red}^o for the half-reactions from Appendix E and calculate E°.   *Solve.*

(a)   Li(s) is oxidized at the anode.

(b)

$$\text{Ag}_2\text{CrO}_4(s) + 2 \text{ e}^- \rightarrow 2 \text{ Ag}(s) + \text{CrO}_4^{2-}(aq) \qquad E_{red}^o = 0.45 \text{ V}$$

$$\underline{2 \left[\text{Li}(s) \rightarrow \text{Li}^+(aq) + 1 \text{ e}^- \right] \qquad\qquad\qquad E_{red}^o = -3.05 \text{ V}}$$

$$\text{Ag}_2\text{CrO}_4(s) + 2 \text{ Li}(s) \rightarrow 2 \text{ Ag}(s) + \text{CrO}_4^-(aq) + 2 \text{ Li}^+(aq)$$

E° = 0.45 V – (–3.05 V) = 3.50 V

(c)   The emf of the battery, 3.5 V, is exactly the standard cell potential calculated in part (b).

(d)   For this battery at ambient conditions, E ≈ E°, so log Q ≈ 0. This makes sense because all reactants and products in the battery are solids and thus present in their standard states. Assuming that E° is relatively constant with temperature, the value of the second term in the Nernst equation is ≈ 0 at 37 °C, and E ≈ 3.5 V.

20.76   (a)   HgO(s) + Zn(s) → Hg(l) + ZnO(s)

(b) $E_{cell}^o = E_{red}^o \text{ (cathode)} - E_{red}^o \text{ (anode)}$

$E_{red}^o \text{ (anode)} = E_{red}^o - E_{cell}^o = 0.098 - 1.35 = -1.25 \text{ V}$

(c) E_{red}^o is different from Zn^{2+} (aq) + 2e$^-$ → Zn(s) (–0.76 V) because in the battery the process happens in the presence of base and Zn^{2+} is stabilized as ZnO(s). Stabilization of a reactant in a half-reaction decreases the driving force, so E_{red}^o is more negative.

20.77   *Analyze/Plan.* (a) Consider the function of Zn in an alkaline battery. What effect would it have on the redox reaction and cell emf if Cd replaces Zn? (b) Both batteries contain Ni. What is the difference in environmental impact between Cd and the metal hydride? *Solve.*

(a) E_{red}^o for Cd (–0.40 V) is less negative than E_{red}^o for Zn (–0.76 V), so E_{cell} will have a smaller (less positive) value.

(b)   NiMH batteries use an alloy such as ZrNi$_2$ as the anode material. This eliminates the use and concomitant disposal problems associated with Cd, a toxic heavy metal.

20.78 (a) $NiO(OH)(s) + H_2O(l) + 1\,e^- \rightarrow Ni(OH)_2(s) + OH^-(aq)$

(b) $Zn(s) + 2\,OH^-(aq) \rightarrow Zn(OH)_2(s) + 2\,e^-$

(c) A nickel-zinc battery will produce about 0.94 V. E^o_{red} for Cd (–0.40 V) is less negative than E^o_{red} for Zn (–0.76 V), so E_{cell} for the nickel-zinc battery will be less positive than that for the nickel-cadmium battery by 0.36 V. That is, the potential of the nickel-zinc battery will be approximately (1.30 V – 0.36 V) = 0.94 V.

(d) Higher. Specific energy density is the amount of energy stored per unit mass of the battery. The potential of the nickel-zinc battery is about 72 % of the nickel-cadmium battery (0.94 V/1.30 V); the molar mass of zinc is about 58 % of the molar mass of cadmium (65.38 g/112.4 g). The reduction in mass by exchanging zinc for cadmium more than compensates for the reduction in cell potential.

20.79 (a) The oxidation states of the elements in $LiCoO_2$ are +1, +3 and –2, respectively. At any point in the operation of the battery, the positive and negative charges must balance. On charging, the oxidation state of some of the cobalt ions increases to +4, which releases some of the Li^+ ions to migrate back to the anode. If approximately half of the Li^+ ions migrate, the same number of Co^{3+} ions are oxidized to the +4 state. When the battery is fully charged, the mole ratios of the elements in the cathode are: 0.5 mol Li^+ to 0.5 mol Co^{3+} to 0.5 mol Co^{4+} to 2 mol O^{2-}. The material in the cathode is $LiCo_2O_4$.

(b) The molar mass of $LiCoO_2$ is 97.87 g/mol.

$$10\text{ g }LiCoO_2 \times \frac{1\text{ mol }LiCoO_2}{97.87\text{ g }LiCoO_2} \times \frac{1\text{ mol }Li}{1\text{ mol }LiCoO_2} = 0.1022 = 0.10\text{ mol }Li^+$$

Half of the Li^+, 0.051 mol, migrates during a full charge.

$$0.051\text{ mol }Li^+ \times \frac{1\text{ mol }e^-}{1\text{ mol }Li^+} \times \frac{96,485\text{ C}}{1\text{ mol }e^-} = 4,921 = 4.9 \times 10^3\text{ C}$$

20.80 (a) $LiMn_2O_4$: FW = 6.941 + 2(54.938) + 4(15.9994) = 180.815 amu

$$\%\text{ Li} = \frac{6.941\text{ g}}{180.815\text{ g}} \times 100 = 3.8387 = 3.839\text{ \%}$$

$LiCoO_2$: FW = 6.941 + 58.9332 + 2(15.9994) = 97.873 amu

$$\%\text{ Li} = \frac{6.941\text{ g}}{97.873\text{ g}} \times 100 = 7.0918 = 7.092\text{ \%}$$

(b) $LiCoO_2$ has almost twice as great a mass percent Li as $LiMn_2O_4$. This definitely explains why $LiMn_2O_4$ cathodes deliver less power on discharging. A Li ion battery produces power by the migration of Li^+, but not all the Li^+ in the cathode material migrates during a full charge. The greater percentage of Li in the cathode material, the more Li^+ that migrates and the greater the power that can be produced upon discharge.

(c) Assume 100.000 g of cathode material. A $LiCoO_2$ cathode has 7.092 g of Li; if half of the Li migrates during charging, 3.546 g Li migrates. A $LiMn_2O_4$ cathode contains 3.839 g of Li. To deliver the same amount of Li to the graphite anode, $\frac{3.546\text{ g}}{3.839\text{ g}} \times 100 = 92.368 = 92\text{ \%}$ of the Li in the $LiMn_2O_4$ cathode would have to migrate.

20.81   *Analyze/Plan.* Write the balanced equation for the spontaneous reaction in a hydrogen fuel cell. Separate the overall reaction into half-reactions in acid solution and basic solution, find the matching standard reduction potentials in Appendix E and calculate standard voltages.   *Solve.*

(a)   The spontaneous reaction in the hydrogen fuel cell is hydrogen gas plus oxygen gas makes water.

(b)
$$O_2(g) + 4\,H^+(aq) + 4\,e^- \rightarrow 2\,H_2O(l) \qquad E^o_{red} = 1.23\ V$$
$$\underline{\quad\quad 2\,H_2(g) \rightarrow 4\,H^+(aq) + 4\,e^- \quad E^o_{red} = 0.00\ V \quad}$$
$$O_2(g) + 2\,H_2(g) \rightarrow 2\,H_2O(l) \qquad E^o = 1.23 - 0.00 = 1.23\ V$$

20.82   (a)   Both batteries and fuel cells are electrochemical power sources. Both take advantage of spontaneous oxidation–reduction reactions to produce a certain voltage. The difference is that batteries are self-contained (all reactants and products are present inside the battery casing), whereas fuel cells require continuous supply of reactants and exhaust of products.

(b)   No. The fuel in a fuel cell must be fluid, either gas or liquid. Because fuel must be continuously supplied to the fuel cell, it must be capable of flow; the fuel cannot be solid.

Corrosion (Section 20.8)

20.83   *Analyze/Plan.* (a) Decide which reactant is oxidized and which is reduced. Write the balanced half-reactions and assign the appropriate one as anode and cathode. (b) Write the balanced half-reaction for $Fe^{2+}(aq) \rightarrow Fe_2O_3 \cdot 3H_2O$. Use the reduction half-reaction from part (a) to obtain the overall reaction.   *Solve.*

(a)   anode: $Fe(s) \rightarrow Fe^{2+}(aq) + 2\,e^-$

cathode: $O_2(g) + 4\,H^+(aq) + 4\,e^- \rightarrow 2\,H_2O(l)$

(b) $2\,Fe^{2+}(aq) + 6\,H_2O(l) \rightarrow Fe_2O_3 \cdot 3H_2O(s) + 6\,H^+(aq) + 2\,e^-$

$O_2(g) + 4\,H^+(aq) + 4\,e^- \rightarrow 2\,H_2O(l)$

(Multiply the oxidation half-reaction by two to balance electrons and obtain the overall balanced reaction.)

20.84   (a)   Calculate E^o_{cell} for the given reactants at standard conditions.

$$O_2(g) + 4\,H^+(aq) + 4\,e^- \rightarrow 2\,H_2O(l) \qquad E^o_{red} = 1.23\ V$$
$$\underline{\quad\quad 2\,[Cu(s) \rightarrow Cu^{2+}(aq) + 2\,e^-] \quad E^o_{red} = 0.34\ V \quad}$$
$$2\,Cu(s) + O_2(g) + 4\,H^+(aq) \rightarrow 2\,Cu^{2+}(aq) + 2\,H_2O(l) \qquad E^o = 1.23 - 0.34 = 0.89\ V$$

At standard conditions with $O_2(g)$ and $H^+(aq)$ present, the oxidation of $Cu(s)$ has a positive E^o value and is spontaneous. $Cu(s)$ will oxidize (corrode) in air in the presence of acid.

(b) Fe^{2+} has a more negative reduction potential (–0.44 V) than Cu^{2+} (+0.34 V), so $Fe(s)$ is more readily oxidized than $Cu(s)$. If the two metals are in contact, $Fe(s)$ would act as a sacrificial anode and oxidize (corrode) in preference to $Cu(s)$; this would weaken the iron support skeleton of the statue. The teflon spacers prevent contact between the two metals and insure that the iron skeleton doesn't corrode when the $Cu(s)$ skin comes in contact with atmospheric $O_2(g)$ and $H^+(aq)$.

20.85   (a)   A "sacrificial anode" is a metal that is oxidized in preference to another when the two metals are coupled in an electrochemical cell; the sacrificial anode has a more negative E°_{red} than the other metal. In this case, Mg acts as a sacrificial anode because it is oxidized in preference to the pipe metal; it is sacrificed to preserve the pipe.

       (b) E°_{red} for Mg^{2+} is –2.37 V, more negative than most metals present in pipes, including Fe ($E^{\circ}_{red} = -0.44$ V) and Zn ($E^{\circ}_{red} = -0.76$ V).

20.86   No. To afford cathodic protection, a metal must be more difficult to reduce (have a more negative reduction potential) than Fe^{2+}. E°_{red} $Co^{2+} = -0.28$ V, E°_{red} $Fe^{2+} = -0.44$ V.

20.87   *Analyze/Plan.* Positive and negative ion charges must balance in neutral compounds. Use that fact along with oxidation numbers to answer the questions.   *Solve.*

       (a)   +3

       (b)   +8/3 or +2.67

       (c)   2 Fe(III) and 1 Fe(II)

20.88   (a)   +1

       (b)   +2

       (c) CuO_2. Peroxide ion is O_2^{2-}, which balances charge with Cu^{2+}.

       (d) Cu_2O_3.

Electrolysis; Electrical Work (Section 20.9)

20.89   (a)   *Electrolysis* is an electrochemical process driven by an outside energy source.

       (b)   Electrolysis reactions are, by definition, nonspontaneous.

       (c) $2\,Cl^-(l) \rightarrow Cl_2(g) + 2\,e^-$

       (d)   When an aqueous solution of NaCl undergoes electrolysis, sodium metal is not formed because H_2O is preferentially reduced to form $H_2(g)$.

20.90   (a)   An *electrolytic cell* is the vessel in which electrolysis occurs. It consists of a power source and two electrodes in a molten salt or aqueous solution.

       (b)   It is the cathode. In an electrolysis cell, as in a voltaic cell, electrons are consumed (via reduction) at the cathode. Electrons flow from the negative terminal of the voltage source and then to the cathode.

       (c)   A small amount of $H_2SO_4(aq)$ present during the electrolysis of water acts as a change carrier, or supporting electrolyte. This facilitates transfer of electrons through the solution and at the electrodes, speeding up the reaction. [Considering $H^+(aq)$ as the substance reduced at the cathode changes the details of the half-reactions, but not the overall E° for the electrolysis. $SO_4^{2-}(aq)$ cannot be oxidized.]

       (d)   If the active metal salt is present as an aqueous solution during electrolysis, water is reduced [to $H_2(g)$] rather than the metal ion being reduced to the metal. This is true for any active metal with an E°_{red} value more negative than –0.83 V.

20.91   *Analyze/Plan.* Follow the logic in Sample Exercise 20.14, paying close attention to units. Coulombs = amps-s; because this is a 3e⁻ reduction, each mole of Cr(s) requires 3 Faradays.   *Solve.*

(a) $7.60 \text{ A} \times 2.00 \text{ d} \times \dfrac{24 \text{ h}}{1 \text{ d}} \times \dfrac{60 \text{ min}}{1 \text{ h}} \times \dfrac{60 \text{ s}}{1 \text{ min}} \times \dfrac{1 \text{ C}}{1 \text{ amp-s}} \times \dfrac{1 \text{ F}}{96,485 \text{ C}}$

$\times \dfrac{1 \text{ mol Cr}}{3 \text{ F}} \times \dfrac{52.00 \text{ g Cr}}{1 \text{ mol Cr}} = 236 \text{ g Cr(s)}$

(b) $0.250 \text{ mol Cr} \times \dfrac{3 \text{ F}}{1 \text{ mol Cr}} \times \dfrac{96,485 \text{ C}}{\text{F}} \times \dfrac{1 \text{ amp-s}}{1 \text{ C}} \times \dfrac{1}{8.00 \text{ h}} \times \dfrac{1 \text{ h}}{60 \text{ min}} \times \dfrac{1 \text{ min}}{60 \text{ s}}$

$= 2.51 \text{ A}$

20.92   Coulombs = amps-s; because this is a 2e⁻ reduction, each mole of Mg(s) requires 2 Faradays.

(a) $4.55 \text{ A} \times 4.50 \text{ d} \times \dfrac{24 \text{ h}}{1 \text{ d}} \times \dfrac{60 \text{ min}}{1 \text{ h}} \times \dfrac{60 \text{ s}}{1 \text{ min}} \times \dfrac{1 \text{ C}}{1 \text{ amp-s}} \times \dfrac{1 \text{ F}}{96,485 \text{ C}}$

$\times \dfrac{1 \text{ mol Mg}}{2 \text{ F}} \times \dfrac{24.31 \text{ g Mg}}{1 \text{ mol Mg}} = 223 \text{ g Mg}$

(b) $25.00 \text{ g Mg} \times \dfrac{1 \text{ mol Mg}}{24.31 \text{ g Mg}} \times \dfrac{2 \text{ F}}{1 \text{ mol Mg}} \times \dfrac{96,485 \text{ C}}{\text{F}} \times \dfrac{1 \text{ amp-s}}{\text{C}} \times \dfrac{1 \text{ min}}{60 \text{ s}} \times \dfrac{1}{3.50 \text{ A}}$

$= 945 \text{ min}$

20.93   *Analyze/Plan.* Follow the logic in Sample Exercise 20.14, paying close attention to units. Take the 85% efficiency into account. Li⁺ is reduced at the anode; Cl⁻ is oxidized at the anode.   *Solve.*

(a)   If the cell is 85% efficient, $\dfrac{96,485 \text{ C}}{\text{F}} \times \dfrac{1 \text{ F}}{0.85 \text{ mol}} = 1.13512 \times 10^5$

$= 1.1 \times 10^5 \text{ C/mol Li is required.}$

$7.5 \times 10^4 \text{ A} \times 24 \text{ h} \times \dfrac{3600 \text{ s}}{1 \text{ h}} \times \dfrac{1 \text{ C}}{1 \text{ amp-s}} \times \dfrac{1 \text{ mol Li}}{1.13512 \times 10^5 \text{ C}} \times \dfrac{6.94 \text{ g Li}}{1 \text{ mol Li}}$

$= 3.962 \times 10^5 = 4.0 \times 10^5 \text{ g Li}$

(b) $E^{\circ}_{cell} = E^{\circ}_{red}(\text{cathode}) - E^{\circ}_{red}(\text{anode}) = -3.05 \text{ V} - (1.36 \text{ V}) = -4.41 \text{ V}$

The minimum voltage required to drive the reaction is the magnitude of E°_{cell}, 4.41 V.

20.94   (a) $7.5 \times 10^3 \text{ A} \times 48 \text{ h} \times \dfrac{3600 \text{ s}}{1 \text{ h}} \times \dfrac{1 \text{ C}}{1 \text{ amp-s}} \times \dfrac{1 \text{ F}}{96,485 \text{ C}} \times \dfrac{1 \text{ mol Ca}}{2 \text{ F}} \times 0.68 \times \dfrac{40.0 \text{ g Ca}}{1 \text{ mol Ca}}$

$= 1.830 \times 10^5 = 1.8 \times 10^5 \text{ g Ca}$

(b) $E^{\circ}_{cell} = E^{\circ}_{red}(\text{cathode}) - E^{\circ}_{red}(\text{anode}) = -2.87 \text{ V} - (1.36 \text{ V}) = -4.23 \text{ V}$

The minimum voltage required to drive the reaction is the magnitude of E°_{cell}, 4.23 V.

20.95   Refer to Table 4.5, "The Activity Series of the Metals." Gold is the least active metal on this table, less active than copper. This means that gold is more difficult to oxidize than copper (and that E_{red}^{o} for Au^{3+} is more positive than E_{red}^{o} for Cu^{2+}). When a mixture of copper and gold is refined by electrolysis, Cu is oxidized from the anode, but any metallic gold present in the mixture is not oxidized, so it accumulates near the anode.

20.96   The standard reduction potential for Te^{4+}, 0.57 V, is more positive than that of Cu^{2+}, 0.34 V. This means the Te^{4+} is "easier" to reduce than Cu^{2+}, but Te is harder to oxidize and less active than Cu. During electrorefining, although Cu is oxidized from the crude anode, Te will not be oxidized. It is likely to accumulate along with other impurities less active than Cu, in the so-called anode sludge.

Additional Exercises

20.97   (a)

$$Ni^{+}(aq) + 1\,e^{-} \rightarrow Ni(s)$$
$$\underline{Ni^{+}(aq) \quad\quad\quad \rightarrow Ni^{2+}(aq) + 1\,e^{-}}$$
$$2\,Ni^{+}(aq) \quad\quad\quad \rightarrow Ni(s) + Ni^{2+}(aq)$$

             (b)

$$MnO_4^{2-}(aq) + 4\,H^{+}(aq) + 2\,e^{-} \rightarrow MnO_2(s) + 2\,H_2O(l)$$
$$\underline{2\,[MnO_4^{2-}(aq) \quad \rightarrow MnO_4^{-}(aq) + 1\,e^{-}]}$$
$$3\,MnO_4^{2-}(aq) + 4\,H^{+}(aq) \rightarrow 2\,MnO_4^{-}(aq) + MnO_2(s) + 2\,H_2O(l)$$

             (c)

$$H_2SO_3(aq) + 4\,H^{+}(aq) + 4\,e^{-} \rightarrow S(s) + 3\,H_2O(l)$$
$$\underline{2\,[H_2SO_3(aq) + H_2O(l) \rightarrow HSO_4^{-}(aq) + 3\,H^{+}(aq) + 2\,e^{-}]}$$
$$3\,H_2SO_3(aq) \quad \rightarrow S(s) + 2\,HSO_4^{-}(aq) + 2\,H^{+}(aq) + H_2O(l)$$

             (d)

$$Cl_2(aq) + 2\,H_2O(l) \rightarrow 2\,ClO^{-}(aq) + 4\,H^{+}(aq) + 2\,e^{-}$$
$$\underline{4\,OH^{-}(aq) \quad\quad\quad\quad + 4\,OH^{-}(aq)}$$
$$Cl_2(aq) + 4\,OH^{-}(aq) \rightarrow 2\,ClO^{-}(aq) + 2\,H_2O(l) + 2\,e^{-}$$
$$\underline{Cl_2(aq) + 2\,e^{-} \rightarrow 2\,Cl^{-}(aq)}$$
$$\underline{1/2\,[2\,Cl_2(aq) + 4\,OH^{-}(aq) \rightarrow 2\,Cl^{-}(aq) + 2\,ClO^{-}(aq) + 2\,H_2O(l)]}$$
$$Cl_2(aq) + 2\,OH^{-}(aq) \rightarrow Cl^{-}(aq) + ClO^{-}(aq) + H_2O(l)$$

20.98   (a)

$$Fe(s) \rightarrow Fe^{2+}(aq) + 2\,e^{-}$$
$$\underline{2\,Ag^{+}(aq) + 2\,e^{-} \rightarrow 2\,Ag(s)}$$
$$Fe(s) + 2\,Ag^{+}(aq) \rightarrow Fe^{2+}(aq) + 2\,Ag(s)$$

$$E_{cell}^{o} = E_{red}^{o}(cathode) - E_{red}^{o}(anode) = 0.80\text{ V} - (-0.44\text{ V}) = 0.36\text{ V}$$

             (b)

$$Zn(s) \rightarrow Zn^{2+}(s) + 2\,e^{-}$$
$$\underline{2\,H^{+}(aq) + 2\,e^{-} \rightarrow H_2(g)}$$
$$Zn(s) + 2\,H^{+}(aq) \rightarrow Zn^{2+}(aq) + H_2(g)$$

$$E_{cell}^{o} = E_{red}^{o}(cathode) - E_{red}^{o}(anode) = 0.00\text{ V} - (-0.76\text{ V}) = 0.76\text{ V}$$

(c)   Cu | Cu^{2+} || ClO_3^-, Cl^- |Pt. Here, both the oxidized and reduced forms of the cathode solution are in the same phase, so we separate them by a comma and then indicate an inert electrode.

$E_{cell}^o = E_{red}^o$ (cathode) $- E_{red}^o$ (anode) $= 1.45$ V $- (0.34$ V$) = 1.11$ V

20.99   We need in each case to determine whether $E°$ is positive (spontaneous) or negative (nonspontaneous).

(a) $I_2(s) + 2$ e$^- \rightarrow 2$ I^-(aq) $E_{red}^o = 0.54$ V

   $\underline{\hspace{1cm} Sn(s) \rightarrow Sn^{2+}(aq) + 2$ e$^-$ $E_{red}^o = -0.14$ V $\hspace{1cm}}$

   $Sn(s) + I_2(s) \rightarrow Sn^{2+}(aq) + 2$ I^-(aq) $E° = 0.54 - (-0.14) = 0.68$ V, spontaneous

(b) $Ni^{2+}(aq) + 2$ e$^- \rightarrow Ni(s)$ $E_{red}^o = -0.28$ V

   $\underline{\hspace{1cm} 2$ I^-(aq) $\rightarrow I_2(s) + 2$ e$^-$ $E_{red}^o = 0.54$ V $\hspace{1cm}}$

   $Ni^{2+}(aq) + 2$ I^-(aq) $\rightarrow Ni(s) + I_2(s)$ $E° = -0.28 - 0.54 = -0.82$ V, nonspontaneous

(c)   2 $[Ce^{4+}(aq) + 1$ e$^- \rightarrow Ce^{3+}(aq)$ $E_{red}^o = 1.61$ V

   $\underline{\hspace{1cm} H_2O_2(aq) \rightarrow O_2(g) + 2$ H^+(aq) $+ 2$ e$^-$ $E_{red}^o = 0.68$ V $\hspace{1cm}}$

   2 $Ce^{4+}(aq) + H_2O_2(aq) \rightarrow 2$ $Ce^{3+}(aq) + O_2(g) + 2$ H^+(aq) $E° = 1.61 - 0.68 = 0.93$ V, spontaneous

(d) $Cu^{2+}(aq) + 2$ e$^- \rightarrow Cu(s)$ $E_{red}^o = 0.34$ V

   $\underline{\hspace{1cm} Sn^{2+}(aq) \rightarrow Sn^{4+}(aq) + 2$ e$^-$ $E_{red}^o = 0.15$ V $\hspace{1cm}}$

   $Cu^{2+}(aq) + S$ $n^{2+}(aq) \rightarrow Cu(s) + Sn^{4+}(aq)$ $E° = 0.34 - 0.15 = 0.19$ V, spontaneous

20.100   (a)   The reduction potential for $O_2(g)$ in the presence of acid is 1.23 V. $O_2(g)$ cannot oxidize Au(s) to Au^+(aq) or Au^{3+}(aq), even in the presence of acid.

(b)   The possible oxidizing agents need a reduction potential greater than 1.50 V. These include Co^{3+}(aq), $F_2(g)$, $H_2O_2(aq)$, and $O_3(g)$. Marginal oxidizing agents (those with reduction potential near 1.50 V) from Appendix E are BrO_3^-(aq), Ce^{4+}(aq), HClO(aq), MnO_4^-(aq), and $PbO_2(s)$.

(c)   4 Au(s) + 8 NaCN(aq) + 2 H_2O(l) + $O_2(g) \rightarrow$ 4 Na[Au(CN)$_2$](aq) + 4 NaOH(aq)

   Au(s) + 2 CN^-(aq) $\rightarrow$ [Au(CN)$_2$]$^-$ + 1 e$^-$

   $O_2(g)$ + 2 H_2O(l) + 4 e$^- \rightarrow$ 4 OH^-(aq)

   Au(s) is being oxidized and $O_2(g)$ is being reduced.

(d)   2 [Na[Au(CN)$_2$](aq) + 1 e$^- \rightarrow$ Au(s) + 2 CN^-(aq) + Na^+(aq)]

   $\underline{\hspace{1cm} Zn(s) \rightarrow Zn^{2+}(aq) + 2$ e$^- \hspace{1cm}}$

   2 Na[Au(CN)$_2$](aq) + Zn(s) $\rightarrow$ 2 Au(s) + $Zn^{2+}(aq)$ + 2 Na^+(aq) + 4 CN^-(aq)

   Zn(s) is being oxidized and [Au(CN)$_2$]$^-$(aq) is being reduced. Although OH^-(aq) is not included in this redox reaction, its presence in the reaction mixture probably causes Zn(OH)$_2$(s) to form as the product. This increases the driving force (and $E°$) for the overall reaction.

20.101 (a)

$$2\,[Ag^+(aq)+1\,e^- \rightarrow Ag(s)] \qquad\qquad E^o_{red}=0.80\ V$$
$$Ni(s) \rightarrow Ni^{2+}(aq)+2\,e^- \qquad\qquad E^o_{red}=-0.28\ V$$

$$2\,Ag^+(aq)+Ni(s)\rightarrow 2\,Ag(s)+Ni^{2+}(aq) \qquad E^o=0.80-(-0.28)=1.08\ V$$

(b) As the reaction proceeds, $Ni^{2+}(aq)$ is produced, so $[Ni^{2+}]$ increases as the cell operates.

(c) $E=E^o-\dfrac{0.0592}{n}\log Q;\ 1.12=1.08-\dfrac{0.0592}{2}\log\dfrac{[Ni^{2+}]}{[Ag^+]^2}$

$-\dfrac{0.04(2)}{0.0592}=\log(0.0100)-\log[Ag^+]^2;\ \log[Ag^+]^2=\log(0.0100)+\dfrac{0.04(2)}{0.0592}$

$\log[Ag^+]^2=-2.000+1.351=-0.649;\ [Ag^+]^2=0.255\ M;\ [Ag^+]=0.474=0.5\ M$

[Strictly speaking, $[E-E^o]$ having only 1 sig fig leads (after several steps) to the answer having only 1 sig fig. This is not a very precise or useful result.]

20.102 (a)

$$I_2(s)+2\,e^- \rightarrow 2\,I^-(aq) \qquad\qquad E^o_{red}=0.54\ V$$
$$2\,[Cu(s)\rightarrow Cu^+(aq)+1\,e^-] \qquad\qquad E^o_{red}=0.52\ V$$

$$I_2(s)+2\,Cu(s)\rightarrow 2\,Cu^+(aq)+2\,I^-(aq) \qquad E^o=0.54-0.52=0.02\ V$$

$E=E^o-\dfrac{0.0592}{n}\log Q=0.02-\dfrac{0.0592}{2}\log[Cu^+]^2\,[I^-]^2$

$E=0.02-\dfrac{0.0592}{2}\log(0.25)^2(3.5)^2=0.02+0.0034=0.0234=0.02\ V$

(b) Because the cell potential is positive at these concentration conditions, the reaction as written in part (a) is spontaneous in the forward direction. Cu is oxidized and Cu(s) is the anode.

(c) Yes. E^o is positive, so Cu is oxidized and Cu(s) is the anode at standard conditions.

(d) $E=0;\ +0.02=\dfrac{0.0592}{2}\log(0.15)^2\,[I^-]^2;\ \dfrac{2(0.02)}{0.0592}=\log(0.15)^2+2\log[I^-];$

$\log[I^-]=1.1617=1.2;\ [I^-]=10^{1.1617}=14.71=1\times10^1\ M\ I^-$

Strictly speaking, E^o having only 1 sig fig leads, after several steps, to the $[I^-]$ having only 1 sig fig. This is not a very precise or useful result. The result does indicate that, at low $[Cu^+]$, the cell potential is relatively insensitive to $[I^-]$.

20.103 Assume that room temperature is 298 K. Use the relationship developed in Solution 20.55 to calculate K from E^o at 298 K. Use data from Appendix E to calculate E^o for the disproportionation.

$$Cu^+(aq)+1\,e^- \rightarrow Cu(s) \qquad\qquad E^o_{red}=0.52\ V$$
$$Cu^+(aq)\rightarrow Cu^{2+}(aq)+1\,e^- \qquad\qquad E^o_{red}=0.15\ V$$

$$2\,Cu^+(aq)\rightarrow Cu(s)+Cu^{2+}(aq) \qquad E^o=0.52\ V-0.15\ V=0.37\ V$$

$E^o=\dfrac{0.0592}{n}\log K,\ \log K=\dfrac{nE^o}{0.0592}=\dfrac{1\times0.37}{0.0592}=6.25=6.3$

$K=10^{6.25}=1.778\times10^6=2\times10^6$

20.104 (a) In discharge, $Cd(s) + 2\,NiO(OH)(s) + 2\,H_2O(l) \rightarrow Cd(OH)_2(s) + 2\,Ni(OH)_2(s)$. In charging, the reverse reaction occurs.

(b) $E° = 0.49\,V - (-0.76\,V) = 1.25\,V$

(c) The 1.25 V calculated in part (b) is the standard cell potential, E°. The concentrations of reactants and products inside the battery are adjusted so that the cell output is greater than E°. Note that most of the reactants and products are pure solids or liquids, which do not appear in the Q expression. It must be $[OH^-]$ that is other than 1.0 M, producing an emf of 1.30 rather than 1.25.

(d) $E° = \dfrac{0.0592}{n}\log K;\ \log k = \dfrac{nE°}{0.0592}$

$\log K = \dfrac{2 \times 1.30}{0.0592} = 43.92 = 43.9;\ K = 8.3 \times 10^{43} = 8 \times 10^{43}$

20.105 (a) The battery capacity expressed in units of mAh indicates the total amount of electrical charge that can be delivered by the battery.

(b) Quantity of electrical charge is measured in coulombs, C. C = A-s

$2850\,mAh \times \dfrac{1\,A}{1000\,mA} \times \dfrac{3600\,s}{h} \times \dfrac{1\,C}{1\,A\text{-}s} = 10,260\,C$

The battery can deliver 10,260 C. Work, electrical or otherwise, is measured in J. J = V × C. If the battery voltage decreases linearly from 1.55 V to 0.80 V, assume an average voltage of 1.175 = 1.2 V.

$10,260\,C \times 1.175\,V = 12,055.5 = 12 \times 10^3\,J = 12\,kJ$

The total maximum electrical work of the battery is 12 kJ.

(This is 3.3×10^{-3} kWh or 3.3 Wh.)

20.106 (a) By analogy to O in hydroxide (OH^-) and in alcohol (R–OH), the oxidation state of S in a thiol (R–SH) is –2.

(b) By analogy to O in peroxide (–O–O–), the oxidation state of S in a disulfide (–S–S–) is –1.

(c) When two thiols react to form a disulfide, the oxidation number of S changes from –2 to –1. The oxidation number of S becomes more positive and the thiols are oxidized.

(d) Converting a disulfide to two thiols is the reverse of the process described in part (c). In order to reduce the disulfide, a reducing agent must be added to the solution.

(e) When two thiols (R–SH) react to form a disulfide (R–S–S–R), the H atoms are probably removed by base before the oxidizing agent does its work.

20.107 (a) Total volume of Cr $= 2.5 \times 10^{-4}\,m \times 0.32\,m^2 = 8.0 \times 10^{-5}\,m^3$

$mol\,Cr = 8.0 \times 10^{-5}\,m^3\,Cr \times \dfrac{100^3\,cm^3}{1\,m^3} \times \dfrac{7.20\,g\,Cr}{1\,cm^3} \times \dfrac{1\,mol\,Cr}{52.0\,g\,Cr} = 11.077 = 11\,mol\,Cr$

The electrode reaction is:

$CrO_4^{2-}(aq) + 4\,H_2O(l) + 6\,e^- \rightarrow Cr(s) + 8\,OH^-(aq)$

Coulombs required $= 11.077\,mol\,Cr \times \dfrac{6\,F}{1\,mol\,Cr} \times \dfrac{96,485\,C}{1\,F} = 6.41 \times 10^6 = 6.4 \times 10^6\,C$

(b) $\quad 6.41 \times 10^6 \, C \times \dfrac{1 \, \text{amp-s}}{1 \, C} \times \dfrac{1}{10.0 \, s} = 6.4 \times 10^5 \, \text{amp}$

(c) $\quad$ If the cell is 65% efficient, $(6.41 \times 10^6/0.65) = 9.867 \times 10^6 = 9.9 \times 10^6 \, C$ is required to plate the bumper.

$$6.0 \, V \times 9.867 \times 10^6 \, C \times \dfrac{1 \, J}{1 \, C\text{-}V} \times \dfrac{1 \, kWh}{3.6 \times 10^6 \, J} = 16.445 = 16 \, kWh$$

20.108   (a)   The standard reduction potential for $H_2O(l)$ is much greater than that of $Mg^{2+}(aq)(-0.83 \, V$ vs. $-2.37 \, V)$. In aqueous solution, $H_2O(l)$ would be preferentially reduced and no $Mg(s)$ would be obtained.

(b) $\quad 97{,}000 \, A \times 24 \, h \times \dfrac{3600 \, s}{1 \, h} \times \dfrac{1 \, C}{1 \, A\text{-}s} \times \dfrac{1 \, F}{96{,}485 \, C} \times \dfrac{1 \, \text{mol Mg}}{2 \, F} \times \dfrac{24.31 \, \text{g Mg}}{1 \, \text{mol Mg}} \times 0.96$

$$= 1.0 \times 10^6 \, \text{g Mg} = 1.0 \times 10^3 \, \text{kg Mg}$$

20.109   *Analyze.* Given mass of aluminum desired, applied voltage, and electrolysis efficiency, calculate kWh of electricity required. *Plan.* Beginning with mass Al and paying attention to units, calculate coulombs required if the process is 100% efficient. Then, take efficiency into account and use V, C, and the relationship between J and kWh to calculate kWh required. *Solve.*

$$1.0 \times 10^3 \, \text{kg Al} \times \dfrac{1000 \, g}{1 \, kg} \times \dfrac{1 \, \text{mol Al}}{26.98 \, \text{g Al}} \times \dfrac{3 \, F}{1 \, \text{mol Al}} \times \dfrac{96{,}485 \, C}{F} = 1.073 \times 10^{10} = 1.1 \times 10^{10} \, C$$

If the cell is 45% efficient, $(1.073 \times 10^{10}/0.45) = 2.384 \times 10^{10} = 2.4 \times 10^{10} \, C$ is required to plate the bumper.

$$4.50 \, V \times 2.384 \times 10^{10} \, C \times \dfrac{1 \, J}{1 \, C\text{-}V} \times \dfrac{1 \, kWh}{3.6 \times 10^6 \, J} = 29{,}801 = 3.0 \times 10^4 \, kWh$$

20.110   (a) $\quad 7 \times 10^8 \, \text{mol } H_2 \times \dfrac{2 \, F}{1 \, \text{mol } H_2} \times \dfrac{96{,}485 \, C}{1 \, F} = 1.35 \times 10^{14} = 1 \times 10^{14} \, C$

(b) $\quad$
$$\begin{array}{ll} 2 \, [2 \, H^+(aq) + 2 \, e^- \rightarrow H_2(g)] & E^o_{red} = 0.00 \, V \\ \underline{2 \, H_2O(l) \rightarrow O_2(g) + 4 \, H^+(aq) + 4 \, e^-} & \underline{E^o_{red} = 1.23 \, V} \\ 2 \, H_2O(l) \rightarrow O_2(g) + 2 \, H_2(g) & E^o = 0.00 - 1.23 = -1.23 \, V \end{array}$$

$P_t = 300 \, \text{atm} = P_{O_2} + P_{H_2}$. Because $H_2(g)$ and $O_2(g)$ are generated in a 2:1 mole ratio, $P_{H_2} = 200 \, \text{atm}$ and $P_{O_2} = 100 \, \text{atm}$.

$$E = E^o - \dfrac{0.0592}{4} \log \, (P_{O_2} \times P^2_{H_2}) = -1.23 \, V - \dfrac{0.0592}{4} \log \, [100 \times (200)^2]$$

$E = -1.23 \, V - 0.100 \, V = -1.33 \, V; \, E_{min} = 1.33 \, V$

(c) $\quad$ Energy $= nFE = 2(7 \times 10^8 \, \text{mol}) \, (1.33 \, V) \dfrac{96{,}485 \, J}{V\text{-mol}} = 1.80 \times 10^{14} = 2 \times 10^{14} \, J$

(d) $\quad 1.80 \times 10^{14} \, J \times \dfrac{1 \, kWh}{3.6 \times 10^6 \, J} \times \dfrac{\$0.85}{kWh} = \$4.24 \times 10^7 = \$4 \times 10^7$

It would cost more than $40 million for the electricity alone.

Integrative Exercises

20.111 $N_2(g) + 3 H_2(g) \rightarrow 2 NH_3(g)$

(a)   The oxidation number of $H_2(g)$ and $N_2(g)$ is 0. The oxidation number of N in NH_3 is –3, H in NH_3 is +1. H_2 is being oxidized and N_2 is being reduced.

(b)   Calculate $\Delta G°$ from ΔG_f^o values in Appendix C. Use $\Delta G° = -RT \ln K$ to calculate K.

$\Delta G° = 2\Delta G_f^o\ NH_3(g) - \Delta G_f^o\ N_2(g) - 3\Delta G_f^o\ H_2(g)$

$\Delta G° = 2(-16.66\ kJ) - 0 - 3(0) = -33.32\ kJ$

$\Delta G° = -RT \ln K,\ \ln K = \dfrac{-\Delta G°}{RT} = \dfrac{-(-33.32 \times 10^3\ J)}{(8.314\ J/mol\text{-}K)(298\ K)} = 13.4487 = 13.45$

$K = e^{13.4487} = 6.9 \times 10^5$

(c) $\Delta G° = -nfE°;\quad E° = \dfrac{-\Delta G°}{nF};\quad n = ?$

2 N atoms change from 0 to –3 or 6 H atoms change from 0 to +1.

Either way, n = 6.

$E° = \dfrac{-(-33.32\ kJ)}{6 \times 96.5\ kJ/V} = 0.05755\ V$

20.112   The redox reaction is: $2\ Ag^+(aq) + H_2(g) \rightarrow 2\ Ag(s) + 2\ H^+(aq)$. n = 2 for this reaction.

$E_{cell}^o = E_{red}^o\ cathode - E_{red}^o\ anode = 0.80\ V - 0.00\ V = 0.80$

$E = E° - \dfrac{0.0592}{n} \log \dfrac{[H^+]^2}{[Ag^+]^2\ P_{H_2}}$

$[H^+]$ in the cell is held essentially constant by the benzoate buffer.

$$C_6H_5COOH(aq) \rightleftharpoons H^+(aq) + C_6H_5COO^-(aq) \quad K_a = ?$$

$K_a = \dfrac{[H^+][C_6H_5COO^-]}{[C_6H_5COOH]}; [H^+] = \dfrac{K_a[C_6H_5COOH]}{[C_6H_5COO^-]} = \dfrac{0.10\ M}{0.050\ M} \times K_a = 2K_a$

Solve the Nernst expression for $[H^+]$ and calculate K_a and pK_a as shown above.

$1.030\ V = 0.80\ V - \dfrac{0.0592}{n} \log \dfrac{[H^+]^2}{(1.00)^2(1.00)}$

$0.23 \times \dfrac{2}{0.0592} = -\log [H^+]^2 = -2 \log[H^+]$

$\dfrac{0.23}{0.0592} = -\log [H^+] = pH;\ pH = 3.885 = 3.9;\ [H^+] = 10^{-3.885} = 1.303 \times 10^{-4} = 1 \times 10^{-4}$

$[H^+] = 2K_a;\ K_a = [H^+]/2 = 6.515 \times 10^{-5} = 7 \times 10^{-5};\ pK_a = 4.186 = 4.2$

Check. According to Appendix D, K_a for benzoic acid is 6.3×10^{-5}.

20.113   (a)   +1

(b)   –1 (While we don't know the oxidation number of N in chloramine, we can assume the most common oxidation number for Cl.)

(c)   Reduced. The oxidation number of Cl becomes more negative, so Cl is reduced.

(d)   +3

(e)   Oxidized. The oxidation number of N goes from –3 to +3, so N is oxidized.

20.114   (a)

$$Ag^+(aq) + e^- \rightarrow Ag(s) \qquad E^o_{red} = 0.80 \text{ V}$$

$$Fe^{2+}(aq) \rightarrow Fe^{3+}(aq) + 1 \text{ e}^- \qquad E^o_{red} = 0.77 \text{ V}$$

$$\overline{Ag^+(aq) + Fe^{2+}(aq) \rightarrow Ag(s) + Fe^{3+}(aq) \qquad E^o = 0.80 \text{ V} - 0.77 \text{ V} = 0.03 \text{ V}}$$

(b) $Ag^+(aq)$ is reduced at the cathode and $Fe^{2+}(aq)$ is oxidized at the anode.

(c) $\Delta G^o = -nFE^o = -(1)(96.5)(0.03) = -2.895 = 3 \text{ kJ}$

$\Delta S^o = S^o \text{ Ag(s)} + S^o \text{ Fe}^{3+}(aq) - S^o Ag^+(aq) - S^o \text{ Fe}^{2+}(aq)$

     $= 42.55 \text{ J} + 293.3 \text{ J} - 73.93 \text{ J} - 113.4 \text{ J} = 148.5 \text{ J}$

$\Delta G^o = \Delta H^o - T\Delta S^o$. Because ΔS^o is positive, ΔG^o will become more negative and E^o will become more positive as temperature is increased.

20.115   (a) $\Delta H^o = 2\Delta H^o \text{ H}_2O(l) - 2\Delta H^o \text{ H}_2(g) - \Delta H^o \text{ O}_2(g) = 2(-285.83) - 2(0) - 0 = -571.66 \text{ kJ}$

$\Delta S^o = 2S^o \text{ H}_2O(l) - 2S^o \text{ H}_2(g) - \Delta S^o \text{ O}_2(g)$

     $= 2(69.91) - 2(130.58) - (205.0) = -326.34 \text{ J}$

(b)   Because ΔS^o is negative, $-T\Delta S$ is positive and the value of ΔG will become more positive as T increases. The reaction will become nonspontaneous at a fairly low temperature, because the magnitude of ΔS^o is large.

(c) $\Delta G = w_{max}$. The larger the negative value of ΔG, the more work the system is capable of doing on the surroundings. As the magnitude of ΔG decreases with increasing temperature, the usefulness of H_2 as a fuel decreases.

(d)   The combustion method increases the temperature of the system, which quickly decreases the magnitude of the work that can be done by the system. Even if the effect of temperature on this reaction could be controlled, only about 40% of the energy from any combustion can be converted to electrical energy, so combustion is intrinsically less efficient than direct production of electrical energy via a fuel cell.

20.116   First, balance the equation:

$$4 \text{ CyFe}^{2+}(aq) + O_2(g) + 4 H^+(aq) \rightarrow 4 \text{ CyFe}^{3+}(aq) + 2 H_2O(l); \text{ E} = +0.60 \text{ V}; n = 4$$

(a)   From Equation 20.12, we can calculate ΔG for the process under the conditions specified for the measured potential E:

$$\Delta G = -nFE = -(4 \text{ mol e}^-) \times \frac{96.485 \text{ kJ}}{1 \text{ V-mol e}^-}(0.60 \text{ V}) = -231.6 = -232 \text{ kJ}$$

(b)   The moles of ATP synthesized per mole of O_2 is given by:

$$\frac{231.6 \text{ kJ}}{O_2 \text{ molecule}} \times \frac{1 \text{ mol ATP formed}}{37.7 \text{ kJ}} = \text{approximately 6 mol ATP/mol } O_2$$

20.117 $AgSCN(s) + e^- \rightarrow Ag(s) + SCN^-(aq)$ $E_{red}^{o} = 0.09 \text{ V}$

$$\frac{Ag(s) \rightarrow Ag^+(aq) + e^- \qquad\qquad E_{red}^{o} = 0.80 \text{ V}}{AgSCN(s) \rightarrow Ag^+(aq) + SCN^-(aq) \quad E^\circ = 0.09 - 0.80 = -0.71 \text{ V}}$$

$E^\circ = \dfrac{0.0592}{n} \log K_{sp}; \; \log K_{sp} = \dfrac{(-0.71)(1)}{0.0592} = -11.993 = -12$

$K_{sp} = 10^{-11.993} = 1.02 \times 10^{-12} = 10^{-12}$

20.118   The reaction can be written as a sum of the steps:

$Pb^{2+}(aq) + 2 \, e^- \rightarrow Pb(s)$ $E_{red}^{o} = -0.13 \text{ V}$

$$\frac{PbS(s) \rightarrow Pb^{2+}(aq) + S^{2-}(aq) \quad \text{``}E^{o}\text{''} = ?}{PbS(s) + 2 \, e^- \rightarrow Pb(s) + S^{2-}(aq) \qquad E_{red}^{o} = ?}$$

"E^{o}" for the second step can be calculated from K_{sp}.

$E^\circ = \dfrac{0.0592}{n} \log K_{sp} = \dfrac{0.0592}{2} \log (8.0 \times 10^{-28}) = \dfrac{0.0592}{2} (-27.10) = -0.802 \text{ V}$

E^{o} for the half-reaction = $-0.13 \text{ V} + (-0.802 \text{ V}) = -0.93 \text{ V}$

Calculating an imaginary E^{o} for a nonredox process like step 2 may be a disturbing idea. Alternatively, one could calculate K for step 1 (5.4×10^{-5}), K for the reaction in question ($K = K_1 \times K_{sp} = 4.4 \times 10^{-32}$), and then E^{o} for the half-reaction. The result is the same.

20.119   The two half-reactions in the electrolysis of $H_2O(l)$ are:

$2 \, [2H_2O(l) + 2 \, e^- \rightarrow H_2(g) + 2 \, OH^-]$

$$\frac{2 \, H_2O(l) \rightarrow O_2(g) + 4 \, H^+ + 4 \, e^-}{2 \, H_2O(l) \rightarrow 2 \, H_2(g) + O_2(g)}$$

4 mol e^-/2 mol $H_2(g)$ or 2 mol e^-/mol $H_2(g)$

Using partial pressures and the ideal-gas law, calculate the mol $H_2(g)$ produced, and the current required to do so.

$P_t = P_{H_2} + P_{H_2O}$. From Appendix B, P_{H_2O} at 25.5 °C is approximately 24.5 torr.

$P_{H_2} = 768 \text{ torr} - 24.5 \text{ torr} = 743.5 = 744 \text{ torr}$

$n = PV/RT = \dfrac{(743.5/760) \text{ atm} \times 0.0123 \text{ L}}{298.5 \text{ K} \times 0.08206 \text{ L-atm/mol-K}} = 4.912 \times 10^{-4} = 4.91 \times 10^{-4} \text{ mol } H_2$

$4.912 \times 10^{-4} \text{ mol } H_2 \times \dfrac{2 \text{ mol } e^-}{\text{mol } H_2} \times \dfrac{96,485 \text{ C}}{1 \text{ mol } e^-} \times \dfrac{1 \text{ amp-s}}{1 \text{ C}} \times \dfrac{1 \text{ min}}{60 \text{ s}} \times \dfrac{1}{2.00 \text{ min}} = 0.790 \text{ amp}$

21 Nuclear Chemistry

Visualizing Concepts

21.1   *Analyze.* Given the name and mass number of a nuclide, decide if it lies within the belt of stability. If not, suggest a process that moves it toward the belt.

Plan. Calculate the number of protons and neutrons in each nuclide. Locate this point on Figure 21.1. If the point is above the belt, beta decay increases protons and decreases neutrons, decreasing the neutron-to-proton ratio. If the point is below the belt, either positron emission or neutron capture decreases protons and increases neutrons, increasing the neutron-to-proton ratio.   *Solve.*

(a) ^{24}Ne: 10 p, 14 n, just above the belt of stability. Reduce the neutron-to-proton ratio via beta decay.

(b) ^{32}Cl: 17 p, 15 n, just below the belt of stability. Increase the neutron-to-proton ratio via positron emission or orbital electron capture.

(c) ^{108}Sn: 50 p, 58 n, just below the belt of stability. Increase the neutron-to-proton ratio via positron emission or orbital electron capture.

(d) ^{216}Po: 84 p, 132 n, just beyond the belt of stability. Nuclei with atomic numbers $\geq$ 84 tend to decay via alpha emission, which decreases both protons and neutrons.

21.2   *Analyze/Plan.* From the diagram, determine the atomic number (number of protons) and mass number (number of protons plus neutrons) of the two nuclides involved. Based on the relationship between the two nuclides, decide whether the reaction is alpha or beta decay, positron emission or electron capture. Complete the nuclear reaction, balancing atomic numbers and mass numbers.

Solve. The two nuclides in the diagram are $^{109}_{46}$Pd and $^{109}_{47}$Ag, so the second product is a beta particle. The balanced reaction is:

$$^{109}_{46}\text{Pd} \rightarrow {}^{109}_{47}\text{Ag} + {}^{0}_{-1}\text{e}$$

Check. Atomic number and mass number balance.

21.3   *Analyze/Plan.* Determine the number of protons and neutrons present in the two heavy nuclides in the reaction. Draw a graph with appropriate limits and plot the two points. Draw an arrow from reactant to product.   *Solve.*

Bi: 83 p, 128 n; Tl: 81 p, 126 n.

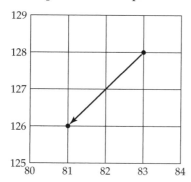

Check. An alpha particle has 2 p and 2 n. The diagram shows a decrease in 2 p and 2 n for the reaction.

21.4   *Analyze/Plan.* Count the protons and neutrons in each particle. Use a periodic table and Table 21.2 to identify the particles. Refer to Sample Exercise 21.4 when writing the reaction using condensed notation. Use Figure 21.1 to determine stability of the product nucleus.

      (a)   From left to right, the particles are $^{10}_{5}B$, $^{1}_{0}n$, $^{7}_{3}Li$, and $^{4}_{2}He$

      (b) $^{10}_{5}B\,(n,\alpha)\,^{7}_{3}Li$

      (c)   The $^{7}_{3}Li$ product nucleus is probably stable. It appears to be in the belt of stability in Figure 21.1. The product nucleus has an odd number of protons but an even number of neutrons. There are 50 stable "odd, even" nuclei.

21.5   *Analyze/Plan.* Use the information in Table 21.3 to identify the type of radioactive decay for each step.

      (a)   (i)   Alpha decay. Mass number decreases by 4 and atomic number decreases by 2.

            (ii) Beta decay. Mass number is the same, atomic number increases by 1.

            (iii) Beta decay. Mass number is the same, atomic number increases by 1.

      (b) $^{228}_{89}Ac$ has the highest activity; it has the shortest half-life.

      (c) $^{232}_{90}Th$ has the lowest activity; it has the longest half-life.

      (d) $^{224}_{88}Rn$. Mass number decreases by 4 and atomic number decreases by 2, relative to the last isotope shown.

21.6   *Analyze/Plan.* Write the balanced equation for the decay. Nuclear decay is a first-order process; use appropriate relationships for first-order processes to determine $t_{1/2}$, k and remaining ^{88}Mo after 12 minutes.   *Solve.*

      (a) $t_{1/2}$ is the time required for half of the original nuclide to decay. Relative to the graph, this is the time when the amount of ^{88}Mo is reduced from 1.0 to 0.5. This time is 7 minutes.

      (b)   For a first-order process, $t_{1/2} = 0.693/k$ or $k = 0.693/t_{1/2}$.

            $k = 0.693/7\ min = 0.0990 = 0.1\ min^{-1}$

(c)   From the graph, the fraction of ^{88}Mo remaining after 12 min is $0.3/1.0 = 0.3$

   Check. $\ln(N_t/N_o) = -kt = -(0.099)(12) = -1.188;\ N_t/N_o = e^{-1.188} = 0.30.$

(d) $^{88}_{42}\text{Mo} \rightarrow {}^{88}_{41}\text{Nb} + {}^{0}_{+1}\text{e}$

21.7   *Analyze/Plan.* Atomic number is number of protons, mass number is (protons + neutrons). Chemical symbol is determined by atomic number. Beta decay increases the number of protons, whereas mass number stays constant. Positron emission decreases the number of protons, whereas mass number stays constant. Half-life, $t_{1/2}$, is the time required to reduce the amount of radioactive material by half.   *Solve.*

(a) $^{10}_{5}\text{B},\ {}^{11}_{5}\text{B};\ {}^{12}_{6}\text{C},\ {}^{13}_{6}\text{C};\ {}^{14}_{7}\text{N},\ {}^{15}_{7}\text{N};\ {}^{16}_{8}\text{O},\ {}^{17}_{8}\text{O},\ {}^{18}_{8}\text{O};\ {}^{19}_{9}\text{F}$

(b)   On the diagram, $^{14}_{6}\text{C}$ is the only radioactive nuclide above and left of the band of stable red nuclides. $^{14}_{6}\text{C}$ will reduce its neutron-to-proton ratio by beta decay.

   $^{14}_{6}\text{C} \rightarrow {}^{14}_{7}\text{N} + {}^{0}_{-1}\text{e}$

(c)   On the diagram, radioactive nuclides right and below the band of stable red nuclides are likely to increase their neutron-to-proton ratio via positron emission. To be useful for positron emission tomography, the nuclides must have a half-life on the order of minutes. (Nuclides with very fast decay disappear before they can be imaged. Those with longer half-lives linger in the patient.) Four radioactive nuclides fit these criteria: $^{11}_{6}\text{C},\ {}^{13}_{7}\text{N},\ {}^{15}_{8}\text{O},\ {}^{18}_{9}\text{F}$

(d)   If 12.5% of the isotope remains, this amounts to three half-lives of decay. If the total decay time is 1 hour, 1/3 of this time is 20 min. The radionuclide with a half-life of 20 min is $^{11}_{6}\text{C}$.

21.8   *Analyze/Plan.* Express the particles in the diagram as a nuclear reaction. Determine the mass number and atomic number of the unknown particle by balancing these quantities in the nuclear reaction.   *Solve.*

(a) $^{239}_{94}\text{Pu} + {}^{1}_{0}\text{n} \rightarrow {}^{95}_{40}\text{Zr} + ? + 2\,{}^{1}_{0}\text{n}$

   The unidentified particle has an atomic number of $(94-40) = 54$; it is Xe. The mass number of the nuclide is $[(239 + 1) - (95 + 2)] = 143$. The unknown particle is $^{143}_{54}\text{Xe}$.

(b) ^{95}Zr: 40 p, 55 n is stable. ^{143}Xe: 54 p, 89 n is above the belt of stability and is not stable; it will probably undergo beta decay.

Radioactivity and Nuclear Equations (Section 21.1)

21.9   *Analyze/Plan.* Given various nuclide descriptions, determine the number of protons and neutrons in each nuclide. The left superscript is the mass number, protons plus neutrons. If there is a left subscript, it is the atomic number, the number of protons. Protons can always be determined from the chemical symbol; all isotopes of the same element have the same number of protons. A number following the element name, as in part (c) is the mass number.   *Solve.*

p = protons, n = neutrons, e = electrons; number of protons = atomic number; number of neutrons = mass number – atomic number

(a) $^{56}_{24}\text{Cr}$: 24p, 32n   (b) ^{193}Tl: 81p, 112n   (c) ^{38}Ar: 18p, 20n

21.10   p = protons, n = neutrons, e = electrons; number of protons = atomic number;
number of neutrons = mass number − atomic number

     (a) $^{129}_{53}$I: 53p, 76n   (b) ^{138}Ba: 56p, 82n   (c) ^{237}Np: 93p, 144n

21.11   *Analyze/Plan.* See definitions in Section 21.1. In each case, the left superscript is mass number, the left subscript is related to atomic number.   *Solve.*

     (a) $^{1}_{0}$n   (b) $^{4}_{2}$He or α (c) $^{0}_{0}\gamma$ or γ

21.12   (a) $^{1}_{1}$p or $^{1}_{1}$H   (b) $^{0}_{-1}$e or β^{-} (c) $^{0}_{+1}$e or β^{+}

21.13   *Analyze/Plan.* Follow the logic in Sample Exercises 21.1 and 21.2. Pay attention to definitions of decay particles in Table 21.2 as well as conservation of mass and charge. *Solve.*

     (a) $^{90}_{37}$Rb $\rightarrow$ $^{90}_{38}$Sr + $^{0}_{-1}$e   (b) $^{72}_{34}$Se + $^{0}_{-1}$e (orbital electron) $\rightarrow$ $^{72}_{33}$As

     (c) $^{76}_{36}$Kr $\rightarrow$ $^{76}_{35}$Br + $^{0}_{+1}$e   (d) $^{226}_{88}$Ra $\rightarrow$ $^{222}_{86}$Rn + $^{4}_{2}$He

21.14   (a) $^{213}_{83}$Bi $\rightarrow$ $^{209}_{81}$Tl + $^{4}_{2}$He   (b) $^{13}_{7}$N + $^{0}_{-1}$e (orbital electron) $\rightarrow$ $^{13}_{6}$C

     (c) $^{98}_{43}$Tc + $^{0}_{-1}$e (orbital electron) $\rightarrow$ $^{98}_{42}$Mo   (d) $^{188}_{79}$Au $\rightarrow$ $^{188}_{78}$Pt + $^{0}_{+1}$e

21.15   *Analyze/Plan.* Using definitions of the decay processes and conservation of mass number and atomic number, work backwards to the reactants in the nuclear reactions. *Solve.*

     (a) $^{211}_{82}$Pb $\rightarrow$ $^{211}_{83}$Bi + $^{0}_{-1}$e   (b) $^{50}_{25}$Mn $\rightarrow$ $^{50}_{24}$Cr + $^{0}_{+1}$e

     (c) $^{179}_{74}$W + $^{0}_{-1}$e $\rightarrow$ $^{179}_{73}$Ta   (d) $^{230}_{90}$Th $\rightarrow$ $^{226}_{88}$Ra + $^{4}_{2}$He

21.16   (a) $^{24}_{11}$Na $\rightarrow$ $^{24}_{12}$Mg + $^{0}_{-1}$e; a beta particle is produced

     (b) $^{188}_{80}$Hg $\rightarrow$ $^{188}_{79}$Au + $^{0}_{+1}$e; a positron is produced

     (c) $^{122}_{53}$I $\rightarrow$ $^{122}_{54}$Xe + $^{0}_{-1}$e; a beta particle is produced

     (d) $^{242}_{94}$Pu $\rightarrow$ $^{238}_{92}$U + $^{4}_{2}$He; an alpha particle is produced

21.17   *Analyze/Plan.* Given the starting and ending nuclides in a nuclear decay sequence, we are asked to determine the number of alpha and beta emissions. Use the total change in A and Z, along with definitions of alpha and beta decay, to answer the question. *Solve.*

The total mass number change is (235–207) = 28. As each alpha particle emission decreases the mass number by four, whereas emission of a beta particle does not correspond to a mass change, there are 7 alpha particle emissions. The change in atomic number in the series is 10. Each alpha particle results in an atomic number lower by two. The 7 alpha particle emissions alone would cause a decrease of 14 in atomic number. Each beta particle emission raises the atomic number by one. To obtain the observed lowering of 10 in the series, there must be 4 beta emissions.

21.18   This decay series represents a change of (232–208 =) 24 mass units. Because only alpha emissions change the nuclear mass, and each changes the mass by four, there must be a total of 6 alpha emissions. Each alpha emission causes a decrease of two in atomic number.

Therefore, the 6 alpha emissions, by themselves, would cause a decrease in atomic number of 12. The series as a whole involves a decrease of 8 in atomic number. Thus, there must be a total of 4 beta emissions, each of which increases atomic number by one. Overall, there are 6 alpha emissions and 4 beta emissions.

Patterns of Nuclear Stability (Section 21.2)

21.19   *Analyze/Plan.* Follow the logic in Sample Exercise 21.3, paying attention to the guidelines for neutron-to-proton ratio.   *Solve.*

(a) $^{8}_{5}B$ - low neutron/proton ratio, positron emission (for low atomic numbers, positron emission is more common than orbital electron capture)

(b) $^{68}_{29}Cu$ - high neutron/proton ratio, beta emission

(c) $^{32}_{15}P$ - slightly high neutron/proton ratio, beta emission

(d) $^{39}_{17}Cl$ - high neutron/proton ratio, beta emission

21.20   (a) $^{3}_{1}H$ - high neutron/proton ratio, beta emission

(b) $^{89}_{38}Sr$ - (slightly) high neutron/proton ratio, beta emission

(c) $^{120}_{53}I$ - low neutron/proton ratio, positron emission

(d) $^{102}_{47}Ag$ - low neutron/proton ratio, positron emission

21.21   *Analyze/Plan.* Use the criteria listed in Table 21.4.   *Solve.*

(a) $^{39}_{19}K$; stable, odd proton, even neutron; 20 neutrons is a magic number.

$^{40}_{19}K$; radioactive, odd proton, odd neutron

(b) $^{209}_{83}Bi$; stable, odd proton, even neutron; 126 neutrons is a magic number.

$^{208}_{83}Bi$; radioactive, odd proton, odd neutron

(c) $^{58}_{28}Ni$; stable, even proton, even neutron

$^{65}_{28}Ni$; radioactive, high neutron/proton ratio

21.22   Use criteria listed in Table 21.4.

(a) $^{40}_{20}Ca$; stable, magic numbers of protons and neutrons

$^{45}_{20}Ca$, radioactive, high neutron/proton ratio.

(b) $^{12}_{6}C$, stable, even proton, even neutron

$^{14}_{6}C$, radioactive, high neutron/proton ratio

(c) $^{206}_{82}$Pb , stable, magic number of protons, even proton, even neutron

      $^{230}_{90}$Th , radioactive, atomic number greater than 84

21.23   *Analyze/Plan.* For each nuclide, determine the number of protons and neutrons and decide if they are magic numbers.   *Solve.*

(a) $^{4}_{2}$He , both   (b) $^{18}_{8}$O has a magic number of protons, but not neutrons

(c) $^{40}_{20}$Ca , both   (d) $^{66}_{30}$Zn has neither   (e) $^{208}_{82}$Pb , both

21.24   Statement (d). Evidence shows that pairs of protons and neutrons have special stability, so elements with even atomic numbers have more abundant isotopes.

Regarding statements (a) and (b), elements with odd atomic numbers can lie above, below, or within the belt of stability. Regarding statement (c), none of the elements on the graph has a magic number of protons.

21.25   Statement (a). Because of its magic numbers, the alpha particle is a very stable emitted particle, which makes alpha emission a favorable process. The proton has no magic numbers, and is an odd proton–even neutron particle. Its instability does not encourage proton emission as a process.

21.26   The criterion employed in judging whether the nucleus is likely to be radioactive is the position of the nucleus on the plot shown in Figure 21.1. If the neutron/proton ratio is too high or low, or if the atomic number exceeds 83, the nucleus will be radioactive.

Radioactive: $^{60}_{27}$Co —odd proton, odd neutron

            $^{92}_{41}$Nb —odd proton, odd neutron

            ^{226}Ra —high atomic number

Stable: $^{58}_{26}$Fe —even proton, even neutron, stable neutron/proton ratio

        ^{202}Hg —even proton, even neutron, stable neutron/proton ratio

Nuclear Transmutations (Section 21.3)

21.27   Statement (b) is the best explanation. Because they are electrically neutral, neutrons are not repelled by the positively charged nucleus being bombarded.

Statement (a) does not apply and statements (c) and (d) are false. Neutrons are not attracted to the nucleus at long distances.

21.28   The element is technetium, Tc. $^{96}_{42}$Mo + $^{2}_{1}$H → $^{98}_{43}$Tc

21.29   *Analyze/Plan.* Determine A and Z for the missing particle by conservation principles. Find the appropriate symbol for the particle.   *Solve.*

(a) $^{252}_{98}$Cf + $^{10}_{5}$B → 3 $^{1}_{0}$n + $^{259}_{103}$Lr   (b) $^{2}_{1}$H + $^{3}_{2}$He → $^{4}_{2}$He + $^{1}_{1}$H

(c) $^{1}_{1}$H + $^{11}_{5}$B → 3 $^{4}_{2}$He   (d) $^{122}_{53}$I → $^{122}_{54}$Xe + $^{0}_{-1}$e

(e) $^{59}_{26}$Fe → $^{0}_{-1}$e + $^{59}_{27}$Co

21.30   (a) $^{14}_{7}N + ^{4}_{2}He \rightarrow ^{17}_{8}O + ^{1}_{1}H$ (b) $^{40}_{19}K + ^{0}_{-1}e \text{ (orbital electron)} \rightarrow ^{40}_{18}Ar$

       (c) $^{27}_{13}Al + ^{4}_{2}He \rightarrow ^{30}_{14}Si + ^{1}_{1}H$ (d) $^{58}_{26}Fe + 2^{1}_{0}n \rightarrow ^{0}_{-1}e + ^{60}_{27}Co$

       (e) $^{235}_{92}U + ^{1}_{0}n \rightarrow ^{135}_{54}Xe + ^{99}_{38}Sr + 2\,^{1}_{0}n$

21.31   *Analyze/Plan.* Follow the logic in Sample Exercise 21.4, paying attention to conservation of A and Z.   *Solve.*

       (a) $^{238}_{92}U + ^{4}_{2}He \rightarrow ^{241}_{94}Pu + ^{1}_{0}n$ (b) $^{14}_{7}N + ^{4}_{2}He \rightarrow ^{17}_{8}O + ^{1}_{1}H$

       (c) $^{56}_{26}Fe + ^{4}_{2}He \rightarrow ^{60}_{29}Cu + ^{0}_{-1}e$

21.32   (a) $^{238}_{92}U + ^{1}_{0}n \rightarrow ^{239}_{92}U + ^{0}_{0}\gamma$ (b) $^{16}_{8}O + ^{1}_{1}H \rightarrow ^{13}_{7}N + ^{4}_{2}He$

       (c) $^{18}_{8}O + ^{1}_{0}n \rightarrow ^{19}_{9}F + ^{0}_{-1}e$

Rates of Radioactive Decay (Section 21.4)

21.33   (a)   True. $k = 0.693/t_{1/2}$. The decay rate constant, k, and half-life, $t_{1/2}$ are inversely related.

       (b)   False. If X is not radioactive, it does not spontaneously decay and its half-life is essentially infinity.

       (c)   True. Changes in the amount of A would be measurable over the 40-year time frame, whereas changes in the amount of X would be very small and difficult to detect.

21.34   (a)   The suggestion is not reasonable. The energies of nuclear states are very large relative to ordinary temperatures. Thus, merely changing the temperature by less than 100 K would not be expected to significantly affect the behavior of nuclei with regard to nuclear decay rates.

       (b)   No. Radioactive decay has no activation energy like a chemical reaction. Activation energy is the minimum amount of energy required to initiate a chemical reaction. Radioactive decay is a spontaneous nuclear transformation from a less stable to a more stable nuclear configuration. Radioisotopes are by definition in a "transition state," prone to nuclear change or decay. Changes in external conditions such as temperature, pressure or chemical state provide insufficient energy to either excite or relax an unstable nucleus.

21.35   *Analyze/Plan.* The half-life is 12.3 yr. Use $t_{1/2}$ to calculate k and the mass fraction (N_t/N_0) remaining after 50 yr.

       Solve. $k = 0.693/\ t_{1/2} = 0.693/12.3 \text{ yr} = 0.056341 = 0.0563 \text{ yr}^{-1}$

       $\ln\dfrac{N_t}{N_0} = -kt; \quad \ln\dfrac{N_t}{N_0} = -0.056341 \text{ yr}^{-1}(50 \text{ yr}) = -2.81707 = -2.8; \quad \dfrac{N_t}{N_0} = 0.05978 = 0.06$

       When the watch is 50 years old, only 6% (or 6.0%) of the tritium remains. The dial will be dimmed by 94%.

21.36 Calculate the decay constant, k, and then $t_{1/2}$. $t = 4$ h 39 min $= 279$ min

$$k = \frac{-1}{t} \ln \frac{N_t}{N_o} = \frac{-1}{279 \text{ min}} \ln \frac{0.25 \text{ mg}}{2.000 \text{ mg}} = 0.007453 = 0.0.0075 \text{ min}^{-1}$$

Using Equation 21.19, $t_{1/2} = 0.693/k = 0.693/0.007453$ min$^{-1} = 92.980 = 93$ min

21.37 *Analyze/Plan.* We are given half-life of cobalt-60, and replacement time when the activity of the sample is 75% of the initial value. Consider the rate law for (first-order) nuclear decay: $\ln(N_t/N_o) = -kt$. *Solve.*

$k = 0.693 / t_{1/2} = 0.693/5.26$ yr $= 0.1317 = 0.132$ yr^{-1}; $N_t/N_o = 0.75$

$$t = \frac{-1}{k} \ln \frac{N_t}{N_o} = -(1/0.1317 \text{ yr}^{-1}) \ln (0.75) = 2.18 \text{ yr} = 26.2 \text{ mo} = 797 \text{ d.}$$

The source will have to be replaced sometime in the summer of 2018, probably in August.

21.38 Follow the logic in Sample Exercise 21.6. In this case, we are given initial sample mass as well as mass at time t, so we can proceed directly to calculate k (Equation 21.19), and then t (Equation 21.20). *Solve.*

$k = 0.693 / t_{1/2} = 0.693/27.8$ d $= 0.02493 = 0.0249$ d^{-1}

$$t = \frac{-1}{k} \ln \frac{N_t}{N_o} = \frac{-1}{0.02493 \text{ d}^{-1}} \ln \frac{0.75}{6.25} = 85.06 = 85 \text{ d}$$

21.39 (a) *Analyze/Plan.* $^{226}_{88}\text{Ra} \rightarrow {}^{222}_{86}\text{Rn} + {}^{4}_{2}\text{He}$

1 alpha particle is produced for each ^{226}Ra that decays. Rate $= kN$. Calculate the number of ^{226}Ra particles in the 10.0 mg sample. Calculate $t_{1/2}$ and k in min, then rate in dis/min, then the number of disintegrations in 5.0 min. *Solve.*

$$10.0 \text{ mg Ra} \times \frac{1 \text{ g}}{1000 \text{ mg}} \times \frac{1 \text{ mol Ra}}{226 \text{ g Ra}} \times \frac{6.022 \times 10^{23} \text{ Ra atoms}}{1 \text{ mol Ra}} = 2.6646 \times 10^{19} = 2.66 \times 10^{19} \text{ atoms}$$

Calculate k in min^{-1}. $1600 \text{ yr} \times \frac{365 \text{ d}}{1 \text{ yr}} \times \frac{24 \text{ h}}{1 \text{ d}} \times \frac{60 \text{ min}}{1 \text{ h}} = 8.410 \times 10^8 \text{ min}^{-1}$

$$k = \frac{0.693}{t_{1/2}} = \frac{0.693}{8.410 \times 10^8 \text{ min}} = 8.241 \times 10^{-10} \text{ min}^{-1}$$

Rate $= kN = (8.241 \times 10^{-10} \text{ min}^{-1})(2.6646 \times 10^{19} \text{ atoms}) = 2.20 \times 10^{10} \text{ atoms/min}$

$(2.20 \times 10^{10} \text{ atoms/min})(5.0 \text{ min}) = 1.1 \times 10^{11} {}^{226}\text{Ra atoms decay in 5.0 min}$

1.1×10^{11} alpha particles emitted in 5.0 min

(b) *Plan.* From part (a), the rate is 2.20×10^{10} disintegrations/min. Change this to dis/s and apply the definition 1 Ci $= 3.7 \times 10^{10}$ dis/s.

$$\frac{2.20 \times 10^{10} \text{ dis}}{1 \text{ min}} \times \frac{1 \text{ min}}{60 \text{ s}} \times \frac{1 \text{ Ci}}{3.7 \times 10^{10} \text{ dis/s}} \times \frac{1000 \text{ mCi}}{\text{Ci}} = 9.891 = 9.9 \text{ mCi}$$

21.40 (a) Proceeding as in Solution 21.39, calculate number of ^{60}Co atoms and k in s^{-1}.

$$3.75 \text{ mg Co} \times \frac{1 \text{ g}}{1000 \text{ mg}} \times \frac{1 \text{ mol Co}}{60 \text{ g Co}} \times \frac{6.022 \times 10^{23} \text{ Co atoms}}{1 \text{ mol Co}} = 3.76375 \times 10^{19} = 3.76 \times 10^{19}$$

$$5.26 \text{ yr} \times \frac{365 \text{ d}}{1 \text{ yr}} \times \frac{24 \text{ h}}{1 \text{ d}} \times \frac{3600 \text{ s}}{1 \text{ h}} = 1.659 \times 10^8 = 1.66 \times 10^8 \text{ s}$$

$$k = \frac{0.693}{t_{1/2}} = \frac{0.693}{1.659 \times 10^8} = 4.178 \times 10^{-9} = 4.18 \times 10^{-9} \text{ s}^{-1}$$

Rate = kN = $(4.178 \times 10^{-9} \text{ s}^{-1})(3.76375 \times 10^{19} \text{ atoms}) = 1.57 \times 10^{11} \text{ atoms/s}$

$(1.57 \times 10^{11} \text{ atoms/s})(600 \text{ s}) = 9.43 \times 10^{13}$ ^{60}Co atoms decay in 600 s

9.43×10^{13} beta particles emitted by a 3.75 mg sample in 600 s

(b) $\dfrac{1.57 \times 10^{11} \text{ dis}}{\text{s}} \times \dfrac{1 \text{ Bq}}{1 \text{ dis/s}} = 1.57 \times 10^{11} \text{ Bq}$

The activity of the sample is 1.57×10^{11} Bq.

21.41   *Analyze/Plan.* Calculate k in yr^{-1} and solve Equation 21.20 for t. N_o = 16.3/min/g, N_t = 9.7/min/g.   *Solve.*

$k = 0.693/t_{1/2} = 0.693/5715 \text{ yr} = 1.213 \times 10^{-4} = 1.21 \times 10^{-4} \text{ yr}^{-1}$

$$t = \frac{-1}{k} \ln \frac{N_t}{N_o} = \frac{-1}{1.213 \times 10^{-4} \text{ yr}^{-1}} \ln \frac{9.7}{16.3} = 4.280 \times 10^3 = 4.3 \times 10^3 \text{ yr}$$

21.42   Follow the logic in Sample Exercise 21.6.

$$t = \frac{-1}{k} \ln \frac{N_t}{N_o}; \quad k = 0.693/5715 \text{ yr} = 1.213 \times 10^{-4} \text{ yr}^{-1}$$

$$t = \frac{-1}{1.213 \times 10^{-4} \text{ yr}^{-1}} \ln \frac{38.0}{58.2} = 3.52 \times 10^3 \text{ yr}$$

21.43   *Analyze/Plan.* Follow the procedure outlined in Sample Exercise 21.6. If the mass of ^{40}Ar is 4.2 times that of ^{40}K, the original mass of ^{40}K must have been 4.2 + 1 = 5.2 times the amount of ^{40}K present now.   *Solve.*

$k = 0.693/1.27 \times 10^9 \text{ yr} = 5.457 \times 10^{-10} = 5.46 \times 10^{-10} \text{ yr}^{-1}$

$$t = \frac{-1}{5.457 \times 10^{-10} \text{ yr}^{-1}} \times \ln \frac{1}{(5.2)} = 3.0 \times 10^9 \text{ yr}$$

21.44   Follow the procedure outlined in Sample Exercise 21.6. The original quantity of ^{238}U is 75.0 mg plus the amount that gave rise to 18.0 mg of ^{206}Pb. This amount is 18.0(238/206) = 20.8 mg.

$k = 0.693/4.5 \times 10^9 \text{ yr} = 1.54 \times 10^{-10} = 1.5 \times 10^{-10} \text{ yr}^{-1}$

$$t = \frac{-1}{k} \ln \frac{N_t}{N_o} = \frac{-1}{1.54 \times 10^{-10} \text{ yr}^{-1}} \ln \frac{75.0}{95.8} = 1.59 \times 10^9 \text{ yr}$$

Energy Changes in Nuclear Reactions (Section 21.6)

21.45   *Analyze/Plan.* Use Equation 21.22 to change 0.1 mg to energy.   *Solve.*

$$\Delta E = c^2 \Delta m = (2.9979246 \times 10^8 \text{ m/s})^2 \times 0.1 \text{ mg} \times \frac{1 \text{ g}}{1000 \text{ mg}} \times \frac{1 \text{ kg}}{1000 \text{ g}} \times \frac{1 \text{ kJ}}{1000 \text{ J}} = 9 \times 10^6 \text{ kJ}$$

21.46   *Analyze/Plan.* Given a particle reaction, calculate Δm and ΔE. An alpha particle is $^{4}_{2}\text{He}$. Use Equation 21.22, $E = mc^2$. Compare the calculated energy to the energy given off by the thermite reaction. *Solve.*

$$2\,^{1}_{1}\text{p} + 2\,^{1}_{0}\text{n} \rightarrow\,^{4}_{2}\text{He}$$

Δm = mass of individual protons and neutrons – mass of $^{4}_{2}\text{He}$

$\Delta m = 2(1.0072765\ \text{amu}) + 2(1.0086649\ \text{amu}) - 4.0015\ \text{amu} = 0.030383 = 0.0304\ \text{amu}$

The mass change for the formation of 1 mol of $^{4}_{2}\text{He}$ can be expressed as 0.0304 g.

$\Delta E = c^2 \Delta m;\ \ 1\ \text{J} = \text{kg-m}^2/\text{s}^2$

$$\Delta E = 0.030383\ \text{g} \times \frac{1\ \text{kg}}{1000\ \text{g}} \times (2.9979 \times 10^8\ \text{m/s})^2 = \frac{2.73 \times 10^{12}\ \text{kg-m}^2}{\text{s}^2} = 2.73 \times 10^{12}\ \text{J}$$

The energy released when one mole of Al_2O_3 is produced is 851.5 kJ or 8.515×10^5 J. The energy released when one mole of $^{4}_{2}\text{He}$ is formed from protons and neutrons is 2.73×10^{12} J. This is 3×10^6 or 3 million times as much energy as the thermite reaction.

21.47   *Analyze/Plan.* Given the mass of an ^{27}Al atom, subtract the mass of 13 electrons to get the mass of an ^{27}Al nucleus. Calculate the mass difference between the ^{27}Al nucleus and the separate nucleons, convert this to energy using Equation 21.22. Use the molar mass of ^{27}Al to calculate the energy required for 100 g of ^{27}Al. *Solve.*

The mass of an electron is 5.485799×10^{-4} amu (inside back cover of the text). The mass of a ^{27}Al nucleus is then 26.9815386 amu $-13(5.485799 \times 10^{-4}$ amu $) = 26.9744071$ amu. $\Delta m = 13(1.0072765\ \text{amu}) + 14(1.0086649\ \text{amu}) - 26.9744071\ \text{amu} = 0.2414960\ \text{amu}$.

$$\Delta E = (2.9979246 \times 10^8\ \text{m/s})^2 \times 0.2414960\ \text{amu} \times \frac{1\ \text{g}}{6.0221421 \times 10^{23}\ \text{amu}} \times \frac{1\ \text{kg}}{1 \times 10^3\ \text{g}}$$

$$= 3.604129 \times 10^{-11} = 3.604129 \times 10^{-11}\ \text{J}/\,^{27}\text{Al nucleus required}$$

If the mass change for a single ^{27}Al nucleus is 0.2414960 amu, the mass change for 1 mole of ^{27}Al is 0.2414960 g.

$$\Delta E = 100\ \text{g}\ ^{27}\text{Al} \times \frac{1\ \text{mol}\ ^{27}\text{Al}}{26.9815386\ \text{g}\ ^{27}\text{Al}} \times \frac{0.241960}{\text{mol}\ ^{27}\text{Al}} \times \frac{1\ \text{kg}}{1000\ \text{g}} \times (2.9979246 \times 10^8\ \text{m/s})^2$$

$$= 8.044234 \times 10^{13}\ \text{J} = 8.044234 \times 10^{10}\ \text{kJ}/\,100\ \text{g}\ ^{27}\text{Al}$$

21.48 Δm = mass of individual protons and neutrons – mass of nucleus

$\Delta m = 10(1.0072765\ \text{amu}) + 11(1.0086649\ \text{amu}) - 20.98846\ \text{amu} = 0.1796189 = 0.17962\ \text{amu}$

$$\Delta E = (2.9979246 \times 10^8\ \text{m/s})^2 \times 0.1796189\ \text{amu} \times \frac{1\ \text{g}}{6.0221421 \times 10^{23}\ \text{amu}} \times \frac{1\ \text{kg}}{1000\ \text{g}}$$

$$= 2.680664 \times 10^{-11} = 2.6807 \times 10^{-11}\ \text{J}/\,^{21}\text{Ne nucleus required}$$

$$2.680664 \times 10^{-11}\ \frac{\text{J}}{\text{nucleus}} \times \frac{6.0221421 \times 10^{23}\ \text{nuclei}}{\text{mol}}$$

$$= 1.6143 \times 10^{13}\ \text{J/mol}\ ^{21}\text{Ne binding energy}$$

21.49   *Analyze/Plan.* Given atomic mass, subtract mass of the electrons to get nuclear mass. Calculate the nuclear binding energy by finding the mass difference between the nucleus and the separate nucleons and converting this to energy using Equation 21.22. Divide by the total number of nucleons to find binding energy per nucleon.   *Solve.*

(a)   Nuclear mass

^{2}H: 2.014102 amu − 1(5.485799 × 10^{-4} amu) = 2.013553 amu

^{4}He: 4.002602 amu − 2(5.485799 × 10^{-4} amu) = 4.001505 amu

^{6}Li: 6.0151228 amu − 3(5.4857991 × 10^{-4} amu) = 6.0134771 amu

(b)   Nuclear binding energy

^{2}H: Δm = 1(1.0072765) + 1(1.0086649) − 2.013553 = 0.002388 amu

$$\Delta E = 0.002388 \text{ amu} \times \frac{1 \text{ g}}{6.022142 \times 10^{23} \text{ amu}} \times \frac{1 \text{ kg}}{1000 \text{ g}} \times \frac{8.987551 \times 10^{16} \text{ m}^2}{\text{s}^2}$$

$$= 3.564490 \times 10^{-13} = 3.564 \times 10^{-13} \text{ J}$$

^{4}He: Δm = 2(1.0072765) + 2(1.0086649) − 4.001505 = 0.030378 amu

$$\Delta E = 0.030378 \text{ amu} \times \frac{1 \text{ g}}{6.022142 \times 10^{23} \text{ amu}} \times \frac{1 \text{ kg}}{1000 \text{ g}} \times \frac{8.987551 \times 10^{16} \text{ m}^2}{\text{s}^2}$$

$$= 4.533636 \times 10^{-12} = 4.5336 \times 10^{-12} \text{ J}$$

^{6}Li: Δm = 3(1.0072765) + 3(1.0086649) − 6.0134771 = 0.0343471 amu

$$\Delta E = 0.0343471 \text{ amu} \times \frac{1 \text{ g}}{6.022142 \times 10^{23} \text{ amu}} \times \frac{1 \text{ kg}}{1000 \text{ g}} \times \frac{8.987551 \times 10^{16} \text{ m}^2}{\text{s}^2}$$

$$= 5.126021 \times 10^{-12} = 5.12602 \times 10^{-12} \text{ J}$$

(c)   Binding energy per nucleon

^{2}H: 3.564490 × 10^{-13} J/2 nucleons = 1.782 × 10^{-13} J/nucleon

^{4}He: 4.533636 × 10^{-12} J/4 nucleons = 1.1334 × 10^{-12} J/nucleon

^{6}Li: 5.126021 × 10^{-12} J/6 nucleons = 8.54337 × 10^{-13} J/nucleon

(d)   This trend in binding energy/nucleon agrees with the curve in Figure 21.12, which shows an irregular increase in binding energy/nucleon up to atomic number 56. The anomalously high value for ^{4}He calculated above is also apparent on the figure.

21.50   Nuclear mass is atomic mass minus mass of electrons. Nuclear binding energy is nuclear mass minus mass of the separate nucleons, converted to energy using Equation 21.22. Divide by the total number of nucleons to find binding energy per nucleon.

(a)   Nuclear mass

^{14}N: 13.999234 − 7(5.485799 × 10^{-4} amu) = 13.995394

^{48}Ti: 47.935878 − 22(5.485799 × 10^{-4} amu) = 47.923809

^{129}Xe: 128.904779 − 54(5.485799 × 10^{-4} amu) = 128.875156 amu

(b)   Nuclear binding energy

^{14}N: $\Delta m = 7(1.0072765) + 7(1.0086649) - 13.995394 = 0.1161959 = 0.116196$ amu

$$\Delta E = 0.1161959 \text{ amu} \times \frac{1 \text{ g}}{6.0221421 \times 10^{23} \text{ amu}} \times \frac{1 \text{ kg}}{1000 \text{ g}} \times \frac{8.987551 \times 10^{16} \text{ m}^2}{\text{s}^2}$$

$$= 1.734127 \times 10^{-11} = 1.73413 \times 10^{-11} \text{ J}$$

^{48}Ti: $\Delta m = 22(1.0072765) + 26(1.0086649) - 47.923809 = 0.4615614 = 0.461561$ amu

$$\Delta E = 0.4615614 \text{ amu} \times \frac{1 \text{ g}}{6.0221421 \times 10^{23} \text{ amu}} \times \frac{1 \text{ kg}}{1000 \text{ g}} \times \frac{8.987551 \times 10^{16} \text{ m}^2}{\text{s}^2}$$

$$= 6.888428 \times 10^{-11} = 6.88843 \times 10^{-11} \text{ J}$$

^{129}Xe: $\Delta m = 54(1.0072765) + 75(1.0086649) - 128.875156 = 1.1676425 = 1.167643$ amu

$$\Delta E = 1.1676425 \text{ amu} \times \frac{1 \text{ g}}{6.0221421 \times 10^{23} \text{ amu}} \times \frac{1 \text{ kg}}{1000 \text{ g}} \times \frac{8.987551 \times 10^{16} \text{ m}^2}{\text{s}^2}$$

$$= 1.742610 \times 10^{-10} \text{ J}$$

(c)   Binding energy per nucleon

^{14}N: 1.73413×10^{-11} J / 14 nucleons $= 1.23866 \times 10^{-12}$ J/nucleon

^{48}Ti: 6.888428×10^{-11} J / 48 nucleons $= 1.43509 \times 10^{-12}$ J/nucleon

^{129}Xe: 1.742610×10^{-10} J/129 nucleons $= 1.350860 \times 10^{-12}$ J/nucleon

21.51   *Analyze/Plan.* Use Equation 21.22 to calculate the mass equivalence of the solar radiation.   *Solve.*

(a) $$\frac{1.07 \times 10^{16} \text{ kJ}}{1 \text{ min}} \times \frac{60 \text{ min}}{1 \text{ h}} \times \frac{24 \text{ h}}{1 \text{ d}} = 1.541 \times 10^{19} \frac{\text{kJ}}{\text{d}} = 1.54 \times 10^{22} \text{ J/d}$$

$$\Delta m = \frac{1.541 \times 10^{22} \text{ kg-m}^2/\text{s}^2/\text{d}}{(2.998 \times 10^8 \text{ m/s})^2} = 1.714 \times 10^5 = 1.71 \times 10^5 \text{ kg/d}$$

(b)   *Analyze/Plan.* Calculate the mass change in the given nuclear reaction, then a conversion factor for g ^{235}U to mass equivalent.   *Solve.*

$\Delta m = 140.8833 + 91.9021 + 2(1.0086649) - 234.9935 = -0.19077 = -0.1908$ amu

Converting from atoms to moles and amu to grams, it requires 1.000 mol or 235.0 g ^{235}U to produce energy equivalent to a change in mass of 0.1908 g.

0.10% of 1.714×10^5 kg is 1.714×10^2 kg $= 1.714 \times 10^5$ g

$$1.714 \times 10^5 \text{ g} \times \frac{235.0 \text{ g} \,^{235}\text{U}}{0.1908 \text{ g}} = 2.111 \times 10^8 = 2.1 \times 10^8 \text{ g} \,^{235}\text{U}$$

(This is about 230 tons of ^{235}U *per day*.)

21.52   First, calculate nuclear masses from atomic masses. Then, the calculated Δm is for one group of single nuclides involved in a reaction, labeled Δm/'atomic reaction.' Multiplying by Avogadro's number changes the quantity to 'mol of reaction.' Because energy is released, the sign of ΔE is negative.

^{1}H: 1.00782 amu $- 1(5.485799 \times 10^{-4}$ amu$) = 1.0072714 = 1.00727$ amu

^{2}H: 2.01410 amu $- 1(5.485799 \times 10^{-4}$ amu$) = 2.0135514 = 2.01355$ amu

^{3}H: 3.10605 amu $- 1(5.485799 \times 10^{-4}$ amu$) = 3.1055014 = 3.10550$ amu

^{3}He: 3.10603 amu $- 2(5.485799 \times 10^{-4}$ amu$) = 3.1049328 = 3.10493$ amu

^{4}He: 4.00260 amu $- 2(5.485799 \times 10^{-4}$ amu$) = 4.0015028 = 4.00150$ amu

(a) $\Delta m = 4.0015028 + 1.0086649 - 3.1055014 - 2.0135514 = -0.1088851 = -0.10889$ amu

$$\Delta E = \frac{-0.1088851 \text{ amu}}{\text{'atomic reaction'}} \times \frac{1 \text{ g}}{6.022 \times 10^{23} \text{ amu}} \times \frac{6.022 \times 10^{23} \text{ 'atomic reaction'}}{\text{mol of reaction}}$$

$$\times \frac{1 \text{ kg}}{10^{3} \text{ g}} \times (2.99792458 \times 10^{8} \text{ m/s})^{2} = -9.7861 \times 10^{12} \text{ J/mol}$$

(b) $\Delta m = 3.1049328 + 1.0086649 - 2(2.0135514) = -0.0864949 = -0.08649$ amu

$\Delta E = -7.774 \times 10^{12}$ J/mol

(c) $\Delta m = 4.0015028 + 1.0072714 - 2.0135514 - 3.1049328 = -0.1097100 = -0.10971$ amu

$\Delta E = -9.8602 \times 10^{12}$ J/mol

21.53   Nucleus (b), ^{51}V, should possess the greatest mass defect per nucleon. Figure 21.12 shows that the binding energy per nucleon (which gives rise to the mass defect) is greatest for nuclei with mass numbers around 50.

21.54   Nuclear mass $= 61.928345$ amu $- 28(5.485799 \times 10^{-4}$ amu$) = 61.912985$

Binding energy $= 28(1.0072765) + 34(1.0086649) - 61.912985 = 0.585363$ amu

$$\Delta E = 0.585363 \text{ amu} \times \frac{1 \text{ g}}{6.0221421 \times 10^{23} \text{ amu}} \times \frac{1 \text{ kg}}{1000 \text{ g}} \times \frac{8.987551 \times 10^{16} \text{ m}^{2}}{\text{s}^{2}}$$

$$= 8.73606 \times 10^{-11} \text{ J}$$

Binding energy/nucleon $= 8.73606 \times 10^{-11}$ J $/ 62 = 1.40904 \times 10^{-12}$ J/nucleon

The value given for iron-56 in Table 21.7 is 1.41×10^{-12} J/nucleon. These values are the same, to three significant figures.

Nuclear Power and Radioisotopes (Sections 21.7, 21.8, and 21.9)

21.55   (a)   NaI is a good source of iodine, because it is a strong electrolyte and completely dissociated into ions in aqueous solution. The I^{-}(aq) are mobile and immediately available for bio-uptake. They do not need to be digested or processed in the body before uptake can occur. Also, iodine is a large percentage of the total mass of NaI.

(b)   After ingestion, I^{-}(aq) must enter the bloodstream, travel to the thyroid and then be absorbed. This requires a finite amount of time. A Geiger counter placed near the thyroid immediately after ingestion will register background, then gradually increase in signal until the concentration of I^{-}(aq) in the thyroid reaches a maximum. Then, over time, iodine-131 decays, and the signal decreases.

(c)   *Analyze/Plan.* The half-life of iodine-131 is 8.02 days. Use $t_{1/2}$ to calculate the decay rate constant, k. Then solve Equation 21.20 for t. $N_{o} = 0.12$ (12% of ingested iodine absorbed); $N_{t} = 1.2 \times 10^{-5}$ (0.01% of the original amount taken up by the thyroid).

Solve.   k = 0.693 t$_{1/2}$ = 0.693/8.02 d = 0.086409 = 0.0864 d^{-1}

ln(N$_t$/N$_o$) = –kt; t = –ln(N$_t$/N$_o$)/k

$$t = \frac{-\ln(1.2\times10^{-5}/0.12)}{0.086409 \text{ d}^{-1}} = 106.59 = 107 \text{ d}$$

Check. N$_t$ is given to 1 sig fig, so 1 × 10^2 days may be a more correct representation of the time frame for decay.

21.56   Radioisotopes used as diagnostic tools are introduced into the body and carried to the point where imaging or some other diagnostic data is needed. We want the decay products of these radioisotopes to leave the body and do as little damage as possible on the way. Gamma rays are penetrating radiation and can escape the body more easily than other radioactive decay products. Also, gamma rays leaving the body can be easily detected using scintillation counters. This is particularly important when imaging is the goal of the procedure.

Alpha emitters are never used as diagnostic tools because alpha particles are ionizing but do not move easily through the body. Trapped inside the body, alpha particles initiate the ionization of water, which ultimately produces free radicals that disrupt the normal operation of cells.

21.57   (a)   Characteristics (ii) and (iv) are required for a fuel in a nuclear power plant.

Two or more neutrons (ii) are required so that a nuclear chain reaction occurs. Fission after absorption of a slow neutron (iv) is the nuclear process that produces energy in all current nuclear power plants. Gamma radiation (i) is produced by most nuclear decay processes and is a non-specific characteristic. Short half-life (iii) would require that fuel be replaced too frequently.

(b) ^{235}U

21.58   Statements (i) and (iv) are true. Natural uranium contains about 0.7% ^{235}U; it must be enriched to about 3–5% for use as a fuel. ^{238}U undergoes neutron-induced fission according to Equation 21.12; no neutrons are produced.

21.59   The *control rods* in a nuclear reactor regulate the flux of neutrons to keep the reaction chain self-sustaining and also prevent the reactor core from overheating. They are composed of materials such as boron or cadmium that absorb neutrons.

21.60   (a)   A moderator slows neutrons, so that they are more easily captured by fissioning nuclei.

(b)   Water is the moderator in a pressurized water generator.

(c)   Graphite is used as a moderator in gas-cooled reactors, and D$_2$O is used in heavy-water reactors.

21.61   (a) $^{2}_{1}\text{H} + ^{2}_{1}\text{H} \rightarrow ^{3}_{2}\text{He} + ^{1}_{0}\text{n}$

(b) $^{239}_{92}\text{U} + ^{1}_{0}\text{n} \rightarrow ^{133}_{51}\text{Sb} + ^{98}_{41}\text{Nb} + 9\,^{1}_{0}\text{n}$

21.62   *Analyze/Plan.* Use conservation of A and Z to complete the equations, keeping in mind the symbols and definitions of various decay products.   *Solve.*

(a) $^{235}_{92}U + ^{1}_{0}n \rightarrow ^{160}_{62}Sm + ^{72}_{30}Zn + 4\ ^{1}_{0}n$

(b) $^{239}_{94}Pu + ^{1}_{0}n \rightarrow ^{144}_{58}Ce + ^{94}_{36}Kr + 2\ ^{1}_{0}n$

21.63   *Analyze/Plan.* At these temperatures, assume the reaction occurs between nuclei rather than atoms. From Table 21.7, the nuclear mass of $^{4}_{2}He$ is 4.00150. The nuclear mass of $^{1}_{1}H$ is simply the mass of a proton, 1.007276467 amu. Note that $^{0}_{+1}e$ is a positron, which has the same mass as an electron, 5.4857991×10^{-4} amu. Calculate the difference in mass between product and reactant nuclei, and the energy released by this mass change. Do the calculation in terms of moles and grams, rather than nuclei and amu. 1 mol amu = 1 g.   *Solve.*

$\Delta m = 4.00150 + 2(5.4857991 \times 10^{-4}) - 4(1.007276467) = -0.0265087 = -0.02651$ g

If the reaction is run with 1 mol of $^{1}_{1}H$, the mass change is $(0.0265087/4) = 0.0066272 = 0.006627$ g

$$\Delta E = c^2 \Delta m = 0.0066272\ \text{g} \times \frac{1\ \text{kg}}{1000\ \text{g}} \times \frac{8.987551 \times 10^{16}\ \text{m}^2}{\text{s}^2} = 5.95621 \times 10^{11}$$

$$= 5.956 \times 10^{11}\ \text{J} = 5.956 \times 10^{8}\ \text{kJ}$$

21.64   (a)   If the spent fuel rods are more radioactive than the original rods, the products of fission must lie outside the belt of stability and be radioactive themselves.

(b)   The heavy (Z > 83) nucleus has a high neutron/proton ratio. The lighter radioactive fission products (for example, barium-142 and krypton-91) also have high neutron/proton ratios, because only 2 or 3 free neutrons are produced during fission. The preferred decay mode to reduce the neutron/proton ratio is beta decay, which has the effect of converting a neutron into a proton. Both barium-142 (86 n, 56 p) and krypton-91 (55 n, 36 p) undergo beta decay.

21.65   (a)   A *boiling water reactor* does not use a secondary coolant.

(b)   A *fast breeder reactor* creates more fissionable material than it consumes.

(c)   A *gas-cooled reactor* uses a gas as a primary coolant.

21.66   (a)   A *heavy water reactor* and a *gas-cooled rector* can use natural uranium as a fuel.

(b)   A *fast breeder reactor* does not use a moderator.

(c)   A *high-temperature pebble-bed reactor* can be refueled without shutting down.

21.67   *Analyze/Plan.* Hydroxyl radical is electrically neutral but has an unpaired electron, • OH. Hydroxide is an anion, OH^-.   *Solve.*

Hydrogen abstraction: $RCOOH + \cdot OH \rightarrow RCOO \cdot + H_2O$

Deprotonation: $RCOOH + OH^- \rightarrow RCOO^- + H_2O$

Hydroxyl radical is more toxic to living systems, because it produces other radicals when it reacts with molecules in the organism. This often starts a disruptive chain of reactions, each producing a different free radical.

Hydroxide ion, OH^-, on the other hand, will be readily neutralized in the buffered cell environment. Its most common reaction is ubiquitous and innocuous:

$H^+ + OH^- \rightarrow H_2O$. The acid–base reactions of OH^- are usually much less disruptive to the organism than the chain of redox reactions initiated by $\cdot OH$ radical.

21.68   X-rays, alpha particles and gamma rays are classified as ionizing radiation.

21.69   *Analyze/Plan.* Use definitions of the various radiation units and conversion factors to calculate the specified quantities. Pay particular attention to units.   *Solve.*

(a) $1\ Ci = 3.7 \times 10^{10}$ disintegrations(dis)/s; $1\ Bq = 1$ dis/s

$14.3\ mCi \times \dfrac{1\ Ci}{1000\ m\ Ci} \times \dfrac{3.7 \times 10^{10}\ dis/s}{Ci} = 5.29 \times 10^8 = 5.3 \times 10^8\ dis/s = 5.3 \times 10^8\ Bq$

(b) $1\ rad = 1 \times 10^{-2}\ J/kg$; $1\ Gy = 1\ J/kg = 100\ rad$. From part (a), the activity of the source is 5.3×10^8 dis/s.

$5.29 \times 10^8\ dis/s \times 14.0\ s \times 0.35 \times \dfrac{9.12 \times 10^{-13}\ J}{dis} \times \dfrac{1}{0.385\ kg} = 6.14 \times 10^{-3} = 6.1 \times 10^{-3}\ J/kg$

$6.1 \times 10^{-3}\ J/kg \times \dfrac{1\ rad}{1 \times 10^{-2}\ J/kg} \times \dfrac{1000\ mrad}{rad} = 6.1 \times 10^2\ mrad$

$6.1 \times 10^{-3}\ J/kg \times \dfrac{1\ Gy}{1\ J/kg} = 6.1 \times 10^{-3}\ Gy$

(c)   rem = rad (RBE); Sv = Gy (RBE) , where 1 Sv = 100 rem

mrem $= 6.14 \times 10^2$ mrad $(9.5) = 5.83 \times 10^3 = 5.8 \times 10^3$ mrem (or 5.8 rem)

Sv $= 6.14 \times 10^{-3}$ Gy $(9.5) = 5.83 \times 10^{-2} = 5.8 \times 10^{-2}$ Sv

21.70   (a) $1\ Ci = 3.7 \times 10^{10}$ dis/s; $1\ Bq = 1$ dis/s

$15\ mCi \times \dfrac{1\ Ci}{1000\ mCi} \times 3.7 \times 10^{10}\ dis/s = 5.55 \times 10^8 = 5.6 \times 10^8\ dis/s = 5.6 \times 10^8\ Bq$

(b) $1\ Gy = 1\ J/kg$; $1\ Gy = 100\ rad$

$5.55 \times 10^8\ dis/s \times 240\ s \times 0.075 \times \dfrac{8.75 \times 10^{-14}\ J}{dis} \times \dfrac{1}{65\ kg} = 1.345 \times 10^{-5} = 1.3 \times 10^{-5}\ J/kg$

$1.3 \times 10^{-5}\ J/kg \times \dfrac{1\ Gy}{1\ J/kg} = 1.3 \times 10^{-5}\ Gy; 1.3 \times 10^{-5}\ Gy \times \dfrac{100\ rad}{1\ Gy} = 1.3 \times 10^{-3}\ rad$

(c)   rem = rad (RBE); Sv = Gy (RBE)

$1.3 \times 10^{-3}\ rad\ (1.0) = 1.3 \times 10^{-3}\ rem \times \dfrac{1000\ mrem}{1\ rem} = 1.3\ mrem$

$1.3 \times 10^{-5}\ Gy\ (1.0) = 1.3 \times 10^{-5}\ Sv$

(d)   The radiation dose (1.3 mrem) is much less than that for a typical mammogram.

Additional Exercises

21.71 *Analyze/Plan.* Atomic number is number of protons; mass number is number of (protons + neutrons). The element symbol is determined by atomic number. Check the list of magic numbers, Figure 21.1 and Table 21.4 to determine which isotope is unstable. Use information in Table 21.3 about positron emission to determine which isotope will produce potassium-39.

(a) (i) $^{38}_{19}K$ (ii) $^{40}_{19}K$ (iii) $^{39}_{20}Ca$ (iv) $^{40}_{20}Ca$

(b) $^{38}_{19}K$ is most likely to be unstable. Both $^{38}_{19}K$ and $^{40}_{19}K$ are odd proton/odd neutron isotopes. $^{38}_{19}K$ is more likely to be unstable because it has a lower neutron/proton ratio. (The two isotopes of Ca have magic numbers of protons and are thus more likely to be stable.)

(c) $^{39}_{20}Ca$ has a magic number of protons. $^{40}_{20}Ca$ has magic numbers of protons and neutrons.

(d) $^{39}_{20}Ca \rightarrow ^{39}_{19}K + ^{0}_{+1}e$

21.72 $^{222}_{86}Rn \rightarrow X + 3\,^{4}_{2}He + 2\,^{0}_{-1}e$

This corresponds to a reduction in mass number of $(3 \times 4 =)$ 12 and a reduction in atomic number of $(3 \times 2 - 2) = 4$. The stable nucleus is $^{210}_{82}Pb$. (This is part of the sequence in Figure 21.1.)

21.73 $^{3}_{2}He + ^{3}_{2}He \rightarrow ^{4}_{2}He + 2\,^{1}_{1}H$. Use nuclear masses calculated in Solution 21.52.

$\Delta m = 4.00150 + 2(1.00727) - 2(3.10493) = -0.19382$ amu

$$\Delta E = \frac{-0.19382 \text{ amu}}{\text{'atomic reaction'}} \times \frac{1 \text{ g}}{6.022 \times 10^{23} \text{ amu}} \times \frac{6.022 \times 10^{23} \text{ 'atomic reaction'}}{\text{mol of reaction}}$$

$$\times \frac{1 \text{ kg}}{10^3 \text{ g}} \times (2.99792458 \times 10^8 \text{ m/s})^2 = -1.7420 \times 10^{13} \text{ J/mol}$$

21.74 (a) $^{36}_{17}Cl \rightarrow ^{36}_{18}Ar + ^{0}_{-1}e$

(b) According to Table 21.4, nuclei with even numbers of both protons and neutrons, or an even number of one kind of nucleon, are more stable. ^{35}Cl and ^{37}Cl both have an odd number of protons *but* an even number of neutrons. ^{36}Cl has an odd number of protons and neutrons (17 p, 19 n), so it is less stable than the other two isotopes. Also, ^{37}Cl has 20 neutrons, a nuclear closed shell.

21.75 $2\,^{1}_{1}p \rightarrow ^{2}_{1}H + ^{0}_{1}e$

21.76 (a) $^{6}_{3}Li + ^{56}_{28}Ni \rightarrow ^{62}_{31}Ga$

(b) $^{40}_{20}Ca + ^{248}_{96}Cm \rightarrow ^{147}_{62}Sm + ^{141}_{54}Xe$

(c) $^{88}_{38}Sr + ^{84}_{36}Kr \rightarrow ^{116}_{46}Pd + ^{56}_{28}Ni$

(d) $^{40}_{20}Ca + ^{238}_{92}U \rightarrow ^{70}_{30}Zn + 4\,^{1}_{0}n + 2\,^{102}_{41}Nb$

21.77   (a)   Z is $^{297}_{117}$Ts; Q is $^{48}_{20}$Ca

        (b)   Isotope Q has 20 protons and 28 neutrons; these are both magic numbers. Even though Q has an unfavorable neutron-to-proton ratio, the special stability associated with magic numbers of protons and neutrons explains its long half-life.

        (c)   The target isotope was $^{248}_{96}$Cm. $^{248}_{96}$Cm + $^{48}_{20}$Ca $\rightarrow$ $^{296}_{116}$Lv

21.78

Time (h)	N_t (dis/min)	ln N_t
0	180	5.193
2.5	130	4.868
5.0	104	4.644
7.5	77	4.34
10.0	59	4.08
12.5	46	3.83
17.5	24	3.18

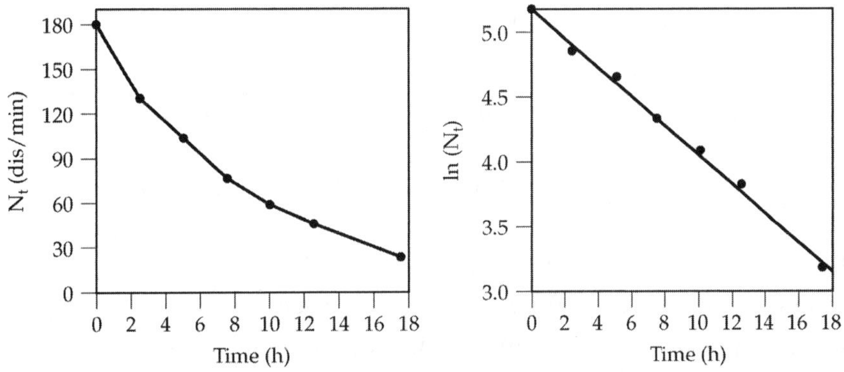

The plot on the left is a graph of activity (disintegrations per minute) vs. time. Choose $t_{1/2}$ at the time where $N_t = 1/2\, N_o = 90$ dis/min. $t_{1/2} \approx 6.0$ h.

Rearrange Equation 21.20 to obtain the linear relationship shown on the right.

$\ln(N_t / N_o) = -kt$; $\ln N_t - \ln N_o = -kt$; $\ln N_t = -kt + \ln N_o$

The slope of this line $= -k = -0.11$; $t_{1/2} = 0.693/0.11 = 6.3$ h.

21.79 1×10^{-6} curie $\times \dfrac{3.7 \times 10^{10} \text{ dis/s}}{\text{curie}} = 3.7 \times 10^4$ dis/s

rate $= 3.7 \times 10^4$ nuclei/s $= kN$

$k = \dfrac{0.693}{t_{1/2}} = \dfrac{0.693}{28.8 \text{ yr}} \times \dfrac{1 \text{ yr}}{365 \times 24 \times 3600 \text{ s}} = 7.630 \times 10^{-10} = 7.63 \times 10^{-10} \text{ s}^{-1}$

3.7×10^4 nuclei/s $= (7.63 \times 10^{-10}/\text{s})$ N; N $= 4.849 \times 10^{13} = 4.8 \times 10^{13}$ ^{90}Sr nuclei

mass ^{90}Sr $= 4.849 \times 10^{13}$ nuclei $\times \dfrac{89.907738 \text{ g Sr}}{6.022 \times 10^{23} \text{ nuclei}} = 7.2 \times 10^{-9}$ g Sr

(mass of ^{90}Sr from webelements.com)

21.80 (a) The C—OH bond of the acid and the O—H bond of the alcohol break in this reaction. Initially, ^{18}O is present in the C—^{18}OH group of the alcohol. In order for ^{18}O to end up in the ester, the ^{18}O—H bond of the alcohol must break. This requires that the C—OH bond in the acid also breaks. The unlabeled O from the acid ends up in the H_2O product.

(b) No. When $TOCH_3$ is used to react with CH_3COOH, T will end up in the H_2O product, regardless of whether the C—OT or O—T bond breaks in the reaction.

21.81 (a) (i) X is $^{11}_{6}C$. The long-hand reaction is $^{14}_{7}N + ^{1}_{1}p \rightarrow ^{11}_{6}C + ^{4}_{2}He$.

(ii) X is $^{1}_{0}n$. The long-hand reaction is $^{18}_{8}O + ^{1}_{1}p \rightarrow ^{18}_{9}F + ^{1}_{0}n$.

(b) (iii) d is $^{2}_{1}H$. The long-hand reaction is $^{14}_{7}N + ^{2}_{1}H \rightarrow ^{15}_{8}O + ^{1}_{0}n$.

It makes sense that "d" represents deuterium, $^{2}_{1}H$, an isotope of hydrogen.

21.82 Because of the relationship $\Delta E = \Delta mc^2$, the mass defect (Δm) is directly related to the binding energy (ΔE) of the nucleus.

^{7}Be: 4p, 3n; $4(1.0072765) + 3(1.0086649) = 7.05510$ amu

Total mass defect = $7.0551 - 7.0147 = 0.0404$ amu

0.0404 amu/7 nucleons = 5.77×10^{-3} amu/nucleon

$$\Delta E = \Delta m \times c^2 = \frac{5.77 \times 10^{-3} \text{ amu}}{\text{nucleon}} \times \frac{1g}{6.022 \times 10^{23} \text{ amu}} \times \frac{1 kg}{1 \times 10^3 \text{ g}} \times \frac{8.988 \times 10^{16} \text{ m}^2}{s^2}$$

$$= \frac{5.77 \times 10^{-3} \text{ amu}}{\text{nucleon}} \times \frac{1.4925 \times 10^{-10} \text{ J}}{1 \text{ amu}} = 8.612 \times 10^{-13} = 8.61 \times 10^{-13} \text{ J/nucleon}$$

^{9}Be: 4p, 5n; $4(1.0072765) + 5(1.0086649) = 9.07243$ amu

Total mass defect = $9.0724 - 9.0100 = 0.06243 = 0.0624$ amu

0.0624 amu/9 nucleons = $6.937 \times 10^{-3} = 6.94 \times 10^{-3}$ amu/nucleon

6.937×10^{-3} amu/nucleon $\times 1.4925 \times 10^{-10}$ J/amu $= 1.035 \times 10^{-12} = 1.04 \times 10^{-12}$ J/nucleon

^{10}Be: 4p, 6n; $4(1.0072765) + 6(1.0086649) = 10.0811$ amu

Total mass defect = $10.0811 - 10.0113 = 0.0698$ amu

0.0698 amu/10 nucleons = 6.98×10^{-3} amu/nucleon

6.98×10^{-3} amu/nucleon $\times 1.4925 \times 10^{-10}$ J/amu $= 1.042 \times 10^{-12} = 1.04 \times 10^{-12}$ J/nucleon

The binding energies/nucleon for ^{9}Be and ^{10}Be are very similar; that for ^{10}Be is slightly higher.

21.83 First, calculate k in s^{-1}

$$k = \frac{0.693}{12.3 \text{ yr}} \times \frac{1 yr}{365 d} \times \frac{1 d}{24 h} \times \frac{1 h}{3600 s} = 1.7866 \times 10^{-9} = 1.79 \times 10^{-9} \text{ s}^{-1}$$

From Equation 21.18, $1.50 \times 10^3 \text{ s}^{-1} = (1.7866 \times 10^{-9} \text{ s}^{-1})(N)$;

$N = 8.396 \times 10^{11} = 8.40 \times 10^{11}$. In 26.00 g of water, there are

$$26.00 \text{ g } H_2O \times \frac{1 \text{ mol } H_2O}{18.02 \text{ g } H_2O} \times \frac{6.022 \times 10^{23} \text{ } H_2O}{1 \text{ mol } H_2O} \times \frac{2 H}{1 H_2O} = 1.738 \times 10^{24} \text{ H atoms}$$

The mole fraction of $^{3}_{1}H$ atoms in the sample is thus

$8.396 \times 10^{11}/1.738 \times 10^{24} = 4.831 \times 10^{-13} = 4.83 \times 10^{-13}$

21.84 (a) $\Delta m = \Delta E/c^2$; $\Delta m = \dfrac{3.9 \times 10^{26} \text{ J/s}}{(2.9979246 \times 10^8 \text{ m/s})^2} \times \dfrac{1 \text{ kg-m}^2/\text{s}^2}{1 \text{ J}} = 4.3 \times 10^9 \text{ kg/s}$

The rate of mass loss is 4.3×10^9 kg/s. (Fewer sig figs in the value of c produce the same result.)

(b) The mass loss arises from fusion reactions that produce more stable nuclei from less stable ones, e.g., Equations 21.26–21.29.

(c) Express the mass lost by the sun in terms of protons per second consumed in fusion reactions like Equations 21.26, 21.27, and 21.29.

$$\dfrac{4.3 \times 10^9 \text{ kg}}{\text{s}} \times \dfrac{1 \text{ proton}}{1.673 \times 10^{-24} \text{ g}} \times \dfrac{1000 \text{ g}}{1 \text{ kg}} = 2.594 \times 10^{36} = 3 \times 10^{36} \text{ protons/s}$$

21.85 $1000 \text{ Mwatts} \times \dfrac{1 \times 10^6 \text{ watts}}{1 \text{ Mwatt}} \times \dfrac{1 \text{ J}}{1 \text{ watt-s}} \times \dfrac{1 \ ^{235}\text{U atom}}{3 \times 10^{-11} \text{ J}} \times \dfrac{1 \text{ mol U}}{6.02214 \times 10^{23} \text{ atoms}}$

$\times \dfrac{235 \text{ g U}}{1 \text{ mol}} \times \dfrac{3600 \text{ s}}{1 \text{ h}} \times \dfrac{24 \text{ h}}{1 \text{ d}} \times \dfrac{365 \text{ d}}{1 \text{ yr}} \times \dfrac{100}{40} \text{(efficiency)} = 1.03 \times 10^6 = 1 \times 10^6 \text{ g U/yr}$

21.86 $2 \times 10^{-12} \text{ curies} \times \dfrac{3.7 \times 10^{10} \text{ dis/s}}{1 \text{ curie}} = 7.4 \times 10^{-2} = 7 \times 10^{-2} \text{ dis/s}$

$\dfrac{7.4 \times 10^{-2} \text{ dis/s}}{75 \text{ kg}} \times \dfrac{8 \times 10^{-13} \text{ J}}{\text{dis}} \times \dfrac{1 \text{ rad}}{1 \times 10^{-2} \text{ J/g}} \times \dfrac{3600 \text{ s}}{\text{h}} \times \dfrac{24 \text{ h}}{1 \text{ d}}$

$\times \dfrac{365 \text{ d}}{1 \text{ yr}} = 2.49 \times 10^{-6} = 2 \times 10^{-6} \text{ rad/yr}$

Recall that there are 10 rem/rad for alpha particles.

$\dfrac{2.49 \times 10^{-6} \text{ rad}}{1 \text{ yr}} \times \dfrac{10 \text{ rem}}{1 \text{ rad}} = 2.49 \times 10^{-5} = 2 \times 10^{-5} \text{ rem/yr}$

Integrative Exercises

21.87 Calculate the molar mass of $NaClO_4$ that contains 29.6% ^{36}Cl. Atomic mass of the enhanced Cl is 0.296(36.0) + 0.704(35.453) = 35.615 = 35.6. The molar mass of $NaClO_4$ is then (22.99 + 35.615 + 64.00) = 122.605 = 122.6. Calculate N, the number of ^{36}Cl nuclei, the value of k in s^{-1}, and the activity in dis/s.

$53.8 \text{ mg NaClO}_4 \times \dfrac{1 \text{ g}}{1000 \text{ mg}} \times \dfrac{1 \text{ mol NaClO}_4}{122.605 \text{ g NaClO}_4} \times \dfrac{1 \text{ mol Cl}}{1 \text{ mol NaClO}_4} \times \dfrac{6.022 \times 10^{23} \text{ Cl atoms}}{\text{mol Cl}}$

$\times \dfrac{29.6 \ ^{36}\text{Cl atoms}}{100 \text{ Cl atoms}} = 7.822 \times 10^{19} = 7.82 \times 10^{19} \ ^{36}\text{Cl atoms}$

$k = 0.693/t_{1/2} = \dfrac{0.693}{3.0 \times 10^5 \text{ yr}} \times \dfrac{1 \text{ yr}}{365 \times 24 \times 3600 \text{ s}} = 7.32 \times 10^{-14} = 7.3 \times 10^{-14} \text{ s}^{-1}$

$\text{rate} = kN = (7.32 \times 10^{-14} \text{ s}^{-1})(7.822 \times 10^{19} \text{ nuclei}) = 5.729 \times 10^6 = 5.7 \times 10^6 \text{ dis/s}$

21.88   Calculate the amount of energy produced by the nuclear fusion reaction, the enthalpy of combustion, $\Delta H°$, of C_8H_{18}, and then the mass of C_8H_{18} required.

Δm for the reaction $4\,{}^1_1H \rightarrow {}^4_2He + 2\,{}^0_{+1}e$ is:

$4(1.00782) - 4.00260$ amu $- 2(5.4858 \times 10^{-4}$ amu$) = 0.027583 = 0.02758$ amu

$$\Delta E = \Delta mc^2 = 0.027583 \text{ amu} \times \frac{1\,g}{6.02214 \times 10^{23}\text{ amu}} \times \frac{1\,kg}{1000\,g} \times (2.9979246 \times 10^8 \text{ m/s})^2$$

$$= 4.11654 \times 10^{-12} = 4.117 \times 10^{-12} \text{ J}/4\,{}^1H \text{ nuclei}$$

$$1.0\,g\,{}^1H \times \frac{1\,{}^1H \text{ nucleus}}{1.00782 \text{ amu}} \times \frac{6.02214 \times 10^{23} \text{ amu}}{g} \times \frac{4.11654 \times 10^{-12} \text{ J}}{4\,{}^1H \text{ nuclei}}$$

$$= 6.1495 \times 10^{11} \text{ J} = 6.1 \times 10^8 \text{ kJ produced by the fusion of } 1.0\,g\,{}^1H.$$

$C_8H_{18}(l) + 25/2\,O_2(g) \rightarrow 8\,CO_2(g) + 9\,H_2O(g)$

$\Delta H° = 8(-393.5 \text{ kJ}) + 9(-241.82 \text{ kJ}) - (-250.1 \text{ kJ}) = -5074.3 \text{ kJ}$

$$6.1495 \times 10^8 \text{ kJ} \times \frac{1 \text{ mol } C_8H_{18}(l)}{5074.3 \text{ kJ}} \times \frac{114.231\,g\,C_8H_{18}}{\text{mol } C_8H_{18}} = 1.384 \times 10^7\,g = 1.4 \times 10^4 \text{ kg } C_8H_{18}$$

14,000 kg $C_8H_{18}(l)$ would have to be burned to produce the same amount of energy as fusion of 1.0 g ${}^1H$.

21.89   Refer to "Chemistry Put to Work: Gas Separations" in Section 10.8.

(a)   The atomic weight of naturally occurring uranium is 238.02891 g/mol. The molar mass of UF_6 is then $[238.02891 + 6(18.998403)] = 352.01933$ g/mol.

$$g = \frac{MM \times RT}{VP} = \frac{352.02\,g}{30.0\,L} \times \frac{0.082058 \text{ L-atm}}{\text{mol-K}} \times \frac{350\,K}{(695/760) \text{ atm}} = 368.522 = 369\,g$$

Check. This seems like a large mass, but 30.0 L is more than the molar volume of a gas at STP, so it is reasonable that more than one mole of UF_6 is in the flask.

(b)   Of the 369 g sample, 0.720% is ${}^{235}UF_6$.

$(368.522 \times 0.00720) = 2.65336 = 2.65\,g\,{}^{235}UF_6$.

The mass of ${}^{235}U$ in 2.65 g ${}^{235}UF_6$ is

$$2.65336\,g\,{}^{235}UF_6 \times \frac{235.044\,g\,{}^{235}U}{349.034\,g\,{}^{235}UF_6} = 1.78681 = 1.79\,g\,{}^{235}U$$

(c)   According to information in "Chemistry Put to Work: Gas Separations," the ratio of effusion rates is 1.0043. That is, ${}^{235}UF_6$ effuses (diffuses) 1.0043 times faster than ${}^{238}UF_6$. That is, the mass of ${}^{235}UF_6$ in the diffused sample is 1.0043 times greater than in the initial sample. The mass of ${}^{235}UF_6$ in the diffused sample is then $(2.65336\,g\,{}^{235}UF_6 \text{ initial} \times 1.0043) = 2.66477 = 2.66\,g\,{}^{235}UF_6$.

$$2.66477\,g\,{}^{235}UF_6 \times \frac{235.044\,g\,{}^{235}U}{349.034\,g\,{}^{235}UF_6} = 1.79449 = 1.79\,g\,{}^{235}U$$

(d)   One more round of enrichment yields $(0.266477 \text{ g } ^{235}UF_6 \times 1.0043) = 2.67623 = 2.68$ g $^{235}UF_6$. The mass % of $^{235}UF_6$ in the sample is then

$$\frac{2.67623 \text{ g } ^{235}UF_6}{368.522 \text{ g sample}} \times 100 = 0.726206 = 0.726 \% \ ^{235}UF_6$$

[This is the same result as twice applying the 1.0043 times enrichment to the natural abundance of ^{235}U. $(0.720\% \times 1.0043 \times 1.0043) = 0.726205 = 0.726\%$]

21.90   (a) $0.18 \text{ Ci} \times \dfrac{3.7 \times 10^{10} \text{ dis/s}}{\text{Ci}} \times \dfrac{3600 \text{ s}}{\text{h}} \times \dfrac{24 \text{ h}}{\text{d}} \times 245 \text{ d} = 1.41 \times 10^{17} = 1.4 \times 10^{17}$ alpha particles

         (b) $P = nRT/V = 1.41 \times 10^{17} \text{ He atoms} \times \dfrac{1 \text{ mol He}}{6.022 \times 10^{23} \text{ atoms}} \times \dfrac{295 \text{ K}}{0.0250 \text{ L}} \times \dfrac{0.08206 \text{ L-atm}}{\text{mol-K}}$

             $= 2.27 \times 10^{-4} = 2.3 \times 10^{-4} \text{ atm} = 0.17 \text{ torr}$

21.91   Calculate N_t in dis/min/g C from 1.5×10^{-2} dis/0.788 g $CaCO_3$. $N_o = 15.3$ dis/min/g C. Calculate k from $t_{1/2}$, calculate t from $\ln(N_t / N_o) = -kt$.

$C(s) + O_2(g) \rightarrow CO_2(g) + Ca(OH_2)(aq) \rightarrow CaCO_3(s) + H_2O(l)$

1 C atom $\rightarrow$ 1 $CaCO_3$ molecule

$$\frac{1.5 \times 10^{-2} \text{ Bq}}{0.788 \text{ g } CaCO_3} \times \frac{1 \text{ dis/s}}{1 \text{ Bq}} \times \frac{60 \text{ s}}{1 \text{ min}} \times \frac{100.1 \text{ g } CaCO_3}{12.01 \text{ g C}} = 9.52 = 9.5 \text{ dis/min/g C}$$

$$k = 0.693/t_{1/2} = 0.693/5.700 \times 10^3 \text{ yr} = 1.216 \times 10^{-4} = 1.22 \times 10^{-4} \text{ yr}^{-1}$$

$$t = -\frac{1}{k} \ln \frac{N_t}{N_o} = \frac{-1}{1.216 \times 10^{-4} \text{ yr}^{-1}} \ln \frac{9.52 \text{ dis/min/g C}}{15.3 \text{ dis/min/g C}} = 3.90 \times 10^3 \text{ yr}$$

21.92   (a) $Ba(NO_3)_2(aq) + Na_2SO_4(aq) \rightarrow BaSO_4(s) + 2 \text{ } NaNO_3(aq)$

         (b) $1.25 \text{ mmol } Ba^{2+} + 1.25 \text{ mmol } SO_4^{2-} \rightarrow 1.25 \text{ mmol } BaSO_4$

Neither reactant is in excess, so the activity of the filtrate is due entirely to $[SO_4^{2-}]$ from dissociation of $BaSO_4(s)$. Calculate $[SO_4^{2-}]$ in the filtrate by comparing the activity of the filtrate to the activity of the reactant.

$$\frac{0.050 \text{ } M \text{ } SO_4^{2-}}{1.22 \times 10^6 \text{ Bq/mL}} = \frac{x \text{ } M \text{ filtrate}}{250 \text{ Bq/mL}}$$

$[SO_4^{2-}]$ in the filtrate $= 1.0246 \times 10^{-5} = 1.0 \times 10^{-5} \text{ } M$

$K_{sp} = [Ba^{2+}][SO_4^{2-}]; \ [SO_4^{2-}] = [Ba^{2+}]$

$K_{sp} = (1.0246 \times 10^{-5})^2 = 1.0498 \times 10^{-10} = 1.0 \times 10^{-10}$

22 Chemistry of the Nonmetals

Visualizing Concepts

22.1   Statement (c) is correct. C_2H_4, the structure on the left, is the stable compound. Carbon, with a relatively small covalent radius owing to its location in the second row of the periodic chart, is able to closely approach other atoms. This close approach enables significant π overlap, so carbon can form strong multiple bonds to satisfy the octet rule. Silicon, in the third row of the periodic table, has a covalent radius too large for significant π overlap. Si does not form stable multiple bonds and Si_2H_4 is unstable.

22.2   (a)   Acid–base (Brønsted)

   (b)   Charges on species from left to right in the reaction: 0, 0, 1+, 1–

   (c) $NH_3(aq) + H_2O(l) \rightleftharpoons NH_4^+(aq) + OH^-(aq)$

22.3   *Analyze.* The structure is a trigonal bipyramid where one of the five positions about the central atom is occupied by a lone pair, often called a see-saw.

   Plan A: Count the valence electrons in each molecule, draw a correct Lewis structure, and count the electron domains about the central atom.

   Plan B: Molecules (a)–(d) each contain four F atoms bound to a central atom through a single bond (F is unlikely to form multiple bonds because of its high electronegativity). This represents 16 electron pairs; the fifth position is occupied by a lone pair, for a total of 17 e^- pairs. A valence e^- count for (a)–(d) will tell us which molecules are likely to have the designated structure. Molecule (e), $HClO_4$, is not exactly of the type AX_4, so a Lewis structure will be required.   *Solve.*

   (a) XeF_4 36 e^-, 16 e^- pairs. Plan B predicts that this molecule *will not* adopt the see-saw structure.

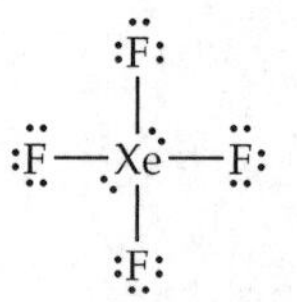

   6 e^- domains about the Xe
   octahedral domain geometry
   square planar structure

   (b) BrF_4^+ 34 e^-, 17 e^- pairs; structure *will* be see-saw

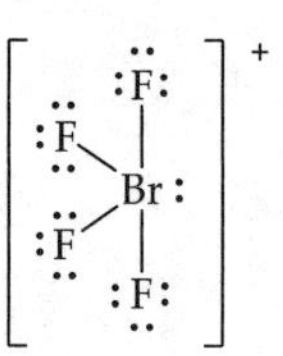

   5 e^- domains about Br trigonal
   bipyramidal domain geometry
   see-saw structure

(c)　SiF$_4$ 32 e$^-$, 16 e$^-$ pairs; structure *will not* be see-saw

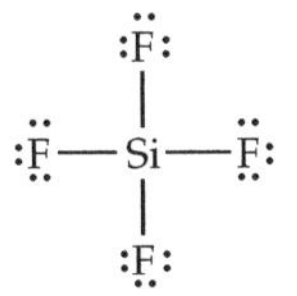

4 e$^-$ domains about Si
tetrahedral domain geometry and structure

(d)　TeCl$_4$ 34 e$^-$, 17 e$^-$ pairs; structure *will* be see-saw

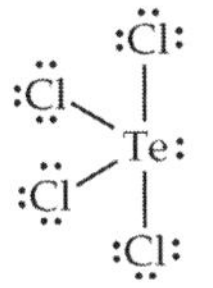

5 e$^-$ domains about Te
trigonal bipyramidal domain geometry
see-saw structure

(e)　HClO$_4$ 32 e$^-$, 16 e$^-$ pairs; structure *will not* be see-saw

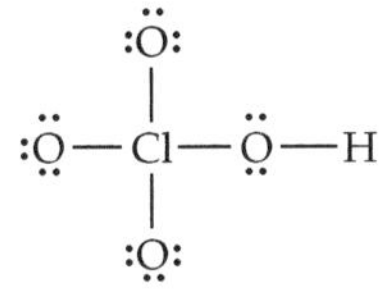

(HClO$_4$ is an oxyacid, so H is bound to O, not Cl.
Other Lewis structures that optimize formal charges
are possible; structure predictions are the same.)

22.4　Both gases are colorless and odorless, so they cannot be distinguished by inspection. (In practice, for safety reasons, one should never work in a lab with unlabeled bottles.) One big difference in the chemical properties of the two gases is their ability to support combustion. Oxygen supports combustion, whereas nitrogen does not. Heat a small piece of steel wool, place it at the mouth of one of the bottles. If the metal flares or glows more strongly, the gas is oxygen. If not, the gas is nitrogen.

22.5　*Analyze.* Given: space-filling models of molecules containing nitrogen and oxygen atoms. Find: molecular formulas and Lewis structures.

Plan. Nitrogen atoms are blue, and oxygen atoms are red. Count the number of spheres of each color to determine the molecular formula. From each molecular formula, count the valence electrons (N = 5, O = 6) and draw a correct Lewis structure. Resonance structures are likely.

Solve. (We list the formulas and Lewis structures from left to right across each row of space-filling models.)

N$_2$O$_5$　　　　40 valence electrons, 20 e$^-$ pairs

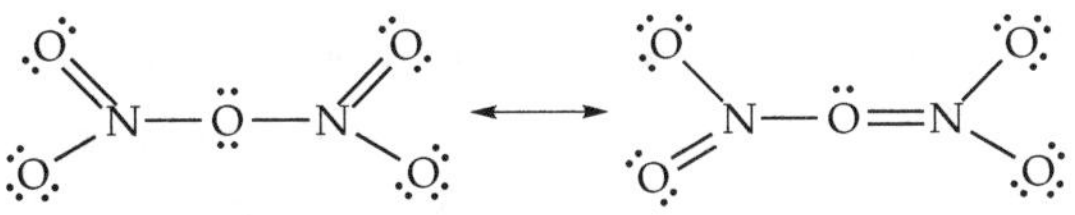

Many other resonance structures are possible. Those with double bonds to the central oxygen (like the right-hand structure above) do not minimize formal charge and are less significant in the net bonding model.

N$_2$O$_3$　　　　28 e$^-$, 14 e$^-$ pairs

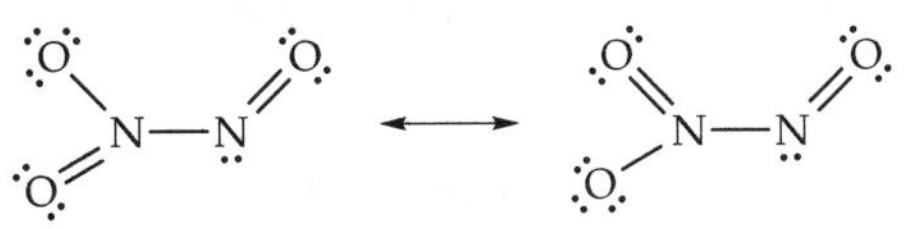

N_2O_4 $34\,e^-,\,17\,e^-$ pairs

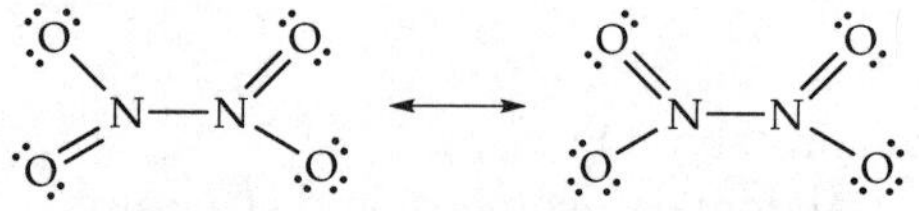

Other equivalent resonance structures with different arrangement of the double bonds are possible.

NO $11\,e^-,\,5.5\,e^-$ pairs

We place the odd electron on N because of electronegativity arguments.

NO_2 $17\,e^-,\,8.5\,e^-$ pairs

We place the odd electron on N because of electronegativity arguments.

N_2O $16\,e^-,\,8\,e^-$ pairs

The right-most structure above does not minimize formal charge and makes smaller contribution to the net bonding model.

22.6   The graph is applicable only to (c) density. Density depends on both atomic mass and volume (radius). Both increase going down a family, but atomic mass increases to a greater extent. Density, the ratio of mass to volume, increases going down the family; this trend is consistent with the data in the figure.

According to periodic trends, (a) electronegativity and (b) first ionization energy both decrease rather than increase going down the family. According to Table 22.5, both (d) X–X single bond enthalpy and (e) electron affinity are somewhat erratic, with the trends decreasing from S to Po, and anomalous values for the properties of O, probably owing to its small covalent radius.

22.7   Statements (a) and (c) are true.

Statement (b) is false. Although nuclear charge does increase going down the group, this effect, taken by itself, would decrease atomic radii. The increase in principle quantum number (n) of the valence electrons dominates and atomic radii increase going down the group. Statement (d) is false because the anion that is the strongest base in water is the conjugate base of the weakest conjugate acid. According to trends in binary hydrides, the acid with the longest X–H bond will be the most readily ionized and the strongest acid. The strongest acid is SeH^- and the weakest is OH^-. Therefore, O^{2-} is the strongest base in water.

22.8   *Analyze/Plan.* Evaluate the graph, describe the trend in data, recall the general trend for each of the properties listed, and use details of the data to discriminate between possibilities.

Solve. The property depicted is (a), first ionization energy. The general trend is an increase in value moving from left to right across the period, with a small discontinuity at S. Considering just this overall feature, both (a) first ionization energy and (c) electronegativity increase moving from left to right, so these are possibilities. (b), Atomic radius, decreases, and can be eliminated. Because Si is a solid and Cl and Ar are gases at room temperature, melting points must decrease across the row; (d), melting point, can be eliminated. According to data in Tables 22.2, 22.5, 22.7, and 22.8, (e), X–X single bond enthalpies, show no consistent trend. Furthermore, there is no known Ar–Ar single bond, so no value for this property can be known; (e) can be eliminated.

Now let's examine trends in (a), first ionization energy, and (c), electronegativity, more closely. From electronegativity values in Chapter 8, we see a continuous increase with no discontinuity at S, and no value for Ar. Values for (a), first ionization energy, from Chapter 7 do match the pattern in the figure. The slightly lower value of I_1 for S is the result of a decrease in repulsion when an electron is removed from a fully occupied orbital. In summary, only (a), first ionization energy, fits the property depicted in the graph.

22.9   The compound on the left, with the strained three-membered ring, will be the most generally reactive. For central atoms with four electron domains*, idealized bond angles are $109°$. From left to right, the bond angles in the three molecules pictured are $60°$, $90°$, and $108°$. The larger the deviation from ideal bond angles, the more strain in the molecule and the more generally reactive it is.

*For the stick structures shown in the exercise, each line represents a C–C single bond and the intersection of two lines is a C atom. To determine the number of electron domains about each atom, visualize or draw the hydrogen atoms and nonbonded electron pairs in each molecule. Alternatively, note that both C and O atoms form only single bonds, so hybridization must be sp^3 and idealized bond angles are $109°$.

22.10   *Analyze/Plan.* The structure shown is a diatomic molecule or ion, depending on the value of n. Each species has 10 valence electrons and 5 electron pairs.   *Solve.*

(a)   Only second row elements are possible, because of the small covalent radius required for multiple bonding. Likely candidates are CO, N_2, NO^+, CN^-, and C_2^{2-}.
       :C≡O: :N≡N: $[:N≡O:]^+$ $[:C≡N:]^-$ $[:C≡C:]^{2-}$

(b)   Because C_2^{2-} has the highest negative charge, it is likely to be the strongest H^+ acceptor and strongest Brønsted base. This is confirmed in Section 22.9 under "Carbides."

Periodic Trends and Chemical Reactions (Section 22.1)

22.11   *Analyze/Plan.* Use the color-coded periodic chart on the front-inside cover of the text to classify the given elements.   *Solve.*

Metals: (b) Sr, (c) Mn, (e) Na;   nonmetals: (a) P, (d) Se, (f) Kr;   metalloids: none

22.12   Metals: (a) Ga, (b) Mo, (f) Ru   nonmetals: (e) Xe   metalloid: (c) Te, (d) As

22.13   *Analyze/Plan.* Follow the logic in Sample Exercise 22.1.   *Solve.*

(a)   O   (b)   Br   (c)   Ba

(d)   O   (e)   Co   (f)   Br

22.14 (a) Cl   (b) K

(c) K in the gas phase (lowest ionization energy), Li in aqueous solution (most positive E° value)

(d) Ne; Ne and Ar are difficult to compare to the other elements because they do not form compounds and their radii are not measured in the same way as other elements. However, Ne is several rows to the right of C and surely has a smaller atomic radius. The next smallest is C.

(e) C

(f) C (graphite, diamond, fullerenes, carbon nanotubes, and graphene)

22.15 Statements (b) and (d) are true.

Statement (a) is false because a nitrogen atom is too small to accommodate five fluorine atoms around it. Statement (c) is false because the reduction potential of Cl_2 is larger than the reduction potential of I_2. The substance with the smaller reduction potential is easier to oxidize.

22.16 Statements (a) and (d) are true.

Statement (b) is false because Si does not readily form π bonds (to itself or other atoms), so Si_2H_4 and Si_2H_2 are not known as stable compounds. Statement (c) is false; the oxidation number of N and P in the two compounds is +5.

22.17 *Analyze/Plan.* Follow the logic in Sample Exercise 22.2. *Solve.*

(a) $NaOCH_3(s) + H_2O(l) \rightarrow NaOH(aq) + CH_3OH(aq)$

(b) $CuO(s) + 2\,HNO_3(aq) \rightarrow Cu(NO_3)_2(aq) + H_2O(l)$

(c) $WO_3(s) + 3\,H_2(g) \overset{\Delta}{\rightarrow} W(s) + 3\,H_2O(g)$

(d) $4\,NH_2OH(l) + O_2(g) \rightarrow 6\,H_2O(l) + 2\,N_2(g)$

(e) $Al_4C_3(s) + 12\,H_2O(l) \rightarrow 4\,Al(OH)_3(s) + 3\,CH_4(g)$

22.18 (a) $Mg_3N_2(s) + 6\,H_2O(l) \rightarrow 2\,NH_3(g) + 3\,Mg(OH)_2(s)$

Because $H_2O(l)$ is a reactant, the state of NH_3 in the products could be expressed as $NH_3(aq)$.

(b) $2\,C_3H_7OH(l) + 9\,O_2(g) \rightarrow 6\,CO_2(g) + 8\,H_2O(l)$

(c) $MnO_2(s) + C(s) \overset{\Delta}{\rightarrow} CO(g) + MnO(s)$ or

$MnO_2(s) + 2\,C(s) \overset{\Delta}{\rightarrow} 2\,CO(g) + Mn(s)$ or

$MnO_2(s) + C(s) \overset{\Delta}{\rightarrow} CO_2(g) + Mn(s)$

(d) $AlP(s) + 3\,H_2O(l) \rightarrow PH_3(g) + Al(OH)_3(s)$

(e) $Na_2S(s) + 2\,HCl(aq) \rightarrow H_2S(g) + 2\,NaCl(aq)$

Hydrogen, the Noble Gases, and the Halogens (Sections 22.2, 22.3, and 22.4)

22.19 *Analyze/Plan.* Use information on the isotopes of hydrogen in Section 22.2 to list their symbols, names, and relative abundances. *Solve.*

(a) ^1_1H-protium; ^2_1H-deuterium; ^3_1H-tritium

(b) The order of abundance is proteum > deuterium > tritium.

(c) $_{1}^{3}\text{H}$-tritium is radioactive.

(d) $_{1}^{3}\text{H} \rightarrow \, _{2}^{3}\text{He} + \, _{-1}^{0}\text{e}$

22.20   Statement (d) is correct. Because deuterium is almost twice as heavy as protium, we expect physical properties influenced by mass to be different for the two isotopes of hydrogen. In fact, D_2O has higher melting and boiling points, and a greater density than H_2O.

22.21   *Analyze/Plan.* Consider the electron configuration of hydrogen and the Group 1A elements. *Solve.*

Like other elements in group 1A, hydrogen has only one valence electron and its most common oxidation number is +1.

22.22   In its standard state, hydrogen is a gas, and thus a nonmetal, like the halogens. Hydrogen can gain an electron to form an anion with a 1– charge. Chemically, hydrogen can combine with group 1A metals to form ionic compounds, where H^- is the anion.

22.23   *Analyze/Plan.* Use information on the descriptive chemistry of hydrogen given in Section 22.2 to complete and balance the equations. *Solve.*

(a) $NaH(s) + H_2O(l) \rightarrow NaOH(aq) + H_2(g)$

(b) $Fe(s) + H_2SO_4(aq) \rightarrow Fe^{2+}(aq) + H_2(g) + SO_4^{2-}(aq)$

(c) $H_2(g) + Br_2(g) \rightarrow 2\,HBr(g)$

(d) $2\,Na(l) + H_2(g) \rightarrow 2\,NaH(s)$

(e) $PbO(s) + H_2(g) \xrightarrow{\Delta} Pb(s) + H_2O(g)$

22.24   (a) $2\,Al(s) + 6\,H^+(aq) \rightarrow 2\,Al^{3+}(aq) + 3\,H_2(g)$

(b) $Mg(s) + H_2O(g) \rightarrow MgO(s) + H_2(g)$

(c) $MnO_2(s) + H_2(g) \rightarrow MnO(s) + H_2O(g)$

(d) $CaH_2(s) + 2\,H_2O(l) \rightarrow Ca(OH)_2(aq) + 2\,H_2(g)$

22.25   *Analyze/Plan.* If the element bound to H is a nonmetal, the hydride is molecular. If H is bound to a metal with integer stoichiometry, the hydride is ionic; with noninteger stoichiometry, the hydride is metallic. *Solve.*

(a)   ionic (metal hydride)

(b)   molecular (nonmetal hydride)

(c)   metallic (nonstoichiometric transition metal hydride)

22.26   (a)   molecular   (b)   ionic   (c)   metallic

22.27   Vehicle fuels produce energy via combustion reactions. The reaction $H_2(g) + 1/2\,O_2(g) \rightarrow H_2O(g)$ is very exothermic, producing 242 kJ per mole of H_2 burned. The only product of combustion is H_2O, a nonpollutant (but like CO_2, a greenhouse gas).

22.28   Electrolysis of water is the cleanest way to produce hydrogen, but it is energy intensive. To make this process sustainable, the energy must come from renewable sources, such as hydroelectric or nuclear power plants, wind generators, or solar cells. The biomass production of fuel for electric power generation would also be a sustainable energy source.

22.29   *Analyze/Plan.* Consider the periodic properties of Xe and Ar.   *Solve.*

Xenon is larger, and can more readily accommodate an expanded octet. More important is the lower ionization energy of xenon; because the valence electrons are a greater average distance from the nucleus, they are more readily promoted to a state in which the Xe atom can form bonds with fluorine.

22.30   Your friend cannot be correct. In general, noble gas elements have very stable electron configurations with complete s and p subshells. They have very large positive ionization energies; they do not lose electrons easily. They have positive electron affinities; they do not attract electrons to themselves. They do not easily gain, lose or share electrons, so they do not readily form the chemical bonds required to create compounds. To date, the only known compounds of noble gases involve Xe and Kr bound to other nonmetals. Specifically, there are no known compounds of Ne.

22.31   *Analyze/Plan.* Follow the rules for assigning oxidation numbers in Section 4.4 and the logic in Sample Exercise 4.8.   *Solve.*

(a) $Ca(OBr)_2$, Br, +1   (b) $HBrO_3$, Br, +5   (c) XeO_3, Xe, +6

(d) ClO_4^-, Cl, +7   (e) HIO_2, I, +3   (f) IF_5; I, +5; F, –1

22.32   (a) ClO_3^-, Cl +5   (b)   HI, I, –1   (c) ICl_3; I, +3; Cl, –1

(d)   NaOCl, Cl, +1   (e) $HClO_4$, Cl, +7   (f) XeF_4; Xe, +4; F, –1

22.33   *Analyze/Plan.* Review the nomenclature rules and ion names in Section 2.8, as well as the rules for assigning oxidation numbers in Section 4.4.   *Solve.*

(a)   iron(III) chlorate, Cl, +5   (b)   chlorous acid, Cl, +3

(c)   xenon hexafluoride, F, –1   (d)   bromine pentafluoride; Br, +5; F, –1

(e)   xenon oxide tetrafluoride, F, –1   (f)   iodic acid, I, +5

22.34   (a)   potassium chlorate, Cl, +5   (b)   calcium iodate, I, +5

(c)   aluminum chloride, Cl, –1   (d)   bromic acid, Br, +5

(e)   paraperiodic acid, I, +7   (f)   xenon tetrafluoride, F, –1

22.35   *Analyze/Plan.* Consider intermolecular forces and periodic properties, including oxidizing power, of the listed substances.   *Solve.*

(a)   Van der Waals intermolecular attractive forces increase with increasing numbers of electrons in the atoms.

(b) F_2 reacts with water: $F_2(g) + H_2O(l) \rightarrow 2\,HF(aq) + 1/2\,O_2(g)$. That is, fluorine is too strong an oxidizing agent to exist in water.

(c)   HF has extensive hydrogen bonding.

(d)   Oxidizing power is related to electronegativity. Electronegativity decreases in the order given.

22.36   (a)   The more electronegative the central atom, the greater the extent to which it withdraws charge from oxygen, in turn making the O–H bond more polar, and enhancing ionization of H^+.

(b)   HF reacts with the silica which is a major component of glass:
$6\,HF(aq) + SiO_2(s) \rightarrow SiF_6^{2-}(aq) + 2\,H_2O(l) + 2\,H^+(aq)$

(c)   If the reaction of NaI and H_2SO_4 did produce HI, it would be a metathesis reaction. If HI is not the product of this reaction, then I^- is probably oxidized by H_2SO_4 to produce I_2.

(d)   The major factor is size; there is not room about Br for the three chlorides plus the two unshared electron pairs that would occupy the bromine valence shell orbitals.

Oxygen and the Other Group 6A Elements (Sections 22.5 and 22.6)

22.37   *Analyze/Plan.* Use information on the descriptive chemistry of oxygen given in Section 22.5 to complete and balance the equations.   *Solve.*

(a) $2\,HgO(s) \xrightarrow{\Delta} 2\,Hg(l) + O_2(g)$

(b) $2\,Cu(NO_3)_2(s) \xrightarrow{\Delta} 2\,CuO(s) + 4\,NO_2(g) + O_2(g)$

(c) $PbS(s) + 4\,O_3(g) \rightarrow PbSO_4(s) + 4\,O_2(g)$

(d) $2\,ZnS(s) + 3\,O_2(g) \xrightarrow{\Delta} 2\,ZnO(s) + 2\,SO_2(g)$

(e) $2\,K_2O_2(s) + 2\,CO_2(g) \rightarrow 2\,K_2CO_3(s) + O_2(g)$

(f) $3\,O_2(g) \xrightarrow{h\nu} 2\,O_3(g)$

22.38   (a) $CaO(s) + H_2O(l) \rightarrow Ca^{2+}(aq) + 2\,OH^-(aq)$

(b) $Al_2O_3(s) + 6\,H^+(aq) \rightarrow 2\,Al^{3+}(aq) + 3\,H_2O(l)$

(c) $Na_2O_2(s) + 2\,H_2O(l) \rightarrow 2\,Na^+(aq) + 2\,OH^-(aq) + H_2O_2(aq)$

(d) $N_2O_3(g) + H_2O(l) \rightarrow 2\,HNO_2(aq)$

(e) $2\,KO_2(s) + 2\,H_2O(l) \rightarrow 2\,K^+(aq) + 2\,OH^-(aq) + O_2(g) + H_2O_2(aq)$

(f) $NO(g) + O_3(g) \rightarrow NO_2(g) + O_2(g)$

22.39   *Analyze/Plan.* Oxides of metals are bases, oxides of nonmetals are acids, oxides that act as both acids and bases are amphoteric, and oxides that act as neither acids nor bases are neutral.   *Solve.*

(a)   acidic (oxide of a nonmetal)

(b)   acidic (oxide of a nonmetal)

(c)   amphoteric

(d)   basic (oxide of a metal)

22.40   (a) Mn_2O_7 (higher oxidation state of Mn)

(b) SnO_2 (higher oxidation state of Sn)

(c) SO_3 (higher oxidation state of S)

(d) SO_2 (more nonmetallic character of S)

(e) Ga_2O_3 (more nonmetallic character of Ga)

(f) SO_2 (more nonmetallic character of S)

22.41 *Analyze/Plan.* Follow the rules for assigning oxidation numbers in Section 4.4 and the logic in Sample Exercise 4.8. *Solve.*

(a) H_2SeO_3, +4   (b) $KHSO_3$, +4   (c) H_2Te, –2   (d) CS_2, –2

(e) $CaSO_4$, +6   (f) CdS, –2   (g) $ZnTe$, –2

Oxygen (a group 6A element) is in the –2 oxidation state in compounds (a), (b), and (e).

22.42 (a) SCl_4, +4   (b) SeO_3, +6   (c) $Na_2S_2O_3$, +2   (d) H_2S, –2

(e) H_2SO_4, +6   (f) SO_2, +4   (g) $HgTe$, –2

Oxygen (a group 6A element) is in the –2 oxidation state in compounds (b), (c), (e) and (f).

22.43 *Analyze/Plan.* The half-reaction for oxidation in all these cases is:

$H_2S(aq) \rightarrow S(s) + 2\,H^+ + 2\,e^-$ [The product could be written as $S_8(s)$, but this is not necessary. In fact it is not necessarily the case that S_8 would be formed, rather than some other allotropic form of the element.] Combine this half-reaction with the given reductions to write complete equations. The reduction in (c) happens only in acid solution. The reactants in (d) are acids, so the medium is acidic. *Solve.*

(a) $2\,Fe^{3+}(aq) + H_2S(aq) \rightarrow 2\,Fe^{2+}(aq) + S(s) + 2\,H^+(aq)$

(b) $Br_2(l) + H_2S(aq) \rightarrow 2\,Br^-(aq) + S(s) + 2\,H^+(aq)$

(c) $2\,MnO_4^-(aq) + 6\,H^+(aq) + 5\,H_2S(aq) \rightarrow 2\,Mn^{2+}(aq) + 5\,S(s) + 8\,H_2O(l)$

(d) $2\,NO_3^-(aq) + H_2S(aq) + 2\,H^+(aq) \rightarrow 2\,NO_2(aq) + S(s) + 2\,H_2O(l)$

22.44 An aqueous solution of SO_2 contains H_2SO_3 and is acidic. Use H_2SO_3 as the reducing agent and balance assuming acid conditions.

(a) $2[MnO_4^-(aq) + 8\,H^+(aq) + 5\,e^- \rightarrow Mn^{2+}(aq) + 4\,H_2O(l)]$

$\underline{5[H_2SO_3(aq) + H_2O(l) \rightarrow SO_4^{2-}(aq) + 4\,H^+(aq) + 2\,e^-]}$

$2\,MnO_4^-(aq) + 5\,H_2SO_3(aq) \rightarrow 2\,MnSO_4(aq) + 3\,SO_4^{2-}(aq) + 3\,H_2O(l) + 4\,H^+(aq)$

(b) $Cr_2O_7^{2-}(aq) + 14\,H^+(aq) + 6\,e^- \rightarrow 2\,Cr^{3+}(aq) + 7\,H_2O(l)$

$\underline{3[H_2SO_3(aq) + H_2O(l) \rightarrow SO_4^{2-}(aq) + 4\,H^+(aq) + 2\,e^-]}$

$Cr_2O_7^{2-}(aq) + 3\,H_2SO_3(aq) + 2\,H^+(aq) \rightarrow 2\,Cr^{3+}(aq) + 3\,SO_4^{2-}(aq) + 4\,H_2O(l)$

(c) $Hg_2^{2+}(aq) + 2\,e^- \rightarrow 2\,Hg(l)$

$\underline{H_2SO_3(aq) + H_2O(l) \rightarrow SO_4^{2-}(aq) + 4\,H^+(aq) + 2\,e^-}$

$Hg_2^{2+}(aq) + H_2SO_3(aq) + H_2O(l) \rightarrow 2\,Hg(l) + SO_4^{2-}(aq) + 4\,H^+(aq)$

22.45 *Analyze/Plan.* For each substance, count valence electrons, draw the correct Lewis structure, and apply the rules of VSEPR to decide electron domain geometry and geometric structure. *Solve.*

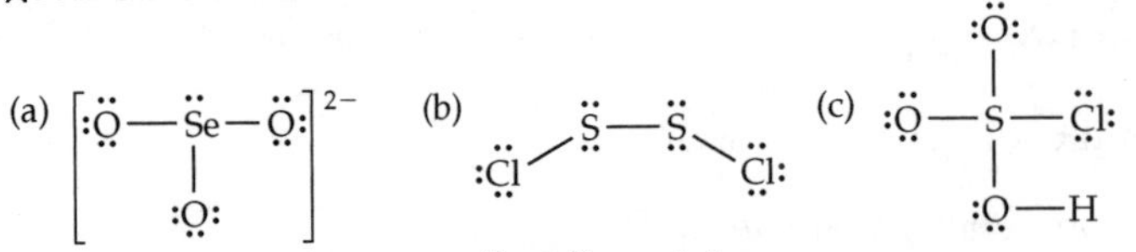

(a) trigonal pyramidal

(b) Bent (free rotation around S–S bond

(c) tetrahedral

670

22.46 SF_4, 34 e^- SF_5^-, 42 e^-

trigonal bipyramidal   octahedral
electron pair geometry   electron pair geometry

see-saw   square pyramidal
molecular geometry   molecular geometry

22.47   *Analyze/Plan.* Use information on the descriptive chemistry of sulfur given in Section 22.6 to complete and balance the equations.   *Solve.*

(a) $SO_2(s) + H_2O(l) \rightarrow H_2SO_3(aq) \rightleftharpoons H^+(aq) + HSO_3^-(aq)$

(b) $ZnS(s) + 2\,HCl(aq) \rightarrow ZnCl_2(aq) + H_2S(g)$

(c) $8\,SO_3^{2-}(aq) + S_8(s) \rightarrow 8\,S_2O_3^{2-}(aq)$

(d) $SO_3(aq) + H_2SO_4(l) \rightarrow H_2S_2O_7(l)$

22.48   (a) $Al_2Se_3(s) + 6\,H^+(aq) \rightarrow 2\,Al^{3+}(aq) + 3\,H_2Se(g)$

         (b) $Cl_2(aq) + S_2O_3^{2-}(aq) + H_2O(l) \rightarrow 2\,Cl^-(aq) + S(s) + SO_4^{2-}(aq) + 2\,H^+(aq)$

Nitrogen and the Other Group 5A Elements (Sections 22.7 and 22.8)

22.49   *Analyze/Plan.* Follow the rules for assigning oxidation numbers in Section 4.4 and the logic in Sample Exercise 4.8.   *Solve.*

       (a) $NaNO_2$, +3   (b) NH_3, –3   (c) N_2O, +1   (d) $NaCN$, –3

       (e) HNO_3, +5   (f) NO_2, +4   (g) N_2, 0   (h) BN, –3

22.50   (a) NO, +2   (b) N_2H_4, –2   (c) KCN, –3

         (d) $NaNO_2$, +3   (e) NH_4Cl, –3   (f) Li_3N, –3

22.51   *Analyze/Plan.* For each substance, count valence electrons, draw the correct Lewis structure, and apply the rules of VSEPR to decide electron domain geometry and geometric structure.   *Solve.*

       (a) $:\ddot{O}=\ddot{N}-\ddot{O}-H \longleftrightarrow :\ddot{O}-\ddot{N}=\ddot{O}-H$

           The molecule is bent around the central oxygen and nitrogen atoms; the four atoms need not lie in a plane. The right-most form does not minimize formal charges and is less important in the actual bonding model. The oxidation state of N is +3.

       (b) $\left[:\ddot{N}=N=\ddot{N}:\right]^- \longleftrightarrow \left[:N\equiv N-\ddot{N}:\right]^- \longleftrightarrow \left[:\ddot{N}-N\equiv N:\right]^-$

           The molecule is linear. The oxidation state of N is –1/3.

(c)

$$\left[\begin{array}{c} \text{H} \quad\quad \text{H} \\ | \quad\quad | \\ \text{H---N---N:} \\ | \quad\quad | \\ \text{H} \quad\quad \text{H} \end{array} \right]^{+}$$

The geometry is tetrahedral around the left nitrogen, trigonal pyramidal around the right. The oxidation state of N is –2.

(d)

$$\left[\begin{array}{c} \text{:Ö:} \\ | \\ \text{:Ö---N=Ö} \end{array} \right]^{-}$$

(three equivalent resonance forms) The ion is trigonal planar. The oxidation state of N is +5.

22.52   (a)

$$\left[\begin{array}{c} \text{H} \\ | \\ \text{H---N---H} \\ | \\ \text{H} \end{array} \right]^{+}$$

The ion is tetrahedral. The oxidation state of N is –3.

(b) $\left[\ddot{\text{O}}=\ddot{\text{N}}-\ddot{\text{O}}: \right]^{-} \longleftrightarrow \left[:\ddot{\text{O}}-\text{N}=\ddot{\text{O}} \right]^{-}$

The ion is bent with a 120° O–N–O angle. The oxidation state of N is +3.

(c) $:\ddot{\text{N}}=\text{N}=\ddot{\text{O}}: \longleftrightarrow :\text{N}\equiv\text{N}-\ddot{\text{O}}: \longleftrightarrow :\ddot{\text{N}}-\text{N}\equiv\text{O}:$

The molecule is linear. Again, the third resonance form makes less contribution to the structure because of the high formal charges involved. The oxidation state of N is +1.

(d) $\ddot{\text{O}}=\dot{\text{N}}-\ddot{\text{O}}: \longleftrightarrow :\ddot{\text{O}}-\dot{\text{N}}=\ddot{\text{O}}:$

The molecule is bent (nonlinear). The odd electron resides on N because it is less electronegative than O. The oxidation state of N is +4.

22.53   *Analyze/Plan.* Use information on the descriptive chemistry of nitrogen given in Section 22.7 to complete and balance the equations.   *Solve.*

(a) $Mg_3N_2(s) + 6\,H_2O(l) \rightarrow 2\,NH_3(g) + 3\,Mg(OH)_2(s)$

Because $H_2O(l)$ is a reactant, the state of NH_3 in the products could be expressed as $NH_3(aq)$.

(b) $2\,NO(g) + O_2(g) \rightarrow 2\,NO_2(g)$, redox reaction

(c) $N_2O_5(g) + H_2O(l) \rightarrow 2\,H^+(aq) + 2\,NO_3^-(aq)$

(d) $NH_3(aq) + H^+(aq) \rightarrow NH_4^+(aq)$

(e) $N_2H_4(l) + O_2(g) \rightarrow N_2(g) + 2\,H_2O(g)$, redox reaction

22.54   (a) $4\,Zn(s) + 2\,NO_3^-(aq) + 10\,H^+(aq) \rightarrow 4\,Zn^{2+}(aq) + N_2O(g) + 5\,H_2O(l)$

(b) $4\,NO_3^-(aq) + S(s) + 4\,H^+(aq) \rightarrow 4\,NO_2(g) + SO_2(g) + 2\,H_2O(l)$

(or $6\,NO_3^-(aq) + S(s) + 4\,H^+(aq) \rightarrow 6\,NO_2(g) + SO_4^{2-}(aq) + 2\,H_2O(l)$)

(c) $2\,NO_3^-(aq) + 3\,SO_2(g) + 2\,H_2O(l) \rightarrow 2\,NO(g) + 3\,SO_4^{2-}(aq) + 4\,H^+(aq)$

(d) $N_2H_4(g) + 5\,F_2(g) \rightarrow 2\,NF_3(g) + 4\,HF(g)$

(e) $4\,CrO_4^{2-}(aq) + 3\,N_2H_4(aq) + 4\,H_2O(l) \rightarrow 4\,Cr(OH)_4^{-}(aq) + 4\,OH^{-}(aq) + 3\,N_2(g)$

22.55   *Analyze/Plan.* Follow the method for writing balanced half-reactions given in Section 20.2 and Sample Exercises 20.2.   *Solve.*

     (a) $HNO_2(aq) + H_2O(l) \rightarrow NO_3^{-}(aq) + 3\,H^{+}(aq) + 2\,e^{-},\ E^{\circ}_{red} = 0.96\ V$

     (b) $N_2(g) + H_2O(l) \rightarrow N_2O(g) + 2\,H^{+}(aq) + 2\,e^{-},\ E^{\circ}_{red} = 1.77\ V$

22.56   (a) $NO_3^{-}(aq) + 4\,H^{+}(aq) + 3\,e^{-} \rightarrow NO(g) + 2\,H_2O(l)$

     (b) $HNO_2(aq) \rightarrow NO_2(g) + H^{+}(aq) + 1\,e^{-}$

22.57   *Analyze/Plan.* Follow the rules for assigning oxidation numbers in Section 4.4 and the logic in Sample Exercise 4.8.   *Solve.*

     (a) H_3PO_3, +3   (b) $H_4P_2O_7$, +5   (c) $SbCl_3$, +3

     (d) Mg_3As_2, −3   (e) P_2O_5, +5   (f) Na_3PO_4, +5

22.58   (a) PO_4^{3-}, +5   (b) H_3AsO_3, +3   (c) Sb_2S_3, +3

     (d) $Ca(H_2PO_4)_2$, +5   (e) K_3P, −3   (f) $GaAs$, −3

22.59   *Analyze/Plan.* Consider the structures of the compounds of interest when explaining the observations.   *Solve.*

     (a)   Phosphorus is a larger atom and can more easily accommodate five surrounding atoms and an expanded octet of electrons than nitrogen can. Also, P has energetically "available" 3d orbitals that participate in the bonding, but nitrogen does not.

     (b)   Only one of the three hydrogens in H_3PO_2 is bonded to oxygen. The other two are bonded directly to phosphorus and are not easily ionized because the P–H bond is not very polar.

     (c) PH_3 is a weaker base than H_2O (PH_4^{+} is a stronger acid than H_3O^{+}). Any attempt to add H^{+} to PH_3 in the presence of H_2O merely causes protonation of H_2O.

     (d)   Refer to the structures of white and red phosphorus in Figure 22.26. White phosphorus consists of P_4 molecules, with P–P–P bond angles of 60°. Each P atom has four VSEPR pairs of electrons, so the predicted electron pair geometry is tetrahedral and the preferred bond angle is 109°. Because of the severely strained bond angles in P_4 molecules, white phosphorus is highly reactive. Red phosphorus is a chain of groups of four P atoms. It has fewer severely strained P–P–P bond angles and is less reactive than white phosphorus.

22.60   (a)   Only two of the hydrogens in H_3PO_3 are bound to oxygen. The third is attached directly to phosphorus, and not readily ionized, because the H–P bond is not very polar.

     (b)   The smaller, more electronegative nitrogen withdraws more electron density from the O–H bond, making it more polar and more likely to ionize.

     (c)   Phosphate rock consists of $Ca_3(PO_4)_2$, which is only slightly soluble in water. The phosphorus is unavailable for plant use.

(d) N_2 can form stable π bonds to complete the octet of both N atoms. Because phosphorus atoms are larger than nitrogen atoms, they do not form stable π bonds with themselves and must form σ bonds with several other phosphorus atoms (producing P_4 tetrahedral or chain structures) to complete their octets.

(e) In solution Na_3PO_4 is completely dissociated into Na^+ and PO_4^{3-}. PO_4^{3-}, the conjugate base of the very weak acid HPO_4^{2-}, has a K_b of 2.4×10^{-2} and produces a considerable amount of OH^- by hydrolysis of H_2O.

22.61 *Analyze/Plan.* Use information on the descriptive chemistry of phosphorus given in Section 22.8 to complete and balance the equations. *Solve.*

(a) $2\ Ca_3(PO_4)_2(s) + 6\ SiO_2(s) + 10\ C(s) \xrightarrow{\Delta} P_4(g) + 6\ CaSiO_3(l) + 10\ CO(g)$

(b) $PBr_3(l) + 3\ H_2O(l) \rightarrow H_3PO_3\ (aq) + 3\ HBr(aq)$

(c) $4\ PBr_3\ (g) + 6\ H_2(g) \rightarrow P_4\ (g) + 12\ HBr(g)$

22.62 (a) $PCl_5(l) + 4\ H_2O(l) \rightarrow H_3PO_4\ (aq) + 5\ HCl(aq)$

(b) $2\ H_3PO_4(aq) \xrightarrow{\Delta} H_4P_2O_7(aq) + H_2O(l)$

(c) $P_4O_{10}(s) + 6\ H_2O(l) \rightarrow 4\ H_3PO_4(aq)$

Carbon, the Other Group 4A Elements, and Boron
(Sections 22.9, 22.10, and 22.11)

22.63 *Analyze/Plan.* Review the nomenclature rules and ion names in Section 2.8. *Solve.*

(a) HCN (b) $Ni(CO)_4$ (c) $Ba(HCO_3)_2$ (d) CaC_2 (e) K_2CO_3

22.64 (a) H_2CO_3 (b) NaCN (c) $KHCO_3$ (d) C_2H_2 (e) $Fe(CO)_5$

22.65 *Analyze/Plan.* Use information on the descriptive chemistry of carbon given in Section 22.9 to complete and balance the equations. *Solve.*

(a) $ZnCO_3(s) \xrightarrow{\Delta} ZnO(s) + CO_2(g)$

(b) $BaC_2(s) + 2\ H_2O(l) \rightarrow Ba^{2+}(aq) + 2\ OH^-(aq) + C_2H_2(g)$

(c) $2\ C_2H_2(g) + 5\ O_2(g) \rightarrow 4\ CO_2(g) + 2\ H_2O(g)$

(d) $CS_2(g) + 3\ O_2(g) \rightarrow CO_2(g) + 2\ SO_2(g)$

(e) $Ca(CN)_2(s) + 2\ HBr(aq) \rightarrow CaBr_2(aq) + 2\ HCN(aq)$

22.66 (a) $CO_2(g) + OH^-(aq) \rightarrow HCO_3^-\ (aq)$

(b) $NaHCO_3(s) + H^+(aq) \rightarrow Na^+(aq) + H_2O(l) + CO_2(g)$

(c) $2\ CaO(s) + 5\ C(s) \xrightarrow{\Delta} 2\ CaC_2(s) + CO_2(g)$

(d) $C(s) + H_2O(g) \xrightarrow{\Delta} H_2(g) + CO(g)$

(e) $CuO(s) + CO(g) \rightarrow Cu(s) + CO_2(g)$

22.67   *Analyze/Plan.* Use information on the descriptive chemistry of carbon given in Section 22.9 to complete and balance the equations.   *Solve.*

(a) $2\ CH_4(g) + 2\ NH_3(g) + 3\ O_2(g) \xrightarrow[\text{cat}]{800°C} 2\ HCN(g) + 6\ H_2O(g)$

(b) $NaHCO_3(s) + H^+(aq) \rightarrow CO_2(g) + H_2O(l) + Na^+(aq)$

(c) $2\ BaCO_3(s) + O_2(g) + 2\ SO_2(g) \rightarrow 2\ BaSO_4(s) + 2\ CO_2(g)$

22.68   (a) $2\ Mg(s) + CO_2(g) \rightarrow 2\ MgO(s) + C(s)$

(b) $6\ CO_2(g) + 6\ H_2O(l) \xrightarrow{h\nu} C_6H_{12}O_6(aq) + 6\ O_2(g)$

(c) $CO_3^{2-}(aq) + H_2O(l) \rightarrow HCO_3^-(aq) + OH^-(aq)$

22.69   *Analyze/Plan.* Follow the rules for assigning oxidation numbers in Section 4.4 and the logic in Sample 4.8.   *Solve.*

(a)	H_3BO_3, +3	(b)	$SiBr_4$, +4	(c)	$PbCl_2$, +2
(d)	$Na_2B_4O_7 \cdot 10H_2O$, +3	(e)	B_2O_3, +3	(f)	GeO_2, +4

22.70

(a)	SiO_2, +4	(b)	$GeCl_4$, +4	(c)	$NaBH_4$, +3
(d)	$SnCl_2$, +2	(e)	B_2H_6, +3	(f)	BCl_3, +3

22.71   *Analyze/Plan.* Consider periodic trends within a family, particularly metallic character, as well as descriptive chemistry in Sections 22.9 and 22.10.   *Solve.*

(a)   Tin; see Table 22.8. The filling of the 4f subshell at the beginning of the sixth row of the periodic table increases Z and Z_{eff} for later elements. This causes the ionization energy of Pb to be greater than that of Sn.

(b)   Carbon, silicon, and germanium; these are the nonmetal and metalloids in group 4A. They form compounds ranging from XH_4 (–4) to XO_2 (+4). The metals tin and lead are not found in negative oxidation states.

(c)   Silicon; silicates are the main component of sand.

22.72   (a)   carbon   (b)   lead   (c)   germanium

22.73   *Analyze/Plan.* Consider the structural chemistry of silicates discussed in Section 22.10 and shown in Figure 22.32.   *Solve.*

(a)   Tetrahedral

(b)   Metasilicic acid will probably adopt the single-strand silicate chain structure shown in Figure 22.32(b). The empirical formula shows 3 O and 2 H atoms per Si atom. The chain has the same Si to O ratio as metasilicic acid. Furthermore, in the chain structure, there are two terminal (not bridging) O atoms on each Si. These can accommodate the 2 H atoms associated with each Si atom of the acid. The sheet structure does not fulfill these requirements.

22.74   Carbon forms carbonates rather than silicates to take advantage of its ability to form π bonds. Because of its relatively compact 2p valence orbitals, carbon can engage in effective π-type overlap and form multiple bonds with itself and other elements. Carbonate, CO_3^{2-}, the anion present in carbonates, takes advantage of this ability to form stable π bonds, and is additionally stabilized by resonance. In silicates, silicon atoms form only single bonds and are tetrahedral.

675

22.75   *Analyze/Plan.* In silicate anions, the oxidation number of silicon is +4 and that of oxygen is –2 (Section 22.10). Determine the charge on the silicate anion, then balance the charges of the cations and anions in the minerals.   *Solve.*

     (a) $x = 2$. The charge on $Si_2O_7^{6-}$ anion is 6–, that on Zn^{2+} cation is 2+. Two Ca^{2+} cations are required to balance charge.

     (b) $x = 2$. The charge on $Si_2O_5^{2-}$ anion is 2–, that on Al^{3+} cation is 3+. Two OH^- anions are required to balance charge.

22.76   (a) $x = 1$. The charge on $Si_3O_8^{4-}$ anion is 4–, that on Al^{3+} cation is 3+. One Na^+ cation is required to balance charge.

     (b) $x = 2$. The charge on $Si_4O_{11}^{6-}$ anion is 6–, that on the cations is 2+. Two OH^- anions are required to balance charge.

22.77   (a)   Diborane (Figure 22.34 and below) has bridging H atoms linking the two B atoms. The structure of ethane shown below has the C atoms bound directly, with no bridging atoms.

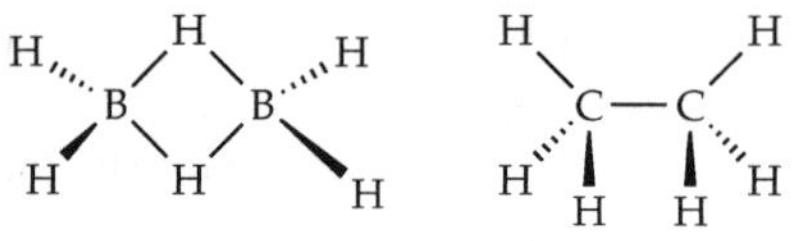

     (b) B_2H_6 is an electron deficient molecule. It has 12 valence electrons, whereas C_2H_6 has 14 valence electrons. The 6 valence electron pairs in B_2H_6 are all involved in B–H σ bonding, so the only way to satisfy the octet rule at B is to have the bridging H atoms shown in Figure 22.34.

     (c)   A hydride ion, H^-, has two electrons, whereas an H atom has one. The term *hydridic* indicates that the H atoms in B_2H_6 have more than the usual amount of electron density for a covalently bound H atom.

22.78   (a) $B_2H_6(g) + 6 H_2O(l) \rightarrow 2 H_3BO_3(aq) + 6 H_2(g)$

     (b) $4 H_3BO_3(s) \xrightarrow{\Delta} H_2B_4O_7(s) + 5 H_2O(g)$

     (c) $B_2O_3(s) + 3 H_2O(l) \rightarrow 2 H_3BO_3(aq)$

Additional Exercises

22.79   (a)   False. $H_2(g)$ and $D_2(g)$ are composed of different isotopes of hydrogen. Allotropes are composed of atoms of a single element bound into different structures.

     (b)   True. An interhalogen is a compound formed from atoms of two or more halogens.

     (c)   False. $MgO(s)$ is a basic anhydride; it is the oxide of a metal.

     (d)   True. $SO_2(g)$ is the oxide of a nonmetal.

     (e)   True. A condensation reaction is the combination of two molecules to form a large molecule and a small one such as H_2O or HCl.

(f)   True. The nucleus of tritium contains one proton and two neutrons.

(g)   False. Disproportionation is an oxidation–reduction process where the same element is both oxidized and reduced. In this reaction, sulfur is oxidized and oxygen is reduced.

22.80   (a) $BrO_3^- (aq) + XeF_2 (aq) + H_2O(l) \rightarrow Xe(g) + 2\,HF(aq) + BrO_4^- (aq)$

        (b) BrO_3^-, +5; BrO_4^-, +7

22.81   (a) $SO_2(g) + H_2O(l) \rightleftharpoons H_2SO_3 (aq)$

        (b) $Cl_2O_7(g) + H_2O(l) \rightleftharpoons 2\,HClO_4(aq)$

        (c) $Na_2O_2(s) + 2\,H_2O(l) \rightarrow H_2O_2(aq) + 2\,NaOH(aq)$

        (d) $BaC_2(s) + 2\,H_2O(l) \rightarrow Ba^{2+}(aq) + 2\,OH^-(aq) + C_2H_2(g)$

        (e) $2\,RbO_2(s) + 2\,H_2O(l) \rightarrow 2\,Rb^+(aq) + 2\,OH^-(aq) + O_2(g) + H_2O_2(aq)$

        (f) $Mg_3N_2(s) + 6\,H_2O(l) \rightarrow 3\,Mg(OH)_2(s) + 2\,NH_3(g)$

        (g) $NaH(s) + H_2O(l) \rightarrow NaOH(aq) + H_2(g)$

22.82   (a) $H_2SO_4 - H_2O \rightarrow SO_3$

        (b) $2\,HClO_3 - H_2O \rightarrow Cl_2O_5$

        (c) $2\,HNO_2 - H_2O \rightarrow N_2O_3$

        (d) $H_2CO_3 - H_2O \rightarrow CO_2$

        (e) $2\,H_3PO_4 - 3H_2O \rightarrow P_2O_5$

22.83   Assume that the reactions occur in acidic solution. The half-reaction for reduction of H_2O_2 is in all cases $H_2O_2 (aq) + 2\,H^+(aq) + 2\,e^- \rightarrow 2\,H_2O(aq)$.

        (a) $N_2H_4(aq) + 2\,H_2O_2 (aq) \rightarrow N_2(g) + 4\,H_2O(l)$

        (b) $SO_2(g) + H_2O_2(aq) \rightarrow SO_4^{2-}(aq) + 2\,H^+(aq)$

        (c) $NO_2^-(aq) + H_2O_2(aq) \rightarrow NO_3^-(aq) + H_2O(l)$

        (d) $H_2S(g) + H_2O_2 (aq) \rightarrow S(s) + 2\,H_2O(l)$

        (e) $$2\,H^+(aq) + H_2O_2(aq) + 2\,e^- \rightarrow 2\,H_2O(l)$$
                $$\underline{2[Fe^{2+}(aq) \rightarrow Fe^{3+}(aq) + e^-]}$$
                $$2\,Fe^{2+}(aq) + H_2O_2(aq) + 2\,H^+(aq) \rightarrow 2\,Fe^{3+}(aq) + 2\,H_2O(l)$$

22.84
$$
\begin{array}{lll}
S(g) + O_2(g) \rightarrow SO_2(g) & \Delta H = -296.9 \text{ kJ} & \text{(a)}\\
SO_2(g) + 1/2\,O_2(g) \rightarrow SO_3(g) & \Delta H = -98.3 \text{ kJ} & \text{(b)}\\
\underline{SO_3(g) + H_2O(l) \rightarrow H_2SO_4(aq)} & \underline{\Delta H = -130 \text{ kJ}} & \text{(c)}\\
S(g) + 3/2\,O_2(g) + H_2O(l) \rightarrow H_2SO_4(aq) & \Delta H = -525.2 = -5.3\times10^2 \text{ kJ} &
\end{array}
$$

$$5000 \text{ lb } H_2SO_4 \times \frac{453.6 \text{ g}}{1 \text{lb}} \times \frac{1 \text{ mol } H_2SO_4}{98.09 \text{ g}} \times \frac{-525 \text{ kJ}}{\text{mol } H_2SO_4} = -1.214\times10^7 = -1\times10^7 \text{ kJ}$$

One mole of H_2SO_4 produces 5.3×10^2 kJ of heat, 5000 lb of H_2SO_4 produces 1×10^7 kJ. (If trailing zeros are not significant, 130 kJ limits the first result to 2 sig figs and 5000 lb limits the second result to 1 sig fig.)

22.85   (a) $PO_4{}^{3-}$, +5; $NO_3{}^-$, +5

       (b)   The Lewis structure for $NO_4{}^{3-}$ would be:

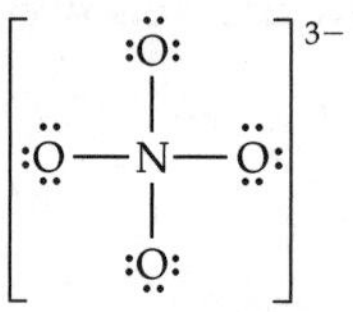

             The formal charge on N is +1 and on each O atom is –1. The four electronegative oxygen atoms withdraw electron density, leaving the nitrogen deficient. Because N can form a maximum of four bonds, it cannot form a π bond with one or more of the O atoms to regain electron density, as the P atom in $PO_4{}^{3-}$ does. Also, the short N–O distance would lead to a tight tetrahedron of O atoms subject to steric repulsion.

22.86   (a)   Although P_4, P_4O_6, and P_4O_{10} all have four P atoms in a tetrahedral arrangement, the bonding *between* P atoms and *by* P atoms is not the same in the three molecules. In P_4, the 4 P atoms are bound only to each other by P–P single bonds with strained bond angles of approximately 60°. In the two oxides, the 4 P atoms are directly bound to oxygen atoms, not to each other. Bonding by P atoms in P_4O_6 and P_4O_{10} is very similar. Each contains the P_6O_6 cage, formed by four P_3O_3 rings that share a P–O–P edge. Phosphorus bonding to oxygen maintains the overall P_4 tetrahedron but allows the P atoms to move away from each other so that the angle strain is relieved relative to molecular P_4. The P–O–P and O–P–O angles in both oxides are near the ideal 109°. In P_4O_6, each P is bound to 3 O atoms and has a lone pair completing its octet. In P_4O_{10}, the lone pair is replaced by a terminal O atom and each P is bound to 3 bridging and 1 terminal O atom.

       (b)

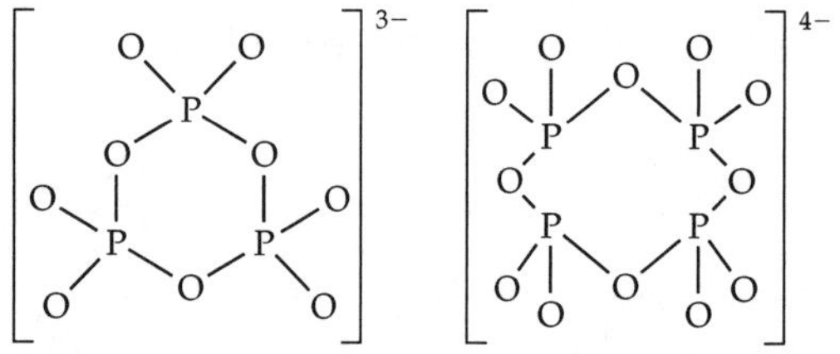

             In both structures there are unshared pairs on all oxygens to give octets and the geometry around each P is approximately tetrahedral.

22.87 $3\,H_3PO_4 \rightarrow H_5P_3O_{10} + 2\,H_2O$

22.88 $GeO_2(s) + C(s) \xrightarrow{\Delta} Ge(l) + CO_2(g)$

       $Ge(l) + 2\,Cl_2\,(g) \rightarrow GeCl_4(l)$

       $GeCl_4(l) + 2\,H_2O(l) \rightarrow GeO_2\,(s) + 4\,HCl(g)$

       $GeO_2(s) + 2\,H_2(g) \rightarrow Ge(s) + 2\,H_2O(l)$

22.89 $KAlSi_3O_8$. The oxidation number of Si is +4, while the oxidation number of Al is +3. If K is present for charge balance, each Al requires one K. That is, (K+Al) replaces Si. The ratio of Si to O in the mineral is 1 : 2. Consider four SiO_2 units, Si_4O_8. If 25% of Si are replaced, that is one out of four Si atoms. Replace one Si with (K+Al). The formula becomes $KAlSi_3O_8$.

22.90 (a) The charge on the aluminosilicate ion shown ($AlSi_3O_{10}$) is 5– (+3 from Al, +12 from Si, –20 from O).

(b) In the silicate ions pictured in Figure 22.32, Si is central and O is either bridging or terminal. We want to select a structure from Figure 22.32 that has the same ratio of central atoms and O atoms as $AlSi_3O_{10}^{5-}$. The ratio of central to O atoms in $AlSi_3O_{10}^{5-}$ is 4 to 10, or 2 to 5. The analogous ion in Figure 22.32 is $Si_2O_5^{2-}$. The $AlSi_3O_{10}^{5-}$ ion will have a sheet structure similar to the one shown in Figure 22.32(c), with ¼ of the Si central atoms replaced by Al central atoms.

Integrative Exercises

22.91 (a) $100.0 \times 10^3 \, g \, FeTi \times \dfrac{1 \, mol \, FeTi}{103.7 \, g \, FeTi} \times \dfrac{1 \, mol \, H_2}{1 \, mol \, FeTi} \times \dfrac{2.016 \, g \, H_2}{1 \, mol \, H_2} = 1944.1 = 1.94 \times 10^3 \, g \, H_2$

(b) $V = \dfrac{1944.1 \, g \, H_2}{2.016 \, g/mol \, H_2} \times \dfrac{0.08206 \, L\text{-}atm}{mol\text{-}K} \times \dfrac{273 \, K}{1 \, atm} = 21,603 = 2.16 \times 10^4 \, L \, H_2$

(c) $2 \, H_2(g) + O_2(g) \rightarrow 2 \, H_2O(l)$

$\Delta H^\circ = 2 \, \Delta H_f^\circ \, H_2O(l) - 2 \, \Delta H_f^\circ \, H_2(g) - \Delta H_f^\circ \, O_2(g)$

$\Delta H^\circ = 2(-285.83) - 2(0) - (0) = -571.66 \, kJ$

$1944.1 \, g \, H_2 \times \dfrac{1 \, mol \, H_2}{2.016 \, g \, H_2} \times \dfrac{-571.66 \, kJ}{2 \, mol \, H_2} = -275,636 = -2.76 \times 10^5 \, kJ$

The minus sign indicates that energy is produced.

22.92 From Appendix C, we need only ΔH_f° for F(g), so that we can estimate ΔH for the process:

$\begin{array}{ll} F_2(g) \rightarrow F(g) + F(g); & \Delta H^\circ = 160 \, kJ \\ \underline{XeF_2(g) \rightarrow Xe(g) + F_2(g)} & \underline{-\Delta H_f^\circ = 109 \, kJ} \\ XeF_2(g) \rightarrow Xe(g) + 2 \, F(g) & \Delta H^\circ = 269 \, kJ \end{array}$

The average Xe–F bond enthalpy is thus 269/2 = 134 kJ. Similarly,

$\begin{array}{ll} XeF_4(g) \rightarrow Xe(g) + 2 \, F_2(g) & -\Delta H_f^\circ = 218 \, kJ \\ \underline{2 \, F_2(g) \rightarrow 4 \, F(g)} & \underline{\Delta H^\circ = 320 \, kJ} \\ XeF_4(g) \rightarrow Xe(g) + 4 \, F(g) & \Delta H^\circ = 538 \, kJ \end{array}$

Average Xe–F bond energy = 538/4 = 134 kJ

$\begin{array}{ll} XeF_6(g) \rightarrow Xe(g) + 3 \, F_2(g) & -\Delta H_f^\circ = 298 \, kJ \\ \underline{3 \, F_2(g) \rightarrow 6 \, F(g)} & \underline{\Delta H^\circ = 480 \, kJ} \\ XeF_6(g) \rightarrow Xe(g) + 6 \, F(g) & \Delta H^\circ = 778 \, kJ \end{array}$

Average Xe–F bond energy = 778/6 = 130 kJ

The average bond enthalpies are: XeF_2, 134 kJ; XeF_4, 134 kJ; XeF_6, 130 kJ. They are remarkably constant in the series.

22.93   (a) $H_2(g) + 1/2\,O_2(g) \rightarrow H_2O(l)$; $\Delta H = -285.83$ kJ/mol H_2

$CH_4(g) + 2\,O_2(g) \rightarrow CO_2(g) + 2\,H_2O(l)$

$\Delta H = 2(-285.83) - 393.5 - (-74.8) = -890.4$ kJ/ mol CH_4

(b)   for H_2: $\dfrac{-285.83\ \text{kJ}}{1\ \text{mol}\ H_2} \times \dfrac{1\ \text{mol}\ H_2}{2.0159\ \text{g}\ H_2} = -141.79$ kJ/g H_2

for CH_4: $\dfrac{-890.4\ \text{kJ}}{1\ \text{mol}\ CH_4} \times \dfrac{1\ \text{mol}\ CH_4}{16.043\ \text{g}\ CH_4} = -55.50$ kJ/g CH_4

(c)   Find the number of moles of gas that occupy 1 m^3 at STP:

$$n = \frac{1\ \text{atm} \times 1\ \text{m}^3}{273\ \text{K}} \times \frac{1\ \text{mol-K}}{0.08206\ \text{L-atm}} \times \left[\frac{100\ \text{cm}}{1\ \text{m}}\right]^3 \times \frac{1\ \text{L}}{10^3\ \text{cm}^3} = 44.64\ \text{mol}$$

for H_2: $\dfrac{-285.83\ \text{kJ}}{1\ \text{mol}\ H_2} \times \dfrac{44.64\ \text{mol}\ H_2}{1\ \text{m}^3\ H_2} = -1.276 \times 10^4$ kJ/m^3 H_2

for CH_4: $\dfrac{-890.4\ \text{kJ}}{1\ \text{mol}\ CH_4} \times \dfrac{44.64\ \text{mol}\ CH_4}{1\ \text{m}^3\ CH_4} = -3.975 \times 10^4$ kJ/m^3 CH_4

22.94   *Analyze/Plan.* $\Delta G° = -RT\ \ln K$. Use $\Delta G°$ for ozone from Appendix C to calculate K for the reaction at 290.0 K, assuming no electrical input.   *Solve.*

The reaction under consideration is: $3\,O_2(g) \rightarrow 2\,O_3(g)$

$\Delta G°$ for ozone at 298 K = 163.4 kJ/mol

$\Delta G°$ for the reaction is then 2(163.4 kJ) − 3(0 kJ) = 326.8 kJ.

$$\ln K = \frac{-\Delta G°}{RT} = \frac{-(326.8 \times 10^3)\,\text{J}}{8.314\ \text{J/K} \times 298.0\ \text{K}} = -131.903 = -131.9;\ \ K = 5 \times 10^{-58}$$

22.95   (a)   First, calculate the molar solubility of Cl_2 in water.

$$n = \frac{1\ \text{atm}\ (0.310\ \text{L})}{\dfrac{0.08206\ \text{L-atm}}{1\ \text{mol-K}} \times 273\ \text{K}} = 0.01384 = 0.0138\ \text{mol}\ Cl_2$$

$$M = \frac{0.01384\ \text{mol}}{0.100\ \text{L}} = 0.1384 = 0.138\ M$$

$[Cl^-] = [HOCl] = [H^+]$ Let this quantity = x. Then, $\dfrac{x^3}{(0.1384 - x)} = 4.7 \times 10^{-4}$

Assuming that x is small compared with 0.1384:

$x^3 = (0.1384)(4.7 \times 10^{-4}) = 6.504 \times 10^{-5}$; x = 0.0402 = 0.040 M

We can correct the denominator using this value, to get a better estimate of x:

$\dfrac{x^3}{0.1384 - 0.0402} = 4.7 \times 10^{-4}$; x = 0.0359 = 0.036 M

One more round of approximation gives x = 0.0364 = 0.036 M. This is the equilibrium concentration of HClO.

(b)   From the equilibrium reaction in part (a), $[H^+] = 0.036\ M$. pH $= -\log[H^+] = 1.4$

The HOCl produced by this equilibrium will ionize slightly to produce additional $H^+(aq)$. However, the K_a value for HOCl is small, 3.0×10^{-8}, and the acid ionization will be suppressed by the presence of $H^+(aq)$ from the solubility equilibrium. $[H^+]$ from ionization of HOCl will be small compared to $0.036\ M$ and will not significantly impact the pH.

22.96   (a) $2\ NH_4ClO_4(s) \overset{\Delta}{\to} N_2(g) + 2\ HCl(g) + 3\ H_2O(g) + 5/2\ O_2(g)$

$NH_4ClO_4(s) \overset{\Delta}{\to} 1/2\ N_2(g) + HCl(g) + 3/2\ H_2O(g) + 5/4\ O_2(g)$

(b) $\Delta H^\circ = \Sigma\ \Delta H_f^\circ\ prod - \Sigma\ \Delta H\ react$

$\Delta H^\circ = \Delta H_f^\circ\ HCl(g) + 3/2\ \Delta H_f^\circ\ H_2O(g) + 1/2\ \Delta H_f^\circ\ N_2(g) + 5/4\ \Delta H_f^\circ\ O_2(g) - \Delta H_f^\circ\ NH_4ClO_4 \Delta H^\circ$

$= -92.30\ kJ + 3/2(-241.82\ kJ) + 1/2\ (0\ kJ) + 5/4\ (0\ kJ) - (-295.8\ kJ)$

$= -159.2\ kJ/mol\ NH_4ClO_4$

(c)   The aluminum reacts exothermically with $O_2(g)$ and HCl(g) produced in the decomposition, providing additional heat and thrust.

(d)   There are $(1/2 + 1 + 3/2 + 5/4) = 4.25$ mol gas per mol NH_4ClO_4 decomposed

$1\ lb\ NH_4ClO_4 \times \dfrac{453.6\ g}{1\ lb} \times \dfrac{1\ mol\ NH_4ClO_4}{117.49\ g\ NH_4ClO_4} \times \dfrac{4.25\ mol\ gas}{1\ mol\ NH_4ClO_4} = 16.408 = 16.4\ mol\ gas$

$V = \dfrac{nRT}{P} = 16.408\ mol\ gas \times \dfrac{0.08206\ L\text{-}atm}{mol\text{-}K} \times \dfrac{273\ K}{1\ atm} = 367.57 = 368\ L$

22.97   (a) $N_2H_4(g) + O_2(g) \to N_2(g) + 2\ H_2O(l)$

(b) $\Delta H^\circ = \Delta H_f^\circ\ N_2(g) + 2\Delta H_f^\circ\ H_2O(l) - \Delta H_f^\circ\ N_2H_4(aq) - \Delta H_f^\circ\ O_2(g)$

$= 0 + 2(-285.83) - 95.40 - 0 = -667.06\ kJ$

(c) $\dfrac{9.1\ g\ O_2}{1 \times 10^6\ g\ H_2O} \times \dfrac{1.0\ g\ H_2O}{1\ mL\ H_2O} \times \dfrac{1000\ mL}{1L} \times 3.0 \times 10^4\ L = 273 = 2.7 \times 10^2\ g\ O_2$

$2.73 \times 10^2\ g\ O_2 \times \dfrac{1\ mol\ O_2}{32.00\ g\ O_2} \times \dfrac{1\ mol\ N_2H_4}{1\ mol\ O_2} \times \dfrac{32.05\ g\ N_2H_4}{1\ mol\ N_2H_4} = 2.7 \times 10^2\ g\ N_2H_4$

22.98   (a) $SO_2(g) + 2\ H_2S(aq) \to 3\ S(s) + 2\ H_2O(g)$ or, if we assume S_8 is the product,

$8\ SO_2(g) + 16\ H_2S(aq) \to 3\ S_8(s) + 16\ H_2O(g)$.

(b)   Assume that all S in the coal becomes SO_2 upon combustion, so that

1 mol S (coal) = 1 mol SO_2; 1 ton = 2000 lb; 760 torr = 1.00 atm

$4000\ lb\ coal \times \dfrac{0.035\ lb\ S}{1\ lb\ coal} \times \dfrac{453.6\ g\ S}{1\ lb\ S} \times \dfrac{1\ mol\ S\ (coal)}{32.07\ g\ S} \times \dfrac{1\ mol\ SO_2}{1\ mol\ S\ (coal)} \times \dfrac{2\ mol\ H_2S}{1\ mol\ SO_2}$

$= 3960 = 4.0 \times 10^3\ mol\ H_2S$

$V = \dfrac{3960\ mol \times (0.08206\ L\text{-}atm/mol\text{-}K) \times 300\ K}{1.00\ atm} = 97,496 = 9.7 \times 10^4\ L$

(c) $3960 \text{ mol H}_2\text{S} \times \dfrac{3 \text{ mol S}}{2 \text{ mol H}_2\text{S}} \times \dfrac{32.07 \text{ g S}}{1 \text{ mol S}} = 1.9 \times 10^5 \text{ g S}$

This is about 210 lb S per ton of coal combusted. (However, two-thirds of this comes from the H_2S, which presumably at some point was also obtained from coal.)

22.99   *Plan.* Vol air → kg air → g H_2S → g FeS. Use the ideal-gas equation to change volume of air to mass of air, (assuming 1.00 atm, 298 K and an average molar mass (MM) for air of 29.0 g/mol. Use (20 mg H_2S/kg air) to find the mass of H_2S in the given mass of air. *Solve.*

$$V_{air} = 12 \text{ ft} \times 20 \text{ ft} \times 8 \text{ ft} \times \dfrac{12^3 \text{ in}^3}{\text{ft}^3} \times \dfrac{2.54^3 \text{ cm}^3}{1^3 \text{ in}^3} \times \dfrac{1 \text{ L}}{1000 \text{ cm}^3} = 5.4368 \times 10^4 = 5 \times 10^4 \text{ L}$$

$$g_{air} = \dfrac{\text{PV MM}}{\text{RT}}; \text{ assume P} = 1.00 \text{ atm}, \text{ T} = 298 \text{ K}, \text{ MM}_{air} = 29.0 \text{ g/mol}$$

$$g_{air} = \dfrac{1.00 \text{ atm} \times 5.4368 \times 10^4 \text{ L} \times 29.0 \text{ g/mol}}{298 \text{ K}} \times \dfrac{\text{mol-K}}{0.08206 \text{ L-atm}} = 64{,}476 = 6 \times 10^4 \text{ g air}$$

$$6.4476 \times 10^4 \text{ g air} \times \dfrac{1 \text{ kg}}{1000 \text{ g}} \times \dfrac{20 \text{ mg H}_2\text{S}}{1 \text{ kg air}} \times \dfrac{1 \text{ g}}{1000 \text{ mg}} = 1.2895 = 1 \text{ g H}_2\text{S}$$

$$\text{FeS(s)} + 2 \text{ HCl(aq)} \rightarrow \text{FeCl}_2\text{(aq)} + \text{H}_2\text{S(g)}$$

$$1.2895 \text{ g H}_2 \times \dfrac{1 \text{ mol H}_2}{34.08 \text{ g H}_2\text{S}} \times \dfrac{1 \text{ mol FeS}}{1 \text{ mol H}_2\text{S}} \times \dfrac{87.91 \text{ g FeS}}{1 \text{ mol FeS}} = 3.3263 = 3 \text{ g FeS}$$

22.100   The reactions can be written as follows:

$H_2(g) + X(\text{std state}) \rightarrow H_2X(g)$	ΔH_f^o
$2 H(g) \rightarrow H_2(g)$	$\Delta H_f^o (H–H)$
$X(g) \rightarrow X(\text{std state})$	ΔH_3
Add: $2 H(g) + X(g) \rightarrow H_2X(g)$	$\Delta H = \Delta H_f^o + \Delta H_f^o (H–H) + \Delta H_3$

These are all the necessary ΔH values. Thus,

Compound	ΔH	D H–X
H_2O	$\Delta H = -241.8 \text{ kJ} - 436 \text{ kJ} - 248 \text{ kJ} = -926 \text{ kJ}$	463 kJ
H_2S	$\Delta H = -20.17 \text{ kJ} - 436 \text{ kJ} - 277 \text{ kJ} = -733 \text{ kJ}$	367 kJ
H_2Se	$\Delta H = +29.7 \text{ kJ} - 436 \text{ kJ} - 227 \text{ kJ} = -633 \text{ kJ}$	317 kJ
H_2Te	$\Delta H = +99.6 \text{ kJ} - 436 \text{ kJ} - 197 \text{ kJ} = -533 \text{ kJ}$	267 kJ

The average H–X bond energy in each case is just half of ΔH. The H–X bond energy decreases steadily in the series. The origin of this effect is probably the increasing size of the orbital from X with which the hydrogen 1s orbital must overlap.

22.101   (a)   MnSi: more than one element, so not metallic; high melting, so not molecular; insoluble in water, so not ionic; therefore covalent network.

(b) $\text{MnSi(s)} + \text{HF(aq)} \rightarrow \text{SiH}_4\text{(g)} + \text{MnF}_4\text{(s)}$

Reduction of Mn(IV) to Mn(II) is unlikely, because F^- is an extremely weak reducing agent. E_{red}^o for $F_2(g) + 2 e^- \rightarrow 2 F^-(aq) = 2.87 \text{ V}$

22.102 $N_2H_5^+(aq) \rightarrow N_2(g) + 5\,H^+(aq) + 4\,e^-$ $E_{red}^{\circ} = -0.23\text{ V}$

Reduction of the metal should occur when E_{red}° of the metal ion is more positive than about -0.15 V. This is the case for (b) Sn^{2+} (marginal), (c) Cu^{2+}, (d) Ag^+ and (f) Co^{3+}

22.103   First write the balanced equation to give the number of moles of gaseous products per mole of hydrazine.

(A) $(CH_3)_2NNH_2 + 2\,N_2O_4 \rightarrow 3\,N_2(g) + 4\,H_2O(g) + 2\,CO_2(g)$

(B) $(CH_3)HNNH_2 + 5/4\,N_2O_4 \rightarrow 9/4\,N_2(g) + 3\,H_2O(g) + CO_2(g)$

In case (A) there are nine moles gas per one mole $(CH_3)_2NNH_2$ plus two moles N_2O_4. The total mass of reactants is $60 + 2(92) = 244$ g. Thus, there are

$$\frac{9\text{ mol gas}}{244\text{ g reactants}} = \frac{0.0369\text{ mol gas}}{1\text{ g reactants}}$$

In case (B) there are 6.25 moles of gaseous product per one mole $(CH_3)HNNH_2$ plus 1.25 moles N_2O_4. The total mass of this amount of reactants is $46.0 + 1.25(92.0) = 161$ g.

$$\frac{6.25\text{ mol gas}}{161\text{ g reactants}} = \frac{0.0388\text{ mol gas}}{1\text{ g reactants}}$$

Thus the methylhydrazine (B) has marginally greater thrust.

22.104   (a) $3\,B_2H_6(g) + 6\,NH_3(g) \rightarrow 2\,(BH)_3(NH)_3(l) + 12\,H_2(g)$

              $3\,LiBH_4(s) + 3\,NH_4Cl(s) \rightarrow 2\,(BH)_3(NH)_3(l) + 9\,H_2(g) + 3\,LiCl(s)$

(b)   30 valence e^-, 15 e^- pairs

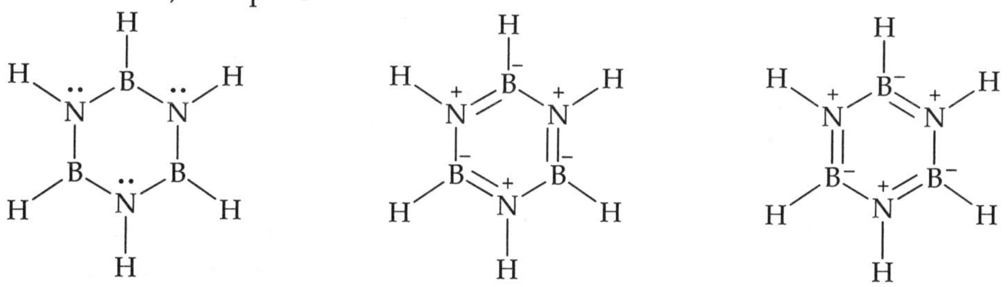

The structure with nonbonded pairs minimizes formal charge, but these electrons are almost certainly delocalized about the six-membered ring, mimicking the bonding in benzene.

(c) $n = \dfrac{PV}{RT} = \dfrac{1.00\text{ atm} \times 2.00\text{ L}}{273\text{ K}} \times \dfrac{\text{mol-K}}{0.08206\text{ L-atm}} = 0.08929 = 8.93 \times 10^{-2}\text{ mol NH}_3$

$0.08929\text{ mol NH}_3 \times \dfrac{2\text{ mol }(BH)_3(NH)_3}{6\text{ mol NH}_3} \times \dfrac{80.50\text{ g }(BH)_3(NH)_3}{1\text{ mol }(BH)_3(NH)_3}$

$= 2.3956 = 2.40\text{ g }(BH)_3(NH)_3$

23 Transition Metals and Coordination Chemistry

Visualizing Concepts

23.1 *Analyze/Plan.* Given graphs of three properties moving across the fourth period of the chart, match each graph to the atomic property it represents. *Solve.*

Graph (a) is the trend in maximum oxidation state. This value corresponds to the maximum number of (4s + 3d) electrons that can be removed from a neutral atom. It increases up to a maximum of +7 for Mn, then decreases. The decrease is partly because of an increase in the attraction of 3d electrons for the nucleus as Z_{eff} increases.

Graph (b) is the trend in effective nuclear charge, Z_{eff}. Moving from left to right in a period, Z_{eff} increases because the increase in Z is not offset by a significant increase in shielding.

Graph (c) is the trend in radius. Increasing Z_{eff} leads to decreasing atomic radius.

23.2 *Analyze.* Given the formula of a coordination compound, draw the structure, determine the coordination number, coordination geometry, oxidation state of the metal, and number of unpaired electrons.

Plan. From the formula, determine the identity of the ligands and the number of coordination sites they occupy. From the total coordination number, decide on a likely geometry. Use ligand and overall complex charges to calculate the oxidation number of the metal. Refer to the d-orbital energy level diagram that corresponds to the structure of the compound and the field strength of the ligands to determine the number of unpaired electrons.

Solve. The ligands are $2Cl^-$, one coordination site each, and en, ethylenediamine, two coordination sites, for a coordination number of 4. This coordination number has two possible geometries, tetrahedral and square planar. Pt is one of the metals known to adopt square planar geometry when CN = 4.

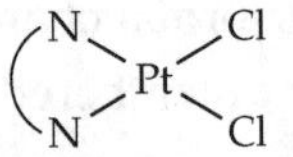

(a) Coordination number is 4

(b) Coordination geometry is square planar

(c) The oxidation state of Pt is +2. $Pt(en)Cl_2$ is a neutral compound, the en ligand is neutral, and the $2Cl^-$ ligands are each –1, so the oxidation state of Pt must be +2, Pt(II).

(d) There are no unpaired electrons. Square planar d^8 complexes are usually low spin, especially with heavier metals like Pt.

23.3 *Analyze.* Given a ball-and-stick figure of a ligand, write the Lewis structure and answer questions about the ligand.

Plan. Assume that each atom in the Lewis structure obeys the octet rule. Complete each octet with unshared electron pairs or multiple bonds, depending on the bond angles in the ball-and-stick model. Black = C, blue = N, red = O, gray = H.

$$\left[\begin{array}{c} \quad\;\; \overset{\displaystyle H \quad H \qquad\quad H \quad :O:}{\underset{\displaystyle H \quad H \quad H \quad H \quad H}{H-\ddot{N}-C-C-\ddot{N}-C-C-\ddot{O}:}} \end{array} \right]^{-}$$

There is a second resonance structure with the double bond drawn to the other O atom.

Check. Write the molecular formula, count the valence electron pairs and see if it matches your structure. $[C_4H_9N_2O_2]^-$ $(16 + 9 + 10 + 12 + 1) = 48$ valence e^-, 24 e^- pair Our Lewis structure also has 24 e^- pairs.

(a) Donor atoms have unshared electron pairs. The potential donors in this structure are the two N and two O atoms.

The ligand is polydentate. If bound to a single metal center, it is likely to be tridentate. Even though there are four possible donor atoms, the structure would be strained if all four were bound to one metal center. It is likely that only one of the two O atoms binds to the same metal as the two N atoms. If it is bound to more than one metal, it could use all four donor atoms.

(b) An octahedral complex has 6 coordination sites. A single ligand has 3 likely donors, so two ligands are needed. From a steric perspective, the likely donors would be the 2 N atoms and 1 of the carbonyl oxygen atoms. The chelate bite of a carboxyl group is relatively small and would require an O—M—O angle of less than 90°.

23.4 *Analyze/Plan.* Given the molecular formula, name the compound. Use the ball-and-stick structures to decide the number of geometric isomers for each molecular geometry. Consider the d-electron configuration of platinum(II) and the appropriate crystal field splitting diagrams (Figures 23.33 and 23.34) to determine magnetic properties for the two molecular geometries. *Solve.*

(a) Diaminodichloroplatinum(II). If the geometry is square planar, the prefix *cis-* would be added.

(b) No, the tetrahedral molecule would not have a geometric isomer.

(c) The tetrahedral molecule would be paramagnetic. Platinum(II) is a d^8 transition metal ion. In a tetrahedral crystal field (Figure 23.33 and Sample Exercise 23.8), there would be two unpaired d-electrons and the molecule would be paramagnetic.

(d) Yes, the square planar molecule would have a geometric isomer, the *trans* isomer.

(e) The square planar molecule would be diamagnetic. A d^8 transition metal ion in a square planar crystal field (Figure 23.34 and Sample Exercise 23.8) has no unpaired electrons.

(f) Yes, you would be able to distinguish the two geometries by determining the number of geometric isomers. The square planar geometry has two possible isomers, cis and trans, while the tetrahedral geometry has one unique structure.

(g) Yes, you would be able to distinguish the two geometries by measuring the molecule's response to a magnetic field. The tetrahedral molecule is attracted to a magnetic field, while the square planar molecule is not.

23.5 *Analyze.* Given 5 structures, visualize which are identical to (1) and which are geometric isomers of (1).

Plan. There are two possible ways to arrange MA_3X_3. The first has bond angles of 90° between all similar ligands; this is structure (1). The second has one 180° angle between similar ligands. Visualize which description fits each of the five structures.

Solve. (1) has all 90° angles between similar ligands.

(2) has a 180° angle between similar ligands (see the blue ligands in the equatorial plane of the octahedron)

(3) has all 90° angles between similar ligands

(4) has all 90° angles between similar ligands

(5) has a 180° angle between similar ligands (see the blue axial ligands)

Structures (3) and (4) are identical to (1); (2) and (5) are geometric isomers.

23.6 *Analyze.* Given four structures, decide which are chiral.

Plan. Chiral molecules have nonsuperimposable mirror images. Draw the mirror image of each molecule and visualize whether it can be rotated into the original molecule. If so, the complex is not chiral. If the original orientation cannot be regenerated by rotation, the complex is chiral. *Solve.*

(1)   (1) mirror

The two orientations are not superimposable and molecule (1) is chiral.

The two orientations are superimposable. Rotate the right-most structure 90° counterclockwise about the B-M-B axis to align the G's; the bidentate ligands then also overlap. Molecule (2) is not chiral.

The two orientations are not superimposable and molecule (3) is chiral.

The two orientations are not superimposable and molecule (4) is chiral.

23.7   *Analyze.* Given the visible colors of two solutions, determine the colors of light absorbed by each solution.

Plan. Apparent color is transmitted or reflected light, absorbed color is basically the complement of apparent color. Use the color wheel in Figure 23.25 to obtain the complementary absorbed color for the solutions.

Solve. Moving from left to right, the solutions appear blue-green (cyan), yellow, green, and red. The solutions absorb red-orange, violet, red, and green.

23.8   *Analyze.* Fit the crystal field splitting diagram to the complex description in each part.

Plan. Determine the number of d-electrons in each transition metal. On the splitting diagrams match the d-orbital splitting patterns to complex geometry and electron pairing to the definition of high-spin and low-spin.

Solve. Octahedral complexes have the 3 lower, 2 higher splitting pattern, whereas tetrahedral complexes have the opposite 2 lower, 3 higher pattern. Low spin complexes favor electron pairing because of large d-orbital splitting. High-spin complexes have maximum occupancy because of small orbital splitting.

(a) Fe^{3+}, 5 d-electrons; weak field: spins unpaired; octahedral: 3 lower, 2 higher d-splitting ∴ diagram (4)

(b) Fe^{3+}, 5 d-electrons; strong field: spins paired; octahedral: 3 lower, 2 higher d-splitting ∴ diagram (1)

(c) Fe^{3+}, 5 d-electrons; tetrahedral: 2 lower, 3 higher d-splitting ∴ diagram (3)

(d) Ni^{2+}, 8 d-electrons; tetrahedral: 2 lower, 3 higher d-splitting ∴ diagram (2)

Check. Diagram (2) was the remaining choice for (d) and it fits the description.

23.9   *Analyze/Plan.* Given the linear diagram and axial labels, answer the questions and predict crystal field splitting. Orbitals with lobes nearest ligand charges (or partial charges) will be highest in energy; orbitals with lobes away from charges are lowest in energy.

Solve. Diagram (c) is the best choice. The d_{z^2} orbital has lobes nearest the charges and is at the highest energy. The $d_{x^2-y^2}$ and d_{xy} orbitals have lobes in the xy-plane farthest from the charges and are lowest in energy. The d_{xz} and d_{yz} orbitals point between the respective axes and are intermediate in energy.

23.10   *Analyze.* Given the colors of two low spin Fe(II) complexes, determine which complex contains the stronger field ligand. *Plan.* We can make this direct comparison because both solutions contain low spin d^6 ions. A solution that appears one color absorbs visible light of the complementary color. Use the color wheel in Figure 23.25 to decide which color and approximate wavelength of visible light is absorbed by the two solutions. The stronger-field ligand causes a larger d-orbital splitting and absorbs light with the shorter wavelength.

Solve. The green solution absorbs red light in the 650 to 750 nm range. The red solution absorbs green light in the 490 to 560 nm range. The complex that produces the red solution has the larger d-orbital splitting and the stronger-field ligand.

The Transition Metals (Section 23.1)

23.11   Trend (c). The lanthanide contraction is the name given to the decrease in atomic size because of the build-up in effective nuclear charge as we move through the lanthanides (elements 57–71) and beyond them. This effect offsets the expected increase in atomic size going from period 5 to period 6 transition elements.

23.12   Trend (b) explains the peak in maximum oxidation state of the transition-metal elements near groups 7B and 8B. As effective nuclear charge increases, d-electrons are more strongly attracted to the nucleus and more difficult to remove from the atom.

23.13   (a) Ti^{2+}, $[Ar]3d^2$ (b) Ti^{4+}, $[Ar]$ (c) Ni^{2+}, $[Ar]3d^8$ (d) Zn^{2+}, $[Ar]3d^{10}$

23.14   Refer to Figure 23.5. Among the period 4 transition metals, only Sc and Zn do not have at least one oxidation state with partially filled 3d orbitals. Sc^{3+} has zero 3d electrons and Zn^{2+} has ten 3d electrons. In Sc^{3+} the 3d orbitals are unoccupied; in Zn^{2+} the 3d orbitals are filled.

23.15   (a) Ti^{3+}, $[Ar]3d^1$ (b) Ru^{2+}, $[Kr]4d^6$ (c) Au^{3+}, $[Xe]4f^{14}5d^8$ (d) Mn^{4+}, $[Ar]3d^3$

23.16   (a) Co^{3+}, $[Ar]3d^6$; 6 valence d-electrons

       (b) Cu^+; $[Ar]3d^{10}$; 10 valence d-electrons

       (c) Cd^{2+}, $[Kr]4d^{10}$; 10 valence d-electrons

       (d) Os^{3+}: $[Xe]\,4f^{14}5d^5$; 5 valence d-electrons

23.17   *Analyze/Plan.* Consider the definitions of paramagnetic and diamagnetic.   *Solve.*

The unpaired electrons in a paramagnetic substance cause it to be weakly attracted into a magnetic field. (A diamagnetic material, where all electrons are paired, is very weakly repelled by a magnetic field.)

23.18   Antiferromagnetic materials cannot be used to make permanent magnets. In an antiferromagnetic material, coupled spins are aligned in opposite directions and the opposing spins exactly cancel.

23.19   *Analyze/Plan.* Consider the orientation of spins in various types of magnetic materials as shown in Figure 23.6.

The diagram shows a material with misaligned spins that become aligned in the direction of an applied magnetic field. This is a paramagnetic material.

23.20   (a) Fe_2O_3 has all Fe atoms in the +3 oxidation state, whereas Fe_3O_4 contains Fe atoms in both the +2 and +3 states. Each Fe_3O_4 formula unit has one Fe(II) and two Fe(III).

        (b)   In an antiferromagnetic material, spins on coupled atoms are oppositely aligned, producing a net spin of zero. This is only possible for Fe_2O_3, where all Fe atoms have the same oxidation state, d-electron configuration and number of unpaired electrons. In Fe_3O_4, Fe(II) and Fe(III) atoms have different d-electron configurations and different numbers of unpaired electrons. Assuming an Fe(II) is coupled to an Fe(III), even if spins on coupled centers are oppositely aligned, their spins do not fully cancel and the material is ferromagnetic.

Transition-Metal Complexes (Section 23.2)

23.21   (a)   Primary valence. In Werner's theory, primary valence is the charge of the metal cation at the center of the complex. "Oxidation state" is a broader term than "ionic charge," but Werner's complexes contain metal ions where cation charge and oxidation state are equal.

        (b)   Coordination number. In Werner's theory, secondary valence is the number of atoms bound or coordinated to the central metal ion.

        (c) NH_3 can serve as a ligand because it has an unshared electron pair, whereas BH_3 does not. Ligands act as a Lewis base in metal-ligand interactions. As such, they must possess at least one unshared electron pair. BH_3, with fewer than 8 electrons about B, has no unshared electron pair and cannot act as a ligand. In fact, BH_3 acts as a Lewis acid, an electron pair acceptor, because it is electron-deficient.

23.22   (a)   Negatively charged ions are more likely to act as ligands, to take advantage of the electrostatic attraction between a positively charged metal cation and a negatively charged ligand.

        (b)   Polar molecules are more likely to act as ligands to take advantage of ion-dipole interactions between a positively charged metal cation and a polar ligand.

23.23 *Analyze/Plan.* Follow the logic in Sample Exercises 23.1 and 23.2. *Solve.*

(a) This compound is electrically neutral, and the NH_3 ligands carry no charge, so the charge on Ni must balance the –2 charge of the 2 Br^- ions. The charge and oxidation state of Ni is +2.

(b) Because there are 6 NH_3 molecules in the complex, the likely coordination number is 6. In some cases Br^- acts as a ligand, so the coordination number could be other than 6.

(c) Assuming that the 6 NH_3 molecules are the ligands, 2 Br^- ions are not coordinated to the Ni^{2+}, so 2 mol AgBr(s) will precipitate.

23.24 (a) $[Cr(H_2O)_6]^{3+}$

(b) Three. Because the Cl^- ions are not in the coordination sphere, all 3 anions react with Ag^+ to form 3 moles of AgCl(s).

(c) If the empirical formula of the compound is $CrCl_3$ and Cr^{3+} has a coordination number of six, some or all of the Cl^- ions must be shared by more than one Cr^{3+} ion. This produces a network solid that is very difficult to dissolve or break down.

23.25 *Analyze/Plan.* Count the number of donor atoms in each complex, taking the identity of polydentate ligands into account. Follow the logic in Sample Exercise 23.2 to obtain oxidation numbers of the metals.

(a) Coordination number = 4, oxidation number = +2

(b) 5, +4

(c) 6, +3

(d) 5, +2

(e) 6, +3

(f) 4, +2

23.26 (a) Coordination number = 6, oxidation number = +3

(b) 4, +2

(c) 6, +4

(d) 6, +3

(e) 6, +3

(f) 5, +2

Common Ligands in Coordination Chemistry (Section 23.3)

23.27 (a) $CH_3CH_2NH_2$, 20 e^-, 10 e^- pr

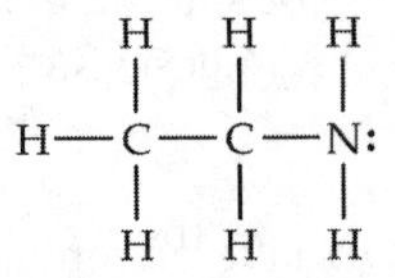

$$
\begin{array}{c}
\quad\ \ H \quad\ \ H \quad\ \ H \\
\quad\ \ | \qquad | \qquad | \\
H - C - C - N: \\
\quad\ \ | \qquad | \qquad | \\
\quad\ \ H \quad\ \ H \quad\ \ H
\end{array}
$$

monodentate ligand, only N atom has a nonbonded pair of electrons

(b) $P(CH_3)_3$, 26 e⁻, 13 e⁻ pr

monodentate ligand, only P atom has a nonbonded pair of electrons

(c) CO_3^{2-}, 24 e⁻, 12 e⁻ pr

either monodentate or bidentate

[All three O atoms are possible bonding sites, but it is not geometrically possible for all three O atoms to be bound to the same metal ion.]

(d) C_2H_6, 14 e⁻, 7 e⁻ pr

unlikely to act as a ligand, no nonbonded pairs of electrons

23.28 (a) 2 coordination sites, 2 N donor atoms

(b) 2 coordination sites, 2 N donor atoms

(c) 2 coordination sites, 2 O donor atoms (Although there are four potential O donor atoms in $C_2O_4^{2-}$, it is geometrically impossible for more than two of these to be bound to a single metal ion.)

(d) 4 coordination sites, 4 N donor atoms

(e) 6 coordination sites, 2 N and 4 O donor atoms

23.29 *Analyze/Plan.* Given the formula of a coordination compound, determine the number of coordination sites occupied by the polydentate ligand. The coordination number of the complexes is probably 4 or 6. Note the number of monodentate ligands and determine the number of coordination sites occupied by the polydentate ligands. *Solve.*

(a) *ortho*-Phenanthroline, *o*-phen, is bidentate. The complex is 6-coordinate, there are 4 monodentate NH_3 ligands, so *o*-phen occupies 2 sites.

(b) Oxalate, $C_2O_4^{2-}$, is bidentate. The complex is 6-coordinate, there are 4 monodentate H_2O ligands, so oxalate occupies 2 coordination sites.

(c) Ethylenediaminetetraacetate, EDTA, is hexadentate. The complex is probably 6-coordinate octahedral with EDTA occupying all six coordination sites.

(d) Ethylenediamine, en, is bidentate. The complex is 4-coordinate and each en ligand occupies 2 coordination sites.

23.30   (a)   6   (b)   6   (c)   6   (d)   6

23.31   *Analyze/Plan.* Anions and polar molecules (with nonbonded electron pairs) are most likely to act as ligands in a metal complex (Solution 23.22).   *Solve.*

(a) CH_3CN, polar molecule with nonbonded electron pair

(b) H^-, anion

(c) CO, polar molecule with a nonbonded electron pair

23.32   (a) Pyridine is a **mono**dentate ligand because it has one N donor atom and therefore occupies one coordination site in a metal complex.

(b) K for this reaction will be smaller than one. Two free pyridine molecules are replaced by one free bipy molecule. There are more moles of particles in the reactants than products, so ΔS is predicted to be negative. Processes with a net decrease in entropy are usually nonspontaneous, have positive ΔG, and values of K less than one. This equilibrium is likely to be spontaneous in the reverse direction.

23.33   False. The ligand shown in the figure does not typically act as a bidentate ligand for a single metal center. The entire molecule is planar; there is no "bend" in the central 6-membered ring that includes the two N atoms. The benzene rings on either side of the two N atoms inhibit their approach in the correct orientation for chelation. It might act as a bridging ligand between two metal centers, but again, the benzene rings would create a significant amount of steric hindrance.

23.34   (a) The complex shown in the exercise has tetrahedral geometry about the silver.

(b) The ligands are neutral molecules and the metal is Ag(I), so the complex will have a 1+ charge.

(c) Yes, one nitrate ion, NO_3^-, will be present in the crystal to provide charge balance for the complex cation.

(d) $[Ag(o\text{-phen})_2]NO_3$

(e) bis(*ortho*-phenanthroline)silver(I) nitrate

Nomenclature and Isomerism in Coordination Chemistry (Section 23.4)

23.35   *Analyze/Plan.* Given the name of a coordination compound, write the chemical formula. Refer to Tables 23.4 and 23.5 to find ligand formulas. Place the metal complex (metal ion + ligands) inside square brackets and the counterion (if there is one) outside the brackets.   *Solve.*

(a) $[Cr(NH_3)_6](NO_3)_3$ (b) $[Co(NH_3)_4CO_3]_2SO_4$ (c) $[Pt(en)_2Cl_2]Br_2$

(d) $K[V(H_2O)_2Br_4]$ (e) $[Zn(en)_2][HgI_4]$

23.36   (a) $[Mn(H_2O)_4Br_2]ClO_4$ (b) $[Cd(bipy)_2]Cl_2$

(c) $K[Co(o\text{-phen})Br_4]$ (d) $Cs[Cr(NH_3)_2(CN)_4]$

(e) $[Rh(en)_3][Co(ox)_3]$

23.37   *Analyze/Plan.* Follow the logic in Sample Exercise 23.3, paying attention to naming rules in Section 23.4.   *Solve.*

(a)   tetraamminedichlororhodium(III) chloride

(b)   potassium hexachlorotitanate(IV)

(c)   tetrachlorooxomolybdenum(VI)

(d)   tetraaqua(oxalato)platinum(IV) bromide

23.38   (a)   dichloroethylenediamminecadmium(II)

(b)   potassium hexacyanomanganate(II)

(c)   pentaamminecarbonatochromium(III) chloride

(d)   tetraamminediaquairidium(III) nitrate

23.39   *Analyze/Plan.*   Consider the coordination number and geometry of each of the complexes, along with the definitions of the various types of isomerism. Use this information to decide which of the complexes could exhibit isomerism of the specified type.   *Solve.*

Complex 1 has a coordination number of 6 and octahedral geometry about the metal. There are 4 monodentate ligands of one kind and two of another.

Complex 2 has a coordination number of 4 and square planar geometry. There are two monodentate ligands of one kind and two of another.

Complex 3 has a coordination number of 6 and octahedral geometry about the metal. There are 2 bidentate ligands and two monodentate ligands.

(a)   Complexes 1, 2, and 3 can have geometric isomers. These are different arrangements of the same set of ligands. All three complexes have cis-trans isomers, where a pair of ligands is either opposite or adjacent to each other.

(b)   Only complex 2 can have linkage isomers. Nitrite ion, NO_2^-, can coordinate through either N or O. It is the only ligand in the three complexes that has this ability.

(c)   Only the cis geometric isomer of complex 3 can have optical isomers. These are isomers, with the same arrangement of bonds, that are mirror images of each other and cannot be superimposed.

(d)   Only complex 1 can have coordination sphere isomers. This is where an anion can either be a ligand or a counterion. Complex 1 is the only example with a counterion that can also be a ligand.

23.40   Complex 1 has a coordination number of 6 and octahedral geometry about the metal. There are 5 monodentate ligands of one kind and one of another.

Complex 2 has a coordination number of 6 and octahedral geometry. There are three monodentate ligands of one kind and three of another.

Complex 3 has a coordination number of 6 and octahedral geometry about the metal. There are 5 monodentate ligands of one kind and one of another.

(a)   Only complex 2 has geometric isomers. It is the only complex with different possible arrangements of the same ligands. There is only one unique way to arrange 5 ligands of one kind and one of another, as in complex 1 and 3.

(b)   Only complex 1 can have linkage isomers. Thiocyanate ion, SCN^-, can coordinate through either N or S. It is the only ligand in the three complexes that has this ability.

(c)   None of the complexes has optical isomers. None of the complexes has bidentate ligands, which are usually present in optically active octahedral complexes.

(d)   Only complex 3 can have coordination sphere isomers. This is where an anion can either be a ligand or a counterion. Complex 3 is the only example with anions that are not specifically written as ligands.

23.41   Yes. A tetrahedral complex of the form MA_2B_2 would have neither structural isomers nor stereoisomers. For a tetrahedral complex, no differences in connectivity are possible for a single central atom, so the terms cis and trans do not apply. No optical isomers with tetrahedral geometry are possible because M is not bound to four different groups. The complex must be square planar with cis and trans geometric isomers.

23.42   Two geometric isomers are possible for an octahedral MA_3B_3 complex (see below). All other arrangements, including mirror images, can be rotated into these two structures. Neither isomer is optically active.

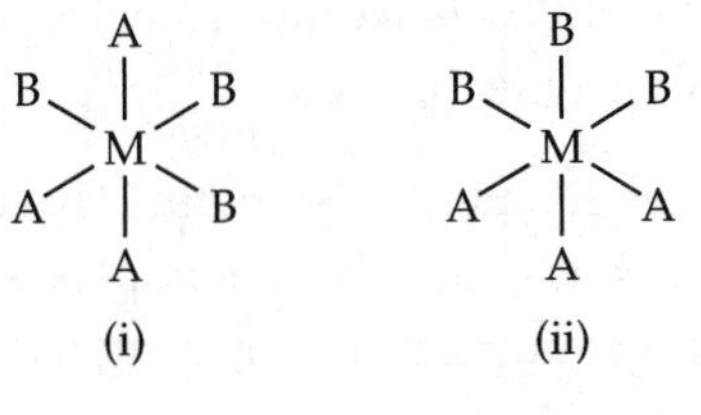

23.43   *Analyze/Plan.* Follow the logic in Sample Exercise 23.4 and 23.5.   *Solve.*

(a)   No geometric isomers

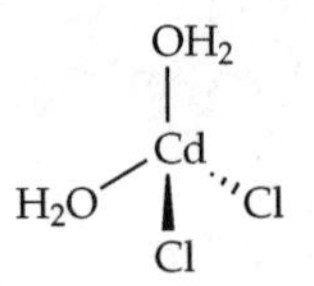

(b)   Two geometric isomers, cis and trans

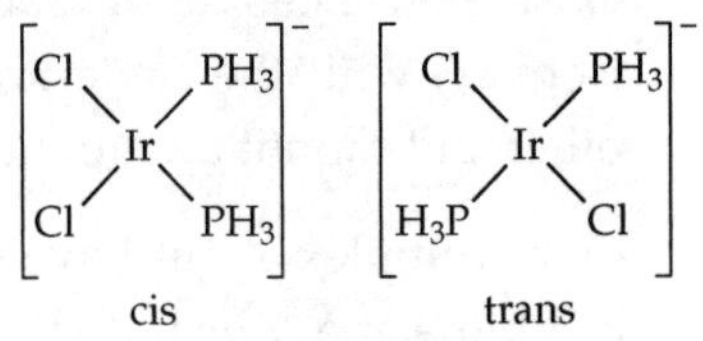

(c)   Three geometric isomers: cis and trans; the cis isomer has enantiomers

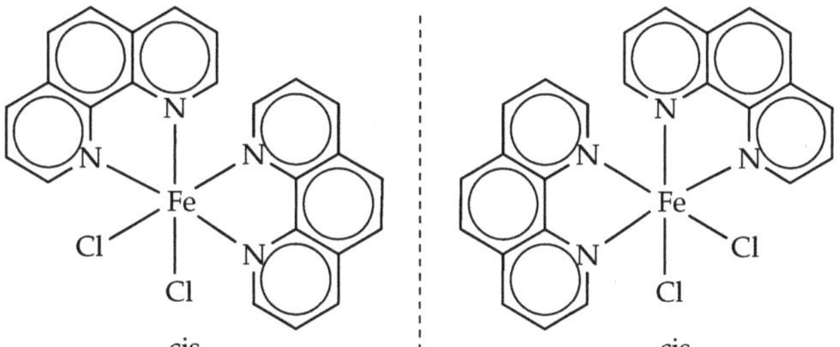

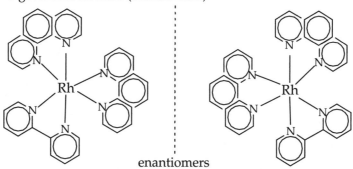

cis   cis

optical isomers

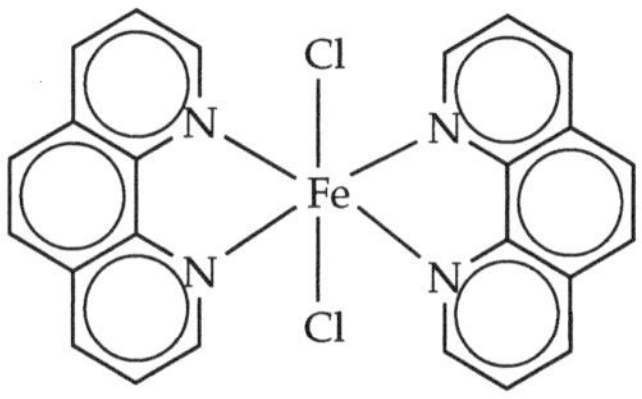

trans

(The three isomeric complex ions in part (c) each have a 1+ charge.)

23.44   (a)   Two geometric isomers (enantiomers)

enantiomers

(b)   Two geometric isomers. In one, the three ammonia ligands are all cis to each other; in the other, one pair of ammonia ligands are trans.

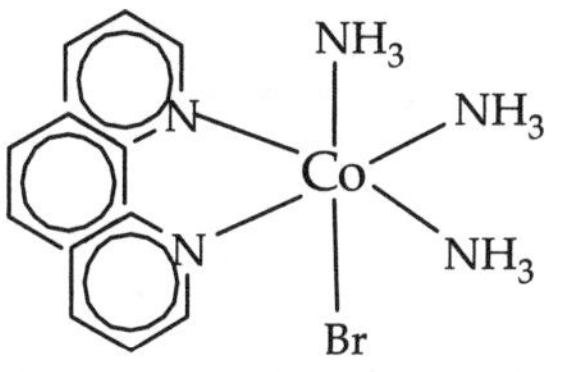

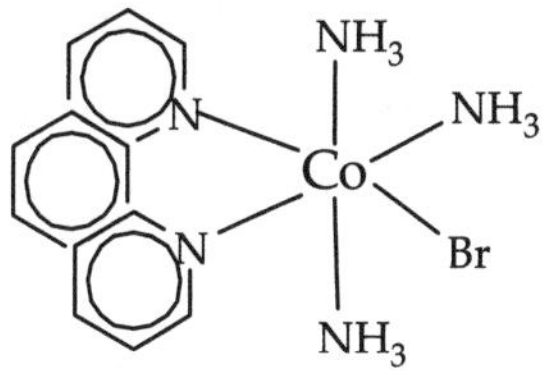

(c)   No geometric isomers

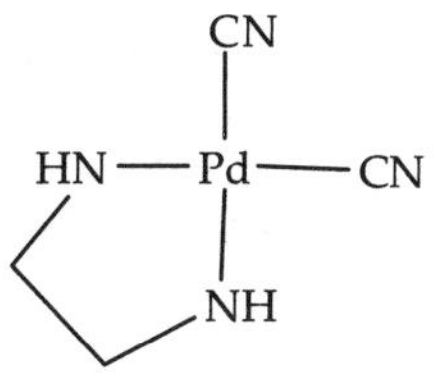

23.45   *Analyze/Plan.* Consider the geometry and arrangement of ligands in each complex. Decide whether the structure has a nonsuperimposable mirror image.   *Solve.*

(a)   Not chiral. The tetrahedral Zn^{2+} is not bonded to four different ligands.

(b)   Not chiral. One mirror image can be superimposed on the other. [See similar trans structure in Solution 23.43 (c)]

(c)   Chiral, has an optical isomer. [See similar cis structures in Solution 23.43 (c)]

23.46   (a)   Not chiral. A square planar complex cannot be chiral.

(b)   Not chiral. One mirror image can be superimposed on the other.

(c)   Chiral. One mirror image cannot be superimposed on the other.

Color and Magnetism in Coordination Chemistry; Crystal-Field Theory (Sections 23.5 and 23.6)

23.47   (a)   Visible light with a wavelength of 610 nm is orange. If the complex absorbs orange light, it will appear blue.

(b) $E(J/photon) = h\nu = hc/\lambda$.

$$E = \frac{6.626 \times 10^{-34} \text{ J-s}}{610 \text{ nm}} \times \frac{2.998 \times 10^8 \text{ m}}{s} \times \frac{1 \text{ nm}}{1 \times 10^{-9} \text{ m}} = 3.257 \times 10^{-19} = 3.26 \times 10^{-19} \text{ J}$$

(c)   Change J/photon to kJ/mol.

$$\frac{3.259 \times 10^{-19} \text{ J}}{photon} \times \frac{1 \text{ kJ}}{1000 \text{ J}} \times \frac{6.022 \times 10^{23} \text{ photons}}{mol} = 196 \text{ kJ/mol}$$

23.48   (a) $E(J/photon) = hc/\lambda$. $\lambda = hc/E$.

$$\lambda = \frac{6.626 \times 10^{-34} \text{ J-s}}{4.51 \times 10^{-19} \text{ J/photon}} \times \frac{2.998 \times 10^8 \text{ m}}{s} \times \frac{1 \text{ nm}}{1 \times 10^{-9} \text{ m}} = 440.46 = 440 \text{ nm}$$

(b)   If the complex absorbs only 441 nm visible light, it absorbs violet and appears yellow.

23.49   *Analyze/Plan.* Given the formula of a coordination complex, determine the oxidation state and electron configuration of the central metal ion. If necessary, use a d-orbital energy level diagram appropriate for the geometry of the complex to decide if the metal ion has unpaired electrons. If so, it is paramagnetic.

(a) Zn^{2+}, $[Ar]3d^{10}$. There are no unpaired electrons, so the complex is diamagnetic.

(b) Pd^{2+}, $[Kr]4d^8$. The complex is probably square planar. Square planar complexes with 8 d-electrons are usually diamagnetic, especially with a heavy metal center like Pd.

(c) V^{3+}, $[Ar]3d^2$. There are 2 d-electrons, so the complex is paramagnetic. The complex is octahedral, but the two electrons would be unpaired in any of the d-orbital energy level diagrams.

(d) Ni^{2+}, $[Ar]3d^8$. The complex is paramagnetic. The geometry is octahedral and there are two unpaired electrons.

23.50 (a) Ag^+, $[Kr]3d^{10}$. The complex is diamagnetic; all electrons are paired.

(b) Cu^{2+}, $[Ar]3d^9$. The complex is paramagnetic. The geometry is square planar, but there would be one unpaired electron in any of the d-orbital energy level diagrams.

(c) Ru^{2+}, $[Kr]3d^6$. The complex is octahedral and bipy is a strong field ligand. The 6 d-electrons are paired in the 3 lower energy d orbitals. The complex is diamagnetic.

(d) Co^{2+}, $[Ar]3d^7$. The complex is paramagnetic. The complex is either tetrahedral or square planar, but in either diagram there will be at least one unpaired electron.

23.51 An electron in a d orbital with lobes that point directly at the ligands will have higher energy than an electron in a d orbital with lobes that do not point directly at the ligands.

23.52 (a) The six ligands in an octahedral arrangement are oriented along the x, y, and z axes of the metal. The d_{xy}, d_{xz}, and d_{yz} metal orbitals point between the x, y, and z axes, and also between the ligands in this arrangement.

(b) The four ligands in a tetrahedral arrangement are oriented between the x, y, and z axes of the metal. The d orbitals along the axes, $d_{x^2-y^2}$ and d_{z^2}, point between the ligands.

23.53 (a)

$$\begin{array}{ll} \text{——} \quad \text{——} & d_{x^2-y^2}, d_{z^2} \\ \\ \\ \text{——} \quad \text{——} \quad \text{——} & d_{xy}, d_{xz}, d_{yz} \end{array}$$

with Δ marked between the upper and lower levels.

(b) The magnitude of Δ and the energy of the d-d transition for a d^1 complex are equal.

(c)
$$\frac{6.626 \times 10^{-34} \text{ J-s}}{545 \text{ nm}} \times \frac{2.998 \times 10^8 \text{ m}}{\text{s}} \times \frac{1 \text{ nm}}{1 \times 10^{-9} \text{ m}} \times \frac{1 \text{ kJ}}{1000 \text{ J}} \times \frac{6.022 \times 10^{23} \text{ photons}}{\text{mol}}$$

$$= 220 \text{ kJ/mol}$$

23.54 (a)
$$\Delta E = hc/\lambda = \frac{6.626 \times 10^{-34} \text{ J-s} \times 2.998 \times 10^8 \text{ m/s}}{500 \times 10^{-9} \text{ m photon}} = 3.973 \times 10^{-19}$$

$$= 3.97 \times 10^{-19} \text{ J/photon}$$

$$\Delta = 3.973 \times 10^{-19} \text{ J/photon} \times \frac{6.022 \times 10^{23} \text{ photons}}{1 \text{ mol}} \times \frac{1 \text{ kJ}}{1000 \text{ J}} = 239.25$$

$$= 239 \text{ kJ/mol}$$

(b) NH_3 is higher than H_2O in the spectrochemical series, an ordering of ligands according to their ability to increase the energy gap, Δ. If H_2O is replaced by NH_3 in the complex, the magnitude of Δ would increase because NH_3 is higher in the spectrochemical series and creates a stronger ligand field.

23.55   *Analyze/Plan.* Determine the oxidation state of the copper ions in each mineral from their molecular formulas. Write the appropriate electron configuration(s). Consider the relationship between d-orbital electron configuration, the color of a complex, the wavelength of absorbed light, and the magnitude of the crystal field splitting Δ.   *Solve.*

(a)   Both minerals contain Cu^{2+} ions. The electron configuration of Cu^{2+} is $[Ar]3d^9$.

(b)   Azurite will probably have the larger Δ. Malachite appears green and absorbs red. Azurite appears blue and absorbs orange. Orange light has a wavelength range of 580 to 650 nm, whereas red light has wavelengths between 650 and 750 nm. The shorter wavelengths of the orange light absorbed by azurite correspond to higher-energy electron transitions and larger Δ values.

23.56   (a)   Red. The $[Ni(bipy)_2\,]^{2+}$ ion absorbs 520 nm green light and appears as the complementary color, red. This fits the absorbed wavelength vs observed color trend shown in Figure 23.30.

(b)   The shorter the wavelength of light absorbed, the greater the value of Δ, and the stronger the ligand field. The order of increasing ligand field strength is the order of decreasing wavelength absorbed.

$H_2O < NH_3 < en < bipy$

23.57   *Analyze/Plan.* Determine the charge on the metal ion, subtract it from the row number (3-12) of the transition metal, and the remainder is the number of d-electrons.   *Solve.*

(a) Ti^{3+}, d^1 (b) Co^{3+}, d^6 (c) Ru^{3+}, d^5

(d) Mo^{5+}, d^1 (e) Re^{3+}, d^4

23.58   (a) Fe^{3+}, d^5 (b) Mn^{2+}, d^5 (c) Ag^+, d^{10}

(d) Cr^{3+}, d^3 (e) Sr^{2+}, d^0

23.59   Yes. A weak-field ligand leads to a small Δ value and a small d-orbital splitting energy. If the splitting energy of a complex is smaller than the energy required to pair electrons in an orbital, the complex is high-spin.

23.60   Octahedral. In an octahedral crystal field, there are more ligand point charges to interact with d orbitals on the metal. And, the point charges point directly at the metal d orbitals so the interaction is greater.

23.61   *Analyze/Plan.* Follow the logic in Sample Exercise 23.7.   *Solve.*

(a)   Mn: $[Ar]4s^2 3d^5$ (b)   Ru: $[Kr]5s^1 4d^7$ (c)   Rh: $[Kr]5s^1 4d^8$
        Mn^{2+}: $[Ar]3d^5$ Ru^{2+}: $[Kr]4d^6$ Rh^{2+}: $[Kr]4d^7$

1 unpaired electron   0 unpaired electrons   1 unpaired electron

23.62 (a) Fe: $[Ar]4s^2 3d^6$ (b) Mo: $[Kr]5s^1 4d^5$ (c) Co: $[Ar]4s^2 3d^7$

Fe^{3+}: $[Ar]3d^5$ Mo^{3+}: $[Kr]4d^3$ Co^{3+}: $[Ar]3d^6$

5 unpaired electrons   3 unpaired electrons   4 unpaired electrons

23.63 *Analyze/Plan.* All complexes in this exercise are six-coordinate octahedral. Use the definitions of high-spin and low-spin along with the orbital diagram from Sample Exercise 23.7 to place electrons for the various complexes. *Solve.*

(a) d^4, high spin   (b) d^5, high spin   (c) d^6, low spin

(d) d^5, low spin   (e) d^3 (f) d^8

23.64 (a) d^2 (b) d^5, high spin   (c) d^5, low spin

(d) d^8 (e) d^8 (f) d^2

23.65 *Analyze/Plan.* Follow the ideas but reverse the logic in Sample Exercise 23.7. *Solve.*

high spin

23.66

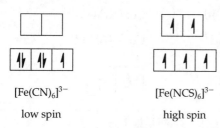

[Fe(CN)$_6$]$^{3-}$
low spin

[Fe(NCS)$_6$]$^{3-}$
high spin

Both complexes contain Fe^{3+}, a d^5 ion. CN$^-$, a strong field ligand, produces such a large Δ that the splitting energy is greater than the pairing energy, and the complex is low spin. NCS$^-$ produces a smaller Δ, so it is energetically favorable for d-electrons to be unpaired in the higher energy d-orbitals. NCS$^-$ is a much weaker-field ligand than CN$^-$. It is probably weaker than NH$_3$ and near H$_2$O in the spectrochemical series.

Additional Exercises

23.67   The paper clip must contain a significant amount of Ni, a ferromagnetic metal. At ambient temperature, the paper clip is below its Curie temperature, behaves ferromagnetically, and is strongly attracted to the permanent magnet. The lighter heats the left paperclip above its Curie temperature (354 $^\circ$C), and it switches from from ferromagnetic to paramagnetic behavior. That is, below its Curie temperature, the spins of the unpaired electrons in Ni are perfectly aligned and the clip is strongly attracted to the permanent magnet. Above the Curie temperature, the unpaired spins become randomly aligned, and the paper clip loses most of its attraction for the permanent magnet.

23.68   We expect radii in a group to increase moving down the periodic table as principle quantum number increases. However, the nuclear build-up associated with filling of the 4f subshell at the beginning of period 6 counteracts this trend. The increased nuclear charge for transition metals of period 6 means that the valence electrons experience a Z$_{eff}$ large enough to offset the increase in principle quantum number. The increased Z$_{eff}$ causes the radii of the metals in group 6 to be smaller than expected, and period 5 and 6 metals in the same group to have very similar radii. This phenomenon is called the lanthanide contraction.

23.69   [Pt(NH$_3$)$_6$]Cl$_4$; [Pt(NH$_3$)$_4$Cl$_2$]Cl$_2$; [Pt(NH$_3$)$_3$Cl$_3$]Cl; [Pt(NH$_3$)$_2$Cl$_4$]; K[Pt(NH$_3$)Cl$_5$]

23.70   (a)

$$\left[\begin{array}{c} \text{Cl} \\ \text{H}_2\text{O} \quad | \quad \text{OH}_2 \\ \text{Ru} \\ \text{H}_2\text{O} \quad | \quad \text{OH}_2 \\ \text{H}_2\text{O} \end{array}\right]^{2+} + 2\,\text{Cl}^-$$

[Ru(H$_2$O)$_5$Cl]Cl$_2$

(b)

$$\left[\begin{array}{c} \text{H}_2\text{O} \\ \text{H}_2\text{O} \quad | \quad \text{OH}_2 \\ \text{Ru} \\ \text{H}_2\text{O} \quad | \quad \text{OH}_2 \\ \text{H}_2\text{O} \end{array}\right]^{3+} + 3\,\text{Cl}^-$$

$\longrightarrow$ [Ru(H$_2$O)$_6$]Cl$_3$

23.71

(a) 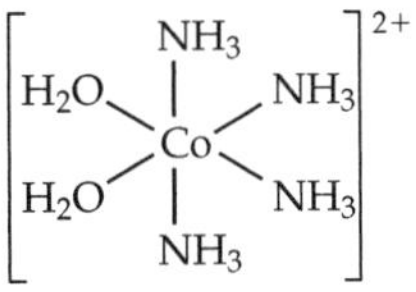
octahedral

(b) 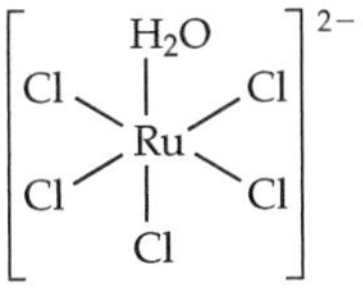
octahedral

(c) 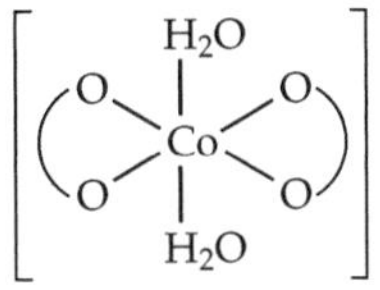
octahedral

(d) 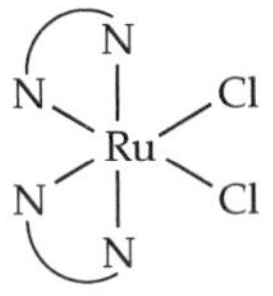
octahedral

   (a)   *cis*-tetraamminediaquacobalt(II) nitrate

   (b)   sodium aquapentachlororuthenate(III)

   (c)   ammonium *trans*-diaquabisoxalatocobaltate(III)

   (d)   *cis*-dichlorobisethylenediammineruthenium(II)

23.72   Only the complex in 23.69(d) has optical isomers. The chelating ethylenediamine ligands in (d) prevent its mirror images (enantiomers) from being superimposable.

23.73   (a)   Valence electrons: $2P + 6C + 16H = 10 + 24 + 16 = 50 \text{ e}^-$, 25 e^- pr

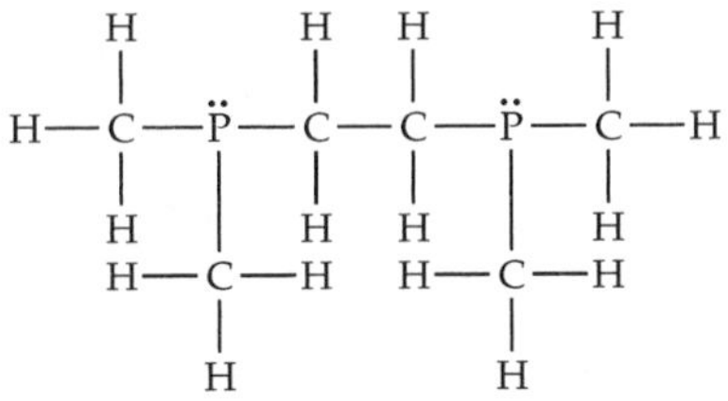

           Both dmpe and en are bidentate ligands. The dmpe ligand binds through P, whereas en binds through N. Phosphorus is less electronegative than N, so dmpe is a stronger electron pair donor and Lewis base than en. Dmpe creates a stronger ligand field and is higher on the spectrochemical series.

           Structurally, P has a larger covalent radius than N, so M–P bonds are longer than M–N bonds. This is convenient because the two –CH_3 groups on each P atom in dmpe create more steric hindrance (bumping with adjacent atoms) than the H atoms on N in en.

    (b)   CO and dmpe are neutral, $2\ CN^- = 2-$, $2\ Na^+ = 2+$. The ion charges balance, so the oxidation state of Mo is zero.

(c)   The symbol P⌒P represents the bidentate dmpe ligand.

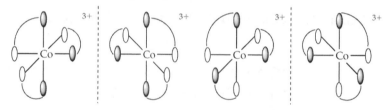

optical isomers

23.74   The *trans* isomer is not observed. In a square planar complex such as [Pt(en)Cl$_2$], if one pair of ligands is trans, the remaining two coordination sites are also trans to each other. Ethylenediamine is a relatively short bidentate ligand that cannot occupy trans coordination sites, so the trans isomer is unknown.

23.75   We will represent the end of the bidentate ligand containing the CF$_3$ group by a shaded oval, the other end by an open oval:

23.76   (a)   Iron. Hemoglobin is the iron-containing protein that transports O$_2$ in human blood.

(b)   Magnesium. Chlorophylls are magnesium-containing porphyrins in plants. They are the key components in the conversion of solar energy into chemical energy that can be used by living organisms.

(c)   Iron. Siderophores are iron-binding compounds or ligands produced by a microorganism. They compete on a molecular level for iron in the medium outside the organism and carry needed iron into the cells of the organism.

(d)   Copper. Hemocyanine is a copper-containing protein responsible for oxygen transport in the blue blood of certain marine animals.

23.77   (a)   Zero. The CO ligand is a neutral molecule and the charge on the complex is zero, so nickel must be present as Ni(0).

         (b)   The electron configuration of a Ni atom in the absence of a ligand field is $[Ar]4s^2 3d^8$. A tetrahedral complex with 8 d-electrons would have two unpaired electrons and be paramagnetic. Because the compound is diamagnetic, the Ni atom must have 10 d-electrons. The electron configuration of Ni in the complex is $[Ar]3d^{10}$.

         (c)   tetracarbonylnickel(0)

23.78   (a)   pentacarbonyliron(0)

         (b)   Because CO is a neutral molecule, the oxidation state of iron must be zero.

         (c) $[Fe(CO)_4CN]^-$ has two geometric isomers. In a trigonal bipyramid, the axial and equatorial positions are not equivalent and not superimposable. One isomer has CN in an axial position and the other has it in an equatorial position.

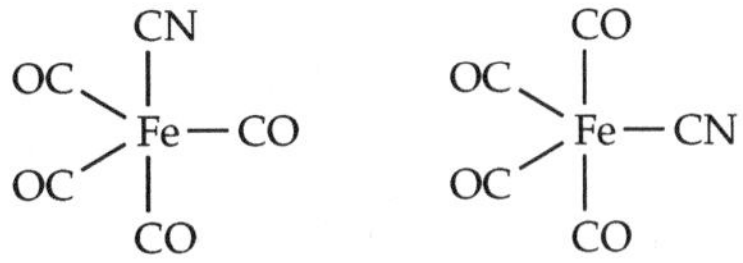

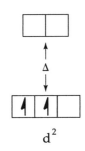

23.79   (a)   left shoe

         (c)   wood screw

         (e)   a typical golf club

23.80   (a)

         (b)   These complexes are colored because the crystal-field splitting energy, Δ, is in the visible portion of the electromagnetic spectrum. Visible light with $\lambda = hc/\Delta$ is absorbed, promoting one of the d-electrons into a higher energy d-orbital. The remaining wavelengths of visible light are reflected or transmitted; the combination of these wavelengths is the color we see.

         (c) $[V(H_2O)_6]^{3+}$ will absorb light with higher energy. H_2O is in the middle of the spectrochemical series, and causes a larger Δ than F^-, a weak-field ligand. Because Δ and λ are inversely related, larger Δ corresponds to higher energy and shorter λ.

23.81 (a) Formally, the two Ru centers have different oxidation states; one is +2 and the other is +3.

(b)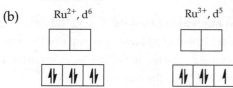

(c) There is extensive bonding-electron delocalization in the isolated pyrazine molecule. When pyrazine acts as a bridging ligand, its delocalized molecular orbitals provide a pathway for delocalization of the "odd" d-electron in the Creutz-Taube ion. The two metal ions appear equivalent because the odd d-electron is delocalized across the pyrazine bridge.

23.82 According to the spectrochemical series, the order of increasing Δ for the ligands is $Cl^- < H_2O < NH_3$. (The tetrahedral Cl^- complex will have an even smaller Δ than an octahedral one.) The smaller the value of Δ, the longer the wavelength of visible light absorbed. The color of light absorbed is the complement of the observed color. A blue complex absorbs orange light (580 to 650 nm), a pink complex absorbs green light (490 to 560 nm), and a yellow complex absorbs violet light (400 to 430 nm). Because $[CoCl_4]^{2-}$ absorbs the longest wavelength, it appears blue. $[Co(H_2O)_6]^{2+}$ absorbs green and appears pink, and $[Co(NH_3)_6]^{3+}$ absorbs violet and appears yellow.

23.83 (a) oxyhemoglobin   deoxyhemoglobin

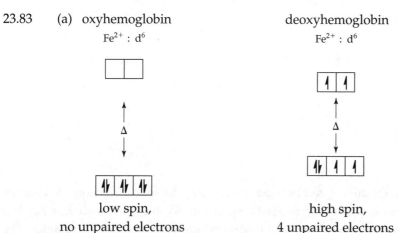

(b) In deoxyhemoglobin, H_2O is bound to Fe in place of O_2.

(c) The two forms of hemoglobin have different colors because they absorb different wavelengths of visible light. They differ by just the H_2O or O_2 ligand, which means that the two ligands have slightly different ligand fields. Oxyhemoglobin appears red and absorbs green light, whereas deoxyhemoglobin appears bluish and absorbs longer wavelength yellow-green light. O_2 has a stronger ligand field than H_2O.

(d) According to Table 18.1, air is 20.948 mole percent O_2. This translates to 209,480 ppm O_2. This is approximately 500 times the 400 ppm concentration of CO in the experiment. If air with a CO concentration 1/500th that of O_2 converts 1/10 of the oxyhemoglobin to carboxyhemoglobin, the equilibrium constant for binding CO is much larger than that for binding O_2.

(e) If CO is a stronger field ligand than O_2, carboxyhemoglobin will absorb shorter wavelengths than hemoglobin. It will absorb blue-green light and appear orange-red.

23.84 (a) The term *isoelectronic* means that the three ions have the same number of electrons.

(b) In each ion, the metal is in its maximum oxidation state and has a d^0 electron configuration. That is, the metal ions have no d-electrons, so there should be no d-d transitions.

(c) A *ligand-metal charge transfer* transition occurs when an electron in a filled ligand orbital is excited to an empty d-orbital of the metal.

(d) Absorption of 565 nm yellow light by MnO_4^- causes the compound to appear violet, the complementary color. CrO_4^{2-} appears yellow, so it is absorbing violet light of approximately 420 nm. The wavelength of the LMCT transition for chromate, 420 nm, is shorter than the wavelength of LCMT transition in permanganate, 565 nm. This means that there is a larger energy difference between filled ligand and empty metal orbitals in chromate than in permanganate.

(e) UV. A white compound indicates that no visible light is absorbed. Going left on the periodic chart from Mn to Cr, the absorbed wavelength got shorter and the energy difference between ligand and metal orbitals increased. The 420 nm absorption by CrO_4^- is at the short wavelength edge of the visible spectrum. It is not surprising that the ion containing V, further left on the chart, absorbs at a still shorter wavelength in the ultraviolet region and that VO_4^{3-} appears white.

23.85 The higher the oxidation state of the metal, the smaller the energy separation between the ligand orbitals and the empty d-orbitals on the metal. The oxidation states of the metals in the tetrahedral oxoanions are: Mn, +7; Cr, +6; V, +5. From Solution 23.84, the energy separation between the ligand orbitals and the empty d-orbitals on the metals increases in the order Mn < Cr < V.

23.86 Application of pressure would result in shorter metal ion–oxide distances. This would have the effect of increasing the ligand-electron repulsions, and would result in a larger splitting in the d-orbital energies. Thus, application of pressure should result in a shift in the absorption to a higher energy and shorter wavelength.

23.87   (a)

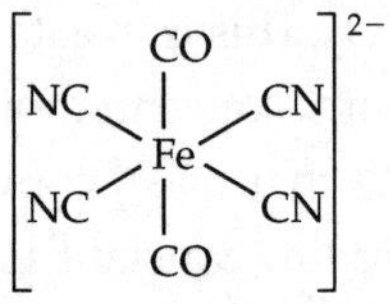

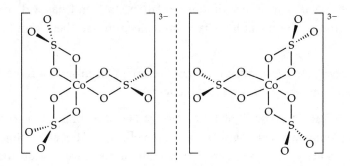

      (b)   sodium dicarbonyltetracyanoferrate(II)

      (c)   +2, 6 d-electrons

      (d)   We expect the complex to be low spin. Cyanide (and carbonyl) are high on the spectrochemical series, which means the complex will have a large Δ splitting characteristic of low spin complexes.

23.88

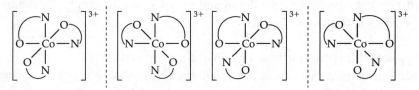

23.89   A large part of the metal-ligand interaction is electrostatic attraction between the positively charged metal and the fully or partially negatively charged ligand. For the same ligand, the greater the charge on the metal or the shorter the M–L separation, the stronger the interaction and the more stable the complex. The greater positive charge and smaller ionic radius of a metal in the 3+ oxidation state means that, for the same ligand, complexes with metals in the 3+ state are more stable than those with metals in the 2+ state.

23.90   (a)   Only one   (b)   Two

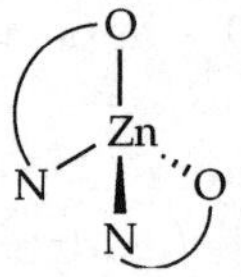

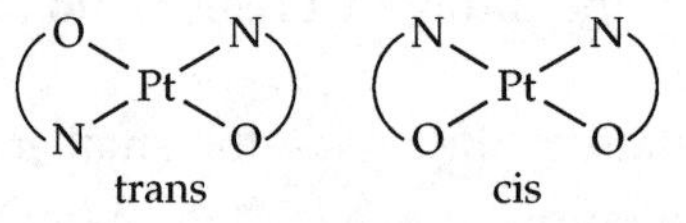

      (c)   Four; two are geometric, the other two are stereoisomers of each of these.

23.91   (a)   Zero. The CO ligand is a neutral molecule and the charge on the complex is zero, so chromium must be present as Cr(0).

      (b)   The electron configuration of a Cr atom in the absence of a ligand field is [Ar]$4s^2 3d^4$. An octahedral complex with 4 d-electrons would have unpaired electrons and be paramagnetic. In order for the complex to be diamagnetic, all valence electrons must be paired. The electron configuration of Cr in the complex is $[Ar]3d^6$. In a strong field complex, the 6 d-electrons will be paired in the three lower energy d orbitals.

(c) A colorless complex has no d-d electron transitions. This indicates a large Δ for the complex, which means CO is a strong-field ligand.

(d) hexacarbonylchromium(0)

Integrative Exercises

23.92 (a)

(b) The pK_a of pure water is 14, that of carbonic anhydrase is 7.5. The active site of carbonic anhydrase is much more acidic than the bulk water. In carbonic anhydrase, the Zn^{2+} ion withdraws electron density from the O atom of water. The electronegative oxygen atom compensates by withdrawing electron-density from the O—H bond. The O—H bond is polarized and H becomes more ionizable, more acidic than in the bulk solvent. This is similar to the effect of an electronegative central atom in an oxyacid such as H_2SO_4.

(c) When the water molecule is deprotonated, the ligand coordinated to water becomes hydroxide ion, OH^-. The three N atoms are unaffected.

(d) In $[Zn(H_2O)_6]^{2+}$, the Zn^{2+} ion has six bound O atoms from which to withdraw electron density. Each O atom donates less electron density than the single O atom in carbonic anhydrase and each O atom withdraws less electron density from its O—H bonds. The O—H bonds in $[Zn(H_2O)_6]^{2+}$ are less polarized and less acidic than those in carbonic anhydrase. $[Zn(H_2O)_6]^{2+}$ is a weaker acid and has a higher pK_a than carbonic anhydrase.

(e) No, we do not expect carbonic anhydrase to have a deep color like hemoglobin. The Zn^{2+} ion is a d^{10} metal center. Its d-orbitals are completely occupied and there is no possibility for the d-d transitions that lead to colored complexes.

23.93 (a) Both compounds have the same general formulation, so Co is in the same (+3) oxidation state in both complexes.

(b) Cobalt(III) complexes are generally inert; that is, they do not rapidly exchange ligands inside the coordination sphere. Therefore, the ions that form precipitates in these two cases are probably outside the coordination sphere. The dark violet compound A forms a precipitate with $BaCl_2(aq)$ but not $AgNO_3(aq)$, so it has SO_4^{2-} outside the coordination sphere and coordinated Br^-, $[Co(NH_3)_5Br]SO_4$.

The red-violet compound B forms a precipitate with $AgNO_3(aq)$ but not $BaCl_2(aq)$ so it has Br^- outside the coordination sphere and coordinated SO_4^{2-}, $[Co(NH_3)_5SO_4]Br$.

Compound A, dark violet   Compound B, red-violet

(c)   Compounds A and B have the same formula but different properties (color, chemical reactivity), so they are isomers. They vary by which ion is inside the coordination sphere, so they are *coordination sphere isomers*.

(d)   Compound A is an ionic sulfate and compound B is an ionic bromide, so both are strong electrolytes. According to the solubility guidelines in Table 4.1, both should be water-soluble.

23.94   Determine the empirical formula of the complex, assuming a 100 g sample.

$$10.0 \text{ g Mn} \times \frac{1 \text{ mol Mn}}{54.94 \text{ g Mn}} = 0.1820 \text{ mol Mn}; 0.182/0.182 = 1$$

$$28.6 \text{ g K} \times \frac{1 \text{ mol K}}{39.10 \text{ g K}} = 0.7315 \text{ mol K}; 0.732 / 0.182 = 4$$

$$8.8 \text{ g C} \times \frac{1 \text{ mol C}}{12.0 \text{ g C}} = 0.7327 \text{ mol C}; 0.733/0.182 = 4$$

$$29.2 \text{ g Br} \times \frac{1 \text{ mol Br}}{79.904 \text{ g Br}} = 0.3654 \text{ mol Br}; 0.365/0.182 = 2$$

$$23.4 \text{ g O} \times \frac{1 \text{ mol O}}{16.00 \text{ g O}} = 1.463 \text{ mol O}; 1.46/0.182 = 8$$

There are 2 C and 4 O per oxalate ion, for a total of two oxalate ligands in the complex. To match the conductivity of $K_4[Fe(CN)_6]$, the oxalate and bromide ions must be in the coordination sphere of the complex anion. Thus, the compound is $K_4[Mn(ox)_2Br_2]$.

23.95   (a) $\Delta G° = -nFE°$. The positive E° values for both sets of complexes correspond to $-\Delta G°$ values. Negative values of $\Delta G°$ mean that both processes are spontaneous. For both *o*-phen and CN^- ligands, the Fe(II) complex is more thermodynamically favorable than the Fe(III) complex.

(b)   The CN^- complex, with the smaller positive E° value, is more difficult to reduce.

(c)   That both the Fe(II) complexes are low spin means that both CN^- and *o*-phen are strong-field ligands. The negatively charged CN^- has a stronger electrostatic interaction with Fe^{3+} than the neutral *o*-phen has. This stabilizes the Fe(III) complex of CN^- relative to the Fe(III) complex of *o*-phen, which reduces the driving force for reduction of $[Fe(CN)_6]^{3-}$ relative to reduction of $[Fe(o\text{-phen})_3]^{3+}$. The E° value and magnitude of $\Delta G°$ for the reduction of the CN^- complex are thus smaller than those values for the *o*-phen complex.

23.96 First determine the empirical formula, assuming that the remaining mass of complex is Pd.

$$37.6 \text{ g Br} \times \frac{1 \text{ mol Br}}{79.904 \text{ g Br}} = 0.4706 \text{ mol Br}; 0.4706/0.2361 = 2$$

$$28.3 \text{ g C} \times \frac{1 \text{ mol C}}{12.01 \text{ g C}} = 2.356 \text{ mol C}; 2.356/0.2361 = 10$$

$$6.60 \text{ g N} \times \frac{1 \text{ mol N}}{14.01 \text{ g N}} = 0.4711 \text{ mol N}; 0.4711/0.2361 = 2$$

$$2.37 \text{ g H} \times \frac{1 \text{ mol H}}{1.008 \text{ g H}} = 2.351 \text{ mol H}; 2.351/0.2361 = 10$$

$$25.13 \text{ g Pd} \times \frac{1 \text{ mol Pd}}{106.42 \text{ g Pd}} = 0.2361 \text{ mol Pd}; 0.2361/0.2361 = 1$$

The chemical formula is $[Pd(NC_5H_5)_2Br_2]$. This should be a neutral square-planar complex of Pd(II), a nonelectrolyte. Because the dipole moment is zero, we can infer that it must be the trans isomer.

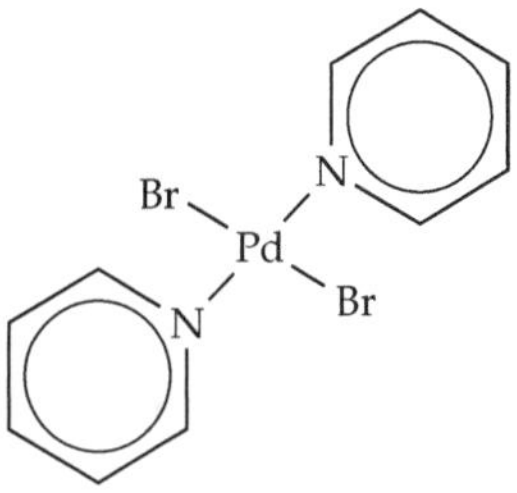

23.97 (a) The reaction that occurs increases the conductivity of the solution by producing a greater number of charged particles, particles with higher charges, or both. It is likely that H_2O from the bulk solvent exchanges with a coordinated Br^- according to the reaction below. This reaction would convert the 1:1 electrolyte, $[Co(NH_3)_4Br_2]Br$, to a 1:2 electrolyte, $[Co(NH_3)_3(H_2O)Br]Br_2$.

(b) $[Co(NH_3)_4Br_2]^+(aq) + H_2O(l) \rightarrow [Co(NH_3)_4(H_2O)Br]^{2+}(aq) + Br^-(aq)$

(c) Before the exchange reaction, there is one mole of free Br^- per mole of complex. mol $Br^- = $ mol Ag^+

$M = $ mol/L; L $AgNO_3 = $ mol $AgNO_3/M$ $AgNO_3$

$$\frac{3.87 \text{ g complex}}{0.500 \text{ L soln}} \times \frac{1 \text{ mol complex}}{366.77 \text{ g complex}} \times 0.02500 \text{ L soln used} =$$

$$5.276 \times 10^{-4} = 5.28 \times 10^{-4} \text{ mol complex}$$

$$5.276 \times 10^{-4} \text{ mol complex} \times \frac{1 \text{ mol Br}^-}{1 \text{ mol complex}} \times \frac{1 \text{ mol Ag}^+}{1 \text{ mol Br}^-} \times \frac{1 \text{ L Ag}^+(aq)}{0.0100 \text{ mol Ag}^+(aq)}$$

$$= 0.05276 \text{ L} = 52.8 \text{ mL AgNO}_3(aq)$$

(d)   After the exchange reaction, there are 2 mol free Br^- per mol of complex. Because M $AgNO_3(aq)$ and volume of complex solution are the same for the second experiment, the titration after conductivity changes will require twice the volume calculated in part (c), 105.52 = 106 mL of 0.0100 M $AgNO_3(aq)$.

23.98   Calculate the concentration of Mg^{2+} alone, and then the concentration of Ca^{2+} by difference. $M \times L = mol$

$$\frac{0.0104 \text{ mol EDTA}}{1 \text{ L}} \times 0.0187 \text{ L} \times \frac{1 \text{ mol } Mg^{2+}}{1 \text{ mol EDTA}} \times \frac{24.31 \text{ g } Mg^{2+}}{1 \text{ mol } Mg^{2+}} \times \frac{1000 \text{ mg}}{\text{g}}$$

$$\times \frac{1}{0.100 \text{ L } H_2O} = 47.28 = 47.3 \text{ mg } Mg^{2+}/L$$

$$0.0104 \text{ } M \text{ EDTA} \times 0.0315 \text{ L} = \text{mol } (Ca^{2+} + Mg^{2+})$$
$$\underline{0.0104 \text{ } M \text{ EDTA} \times 0.0187 \text{ L} = \text{mol } Mg^{2+}}$$
$$0.0104 \text{ } M \text{ EDTA} \times 0.0128 \text{ L} = \text{mol } Ca^{2+}$$

$$0.0104 \text{ } M \text{ EDTA} \times 0.0128 \text{ L} \times \frac{1 \text{ mol } Ca^{2+}}{1 \text{mol EDTA}} \times \frac{40.08 \text{ g } Ca^{2+}}{1 \text{ mol } Ca^{2+}} \times \frac{1000 \text{ mg}}{\text{g}} \times \frac{1}{0.100 \text{ L } H_2O}$$

$$= 53.35 = 53.4 \text{ mg } Ca^{2+}/L$$

23.99   Use Hess' law to calculate $\Delta G°$ for the desired equilibrium. Then $\Delta G° = -RT \ln K$ to calculate K.

$$Hb + CO \rightarrow HbCO \qquad\qquad \Delta G° = -80 \text{ kJ}$$
$$\underline{HbO_2 \rightarrow Hb + O_2 \qquad\qquad\quad \Delta G° = \;\;70 \text{ kJ}}$$
$$HbO_2 + Hb + CO \rightarrow HbCO + Hb + O_2$$
$$HbO_2 + CO \rightarrow HbCO + O_2 \qquad \Delta G° = -10 \text{ kJ}$$

$$\Delta G° = -RT \ln K, \; \ln K = \frac{-\Delta G°}{RT} = \frac{-(-10 \text{ kJ})}{8.314 \text{ J/K-mol} \times 298 \text{ K}} \times \frac{1000 \text{ J}}{\text{kJ}} = 4.036 = 4.04$$

$$K = e^{4.04} = 56.61 = 57$$

23.100 $$\frac{182 \times 10^3 \text{ J}}{1 \text{ mol}} \times \frac{1 \text{ mol}}{6.022 \times 10^{23} \text{ molecules}} = 3.022 \times 10^{-19} = 3.02 \times 10^{-19} \text{ J/photon}$$

$$\Delta E = h\nu = 3.02 \times 10^{-19} \text{ J}; \; \nu = \Delta E/h$$

$$\nu = 3.022 \times 10^{-19} \text{ J}/6.626 \times 10^{-34} \text{ J-s} = 4.561 \times 10^{14} = 4.56 \times 10^{14} \text{ s}^{-1}$$

$$\lambda = \frac{2.998 \times 10^8 \text{ m/s}}{4.561 \times 10^{14} \text{ s}^{-1}} = 6.57 \times 10^{-7} \text{ m} = 657 \text{ nm}$$

We expect that this complex will absorb in the visible, at around 660 nm. It will thus exhibit a blue-green color (Figure 23.25).

23.101 The process can be written:

$$H_2(g) + 2\,e^- \rightarrow 2\,H^+(aq) \qquad\qquad E^o_{red} = 0.0\ V$$

$$Cu(s) \rightarrow Cu^{2+} + 2\,e^- \qquad\qquad E^o_{red} = 0.337\ V$$

$$\underline{Cu^{2+}(aq) + 4\,NH_3(aq) \rightarrow [Cu(NH_3)_4]^{2+}(aq) \qquad\qquad "E^o_f" = ?}$$

$$H_2(g) + Cu(s) + 4\,NH_3(aq) \rightarrow 2\,H^+(aq) + [Cu(NH_3)_4]^{2+}(aq) \qquad E = 0.08\ V$$

$$E = E^o - RT\ln K;\ K = \frac{[H^+]^2[Cu(NH_3)_4^{2+}]}{P_{H_2}[NH_3]^4}$$

$$P_{H_2} = 1\ atm,\ [H^+] = 1\ M,\ [NH_3] = 1\ M,\ [Cu(NH_3)_4]^{2+} = 1\ M,\ Q = 1$$

$$E = E^o - RT\ln(1);\ E = E^o - RT(0);\ E = E^o = 0.08\ V$$

Because we know E^o values for two steps and the overall reaction, we can calculate "E^o" for the formation reaction and then K_f, using $E^o = \dfrac{0.0592}{n}\log K_f$ for the step.

$$E_{cell} = 0.08\ V = 0.0\ V - 0.337\ V + "E^o_f" \qquad "E^o_f" = 0.08\ V + 0.337\ V = 0.417\ V = 0.42\ V$$

$$"E^o_f" = \frac{0.0592}{n}\log K_f;\ \log K_f = \frac{n(E^o_f)}{0.0592} = \frac{2(0.417)}{0.0592} = 14.0878 = 14$$

$$K_f = 10^{14.0878} = 1.2 \times 10^{14} = 10^{14}$$

24 The Chemistry of Life: Organic and Biological Chemistry

Visualizing Concepts

24.1 *Analyze/Plan.* Follow the logic in Sample Exercise 24.1 to name each compound. Decide which structures are the same compound. *Solve.*

(a) 2,2,4-trimethylpentane

(b) 3-ethyl-2-methylpentane

(c) 2,3,4-trimethylpentane

(d) 2,3,4-trimethylpentane

Structures (c) and (d) are the same molecule.

24.2 *Analyze/Plan.* Given structural formulas, specify which molecules are unsaturated. Consider the definition of *unsaturated* and apply it to the molecules in the exercise. *Solve.*

Unsaturated molecules contain one or more multiple bonds. Saturated molecules contain only single bonds. Molecules (c) and (d) are unsaturated.

24.3 *Analyze/Plan.* Given structural formulas, decide which molecule will undergo addition. Consider which functional groups are present in the molecules, and which are most susceptible to addition. *Solve.*

(a) Molecule (iii), an alkene, will readily undergo addition. Addition reactions are characteristic of alkenes. [Molecule (i) will not typically undergo addition, because its delocalized electron cloud is too difficult to disrupt. Molecules (b) and (d) contain carbonyl groups (actually carboxylic acid groups) that do not typically undergo addition, except under special conditions.]

(b) Molecule (i) is an aromatic hydrocarbon.

(c) Molecule (i) most readily undergoes a substitution reaction.

24.4 *Analyze/Plan.* Given condensed structural formulas, predict which molecule will have the highest boiling point. Boiling point is determined by strength of intermolecular forces; for neutral molecules with similar molar masses, the strongest intermolecular force is hydrogen bonding.

We are also asked to identify various functional groups, given condensed structural formulas. Refer to Table 24.6 for information on functional groups. *Solve.*

(a) Molecule (ii), an alcohol, forms hydrogen bonds with like molecules; it has the highest boiling point.

(b) Molecule (iv) is most oxidized; it has the most oxygen atoms.

(c) None of the compounds is an ether. Ethers are characterized by an oxygen atom bound to two carbon atoms. Compound (iv) is an ester, which has an ether linkage.

(d) Compound (iv) is an ester.

(e) None of the compounds is a ketone. Ketones are characterized by a carbonyl group bonded to two alkyl groups. Compound (i) is an aldehyde; the carbonyl group is bonded to one alkyl group and one hydrogen atom.

24.5 *Analyze.* Given structural formulas, name the compound, identify the functional groups present, and determine if the molecule is chiral. *Plan.* Apply principles of organic nomenclature in Section 24.2. Identify functional groups using Table 24.6. Assess chirality by looking for one or more carbon atoms that are bonded to four different groups. *Solve.*

(a) Valine. This molecule is an amino acid shown in the zwitterion form. Use Figure 24.16 to name the amino acid. In the neutral amino acid, there is a carboxlyic acid group and an amine group. (In the zwitterion, these become a carboxyl group and an ammonium group.) The amino acid is chiral. The C atom to which the $-NH_3^+$ group is bound is a chiral center, because it is bound to four different groups. All amino acids except glycine are chiral.

(b) 3-chlorobenzoic acid. The functional groups are a carboxylic acid group and an (aryl) halide. The molecule is not chiral; no C atom is bound to four different groups.

(c) 2-pentene. The molecule contains an alkene functional group and is not chiral.

(d) Propane. The molecule is an alkane, there are no functional groups, and it is not chiral.

24.6 *Analyze/Plan.* Given ball-and-stick models, select the molecule that fits the description given. From the models, decide the type of molecule or functional group represented.

Solve. Molecule (i) is a sugar, (ii) is an ester with a long hydrocarbon chain, (iii) is an organic base and a component of nucleic acids, (iv) is an amino acid, and (v) is an alcohol.

(a) Molecule (i) is a disaccharide composed of galactose (left) and glucose (right); it can be hydrolyzed to form a solution containing glucose. Because it is the only sugar molecule depicted, it was not necessary to know the exact structure of glucose to answer the question.

(b) Amino acids form zwitterions, so the choice is molecule (iv).

(c) Molecule (iii) is an organic base present in DNA (again, the only possible choice).

(d) Molecule (v) because alcohols react with carboxylic acids to form esters.

(e) Molecule (ii), because it has a long hydrocarbon chain and an ester functional group.

Introduction to Organic Compounds; Hydrocarbons
(Sections 24.1 and 24.2)

24.7 (a) False. Butane is an alkane; it contains carbon atoms that are sp^3 hybridized.

(b) False. Cyclohexane is a saturated hydrocarbon, whereas benzene is aromatic.

(c) True.

(d) False. Olefin is another name for alk**ene**.

24.8 (a) False. Hexane contains six carbon atoms and pentane contains only five.

(b) True. For molecules with similar structures, the larger the molecule, the stronger their dispersion forces and the higher the boiling point.

(c) True. The carbon atoms involved in an alkyne group are sp hybridized.

(d) False. There is only one way to arrange the three carbon atoms in propane.

24.9 *Analyze/Plan.* Given a condensed structural formula, determine the bond angles and hybridization about each carbon atom in the molecule. Visualize the number of electron domains about each carbon. State the bond angle and hybridization based on electron domain geometry.   *Solve.*

C1 has trigonal planar electron domain geometry, 120° bond angles, and sp^2 hybridization. C3 and C4 have linear electron domain geometry, 180° bond angles, and sp hybridization. C2 and C5 both have tetrahedral electron domain geometry, 109° bond angles, and sp^3 hybridization.

24.10

(a) C2, C5, and C6 have sp^3 hybridization (4 e⁻ domains around C)

(b) C7 has sp hybridization (2 e⁻ domains around C)

(c) C1, C3, and C4 have sp^2 hybridization (3 e⁻ domains around C)

24.11 *Analyze/Plan.* For each molecule, count the number of carbon atoms in the root name and in the substituent group(s).   *Solve.*

(a) One. The root "meth" indicates one carbon atom. Methane is the simplest alkane.

(b) Ten. The root "dec" indicates 10 carbon atoms.

(c) Seven. The root "hex" indicates 6 carbon atoms, plus 1 in the "methyl" substituent.

(d) Five. Although this is a common name (not an IUPAC name), the root "pent" still indicates 5 total carbon atoms.

(e) Two. Again a common name, acetylene is the simplest two-carbon alkyne. The IUPAC name is ethyne.

24.12   True. Note that we are comparing enthalpies for only single bonds in the molecule.

24.13   (a)   True.

   (b)   True.

   (c)   False. Alkenes contain carbon-carbon double bonds.

   (d)   False. Alkynes contain carbon-carbon triple bonds.

   (e)   True.

   (f)   False. Cyclohexane is a saturated hydrocarbon; benzene is aromatic.

   (g)   True.

24.14   All the classifications listed are hydrocarbons; they contain only the elements hydrogen and carbon.

   (a)   *Alkanes* are hydrocarbons that contain only single bonds.

   (b)   *Cycloalkanes* contain at least one ring of three or more carbon atoms joined by single bonds. Because it is a type of alkane, all bonds in a cycloalkane are single bonds.

   (c)   *Alkenes* contain at least one C=C double bond.

   (d)   *Alkynes* contain at least one C≡C triple bond.

   (e)   A *saturated hydrocarbon* contains only single bonds. Alkanes and cycloalkanes fit this definition.

   (f)   An *aromatic hydrocarbon* contains one or more planar, six-membered rings of carbon atoms with delocalized π-bonding throughout the ring.

24.15   *Analyze/Plan.* Follow the rules for naming alkanes given in Section 24.2 and illustrated in Sample Exercise 24.1.   *Solve.*

   (a)   2-methylhexane

   (b)   4-ethyl-2,4-dimethyldecane

   (c)
$$CH_3CH_2CH_2CH_2-CH_2-\overset{\overset{\displaystyle CH_3}{|}}{CH}-CH_3$$

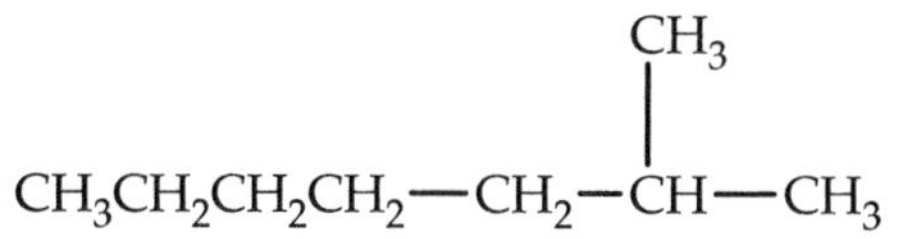

   (d)
$$CH_3CH_2CH_2CH_2-\overset{}{\underset{\underset{\displaystyle CH_2CH_3}{|}}{CH}}-\overset{\overset{\displaystyle CH_3}{|}}{CH}-\overset{}{\underset{\underset{\displaystyle CH_3}{|}}{CH}}-CH_3$$

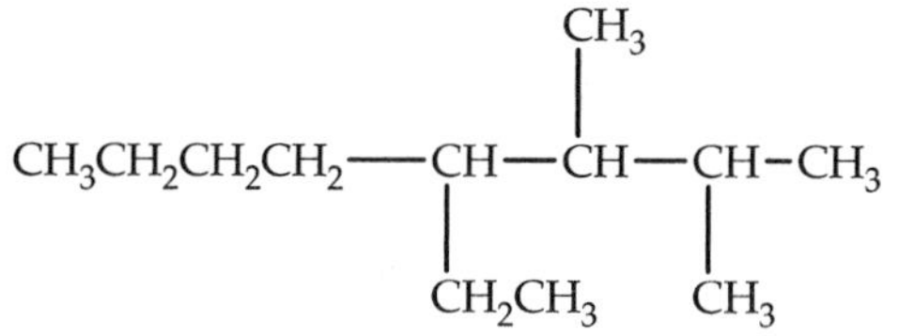

   (e)

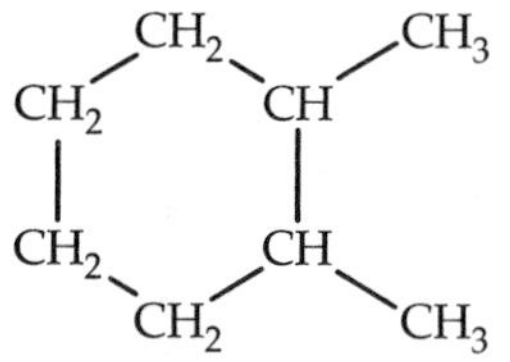

24.16   (a)   3,3,5-trimethylheptane

        (b)   3,4,4-trimethylheptane

        (c)

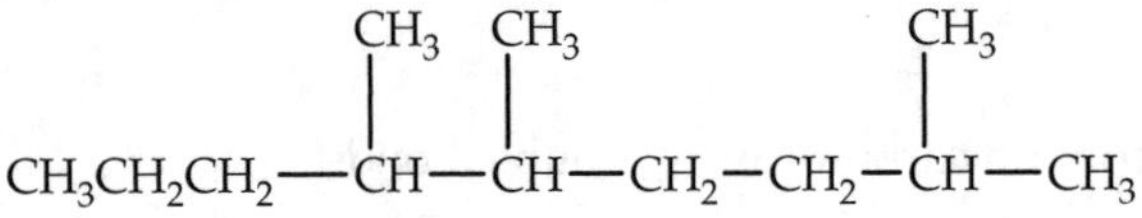

$$\begin{array}{ccccccc} & & CH_3 & CH_3 & & & CH_3 \\ & & | & | & & & | \\ CH_3CH_2CH_2 & - & CH & - CH & -CH_2-CH_2- & CH & -CH_3 \end{array}$$

        (d)

$$\begin{array}{ccccc} & & CH_3 & CH_3 & \\ & & | & | & \\ CH_3CH_2CH_2 & -CH & -CH & -CH & -CH_2CH_2CH_2CH_2CH_2CH_3 \\ & | & & & \\ & CH_3-CH_2 & & & \end{array}$$

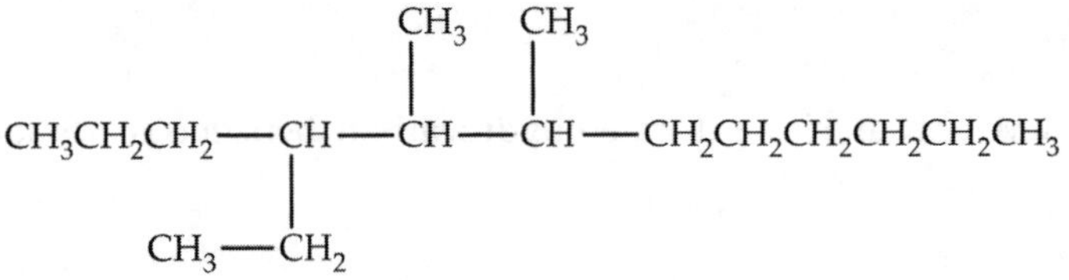

        (e)

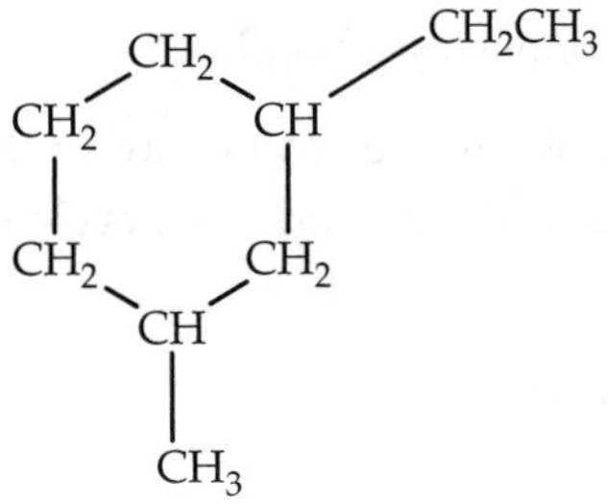

24.17   *Analyze/Plan.* Follow the rules for naming alkanes given in Section 24.2 and illustrated in Sample Exercise 24.1.   *Solve.*

        (a)   2,3-dimethylheptane

        (b)

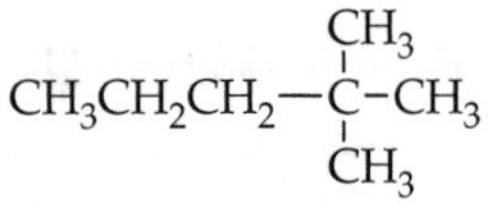

$$\begin{array}{c} CH_3 \\ | \\ CH_3CH_2CH_2-C-CH_3 \\ | \\ CH_3 \end{array}$$

        (c)

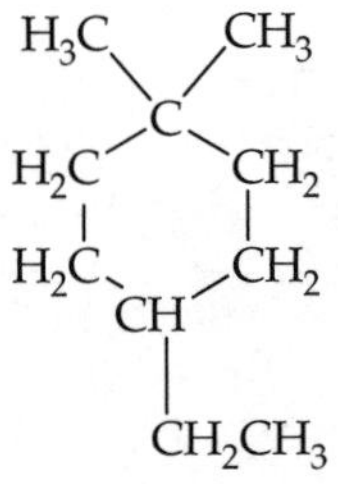

        (d)   2,2,5-trimethylhexane

        (e)   3-ethylheptane

24.18   (a)

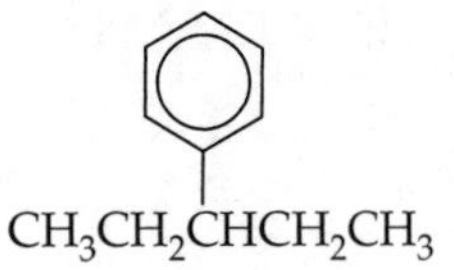

$$CH_3CH_2CHCH_2CH_3$$

(b) 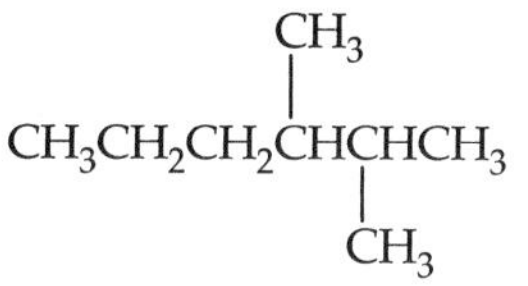 $CH_3CH_2CH_2\overset{\overset{\displaystyle CH_3}{|}}{C}H\overset{\overset{}{}}{C}H\underset{\underset{\displaystyle CH_3}{|}}{}CH_3$

(c) $CH_3CH_2CH_2CH_2CH_2\overset{\overset{\displaystyle CH_3}{|}}{\underset{\underset{\displaystyle CH_2CH_3}{|}}{C}}CH_3$

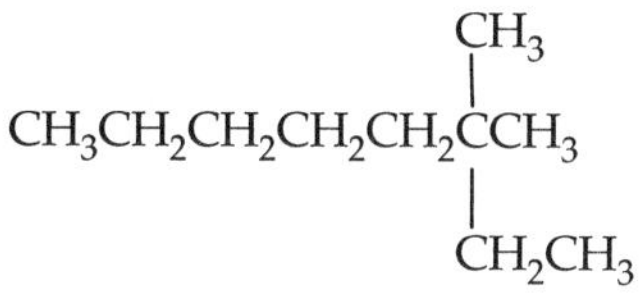

(d)   2,4-dimethylhexane

(e)   methylcyclobutane

24.19   Assuming that each component retains its effective octane number in the mixture (and this isn't always the case), we obtain: octane number = 0.35(0) + 0.65(100) = 65.

24.20   Octane number can be increased by increasing the fraction of branched-chain alkanes or aromatics, because these have high octane numbers. This can be done by cracking. The octane number also can be increased by adding an anti-knock agent such as tetraethyl lead, $Pb(C_2H_5)_4$ (no longer legal); methyl t-butyl ether (MTBE); or an alcohol, methanol, or ethanol.

Alkenes, Alkynes, and Aromatic Hydrocarbons (Section 24.3)

24.21   (a) C_4H_6 is an unsaturated hydrocarbon. The maximum number of hydrogen atoms for 4 C atoms in a saturated alkane is [(2 × 4) + 2] = 10. C_4H_6 does not contain the maximum possible hydrogen atoms and is unsaturated.

(b)   Yes, all alkynes are unsaturated. The presence of a triple bond means that the alkyne carbon atoms are not bound to the maximum possible number of hydrogen atoms.

24.22   (a)   The molecule $CH_3CH=CH_2$ is unsaturated because it contains a double bond. It is possible to add more hydrogen to the molecule.

(b)   The formula $CH_3CH_2CH=CH_3$ has too many H atoms bound to the right-most C atom; the formula as it stands implies 5 bonds to this atom. A correct formula is $CH_3CH_2CH=CH_2$, with 2 H atoms on the right-most C atom.

24.23   *Analyze/Plan.* Consider the definition of the stated classification and apply it to a compound containing five C atoms.   *Solve.*

(a) $CH_3CH_2CH_2CH_2CH_3$, C_5H_{12}

(b) 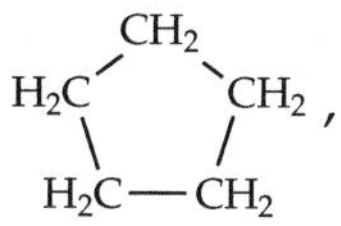 $\begin{array}{c} CH_2 \\ H_2C \qquad CH_2 \\ H_2C - CH_2 \end{array}$, C_5H_{10}

(c) $CH_2=CHCH_2CH_2CH_3$, C_5H_{10}

(d) $HC\equiv CCH_2CH_2CH_3$, C_5H_8

24.24   cyclic alkane, H_2C—CH_2, C_6H_{12}

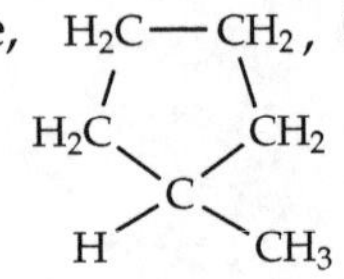

cyclic alkene, HC=CH, C_6H_{10}

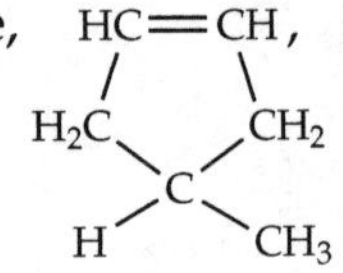

alkyne, CH_3–CH_2–C≡C–CH_2–CH_3, C_6H_{10}

aromatic hydrocarbon, H—C ⬡ C—H, C_6H_6

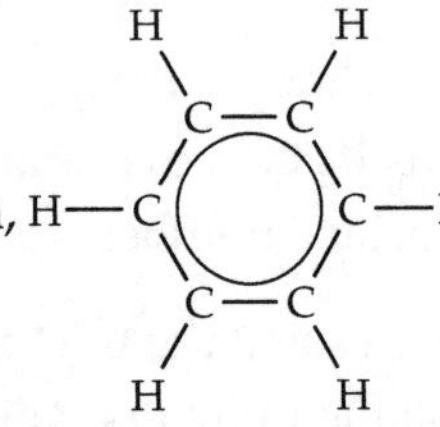

24.25   *Analyze/Plan.* We are given the class of compounds "enediyne." Based on organic nomenclature, determine the structural features of an enediyne. Construct a molecule with 6 C atoms in a row that has these features.   *Solve.*

The term "enediyne" contains the suffixes –ene and –yne. The suffix –ene is used to name alkenes, molecules with one double bond. The suffix –yne is used to name alkynes, molecules with one triple bond. An enediyne then features one double and two triple bonds. Possible arrangements of these bonds involving 6 C atoms in a row are:

$$CH_2=CH–C≡C–C≡CH \qquad CH≡C–CH=CH–C≡CH$$

Check. The formula of a saturated alkane is C_nH_{2n+2}. For each double bond subtract 2 H atoms, for each triple bond subtract 4 H atoms. A saturated 6 C alkane has 14 H atoms. For the enediyne, subtract (2 + 4 + 4 =) 10 H atoms. The molecular formula is C_6H_4. That is the molecular formula of each structure above.

24.26 C_nH_{2n-2}

24.27   *Analyze/Plan.* Follow the logic in Sample Exercise 24.3.

Solve. There are many correct structures that are alkenes or alkynes and have the molecular formula C_6H_{10}. The molecule will have two points of unsaturation. Molecules can include various arrangements of one alkyne group, two alkene groups, or one cyclic mono-alkene. A few of the possibilities are shown below.

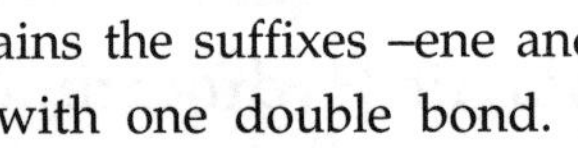

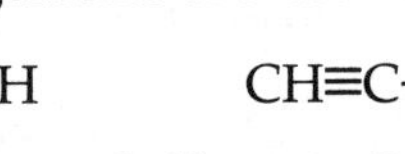

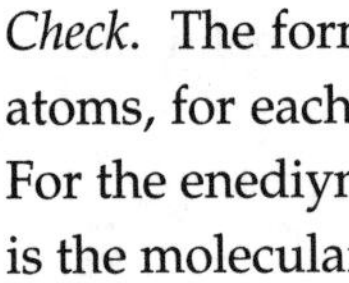

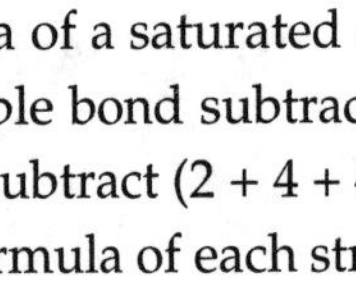

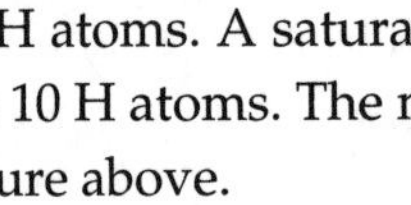

24.28 $CH_3-CH_2-CH_2-CH=CH_2$

pentene

$CH_3-CH_2-CH=CH-CH_3$

2-pentene

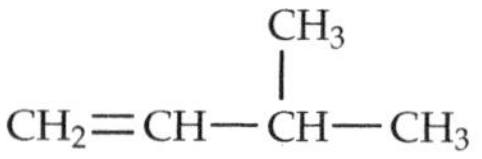

$$CH_2=CH-\underset{\underset{CH_3}{|}}{CH}-CH_3$$

3-methyl-1-butene

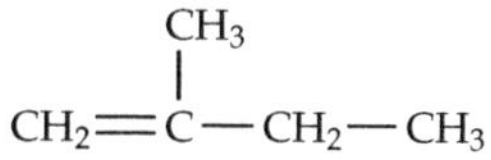

$$CH_2=\underset{\underset{CH_3}{|}}{C}-CH_2-CH_3$$

2-methyl-1-butene

$$CH_3-\underset{\underset{CH_3}{|}}{C}=CH-CH_3$$

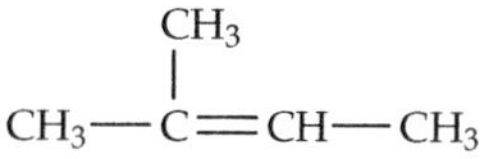

2-methyl-2-butene

24.29   *Analyze/Plan.* Follow the logic in Sample Exercises 24.1 and 24.4.   *Solve.*

(a)

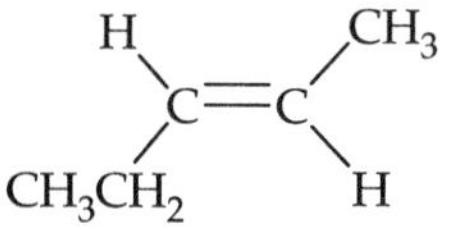

(b)

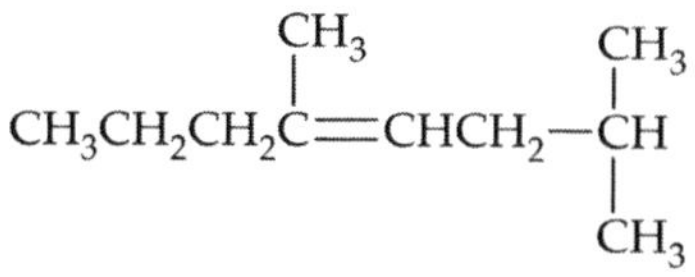

(c)   *cis*-6-methyl-3-octene

(d)   *para*-dibromobenzene (or 1,4-dibromobenzene)

(e)   4,4-dimethyl-1-hexyne

24.30   (a)

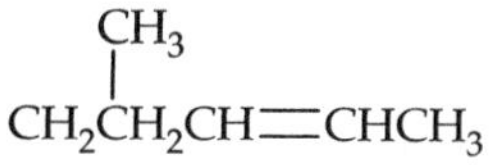

(b)

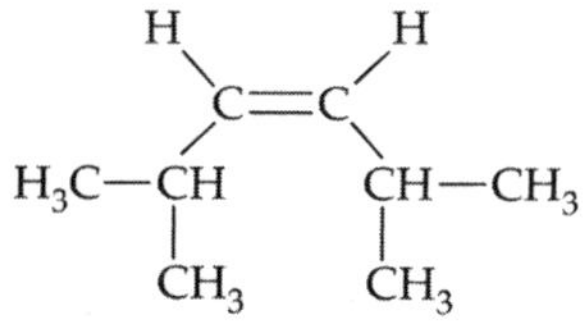

(c)

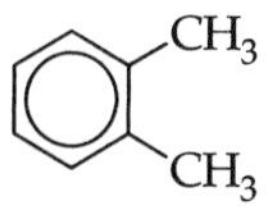

(d)   1-butyne

(e)   *trans*-2-heptene

24.31   (a)   True

(b)   True. (Alkenes *can* have cis and trans isomers, but whether they *do* have them depends on the groups bonded to the sp^2 C atoms of the alkene.)

(c)   False. The geometry of the alkyne functional group is linear.

24.32   Butene is an alkene, C_4H_8. There are two possible placements for the double bond:

$CH_2=CHCH_2CH_3$ or $CH_3CH=CHCH_3$

1-butene   2-butene

These two compounds are *structural isomers*.

For 2-butene, there are two different, noninterchangeable ways to construct the carbon skeleton (owing to the absence of free rotation around the double bond). These two compounds are *geometric isomers*.

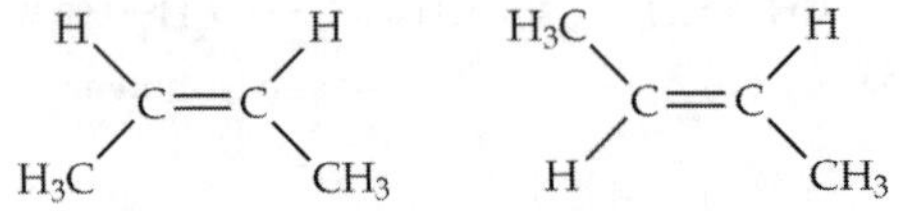

cis-2-butene   trans-2-butene

24.33   *Analyze/Plan.* In order for geometrical isomerism to be possible, the molecule must be an alkene with two different groups bound to each of the alkene C atoms.   *Solve.*

(a) 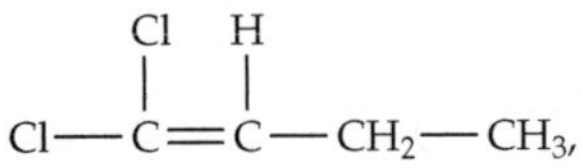 $Cl-C=C-CH_2-CH_3$, no

(b)

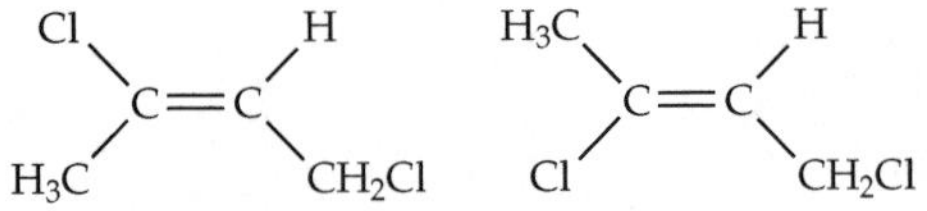

(c)   no, not an alkene

(d)   no, not an alkene

24.34

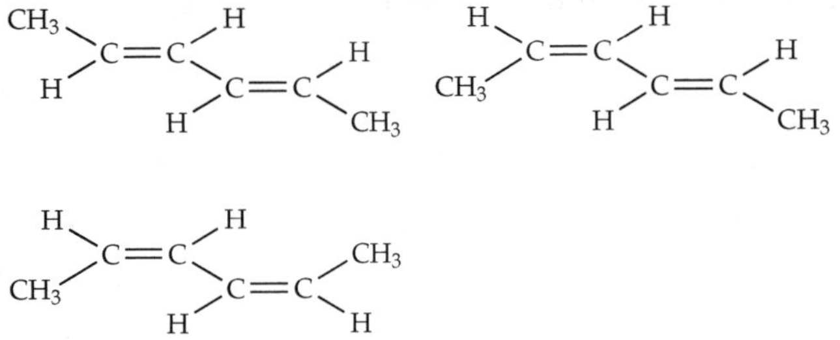

24.35   (a)   True

(b)   *Plan.* Draw the condensed structural formula of 2-pentene. Consider the part of the molecule that is likely to react with Br_2.   *Solve.*

$CH_3CH_2CH=CH-CH_3 + Br_2 \rightarrow CH_3CH_2CH(Br)CH(Br)CH_3$

2-pentene   2,3-dibromopentane

This is an addition reaction. The π bond is broken and a Br atom adds to each of the C atoms involved in the π bond.

(c)   *Plan.* Draw the structure of benzene. The term *para* means the Cl atoms will be opposite each other across the benzene ring in the product.   *Solve.*

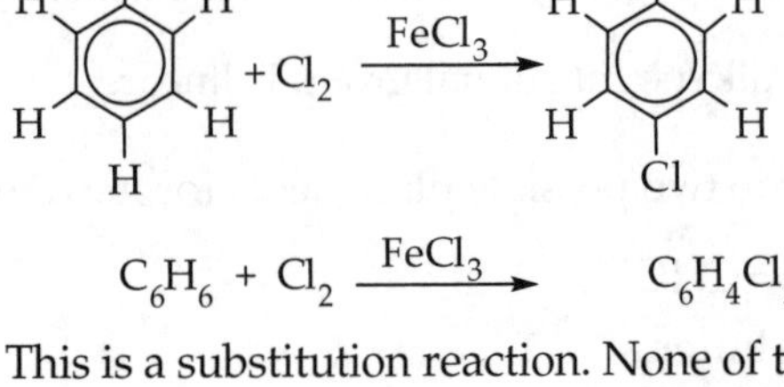

$C_6H_6 + Cl_2 \xrightarrow{FeCl_3} C_6H_4Cl_2$

This is a substitution reaction. None of the double bonds in the benzene ring are broken. Each Cl atom has replaced an H atom on the the ring. The Cl atoms have substituted for two of the H atoms that are opposite each other across the ring.

24.36 (a)

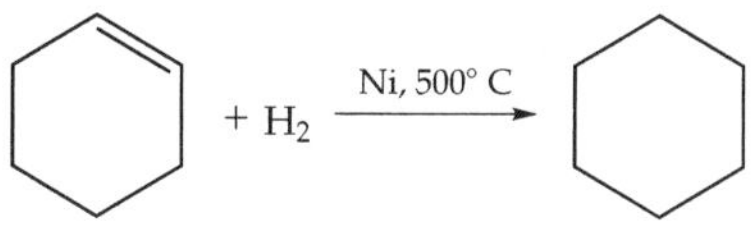

(b)

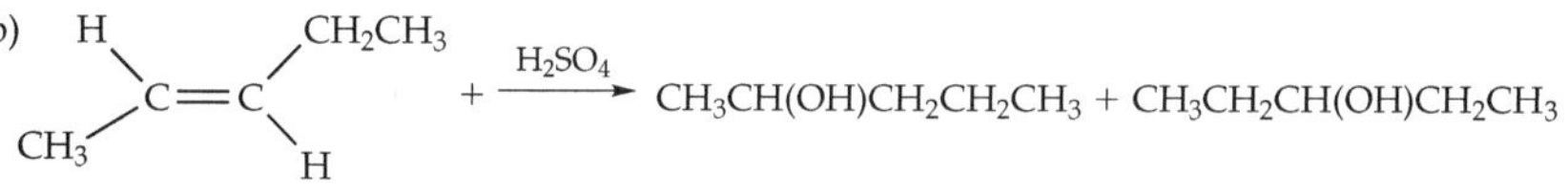

(c)

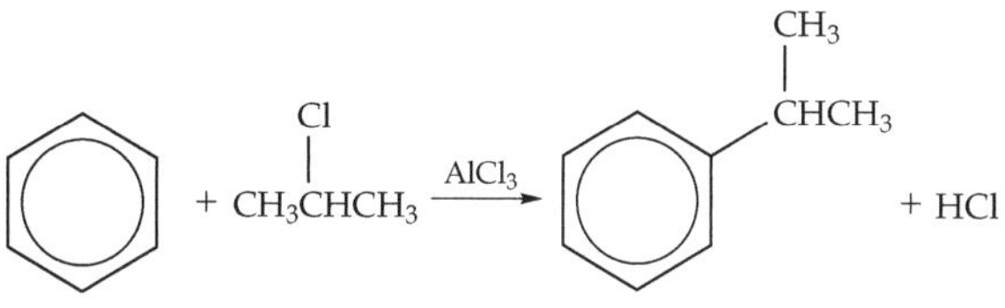

24.37 (a) *Plan.* Consider the structures of cyclopropane, cyclopentane, and cyclohexane. *Solve.*

The small 60° C–C–C angles in the cyclopropane ring cause strain that provides a driving force for reactions that result in ring opening. There is no comparable strain in the five- or six-membered rings.

(b) *Plan.* First form an alkyl halide: $C_2H_4(g) + HBr(g) \rightarrow CH_3CH_2Br(l)$; then carry out a Friedel-Crafts reaction. *Solve.*

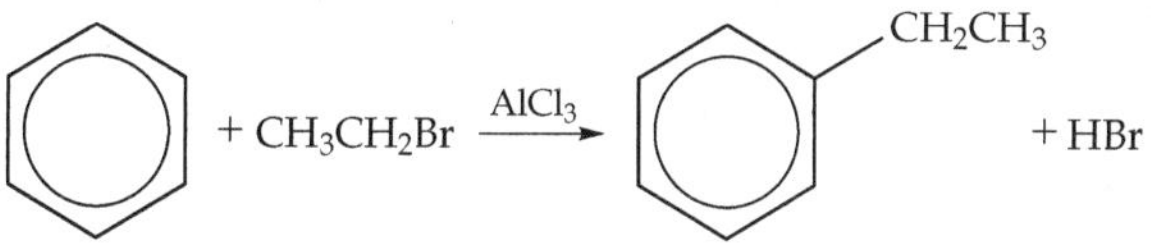

24.38 (a) The reaction of Br_2 with an alkene to form a colorless halogenated alkane is an addition reaction. Aromatic hydrocarbons do not readily undergo addition reactions, because their π-electrons are stabilized by delocalization.

(b) *Plan.* Use a Friedel-Crafts reaction to substitute a $-CH_2CH_3$ onto benzene. Do a second substitution reaction to get *para*-bromoethylbenzene. *Solve.*

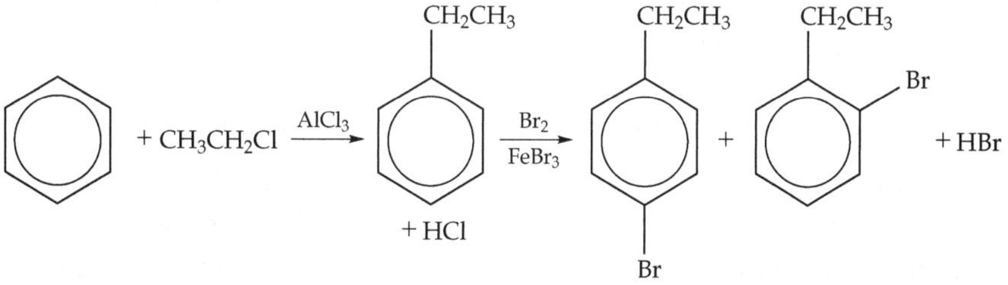

It appears that ortho, meta, and para geometric isomers of bromoethylbenzene would be possible. However, because of electronic effects beyond the scope of this chapter, the ethyl group favors formation of ortho and para isomers, but not the meta. The ortho and para products must be separated by distillation or some other technique.

24.39 Yes, this information suggests (but does not prove) that the reactions proceed in the same manner. That the rate laws are both first order in both reactants and second order overall indicates that the activated complex in the rate-determining step in each mechanism is bimolecular and contains one molecule of each reactant. This is usually an indication that the mechanisms are the same, but it does not rule out the possibility of different fast steps, or a different order of elementary steps.

24.40 The partially positive end of the hydrogen halide, $\overset{\delta^+ \ \ \delta^-}{H-X}$, is attached to the π electron cloud of the alkene cyclohexene. The electrons that formed the π bond in cyclohexene form a sigma bond to the H atom of HX, leaving a halide ion, X^-. The intermediate is a carbocation; one of the C atoms formerly involved in the π bond is now bound to a second H atom. The other C atom formerly involved in the π bond carries a full positive charge and forms only three sigma bonds, two to adjacent C atoms and one to H.

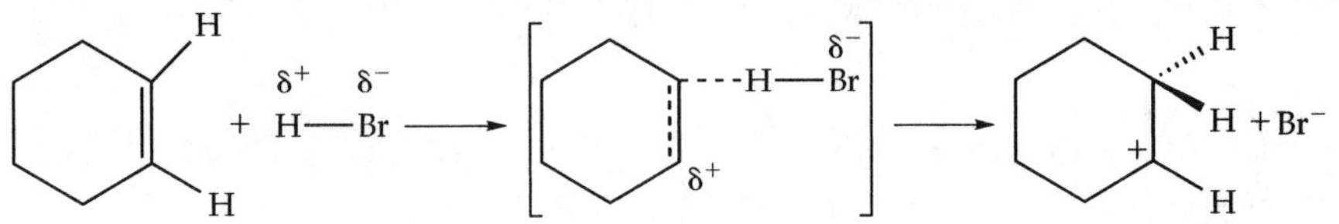

24.41 *Analyze/Plan.* Both combustion reactions produce CO_2 and H_2O:

$$C_3H_6(g) + 9/2\,O_2(g) \rightarrow 3\,CO_2(g) + 3\,H_2O(l)$$

$$C_5H_{10}(g) + 15/2\,O_2(g) \rightarrow 5\,CO_2(g) + 5\,H_2O(l)$$

Thus, we can calculate the ΔH_{comb} / CH_2 group for each compound.   *Solve.*

$$\frac{\Delta H_{comb}}{CH_2\ group} = \frac{2089\ kJ/mol\ C_3H_6}{3\ CH_2\ groups} = \frac{696.3\ kJ}{mol\ CH_2}\ ;\ \frac{3317\ kJ/mol\ C_5H_{10}}{5\ CH_2\ groups} = 663.4\ kJ/mol\ CH_2$$

$\Delta H_{comb}/CH_2$ group for cyclopropane is greater because C_3H_6 contains a strained ring. When combustion occurs, the strain is relieved and the stored energy is released during the reaction.

24.42

	ΔH
$C_{10}H_8(l) + 12\,O_2(g) \rightarrow 10\,CO_2(g) + 4\,H_2O(l)$	-5157 kJ
$-[C_{10}H_{18}(l) + 29/2\,O_2(g) \rightarrow 10\,CO_2(g) + 9\,H_2O(l)$	$-(-6286)$ kJ
$C_{10}H_8(l) + 5\,H_2O(l) \rightarrow C_{10}H_{18}(l) + 5/2\,O_2(g)$	$+1129$ kJ
$5/2\,O_2(g) + 5\,H_2(g) \rightarrow 5\,H_2O(l)$	$5(-285.8)$ kJ
$C_{10}H_8(l) + 5\,H_2(g) \rightarrow C_{10}H_{18}(l)$	-300 kJ

Compare this with the heat of hydrogenation of ethylene:

$C_2H_4(g) + H_2(g) \rightarrow C_2H_6(g)$; $\Delta H = -84.7 - (52.3) = -137$ kJ. This value applies to just one double bond. For five double bonds, we would expect about -685 kJ. The fact that hydrogenation of napthalene yields only -300 kJ indicates that the overall energy of the napthalene molecule is lower than expected for five isolated double bonds. The resonance energy is then $[-685$ kJ $- (-300$ kJ$)]$ or -385 kJ. Resonance has stabilized or lowered the overall energy of naphthalene by 385 kJ.

Functional Groups and Chirality (Sections 24.4 and 24.5)

24.43 *Analyze/Plan.* Match the structural features of various functional groups shown in Table 24.6 to the molecular structures in this exercise.   *Solve.*

(a)   (iii)

(b)   (i)

(c)   (ii) Amines are organic bases; they are H⁺ acceptors because of the the lone pair of electrons on the N atom.

(d)   (iv)

(e)   (v)

24.44   (a) $-\overset{\overset{O}{\|}}{C}-O$, ester

(b)   –Cl, halocarbon; –OH, alcohol (aromatic alcohols are phenols)

(c) $-\overset{\overset{O}{\|}}{C}-NH-$, amide   (d) alkane

(e)   —C=C—, alkene; $-\overset{\overset{O}{\|}}{C}H$, aldehyde   (f) $-\overset{\overset{O}{\|}}{C}-$, ketone

24.45   *Analyze/Plan.* Given the name of a compound, write its molecular formula. Identify the structure of the functional group present in the isomer. Finally, draw the structural formula of a molecule that contains the new functional group and has the same molecular formula as the parent compound.   *Solve.*

(a)   The formula of acetone is C_3H_6O. An aldehyde contains the group $-C\overset{\nearrow O}{\underset{\searrow H}{}}$

An aldehyde that is an isomer of acetone is propionaldehyde (or propanal):

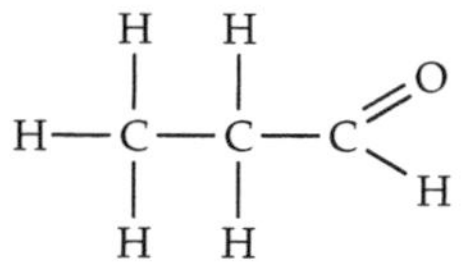

(b)   The formula of 1-propanol is C_3H_8O. An ether contains the group –O–. An ether that is an isomer of 1-propanol is ethylmethyl ether:

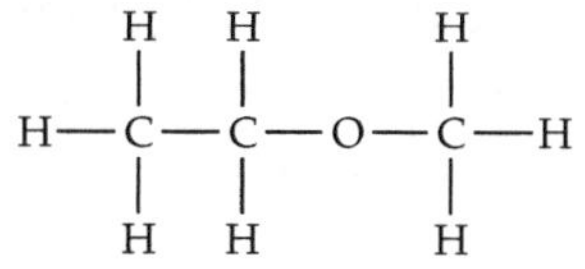

24.46   (a) C_4H_8O,

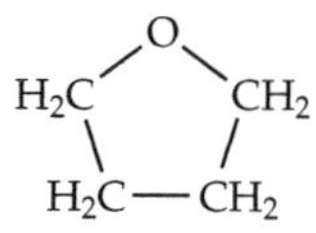

(b) $CH_3CH_2\overset{\overset{O}{\|}}{C}CH_3$, $CH_3CH_2CH_2\overset{\overset{O}{\|}}{C}-H$

$CH_2=CH_2CH_2CH_2OH$, $CH_3CH=CHCH_2OH$,   (cis and trans)
$CH_2=CHCH(OH)CH_3$ (enantiomers)

(Structures with the —OH group attached to an alkene carbon atom are not included. These molecules are called "vinyl alcohols" and are not the major form at equilibrium.)

24.47   *Analyze/Plan.* From the hydrocarbon name, deduce the number of C atoms in the acid; one carbon atom is in the carboxyl group.   *Solve.*

(a)   meth = 1 C atom

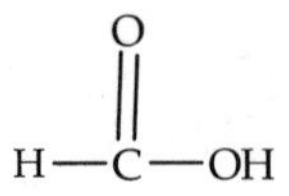

$$\underset{\text{H}-\overset{\displaystyle \text{O}}{\overset{\|}{\text{C}}}-\text{OH}}{}$$

(b)   pent = 5 C atoms

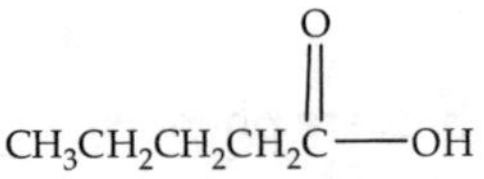

$$\text{CH}_3\text{CH}_2\text{CH}_2\text{CH}_2\overset{\displaystyle \text{O}}{\overset{\|}{\text{C}}}-\text{OH} \quad \text{or} \quad$$

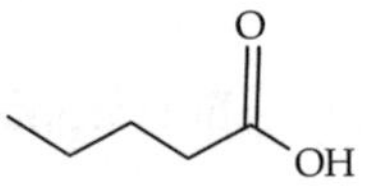

(c)   dec = 10 C atoms in backbone

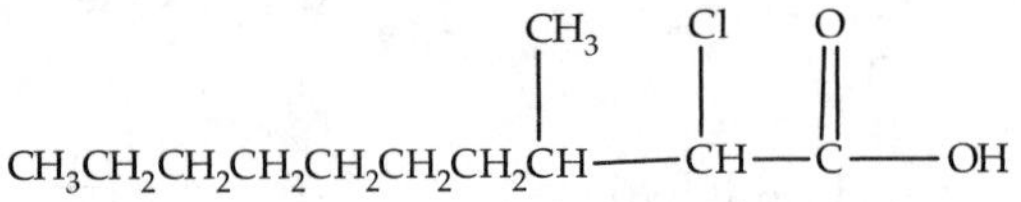

$$\text{CH}_3\text{CH}_2\text{CH}_2\text{CH}_2\text{CH}_2\text{CH}_2\text{CH}_2\underset{\text{CH}_3}{\text{CH}}-\underset{\text{Cl}}{\text{CH}}-\overset{\displaystyle \text{O}}{\overset{\|}{\text{C}}}-\text{OH}$$

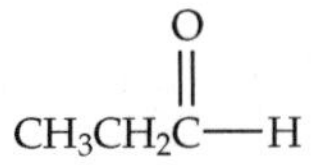

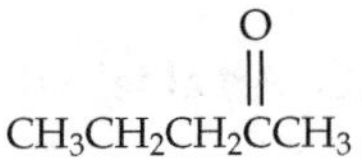

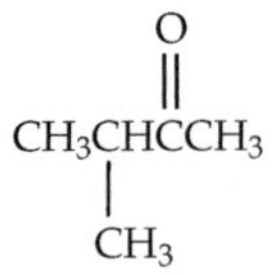

or

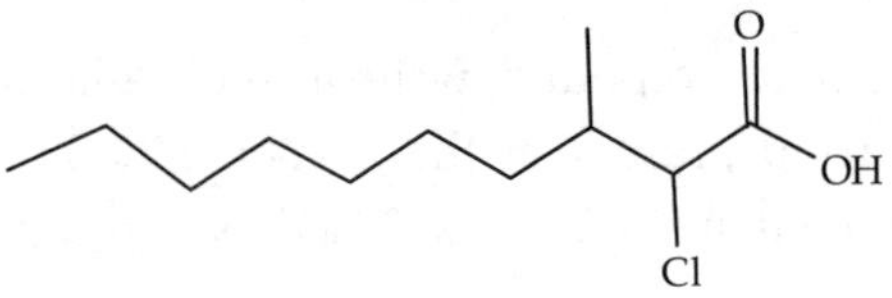

24.48   (a) $\text{CH}_3\text{CH}_2\overset{\displaystyle \text{O}}{\overset{\|}{\text{C}}}-\text{H}$ (b) $\text{CH}_3\text{CH}_2\text{CH}_2\overset{\displaystyle \text{O}}{\overset{\|}{\text{C}}}\text{CH}_3$

(c) $\text{CH}_3\underset{\text{CH}_3}{\text{CH}}\overset{\displaystyle \text{O}}{\overset{\|}{\text{C}}}\text{CH}_3$ (d) $\text{CH}_3\text{CH}_2\underset{\text{CH}_3}{\text{CH}}\overset{\displaystyle \text{O}}{\overset{\|}{\text{C}}}-\text{H}$

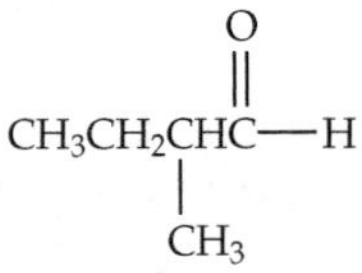

24.49   *Analyze/Plan.* In a condensation reaction between an alcohol and a carboxylic acid, the alcohol loses its –OH hydrogen atom and the acid loses its –OH group. The alkyl group from the acid is attached to the carbonyl group and the alkyl group from alcohol is attached to the ether oxygen of the ester. The name of the ester is the alkyl group from the alcohol plus the alkyl group from the acid plus the suffix *-oate.*   *Solve.*

(a)

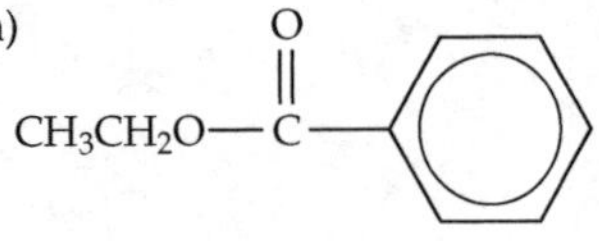

ethylbenzoate

(b)

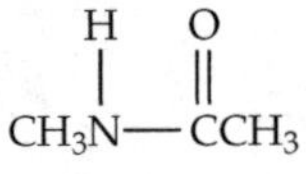

N-methylethanamide
or N-methylacetamide

(c)

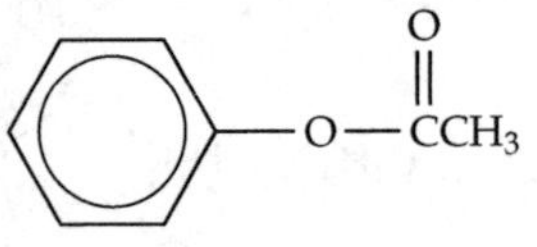

phenylacetate

24.50   (a)

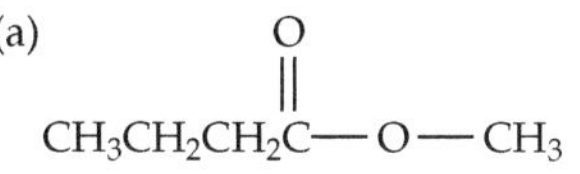

$$CH_3CH_2CH_2\overset{\overset{\displaystyle O}{\|}}{C}-O-CH_3$$

methylbutanoate

(b)

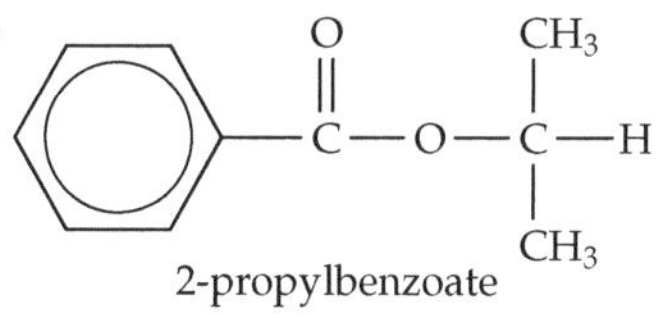

2-propylbenzoate

(c)

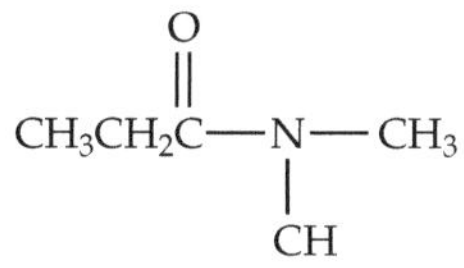

$$CH_3CH_2\overset{\overset{\displaystyle O}{\|}}{C}-\underset{\underset{\displaystyle CH}{|}}{N}-CH_3$$

N, N-dimethylpropanamide

24.51   *Analyze/Plan.* Follow the logic in Sample Exercise 24.6.   *Solve.*

(a)

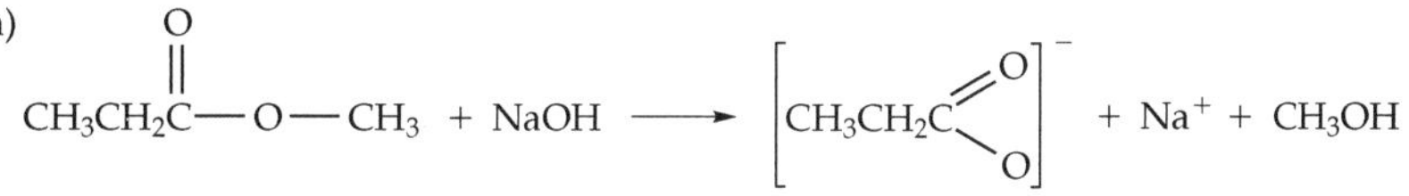

$$CH_3CH_2\overset{\overset{\displaystyle O}{\|}}{C}-O-CH_3 + NaOH \longrightarrow \left[CH_3CH_2C\overset{\diagup O}{\diagdown O}\right]^- + Na^+ + CH_3OH$$

(b)

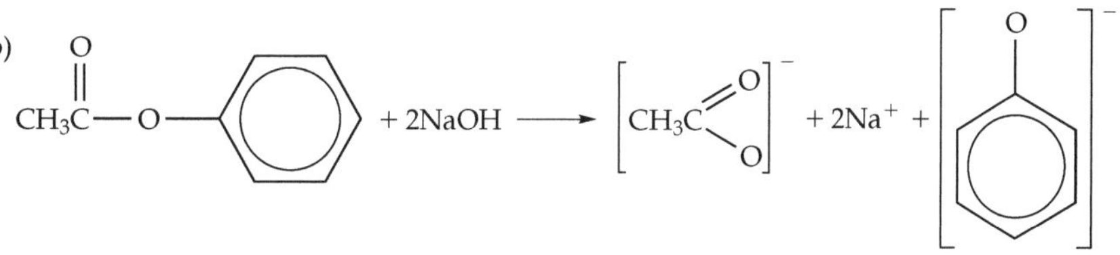

24.52   (a)

$$CH_3CH_2CH_2CH_2OH + HO\overset{\overset{\displaystyle O}{\|}}{C}CH_2CH_3 \longrightarrow CH_3CH_2CH_2CH_2O\overset{\overset{\displaystyle O}{\|}}{C}CH_2CH_3$$

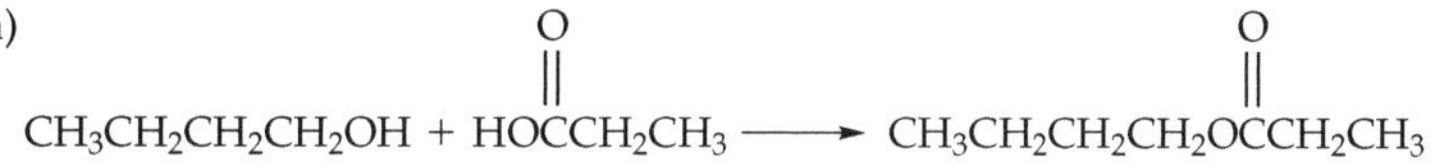

1-butanol   propionic acid   butyl proprionate
              (propanoic acid)

(b)

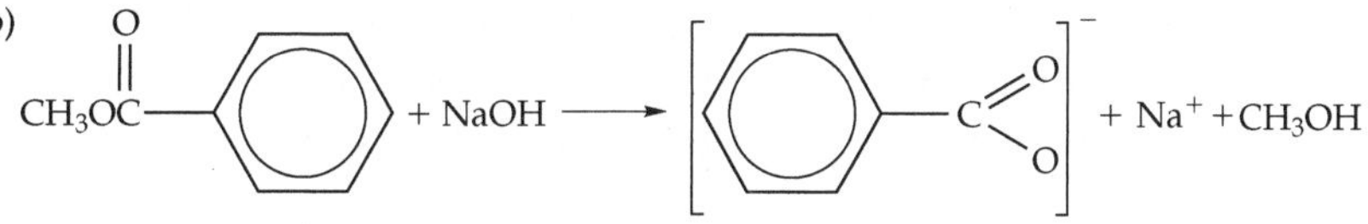

24.53   High melting and boiling points are indicators of strong intermolecular forces in the bulk substance. The strongest intermolecular force among neutral covalent molecules is hydrogen bonding. The carboxyl group of acetic acid has —OH, which acts as a donor, and —C=O, which acts as an acceptor in hydrogen bonding. We expect acetic acid to be a strongly hydrogen-bonded substance, as shown by its physical properties. The melting and boiling points of acetic acid are somewhat higher than those of water, another substance that experiences strong hydrogen bonding.

24.54 $$2\ CH_3COOH(l) \longrightarrow CH_3\overset{\overset{\displaystyle O}{\|}}{C}O\overset{\overset{\displaystyle O}{\|}}{C}CH_3(l) + H_2O(l)$$

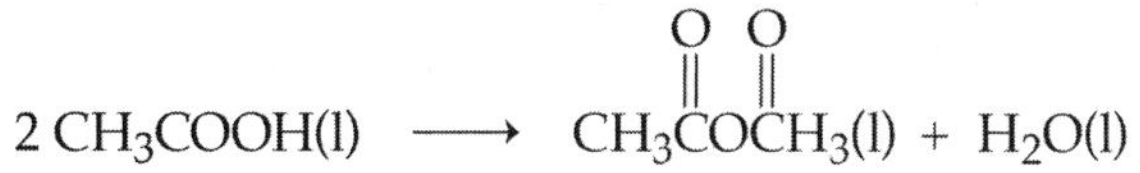

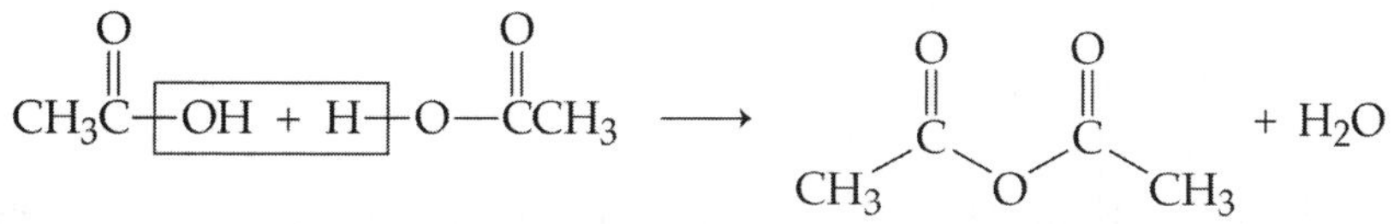

24.55   *Analyze/Plan.* Follow the logic in Sample Exercise 24.2, incorporating functional group information from Table 24.6.   *Solve.*

(a) $CH_3CH_2CH_2CH(OH)CH_3$ (b) $CH_3CH(OH)CH_2OH$

(c) $CH_3\overset{\overset{O}{\|}}{C}OCH_2CH_3$ (d)

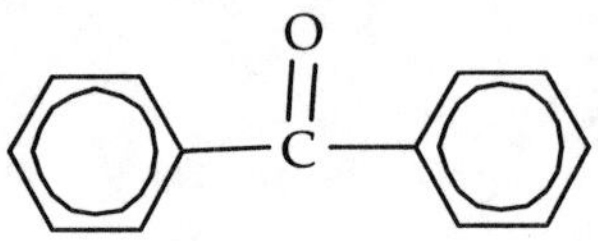

(e) $CH_3OCH_2CH_3$

24.56   (a) $CH_3CH_2CH_2CH_2CH(C_2H_5)CH_2OH$

(b) 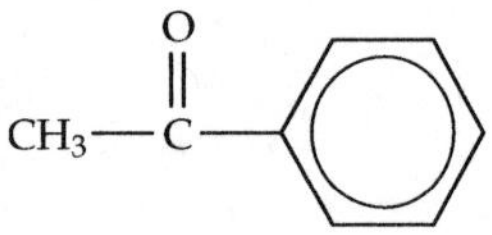   (c)

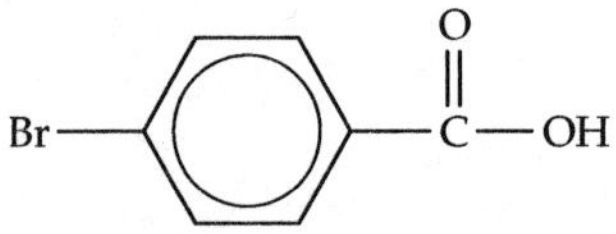

(d) $CH_3CH_2OCH_2CH_2CH_2CH_3$

(e)

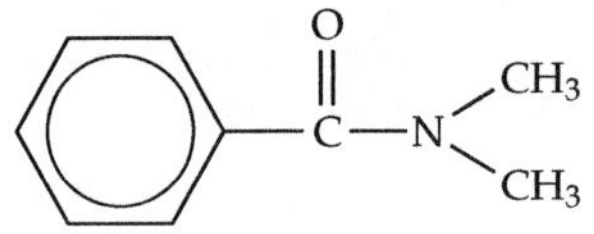

24.57   *Analyze/Plan.* Review the rules for naming alkanes and haloalkanes; draw the structure. That is, draw the carbon chain indicated by the root name, place substituents, fill remaining positions with H atoms. Each C atom attached to four different groups is chiral.   *Solve.*

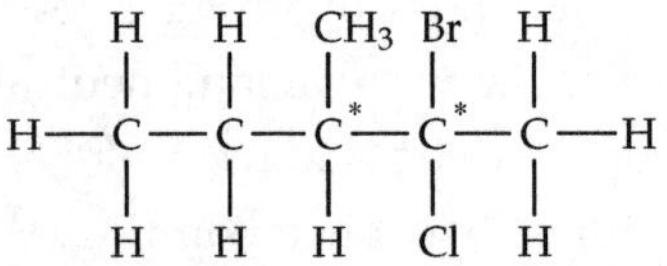

The correct choice is (c); there are two chiral carbon atoms in the molecule. C2 is obviously attached to four different groups. C3 is chiral because the substituents on C2 render the C1-C2 group different than the C4-C5 group.

24.58

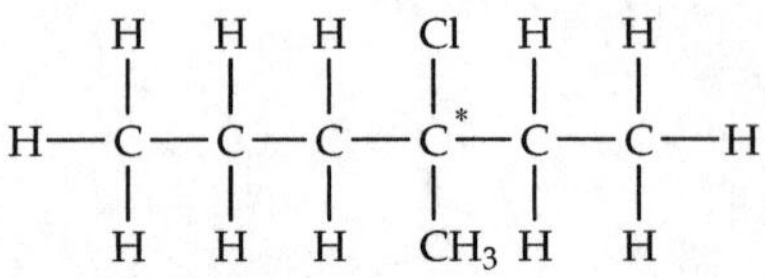

Yes, the molecule is chiral. The chiral carbon atom is attached to chloro, methyl, ethyl, and propyl groups. (If the root was a 5-carbon chain, the molecule would not have optical isomers because two of the groups would be ethyl groups.)

Introduction to Biochemistry; Proteins (Sections 24.6 and 24.7)

24.59   (a)

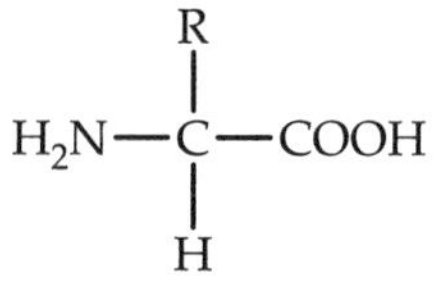

$$H_2N-\underset{\underset{H}{|}}{\overset{\overset{R}{|}}{C}}-COOH$$

(b)   In forming a protein, amino acids undergo a condensation reaction between the amino group and carboxylic acid:

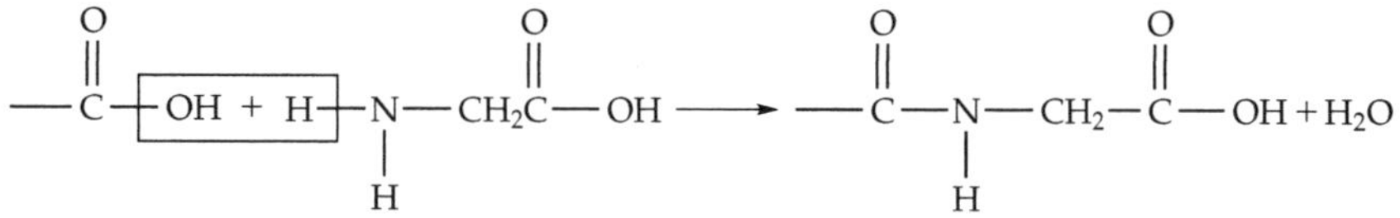

(c)   The bond that links amino acids in proteins is called the *peptide* bond.

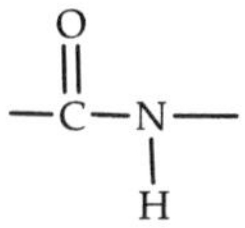

24.60   (a)   True.

(b)   True. Lysine is in the group of *basic* amino acids. Near pH 7, these amino acids have more positively charged functional groups than negatively charged groups.

(c)   False. There is one amide group, but other N-containing group is an amine; it is separated from the carbonyl group by one C atom.

(d)   False. Leucine and isoleucine are structural isomers, but not enantiomers.

(e)   False. The R group of valine is hydrophobic; it is a nonpolar hydrocarbon. The R group of arginine is hydrophilic; although it is larger than that of valine, it has one imine and two amine groups capable of hydrogen-bonding interactions with water molecules.

24.61   *Analyze/Plan.* Two dipeptides are possible. Either peptide can have the terminal carboxyl group or the terminal amino group.   *Solve.*

histadylaspartic acid, His-Asp or HD

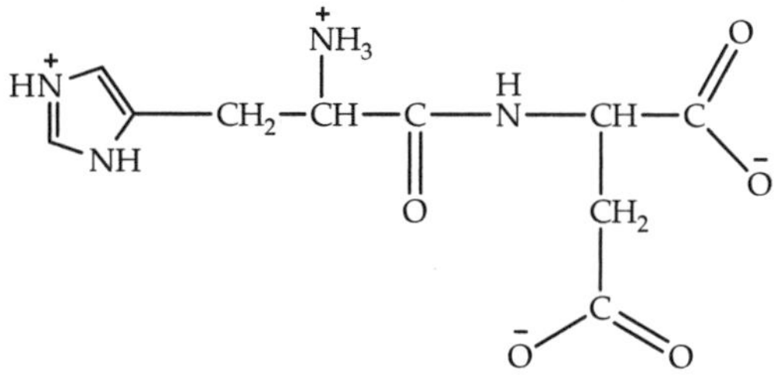

aspartylhistidine, Asp-His or DH

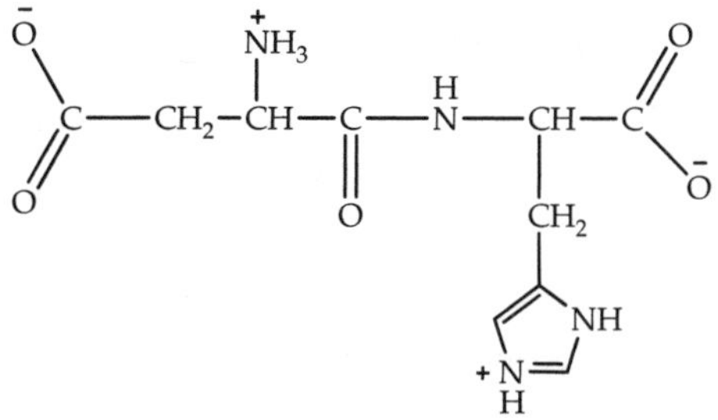

24.62

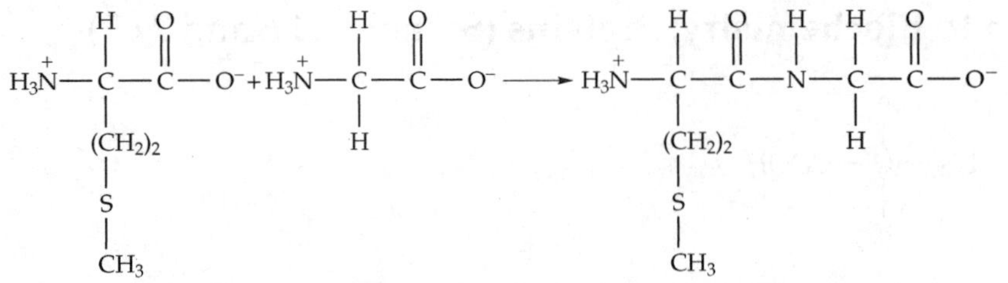

methionine   glycine   methionylglycine

24.63   *Analyze/Plan.* Follow the logic in Sample Exercise 24.7.   *Solve.*

(a)

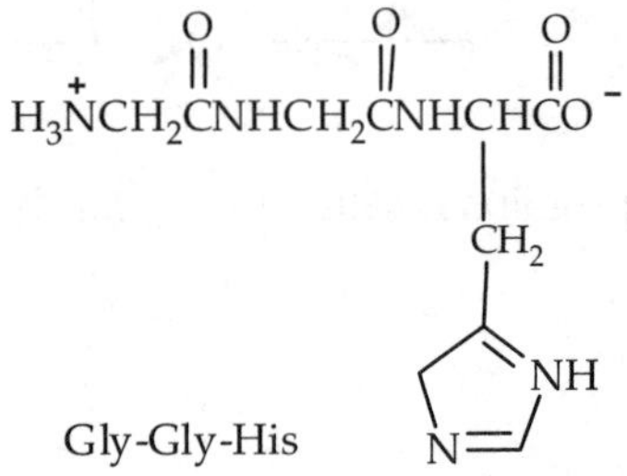

Gly-Gly-His

(b)   Three tripeptides are possible: Gly-Gly-His, GGH; Gly-His-Gly, GHG; His-Gly-Gly, HGG

24.64   (a)   Valine, serine, glutamic acid

(b)   Six (assuming the tripeptide contains all three amino acids):

Gly-Ser-Glu, GSE; Gly-Glu-Ser, GES; Ser-Gly-Glu, SGE; Ser-Glu-Gly, SEG; Glu-Ser-Gly, ESG; Glu-Gly-Ser, EGS

24.65   (a)   True.

(b)   False. Alpha helix and beta sheet are examples of secondary structure. They are different configurations of the protein chain.

(c)   False.

24.66   (a)   False. In an alpha helix, hydrogen bonding occurs between carbonyl groups and amide hydrogen atoms along the protein backbone.

(b)   False. In a beta sheet, hydrogen bonding occurs between carbonyl groups and amide hydrogen atoms along two different protein chains. This interaction zips the two chains together into a beta sheet.

Carbohydrates and Lipids (Sections 24.8 and 24.9)

24.67   (a)   True. A disaccharide is composed of two sugar units. Sugars are carbohydrates.

(b)   False. Sucrose is a disaccharide.

(c)   True. Well, most carbohydrates have this general formula.

24.68   (a)   No. Carbon 1 in the cyclic form of glucose is chiral. The configuration at C1 in alpha glucose is the opposite of that in beta glucose. However, there are four other chiral centers in each of the two forms that do not have opposite configurations. In order for molecules to be enantiomers, they must be mirror images of each other; all analogous chiral centers must have opposite configurations.

(b)

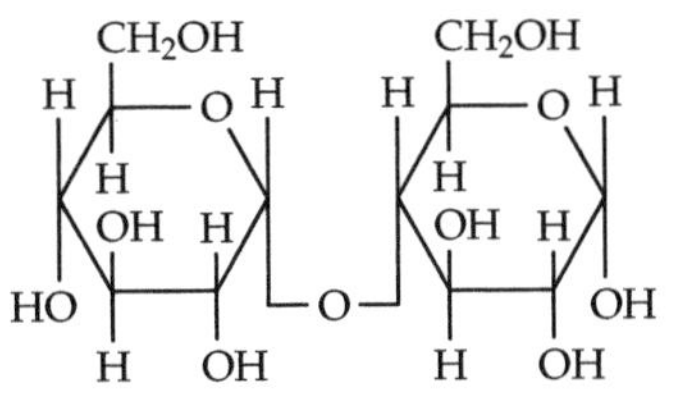

α-linkage

(c)

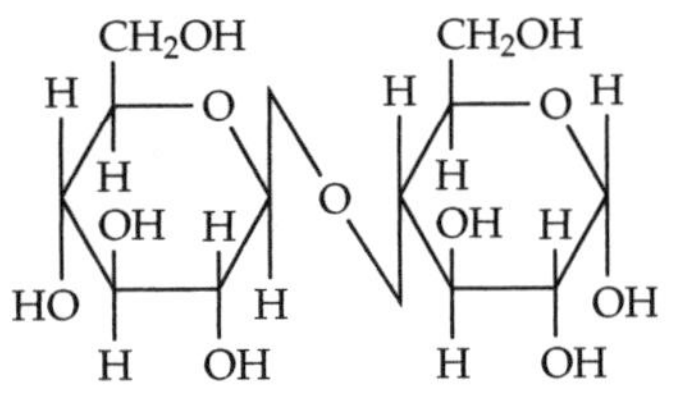

β-linkage

24.69   (a)   The empirical formula of cellulose is $C_6H_{10}O_5$.

(b)   The six-membered ring form of glucose forms the monomer unit that is the basis of the polymer cellulose. In cellulose, glucose monomer units are joined by β linkages.

(c)   Ether linkages connect the glucose monomer units in cellulose.

24.70   (a)   The empirical formula of starch is $C_6H_{10}O_5$.

(b)   The six-membered ring form of glucose is the unit that forms the basis of starch. The glucose monomer units are joined by α linkages.

(c)   Ether linkages connect the glucose monomer units in starch.

24.71   (a)   Yes, D-mannose is a sugar; it is a polyhydroxy aldehyde.

(b)   In the linear form of mannose, there are four chiral carbon atoms, C2, C3, C4, and C5. The two terminal carbon atoms, C1 and C6, are not chiral.

(c)   Both the α (left) and β (right) forms are possible.

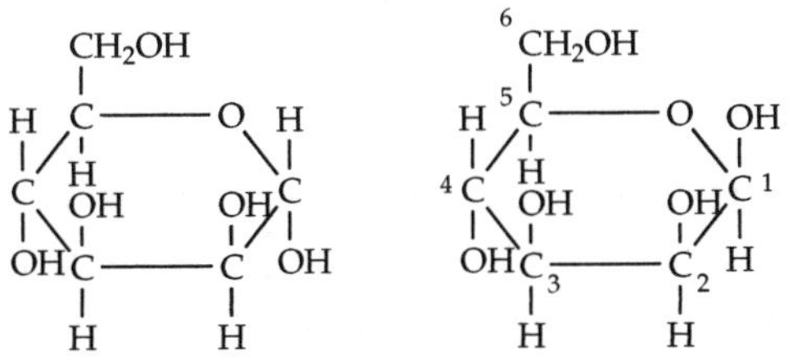

24.72   (a)   Galactose is a sugar; it is a polyhydroxy aldehyde.

(b)   In the linear form of galactose, there are four chiral carbon atoms, C2, C3, C4, and C5. The two terminal carbon atoms, C1 and C6, are not chiral.

(c)   The structure is best deduced by comparing galactose with glucose, and inverting the configurations at the appropriate carbon atoms. Recall from Solution 25.71 that both the β-form (shown here) and the α-form (OH on carbon 1 on the opposite side of ring as the CH_2OH on carbon 5) are possible.

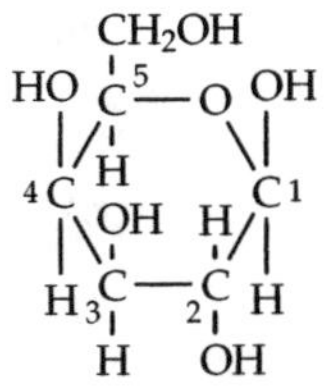

galactose

24.73　(a)　False. Fat molecules contain ester functional groups.

(b)　True. See Figure 24.24.

(c)　True. See Figure 24.24

24.74　(a)　True. Consider the fuels ethane, C_2H_6, and ethanol, C_2H_5OH, where one C–H bond has been replaced by C–O–H, a C–O and an O–H bond. Combustion reactions for the two fuels follow.

$$C_2H_6 + 7/2\,O_2 \rightarrow 2\,CO_2 + 3\,H_2O;\ \ C_2H_5OH + 3\,O_2 \rightarrow 2\,CO_2 + 3\,H_2O$$

More energy is required to break bonds in the combustion of one mole of ethanol. The reaction is less exothermic overall than the combustion of ethane. As ethane has the more exothermic combustion reaction, we say that more energy is "stored" in C_2H_6 than in C_2H_5OH.

(b)　False. Trans fats contain at least one double bond, with substituents in the trans orientation.

(c)　True.

(d)　False. Monounsaturated fatty acids have one carbon-carbon double bond in the chain, whereas the rest are single bonds.

Nucleic Acids (Section 24.10)

24.75　Dispersion forces increase as molecular size (and molar mass) increases. The larger purines (2 rings vs. 1 ring for pyrimidines) have larger dispersion forces.

24.76

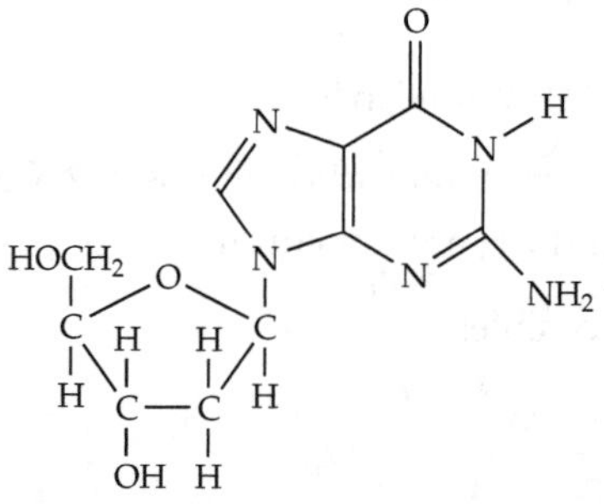

24.77　*Analyze/Plan.* Consider the structures of the organic bases in Section 24.10. The first base in the sequence is attached to the sugar with the free phosphate group in the 5′ position. The last base is attached to the sugar with a free –OH group in the 3′ position.   *Solve.*

The DNA sequence is 5′–TACG–3′.

24.78　*Analyze/Plan.* Refer to Figures 24.26 and 24.27 for details of DNA structure. In the DNA double strand, each adenine is hydrogen-bonded to a thymine and each cytosine is hydrogen-bonded to a guanine. Each base in a DNA strand is associated with a phosphate group with a 1– charge.

(a)　One adenine to one thymine

(b)　One cytosine to one guanine

(c)　24 sodium ions per dodecamer. The double-stranded dodecamer has 24 bases, 12 in each strand, for an overall 24– charge. 24 sodium ions per dodecamer are required to achieve charge balance.

24.79   *Analyze/Plan.* Recall that there is complimentary base pairing in nucleic acids because of hydrogen bond geometry. The DNA pairs are A–T and G–C. (The RNA pairs are A–U and G–C.)

Solve. From the single strand sequence, formulate the complimentary strand. Note that 3′ of the complimentary strand aligns with 5′ of the parent strand.

5′–GCATTGGC–3′

3′–CGTAACCG–5′

24.80   Statement (d) best explains the chemical differences between DNA and RNA. Statement (b) is true, but it does not prevent hydrogen bonding between complimentary base pairs.

Additional Exercises

24.81

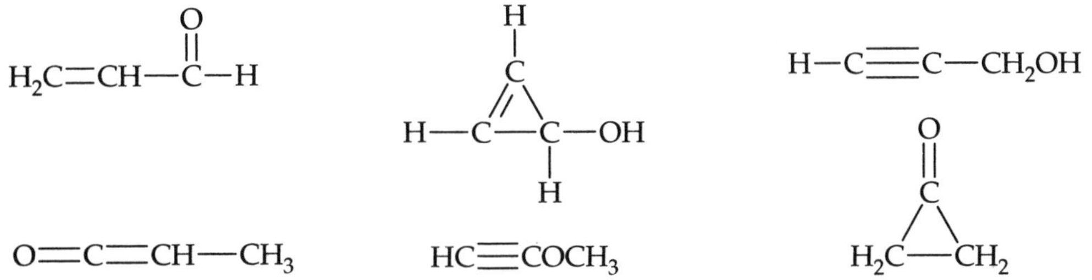

Structures with the –OH group attached to an alkene carbon atom are not included. These molecules are called "vinyl alcohols" and are not the major form at equilibrium.

24.82   *Analyze/Plan.* We are asked the number of structural isomers for two specified carbon chain lengths and a certain number of double bonds. Structural isomers have different connectivity. Since the chain length is specified, we can ignore structural isomers created by branching. We are not asked about geometrical isomers, so we ignore those as well. The resulting question is: How many ways are there to place the specified number of double bonds along the specified C chain?   *Solve.*

5 C chain with one double bond: 2 structural isomers

C=C—C—C—C   C—C=C—C—C

6 C chain with two double bonds: 6 structural isomers

C=C—C=C—C—C   C=C—C—C=C—C   C=C—C—C—C=C
C—C=C—C=C—C   C=C=C—C—C—C   C—C=C=C—C—C

24.83   (a)

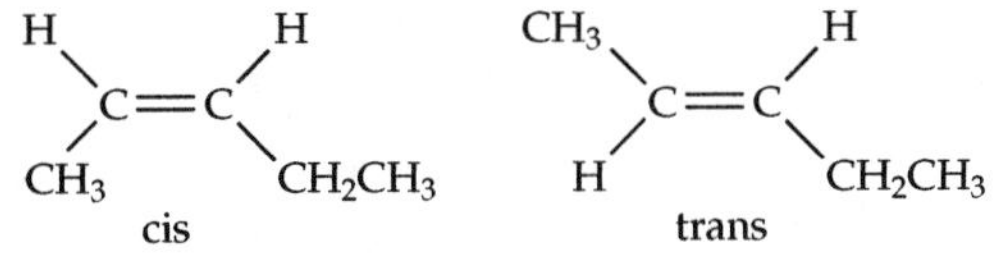

(b)   Cyclopentene does not show cis-trans isomerism because the existence of the ring demands that the C–C bonds be cis to one another.

(c)   1-pentyne does not have enantiomers because the geometry about the alkyne group is linear.

24.84   The suffix –ene signifies an alkene, –one a ketone. The molecule has alkene and ketone functional groups.

731

24.85 (a)

$\overset{O}{\overset{\|}{-CH}}$, aldehyde; 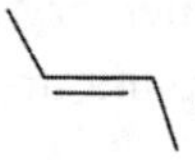 , *trans*-alkene; 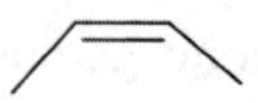 , *cis*-alkene

(b) $-O-$, ether; $-OH$, alcohol; $C=C$, alkene;

$-\overset{|}{N}-$, amine (two of these, one aliphatic and one aromatic);

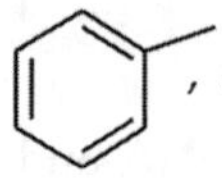

, aromatic (phenyl) ring

(c)

$-\overset{O}{\overset{\|}{C}}-$, ketone (2 of these); $-\overset{|}{N}-$, amine (2 of these)

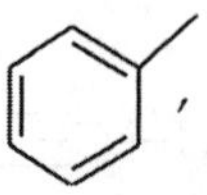

, aromatic (phenyl) ring (2 of these)

(d)

$-\overset{O}{\overset{\|}{C}}-\overset{|}{N}-$, amide; 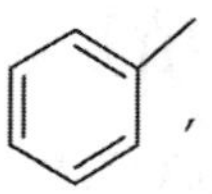 , aromatic (phenyl) ring;

$-OH$, alcohol (aromatic alcohol, phenol)

24.86 (a) Quinine, molecule (b), and indigo, molecule (c), both contain amine functional groups and would produce basic solutions if dissolved in water.

(b) Acetaminophen, molecule (d), would produce an acidic solution if dissolved in water. None of the four molecules is a carboxylic acid. Acetaminophen is both an amide and a phenol. Both of these functional groups have ionizable H atoms and produce very weakly acidic aqueous solutions.

(c) Acetaminophen, molecule (d), is probably most water soluble. Both (c) and (d) can form hydrogen bonds with water, but acetaminophen has a lower molecular weight and smaller nonpolar portion to interfere with interactions between solute and solvent.

24.87 (a)

$CH_3CH_2CH_2\overset{O}{\overset{\|}{C}}OH$ or $(CH_3)_2CH\overset{O}{\overset{\|}{C}}OH$

(b)

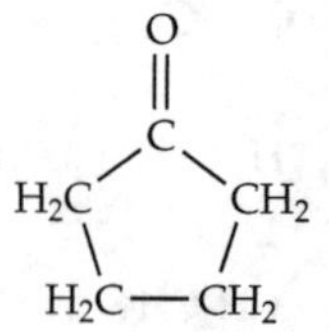

(c)

$CH_3-\overset{\overset{OH}{|}}{CH}-\overset{\overset{OH}{|}}{CH_2}$ or $\overset{\overset{OH}{|}}{CH_2}-CH_2-\overset{\overset{OH}{|}}{CH_2}$

(d)

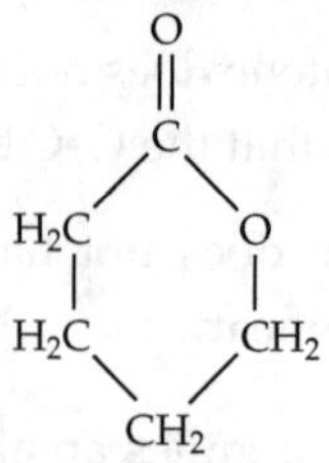

24.88   (a)   nitroglycerin, 1,2,3-trinitroxypropane

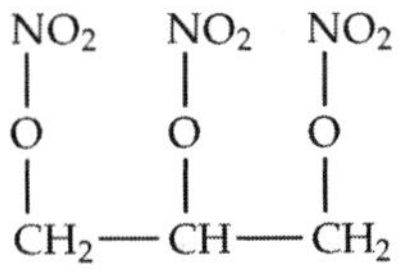

$$NO_2 \quad NO_2 \quad NO_2$$
$$| \qquad | \qquad |$$
$$O \qquad O \qquad O$$
$$| \qquad | \qquad |$$
$$CH_2 — CH — CH_2$$

(b)   Putrescine, 1,4-diaminobutane

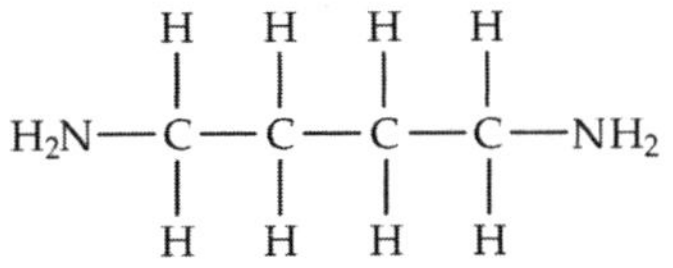

$$\begin{array}{cccc} H & H & H & H \\ | & | & | & | \\ H_2N—C—C—C—C—NH_2 \\ | & | & | & | \\ H & H & H & H \end{array}$$

(c)   Cyclohexanone, a precursor to Nylon

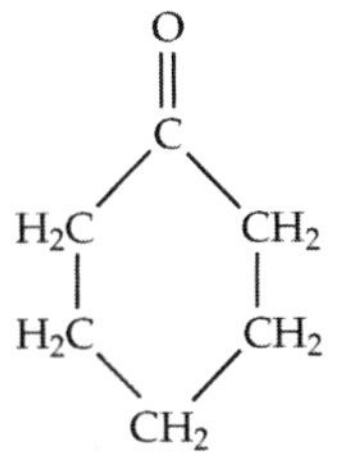

$$\begin{array}{c} O \\ \| \\ C \\ H_2C \qquad CH_2 \\ | \qquad | \\ H_2C \qquad CH_2 \\ CH_2 \end{array}$$

(d)   1,1,2,2-tetrafluoroethene, precursor to Teflon

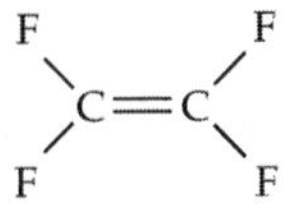

$$\begin{array}{cc} F & F \\ \diagdown \quad \diagup \\ C=C \\ \diagup \quad \diagdown \\ F & F \end{array}$$

(e)   Oleic acid, *cis*-9-octanedecenoic acid

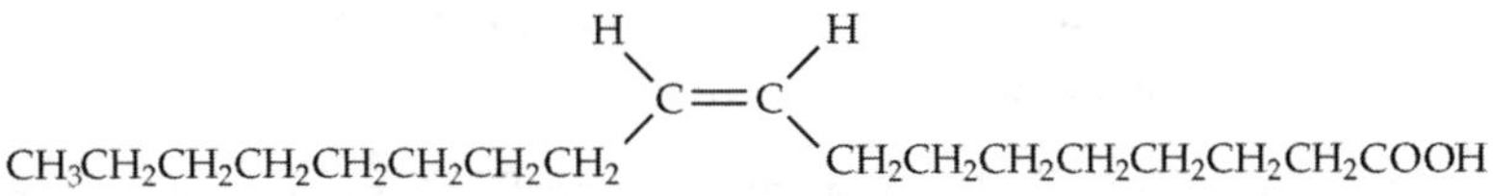

$$\begin{array}{cc} H & H \\ \diagdown \quad \diagup \\ C=C \\ \end{array}$$
$$CH_3CH_2CH_2CH_2CH_2CH_2CH_2CH_2 \qquad CH_2CH_2CH_2CH_2CH_2CH_2CH_2COOH$$

24.89   In order for indole to be planar, the N atom must be sp^2 hybridized. The nonbonded electron pair on N is in a pure p orbital perpendicular to the plane of the molecule. The electrons that form the π bonds in the molecule are also in pure p orbitals perpendicular to the plane of the molecule. Thus, each of these p orbitals is in the correct orientation for π overlap; the delocalized π system extends over the entire molecule and includes the "nonbonded" electron pair on N. The reason that indole is such a weak base (H^+ acceptor) is that the nonbonded electron pair is delocalized and an H^+ ion does not feel the attraction of a full localized electron pair.

24.90   (a)   The molecule has one ketone and two alcohol functional groups. There are no chiral centers, no carbon atoms attached to four different groups.

(b)   The molecule has one ketone and three alcohol functional groups. There is one chiral center. The carbon bearing the secondary –OH has four different groups attached, and is thus chiral.

(c)   The molecule has a carboxylic acid and an amine functional group. It is an amino acid, shown in its neutral (not zwitterion) form. There are two chiral centers. The carbon bearing the –NH₂ group and the carbon bearing the –CH₃ group are both chiral.

24.91   In the zwitterion form of a tripeptide present in aqueous solution near pH 7, the terminal carboxyl group is deprotonated and the terminal amino group is protonated, resulting in a net zero charge. The molecule has a net charge only if a side (R) group contains a charged (protonated or deprotonated) group. The tripeptide is positively charged if a side group contains a protonated amine. According to Figure 24.16, the only amino acids with protonated amines in their side groups are histidine (His), lysine (Lys), and arginine (Arg). Of the tripeptides listed, only (a) Gly-Ser-Lys will have a net positive charge at pH 7. [Note that aspartic acid (Asp) has a deprotonated carboxyl in its side group, so (c) Phe-Tyr-Asp will have a net negative charge at pH 7.]

24.92   Glu-Cys-Gly is the only possible order. Glutamic acid has two carboxyl groups that can form a peptide bond with cysteine, so there are two possible structures.

24.93   Both glucose and fructose contain six C atoms, so both are hexoses. Glucose contains an aldehyde group at C1, so it is an aldohexose. Fructose has a ketone at C2, so it is a ketohexose.

24.94   DNA and RNA have the bases guanine, cytosine, and adenine in common, but DNA contains thymine, whereas RNA contains uracil. Thymine and uracil differ by a single methyl group, so both have a similar hydrogen bonding pattern with the complementary base adenine. This means that there is the possibility of a DNA strand binding to a "complementary" RNA strand.

Given this possibility, RNA is not involved in DNA replication. However, during the process of transcription and in the presence of the enzyme RNA polymerase, a complementary strand of mRNA is assembled along a segment of the backbone of a single DNA strand. During transcription, when there is adenine (A) in the DNA strand, uracil (U) is added to the RNA strand.

Integrative Exercises

24.95

methane   difluoromethane

CH$_3$CH$_2$OH   CH$_3$—O—CH$_3$
ethanol   dimethyl ether

(a)   Methane boils at –128 °C. Methane is nonpolar and has the smallest molar mass. It experiences only weak dispersion forces and has the lowest boiling point.

(b)   Difluoromethane boils at –52 °C. It is somewhat polar and experiences weak dipole-dipole and dispersion forces. Although difluoromethane has a larger molar mass than dimethyl ether, it has approximately spherical shape, which reduces the strength of its dispersion forces.

(c) Dimethyl ether boils at –25 °C. It is somewhat polar and experiences weak dipole-dipole and dispersion forces. Its shape is bent, which enables stronger dispersion forces than the distorted spherical shape of difluoromethane. [This explanation of the relative boiling points of dimethyl ether and difluoromethane is a rationalization based on the actual boiling points of the compounds. It is difficult to distinguish the two based on simple principles of relative strengths of intermolecular forces.]

(d) Ethanol boils at 78 °C. It is an alcohol that participates in hydrogen bonding. This is the strongest kind of intermolecular force and explains the highest boiling point.

24.96 Determine the empirical formula of the unknown compound and its oxidation product. Use chemical properties to propose possible structures.

$$68.1 \, g \, C \times \frac{1 \, mol \, C}{12.01 \, g \, C} = 5.6703; \, 5.6703 \, / \, 1.1375 = 4.98 \approx 5$$

$$13.7 \, g \, H \times \frac{1 \, mol \, H}{1.008 \, g \, H} = 13.5913; \, 13.5913 \, / \, 1.1375 = 11.95 \approx 12$$

$$18.2 \, g \, P \times \frac{1 \, mol \, O}{16.00 \, g \, O} = 1.1375; \, 1.1375 \, / \, 1.1375 = 1$$

The empirical formula of the unknown is $C_5H_{12}O$.

$$69.7 \, g \, C \times \frac{1 \, mol \, C}{12.01 \, g \, C} = 5.8035; \, 5.8035 \, / \, 1.1625 = 4.99 \approx 5$$

$$11.7 \, g \, H \times \frac{1 \, mol \, H}{1.008 \, g \, H} = 11.6071; \, 11.6071 \, / \, 1.1625 = 9.99 \approx 10$$

$$18.6 \, g \, O \times \frac{1 \, mol \, O}{16.00 \, g \, O} = 1.1625; \, 1.1625 \, / \, 1.1625 = 1$$

The empirical formula of the oxidation product is $C_5H_{10}O$.

The compound is clearly an alcohol. Its slight solubility in water is consistent with the properties expected of a secondary alcohol with a five-carbon chain. The fact that oxidation results in a ketone, rather than an aldehyde or a carboxylic acid, tells us that it is a secondary alcohol. Some reasonable structures for the unknown secondary alcohol are:

$$\underset{\underset{OH}{|}}{CH_3CHCH_2CH_2CH_3} \quad \underset{\underset{OH}{|}}{CH_3CHCHCH_2CH_3} \quad \underset{\underset{OH}{|}}{CH_3CHCH(CH_3)_2}$$

24.97 Determine the empirical formula, molar mass, and thus molecular formula of the compound. Confirm with physical data.

$$66.7 \, g \, C \times \frac{1 \, mol \, C}{12.01 \, g \, C} = 5.554 \, mol \, C; \, 5.554 \, / \, 1.381 = 4.021 = 4$$

$$11.2 \, g \, H \times \frac{1 \, mol \, H}{1.008 \, g \, H} = 11.11 \, mol \, H; \, 11.11 \, / \, 1.381 = 8.043 = 8$$

$$22.1 \, g \, O \times \frac{1 \, mol \, O}{16.00 \, g \, O} = 1.381 \, mol \, O; \, 1.381 \, / \, 1.381 = 1$$

The empirical formula is C_4H_8O. Using Equation 10.11 (MM = molar mass):

$$MM = \frac{(2.28 \text{ g/L})(0.08206 \text{ L-atm/mol-K})(373 \text{ K})}{0.970\text{-atm}} = 71.9 \text{ g/mol}$$

The formula weight of C_4H_8O is 72, so the molecular formula is also C_4H_8O. Because the compound has a carbonyl group and cannot be oxidized to an acid, the only possibility is 2-butanone.

$$CH_3\overset{\overset{O}{\|}}{C}CH_2CH_3$$

The boiling point of 2-butanone is 79.6 °C, confirming the identification.

24.98   Determine the empirical formula, molar mass, and thus molecular formula of the compound. Confirm with physical data.

$$85.7 \text{ g C} \times \frac{1 \text{ mol C}}{12.01 \text{ g C}} = 7.136 \text{ mol C}; 7.136/7.136 = 1$$

$$14.3 \text{ g H} \times \frac{1 \text{ mol H}}{1.008 \text{ g H}} = 14.19 \text{ mol H}; 14.19/7.136 \approx 2$$

Empirical formula is CH_2. Using Equation 10.11 (MM = molar mass):

$$MM = \frac{(2.21 \text{ g/L})(0.08206 \text{ L-atm/mol-K})(373 \text{ K})}{(735/760) \text{ atm}} = 69.9 \text{ g/mol}$$

The molecular formula is thus C_5H_{10}. The absence of reaction with aqueous Br_2 indicates that the compound is not an alkene, so the compound is probably the cycloalkane cyclopentane. According to the *Handbook of Chemistry and Physics*, the boiling point of cyclopentane is 49 °C at 760 torr. This confirms the identity of the unknown.

24.99   The reaction is: $2 \text{ NH}_2\text{CH}_2\text{COOH(aq)} \rightarrow \text{NH}_2\text{CH}_2\text{CONHCH}_2\text{COOH(aq)} + \text{H}_2\text{O(l)}$
$\Delta G° = (-488) + (-237.13) - 2(-369) = 12.87 = 13 \text{ kJ}$

24.100 (a)   At low pH, the amine and carboxyl groups are protonated.   At high pH, the amine and carboxyl groups are deprotonated.

(b) $CH_3\overset{\overset{O}{\|}}{C}\text{—OH(aq)} \longrightarrow CH_3\overset{\overset{O}{\|}}{C}\text{—O}^-\text{(aq)} + H^+\text{(aq)}$

$K_a = 1.8 \times 10^{-5}$, $pK_a = -\log(1.8 \times 10^{-5}) = 4.74$

The conjugate acid of NH_3 is NH_4^+.

$NH_4^+\text{(aq)} \rightleftharpoons NH_3\text{(aq)} + H^+\text{(aq)}$

$K_a = K_w/K_b = 1.0 \times 10^{-14}/1.8 \times 10^{-5} = 5.55 \times 10^{-10} = 5.6 \times 10^{-10}$

$pK_a = -\log(5.55 \times 10^{-10}) = 9.26$

In general, a –COOH group is a stronger acid than a $-NH_3^+$ group. The smaller pK_a value for amino acids is for the ionization (deprotonation) of the –COOH group and the larger pK_a is for the deprotonation of the $-NH_3^+$ group.

$$pK_a = 9.47$$

(c)

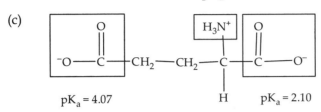

$$pK_a = 4.07 \qquad\qquad H \qquad pK_a = 2.10$$

By analogy to serine, the carboxyl group near the amine will have pK_a ~2 and the amino group will have pK_a ~9. By elimination, the carboxyl group in the side chain has pK_a ~4.

(d) From the titration curve, the unknown amino acid has three pK_a values: 2.0, 4.0, and 10.0. The unknown will be an acidic amino acid, because two of the pK_a values are significantly less than 7. The two acidic amino acids are glutamic acid and aspartic acid. From part (c), the pK_a values of glutamic acid are close to but not an exact match to the ones on the titration curve. The most likely candidate is aspartic acid.

Asparagine and glutamine are the amide forms of aspartic and glutamic acid. The "middle" pK_a value for these two amino acids represents ionization of the amide hydrogen and will be large than 4.0.

24.101   (a) Because the native form is most stable, it has a lower, more negative free energy than the denatured form. Another way to say this is that ΔG for the process of denaturing the protein is positive.

(b) ΔS is negative in going from the denatured form to the folded (native) form; the native protein is more ordered.

(c) The four S–S linkages are strong covalent links holding the chain in place in the folded structure. A folded structure without these links would be less stable (higher G) and have more motional freedom (more positive entropy).

(d) After reduction, the eight S–S groups will form hydrogen-bond-like interactions with acceptors along the protein backbone, but these will be weaker and less specifically located than the S–S covalent bonds of the native protein. Overall, the tertiary structure of the reduced protein will be looser and less compact because of the loss of the S–S linkages; the entropy will be higher.

(e) The amino acid cysteine must be present in order for –SH bonds to be found in ribonuclease A. (Methionine contains S, but no –SH functional group.)

24.102 $AMPOH^-(aq) \rightleftharpoons AMPO^{2-}(aq) + H^+(aq)$

$$pK_a = 7.21; K_a = 10^{-pK_a} = 6.17 \times 10^{-8} = 6.2 \times 10^{-8}$$

$$K_a = \frac{[AMPO^{2-}][H^+]}{[AMPOH^-]} = 6.2 \times 10^{-8}. \text{ When } pH = 7.40, [H^+] = 3.98 \times 10^{-8} = 4 \times 10^{-8}.$$

$$\text{Then } \frac{[AMPOH^-]}{[AMPO^{2-}]} = 3.98 \times 10^{-8} / 6.17 \times 10^{-8} = 0.6457 = 0.6$$

NOTES